12th International Workshop on Ship and Marine Hydrodynamics (IWSH-2023)

IOP Conference Series: Materials Science and Engineering
Volume 1288

Espoo, Finland
28 August - 1 September 2023

Editors:

Spyros Hirdaris
Decheng Wan

ISBN: 978-1-7138-7728-8
ISSN: 1757-8981

Printed from e-media with permission by:

Curran Associates, Inc.
57 Morehouse Lane
Red Hook, NY 12571

Some format issues inherent in the e-media version may also appear in this print version.

Institute of Physics
Dirac House, Temple Back
Bristol BS1 6BE UK

Phone: 44 1 17 929 7481
Fax: 44 1 17 920 0979

techtracking@iop.org

Additional copies of this publication are available from:

Curran Associates, Inc.
57 Morehouse Lane
Red Hook, NY 12571 USA
Phone: 845-758-0400
Fax: 845-758-2633
Email: curran@proceedings.com
Web: www.proceedings.com

TABLE OF CONTENTS

IOP Conf. Series: Materials Science and Engineering 1288 (2023) 011001 doi:10.1088/1757-899X/1288/1/011001

The 12th International Workshop on Ship and Marine Hydrodynamics

IWSH 2023

Aalto University
School of Engineering

28 August to 1 September 2023

Aalto University, Espoo, Finland

Editors

Eur.Ing. Dr. Spyros Hirdaris CEng FRINA, FSNAME

Aalto University, Finland

AND

Professor Decheng Wan

Shanghai Jiao Tong University, China

IOP Conf. Series: Materials Science and Engineering 1288 (2023) 011001 doi:10.1088/1757-899X/1288/1/011001

Preface

The 12th International Workshop on Ship and Marine Hydrodynamics (IWSH 2023) was held from the 28th of August to the 1st of September 2023 at Aalto University, Finland. Aalto University was established in 2010 as a merger the Helsinki University of Technology (HUT), the Helsinki School of Economics and the University of Art and Design of Helsinki. Marine and Arctic Technology has been solid part of the well cherished traditions of HUT dating back to the 19th century.

IWSH inaugurated by Prof. Zaojian Zou of Wuhan University of Technology in 1999. Early national events were organised by Wuhan University of Technology (2001,2003), Shanghai Jiao Tong University (2005), Jiangshu University of Science and Technology (2007) and Harbin Engineering University (2009).

In 2011 Professor Decheng Wan organised the first international version of the workshop at Shanghai Jiao Tong University. Since then, the event has been organised by Seoul National University - South Korea (2013), the University of Strathclyde - UK (2015), the National Taiwan Ocean University (2017) and Hamburg University of Technology - Germany (2019). The 2021 edition of IWSH did not take place because of the COVID-19 pandemic. The "all-inclusive" nature of the 12th sequel of the workshop held in 2023 included scientific and social activities, fostering opportunities for networking.

The workshop focused on publications of relevance to ship dynamics, hydrodynamics, wave mechanics, sea loads and responses, fluid structure interactions, ship resistance and propulsion, etc. Notably, emerging multiphysics methods with application on both ships and ocean engineering including renewable energies were evident. Six keynotes were delivered by Dr. Nuno Fonseca of SINTEF (Norway), Dr. Sime Malenica of Bureau Veritas (BV, France), Alessandro Iafrati of Istituto di iNgegneria del Mare (INM, Italy), Dr. Michel Visonneau of Centre National de la Recherche Scientifique (CNRS, France), Professor Stefano Brizzolara of Virginia Technological Institute and State University (USA) and Professor Yonghwan Kim of Seoul National University (South Korea), were invited to foster top level discussion on relevant subjects. As part of the workshop proceedings the main event sponsor, SINTEF Ocean (Norway) hosted a special session of their International Collaboration Network in Marine Technology.

The oral technical program was organized in parallel sessions. In these proceedings, the papers are organized by major themes. The themes are: (1) Ship dynamics, (2) Wave mechanics, (3) Fluid Structure Interactions, (4) Ship motions and sea loads, (5) Ocean Engineering, (6) Ship resistance, (7) Marine propulsion.

The workshop confirmed that the complexity of the subject as related to conventional and emerging technologies attracts a great deal of interest of relevance to engineering science and technological innovation. The field of ship and marine hydrodynamics remains more than relevant for the safety and sustainability of the ship and ocean engineering sectors.

Dr. Spyros Hirdaris CEng FRINA FSNAME

Aalto University, Espoo, Finland, September 2023

Sponsors

The organizing committee of IWSH 2023 would like to thank SINTEF Ocean for supporting this conference.

SINTEF Ocean conducts research and innovation related to ocean space for national and international industries. The organisation's ambition is to continue Norway's leading position in marine technology and bio-marine research. Many of the challenges of modern society can be solved through sustainable use of the ocean. Transport, food and energy production represent the backbone of ocean-based industries and are also core research areas for SINTEF. In addition, there is focus on environmental technology, with one of the world's leading professional environments in marine environmental technology. Through cooperation in the group, SINTEF integrate their own expertise with expert technological knowledge from other industry sectors.

IOP Conf. Series: Materials Science and Engineering 1288 (2023) 011001 doi:10.1088/1757-899X/1288/1/011001

Reviewers acknowledgement

The organizing committee of IWSH 2023 would like to thank the following academics for supporting the review of papers presented in this special volume.

Abbas Dashtimanesh, KTH Royal Institute of Technology, Sweden

Atilla Incecik, Strathclyde University, UK

Changhong Hu, Kyushu University, Japan

Decheng Wan, Shanghai Jiao Tong University, China

Ghalin Humayun Taimuri, Aalto University and NAPA Ltd., Finland

Jerzy Matusiak, Aalto University, Finland

Liangliang Lu, Aalto University, Finland

Mingyang Zhang, Aalto University, Finland

Mikko Suominen, Aalto University, Finland

Pekka Ruponen, Aalto University and NAPA Ltd., Finland

Tommi Mikkola, Aalto University, Finland

Robinson Peric, TUHH, Germany

Sasan Tavakoli, Aalto University, Finland

Victor Bolbot, Aalto University, Finland

Youjiang Wang, Shanghai Jiao Tong University, China

IOP Conf. Series: Materials Science and Engineering 1288 (2023) 011001 doi:10.1088/1757-899X/1288/1/011001

Aalto University's IWSH 2023 local organizing committee

Spyros Hirdaris, Associate Professor

Tommi Mikkola, Senior Lecturer

Pekka Ruponen, Professor of Practice

Pentti Kujala, Professor em.

Sasan Tavakoli, Post Doctoral Scholar

Liangliang Lu, Post Doctoral Scholar

Ghalib Taimuri, Post Doctoral Scholar

Mingyang Zhang, Post Doctoral Scholar

Aleksandr Kondratenko, Post Doctoral Scholar

Victor Bolbot, Post Doctoral Scholar

Dongni Yan, Research student

Aaro Karola, Research student

Zongyu Jiang, Research student

IWSH 2023 International Standing Committee

Moustafa Abdel-Maksoud, Technical University of Hamburg – Harburg, Germany

Jiahn-Horng Chen, National Taiwan Ocean University, Taiwan

Atilla Incecik, University of Strathclyde, UK

Masashi Kashiwagi, Osaka University, Japan

Yonghwan Kim, Seoul National University, South Korea

Decheng Wan, Shanghai Jiao Tong University, China

Spyros Hirdaris, Aalto University, Finland

IOP Conf. Series: Materials Science and Engineering 1288 (2023) 011001 doi:10.1088/1757-899X/1288/1/011001

IWSH 2023 International Scientific Committee

Moustafa Abdel-Maksoud, TUHH, Germany

Jiahn-Horng Chen, National Taiwan University, Taiwan, China

Sören Ehlers, TUHH, Germany

Pierre Ferrant, Ecole Centrale Nantes, France

Ismail Hakki Helvacioglu, ITU, Turkey

Chin-Yeh Hsin, National Taiwan Ocean University, Taiwan, China

Changhong Hu, Kyushu University, Japan

Tomoki Ikoma, Nihon University, Japan

Atilla Incecik, University of Strathclyde, UK

Masashi Kashiwagi, Osaka University, Japan

Yonghwan Kim, Seoul National University, Korea

Dezhi Ning, Dalian University of Technology, China

Lian Shen, University of Minnesota, USA

Sverre Steen, NTNU, Norway

Yougang Tang, Tianjin University, China

Stephen Turnock, University of Southampton, UK

Marc Vantorre, Ghent University, Belgium

Decheng Wan, Shanghai Jiaotong University, China

Chi Yang, George Mason University, USA

Zaojian Zou, Shanghai Jiao Tong University, China

Spyros Hirdaris, Aalto University, Finland.

IOP Conf. Series: Materials Science and Engineering 1288 (2023) 011002 doi:10.1088/1757-899X/1288/1/011002

Peer Review Statement

All papers published in this volume have been reviewed through processes administered by the Editors. Reviews were conducted by expert referees to the professional and scientific standards expected of a proceedings journal published by IOP Publishing.

- **Type of peer review:** Single Anonymous
- **Conference submission management system:** Morressier
- **Number of submissions received:** 83
- **Number of submissions sent for review:** 83
- **Number of submissions accepted:** 66
- **Acceptance Rate (Submissions Accepted / Submissions Received × 100):** 79.5
- **Average number of reviews per paper:** 1
- **Total number of reviewers involved:** 16
- **Contact person for queries:**
  Name: Dr. Spyros Hirdaris
  Email: spyros.hirdaris@aalto.fi
  Affiliation: Associate Professor of Marine Technology, Aalto University, Finland (IWSH organiser and proceedings editor)

IOP Conf. Series: Materials Science and Engineering 1288 (2023) 012001 doi:10.1088/1757-899X/1288/1/012001

Maritime Emergency Preparedness – Drifting Ships

T E Berg[1], Ø Selvik[2], R Indergård[3] and E Ringen[4]

[1] Scientific Advisor, SINTEF Ocean, Trondheim, Norway
[2] Senior Advisor, SINTEF Ocean, Trondheim, Norway
[3] Senior Advisor, SINTEF Ocean, Trondheim, Norway
[4] Senior Software Developer, SINTEF Ocean, Trondheim, Norway

E-mail: Tor.berg@sintef.no

Abstract. The Norwegian Coastal Administration registers maritime incidents in Norwegian waters. Drifting ships and drifting objects may cause serious incidents and possible disasters. By analysing ship traffic in Norwegian waters, it is documented that the frequency of drifting ships has been approximately 0.15% of the registered ship transits, or 150–165 drifting cases annually. In addition to a possible grounding when a vessel reaches the coastline, the increased use of the ocean space for wind farms, floating solar installations, and fish farming structures (moving from sheltered coastal sites to exposed locations) increase the risk of collision with different types of ocean structures, including offshore oil and gas installations. The need for adequate emergency preparedness to prevent drifting ships from colliding with different types of ocean structures or a final grounding can be solved through different types of mitigating measures. One measure could be to have dedicated emergency towing vessels for handling vessels drifting towards ocean structures or on a course leading to a possible grounding. A reliable tool for predicting a drifting vessel's path is important for planning an emergency towing operation. Systems to provide a fast and safe setup of an emergency towing connection under different environmental conditions are needed, while the competence of the crew on the emergency towing vessel and the ship to be towed is a critical factor for a successful emergency towing operation under adverse conditions.

1. Introduction

The Norwegian Coastal Administration (NCA) has the national responsibility for pollution preparedness from all ships within the Norwegian Economic Exclusive Zone (EEZ). Vardø Vessel Traffic Centre (NOR VTS) is responsible for monitoring vessel traffic in Northern Norway. One major task is to identify vessels with suspicious tracks and find out the cause of this behaviour. Special care is given to tracks deviating from the vessel's voyage plan. By contacting the vessels, the cause is identified. It may be that a vessel is performing planned maintenance during the voyage, but in some cases, a vessel is drifting due to a blackout or problems with the steering system. If a drifting vessel approaches the 12 nm zone off the Norwegian coast and there is a possibility that the vessel will cause pollution, the NCA may scramble the Norwegian Emergency Towing System (NETS) to assist the disabled drifting vessel using one of the Norwegian Coast Guard vessels allocated for NETS. Some brief descriptions of the organization of NETS and actual emergency towing incidents are prepared by SINTEF Ocean in collaboration with the Norwegian Coastal Administration and Norwegian Coast Guard [1]–[3]

In addition to grounding, the possibility of collision with different ocean structures is increasing as more such structures are placed in more exposed coastal waters (different types of fish farms). Within

IOP Conf. Series: Materials Science and Engineering 1288 (2023) 012001 doi:10.1088/1757-899X/1288/1/012001

the Norwegian EEZ, there are a number of oil and gas installations and a growing number of offshore windmills/wind farms that need protection from drifting ships.

Early observation of drifting vessels, good models for calculation of drifting paths in all types of weather, combined with available emergency towing vessels positioned to reduce transit time to a disabled drifting vessel.are all part of the solution to this problem.

2. Drifting ship statistics

The Norwegian Coastal Administration gathers information on all types of incidents for vessels sailing within the Norwegian EEZ. They produce annual reports as shown in Fig. 1. The figure compares incident types for the years 2020–2022. The two largest incident types are drifting vessels and drifting objects. From the analyses of vessel traffic in Norwegian waters, it is documented that the frequency of drifting ships has been approximately 0.15% of the registered ship transits, or 150–165 drifting cases annually. In addition to a possible grounding when a vessel reaches the coastline, the increased use of the ocean space for wind farms, floating solar installations, and fish farming structures (moving from sheltered coastal sites to exposed locations) increases the risk of collision with different types of ocean structures, including offshore oil and gas installations.

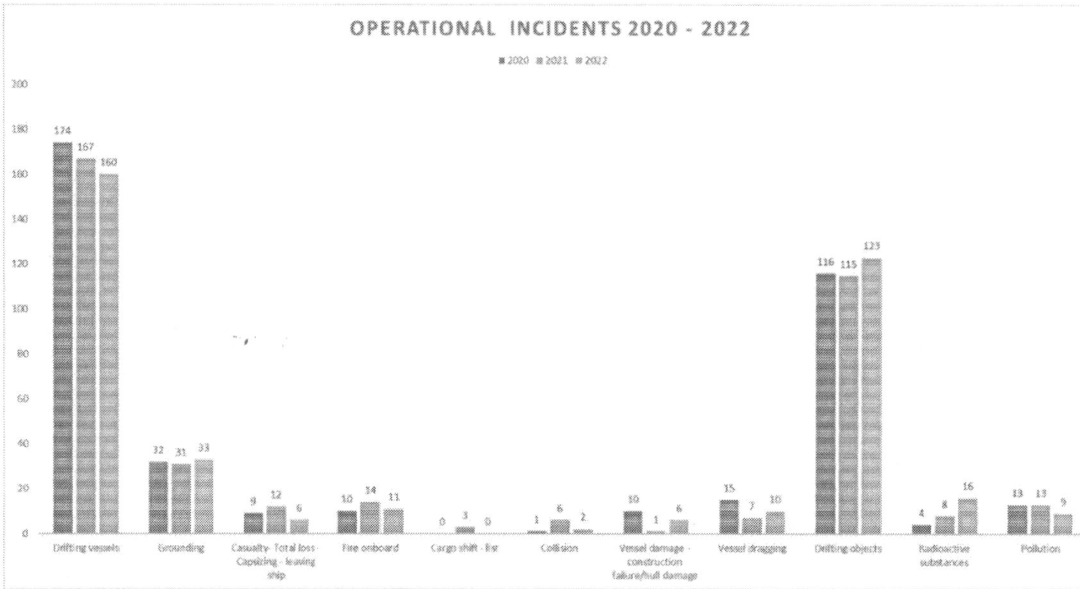

Fig. 1. Incident statistics collected by the Norwegian Coastal Administration's Ocean and Coastal Vessel Monitoring Vardø division [4].

3. Drifting ships – emergency towing cases

3.1. Cases in Norwegian waters – 2021–2023

3.1.1. MV Melinda. The bulk vessel MV Melinda (LOA 190m, GT 32 839) was in transit from Riga to Murmansk. As the ship sailed through the Norwegian EEZ, it had reported to and was in the Barents Ship Reporting System [5]. On January 17, 2022, at 14.45 hours, the ship called NOR VTS using VHF, stating that the ship had some problems with the rudder and steering system. The master informed NOR VTS that they needed some time to test the system. Two hours later, he called for assistance. At that time, the vessel was in the TSS 37 nm off Andøya (see Fig. 2). The weather was deteriorating and the risk of drifting into the coastline was increasing. Using the BarentsWatch system, the VTS got information on other vessels close to MV Melinda. From the system, it was found that Norwegian Coast

IOP Conf. Series: Materials Science and Engineering 1288 (2023) 012001 doi:10.1088/1757-899X/1288/1/012001

Guard Vessel KV Harstad was the closest vessel. This Coast Guard vessel happens to be one of the vessels in the Norwegian Emergency Towing Services [6].

Based on available information, the Norwegian Coastal Administration declared the MV Melinda drifting as a possible case for a major pollution incident. Resources from the Joint Rescue Coordination Centre (JRCC in Bodoe), Norwegian Coast Guard, Norwegian Coastal Administration, and Norwegian Navy were called to set up an emergency response group. To give the response group background information for decision-making about a possible rescue and pollution prevention, NOR VTS performed a drifting path estimation using a tool found in the restricted part of BarentsWatch. More information on this tool is presented in Section 4.2. The results are shown in Fig. 3. Based on the outcomes from the drifting path study, it was seen that the vessel could drift into the Norwegian Coast and result in a need for a SAR operation and finally a major pollution accident.

The ship owner, after discussions with the master, asked a UK salvage company to assist the drifting vessel. A local tug company in Tromsoe (with a small ocean-going tug), got a contract to take control of the drifting ship. However, this tug had an expected ETA of 10 hours, which was estimated to be too long before an attempt to establish an emergency towing line could start. The governmental emergency response team then decided that a government-led operation had to be initiated. NCG KV Harstad was given the responsibility to be the On-Scene Commander. During the first contact between the two masters, the MV Melinda master rejected a towing line from the Coast Guard vessel. NOR VTS then contacted the UK Salvage company with the information that NCG KV Harstad was representing NETS. In the following contact between the two masters, the one on MV Melinda agreed to set up an emergency towing line to the Coast Guard vessel.

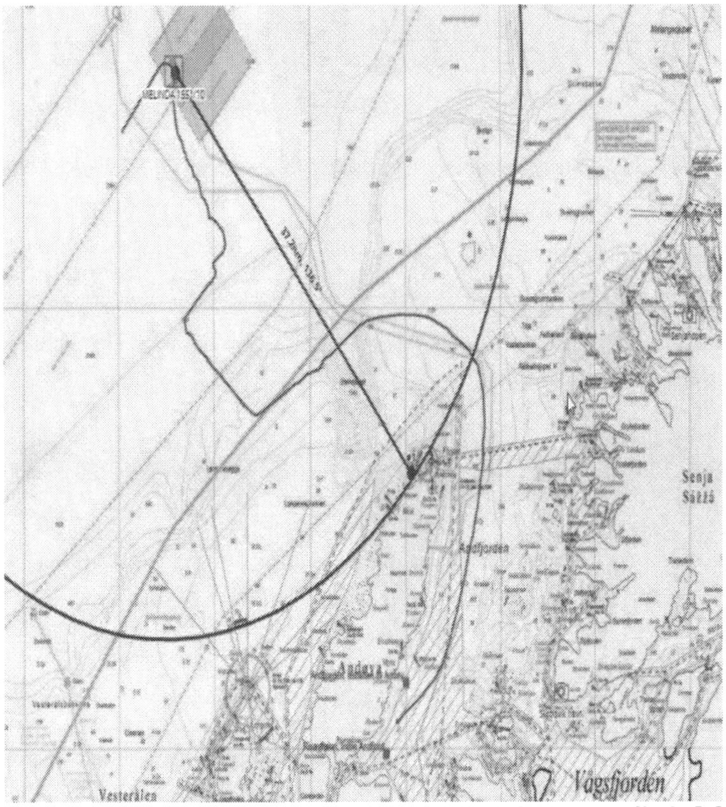

Fig. 2. Starting position, drifting track and towed track (red line) for MV Melinda [7].

IOP Conf. Series: Materials Science and Engineering 1288 (2023) 012001 doi:10.1088/1757-899X/1288/1/012001

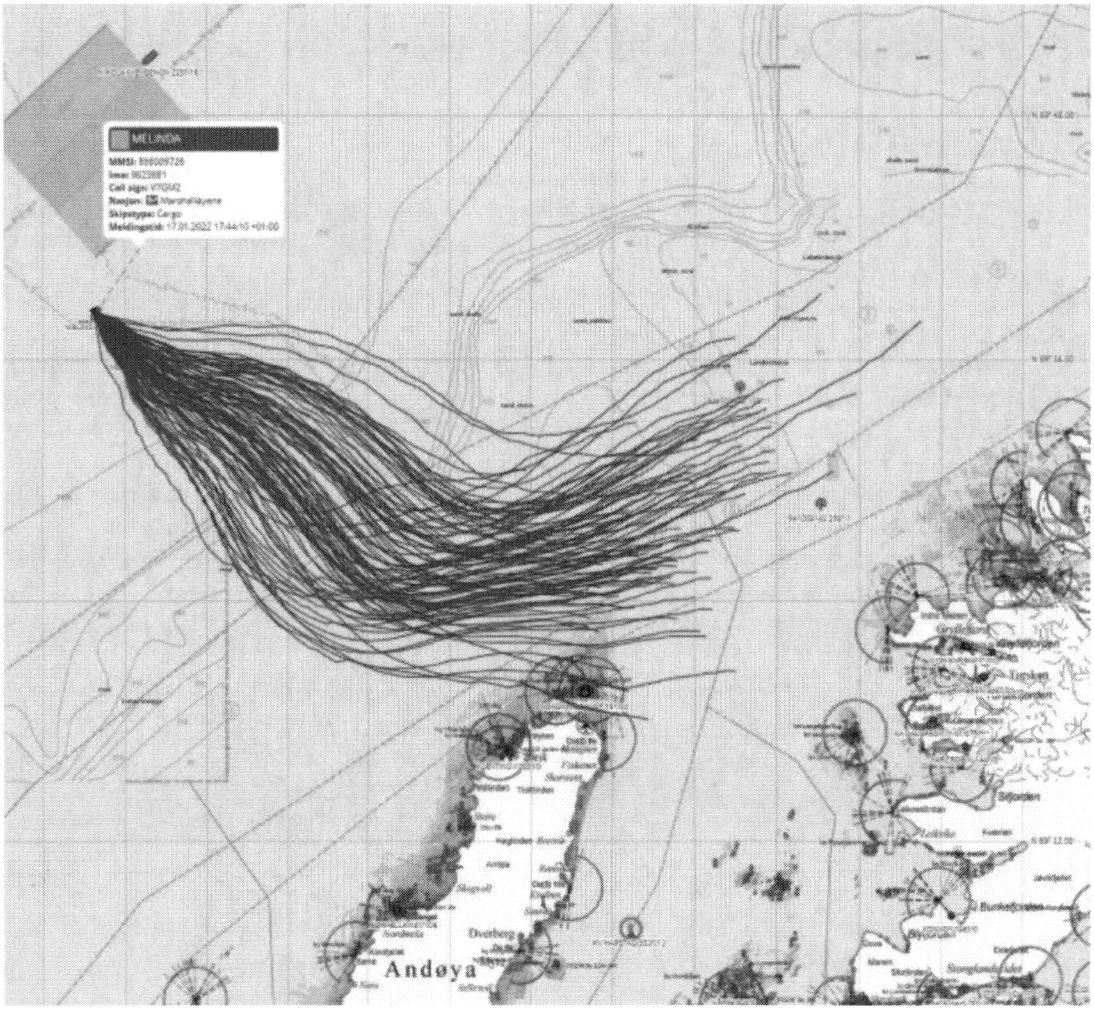

Fig. 3. Drifting path ensemble – varying initial conditions and weather parameters (courtesy BarentsWatch/Norwegian Coastal Administration) [7].

The weather conditions when the emergency towing line was set up:

- Wind: NNW 12–17 m/s
- Sea: 4.5–5 m waves.
- Snow showers

while MV Melinda was drifting SSE at a speed of 3–3.5 knots. Fig. 4 illustrates the preparation of the emergency towing gear on KV Harstad prior to the transfer of the emergency towing connection.

A detailed description of the MV Melinda incident was presented in a meeting arranged by SINTEF Ocean on behalf of a national competence network (Kompetanseforum for krevende fartøysoperasjoner) in June 2022 [7]. The first attempt to transfer an emergency towing line from KV Harstad failed. The gun line was shot to MV Melinda, but the line broke when the transfer of the messenger line started. The second attempt was successful, the uncontrolled drifting stopped, and the disabled vessel was towed into sheltered waters. During this part of the operation, it was reported that the rudder on the towed vessel was locked 15 degrees starboard, leading to a shearing motion with rather high tension in the towing line. The towing speed was reduced; then, the towed vessel lined up with the towing vessel. More information, including pictures and video clips, has been published by the Norwegian Coastal Administration [8] and Norwegian Broadcasting Corporation [9].

Fig. 4. Preparing the emergency towing gear on KV Harstad (Photo: Bjørn Amundsen, Norwegian Coast Guard).

3.1.2. MT Bergstraum. In November 2022, the coastal tanker MV Bergstraum (LOA 123 m) was sailing along the western coast of Norway, heading for the port at Elnesvågen. Early in the morning, the vessel had a technical failure leading to an engine blackout. The vessel had a position approximately 0.6 nm off the coast (see Fig. 5). Luckily, the weather was calm (including a low current speed) when the blackout happened. A Coast Guard vessel (KV Njord) was close by and able to swiftly establish an emergency towing line, assisted by the crew on the tanker. Towing into the port met no problems.

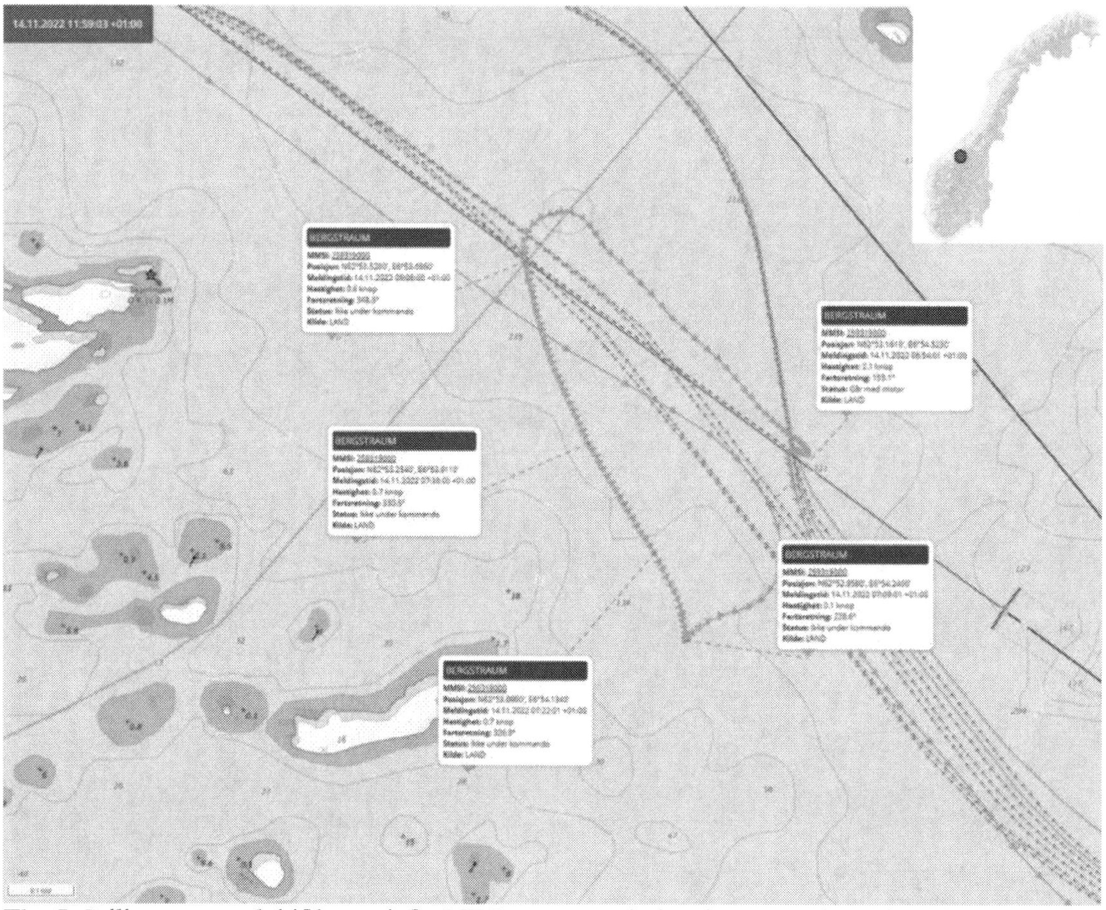

Fig. 5. Sailing route and drifting path for MT Bergstraum during the approach to Elnesvågen. The vessel goes from the north-west along the orange path with triangles. At 7.09 am the vessel was freely drifting until 8.06 am, when the daughter craft from KV Njord arrived.

3.2. An international case

Early in the morning on January 31, 2022, the tanker PECHORA STAR collided with bulk carrier JULIETTA D at Ijmuiden outer anchorage, North Sea. The bulk carrier was damaged, resulting in water ingress in the engine room. The vessel started drifting in harsh weather conditions. To prevent a possible emergency rescue operation, JULIETTA D's crew were lifted off by helicopters. The vessel drifted and made contact with a wind farm grid transformer platform, and her hull was damaged for the second time. In the morning the following day, the OSV SOVEREIGN was able to establish an emergency towline on the unmanned bulk carrier. It was safely towed to Rotterdam, for more information see [10].

A result of this and other incidents was that the Netherlands decided to allocate two Emergency Towing Vessels, primarily to prevent drifting ships from entering sea areas with offshore wind farms.

4. Numerical models for calculation of drifting paths for ships

4.1. Review of existing models

Important parameters for vessel drift calculation:

Hydrodynamic vessel model – draught/trim, bow/stern shape, wave and current forces.

Aerodynamic vessel model – Superstructure volume and distribution, longitudinal/transverse projected area, heel angle, drift angle, wind forces.

Quality of forecasted environmental parameters – current, waves, and wind.

4.2. SINTEF Ocean models

4.2.1. Quasistatic model. The ShipDrift code is a 3 DOF static solver where environmental forces (wind, wave drift, and current) are balanced against the hull forces caused by moving the vessel through the water. The solver calculates the steady-state scenario and returns the ship's heading, steady-state drift velocity, and direction. Environmental directions/magnitude (wind speed, wave height, etc.) can be changed whenever the user likes, and the program will recalculate the new steady state and continue drifting. The program solves the equilibrium calculations quickly, making it very well suited for a large number of drifting scenarios to generate data for the user to take a statistical approach to where the ship will end up after drifting for a given time in a changing environment.

Environmental forces are calculated based on pre-calculated force coefficients scaled to forces using the actual main dimensions of the vessel. The program contains a predefined set of force coefficients for a typical vessel and can be extended with other force coefficients from different sources.

The quasistatic model can also tune the drifting paths based on actual drift if the user has real-time or historical drifting data from the ship. This functionality is useful for emergency preparedness to best use available data to predict the drifting path of a disabled vessel.

4.3. Wind forces

The static wind forces and moments are calculated as follows:

Surge: $$F_x = C_x(\alpha) \cdot \frac{1}{2} \cdot \rho_{air} \cdot U^2 \cdot A_{transverse}$$

Sway: $$F_y = C_y(\alpha) \cdot \frac{1}{2} \cdot \rho_{air} \cdot U^2 \cdot A_{lateral}$$

Yaw: $$M_n = C_n(\alpha) \cdot \frac{1}{2} \cdot \rho_{air} \cdot U^2 \cdot A_{lateral} \cdot L_{OA}$$

The symbols are defined in the following table.

Symbol	Description	Unit
F_x	Wind force in surge for the relative wind heading.	N
F_y	Wind force in sway for the relative wind heading.	N
M_n	Wind moment in yaw for the relative wind heading.	Nm
C_x	Wind coefficient in surge for the relative wind heading.	-
C_y	Wind coefficient in sway for the relative wind heading.	-
C_n	Wind coefficient in yaw for the relative wind heading.	-
α	Wind direction relative to the ship's heading.	deg
U	Wind velocity.	m/s
ρ_{air}	Density of air. Defined to be 1.225.	kg/m^3
$A_{transverse}$	Transverse area of ship superstructure.	m^2
$A_{lateral}$	Lateral area of ship superstructure.	m^2
L_{OA}	Length overall.	m

4.4. *Current forces*

The static current forces and moments are calculated as follows:

Surge: $$F_x = C_x(\alpha)\cdot\frac{1}{2}\cdot\rho\cdot U^2\cdot B\cdot T$$

Sway: $$F_y = C_y(\alpha)\cdot\frac{1}{2}\cdot\rho\cdot U^2\cdot L_{PP}\cdot T$$

Yaw: $$M_n = C_n(\alpha)\cdot\frac{1}{2}\cdot\rho\cdot U^2\cdot L_{PP}^2\cdot T$$

The symbols are defined in the following table.

Symbol	Description	Unit
F_x	Current force in surge for the relative current heading.	N
F_y	Current force in sway for the relative current heading.	N
M_n	Current moment in yaw for the relative current heading.	Nm
C_x	Current coefficient in surge for the relative current heading.	-
C_y	Current coefficient in sway for the relative current heading.	-
C_n	Current coefficient in yaw for the relative current heading.	-
α	Current direction relative to the ship's heading.	deg
U	Current velocity.	m/s
ρ	Density of water.	kg/m^3
B	Beam at midships.	m
T	Draught at midships.	m
L_{PP}	Length between ship perpendiculars.	m

The wind-generated current velocity vector and the tidally generated current velocity vector are added, and the resultant velocity vector is used when calculating current forces.

4.5. *Wave forces*

The static wave drift forces and moments are calculated as follows:

Surge: $$F_x = 2\cdot\frac{\rho\cdot g\cdot B^2}{L_{PP}}\cdot\int_0^\infty S(\omega)\cdot C_x(\omega,\alpha)\cdot d\omega$$

Sway: $$F_y = 2\cdot\frac{\rho\cdot g\cdot B^2}{L_{PP}}\cdot\int_0^\infty S(\omega)\cdot C_y(\omega,\alpha)\cdot d\omega$$

Yaw: $$M_n = 2\cdot\rho\cdot g\cdot B^2\cdot\int_0^\infty S(\omega)\cdot C_n(\omega,\alpha)\cdot d\omega$$

Where

$$C_x(\omega,\alpha)=\frac{\overline{F}_x(\omega,\alpha)}{\dfrac{\rho\cdot g\cdot B^2\cdot\zeta^2}{L_{PP}}}$$

$$C_y(\omega,\alpha)=\frac{\overline{F}_y(\omega,\alpha)}{\dfrac{\rho\cdot g\cdot B^2\cdot\zeta^2}{L_{PP}}}$$

$$C_n(\omega,\alpha)=\frac{\overline{M}_n(\omega,\alpha)}{\rho\cdot g\cdot B^2\cdot\zeta^2}$$

The symbols are defined in the following table.

Symbol	Description	Unit
F_x	Wave drift force in surge for the relative wave heading.	N
F_y	Wave drift force in sway for the relative wave heading.	N
M_n	Wave drift moment in yaw for the relative wave heading.	Nm
C_x	Wave drift coefficient in surge for the relative wave heading.	-
C_y	Wave drift coefficient in sway for the relative wave heading.	-
C_n	Wave drift coefficient in yaw for the relative wave heading.	-
α	Wave direction relative to the ship's heading.	deg
S	Wave spectrum.	-
ρ	Density of water.	kg/m^3
g	Gravity.	m/s^2
ζ	Wave amplitude.	m
B	Beam at midships.	m
L_{PP}	Length between ship perpendiculars.	m

4.5.1. Use of model on a drifting ship – general cargo ship. The quasistatic model described above is implemented and in use by the Norwegian Coastal Administration to plan emergency response and as a tool during incidents to predict the drifting path of the vessel. When used to predict the drifting path of a vessel, the model is tuned based on the automatic identification system (AIS) of the initial phase of drifting. The use of this model when Eemslift Hendrica started drifting on April 5, 2021, 60 nm west of Ålesund, Norway, is described by Berg et al. [3].

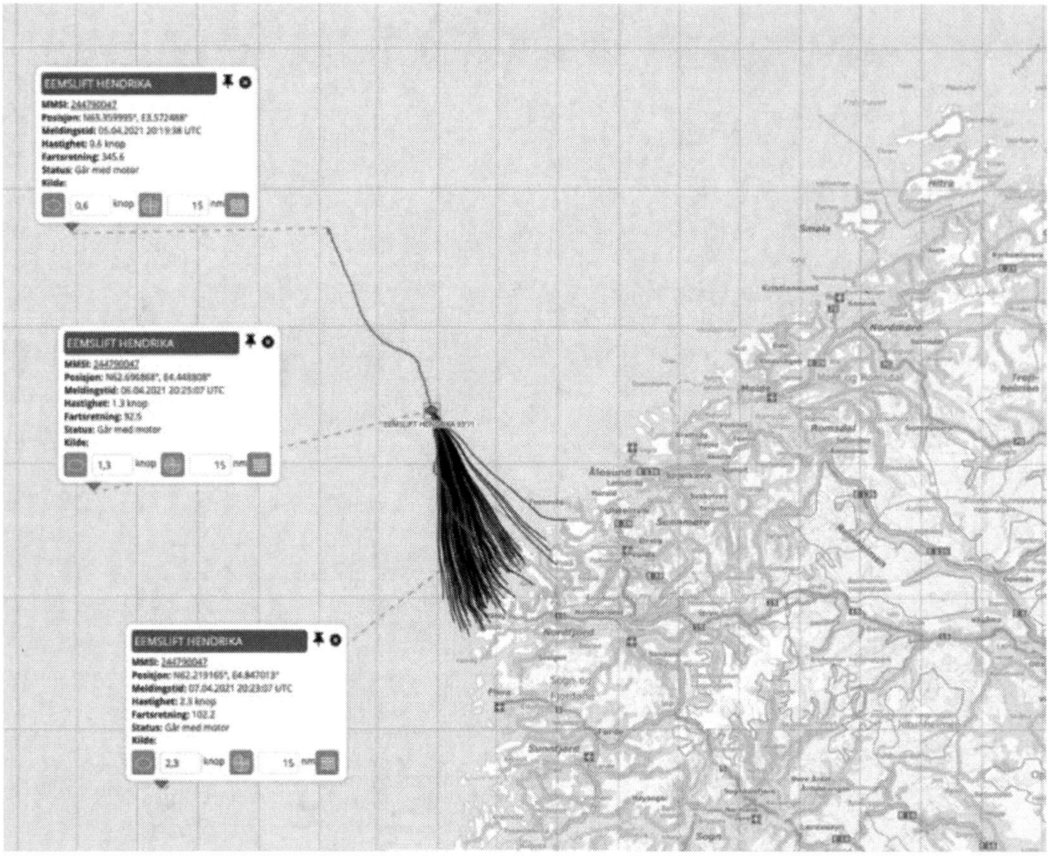

Fig. 6. Sailing actual drift and predicted drift for Eemslift Hendrica (Figure from [3]).

5. Application of drifting path models in the Norwegian governmental maritime emergency preparedness system

NOR VTS is responsible for monitoring ship traffic in the Norwegian Exclusive Economic Zone (EEZ). The incident statistics in Fig. 1 are compiled by their «Ocean and Coastal Vessel Monitoring Vardø» division. Part of their work is to identify drifting vessels and contact them to find out if they need assistance. Often the response is that the situation onboard is under control. However, as shown in figure 1, there is a significant number of vessels needing assistance to prevent them from drifting ashore or into areas with different types of ocean structures. If NOR VTS evaluate that the risk for an incident is high, they may initiate a governmental emergency response action. When there is a need to use state towage service, NOR VTS notifies the Norwegian Coast Guard. The Coast Guard has since 2020 been the operational part of the Norwegian Emergency Response Service (NETS) and will then dedicate vessels to the mission.

A part of the agreement between NCA and the Coast Guard includes a risk-based approach to position and mobilize the state towing capacities. Six vessels have been dedicated to NETS. Two of the vessels are anchor handling tugs (approximately 270 tonnes bollard pull) and four are ordinary Coast Guard vessels (approximately 100 tonnes bollard pull). Based on weather forecasts, NOR VTS and the Coast Guard agree on the position the emergency towing vessels to optimize their position with respect to the traffic pattern and specific vessels that have a large pollution potential in case of an accident.
When a critical vessel drift scenario develops, the emergency response team applies a dedicated tool for the study of possible drifting tracks. SINTEF Ocean has in collaboration with NCA, and the Coast Guard developed a tool for prediction of possible drifting paths using the quasistatic tool described in subsection 4.2. Through variation of the starting condition for the simulation tool, and different

environmental parameters (wind, waves and current), an envelope of predicted paths is produced. The outcome is then used by the master on the emergency towing vessel to optimize the sailing route to the drifting vessel.

6. Emergency towing of disabled drifting vessels

6.1. International requirements – onboard equipment for emergency towing

IMO resolution MSC.35(63) [10], adopted on May 20, 1994, requires an emergency towing arrangement only for oil, chemical, and gas tankers of 20 000 tonnes deadweight and above. The arrangements shall be fitted both forward and aft of the vessel and should always be capable of rapid deployment in the absence of main power on the ship to be towed and easy connection to the towing vessel. Strength calculation for the strongpoint and fairleads is to be based on IMO MSC/Circ. 1175 [11].

For all other cargo ships above 500 GT, new requirements were included in SOLAS based on two IMO documents [12], [13]. For such ships, a ship-specific emergency towing procedure should be developed. It has to be carried aboard the ship, and the procedure is to be based on the ship's existing equipment. The procedure contains a section on onboard training for personnel responsible for deploying the emergency towing gear.

6.2. Calculation of required towing force for a disabled vessel

The required towing force (RTF) is the force necessary to hold or tow the vessel in the environmental conditions of waves, wind, and current. The RTF can be explained by the following equation:

$$RTF = R_{Calm} + R_{Wind} + R_{Wave}$$

where

R_{Calm} = Hull resistance in calm water
R_{Wind} = Wind resistance
R_{Wave} = Hull resistance in waves

Calm water resistance is what a vessel would experience in the event of totally calm weather conditions, i.e., no waves excluding the waves created by the ship. The calm water hull resistance typically consists of frictional resistance due to the viscosity of the water, which creates friction with the hull of the ship and depends, among other things, on the cleanness of the hull. There is also resistance from the waves made by the vessel, depending highly on the hull shape. The resistance from the propeller and eventually the rudder can also be considered as calm water resistance. The resistance depends on whether the propeller or rudder is installed or not, locked, or free to rotate. A factor that tends to influence water resistance is the depth of the water, as in shallow waters, the water has greater difficulty moving aftwards.

The wind resistance during emergency towing of a disabled vessel depends, among other things, on the wind velocity and projected area. The wind resistance may be significant depending on the size of the disabled vessel and the severe wind conditions often experienced by drifting vessels.

The towed vessel is, in addition to the wind resistance and calm water resistance, subject to wave drift forces. The mean wave drift forces can be estimated through either direct pressure integration or conservation of fluid momentum. The wave drift force is often calculated without forward speed; however, forward speed will increase the forces. The mean wave drift force is proportional to the wave amplitude squared and linear with the area subjected to the waves.

The tow force required to tow a containership under emergency conditions was studied by Shigunov and Schellin [14]. Five containerships of different sizes were considered in steady-state wave, wind, and current conditions. Tow forces were obtained for the wind and current aligned with the primary direction of the waves, wind deviating 30 degrees from the wave direction, and current deviating up to 90 degrees from the waves. The direction of the tow force deviated up to 90 degrees from the vessel heading. A formulation estimating the maximum towing force as a function of the significant wave height and the ship length between particulars, Lpp, was proposed. The formulation was based on the equilibrium equations of the towed vessel in the horizontal plane.

6.3. Effective bollard pull towing a disabled vessel

The estimated required towing force can be compared to the tug's bollard pull. A tug's bollard pull is normally measured in a test in calm water and limited current and wind conditions. During the test, the tug's trim may also have been optimized, and sufficient depth and clearance for a free propeller stream are ensured. The calm weather, optimized trim, and sufficient clearance for a free propeller stream cannot be achieved in a real operation and need to be considered when comparing the required towing force with the strength of the tug.

During an emergency tow, the weather may be harsh, i.e., the power required for the tug itself to achieve forward speed in the waves can be considerable. Further, the distance between the towed object and the tug may be short, and the tug's propeller may induce propeller wash at the towed object, increasing the towing resistance.

6.4. Connecting to a disabled vessel

The Norwegian Coast Guard's 2022 Presentation of the MV Melinda incident [7] explained the emergency towing tug's arrival procedure at the location of the disabled vessel MV Melinda. When the tug arrived at the vessel's location, the tug circled around the ship to check its condition, draughts, and trim. Different ships will lay to weather, seas, and current differently and will drift to the lee side at different rates, so the tug laid ahead and astern of the ship to see how the ship and tug drifted and laid relative to each other at each location. The tug decided to lay to the side of MV Melinda's bow to connect the emergency towing equipment to the bow of MV Melinda. If the weather is heavy, the disabled vessel is not in immediate danger of going aground, and it is in the interest of safety, the tug may choose to delay the tow connection until weather and sea conditions improve.

Systems to provide a fast and safe setup of an emergency towing connection under different environmental conditions are required only for oil, chemical, and gas tankers of 20 000 tonnes deadweight and above [15]. The tug's connection to the disabled vessel will depend on the towing arrangement identified in the disabled vessel's Emergency Towing Book (ETB) if existing. Preferably prior to departure, the tug should obtain a copy of the ship's ETB and talk to the vessel master to ensure that the towing gear is ready to be deployed and determine how to transfer the equipment to the tug. Alternatively, the tug may use its own equipment. How and where the disabled vessel's towing equipment is stored, the condition of the equipment, and the time taken by the disabled vessel to assemble the equipment are unknown. For the MV Melinda incident [7], the tug used its own towing equipment to be connected to the disabled vessel.

Since emergency ship towage varies, it is difficult to outline exactly how the tug should connect to a disabled vessel, and this is one of the reasons why the competence of the crew on the emergency towing vessel and the ship to be towed is a critical factor during the transfer of the towing equipment.

The transfer of the towing line normally starts with the tug or the disabled vessel receiving a messenger line. The messenger line can be thrown manually, or by use of a line-throwing gun. The messenger line needs to be passed through the closed chock at the bow or stern of the disabled vessel and not on top of the bulwark. At the end of the messenger line, the towing line or the bridle is connected. The towing line or bridle is connected to the disabled vessel's mooring system or similar on the deck and to the tug's main towing line. The messenger line is removed. Preferably, winch assisting is available to transfer the equipment. If winch assisting is not available on the disabled vessel, the transfer can be performed manually by the crew onboard the disabled vessel; eventually, the messenger line can be passed from the tug back to the disabled vessel via one of the bollards on the disabled vessel so the transfer can be performed by one of the winches onboard the tug. Another alternative if the disabled vessel's towing arrangement is used is to connect a floating buoy at the end of the messenger line to be picked up by the tug. The tug then transfers the towing arrangement on the deck of the tug and connects to its towing wire.

If there is no crew onboard the disabled vessel, the connection of the towing arrangement is more challenging, especially if the crew members have not deployed the emergency towing arrangement prior to abandoning the vessel.

The connection to the disabled vessel can either be a bridle connection or a single line. From a towing direction stability point of view, a bridle connected to the bow of the vessel is preferred.

Communication between the tug and the disabled vessel is crucial for a successful connection. Often, ships route all external communication through the ship's bridge and then on to the working deck crew. The master onboard the tug for the Melinda incident [7] also pointed out language barriers to communication.

6.5. Forces on the towing line when performing an emergency tow

The total towline tension during towing can be split into static tension and dynamic tension. The static tension can be estimated by parabolic equations for cable slope based on the horizontal towing resistance, the distance between the tug and the towed vessel, and the distributed submerged weight of the towing cable. Estimation of the dynamic tension requires a mathematical model of the complete towing system considering independent motion analysis of the tug and towed object in a given sea state to achieve the resulting towline forces.

The stiffness of the towline controls the dynamic tensions in the towing line in addition to the slow-varying motions between the tug and the towed object. The dynamic forces on the towing line are highly dependent on the towline stiffness. The stiffness can be split into two components, one due to elastic elongation of the line and one due to the change of geometry of the towline catenary. Elastic elongation is defined by the modulus of elasticity, which depends on the material of the towing line. During towing, it is desirable that geometrical stiffness dominates, i.e., the towline characteristics are weight-dominated by ensuring a parabolic shape of the towing line. With an increased distance between the tug and towed object for a towing line with constant length, the dynamic tension increases in the towing line, so the working line's characteristics get more and more dominated by the elastic elongation. A towing line without sag is only dominated by the elastic elongation and may give high peak tensions in the towing line.

For emergency towing, the towing line material is often chosen based on a handling point of view, not the towing line's characteristics. The synthetic ropes used to connect to disabled vessels provide no or little geometrical stiffness. To achieve geometrical stiffness and sufficient distance between the tug and towed vessel, it is recommended to connect a steel towing wire to the synthetic rope if the rope used to connect to the disabled vessel is used during the towing operation.

6.6. Strength loss in the towing equipment during towage

In comparing the strength of the towing equipment with the dynamic towing force, the strength loss of the ropes when loaded around a curved surface, typical towing connection points, Panama chocks, and connection between two towing lines need to be considered. More severe bends, i.e., small surface radii, can result in great strength loss.

The strength of tow assemblies can also be reduced due to heating up when in use due to internal and external friction. Due to their elastic properties, synthetic ropes will heat up more and faster than wire ropes. Internal friction is the force-resisting motion between the fibres in a line while it undergoes deformation due to bending of the line over a curved surface and due to the cycling tension load. External friction is the chafing between a line and a surface when these move against each, e.g., the towing line through a fairlead or at the connection point between components in the towing line assembly. All synthetic fibre towing assembly components under tension can become damaged if exposed to contact surfaces. Damage to steel wire is less likely, but broken strands are still possible. Chain is the preferred material through areas where chafing can occur, e.g., towing connection points, fairleads, and the deck edge. However, chain is often not suitable for emergency towing due to its heavy weight for handling and is replaced with synthetic line for emergency towing. The synthetic lines need to be protected by using a protective line cover, sleeve, or similar in chafing areas. The total towline tension during towing can be split into static tension and dynamic tension.

7. Summary and recommendations

The number of drifting ships in the Norwegian EEZ is approximately 0.15% of the vessels sailing in that area. Industrialization of coastal waters through installation of ocean wind farms, solar power plants, and exposed fish farms requires improved maritime emergency preparedness. National and regional capability to handle drifting vessels is important.

As part of the Norwegian maritime emergency preparedness system, SINTEF Ocean has delivered a tool for calculating the path of drifting vessels. The tool is implemented in the closed part of the BarentsWatch portal. In case of an incident with a drifting vessel, the tool generates an ensemble of paths from a variety of start conditions for the drifting vessel, a generic vessel category, and the available weather forecast.

Emergency towing to prevent a disabled vessel from colliding with ocean structures or grounding in coastal waters is a challenging operation in harsh weather. The lack of an international requirement for an onboard emergency towing system for SOLAS ships (except tankers over 20 000 dwt) complicates the work to establish an emergency towing line, especially in harsh weather situations.

The authors recommend the following actions:

- Increased international sharing of knowledge and experience related to emergency towing operations.
- Further development of ship-specific models for prediction of paths for a drifting vessel.
- Follow-up of onboard training according to the vessel's Emergency Towing Procedures/Emergency Towing Book.
- International collaboration work to develop input for guidelines on Emergency Towing Systems on ships not presently covered by IMO requirements.

References

[1] T. E. Berg, Ø. Selvik, and O. K. Jordheim, "Norwegian Emergency Towing Service – Past – Present and Future," *TransNav*, vol. 14, no. 1, pp. 83–88, 2020, doi: 10.12716/1001.14.01.09.

[2] T. E. Berg and Ø. Selvik, "Changes in and Recent Experiences from Norwegian Emergency Towing Service (NETS)," *TransNav*, vol. 15, no. 1, pp. 203–208, 2021, doi: 10.12716/1001.15.01.21.

[3] T. E. Berg, Ø. Selvik, K. Aasebø, and S. Sveinungsen, "Norwegian Tools for Maritime Emergency Response, Planning and Operation Phases," presented at the 1st International Conference on the Stability and Safety of Ships and Ocean Vehicles, Glasgow, Scotland, UK, Jun. 2021.

[4] "Annual operational incident report 2022," Norwegian Coastal Administration, Vardø, Norway, 2023.

[5] "Barents Ship Reporting System." Norwegian Coastal Administration, Apr. 25, 2023. Accessed: Apr. 25, 2023. [Online]. Available: https://kystverket.no/en/sea-transport-and-ports/safeseanet-norway/barents-srs

[6] "Norwegian Emergency Towing Service." Norwegian Coastal Administration, Apr. 25, 2023. Accessed: Apr. 25, 2023. [Online]. Available: https://kystverket.no/en/preparedness-and-emergency-response-against-acute-pollution/responsibilities-roles-and-resources/norwegian-emergency-towing-service/

[7] B. Amundsen, "Presentation of the MV Melinda incident (in Norwegian)," presented at the Kompetanseforum for krevende fartøysoperasjoner: Krevende nødslepoperasjoner, Digital meeting, Jun. 16, 2022. Accessed: Apr. 30, 2023. [Online]. Available: https://www.sintef.no/arrangementer-og-kurs/arkiv/2022/kompetanseforum-for-krevende-fartoysoperasjoner-nod-og-havslep/

[8] "Incident description - MV Melinda." Apr. 25, 2023. Accessed: Apr. 25, 2023. [Online]. Available: https://www.kystverket.no/nyheter/2020/fartoy-pa-190-meter-i-drift--ulykke-avverget/
[9] "Incident review." Norwegian Broadcasting Corporation, Jan. 20, 2022. Accessed: Apr. 25, 2023. [Online]. Available: https://www.nrk.no/nordland/fartoyet-melinda-pa-190-meter-mistet-styringen-_-ble-slept-i-land-1.15817883
[10] "Bulk carrier JULIETTA D towed to safety, was abandonment justified? UPDATES," *Maritime Bulletin*, Feb. 07, 2022. https://www.maritimebulletin.net/2022/01/31/bulk-carrier-abandoned-sinking-after-she-was-struck-by-tanker-netherlands/ (accessed May 01, 2023).
[11] "Revised guidance on shipboard towing and mooring equipment MSC.1/Circ.1175/Rev.1." International Maritime Organization, 2020. Accessed: Apr. 29, 2023. [Online]. Available: https://wwwcdn.imo.org/localresources/en/Documents/MSC.1-Circ.1175-Rev.1.pdf
[12] "Adoption of amendments to the international convention for safety at sea 1974 as amended MSC.256(840)." International Maritime Organization, 2008. Accessed: Apr. 29, 2023. [Online]. Available: https://wwwcdn.imo.org/localresources/en/KnowledgeCentre/IndexofIMOResolutions/MSCResolutions/MSC.256(84).pdf
[13] "Guidelines for owners/operators on preparing emergency towing procedures MSC.1/Circ.1255." International Maritime Organization, 2008.
[14] V. Shigunov and T. E. Schellin, "Tow Forces for Emergency Towing of Containerships," 2014, doi: 10.13140/2.1.1389.7603.
[15] "Adoption of guidelines for emergency towing arrangement on tankers. IMO resolution MSC.35(63)." International Maritime Organization, 1994. Accessed: Apr. 29, 2023. [Online]. Available: https://wwwcdn.imo.org/localresources/en/KnowledgeCentre/IndexofIMOResolutions/MSCResolutions/MSC.35(63).pdf

Numerical PMM test in shallow water using CFD method

Zhang Zhu[1,2], **Yonghwan Kim**[1], **Jun-Hoi Choi**[1]

[1] Department of Naval Architecture and Ocean Engineering, Seoul National University, 1 Gwanak-ro, Gwanak-gu, Seoul, 08826, Republic of Korea
[2] Corresponding author, zhuzhang@snu.ac.kr

Abstract. This study aims to the estimation of the ship manoeuvrability in shallow water by using the numerical PMM approach, which analyses the hull hydrodynamic responses and manoeuvring performance from the viewpoint of coefficients. The test model is a scaled KVLCC2 tanker operating at a low Froude number of 0.0514. Static drift test and pure yaw test are considered, and the water depths vary from deep water to very shallow water of 20% under-keel clearance. The numerical tank is built based on an OpenFOAM+ package. The free surface is modelled by volume-of-fraction method, and the k-ω SST model associating with wall functions is applied for turbulence. After the convergence tests and validation works, the numerical tank is employed to investigate the wave-depth effect on hydrodynamic forces and manoeuvring coefficients. The present results show that the forces of static drift test, particularly sway force, can be significantly changed as the water depth decreases.

1. Introduction

The navigational challenge posed by the shallow water problem in harbours or inland waterways is an ever-present concern for maritime industry. The confined water regions encountered in these environments can significantly obstruct a ship's manoeuvrability, which is a crucial performance metric for vessels. As modern merchant vessels continue to grow in size, the impact of the shallow water problem is becoming even more pronounced. Furthermore, the limitation of harbour infrastructure further underscores the significance of assessing the ship manoeuvrability in confined water regions.

Many recent studies have been carried out to evaluate the manoeuvring performance of ships in confined water using experiment, empirical formulae, data-driven analysis, and numerical computation [1, 2, 3, 4]. Toxopeus [5] presented a benchmark study on the PMM tests of KVLCC2 tanker in various water depths. Several RANS solvers have been employed for comparing the body forces and flow fields with the experimental PMM tests in towing tank. The recent SIMMAN workshop [6] also announced the benchmark work on the ship manoeuvrability in shallow water, which involves the hull's manoeuvring coefficients and hull-propeller-rudder interaction of multiple vessels in various shallow-water depths. KVLCC2 tanker is one of the benchmark models used for the workshop.

In this study, a numerical tank that is built based on an OpenFOAM+ package is employed to carry out the numerical PMM tests in finite depths, with a particular focus on shallow water. The test model is the KVLCC2 tanker manoeuvring at low speed. Both static drift test and pure yaw test are carried out for water depths ranging from deep water to very shallow water. Prior to investigating the water-depth effect, the numerical convergence tests and validation works in shallow-water conditions have been carried out for the CFD solver. Then, a series of PMM tests are conducted to observe the effect of water depth on hydrodynamic forces and manoeuvring coefficients.

2. Test overview

2.1. PMM tests

Within the inertia coordinate system o_e–$x_ey_ez_e$, the body coordinate system o–xyz of right-hand rule is used to specify ship PMM motions and hydrodynamic forces. Figure 1 describes two types of PMM tests considered in this study. The origin of body coordinate system is located at the midship for virtual towing motion and the calculation of yaw moment. Body coordinate system is fixed to the moving body. The positive x axis denotes the forward motion direction, the positive y axis indicates starboard side, and the positive z axis is vertically downwards.

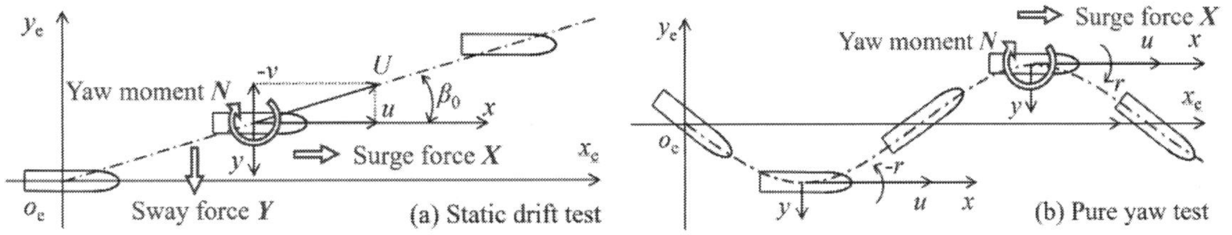

Figure 1. Description of PMM tests.

Two types of PMM tests are involved in this study. One is the static drift test which is characterized by the drift angle β_0, as shown in Figure 1 (a). Equation (1) defines the drift angle based on body surge velocity u_0 and sway velocity v_0. The other is the pure yaw test which is characterized by a harmonic yawing motion, as shown in Figure 1 (b). The yawing ship has a sine yaw rate as Equations (2) and (3), and the maximum yaw rate r_0 is determined by the maximum orientation ψ_0 and harmonic frequency ω_0. In this study, the harmonic frequency is fixed to be 0.3142 for all pure yaw tests.

$$\beta_0 = \tan^{-1}\left(\frac{-v_0}{u_0}\right) \tag{1}$$

$$r = -r_0 \sin(\omega_0 t) \tag{2}$$

$$r_0 = \psi_0 \omega_0 \tag{3}$$

2.2. Ship model

The test model is blunt KVLCC2 tanker, as shown in Figure 2. Table 1 lists the principal particulars of KVLCC2, and the present test uses the 7-m model based on the benchmark work on shallow-water problems [5, 6]. Since the towing speed is low and the rudder deflection angle is zero, the rudder has a negligible interactive effect on the force/moment of hull. Additionally, ship sinkage, heel, and trim are fixed during the PMM tests.

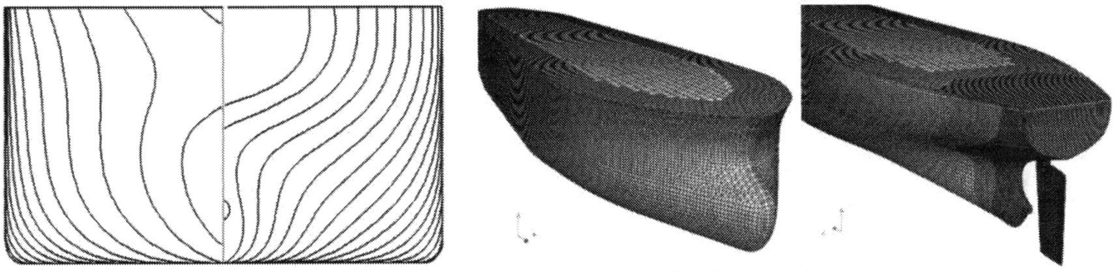

(a) Body plan (b) Discretized surface

Figure 2. KVLCC2 model.

Table 1. Principle particulars of KVLCC2 hull.

Particulars	Full scale	Present model
Ship length L_{pp} (m)	320	7.0000
Ship breadth B_{wl} (m)	58.0	1.2688
Ship draught T (m)	20.8	0.4550
Horizontal aspect ratio L/B	5.5170	5.5170
Block coefficient C_b	0.8098	0.8098
Displacement ∇ (m^3)	312622	3.2724
Wetted surface w/o rudder (m^2)	27194	13.0129
Froude number Fr	-	0.051/0.064
Reynolds number Re ($\times 10^6$)	-	2.617/3.277

2.3. Test matrix

To obtain the manoeuvring coefficients in different water depths, various motion conditions are arranged as listed in Table 2. Four different water depths that range from very shallow-water condition (H/T = 1.2) to deep-water condition ($H/T = \infty$) are considered, where H and T denote the absolute water depth and ship draught, respectively.

For the static drift test, the drift angle increases from 0° to 18°. The mean force/moment $X^{(0)}$, $Y^{(0)}$, $N^{(0)}$ are measured for obtaining the manoeuvring coefficients X_{vv}, X_{vvvv}, Y_v, Y_{vvv}, N_v, N_{vvv}. For the pure yaw test that the maximum yaw rate r_0' increases from 0.1 to 0.5. The mean surge force $X^{(0)}$ is measured for manoeuvring coefficient X_{rr}. The first-order sine component of sway force and yaw moment ($Y^{(1,\sin)}$ and $N^{(1,\sin)}$), which is related to the maximum yaw rate, are measured for manoeuvring coefficients Y_r, Y_{rrr}, N_r, N_{rrr}. All the notations are typical ones of ship manoeuvring field, and the same with those of Zhu et al. [8].Yaw rate r_0 is normalized by $U_0(L)^{-1}$. Force is normalized by $0.5\rho U_0^2TL$ and moment is normalized by $0.5\rho U_0^2TL^2$. The symbol L simply denotes the ship length between perpendiculars L_{pp} in this study.

Table 2. Test matrix.

Type of test	Static drift test	Pure yaw test
Motion range	$\beta_0 = 0°, 2°, 4° \ldots 18°$	$r_0' = 0.1, 0.2 \ldots 0.5$
Water depth H/T	1.2, 1.5, 2.0, ∞	1.2, 1.5, 2.0, ∞
Target force components	$X^{(0)}$, $Y^{(0)}$, $N^{(0)}$	$X^{(0)}$, $Y^{(1,\sin)}$, $N^{(1,\sin)}$
Manoeuvring coefficients	X_{vv}, X_{vvvv}, Y_v, Y_{vvv}, N_v, N_{vvv}	X_{rr}, Y_r, Y_{rrr}, N_r, N_{rrr}

3. Numerical method

3.1. Governing equations

The Reynolds-averaged Navier–Stokes (RANS) equations of incompressible and immiscible fluids are solved using the snuMHLFoam solver. It is a CFD package that the Marine Hydrodynamic Lab (MHL) of Seoul National University (SNU) has developed based on the OpenFOAM+ v1912 platform, and its validations and applications on ship seakeeping and manoeuvring can be found in these works [7, 8, 9]. Following equations are the continuity equation and momentum equation.

$$\nabla \cdot \vec{u} = 0 \tag{4}$$

$$\frac{\partial(\rho\vec{u})}{\partial t} + \nabla \cdot (\rho\vec{u}\vec{u}) = -\nabla p_{\rho gh} + \nabla \cdot \vec{T} + \rho \vec{f}_{ext} \tag{5}$$

The present solver achieves the prescribed PMM motions by moving the global computation domain, the $\vec{u}$ indicates the pure fluid velocity without the velocity of domain movement. The ρ and $p_{\rho gh}$ are the density and pressure, respectively. The $\vec{T}$ is the deviatoric stress tensor of incompressible flow and $\vec{f}_{ext}$ is the external forces.

To discretize the governing equations, the implicit Euler method is used for the transient term, and the Gauss linear scheme is applied for discretized gradient terms. The discretization of divergent terms depends on the transported quantities: the total variation diminishing (TVD) scheme and van Leer limiter are used for the volume of fraction, while the central difference scheme and linear upwind scheme are selected for the velocity and turbulent quantities, respectively. For the Laplacian term of the momentum equation, the diffusion coefficient is transformed from the cell to the cell faces through linear interpolation. The gradient evaluated at the center of two neighbouring grids has a normal component, and this surface-normal component is discretized by central difference scheme.

The coupling of pressure and velocity employs the PIMPLE algorithm, which combines the Pressure Implicit with Splitting of Operators (PISO) algorithm and the Semi-Implicit Method for Pressure Linked Equations (SIMPLE) algorithm. Two inner iterations of SIMPLE part is applied, and two outer iterations of pressure correction is used. The free surface is modelled using the Volume of Fraction (VOF) method and flux-corrected-transport method, and a leapfrog method of three sub-steps is used to integrate the transport equation of VOF.

The k-ω SST turbulence model is applied, with the additional buoyancy term and artificial viscosity limiter added for improving the computational stability. The details can be found in Zhu et al. [8]. The blending wall functions are used for near-wall treatment, as given in the following equations. The empirical constants β_1, C_μ, κ, and E use the default values provided by OpenFOAM, and the y^+_{lam} that represents the intersection between the viscous layer and log layer is estimated at the beginning of computation.

$$\omega = \begin{cases} \dfrac{6\nu}{\beta_1 y^2} & \text{in viscous layer, } y^+ < y^+_{lam} \\ \dfrac{k^{1/2}}{C_\mu^{\ 1/4}\kappa y} & \text{in log layer, } y^+ > y^+_{lam} \end{cases} \tag{6}$$

$$\nu_t = \begin{cases} 0 & \text{in viscous layer, } y^+ < y^+_{lam} \\ \nu\left\{\dfrac{\kappa y^+}{\ln\left[\max(Ey^+,1)\right]} - 1\right\} & \text{in log layer, } y^+ > y^+_{lam} \end{cases} \tag{7}$$

3.2. Numerical tank

To carry out the numerical PMM tests, a cylindrical computation domain is used as shown in Figure 3. The radius of domain is set to $2L$ to eliminate the lateral blockage effect of the side wall. The downward depth of the domain is adjustable depending on the target water depth. The domain is divided into four major boundaries: *top*, *inlet*, *bottom*, and *hull*. To maintain the sharpness of hull geometry, the *hull* boundary is further divided into several feature patches, e.g. *starboard* patch, *port* patch, and *deck* patch.

The zero-gradient boundary condition is generally used for the flow fields, such as volume fraction, velocity, pressure, and turbulence. For the *top* boundary, the pressure field is fixed at atmospheric pressure, while zero-gradient condition is applied all other fields. The *inlet* boundary is set to the zero-gradient condition for all fields. The *bottom* boundary is applied with the wall functions for turbulence field, and no-slip condition is used for velocity. The *hull* boundary is also assigned with wall functions and no-slip condition, and its velocity is the prescribed PMM velocity which is updated at every time step. The overall computation domain moves according to the updated velocity condition of *hull*. The pressure and volume fraction of *hull* boundary is of zero-gradient condition.

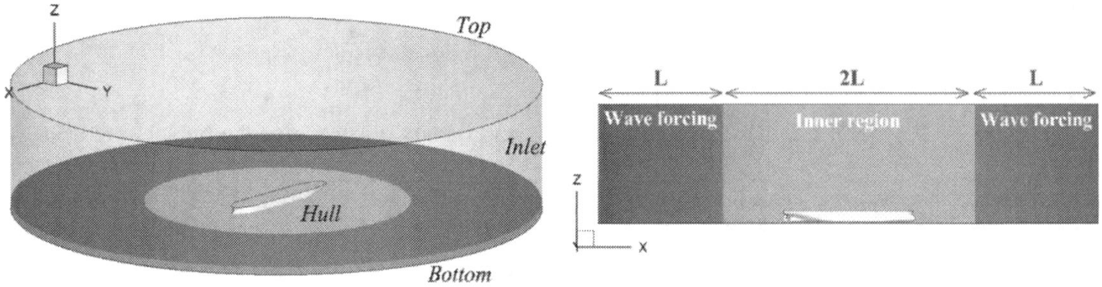

(a) Overall domain (b) Sectional dimension of domain

Figure 3. Computational domain with circular wave forcing region (case: $H/T = 1.2$).

Due to the very shallow depth, the flow fields beneath free surface have very intensive variation during the numerical computation. To damp the undesired oscillation or reflection of flow fields, the wave forcing function of *waves2Foam* toolkit [10] is added to the present computation domain. As marked in Figure 3 (b), the forcing is performed in a circular region of radius L where the wave elevation, velocity, and pressure are gradually forced to be zero. The "inner region" indicates the computational region without any forcing function. The exponential forcing weight depending on radial distance is given by Equations (8) and (9), where the empirical forcing constant f_0 is 3.5 and the relative distance σ is calculated by the radial distance r_f. The wave forcing function suits well with the zero-gradient condition of *inlet* boundary.

$$w_R = \frac{\exp\left(\sigma^{f_0}\right) - 1}{\exp(1) - 1} \tag{8}$$

$$\sigma = \frac{r_f - L}{L} \tag{9}$$

The computation domain is discretized into an unstructured mesh system with local refinements near the free surface, ship, and under-keel clearance. Figure 4 shows an example of the mesh structure for the very shallow-water case. Very thin layer grids are used to handle with the boundary flow near both hull and bottom wall. Totally 8 layers of grids are applied with a constant expansion ratio of 1.15. The thickness of the first layer grid (Δn) is 1/8000 of ship length L, which empirically provides a y^+ value of 17.6 according to Equation (10) suggested by ITTC [11]. The C_f denotes ship friction coefficient. Rudder interactive effect is negligible for the force/moment of hull in the present study, the ship stern region has a high grid resolution mainly for computational stability. For the numerical model, the overall number of grid is 6.7 million, and the time step is 0.0043 second.

$$y^+ = \Delta n \frac{\mathrm{Re}\left(0.5C_f\right)^{1/2}}{L} \tag{10}$$

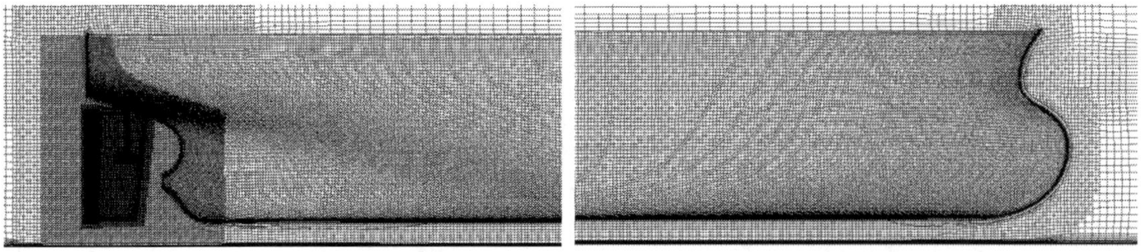

Figure 4. Mesh structure near ship (case: $H/T = 1.2$).

4. Numerical validation works

4.1. Numerical convergence tests

In previous work [8], the numerical uncertainties of the present PMM solver have been evaluated with respect to surface y^+, grid solution, time step, and the time window of data analysis. Compared with the deep-water case, the present concern is the numerical discretization of ship under-keel clearance. Therefore, two sets of numerical convergence tests are conducted, as listed in Table 3. The first set is about the grid size Δx which is selected for the grid resolution of under-keel clearance, namely $(H-T)(\Delta x)^{-1}$. The hull surface is also discretized by the grid size Δx. The second set of numerical convergence tests is about the time step Δt which is selected based on the ideal CFL number of under-keel-clearance region. The ideal value of CFL_{UKC} is calculated by $U_0 \Delta t(\Delta x)^{-1}$, where U_0 is the prescribed towing velocity. Constant time step is used for computation, and the real CFL number fluctuates during computation. The test case is the static drift test of 18°, and water depth H/T is 1.2. The thickness of boundary layer mesh is constant for all cases, with the details given in previous section 3.2.

Table 3. Matrix of numerical convergence tests (Case: $\beta_0 = 18°$, $H/T = 1.2$, $Fr = 0.064$).

		Grid resolution $(H-T)(\Delta x)^{-1}$		
Case #	$(H-T)(\Delta x)^{-1}$	$L(\Delta x)^{-1}$	CFL_{UKC}	Total number of grids
1.1	2	154	0.2	4.92 million
1.2	4	308	0.2	5.31 million
1.3	6	462	0.2	5.93 million
1.4	8	615	0.2	6.73 million
		Time resolution CFL_{UKC}		
Case #	CFL_{UKC}	Δt	$(H-T)(\Delta x)^{-1}$	Total number of grids
2.1	0.60	0.0128 s	8	6.73 million
2.2	0.47	0.0100 s	8	6.73 million
2.3	0.33	0.0070 s	8	6.73 million
2.4	0.20	0.0043 s	8	6.73 million

Figure 5 summarizes the results of numerical convergence tests. Surge force $X^{(0)}$, sway force $Y^{(0)}$, and yaw moment $N^{(0)}$ are observed with respect to the grid and time resolutions. Generally, a convergent tendency is observed from the results. Although it is a steady test, the hydrodynamic forces are found be to slightly overestimated when CFL_{UKC} is larger than 0.33. The reason is the accelerated flow passing under-keel clearance, which also induces strong shedding of bilge vortices.

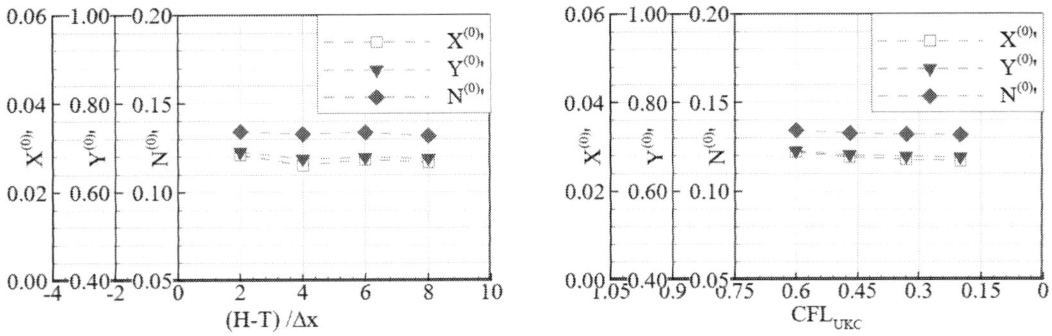

Figure 5. Numerical convergence tests of static drift test (Case: $\beta_0 = 18°$, $H/T = 1.2$, $Fr = 0.064$).

IOP Conf. Series: Materials Science and Engineering 1288 (2023) 012002 doi:10.1088/1757-899X/1288/1/012002

4.2. Validation on static drift test

The validation is conducted by comparing with the benchmark work of the NATO AVT group [5], in which a 7-m KVLCC2 hull has been used for the PMM tests in shallow waters. In the benchmark, INSEAN's experiment is involved, and several CFD solvers of different research groups have been employed for comparison. One validation case is the static drift test with a small drift angle of 4°, and the body force/moment is compared in Figure 6. In the legend box of Figure 6, the terms "FS" and "DB" denote the free-surface model and the double-body model, respectively. Due to the low speed, the free-surface effect can be insignificant if the side-wall effect of tank is negligible. The term "tank" means that the numerical domain has the same lateral size as that of the INSEAN's towing tank facility, which may exist the lateral blockage effect by the side walls. By comparison, the term "wide" indicates a large computation domain that its side wall generates negligible blockage effect.

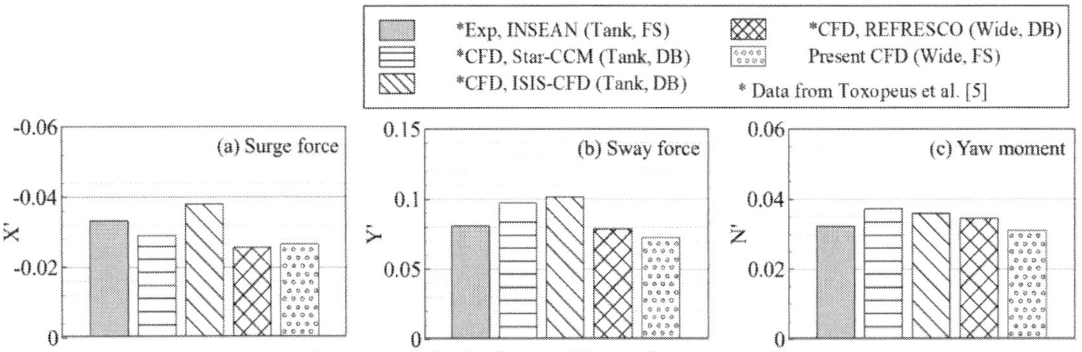

Figure 6. Comparison on the body forces (Case: $\beta_0 = 4°$, $H/T = 1.2$, $Fr = 0.064$).

Compared with the other results that may contain side-wall effect, both the ReFRESCO solver and present solver that apply a wide domain have relatively smaller surge force and sway force. Compared with ReFRESCO's prediction, the present solver has slightly lower sway force and yaw moment. Both two solvers use midship as the longitudinal origin of body coordinate system. Furthermore, Figure 7 shows that the two solvers have the similar profiles of axial velocity. Primary stern vortex is observed around the hub, and the bilge vortex induced by static drift motion is also well captured by the present solver.

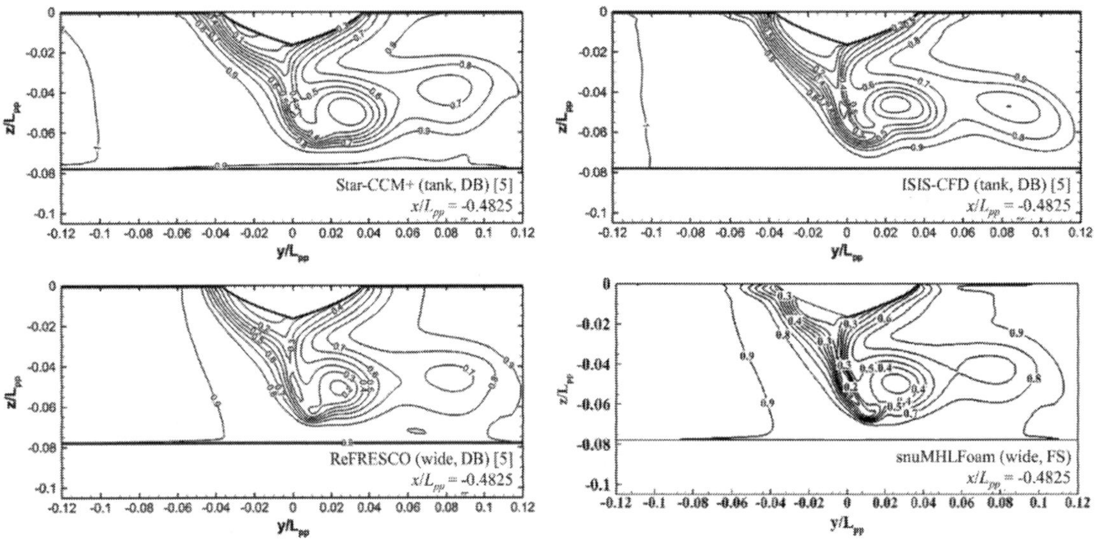

Figure 7. Comparison on axial velocity (Case: $\beta_0 = 4°$, $H/T = 1.2$, $Fr = 0.064$).

IOP Conf. Series: Materials Science and Engineering 1288 (2023) 012002 doi:10.1088/1757-899X/1288/1/012002

4.3. Validation on pure yaw test

The second validation case is the pure yaw test of yaw rate 0.4. Figure 8 presents the raw signals of force and moment generated by the present CFD solver. The harmonic signal of yaw rate is added in the figure for understanding the interaction between yaw rate and body force/moment. The yaw rate r_0' is 0.4, and the harmonic motion period is 20 seconds. Ship sinkage and trim are always fixed as the experiment. The signal of surge force has an oscillatory behaviour due to the numerical integration of unsteady solution, but the oscillation tends to become steady as computation runs. Both the sway force and yaw moment show smooth and periodical signals, and the sway force is observed to contain more nonlinear components. The phase difference is found between the yaw motion and body forces, showing that both added mass component and damping component are significantly increased in very shallow water.

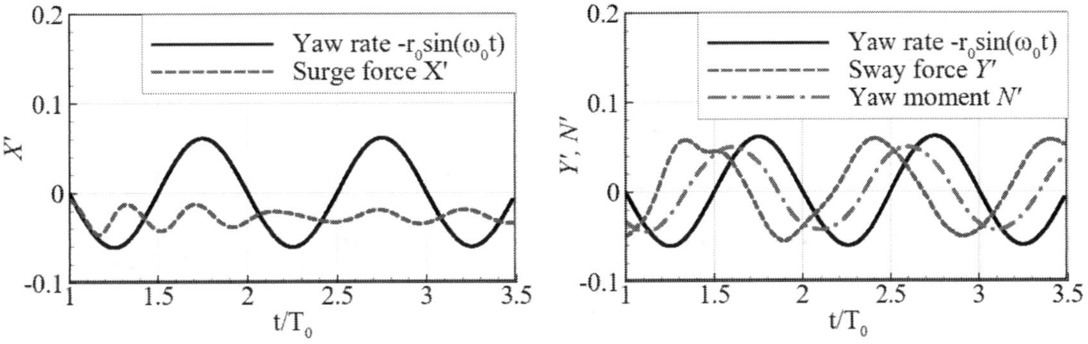

Figure 8. Force signals of present solver (Case: $r_0' = 0.4$, $H/T = 1.2$, $Fr = 0.064$).

Table 4 compares the mean surge force $X^{(0)\prime}$ and the 1st-order damping components of sway force and yaw moment ($Y^{(1,\sin)\prime}$ and $N^{(1,\sin)\prime}$) in three different water-depth conditions. The experimental measurement of INSEAN and ReFRESCO's CFD results [5] are available for comparison. The blockage effect may be involved in the experimental data, but it is ignorable for the two CFD solvers that use a wide domain. The present solver has a much smaller sway force component than that of ReFRESCO. The purely yawing ship has a low speed and very small magnitude of sway force, so the reason for the discrepancy is hard to be explained yet. However, for the yaw moment that is mainly determined by the lateral force on bow, two CFD solvers have shown an agreement in different water-depth cases.

Table 4. Comparison on the body forces (Case: $r_0' = 0.4$, $H/T = 1.2$, $Fr = 0.064$).

Water depth	Method	FS/DB	Domain	$X^{(0)\prime}$	$Y^{(1,\sin)\prime}$	$N^{(1,\sin)\prime}$
	Exp.	FS	Tank	-0.0156	0.0193	-0.0159
H/T = deep	ReFRESCO	DB	Wide	-0.0157	0.0158	-0.0177
	Present CFD	FS	Wide	-0.0172	0.0141	-0.0169
	Exp.	FS	Tank	-0.0264	-	-0.0264
H/T = 1.5	ReFRESCO	DB	Wide	-0.0209	0.0230	-0.0215
	Present CFD	FS	Wide	-0.0226	0.0186	-0.0195
	Exp.	FS	Tank	-0.0258	-	-0.0389
H/T = 1.2	ReFRESCO	DB	Wide	-0.0253	0.0489	-0.0289
	Present CFD	FS	Wide	-0.0257	0.0269	-0.0274

IOP Conf. Series: Materials Science and Engineering 1288 (2023) 012002 doi:10.1088/1757-899X/1288/1/012002

5. Results and discussions

Based on the present numerical tank, static drift test and pure yaw test are carried out to examine the water-depth effect on velocity-dependent manoeuvring coefficients. Four water depths that vary from deep water to very shallow water are considered. The Froude number is 0.051, which is slightly lower than that of the previous validation cases.

5.1. Static drift test

Figure 9 presents a comparison of the flow fields during a static drift test for two different water depths. The figures display the shear stress of hull surface and the Q-criterion field near the hull, for a large drift angle of 18°. The Q criterion is a mathematical value used to identify vortical structures. For the very shallow-water case shown in Figure 9 (b), the flow passing beneath the keel is compressed and accelerated due to the confined region, leading to the formation of strong vortex shedding near the bilge on both sides of the ship. Meanwhile, the accelerated flow results in a noticeable increase in the wall shear stress near the bow, as shown in Figure 9 (b). More importantly, a large drop of dynamic pressure occurs in the region of increased wall shear stress, as evident from Figure 10 (b). The magnitude of pressure drop is much larger than that of the shear stress, making the pressure drop on the leeward side the main cause of the force change in very shallow water.

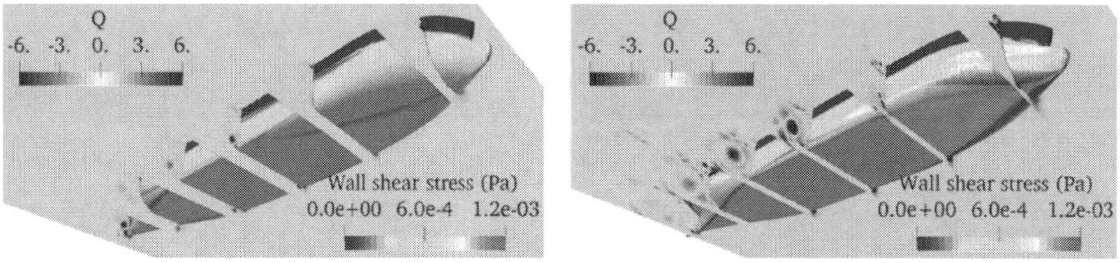

(a) H/T = deep (b) H/T = 1.2

Figure 9. Water-depth effect on vortices and surface shear stress (Case: β_0 = 18°, Fr = 0.051).

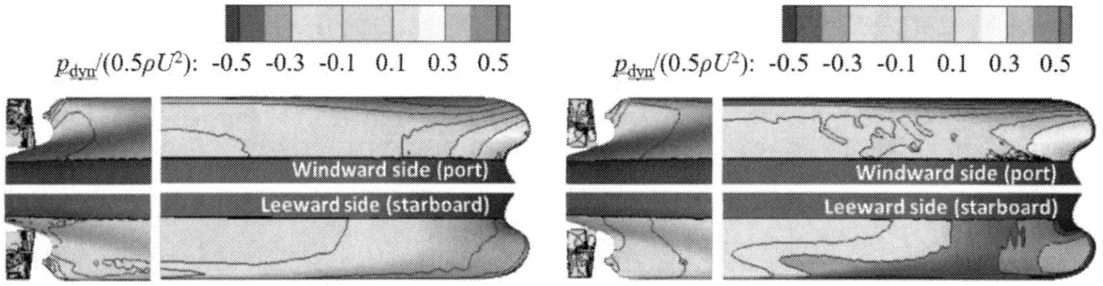

(a) H/T = deep (b) H/T = 1.2

Figure 10. Water-depth effect on the dynamic pressure of hull surface (Case: β_0 = 18°, Fr = 0.051).

Figure 11 summarizes the body force/moment of the static drift tests in 4 different water depths. The dashed lines are the curve-fitting lines based on the present computational data. Generally, the decreased water depth has caused a significant change of the hydrodynamic force/moment. The surge force $X^{(0)\prime}$ shows a remarkable difference in the very shallow-water case. Without drift angle, the finite depth induces more surge force, namely a stronger resistance. However, an opposite tendency can be observed when the drift angle becomes high. The surge force becomes positive at a large drift condition, which means the hull is being pushing forward in its surge direction. This is due to the negative pressure region of leeward side observed in Figure 10 (b). Meanwhile, both sway force $Y^{(0)\prime}$ and yaw moment $N^{(0)\prime}$ show a gradually increasing tendency with respect to the reducing water depth and increasing drift angle.

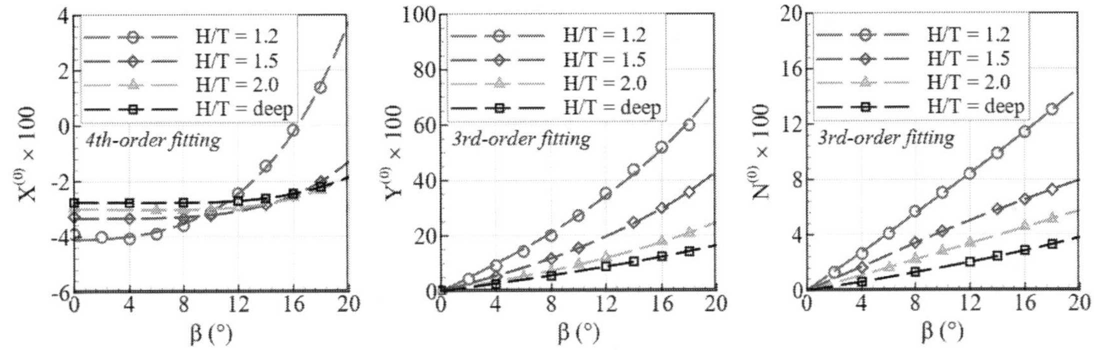

Figure 11. Water-depth effect on the body force/moment of static drift test.

The manoeuvring coefficients of sway velocity are summarized in Figure 12. The coefficients have shown an inversely proportional tendency with respect to the water depth *H/T*. The relative increase of coefficients can be higher than 300% due to the very shallow water, even the nonlinear coefficients. It indicates that the hull becomes more resistant to sway motion, and the hull's manoeuvrability has a significant drop.

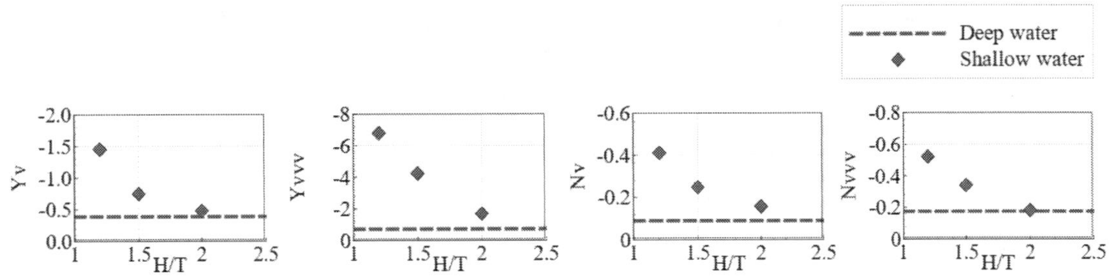

Figure 12. Water-depth effect on sway damping coefficients Y_v, Y_{vvv}, N_v and N_{vvv}.

5.2. *Pure yaw test*

Figure 13 presents the hydrodynamic characteristics of the pure yaw test. The presented case has a yaw rate of 0.5. Compared with the static drift motion, the yawing motion induces weaker vortices around hull, which can be observed in Figure 13 (a). No significant vortices are observed near bow or bilge. Meanwhile, the longitudinal distributions of lateral force are given in Figure 13 (b). Positive *x/L* indicates the fore body, and negative *x/L* is the aft body. The lateral force is dominated by the pressure effect, and it can be observed that the bow is still the main source of the pressure changes caused by finite water depth.

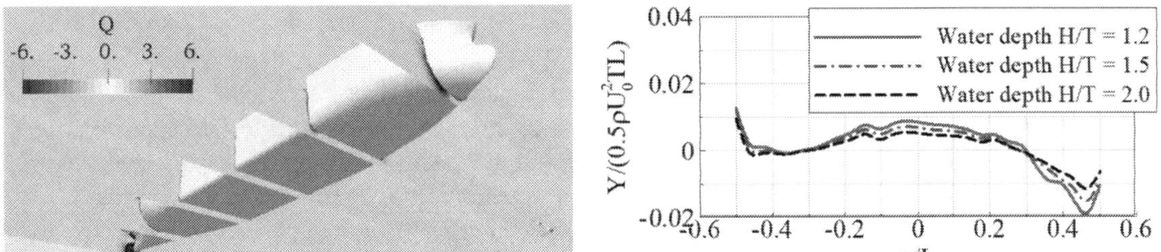

(a) *Q*-criterion field (Case: *H/T* = 1.2) (b) Water-depth effect on distribution of lateral force

Figure 13. Hydrodynamic characteristics of pure yaw test (Case: r_0' = 0.5, Fr = 0.051).

Figure 14 summarizes the body force/moment of the pure yaw tests in 4 different water depths, and the yaw rate is up to 0.5. The mean surge force and the 1st-order sine component (damping component) of sway force and yaw moment are shown in the figures. The mean surge force $X^{(0)}$ shows a main dependency on water depth rather than yaw rate. By comparison, both sway force and yaw moment show a significantly increase only in the very shallow water. When the water depth H/T is 1.5 or 2.0, both sway force and yaw moment are very close to the values of deep-water condition. It indicates that only the very shallow-water condition gives the ship a highly increased damping forces to resist the rotational motion.

Compared with the manoeuvring coefficients of sway velocity, yaw damping coefficients have a less increasing ratio in very shallow-water condition, as shown in Figure 15. The nonlinear coefficients have small magnitudes and they do not dramatically change the ship manoeuvrability.

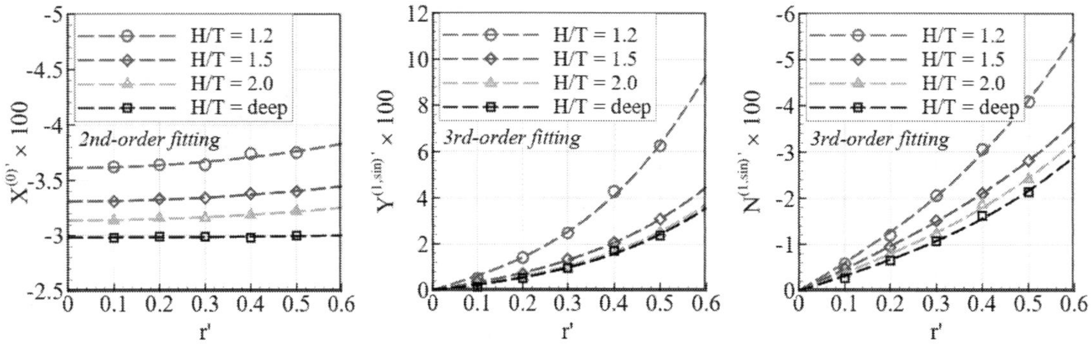

Figure 14. Water-depth effect on the body force/moment of pure yaw test.

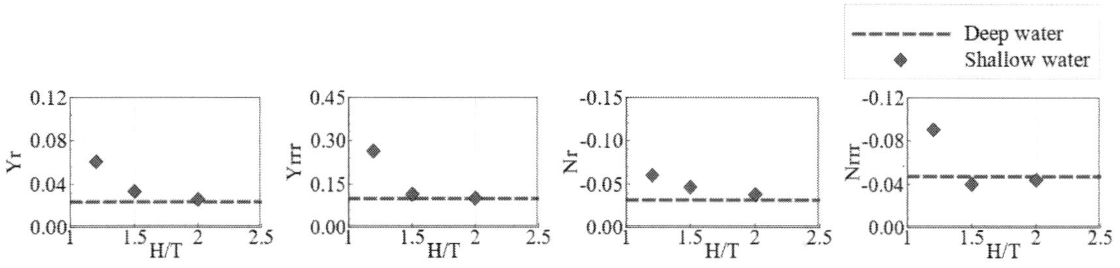

Figure 15. Water-depth effect on yaw damping coefficients Y_r, Y_{rrr}, N_r and N_{rrr}.

6. Conclusions

This study applies a numerical tank for the PMM tests of KVLCC2 model in shallow water. Both static drift test and pure yaw test have been carried out, and the effect of different water depths on manoeuvring coefficients have been estimated. The following conclusions are drawn:

- Based on a static drift test in very shallow water, the numerical convergence tests have been conducted with respect to the grid size and time step of the small under-keel clearance. A convergent tendency is observed for the present numerical solver.
- The numerical validation tests have been carried out for both static drift test and pure yaw test. Compared with the experimental measurement and the CFD predictions of benchmark work, the present solver has provided an acceptable prediction on body forces and flow fields.
- It is shown that the water depth provides significant changes to the manoeuvring coefficients of sway velocity, and the relative increase can be higher than 300% in very shallow water. The direct reason is the significant pressure drop of ship leeward side during static drift motion.
- By comparison, the manoeuvring coefficients of yaw rate show less increasing ratio with respect to finite water depth. The significant changes of yaw damping coefficients are mainly observed in very shallow water (H/T = 1.2).

Acknowledgments
This study was supported by the Lloyd's Register Foundation (LRF)-Funded Research Center (LRFC) at Seoul National University, South Korea, grant number GA100050. Their support is greatly acknowledged.

References
[1] Taimuri G, Matusiak J, Mikkola T, Kujala P and Hirdaris S 2020 A 6-DoF maneuvering model for the rapid estimation of hydrodynamic actions in deep and shallow waters *Ocean Eng.* vol 218 pp 108103
[2] Liu Y, Zou L, Zou Z J, Lu T C and Liu J X 2017 Numerical predictions of hydrodynamic forces and squat of ships in confined waters *In Proc. the 8th Int. Conf. on Computational Methods (Guilin, China)*
[3] Eloot K and Vantorre M 2011 Ship behaviour in shallow and confined water: an overview of hydrodynamic effects through EFD *NATO Res. and Tech. Organisation Report (Portsdown, UK)*
[4] Mucha P, Deng G, Gourlay T and Moctar O el 2016 Validation studies on numerical prediction of ship squat and resistance in shallow water *In Proc. the 4th Int. Conf. on Ship Manoeuvring in Shallow and Confined Water (Hamburg, Germany)*
[5] Toxopeus S L, Simonsen C D, Guilmineau E, Visonneau M, Xing T and Stern F 2013 Investigation of water depth and basin wall effects on KVLCC2 in manoeuvring motion using viscous-flow calculations *J. Mar. Sci. and Tech.* vol 18 pp 471-496
[6] SIMMAN 2020 *Workshop on Verification and Validation of Ship Manoeuvring Simulation Methods (Songdo, Korea)*
[7] Kim B S, Park D M and Kim Y 2022 Study on nonlinear heave and pitch motions of conventional and tumblehome hulls in head seas *Ocean Eng.* vol 247 pp 110671
[8] Zhu Z, Kim B S, Wang S and Kim Y 2022 Study on numerical PMM test and its application to KCS hull *Applied Ocean Res.* vol 127 pp 103327
[9] Wang S G, Kim B S, Zhu Z and Kim Y 2023 Numerical course-keeping tests of ONR tumblehome in waves with different rudder control strategies *Ocean Eng.* vol 269 pp 113584
[10] Jacobsen N G, Fuhrman D R and Fredsøe J 2012 A wave generation toolbox for the open-source CFD library: OpenFoam® *Int. J. forNum. Methods in Fluids* vol 70(9) pp 1073-1088
[11] ITTC Specialist Committee 2014 Recommended procedures and guidelines: practical guidelines for ship CFD applications *In 27th Int. Towing Tank Conf. (Copenhagen, Denmark)*

Numerical studies on tugboat performance during pushing operations

Z Abdelghafor[1], G Taimuri[1,¥], P Kujala[1], Y Wang[2] and S Hirdaris[1]

1Aalto University, School of Engineering, Marine Technology Group, Espoo, Finland
2 Shanghai Jiao Tong University, School of Naval Architecture, Ocean & Civil Engineering, China
¥ Corresponding author: ghalib.taimuri@aalto.fi

Abstract. This paper introduces a RANS CFD methodology for the evaluation of tugboat dynamics during pushing operations. Two- and three- dimensional methods that respectively utilize "Dynamic Fluid Body Interactions - (DFBI)" and "Tug Force Equilibrium kinematics- (TFE)" are assessed and compared with the aim to better understand the influence of fluid modelling on ship dynamics. For the DFBI method, an unsteady RANS solver comprising of a dynamic fluid body interaction module and a contact mechanics coupling algorithm are used to predict the forces between a tugboat and an assisted ship. For the TFE method, a steady RANS method is applied and contact actions are calculated as a sum of the hydrodynamic forces on the hull and the propeller. Whereas DFBI accounts for the time variation of the contact forces, the TFE is more rapid and could be used to derive operational decision support criteria. To demonstrate the latter the TFE method is used to derive the pushing forces based on a set of 16 numerical simulations. It is concluded that irrespective of the model used the tugboat speed and orientation may amplify the pushing forces. This effect could be prominent, especially at slow speeds for which the sway force acts in opposite direction to the tug.

1. Introduction

Tugboats serve a range of berthing, salvage and ice-breaking shipping operations (see Figure 1) [1]. Yet, accidents that involve tugboat operations in confined waters and fairways are broadly associated with wrong decision making [2,3]. A review of accidents reported in the Baltic Sea area during the past two decades [4,5] reveals that poor communication during tug operations and the lack of pilotage assistance are the major causes of accidents between 1997 and 2011 [6]. To assess tug performance and safety during ship handling operations, it is essential to understand the influence of operational conditions on ship dynamics. This is particularly relevant for tug pulling and escorting operations for which the risk of accidents due to tug-ship hydrodynamic interactions should be minimized [7].

Tugboat maneuvering dynamics have been tackled using numerical and experimental methods with an eye on the hydrodynamic forces that act on the tugboat hull and appendages, the tug motions, and trajectories at different speeds [8]. Well validated RANS CFD methods could enable the broad derivation of improved hydrodynamic derivatives for use by modular maneuvering models [9]. An example of these models is the Mathematical Model Group (MMG) presented by Piaggio [10]. RANS CFD [11], potential flow (PF) hydrodynamics [12] or captive model tests [13,14] are distinct approaches that may be used to better understand ship - tug dynamics. Recently, [15,16] explored uncertainties of simulations as compared to experiments for the case where parallel and drifting tug operations interplay at lateral and longitudinal positions of a tanker.

(a) (b)

Figure 1. Tugboats operations near ports and canals. (a) Tugboats assisting Ever Given [17]. (b) berths clearing from ice [18]

To date, most of the studies conducted to model the direct contact between a tugboat and an assisted vessel rely on the mathematical modular maneuvering models [19,20]. These models evaluate the hull, propeller, rudder, and other devices separately in addition to the external forces e.g., wind and waves. Ship-tug, propeller-hull, and propeller-rudder interactions are usually ignored. This paper pushes the state of the art by introducing the potential use of CFD methods for the evaluation of the contact forces applied on tugs during pushing mode assistance operations. Two- and three- dimensional methods that, respectively utilize "Dynamic Fluid Body Interactions - (DFBI)" and "Tug Force Equilibrium kinematics- (TFE)" are assessed and compared with the aim to understand the influence of fluid modelling on ship dynamics. Throughout the simulations, it is assumed that the behavior of the tugboat in oblique flow resembles the flow around a wet transom.

2. Theory

STARCCM+ was used for the numerical simulations [21]. The motions of the tugboat were simulated using the DFBI integrated into the equations of motion for all forces and moments. The aim of this process was to calculate accelerations in each time step.

2.1. *Equations of Fluid Flow and Turbulence models*

The flow was assumed turbulent, incompressible, and isothermal. The RANS method was adopted. The velocity and pressure were decomposed into mean ($\bar{u}$ and $\bar{p}$) and fluctuating (u' and p') parts. The mean flow velocities and pressures were obtained from the continuity and momentum equations

$$\frac{\partial \bar{u}_i}{\partial x_i} = 0 \tag{1}$$

$$\rho\frac{\partial \bar{u}_i}{\partial x_j} + \rho\frac{\partial}{\partial x_j}(\bar{u}_i\bar{u}_j) = -\frac{\partial \bar{p}}{\partial x_j} + \mu\frac{\partial^2 \bar{u}_i}{\partial x_j x_j} - \frac{\partial \tau_{ij}}{\partial x_j} + f_i \tag{2}$$

where $\rho\ [kg/m^3]$ is the density, $\mu\ [kg/m.s]$ is the molecular viscosity, $u_i\ [m/s]$ is the velocity, $p\ [N/m^2]$ is the pressure, $\tau_{ij} = \rho\left(\overline{u_i'u_j'}\right)\ [N/m^2]$ is the Reynolds stress tensor, and $f_i\ [N]$ is the body force term. The appearance of the Reynolds stress term introduces nine additional unknowns to the equations for which additional equations must be obtained. In this study, the k-omega-SST turbulence model was used. This model is based on the combination of the k-ω and the k-ε models. The former model was activated in the near wall region. The latter was activated in the outer region of the boundary layer and the free shear layers. Implementation of the k-ω-SST turbulence introduced two additional equations for the turbulent kinetic energy $k\ [m^2/s^2]$ and the dissipation rate $\omega\ [m^2/s^3]$. The Reynolds stress term was replaced with the mean velocity gradients and Eddy viscosity $\mu_T\ [kg/m.s]$ as follows

$$\rho\frac{\partial k}{\partial t} + \rho\bar{u}_i\frac{\partial k}{\partial x_j} = \tau_{ij}\frac{\partial \bar{u}_i}{\partial x_j} - \beta^*\rho k\omega + \frac{\partial}{\partial x_j}[(\mu + \sigma^*\mu_T)\frac{\partial k}{\partial x_j}] \tag{3}$$

$$\rho\frac{\partial \omega}{\partial t}+\rho\bar{u}_i\frac{\partial \omega}{\partial x_j}=\frac{\gamma}{v_T}\tau_{ij}\frac{\partial \bar{u}_i}{\partial x_j}-\chi\rho\omega^2+\frac{\partial}{\partial x_j}\left[(\mu+\sigma_\omega\mu_T)\frac{\partial \omega}{\partial x_j}\right]+2\rho(1-F_1)\sigma_{\omega 2}\frac{1}{\omega}\frac{\partial k}{\partial x_j}\frac{\partial \omega}{\partial x_j} \quad (4)$$

In the above equations F_1 represents a blending function, which ensures that a k-ω formulation is used in the inner parts of the boundary layer and a k-ε behavior is achieved in the free-stream. The terms β^*, σ^*, γ, χ, σ_ω, $\sigma_{\omega 2}$ are model coefficients and may also depends on F_1.

In turbulent flows, the presence of viscosity affects the flow significantly near wall boundaries that may be sources of perturbations and vortices. The inner boundary layer region is further divided into three sublayers namely (a) viscous, (b) buffer, and (c) logarithmic. To capture the velocity gradient in the viscous sublayer a $y+\leq 1$ is needed. Thus, solving the flow through the viscous sub-layer is computationally expensive. This challenge is overcome by using wall functions, i.e., by setting a $y+$ value for the first cell next to the wall (i.e., in way of either the viscous or the logarithmic sublayers) to provide sufficiently accurate results. STAR-CCM+ utilizes the following blended wall function

$$u_* = \delta\sqrt{\frac{\mu|\vec{V}_{tangential}|}{\rho y}}+(1-\delta)C_\mu^{1/4}k^{1/2} \quad (5)$$

where u_* is the frictional velocity, δ is the blending function, $\vec{V}_{tangential}$ is the wall-tangential velocity vector, y is the wall distance, and C_μ is the turbulence model coefficient. In this study $30 < y+ < 300\sim500$ and $y+ < 1$ were set in way of the log-layer and the viscous sublayer, respectively [22].

2.2. Computations of Motions and Contact Forces

Unless constraints to the body motions are defined, STAR-CCM+ solves the translation and rotation of the center of mass of the body at the earth-fixed coordinate system in all degrees of freedom. The equation of motion for translation and rotation in the earth-fixed coordinate system is defined as

$$m\frac{d\vec{V}_b}{dt}=\vec{F} \text{ and } M\frac{d\vec{\varpi}}{dt}+\vec{\varpi}\times M\vec{\varpi}=\vec{H} \quad (6)$$

where $\vec{V}_b$ $[m/s]$ is the velocity vector of the body, m $[kg]$ is mass of the tugboat, and $\vec{F}$ $[N]$ is external force vector acting on tugboat and t is the time in seconds, $\vec{\varpi}$ $[rad/s]$ the angular velocity vector, M $[kg.m^2]$ is the inertia matrix of the body and $\vec{H}$ $[N.m]$ is the external moment vector.

The contact force between the tugboat and the assisted ship is calculated with the contact coupling mechanism embedded in the DFBI module of STAR-CCM+. For the model to resemble a real contact, the contact force depends on the distance between the tugboat and the boundary (ship hull) at which the contact should be well idealized. The contact force is activated once this user-defined distance i.e., the effective range, is reached (Figure 2). As the tugboat gets near to the ship hull, the contact force increases until reaches a maximum value which stops the tugboat from moving further and avoid penetration. The contact force is applied on the mesh faces of the tugboat and the ship hull and may be calculated as

$$\vec{F}_f = F_n\hat{\imath}+F_t\hat{\jmath}\text{ , } F_n=\vec{a}_f\left[k_1(d_0-d_f)-k_2\dot{d}_f\right]\ .\ \vec{n}_{bf}\text{ , and } F_t=-\mu|F_n|\tanh(k_tV_t) \quad (7)$$

where F_n $[N]$ and F_t $[N]$ is the normal and tangential component contact force, respectively; $\vec{a}_f$ $[m^2]$ is the face area vector, k_1 $[N/m]$ is the elastic coefficient, k_2 $[N.s/m]$ is the damping coefficient, d_0 $[m]$ is the effective range, d_f $[m]$ is the distance between the boundary face and the opposing boundary, $\vec{n}_{bf}$ is the normal vector to the boundary face, μ is the friction coefficient, k_t is the '*tanh*' coefficient, and V_t $[m/s]$ is the tangential velocity of the face f.

Two main defects in the STARCCM+ contact coupling mechanism should be overcome to resemble contact. At first, the contact force applied on the tugboat from the fluid cells is sensitive to the user-defined effective range. A considerably large effective range may lead to excessive rebounding of the tugboat after the first contact with the assisted ship. Thus, it should be adjusted at the start of the simulation, to avoid the magnification of the contact force. The other defect is that the software cannot model static friction when two bodies are in contact. Hence, regardless of the value of the tangential component of the contact force between the two bodies, they may slide over each other and wrongly idealize the loss of contact. To overcome the latter, a force is added at the center of gravity of the tug-

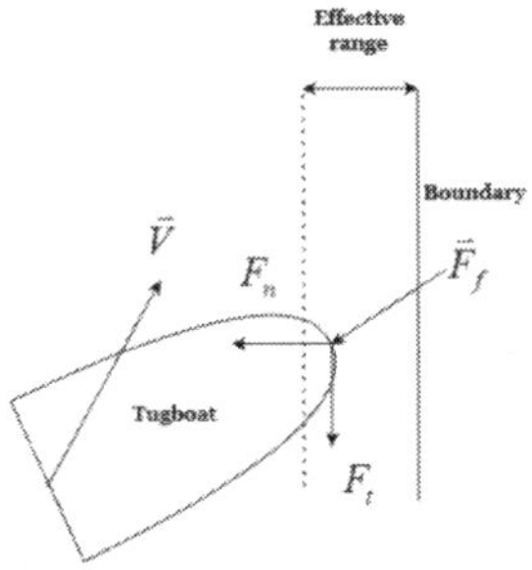

Figure 2. Contact coupling mechanism.

body which has the same magnitude, but opposite direction of the contact force, together with a counter moment to oppose the contact moment generated on the tugboat. These forces and moments are activated only when the contact force value is under the static friction value defined as

$$\vec{F}_{Ext} = \begin{cases} -\vec{F}_f & for\ F_t \leq \eta_{static} \cdot F_n \\ 0 & for\ F_t > \eta_{static} \cdot F_n \end{cases}, \text{and } M_{Ext} = \begin{cases} -M_f & for\ F_t \leq \eta_{static} \cdot F_n \\ 0 & for\ F_t > \eta_{static} \cdot F_n \end{cases} \quad (8)$$

where $\vec{F}_{Ext}$ and M_{Ext} are the counter force and moment respectively, and η_{static} is the static friction coefficient between the tugboat fenders and the assisted ship side shell is assumed as steel.

2.3. *Propeller idealization*

To avoid high computational costs, the propeller was substituted by a virtual disk model. The virtual disk force is inserted as a source term in the momentum equations. As defined in STAR - CCM+ the source term comprises of a volume force $\vec{F}_b$ distributed over the cylindrical virtual disk, with an axial component f_{bx} and radial component $f_{b\theta}$ following the Goldstein optimum method [23] according to

$$f_{bx} = A_x r^* \sqrt{1-r^*}, \text{and } f_{b\theta} = A_\theta \cdot \frac{r^*\sqrt{1-r^*}}{r^*(1-r_h')+r_h'}; \text{ where } \{r^* = \frac{r'-r'}{1-r_h'}\ ;\ r_h' = \frac{R_H}{R_P} \text{ and } r' = \frac{r}{R_P}\} \quad (9)$$

where r is the radial coordinate, R_H is the hub radius, R_P is the propeller tip radius, and the radial distribution of the body forces is r^*. Both A_x and A_θ are constants computed from the propeller thrust T and torque Q as $A_x = \frac{105}{8} \cdot \frac{T}{\pi\Delta(3R_H+4R_P)(R_P-R_H)}$ *and* $A_\theta = \frac{105}{8} \cdot \frac{Q}{\pi\Delta R_P(3R_H+4R_P)(R_P-R_H)}$ where Δ represents the virtual disk thickness.

2.4. *Tug Force Equilibrium Model (TFE)*

The TFE is a model that assumes that the tug while pushing an assisted vessel reaches a steady equilibrium. Based on this idealization the contact force may be calculated as a summation of the hydrodynamic forces acting on the tugboat as follows

$$\begin{aligned} X_H + X_P + X_r &= X_c \\ Y_H + Y_P + Y_r &= Y_c \\ N_H + N_P + N_r &= N_c \end{aligned} \quad (10)$$

where X, Y respectively represent the surge and sway forces, N is the yaw moment, subscripts H, P, R, and C represent the forces of the hull, propeller, rudder and contact, respectively. With the TFE model, the body is fixed and a steady CFD RANS simulation is conducted. Such idealization allows for reduced computational costs in comparison to the DFBI simulation that is transient.

Results presented in this paper correspond to the case of a conventional tugboat with single screw propeller and rudder configuration. The general particulars of the vessel are shown in Table 1. In 2D, the model scale $\lambda = 1/25$ and $Fn = 0.0602$ were used to reduce the computational effort. The data on the propeller geometry and characteristics were limited to the propeller hub/tip diameters and the number of blades. The propeller type chosen was a Wageningen B-series of an expanded area ratio $A_E/A_0 =$ 0.8 and pitch to diameter ratio $P/D = 1.1$. The performance curves used were based on [24] and are shown in Figure 3. The pushing operational scenario shown in Figure 4 assumes the case whereby two tugboats are pushing the assisted vessel with the same force. Accordingly, the assisted vessel has a constant heading. Both tugboats are moving with a constant velocity and a heading like the assisted vessel.

Table 1. Tugboat general particulars.

Dimension	Value	Unit
Length overall (LOA)	32.9	m
Length between perpendiculars (L_{PP})	30	m
Breadth molded	10	m
Depth molded	5.72	m
Maximum draft	4.75	m
Engine power	2880	KW
Bollard pull	39(382.5)	tons (KN)
Maximum speed	14 (7.2)	Knots (m/s)
Estimated operational speed during pushing	2(1.03)	Knots (m/s)

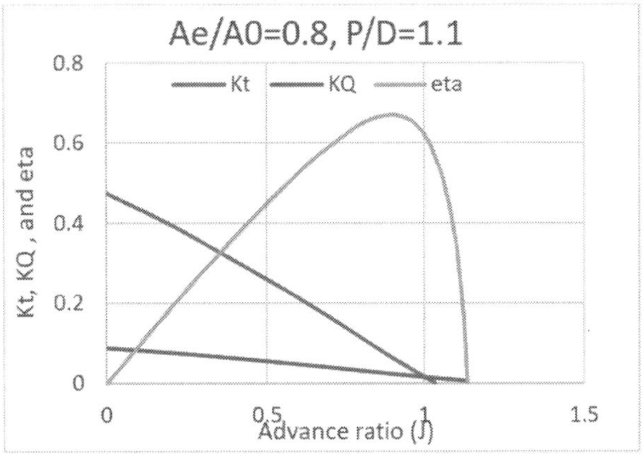

Figure 3. Propeller characteristics

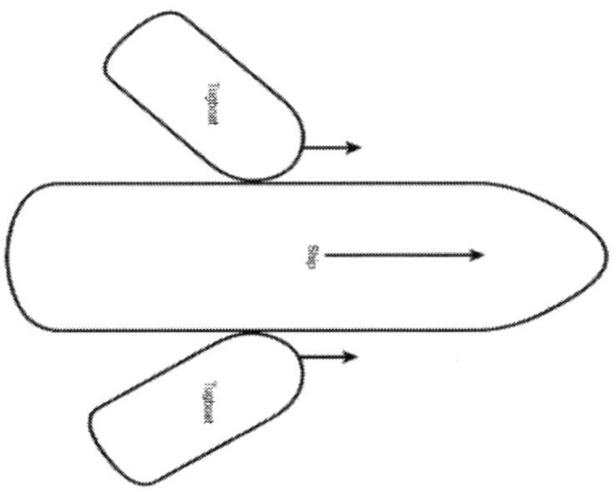

Figure 4. Pushing assistance scenario

Three cases of tugboat pushing operations were considered. In all cases, the influence of free surface hydrodynamics was disregarded as vessel and tug speeds were low. For the first two cases, results were based on a 2D model for which the assisted vessel was replaced by a rectangle and the propeller was idealized by a user-defined thrust force. The purpose of the exercise was to test the ability of the proposed DFBI model to predict the contact force as compared to a traditional equilibrium model. Figure 5 shows the domain dimensions for case 1 that utilizes the DFBI model in 2D. The tugboat overset region is set as a 2 degrees of freedom (DOF) body while the plate overset region representing the contact shell of the assisted ship is set as an 1 DOF body. The background region is assumed stationary. Case 2 applies the TFE model in which both the ship and the tugboat are assumed stationary. The fluid is flowing from the inlet and goes around the two bodies (Figure 6). Case 3 is based on a 3D idealization for which a TFE model is used to predict the contact forces on the tugboat at different velocities and drift angles (Figure 7). The tugboat has a zero heading while the inlet flow is set with an angle of attack equal to the drift angle required for a simulation. Such modelling approach helps simplify the meshing as compared to the case when the tugboat is rotated to a certain drift angle. Various boundary conditions were set along the lines for 2D models and on surfaces in 3D models. Tables 2, show the boundary conditions for each case. It is noted that $\vec{V}_\infty$ and θ_{spec} are user-defined values of the flow velocity and its direction, respectively. During modelling it was assumed that there is nothing predefined in way of the overset boundaries.

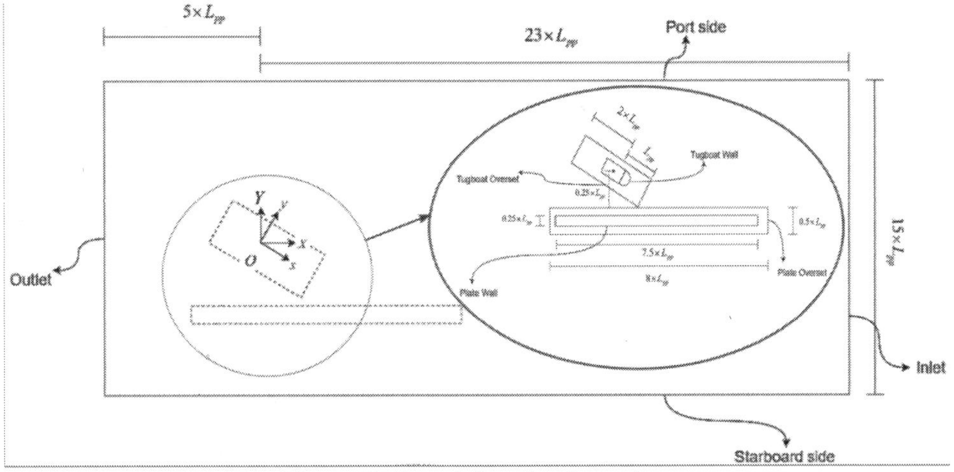

Figure 5. Case 1 domain dimensions

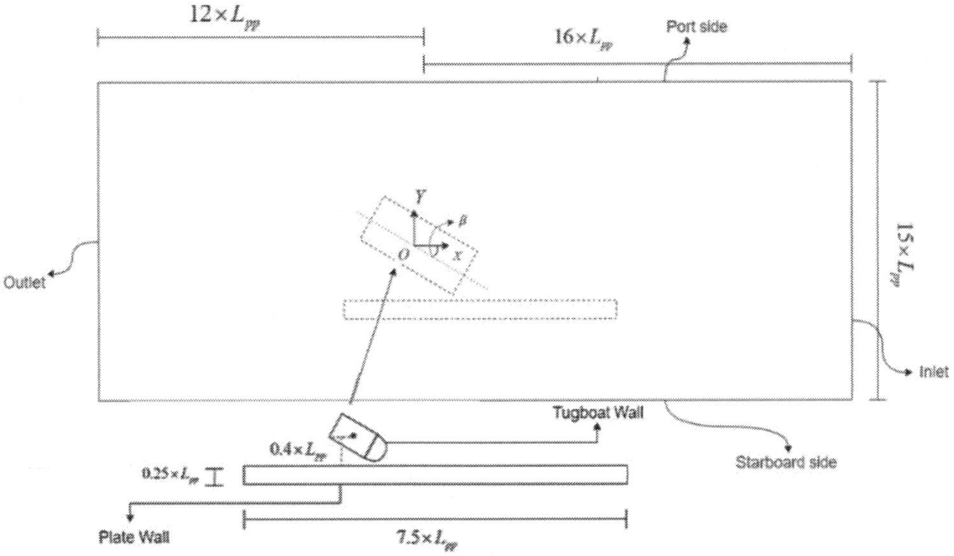

Figure 6. Case 2 domain dimensions

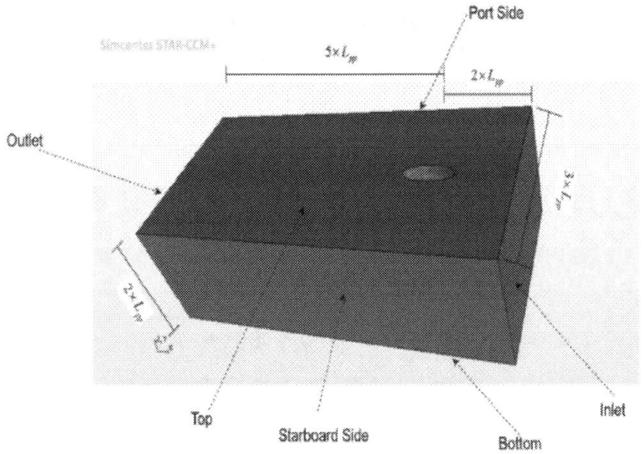

Figure 7. Case 3 domain dimensions

Table 2. : Boundary conditions

Region	Line/surface name	Velocity	Pressure	Motion specification
Case 1				
Background	Inlet	$V = 0$	Zero Gradient	Stationary
	Outlet	-	$P = 0$	
	Starboard Side	Symmetry	Symmetry	
	Port Side	Symmetry	Symmetry	
Tugboat	Tug Overset	-	-	DFBI translation (2DOF) (X, Y)
	Tugboat wall	$V = \vec{V}_{wall}$	-	
Plate	Plate Overset	-	-	Translation 1 DOF X
	Plate Wall	$V = \vec{V}_{wall}$	-	
Case 2				
Domain	Inlet	$V = \vec{V}_{\infty}$	Zero Gradient	Stationary
	Outlet	-	$P = 0$	
	Starboard Side	Symmetry	Symmetry	
	Port Side	Symmetry	Symmetry	
	Tugboat wall	$V = \vec{V}_{wall}$	Zero Gradient	
	Plate Wall	$V = \vec{V}_{wall}$	Zero Gradient	
Case 3				
Domain	Inlet	$V = \vec{V}_{\infty} \cdot \theta_{spec}$	Zero Gradient	Stationary
	Outlet	-	$P = 0$	
	Starboard Side	$V = \vec{V}_{\infty} \cdot \theta_{spec}$	Zero Gradient	
	Port Side	$V = \vec{V}_{\infty} \cdot \theta_{spec}$	Zero Gradient	
	Tugboat wall	$V = V_{wall}$	Zero Gradient	
	Bottom	$V = \vec{V}_{\infty} \cdot \theta_{spec}$	Zero Gradient	
	Top	Symmetry	Symmetry	

Polyhedral and hexahedral volume meshes were considered. The polyhedral mesh consists of arbitrary polyhedral-shaped cells. In this case, each individual cell has multiple neighbors. Hence, gradients may be accurately approximated. Polyhedral idealizations provide smooth transition between the prism layers and the rest of the volume mesh, albeit at the cost of larger number of cells as compared to hexahedral meshes [25]. A polyhedral mesh allows for the affordable use of many cells and may therefore be considered more appropriate for the 2D cases Figure 8. A hexahedral mesh is a template mesh comprising of cells that cut through the input geometry and may be more useful in 3D idealizations (Figures 9-10). Irrespective to the type of mesh used near to boundary walls layers of relatively small orthogonal prismatic cells should be formed. In this study the number of prismatic layers, their thickness and thickness distributions were controlled to achieve the desired range of $y+$ values as much as practically possible. The overlapping mesh technique was employed for the DFBI model. The moving bodies in such cases have an independent mesh domain that overlaps the main background domain. The outer region of the overlapping domain may exchange the flow characteristics with the background domain present at every step of the simulation.

For cases 1,2 there is a refined region along the length and width of $5 \times L_{pp}$ of the domain. The cell size in the refined region was set to be near or equal to the size of the overlapping mesh. In case 3, two sub-regions were refined namely (a) a region along the length of the domain with width of $5 \times L_{pp}$ and a depth of $Max. draft + 1m$ (same as in 2D cases) and (b) a cylindrical domain before and beyond the location of the virtual disk. The latter idealization accounts for abrupt flow changes.

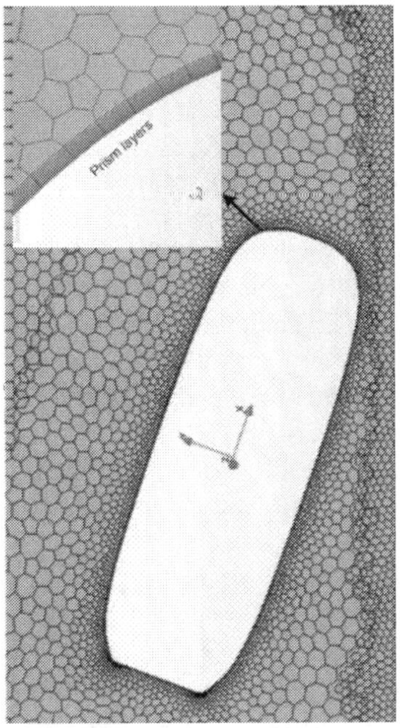

Figure 8. The prismatic layers around the hull for case 1 and case 2.

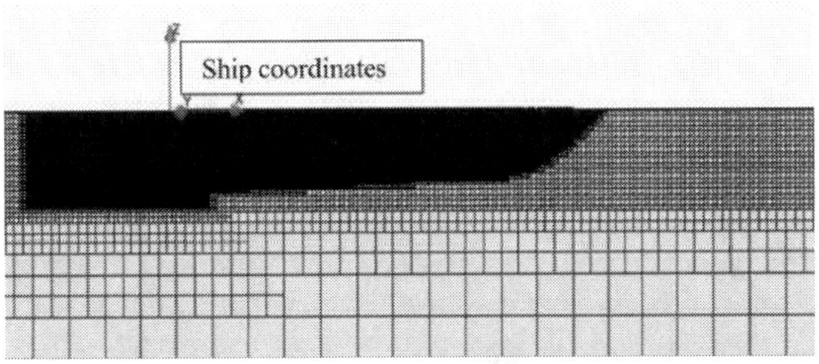

Figure 9. Case 3 refined regions along the depth

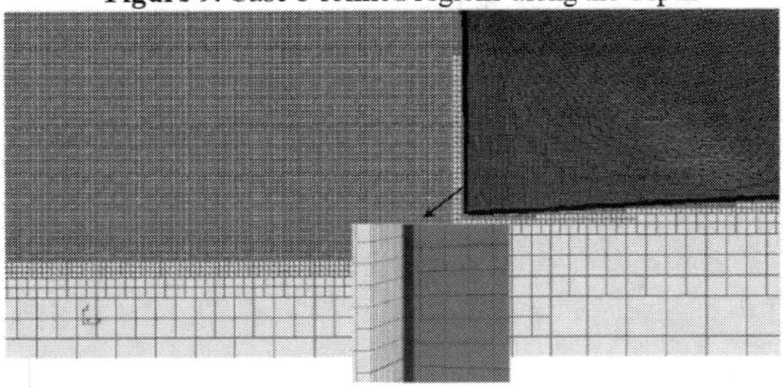

Figure 10. Case 3 virtual disk refined region and prism layers

3. Verification of computational models

In cases 1,2 the behavior of the DFBI and the TFE models was assessed and compared with the aim to confirm that the TFE model idealizes accurately the contact forces. Hence, the results are discussed only qualitatively and have not been compared to the 3D results. For case 3, the results are discussed qualitatively and quantitatively. The main sources of error and hence uncertainty consisted of the iteration error of the discretized equations, the grid resolution and the time step error for unsteady simulations, and other residual sources of error. The study focused on verifying the model in terms of the truncation error, i.e., the model is verified by assessing the mesh uncertainty of the simulations. The tool applied made use of the Richardson extrapolation method [26]. A convergence study using multiple solutions of different mesh resolutions with a uniform refinement ratio of $\sqrt{2}$ was carried out. Drift angle and velocity were kept constant at 15 degrees and 2 knots, respectively. The solution parameter on which the convergence study was performed has been the drag value. Three meshes were used with number of cells and drag solutions as shown in Table 3.

Table 3. Mesh uncertainty data

Refinement ratio	Base Size (m)	Number of cells	Drag	Unit	Solution symbol
	4.5	14.1 million	2395	N	S_1
$\sqrt{2}$	6.363	7.7 million	2450	N	S_2
	9	4.1 million	2930	N	S_3

The error estimate between the solution at the finest grid and the estimated exact solution S_0 at the infinitesimal base-size was defined as

$$\delta_{RE} = S_1 - S_0 = \frac{S_2 - S_1}{r^p - 1} = 7.1\ N \tag{11}$$

where r is the refinement ratio, and $p = 6.25$ is the order of the curve. The error estimate was eventually multiplied by a factor of safety of 1.25 to limit the simulation error to be $U_G = C\delta_{RE} = 1.25\delta_{RE} =$

8.89N. Although the error at the finest mesh is minimal, a medium mesh resolution (base size=6.36) was used for the simulations to reduce the computation time.

4. Results and discussion

This section presents the results of the hull hydrodynamic, thrust and contact forces acting on the tugboat for each studied case. Then, results of case 3 are used to form decision-making charts to be used by tug masters. For cases 1,2, the simulation time was 100 seconds with a time step of 0.05 seconds. The simulation scenario assumed that the tugboat is initially positioned at some distance from the ship. During the first impact, the contact force dramatically increased. Eventually, the tugboat moved out of the effective range due to rebounding, so the contact force diminished. The bounding and rebounding actions lasted for 20 seconds until the tugboat reached a steady contact with the plate and almost constant contact force values. After another 20 seconds, the contact forces started to change and fluctuate about a mean value. The force summation (Figure 11) shows that despite the variation of the contact and hydrodynamic forces the thrust force remained constant at zero. This means that any increase in the hydrodynamic force can be followed by a contact force decrement in the same direction. This observation supports the use of the TFE model in predicting contact forces. It also suggests that the variation of the contact force can be linked directly to the change in fluid actions on the tugboat.

The contact force variation in both cases follows the same trend. This reflects the strong dependency on hull forces. The mean forces show a difference of less than 2.5% (Figures 12, 13). The Y components of contact forces from case 1 lag those of case 2 (Figure 13). This reflects ship dynamics during the unsteady bounding / rebounding initial phase of contact that lasted for 20 seconds. For case 3, a set of 16 steady simulations over 2-5 knots and drift angles from 15 to 60 degrees were considered. Case 3 results are shown in Figure 14–18. A plot of the drag force against the drift angle for a range of velocities is depicted in Figure14. The drag increased with increasing velocity. The differences in the drag forces for the speed range considered at low drift angles was not significant. As the drift angle increased, this difference was dramatically raised. The speed increased proportionally to drag.

Figure 15 displays the variation of the sway force with the drift angle at different velocities. The sway force generally increased until the drift angle reached 45 degrees. After this point it decreased as the speed increased. When the sway force acted in the negative direction, the pushing force increased. Although the increase in the speed might cause a rise in the drag force and reduce pushing, it led to an increase of sway forces in the negative direction. This means that from an operational perspective the speed and drift angles should be kept at optimum levels to reach an ideal pushing force. The thrust coefficient variation against the drift angle is shown in Figure 16. As expected, it declined by increasing the speed, whereas it was slightly varied at the same speed with changing the drift angle. To get the maximum thrust available for pushing, the speed must be kept to a minimum. Figures 17 and 18 demonstrate the variation of the components of the pushing forces.

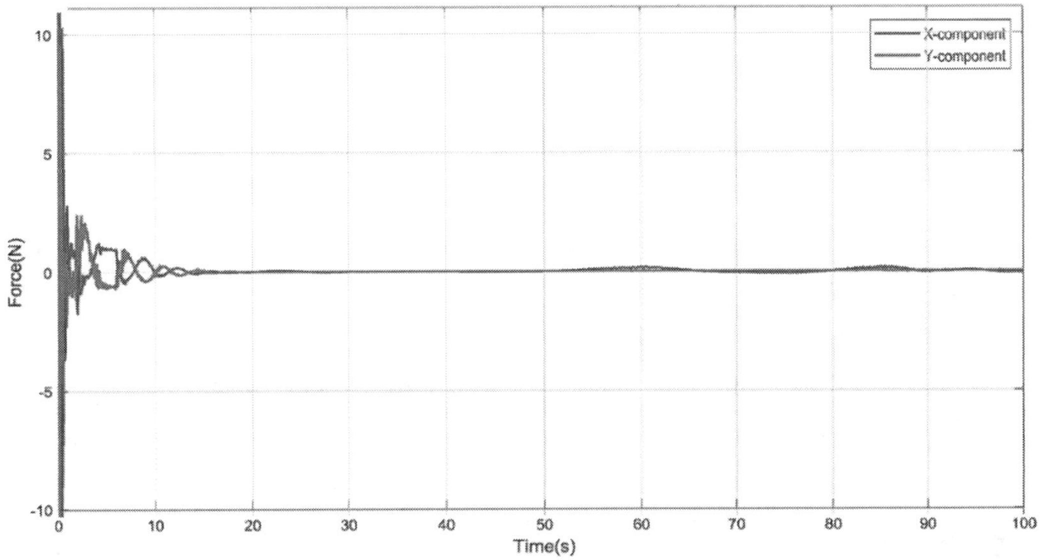

Figure 11. Force summation in the X, and Y directions.

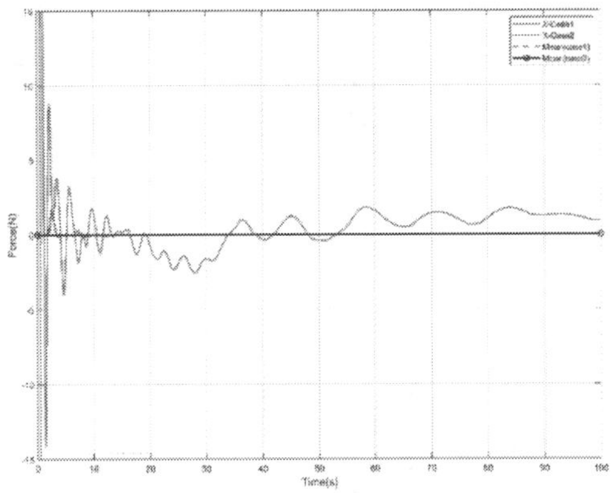

Figure 12. X component of the contact force in case 1 and case 2.

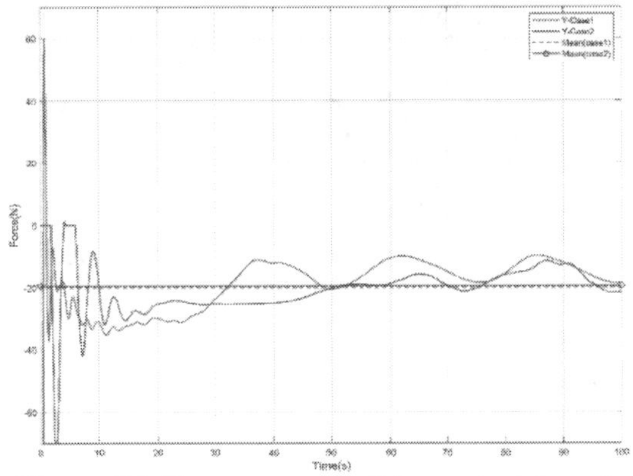

Figure 13. Y components of the contact force in case 1 and case 2.

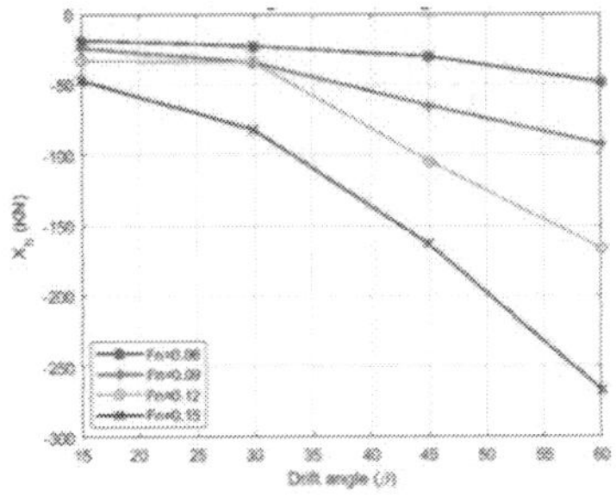

Figure 14. Drag versus drift angle

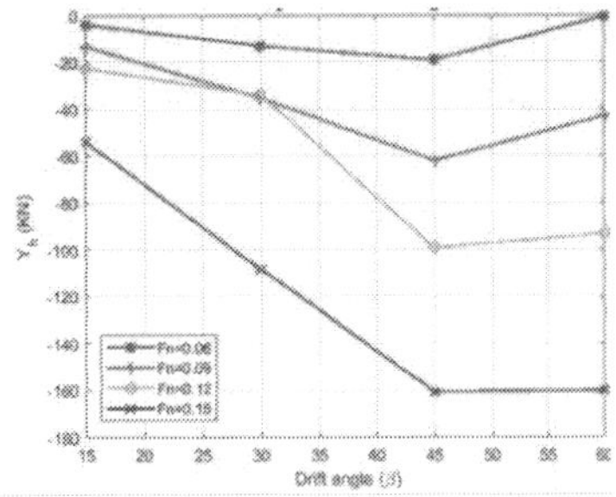

Figure 15. Sway force versus drift angle

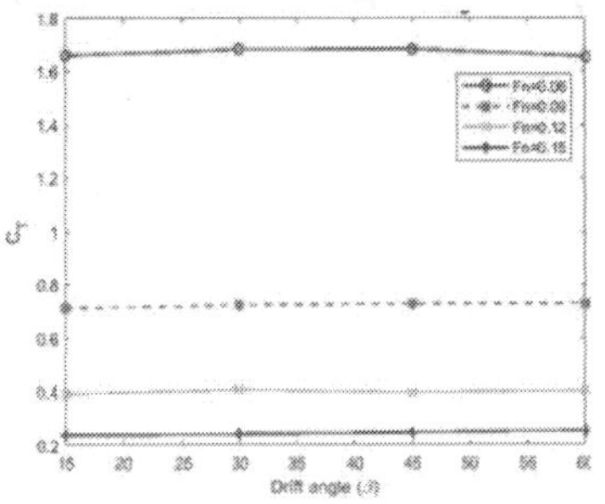

Figure 16. Thrust coefficient vs drift angle

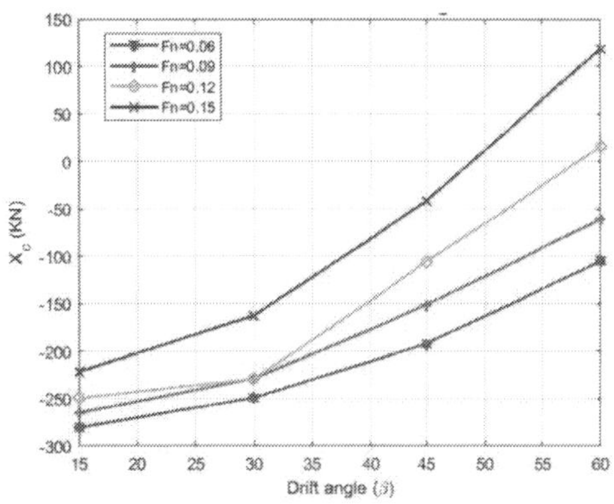

Figure 17. X component of the contact force

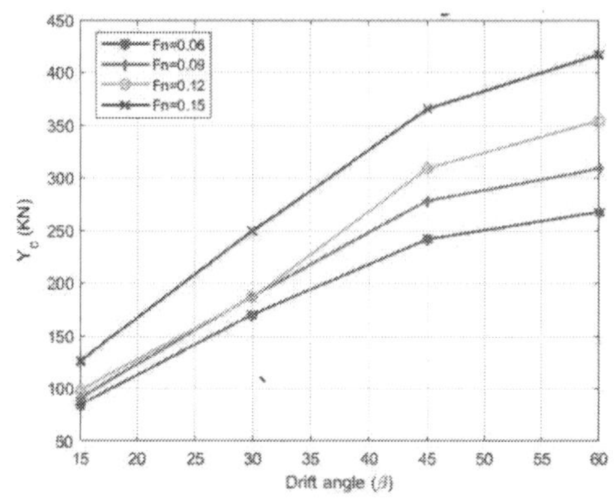

Figure 18. Y component of the contact force

4.1. Stability Against Sliding

Stability against sliding was determined by studying (i) whether the contact force on the tugboat is enough for sliding to occur and (ii) observing the variation of the critical angle at each velocity before sliding. The sliding condition states that the x-component of the contact force must be lower than the y-component multiplied by the static friction coefficient to avoid sliding $X_C \leq \eta \cdot Y_C$. Figures 19-22 demonstrate the range of stability against sliding using three friction coefficient values. The lower the angle at the intersection of the sliding curve with the x-axis, the higher the stability range without sliding. At low drift angles, the tugboat was unstable for the whole range of velocities. However, as the drift angle increased, the Y-component contact force increased, and the X-component contact force decreased until the critical angle of 45 degrees. Generally, speed increase widened the stability range against sliding. This could be attributed to the fact that the critical angle decreased.

4.2. Tug Performance Charts

The pushing forces and sliding curves were used to generate the tug performance charts (see Figure 23-26). The pushing force is presented in the polar coordinates (r,θ) instead of the cartesian coordinates to facilitate the decision making. At any point on the pushing force line, the r-component represents the pushing force value in tons, while the θ-coordinate represents the drift angle. The straight lines shown indicate whether the condition is stable against sliding or not. There are three different straight lines for each of the friction coefficient values. The region on the right side of the straight line represents the unstable area in which sliding will occur, and the tug master is recommended to use the rudder to return the tugboat to its required position. The left side of the friction coefficient line is expected to be stable; there is no need for the tug master to involve. The intersection point between friction coefficient lines and the pushing force line represents the critical angle.

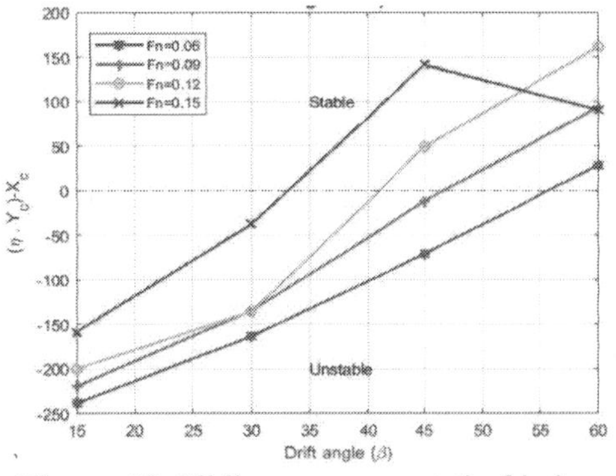

Figure 19. Sliding curves at static friction coefficient = 0.5

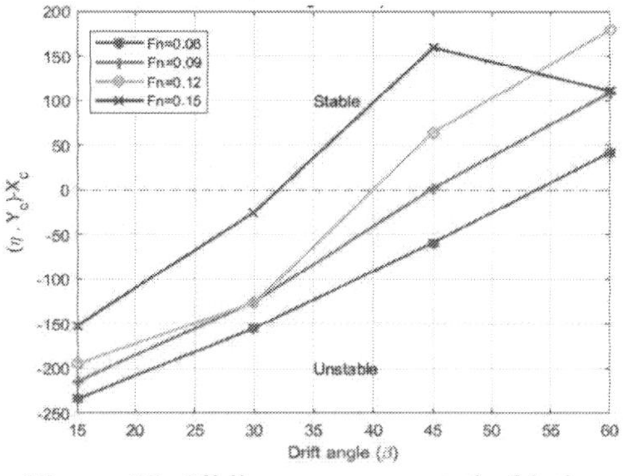

Figure 20. Sliding curves at static friction coefficient = 0.55

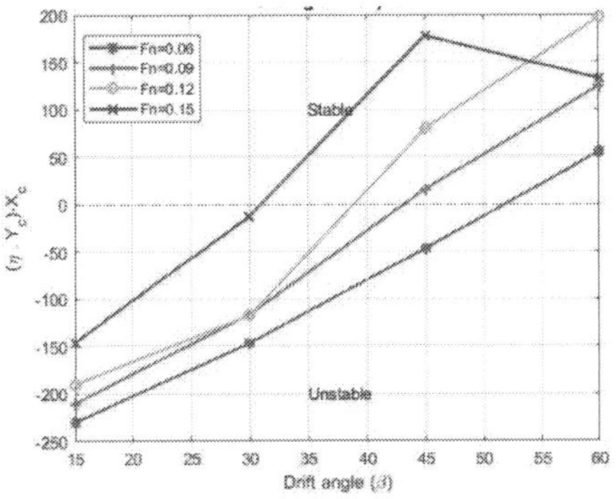

Figure 21. Sliding curves at static friction = 0.6.

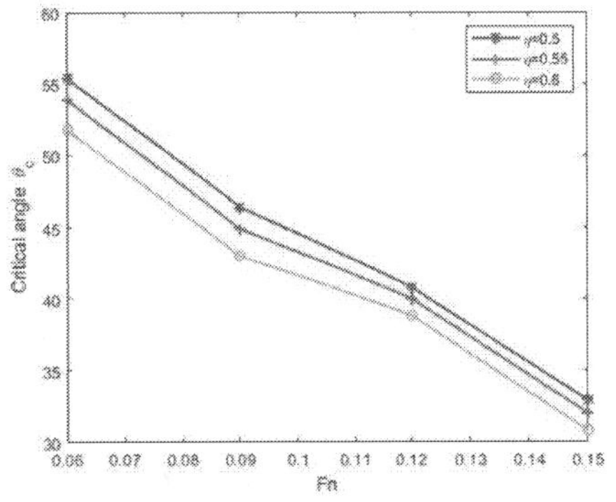

Figure 22. Critical angle against velocity.

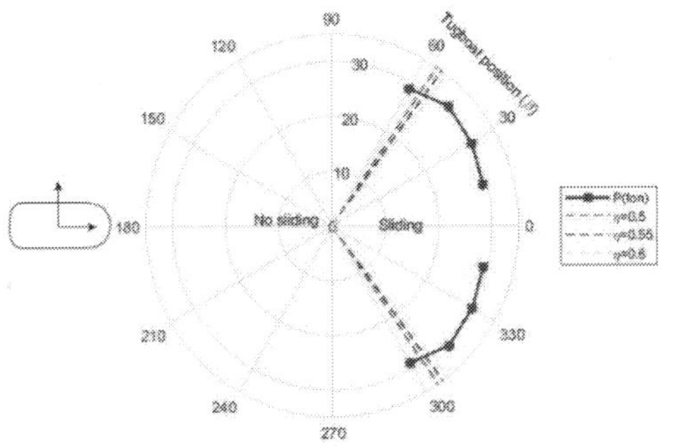

Figure 23. Pushing performance chart at 2 knots

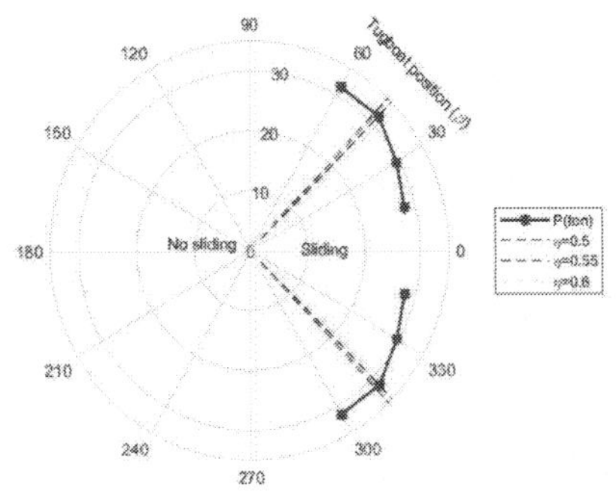

Figure 24. Pushing performance at 3 knots.

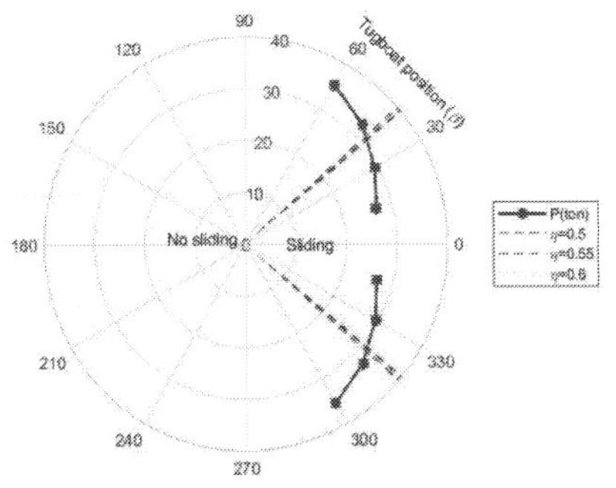

Figure 25. Pushing performance at 4 knots.

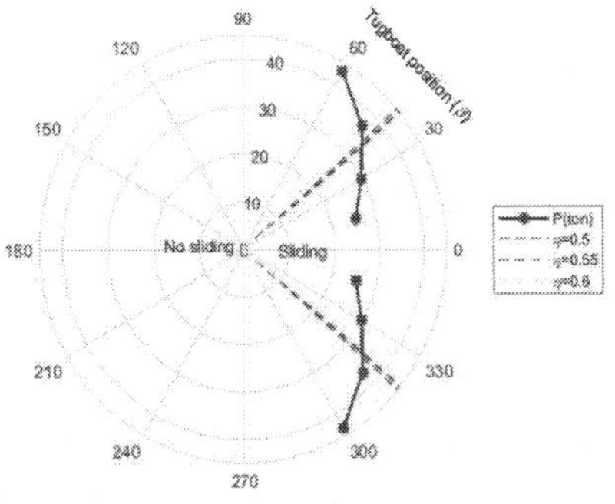

Figure 26. Pushing performance at 5 knots

5. Conclusions

This paper presented an investigation on the pushing performance of tugboats during ship assistance operations using CFD. The contact forces that act on the tugboat were predicted using two models. The first model applied an unsteady RANS CFD method with DFBI module and the contact mechanics coupling module in the STARCCM+ solver to simulate the contact between the two vessels and the tug motions. The second model utilized the same solver with either a steady or unsteady RANS method but without the DFBI and the contact mechanics coupling modules. A comparison of the contact forces predicted between the two models showed that the DFBI might be more elaborative in capturing the time variations of the contact forces. However, the tug force equilibrium model is more economic.

A set of 16 steady RANS simulations was conducted with the aim to predict the mean values of the contacts force acting on the tugboat and to develop decision-making charts. The simulations were conducted for a range of drift angles from 15 to 60 degrees and a speed range from 2 to 5 knots. Overall, the results demonstrated that the proper choice of orientation and speed of the tugboat might increase the pushing force to levels of its bollard pull and even more as the hull forces in such cases are supplementing the propeller forces. Comparison of results suggests that the DFBI model is able to capture well the contact coupling mechanism and could be employed for the derivation of more detailed decision-making charts for use in ship handling operations.

6. Acknowledgements

The authors wish to acknowledge CSC – IT Center for Science, Finland, for the provision of parallel computing computational resources used in this study and Alfons Hakans tug operators for the useful discussions during the preparation of this manuscript.

References

[1] Viran A and Mentes A 2021 Assessment of propulsion systems performance in tugboat *GİDB Dergi (Istanbul Tech. Univ.)* 53 – 66
[2] Taguchi H, Haraguchi T, Minami M and Houtani H 2015 An investigation into the capsizing accident of a pusher tug boat *12th International Conference on the Stability of Ships and Ocean Vehicles STAB 2015* (Glasgow) pp 903–910
[3] Lahtinen J, Valdez Banda O A, Kujala P and Hirdaris S 2020 Remote piloting in an intelligent fairway – A paradigm for future pilotage *Saf. Sci.* **130**
[4] HELCOM – Baltic Marine Environment Protection Commission 2011 *Report on shipping accidents in the Baltic Sea area during 2011.* (Helsinki, Finland)
[5] HELCOM – Baltic Marine Environment Protection Commission 2018 *Report on shipping accidents in the Baltic Sea from 2014 to 2017.* (Helsinki, Finland)
[6] Butt N, Johnson D, Pryce-roberts N and Vigar N 2015 15 Years of Shipping Accidents : A review for WWF Southampton Solent University *Shipp. Accid. Rep.* **44** 37–63
[7] Vantorre M, Eloot K, Delefortrie G, Lataire E, Candries M and Verwilligen J 2017 Maneuvering in shallow and confined Water *Encyclopedia of Maritime and Offshore Engineering* (John Wiley & Sons, Ltd) pp 1–17
[8] Piaggio B, Villa D and Viviani M 2020 Numerical analysis of escort tug manoeuvrability characteristics *Appl. Ocean Res.* **97** 102075
[9] Figari M, Martinelli L, Piaggio B, Enoizi L, Viviani M and Villa D 2020 An All-Round Design-to-Simulation Approach of a New Z-Drive Escort Tug Class *J. Offshore Mech. Arct. Eng.* **142**
[10] Piaggio B, Viviani M, Martelli M and Figari M 2019 Z-Drive Escort Tug manoeuvrability model and simulation *Ocean Eng.* **191** 106461
[11] Simonsen C D, Nielsen C K, Otzen J F and Agdrup K 2011 CFD based prediction of ship-ship interaction forces on a TUG beside a tanker *RINA, Royal Institution of Naval Architects - 2nd International Conference on Ship Manoeuvering in Shallow and Confined Water: Ship to Ship Interactions 2011* pp 329–38
[12] Sutulo S and Guedes Soares C 2009 Simulation of close-proximity maneuvers using an online 3D potential flow method *MARSIM '09 Conference, Proceedings* pp M-9-1/10
[13] Lataire E, Vantorre M and Delefortrie G 2009 Captive model testing for ship to ship operations *MARSIM 09 Conference Proceedings* pp M-14-1/10
[14] Falter J 2010 *Validation of a potential flow code for computation of ship-ship interaction forces with captive model test results*
[15] Jayarathne N, Ranmuthugala D, Leong Z, Fei J, Chai S and Chen H M 2019 Numerical and experimental prediction of hydrodynamic interaction effects acting on tugs during ship manoeuvres *J. Mar. Sci. Technol.* **27** 185–201

[16] Jayarathne B, Ranmuthugala D, Chai S and Fei J 2014 Estimation of Drag Force on Tug Hulls at Different Drift Angles *3rd International Symposium of Maritime Sciences* pp 10–4
[17] Stickings T, Griffith K, Mitchell C, Wright J and Patel B 2021 MailOnline *dailymail*
[18] Håkans A 2020 Icebreaking
[19] Xu D, Zhang X and Yin Y 2016 The computer simulation study of tug pushing operation system *Chinese Control Conference, CCC* vol 2016-Augus (Chengdu, China: IEEE) pp 2074–80
[20] Artyszuk J 2013 Simplified steady-state analytical model for manoeuvring and control of ASD tug in escort push operations *Zesz. Nauk. / Akad. Morska w Szczecinie* **nr 36 (108** 5–14
[21] Menter F R 1994 Two-equation eddy-viscosity turbulence models for engineering applications *AIAA J.* **32** 1598–605
[22] Kundu P K and Cohen I M 2002 *Fluid Mechanics* (California: Elsevier)
[23] S. Goldstein 1929 *On the vortex theory of screw propellers* (London)
[24] Bernitsas M M, Ray D and Kinley P 1981 *KT, KQ and efficiency curves for the Wageningen B-series propellers*
[25] Sosnowski M, Krzywanski J, Grabowska K and Gnatowska R 2018 Polyhedral meshing in numerical analysis of conjugate heat transfer *EPJ Web Conf.* **180**
[26] Resistance Committee of 23rd International Towing Tank Conference 2002 *ITTC - Recommended Procedures and Guidelines Uncertainty Analysis in CFD , Verification and Validation Methodology and Procedures*

URANS assessment of ship extreme roll event in irregular stern quartering sea

X Chen[3], L W Liu[1, 2]*, Z G Zhang[1, 2], X Z Wang[1, 2], D K Feng[1, 2]

[1] School of Naval Architecture and Ocean Engineering, Huazhong University of Science and Technology, Wuhan, China
[2] Key Laboratory of Ship and Ocean Hydrodynamics of Hubei Province, Wuhan, China
[3] China Ship Development and Design Center (Shanghai Branch), Shanghai, China

*Email : llw@hust.edu.cn

Abstract. In this study, the extreme roll events of a free-running ONR Tumblehome model in stern quartering seas are predicted by an unsteady Reynolds-averaged Navier–Stokes (URANS) solver coupled with a dynamic overset grid approach. The validation study is first performed for the prediction of ship extreme roll amplitude in regular waves, indicating satisfactory agreement with an average difference of less than 3% between the simulation and experimental results. The URANS approach is then applied to the assessment of the extreme roll events as the ship is sailing in irregular seas. The ship is free to 6 degrees of freedom (DOF) motions with a course-keeping control. It is concluded that the ship's yaw motion plays an important role in the occurrence of ship extreme roll events in an irregular stern quartering sea. A rapid increase of yaw angle is observed for the ship coupled with the occurrence of the extreme roll, even leading to the capsizing.

1. Introduction

As a ship is sailing in stern quartering seas, the encounter frequency may be low due to the Doppler effect. This situation may lead to the wave crest passing the midship section of a ship with a speed higher than the ship's forward speed, causing the relatively small restoring arm and this state exists long enough. The decrease in roll stiffness can be large enough to make the dynamical system very soft, creating a very large roll angle or even capsizing of the ship. The loss of stability of a ship in stern quartering seas is becoming an important reason for marine accidents and it has been considered one of the dynamic stability failure modes in the second-generation intact stability criteria, proposed by the International Maritime Organization (IMO), with the name of "pure loss of stability".

With the occurrence of the stability failure mode, the ship will experience the large-amplitude extreme roll, which causes potential danger for the high-speed ship. Thus, the predicting methods for the occurrence of the stability failure mode in waves are required urgently as stated by IMO [1], along with the establishment of multitiered criteria, including the level 1 and level 2 vulnerability criteria and level 3 direct stability assessment. If the vulnerability criteria are not met, the direct stability assessment should be adopted using a numerical simulation tool.

As for the development of the direct stability assessment method, plenty of research efforts have been carried out to reproduce these phenomena in following and stern quartering seas to understand their dynamics in regular waves, with emphasis on the estimation of the restoring arm variation. Hashimoto

[2] and Lilienthal et al. [3] adopted a surge-roll 2 degrees of freedom (DOF) model. Lu et al. [4] predicted the nonlinear ship motions with a surge-heave-roll-pitch 4DOF model. Lu et al. [5] conducted the predictions of the large-amplitude ship motions in stern quartering seas by a 6DOF mathematical model.

With the development of high-performance computational techniques, the computational fluid dynamics (CFD) method makes it possible to predict the ship's nonlinear motion directly without the need for inputs such as the hydrodynamic coefficients, damping coefficients, restoring coefficients, coupling term coefficients and maneuvering coefficients, which can consider all relevant physics in an accurate way. CFD study efforts have been performed for the predictions of ship motions with large amplitudes in waves. Sadat-Hosseini et al. [6], Zhou et al. [7], and Liu et al. [8] carried out CFD simulations of ship parametric roll in head waves, and the numerical results agreed well with the experimental data. Carrica et al. [9], Sadat-Hosseini et al. [10], and Begovic et al. [11] performed studies on the surf-riding and broaching phenomena of the ship in following sea and stern quartering sea with the CFD approaches. Overall, it is concluded that the application of the direct CFD method is possible for the evaluation of the nonlinear stability failure modes of the ship in waves at the design stage.

Above the literature reviews, the CFD application on the estimation of ship extreme roll in a realistic sea is still limited. Therefore, a further application of the CFD method, an in-house unsteady Reynolds-averaged Navier–Stokes (URANS) solver coupled with the dynamic overset approach, has been studied for a free-running ONR Tumblehome model in an irregular stern quartering sea. The ship is free to 6DOF motions with a course keeping control at nominal Froude number (Fn) of 0.40, corresponding to a high-speed condition, and the occurrence of the extreme roll event for the free-running ship is predicted, along with the assessment of the force and moment acting on the ship. Discussions of simulation results are presented and some conclusions are drawn finally.

2. Ship model and simulation conditions

The ONR Tumblehome model 5613, a preliminary design for a modern concept of a surface combatant, is used as the study object. Figure 1 shows the geometry of the ship model. The 1/40.526 scaled ship model is appended with skeg, bilge keels, twin rudders, and twin propellers but without the superstructure, which is consistent with the setup in the model test performed by Lu et al. [5]. They carried out the free-running model test in the seakeeping basin (69m length, 46m breadth, 4m depth) of China Ship Scientific Research Center. During the experiment, the extreme roll event occurs for the free-running ship model in regular stern quartering waves, as well as the capsizing, which are used for the validation of the unsteady RANS solver. The details of the main parameters are shown in table 1.

Figure 1. Geometry of ONR Tumblehome.

Table 1. Main parameters of ONR Tumblehome model.

Description	Symbol	Units	Value
Scale		-	40.526
Length of waterline	L_{WL}	m	3.800
Beam	B	m	0.463
Draft	T	m	0.136
Freeboard	F	m	0.222
Displacement	Δ	kg	127.8
Block coefficient	C_B	-	0.535
Vertical center of gravity	$\overline{KG}$	m	0.203
Metacentric height	$\overline{GM}$	m	0.037
Natural roll period	T_ϕ	s	2.199
Radius of gyration along the y-axis	K_{yy}	m	0.95
Radius of gyration along the z-axis	K_{zz}	m	0.95
Rudder			
Type		Spade	
Rudder area	A_R	m^2	0.0145
Maximum rudder angle	δ_{max}	°	±35
Propeller			
Type		Fixed pitch	
No. of blades	-	-	4
Propeller diameter	D_p	m	0.129
Hub ratio	D_h/D_p	-	0.20
Rotation		Inwards	

The unsteady simulation of the ship model is carried out for the ship advancing at Fn = 0.40, corresponding to the advancing speed of 2.44 m/s, where Fn is the Froude number defined as follows:

$$Fn=\frac{U}{\sqrt{gL_{WL}}} \tag{1}$$

where U is the ship sailing speed and g is the acceleration due to gravity. It is noted that the specified propeller rotational speed is kept constant throughout the simulation, and set to be the same with the corresponding propeller rate, which can drive the ship model to achieve the nominal Fn in calm water. During the simulation, the ship model is free to undergo 6DOF motions with the course keeping control in a stern quartering wave with heading angle χ = 30° (wave heading from starboard quarter).

The irregular wave composed of wind sea and swell components is considered as the wave condition to assess ship extreme roll behavior in a realistic sea. The well-known Ochi-Hubble bimodal spectrum, proposed by Ochi and Hubble [12], is applied to model the mixed sea. This spectrum is composed of two components, the high-frequency part of wind-sea and the low-frequency part of swell, which can model the mixed sea properly, expressed as follows:

$$S(\omega)=\frac{1}{4}\sum_{j}\frac{\left(\frac{4\lambda_j+1}{4}\omega_{pj}^4\right)^{\lambda_j}}{\Gamma(\lambda_j)}\frac{H_j^2}{\omega^{4\lambda_j+1}}\exp\left(-\left(\frac{4\lambda_j+1}{4}\right)\left(\frac{\omega_{pj}}{\omega}\right)^4\right) \tag{2}$$

where H_j is the significant wave height; ω_{pj} is the angular peak frequency; λ_j is the spectral shape factor; and j=1, 2 represents the swell and wind-sea components respectively. The wave parameters are shown in table 2, following the previous study [13], where $H_{1/3}$ is the significant height and T is the mean period, corresponding to the sea state 6. Figure 2 shows the bimodal spectrum of the mixed sea.

Table 2. Wave parameters for the mixed sea modeled by Ochi-Hubble bimodal spectra.

Swell			Wind-sea			$H_{1/3}$ (m)	T (s)
H_1 (m)	ω_{p1} (rad/s)	λ_1	H_2 (m)	ω_{p2} (rad/s)	λ_2		
4.58	0.45	3.19	3.64	0.88	1.55	5.85	9.08

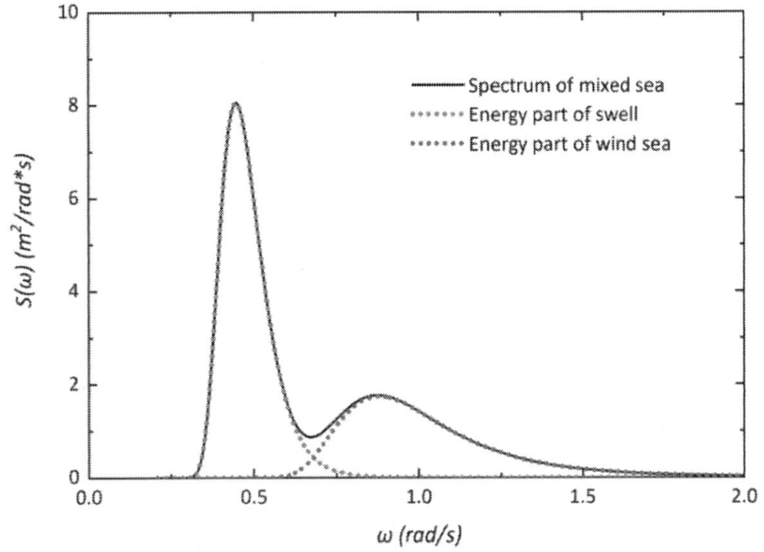

Figure 2. Bimodal spectrum of the mixed sea composed of wind sea and swell.

3. Computational method

The inhouse viscous CFD code is developed for ship hydrodynamics and CFD simulations. URANS equations are shown as follows, which is solved in liquid phase of free surface flow.

$$\frac{\partial U_i}{\partial x_i} = 0 \tag{3}$$

$$\frac{\partial U_i}{\partial t} + \frac{\partial U_i U_j}{\partial x_j} = -\frac{1}{\rho}\frac{\partial P}{\partial x_i} + \frac{1}{\rho}\frac{\partial}{\partial x_j}\left(\mu \frac{\partial U_i}{\partial x_j} - \rho\overline{u_i u_j}\right) \tag{4}$$

where U_i and U_j (i, j =1, 2, 3) represent time-averaged velocity components; ρ is density; $\bar{p}$ is time-averaged pressure; μ is dynamic viscosity and $\rho\overline{u_i u_j}$ is Reynolds stress term. This term can be solved by using *SST k-ω* turbulence model to close the governing equations.

URANS equations are discretized using the finite difference method at the structured grid points. The projection algorithm is employed for the solution of pressure-velocity coupling effectively. The free surface of the flow is captured using the level-set method. Based on the velocity-pressure flow field computation, the forces and moments acting on the ship hull are computed with respect to the inertial coordinate system and projected into the non-inertial coordinate system, then inputted to the rigid-body motion equations. The translational and rotational accelerations are obtained by solving the rigid-body motion equations to predict the 6DOF motions of the ship at each time step. The proportional–integral controller is used for the free-running ship simulation. The number of revolutions per second (rps) of the propeller is controlled to achieve the ship's target speed. The rudder angle can be executed by the feedback controller following the deviation of the ship's yaw angle from the target heading angle, to achieve the course-keeping control for the ship. The body force model is used to simplify the propeller as an actuator disk. The local instantaneous velocity distribution is first obtained at the propeller plane during the unsteady RANS simulation and directly inputted to the body force model. The lift and drag distributions are then calculated based on the simplified quasi-steady two-dimensional blade element momentum theory in each iteration step, where the propeller blade is equally divided in the radial direction and considered as an airfoil segment, as proposed by Tokgoz [14].

Dynamic overset grid approach is applied for the connection between the computational domain, ship hull, and rudders, allowing the large-amplitude motion of the ship and relatively rotational motion of the propeller and rudder deflection with respect to the ship-fixed coordinate system. The overset grid system is updated dynamically at each time step following the predicted or prescribed motions, along with the real-time computations of the domain connectivity information, thereby realizing the relative movement among the neighboring grids without regenerating the grids. In addition, as a result of the application of the overset grid approach, the computational grid model can be generated independently for each grid part, including the computational domain, ship hull, and different appendages. The Cartesian grid is generated for the computational domain. The body-fitted structured grids are generated for the ship hull and appendages and the overset grid system generated by assembling different grids together is presented in figure 3. The details of the URANS solver and overset treatment are presented in Liu et al. [13, 15, 16].

The extreme roll results of the free-running model in a regular stern quartering wave from Lu et al. [5] are used as the benchmark data and compared with the CFD simulation results for the validation study. The CFD predictions have been carried out in earlier work [15] for the model under different speed conditions with the course-keeping control. Satisfactory agreements among the results of the extreme roll angle are observed between the CFD and experimental data, with an average difference of less than 3%. In addition, the threshold for the occurrence of ship capsizing is observed at the speed condition of $Fn = 0.3$ during the CFD simulations, which is consistent with the experimental results.

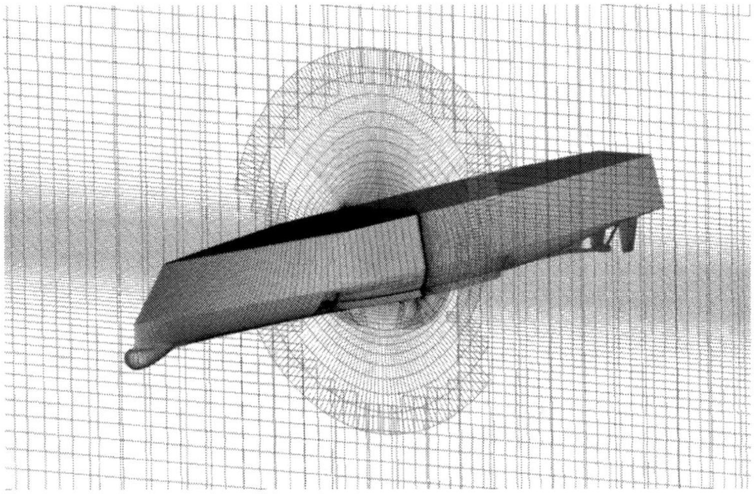

Figure 3. Total overset grid system for the numerical simulation.

4. Result and discussion

The CFD simulation is carried out for the free-running ship with course-keeping control in an irregular stern quartering sea. The results of the heave, roll, pitch, and yaw motions are shown in figure 4. After several wave encounters, the ship roll angle is observed increasing rapidly towards portside at 93 s and the ship eventually experiences the capsizing event. In addition, together with the increase of the roll angle, the rapid increase occurs for the ship's yaw motion towards starboard. As the ship is rolling 90° towards portside at 95.4 s, the yaw angle reaches 22° towards starboard, indicating the strong correlation between the yaw and roll motions in stern quartering seas.

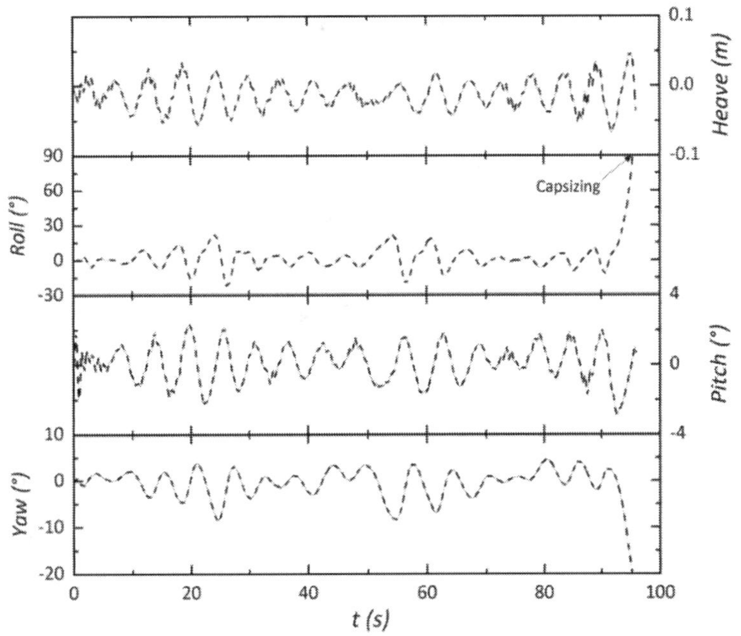

Figure 4. Time histories of the heave, roll, pitch, and yaw motions for the free-running model in irregular stern quartering sea.

Since the extreme roll event (capsizing) occurs for the 6DOF ship model with the course-keeping control, the snapshots of the CFD simulations from 93 s to 94.9 s are presented in figure 5. It is observed that a wave crest propagating from starboard quarter reaches the ship stern at 93 s, creating a sudden turning motion for the ship towards starboard. The stern quartering wave is then propagating through the ship (from 93 s to 94.9 s). During this stage, the ship roll angle increases due to the coupling effect of the centrifugal effect and wave-induced load. At 94.9 s the ship is rolling 65° to portside with a yaw angle of 18° to starboard. In addition, the starboard rudder is observed totally out of the water at the same time.

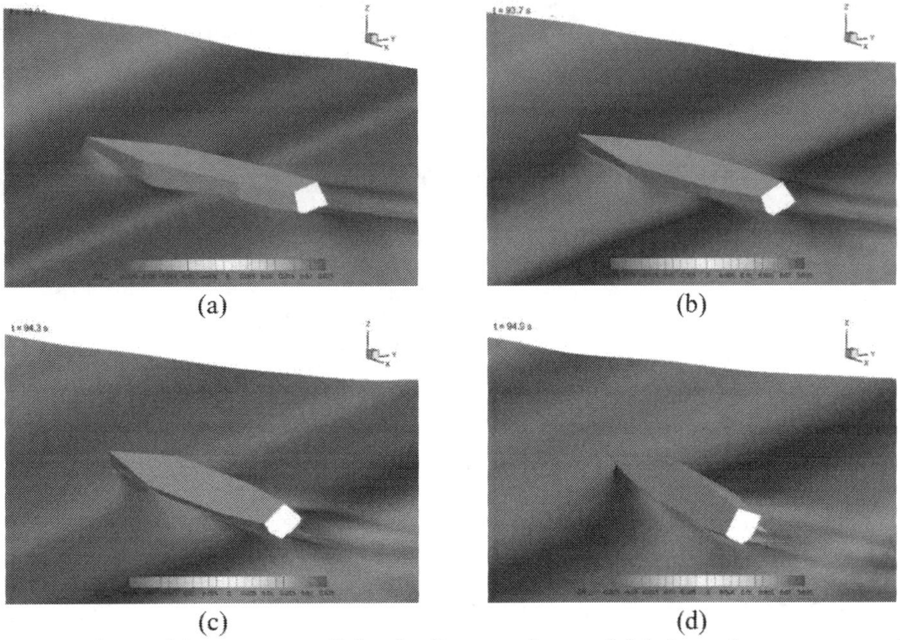

Figure 5. Snapshots of the extreme roll for the free-running model in irregular stern quartering sea.

Figure 6 shows the time histories of the roll moment M_x and yaw moment M_z acting on the free-running model in an irregular stern quartering sea. It is noted that the ship yaw moment acting on the ship under 6DOF mode is mainly induced by two effects, one the wave load and the other the rudder action. Therefore, the yaw moment is decomposed into wave-induced component M_{z_wave}, the moment due to portside rudder action $M_{z\_portside\ rudder}$, and starboard rudder action $M_{z\_starboard\ rudder}$. With the variation of the ship yaw angle, the rudders are deflected by the course keeping control to provide a reactive yaw moment. At 92.7 s the ship is turning towards starboard with the rudder deflection towards portside, thereby causing a lateral force towards starboard, in consequence creating a reactive yaw moment towards portside, which is a significant contributor to the total yaw moment (see Fig. 6c). However, owing to the larger wave-induced moment, the ship still turns towards starboard with an increase of the yaw angle. In addition, the lateral force induced by the rudder deflection creates a positive roll moment (towards portside) that reaches 4.2 N.m at 94 s, thus contributing to the roll motion towards portside (see Fig. 6b). Therefore, the ship yaw motion and corresponding rudder action are the important contributors to the occurrence of the extreme roll event.

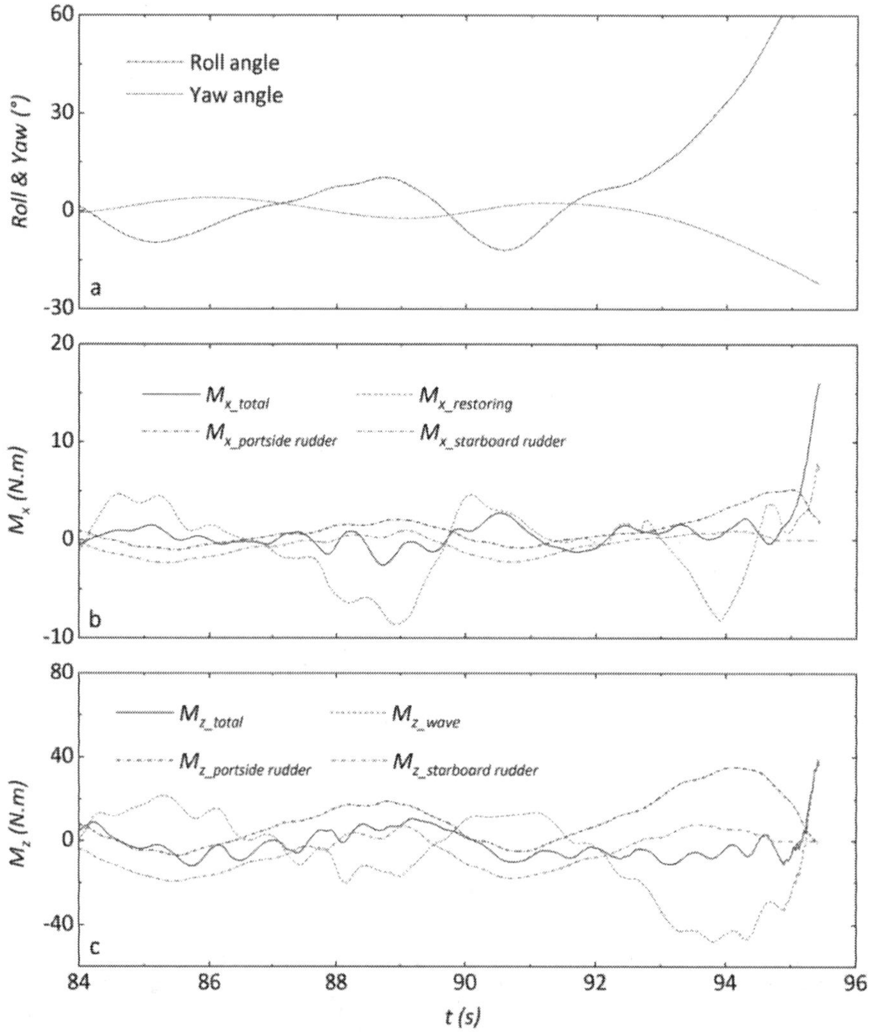

Figure 6. Time histories of the ship roll and yaw angle, roll moment, and yaw moment for the free-running model in irregular stern quartering sea.

5. Conclusion

In this study, a URANS solver coupled with a dynamic overset grid approach is applied to the prediction of the extreme roll events of a free-running ONR Tumblehome model in irregular stern quartering seas of $\chi = 30°$. The ship is free to 6DOF motions with a course keeping control at nominal Fn of 0.40, corresponding to a high-speed condition. The irregular wave composed of wind sea and swell components, modeled by the Ochi-Hubble bimodal spectrum, is considered as the wave condition to assess the ship's extreme roll behavior in a realistic sea.

The validation study is first performed for the prediction of ship extreme roll amplitude in regular waves, indicating satisfactory agreement with an average difference of less than 3% between the simulation and experimental results at different speed conditions.

The results of the irregular-wave simulation show the occurrence of the capsizing for the free-running ship in a stern quartering sea. It is concluded that the ship's yaw motion and rudder action are important contributors to the occurrence of the extreme roll event. A rapid increase of yaw angle is observed for the ship coupled with the occurrence of the extreme roll, even leading to the capsizing.

References

[1] IMO 2017 Finalization of second generation intact stability criteria *Ship Design and Construction (SDC) 4th Session* WP P 4.
[2] Hashimoto H 2009 Pure loss of stability of a tumblehome hull in following seas *Proceedings of the 19th International Offshore and Polar Engineering Conference* Osaka Japan
[3] Lilienthal T, Matsuda A, and Thomas G 2007 Dynamic stability in following seas: predictive and experimental approaches *Journal of Marine Science and Technology* 12(2) 111-118
[4] Lu J, Gu M and Boulougouris E 2019 Model experiments and direct stability assessments on pure loss of stability of the ONR Tumblehome in following seas *Ocean Engineering* 194 106640
[5] Lu J, Gu M and Boulougouris E 2020 Model experiments and direct stability assessments on pure loss of stability in stern quartering waves *Ocean Engineering* 216 108035
[6] Sadat-Hosseini H, Stern F, Olivieri A, Campana E F, Hashimoto H, Umeda N, Bulian G and Francescutto A 2010 Head-wave parametric rolling of a surface combatant *Ocean Engineering* 37(10) 859-878
[7] Zhou Y, Ma N, Lu J and Gu X 2015 A study on applicability of CFD approach for predicting ship parametric rolling *Proceedings of the 12th International Conference on the Stability of Ships and Ocean Vehicles* Glasgow UK
[8] Liu L W, Chen M X, Wang X Z, Zhang Z G, Yu J W, and Feng D K 2021 CFD prediction of full-scale ship parametric roll in head wave *Ocean Engineering* 233 109180
[9] Carrica P M, Paik K, Sadat-Hosseini H and Stern F 2008 URANS analysis of a broaching event in irregular quartering seas *Journal of Marine Science and Technology* 13(4) 395-407
[10] Sadat-Hosseini H, Carrica P, Stern F, Umeda N, Hashimoto H, Yamamura S and Mastuda A 2011 CFD, system-based and EFD study of ship dynamic instability events: surf-riding, periodic motion, and broaching *Ocean Engineering* 38(1) 88-110
[11] Begovic E, Gatin I, Jasak H and Rinauro B 2020 CFD simulations for surf-riding occurrence assessment *Ocean Engineering* 218 107975
[12] Ochi M K, Hubble E N 1976 Six-parameter wave spectra *Coastal Engineering Proceedings* 15(1) 301-328
[13] Liu L, Chen M, Yu J, Feng D and Wang X 2022 Numerical study of swell impact on seakeeping performance of surface combatant *Ocean Engineering* 249 110827
[14] Tokgoz E 2015 A CFD study on the propeller-hull interaction flow in waves using body-force distribution model *Osaka University* Suita Japan
[15] Liu L, Feng D, Wang X, Zhang Z, Yu J and Chen M 2022 Study on extreme roll event with capsizing induced by pure loss of stability for the free-running ONR tumblehome *Ocean Engineering* 257 111656
[16] Liu L, Yao C, Feng D, Wang X, Yu J and Chen M 2022 Numerical study of the interaction between the pure loss of stability, surf-riding, and broaching on ship capsizing *Ocean Engineering* 266 112868

IOP Conf. Series: Materials Science and Engineering 1288 (2023) 012005 doi:10.1088/1757-899X/1288/1/012005

A numerical investigation on ship-ship overtaking interactions in confined waterways

Z Q Zheng[1], **L Zou**[1,2] **and Z J Zou**[1,2]

[1] School of Naval Architecture, Ocean and Civil Engineering, Shanghai Jiao Tong University, Shanghai 200240, China
[2] State Key Laboratory of Ocean Engineering, Shanghai Jiao Tong University, Shanghai 200240, China

E-mail: luzou@sjtu.edu.cn

Abstract. The whole process of a KCS container ship overtaking a low-speed Neo-Panamax container ship is numerically simulated by solving the unsteady Reynolds Averaged Navier-Stokes (URANS) equations. The dynamic overset mesh technique is adopted to simulate the relative motions between the two ships and the ship attitude variations. The free-surface elevations, ship attitudes, as well as hydrodynamic forces and moments on the two ships are predicted. In order to discuss the effects of different navigation parameters on the ship-ship interactions, a set of ship speeds, lateral distances and water depths are simulated. The results illustrate that the overtaken ship is more significantly impacted by the interaction than the overtaking ship, and the drawdown generated by the overtaking ship is the key factor during the overtaking process. The increase of ship speed, the decrease of lateral distance or water depth can lead to more intense drawdown, and hence even more significant ship-ship interaction effects.

1. Introduction

The ship-ship interaction problems often occur in confined waterways, such as ship meeting and overtaking. These interactions may induce significant impacts on ship motions, stability, and navigation safety. For example, a large container ship overtook a feeder vessel on the Elbe river in 2004, resulting in a collision accident [1]. As the shipping industry continues to develop and the ship sizes tend to be increasingly larger, ship-ship interactions become more frequent and intense. Therefore, the accurate prediction of ship-ship interactions in confined waterways is a long-term concern in the engineering community. Studying this problem can provide valuable references for evaluating navigation safety in confined waterways.

At present, the main methods to investigate ship-ship interaction problems include model tests, potential flow methods with the inviscid-fluid simplification and Computational Fluid Dynamics (CFD) methods considering the fluid viscosity. The model tests are regarded as the most reliable method [2]. Remery [3], Vantorre et al. [4], Gronarz [5] and Mousaviraad et al. [6] carried out extensive captive model tests to investigate the influences of different navigation parameters on ship-ship interactions. However, due to the high costs and long periods of model tests, the numerical methods gradually become another important tool to explore the ship-ship interaction problems. With the advantage of rapid calculations, the potential flow method was firstly widely applied in this field. Using two potential flow codes, Pinkster and Ruijter [7] calculated the interaction forces on a moored ship due to the passing-ship effect. The main difference between these two codes was whether the free surface elevation was taken into account in the simulations. The results illustrated that the free-surface effects should be considered in the cases with complex harbour geometry. Sutulo et al. [8-10] and Degrieck et al. [11]

developed real-time calculation codes to investigate ship-ship interactions. It was found that the potential flow methods were normally suitable for simple cases like open-water or low-speed situations, but there were still errors in the cases with close ship-ship or ship-bank distances due to the neglect of fluid viscosity. Therefore, the CFD method, especially by solving Reynolds-averaged Navier-Stokes (RANS) equations, becomes a more appropriate choice for investigating the ship-ship interaction problems. Applying the CFD method, Nandhini and Nallayarasu [12] discussed the effects of different parameters on a ship-ship interaction problem. Zhou et al. [13] mainly explored the effects of a high-speed frigate passing a moored container ship with the focus on the Kelvin wave system. Applying the benchmark test cases of the 6th International Conference on Ship Manoeuvring in Shallow and Confined Water (MASHCON2022) [14], Zheng et al. [15-17] established a reliable numerical method to predict free-surface elevations, unsteady ship motions, as well as hydrodynamic forces and moments in the passing-ship problem and to investigate the physical mechanisms of passing-ship effects on a moored ship with high blockage effects. On this basis, this paper aims to extend the study to the ship-ship overtaking problems, where both ships are moving but with speed differences.

To obtain reliable predictions of unsteady flow features during the overtaking process, the entire process of a ship at higher speeds overtaking another slower ship is numerically simulated by the unsteady Reynolds Averaged Navier-Stokes (URANS) method. Dynamic overset mesh technique is utilized to simulate ship motions and changes of the ship attitudes. Effects of ship speeds, water depths and lateral distances between the two ships are considered in the present study, and the free-surface elevations around the two ships, ship attitudes, hydrodynamic forces and moments on the two ships are predicted to explore the effects of these navigation parameters on the ship-ship overtaking interactions. The analyses based on the predicted quantities and flow field results are conducted to evaluate the mechanism of overtaking interactions in confined waterways.

2. Problem description

The container ship KCS is selected as the overtaking ship, and its forward speed is U_1; the Neo-Panamax container ship with the speed U_2 is selected as the overtaken ship. Both ships are without appendages, and their geometries and main particulars are shown in Figure 1 and Table 1.

Figure 1. Ship geometries (left: KCS; right: Neo-Panamax).

Table 1. Main particulars of the two ships [14].

Main particulars	KCS	Neo-Panamax
Scale ratio (-)	1/80	1/80
L_{pp} (m)	4.367	4.350
B (m)	0.611	0.610
T (m)	0.190	0.190
m (kg)	320.6	326.2
x_G (m)	−0.048	−0.114
z_G (m)	0.003	−0.002
I_{xx} (kgm^2)	11.9	11.2
I_{yy} (kgm^2)	367.4	396.6
I_{zz} (kgm^2)	385.8	376.1

Figure 2 illustrates a schematic diagram of the present overtaking problem. The two ships move parallelly with different speeds in the waterway, where the waterway width W is only 6.11 m. Three right-handed coordinate systems are applied in the numerical simulations: an earth-fixed coordinate system and two ship coordinate systems. The earth-fixed coordinate system $O_0x_0y_0z_0$ is fixed at the waterway. The O_0-x_0y_0 plane is placed on the undisturbed water-surface, with x_0-axis pointing toward the moving direction of the two ships, y_0-axis perpendicular to the solid wall of the waterway and z_0-

axis pointing upwards. The two ship coordinate systems $O_1x_1y_1z_1$ and $O_2x_2y_2z_2$ are fixed at the overtaking ship (denoted by the subscript '1') and the overtaken ship (denoted by the subscript '2'), respectively. Their x-axes point toward the bows, y-axes are positive to the ports, and z-axes are positive upward. In this paper, the non-dimensional parameter ξ is adopted to describe the longitudinal distance between the two ships, which is defined as $\xi = 2(x_1 - x_2)/(L_{pp,1} + L_{pp,2})$, where x_1 and x_2 are the longitudinal coordinates of the two ships in the $O_0x_0y_0z_0$ coordinate system, and $L_{pp,1}$ and $L_{pp,2}$ represent their lengths between the perpendiculars, respectively. In addition, the relatively lateral distance between the two ships is described by the parameter d_{pas} (see Figure 2); the parameter d denotes the distance between the starboard side of the overtaken ship and the waterway wall, which is fixed at 1.325 m in the present study.

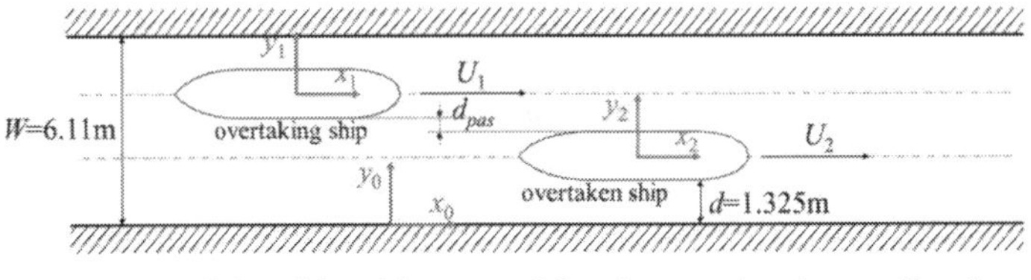

Figure 2. Diagram of the ship-ship overtaking interaction in confined waterway.

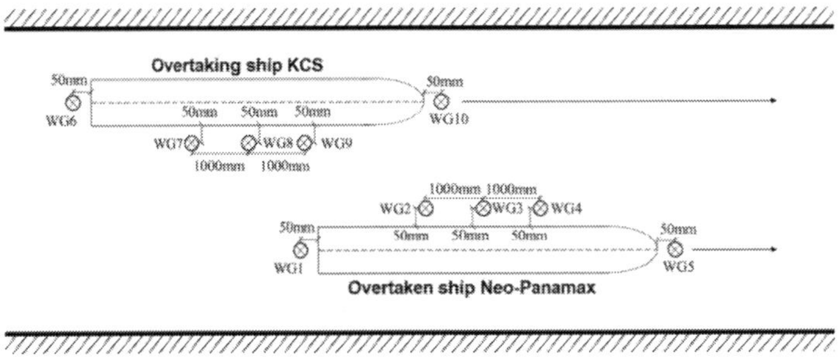

Figure 3. Measurement locations of free-surface elevations.

In the simulations, both ships are free to heave and pitch, but are restrained in 4DOFs (degrees of freedom): surge, sway, yaw and roll. The interaction effects are measured by hydrodynamic forces and moments, namely, longitudinal force X, lateral force Y, roll moment K and yaw moment N on the two ships. Besides, the free-surface elevations at five locations around each ship are measured to illustrate the effects by the ship-generated waves during the overtaking process, as shown in Figure 3. WG1-5 are placed around the overtaken ship, while WG6-10 are placed around the overtaking ship. For each ship, two locations are respectively located in front of and behind the ship, while the other three are placed on the side close to the other ship. All these locations are only 0.05 m from each hull.

Table 2. Simulation conditions.

Item	Case No.	U_1 (m/s)	Fr_1 (-)	$Fr_{h,1}$ (-)	U_2 (m/s)	Fr_2 (-)	$Fr_{h,2}$ (-)	h (m)	UKC (-)	d_{pas} (-)	d_{pas}/B_1 (-)
speed	1	0.69	0.105	0.413	0.1	0.015	0.060	0.285	50%	1.395	2.28
	2	0.8	0.122	0.498	0.1	0.015	0.060	0.285	50%	1.395	2.28
	3	0.9	0.138	0.538	0.1	0.015	0.060	0.285	50%	1.395	2.28
	4	0.69	0.105	0.413	0.15	0.023	0.090	0.285	50%	1.395	2.28
	5	0.69	0.105	0.413	0.2	0.120	0.120	0.285	50%	1.395	2.28
water depth	6	0.69	0.105	0.382	0.15	0.023	0.078	0.38	75%	1.395	2.28
	7	0.69	0.105	0.435	0.15	0.023	0.094	0.257	35%	1.395	2.28
lateral distance	8	0.69	0.105	0.413	0.15	0.023	0.090	0.285	50%	1.895	3.10
	9	0.69	0.105	0.413	0.15	0.023	0.090	0.285	50%	1.095	1.79

In this paper, different ship speeds, water depths and lateral distances between the two ships are considered to evaluate their effects on the ship-ship overtaking interactions. The simulation conditions are given in Table 2, where $Fr = U/\sqrt{gL_{pp}}$ and $Fr_h = U/\sqrt{gh}$ are the Froude numbers based on ship length and water depth, respectively; h is the water depth; UKC represents the under-keel clearance of the overtaking ship; B_1 is the breadth of the overtaking ship.

3. Governing equations and numerical methods

3.1. Governing equations and turbulence model

The viscous flow during the whole overtaking process is numerically simulated by the software STAR-CCM+, where the fluid is assumed to be incompressible. To appropriately describe the unsteady flows, the URANS equations are adopted as the governing equations:

$$\begin{cases} \dfrac{\partial \overline{u}_i}{\partial x_i} = 0 \\ \dfrac{\partial \overline{u}_i}{\partial t} + \dfrac{\partial}{\partial x_j}\left(\overline{u}_i \overline{u}_j\right) = -\dfrac{1}{\rho}\dfrac{\partial \overline{p}}{\partial x_i} + \dfrac{1}{\rho}\dfrac{\partial}{\partial x_j}\left(\overline{\sigma}_{ji} - \rho \overline{u'_i u'_j}\right) + f_i \end{cases} \tag{1}$$

where $\overline{u}_{i(j)}$, $\overline{p}$ and $\overline{\sigma}_{ji}$ denote the time-averaged velocity, pressure and stress tensor components (i,j=1, 2, 3); f_i is the body force, which is regarded as a constant term; ρ is the fluid density; $-\rho\overline{u'_i u'_j}$ are the Reynolds stresses resulted from the time-averaging process, which should be solved by turbulence modeling. In the present study, the realizable k-ε turbulence model [18] is used to close the governing equations, which was proved to be suitable for ship-ship interaction problems by Wnęk et al. [19], Zhou et al. [13], Zheng et al. [15-17].

3.2. Dynamic overset mesh technique

The dynamic overset mesh technique is suitable for the problem involving relative motions among different bodies [20]. This paper uses this technique to simulate movements and attitude changes of the two ships. The computational domain is divided into a background domain and two overset domains for the two ships. The meshes on the interface among the background domain and the overset domains are the acceptor meshes, which are used to exchange flow information through interpolation. The two ships are constantly moving during the overtaking process, thus the interfaces between the background and the overset domains are also accordingly changing constantly. Besides, the Dynamic Fluid-Body Interaction (DFBI) module in STAR-CCM+ is combined with the dynamic overset mesh technique to solve the attitude changes of the two ships.

3.3. Numerical settings

Figure 4 shows the computational domain in the numerical simulations. As mentioned above, the whole computational domain is divided into the background domain and the two overset domains for the two ships. To ensure that the overtaking process is not affected by reflected waves from the left and right boundaries, the computation domain is set up with a long longitudinal length. The domain width is corresponding to the waterway width W (Figure 2). The height between the undisturbed water-surface and the top plane (approximately equal to $L_{pp,1}$) is sufficient to reduce numerical interferences, and the depth from the undisturbed water-surface to the domain bottom depends on the simulation conditions in Table 2. To appropriately describe the flow characteristics, the no-slip wall condition is applied on the hull surfaces of the two ships, the two side boundaries and the bottom of the computational domain; the velocity inlet domain is set on the left and top boundaries; while the pressure outlet is adopted on the right boundary.

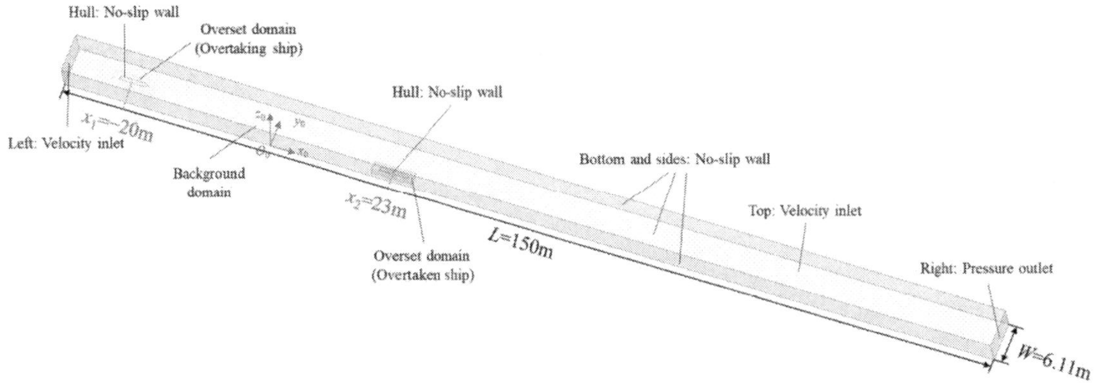

Figure 4. Computational domain with boundary conditions.

Different form steady flow simulations, the mesh generations of the present ship-ship interaction problem are more complex. Trimmed meshes and prism layer meshes are applied to discretize the fluid domain, and the mesh distributions are shown in Figure 5. Firstly, the prism layer meshes are utilized to generate orthogonal prismatic cells next to the hull surfaces, and the all-y^+ wall treatment with $y^+<1$ is implemented for the near-wall flow simulation, as shown in Figure 6. Secondly, the local mesh refinements by volumetric and surface controls are applied for the regions with complex flows, such as the bow- and the stern-regions. In addition, to ensure interpolation accuracy, the mesh sizes on the interface between the two domains should be similar, therefore the meshes in the moving track regions are also refined to maintain similar sizes.

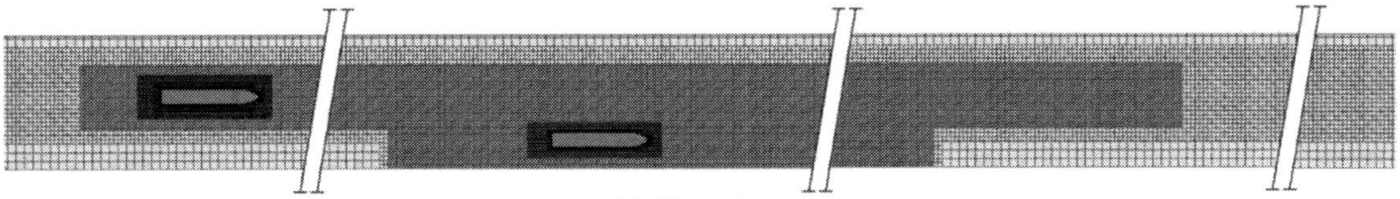

(a) Top view

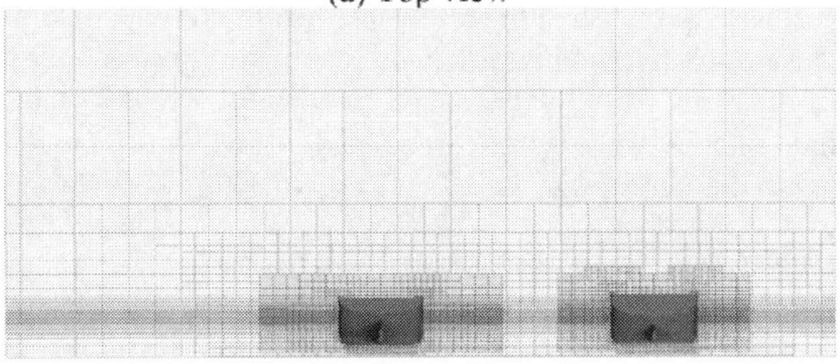

(b) Stern view

Figure 5. Mesh distributions.

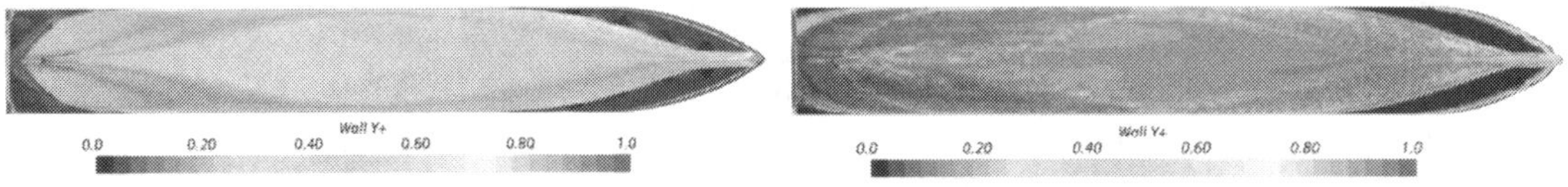

Figure 6. y^+ distributions on the hull surfaces (left: KCS; right: Neo-Panamax).

The Volume of Fluid (VOF) method is used to capture the free-surface elevations. To increase the accuracy of free-surface captures, the High-Resolution Interface Capturing (HRIC) scheme is adopted coupled with the VOF method. A verification and validation study is commonly necessary to make sure that the numerical model is reliable. By applying the MASHCON2022 benchmark test cases [14], Zheng et al. [15, 17] evaluated the accuracy in spatial and temporal discretizations for the passing-ship problems. The results demonstrated that the numerical solutions converged as the mesh and time-step reduced, and good agreements between the numerical predictions and test data were achieved, indicating a high accuracy of the model. Based on this, this numerical model is adopted to simulate the ship-ship overtaking problem in this paper since there are relative motions between the two ships as well. Similarly, a mesh basic size of 0.1 m with a 0.04 s time-step is used in the present study.

4. Analyses of numerical results

4.1. Effect of ship speed

The ship speed is one of the important factors in ship-ship interaction problems. In this section, the numerical results of Case1-5 are compared to analyse its effects and the corresponding physical mechanisms.

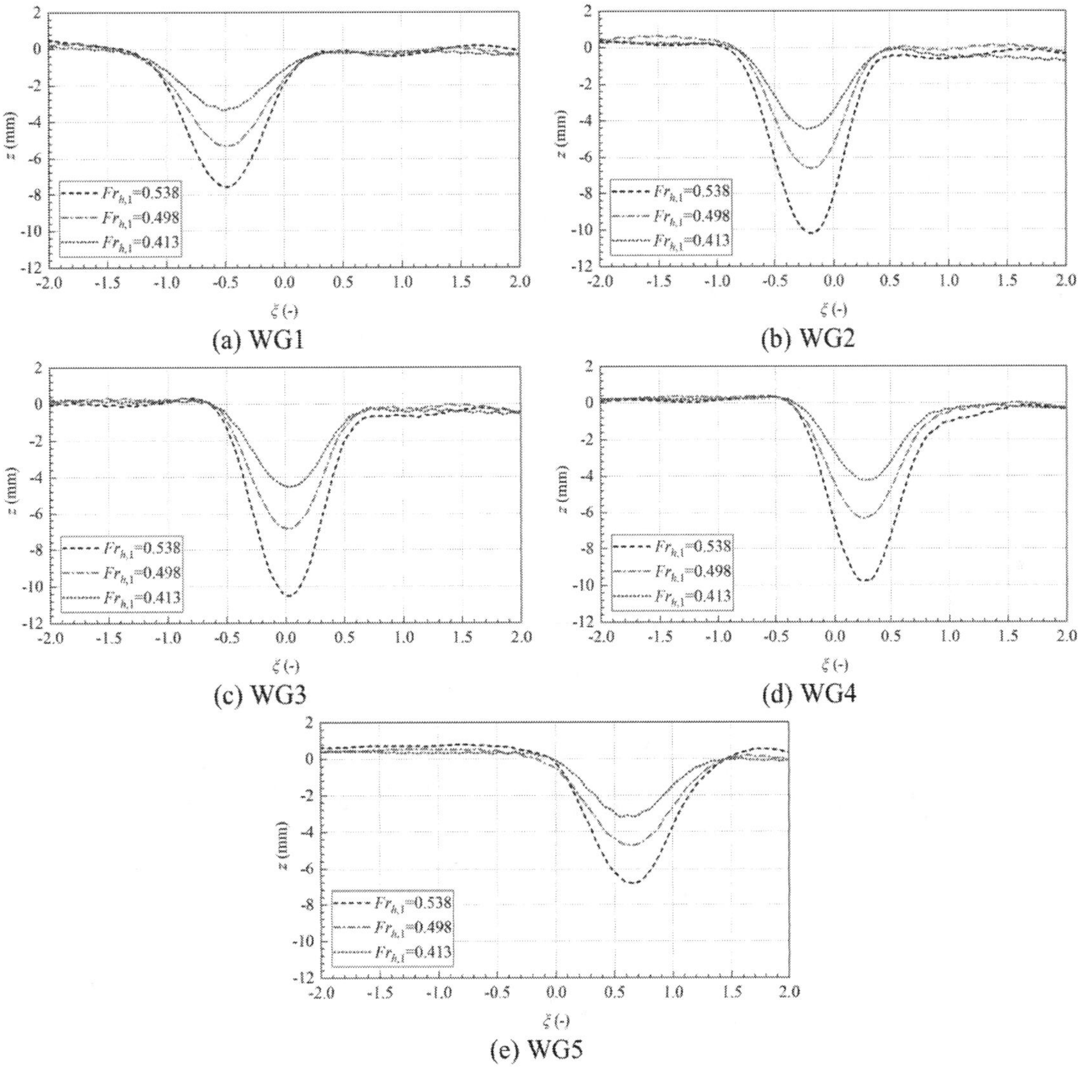

Figure 7. Free-surface elevations around the overtaken ship against overtaking-ship speeds.

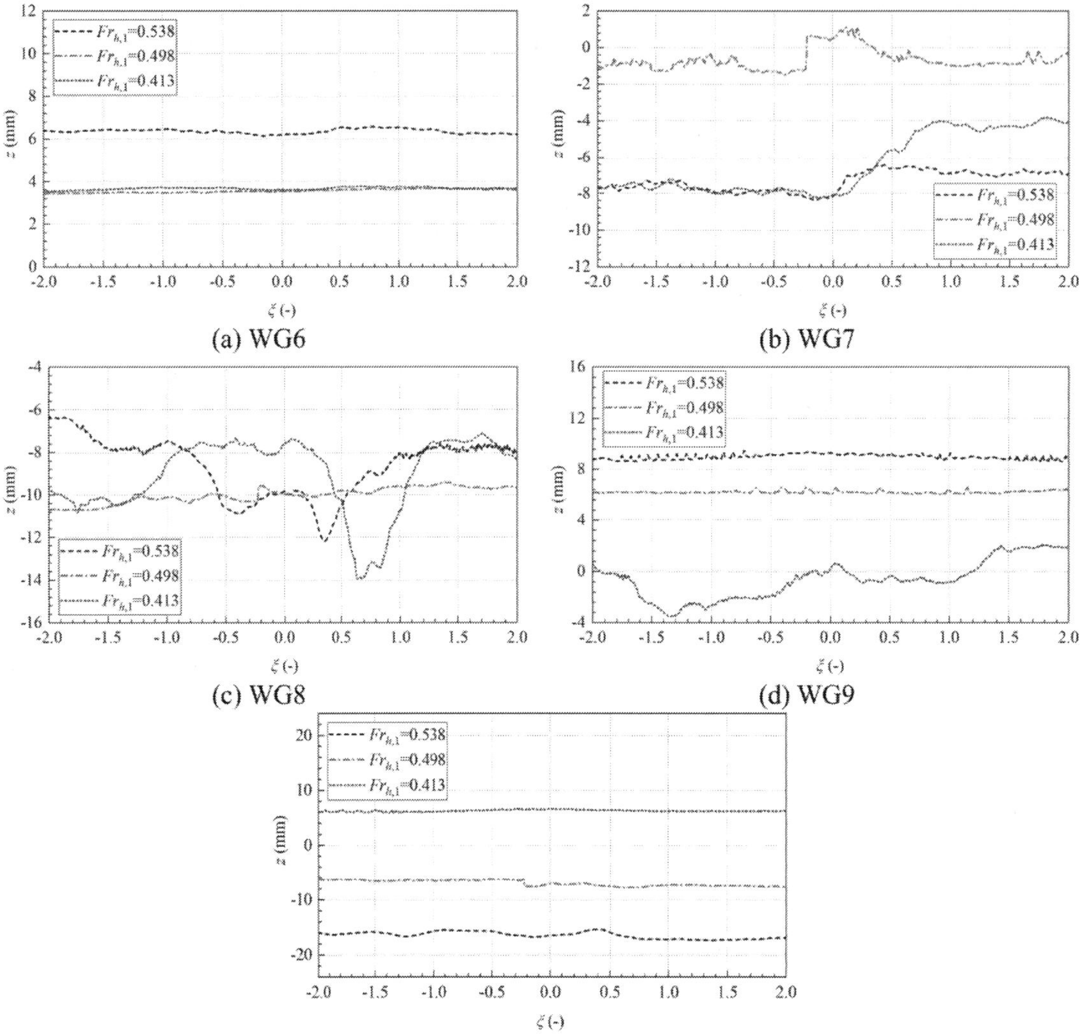

(a) WG6 (b) WG7 (c) WG8 (d) WG9 (e) WG10

Figure 8. Free-surface elevations around the overtaking ship against overtaking-ship speeds.

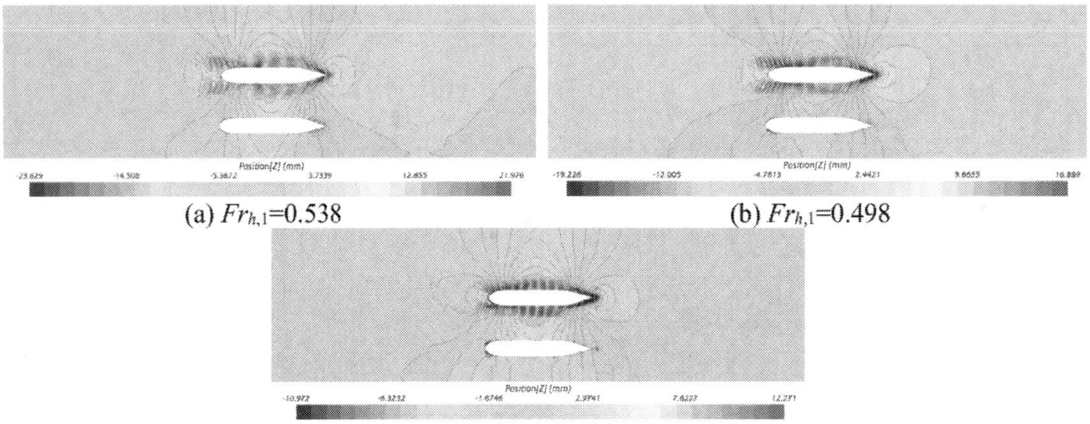

(a) $Fr_{h,1}$=0.538 (b) $Fr_{h,1}$=0.498 (c) $Fr_{h,1}$=0.413

Figure 9. Wave patterns at different passing speeds when the two ships are side by side.

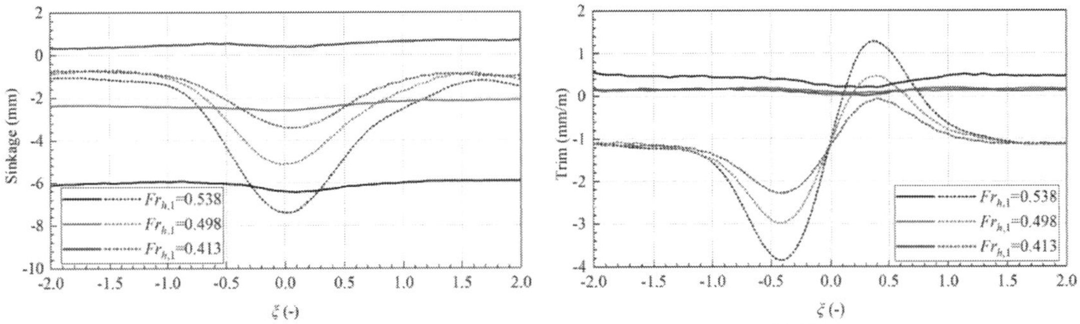

(a) Sinkage (b) Trim

Figure 10. Attitudes of the two ships against overtaking-ship speeds (Solid curves: overtaking ship; dotted curves: overtaken ship).

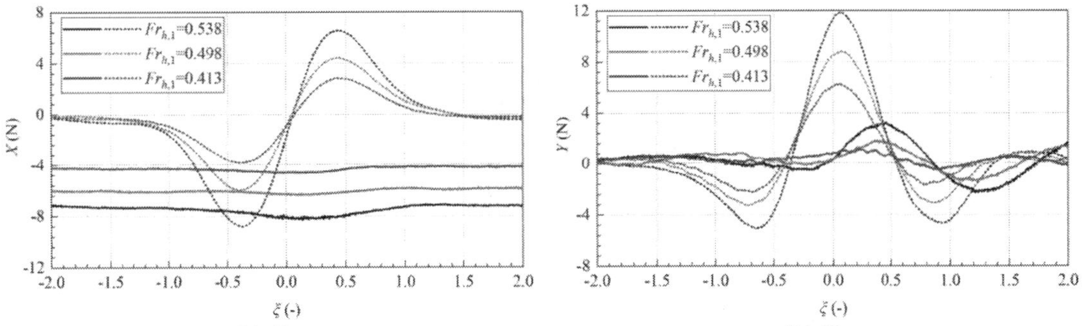

(a) X (b) Y

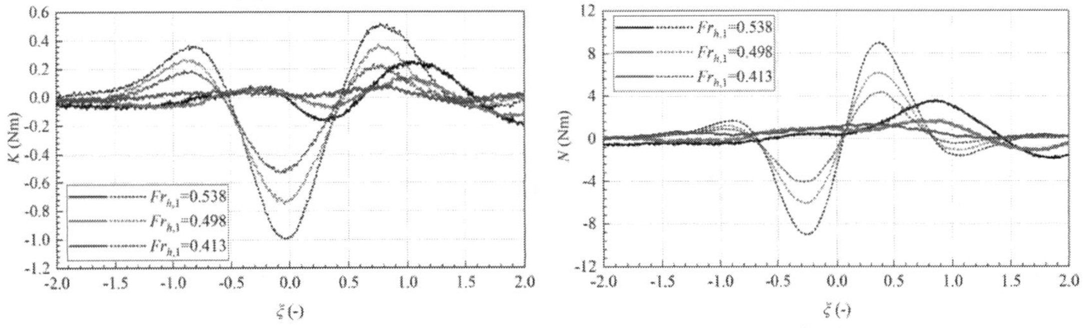

(c) K (d) N

Figure 11. Hydrodynamic quantities of the two ships against overtaking-ship speeds (Solid curves: overtaking ship; dotted curves: overtaken ship).

Firstly, the speed of the overtaken ship is fixed at 0.15 m/s, and the impact of the overtaking ship speeds (Case1-3) on the ship-ship interactions is investigated. The free-surface elevations around the two ships at different speeds are shown in Figures 7 and 8. It can be found that there is an obvious water-level drop around the overtaken ship when the overtaking ship is close to it, and the amplitude of this drop is positively related to the overtaking-ship speed, i.e., the larger the overtaking-ship speed, the more drastic change in the water level around the overtaken ship. However, the free-surface elevations around the overtaking ship are more unstable. The wave patterns around the two ships when they are side by side (ξ=0) are presented in Figure 9, to indicate the characteristics of ship-generated waves during the overtaking process. The primary wave system, especially the drawdown generated by the overtaking ship, dominates during the whole overtaking process, but the waves generated by the overtaken ship are not obvious. This disparity arises due to the significantly lower speed of the overtaken ship compared to the overtaking ship. Overall, the overtaken ship is more sensitive to the ship-ship interaction effects. Therefore, the following discussions and analyses mainly focus on the overtaken ship. Secondly, Case1, 4 and 5 are selected to discuss the effects of the overtaken-ship speed. Figures 12 and 13 show the ship attitudes, hydrodynamic forces and moments of the overtaken ship at different speeds. Except for sinkage, all the numerical results at different speeds are similar. It demonstrates that the effects of the small-value overtaken-ship speed on the overtaking process are not pronounced.

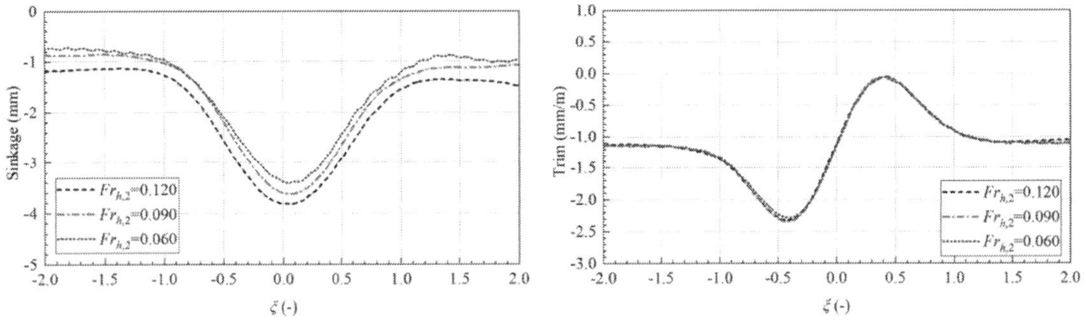

(a) Sinkage (b) Trim

Figure 12. Attitudes of the overtaken ship against overtaken-ship speeds.

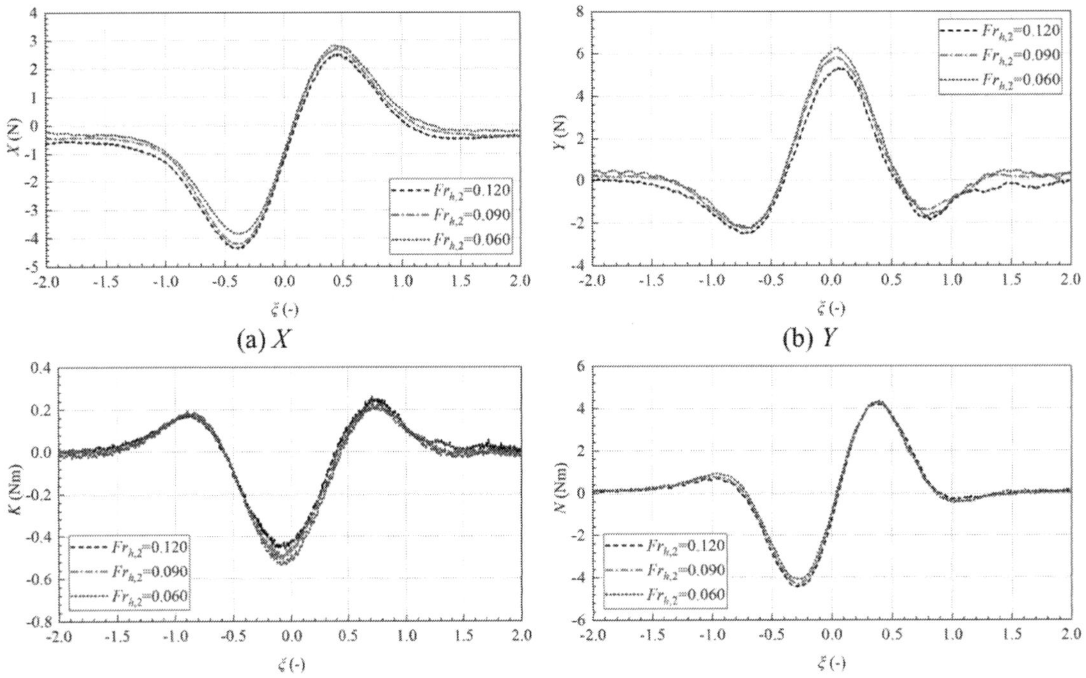

(a) X (b) Y

(c) K (d) N

Figure 13. Hydrodynamic quantities of the overtaken ship against overtaken-ship speeds.

4.2. Effect of water depth

The predicted results of Case3, 6 and 7 are compared to illustrate the effects of water depth (represented by UKC) on the ship-ship interaction. Figure 14 shows the free-surface elevations around the overtaken ship. It can be seen that with smaller water depth, there are larger variations in free-surface elevations. Combined with the simulated wave patterns in Figure 15, it illustrates that as the water depth decreases, the overtaking ship results in a more significant drawdown affecting the overtaken ship. Figures 16 and 17 illustrate the predicted attitudes and hydrodynamic quantities of the overtaken ship under the interaction impacts. The numerical results demonstrate that the more intense drawdown gives rise to more pronounced hydrodynamic interactions.

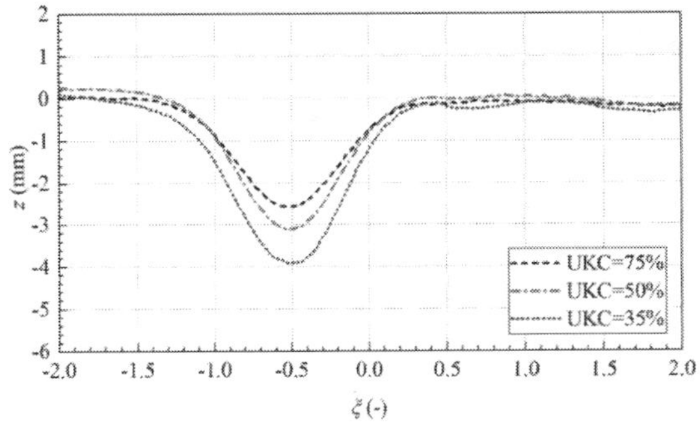

(a) WG1

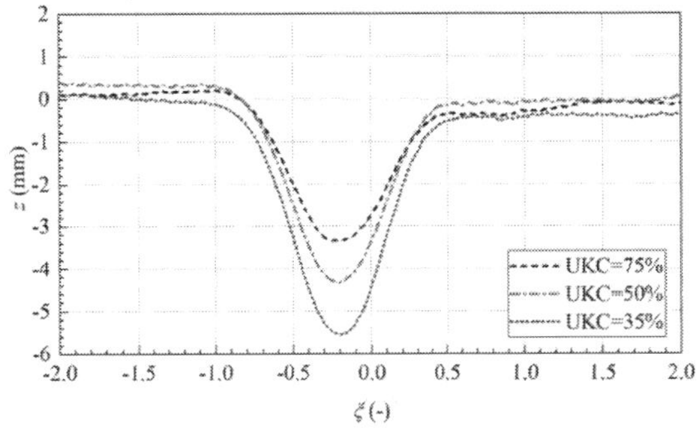

(b) WG2

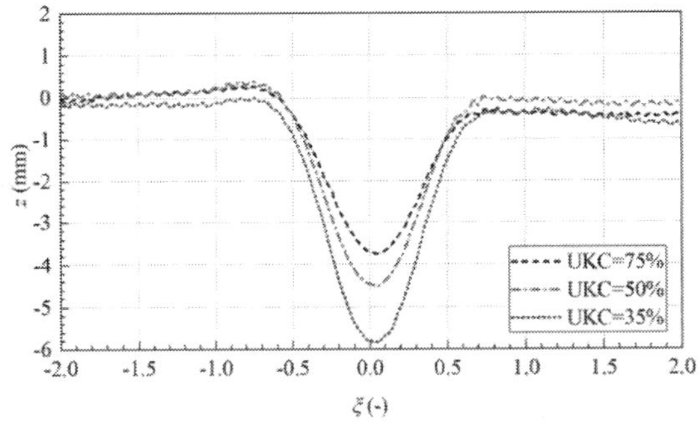

(c) WG3

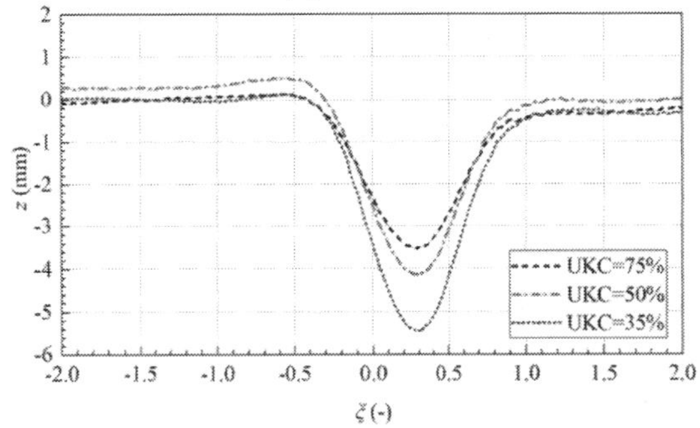

(d) WG4

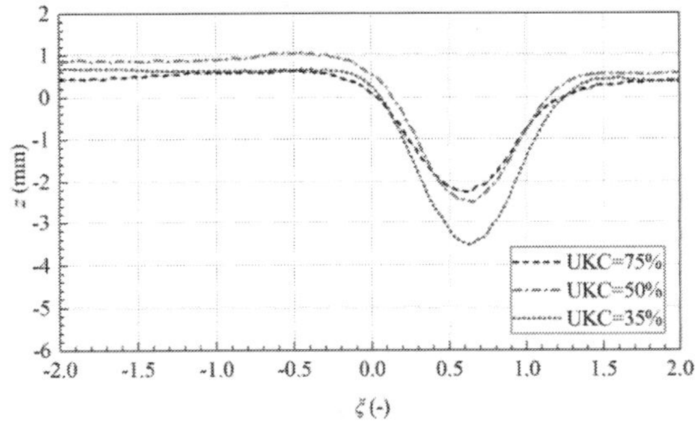

(e) WG5

Figure 14. Free-surface elevations around the overtaken ship at different water depths.

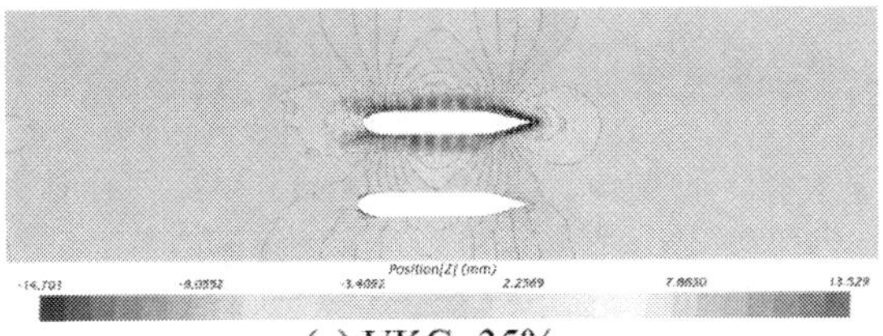
(a) UKC=35%

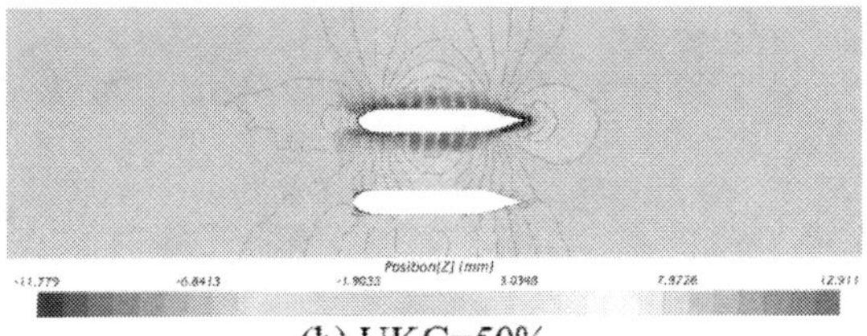
(b) UKC=50%

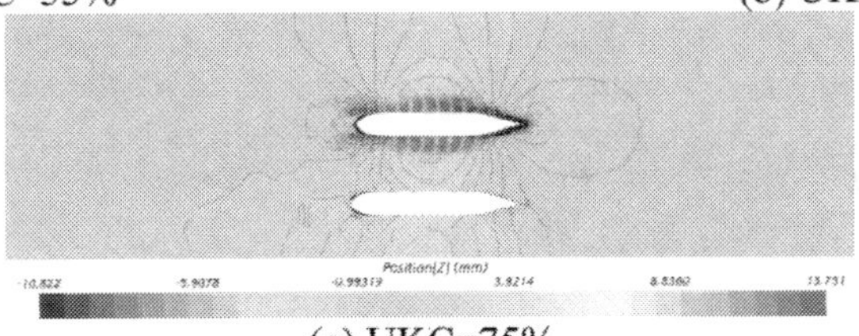
(c) UKC=75%

Figure 15. Wave patterns at different water depths when the two ships are side by side.

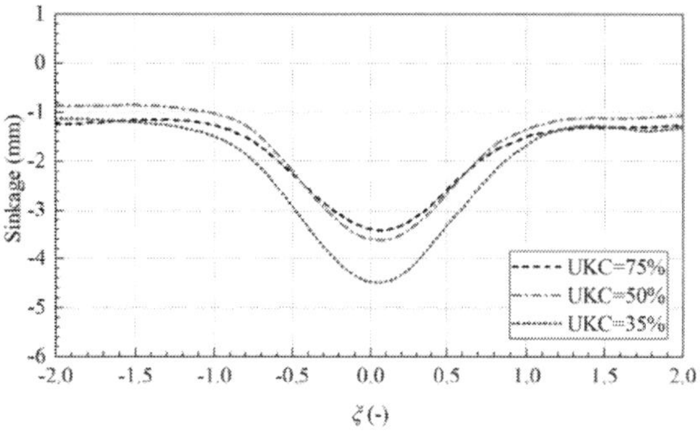

(a) Sinkage

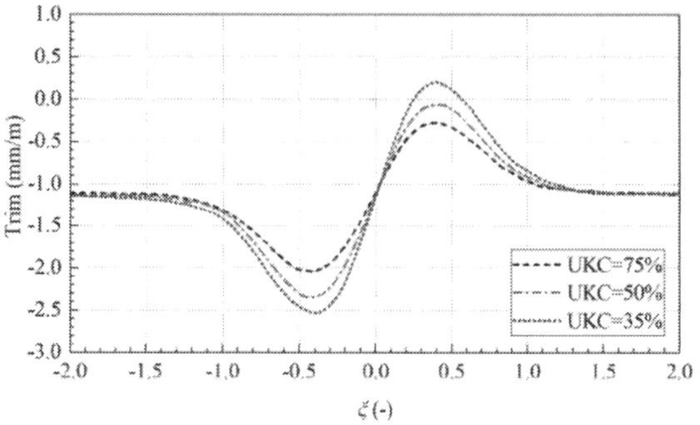

(b) Trim

Figure 16. Attitudes of the overtaken ship at different water depths.

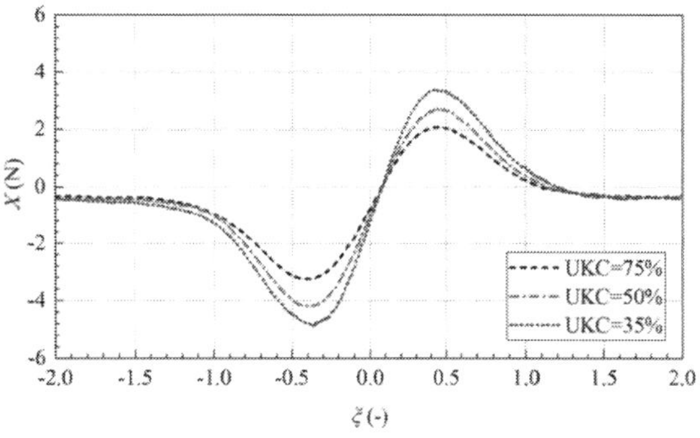

(a) X

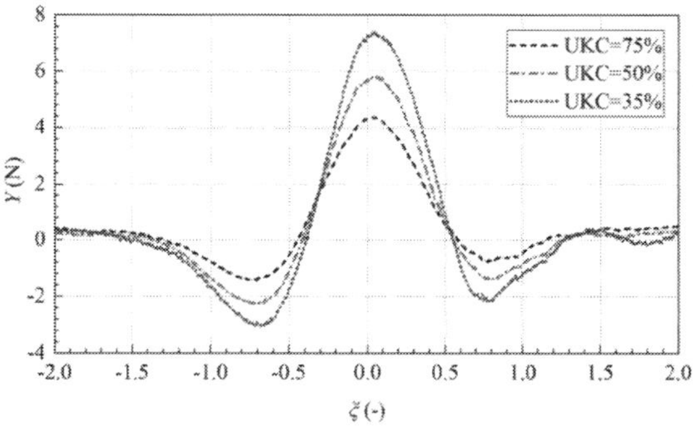

(b) Y

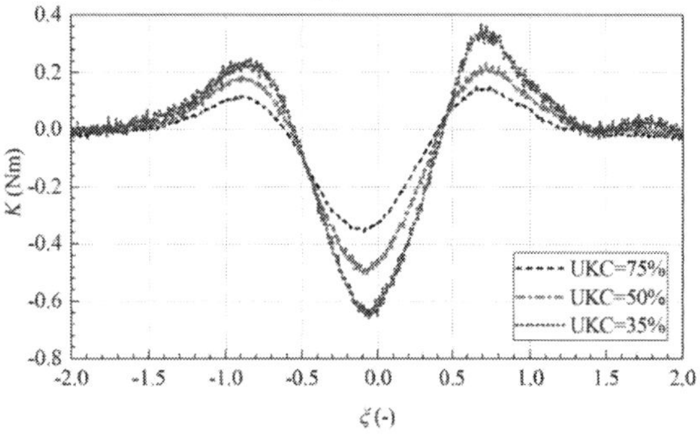

(c) K

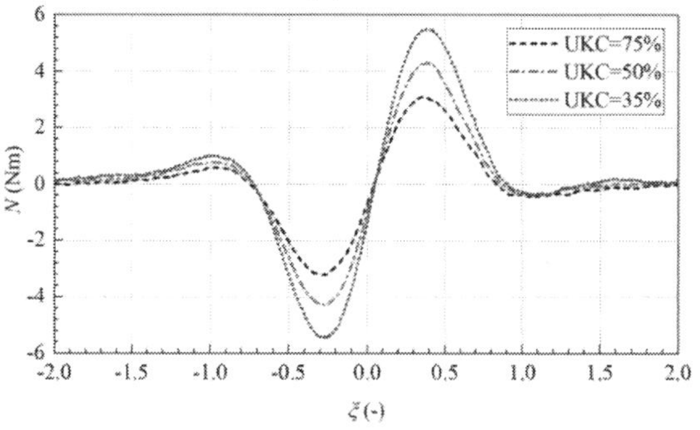

(d) N

Figure 17. Hydrodynamic quantities of the overtaken ship at different water depths.

4.3. Effect of lateral distance

The numerical results of Case3, 8 and 9 are compared to investigate the effect of lateral distance between the two ships (represented by d_{pas}/B_1). The free-surface elevations around the overtaken ship with different lateral distances are presented in Figure 18. As the lateral distance decreases, the free-surface elevations around the overtaken ship vary more dramatically, resulting in more significant ship motions, as well as more pronounced hydrodynamic forces and moments on the overtaken ship, as shown in Figures 19 and 20. It indicates that the ship-ship interaction effects become more significant when the lateral distance decreases.

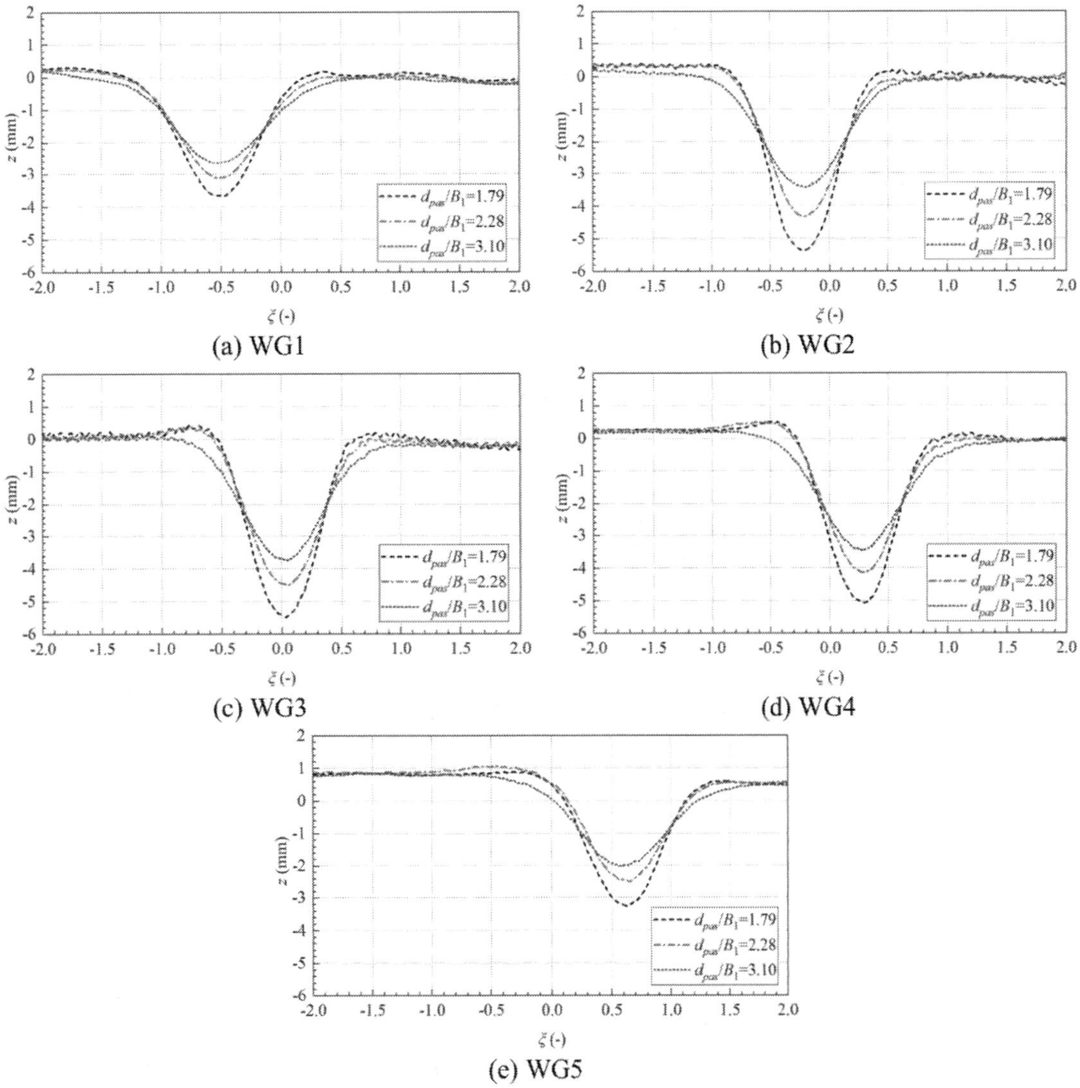

Figure 18. Free-surface elevations around the overtaken ship with different lateral distances.

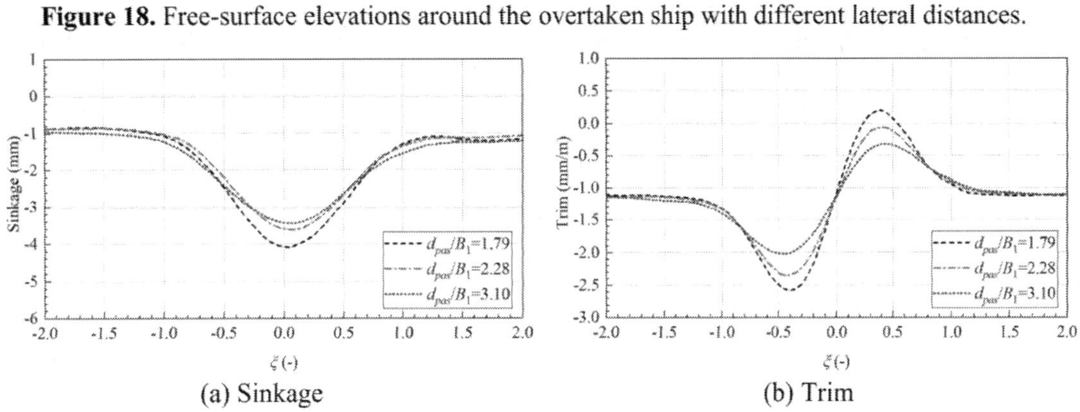

Figure 19. Attitudes of the overtaken ship with different lateral distances.

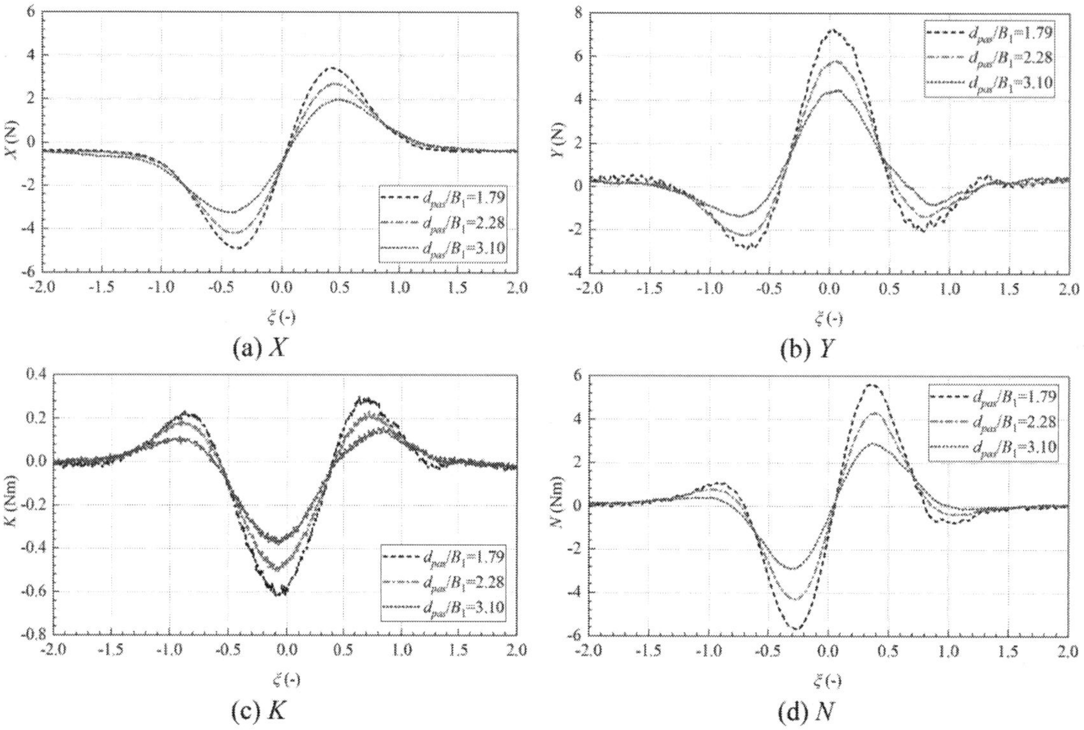

Figure 20. Hydrodynamic quantities of the overtaken ship with different lateral distances.

5. Conclusions

This paper investigates ship-ship overtaking interactions in confined waterways by conducting numerical simulations of a KCS container ship overtaking a slower Neo-Panamax container ship using the URANS method. The dynamic overset mesh technique is utilized to simulate the relative motions between the two ships and their attitude variations during the entire overtaking process. The free-surface elevations around the ships, ship attitudes, as well as hydrodynamic forces and moments on both ships are predicted. Three navigation parameters, ship speed, water depth and lateral distance between the ships, are considered to explore their effects on ship-ship overtaking interactions. The results demonstrate that the overtaken ship is more significantly impacted by the interactions compared to the overtaking ship. The drawdown generated by the overtaking ship is shown to be a key factor during the overtaking process in confined waterways. Furthermore, the results indicate that overtaking-ship speed, lateral distance, and water depth are important factors influencing ship-ship interactions. The increase of overtaking-ship speed, the decrease of lateral distance or water depth will result in a more pronounced drawdown, and hence even more significant ship-ship interaction effects on the overtaken ship.

6. Acknowledgements

This work is supported by the National Natural Science Foundation of China (Grant No. 51979164) and the Fundamental Research Funds for the Central Universities (Grant No. AF0100142/005/002).

7. References

[1] Söding H and Conrad F 2005 Analysis of overtaking manoeuvres in a narrow waterway *Ship Technol. Res.* **52** 189–93

[2] ITTC Manoeuvring Committee 2021 *Final report and recommendations to the 29th ITTC*

[3] Remery G F 1974 Mooring forces induced by passing ships *Proc. of the Offshore Technology Conf. (May Texas: Houston)*

[4] Vantorre M, Verzhbitskaya E and Laforce E 2002 Model test based formulations of ship-ship interaction forces *Ship Technol. Res.* **49** 124–41

[5] Gronarz A 2006 Ship-ship interaction: overtaking and encountering of inland vessels on shallow water *Proc. of the Int. Conf. on Marine Simulation and Ship Manoeuvrability (June 25-30 The Netherlands: Terschelling)* pp M–1–1–5

[6] Mousaviraad S M, Sadat-Hosseini S H and Stern F 2016 Ship-ship interactions in calm water and waves. Part 1: Analysis of the experimental data *Ocean Eng.* **111** 615–26

[7] Pinkster J A and Ruijter M N 2004 The influence of passing ships on ships moored in restricted waters *Proc. of the Offshore Technology Conf. (May Texas: Houston)*

[8] Sutulo S and Guedes Soares C 2008 Simulation of the hydrodynamic interaction forces in close-proximity manoeuvring *Proc. of the Int. Conf. on Offshore Mechanics and Arctic Engineering* pp 839–48

[9] Sutulo S and Guedes Soares C 2009 Simulation of close-proximity maneuvers using an online 3d potential flow method *Proc. of the Int. Conf. on Marine Simulation and Ship Manoeuvrability (August 17-20 Panama: Panama City)* pp M–9–1–10

[10] Sutulo S, Guedes Soares C and Otzen J F 2012 Validation of potential-flow estimation of interaction forces acting upon ship hulls in parallel motion *J. Ship Res.* **56** pp 129–45

[11] Degrieck A, Uyttersprot B, Sutulo S, Guedes Soares C, Van Hoydonck W, Vantorre M and Lataire E 2021 Hydrodynamic ship-ship and ship-bank interaction: A comparative numerical study *Ocean Eng.* **230** 108970

[12] Nandhini V and Nallayarasu S 2020 CFD simulation of the passing vessel effects on moored vessel *Ships Offshore Struct.* **15** 184–99

[13] Zhou L, Abdelwahab H S and Guedes Soares C 2021 Experimental and CFD investigation of the effects of a high-speed passing ship on a moored container ship *Ocean Eng.* **228** 108914

[14] Van Zwijnsvoorde T, Delefortrie G and Lataire E 2022 Passing ship effects in shallow and confined water: open model test data for validation purposes *Proc. of the 6th MASHCON Int. Confe. on Ship Manoeuvring in Shallow and Confined Water (May UK: Glasgow)* pp 255–83

[15] Zheng Z Q, Zou L and Zou Z J 2023 A numerical study on the effects of ship-generated waves on a moored ship in restricted waterways considering initial acceleration process *J. Mar. Sci. Eng.* **11** 483

[16] Zheng Z Q, Zou L and Zou Z J 2022 A numerical investigation on the ship-ship interaction between a passing ship and a moored ship in restricted waterways *Proc. of the 6th MASHCON Int. Conf. on Ship Manoeuvring in Shallow and Confined Water (May UK: Glasgow)* pp 344–55

[17] Zheng Z Q, Zou L and Zou Z J 2023 A numerical study of passing ship effects on a moored ship in confined waterways with new benchmark cases *Ocean Eng.* **280** 114643

[18] Shih T H, Liou W W, Shabbir A, Yang Z and Zhu J 1995 A new k-ε eddy viscosity model for high reynolds number turbulent flows *Comput. Fluids* **24** 227–38

[19] Wnęk A D, Sutulo S and Guedes Soares C 2018 CFD analysis of ship-to-ship hydrodynamic interaction *J. Mar. Sci. Appl.* **17** 21–37

[20] Hadzic H 2005 *Development and Application of Finite Volume Method for the Computation of Flows Around Moving Bodies on Unstructured, Overlapping Grids* (Hamburg: Hamburg University of Technology)

IOP Conf. Series: Materials Science and Engineering 1288 (2023) 012006 doi:10.1088/1757-899X/1288/1/012006

Influence of interface-sharpening on the ship's wave system

Robinson Perić[1,*], Jannes Berndt[1], Moustafa Abdel-Maksoud[1]

[1]Hamburg University of Technology, Institute for Fluid Dynamics and Ship Theory
Am Schwarzenberg-Campus 1, 21073 Hamburg, Germany

E-mail: robinson.peric@tuhh.de

Abstract. This article investigates the influence of the interface-sharpening algorithm on the wave height near ships in flow simulations. The focus is on finite-volume-based flow simulations with the volume of fluid (VOF) method to account for the two phases, gaseous air and liquid water. A common approach to obtain a comparatively sharp interface between the water phase and the air phase is to use interface-capturing approaches combined with interface-sharpening schemes. The present work demonstrates that the interface-sharpening schemes can contain case-dependent parameters, and that the optimum values for these parameters may not be known a priori. Therefore, if the parameters are not set correctly, the wave height at some distance to the ship may be unrealistically small or large. To demonstrate this, flow simulations of calm-water resistance of the KCS ship are performed with different discretizations and different parameters of the interface-sharpening scheme. The present simulation results are compared to experimental results to provide recommendations on the optimum choice of interface-sharpening parameters and guidelines for performing such flow simulations, to improve predictions of the ship's wave system.

1. Introduction

Computational fluid dynamics (CFD) is increasingly used to predict ship resistance and the ship's wave field, with a trend towards automatizing CFD simulations.

The ship's calm-water resistance can be well predicted via CFD simulations: For example, in the Tokyo 2015 workshop, the mean absolute comparison error in all ship-resistance computations was 2% [2]. Comparison errors were below 4% when at least 10^7 grid cells were used. The number of cells necessary to obtain accurate results may be reduced by using adaptive grid refinement [8].

However, the prediction of the ship's wave field via CFD remains challenging: Although wave profiles along the ship hull matched experimental data closely in the Tokyo 2015 workshop, there was more discrepancy relative to measurements for two wave cuts at some distance to the ship. Thus, already at a distance in the order of one ship breadth, the wave elevation may be substantially smaller or even larger than the physically correct solution. So far, literature does neither provide an explanation nor a solution for this problem.

Although often in practice only the calm-water resistance is of interest (e.g. when optimizing the hull geometry), accurate prediction of the ship's wave fields is also important for various applications, e.g. to determine planing vessel behavior next to a larger craft or to predict ship-ship interactions such as refueling or launching and recovery of life boats from moving ships.

Furthermore, approaches to determine the wave resistance from wave cuts also rely on correct prediction of wave elevations.

The aim of this work is to investigate via flow-simulations based on OpenFOAM why predicting the ship's wave fields is challenging for common CFD approaches and which aspects should be taken into account to accurately perform such flow simulations.

For this, the present work investigates the hypothesis that *the root of these problems are the interface-capturing and interface-sharpening algorithms*, in particular that these contain case-dependent parameters and are sensitive to the grid size and cell aspect-ratios near the free surface.

2. Governing equations

The governing equations for fluid flow considered in this work are

$$\frac{\mathrm{d}}{\mathrm{d}t}\int_V \rho \, \mathrm{d}V + \int_S \rho\mathbf{v}\cdot\mathbf{n}\, \mathrm{d}S = 0 \quad , \tag{1}$$

$$\frac{\mathrm{d}}{\mathrm{d}t}\int_V \rho u_i \, \mathrm{d}V + \int_S \rho u_i\mathbf{v}\cdot\mathbf{n}\, \mathrm{d}S = \int_S (\tau_{ij}\mathbf{i}_j - p\mathbf{i}_i)\cdot\mathbf{n}\, \mathrm{d}S + \int_V \rho\mathbf{g}\mathbf{i}_i \, \mathrm{d}V + Q_i \quad , \tag{2}$$

$$\frac{\mathrm{d}}{\mathrm{d}t}\int_V \alpha \, \mathrm{d}V + \int_S \alpha\mathbf{v}\cdot\mathbf{n}\, \mathrm{d}S = Q_\alpha \quad , \tag{3}$$

with volume V of control volume (CV) bounded by the closed surface S, fluid velocity vector $\mathbf{v}$ with the Cartesian components u_i, unit vector $\mathbf{n}$ normal to S and pointing outwards, time t, pressure p, fluid density ρ, components τ_{ij} of the viscous stress tensor, unit vector $\mathbf{i}_j$ in direction x_j, and volume fraction α of water.

Source terms Q_i and Q_α can be used to implement forcing zones for wave damping. Although this can be advantageous for ship resistance computations (cf. [6]), e.g. to improve convergence towards a steady solution, this is not mandatory and often not applied. Therefore, in this work no forcing zones will be used, thus $Q_i = 0$.

Interface-sharpening is applied in interface-capturing schemes to avoid smearing of the air-water interface over several cells, illustrated in Fig. 1. Interface-capturing schemes can be classified into nondimensionalized-variable-diagram (NVD) schemes and artificial compressive source terms. This work focuses on artificial compressive source terms as implemented in OpenFOAM.

The artificial compressive source terms are implemented in the transport equation for volume fraction α, Eq. (3), i.e.:

$$Q_\alpha = -\int_S c_\alpha \alpha(1-\alpha)|\mathbf{v}|\mathbf{n}_\alpha\cdot\mathbf{n}\, \mathrm{d}S \quad , \tag{4}$$

where $\mathbf{n}_\alpha$ is the normal vector to the air-water interface and the source term magnitude is regulated by user-defined parameter c_α.

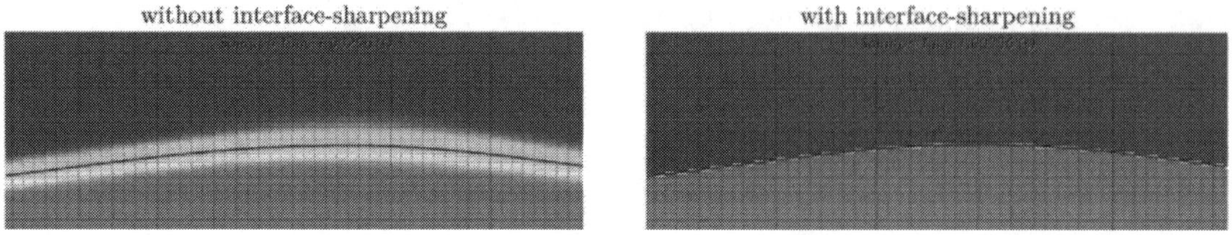

Figure 1. Visualization of smearing of the air-water interface without interface-sharpening (left) and maintaining a sharp air-water interface with interface-sharpening (right); with volume fractions of air (blue) and water (red), computational grid and air-water interface (thick black curve)

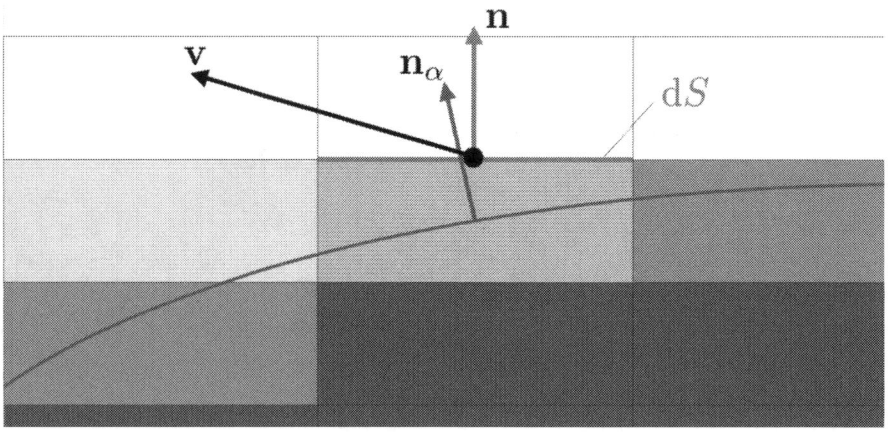

Figure 2. Visualization of the vectors from Eq. (4): velocity vector $\mathbf{v}$ and face-normal vector $\mathbf{n}$ at centroid of face with surface area $\mathrm{d}S$ as well es interface-normal vector $\mathbf{n}_\alpha$ belonging to the same cell

3. Simulation setup

The testcase was the KCS ship in model scale $1 : 31.599$ at Froude number $\mathrm{Fr} = 0.26$ from the Gothenburg 2010 workshop [4], with main parameters shown in Table 1. The ship was held fixed in its floating position at zero speed (even keel) as in the experiment [7]. By exposing the ship to uniform inflow velocity $u_\mathrm{m} = 2.196\,\mathrm{m/s}$, the ship's wave field develops and its calm-water resistance can be computed.

CFD simulations were performed with flow solver OpenFOAM [9] version 2212 (ESI Group, website: `https://www.openfoam.com/`) based on the finite-volume method with volume of fluid (VOF) method to account for the air and water phases. The intention for the present CFD simulations was to start from widely used solver settings. Therefore, most solver settings (cf. Table 2) correspond to those from the OpenFOAM tutorial located in the folder *tutorials/multiphase/interFoam/RAS/DTCHullMoving*.

The simulation domain was box-shaped with dimensions $-1.648L_\mathrm{pp} \leq x \leq 2.06L_\mathrm{pp}$, $0 \leq y \leq 1.648L_\mathrm{pp}$ and $-1.373L_\mathrm{pp} \leq z \leq 1.09L_\mathrm{pp}$ (cf. Fig. 3). The coordinate system was located on the calm-water level, in the ship's symmetry plane with normal vector $(0, 1, 0)^\mathrm{T}$ and on x-position corresponding to $0.024605L_\mathrm{oa} \approx 0.189\,\mathrm{m}$ distance from the stern, with x-axis pointing in direction of the ship's speed and z-axis pointing upwards.

Table 1. Parameters of the KCS ship in model scale for the present flow simulations

Main particulars	full scale (1 : 1)	model scale (1 : 31.599)
Length between perpendiculars L_{pp}	230.0 m	7.2786 m
Maximum beam of waterline B	32.2 m	1.019 m
Draft T	10.8 m	0.3417 m
Displacement Δ	52030 m^3	1.649 m^3
Wetted area w/o rudder S_0	9424 m^2	9.438 m^2
Block coefficient C_{B}	0.6505	0.6505
Speed u	12.34 m/s	2.196 m/s
Froude number	0.26	0.26

Only half of the ship was simulated by applying a symmetry boundary condition at $y = 0$. At the inlet boundary $x = 2.06L_{\text{pp}}$, the inflow velocity was set to u_{m}. The bottom boundary $z = -1.373L_{\text{pp}}$ was set to slip wall. At the top boundary $z = 1.09L_{\text{pp}}$, atmospheric pressure was prescribed. The side boundary $y = 1.648L_{\text{pp}}$ was set to symmetry plane. The outlet boundary $x = -1.648L_{\text{pp}}$ was set to pressure outlet. Apart from the symmetry boundaries, the volume fraction α at the domain boundaries was set to calm-water solution, i.e. $\alpha = 1$ (water) for $z \leq 0$ and $\alpha = 0$ (air) everywhere else. The ship was held in fixed position (i.e. zero degrees of freedom, '0DOF').

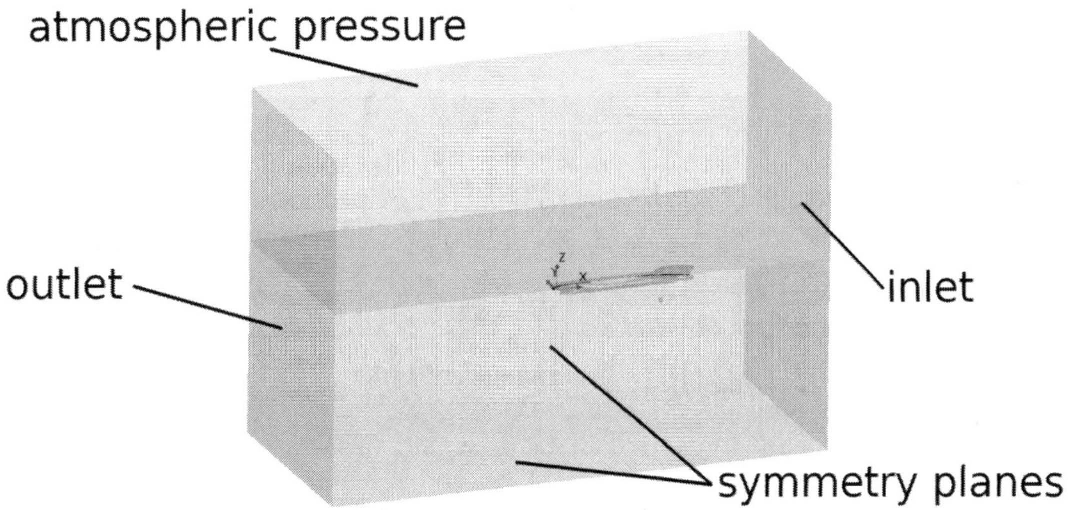

Figure 3. Visualization of simulation domain, boundary conditions and coordinate system

The k-ω-SST turbulence model [5] was used with default settings. In all simulations, the average y^+-value for the ship hull lay within the log-law range, mostly $50 \lesssim y^+ \lesssim 100$. The solvers used were *GAMG* for pressure p and *smoothSolver* with smoother *symGaussSeidel* for velocities $\mathbf{u}$ and volume fraction α. For interface-sharpening, the Multidimensional Universal Limiter with Explicit Solution (*MULES*) was applied with the artificial compressive source term in Eq. (4).

The flow simulations were unsteady with constant time-step size Δt. The simulations were started at $t = 0$ s and a converged quasi-steady solution was obtained at time $t_{\text{end}} = 90$ s.

Table 2. Finite-volume schemes used in the simulations with OpenFOAM naming convention

Time derivative (`ddtSchemes`)	
	Euler
Divergence operator (`divSchemes`)	
div(rhoPhi,U)	Gauss linearUpwind grad(U)
div(phi,alpha)	Gauss vanLeer
div(phirb,alpha)	Gauss linear
div(phi,k)	Gauss linearUpwind limitedGrad
div(phi,omega)	Gauss linearUpwind limitedGrad
div(((rho*nuEff)*dev2(T(grad(U))))) Gauss linear	
Laplacian operator (`laplacianSchemes`)	
	Gauss linear corrected
Gradient operator (`gradSchemes`)	
default	Gauss linear
limitedGrad	cellLimited Gauss linear 1
Surface-normal gradient operator (`snGradSchemes`)	
	corrected
Interpolation (`interpolationSchemes`)	
	linear
Wall distance (`wallDist`)	
method	meshWave

The computational grid was generated via Simcenter STAR-CCM+ version 18.02.008-R8 by Siemens. The grid was rectilinear with localized grid refinement near the ship and in the vicinity of the free surface as shown in Fig. 4 and also in later Figures. As reference grid, the grid from a previous computation [6] was selected, where close agreement between results from experiments and CFD simulation was achieved. The grid was stored as ccm-file and imported into OpenFOAM by the utility *ccmToFoam*. Therefore, the grid was considered satisfactory, so that with suitable solver settings it should be possible to compute the ship's calm-water resistance with satisfactory accuracy.

Parameter studies were performed for different interface-sharpening settings and different discretization:

(i) Each CFD simulation was repeated with different interface-sharpening settings, with c_α **varied between** 0 **and** 10.

(ii) Simulations were repeated with the cell size in vicinity of the free surface systematically refined, by **varying horizontal cell sizes** Δx **and** Δy **and vertical cell size** Δz by factors 2.0, 1.0 and 0.5. Investigations were performed with varied base size, i.e. changing the the cell sizes in the whole grid by factors 2.0 (finer grid) and 4.0 (coarser grid). This resulted in grids with 10^5 to 10^6 cells.

Tables 3 and 4 and Figures 5 and 6 illustrate the cell-aspect ratios resulting from refining the grid near the free surface in horizontal and/or vertical direction.

The present work will look in detail at the surface elevation at the wave cut at $y = 3.01715B$. There, the free surface was discretized by cell sizes $\Delta x_{\mathrm{ref}} = 0.2\,\mathrm{m}$ and $\Delta z_{\mathrm{ref}} = 0.025\,\mathrm{m}$ (coarser grid) and $\Delta x_{\mathrm{ref}} = 0.1\,\mathrm{m}$ and $\Delta z_{\mathrm{ref}} = 0.0125\,\mathrm{m}$ (finer grid) for the reference simulation.

To judge whether the discretization can be considered sufficient to resolve the ship's wave field, first, the aforementioned cell sizes will be compared to the range of wave heights occurring in the whole domain and to the wavelength $\lambda_{\mathrm{t}} \approx 3.1\,\mathrm{m}$ of the transverse waves generated by the ship; λ_{t} can be computed from linear wave theory as the wavelength of the regular waves

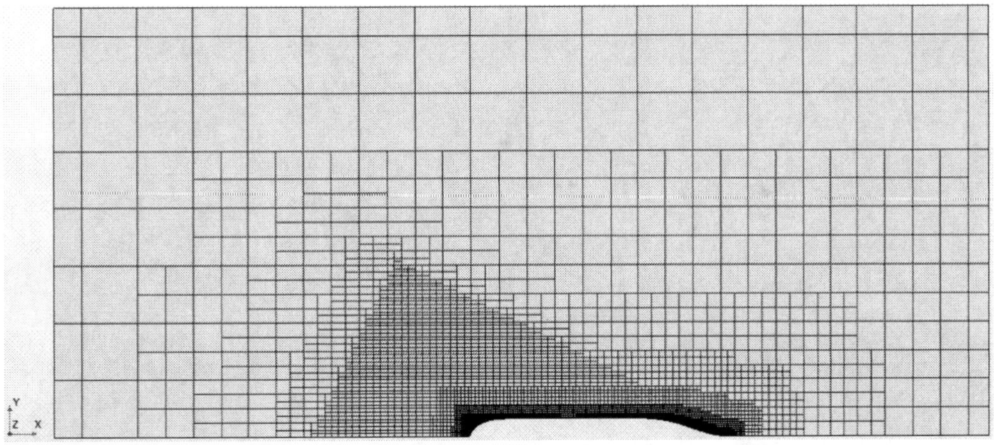

Figure 4. Computational grid for horizontal cut through domain at calm water level $z = 0$, illustrating the local grid refinement near the ship hull and for the ship's wave system; for reference simulation on *coarser grid*

Table 3. For simulations with different refinement of the horizontal (Δx, $= \Delta y$) or vertical (Δz) cell sizes in the vicinity of the free surface: Aspect ratio $\Delta z/\Delta x$ of cells in the vicinity of the free surface, total number of grid cells, time-step Δt, clock time for simulation to complete in parallel on 10 cores; for *coarser grid*

refinement	$\Delta z/\Delta x$	cells	Δt	clock time
$\Delta x = 2.0\Delta x_{\text{ref}}$	1 : 16	$1.8 \cdot 10^5$	0.005 s	1.8 h
$\Delta x = 1.0\Delta x_{\text{ref}}$	1 : 8	$1.9 \cdot 10^5$	0.0035 s	2.6 h
$\Delta x = 0.5\Delta x_{\text{ref}}$	1 : 4	$2.2 \cdot 10^5$	0.0025 s	4.4 h
$\Delta z = 2.0\Delta z_{\text{ref}}$	1 : 4	$1.1 \cdot 10^5$	0.01 s	0.6 h
$\Delta z = 1.0\Delta z_{\text{ref}}$	1 : 8	$1.9 \cdot 10^5$	0.005 s	1.9 h
$\Delta z = 0.5\Delta z_{\text{ref}}$	1 : 16	$3.6 \cdot 10^5$	0.0025 s	7.6 h

Table 4. For simulations with different refinement of the horizontal (Δx, $= \Delta y$) or vertical (Δz) cell sizes in the vicinity of the free surface: Aspect ratio $\Delta z/\Delta x$ of cells in the vicinity of the free surface, total number of grid cells, time-step Δt, clock time for simulation to complete in parallel on 10 cores; for *finer grid*

refinement	$\Delta z/\Delta x$	cells	Δt	clock time
$\Delta x = 2.0\Delta x_{\text{ref}}$	1 : 16	$4.8 \cdot 10^5$	0.0025 s	11.9 h
$\Delta x = 1.0\Delta x_{\text{ref}}$	1 : 8	$5.3 \cdot 10^5$	0.00176 s	16.2 h
$\Delta x = 0.5\Delta x_{\text{ref}}$	1 : 4	$7.4 \cdot 10^5$	0.00125 s	26.1 h
$\Delta z = 2.0\Delta z_{\text{ref}}$	1 : 4	$3.5 \cdot 10^5$	0.005 s	4.5 h
$\Delta z = 1.0\Delta z_{\text{ref}}$	1 : 8	$5.3 \cdot 10^5$	0.0025 s	12.8 h
$\Delta z = 0.5\Delta z_{\text{ref}}$	1 : 16	$10.2 \cdot 10^5$	0.00125 s	51.5 h

following the ship at phase velocity equals ship speed. This results in ca. 5 (coarser grid) or 10 (finer grid) cells per maximum elevations and 15.5 (coarser grid) or 31 (finer grid) cells per

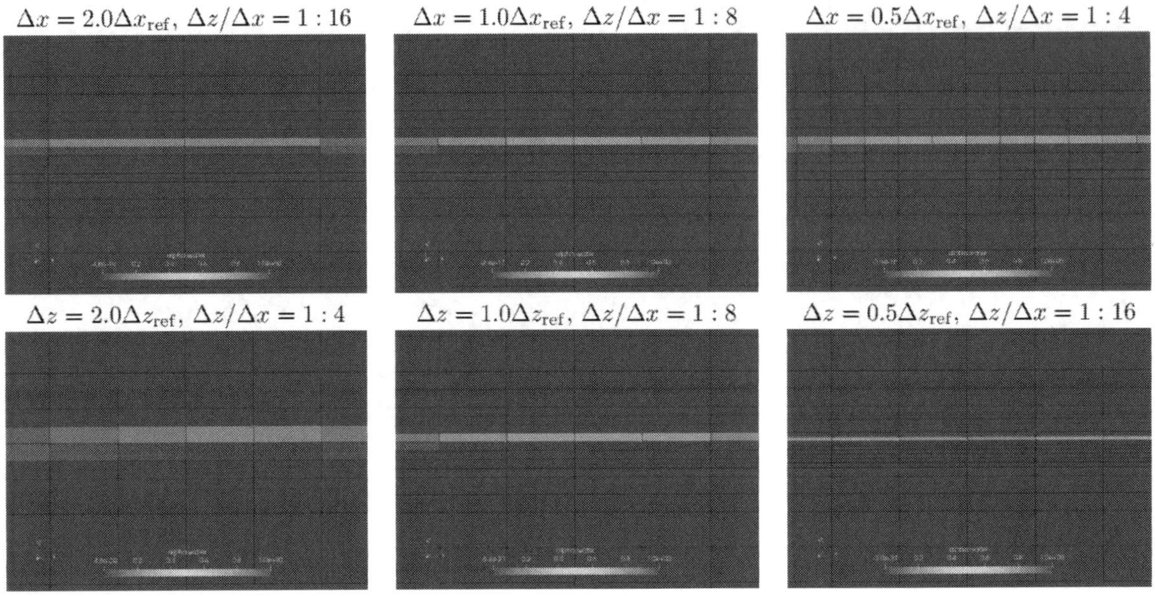

Figure 5. Visualization of aspect ratio $\Delta z/\Delta x$ of the cells near the water-air interface in a close-up of the ship's wave field for the wave cut at $y = 3.01715B$ and the *coarser grid*

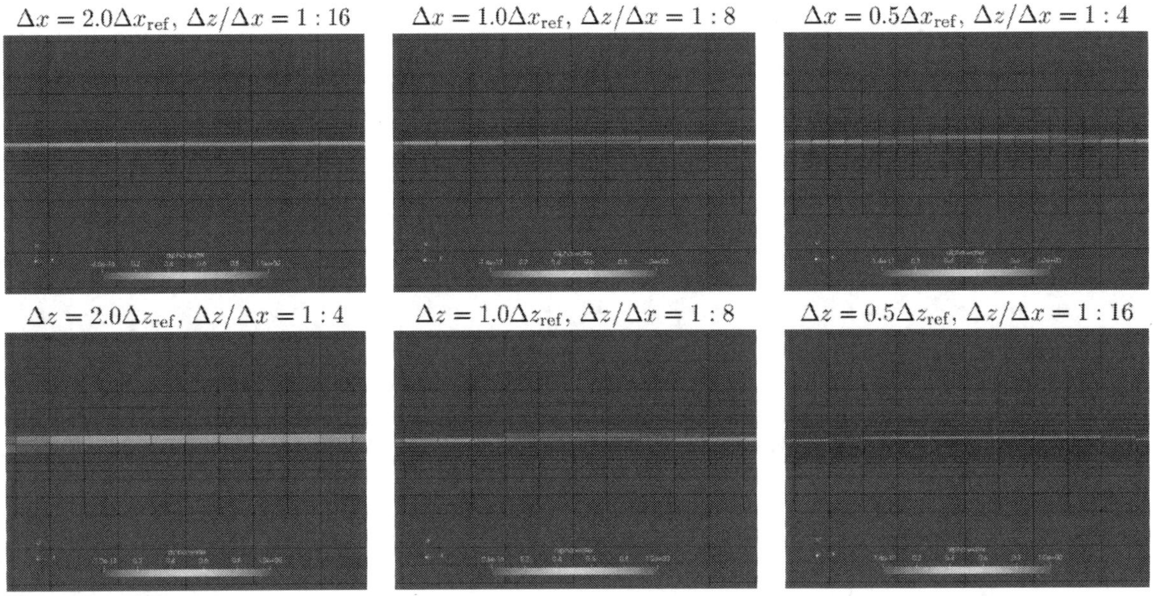

Figure 6. Visualization of aspect ratio $\Delta z/\Delta x$ of the cells near the water-air interface in a close-up of the ship's wave field for the wave cut at $y = 3.01715B$ and the *finer grid*

wavelength λ_t. This corresponds to an acceptable resolution, which could also be used when simulating wave propagation over distances of a few wavelengths.

However, the ship's wave field also contains finer structures. It is customary that these are not resolved as accurately. For example, in relation to the largest wave height at the wave cut at $y = 3.01715B$ (from experiments, largest wave height $H = 0.02785\,\mathrm{m}$ and corresponding wavelength $\lambda = 1.1\,\mathrm{m}$, measured between wave peak at $x = -0.70\,\mathrm{m}$ and wave trough at $x = -1.80\,\mathrm{m}$), the discretization corresponds to only 1.1 (coarser grid) or 2.2 (finer grid) cells per wave height and 5.5 (coarser grid) or 11 (finer grid) cells per wavelength.

The total resistance coefficient for the ship was evaluated as

$$C_\mathrm{T} = \frac{R_\mathrm{T}}{0.5\rho S_0 u_\mathrm{m}^2} \quad , \tag{5}$$

with ship resistance $R_\mathrm{T} = -2F_x$, x-component F_x of the force acting on the ship, density of water $\rho = 998.8\,\mathrm{kg/m^3}$, wetted surface $S_0 = 9.4379\,\mathrm{m^2}$ of the hull without rudder at zero speed (even keel) and ship speed $u_\mathrm{m} = 2.196\,\mathrm{m/s}$.

4. Results

4.1. Influence of discretization and interface-sharpening on ship resistance

Figures 7 and 8 demonstrate that the cell aspect ratio and the magnitude of the compressive source term c_α have a comparatively small influence on the computed value for ship resistance ($< 2\%$ for most simulations). Differences were smaller for the finer grid. The CFD results for the coefficient of total ship resistance, C_T, were ca. 7% to 13% larger than experimental results on the coarser grid, and ca. 6% to 7% larger than experimental results on the finer grid. The reason for these differences was not found, but it is expected that refining the grid or improving the numerical setup could produce closer agreement to experimental results.

Whereas on the coarser grid all simulations completed successfully, on the finer grid some simulations experienced stability problems and blew up. This was attributed to the presence of a few low-quality cells in the grid, which caused problems with the present simulation setup. These problems occurred at different times in the simulation and for different discretizations and for different values of c_α.

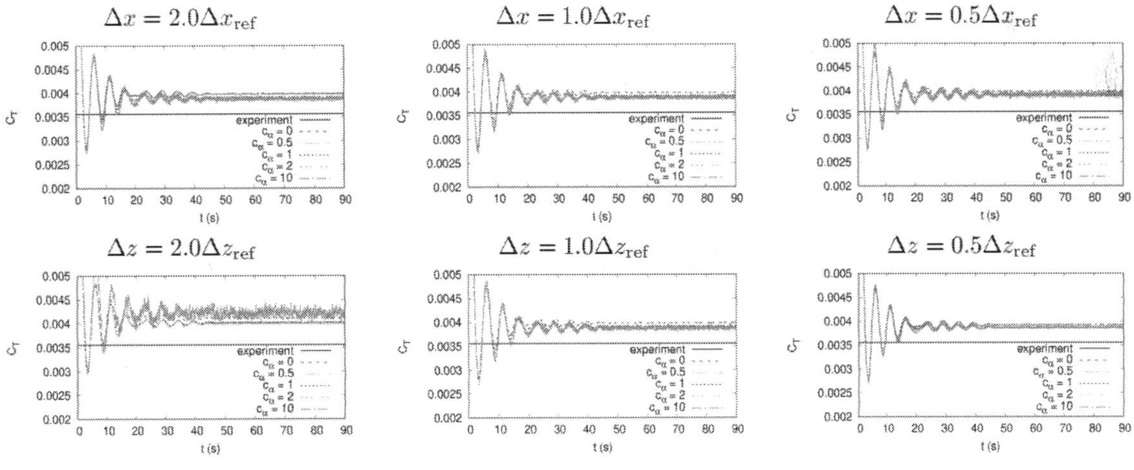

Figure 7. Ship resistance coefficient C_T as a function of time t; for different values of compressive source term magnitude c_α and for refinements of the grid sizes Δx and Δz near the free surface, compared to reference grid size Δx_ref and Δz_ref; for *coarser grid*

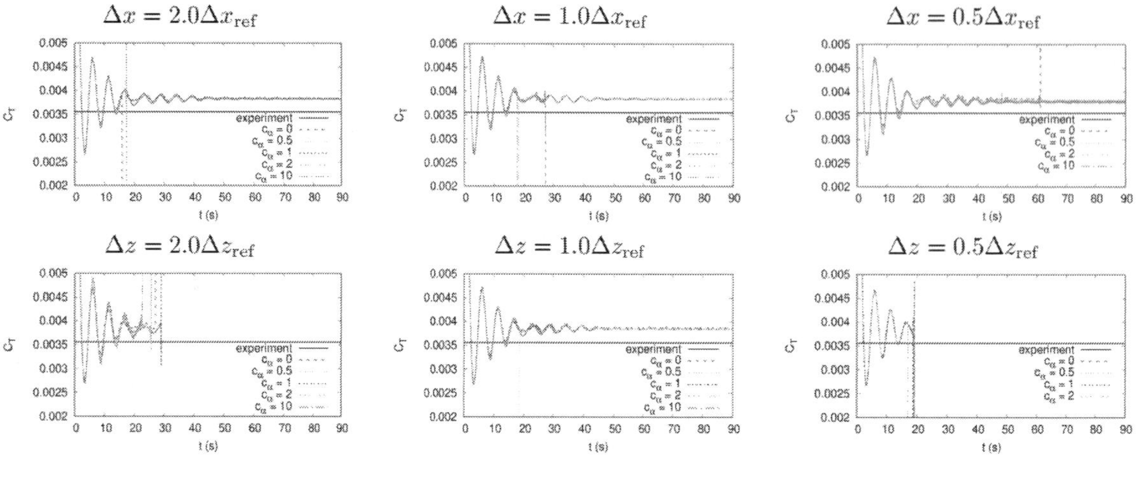

Figure 8. Ship resistance coefficient C_T as a function of time t; for different values of compressive source term magnitude c_α and for refinements of the grid sizes Δx and Δz near the free surface, compared to reference grid size Δx_ref and Δz_ref; for *finer grid*

4.2. Influence of discretization and interface-sharpening on the ship's wave fields

Figures 9 to 16 demonstrate that the cell aspect ratio and the manitude of the compressive source term c_α can have a large influence on the ship's wave field.

Figure 9 shows that uniform grid refinement leads to a better resolution of the ship's wave field.

Figure 11 shows that increasing c_α can substantially increase the wave elevation in the ship's wave field. No significant differences are observed next to the hull, but at some distance (e.g. the highlighted wave cut) the influence is substantial.

Figure 11 also shows that for $c_\alpha = 0$ (no compressive source term), the air-water interface was smeared over ca. 3 cells (coarser grid) or over ca. 2 cells (finer grid, cf. Fig. 9) in z-direction. Already for $c_\alpha = 1$, the air-water interface extended mostly only over 1 cell (i.e. there was only one cell with $0 < \alpha < 1$ between the cells with volume fraction $\alpha = 0$ (air) and $\alpha = 1$ (water).

For $c_\alpha = 10$ (comparatively large compressive source term), disturbances occurred in the free surface in the form of 'ripples' or short-wavelength waves, particularly in the finer grid near the ship hull.

Figures 11 and 12 demonstrates that the influence of compressive source-term magnitude c_α can strongly depend on the aspect ratio $\Delta z/\Delta x$ of the cells near the water-air interface.

For flat cells ($\Delta z/\Delta x = 1 : 16$), varying c_α had a negligible effect on the wave elevation, except for the generation of unphysical ripples for $c_\alpha = 10$.

The smaller the cell aspect-ratio $\Delta z/\Delta x$, the larger was the influence of c_α on the wave elevation, with larger c_α-values leading to a greater increase in wave height in the ship's wave field.

For a given value of c_α, e.g. default setting $c_\alpha = 1$, changing the cell aspect ratio could substantially change the wave height in the ship's wave field.

Figures 13 to 16 show the previous observations more quantitatively. The comparison to experimental data shows that certain combinations of cell aspect-ratio $\Delta z/\Delta x$ and compressive source-term magnitude c_α can lead to an unphysical increase in wave height, i.e. the ship would generate larger and more energetic waves than physically possible for its speed.

IOP Conf. Series: Materials Science and Engineering 1288 (2023) 012006 doi:10.1088/1757-899X/1288/1/012006

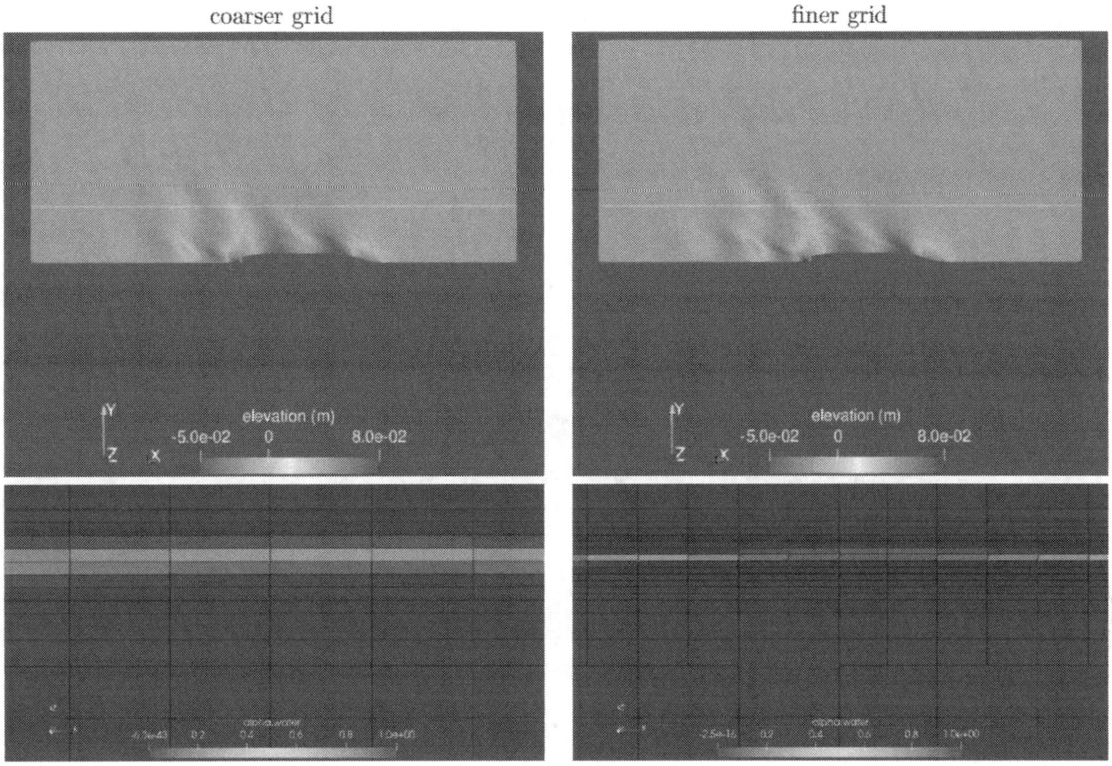

Figure 9. Ship wave fields in whole domain for finished simulations; the light line illustrates the wave cut at $y = 3.01715B$, at which the surface elevations will be compared; for compressive source term magnitude $c_\alpha = 1$, reference grids with $\Delta z = 1.0\Delta z_{\text{ref}}$ and coarser and finer grid

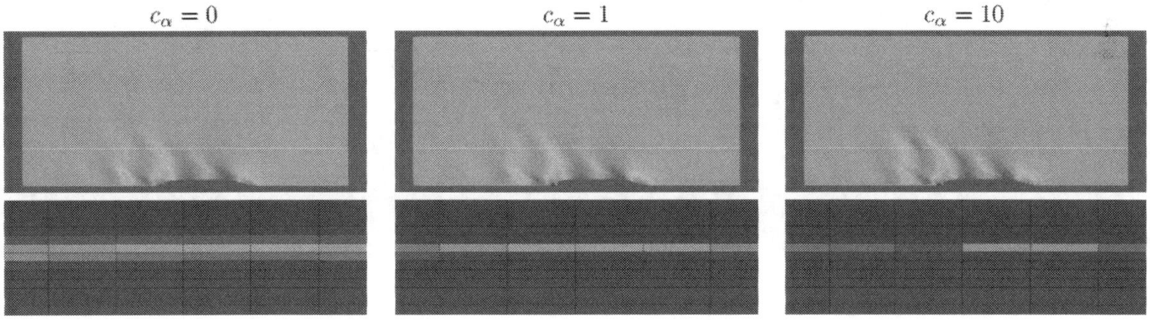

Figure 10. As Fig. 9 top left, except for different values of compressive source term c_α

Changing c_α from 0 to 10 could more than double the local wave elevation ($\Delta z/\Delta x = 1 : 4$) or have no effect at all ($\Delta z/\Delta x = 1 : 16$). Thus, the cell aspect ratio and the magnitude of the compressive source term c_α can have a large influence on the ship's wave field.

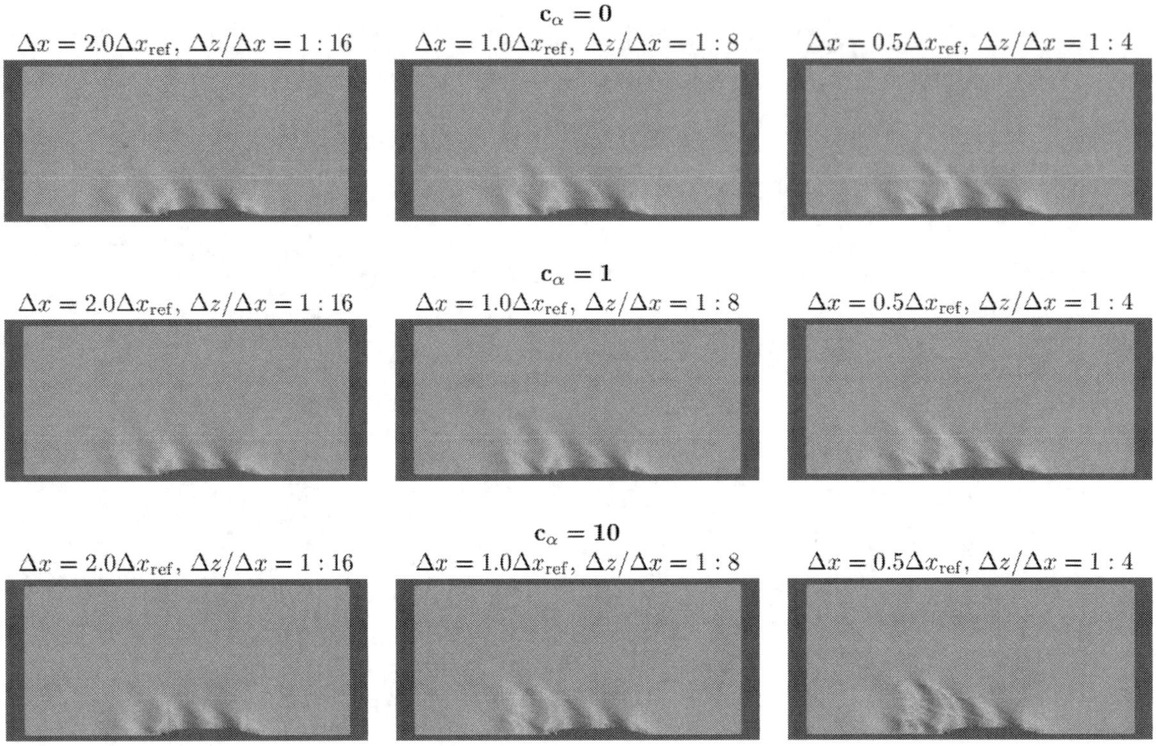

Figure 11. As Fig. 9 top left (coarser mesh), except for different values of compressive source term c_α and different refinement of horizontal cell size Δx, resulting in different aspect ratios $\Delta z/\Delta x$ for the cells near the water-air interface

IOP Conf. Series: Materials Science and Engineering 1288 (2023) 012006 doi:10.1088/1757-899X/1288/1/012006

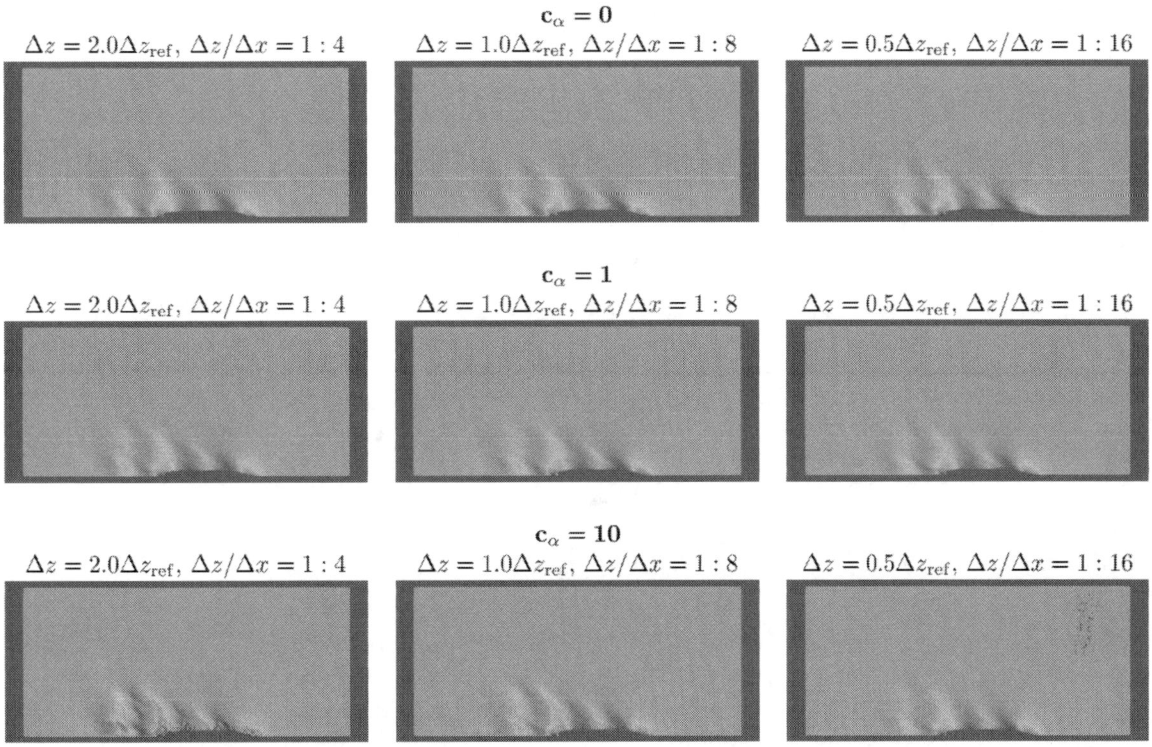

Figure 12. As Fig. 9 top left (coarser mesh), except for different values of compressive source term c_α and different refinement of vertical cell size Δz, resulting in different aspect ratios $\Delta z/\Delta x$ for the cells near the water-air interface

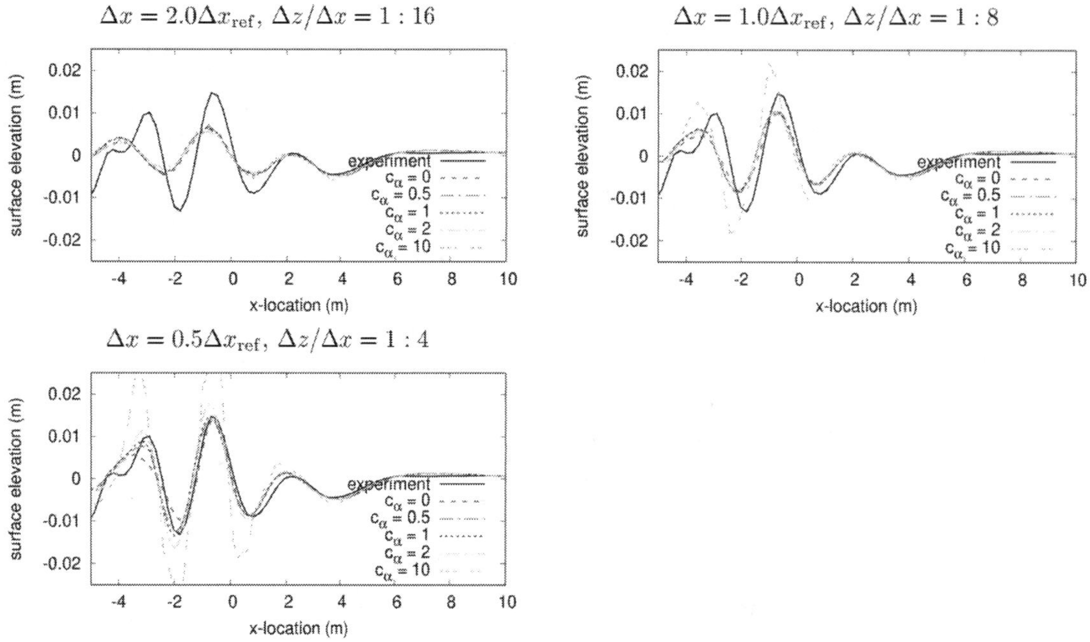

Figure 13. Simulation and experimental results of wave elevation along cut at $y = 3.01715B$; the KCS model is located roughly between $0 < x < 7.5\,\mathrm{m}$; for different horizontal cell size Δx, different compressive source-term magnitude c_α and *coarser grid*

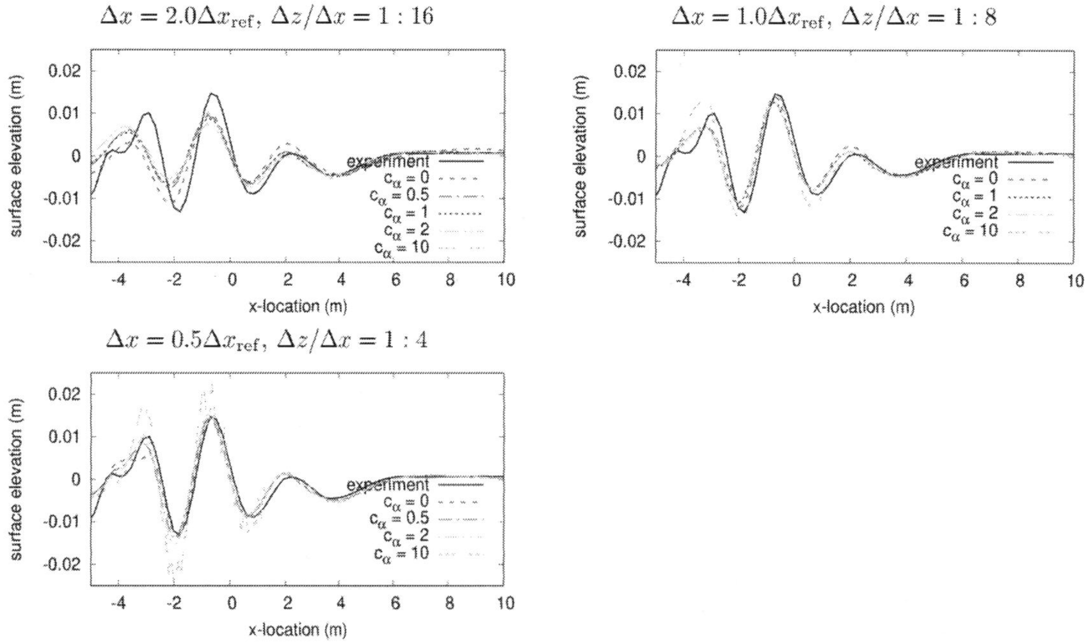

Figure 14. As Fig. 13, except for *finer grid*

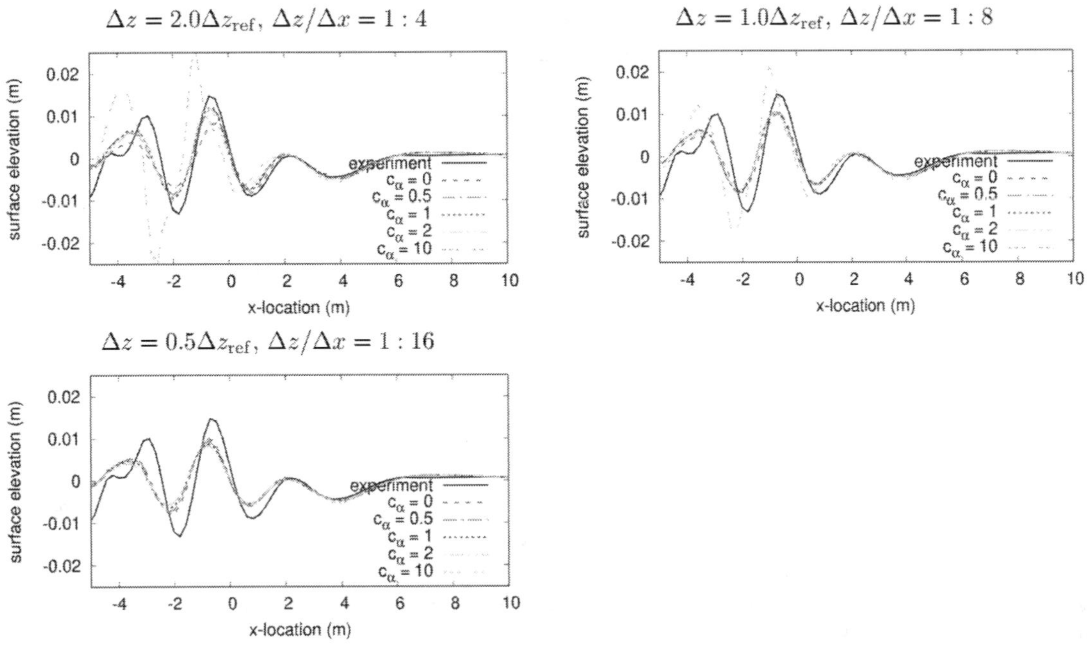

Figure 15. As Fig. 13, except for grid refinement in z-direction via Δz

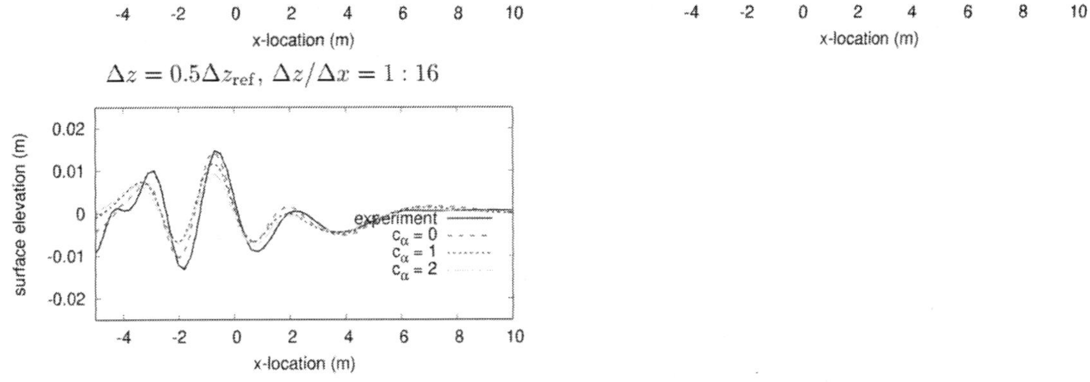

Figure 16. As Fig. 15, except for *finer grid*

5. Discussion

The present results demonstrate how the magnitude of the artificial compressive source terms, regulated by c_α in Eq. (4), influences the solution for the ship's wave field:

For a too small value of c_α, the interface may smear over too many cells, which can cause a loss of wave height.

For too large values of c_α, the wave height can increase unphysically, i.e. the source term can introduce energy into the flow. Furthermore, ripples may occur in the free surface, and for very large source terms, the simulation may become unstable and blow up.

The main conclusion from the present results is that the optimum value of c_α is case-dependent[1]: When refining the discretization, a different value of c_α may give the most correct results.

However, the correct scaling of c_α is unknown; the present results do not suffice to deduce its dependence. Consequently, c_α poses a challenge to any grid- and time-step-dependence study of flows with free surfaces, by preventing comparison of results on coarse and fine mesh. For example, when all simulations are performed with same c_α, then it could be that on the coarse grid, c_α may be too small, while on the fine grid, it could be too large (cf. e.g. the curve for $c_\alpha = 2$ in Figs. 15 top right and 16 top right).

When uniformly refining the grid, the present results indicate that the influence of c_α on the ship's wave field will eventually decrease. Therefore, when using particularly fine grids, as done e.g. in Large Eddy Simulations (LES) with hundreds of millions of cells, then the choice of c_α may become less relevant.

However, for efficient CFD setups as used frequently in industrial practice, the influence of cell aspect-ratio $\Delta z/\Delta x$ and compressive source-term magnitude c_α can be substantial.

Therefore, at present it is not possible to determine the correct wave field of a ship from CFD simulation results alone, when artificial compressive source-terms are used for interface-sharpening, unless exceptional computational effort is invested.

Apart from that finding, an interesting observation was that refining the grid in z-direction could even reduce the wave height in the ship's wave field. This is counter-intuitive, because smaller vertical cell sizes would typically be associated with smaller vertical discretization errors, so a decrease in numerical diffusion and consequently an increase in wave height would be expected. This has particular relevance for adaptive grid refinement techniques:

First, to ensure that adaptive grid refinement yields a more correct solution than the coarser grid, the grid must be refined uniformly, i.e. the cell aspect-ratio $\Delta z/\Delta x$ must not change near the water-air interface.

Second, the necessary refinement to obtain a good solution for the ship's wave field can depend substantially on the cell aspect-ratio $\Delta z/\Delta x$ for the base grid, to which adaptive grid refinement is applied: For example, in the present simulations, the coarser grid with $\Delta x = 0.5\Delta x_{\text{ref}}$ ($2.2 \cdot 10^5$ cells, aspect ratio $\Delta z/\Delta x = 1:4$, $c_\alpha = 1$) produced close agreement to experimental wave elevations, whereas the finer grid with $\Delta z = 0.5\Delta z_{\text{ref}}$ ($10.2 \cdot 10^5$ cells, aspect ratio $\Delta z/\Delta x = 1:16$, $c_\alpha = 1$) produced much worse results in spite of the ca. 7 times larger computational effort.

Based on the present results, cell aspect ratios in the range of $\Delta z/\Delta x = 1:4$ appear to be preferable, whereas substantially larger cell-aspect ratios like $\Delta z/\Delta x > 1:8$ should be avoided, because these could result in unphysically low wave height in the ship's wave field.

Because the cell aspect-ratio $\Delta z/\Delta x$ can have such a pronounced effect, it can be estimated that the rotation of the cells (not investigated in this work) may also have a substantial influence. Therefore, mesh deformation should not be used or at least should be used with caution, because

[1] For propagating free-surface waves, [3] observed that the choice of c_α, the time-step and the numerical schemes used can affect the loss of wave height the further waves travel through the domain. Furthermore, [1] found that interface-sharpening approaches based on the NVD-diagram contain case-dependent parameters as well.

a rotation of the cells relative to the horizontal axis might cause a change in the local wave height, as well. The investigation of this hypothesis is left to future work.

Furthermore, it was observed that the interface was already comparatively sharp without compressive source terms (i.e. $c_\alpha = 0$), where the change from water ($\alpha = 1$) to air ($\alpha = 0$) occurred over only two to three cells. This is expected to be due to the MULES limiter applied to the discretization of the advection term of the volume fraction in Eq. (3). Thus, MULES might be interpreted as an interface-sharpening approach in its own right. Future work will investigate this aspect in more detail.

Finally, another topic for future work could be to improve the OpenFOAM solver settings: For some grids, OpenFOAM simulations blew up and terminated with a segmentation fault, mostly at a nearly converged state of the simulation (between simulation times $10\,\mathrm{s} \lesssim t \lesssim 60\,\mathrm{s}$). The reason for this was not found, but from previous experience it was expected that sensitivity of OpenFOAM to the presence of a few cells with low quality may be the reason for this. Therefore, improving OpenFOAM's resilience to the presence of low-quality cells would also be an important topic for future research.

6. Conclusion

The present work demonstrates that, at present, it is not possible to determine the correct wave field of a ship from CFD simulation results alone, when artificial compressive source-terms are used for interface-sharpening, unless exceptional computational effort is invested.

The reason for this is that the optimum magnitude c_α of the artificial compressive source-terms was found to be case-dependent: Both uniform and non-uniform mesh refinement were found to substantially influence its optimum value.

It was found that if c_α is selected larger than optimal, it can lead to an unphysical increase in wave height and thus also wave energy in the ship's wave field. At present, it is an open research question how to determine the optimum value of c_α for a given discretization.

However, the present findings provide recommendations for the choice of aspect-ratio of cells near the water-air interface and for selection of suitable c_α-values. The results are also relevant for adaptive grid refinement techniques, indicating that grid refinement near the water-air interface should be uniform and start from a suitable cell aspect ratio.

Further research is necessary to investigate the dependency of c_α on the discretization to predict its optimum value before performing the flow simulations. This would be necessary to correctly predict the ship's wave field from CFD simulations using artificial compressive source terms.

Another topic of future research could be improving the numerical stability to increase resilience of OpenFOAM to the presence of low-quality cells.

References

[1] Berndt, J. C., Perić, R. and Abdel-Maksoud, M., 2021. "Improved simulation of flows with free-surface waves by optimizing the angle factor in the HRIC interface-sharpening scheme." *Journal of Applied Fluid Mechanics*, 14, 3, 909-920. DOI: `https://doi.org/10.47176/jafm.14.03.32062`

[2] Hino, T., Stern, F., Larsson, L., Visonneau, M., Hirata, N. and Kim, J. (Eds.), 2020. "Numerical ship hydrodynamics: an assessment of the Tokyo 2015 Workshop (Vol. 94)." *Springer Nature*. DOI: `https://doi.org/10.1007/978-3-030-47572-7`

[3] Larsen, B. E., Fuhrman, D. R. and Roenby, J., 2019. "Performance of interFoam on the simulation of progressive waves." *Coastal Engineering Journal*, 61(3), 380-400. DOI: `https://doi.org/10.1080/21664250.2019.1609713`

[4] Larsson, L., Stern, F. and Visonneau, M. (Eds.). (2013). "Numerical ship hydrodynamics: an assessment of the Gothenburg 2010 workshop." *Springer Science & Business Media*. DOI: `https://doi.org/10.1007/978-94-007-7189-5`

[5] Menter, F. R., 1992. "Improved two-equation k-omega turbulence models for aerodynamic flows (No. A-92183)". Link: `https://ntrs.nasa.gov/citations/19930013620`

[6] Perić, R. and Abdel-Maksoud, M. (2016). Reliable damping of free-surface waves in numerical simulations. Ship Technology Research, 63(1), 1-13. DOI: https://doi.org/10.1080/09377255.2015.1119921

[7] Van, S. H., Kim, W. J., Yim, G. T., Kim, D. H., and Lee, C. J., 1998: "Experimental Investigation of the Flow Characteristics Around Practical Hull Forms." *Proc. 3rd Osaka Colloquium on Advanced CFD Applications to Ship Flow and Hull Form Design*, Osaka, Japan.

[8] Wackers, J., Deng, G., Raymond, C., Guilmineau, E., Leroyer, A., Queutey, P. and Visonneau, M., 2022. "Adaptive grid refinement for ship resistance computations." *Ocean Engineering*, 250, 110969. DOI: https://doi.org/10.1016/j.oceaneng.2022.110969

[9] Weller, H. G., Tabor, G., Jasak, H. and Fureby, C., 1998. "A tensorial approach to computational continuum mechanics using object-oriented techniques." *Computers in physics*, 12(6), 620-631. DOI: https://doi.org/10.1063/1.168744

A deep learning method for the prediction of focused waves in a wave flume

Mingyang Zhang, Sasan Tavakoli * and Spyros Hirdaris

Department of Mechanical Engineering, Aalto University. Espoo, Finland
(Corresponding author: sasan.tavakoli@aalto.fi)

Abstract. Rogue waves pose a significant risk to marine safety, emphasizing the need to accurately predict their occurrence in the open ocean. However, the complexity of their evolution, which may involve nonlinear physical phenomena such as wave-wave interaction and modulation instability, makes this task challenging. Currently the reconstruction of rogue waves involves generating focused waves through the superposition of different spectral components of irregular waves that are in phase at the focusing point. Despite its effectiveness, this approach is limited. The paper introduces a deep learning method based on Long short-term memory (LSTM) to predict focused waves generated in a Computational Fluid Dynamics (CFD) flume in the time domain. The model is trained on 60% of the generated wave time series, with the remaining 40% used for both validation and testing. The results demonstrate that the proposed method can assist with the prediction of focused waves at various observation points, indicating its potential as a promising approach for predicting rogue wave behaviour in the ocean.

1 Introduction

Rouge waves can be a potential natural source of danger for ships and offshore structures as they can result in very large sea loads [1,2]. Over last decades, several ship accidents have been found to be triggered by such waves [3]. Forecasting and simulating such waves in physical and virtual wave flumes is an important task, that make it possible to predict their evolution, study their physics and understand the way they interact with ships and floating bodies locating in their paths. Research in this area may lead to the development of advanced ship seakeeping and navigation criteria decision for proactive ship safety management.

In any tank, either a wave flume or a virtual one (e.g., numerical), freak waves can be formed using different mechanisms depending on the aim of the study. One of the most common mechanisms used to trigger freak waves is the modulation instability, which unstabilises the wave growth over the tank. The phenomenon relates to the fact that different components of a random wave may interact with each other [4]. This may eventually lead to the generation of rouge waves along the tank. Soliton like breathers during wave making, may be also localized in time and space leading to extreme wave events [5, 6]. Breathers are found by solving the Nonlinear Schrödinger Equation (NLS).

Rouge waves can be reconstructed by forcing the components of random waves to be in phase at a focusing point [7]. This method is popular in terms of reconstructing freak waves numerically and physically as (i) the required physical time for the evolution of a freak wave is relatively short as compared to the way random freak waves are generated using the other dispersive methods and (ii) a random wave pattern is generated at the wave maker, which is less challenging as compared to the way breathers are generated at a wavemaker.

In the realm of water wave dynamics, focused waves can be generated in physical and numerical tanks [8,9]. However, the advent of artificial intelligence, particularly machine learning, presents an opportunity to advance wave forecasting capabilities. Recent advances in ocean and coastal engineering confirm that researchers made strides in forecasting water waves based on available data, although not with a specific focus on the formation of focused waves. For example, Artificial Neural Networks (ANN) have been used to predict wave heights using the buoy observation data [10-12]. The results indicate that ANN outperform physical model [13] and traditional regression models [14,15]. More recently, Convolutional Neural Networks (CNN) have been applied to predict short-term wave power. Comparisons indicate that CNN outperform ANN in terms of predicting long wave patterns [16]. Recurrent Neural Networks (RNN) may outperform traditional ANN and CNN due to their ability to handle sequential data [13,17]. Long Short-Term Memory (LSTM) is a variant of Recurrent Neural Network (RNN) that is specifically designed to handle the learning of long-term dependencies, making it more effective than traditional RNNs in certain tasks. LSTM is equipped with a specialized memory unit which enables the network to selectively retain or forget information over time. This capability allows LSTM to effectively handle long sequences of data, such as waveforms.

This paper aims to develop an LSTM-based model for the prediction of focused waves. These waves are numerically generated in a virtual tank using CFD, i.e., for fully nonlinear conditions. The paper is structured as follows: Section 2 presents the steps taken to develop the deep learning method. The first step is to design the CFD tank and generate water waves. The second step is to break down the time history of waves recorded in the CFD tank into input and output using a method called autocorrelation. The third step is to develop the deep learning model. Section 3 presents the results, including CFD data and the predictions of the deep learning model. Conclusions and future research are given in Section 4.

2 Methods

The deep learning method used for the prediction of focused waves as presented in this paper accounts for temporal and spatial features. The approach presented comprises the following three steps namely (1) **Step I** where CFD modelling by OpenFOAM is used for wave generation. A two-dimensional wave flume with a length of 30 m and height of 2.2 m is designed to generate waves at various observation points; (2) **Step II** that focuses more on the analysis of the autocorrelation of waves. To capture the lag feature of waves, an autocorrelation function is used. The waves are segmented by utilizing a sliding window approach considering the wave lag feature. Subsequently, input and output datasets are identified, which are used to train the deep learning model. The lag value is then utilized to design the LSTM model; (3) **Step III** where an LSTM based model is developed to forecast focused waves. The model is trained using 60% of the available large-scale data streams, and predictions of waves at different observation points are obtained based on the outcomes of Steps I and II. Validation and testing make use of the remaining 40% of the total dataset. The trained model is utilized to predict the focused wave at a new observation point.

2.1. CFD model for wave generation

Consider a two-dimensional wave flume of length 30 m, height 2.2 m and water depth 1.5 m. The air-water flow is assumed to be viscous. This allows to solve the problem under fully nonlinear conditions that are more likely to arise during the propagation of freak waves. As such, Navier-Stokes equations govern the fluid motion of mixed air-water flow. A Volume of Fluid (VoF) model is used to capture the temporal and spatial evaluation of free surface over the tank, which is governed by a conservation equation [18]. Random waves are generated by a wavemaker located on the left side of the numerical tank. Waves are formulated using linear random wave theory [19] described by the function

$$\eta = \sum_{i=1}^{N} a_i \cos(\omega_i t - k_i x + \varphi_i), \tag{1}$$

where a_i, ω_i, k_i and φ_i are respectively wave amplitude, wave angular frequency, wavenumber and wave phase corresponding to i_{th} component.

A JONSWAP (Joint North Sea Wave Project) wave spectrum is used to generate random waves [19]. The peak period of random waves is set to be 2 s, and the corresponding wavelength is ≈ 6 m. This can be found using the dispersion relationship namely $\omega^2 = gk \tanh kD$, (g is gravitational acceleration, ω

is the wave frequency, k is wavenumber and D is the water depth). The significant wave height, H_s, is prescribed to be 0.05 m. This leads to a steepness of $k_0 H_S/2 \approx 0.026$, which is below the wave breaking limit. The height of the focused wave is expected to be $4.5H_s$ and thus the temporal wave steepness at focusing would be lower than 0.12; a value that is still below the wave breaking limit. This has been done purposely to avoid any wave breaking and resulting wave turbulence, which may lead to complicated physical phenomena (e.g., energy dissipation). Note that the aim of this step is to generate wave data for the development of a deep learning method for the prediction of focused waves only. Modelling of wave turbulence, wave breaking, or any other physical phenomena is considered to be out of the scope of present study.

The numerical tank is generated using OpenFOAM (see Figure 1) utilising the Finite Volume Method (FVM). The interFOAM solver which is capable of solving the VOF (Volume of Fluid) equations [20] is utilised. Waves are generated using the waves2foam library [21]. A 6 m long relaxation zone (corresponding to a wavelength) is placed in front of the wavemaker, and a 9 m long relaxation zone (corresponding to a wavelength and half long) is positioned in the vicinity of the other end of the tank to cancel out the wave reflection from the tank boundaries. Hence, waves propagate over a length of 15 m, which is 2.5 times of the wavelength of the peak period. Waves are set to be in phase in way of the middle of the tank $x = 15$ m. This is longer than 2 wavelengths far from the wave maker, and more than 1 wavelength far from the relaxation zone.

A zero-speed condition is prescribed for the bottom of the tank and an open-atmosphere boundary condition is set for the upper patch of the domain. A structured mesh is generated, as shown in close-up views in Figure 1. The cell size around the water surface is 0.003×0.002 (dense mesh zone). Hence, waves with a height of H_s would be covered by $\approx$ 25 cells, and each wavelength would fit in $\approx$ 2000 cells. Note that the high resolution of the cell size along the tank is chosen to avoid 1) large aspect ratio and 2) energy loss of the short waves (the shortest wave is 0.7 m long and is covered by $\approx$ 233 cells).

To run the problem, courant number is forced to be 0.5. At the present stage, a laminar fluid flow pattern is used as no wave breaking and modulation instability occurs. Within the domain of interest ($1 < x/\lambda < 3$), water surface elevation, pressure and velocity components are sampled at 7 different water columns which are located within 2 m distance intervals (a third of the wavelength). Such setup is designed based on the experience and recommendations [20].

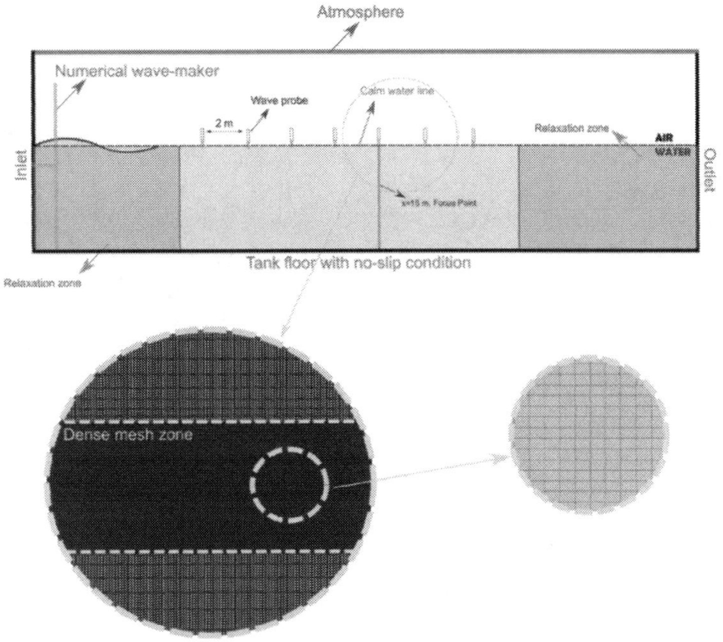

Figure 1. A schematic of the numerical wave tank and the generated mesh used for simulation of focused waves at various wave probe.

2.2. Autocorrelation for time series of waves

Autocorrelation analysis is a valuable tool for time series (waves) predictions, particularly in the context of LSTM models. Autocorrelation refers to the degree of similarity between a time series and a lagged version of itself and can be measured using the Autocorrelation Function (ACF) [22]. ACF measures the correlation between a time series and its own past values at different time lags. A typical ACF is defined as

$$R\ (lag) = \frac{E[(X_i-\mu_i)(X_{i+lag}-\mu_{i+lag})]}{\sigma_i\sigma_{i+lag}}, \tag{2}$$

where, $E[\cdot]$ is the expected value, μ is the mean value, and σ is the variance. The $R\ (lag)$ denotes the correlation between a time series X_i and its own past values at $i + lag$.

A sliding window model [23] is used to split the time series of generated waves into inputs and outputs, and autocorrelation analysis is applied to evaluate the correlation between the splitting waves in various domain, as shown in Figure 2. Therefore, the optimal sequence length can be determined for the development of LSTM model.

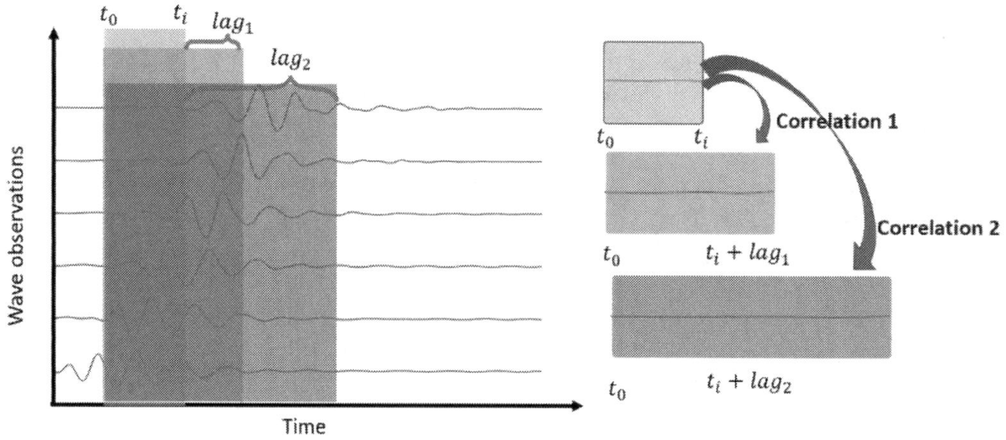

Figure 2. Autocorrelation analysis of waves.

2.3. LSTM based deep learning model

In this paper, an LSTM RNN approach is applied in Figure 3. The LSTM unit comprises of four interconnected elements, including the input and control signals for the input, forgetting, and output gates. These components work together to regulate the memory storage, retention, and output, the details of which can be found in [24, 25]. Figure 4 provides an illustrative depiction of the internal structure of the LSTM unit, where Z denotes the input; Z_i, Z_f, and Z_o represent the control signals of the input gate, forgetting gate, and output gate, respectively. The $f(x)$ function is typically a sigmoid (see Equation (3)), which controls the degree of opening of the gate within the range of [0, 1]; $g(x)$ and $h(x)$ are activation functions.

IOP Conf. Series: Materials Science and Engineering 1288 (2023) 012007 doi:10.1088/1757-899X/1288/1/012007

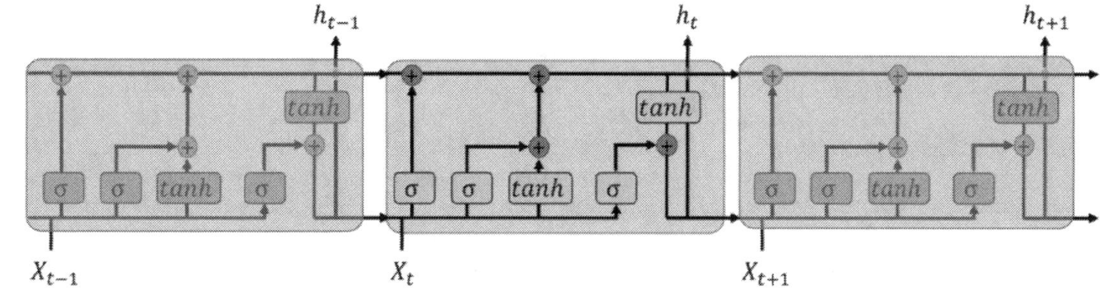

Figure 3. The structural unit diagram of LSTM.

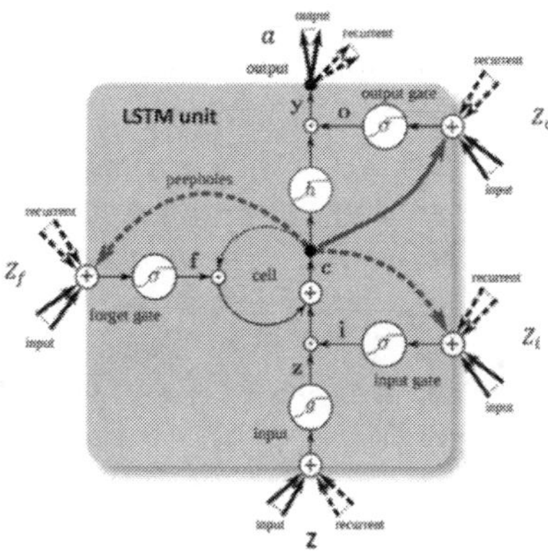

Figure 4. The internal structure of the LSTM unit [27].

To activate Z, the unit first obtains $g(Z)$. Then, Z_i is activated using the sigmoid function to obtain $f(Z_i)$, which is then multiplied by $g(Z)$ to yield $g(Z)\,f(Zi)$. Z_f is obtained by applying the sigmoid function to Z_f, resulting in $f(Z_f)$. The value of the memory unit existing at the previous time step is then multiplied by $f(Z_f)$, which yields $cf(Z_f)$. The memory unit is updated using Equation (4). After activating Z, $h(c')$ is obtained. Finally, Z_o is activated using the sigmoid function to obtain $f(Z_o)$, which is then multiplied by $h(\cdot)$ to obtain the output, denoted by a and given by Equation (5), as

$$f(x) = \frac{1}{1+e^{-x}}, \tag{3}$$

$$c' = g(Z)f(Z_i) + cf(Z_f), \tag{4}$$

$$a = h(c')f(Z_o). \tag{5}$$

The input x_t at time t and the output h_{t-1} of the hidden layer neuron at time $t-1$ are the joint inputs to the hidden layer at time t. These inputs are multiplied by distinct weight vectors, and upon application of the activation function, the control signals Z_f, Z_i, Z_o, and the input value Z of the three gates are generated. This process is represented by Equations (6-9), where the weight vector is denoted by w and the activation amount by b. The biases for different connection weights are represented by b_f, b_i, b_o, and b_x. Subsequently, following the operation of the LSTM unit, the value of the memory unit c is updated, as stated in Eq. (5), which is represented as Equation (10). Accordingly, the output of hidden layer neurons can be presented as shown in Equation (11).

$$Z_f = w_f[h_{t-1}, x_t] + b_f, \tag{6}$$

$$Z_i = w_i[h_{t-1}, x_t] + b_i, \tag{7}$$

$$Z_o = w_o[h_{t-1}, x_t] + b_o, \tag{8}$$

$$Z = w_x[h_{t-1}, x_t] + b_x, \tag{9}$$

$$c' = g(w_x[h_{t-1}, x_t] + b_x)f(w_i[h_{t-1}, x_t] + b_i) + cf(w_f[h_{t-1}, x_t] + b_f), \tag{10}$$

$$h_t = h(c')f(Z_o) = h\Big(g(w_x[h_{t-1}, x_t] + b_x) \cdot f(w_i[h_{t-1}, x_t] + b_i) + cf\big(w_f[h_{t-1}, x_t] + b_f\big)\Big) \cdot f(w_o[h_{t-1}, x_t] + b_o). \quad (11)$$

The input to an LSTM includes not only the output (h_{t-1}) of the neuron in the hidden layer at the previous time step but also the value of the memory unit in the LSTM unit. Because of this design, the LSTM network can effectively prevent the occurrence of gradient disappearance (which is a common issue happening when training traditional ANNs). Thus, the LSTM may effectively remember long-term historical information, and hence it can more accurately model long-term time series data. This is the main reason why the LSTM neural network is utilized for predicting the long-time history of the focused wave in this paper.

This study proposes a deep learning method based on LSTM for wave forecasting, as depicted in Figure 5. As explained in Section 2.2.1, the wave generation is carried out in a wave tank using OpenFOAM, and wave data are collected from several wave probes (Figure 1). The lag length of generated data is determined using ACF (which was introduced in Section 2.2.2). A sliding window model is used to perform the calculation for historical data lying in the time interval $[t, t + t_W]$ (t_W is the window length). This may increase the accuracy level of trained model. The deep learning model is designed to have three different layers. The first and second layers contain 64 and 32 units (see Figures 5). The third layer is termed *dense output* which contains six units corresponding to the input data features. The trained LSTM model is compiled with the Mean Squared Error (MSE) loss function. Further, it is equipped with the Adam optimizer, which is a prevalent optimization algorithm in deep learning used for calculation of gradients [26].

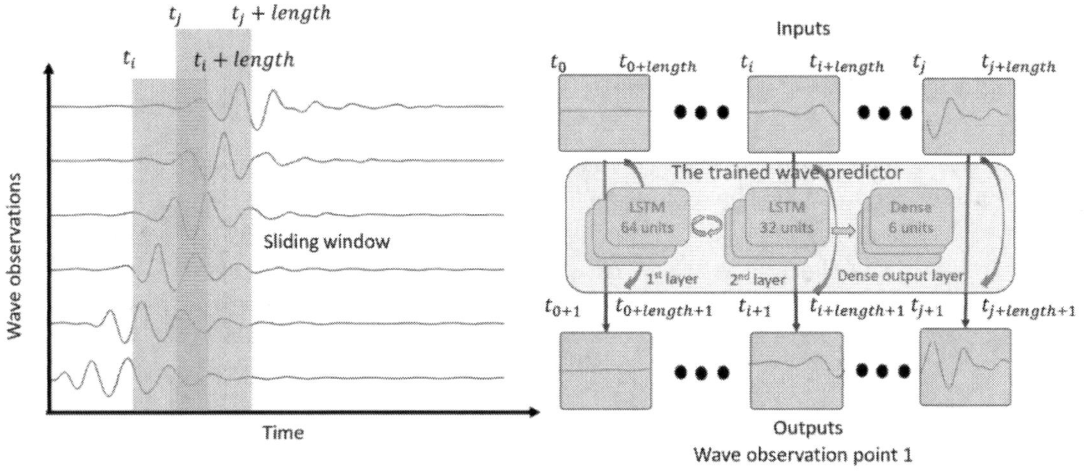

Figure 5. Focused waves predicting from autoregression theory perspectives.

To calculate the errors between the real and the predicted waves, Root Mean Square Error (RMSE), R^2 (Coefficient of Determination), and error rate e_n are used as shown in Equations (12-14).

$$RMSE = \sqrt{\frac{1}{N}\sum_{n=1}^{N}(y_n - \hat{y}_n)^2}, \quad (12)$$

$$R^2 = 1 - \frac{\sum_{n=1}^{N}(y_n-\hat{y}_n)^2}{\sum_{n=1}^{N}(y_n-\bar{y}_n)^2}, \quad (13)$$

$$e_n = (y_n - \hat{y}_n)/(y_n), \quad (14)$$

where, y_n is the actual value, $\hat{y}_n$ denotes the predicted value, $\bar{y}_n$ is the mean value.

IOP Conf. Series: Materials Science and Engineering 1288 (2023) 012007 doi:10.1088/1757-899X/1288/1/012007

3 Results

3.1. Waves from CFD model

The results of CFD model (time series of recorded waves) are shown in Figure 6. Each curve represents the time history of the recorded wave at one of the probes. At every station, the wave is seen to reach a very sharp crest. This crest grows as it travels along the tank and reaches its maximum value at the focus point ($x = 15$ m). As the wave passes the focus point, the crest height decreases, and a random wave pattern starts to develop. As seen, the crest at which the energy is focused travels a distance of 8 m over a time frame of ≈ 5 s. The wave phase speed and wave group speed of the peak frequency are respectively 3 m/s and 1.5 m/s. This suggests that the wave crest travels at the speed of a wave group speed and simulations match with actual physics. In the present paper, waves recorded at probes located at $x = 7$, 5, 9, 11, 15 and 17 m are used to train the LSTM based model, and the wave recorded at $x = 19$ m is used to validate the generalisation of the trained model.

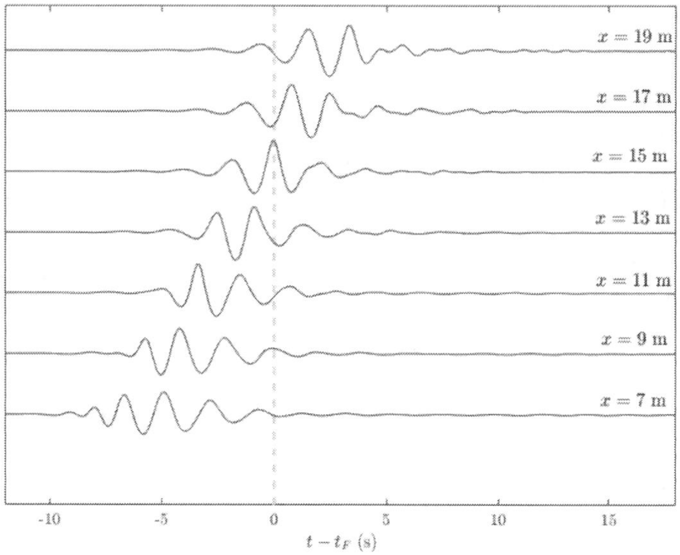

Figure 6. Wave growth along the CFD tank. The dashed grey line marks the focusing.

3.2. Autocorrelation analysis

The pattern of the sliding window model used for splitting the time series into data lying in time interval of $[t, t + t_W]$ is shown in Figure 5. When the sliding window model is used, the sequence length (the length of the data sequence for training the model) is believed to be a critical factor that requires careful consideration. This is because a very long sequence may reduce the accuracy of the trained model, while a very short sequence length may not adequately capture the features of the generated data. Therefore, autocorrelation analysis is employed to help us select the most suitable sequence length (t_W).

To perform the autocorrelation analysis, the sequence length is set to vary from 0 to 250, as illustrated in Figure 7. The correlation coefficient for the given sequence length is calculated, ranging from 0 to 1. We are aiming to find a sequence length that is not too short, but still yields a correlation coefficient close to 1.0. Our findings suggest that autocorrelation corresponding to a sequence length of 98 is ≈ 0.85. This sequence length appears sufficient for capturing features of the CFD data and training the LSTM model.

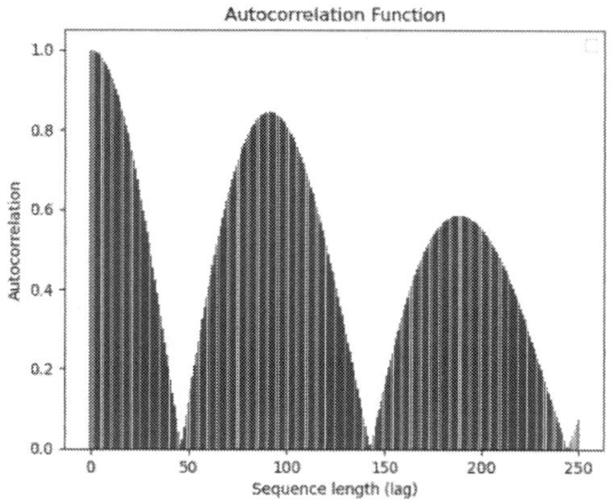

Figure 7. ACF values found for various lags.

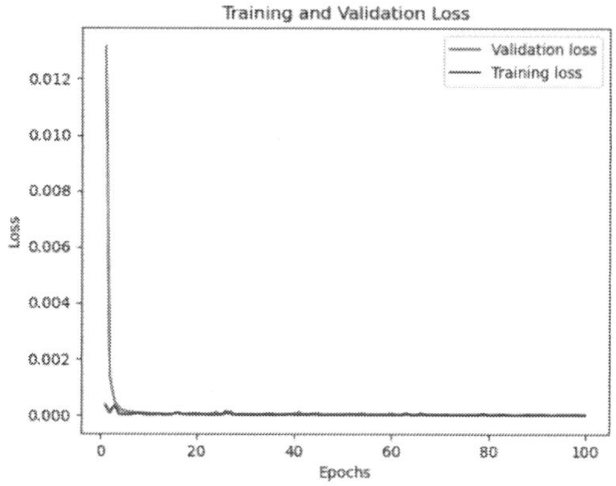

Figure 8. The model performance evaluation.

3.3. Waves prediction using LSTM

The LSTM model was trained using time series recorded at six waves probes located at x = 7, 5, 9, 11, 15 and 17 m of the CFD tank, see more in Figure 1. 60% of each time series was used for training of the model, 20% was used for validation, and the rest was used for testing.

To obtain the optimal LSTM model, 100 epochs were defined. The training and validation losses were calculated based on respective datasets for these epochs. Figure 8 shows the loss values versus epochs. It is observed that the LSTM-based model exhibits an optimal fit without over- or underfitting. It is noteworthy that water waves obtained through CFD simulations are not complex, as they do not exhibit wave breaking, turbulence, or noticeable wave-wave interaction. This simplified nature of the water waves makes it easier for the deep learning model to capture the physical properties of the waves. Thus, the model showed quick convergence and reached its optimal performance at epoch 12.

After training the deep learning model, it was used to predict the time history of waves at x = 7, 5, 9, 11, 15, and 17 m of the CFD tank. The predictions of the deep learning model (green curve) are shown in Figure 9, along with the CFD data (red curve) plotted. The RMSE of the deep learning model in predicting the wave is found to be 0.0018 (see Equation 12). This value provides evidence of the high accuracy level of the developed deep learning model. As observed, the waves predicted by the deep learning model closely match those recorded in the CFD tank. There is no phase shift between the predictions of the deep learning model and the ground truth curve (CFD simulations) in all six panels of Figure 9. This indicates that the dispersion process of waves generated in the CFD tank is accurately captured by the deep learning model.

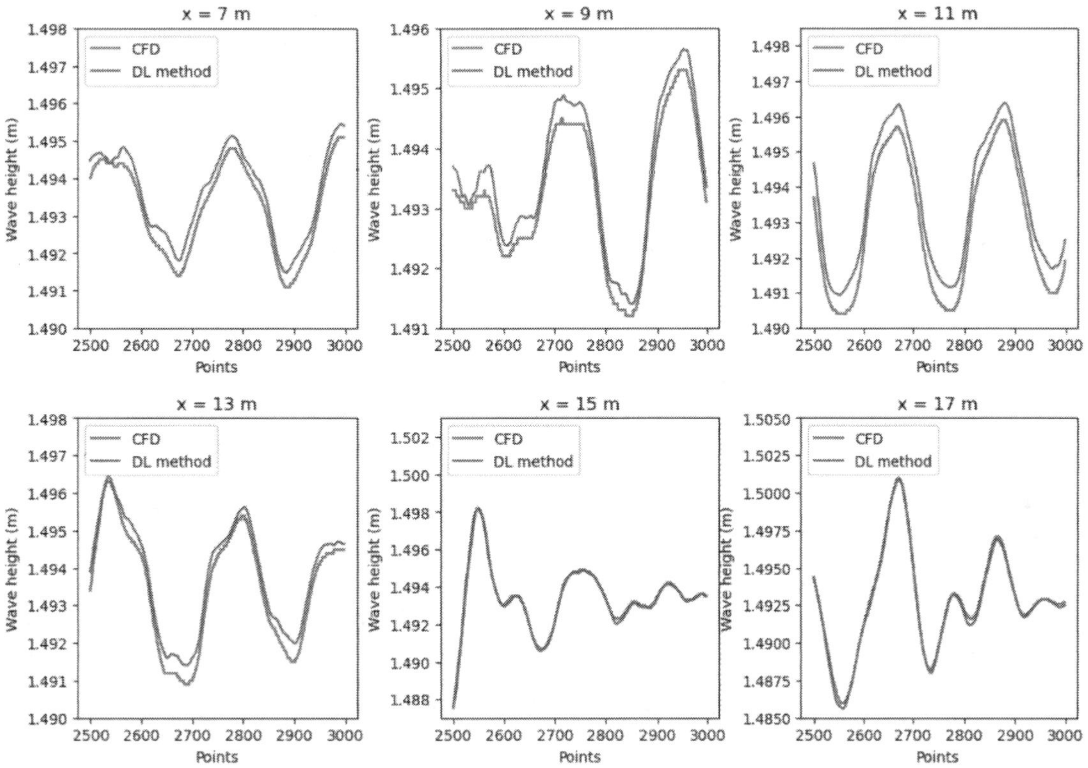

Figure 9. A part of time series of focused waves predicted by the DL method in green and recorded in the CFD tank in red.

The wave at x=19 m was not included in the training data, as mentioned earlier. The deep learning model is utilized to predict the time history of the wave at this point, and its results, along with the CFD simulations, are presented in Figure 10. The error rate is calculated by Equation (14) and is plotted in lower panel of Figure 10. This rate is observed to be lower than 2%, which further validates the reliability of the developed deep learning model in predicting waves at an untrained point (which is x=19 m). In other words, the deep learning model exhibits fair performance in handling new data and making accurate predictions. It is noteworthy that the predicted wave crest and trough are lower than the CFD values. This could be attributed to the fact that focusing has already occurred at x=15 m, and the different components of the wave are not in phase at x=19 m. To address this issue in the future, wave propagation in a longer CFD tank with more observation points may need to be considered. This would yield more data, which can enhance the accuracy level of the deep learning model.

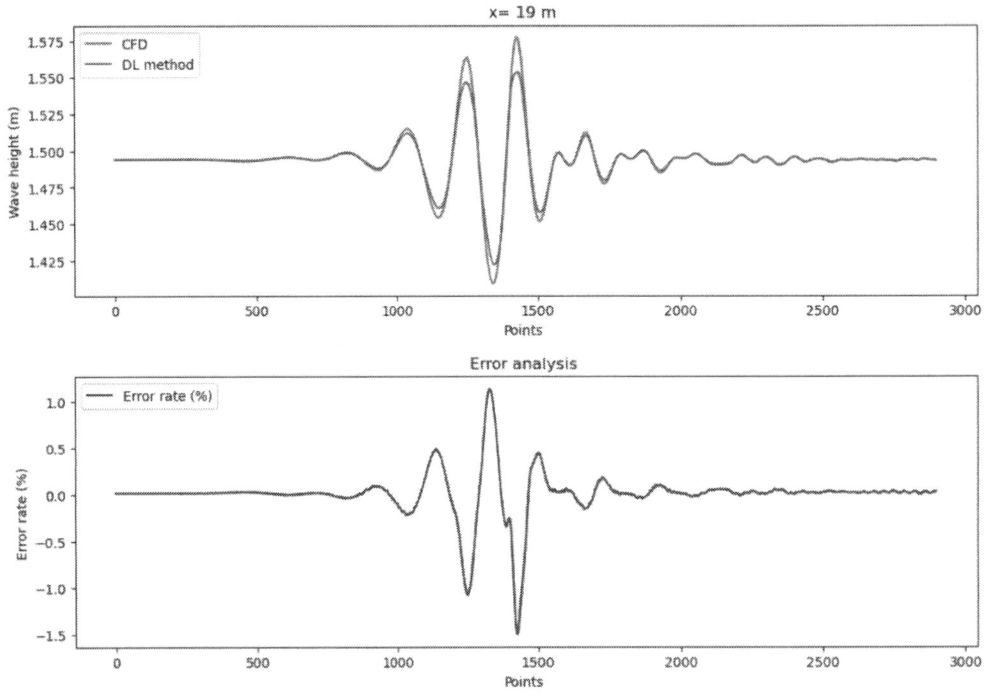

Figure 10. The predicted wave in green and CFD records in red at new observation point, and error analysis in blue.

4 Conclusion

This paper presents a deep learning model based on LSTM for predicting focused waves generated in a CFD tank. Propagation of waves in a virtual tank using CFD was reconstructed using the interFOAM solver, which decomposes fluid equations using the FVM approach. The proposed deep learning model utilizes a sliding window to divide the time series of the recorded waves into intervals. The optimal sequence length of these intervals is selected using an ACF. The recorded waves at six observation points in the tank are used for training the deep learning model. The results demonstrate that the latter effectively captures the physical characteristics of waves and predicts their behaviour at various observation points (wave probes) in the wave tank. Further development of the present model using different spectra is necessary to enhance its accuracy and applicability in practice.

Acknowledgment

The research presented has received funding from the Merenkulun Säätiö, which is gratefully acknowledged.

References

[1] Clauss G, and Klein M 2011 The New Year Wave in a seakeeping basin: generation, propagation, kinematics and dynamics *Ocean Engineering* **38** pp 1624-1639.

[2] Tavakoli S, Khojasteh D, Haghani M, and Hirdaris S 2023 A review on the progress and research directions of ocean engineering *Ocean Engineering* **272** pp 113617.

[3] Zhang Z, and Li X 2017 Global ship accidents and ocean swell-related sea states *Natural Hazards and Earth System Sciences* **17** pp 2041–2051.

[4] Kirezci C, Babanin AV, and Chalikov D 2021 Probabilistic assessment of rogue wave occurrence in directional wave fields *Ocean Dynamics* **71** pp 1141-1166.

[5] Chabchoub A, Hoffmann NP, and Akhmediev N 2011 Rogue wave observation in a water wave tank, *Phys. Rev. Lett.* **106** pp 204502.

[6] Dudley JM, Genty G, Mussot A, Chabchoub A, and Dias F 2019 Rogue waves and analogies in optics and oceanography *Nature Reviews Physics* **1** (11) pp 675 – 689.
[7] Adcock T, and Taylor P 2016 Non-linear evolution of uni-directional focussed wave-groups on a deep water: a comparison of models *Appl. Ocean Res.* **59** pp 147-152.
[8] Wang W, Kamath A, Pakozdi C, and Bihs H 2019 Investigation of focusing wave properties in a numerical wave tank with a fully nonlinear potential flow model *Journal of Marine Science and Engineering* **7**(10) p 375.
[9] Bredmose H, Larsen S E, Matha D, Rettenmeier A, Marino E, and Saettran L 2012 Collation of offshore wind-wave dynamics: Marine Renewables Infrastructure Network for Emerging Energy Technologies D2. 4.
[10] Deo MC, and Naidu CS 1998 Real time wave forecasting using neural networks *Ocean Engineering* **26**(3) pp 191-203.
[11] Makarynskyy O 2004 Improving wave predictions with artificial neural networks *Ocean Engineering* **31**(5-6) pp 709-724.
[12] Lond SN, Panchang V 2006 One-Day Wave Forecasts Based on Artificial Neural Networks *Journal of Atmospheric and Oceanic Technology* **23**(11) pp 1593-1603.
[13] Minuzzi FC, and Farina L 2023 A deep learning approach to predict significant wave height using long short-term memory *Ocean Modelling* **181** pp 102151.
[14] Mahjoobi J, and Mosabbeb E 2009 Prediction of significant wave height using regressive support vector machines *Ocean Engineering* **36**(5) pp 339-347.
[15] Cornejo-Bueno L, Garrido-Merchán EC, Hernández-Lobato D, and Salcedo-Sanz S 2018 Bayesian optimization of a hybrid system for robust ocean wave features prediction *Neurocomputing* **275** pp 818-828.
[16] Bento P, Pombo J, Calado MDR, Mariano S 2021 Ocean wave power forecasting using convolutional neural networks *IET Renewable Power Generation* **15**(14) pp 3341-3353
[17] Sadeghifar T, Nouri Motlagh M, Torabi Azad M, and Mohammad Mahdizadeh M 2017 Coastal wave height prediction using Recurrent Neural Networks (RNNs) in the south Caspian Sea *Marine Geodesy* **40**(6) pp 454-465.
[18] Hirt CW, Nichols ND 1981 Volume of fluid (VOF) method for the dynamics of free boundaries *J. Comput. Phys.* **39** pp 201-225.
[19] Dean RG, Dalrymple RA 1991 Water Wave Mechanics for Engineers and Scientists *world scientific publishing company.*
[20] Huang L, Li Y, Benites-Munoz D, Windt CW, Feichtner A, Tavakoli S, Davidson J, Paredes R, Quintuna T, Ransley E, Colombo M, Li M, Cardiff P, and Tabor G 2022 A review on the modelling of wave-structure interactions based on OpenFOAM *OpenFOAM® Journal* **2** pp 116–142.
[21] Jacobsen NG, Fuhrman DR, and Fredsøe J 2012 A wave generation toolbox for the open-source cfd library: Open-foam® *International Journal for Numerical Methods in Fluids* **70** pp 1073–1088.
[22] Podulka P 2022 Resolving selected problems in surface topography analysis by application of the autocorrelation function *Coatings* **13**(1) pp 74.
[23] Chou JS, and Ngo NT 2016 Time series analytics using sliding window metaheuristic optimization-based machine learning system for identifying building energy consumption patterns. *Applied energy* **177** pp 751-770.
[24] Hewamalage H, Bergmeir C, and Bandara K 2021 Recurrent neural networks for time series forecasting: Current status and future directions *International Journal of Forecasting* **37**(1) 388-427
[25] Ju J, Liu KN, Liu FA 2022 Prediction of SO2 Concentration Based on AR-LSTM Neural Network *Neural Processing Letters* pp 1-19.
[26] Haji SH, and Abdulazeez AM 2021 Comparison of optimization techniques based on gradient descent algorithm: A review. *PalArch's Journal of Archaeology of Egypt/Egyptology* **18**(4) pp 2715-2743.
[27] Greff K, Srivastava RK, Koutník J, Steunebrink BR, and Schmidhuber J 2016 LSTM: A search space odyssey. *IEEE transactions on neural networks and learning systems* **28**(10) 2222-2232.

Hydrodynamic response of dual obliquely submerged non-uniform flexible thin plates

Mansi Singh and R Gayen

Department of Mathematics, Indian Institute of Technology Kharagpur, Kharagpur 721302, West Bengal, India

E-mail: mansi.maths.iitkgp@gmail.com; rupanwita.gayen@gmail.com

Abstract. Examining inclined flexible plates with variable thickness becomes crucial when it comes to optimizing or achieving controlled reflection and transmission of waves, especially for the construction of breakwaters. Unlike horizontal breakwaters, inclined barriers have the ability to penetrate through numerous layers of the fluid, having different particle velocities and foster their interactions. This causes wave breaking which leads to loss of wave energy. Also, despite exhibiting a similar behavior, vertical structures do not attenuate waves as effectively as inclined ones do. Additionally, the resonant motion of the fluid trapped between two structures proves to enhance the attenuation, thus it is recommended to include dual structures in the model rather than just one. Therefore in the present study, we examine the water wave scattering phenomenon by a pair of symmetric flexible thin plates with non-uniform thickness obliquely submerged in deep water. We use the linear water wave theory and Kirchhoff's thin plate theory to model the physical problem. The boundary value problem is converted into a system of coupled integral equations using repeated integration and Green's integral theorem. Using appropriate approximations, this system is solved and its solutions are used to determine numerical values of different hydrodynamic quantities. Results of two horizontal as well as two vertical plates could be obtained from the present model, thus it is a very general model. Also results are illustrated to analyze the contribution of the thickness variation and the inclinations of the two flexible plates towards the wave scattering process and some related physical quantities.

1. Introduction

The construction of optimally designed breakwaters made of different materials to shelter shorelines, safeguard coastal lives, control soil erosion, extract wave energy has been a topic of great interest since the past many decades. Effectively constructed breakwater systems can help to achieve desired results along with increasing the durability of the structure. An efficient way to regulate the reflection and transmission of waves is through the use of breakwater systems comprised of an array of inclined flexible plates arranged asymmetrically. Flexible structures have proved to be more advantageous over rigid structures due to their property of withstanding extreme loads without breaking, and thus increasing the life of the structure. Further, adding the property of inclinations to the flexible plate models will not only enhance the interaction of the plate with the different layers of water, but will also allow to achieve controlled physical outputs as per need just by adjusting the inclinations. This general model will serve the purpose of studying the hydrodynamic behaviours of both the horizontal and the vertical plates, without the need to investigate them separately.

Studies on inclined structures in the context of linear water wave theory have been performed by many researchers. To list a few, we have the work of Parsons and Martin [1] for a rigid plate, Cho and Kim [2] and Gayen and Mondal [3] for a porous plate, Kundu *et. al.* [4] for an elastic plate and Singh *et. al.* [5] for a porous elastic plate. In all these previously stated works, the authors have considered wave interaction with a single structure. However, investigations show that use of multiple structures is more beneficial due to their capability of trapping waves and thus exhibiting resonant behaviour. Investigations on water wave scattering by two structures submerged in water can be found in the works of Levine and Rodemich [6], Jarvis [7], Evans and Morris [8], Roy et al. [9], Gayen and Mondal [10], Kundu *et. al.* [11] and many more. In each of these works or for most of the works present in literature, the focus is on the symmetric or asymmetric arrangement of the two barriers.

Various mathematical techniques like the boundary element method, domain decomposition method, matched eigenfunction expansion method, etc. have been employed to solve different problems pertaining to wave–structure interactions. However, the method of hypersingular integral equations that arise due to application of Green's integral theorem to boundary value problems involving Neumann conditions, along with suitable numerical techniques is found to be comparatively convenient. This is because this approach can easily tackle barriers with arbitrary shapes and orientations. Additionally, properties like faster convergence and computational efficiency, makes the method more preferable over the formerly mentioned conventional and popular methods. Thus, the use of hypersingular integral equations can be seen in various works, for example, Parsons and Martin [1], Martin and Farina [12], Davydov *et. al.* [13], Renzi and Dias [14], Kundu *et. al.* [11], Singh *et. al.* [15], etc.

It is also interesting to note that introducing variations in the thickness of the flexible structures can also lead to interesting and important hydrodynamical results. The thickness variations can also significantly effect various physical quantities. This fact has been established in the work of Singh and Gayen [16], where they investigate the effect of the thickness variations of a flexible plate vertically submerged in water in frequency domain. Recently, Singh *et. al.* [15] extended the work of Singh and Gayen [16] for a floating and an obliquely submerged non-uniform flexible plate in time domain. Thus, in the present work we extend their methodology to examine the behaviours of two obliquely submerged flexible plates with variable thickness to incoming water waves. Further, we also try to obtain the results for two horizontal and two vertical non-uniform flexible plates from this general model.

The outline of the paper is as follows. In Section 2, we elaborate the mathematical formulation of the physical problem in consideration. In Section 3, we briefly discuss the solution procedure to solve the boundary value problem obtained in Section 2. Further, in Section 4, we first validate the correctness of the method used in the present work and then illustrate some interesting graphical results related to the reflection coefficient and the deflection of the two inclined flexible plates. Finally, in Section 5, we state the concluding remarks and certain important observations based on the analysis of the present problem.

2. Mathematical formulation

The fluid domain is described by a Cartesian coordinate system (x, y) where the positive $y-$axis points vertically downwards into the fluid domain and the $x-$axis represents the mean free surface. We consider two symmetric thin flexible plates Γ_1 and Γ_2 each of length $2l$ and having non-uniform thickness, to be obliquely submerged in water of infinite depth. The latter meaning that the plates have arbitrary inclinations θ_1 and θ_2 with the vertical direction. Both the plates are equally separated from the $y-$axis on its either sides, so that the total distance between their midpoints is $2d$. The depths of the midpoints of the plates from the $x-$axis are h_1 and h_2. This description of the dual plate system is depicted through a schematic diagram shown in Figure 1. The position of the two plates in this setup can be given by

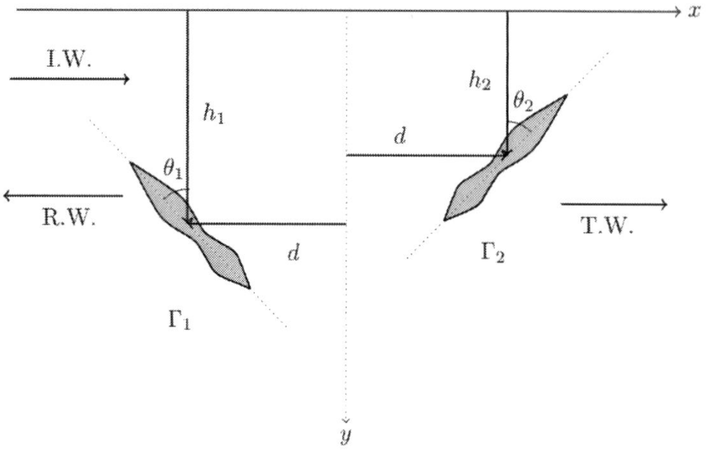

Figure 1: Schematic diagram of the obliquely submerged non-uniform flexible plate system.

$\Gamma_i = \{\, (x, y) \,|\, x = (-1)^i(d - u\, l \sin\theta_i),\ y = h_i + u\, l \cos\theta_i;\ -1 \le u \le 1\}$ for $i = 1, 2$, following the technique of Singh and Gayen [16]. Assuming linear theory of water waves and potential fluid flow, the fluid motion can be represented by the velocity potential in time harmonic form as

$$\Phi(x, y, t) = \mathrm{Re}\left\{ \frac{g^2}{\omega^3}\, \phi(x, y)\, \mathrm{e}^{-\mathrm{i}\omega t} \right\}, \tag{1}$$

where g is the acceleration due to gravity , ω is the angular frequency and t is the time. Here the time-independent potential function $\phi(x, y)$ satisfies:
the two-dimensional Laplace equation

$$\frac{\partial^2 \phi}{\partial x^2} + \frac{\partial^2 \phi}{\partial y^2} = 0, \text{ in the fluid domain;} \tag{2}$$

linear free surface condition

$$K\phi + \frac{\partial \phi}{\partial y} = 0, \quad \text{on } y = 0, \quad \text{with } K = \frac{\omega^2}{g}; \tag{3}$$

the condition of no flow across the domain bottom

$$|\nabla \phi| \to 0, \quad \text{as} \quad y \to \infty; \tag{4}$$

and the radiation conditions

$$\phi(x, y) \to \begin{cases} T\phi^{inc}(x, y) & \text{as } x \to \infty, \\ \phi^{inc}(x, y) + R\phi^{inc}(-x, y) & \text{as } x \to -\infty, \end{cases} \tag{5}$$

where R and T denote the unknown reflection and transmission coefficients and $\phi^{inc}(x, y)$ denotes the incident wave potential given by

$$\phi^{inc}(x, y) = \mathrm{e}^{-Ky + \mathrm{i}Kx}. \tag{6}$$

Let the two flexible plates be normally deflected by the incoming water waves with small displacements

$$W_i(p_i,t) = \mathrm{Re}\left\{\sigma_i(p_i)\, \mathrm{e}^{-\mathrm{i}\omega t}\right\},\ p_i \in \Gamma_i,\ i=1,2, \tag{7}$$

where $\sigma_i(p_i)$ represents the complex-valued displacement amplitude of the inclined flexible plates. Here $p_i = (x_i, y_i)$.
Thus, the linearised kinematic condition on the two inclined flexible plates Γ_1 and Γ_2 is obtained as

$$\frac{\partial \phi}{\partial n_{p_i}} = -\mathrm{i}K^2\ \sigma_i(p_i), \quad p_i \in \Gamma_i,\ i = 1,2. \tag{8}$$

Further, we assume that each plate per unit width behaves as a one-dimensional flexible beam with small displacements. Thus, based on Kirchhoff-Love beam theory, the motion of each flexible plate due to the fluid pressure is analysed. The resulting force balance equation gives a linearised boundary condition on the two inclined plates in terms of $\phi_{n_i}(p_i)$ as

$$\frac{\partial^2}{\partial s_i^2}\left[D_i(p_i)\ \frac{\partial^2 \phi_{n_i}}{\partial s_i^2}\right] - \beta K\ b_i(p_i)\ \phi_{n_i} = -K[\phi](p_i), \quad p_i \in \Gamma_i,\ i=1,2, \tag{9}$$

where the first term on the left hand side represents the force due to the plate flexure, the second term represents the inertial force, while the term on the right hand side represents the force exerted by fluid. Further, in Eq. (9), $[\phi](p_i) = \phi(p_i^+) - \phi(p_i^-)$ represents the potential difference across the plate Γ_i, $\beta = \frac{\rho'}{\rho}$, and

$$D_i(p_i) = \frac{E\ b_i^3(p_i)}{12\rho g(1-\nu^2)},$$

is the flexural rigidity parameter with E as the Young's modulus, ν as the Poisson's ratio, ρ' as the material density and ρ as the density of water. Further, $b_i(p_i) = b_0 + \epsilon\hat{b}_i(p_i)$ denotes the non-uniform thickness of the two inclined flexible plates varying along their lengths. In this expression of thickness, ϵ and b_0 are small constant parameters, while $\hat{b}_i(p_i)$ describes the outer shape of the thickness of the two plates. The details regarding this choice of thickness can be found in the work of Singh and Gayen [16].

Further, we allow the top ends of the flexible plates that are towards the free surface to move freely and fix their bottom ends, thus, we have the conditions at the ends of the plates as

$$\frac{\partial^2 \phi_{n_i}}{\partial s_i^2} = 0 = \frac{\partial^3 \phi_{n_i}}{\partial s_i^3}, \quad \text{at the top end of } \Gamma_i \tag{10}$$

and

$$\phi_{n_i} = 0 = \frac{\partial \phi_{n_i}}{\partial s_i}, \quad \text{at the bottom end of } \Gamma_i, \tag{11}$$

for $i = 1, 2$. These kind of edge conditions are considered for piled breakwaters, built near coastlines or beaches to reflect waves. Also, they are relevant to oscillating surge wave energy converters (Oyster type) or point absorbers (Archimedes Wave Swing) installed near shore (see [17]).

Further, at the tips of the plates, the behaviour of ϕ is governed as

$$\phi \sim O(r^{1/2}) \text{ as } r \to 0, \tag{12}$$

where r is the distance between a point in the fluid and either end of the plates. Thus, the mathematical formulation of the problem of water wave scattering by dual obliquely submerged non-uniform flexible plates is now complete. In the next section we elaborate on the procedure to solve the obtained boundary value problem.

3. Method of solution

For the ease of mathematical calculations, we first transfer the two inclined plates into two new coordinate systems (X_i, Y_i), $i = 1,2$. In the new system, the plates lie along the Y_i-axis and the X_i-axis is perpendicular to the Y_i-axis. The mid-points of the two plates are the origins of these new coordinate systems. Thus, the relation established between the systems (X_i, Y_i) and (x, y) is given as

$$X_i = \big((-1)^i x - d\big)\cos\theta_i + (y - h_i)\sin\theta_i, \tag{13}$$

$$Y_i = \big((-1)^{i+1} x + d\big)\sin\theta_i + (y - h_i)\cos\theta_i, \tag{14}$$

where $0 \le \theta_i \le \pi/2$, $i = 1,2$.
Thus, the boundary conditions (9) on the inclined plates can be rewritten in these new coordinate systems as

$$\frac{\partial^2}{\partial Y_i^2}\left[D_i(Y_i)\ \frac{\partial^2 \phi_{X_i}}{\partial Y_i^2}(0, Y_i)\right] - \beta K\ b_i(0, Y_i)\ \phi_{X_i}(0, Y_i) = -K[\phi](0, Y_i), \quad -l \le Y_i \le l, \tag{15}$$

and the kinematic condition (8) can be written as

$$\phi_{X_i}(0, Y_i) = \frac{\partial \phi}{\partial X_i}(0, Y_i) = -\mathrm{i}K^2\ \sigma_i(0, Y_i), \quad -l \le Y_i \le l, \tag{16}$$

for $i = 1,2$.
Now we proceed to further solve the problem for which we first rewrite the transformed plate boundary condition (15) in the form

$$\frac{\partial^2}{\partial Y_i^2}\left[b_i^3(Y_i)\ \frac{\partial^2 \phi_{X_i}}{\partial Y_i^2}(0, Y_i)\right] - \frac{\beta K}{\alpha}\ b_i(Y_i)\ \phi_{X_i}(0, Y_i) = -\frac{K}{\alpha}\ [\phi](0, Y_i), \quad -l \le Y_i \le l, \tag{17}$$

$i = 1,2$, where $\alpha = \frac{E}{12\rho g(1-\nu^2)}$ and thus $D_i(Y_i) = \alpha b_i^3(Y_i)$.

Next we integrate the above equations (17) repeatedly four times so that the derivatives vanish. This will result in two integral equations as

$$b_i^3(Y_i)\ \phi_{X_i}(0, Y_i) + \int_{-l}^{l} K_i(Y_i, Z_i)\ \phi_{X_i}(0, Z_i)\ \mathrm{d}Z_i + \int_{-l}^{l} J_i(Y_i, Z_i)\ [\phi](0, Z_i)\ \mathrm{d}Z_i = 0,$$
$$-l \le Y_i \le l, \quad i = 1,2, \tag{18}$$

or, in the original Cartesian coordinate system as

$$b_i^3(p_i)\ \frac{\partial \phi}{\partial n_{p_i}} + \int_{\Gamma_i} K_i(p_i, q_i)\ \frac{\partial \phi}{\partial n_{q_i}}\ \mathrm{d}s_{q_i} + \int_{\Gamma_i} J_i(p_i, q_i)\ [\phi](q_i)\ \mathrm{d}s_{q_i} = 0, \quad p_i \in \Gamma_i,$$
$$i = 1,2, \tag{19}$$

The four kernels $K_i(p_i, q_i)$, $J_i(p_i, q_i)$, $(i = 1,2)$ appearing in the above integral equations are all known regular kernels obtained with the help of the conditions (10) and (11) at the ends of the two inclined plates Γ_1 and Γ_2.

Further, we apply Green's integral theorem to the scattered potential $\phi(x, y) - \phi^{inc}(x, y)$ and the fundamental potential function $\mathcal{G}(x, y; \xi, \eta)$ for deep water (cf. [18]), to obtain

$$\phi(p) - \phi^{inc}(p) = -\sum_{i=1}^{2} \int_{\Gamma_i} [\phi](q_i)\ \frac{\partial \mathcal{G}}{\partial n_{q_i}}(p; q_i)\ \mathrm{d}s_{q_i}. \tag{20}$$

Now taking normal derivative on both sides of Eq. (20) at points p_i on Γ_i $(i = 1, 2)$, we obtain a set of two of coupled integral equations, combinedly written as

$$\frac{\partial \phi}{\partial n_{p_i}} + \int\!\!\!\!\!\!\times_{\Gamma_i} [\phi](q_i) \frac{\partial^2 \mathcal{G}}{\partial n_{p_i} \partial n_{q_i}} (p_i; q_i) \, \mathrm{d}s_{q_i} + \int\!\!\!\!\!\!\times_{\Gamma_j} [\phi](q_j) \frac{\partial^2 \mathcal{G}}{\partial n_{p_i} \partial n_{q_j}} (p_i; q_j) \, \mathrm{d}s_{q_j} = \frac{\partial \phi^{inc}}{\partial n_{p_i}}, \quad p_i \in \Gamma_i, \tag{21}$$

where $j = 3 - i$ for $i = 1, 2$.

It should be noted that in the above expression the integrals having the cross sign should be treated as Hadamard finite part integrals since they contain hypersingular kernels, thus the integration would not be the usual one.

Thus, the Eqs. (19) and (21) form a set of four coupled integral equations with four unknowns, *viz.*, $[\phi](p_i)$ and $\frac{\partial \phi}{\partial n_{p_i}}$, $i = 1, 2$. Now, we proceed to solve this consistent system to determine the four unknowns numerically. Firstly, to avoid complex calculations, we make use of the parameterization of the points $p_i(x_i, y_i)$ and $q_i(\xi_i, \eta_i)$ as

$$(x_i, y_i) = \left((-1)^i (d - lu \sin\theta_i), h_i + lu \cos\theta_i\right), \quad -1 \le u \le 1, \quad i = 1, 2, \tag{22}$$

$$(\xi_i, \eta_i) = \left((-1)^i (d - lv \sin\theta_i), h_i + lv \cos\theta_i\right), \quad -1 \le v \le 1, \quad i = 1, 2. \tag{23}$$

Thus, by virtue of the above parameterization and setting

$$f_i(v) \equiv [\phi](q_i) \quad \text{and} \quad g_i(u) \equiv \frac{\partial \phi}{\partial n_{p_i}}, \quad i = 1, 2,$$

Eqs. (19) and (21) get transformed to

$$b_i^3(u) \, g_i(u) \; + l \left[\int_{-1}^{1} K_i(u, v) \, g_i(v) \, \mathrm{d}v + \int_{-1}^{1} J_i(u, v) \, f_i(v) \, \mathrm{d}v \right] = 0, \quad i = 1, 2 \tag{24}$$

and

$$g_i(u) - \frac{1}{2\pi l} \int\!\!\!\!\!\!\times_{-1}^{1} \frac{f_i(v)}{(u - v)^2} \, \mathrm{d}v + \sum_{j=1}^{2} \int_{-1}^{1} \mathcal{R}_{ij}(u, v) \, f_j(v) \, \mathrm{d}v = \frac{\partial \phi^{inc}}{\partial n_{p_i}}, \quad i = 1, 2, \tag{25}$$

where, $\mathcal{R}_{ij}(u, v)$, $(i, j = 1, 2)$ are known bounded regular kernels.

Now due to Eq. (12) we have the behaviours of f_i's and g_i's as $f_i \sim O(r^{1/2})$ and $g_i \sim O(r^{-1/2})$, respectively, $(i = 1, 2)$ as $r \to 0$. Also since there is no potential difference at the tips of the two elastic plates, thus we have that $f_i(\pm 1) = 0$. Based on these conditions, we approximate the unknown f_i's and g_i's as series expansions in terms of Chebyshev polynomials of the first kind $T_n(u)$ and second kind $U_n(u)$, multiplied with appropriate weight functions as follows:

$$f_i(v) = \sqrt{1 - v^2} \sum_{n=0}^{N} F_n^i \, U_n(v), \quad i = 1, 2, \tag{26}$$

$$g_i(v) = \frac{1}{\sqrt{1 - v^2}} \sum_{n=0}^{N} G_n^i \, T_n(v), \quad i = 1, 2. \tag{27}$$

Here F_n^i's and G_n^i's $(i = 1, 2)$ in the above series expansions are unknown coefficients to be determined. Further, by using the technique of expansion-cum-collocation, with an appropriate choice of the collocation points $u = u_n (n = 0, ..., N)$, we obtain a system of $4N + 4$ linear equations whose matrix representation is

$$\begin{bmatrix} A1 & A2 & A3 & 0 \\ B1 & B2 & 0 & B3 \\ C1 & 0 & C2 & 0 \\ 0 & D1 & 0 & D2 \end{bmatrix} \begin{bmatrix} F_n^1 \\ F_n^2 \\ G_n^1 \\ G_n^2 \end{bmatrix} = \begin{bmatrix} E1 \\ E2 \\ O \\ O \end{bmatrix} \tag{28}$$

where the coefficient matrices $[Aj], [Bj]$ $(j = 1, 2, 3)$, $[Cj], [Dj]$ $(j = 1, 2)$ have order $(N + 1) \times (N + 1)$, while the unknown column matrices $[F_n^j], [G_n^j] (j = 1, 2)$, and the known column matrices $[Ej]$ $(j = 1, 2, 3)$ have order $(N+1) \times 1$. Further, $[O]$ represents the zero vector of order $(N+1) \times 1$. Finally solving this matrix system (28) using any system solving algorithm will produce the numeric estimates of the coefficients F_n^i and G_n^i $(i = 1, 2)$, $(n = 0, ..., N)$ appearing in the expansions (26) and (27).

3.1. *Reflection and transmission coefficients*

Now letting $x \to \mp\infty$ in Eq. (21) and simultaneously using the radiation condition (5), along with the asymptotic behaviour of the fundamental potential function $\mathcal{G}(x, y; \xi, \eta)$, we can obtain the expressions for the reflection coefficient R and the transmission coefficient T as

$$R = (Kl) \sum_{i=1}^{2} \sum_{n=0}^{N} \int_{-1}^{1} (-1)^i \ F_n^i \left\{ \sqrt{1 - v^2} \ U_n(v) \ \exp\left(-K\eta_i(v) + \mathrm{i}K\xi_i(v) + (-1)^i \mathrm{i}\theta_i\right) \right\} \mathrm{d}v \tag{29}$$

and

$$T = 1 + (Kl) \sum_{i=1}^{2} \sum_{n=0}^{N} \int_{-1}^{1} (-1)^i \ F_n^i \left\{ \sqrt{1 - v^2} \ U_n(v) \ \exp\left(-K\eta_i(v) - \mathrm{i}K\xi_i(v) + (-1)^{i+1} \mathrm{i}\theta_i\right) \right\} \mathrm{d}v. \tag{30}$$

3.2. *Plate deflection*

The plate deflection amplitudes $\sigma_i (i = 1, 2)$ of the two inclined non-uniform flexible plates Γ_1 and Γ_2 respectively can be directly determined from the kinematic condition (8) as

$$\sigma_i(p_i) = \frac{\mathrm{i}}{K^2} \frac{\partial \phi}{\partial n_{p_i}}, \quad i = 1, 2, \tag{31}$$

$$\hat{\sigma}_i(u) \equiv \sigma_i(x(u), y(u)) = \frac{\mathrm{i}}{K^2} \sum_{n=0}^{N} \frac{1}{\sqrt{1 - u^2}} G_n^i \ T_n(u), \quad -1 < u < 1, \quad i = 1, 2, \tag{32}$$

where the last expression which is the parameterised form of the plate deflection of the two plates, comes from the expansion (27).

4. Results and discussion

By implementing a code in MATLAB, the numerical values of different hydrodynamic quantities whose expressions were derived in the previous section are computed and then presented graphically. Firstly to validate the correctness of the method used in the present work, numerical values of the reflection coefficient $|R|$ determined using the current method are compared with the results available in literature for the limiting case of our model. Figures 2(a) and 2(b) compare the values of $|R|$ obtained using our method with the results of Evans and Morris [8] and Gayen and Mondal [10] respectively. The former considers the case of two surface piercing vertical rigid symmetric plates, while the latter assumes the case of two rigid thin plates both inclined at angles of $\pi/4$, both with constant thickness. The present model is appropriately calibrated to match the two existing models for proper comparison. It should be noted that letting the flexural rigidity parameter $\frac{\alpha}{2l}$ to be sufficiently large, makes the flexible plates tend to behave like rigid plates. Further, assuming the plate thickness shape function $\hat{b_i}(p_i) = 0, (i = 1, 2)$, we can restrict the two plates to have constant thickness. So, the perfect matching of the two results as seen from Figure 2 proves the correctness of the present analysis and assures its ability to render promising results.

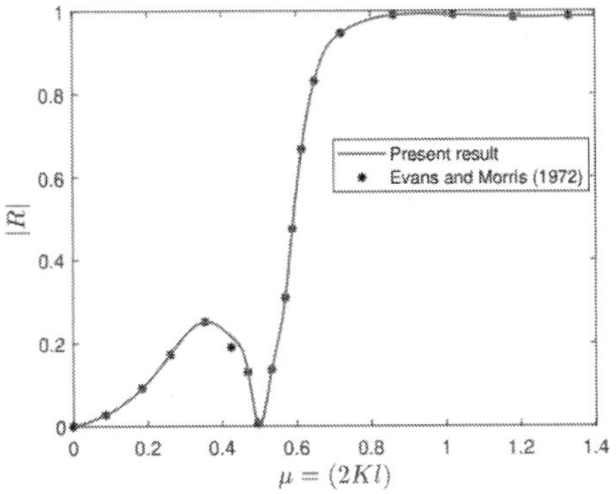

(a) Comparison with Evans and Morris [8] by considering dual vertical surface piercing rigid plates with constant thickness. Parametric values are set as $\frac{d}{2l} = 1$, $\theta_1 = \theta_2 = 0$, $\frac{b_0}{2l} = 0.005$, $\hat{b_1}(Y_1) = \hat{b_2}(Y_2) = 0$, $\frac{\alpha}{2l} = 10^9$ and $\frac{h_1}{2l} = \frac{h_2}{2l} = 0.5$.

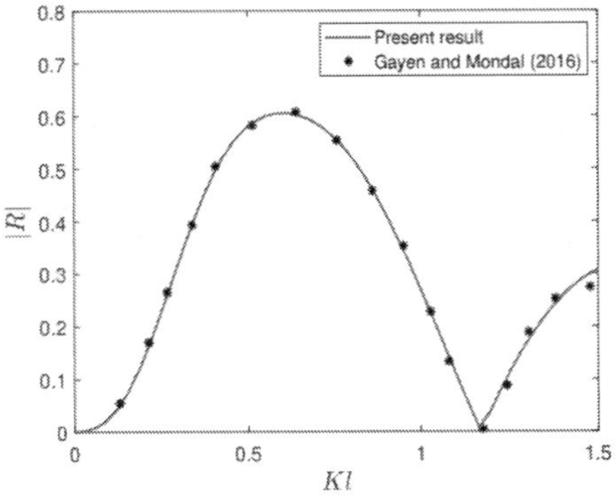

(b) Comparison with Gayen and Mondal [10] by considering dual inclined submerged rigid plates with constant thickness. Parametric values are set as $\frac{d}{l} = 1$, $\theta_1 = \theta_2 = \frac{\pi}{4}$, $\frac{b_0}{l} = 0.005$, $\hat{b_1}(Y_1) = \hat{b_2}(Y_2) = 0$, $\frac{\alpha}{l} = 10^9$ and $\frac{h_1}{l} = \frac{h_2}{l} = 1.15$.

Figure 2: Graphical Validation - Comparison of values of $|R|$ obtained using present method with those already existing in literature.

For all the figures from here onwards we fix the shape function of the thickness of the two inclined flexible plates by $\hat{b_i}(Y) = \frac{(Y+l)(l-Y)}{2l}$, $i = 1, 2$ and $\epsilon = 0.001$. Figures 3(a) and 3(b) depict the variation of $|R|$ corresponding to different values of the submergence depth of plate Γ_2 while fixing the submergence depth of plate Γ_1, in case of two horizontal and two vertical non-uniform flexible plates, respectively. It is observed that the two vertical plate system shows greater reflection of wave energy than the two horizontal plate system, when the submergence depth of the second plate Γ_2 is small.

Figure 4 illustrates the effect of changing plate inclination of plate Γ_2 on the reflection coefficient, when plate inclination of plate Γ_1 is fixed with $\theta_1 = \pi/4$. It is noted that as the inclination of plate Γ_2 grows from $\theta_2 = 0$ to $\theta_2 = \pi/4$, the peak values of the curves shift towards lower values of the wavenumber. Interestingly, it can be analysed from Figures 3 and 4 that only for a symmetric arrangement of the two non-uniform flexible plates, indicated by the red curves in these graphs, zero reflection is achieved for more than one value of the wavenumber.

Figure 5 depicts the effect of the separation distance between the two plates on the reflection coefficient. It is clearly visible from the graph that when the two plates are separated by a larger distance, higher resonance in the reflection curves is achieved even for smaller wavenumbers, *i.e.*,

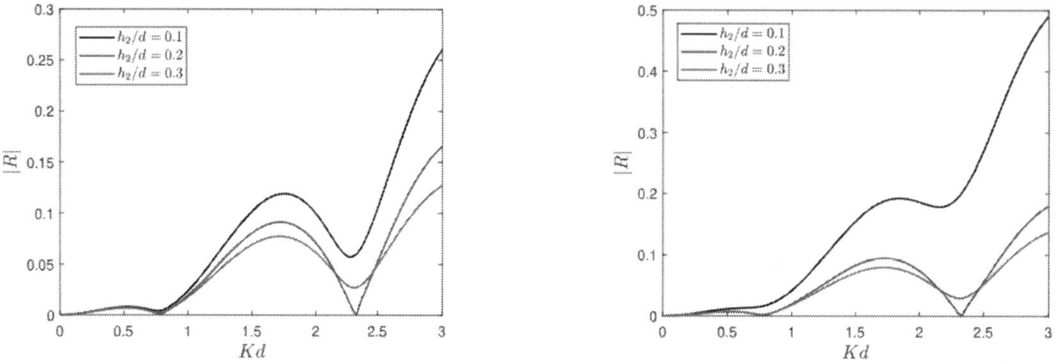

(a) Parameters are set as $\frac{l}{d} = 0.1$, $\theta_1 = \theta_2 = \frac{\pi}{2}$, $\frac{b_0}{d} = 0.005$, $\frac{\alpha}{d} = 10^6$ and $\frac{h_1}{d} = 0.2$.

(b) Parameters are set as $\frac{l}{d} = 0.1$, $\theta_1 = \theta_2 = 0$, $\frac{b_0}{d} = 0.005$, $\frac{\alpha}{d} = 10^6$ and $\frac{h_1}{d} = 0.2$.

Figure 3: Graphs of Reflection coefficient $|R|$ against non-dimensional wavenumber Kd for the cases of (a) dual horizontal non-uniform flexible plates and (b) dual vertical non-uniform flexible plates, when the submergence depth $\frac{h_2}{d}$ of plate Γ_2 is varied but $\frac{h_1}{d} = 0.2$ is fixed.

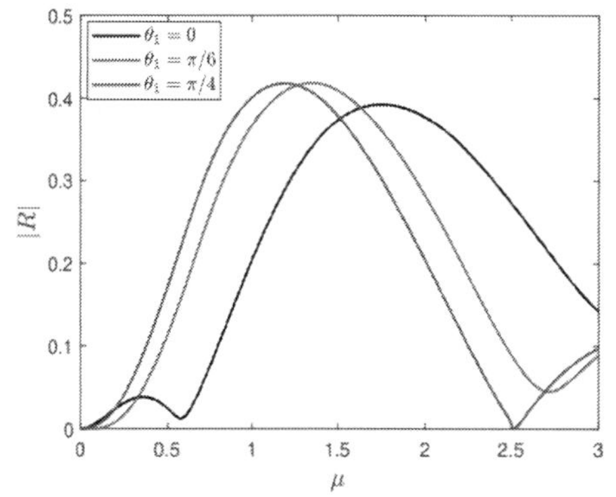

Figure 4: Graph of $|R|$ against non-dimensional wavenumber $\mu = 2Kl$ for varying values of θ_2. Parameters are set as $\frac{d}{2l} = 0.5$, $\theta_1 = \frac{\pi}{4}$, $\frac{b_0}{2l} = 0.005$, $\frac{\alpha}{2l} = 10^6$ and $\frac{h_1}{2l} = \frac{h_2}{2l} = 0.6$.

the number of peaks in the curves increase. This is indicative of the fact that even for lower frequencies higher reflection can be attained multiple times by positioning the two plates far from each other.

The variations in the plate deflection amplitudes of the two non-uniform flexible plates along their lengths, corresponding to different values of the separation distance between them are shown in Figure 6. It is observed from the two graphs that increasing the distance between the two plates only affects the deflection of the second flexible plate Γ_2, while the plate Γ_1 with whom the incoming waves initially interact experiences negligible impact.

5. Conclusion

In the present work the hydrodynamic behaviour of two obliquely submerged thin flexible plates of variable thickness to incoming water waves is investigated. The problem is considered in a two-dimensional coordinate system. The concepts of small amplitude wave theory and thin plate theory are used to formulate the physical problem. The problem is solved semi-analytically using

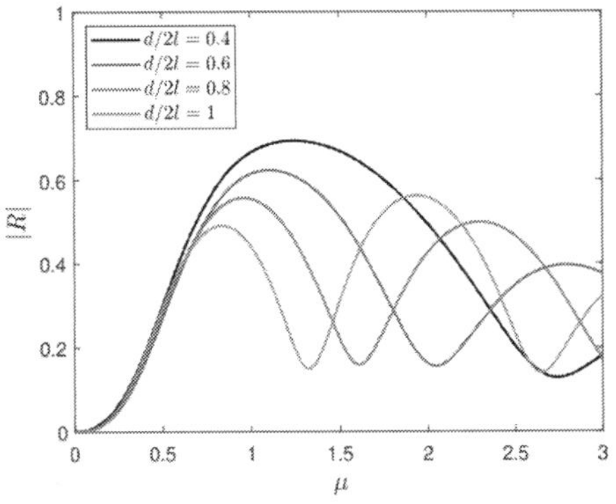

Figure 5: Graph of $|R|$ against $\mu = 2Kl$ for varying values of the separation distance $\frac{d}{2l}$ between the two plates. Parameters are set as $\theta_1 = \theta_2 = \frac{\pi}{4}$, $\frac{b_0}{2l} = 0.005$, $\frac{\alpha}{2l} = 10^6$, $\frac{h_1}{2l} = 0.4$ and $\frac{h_2}{2l} = 0.5$.

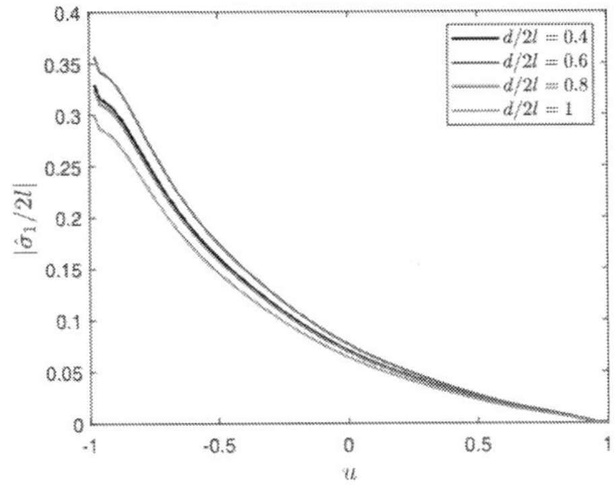

(a) Deflection of plate Γ_1.

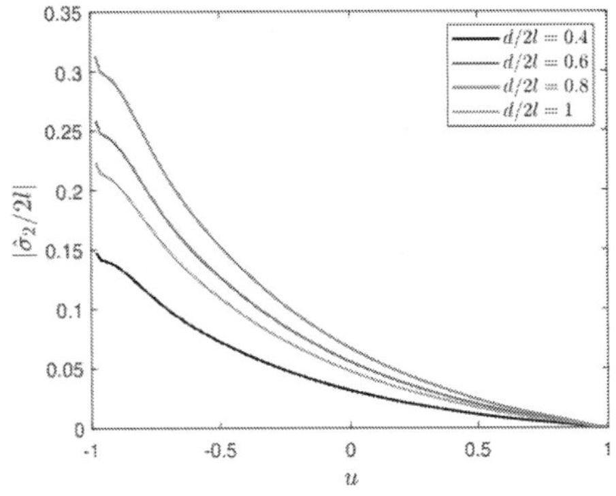

(b) Deflection of plate Γ_2.

Figure 6: Graphs of plate deflections $|\hat{\sigma}_i/2l|, i = 1, 2$ against parameterised plate length u for varying values of the separation distance $\frac{d}{2l}$ between the two plates. Parameters are set as $\theta_1 = \theta_2 = \frac{\pi}{4}$, $\frac{b_0}{2l} = 0.005$, $\frac{\alpha}{2l} = 10^6$, $Kl = 1$, $\frac{h_1}{2l} = 0.4$ and $\frac{h_2}{2l} = 0.5$.

a set of coupled Fredholm integral equations. Numerical values of hydrodynamic quantities like reflection coefficient and plate deflection are obtained and presented graphically. A particular thickness profile for the two flexible plates is considered and then by varying the geometric configurations like submergence depths, inclination angles and the gap between them, some interesting and crucial results are determined and illustrated. Results for two horizontal plates and two vertical plates could be efficiently obtained using the present general model of the dual flexible plate system. Zero reflection is seen at certain wavenumbers only in case of symmetric setup of the two flexible plates. Further, higher resonance of waves is observed towards lower wavenumbers when the two plates are situated quite far apart. The present study will be beneficial for engineers and scientists to optimally design and construct durable breakwaters to protect coastlines and other marine structures. The breakwater system can also be integrated with wave energy devices to enhance their energy absorption efficiency. The present work can be extended by including the uneven sea bottom topography and also by analysing the problem in time domain to visualize concealed hydrodynamic behaviours.

Acknowledgments

This work is partially supported by the Council of Scientific and Industrial Research (CSIR), India (25(0311)/20/EMR-II) and by Science and Engineering Research Board (SERB), DST (TAR/2022/000107) through research grants to **R. G.**. Also, **M. S.** is thankful to PMRF, an initiative of the Government of India, for providing support to her as a PhD student in the present research work.

References

[1] Parsons N F and Martin P A 1992 Scattering of water waves by submerged plates using hypersingular integral equations *Appl. Ocean Res.* **14(5)** 313–321

[2] Cho I H and Kim M H 2008 Wave absorbing system using inclined perforated plates *J. Fluid Mech.* **608** 1–2

[3] Gayen R and Mondal A 2014 A hypersingular integral equation approach to the porous plate problem *Appl. Ocean Res.* **46** 70–78

[4] Kundu S, Gayen R and Datta R 2018 Scattering of water waves by an inclined elastic plate in deep water *Ocean Eng.* **167** 221–228

[5] Singh M, Gayen R and Kundu S 2022 Linear water wave propagation in the presence of an inclined flexible plate with variable porosity *Arch. Appl. Mech.* **92(9)** 2593–2615

[6] Levine H and Rodemich E 1958 *Scattering of Surface Waves on an Ideal Fluid* (Technical report no. 78 Stanford University California)

[7] Jarvis J 1971 The scattering of surface waves by two vertical plane barriers *J. Inst. Math. Appl.* **7** 207–215

[8] Evans D V and Morris C A N 1972 Complementary approximations to the solution of a problem in water waves *IMA J. Appl. Math.* **10(1)** 1–9

[9] Roy R, Basu U and Mandal B N 2016 Water-wave scattering by two submerged thin vertical unequal plates *Arch. Appl. Mech.* **86(9)** 1681–1692

[10] Gayen R and Mondal A 2016 Water wave interaction with two symmetric inclined permeable plates *Ocean Eng.* **124** 180–191

[11] Kundu S, Gayen R and Gupta S 2021 Propagation of surface waves past asymmetric elastic plates *J. Eng. Math.* **126(1)** 1–24

[12] Martin P A and Farina L 1997 Radiation of water waves by a heaving submerged horizontal disc *J. Fluid Mech.* **337** 365-379

[13] Davydov A G, Zakharov E V and Pimenov Y V 2006 Hypersingular integral equations for the diffraction of electromagnetic waves on homogeneous magneto-dielectric bodies *Comput. Math. Model.* **17(2)** 97–104

[14] Renzi E and Dias F 2012 Resonant behaviour of an oscillating wave energy converter in a channel *J. Fluid Mech.* **701** 482–510

[15] Singh M, Meylan M H and Gayen R 2023 Time-domain motion of a floating or obliquely submerged non-uniform elastic plate *Phys. Fluids* **35(4)** 047117

[16] Singh M and Gayen R 2023 Mathematical study on the potential flow past a vertical submerged flexible plate of non-uniform thickness *J. Fluids Struct.* **116** 103795

[17] Fadaeenejad M, Shamsipour R, Rokni S D and Gomes C 2014 New approaches in harnessing wave energy: With special attention to small islands *Renew. Sust. Energ. Rev.* **29** 345-354

[18] Mandal B N and Chakrabarti A 2000 *Water Wave Scattering by Barriers* (Wit Pr/Computational Mechanics)

IOP Conf. Series: Materials Science and Engineering 1288 (2023) 012009 doi:10.1088/1757-899X/1288/1/012009

Towards accurate stereo-video based free-surface reconstruction for wave tank experiments

Sacha Le Page[1], Alan Tassin[1], Julien Caverne[1], Yves Le Gall[1], Benoit Gomez[1], Guillaume Ducrozet[2]

[1]Ifremer, RDT Research and Technological Development, F-29280 Plouzané, France
[2]Nantes Université, École Centrale Nantes, CNRS, LHEEA, UMR 6598, F-44000 Nantes, France

E-mail: slepage@ifremer.fr

Abstract. Measuring accurately the shape of the free surface in medium-scale hydrodynamic testing facilities is challenging with conventional wave gauges. To overcome this problem, it is necessary to go towards higher resolution measurement systems. For this purpose, we investigate the ability of stereo-video acquisition systems to reconstruct free surface waves in laboratory conditions. This technique has been proven to give satisfactory results in open sea conditions, although it is dependent on the environmental conditions (weather, solar incidence angle). At sea, the stereo-video reconstruction algorithm makes profit of the short waves generated by the wind in order to correlate the points from the left and right images. The main challenge of adapting this technique to laboratory conditions, *i.e.* in absence of sun and wind, is to overcome the absence of texture on the free surface. This paper presents recent work aiming at developing a stereo-vision measurement system for laboratory conditions. Our efforts have been devoted to finding an adapted lighting system and generating a texture suitable for the reconstruction algorithm. Different lighting configurations have been tested in order to understand its impact on the reconstructions. For the texture generation, we have investigated different means of generating short surface waves (water droplets, air blowing, circulation current, underwater acoustic emissions) in different experimental facilities. Our conclusions show that it is possible to reconstruct the shape of surface gravity waves in the presence of short surface waves and that the effects of lighting and texture on the quality of the reconstruction are strongly coupled.

1. Introduction

Wave-structure interaction is still a challenging field of research with stimulating engineering applications. There has been a lot of efforts dedicated to the development of non-linear wave-wave interaction models and non-linear wave-structure interaction models in the last decades. From an experimental point of view, these phenomena are also challenging to investigate in wave tanks, as they require innovative developments in terms of measurements. Indeed, one may be interested in having a better spatial description of the free-surface profile, tracking the evolution of the wetted surface and measuring high-frequency force components induced by higher-order wave-structure interactions. Regarding the measurement of the free-surface, there has been several attempts to improve the resolution of the measurements. Baldock & Swan [1] proposed a non-linear reconstruction method based on a punctual free-surface elevation measurement and a double Fourier series expansion of the velocity potential and of the free-surface elevation. This method has been revisited recently by Carciunescu & Christou [2], showing interesting results up to mildly breaking waves. Until recently, the only way of getting a refined measurement of the wave field was to deploy tens of wave gauges and to repeat the experiments with the wave gauge rack at different locations, as performed by Swan & Sheikh [3] who have measured the wave elevation around a vertical cylinder at 1400 locations. In narrow wave flumes, it is possible to visualize the free-surface profile with a video camera by tracking the intersection at the wall [4] or by a side view through a glass wall [5], but the wave profile along the wall is affected by the presence of the wall. For more complex wave fields, it is necessary to use set-ups with two or more video cameras. Aubourg *et al.* [6] used a stereo-PIV set-up with 3 cameras making it possible to measure the wave field and the velocity field at the free surface. A stereo-refraction approach was used by Gomit *et al.* [7] to measure the wave field while a stereo-thermal PIV approach was used by Savelyev & Fuchs [8] to measure the free-surface elevation and the velocity field at the free surface. The stereo-PIV method requires seeding in the fluid with floating particles which may be missing in certain areas (*e.g.* wake of a ship, behind the crest of a breaking wave, or due to a segregation process). Stereo-refraction is very sensitive to small deformations of the free-surface but is not adapted to steep waves. Stereo-thermal PIV requires infra-red cameras and a powerful infra-red laser, which is expensive and presents safety issues for users. In the present paper, we investigate the feasibility of stereo-video reconstruction in a wave tank using short waves at the free-surface. This approach tends to reproduce the principles of stereo-video wave measurement at sea using WASS, an open-source stereo processing pipeline developed by Bergamasco *et al.* [9] and Benetazzo *et al.* [10, 11]. We report here on different experiments which have been carried out for that purpose. The paper is organised as follows: **section 2** presents a first attempt to reconstruct a free-surface with short waves, **section 3** investigates the effect of the lighting set-up on the reconstruction and **section 4** investigates the feasibility of using underwater ultrasonic transducers to generate short free-surface waves. Finally, some conclusions are drawn in **section 5**.

2. Reconstruction of the free-surface from shortwaves: influence of the texture generation

This section aims to show the influence of the texture generation on free-surface reconstruction via observations made during an experimental campaign at the wave and current flume tank of IFREMER located in Boulogne-sur-Mer (Pas-de-Calais, France).

2.1. Experimental Set-up

The working section of the facility is 4m wide, 18m long and 2m deep. The facility is filled with freshwater, it can generate current up to 2m/s and waves can be superimposed to the current. In our case, the wave maker was located upstream, generating waves propagating in the same direction as the current. The tank is equipped with a 4m by 4m false floor that can be lifted up to a few centimeter above the free surface. Images of the experimental setup are shown in **Fig. 1** and a side view sketch is shown in **Fig. 2**.

Two high-speed cameras (PHOTRON Mini AX, 1024x1024 pixel) with 35mm lenses are installed 1.74m above free surface, in the median axis of the flume, with a 33° angle to the horizontal. The baseline, *i.e.*, the distance between both cameras optic center, is about 577 mm.

A large translucent screen (a polyane tarpaulin) is hung over an overhead crane positioned 7m ahead of the cameras, both layers are spaced from 20cm. With a 3.6m height and a 6m width, the tarpaulin provides a large screen to uniformly illuminate the free surface. To prevent direct lighting on the screen, (unfavorable to stereo-video reconstruction) the latter is lit by retro-lighting, *i.e.*, two LED lighting heads are placed behind the screen and oriented towards the white ceiling.

The stereo camera calibration is done with a 405x360mm chessboard using MATLAB Stereo Camera Calibrator [12], inspired by Bouguet [13], Zhang [14], and Heikkila et al [15]. The

Figure 1. Boulogne-sur-Mer experimental campaign setup. On the left: Experimental setup on the cameras side. On the right: Setup, behind the translucent screen

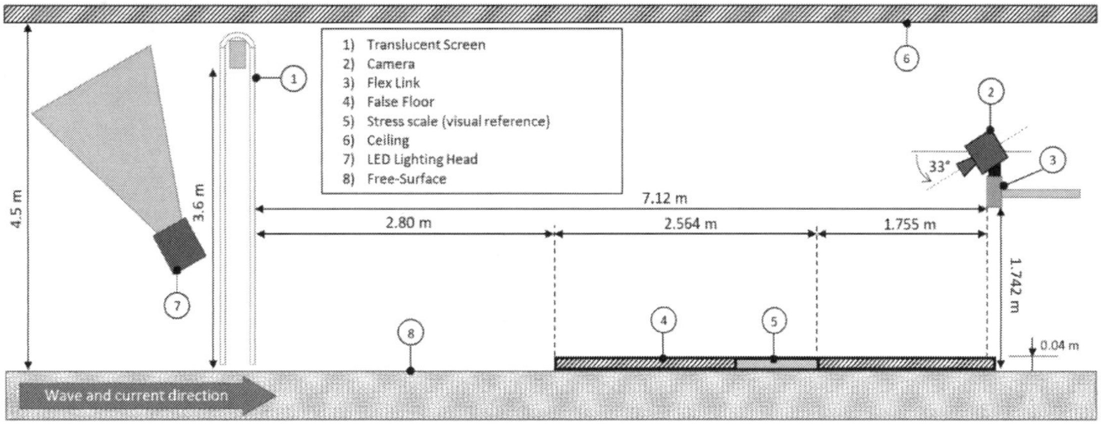

Figure 2. Boulogne-sur-Mer experimental campaign setup. Side view scheme

results show that both cameras have low distortion and the reprojection error is very low (about 0.04 px). In order to assess the accuracy of the stereo-video reconstruction, the false floor is reconstructed using the WASS pipeline. As the false floor is allegedly planar, a comparison of the altitudes of the reconstructed points with a median plane gives an estimate of the accuracy of the setup. **Fig. 3** shows the rectified pair of images used for the reconstruction of the false bottom. Some epipolar lines are displayed in red over the pair of images. **Fig. 4** shows the 3D reconstruction of the false floor. Each calculated point is given the "color" (in gray scale) of the associated pixel on the pair of pictures, which gives a rather realistic render to the reconstruction. Removing outliers, the point cloud can easily be compared to a median plane. The analysis of the altitude of the points around this plane gives an error of standard deviation from the median plane of $2.7mm$.

In order to generate texture, we take advantage of the short waves induced by the water circulation in the flume. After testing several current velocities (from 0.5 m/s to 1 m/s), it appears that a water circulation of 0.7 m/s gives the best results in terms of calm (no waves) free-surface reconstruction without altering the free-surface shape too much. Unfortunately, when regular waves are generated in the flume, huge gaps remain in the middle of the reconstructed

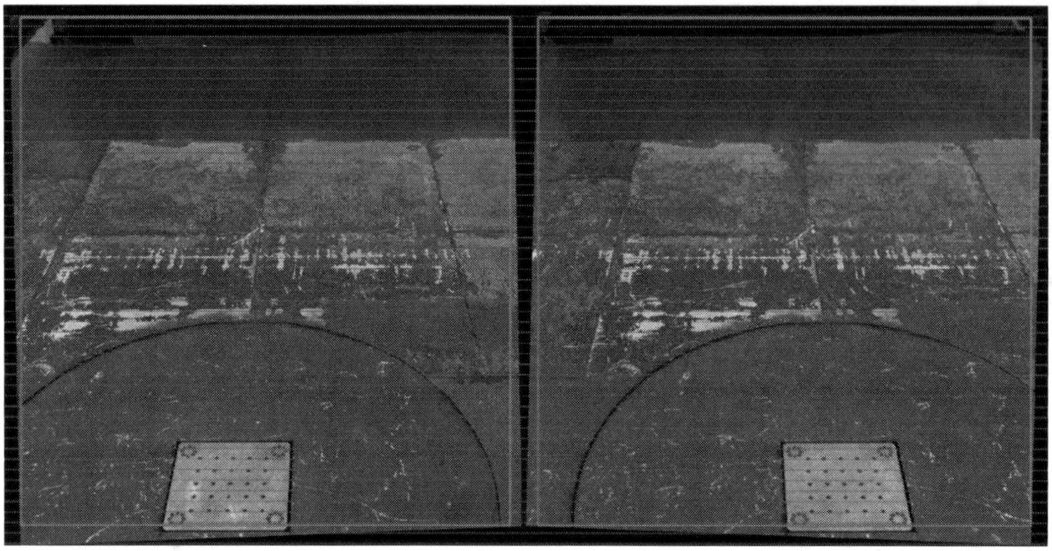

Figure 3. Rectified pair of images of the false floor, used for 3D reconstruction

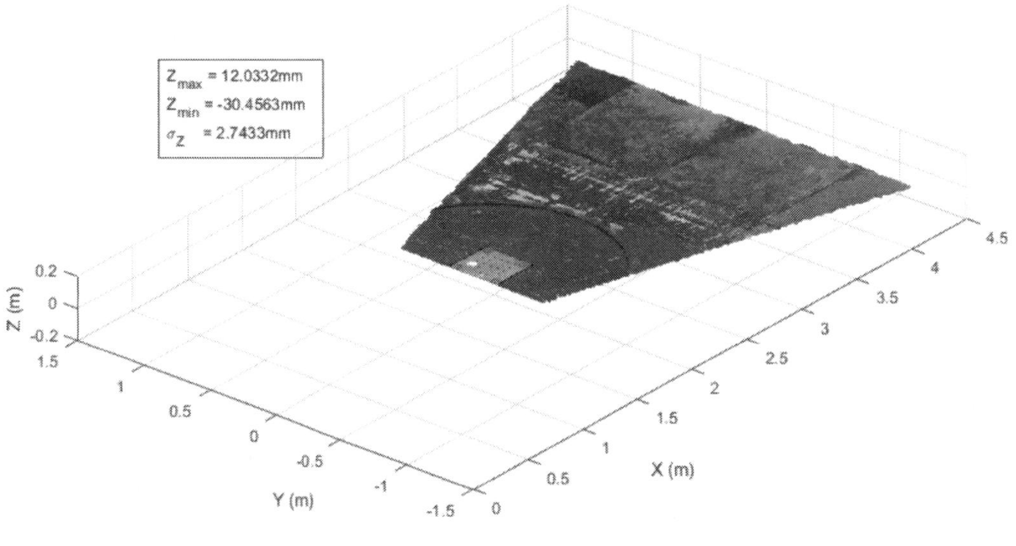

Figure 4. 3D reconstruction of the false floor (point cloud). The points beyond X=4.3m are considered as outliers and have been removed (X=0 is the position of the cameras)

zone indicating that the texture generated by the water circulation is not enough for accurate free-surface reconstruction with waves.

With a view to extend the surface of reconstruction and to improve the accuracy of the reconstruction, three additional means of generating texture at the free-surface are investigated. Firstly, a rope is tied above the flume, flicking the free-surface, so the impact between the rope and the circulating water generates short waves. Secondly, water droplets are projected over the surface by means of a hose pipe. Finally, through the use of a blow gun, compressed air is projected on the free-surface.

The addition of compressed air projection to water circulation gives the best results in terms of reconstructed surface extension and accuracy increase. **Fig.5** shows the reconstruction of regular waves (crest-to-trough wave height H=0.22m, wave period T=1.33s) at four different instants over a period. For this reconstruction, the video is acquired with a 10 Hz frame rate. Reconstructions produced with those conditions allow to observe an entire wavelength ($\lambda = 2.76m$) at different time. The free-surface reconstructions are shown as point clouds directly generated by WASS (same as **Fig. 4**). One observes that almost all the field covered by the stereo-rig is fully reconstructed, except the furthest points to the cameras (about 4.6m from the cameras) that are not well reconstructed. This is due to the fact that the translucent screen does not fall up to the free surface, hence there is a band on the free surface that is not illuminated from the point of view of the cameras. The results is quite satisfactory as it confirms that free surface reconstruction with waves is possible with fine texture generated by short waves. Moreover, the reconstruction seems accurate as the measured crests to troughs distance is of the same magnitude as the set value.

2.2. Tests with compressed air blowing only

Following the positive results obtained with the texture generated by current and compressed air blowing, another test is done on a calm free surface, *i.e.*, without water circulation nor waves. **Fig.6** shows on the right image a left camera view of the free surface, so one can observe the short waves generated by the compressed air blowing. On the left image, the green area refers to the reconstructed points. The right part of the surface is not reconstructed, nor is the top part, due to the lack of texture. The left part is not reconstructed because it is out of the right camera field. The bottom part lacks lighting (the screen does not go high enough) so texture is harder to distinguish letting the floor of the tank visible. As it is difficult to uniformly generate short waves on the measured free-surface using a single blow-gun from the edge of the flume, other reconstructions are done after changing the location of the impact of the air projection. For example, we managed to reconstruct the non-reconstructed area on the right in **Fig.6** by moving the air impact closer to this zone, to the detriment of the reconstruction of the left side. That is to say that with a correct lighting one can sensibly advocate that the more dense and shorter the short waves are, in a specific area, the better the reconstruction will be in this area. Adding texture on free-surface allows to "hide" the bottom of the tank that is seen through the water due to the angle of incidence of the cameras (especially close to the cameras).

After a few outliers removal on the reconstruction obtained with pictures from **Fig.6**, the standard deviation of the altitude (z-coordinate) around a horizontal median plane drops to $\sigma_Z = 7.38mm$. This error on the altitude is quite encouraging and shows the free-surface reconstruction is rather flat, given that the short waves generated by the pressurized air blown are a few millimeters high.

2.3. Regular waves reconstruction (texture generated by compressed air projection)

To conclude the experimental campaign performed in the Boulogne-sur-Mer facility to investigate the feasibility of free-surface reconstruction using short waves as texture, the reconstruction of three different regular waves condition has been tested: H=0.048m, T=0.67s ; H=0.05m, T=0.56s and H=0.048m, T=0.5s. The texture is once again generated by compressed air projection on the free surface. The air flow is located 2m away from the camera as it seems to be the best compromise to have both a large surface reconstruction and a dense point cloud. All three reconstructions cover almost all of the filmed free surface, *i.e.*, a 3m high trapezoid with bases going from 1.5m to 3m.

It is noticed that the shorter the period is, the less troughs are seen by the cameras hence reconstructed. As the period decreases, the wave curvature increases and favours the troughs masking by the crests. On **Fig. 7**, showing the reconstruction of (H=0.05m, T=0.56s), the first

IOP Conf. Series: Materials Science and Engineering 1288 (2023) 012009 doi:10.1088/1757-899X/1288/1/012009

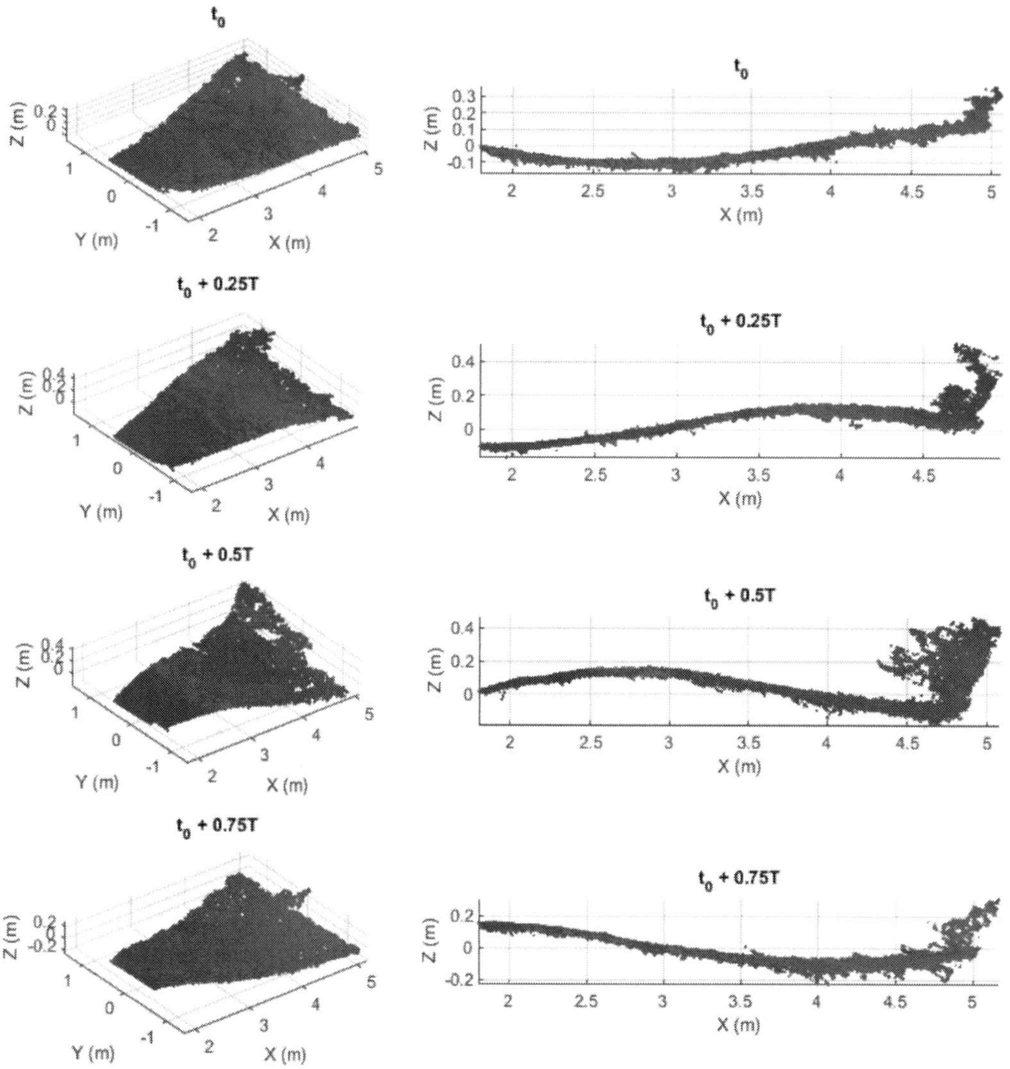

Figure 5. 3D reconstructions (point cloud) of regular waves (H=0.22m, T=1.33s) with current ($V_c = 0.7m/s$) and compressed air projection

two wave lengths, extending along 1.5m, are entirely reconstructed. Then, the crest masking effect takes precedence and troughs are no longer reconstructed. On the side view (**Fig. 7** right image), it seems that all the observed crests and troughs are fully reconstructed, but a change of viewing angle (left image) shows gaps between the further crests.

Nonetheless, the results of those tests are promising and show that with appropriate lighting and fine texture, free-surface reconstruction and waves measurement in test tanks are possible on an approximately 3m by 3m surface. However, the tests also showed some limitations in our process, namely the troughs masking, but they might be avoided by setting the cameras perpendicularly to the waves principal direction.

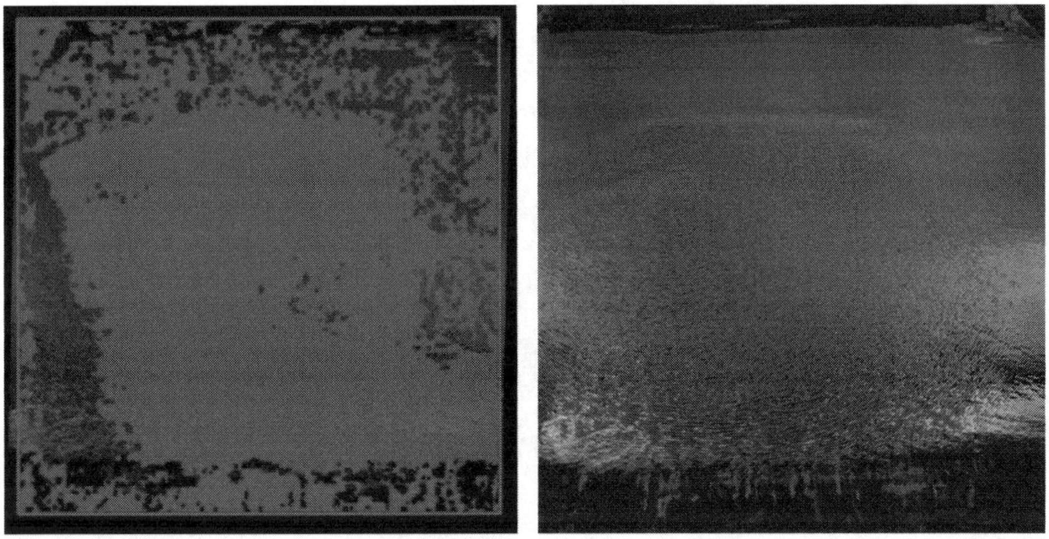

Figure 6. On the right: Calm free surface textured by compressed air blowing (left camera view), on the left: reconstructed zone

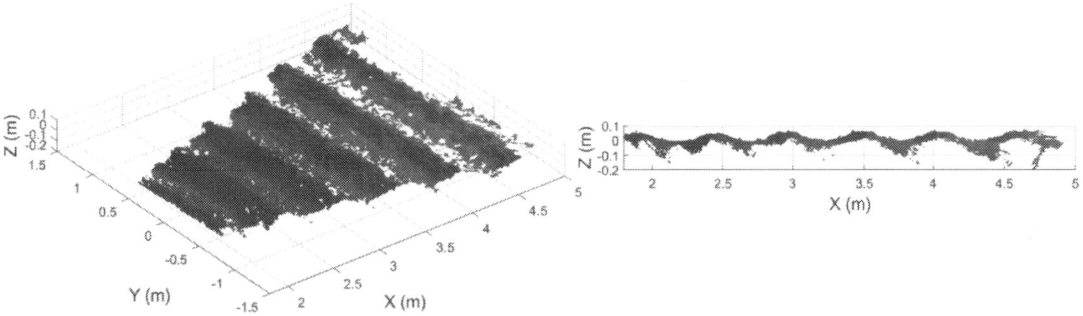

Figure 7. Regular waves reconstruction (H=0.05m, T=0.56s)

3. Reconstruction of the free-surface from shortwaves: influence of the lighting setup

Following an experimental campaign principally focused on texture generation and its influence on stereo video based free-surface reconstruction, a second experimental campaign was conducted at the wave flume of IFREMER located in Plouzané (Brittany, France), to have a better understanding of the influence of the lighting set-up. The main idea for this campaign is to create, over the flume, an artificial sky (AS) that would deliver a uniform and diffuse lighting on the observed free-surface.

3.1. Experimental Set-up

The flume is 50m long, 4m large and 3 meter deep. The side walls go about 1 m high above the free-surface. For this campaign, the same cameras and lenses were used and placed in a similar way as previously. The cameras are installed in the median axis of the flume, 1.9m above the still free-surface, on a footbridge. They are oriented towards the free surface with a 38° angle to

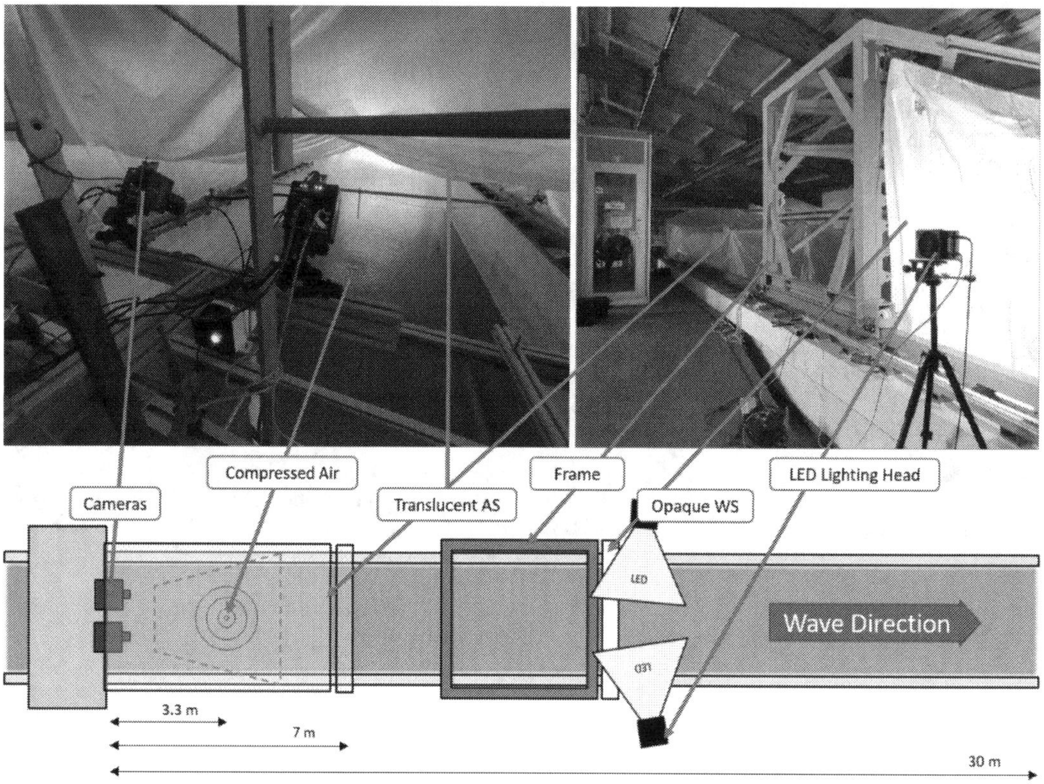

Figure 8. Experimental set-up with translucent artificial sky (AS) and opaque white screen (WS) (Config. 2)

the horizontal and the baseline is again 577 mm. The stereo camera calibration is done with the same technique and gives once again a very low reprojection error (0.04 px). As compressed air projection on the surface has been proven to be our best way to generate short waves, hence fine texture on the free surface, a small flexible pipe ($\o 6mm$) supplied with compressed air is placed above the flume, 3.3m away from the cameras. The pressure of the compressed air distribution network is lower than at the facilty in Boulogne-sur-Mer hence the pipe was installed vertically, about 1m above the free-surface. It creates a source point, right under the pipe, from which short waves propagates. This installation compromises between having a decent texture and keeping the pipe out of the cameras' field, nonetheless, the generated texture is less fine than the one obtained in Boulogne-sur-Mer. **Fig.8** shows two pictures and a top view scheme of the set-up.

The previously mentionned artificial sky is set-up either by hanging the same translucent polyane tarpaulin used in Boulogne-sur-Mer (**Config. 1 and 2**, see **Tab.1**), or more opaque white screens coming from boat sails (**Config. 3**) over the flume. The artificial sky is hung by means of ropes, tensed from the previously mentioned footbridge to a large frame that can be moved along the flume. It goes over the cameras and extends to 7m in front of them. Both sides are also covered as the artificial sky drops to the flume walls. The artificial sky hence forms a tunnel, however, due to its own weight and accessibility issues it cannot be properly taut and many folds appear, as it can be seen on **Fig.8**. The end of the tunnel is "closed" on the opposite side of the cameras to prevent direct lighting on the surface, highly unfavorable to stereo-video reconstructions.

In total, reconstructions with 88 different configurations have been tested by modifying the artificial sky set-up and by playing with the lighting conditions (adding a diffusive screen, changing the location and orientation of the lighting heads). Those configurations can be summarised in three main set-up configurations. In the first configuration, the artificial sky is entirely composed with translucent tarpaulins. The addition of a large opaque white screen (WS) to diffuse the light, hung across the width of the flume in front of the artificial sky, forms the second configuration. It is the configuration represented on the pictures and scheme in **Fig.8**. The third and last configuration consists in an artificial sky only made with opaque white screens only (same material as the diffusing screen). **Tab. 1** sums up the main set-up configurations to which this paper will refer to.

Table 1. Main set-up configurations for lighting influence studies

	Artificial Sky (AS)	Light Diffusion Screen
`Config.1`	Translucent	No
`Config.2`	Translucent	Yes: Opaque White Screen
`Config.3`	Opaque	No

3.2. Influence of the artificial sky on the reconstruction

The artificial sky is supposed to have two main purposes: to diffuse the light and to prevent nonuniform reflections on the free-surface. Free-surface reconstruction attempts without any artificial sky gives no exploitable result, no matter the lighting conditions. When the **Config. 1** is fully set up, the reconstruction becomes possible using the natural surrounding light, but also using LED lighting heads in certain conditions (see **Section 3.3**). This brief study shows how the artificial sky works as a filter for surrounding parasite lights. In the wave and current facility of Boulogne-sur-Mer, less parasite lights were present as the room could go almost totally dark.

3.3. Influence of the direct/diffuse lighting

Now that the effects of the artificial sky are proven to be beneficial for the reconstruction, one can focus on the lighting set-up. Firstly, **Config. 1** is set up. With a will of lighting the free surface, from above, using the artificial sky, two LED lighting heads are set on tripods and oriented downwards, towards the artificial sky. Even with the largest aperture, the light is not diffused enough, and despite the translucent artificial sky, the lighting is not uniform on the free surface, creating shades differently seen from both cameras leading to low quality reconstructions. Moreover, no matter the location and orientation/angle of the lighting heads, when they are positioned only a few meters away from the artificial sky, the obtained reconstructions are not dense enough.

From these observations, we tried to create a more diffuse light by relocating both lighting heads further, at the end of the flume, about 35m from the artificial sky. **Fig. 9** shows a comparison of still free-surface reconstructions obtained with three different lighting set-ups, and schemes of the set-ups. In **Fig. 9.a**, both lighting heads are oriented horizontally with large aperture towards the artificial sky. The resulting reconstructions is completely unusable. Inspired by the lighting set-up used during the experimental campaign in Boulogne-sur-Mer, in **Fig. 9.b** the lighting heads are oriented in the opposite direction, to light a large reflective and diffusive surface, *i.e.*, a 5m x 3m metallic door. When surrounding lighting is strong enough, the free-surface is reconstructed, however, the point density is still not high and a huge gap appears

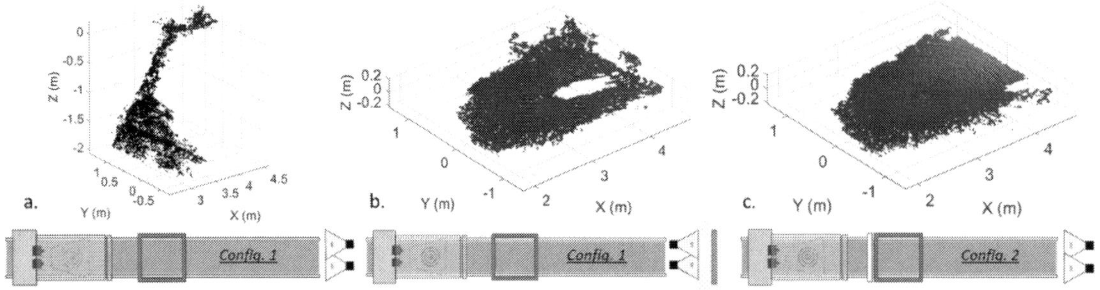

Figure 9. Still free-surface reconstruction comparison. a. Direct lighting. b. Indirect light (retro-lighting). c. WS illuminated by LED at the end of the flume.

in the middle of the reconstructed surface. As light intensity seems to be rather important for the reconstruction, we decided to bring the light source closer to the translucent artificial sky whilst keep a diffuse lighting. Consequently, **Config. 2** is set-up, and the opaque white screen is illuminated by both LED lighting heads from the flume end (see **Fig. 9.c**). The addition of the white screen creates a closer and more intense diffuse lighting, resulting in a free-surface reconstruction with a far better resolution and no gaps.

From this comparison one can sensibly assume that "direct" lighting is definitely to be avoided in contrast to diffuse lighting either using reflective lighting (like during the Boulogne-sur-mer campaign), or by means of the intermediary diffusive screen (**Config. 2**).

On **Fig. 9.b and 9.c**, it can be seen that reconstructions are less dense closer to the cameras. Indeed, short waves dissipate with distance from the source point, making the texture poorer. Moving the source point closer to the cameras allows to reconstruct an area that is closer to the cameras at the expense of the surface located further, confirming what have been observed in **Section 2.2**.

3.4. Translucent v.s. *Opaque artificial sky*

Now that the benefits of diffusing the light through an opaque white screen have been shown, one can focus on the comparison between regular waves reconstructions obtained when a translucent artificial sky with an opaque screen ahead (**Config. 2**) is set and when the artificial sky is made of opaque white screens (**Config. 3**).

Fig. 10 shows the reconstruction of regular waves with parameters H=0.048m, T=0.67s. The set-up for this reconstruction is exactly the same as the one pictured on **Fig. 9.c**, *i.e.*, two lighting head positioned at the end of the flume, illuminating the white screen in front of the translucent artificial sky. The left image on **Fig. 10** presents the raw reconstruction (untreated point cloud), while the right image is a reconstruction obtained by linear interpolation on a regular mesh ($\Delta x, \Delta y = 10mm$) and then the application of a z-coordinates median filter. The reconstruction is comparable to what was obtained at the end of the experimental campaign in Boulogne-sur-Mer (see **Section 2.3**) as the reconstructed surface is rather dense and large.

Fig. 11 shows a reconstruction made with **Config. 3**. It has the same wave parameters and same lighting heads configuration as used for **Fig. 10**. Nonetheless, less surface is reconstructed, and gaps appear behind the furthest crest while this phenomenon does not append in **Fig. 10**. It is due to an overexposure caused by direct lighting coming from a small gap between the white screen and the free-surface. In the case of (**Config. 2**), the translucent artificial sky filters this "direct light". Finally, the reconstruction seems laterally cropped compared to **Fig. 10**. The artificial sky being more opaque prevents surrounding light to brighten the surface, making the

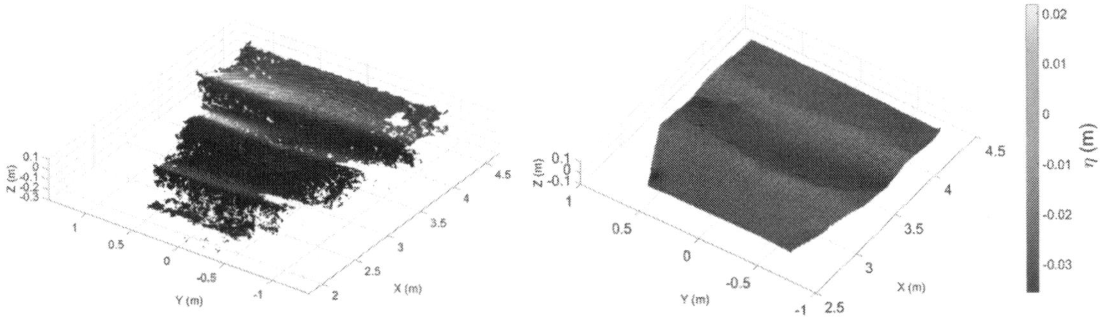

Figure 10. Regular Waves Reconstruction (H=0.048 , T=0.67s) with Config. 2. Left Image: Raw point cloud. Right Image: Linear Interpolation

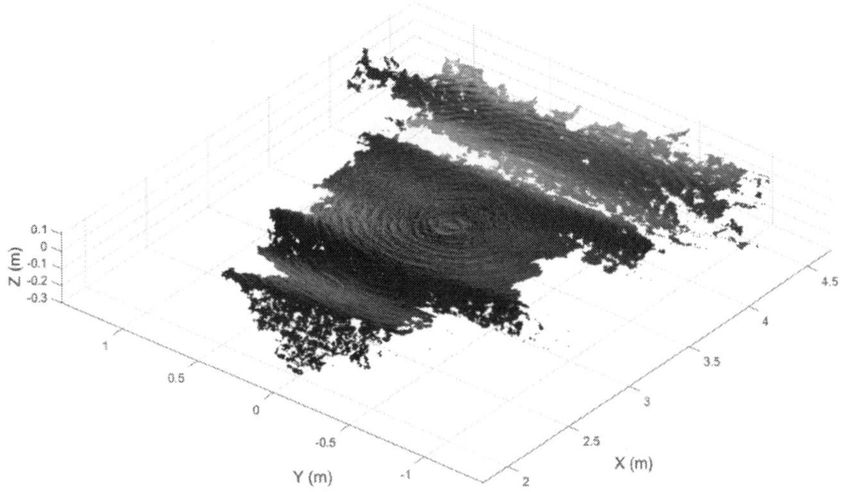

Figure 11. Regular Waves Reconstruction (H=0.048 , T=0.67s) with Config. 3.

front screen the only source of light and intensify shades from the side walls.

In contrast, although only one translucent screen was used to diffuse the light during the first experimental campaign in Boulogne-sur-Mer, the screen was wider than the flume, decreasing shades on the sides hence maximizing the reconstructed area.

To conclude this section, one can say that **Config. 2** seems to be the best configuration for waves reconstruction. The association of a translucent artificial sky and a diffusive white screen allows to have a diffuse and filtered lighting of the free-surface.

4. Generation of texture using Underwater Acoustics (UA)

With the purpose of generating texture at the free-surface without disturbing the generated waves (unlike compressed air projection), other texture generation ways have been investigated. UA is one of the considered options. As mentioned in [16],[17] and [18], ultrasonic acoustic emissions may generate short waves at the free-surface.

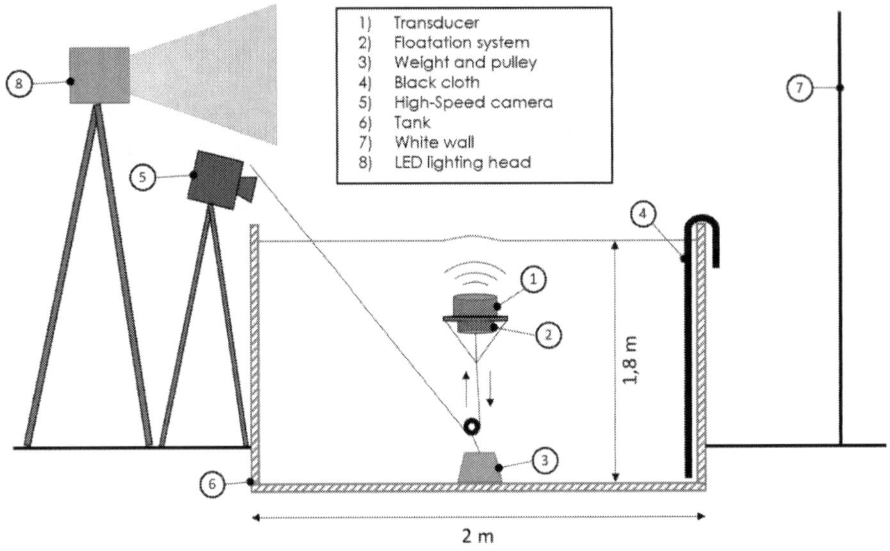

Figure 12. Experimental set-up for texture generation with ultrasonic transducers tests

4.1. Short waves generation using ultrasonic transducers

Table 2. Characteristics of tested SIMRAD® transducers

	ES 200-7C	ES 333-7C	ES 38-10
Nominal frequency	200 kHz	333 kHz	38 kHz
Frequency range	160-220 kHz	280-450 kHz	35-45 kHz
Beamwidth	7°	7°	10°
Dimensions (HxØ)	75x120 mm	75x120 mm	150x340 mm
Max. Input Power	1000 W	200 W	1500 W
Acoustic Power	227 dB	220 dB	227 dB
Max. Pulse Length	16 ms	16 ms	16 ms

To study this alternative, three underwater ultrasonic transducers have been tested (the characteristics of each transducers are described in **Tab.2**). The transducers can emit pulsed signals of adjustable duration. The period between each pulse is also adjustable but is of a minimum duration of about 100 times the pulse duration (amplifier's limit). Transducers can work in continuous wave mode (CW), emitting a signal with constant amplitude and frequency, and in frequency modulation mode (FM) where the frequency varies along a range over a pulse. **Fig. 12** shows a scheme of the experimental set-up used to observe free-surface short waves in a small tank (4m x 2m, 1.8m deep). Each transducer is plunged underwater with a system enabling depth change. In line with the previous stereo-video campaign, the lighting consist in illuminating a white wall, located on the opposite side of the cameras so the reflection contrasts the short waves generated by the transducers.

The first trials at a depth of 46cm show that both ES200-7C and ES333-7C emitting in CW mode at nominal frequency can generate a source point at the free-surface inducing gravity waves

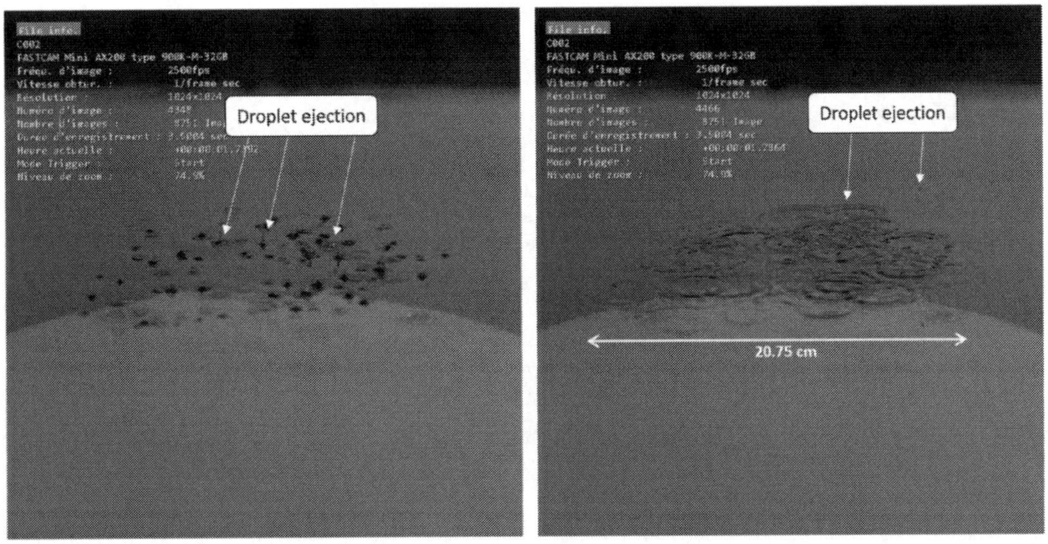

Figure 13. Shortwaves generated with ES38-10 (Depth: 21cm, Pulse Length: 8.192ms, Inter-Pulse: 2400ms, Mode: FM, Freq. Range: 35-45 kHz). Pictures taken 47 ms apart (from left to right)

with the period equals to the inter-pulse duration (time between two consecutive pulses). The amplitude of the generated gravity waves are higher with the ES200-7C than with the ES333-7C suggesting that more acoustic power creates more pressure under the surface. Limited by the amplifier capacity, one cannot set shorter inter-pulse duration, that would generate a finer texture. The generated gravity waves are not interesting since they propagate and disturb the free-surface too much.

When transducers are set at 21cm from the free-surface, one can observe an interesting shorter wave structure of about 5cm wide above the ES200-7C that is not present with the ES333-7C, confirming the idea that acoustic power has an impact on surface excitation. When ES38-10 is set-up 20cm deep, a similar but wider short wave structure (about 20 cm wide) is observable and shown in **Fig. 13**. Filming with high frame rate allows to observe first the appearance of many punctual free-surface elevation, energetic enough to expel droplets in the air (left image) and then in a second time, the appearance of short waves (right image). The observed short wave structure would make a decent texture if one could extend the affected surface. With few other tests it could be noticed that using FM mode, instead of CW mode, but also increasing the pulse duration make the short waves density increase. At 67cm deep, short waves are still present but not dense enough to be used for stereo-video. At 1.56m, no free-surface perturbation is observed.

These tests show that short waves generated by ultrasonic signal emissions seem highly dependant on the acoustic power of the used transducer and energy-intensive. The radius of the excited surface seems also to be proportional to the radius of the transducer. Unfortunately, the transducers used during these tests were only powerful enough to create interesting short waves at low depth (around 20cm), making their use only suitable for coastal test tanks. At reasonable depth for deeper wave tanks, from 1m, no short waves were discernible. It seems that more powerful with larger diameter transducers would be necessary. Knowing that the levels of emission used during the tests were already high, it raises the question about the feasibility of using short waves generated by ultrasonic signals as texture for stereo-video wave measurements.

4.2. "Stains" appearance using ultrasonic transducers

While testing short waves generation with the ES38-10, we observed the presence of "white stains" before the appearance of the punctual free-surface elevations, at depths of 20 cm, 67cm and even 1.56m (under different lighting conditions). On **Fig. 14**, one can observes the appearance and disappearance of the mentioned stains. Using a trigger on the signal emission to turn the camera recording on, one can notice that the stains appear right after a pulse emission and their appearance duration corresponds exactly to the emitted pulse duration. Continuously played, this record shows some stains moving, leading one to think that the stains could actually be bubbles.

Many articles mention the use of ultrasonic transducers to generate acoustic cavitation. Acoustic cavitation is described by Ashokkumar [19] as the growth and collapse of preexisting micro-bubbles under the excitation of an ultrasonic field. In fact, non degassed liquids contains microscopic dissolved gas nuclei that can grow and collapse or implode. According to Gugulothu *et al* [20], the use of piezoelectric transducer is the easiest way to yield acoustic cavitation. Frequencies associated to acoustic cavitation would range between 20 and 60 kHz which corresponds to the range swept through during tests with ES38-10. Furthermore, apparently higher frequencies can increase the bubble activity within the water column. High frequencies generated bubbles could act as nuclei for lower frequencies hence favoring acoustic cavitation. Liu & Hsieh [21] have experimentally demonstrated this theory showing that bubbles generated by acoustic cavitation occupy a larger space in the water column when stimulated at 83 kHz and 241 kHz simultaneously rather than separated. On our side, further observations show that the stain density increases with the pulse duration, along with the use of the FM mode instead of CW mode, which could correlate Liu & Hsieh [21] experience. Another explanation could be the fact that sweeping a range of frequencies would make bubble nuclei of difference sizes resonate hence increase the bubble activity in contrast to only one frequency making only bubbles of a certain size resonate.

So far, it is impossible to know whether the observed stains are actually bubbles generated by acoustic cavitation or not. Thus, it is not known if the observed phenomenon happens at, or close to, the free-surface, which, in this case, could make it attractive as a texture generation mean for stereo-video measurements.

5. Conclusion and prospects

Based on the results obtained during both stereo-video experimental campaigns, we have shown that it is possible to reconstruct the free-surface, indoor, using short waves as texture. We also showed that fine texture and correct lighting are indissociable to produce dense and accurate free-surface reconstructions. About the lighting, the campaign showed that direct lighting is to be avoided at all cost. The use of a translucent artificial sky over the reconstructed area to filter and diffuse surrounding lighting, combined with a large illuminated diffusive white screen, seems to be favorable to good reconstructions. However, with a very good texture, it is almost possible to reconstruct the free-surface everywhere, even with a lighting set-up with a low level of intensity.

Short-wave generation using ultrasonic transducers is a seducing approach as it is potentially less intrusive than compressed air blowing. However, it requires very high levels of emission to obtain exploitable texture at larger depths and on larger areas. In a near future, other tests will be performed on ultrasonic transducers in order to observe the location of the stains in the water column and see if they can be correlated to the punctual free-surface elevations responsible for the short waves generation.

Other ways of generating texture are to be investigated, as it is the most challenging aspect to improve in order to develop a robust and accurate stereo-video based free-surface reconstruction for wave tank experiments.

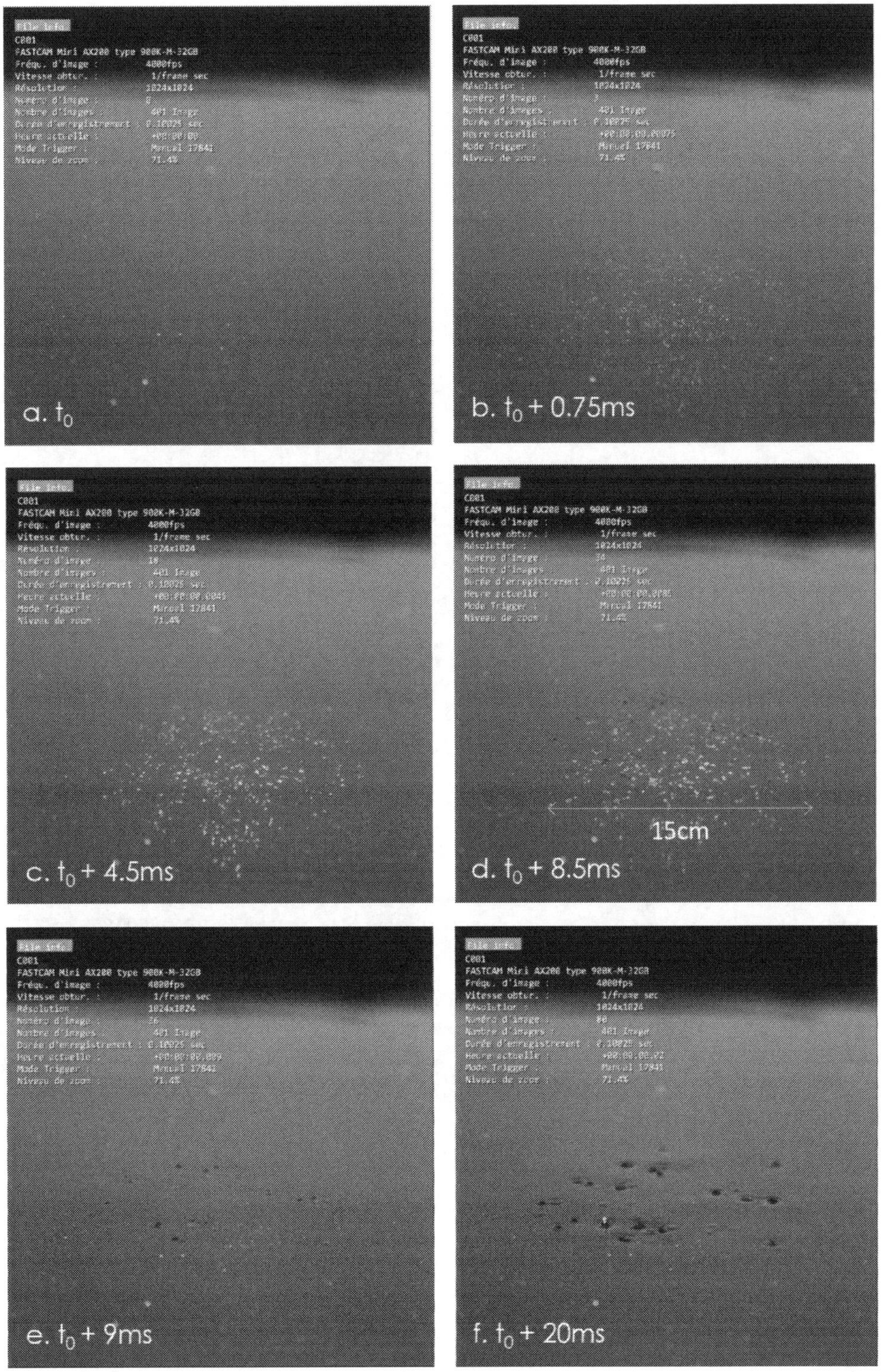

Figure 14. Free-surface evolution above ES38-10 from pulse emission (a. t_0) to short waves appearance (f. t_0+20ms) (Depth: 67 cm, Pulse Duration: 8.192 ms, Inter-pulse: 2400ms, Mode: FM, Freq. Range: 35-45 kHz)

Acknowledgments

This work has received funding from the Carnot Marine Engineering Research for Sustainable, Safe and Smart Seas Institute.

The authors would like to express their gratitude for help in anyway to: Tristan Adam, Thomas Baccheti, Sebastien Chalony, Benoit Gaurier, Jean-Valery Facq, Marie Ponchart, Naïg Le Bouffant and Laurent Berger. Special thanks are addressed to *ALL PURPOSE Roscoff* for the boat sails donation.

References

[1] Baldock T E and Swan C 1994 *Applied Ocean Research* **16** 101–112 ISSN 0141-1187 URL `https://www.sciencedirect.com/science/article/pii/014111879490006X`

[2] Craciunescu C C and Christou M 2019 Identifying Breaking Waves from Measured Time Traces (OnePetro) URL `https://onepetro.org/ISOPEIOPEC/proceedings-abstract/ISOPE19/All-ISOPE19/21000`

[3] Swan C and Sheikh R 2015 *Philosophical Transactions of the Royal Society A: Mathematical, Physical and Engineering Sciences* **373** 20140114 publisher: Royal Society URL `https://royalsocietypublishing.org/doi/10.1098/rsta.2014.0114`

[4] Hulin F, Tassin A, Filipot J F and Jacques N 2022 Experimental investigation of the hydrodynamic loads induced by breaking wave impacts on a SPAR-type floating offshore wind turbine URL `https://jh2022.sciencesconf.org/420514`

[5] Lubin P, Kimmoun O, Véron F and Glockner S 2019 *European Journal of Mechanics - B/Fluids* **73** 144–156 ISSN 0997-7546 URL `https://www.sciencedirect.com/science/article/pii/S0997754617302881`

[6] Aubourg Q, Campagne A, Peureux C, Ardhuin F, Sommeria J, Viboud S and Mordant N 2017 *Physical Review Fluids* **2** 114802 publisher: American Physical Society URL `https://link.aps.org/doi/10.1103/PhysRevFluids.2.114802`

[7] Gomit G, Chatellier L, Calluaud D and David L 2013 *Experiments in Fluids* **54** 1540 ISSN 1432-1114 URL `https://doi.org/10.1007/s00348-013-1540-4`

[8] Savelyev I and Fuchs J 2018 *Frontiers in Mechanical Engineering* **4** ISSN 2297-3079 URL `https://www.frontiersin.org/articles/10.3389/fmech.2018.00001`

[9] Bergamasco F, Torsello A, Sclavo M, Barbariol F and Benetazzo A 2017 *Computers & Geosciences* **107** 28–36 ISSN 0098-3004 URL `https://www.sciencedirect.com/science/article/pii/S0098300417304302`

[10] Benetazzo A 2006 *Coastal Engineering* **53** 1013–1032 ISSN 0378-3839 URL `https://www.sciencedirect.com/science/article/pii/S0378383906000962`

[11] Benetazzo A, Fedele F, Gallego G, Shih P C and Yezzi A 2012 *Coastal Engineering* **64** 127–138 ISSN 0378-3839 URL `https://www.sciencedirect.com/science/article/pii/S0378383912000166`

[12] MATLAB 2023 Estimate geometric parameters of a stereo camera - MATLAB - MathWorks France URL `https://fr.mathworks.com/help/vision/ref/stereocameracalibrator-app.html`

[13] Bouguet J Y 2008 Camera Calibration Toolbox for Matlab URL `http://robots.stanford.edu/cs223b04/JeanYvesCalib/`

[14] Zhang Z 2000 *IEEE Transactions on Pattern Analysis and Machine Intelligence* **22** 1330–1334 ISSN 1939-3539 conference Name: IEEE Transactions on Pattern Analysis and Machine Intelligence

[15] Heikkila J and Silven O 1997 A four-step camera calibration procedure with implicit image correction *Proceedings of IEEE Computer Society Conference on Computer Vision and Pattern Recognition* pp 1106–1112 iSSN: 1063-6919

[16] Krutyansky L, Preobrazhensky V, Makalkin D, Brysev A and Pernod P 2020 *Ultrasonics* **100** 105972 ISSN 0041-624X URL `https://www.sciencedirect.com/science/article/pii/S0041624X19300344`

[17] Boubekri R, Gross M, In M, Diat O, Nobili M, Möhwald H and Stocco A 2016 *Langmuir: the ACS journal of surfaces and colloids* **32** 10177–10183 ISSN 1520-5827

[18] Prokhorov V E and Chashechkin Y D 2009 *Izvestiya, Atmospheric and Oceanic Physics* **45** 495–502 ISSN 1555-628X URL `https://doi.org/10.1134/S0001433809040094`

[19] Ashokkumar M 2011 *Ultrasonics Sonochemistry* **18** 864–872 ISSN 1350-4177 URL `https://www.sciencedirect.com/science/article/pii/S1350417710002312`

[20] Gugulothu S K, Kumar P R V and Deekshith P 2012 *Procedia Engineering* **38** 154–164 ISSN 1877-7058 URL `https://www.sciencedirect.com/science/article/pii/S1877705812019352`

[21] Liu H L and Hsieh C M 2009 *Ultrasonics Sonochemistry* **16** 431–438 ISSN 1350-4177 URL `https://www.sciencedirect.com/science/article/pii/S1350417708001508`

Hydroelastic analysis of wave diffraction by multiple flexible porous plates

Anjan Sasmal and Soumen De*

Department of Applied Mathematics, University of Calcutta, 92, A.P.C. Road, Kolkata-700 009, India.

*Corresponding author, email: soumanisi@gmail.com

Abstract. The study investigates the phenomenon of wave diffraction by multiple poroelastic plates in infinitely deep water using the two-dimensional linearized potential theory. The study focuses on numerically computing the reflection and transmission coefficients using Galerkin approximations specifically designed for two nonidentical vertical poroelastic plates. By varying the separation lengths, porosity, and flexural rigidity of the plates, the study explores the hydrodynamic quantities associated with the wave-plate interaction. These quantities are then represented graphically to visualize their behavior under different conditions. The results of the study show that the flexural rigidity of the porous plates has a significant effect on wave reflection and transmission. The correctness of the method used in the study is confirmed by comparing the results available in the literature. Overall, the study provides valuable insights into the behavior of waves interacting with multiple poroelastic plates and highlights the importance of considering the flexural rigidity of the plates in such systems. The numerical approach used in the study can also be applied to other similar problems and can provide a useful tool for predicting the behavior of waves in complex systems.

1. Introduction

The design of hydroelastic structures can be optimized to achieve the desired level of wave energy absorption and reduction of wave impact on offshore facilities. The use of such structures can be especially important in regions where the impact of waves is high, such as in coastal areas with a high incidence of storms, hurricanes, or tsunamis. In addition to their protective function, hydroelastic structures can also be used to improve the performance and stability of offshore facilities. For example, they can be used to reduce the motion and vibration of floating structures, which can improve the safety and comfort of workers on these structures. This is especially important in harsh weather conditions where the motion of the structure can be extreme and may cause damage or injury. Overall, the use of hydroelastic structures as breakwaters is an important area of research in coastal engineering, with many potential applications in protecting and improving offshore facilities.

It is true that many researchers have investigated the diffraction of surface water waves by single or multiple submerged thin vertical barriers in deep or finite-depth water using mathematical techniques. These studies aim to optimize the design and performance of hydroelastic structures as breakwaters and to understand how they interact with waves. Various mathematical techniques have been applied to solve the problem of water wave scattering by submerged barriers, including complex variable techniques and integral equation formulations. Researchers like Dean [1], Ursell [2], Evans [3], Porter [4], and Banerjea et al. [5] have contributed to this area of research by using these techniques to investigate different aspects of wave interaction with submerged barriers, including wave diffraction, reflection, and scattering. In addition, the interaction of waves with multiple submerged barriers has also been studied by researchers such as Evans and Morris [6], Porter and Evans[7], and Roy et al. [8]. These studies have contributed to our understanding of how multiple barriers can be used together to protect offshore structures from wave impact.

While previous studies have investigated the interaction of water waves with submerged barriers, few have accounted for the porosity of the barriers. However, porosity is an important factor to consider because it affects the hydrodynamic pressure on both sides of the barrier and can make the breakwater more efficient in dissipating wave energy as mentioned by Chwang and Chan [9]. There have been some studies that have investigated the reflection and transmission coefficients for certain types of porous structures, but there is still much research to be done in this area to fully understand the feasibility of using porous barriers as breakwaters. The analytical solution of wave interaction through porous breakwater was developed by Sollitt and Cross [10]. Macaskill [11] obtained the solution of normally incident wave scattering involving a permeable thin barrier in the water of infinite depth. The porous wave maker theory in the water of finite depth water was developed by Chwang [12]. Yu [13] used an approximate method to solve the problem of diffraction of surface water waves by a semi-infinite porous breakwater employing a boundary condition based on the formulation of Sollitt and Cross [10]. The analytical solution for normally incident wave interaction with a partially immersed porous barrier was found by Isaacson [14]. An accurate solution to water wave scattering by vertical thin porous barriers using multi-term Galerkin approximation was obtained by Li et al. [15]. The two-dimensional problem of scattering of surface water waves by two symmetric vertical thin porous plates is investigated by Gayen and Mondal [16] using the hypersingular integral equation method. Manam and Sivanesan [17] found the solution for deep water wave motion using an analytic approach over a surface piercing or a bottom-standing porous barrier. Sasmal et al. [18] investigated the problem of water wave scattering by two unequal vertical partially immersed porous barriers and two unequal totally submersed porous barriers extended to infinite depth. Sasmal and De [19] investigated the problem of oblique water wave scattering by two vertical porous barriers with submerged gaps of different widths in infinitely deep water. Recently, Sasmal and De [20] studied the effect of multiple arbitrary porous barriers in deep water, in mitigating the wave effects.

On the other hand, hydro elasticity theories for structures like floating elastic plates of arbitrary shape and disks have recently been reviewed by many researchers. For example, the problem of water wave scattering by a vertical elastic sheet is studied by Meylan[21] using Green's function. Meylan and Peter [22] presented the solution to the water-wave interaction with a submerged elastic plate of negligible thickness by the eigenfunction-matching method. Williams and Meylan [23] investigated the problem of normally incident linear waves interaction with a submerged elastic plate of semi-infinite extent by applying WienerHopf (WH) and residue calculus (RC) methods. Zheng et al. [24] developed a semi-analytical model based on linear potential flow theory and an eigenfunction expansion method to study wave scattering by a floating porous elastic plate with an arbitrary shape, considering three different types of edge conditions, i.e., a clamped edge, a simply supported edge, and a free edge. The problem of water-wave scattering by a semi-infinite submerged thin elastic plate that is either porous or non-porous is investigated by Smith et al. [25] using the Wiener-Hopf technique. Ashok et al [26] presented explicit solutions to the problem of scattering of normally incident waves by a submerged flexible porous thin plate.

This paper focuses on investigating the phenomenon of oblique wave diffraction by two

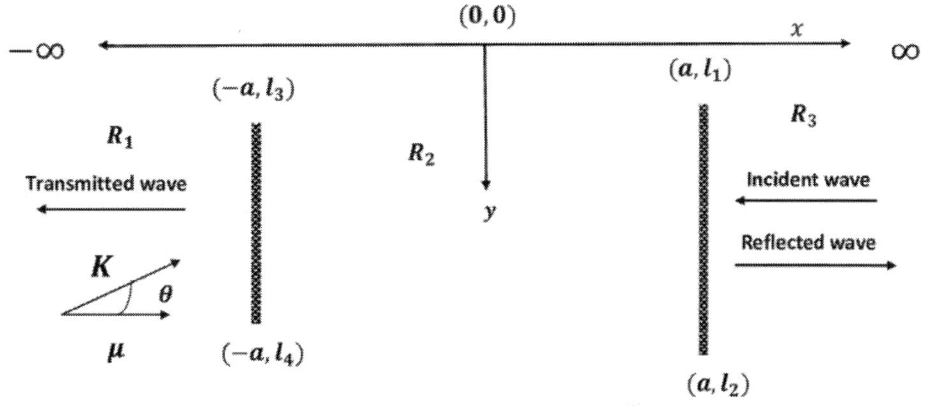

Figure 1. Definition diagram

nonidentical vertical poroelastic plates that are submerged beneath the free surface in infinitely deep water. As far as the author's knowledge no previous research has explored the scattering of water waves by such a configuration of nonidentical plates. The problem is divided into three parts, each involving a potential function that describes the resulting motion in the fluid region. The Havelock inversion formula is then employed, followed by a matching process to transform the potential function into linear integral equations using eigenfunction expansions. In the final stage, we use a single-term Galerkin approximation to approximately solve the integral equations and determine the reflection and transmission coefficients. Through this analysis, we find that both porosity and flexural rigidity have a significant impact on the reflection coefficient.

2. Mathematical Formulation

Let us consider a Cartesian coordinate system in which the y axis is taken vertically downwards along the fluid region $y \geq 0$ extended to infinite depth and the mean free surface as $y = 0$. The two vertical porous plates are situated at $x = -a$, $y \in L_1 \equiv (l_1, l_2)$(first porous plate) and $x = a$, $y \in L_2 \equiv (l_3, l_4)$(second porous plate) respectively(see Fig 1). The fluid region is divided into three subregions, namely $R_1 \equiv (-\infty < x < -a, 0 < y < \infty)$; $R_2 \equiv (-a < x < a, 0 < y < \infty)$ and $R_3 \equiv (a < x < \infty, 0 < y < \infty)$, respectively. The fluid is assumed to be incompressible, homogeneous, inviscid and the motion is irrotational and simple harmonic in time. Assuming linear theory, the time harmonic motions in the fluid are described by the velocity potential $\Phi_j(x, y, z, t) = Re\{\phi_j(x, y)e^{i\nu z - i\omega t}\}$ where Re denotes the real part, ω is the angular frequency of the incident wave and ν is the component of the wave number along the z-axis. Here the component ν is defined by $\nu = K \sin\theta$, where $K = \frac{\omega^2}{g}$, g being the acceleration due to gravity and θ is the angle of the incident wave to the x-axis. The spatial velocity potential $\phi_j(x, y)$ for $j = 1, 2, 3$ in each sub-region satisfies the governing partial differential equation given by

$$(\nabla^2 - \nu^2)\phi_j = 0 \quad \text{in the fluid region.} \tag{1}$$

where, $\nabla^2 = \partial^2/\partial x^2 + \partial^2/\partial y^2$. The free surface boundary condition is given by

$$K\phi_j + \frac{\partial \phi_j}{\partial y} = 0 \quad \text{on } y = 0, \tag{2}$$

the bottom boundary condition,

$$|\nabla \phi_j| \to 0, \quad \text{as } y \to \infty, \tag{3}$$

IOP Conf. Series: Materials Science and Engineering 1288 (2023) 012010 doi:10.1088/1757-899X/1288/1/012010

where $\nabla = (\partial/\partial x, \partial/\partial y)$. The plate is deflected horizontally with the displacement $\eta_j(y,t) = Re[v_j(y)e^{(-1)^j i\omega t}]$ $j = 1,2$, where $v_j(y)$, $j = 1,2$ is the complex quantity whose absolute value is assumed to be small.

The conditions on the porous plates are given by

$$\frac{\partial \phi_j}{\partial x} = \frac{\partial \phi_{j+1}}{\partial x} = -iKG_j(\phi_{j+1} - \phi_j) - i\omega v_j(y), \quad x = x_j, \quad y \in L_j, \; j = 1,2, \tag{4}$$

where, $G_j = G_j^r + iG_j^i$, $j = 1,2$ are the dimensionless porous effect parameters of poroelastic plates.

The complex porous effect parameters can be written as

$$G_j = \frac{\varepsilon_j(f_j + iS_j)}{K\alpha_j(f_j^2 + S_j^2)}$$

in which ε_j are the porosity constant, f_j are the resistance force coefficient, S_j are the inertial force coefficient, α_j are the thickness of the porous medium. The real part G_j^r represents the resistance effect of the porous material against the flow, while the imaginary part G_j^i represents the inertia effect of the fluid inside the porous material.

The displacement amplitude $v_j(y)$, $j = 1,2$ satisfies

$$\frac{\partial^4 v_j}{\partial y^4} - \lambda^2 v_j = -\frac{i\omega\rho}{EI}(\phi_{j+1} - \phi_j), \quad y \in L_j, \; j = 1,2, \tag{5}$$

where $\lambda = \omega\sqrt{m_b/EI}$, EI is uniform flexural rigidity of porous plate, m_b is uniform mass per unit length and ρ is the density of water. The ends of the plates can be clamped or simply supported or allowed to be free with the mathematical conditions $v_j = \frac{\partial v_j}{\partial y} = 0$ or $v_j = \frac{\partial^2 v_j}{\partial y^2} = 0$ or $\frac{\partial^2 v_j}{\partial y^2} = \frac{\partial^3 v_j}{\partial y^3} = 0$ at the end points, respectively(see Ashok et al. [26]). Using these end conditions, one can find the displacement amplitude $v_j(y)$ solving the 4th order differential equation (5).

In the vicinity of sharp submerged edges velocity potential behaves as

$$|\nabla \phi_j| = O(r^{-1/2}) \quad \text{as} \quad r \to 0, \tag{6}$$

where r is the distance from the edges of plates.

The conditions at infinity are given by

$$\begin{aligned} \phi_1(x,y) &= \phi_0(y)Te^{-i\mu(x-a)} \quad \text{as} \quad x \to -\infty \\ \phi_3(x,y) &= \phi_0(y)\{e^{-i\mu(x-a)} + Re^{i\mu(x+a)}\} \text{ as } x \to \infty \end{aligned} \tag{7}$$

where R and T are the unknown complex reflection and transmission coefficients to be determined and $\phi_0(y) = e^{-Ky}$, $\mu = K\cos\theta$.

At the common boundary of the region velocity potential ϕ_j and gradient of velocity potential $\partial\phi_j/\partial x$ are continuous across the gap. Let us consider

$$p_j(y) = \frac{\partial \phi_j(x_j,y)}{\partial x} = \frac{\partial \phi_{j+1}(x_j,y)}{\partial x}, q_j(y) = \phi_{j+1}(x_j,y) - \phi_j(x_j,y), \tag{8}$$

for $j = 1,2$.

Eliminating $v_j(y)$, $j = 1,2$ from Eqs.(4) and (5) along with the Eqs.(8), we get

$$\begin{aligned} &\left(\frac{\partial^4}{\partial y^4} - \lambda^2\right)\frac{\partial \phi_j}{\partial x} = \left(\frac{\partial^4}{\partial y^4} - \lambda^2\right)\frac{\partial \phi_{j+1}}{\partial x} = \\ &-iKG_j\frac{\partial^4 q_j(y)}{\partial y^4} + \lambda^2\left(iKG_j - \frac{\rho}{m_b}\right)q_j(y), \quad y \in L_j, \; j = 1,2, \end{aligned} \tag{9}$$

3. Method of Solution

The velocity potentials $\phi_j(x,y)$ satisfies the Helmholtz equation in the fluid domain as in Eq.(1) along with the boundary conditions Eq.(2) and Eq.(3). Using the Havelock's expansion formula, the velocity potentials $\phi_j(x,y)$ for each of the three regions are given by

$$\phi_1(x,y) = T\phi_0(y)e^{-i\mu(x+a)} + \int_0^\infty \frac{A_1(k)Z(k,y)}{l(k^2+K^2)}e^{l(x+a)}dk. \tag{10}$$

$$\phi_2(x,y) = \phi_0(y)(be^{i\mu x} + de^{-i\mu x}) + \int_0^\infty \frac{\{A_2(k)e^{lx} + A_3(k)e^{-lx}\}Z(k,y)}{l(k^2+K^2)}dk. \tag{11}$$

$$\phi_3(x,y) = \phi_0(y)\{e^{-i\mu(x-a)} + Re^{i\mu(x-a)}\} + \int_0^\infty \frac{A_4(k)Z(k,y)}{l(k^2+K^2)}e^{-l(x-a)}dk. \tag{12}$$

where $l = \sqrt{k^2+\nu^2}$, $Z(k,y) = k\cos ky - K\sin ky$ and b, d are arbitrary constants and $A_1(k)$, $A_2(k)$, $A_3(k)$, $A_4(k)$ are unknown functions of k.

Applying Havelock's inversion formula along with the expression of the velocity potential for three regions in Eqs.(10) – (12) and with the expression in the Eqs.(8), for the gap, we obtain

$$Te^{2i\mu a} = de^{i\mu a} - be^{-i\mu a} = \frac{2i}{\cos\theta}\int_{\overline{L}_1} p_1(y)\phi_0(y)dy. \tag{13}$$

$$1 - Re^{2i\mu a} = de^{-i\mu a} - be^{i\mu a} = \frac{2i}{\cos\theta}\int_{\overline{L}_2} p_2(y)\phi_0(y)dy. \tag{14}$$

$$\left.\begin{aligned} A_1(k) = A_2(k)e^{-al} - A_3(k)e^{al} &= \frac{2}{\pi}\int_{\overline{L}_1} p_1(y)Z(k,y)dy, \\ A_4(k) = A_2(k)e^{al} - A_3(k)e^{-al} &= \frac{2}{\pi}\int_{\overline{L}_2} p_2(y)Z(k,y)dy \end{aligned}\right\} \tag{15}$$

$$be^{-i\mu a} + de^{i\mu a} - Te^{2i\mu a} = 2K\int_{L_1} q_1(y)\phi_0(y)dy. \tag{16}$$

$$1 + Re^{2i\mu a} - be^{i\mu a} - de^{-i\mu a} = 2K\int_{L_2} q_2(y)\phi_0(y)dy. \tag{17}$$

$$\left.\begin{aligned} A_2(k)e^{-al} + A_3(k)e^{al} - A_1(k) &= \frac{2l}{\pi}\int_{L_1} q_1(y)Z(k,y)dy, \\ A_4(k) + A_2(k)e^{al} + A_3(k)e^{-al} &= -\frac{2l}{\pi}\int_{L_2} q_2(y)Z(k,y)dy \end{aligned}\right\} \tag{18}$$

Using the first two terms of the Eqs.(13), (14), the constants b and d can be found in terms of R and T, as follows

$$\left.\begin{aligned} b &= \frac{i}{2}\csc(2\mu a)e^{i\mu a}\{1 - Re^{2i\mu a} - T\}, \\ d &= \frac{i}{2}\csc(2\mu a)e^{-i\mu a}\{1 - Re^{2i\mu a} - Te^{4i\mu a}\} \end{aligned}\right\} \tag{19}$$

Now we find the expression of $A_2(k)$ and $A_3(k)$ in terms of $A_1(k)$ and $A_4(k)$ using the first two terms of 1st and 2nd equation of (15) as follows

$$(A_2(k),\ A_3(k)) = \left(\frac{A_4(k)e^{al} - A_1(k)e^{-al}}{2\sinh(2al)},\ \frac{A_4(k)e^{-al} - A_1(k)e^{al}}{2\sinh(2al)}\right). \tag{20}$$

Using the expression of b and d from the Eq.(19), the Eq.(16) and Eq.(17) reduce to

$$i\csc(2\mu a)\{1 - Re^{2i\mu a} - T\} = 2K\int_{L_1} q_1(y)\phi_0(y)dy. \tag{21}$$

$$i\csc(2\mu a)\{R - e^{2i\mu a} + Te^{2i\mu a}\} = 2K\int_{L_2} q_2(y)\phi_0(y)dy. \tag{22}$$

Using the expression of $A_2(k)$ and $A_3(k)$ from the Eq.(20), the two Eqs. of (18) reduce to

$$\left(\frac{A_4(k) - A_1(k)e^{2al}}{\sinh(2al)},\ \frac{A_4(k)e^{2al} - A_1(k)}{\sinh(2al)}\right) = \frac{2l}{\pi}\left(\int_{L_1} q_1(y)Z(k,y)dy,\ -\int_{L_2} q_2(y)Z(k,y)dy\right) \tag{23}$$

Now we simplify Eqs.(23) to find the expressions of $A_1(k)$, $A_4(k)$ in terms of the functions $q_j(y)$; $(j = 1, 2)$ as given by

$$\begin{pmatrix} A_1(k) \\ A_4(k) \end{pmatrix} = -\frac{l}{\pi}\begin{pmatrix} \int_{L_1} q_1(y)Z(k,y)dy + \int_{L_2} e^{-2al}q_2(y)Z(k,y)dy \\ \int_{L_1} e^{-2al}q_1(y)Z(k,y)dy + \int_{L_2} q_2(y)Z(k,y)dy \end{pmatrix} \tag{24}$$

Now the expressions of $A_1(k)$, $A_4(k)$ from the Eqs(24), substituting in the expression of $p_1(y)$ and $p_2(y)$ and using the relation (9), we obtain the following integral equations in $q_j(y)$ $(j = 1, 2)$:

$$\sum_{m=1}^{2}\int_{L_m} q_m(t)\mathcal{F}_{nm}(y,t)dt = -\frac{\lambda^2\left(iKG_n - \frac{\rho}{m_b}\right)}{K^4 - \lambda^2}q_n(y) + H_n(y), n = 1, 2 \tag{25}$$

where, $H_1(y) = -i\mu Te^{2i\mu a}\phi_0(y)$; $H_2(y) = -i\mu(1 - Re^{2i\mu a})\phi_0(y)$ and

$$\left.\begin{aligned} \mathcal{F}_{nn}(y,t) &= \frac{1}{\pi(K^4 - \lambda^2)}\int_0^\infty \left(\frac{l(k^4 - \lambda^2)}{k^2 + K^2} - \frac{2iKG_nk^4}{k^2 + K^2}\right)Z(k,y)Z(k,t)dk \\ &\quad - \frac{2iK^6G_n}{(K^4 - \lambda^2)}\phi_0(y)\phi_0(t), n = 1, 2, \\ \mathcal{F}_{mn}(y,t) &= \mathcal{F}_{nm}(y,t) = \frac{1}{\pi(K^4 - \lambda^2)}\int_0^\infty \frac{l(k^4 - \lambda^2)}{k^2 + K^2}e^{-2al}Z(k,y)Z(k,t)dk, ; m, n = 1, 2. \end{aligned}\right\} \tag{26}$$

Let us consider potential difference function $q_m(t)$ as

$$q_m(t) = -i\mu Te^{2i\mu a}O_{m1}(t) - i\mu(1 - Re^{2i\mu a})O_{m2}(t), m = 1, 2. \tag{27}$$

After using the expressions of $q_m(t)$ from Eq. (27), the integral equations in Eq. (25) can be reduced as

$$\sum_{m=1}^{2}\int_{L_m} O_{mj}(t)\mathcal{F}_{nm}(y,t)dt + \frac{\lambda^2\left(iKG_n - \frac{\rho}{m_b}\right)}{K^4 - \lambda^2}O_{nj}(y) = \delta_{nj}\phi_0(y), y \in L_n, n, j = 1, 2, \tag{28}$$

where δ_{nj} is the Kronecker delta symbol.

3.1. Solutions of integral equations using Galerkin's approximation technique:
The integral equations given by Eq. (28) have been solved using the single-term Galerkin technique. The unknown functions $O_{ij}(y)$ are chosen in the following form

$$O_{mj}(t) = c_{mj}\ \mathcal{S}_m(t), \qquad m, j = 1, 2, \tag{29}$$

where $\mathcal{S}_j(t)$, are suitably chosen basis functions, and c_{mj} unknown constants are to be found.

In order to deduce the unknown constants c_{mj} appearing in Eq. (29), we substitute the expressions of $O_{mj}(t)(m, j = 1, 2,)$ from Eq. (29) into the integral equations in (28), then multiplying the m-th integral equation in (28) by $\mathcal{S}_m(t)$ and integrating over L_m. This yields the following linear systems of equations in $c_{mj}(m, j = 1, 2)$:

$$\sum_{m=1}^{2} c_{mj}\mathcal{A}_{nm} = \delta_{nj}\mathcal{D}_n, \quad n, j = 1, 2, \tag{30}$$

where

$$\left.\begin{aligned}
\mathcal{A}_{nn} &= \int_{L_n} \mathcal{S}_n(y)\left\{\int_{L_n} \mathcal{S}_n(t)\mathcal{F}_{nn}(y,t)dt\right\}dy + \frac{\lambda^2\left(iKG_n - \frac{\rho}{m_b}\right)}{K^4-\lambda^2}\int_{L_n}[\mathcal{S}_n(y)]^2dy,\\
\mathcal{A}_{nm} = \mathcal{A}_{mn} &= \int_{L_n} \mathcal{S}_n(y)\left\{\int_{L_m} \mathcal{S}_m(t)\mathcal{F}_{nm}(y,t)dt\right\}dy, \quad n \neq m,\\
\mathcal{D}_n &= \int_{L_n} \mathcal{S}_n(y)\phi_0(y)dy, \qquad n, m = 1, 2.
\end{aligned}\right\} \tag{31}$$

We choose the basis function $\mathcal{S}_m(y), m = 1, 2$ as follows

$$\mathcal{S}_1(y) = e^{-Ky}\left[\int_{l_1}^{y} \frac{(d_1^2 - z^2)e^{Kz}}{\sqrt{(z^2-l_1^2)(l_2^2-z^2)}}dz\right], \quad l_1 < y < l_2 \tag{32}$$

$$\mathcal{S}_2(y) = e^{-Ky}\left[\int_{l_3}^{y} \frac{(d_2^2 - z^2)e^{Kz}}{\sqrt{(z^2-l_3^2)(l_4^2-z^2)}}dz\right], \quad l_3 < y < l_4 \tag{33}$$

where

$$d_1^2 = \frac{\displaystyle\int_{l_1}^{l_2} \frac{z^2e^{Kz}}{\sqrt{(z^2-l_1^2)(l_2^2-z^2)}}dz}{\displaystyle\int_{l_1}^{l_2} \frac{e^{Kz}}{\sqrt{(z^2-l_1^2)(l_2^2-z^2)}}dz}, \quad d_2^2 = \frac{\displaystyle\int_{l_3}^{l_4} \frac{z^2e^{Kz}}{\sqrt{(z^2-l_3^2)(l_4^2-z^2)}}dz}{\displaystyle\int_{l_3}^{l_4} \frac{e^{Kz}}{\sqrt{(z^2-l_3^2)(l_4^2-z^2)}}dz}$$

Substituting the expressions of $\mathcal{S}_m(y)$ from Eq. (32), (33) and $\mathcal{F}_{mn}(y,t)$ from Eq. (26) into the Eq. (31), we have

$$\begin{aligned}
\mathcal{A}_{nn} &= \frac{1}{\pi(K^4-\lambda^2)}\int_0^\infty \left(\frac{l(k^4-\lambda^2)}{k^2+K^2} - \frac{2iKG_nk^4}{k^2+K^2}\right)\left[\int_{L_n} \frac{(d_n^2-z^2)\sin kz}{\sqrt{(z^2-l_{2n-1}^2)(l_{2n}^2-z^2)}}dz\right]^2 dk\\
&- \frac{iK^4G_n}{2(K^4-\lambda^2)}\left[\int_{L_n} \frac{(d_n^2-z^2)e^{-Kz}}{\sqrt{(z^2-l_{2n-1}^2)(l_{2n}^2-z^2)}}dz\right]^2 + \frac{\lambda^2\left(iKG_n - \frac{\rho}{m_b}\right)}{K^4-\lambda^2}\int_{L_1} \mathcal{S}_n(t)dt, \quad n = 1, 2,
\end{aligned} \tag{34}$$

$$\mathcal{A}_{mn} = \mathcal{A}_{nm} = \frac{1}{\pi\left(K^4-\lambda^2\right)}\int_0^{\infty}\frac{l\left(k^4-\lambda^2\right)e^{-2al}}{k^2+K^2}$$
$$\left[\int_{L_n}\frac{(d_n^2-z^2)\sin kz}{\sqrt{(z^2-l_{2n-1}^2)(l_{2n}^2-z^2)}}dz\right]\left[\int_{L_m}\frac{(d_m^2-z^2)\sin kz}{\sqrt{(z^2-l_{2m-1}^2)(l_{2m}^2-z^2)}}dz\right]dk, m,n=1,2,\ (n\neq m) \tag{35}$$

$$\mathcal{D}_n = \frac{1}{2K}\left[\int_{L_n}\frac{(d_n^2-z^2)e^{-Kz}}{\sqrt{(z^2-l_{2n-1}^2)(l_{2n}^2-z^2)}}dz\right],\quad n=1,2, \tag{36}$$

3.2. Determination of Reflection and Transmission Coefficients:
In order to determine the coefficients R and T, we substitute the expression from Eq. (27) into Eqs. (21) – (22), then these equations can be written in the matrix notation as

$$WX = Y \tag{37}$$

where,

$$W = \begin{bmatrix} \csc(2\mu a) - 2\mu K e^{2i\mu a}V_{11} & (\csc(2\mu a) + 2\mu K V_{12})e^{2i\mu a} \\ (\csc(2\mu a) + 2\mu K V_{21})e^{2i\mu a} & \csc(2\mu a) - 2\mu K V_{22}e^{2i\mu a} \end{bmatrix},$$
$$X = \begin{bmatrix} T \\ R \end{bmatrix}, Y = \begin{bmatrix} \csc(2\mu a) + 2\mu K V_{12} \\ \csc(2\mu a)e^{2i\mu a} - 2\mu K V_{22} \end{bmatrix}.$$

where, $V_{mj} = \int_{B_m} O_{mj}(t)e^{-Kt}dt,\quad m,j=1,2.$

Once the unknowns constants $c_{mj}, m,j = 1,2$ are obtained after solving the system of equations (30), then $O_{mj}, m,j = 1,2$ from relations of Eq. (29) can be obtained. Hence, R and T can be determined by solving the Eq. (37).

4. Numerical Results and Discussion

In this section, the numerical results for the reflection and transmission coefficients of the poroelastic plates are presented. These coefficients are studied as a function of the non-dimensional frequency, defined as Kl_2. The numerical results are plotted assuming the dimensionless parameters $M = \frac{m_b}{\rho l_2}$, $F_r = \frac{EI}{\rho g l_2^4}$ for the elastic porous plates. Throughout the numerical computations, certain parametric values are kept constant. The fluid density is set to $\rho = 1025.0\ kg/m^3$, and the gravitational acceleration is taken as $g = 9.81\ m/s^2$.

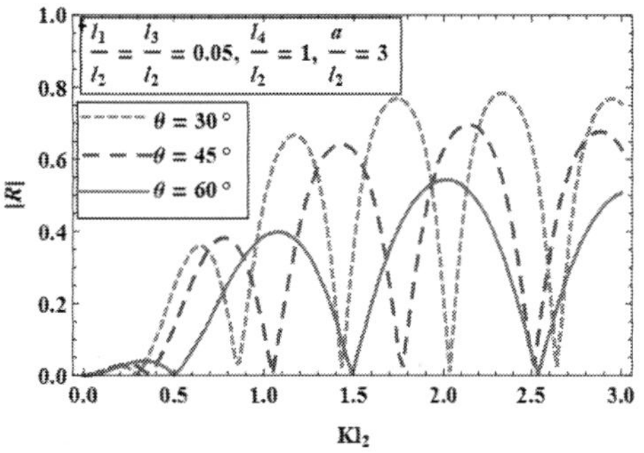

Figure 2. $|R|$ against Kl_2 for various values of angle of incident θ.

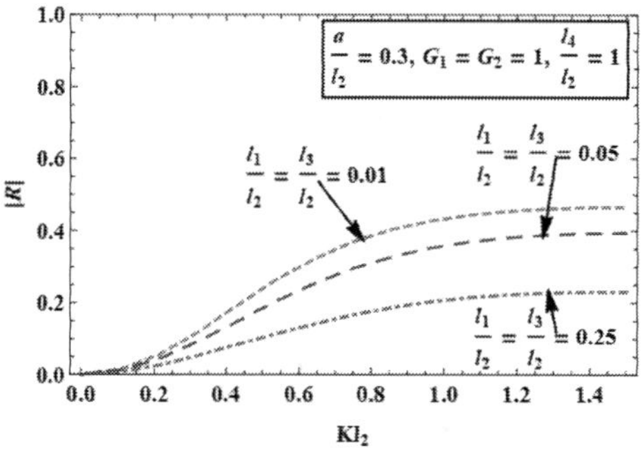

Figure 3. $|R|$ against Kl_2 for different values of length of the plates.

4.1. Comparison with existing results

In order to validate the current method, the comparison is made with the corresponding cases available in the literature. The reflection coefficients $|R|$ are depicted graphically against Kl_2 for three different values of angle of incidence $\theta = 30^o, 45^o, 60^o$ and fixed values of $G_1 = G_2 = 0; \frac{a}{l_2} = 3$ and $\frac{l_1}{l_2} = \frac{l_3}{l_2} = 0.05$, $\frac{l_4}{l_2} = 1, M = 0.01, F_r = 0.999$ in the Fig.2. The curves in this figure almost agree with the corresponding curves plotted by Roy et al. [8] for two equal submersed plates. This comparison with the results available in the literature provides validation for the current method.

To further verify the results obtained for permeable plates in this study, we compare our results with those reported by Gayen and Mondal [16], who used a different method for analyzing two equal plates. Gayen and Mondal [16] plotted $|R|$ against dimensionless wave number $K(d+b)$ ($\equiv Kl_2$ here) for three different values of $\mu = \frac{d-b}{d+b}$ ($\equiv \frac{l_1}{l_2} = \frac{l_3}{l_2}$ here) and fixed value of the length of separation of plates and porosity and normal incident wave in their Fig.4. In this paper, $|R|$ is plotted against Kl_2 for the same values of $\frac{a}{l_2} = 0.3$, $\frac{l_4}{l_2} = 1$, $G_1 = G_2 = 1$ and $\frac{l_1}{l_2} = \frac{l_3}{l_2} = 0.01, 0.05, 0.25$ and $\theta = 0^o, M = 0.01, F_r = 0.999$ in Fig.3. The corresponding curves of Gayen and Mondal [16] for two equal plates almost coincide with the corresponding curves in Fig.3.

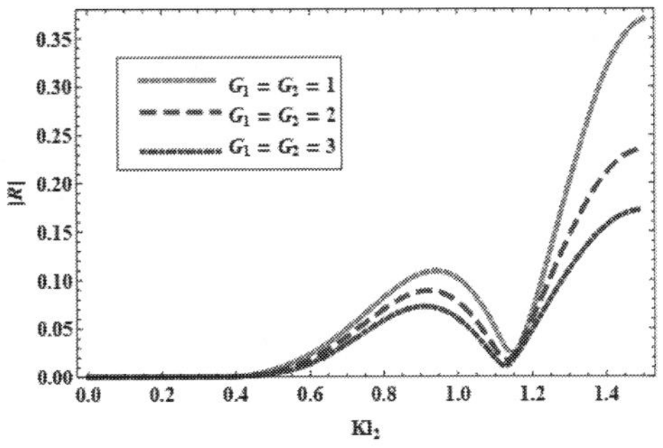

Figure 4. Reflection coefficient $|R|$ against Kl_2 for different values of porosity.

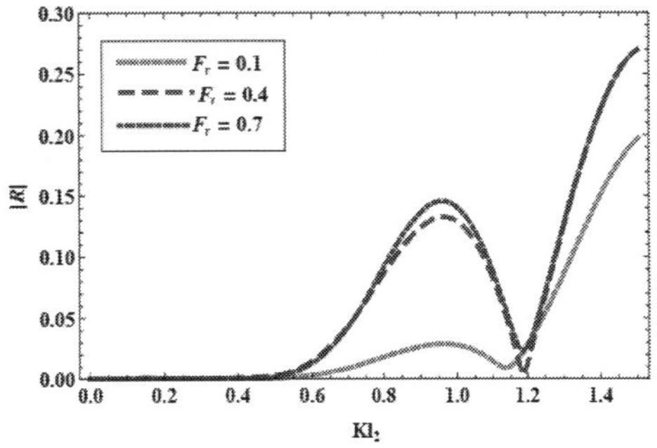

Figure 5. Reflection coefficient $|R|$ against Kl_2 for different values of flexural rigidity.

4.2. Results

The effect of porosity on $|R|$ for two equal plates is demonstrated in Fig.4. Fig. 4 is plotted for different values of $G_1 = G_2 = 1; G_1 = G_2 = 2; G_1 = 1, G_2 = 3$ and fixed values of $\frac{a}{l_2} = 3$ and $\frac{l_1}{l_2} = \frac{l_3}{l_2} = 0.05$, $\frac{l_4}{l_2} = 1, M = 0.01, F_r = 0.4, \theta = 45^o$. From this figure, it is clear that the peak value of $|R|$ decreases as the value of porous effect parameters increases, due to the increased wave energy dissipation caused by higher porosity. Additionally, it is observed that the reflection coefficients are oscillatory in nature for larger wave numbers $Kl_2 > 0.6$. This is regarded as a very important phenomenon in coastal engineering since the porous structures cause smaller surface fluctuation due to the low reflection.

Figure 5 shows the effect of flexural rigidity on the reflection coefficient $|R|$ for different values of F_r. Figure 5 is plotted for different values of $F_r = 0.1; F_r = 0.4; F_r = 0.7$ and fixed values of $\frac{a}{l_2} = 3$ and $\frac{l_1}{l_2} = \frac{l_3}{l_2} = 0.05$, $\frac{l_4}{l_2} = 1, M = 0.01, G_1 = G_2 = 1, \theta = 45^o$. The figure demonstrates that the values of $|R|$ increase with increasing values of flexural rigidity F_r. This suggests that the presence of flexible porous barriers enhances the reflection to some extent. These results provide important insights into the behavior of flexible porous barriers under different flexural rigidity values, which can be useful for designing and optimizing these barriers for various engineering applications.

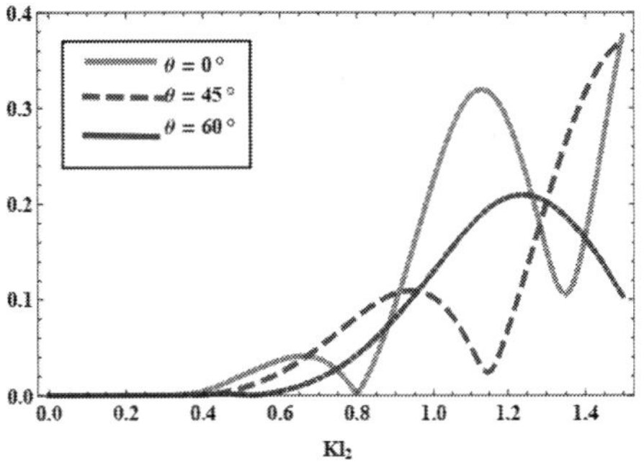

Figure 6. Reflection coefficient $|R|$ against Kl_2 for different values of incident angles θ.

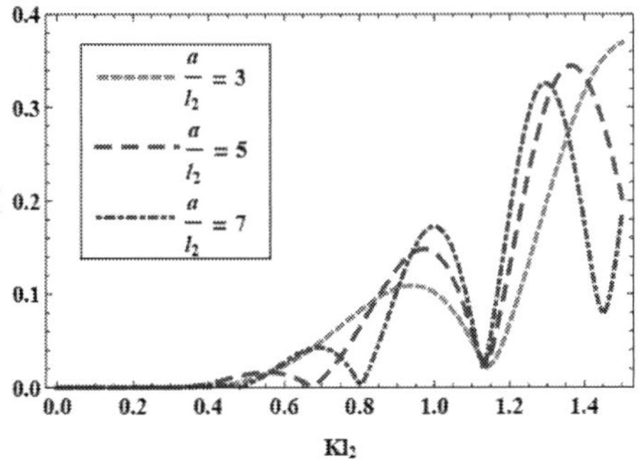

Figure 7. Reflection coefficient $|R|$ against Kl_2 for different values of length of separation.

The reflection coefficient $|R|$ is depicted against Kl_2 for different values of incident angle $\theta = 0^o, 45^o, 60^o$ and fixed values of $\frac{a}{l_2} = 3$ and $\frac{l_1}{l_2} = \frac{l_3}{l_2} = 0.05$, $\frac{l_4}{l_2} = 1, M = 0.01, G_1 = G_2 = 1, F_r = 0.4$ in Fig. 6. It is observed that the peak value of the reflection coefficient decreases with an increase in the angle of incidence. This is due to the interaction between the incident and reflected waves within the barriers. The maximum and minimum values of $|R|$ are shifted from left to right as the angle of incidence increases. The reflection coefficients are found to be oscillatory in nature, and higher oscillation occurs for smaller values of θ.

Figure 7 show the variation of the reflection coefficient $|R|$ versus Kl_2 for different values of plate separation length $\frac{a}{l_2} = 3, 5, 7$, and fixed values of $\frac{l_1}{l_2} = \frac{l_3}{l_2} = 0.05$, $\frac{l_4}{l_2} = 1, M = 0.01, G_1 = G_2 = 1, F_r = 0.4, \theta = 45^o$. This figure clearly demonstrates that the values of $|R|$ shift to the right as the separation length decreases. It is also observed that higher oscillation occurs for larger values of plate separation length. This is because incident and reflected waves interfere constructively and/or destructively within the flexible porous plates, leading to oscillations in the reflection coefficient.

5. Conclusion

This study focuses on investigating the diffraction of oblique water waves by two unequal thin parallel porous elastic plates in an infinitely deep and infinite channel, using linear water wave theory. The study proposes an approximate method based on a one-term Galerkin approximation to compute the reflection and transmission coefficients. The obtained results are then presented graphically, taking into account the porous effect parameter and flexural rigidity. The study finds that the derived result for the reflection coefficient coincides graphically with the results already present in the literature. This agreement provides validation for the proposed method used in the study. The computational results reveal that, for specific values of the wave number, the peak value of the reflection coefficient decreases as the porosity of the plates increases. Additionally, an oscillatory nature is observed in the curve of the reflection coefficient. These findings indicate the significance of porosity in wave energy dissipation and surface fluctuation reduction. Furthermore, the study demonstrates that the flexural rigidity of the plates has a noticeable impact on the hydrodynamic coefficients. Higher values of flexural rigidity correspond to higher reflection coefficients. Overall, this study offers valuable insights into the hydrodynamics of oblique water wave diffraction by two unequal thin parallel porous elastic plates. The proposed approximate method provides a practical approach for calculating the reflection and transmission coefficients. The study also highlights the potential for further research in this area, particularly regarding water wave scattering by multiple thin vertical porous plates.

6. Acknowledgments
This work is partially supported by a SERB, DST [grant number TAR/2022/000107].

7. References
[1] Dean WR 1945 *Math. Proc. Camb. Phil. Soc.* **41** 231-238
[2] Ursell F 1947 *Math. Proc. Camb. Phil. Soc.* **43** 374-382
[3] Evans DV 1970 *J. Fluid Mech.* **40** 433-451
[4] Porter D 1972 *Math. Proc. Camb. Phil. Soc.* **71** 411-422
[5] Banerjea S, Kanoria M, Dolai DP and Mandal BN 1996 *Appl. Ocean Res.* **18** 319-327
[6] Evans DV and Morris CAN 1972 *J. Inst. Math. Applications* **10(1)** 1-9
[7] Porter R and Evans DV 1995 *J. Fluid Mech.* **294** 155-180
[8] Roy R, Basu U and Mandal BN 2016 *Arch. Appl. Mech.* **86** 1681-1692
[9] Chwang AT and Chan AT 1998 *Ann. Rev. Fluid Mech.* **30** 53-84
[10] Sollitt CK and Cross RH 1972 *Coast. Eng. Proc.* **1** 1827-1846
[11] Macaskill C 1979 *J. Fluid Mech.* **75** 141-157
[12] Chwang AT 1983 *J. Fluid Mech.* **132** 395-406
[13] Yu X 1995 *J. Waterw Port Coast Ocean Eng.* **121** 275-282
[14] Isaacson M, Premasiri S and Yang G 1998 *J. Waterw Port Coast Ocean Eng.* **124** 118-126
[15] Li AJ, Liu Y and Li HJ 2015 *Math. Probl. Eng.* **2015** 1-11
[16] Gayen R and Mondal A 2015 *Geophysical and Astrophysical Fluid Dynamics* **109(5)** 480-496
[17] Manam SR and Sivanesan M 2016 *Wave Motion* **67** 89-101
[18] Sasmal A, Paul S and De S 2019 *J. Mar. Sci. Appl.* **18** 417-432
[19] Sasmal A and De S 2019 *Meccanica* **54 (10)** 1525-1544
[20] Sasmal A and De S 2021 *Waves in Random and Complex Media* **DOI: 10.1080/17455030.2021.1915514**
[21] Meylan MH 1995 *International Journal of Offshore and Polar Engineering* **5(2)** 422-427
[22] Meylan MH and Peter MA 2009 *The Quarterly Journal of Mechanics and Applied Mathematics* **62(3)** 321-344
[23] Williams TD and Meylan MH 2012 *Journal of Engineering Mathematics* **75** 81-106
[24] Zheng S, Meylan MH, Fan L, Greaves D and Iglesias G 2020 *Journal of Fluids and Structures* **92** 102827
[25] Smith MJA, Peter MA, Abrahams ID and Meylan MH 2020 *Proceedings of the Royal Society A* **476** 2242
[26] Ashok R, Gunasundari C and Manam SR 2020 *Journal of Fluids and Structures* **99** 103149

IOP Conf. Series: Materials Science and Engineering 1288 (2023) 012011 doi:10.1088/1757-899X/1288/1/012011

Nonlinear computation for focused wave groups on abrupt depth transitions

Q Wu[1,2], Xingya Feng[2,*] and Y Dong[1]

[1]Department of Civil and Environmental Engineering, The Hong Kong Polytechnic University, Hong Kong 999077, China

[2]Department of Ocean Science and Engineering, Southern University of Science and Technology, Shenzhen, 518055, China

E-mail: fengxy@sustech.edu.cn

Abstract. Focused wave groups in intermediate water depth propagating over varying abrupt depth transitions are studied numerically and experimentally. The experiments well validate the fully nonlinear model by a conformal mapping method. The generation of second-order and third-order harmonics is observed near the top of the submerged step, as well as the decrease of the main (first-order) components propagating along with the step. In addition, an enhancement of kurtosis occurs just after the step with a high-trailed profile, revealing a considerable asymmetry of the surface elevations. In addition, a wave group passing two or three steps is also investigated numerically. More submerged obstacles effectively strengthen the reflection capability in terms of the reflection coefficients computed with the amplitude of wave envelopes. Although the harmonics manifest a higher risk of extreme waves on abrupt depth transitions, the increase of water depth changes might not further enhance the maximum skewness and kurtosis over the varying bottom.

1. Introduction

Extreme waves have attracted the attention of researchers in recent decades due to the potential for serious accidents that can cause significant damage to coastal structures and pose a high risk to human safety [1]. To investigate the nonlinear evolution of extreme waves, focused wave groups are commonly used as a representation of extreme waves in studies [2, 3]. These focused wave groups consist of a range of sinusoidal waves with different wave speeds and are designed to reach a peak at a specific position and time. This characteristic effectively describes the sudden occurrence of extreme waves in the ocean. While the formation mechanism of extreme waves has not yet been confirmed, an abrupt depth transition (ADT) has been identified as a potential mechanism that increases the probability of extreme wave occurrence [4].

The parameters of skewness and kurtosis, which describe the statistical distribution of surface elevations, are widely used to determine the occurrence of extreme waves. These parameters refer to the asymmetry of wave profiles and are related to the third and fourth powers of surface elevations, respectively. In 2020, Trulsen et al. [4] discovered a local increase in statistical distribution at ADTs, indicating a higher probability of extreme wave occurrence. The propagation of wave groups over varying bottom topography was then studied both theoretically

Table 1. Case parameters for experiments.

$No.$	f_p	a	kh	k_sh_s	ka
1*	0.94	0.0388	1.442	0.739	0.155
2	–	0.0475	—	—	0.190
3	–	0.0540	—	—	0.216
4*	1.00	0.0468	1.578	0.791	0.205
5*	1.41	0.0292	2.892	1.229	0.235

and experimentally by Li et al. [5]. However, the effect of ADTs on the nonlinearity of extreme waves has not been extensively studied in previous research. Understanding the enhancement of ADTs on focused wave groups is important for forecasting and protecting against extreme waves. The conformal mapping method is an effective method for solving a potential model. It transforms the Euler equations in the physical domain into a simplified mapped domain governed by two parallel lines. This method greatly reduces computational effort. Besides, any complex physical bathymetry can be directly applied in the model, which overcomes limitations in different water depth regions and achieves a fully nonlinear numerical simulation [6].

This study aims to investigate the nonlinear evolution of focused wave groups over ADTs using fully nonlinear numerical simulations. Good agreements between the experimental and numerical methods are found. Section 2 describes the experimental setup and the numerical model. Section 3 presents the validation of the numerical model and discusses the elevation superharmonics induced by ADTs. Finally, conclusions are drawn in section 4.

1.1. *Experimental setup*

Experiments were conducted in the Hydraulic laboratory at the Hong Kong Polytechnic University with a length of 27 meters, a width of 0.75 meters and a depth of 1.5 meter. It is equipped with a piston-type wave generator and a wave absorber to minimize wave reflection. Seven captive-type wave gauges are positioned along the wave tank, with four gauges primarily located near the first ADT, as shown in figure 1. A submerged step constructed from a stainless steel frame and acrylic plates and measuring 2.4 meters, is placed 6 meters in front of the generator. All wave conditions refer to intermediate water regions where the deeper water depth h is 0.36 meters and the shallower depth h_s is 0.13 meters. It should be noticed that the amplitude a refers to the envelope amplitude of linear harmonics. Wave numbers are different in the deeper region k and shallower region k_s. A total of five wave conditions were measured in the laboratory tank for both non-step and step cases, with case parameters shown in table 1 where * represents the cases that are repeated three times.

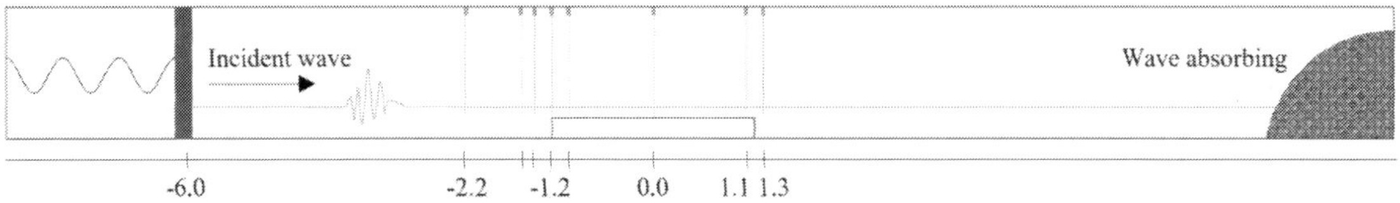

Figure 1. Sketch of the numerical tank.

2. Method

2.1. Numerical model

A fully nonlinear numerical model has been developed using the conformal mapping method to solve the exact Euler equation. The free surface waves and bathymetry in the physical plane are transformed into two parallel lines in the mathematical plane, as shown in figure 2. An analytic equation for this transformation was proposed by Camassa in 1999 [6], resulting in a simplified Euler equation in the mathematically mapped plane. The governing equations in the physical plane are written as follows

$$\nabla^2\phi = 0 \quad b(x) \leq y \leq \eta(x), \tag{1}$$

$$\eta_t + \phi_x\eta_x - \phi_y = 0 \quad y = \eta(x), \tag{2}$$

$$\phi_t + \frac{1}{2}\left({\phi_x}^2 + {\phi_y}^2\right) + g\eta = 0 \quad y = \eta(x), \tag{3}$$

$$\phi_y = 0 \quad y = b(x), \tag{4}$$

where ∇ represents the Laplace operator, $b(x)$ describes the seabed and $\eta(x)$ represents the free surface elevations. Equations 2-3 refer to the free surface kinematic and dynamic boundary conditions, respectively. The symbol g is the gravitational acceleration. The transformed Euler equation in the mathematical plane is then solved using the pseudo-spectral method. The results can then be transformed back into the physical plane. To validate the numerical model, comparisons were made between the measured and the numerical results for both non-step and step cases.

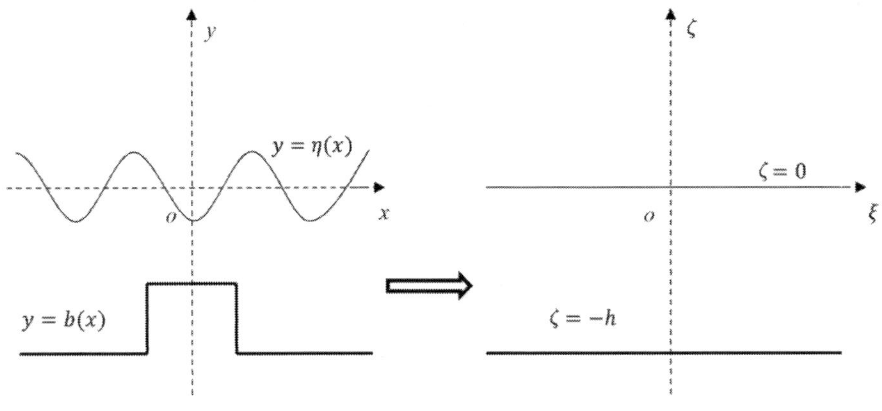

Figure 2. A transformation of boundary conditions between the physical and mathematical planes.

In the fully nonlinear numerical model, the number of Fourier discrete points is chosen as 60 per wavelength. The time series in a single run is about 40 wave periods with a time step of 0.02 seconds. This appears to be sufficient for satisfactory convergence of all statistical quantities considered in this study. Two parameters describing statistical distributions are introduced, skewness and kurtosis, which are written as

$$skewness = \frac{\frac{1}{n}\sum_{i=1}^{n}(\eta-\overline{\eta})^3}{(\sqrt{\frac{1}{n}\sum_{i=1}^{n}(\eta-\overline{\eta})^2})^3} \quad kurtosis = \frac{\frac{1}{n}\sum_{i=1}^{n}(\eta-\overline{\eta})^4}{(\sqrt{\frac{1}{n}\sum_{i=1}^{n}(\eta-\overline{\eta})^2})^4}, \tag{5}$$

where η denotes the average value of surface elevations. It is known that the skewness and kurtosis are fixed for a Gaussian random wave, as 0 and 3, respectively.

3. Results and discussion

To validate the fully nonlinear model, surface elevations along the tank were compared with the measured data, as shown in figure 3. Good agreement was found between the measured and the numerics for both non-step (figure 3a) and step conditions (figure 3b). The free surface elevations were normalized by the envelope amplitude of the first-order component. Surface elevations near the focused positions (WG 5) for both non-step and step conditions are presented in figure 4. The NewWave theory is widely accepted as the linear result of focused wave groups with a superposition of individual monochromatic components. In figure 4a, the differences between experimental, numerical, and theoretical results can be considered negligible in the non-step condition. However, the numerical results are much closer to the experimental data than the linear results in the step condition, particularly in terms of wave crest and trough (figure 4b). Additionally, the focused wave elevations have sharper crests than those predicted by the NewWave theory. This indicates that the numerical model using the conformal mapping method can accurately describe the nonlinear propagation of focused wave groups over a varying bottom.

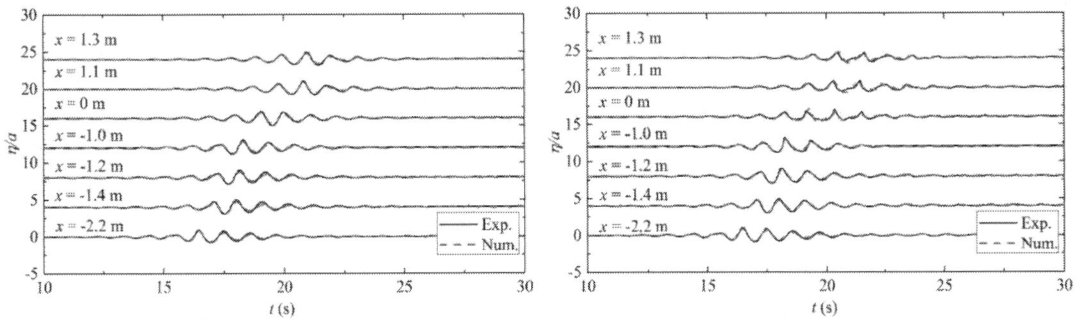

Figure 3. Surface elevations at different seven positions. (a) non-step condition and (b) step condition.

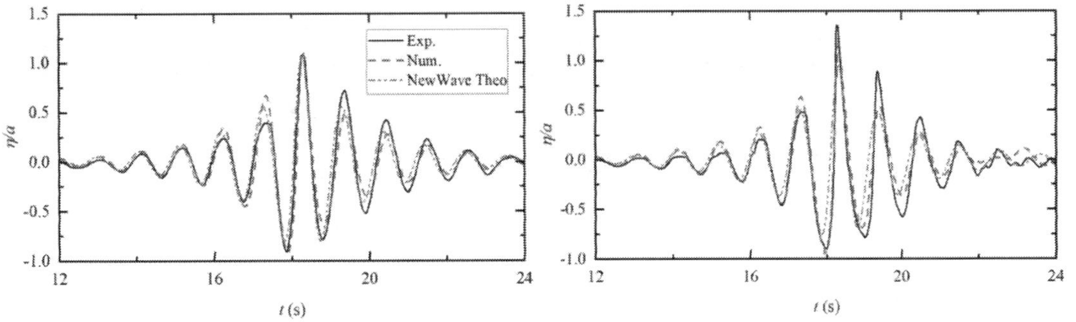

Figure 4. Focused wave profiles at the $x = -1$ m. (a) non-step condition and (b) step condition.

To examine the presence of superharmonics, both linear and superharmonic components were analyzed at four positions, as depicted in Figure 5. Superharmonics were separated using Fast Fourier Transform (FFT) on the surface elevations. A stronger nonlinearity of wave profiles

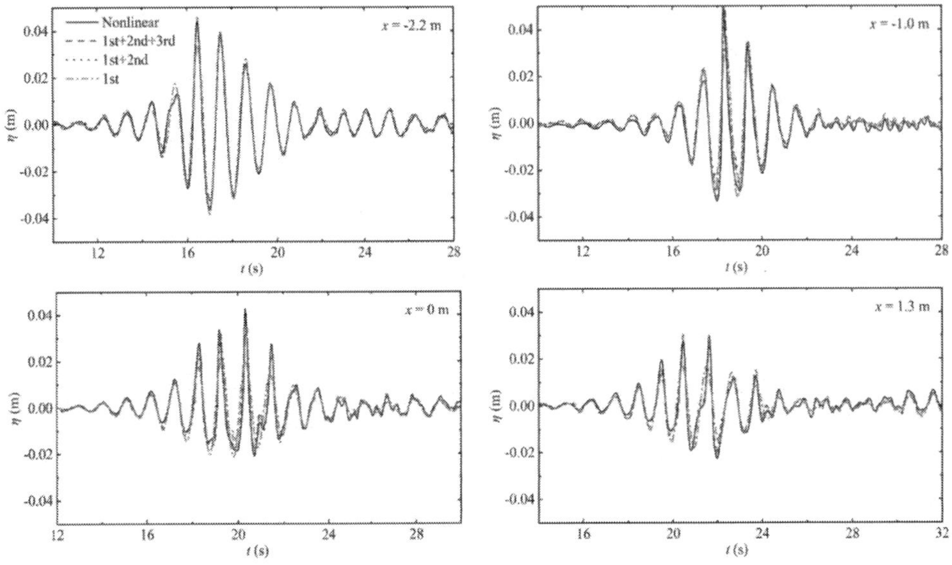

Figure 5. Comparison of nonlinear wave profiles and the sum of first three superharmonics at four positions.

was observed in shallower regions ($x = -1.0, 0, 1.3$ m) with flatter wave troughs and sharper crests. Notably, the superposition of the first two harmonics is not sufficient to describe the nonlinear surface profiles, but the inclusion of the first three components resulted in a closer approximation to the nonlinear result. This indicates that the third harmonics are also significant in the nonlinear evolution. In addition, the statistical distribution of surface elevations in both non-step and step conditions are displayed in figure 6. In the non-step case, both parameters reach their maximum values at $x = -1.0$ m, which is the prescribed focused position. This was particularly evident for kurtosis, which is more representative of extreme waves. In contrast, in the step case, the locations of greatest asymmetry were observed near $x = -0.43$ m. Even in the linear case ($a = 0.001$ m) with wave propagation over ADTs, the position of superharmonic occurrence shifted backwards.

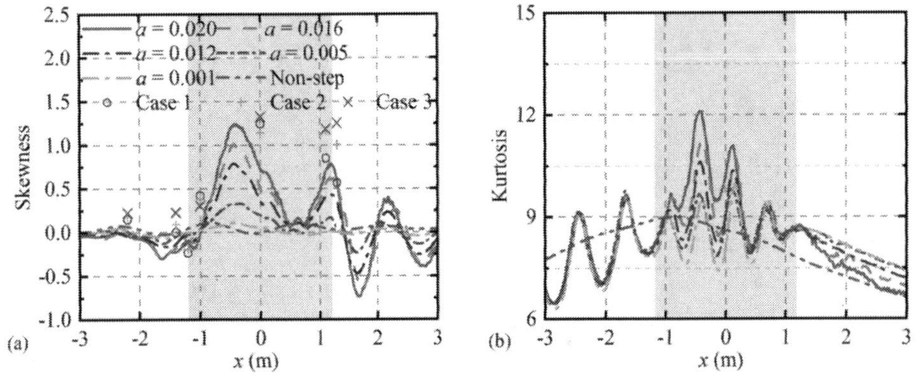

Figure 6. Statistical distribution along the tank. (a) skewness and (a) kurtosis.

The effect of incident amplitude ka is presented in Figure 7. Both skewness and kurtosis rise with the increased amplitude. This suggests that the presence of ADTs induces higher nonlinearity of focused waves in shallower regions and results in stronger superharmonics ahead of the focused position. Additionally, an increase in both skewness and kurtosis was observed at the second ADT, although the latter was smaller in magnitude (in figure 6). This suggests that the presence of the second Abrupt Depth Transition (ADT) results in a greater enhancement of asymmetry along the horizontal axis compared to the vertical axis. In other words, the sudden drop in water depth at the second ADT induces a higher degree of skewness but only marginally increases the likelihood of extreme wave occurrence.

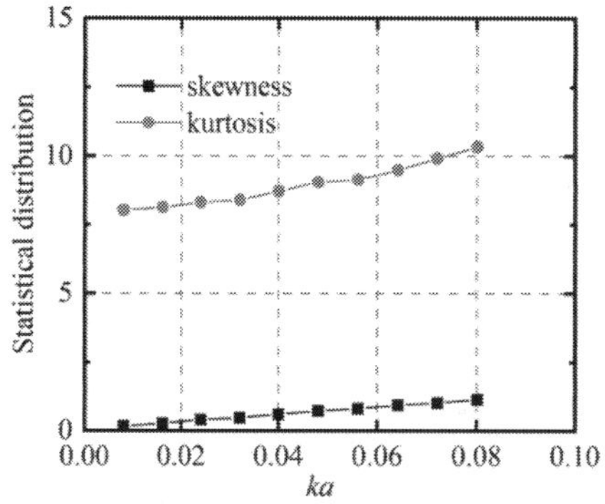

Figure 7. Statistical distribution with the increasing incident amplitude ka at $x = -0.43$ m.

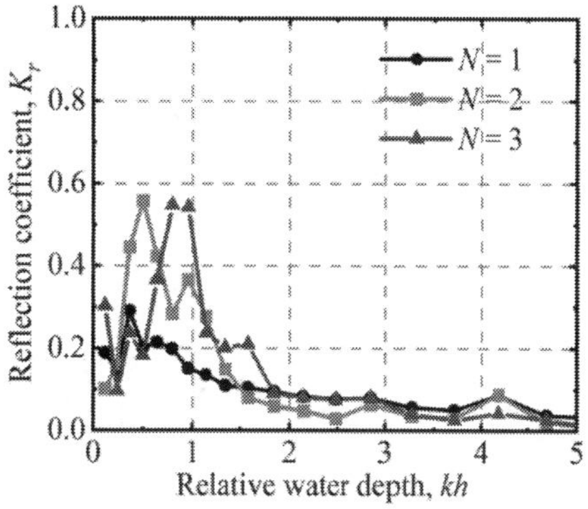

Figure 8. Reflection coefficients of focused wave groups with one, two and three submerged steps.

In addition to the one-step case, cases with two and three steps were also tested for wave reflection, as shown in figure 8. The reflection coefficient is defined as the ratio of the maximum amplitude of the reflected wave to that of the focused wave group. Reflected waves were measured at a position 3.5 times the peak wavelength before the first ADT in order to capture steady reflected waves. It was observed that the maximum reflection coefficients occurred at different relative water depths kh for each step case, which is inconsistent with Bragg resonance predictions for regular waves [7]. The maximum reflection coefficients are closed for both two-step and three-step conditions. As the number of ADTs increased, the relative water depth kh at which the peaks of the reflection coefficients occurred also increased. This suggests that a greater number of ADTs may enhance the effects on wave components with higher wave numbers k. Meanwhile, more ADTs effectively weaken the propagation of extreme waves. Furthermore, an increasing number of ADTs effectively weakened the propagation of extreme waves. As there is no accepted computation of reflection coefficients for focused wave groups, further investigation of individual wave components may be significant.

4. Conclusions

A fully nonlinear model was developed using the conformal mapping method to study the effect of ADT on focused wave groups. Experimental data validated the numerical simulation for both non-step and step conditions.The presence of a submerged step induces higher superharmonics, with third harmonics also playing a significant role in shaping surface wave profiles in shallower regions. Furthermore, ADTs were observed to result in peak superharmonics occurring ahead of the prescribed focused position. Increased skewness and kurtosis further demonstrated stronger asymmetry occurring at the ADT. Stronger reflection occurs with an increasing number of ADTs, with the corresponding relative water depth kh being higher. This investigation into the effects of ADTs on focused wave groups provides insights into potential mechanisms for strong extreme waves in coastal areas and enhances our understanding of topography effects on extreme waves. The wave propagation at the second ADT is complicated due to wave reflection, wave transmission, and wave dispersion. Thus, it is meaningful to investigate wave nonlinearity on the second ADT as well as potential Bragg resonance during wave-bottom interaction.

References

[1] Didenkulova E 2020 *Ocean & Coastal Management* **188** 105076
[2] Baldock T, Swan C and Taylor P 1996 *Phil. Trans. R Soc. Lond.* A **354** 649–676
[3] Westphalen J, Greaves D, Williams C, Hunt-Raby A and Zang J 2012 *Ocean Eng.* **45** 9-21
[4] Trulsen K, Raustøl A, Jorde S and Bæverfjord Rye L 2020 *J. Fluid Mech.* **882**
[5] Li Y, Draycott S, Adcock T A A and Van Den Bremer T S 2021 *J. Fluid Mech.* **915** A72
[6] Choi W and Camassa R 1999 *J. Engng Mech.* **125** 756
[7] Cho Y S, Lee C 2000 *Journal of Coastal Research* **16** 870-876

IOP Conf. Series: Materials Science and Engineering 1288 (2023) 012012 doi:10.1088/1757-899X/1288/1/012012

Nonlinear wave propagation in a vegetated sandbar-lagoon system

X Cong, C P Kuang*, H Y Li, W Xing, and J L Chen

College of Civil Engineering, Tongji University, Shanghai, China

*cpkuang@tongji.edu.cn; congxin@tongji.edu.cn

Abstract. The sandbar-lagoon system presents a rich variety of values. At the same time, the degradation of ecosystems and coastal erosion caused by human activities and climate change are even more severe. A series of ecological conservation and restoration projects focused on vegetation rehabilitation have been implemented. However, there is insufficient scientific evidence to guide coastal engineering in the presence of vegetation. In this paper, ten tests were presented to experimentally investigate nonlinear wave propagation in a vegetated sandbar-lagoon system. The following conclusions can be obtained. In the absence of vegetation, the low-frequency wave energy proportion gradually increases with the rise of incident wave height. Double-peaked and even multi-peaked spectra are notable throughout the propagation in tests with larger incident wave heights. Wave height decay and wave energy dissipation due to vegetation effects increase as the incident wave height increases. The dependence of the wave energy dissipation induced by vegetation on the incident wave energy (the square of wave height) presents a linear relationship. Vegetation plays an important and complex role in both sea-swell wave and infragravity wave attenuation.

1. Introduction

Ecosystem degradation due to human disturbance and climate change has become a worldwide problem. Nature-based solutions were once known as a popular and accepted approach to coastal restoration. Vegetation not only provides excellent natural conditions for flora and fauna to inhabit and multiply, but also protects coastal systems from erosion and storms. The lagoon coast is one of the most productive and valuable ecosystems on earth, while global climate change has a significant impact on such a coastal zone [1]. Waves, especially storm waves, are one of the main motivating factors for coastal development and have been considered in most coastal projects. Wave transformations induced by submarine geomorphology and coastal structures lead to changes in the wave frequency distribution. Clearly, nonlinear wave propagation is a significant point in sandbar-lagoon systems, especially in the presence of vegetation.

Vegetation has significant wave attenuation capabilities. In the past decades, considerable laboratory studies have been concentrated on the wave attenuation by rigid and flexible vegetation [2]. Ma et al. [3-5] studied experimentally the effect of artificial reef on wave spectra induced by regular and irregular waves. This facilitates the identification of detailed wave decay mechanisms. Spectral analysis can extract the wave components at different frequencies, namely the dominant and generated higher or lower harmonics. Low-frequency wave (long wave) separated by spectral analysis is one of identified important drivers for port operation efficiency [6], wave run-up or overtopping [7], and sediment transport [8, 9]. Infragravity (hereafter IG) waves are surface ocean waves with frequencies below those of wind-generated short waves. They are generated by two widely acknowledged mechanisms: bound wave and moving breakpoint [10]. Actually, they are the mutual transformation or joint action of multiple mechanisms in many cases [11] instead of a single mechanism. Moreover, Bertin et al. [12] presented another generation mechanism: bore merging. Low-frequency waves are influenced by many factors, which not simply depend on local submergence, but also the history of wave spatial evolution, especially the generation region of maximum storm [13]. The energy conversion analysis showed that the sandbar plays a key role in modulating the IG wave [14]. Saprykina [15] found that the IG wave height depends on the gravity wave height within a large range and has a quadratic correlation on average, and the dependence is stronger for the breaking wave and the wave after breaking.

Based on the wave data measured by physical experiment in a sandbar-lagoon system, previous studies have concentrated on wave attenuation and beach profile changes with emergent vegetation of bamboo stick [16], real aquatic vegetation of *dracaena sanderiana* [17], and submerged vegetation made of optical fibre [18, 19], this study focuses on nonlinear wave propagation in terms of wave energy and different wave frequency bands. The importance of this research lies in the study of erosion, sediment transport, and regional wave modelling with high resolution, where the mechanism of nonlinear wave propagation affected by vegetation is required.

2. Method

2.1. Experimental set-up

The experiment was conducted in the flume of 50 m×0.8 m×1.2 m in the Laboratory of Hydraulic and Harbor Engineering, Tongji University. The specific experimental setup is illustrated in Figure 1. A resin sand with a median grain size of 0.15 to 0.18 mm and a density of 1.40 to 1.45 g/cm^3 was selected to construct the sandbar-lagoon-dune system. The both sandbar and dune have a slope of 1:2 and the coordinate system is marked in Figure 1. Seven capacitive wave gauges and four acoustic wave gauges were positioned along the flume. Ten different experimental tests are given in Table 1. The length scale was confirmed as 1:10 according to the Shields, Rouse and Froude similitude scaling [5, 20, 21]. The still water depth of 0.55 m and five incident wave heights were considered. JONSWAP spectrum was used to generate the incident irregular wave. The theoretical and obtained wave spectra in the flume without sandbar-lagoon system are shown in Figure 2. It shows that the obtained spectral peak at the highest incident wave height of 0.21 m is lower than the theoretical value and its low frequency band increases, presumably due to confinement of the wavemaker and increased reflection signal induced by wave absorption. A bamboo stick with a diameter of 4 mm was used to mimic vegetation. They were planted in the sea-side sandbar crest with a cover width of 0.5 m. The vegetation density was set to 2500 units/m^2 and each plant was placed in a rectangular arrangement with a spacing of 2 cm.

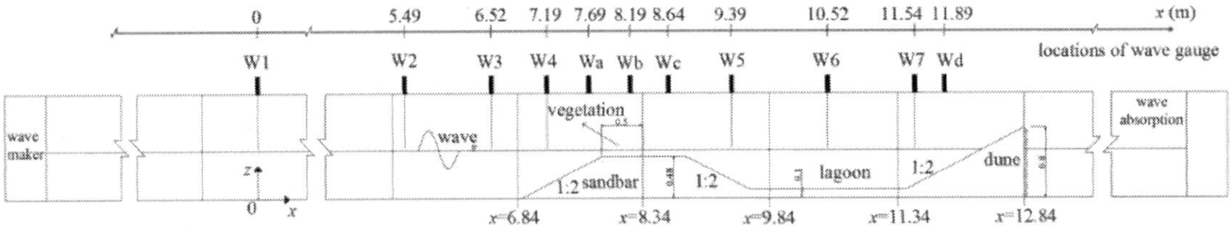

Figure 1. Experimental setup.

2.2. Data analysis

Spectral analysis was used to observe the energy dissipation process of wave propagation. The sea-swell and IG wave heights can be obtained by separating the different frequency bands. The following formulas were adopted to separate the waves.

$$H_{ig} = 4\left(\int_{f_1}^{f_2} S_f df\right)^{1/2} \tag{1}$$

$$H_{ss} = 4\left(\int_{f_2}^{\infty} S_f df\right)^{1/2} \tag{2}$$

where, f_1 and f_2 are the cutoff frequencies; H_{ig} is the IG wave height ($f_1 < f < f_2$); H_{ss} is the sea-swell wave height ($f > f_2$). The cutoff frequency of IG wave is slightly different in different studies [7, 10]. The high cutoff frequency is mostly 0.04 or 0.05 Hz, and the low cutoff frequency is mostly 0.004 or 0.005 Hz. In this study, the IG wave ($f < 0.16$ Hz) and the sea-swell wave ($f \geqslant 0.16$ Hz) were confirmed based on the physical model scale (1:10) and proto cutoff frequency 0.05 Hz. The wave energy is defined as the area between the wave spectrum and the horizontal axis.

Table 1. Experimental tests.

Test	Water depth d (m)	Incident sea-swell wave height H (m)	Wave period (s)	Vegetation (Yes or No)
T1	0.55	0.06	1.33	No
T2	0.55	0.06	1.33	Yes
T3	0.55	0.10	1.61	No
T4	0.55	0.10	1.61	Yes
T5	0.55	0.13	1.68	No
T6	0.55	0.13	1.68	Yes
T7	0.55	0.16	1.80	No
T8	0.55	0.16	1.80	Yes
T9	0.55	0.21	2.12	No
T10	0.55	0.21	2.12	Yes

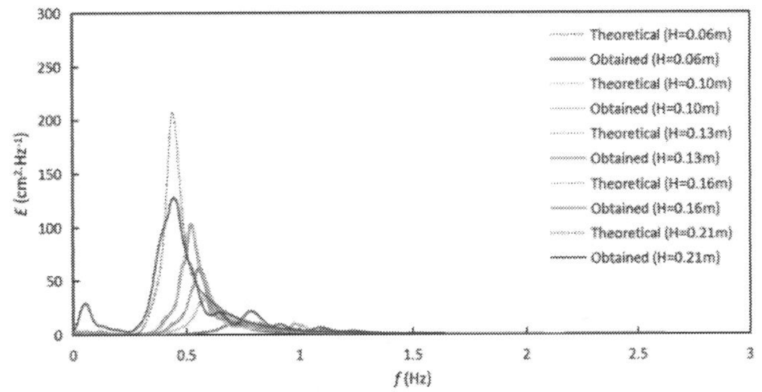

Figure 2. Theoretical and obtained wave spectra in a flume without sandbar-lagoon system, where H is the incident wave height.

3. Results and discussions

3.1. Wave spectra

Wave energy spectra were obtained to observe frequency band variability. Figure 3a, 4a, 5a, 6a and 7a show the energy spectra characteristics of different incident wave heights along the sandbar-lagoon system without vegetation. In the absence of vegetation, the low-frequency wave energy proportion gradually increases as the incident wave height increases. A wave with two-order harmonics is generated after the vegetation region when the incident wave height is 0.06 m (Figure 3a), which is similar with the study by Gong et al. [22] that the two-order harmonics occur within the vegetation field under regular wave condition. Moreover, Double or even multiple peaks occur remarkably throughout the whole propagation process especially for the tests with incident wave height of 0.13m, 0.16m, and 0.21 m in Figure 5a, 6a and 7a. It follows that low-frequency wave plays a more important role due to increasing proportion in the tests of higher incident energy. Moreover, the sandbar significantly reduces the wave energy in all tests. Figure 3b, 4b, 5b, 6b and 7b present wave spectra influenced by vegetation. The vegetation induces the frequency dispersion (e.g. Figure 3a and b, Wa), the reducing peak frequency energy (e.g. Figure 4a and b, Wa), and even the increasing peak frequency energy (Figure 7a and b, Wa, this reason can be found in subsection 3.2). More low-frequency wave energy is reduced obviously by vegetation in the test with the incident wave height of 0.21 m (Figure 7b). In this way, the effects of vegetation are diversified and are combined with the nonlinear wave propagation in the mobile sandbar-lagoon system. In general, vegetation has a prominent weakening effect on wave energy in the area after the cover zone of vegetation (Wc~Wd and W5~W7).

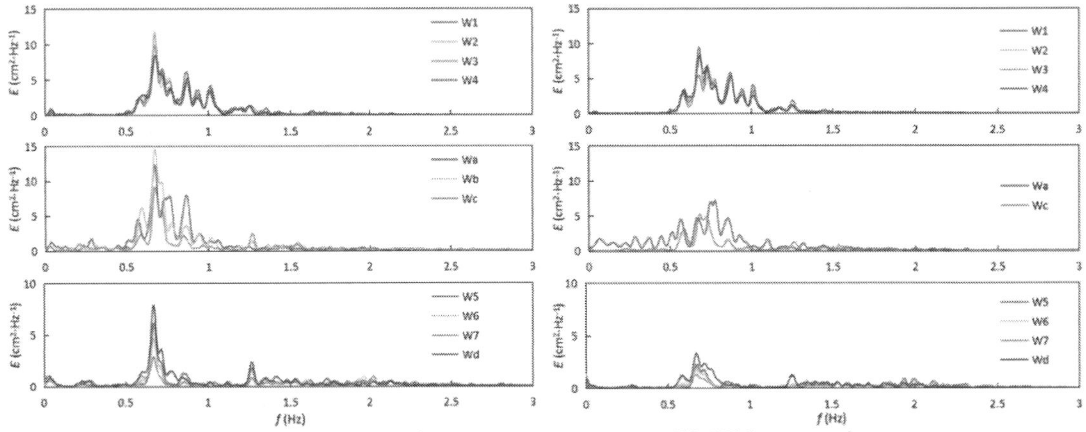

(a) Without vegetation (b) With vegetation

Figure 3. Comparison of wave spectra with the incident wave height H of 0.06 m.

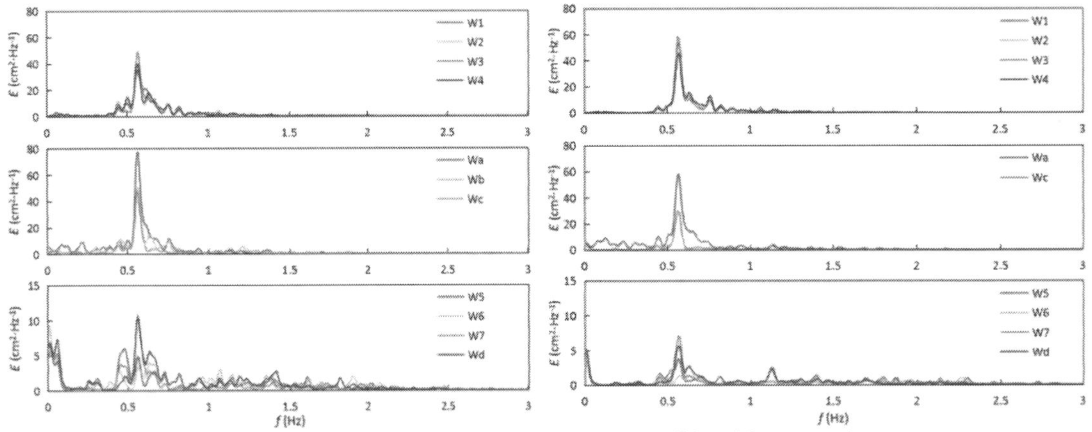

(a) Without vegetation (b) With vegetation

Figure 4. Comparison of wave spectra with the incident wave height H of 0.10 m.

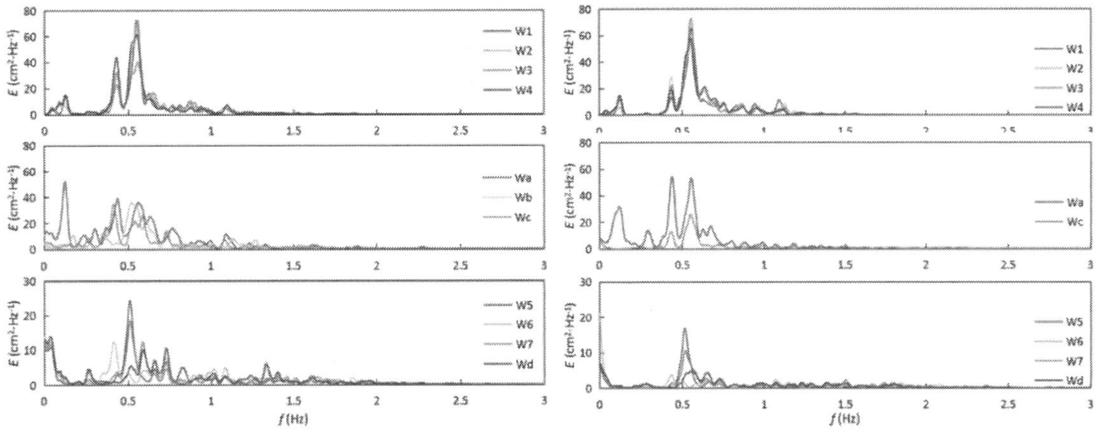

(a) Without vegetation (b) With vegetation

Figure 5. Comparison of wave spectra with the incident wave height H of 0.13 m.

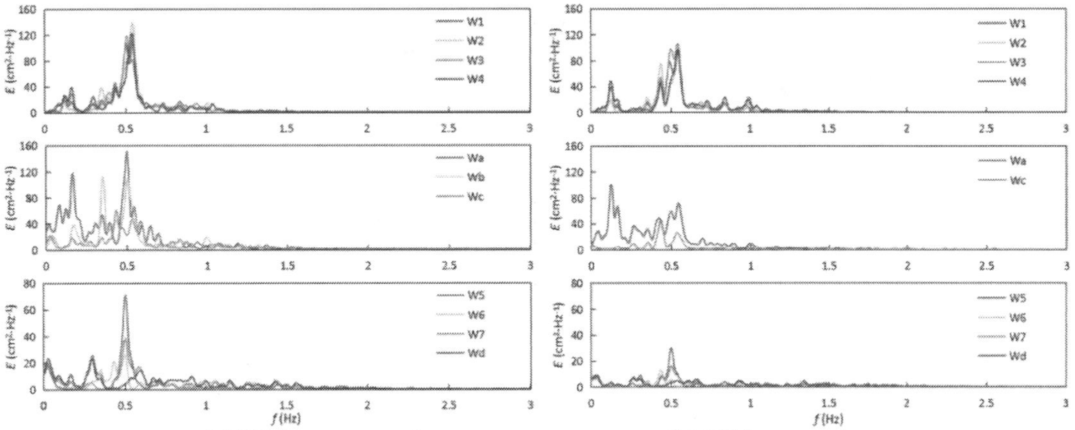

(a) Without vegetation (b) With vegetation

Figure 6. Comparison of wave spectra with the incident wave height H of 0.16 m.

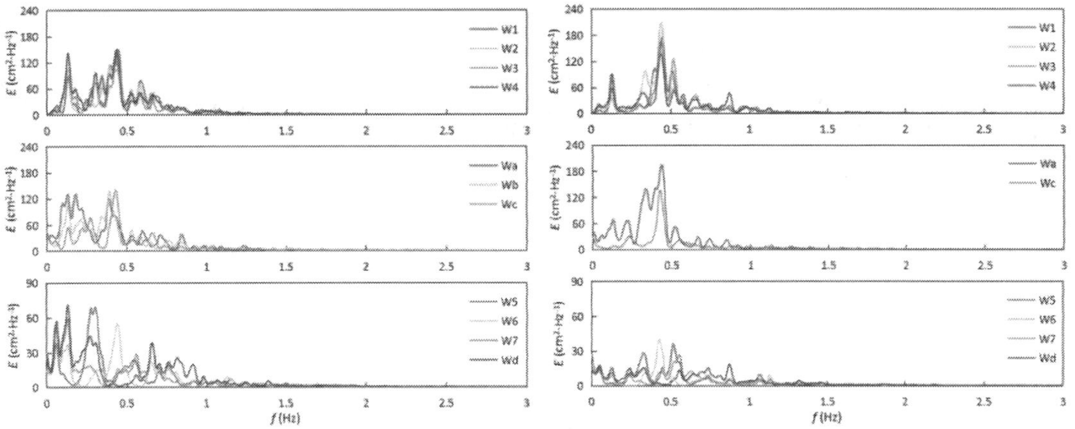

(a) Without vegetation (b) With vegetation

Figure 7. Comparison of wave spectra with the incident wave height H of 0.21 m.

3.2. Wave energy evolution

To quantify the evolution of the wave energy, Figure 8 depicts the process of wave energy adjustment along the flume. Overall, the curve first drops slightly and then rises in the slope of the seaward sandbar, but the wave energy is significantly reduced by the vegetation behind the vegetation covered area. In the foredune, the curve rises slightly again. From an overall perspective, the wave energy dissipation increases in magnitude with the rise of incident wave height and vegetation has a negligible effect on the position of wave breaking point. However, when the incident wave height is 0.21 m, vegetation reduces wave energy in the margin of surf zone by 11.1 cm^2 (W4) but have no obvious effect on the position of wave breaking point (Wa). By contrast, the energy at the position of wave breaking point decreases by 9.6 cm^2 (Wa) but it slightly decreases in the margin of surf zone (W4) with the incident wave height of 0.16 m. By observing the video recording, this is most likely related with an excess of wave energy at the fore-dune in the vegetation-free test with the largest incident wave height, leading to an additional loss of wave energy through over-topping. And this will decrease the reflection energy from the fore dune.

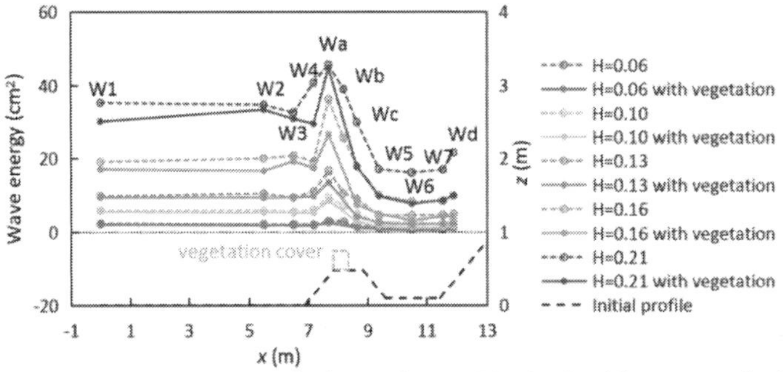

Figure 8. Wave energy evolution, where *H* is the incident wave height.

The mean wave energy dissipation E_d among 5 gauges (Wc, W5, W6, W7, and Wd) is calculated by Equation 3 below. It increases with the rise of incident wave height and the largest value E_d of 9.5 cm^2 is observed with the incident wave height of 0.21 m. Thus, vegetation has a negligible effect on the position of wave breaking point and plays a significant role in wave damping in the break region and behind it.

$$E_{d} = n^{-1}\sum_{i=1}^{n}(E_{ni} - E_{pi}) \tag{3}$$

where, E_d is the mean wave energy dissipated after vegetation cover; E_{ni} and E_{pi} are the *i*th wave energy without and with vegetation, respectively; *n* is the number of wave gauges used.

Localized energy dissipation by vegetation creates a shadow region of low wave energy, which may have important implications for neighboring coastal area [23]. Based on the rate of energy change for dislocating sediment particles proposed by Türker and Kabdasli [24], a linear correlation was depicted [25] between the wave energy dissipated by vegetation cover and the energy to erode coastal dune. In this paper, the relationship relating the mean wave energy dissipation to the incident wave height and the incident wave energy (the square of incident wave height) is depicted in Figure 9. According to the linear fitting in Figure 9a, the determination coefficient R^2 is 0.96. However, the intercept of this line is not zero, which is not consistent with objective reality. The determination coefficient R^2 reaches 0.99 with power fitting and the exponent is up to 2.71 (Figure 9a). Moreover, the exponential curve pass through the origin in corresponding with the actual environment. The wave energy dissipation induced by vegetation appears to increase exponentially as the incident wave height increases. However, biological features such as its uprooting, broken, and seasonal changes are also important factors leading to a decrease in the wave energy dissipation. The results presented in Figure 9b show that the energy dissipation is proportional to the square of the incident wave height, with a larger R^2 of 0.97 compared to the linear fitting in Figure 9a. The reason is that in Figure 9a the energy dissipation is plotted against the wave height, whereas in Figure 9b, it is plotted against the square of the wave height. That is, Figure 9a violates a physical law of the dimensional unity. In addition, the trend line (Figure 9b) through the origin is more grounded in reality. These relationships can be used to scientifically guide ecological restoration projects. Certainly, this relationship requires broader data support and more experimental tests considering vegetation density and water depth, due to their significance in practical coastal restoration projects.

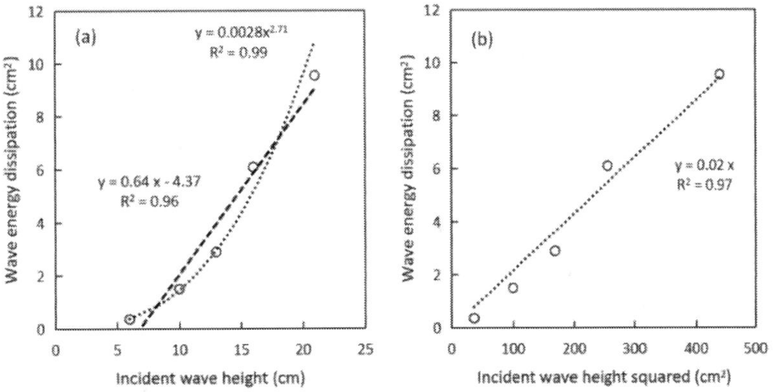

Figure 9. Relationship linking the incident wave height (a) and the incident wave height squared (b) to the mean wave energy dissipation.

3.3. *Wave height evolution*

Low-frequency waves play a key role in various coastal systems, including reef or lagoon systems and coastal engineering. To identify different contribution of vegetation on different wave frequency bands, the sea-swell (high-frequency) and IG (low-frequency) wave heights evolution along the flume are illustrated in Figure 10. The trend of the sea-swell wave height is the same as the wave energy (Figure 8). As indicated by the IG wave height (Figure 10b), there is a continuous rise rather than a decrease of the sea-swell wave height in the front of the sandbar fore-slope (Figure 10a). In general, the wave height decay, including the sea-swell and IG waves, increases with the rise of incident wave height. However, at an incident wave height of 0.16 m, the sea-swell wave height affected by vegetation decreases more than other tests before the wave breaks and the IG wave height decreases only after the wave breaks. For an incident wave height of 0.21 m, the difference is that the IG wave height decays more along the whole flume, and vegetation results in larger sea-swell wave breaking heights. As mentioned in the subsection 3.2, this is most likely due to an additional loss of wave energy through over-topping in the vegetation-free test with the incident wave height of 0.21 m. At the same time, the IG wave height associated with sediment transport and coastal erosion can be significantly reduced by vegetation with an incident wave height of 0.21 m. This indicates that vegetation plays an important and complex role in coastal restoration projects, especially in the presence of strong wave conditions.

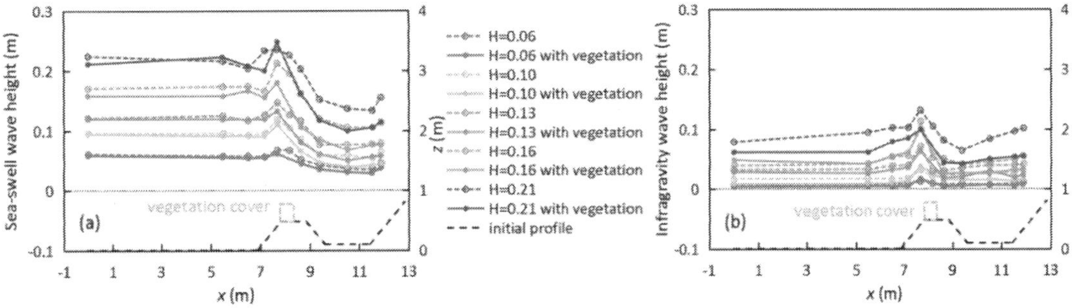

Figure 10. Sea-swell (a) and IG (b) wave height evolution, where *H* is the incident wave height.

4. Conclusions

Ten tests were presented to experimentally study the propagation of nonlinear waves in a vegetated sandbar-lagoon system. The following conclusions can be drawn. In the absence of vegetation, the low-frequency wave energy proportion gradually increases as the incident wave height increases. In general, the wave energy and sea-swell wave height decrease slightly before breaking and then increase with the decreasing water depth around the seaward sandbar, but they are significantly weakened by vegetation behind the covered area. In contrast, the IG wave height increases continuously before the wave breaks. The wave energy dissipation magnitude, as well as the wave height attenuation of the vegetation-influenced sea-swell and IG waves gradually increase with the rise of incident wave height. Double-peaked and even multi-peaked spectra are notable throughout the propagation in tests at high incident wave heights. Vegetation has a negligible effect on the position of wave breaking point, but a considerable effect on the wave energy dissipation in the front of and behind the vegetation cover. The dependence of the wave energy dissipation induced by vegetation on the incident wave energy (the square of the incident wave height) presents a linear relationship and the determination coefficient is up to 0.97. Vegetation plays an important role in coastal restoration projects, especially during strong wave conditions. This makes sense for wave propagation mechanisms that include different frequency bands affected by coastal vegetation.

Acknowledgments

This study was supported by the National Natural Science Foundation of China (No. 41976159) and the National Key Research and Development Program of China (No. 2022YFC3106205).

References

[1] Hanley M E, Bouma T J and Mossman H L 2020 The gathering storm optimizing management of coastal ecosystems in the face of a climate-driven threat *Ann. Bot.* **125** pp 197-212
[2] Kuang C P, Cong X, Fan J D and Li H Y 2023 Review on physical model experiments on ecological protection mechanism of coastal vegetation *J. Tongji Univ. (Natual Sci.)* (*Preprint*)
[3] Ma Y, Kuang C P, Han X J and Dong B L 2018 Wave attenuation mechanism of the artificial reef in Beidaihe, China *The 28th Int. Ocean Polar Eng. Conf.* OnePetro pp 1446-1452
[4] Ma Y, Kuang C P, Han X J and Zhu L 2018 Experimental investigation on regular wave process through an artificial reef *The 13th ISOPE Pac./Asia Offshore Mech. Symp.* OnePetro pp 96-104
[5] Ma Y, Kuang C P, Han X J, Niu H B, Zheng Y H and Shen C 2020 Experimental study on the influence of an artificial reef on cross-shore morphodynamic processes of a wave-dominated beach *Water* **12** p 2497
[6] Guo L, Ma X and Dong G 2021 Performance accuracy of surfbeat in modeling infragravity waves near and inside a harbor *J. Mar. Sci. Eng.* **9** p 918
[7] Lashley C H, Jonkman S N, van der Meer J, Bricker J D and Vuik V 2022 The influence of infragravity waves on the safety of coastal defences: a case study of the Dutch Wadden Sea *Nat. Hazards Earth Syst. Sci.* **22** pp 1-22
[8] Pomeroy A W M, Storlazzi C D, Rosenberger K J, Lowe R J, Hansen J E and Buckley M L 2021 The contribution of currents, sea-swell waves, and infragravity waves to suspended-sediment transport across a coral reef-lagoon system *J. Geophys. Res.: Oceans* **126** p e2020JC017010
[9] Grossmann F, Hurther D, van der Zanden J, Caceres I, Sanchez-Arcilla A and Alsina J M 2022 Near-Bed sediment transport during offshore bar migration in large-scale experiments *J. Geophys. Res.: Oceans* **127** p e2021JC017756
[10] Zhu G, Ren B, Dong P, Wang G Y and Chen W D 2021 Experimental investigation on infragravity wave on different reef systems under irregular wave action *Ocean Eng.* **226** 108851
[11] Liu Y, Yao Y, Liao Z L, Li S W, Zhang C and Zou Q P 2023 Fully nonlinear investigation on energy transfer between long waves and short-wave groups over a reef *Coast. Eng.* **179** p 104240

[12] Bertin X, et al. 2018 Infragravity waves: From driving mechanism to impacts *Earth-Sci. Rev.* **177** pp 774-799

[13] Rijnsdorp D P, Reniers A J H M and Zijlema M 2021 Free infragravity waves in the North Sea *J. Geophys. Res.: Oceans* **126** p e2021JC017368

[14] Matsuba Y, Shimozono T and Tajima Y 2021 Tidal modulation of infragravity wave dynamics on a reflective barred beach *Estuar. Coast. Shelf Sci.* **261** p 107562

[15] Saprykina Y V 2022 *Role of nonlinear processes in infragravity waves evolution* Springer pp 1-10

[16] Kuang C P, Huang G W, Cong X, Liu H X and Zhu L 2021 Experimental study on wave attenuation by a sandbar with emerged rigid vegetation *The 31st Int. Ocean Polar Eng. Conf.* pp 2211-2215

[17] Cong X, Kuang C P, Han X J, Gong L X and Shen C 2021 Experimental observation on wave and profile changes in a sandbar-lagoon system with emergent aquatic plants on the sandbar crest *The 31st Int. Ocean Polar Eng. Conf.* pp 2216-2222

[18] Cong X, Kuang C P, Wu Y L and Xia Z L 2021 Study on erosion and deposition in a sandbar-lagoon system influenced by submerged vegetation under erosion wave conditions *J. Trop. Oceanogr.* **41** pp 31-37

[19] Cong X, Kuang C P, Liu H X, Zhu L and Shen C 2022 Experimental observation on wave and profile changes in a sandbar-lagoon system by submerged vegetation on the sandbar slope *Proc. Coasts Ports 2021 Conf.* pp 1014-1019

[20] Yin S, Pan Y and Chen Y P 2017 Scale design based on local curve fitting method for low-energy sandy beach *Hydro-Sci. Eng.* **4** pp 43-51

[21] Kuang C P, Ma Y, Han X J, Pan S Q and Zhu L 2020 Experimental observation on beach evolution process with presence of artificial submerged sand bar and reef *J. Mar. Sci. Eng.* **8** p 1019

[22] Gong S P, Chen J, Jiang C B, Mao Y Q and He F 2020 Experimental study on wave elimination of vegetation generalized model under regular wave action *Chin. J. Hydrodyn. (A)* **35** pp 213-221

[23] Dalrymple R A, Kirby J T and Hwang, P A 1984 Wave diffraction due to areas of energy dissipation *J. Waterw. Port Coast. Ocean Eng.* **110** pp 67-79

[24] Türker U and Kabdasli M S 2004 Average sediment dislocation analysis for barred profiles *Ocean Eng.* **31** pp 1741-1756

[25] Türker U, Yagci O and Kabdasli M S 2019 Impact of nearshore vegetation on coastal dune erosion: assessment through laboratory experiments *Environ. Earth Sci.* **78** p 584

IOP Conf. Series: Materials Science and Engineering 1288 (2023) 012013 doi:10.1088/1757-899X/1288/1/012013

Absorption of wave reflections in way of an inlet boundary

A Karola, S Tavakoli, T Mikkola, J Matusiak, S Hirdaris

Marine and Arctic Technology Group, Aalto University, Espoo, Finland

E-mail: aaro.karola@aalto.com

Abstract. In recent years, Computational Fluid Dynamics (CFD) is increasingly becoming more attractive for the analysis of ship seakeeping. This is because of the increasing efficiency of computers and the accuracy of solvers (e.g. Reynold Navier Stokes Averaged – RANS) that may be more accurate in terms of idealizing nonlinear phenomena in comparison to potential flow methods. Notwithstanding this, RANS methods remain sensitive to accurate wave modelling across the fluid domain as well as the reflection of domain boundaries. This paper studies the effective absorption of the reflected waves from a structure heading back towards the inlet boundary with a relaxation zone method. This is achieved by applying the RANS CFD solver OpenFOAM-v2206 and wave generation toolbox waves2Foam. The modelled case is a 2D box in waves with different wave lengths. The distance of the relaxation zone and the distance after the relaxation zone until the box varies and their effect to excitation forces is studied. It is demonstrated that longer distances reduce the problems caused by wave reflections. The effects are reduced with coarser meshes as the mesh itself dampens the waves.

1. Introduction

Since 2010s, the interest in Computational Fluid Dynamics (CFD) methods for use in ship seakeeping analysis has been increasing due to increasing computational efficiency. Reynolds-Averaged Navier-Stokes (RANS) based CFD methods allow for more accurate modelling of nonlinear effects, such as wave breaking or water on decks, in comparison to nonlinear potential flow methods. However, RANS methods have their own limitations which are often related to high sensitivity to the modelling choices. These difficulties are highlighted in complex simulation cases such as wave-structure interaction modelling [1].

One of the problems in wave modelling with CFD is the waves and wave reflections reaching the boundaries of the computational domain. Waves can be reflected from the domain boundaries which increases inaccuracies in the analysis. There are multiple methods to absorb waves and reflections heading towards the boundaries. Popular methods are coarsening the mesh towards the boundaries [2], numerical or modelled beaches [3] and relaxation zone methods [4]. Windt et al. carried out an extensive study of the different methods that absorb waves close to boundaries. The relaxation zone proved to have overall good performance [5].

Among wave reflections, those heading back towards the inlet boundary are especially problematic. This is because reflected waves could be re-reflected. The mesh can not be coarsened towards the inlet boundary as dense mesh is required for accurate wave modelling. A solution is to stop the simulation before reflected waves reach the inlet boundary [6]. However, in many cases long simulation times are required to achieve steady state or quasi-static conditions.

Therefore, numerical damping with relaxation zones is popular method to dampen the waves in wanted directions.

In OpenFOAM simulations, often applied waves2Foam toolbox includes a relaxation zone method to absorb waves and reflections towards boundaries [7]. Various numerical wave modelling related studies performed with OpenFOAM also include waves2Foam toolbox. Study of wave-structure interactions for a semi-immersed cylinder by Chen et al. showed that inclusion of relaxation zones reduced reflections significantly [8]. Zhou et al. applied relaxation zones in their breaking wave studies. [9]. Liu et al. applied waves2Foam for their ship motion and wave forces study [10].

Inclusion of relaxation zones increases the size of the computational domain and hence demands more computational resources [5]. It is important to find balance between long enough relaxation zones so that they work properly but their length should not be unnecessary extensive. This paper studies the effect of relaxation zone length and also the length between the end of the relaxation zone and the structure where waves are reflected. This is achieved by applying OpenFOAM-v2206 [11] and waves2Foam toolbox [7] to perform a parametric study of a box structure in waves and in a 2D domain by varying the two mentioned distances. Two different wave lengths are studied to study the effect of short and long waves. Additionally, two mesh densities are studied to understand how fluid domain discretisation may affect the results.

2. Theory

The governing equations in RANS CFD analyses are the Navier-Stokes equation and the continuity equation [12]. For an incompressible fluid, the Navier-Stokes equation becomes

$$\frac{\partial u_i}{\partial t} + u_i \frac{\partial u_j}{\partial x_i} = -\frac{1}{\rho}\frac{\partial p}{\partial x_i} + \nu \frac{\partial^2 u_i}{\partial x^2} + g_i, \tag{1}$$

where u_i is the velocity, x_i is coordinate axis direction, p is the pressure, g_i is the gravitational acceleration and ρ and ν are the fluid density and kinematic viscosity respectively. The incompressible form of continuity equation is

$$\frac{\partial u_i}{\partial x_i} = 0. \tag{2}$$

OpenFOAM applies the Volume of Fluid (VOF) method, where the volume fraction of water, α, is defined for each cell in the computational domain [11]; $\alpha = 1$ corresponds the cell being completely filled with water and $\alpha = 0$ full of air. Thus the free surface surface is found in location where $\alpha = 0.5$. The volume fraction is transported inside the domain according to the transport equation

$$\frac{\partial \alpha}{\partial t} + \frac{\partial \alpha u_i}{\partial x_i} = 0. \tag{3}$$

The waves2Foam toolbox applies relaxation zones to absorb waves. Inside the relaxation zone, the computational variable ϕ, which can be either α or u_i, is corrected by a weighted sum

$$\phi = (1 - \omega_R)\phi_{target} + \omega_R \phi_{computed}, \tag{4}$$

where ϕ_{target} is the theoretical value and $\phi_{computed}$ is simulated value; $\omega_R \in [0, 1]$ is the weighting function which is a function of the local coordinate inside the relaxation zone [7].

3. Numerical methods

OpenFOAM applies the Finite Volume Method (FVM) to discretise the computational domain. Accurate excitation force predictions require that the waves propagate accurately from the inlet boundary towards the structure. Therefore, applied numerical schemes (table 1) mostly follow results presented by Larsen et al. for accurate and stable wave modelling [13]. Crank-Nicholson scheme was applied for time derivatives, linear scheme for gradients and linear upwind scheme for convection terms. In simulation the time step is controlled by the Courant number. The maximum value inside the domain is set to be 0.3 for balancing between accurate wave modelling and fast simulations [13]. The governing equations are solved with OpenFOAM's interFoam solver and PIMPLE algorithm which is combination of SIMPLE and PISO algorithms [11]. The main numerical set up is shown in table 1.

Table 1. Main numerical set up for the simulations with OpenFOAM terminology.

ddt	CrankNicholson 0.645
grad	Gauss linear
div(rhoPhi,U)	Gauss linearUpwind grad(U)
div(phi,alpha)	Gauss vanLeer
div(phi,k)	Gauss linearUpwind limitedGrad
div(phi,omega)	Gauss linearUpwind limitedGrad
laplacian	Gauss linear corrected
interpolation	linear
PIMPLE nOuterCorrectors	8
PIMPLE nCorrectors	2
Relaxation zone shape	rectangular
Relaxation zone weight	exponential [7]

Mesh density is also an important factor for accurate wave modelling. Previous studies show that a number of cells per wave height is more crucial factor than cells per wave length and more than 10 cells per wave height ($H/\Delta z$) is recommended to be used for accurate wave modelling [13] [14] [15]. Therefore, in this study 15 cells per height were used to idealise dense mesh cases and 10 cells per wave height were used for coarse meshes. Previous studies recommend to use cell aspect ratio of 1 [15] [16]. However, this could lead to very large total number of cells. Therefore, the aspect ratio inside the wave modelling area was maintained at 4.

4. Case study

The effects of relaxation zone length and the length between the end of relaxation zone and the structure were studied for the 2D case of a static box floating in deep waters. The model corresponds to the simplified case of a long barge restricted in motion in beam seas. The box is $37.0m$ wide and $9.92m$ deep. Figure 1 depicts the computational domain. A parametric study is performed by varying the length of the relaxation zone (parameter A times the wave length) and the distance from the end of the relaxation zone until the box (parameter B times the wave length). Thus, the domain size varies in horizontal direction for each case, but in the vertical direction the domain size is constant. The water depth is $290m$ and the top boundary is at $163m$.

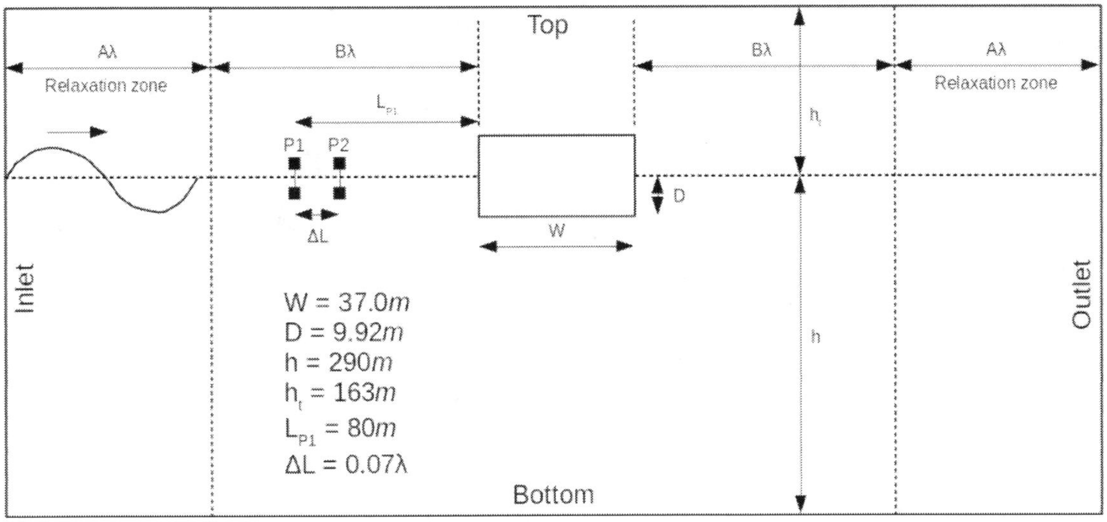

Figure 1. Schematic picture of the computational domain. P1 and P2 are two wave probes measuring wave amplitude.

The waves2Foam toolbox handles the boundary conditions for velocity and volume fraction α at the inlet and outlet boundaries with zero gradient boundary condition for pressure. At the bottom boundary, a symmetryPlane condition is applied for deep water conditions. At the top, an atmospheric boundary is applied. For the turbulence terms, k, ω, ν_t, fixed values are applied at the inlet boundary. At the outlet and top boundaries, the boundary conditions for k and ω vary based on inflow (fixed value) or outflow (zero gradient) while ν_t has a zero gradient [17]. Wall functions are applied at the box surface.

Two different wave lengths were studied namely, $\lambda = 174m$ and $\lambda = 348m$, while the wave height is kept constant at $H = 2m$. Two mesh densities are considered namely, a dense mesh that has $H/\Delta z = 15$ and the coarse mesh that has $H/\Delta z = 10$. Figure 2 shows an example of the mesh in the computational domain and around the box for case B1 with dense mesh. The dense mesh area from inlet to the box is created for accurate wave modelling and towards top, bottom and outlet boundaries the mesh is coarsened to reduce the number of cells. All the cases with different relaxation zone layouts are described in table 2. The simulation time in all cases where set to be 20 wave periods after the wave group has reached the box to achieve steady state conditions.

Table 2. Case study set ups.

Case	A [$\times\lambda$]	B [$\times\lambda$]	N_{cells} $\lambda = 174m$ (coarse/dense)	N_{cells} $\lambda = 348m$ (coarse/dense)
A1	0.5	2	21000 / 400000	37000 / 69000
A2, B3	1	2	24000 / 46000	43000 / 82000
A3	2	2	30000 / 57000	56000 / 107000
B1	1	0.5	15000 / 28000	24000 / 46000
B2	1	1	19000 / 34000	30000 / 57000
B4	1	3	30000 / 57000	56000 / 107000

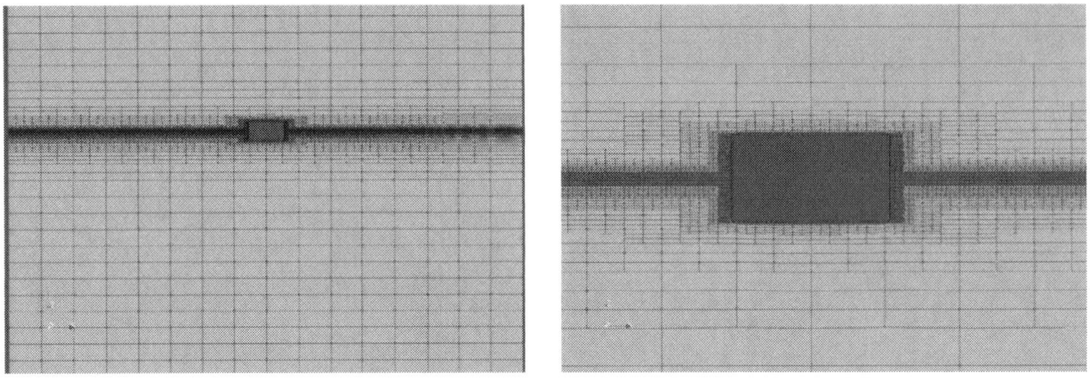

Figure 2. The mesh in the domain and close up around the box.

5. Results and discussion

Simulation results focus on the accuracy of wave modelling for each case. Then, the effect of the relaxation zone length and the length between the end of the relaxation zone and the box to the excitation forces are presented. Finally, the effect of mesh density to the excitation forces is discussed.

5.1. Parametric study on wave amplitudes

Two wave probes were set to measure wave amplitudes in front of the box (see figure 1). Following the method described by Goda and Suzuki, the measured wave amplitude can be divided into incident and reflected wave components [18]. Table 3 shows the incident and reflected wave amplitudes for all cases.

Table 3. Incident / reflected wave amplitudes [m], input amplitude is $1m$.

Case	A1	A2, B3	A3	B1	B2	B4
λ_{174}, coarse	0.76 / 0.63	0.67 / 0.55	0.69 / 0.57	0.9 / 0.9	0.88 / 0.88	0.5 / 0.4
λ_{174}, dense	0.97 / 0.81	0.89 / 0.77	0.87 / 0.75	0.8 / 0.8	0.86 / 0.86	0.78 / 0.66
λ_{348}, coarse	0.91 / 0.49	0.81 / 0.42	0.77 / 0.46	0.96 / 0.63	0.9 / 0.46	0.62 / 0.43
λ_{348}, dense	1.02 / 0.5	0.99 / 0.5	0.83 / 0.58	1.01 / 0.67	0.97 / 0.47	0.85 / 0.5

The results show that a coarse mesh dampens the wave amplitude more than the dense mesh in all cases. The only exception is case B2. Dense mesh results show amplitudes close to $1m$. This could be attributed to numerical inaccuracy in the wave decomposition method. In longer domains the incident wave amplitude is reduced more (see cases A3 and B4 where the incident wave amplitudes have the lowest values). The latter could be attributed to numerical damping i.e., it takes longer time to affect the waves generated at the inlet boundary before they reach the box. Overall, the simulation results agree with published wave modelling studies [13] [14] [15].

Reflections increase in shorter waves. In very short domains, such as those presented in cases B1 and B2, the wave seems to completely reflect from the box. In other cases, the reflected wave amplitude is more than half of the incident wave amplitude. In longer waves reflections reduce as in most cases the reflected wave amplitude is about half of the incident wave amplitude. An

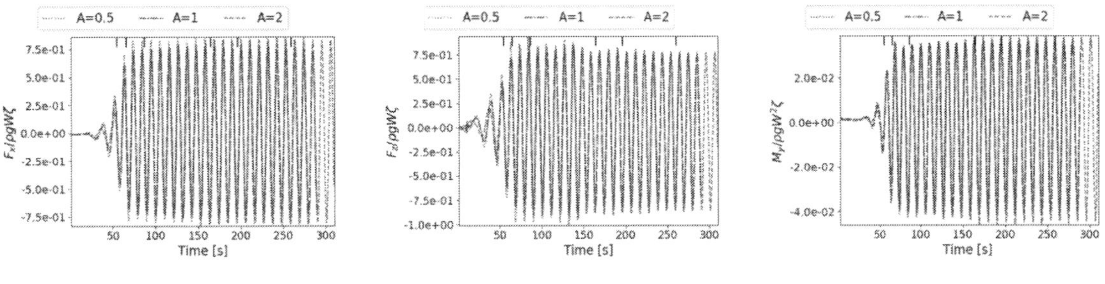

Figure 3. Time histories for cases A1-A3 with $\lambda = 174m$ and coarse mesh showing the effect of relaxation zone length to Fx, Fz and My.

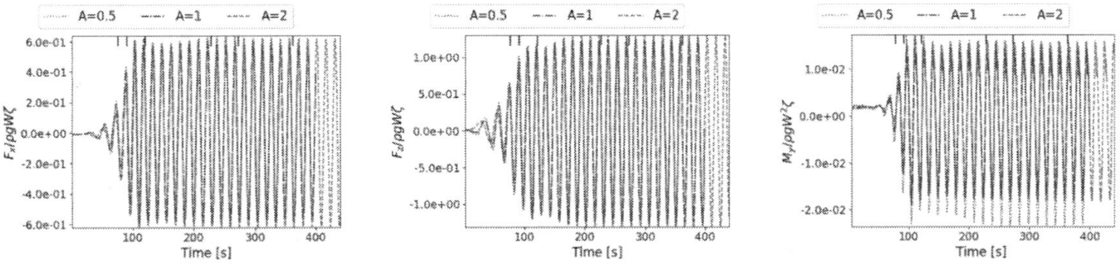

Figure 4. Time histories for cases A1-A3 with $\lambda = 348m$ and coarse mesh showing the effect of relaxation zone length to Fx, Fz and My.

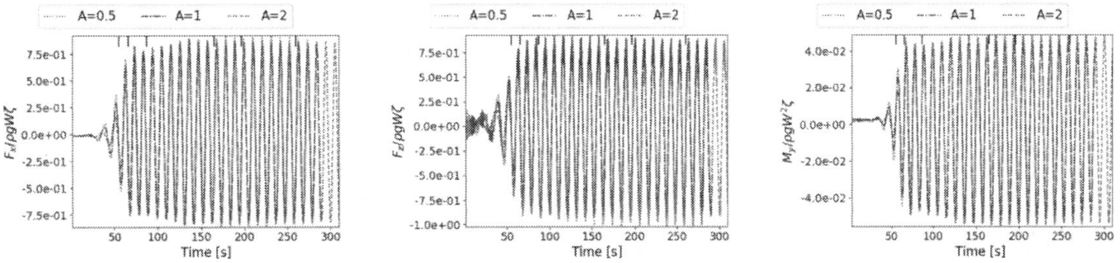

Figure 5. Time histories for cases A1-A3 with $\lambda = 174m$ and dense mesh showing the effect of relaxation zone length to Fx, Fz and My.

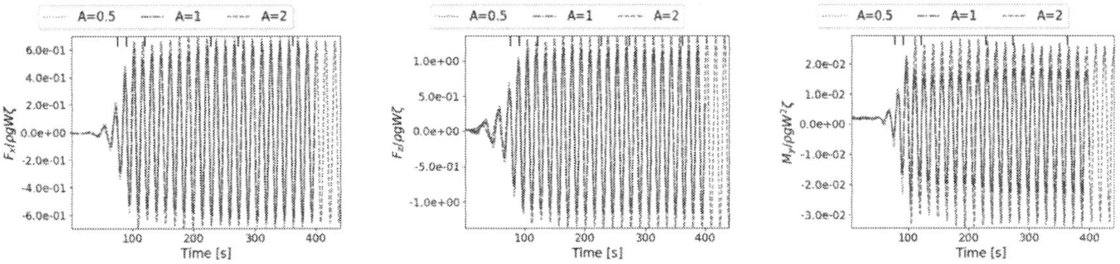

Figure 6. Time histories for cases A1-A3 with $\lambda = 348m$ and dense mesh showing the effect of relaxation zone length to Fx, Fz and My.

exception to this are cases A3, B1 and B4 where the reflected wave amplitude is more than half of the incident wave amplitude. For cases A3 and B4 this could be due to the incident wave amplitude being damped more due to long domain size which alters the wave profile and could increase reflections. For case B1, the short distance between the end of relaxation zone and the box could affect to the modelling of the wave.

5.2. Relaxation zone length (parameter A)

Figures 3 - 6 display the effect of the relaxation zone length to horizontal (Fx) and vertical (Fz) excitation forces and roll moment (My) by comparing cases A1-A3 against each other in nondimensional format. The decomposed incident wave amplitudes in table 3 is applied in nondimensional format so as to ignore the numerical damping of wave amplitude. Each figure has six black marks at the top of the graph. The three first marks present the time instances when the wave group reaches the box after the start of the simulation for cases A1-A3 respectively. The latter three marks correspond to three times longer time instances when possibly re-reflected waves from the inlet boundary reaches the box. In all cases, it is visible that the force and moment values increase significantly when the wave group reaches the box and soon after that the forces and moment reaches steady state.

In coarse mesh cases (figures 3 and 4), all cases result in very similar loading. An exception to this is seen in roll moment results for the longer wave length where case A1 ($A = 0.5$) shows larger negative peaks. Similar trend is visible for dense mesh cases (figures 5 and 6), except in the longer wave length results. In these results, case A3 ($A = 2$) seems to predict larger vertical force and the roll moment case A2 ($A = 1$) shows lower loading. This could be attributed to significantly lower incident wave amplitude used for nondimesionalising case A3 in (see table 3). Therefore, the force and moment values for case A3 are increased in comparison to case A2 while case A1 shows larger loading similarly to coarse mesh cases.

When a dense mesh is used for case A1 the peak values of horizontal force and roll moment seem to slightly increase after the time instance corresponding to re-reflected waves from the inlet boundary reaching the box. This is expected since the initially reflected wave is not fully relaxed, i.e. the reflection is assumed to reach the inlet boundary and being partly re-reflected. This reflection then reaches the box at the marked time point and increases the horizontal force. This effect is not visible in coarser meshes due to the waves being numerically dampened more in comparison to dense mesh cases. In conclusion, the results show that too short relaxation zone ($A = 0.5$) does not work properly and the length of the zone should be at least one wave length long.

5.3. The influence of length between the end of the relaxation zone and the box (parameter B)

The effect of the length between the end of relaxation zone and the box is shown in figures 7- 10 where cases B1-B4 are compared. Similarly to the results shown in figures 3 - 6, incident wave amplitudes from table 3 are used for nondimensionalising and the black marks show the time instances when wave group first reaches the box and when possible re-reflected waves from the inlet boundary reach the box.

In short wave and coarse mesh cases (figure 7) all results match well. However, in longer wave larger variations are visible. Horizontal forces displayed in cases B1 and B2 ($B = 0.5$ and $B = 1$) show that the force reduces quickly and then finds steady state after the wave group has reached the box. Therefore, the force peaks are lower in comparison to cases B3 and B4 ($B = 2$ and $B = 3$). Similar trend is visible also for roll moment results where case B3 shows slight reduction in moment peak values after wave group has reached the box. In vertical force results, case B1 shows larger loading than the other three cases. These results show that too short distance after the relaxation zone until the structure affects results greatly.

Dense mesh results show the effect of short distance between the end of the relaxation zone and the structure even more clearly. In short waves (figure 9), cases B1 and B2 show larger

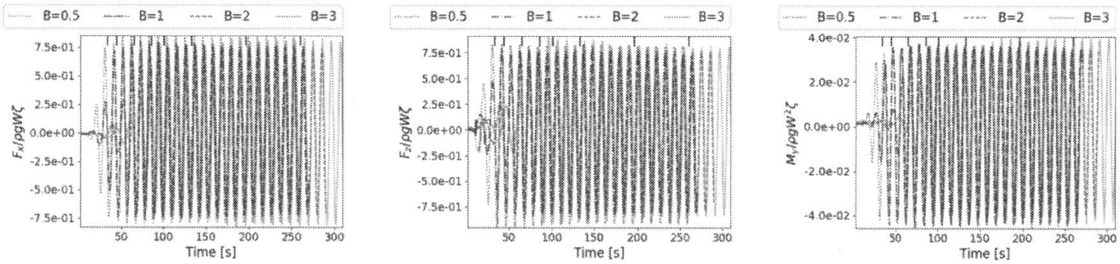

Figure 7. Time histories for cases B1-B4 with $\lambda = 174m$ and coarse mesh showing the effect of the distance to the box after relaxation zone to Fx, Fz and My.

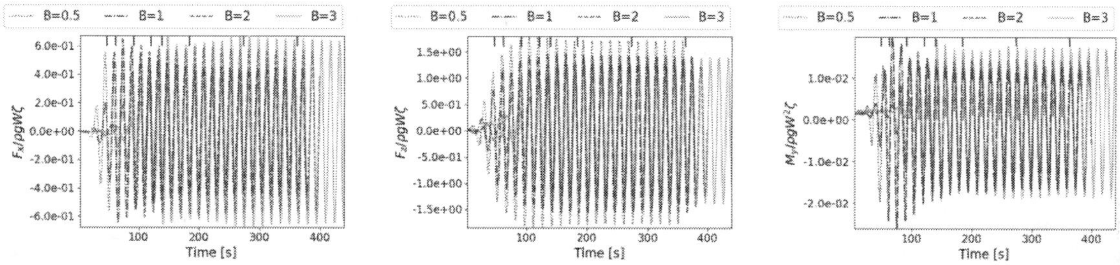

Figure 8. Time histories for cases B1-B4 with $\lambda = 348m$ and coarse mesh showing the effect of the distance to the box after relaxation zone to Fx, Fz and My.

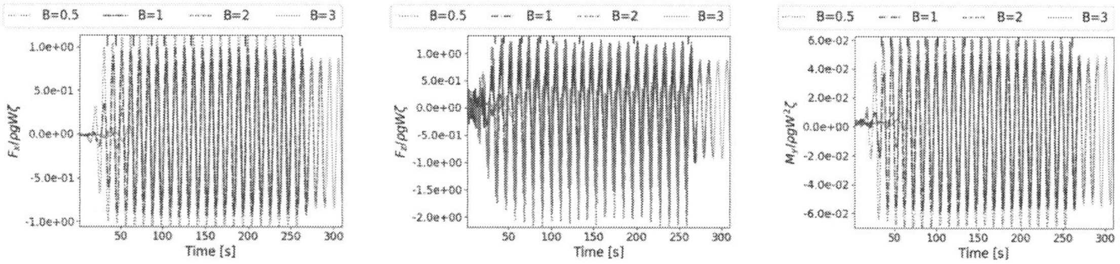

Figure 9. Time histories for cases B1-B4 with $\lambda = 174m$ and dense mesh showing the effect of the distance to the box after relaxation zone to Fx, Fz and My.

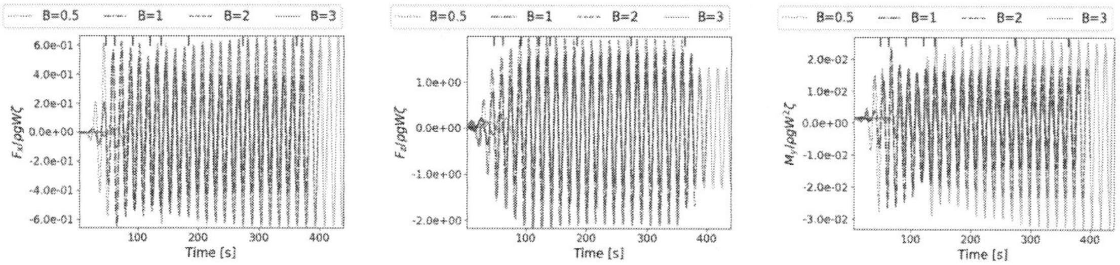

Figure 10. Time histories for cases B1-B4 with $\lambda = 348m$ and dense mesh showing the effect of the distance to the box after relaxation zone to Fx, Fz and My.

loading for forces than the other two cases and in roll moment case B1 predicts larger moments.

In long wave results the trend of reducing force in time from coarse mesh is highlighted as cases B1 and B2 show reducing horizontal force values after reaching initially maximum forces. Similarly, cases B1-B3 have reducing moment values. In vertical force results, cases B1 and B2 show larger loading in comparison to the other two cases. In conclusion, the distance between the end of relaxation zone and the structure has to be large enough and at least two wave lengths is recommended.

5.4. The influence of mesh density to the excitation forces and moments

Table 4 shows the average between minimum and maximum force and moment peaks for each case after steady state is reached. In general, dense mesh results in slightly higher force peaks with some exceptions in $\lambda = 348m$ cases such as vertical excitation force in cases A1 and A2. Higher forces are due to more accurate modelling of the box and more detailed modelling of the flow field around the hull. Cases B1 and B2 show larger differences between coarse and dense mesh forces which is associated with too short distance between the end of the relaxation zone and the box; thus causing unexpected wave forces.

Similarly to forces, roll moment values are also larger with dense mesh results. However, the difference is generally larger than the one observed for force values and in some long wave cases (cases A3 and B4) dense mesh results show almost twice as large loading peak value. This could relate to more accurate modelling of the flow field around around the box corners. The effect is depicted in figure 11 which shows flow velocity magnitude around the box for case A3 with $\lambda = 348m$. It is seen that the large velocity magnitude areas close to the corners of the box are larger with dense mesh as a coarse mesh has larger numerical diffusivity. The large local difference in the flow field has small effect in the forces. However, in this case the effect on moments is large as the corners are far away from the box center of gravity.

Table 4. Coarse / dense mesh average loading peak amplitudes in nondimensional format.

	$\lambda = 174m$			$\lambda = 348m$		
Case	Fx	Fz	My	Fx	Fz	My
A1	0.84 / 0.89	0.85 / 0.92	0.039 / 0.052	0.63 / 0.69	1.29 / 1.15	0.02 / 0.03
A2, B3	0.75 / 0.83	0.79 / 0.90	0.039 / 0.047	0.57 / 0.64	1.24 / 1.17	0.016 / 0.021
A3	0.82 / 0.84	0.82 / 0.90	0.042 / 0.051	0.62 / 0.66	1.3 / 1.36	0.017 / 0.029
B1	0.82 / 1.09	0.87 / 1.69	0.04 / 0.066	0.33 / 0.25	1.78 / 2.05	0.01 / 0.011
B2	0.75 / 0.97	0.79 / 1.14	0.04 / 0.055	0.45 / 0.42	1.51 / 1.7	0.015 / 0.015
B4	0.82 / 0.86	0.83 / 0.90	0.042 / 0.052	0.65 / 0.65	1.39 / 1.31	0.018 / 0.03

6. Conclusions

This paper studied the effect of the relaxation zone length and the length between the end of the relaxation zone and a 2D box like floating structure. This was achieved by performing a parametric study with OpenFOAM-v2206 and waves2Foam toolbox for the case of a box in deep water waves while varying the two mentioned distances (see figure 1 and table 2). Two different wave lengths and two mesh densities were considered.

The results show that a relaxation zone is able to effectively absorb the reflected waves when its length and the distance between the end of the zone and the structure is long enough. Too short relaxation zone length (half of wave length) increases induced forces and moments and

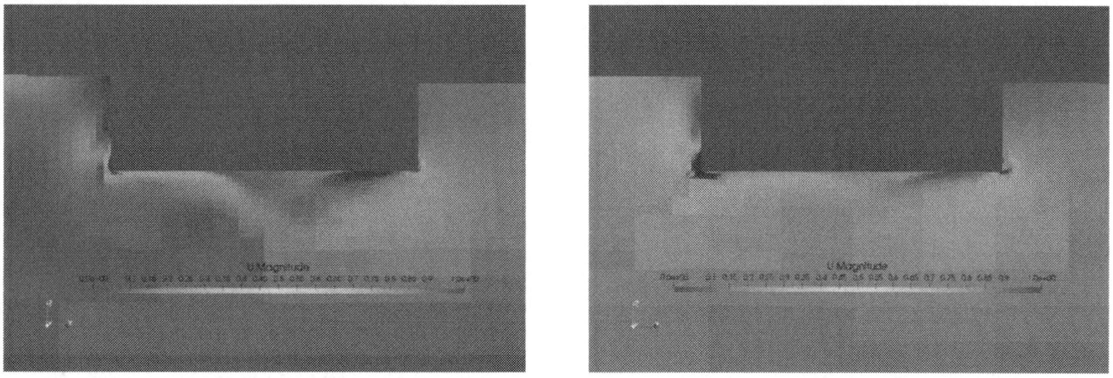

Figure 11. Flow field with velocity magnitude around the box in case A3 with long wave for coarse (left) and dense (right) mesh.

may not be able to completely absorb reflections while re-reflections occur. Similarly, too short distances between the end of the relaxation zone and the box (less than two wave lengths) resulted in unexpected behaviour when force and moment peak values reduced in time. Based on the results, the recommended length of the relaxation zone is at least one wave length and the distance between the end of the relaxation zone and the structure at least two wave lengths.

Relaxation zones increase the domain size which could cause very large domain sizes in 3D simulations with long wave lengths which is often the case with ship seakeeping studies. However, inclusion of motions into the simulation reduces the reflections especially in long waves. Therefore, it could be possible in these cases to reduce the recommended relaxation zone length and the distance to the structure; thus reducing the domain size. Future study will focus on this.

Acknowledgments
The authors wish to acknowledge CSC – IT Center for Science, Finland, for computational resources.

References
[1] Huang L *et al.* 2022 *OpenFOAM® Journal* **2** 116–142
[2] Yao J, Cheng X, Song X, Zhan C and Liu Z 2021 *Journal of Marine Science and Engineering* **9** 1176
[3] Schmitt P, Windt C, Davidson J, Ringwood J V and Whittaker T 2020 *Journal of ocean engineering and marine energy* **6** 277–292
[4] Islam H, Mohapatra S, Gadelho J and Soares C G 2019 *Ocean Engineering* **193** 106532
[5] Windt C, Davidson J, Schmitt P and Ringwood J V 2019 *Journal of Marine Science and Engineering* **7** 47
[6] Devolder B, Rauwoens P and Troch P 2017 *Coastal Engineering* **125** 81–94
[7] Jacobsen N G, Fuhrman D R and Fredsøe J 2012 *International Journal for Numerical Methods in Fluids* **70** 1073–1088
[8] Chen L, Zang J, Hillis A J, Morgan G C and Plummer A R 2014 *Ocean Engineering* **88** 91–109
[9] Zhou Z, Hsu T J, Cox D and Liu X 2017 *Journal of Geophysical Research: Oceans* **122** 207–235
[10] Liu C, Wang J and Wan D 2018 *International journal of offshore and polar engineering* **28** 154–163
[11] OpenFOAM Openfoam user guide `https://www.openfoam.com/documentation/guides/latest/doc/index.html` [Accessed: 4/2023]
[12] el Moctar B O, Schellin T E and Söding H 2021 *Numerical Methods for Seakeeping Problems* (Springer)
[13] Larsen B E, Fuhrman D R and Roenby J 2019 *Coastal Engineering Journal* **61** 380–400
[14] Connell K O and Cashman A 2016 *2016 International Conference on Electrical, Electronics, and Optimization Techniques (ICEEOT)* (IEEE) pp 3008–3012
[15] Roenby J, Larsen B E, Bredmose H and Jasak H 2017 *MARINE VI: Proceedings of the VI International Conference on Computational Methods in Marine Engineering* (CIMNE) pp 266–277

[16] Jacobsen N G, Fuhrman D R and Fredsøe J 2012 *International Journal for numerical methods in fluids* **70** 1073–1088
[17] NASA Nasa langley turbulence model resource `https://turbmodels.larc.nasa.gov/sst.html` [Accessed: 4/2023]
[18] Goda Y and Suzuki Y 1976 *Coastal Engineering 1976* pp 828–845

A two-way coupled FSI ship hard grounding dynamics model

G Taimuri[1,¥], T Mikkola[1], P Kujala[1] and S Hirdaris[1]

[1] Department of Mechanical Engineering, Aalto University, Espoo, Finland
[¥] Corresponding author: ghalib.taimuri@aalto.fi

Abstract. In the maritime sector, ship grounding incidents continue to be a serious problem that may lead to oil spills, capsizing, severe property damage, and even fatalities. Grounding accidents are currently understood utilizing inadequate statistical datasets, probabilistic grounding scenarios, and deterministic computational crashworthiness methodologies due to a lack of practical tools and techniques. The implementation of multiphysics assessment techniques is essential for the development of improved ship safety standards. This involves the use of rapid models for structural integrity rules and damage stability regulations. Taimuri et al., (2022) [1] introduced a rapid two-way coupled fluid-structure interaction (FSI) model that seeks to efficiently examine accidental loads following a ship hard grounding event to fulfil such a requirement. The current study offers an overview of the algorithm and its limitations.

1. Introduction

To meet safe and sustainable shipping targets, it is essential to proactively reduce the cost of accidents to human life, assets, and the marine environment. Today, despite improvements in technology and crew standards, operational accidents still account for most ship accidents. According to Allianz Global Corporate & Specialty (AGCS) [2] collisions, fires, and groundings are the main causes of catastrophic marine accidents. Navigational accidents caused by collisions, contacts, and groundings continue to dominate ship fatality statistics, making up 43% of all ship-related occurrences between 2014 and 2020, according to European Maritime Safety Agency (EMSA) [3], Figure 1.

The accidental distribution of international marine casualty incidents for various ship types during the period of 2002-2022 has recently been reviewed by Zhang et al. 2023 [4]. According to these findings, the top 5 ship accident records are dominated by container ships, passenger ships, dry cargo vessels (e.g., bulk carriers), tankers, and fishing vessels, Figure 2. Technical problems, routing errors, transportation complications, power failure, loss of ship command, and weather situations are all causes that contribute to ship grounding accidents. In recent years we witnessed a crucial need for increased knowledge of vessel responses following grounding events, especially due to the life cost, the environmental contamination and the expense of these catastrophic events. The Exxon Valdez [5] and Wakashio [6] disasters are examples of ship grounding that polluted the ecosystem. Ship and life loss relate with the accident cases of Express Samina [7], Sea Diamond [8], and Costa Concordia [9]. Recently, the Ever Given accident at the Suez Canal [10], demonstrated that grounding accidents can potentially interrupt marine activities.

To date finite element analysis methods, empirical, experimental, and analytical approaches have been used in consequence modelling. These techniques, however, do not idealise ship maneuverability and the influence of surrounding water on ship dynamics [11]. Ship grounding dynamics are influenced by actions associated with both internal and external mechanics. The former may be used to describe the ship's 6-DoF rigid body motions in response to external inputs and the surrounding water. In hard grounding accidents energy is dissipated when the hull bottom encounters a hard rock leading to structural deformation. Thus, internal mechanics can be used to idealise structural response that results in energy dissipation following contact with the rock.

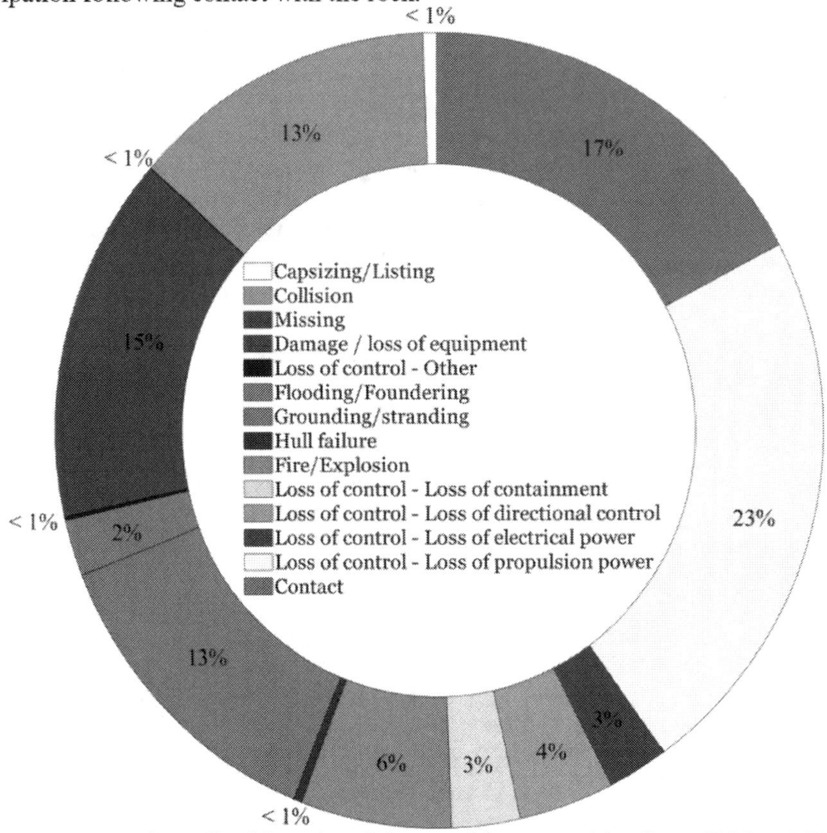

Figure 1. Proportion of incidents involving maritime casualties from 2014 to 2020 [3]

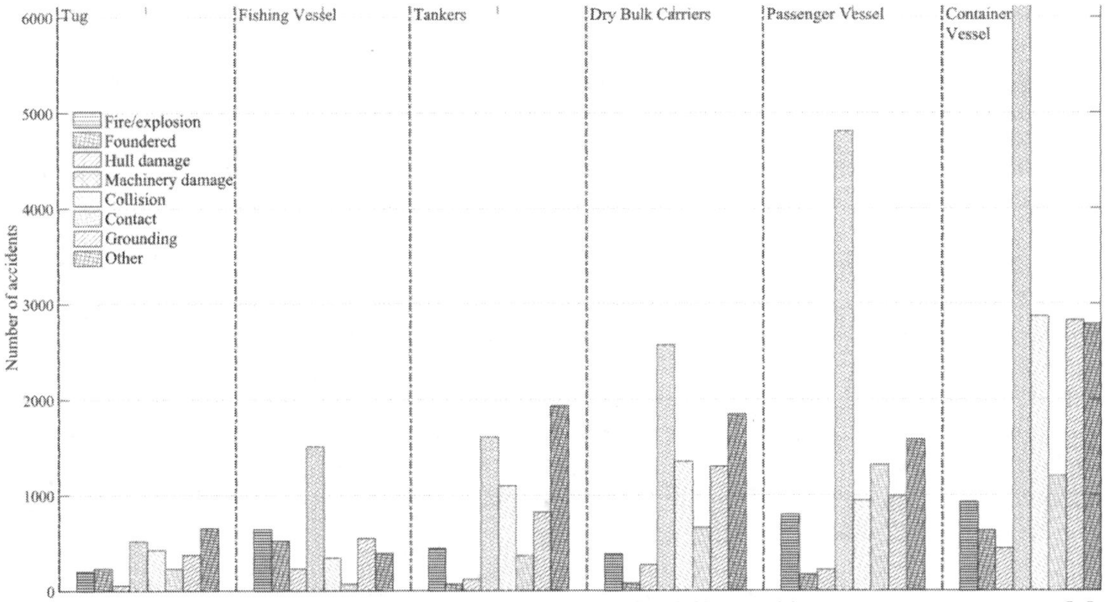

Figure 2. Distribution of global maritime casualty events for different ship types over 2002-2022. [4]

Figure 3 illustrates a review of ship grounding evaluation techniques. The methodologies used to estimate the likelihood of accidents using traffic distributions, historical records, and related risks make up the probabilistic risk evaluation [12]. Using probabilistic statistics to create future predictions may be challenging [13].

Analytical approaches are often employed to assess the crashworthiness of ship structures. The upper-bound theorem is used in these approaches to quantify how much energy is dissipated by major structural components based on the penetration of a rigid object (such as a rock) into a ship's hull. These methods can predict deformations. To analyse internal mechanics analytical formulae are available [14–20]. Yet, the most extensively used and dependable structural response assessment approach is the Finite Element Method (FEM). Despite time consuming FEM is still used in the majority of crashworthiness simulations [20]. Experiments on a big and small scale have been conducted to date [21–27], which are prohibitively expensive, labour-intensive, and may provide findings that are difficult to explain on a full-scale vessel groundings.

Traditionally, external mechanics may be addressed by decoupled methods, which ignore the impact of ship motions on structural deformation. Since these techniques do not adequately capture the consequences of the dynamic impact on a hull surrounded by a fluid, coupled approaches are necessary [28].

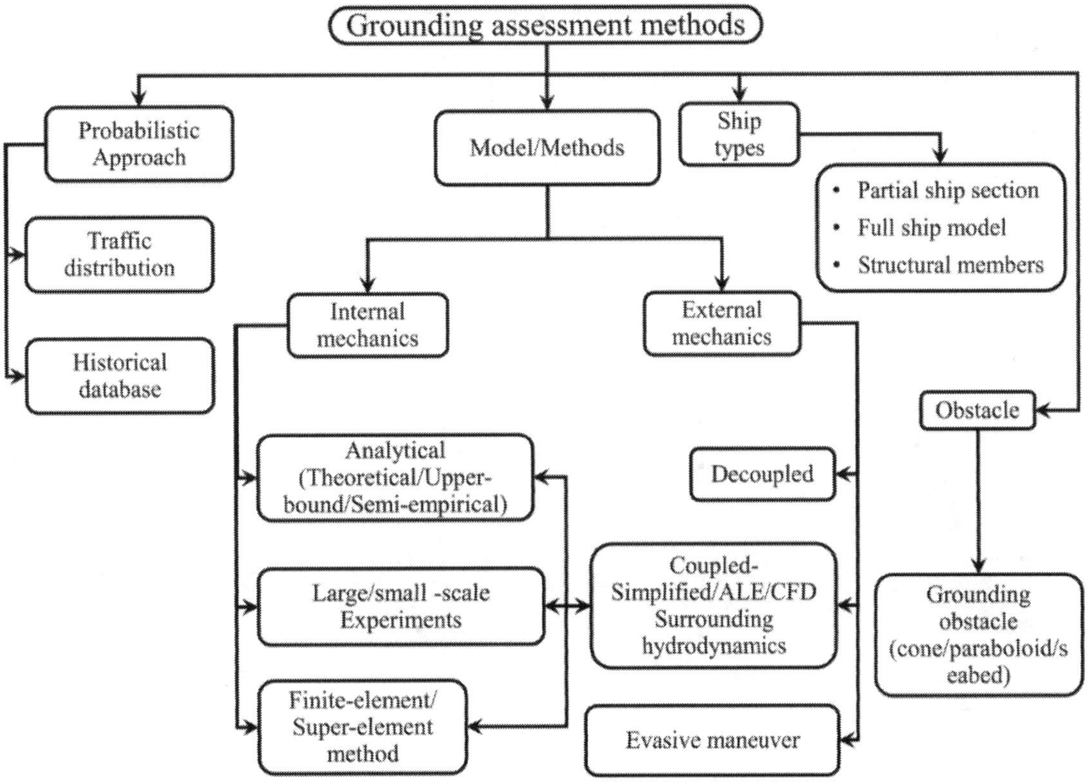

Figure 3. Traditional ship collision and grounding assessment procedure. [4]

The key features and limitations of ship grounding assessment models that have been published during the period of 1997-2022 are depicted in Figure 4. The evaluation is based on internal and external mechanics, investigated vessel types, structural model features, rock forms, and their evaluation. The review demonstrates how little attention has been paid to rapid multiphysics models that idealize coupled ship grounding dynamics [1].

IOP Conf. Series: Materials Science and Engineering 1288 (2023) 012014 doi:10.1088/1757-899X/1288/1/012014

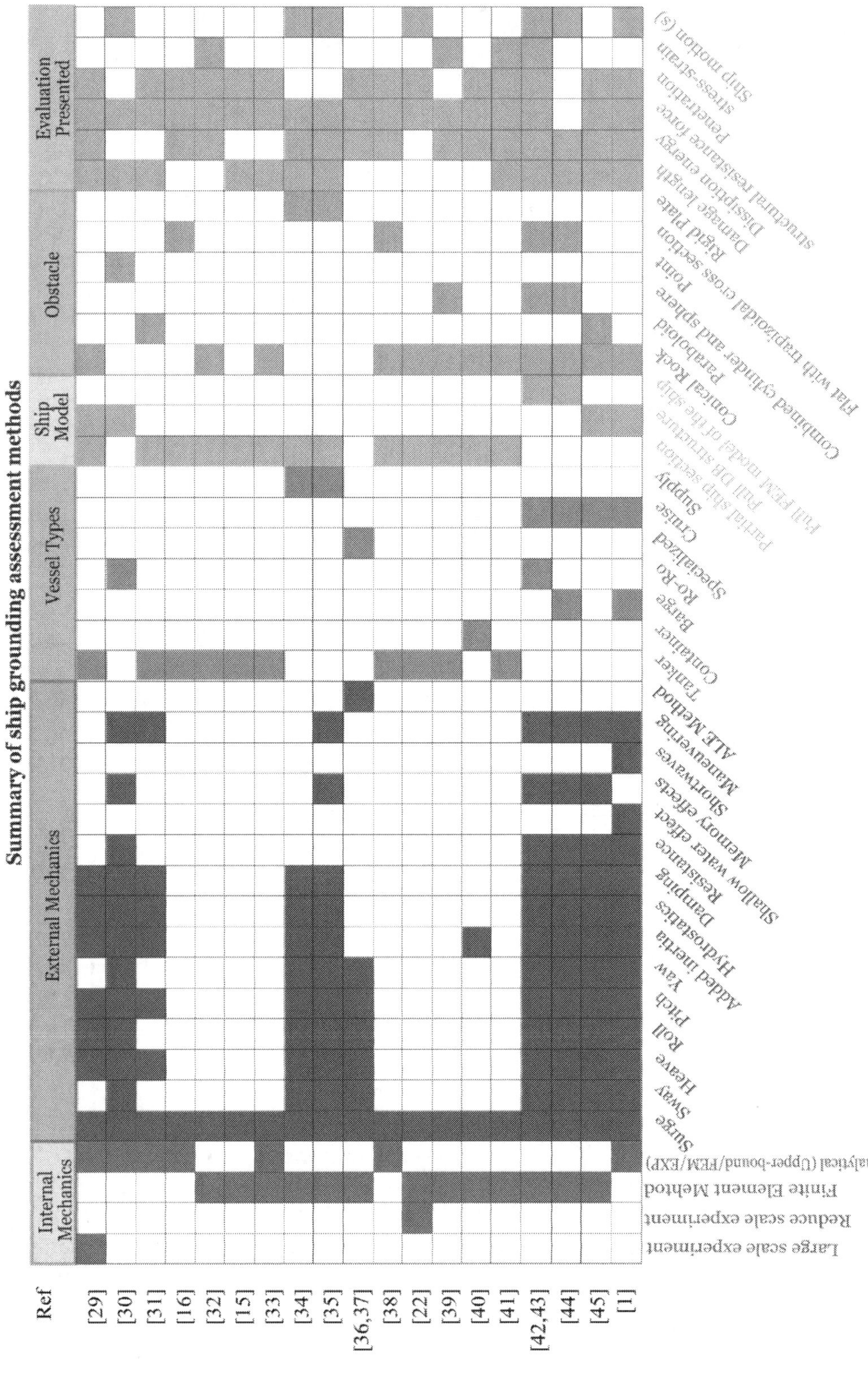

Figure 4. Ship hard grounding assessment methods.

2. Two-way coupled approach

Figure 5 depicts the methods utilized to couple internal and external mechanics. The external mechanics accounts also for possible maneuvering actions. The modelling of a ship's 6-DoF motion during grounding begins with the specification of essential inputssuch as panel mesh of the ship hull, hull hydrodynamic coefficients, rudder-propeller details, and environmental variables. A reference approach for evaluating hydrodynamic hull coefficients is devised for twin screw vessels. For the structural model the rock specification (profile and locations) and ship structural components such as girders, floors, plates, stiffeners, and bulkheads are used. The ship's motions are solved using a set of 2nd-order differential equations, with the ship modelled as a rigid structure navigating in realistic conditions. The contact detection algorithm searches for the interface between the ship panel and the idealized rock After identifying the contact, the structure deforms using an energy approach known as the "upper-bound theorem of plasticity" [40],[46]. The method implies that deformations are influenced by ship motions. Seakeeping actions are considered until the simulation is over, assuming that there is no contact between the rock and hull. Otherwise, the structural deformations caused by the rock and ship bottom contact are incorporated in the ship's 6-DoF motions at each time step. This considers the influences of the sea environment.

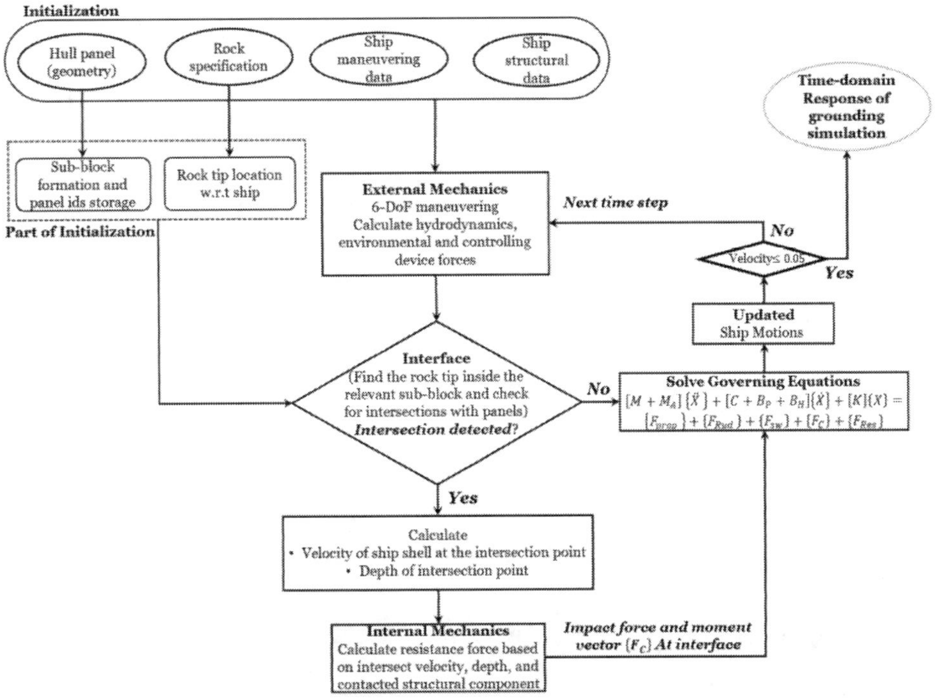

Figure 5. Two-way coupled approach of ship grounding assessment.

3. *Model for coupling ship grounding*

Figure 6 illustrates the coordinate systems utilized for the idealization of a hard grounding event over a rounded-tip conical rock. Using the Earth-fixed coordinate system $O - XYZ$, the vessel's trajectory is determined. The ship's body-fixed origin is measured with reference to the Earth-fixed coordinate system at a distance namely, $\vec{r} = [X_0, Y_0, Z_0]$. The origin of the body-fixed system $(o - xyz)$ is located amidships on an xz-plane of symmetry and at a draft level, where the x-axis points towards the bow, the y-axis towards the starboard side, and the z-axis points downwards. A position vector $\vec{r}_G = [x_G, y_G, z_G]$ is used to define the difference between the center of gravity and the body-fixed origin. Hereby x_G is considered positive towards the bow and z_G is positive towards the base of the ship. The rock location $\vec{r}_{Gr} = [x_{Gr}, y_{Gr}, z_{Gr}]$ is defined as a point node in Earth-fixed coordinate system.

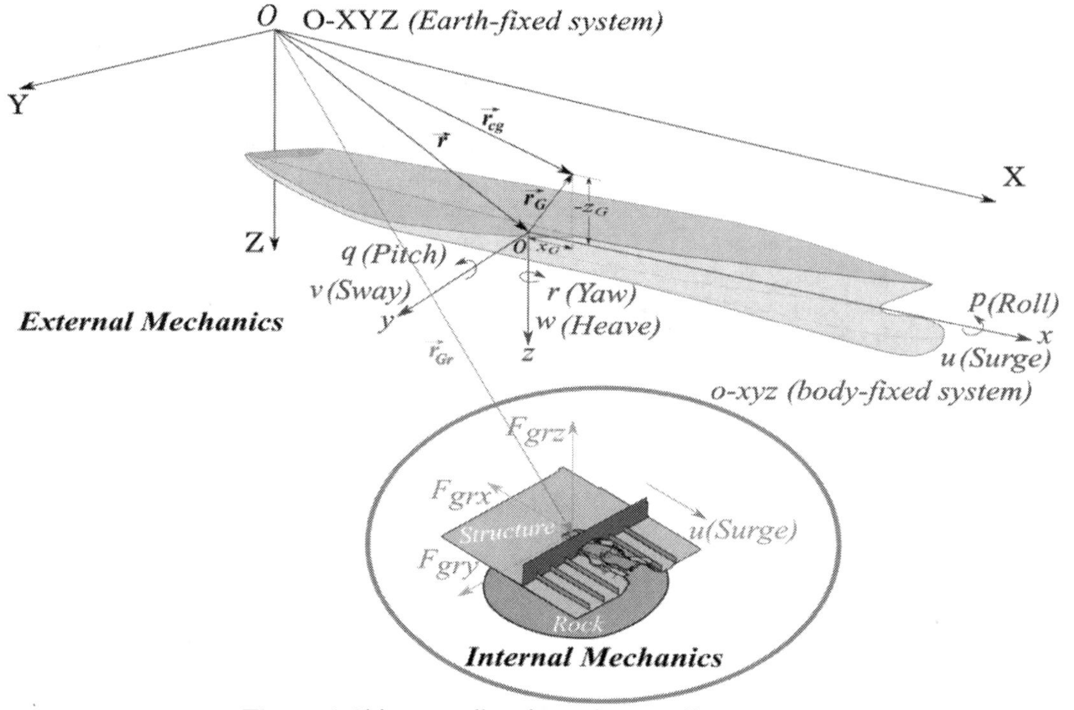

Figure 6. Ship grounding dynamics coordinate system

Assuming a body-fixed system, Figure 6 shows the translational velocity components surge u, sway v, and heave w as well as the rotational velocity components of roll p, pitch q, and yaw r. The fundamental equations describing the mechanics of ship grounding are expressed in Equation (1).

$$
\begin{aligned}
&\text{Surge} && (m - X_{\dot{u}})\dot{u} + (mz_G - 0.5X_{\dot{u}}T) \\
& && \quad = m(rv + x_G(r^2 + q^2) - wq - z_G pr) + \rho g A_{\wp}(\Delta z)sin(\theta) + X_{Hull} \\
& && \quad + X_{Res}(1 - t) + X \quad + X_{Rud} + X_{SW} + X_{C,GRD} \\
&\text{Sway} && (m - Y_{\dot{v}})\dot{v} - (mz_G + Y_{\dot{p}})\dot{p} + (mx_G - Y_{\dot{r}})\dot{r} \\
& && \quad = m(\quad - ru - z_G qr - x_G pq) - \rho g A_{\wp}(\Delta z)sin(\phi)cos(\theta) + Y_{Hull} \\
& && \quad + Y_{Rud} + Y_{SW} + Y_{C,GRD} \\
&\text{Heave} && (m - Z_{\dot{w}})\dot{w} - (mx_G + Z_{\dot{q}})\dot{q} \\
& && \quad = m(uq - vp + z_G(p^2 + q^2) - x_G rp) - \rho g A_{\wp}(\Delta z)cos(\phi)cos(\theta) \\
& && \quad + \rho g A_{\wp} x_f \theta - 2\zeta_w \omega_w w(m - Z_{\dot{w}}) + Z_{C,GRD} \\
&\text{Roll} && -(mz_G + K_{\dot{v}})\dot{v} + (I_X - K_{\dot{p}})\dot{p} - (mx_G z_G + K_{\dot{r}})\dot{r} \\
& && \quad = m(z_G ur - z_G \wp + x_G z_G pq) + (I_Y - I_Z)qr \\
& && \quad - \rho g \nabla GM_T sin(\phi)cos(\phi)cos(\theta) + K_{Hull} + Z_R Y_{Rud} + K_{SW} \\
& && \quad - 2\zeta_\phi \omega_\phi p(I_X - K_{\dot{p}}) + K_{C,GRD} \\
&\text{Pitch} && (mz_G - 0.5X_{\dot{u}}T)\dot{u} - (mx_G + Z_{\dot{q}})\dot{w} + (I_Y - M_{\dot{q}})\dot{q} \\
& && \quad = m\left(z_G vr - z_G \left| \quad \right| wq + x_G vp - x_G uq - x_G z_g(p^2 - r^2)\right) \\
& && \quad + (I_Z - I_X)pr - \rho g \nabla GM_L sin(\theta)cos(\phi)cos(\theta) + \rho g A_{\wp} x_f(\Delta z) \\
& && \quad + Z_R X_{Rud} - 2\zeta_\theta \omega_\theta q(I_Y - M_{\dot{q}}) + M_{C,GRD} \\
&\text{Yaw} && (mx_G - N_{\dot{v}})\dot{v} - (mx_G z_G + K_{\dot{r}})\dot{p} + (I_Z - N_{\dot{r}})\dot{r} \\
& && \quad = m(x_G \wp - x_G ur - x_G z_G qr) + (I_X - I_Y)pq \\
& && \quad - \rho g \nabla(-GM_L cos(\theta) + GM_T)sin(\phi)sin(\theta) + N_{Hull} + X_R Y_{Rud} \\
& && \quad + N_{SW} + N_{C,GRD}
\end{aligned}
\tag{1}
$$

The inertia of the ship, including added masses and moments of inertia, appears on the left side of Equation (1). Semi-empirical relationships or strip theory can be used to estimate these. Coriolis and centripetal forces caused by the rotation of the moving body-fixed system about the Earth-fixed coordinate system are the mass multiplier terms on the right-hand side of Equation (1). The quantities that appear as multiples of acceleration of gravity "g" describe the hydrostatic forces acting on a ship.

The damping of a vessel during maneuvering is represented by in-plane (surge X_{Hull}, sway Y_{Hull}, and yaw N_{Hull}) hydrodynamic damping coefficients. This idealization is based on linear and non-linear semi-empirical relationships. The semi-empirical expression is the outcome of a regression analysis of captive model test (CMT) data. The out-of-plane (heave, roll, and pitch) radiation damping is determined by the ship's natural period (ω_W, ω_ϕ, and ω_θ). The critical damping ratios $\zeta_W, \zeta_\phi \wedge \zeta_\theta$ are calculated using a nonlinear time-domain 6-DoF solver and a numerical decay test. The subscripts "*Res*", "*Prop*", "*Rud*", and "*SW*" represent ship resistance, propeller thrust, control forces, and short-waves forces, respectively.

The behavior of the structure when a ship runs aground over a seabed obstacle is analyzed using the closed-form formulation of structural component resistance forces, represented with a subscript "*C,GRD*" in Equation (1). The depth of penetration of the rock into the hull bottom governs these resistance forces. The procedure considers the bottom structural deformation model introduced by Simonsen [40] and Sun et al. [46]. The formulation is based on the so-called upper bound method of the plastic limit theorem presented in Equation (2).

$$F_H U = F_P . U + \int_S p\mu U_{rel} dS \tag{2}$$

where, F_H is the resistance force of the structure in the direction of relative velocity U between the ship and a rock, F_P is the plastic resistance which includes both plasticity and fracture; μ is the Coulomb coefficient of friction and the last term represents the sliding energy dissipation due to normal and frictional forces on the contact surface of the rock and the plate; U_{rel} is the relative velocity between rock and plate [40].

A conical rock with a rounded tip specified by a radius R_R and semi-apex angle ϕ is used to idealise grounding over a rigid object. The structure consists of the hull's inner and outer shells, longitudinal girders and stiffeners, as well as transverse floors. The forces generated by the deformation of different structural elements are summed together to produce the resultant contact force, which is applied as external forces at the location of the contact, this is represented as a subscript "C,GRD" in Equation (1). Figure 7 depicts the rock profile, structural model, and plate splitting angle after grounding. As the bottom of the hull encounters a rock, the plate tears in a fashion that forms a splitting angle as shown in Figure 7. The magnitude of the plate deformation in front of the conical rock is indicated by this angle. As the energy associated with structural deformation depends on a precise calculation of this angle, it is essential to make a correct assumption in defining the splitting angle. In the previous studies [40,44,46,47], the plate tearing angle was assumed constant throughout the grounding simulation and was precomputed in such a way that it gave the least overall structural resistance force. In other words, the split angle that offers the least amount of resistance is chosen after total structural resistance forces are computed over a range of plate split angles from 0 degrees to 90 degrees. However, this is not the case when a FEM simulation was run, and it was figured out that plate splitting angle depends on the amount of the rock penetrated into the hull, ship motions and the profile of the rock [1]. Taimuri et al. [1], proposed a unique technique to estimate plate splitting angle by utilizing FEM simulation and a curve fitting methodology to develop a plate splitting angle as a function of rock profile and penetration of the rock into the hull. Due to the dynamic nature of the grounding model [1], changes in ship motion will also have an impact on the plate splitting angle.

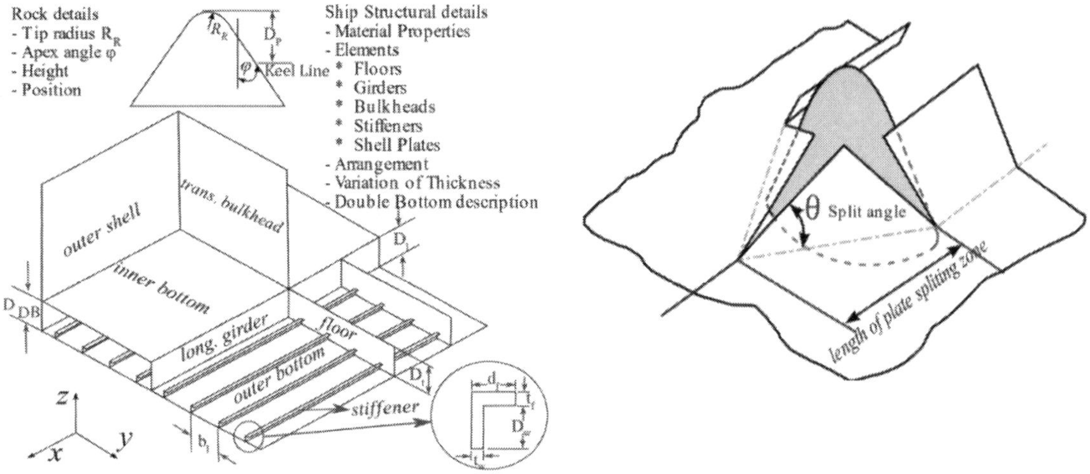

Figure 7. Idealization of ship bottom structure, conical rock, and plate failure mode.

4. *Coupling of internal and external mechanics*

An ideal vertical line from the rock's tip to the seabed is used to evaluate the contact between the rock's tip and the hull. This results in the displacement of the rock-tip inside the hull namely, δ_{out}. This displacement allows for the calculation of the contact force at the interface, using analytical equations of structural resistance force.

Triangular panels are used to idealise the hull model. Each panel's coordinates are defined in relation to the ship's origin. These panels are saved in the user-defined sub-blocks. The 6-DoF time-domain maneuvering model [1] evaluates the relative location of the hull and the position of the rock tip at each time step. The search for the contact of the rock and the panel starts when the rock tip enters a sub-block. Figure 8 depicts the implementation phases, summed up as follows:

a) Based on the ship's main dimensions, a cartesian grid that surrounds the ship is created. The grid is made up of identical 3D sub-blocks along the length, beam, and depth of the ship.
b) In each 3D sub-bock panels are stored depending on the location of the panels with respect to the 3D sub-block position. The initialization steps under(a) and (b) occur only once.
c) To search for the rock inside the sub-block, the rock tip $\vec{r}_{Gr}$ is changed from an inertial coordinate system to a ship-fixed coordinate system.
d) The vector $\vec{r}_{Gr}$ (Figure 6) is used to identify whether the rock is in the bounding cartesian grid of the ship. If the rock tip is within the cartesian grid, a 3D sub-block containing the rock tip is detected Figure 8. If not, the rock has no effect on the ship structure and the ship will perform her intended movement.
e) A vertical vector is formed, which points towards the base of the seabed from the tip of the rock. If the rock tip is within the 3D sub-block then, this vector is used to search the panel. If a 3D sub-block does not contain panels, then neighbouring block in the direction of vector towards seabed is searched to locate panels. This will follow until the vector reached the edge of the last sub-block in the direction of the vector. If an intersection is found in between above steps the intersection point is stored and the penetration is calculated using a Ray-tracing method [48].
f) The contact coupling model outputs the penetration into the hull following the identification of an intersection. After that, the internal mechanics model receives this penetration and uses it to determine the structural resistance force.

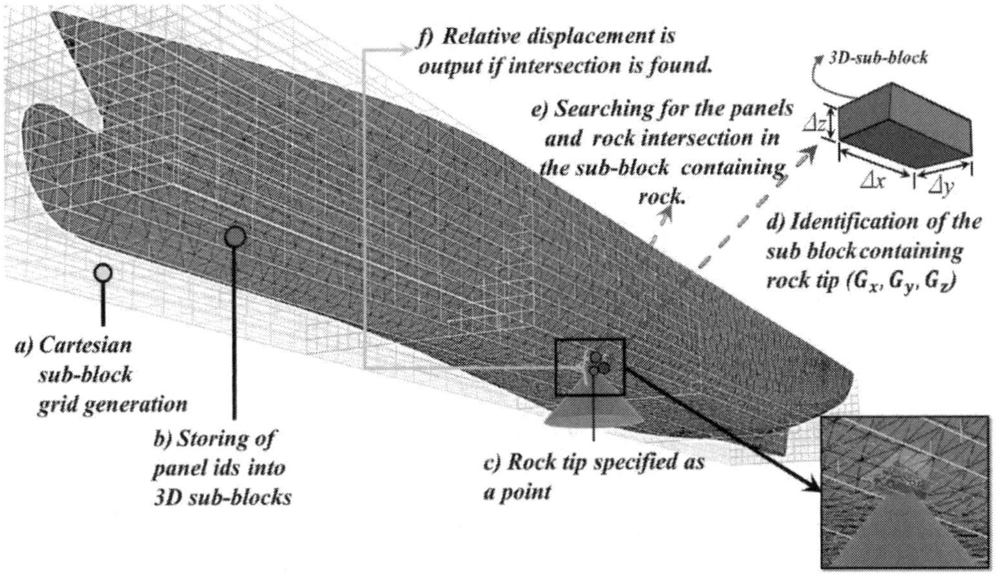

Figure 8. Coupling algorithm details

5. Limitations

The detailed modelling and validation results of the 6-DoF maneuvering model under the influence of the surrounding environment and the comparison coupling algorithm with commercial software LSDYNA can be found in [49] and [1], respectively. The limitation of the models is summarized as follows:

- The hydrodynamic coefficients were evaluated and found to yield satisfactory results for four distinct vessel types. Two different twin-screw and twin-rudder vessels have been validated using the reference approach.
- Drifting of a vessel is considered in shortwaves. As a result, the effect of the waves on heave, pitch, and roll motions is not taken into consideration.
- Yaw-roll hydrodynamic coupling is not taken into consideration. Such modelling could be necessary for a turn at a high speed.
- Pitch motions caused by ship roll are not taken into account. This might result in an underestimation of pitch oscillations during maneuvers. However, the validation of ship off-centre grounding against LS-DYNA MCOL simulations yields satisfactory pitch motions, but the roll motion was overpredicted. These uncertainties need further investigation from the perspective of hydrostatics, and lateral deformation forces of longitudinal members.
- In-plane (hydrodynamic coefficients) and out-of-plane (linear radiation) damping formulations based on regression inherit the memory effects. Yet, memory effects are not specifically considered.
- The plate deformation follows a single clean curling cut. The concertina tearing (plate folding back and forth) and braided cuts (flaps folding back and forth) are not considered. Furthermore, grounding over a conical rock may result in two stable cuts, which is not considered.
- The model overestimates the vertical force in a refined geometric region such as bulbous bow and bilge curves. The model is ideally suited for flat plate interactions.
- The impact of lateral forces on longitudinal girders secured between rocks is not considered.
- The very first contact does not occur between the slant height of the cone and the ship. Instead, the initial contact occurs with the tip of the rock (some distance above the keel base).
- The developed model idealises a rounded tip conical rock. Other shapes are ignored.
- The effect of flooding during an event of grounding and post-grounding is not considered.

6. Conclusions and future study

The goal of developing a two-way coupled rapid FSI grounding dynamics model was to propose a novel and efficient method of assessing ship accidents from a multiphysics viewpoint, as existing numerical methods are time-consuming and large-scale experimentation is impractical. The research establishes the foundation for a computationally efficient technique for coupling external mechanics with internal mechanics. In comparison to LS-DYNA MCOL simulations, which required 4 days of supercomputer time for the simulation, the current FSI model calculation time was reported as being less than 1 second for the same simulation [50]. It is established after successful validation that:

- It is possible to predict the manoeuvring trajectories of both new and existing vessels using the reference technique. The same method can be used to estimate the hydrodynamic activities that will occur in the event of a grounding contact.
- Based on comparisons against experimental data, shallow water and short waves assumptions may be considered acceptable. The same holds for damaged length, breadth, hull penetration, and total deformation energy predicted.
- The assumption of a constant plate split angle approach should be discarded, and structural resistance forces should be evaluated using motion-dependent plate split angles.
- The assumption of linear out-of-plane motions (heave, roll, and pitch) is appropriate for examining grounding dynamics.
- To accurately forecast structural deformation and ship motions, ship restoring forces, damping, and 6-DoF rigid body dynamics must be included. Resistance and shallow water effects have minimal effect on structural resistance force and ship dam-age extents.

Future work should consider hydrodynamic coefficients and their combinations for various hull forms. The lateral loading of the longitudinal elements (girders and stiffeners) is not properly modelled, which results in an underestimation of the lateral forces in an assumed straight-course operation. The severe curvature of the hull's extremities (e.g., bulbous bow region) causes an overestimation of vertical forces. Thus, the model primarily addressed the impact of conical rock profiles and further development of structural resistance force for different rock type are required.

References

[1] Taimuri G, Kim S J, Mikkola T and Hirdaris S 2022 A two-way coupled FSI model for the rapid evaluation of accidental loads following ship hard grounding *J. Fluids Struct.* **112** 103589

[2] Allianz Global Corporate & Specialty 2022 *Safety and Shipping Review 2022 The safety and shipping review from Allianz Global Corporate & Specialty (AGCS) is an annual review of trends and developments in shipping losses, risk challenges and safety*

[3] European Maritime Safety Agency 2021 *Annual overview of marine casualties and incidents - EMSA - European Maritime Safety Agency*

[4] Zhang M, Taimuri G, Zhang J, Zhang D, Yan X and Hirdaris S Ship collision and grounding assessment: State of the art and future perspectives (under review) *Engineering*

[5] Peterson C H, Rice S D, Short J W, Esler D, Bodkin J L, Ballachey B E and Irons D B 2003 Long-term ecosystem response to the Exxon Valdez oil spill *Science (80-.).* **302** 2082–6

[6] Seveso D, Louis Y D, Montano S, Galli P and Saliu F 2021 The Mauritius oil spill: what's next? *Pollutants* **1** 18–28

[7] Papanikolaou A, Spanos D, Boulougouris E, Eliopoulou E and Alissafaki A 2004 Investigation into the sinking of the RO-RO passenger ferry Express Samina *Int. Shipbuild. Prog.* **51** 95–120

[8] Dimitrakakis E, Hahladakis J and Gidarakos E 2014 The "Sea Diamond" shipwreck: Environmental impact assessment in the water column and sediments of the wreck area *Int. J. Environ. Sci. Technol.* **11** 1421–32

[9] Schröder-Hinrichs J U, Hollnagel E and Baldauf M 2012 From Titanic to Costa Concordia—a century of lessons not learned *WMU J. Marit. Aff.* **11** 151–67

[10] Stickings T, Griffith K, Mitchell C, Wright J and Patel B 2021 MailOnline (Suez Canal mega-

ship is on the move) *dailymail*

[11] Bakdi A, Glad I K, Vanem E and Engelhardtsen Ø 2020 AIS-based multiple vessel collision and grounding risk identification based on adaptive safety domain *J. Mar. Sci. Eng.* **8(1)** 1–19

[12] Mazaheri A, Montewka J and Kujala P 2014 Modeling the risk of ship grounding—a literature review from a risk management perspective *WMU J. Marit. Aff.* **13** 269–97

[13] Zhang S, Pedersen P T and Villavicencio R 2019 Probability of ship collision and grounding *Probability and mechanics of ship collision and grounding* ed Butterworth-Heinemann (Kidlington: Matthew Deans) pp 1–61

[14] Haris S and Amdahl J 2013 Analysis of ship–ship collision damage accounting for bow and side deformation interaction *Mar. Struct.* **32** 18–48

[15] Heinvee M and Tabri K 2015 A simplified method to predict grounding damage of double bottom tankers *Mar. Struct.* **43** 22–43

[16] Hong L and Amdahl J 2012 Rapid assessment of ship grounding over large contact surfaces *Ships Offshore Struct.* **7** 5–19

[17] Hong L and Amdahl J 2008 Crushing resistance of web girders in ship collision and grounding *Mar. Struct.* **21** 374–401

[18] Cerup-Simonsen B, Törnqvist R and Lützen M 2009 A simplified grounding damage prediction method and its application in modern damage stability requirements *Mar. Struct.* **22** 62–83

[19] Song M, Kim E, Amdahl J, Ma J and Huang Y 2016 A comparative analysis of the fluid-structure interaction method and the constant added mass method for ice-structure collisions *Mar. Struct.* **49** 58–75

[20] Zhang S, Pedersen P T and Villavicencio R 2019 Internal mechanics of ship collision and grounding *Probability and mechanics of ship collision and grounding* pp 147–270

[21] Calle M A G, Salmi M, Mazzariol L M and Kujala P 2020 Miniature reproduction of raking tests on marine structure: Similarity technique and experiment *Eng. Struct.* **212** 110527

[22] Calle M A G, Oshiro R E and Alves M 2017 Ship collision and grounding: Scaled experiments and numerical analysis *Int. J. Impact Eng.* **103** 195–210

[23] Kitamura O and Kuriowa T 1996 Large-scale grounding experiments and numerical simulations *Sh. Technol. Res.* **43** 62–9

[24] Lemmen P M, Vredeveldt W and Pinkster J A 1996 Design analysis for grounding experiments *International Conference on Designs and Grounding Protection of Ships, San Francisco, California, August 22-23* p 6

[25] Rodd J L and Sikora J P 1995 Double hull grounding experiments *Proceedings of the International Offshore and Polar Engineering Conference* pp 446–56

[26] Rodd J L 1996 Large scale tanker grounding experiments *Proceedings of the International Offshore and Polar Engineering Conference* vol 4 pp 483–94

[27] Tabri K, Määttänen J and Ranta J 2008 Model-scale experiments of symmetric ship collisions *J. Mar. Sci. Technol.* **13** 71–84

[28] Ehlers S 2011 A review of collision and grounding damage assessment methods *Mar. Syst. Ocean Technol.* **6** 5–15

[29] Yu Z, Shen Y, Amdahl J and Greco M 2016 Implementation of linear potential-flow theory in the 6DOF coupled simulation of ship collision and grounding accidents *J. Sh. Res.* **60** 119–44

[30] Lee S G, Zhao T and Nam J H 2013 Structural safety assessment of ship collision and grounding using FSI analysis technique *Collision and Grounding of Ships and Offshore Structures - Proceedings of the 6th International Conference on Collision and Grounding of Ships and Offshore Structures, ICCGS 2013* (Taylor and Francis - Balkema) pp 197–204

[31] Lee S G, Lee J S, Lee H S, Park J H and Jung T Y 2017 Full-scale ship collision, grounding and sinking simulation using highly advanced M&S system of FSI analysis technique *Procedia Engineering* vol 173 (Elsevier Ltd) pp 1507–14

[32] Song Z and Hu Z 2017 An integrated analytical tool on predicting structural responses of ships under collision and grounding scenarios *Proceedings of the International Conference on*

Offshore Mechanics and Arctic Engineering - OMAE, 57656

[33] Prabowo A R, Cao B, Sohn J M and Bae D M 2020 Crashworthiness assessment of thin-walled double bottom tanker: Influences of seabed to structural damage and damage-energy formulae for grounding damage calculations *J. Ocean Eng. Sci.* **5** 387–400

[34] Liu B, Villavicencio R, Pedersen P T and Guedes Soares C 2021 Analysis of structural crashworthiness of double-hull ships in collision and grounding *Mar. Struct.* **76** 102898

[35] Brubak L, Hu Z, Kõrgesaar M, Schipperen I and Tabri K 2021 Numerical simulations of grounding scenarios – benchmark study on key parameters in FEM moddling *Lecture Notes in Civil Engineering* pp 257–69

[36] Kim S J, Kõrgersaar M, Ahmadi N, Taimuri G, Kujala P and Hirdaris S 2021 The influence of fluid structure interaction modelling on the dynamic response of ships subject to collision and grounding *Mar. Struct.* **75** 102875

[37] Kim S J, Kõrgersaar M, Taimuri G, Kujala P and Hirdaris S 2020 A quasi-dynamic approach for the evaluation of structural response in ship collisions and groundings *Proceedings of the International Offshore and Polar Engineering Conference* vol 2020-Octob (International Society of Offshore and Polar Engineers) pp 3174–80

[38] Kim S J, Sohn J M, Kujala P and Hirdaris S 2022 A simplified fluid structure interaction model for the assessment of ship hard grounding *J. Mar. Sci. Technol.* **27** 695–711

[39] Kim S J, Taimuri G, Kujala P, Conti F, Le Sourne H, Pineau J P, Looten T, Bae H, Mujeeb-Ahmed M P, Vassalos D, Kaydihan L and Hirdaris S 2022 Comparison of numerical approaches for structural response analysis of passenger ships in collisions and groundings *Mar. Struct.* **81** 103125

[40] Simonsen B C 1997 *Mechanics of ship grounding. Department of Naval Architecture and Offshore Engineering* (Technical University of Denmark (DTU))

[41] Matusiak J and Varsta P 2002 Transient motion of ship during hard grounding *Proceedings of the 6th International Ship Stability Workshop, Webb Institute, Proceedings. Paper: P2002-1 Proceedings.*

[42] Nguyen T H, Amdahl J, Garrè L and Leira B J 2011 A study on dynamic grounding of ships *Advances in Marine Structures - Proceedings of the 3rd International Conference on Marine Structures, MARSTRUCT 2011* pp 373–80

[43] Abubakar A and Dow R S 2013 Simulation of ship grounding damage using the finite element method *Int. J. Solids Struct.* **50** 623–36

[44] Zeng J, Hu Z and Chen G 2016 A steady-state plate tearing model for ship grounding over a cone-shaped rock *Ships Offshore Struct.* **11** 245–57

[45] Yu Z and Amdahl J 2016 Full six degrees of freedom coupled dynamic simulation of ship collision and grounding accidents *Mar. Struct.* **47** 1–22

[46] Sun B, Hu Z and Wang J 2017 Bottom structural response prediction for ship-powered grounding over rock-type seabed obstructions *Mar. Struct.* **54** 127–43

[47] Friis-Hansen P and Simonsen B C 2002 GRACAT: Software for grounding and collision risk analysis *Mar. Struct.* **15**

[48] Moller T and Trumbore B 1997 Fast, minimum storage ray-triangle intersection *J. Graph. Tools* **2** 21–8

[49] Taimuri G, Matusiak J, Mikkola T, Kujala P and Hirdaris S 2020 A 6-DoF maneuvering model for the rapid estimation of hydrodynamic actions in deep and shallow waters *Ocean Eng.* **218** 108103

[50] Taimuri G H 2022 *A fully coupled Fluid Structure Interaction model for assessing ship hard grounding dynamics* (Aalto University)

IOP Conf. Series: Materials Science and Engineering 1288 (2023) 012015 doi:10.1088/1757-899X/1288/1/012015

Coupled Lattice Boltzmann and Discrete Element Simulations of Ship-Ice Interactions

Seiya Watanabe[1], **Changhong Hu**[1] **and Takayuki Aoki**[2]

[1] Research Institute for Applied Mechanics, Kyushu University, 6-1 Kasuga-koen Kasuga-city, Fukuoka, Japan

[2] Global Scientific Information and Computing Center, Tokyo Institute of Technology, 2-12-1 i7-3 O-okayama, Meguro-ku, Tokyo, Japan

E-mail: swatanabe@riam.kyushu-u.ac.jp

Abstract. Evaluating ice loads acting on ships is essential for the safety of ships navigating in ice-covered seas. In this study, we develop a CFD method to handle ship, ice, and fluid interaction. The lattice Boltzmann method, capable of large-scale calculations, is applied to the simulation of free-surface fluids. The ice motion is computed by solving the equations of motion of a rigid body, and the discrete element method models the ice-ice and ice-ship contact interactions. A momentum exchange scheme couples the lattice Boltzmann method and particle-based rigid body simulation. We introduce tree-based adaptive mesh refinement and multiple GPU computing to improve grid resolution and computational time. The proposed method is applied to model scale simulations of ship navigation in a brash ice channel. Simulations were performed for various conditions with different ice concentrations and ship velocities, and we observed that ice resistance increased with the ice concentration and the ship velocity increased. The ice motions and resistances obtained from our simulations are reasonable compared to model equations of Finnish-Swedish ice class rules (FSICR) and numerical analyses of a previous study.

1. Introduction

With the shrinking of ice due to global warming, the Arctic Ocean has become more attractive in recent years, and the number of ocean-going commercial vessels using the Northern Sea Route has increased. These ships select routes where it is possible to navigate without ice-breaking capabilities. Since ships collide with ice floes, it is essential to evaluate the ice force acting on the ship, which directly affects the safety of navigation, by experiments and numerical models.

An ice tank experiment using a model ship is an effective measure to predict ice resistance performance. However, facilities with ice tanks are limited to a few places worldwide, and it takes time to fabricate ice channels. Thus ice tank experiments are time-consuming and costly.

Numerical simulation tools that can accurately and quickly predict ice resistance are required to reproduce ice tank experiments. Since interactions between ice, fluid, free surface, and ship are very complex physical phenomena, many studies proposed simulation methods deal only with the ship-ice interaction [1, 2]. Flow fields and free interfaces are not calculated, and floating ice floes are subjected to drag and buoyancy forces modeled by empirical equations. Computational models that handle only ice-ship interactions efficiently estimate ice resistance in broken ice and brash ice channels [3, 4]. However, if the flow field around an ice floe affects other ice floes; for example, if the ice floes move densely together, the accuracy of the ice force estimate is reduced.

Recent performance improvements in computational hardware have enabled computational fluid dynamics (CFD) to calculate even the flow field around ice floes and to treat ice-fluid interactions without empirical modeling. Shafiul Mintu and David Molyneux [5] developed a fluid-solid interaction model combining the smoothed particle hydrodynamics (SPH) and the discrete element method (DEM) for ship-ice interaction in waves. The authors of [6] have developed simulation software that combines the OpenDynamics Engine (ODE) with the lattice Boltzmann method (LBM)-based software ELBM. The GPU, which is high-performance hardware, was used to accelerate the fluid calculations, and their ODE-ELBM software simulates ship-fluid-rigid-ice interactions. The high computational cost of direct ice-fluid interactions still needs to be addressed, limiting the number of floating ice floes in simulations.

In this study, we develop a high-performance CFD tool for ship-fluid-ice interactions using the LBM and the DEM to increase the number of ice pieces handled in simulations. To solve the high computational cost of direct coupling of fluid and ice motions, we introduce the adaptive mesh refinement (AMR) method, which locally assigns a high-resolution grid, and perform parallel computation using multiple GPUs on a supercomputer. The ice floe geometry is represented by a simple model of rigidly-connected micro DEM particles, and all DEM calculations are performed on GPUs. In this paper, we report simulation results of the navigation of a model ship in a brash ice channel composed of small ice cubes. The motion of the ice floes and the ice force acting on the ship are compared with simulations of previous studies and the FSICR formula for validation.

2. Numerical method

We develop a robust and fast computational method for fluid-ship-ice interactions. The single-phase lattice Boltzmann method calculates hydrodynamics with free-surface boundaries. The phase-field method simulates the advection and deformation of the free surface. We apply particle-based rigid body dynamics for the ice simulation and do not consider the fracture and consolidation of the ice floes. A mass-spring-damper system of the DEM computes ice-ice and ice-ship contact interactions. The block-based adaptive mesh refinement method and parallel computing on multiple GPUs speed up simulations.

Our previous studies [7][8] describe the details of the computational method and validation results on benchmark problems. We briefly explain the methodology in this section.

2.1. Lattice Boltzmann method

The lattice Boltzmann method (LBM) assumes that the fluid is virtual particles that collide and stream on a grid. The statistical behavior of the virtual particles follows the Boltzmann equation. By discretizing the Boltzmann equation with the finite velocity direction $\boldsymbol{\xi}_{ijk}$, the lattice spacing Δx, and the time step Δt, the lattice Boltzmann equation is obtained as

$$f_{ijk}\left(\boldsymbol{x}+\boldsymbol{\xi}_{ijk}\Delta t, t+\Delta t\right) = f_{ijk}\left(\boldsymbol{x}, t\right) + \Omega_{ijk}\left(\boldsymbol{x}, t\right). \tag{1}$$

The subscript ijk indicates the discretized velocity direction, which follows the D3Q27 velocity model in this study. A stable and accurate cumulant LBM model [9] is introduced for the collision term Ω to calculate the flow around a ship at high Reynolds numbers.

The fluid density ρ and the flow velocity $\boldsymbol{u}$ are calculated as

$$\rho\left(\boldsymbol{x}, t\right) = \sum_{ijk} f_{ijk}(\boldsymbol{x}, t), \tag{2}$$

$$\rho\boldsymbol{u}\left(\boldsymbol{x}, t\right) = \sum_{ijk} \boldsymbol{\xi}_{ijk} f_{ijk}(\boldsymbol{x}, t). \tag{3}$$

Under the weakly compressible fluid assumption, the fluid pressure is proportional to density as $p = c_s^2\rho$, where c_s is the pseudo sound speed.

2.2. Free-surface model

The free surface shape is captured by a phase-field method that defines order parameter ϕ identifying the liquid and gas phases. We define $\phi = 1$ for the liquid phase and $\phi = 0$ for the gas phase, and the gas-liquid interface varies ϕ smoothly over three lattice widths. The time evolution of the phase-field variable ϕ follows the conservative Allen-Cahn equation,

$$\frac{\partial \phi}{\partial t} + \nabla \cdot (\boldsymbol{u}\phi) = \nabla \cdot \left[M \left\{ \nabla \phi - \frac{1 - 4\left(\phi - \phi^{\mathrm{ave}}\right)^2}{W} \boldsymbol{n} \right\} \right], \tag{4}$$

where, W is the interface thickness, and the M parameter sets the interface mobility. ϕ^{ave} is the average value of ϕ between the liquid and gas phases, which in this study is 1/2.

To better handle moving solid boundaries such as a ship and ice floes, we use a phase-field LBM [10] to solve the conservative Allen-Chan equation. For the order parameter ϕ, we define the velocity distribution function h as

$$\phi\left(\boldsymbol{x}, t\right) = \sum_{ijk} h_{ijk}(\boldsymbol{x}, t). \tag{5}$$

h is time updated with a simple collision term using a single relaxation time parameter τ_ϕ, as

$$h_{ijk}\left(\boldsymbol{x} + \boldsymbol{\xi}_{ijk}\Delta t, t + \Delta t\right) = h_{ijk}\left(\boldsymbol{x}, t\right) - \frac{1}{\tau_\phi}\left(h_{ijk}\left(\boldsymbol{x}, t\right) - h_{ijk}^{\mathrm{eq}}\left(\boldsymbol{x}, t\right)\right). \tag{6}$$

h_{ijk}^{eq} is the distribution function of the equilibrium state, given by

$$h_{ijk}^{\mathrm{eq}} = \phi \Gamma_{ijk} + \theta w_{ijk} \boldsymbol{\xi}_{ijk} \cdot \boldsymbol{n}, \tag{7}$$

$$\Gamma_{ijk} = w_{ijk}\left\{1 + \frac{\boldsymbol{\xi}_{ijk} \cdot \boldsymbol{u}}{c_{\mathrm{s}}^2} + \frac{\left(\boldsymbol{\xi}_{ijk} \cdot \boldsymbol{u}\right)^2 - c_{\mathrm{s}}^2 \left|\boldsymbol{u}\right|^2}{2c_{\mathrm{s}}^4}\right\}, \tag{8}$$

$$\theta = \frac{M}{c_{\mathrm{s}}^2}\left\{\frac{1 - 4\left(\phi - \phi^{\mathrm{ave}}\right)^2}{W}\right\}, \tag{9}$$

where, w_{ijk} is the weight parameter of the D3Q27 model. The relaxation parameter τ_ϕ is determined from the interface mobility M as

$$\tau_\phi = \frac{1}{2} + \frac{M}{c_s^2 \Delta t}. \tag{10}$$

The cumulant LBM calculates only the water domain ($\phi \geq \phi^{\mathrm{ave}}$) and imposes boundary conditions on the free surface for computational stability. The velocity distribution function f for the gas ($\phi < \phi^{\mathrm{ave}}$)is undefined. At grid points near the free surfaces, missing distribution functions from the gas phase are given as

$$f_{ijk}\left(\boldsymbol{x}, t + \Delta t\right) = f_{ijk}^{\mathrm{eq}}\left(\rho_0, \boldsymbol{u}\right) - \left[f_{\overline{ijk}}\left(\boldsymbol{x}, t\right) - f_{\overline{ijk}}^{\mathrm{eq}}\left(\rho_0, \boldsymbol{u}\right)\right], \tag{11}$$

$$f_{ijk}^{\mathrm{eq}}\left(\rho_0, \boldsymbol{u}\right) = w_{ijk}\rho_0\left[1 + 3\boldsymbol{\xi}_{ijk} \cdot \boldsymbol{u} + \frac{9}{2}\left(\boldsymbol{\xi}_{ijk} \cdot \boldsymbol{u}\right)^2 - \frac{3}{2}\boldsymbol{u} \cdot \boldsymbol{u}\right], \tag{12}$$

where, $f_{\overline{ijk}}$ is the distribution function in the opposite direction of f_{ijk}, and ρ_0 is the reference density.

IOP Conf. Series: Materials Science and Engineering 1288 (2023) 012015 doi:10.1088/1757-899X/1288/1/012015

2.3. Ice model

A particle-based rigid body simulation method [11] calculates ice motions and ice-ice, ice-ship contact interactions. A rigid body consisting of many small spherical particles represents the shape of a ship or floating ice floes. The particles, placed on a solid surface, detect collisions with other objects and calculate the contact force based on the spring and dashpot models of the DEM [12]. The DEM has been applied in many numerical studies of ice-structure interactions[13]. The spring applies a repulsive force proportional to the penetration depth of the particles, and the dashpot adds a damping force that depends on the relative velocity of the particles. The friction between the particles is expressed by setting an upper bound on the tangential force according to Coulomb's law of friction. Ice accretions are not considered. A sum of contact forces $\boldsymbol{F}_i^{\mathrm{DEM}}$ acting on i-th spherical particle gives the total force acting on the rigid body and the torque around the center of gravity.

The motion of a rigid body is calculated from two parts: translation and rotation. The equation of motion for the translation of a rigid body is

$$m_s \frac{\mathrm{d}\boldsymbol{v}_s}{\mathrm{d}t} = \sum_i \boldsymbol{F}_i^{\mathrm{DEM}} + \boldsymbol{F}_s^{\mathrm{fluid}} + m_s \boldsymbol{g}, \tag{13}$$

where m_s and $\boldsymbol{v}_s$ are the rigid body's mass and translation velocity, respectively. The equation of motion for rigid body rotation is

$$\frac{\mathrm{d}\boldsymbol{L}_s}{\mathrm{d}t} = \sum_i \boldsymbol{r}_i \times \boldsymbol{F}_i^{\mathrm{DEM}} + \boldsymbol{M}_s^{\mathrm{fluid}}, \tag{14}$$

$$\boldsymbol{\omega}_s(t) = \boldsymbol{I}_s(t)^{-1} \boldsymbol{L}_s(t), \tag{15}$$

where, $\boldsymbol{L}_s$ is the angular momentum of the rigid body, $\boldsymbol{\omega}_s$ is the angular velocity, and $\boldsymbol{I}_s$ is the inertia tensor. $\boldsymbol{r}_i$ is the relative position of the i-th particle to the center of gravity of the rigid body. $\boldsymbol{F}_s^{\mathrm{fluid}}$ and $\boldsymbol{M}_s^{\mathrm{fluid}}$ are the force and torque given by the LBM fluid and are explained in the next section. These equations of motion are integrated over time using the first-order Euler method.

2.4. Fluid-structure interaction

The mesh for the LBM calculations is fine enough for the size of the ice floes, and the fluid-ice interactions are calculated directly from the flow field around the ice. We use a coupling method based on momentum exchange between objects and fluid. The motion of the ice pieces is treated as a moving boundary condition in the LBM. The ice shape is implicitly represented by a signed distance function, the distance from the ice surface to the grid point. A non-slip moving boundary condition based on the interpolated bounce-back scheme [14] is used at the ice-fluid interface. According to the normalized signed distance function q, the velocity distribution function bouncing off the object surface is calculated as

$$f_{ijk}(\boldsymbol{x}, t+\Delta t) = \begin{cases} 2q f_{\overline{ijk}}(\boldsymbol{x}, t) + (1-2q) f_{\overline{ijk}}(\boldsymbol{x} + \boldsymbol{\xi}_{ijk}\Delta t, t) + \\ \quad \dfrac{6\omega_{ijk}\rho(\boldsymbol{\xi}_{ijk} \cdot \boldsymbol{u}_{\mathrm{wall}})}{c^2} & \text{if } 0 < q \le 0.5 \\ \dfrac{1}{2q}\left\{ f_{\overline{ijk}}(\boldsymbol{x}, t) + \dfrac{6\omega_{ijk}\rho(\boldsymbol{\xi}_{ijk} \cdot \boldsymbol{u}_{\mathrm{wall}})}{c^2} \right\} + \\ \quad \dfrac{(2q-1)}{2q} f_{ijk}(\boldsymbol{x}, t) & \text{if } 0.5 < q \le 1.0, \end{cases} \tag{16}$$

where, $\boldsymbol{u}_{\mathrm{wall}}$ is the wall velocity.

The force exerted on an object by a distribution function bouncing off an object's surface is calculated as

$$\boldsymbol{F}_{ijk}(\boldsymbol{x},t) = (\boldsymbol{\xi}_{ijk} - \boldsymbol{u}_{\text{wall}}) f_{ijk}(\boldsymbol{x},t) - \left(\boldsymbol{\xi}_{\overline{ijk}} - \boldsymbol{u}_{\text{wall}}\right) f_{\overline{ijk}}(\boldsymbol{x},t+\Delta t). \tag{17}$$

This method calculates the momentum exchange considering the object velocity to satisfy Galilean invariance [15]. By summing $\boldsymbol{F}_{ijk}$ at the object surface region Ω_s, the total fluid force $\boldsymbol{F}_s$ and torque $\boldsymbol{M}_s$ is obtained as

$$\boldsymbol{F}_s(t) = \sum_{\boldsymbol{x}\in\Omega_s}\sum_{ijk\in q^*} \boldsymbol{F}_{ijk}(\boldsymbol{x},t), \tag{18}$$

$$\boldsymbol{M}_s(t) = \sum_{\boldsymbol{x}\in\Omega_s}\left\{(\boldsymbol{x}-\boldsymbol{x}_s)\times\sum_{ijk\in q^*}\boldsymbol{F}_{ijk}(\boldsymbol{x},t)\right\}, \tag{19}$$

where, q^* denotes the set of distribution functions that bounced off the surface. To stabilize the computation of fluid-solid couplings, $\boldsymbol{F}_s^{\text{fluid}}$ and $\boldsymbol{M}_s^{\text{fluid}}$, which are substituted into the equations of motion (13)(14), are the average values of $\boldsymbol{F}_s$ and $\boldsymbol{M}_s$ of the previous and current time steps [16] as

$$\boldsymbol{F}_s^{\text{fluid}} = \frac{1}{2}\{\boldsymbol{F}_s(t) + \boldsymbol{F}_s(t-\Delta t)\}, \tag{20}$$

$$\boldsymbol{M}_s^{\text{fluid}} = \frac{1}{2}\{\boldsymbol{M}_s(t) + \boldsymbol{M}_s(t-\Delta t)\}. \tag{21}$$

Moving boundary conditions also apply to the ship-fluid boundaries as well as to the ice. In this study, the ship velocity is constant, so the effect is one-way from the ship to the fluid.

2.5. Adaptive mesh refinement (AMR)

To reduce the computational cost (corresponding to the number of computational grids), we introduce an adaptive mesh refinement method (AMR method) in the LBM and the phase-filed computation. The AMR method allocates a high-resolution grid in a local part of the computational domain. Figure 1 shows that the octree data structure recursively divides the computational domain into cubic regions called blocks. Our AMR strategy places fine blocks near ice pieces, ships, water surfaces, and eddies. Coarse blocks are away from the refinement area. A tree structure corresponds to a cubic region; therefore, the forest of octree data structure, which arranges multiple octree structures, subdivides non-cubic computational domains.

In our simulation code, almost all computational subroutines, such as the LBM, the interface capture, and the DEM, are executed on the GPU; the CPU handles mesh generation and data output. To efficiently parallelize threads on the GPU, a block-structured mesh allocates 8 × 8 × 8 computational cells to each block divided by an octree structure. The calculation points of the LBM are placed at the centers of the computational cells, so-called cell-centered meshes.

At the boundaries where the resolution changes, the velocity distribution function is coarsened and refined using trilinear interpolation. The non-equilibrium components are not rescaled.

Multi-GPU computing is necessary for faster and larger-scale computations. For efficient parallel computing, load balancing is essential to equalize the computational cost of each GPU as much as possible. In this study, we use the Morton curve, one of the space-filling curves, to address the load balancing problem. A Morton curve divides the computational domain to equal the number of computational blocks. A single MPI process and its corresponding GPU handle the computation in one of the divided subdomains.

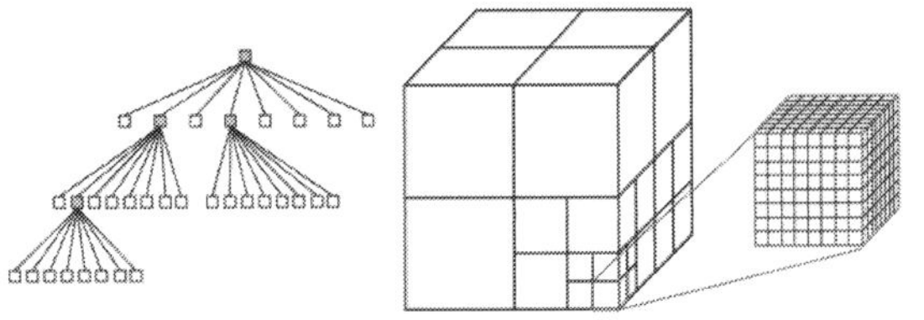

Figure 1. Block-structured mesh based on octree data structure. Each block has $8 \times 8 \times 8$ cells.

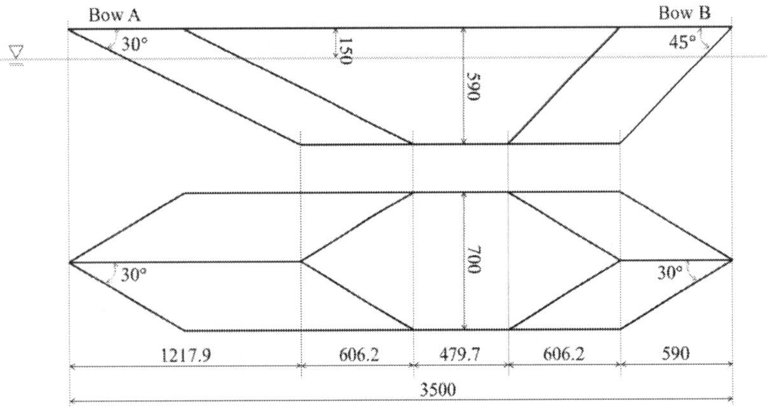

Figure 2. Geometry of model ship (Upper: side view, Lower: bottom view).

Calculation points near the subdomain boundaries access data on the GPU handling the neighboring subdomain. Therefore, data communication with the neighboring GPUs occurs every time step. Some computer systems allow direct data transfer between GPUs via CUDA-aware MPI or NVLink. To ensure that the simulation code does not depend on a specific GPU system, however, in our code, the data transfer between GPUs is realized by the CPU.

3. Simulation setup

We evaluate the ice resistance of ship navigation in a brash ice channel by model-scale simulations. The computational conditions were set up regarding numerical tests conducted by Konno et al. [4]. The model ship B-063 of the National Maritime Research Institute was used. This model ship has a simple shape for ice-breaking experiments. As shown in Figure 2, it is a hexagonal pyramidal trapezoidal shape with a total length of 3.5 m and a width of 0.7 m. The bow at both ends of the hull can be used for experiments. One side has a bow inclination of 30 degrees (Bow A), and the other has a bow inclination of 45 degrees (Bow B).

The ice class was 1A according to the Finnish-Swedish Ice Class Rules (FSICR) [17]. The scale ratio of the model to the full-scale ship was set to 25. The average thickness of the brash ice channel in the model scale was 41 mm (corresponding to 1 m at full-scale) based on the FSICR guidelines.

The computational domain was 11.15 m × 1.39 m × 1.39 m, and the water depth was set at 1.11 m. A brash ice channel was modeled by floating 41 mm ice cubes at equal intervals. The width and length of the brash ice channel were 1.39 m and 5.56 m, respectively. Figure 3 shows the arrangement of the model ship and the brash ice channel in the computational domain. The ice density was 950 kg/m^3. We investigated the effect of ice concentration on ice resistance in three cases with ice concentrations of 50%, 68%, and 88%, corresponding to 2034, 3136, and 4096 ice cubes, respectively.

The full-scale ship velocity in a brash ice channel was 5 knots according to FSICR, and the model-scale ship velocity was determined by the similarity rule to be 1 knot = 0.514 m/s. For the two brash ice channels with ice concentrations of 50% and 88%, the ship velocity was set

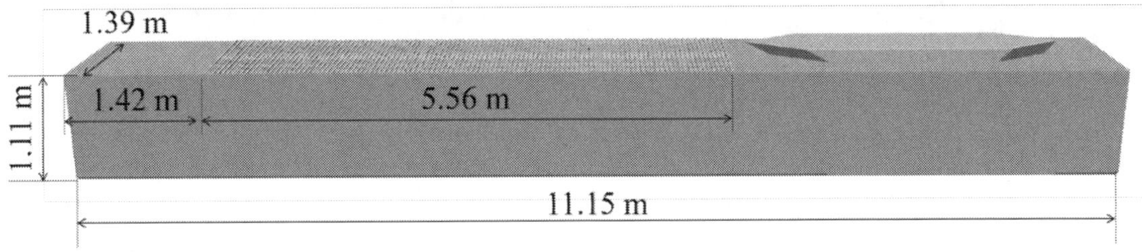

Figure 3. Initial arrangement of model ship and brash ice channel in computational domain.

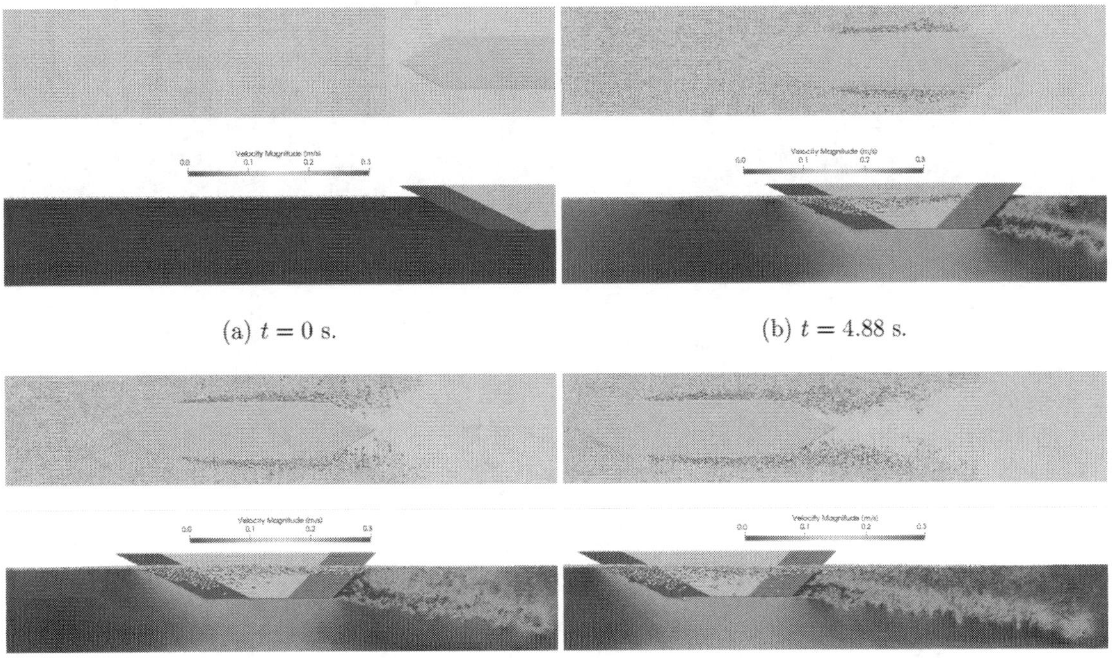

(a) $t = 0$ s.

(b) $t = 4.88$ s.

(c) $t = 7.32$ s.

(d) $t = 9.75$ s.

Figure 4. Snapshots of ship-ice interaction simulation with navigation speed 0.514 m/s and ice concentration 68%. Bottom panels show velocity profile.

to 1 knot. Three cases with different ship velocities of 1, 1.5 and 2 knots in model scale were run for the 68% concentration to study the effect of ship velocity on the ice resistance and the motion of ice cubes.

The finest mesh had a resolution of 2.72 mm and was placed near the free surface, the ship, the ice cubes, and the eddies. Approximately 15 meshes were per side of 41 mm of the ice cube. The coarsest grid was 43.4 mm. At initialization, the number of grid points was about 160 million, about 15% of that using a uniform mesh. The calculations were performed on 32 GPUs of NVIDIA Tesla P100 and were completed within 24 hours in each case.

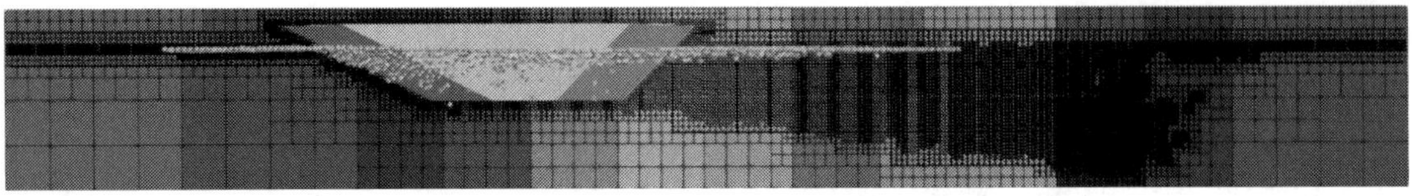

Figure 5. Computational blocks for ship-ice interaction simulation. The color indicates domain decomposition for multi-GPU computation.

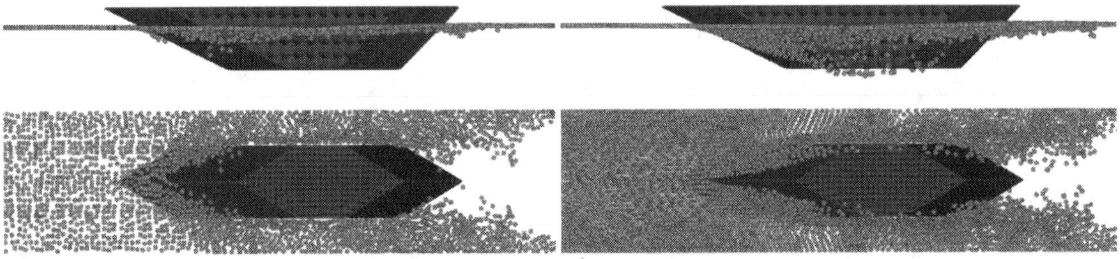

(a) 50%. (b) 88%.

Figure 6. Comparison of ice motions with different ice concentrations of channel.

4. Results and discussion

4.1. Ice motion

As an example of the simulation results, Figure 4 shows simulation snapshots under the ship velocity of 0.514 m/s and the ice concentration of 68%. The top panels are a top view visualizing the free surface, ice, and ship, and the bottom panels are a side view showing the ice, ship, and velocity profile in the center of the channel. Ice cubes colliding with the ship were either pushed to the side of the hull or sank to the bottom. The sinking ice cubes floated to the side of the ship. An ice-free channel was formed in the ship trace. Simulations of previous studies [4] have confirmed similar ice behavior. In addition, the ship's navigation generated waves that induced the heaving motion of the ice pieces. Figure 5 shows the computational block. Due to the eddy, the blocks were divided into smaller blocks in the ship wake. The colors indicate the domain partitioning for multiple GPU computations.

Figure 6 shows the simulated ice motions for different ice concentrations in the brash ice channel. As the ice concentration increased, more ice pieces sank deeper into the lower part of the hull. The width of the ice-free channel narrowed as the ice concentration increased.

Figure 7 shows ice motion differences due to ship velocity. The ship velocity did not significantly affect the sinking motion of the ice cubes. The width of the ice-free channel formed behind the ship narrows with increasing velocity. Faster ship velocity generated higher ship waves; thus, the heaving motion of ice pieces tended to increase with the increase in ship velocity.

Finally, we check the effect of bow shape on ice motion. Figure 8 shows the simulation results for two bow shapes, Bow A and Bow B, at a ship speed of 0.514 m/s and 68% concentration. Bow A, which has a shallower bow angle, has more ice pieces under the hull than Bow B. Bow B has a wider ice-free channel behind the ship.

4.2. Ice resistance

We evaluate ice resistance in a brash ice channel at model ship scale. Ice resistance was calculated by summing the contact forces between the ice cubes and the ship calculated by the DEM model

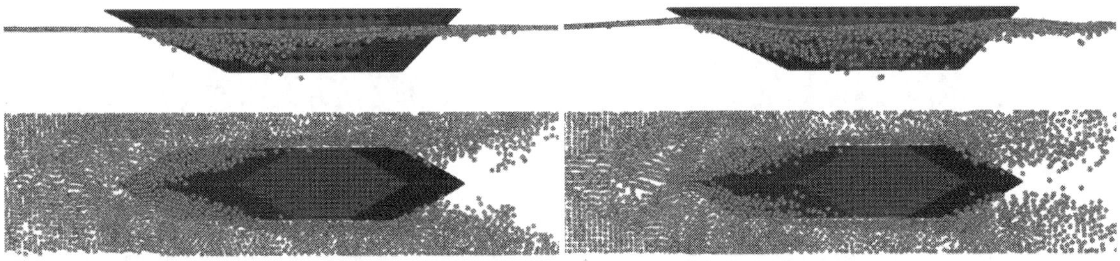

(a) 0.512 m/s (1 knots). (b) 1.024 m/s (2 knots).

Figure 7. Comparison of ice motions with different ship velocities.

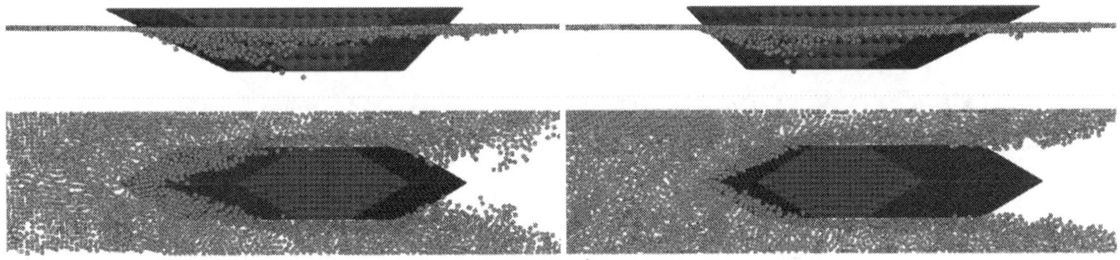

(a) Bow A. (b) Bow B.

Figure 8. Comparison of ice motions with different bow angles.

at the ship's surface. Ice resistance did not contain water resistance.

The time histories of ice force for different ice concentration levels are shown in Figure 9. The ship velocity was 0.512 m/s. Figure 9 (a), (b), and (c) show the ice forces on a log scale, with the horizontal axis representing the ship position. The ice forces varied widely on the order of 10^{-3} N to 10^{2} N. In previous studies [6], such oscillations in ice force have been reported in CFD simulations. The ice load oscillations were smaller for higher ice concentrations. The number of ice pieces in contact with the ship affects the load oscillation. The higher the concentration, the smaller the variation in the number of pieces in contact and, therefore, the smaller oscillations in ice force. The red lines in Figure 9 (a), (b), and (c) show the time-averaged ice force, and Figure 9 (d) shows the relationship between ice concentration and time-averaged ice resistance. It was observed that the ice force increased with increasing ice concentration.

The time series of ice forces in a channel with 68% ice concentration with different ship velocities are shown in Figure 10. The ice force tends to increase as the ship velocity increases. A gradual load change can be observed at the velocity of 2 knots. We consider that this is due to the heave motions of the ice pieces caused by the ship waves and the change in the ice contacting area on the ship.

Figure 9 (d) and Figure 10 (d) compare the ice resistance using Bow A and Bow B. The ice resistance was larger when using Bow B, which has a greater bow angle, than Bow A.

The real-scale ice resistances were estimated based on the cubic law (scale ratio was 25) and

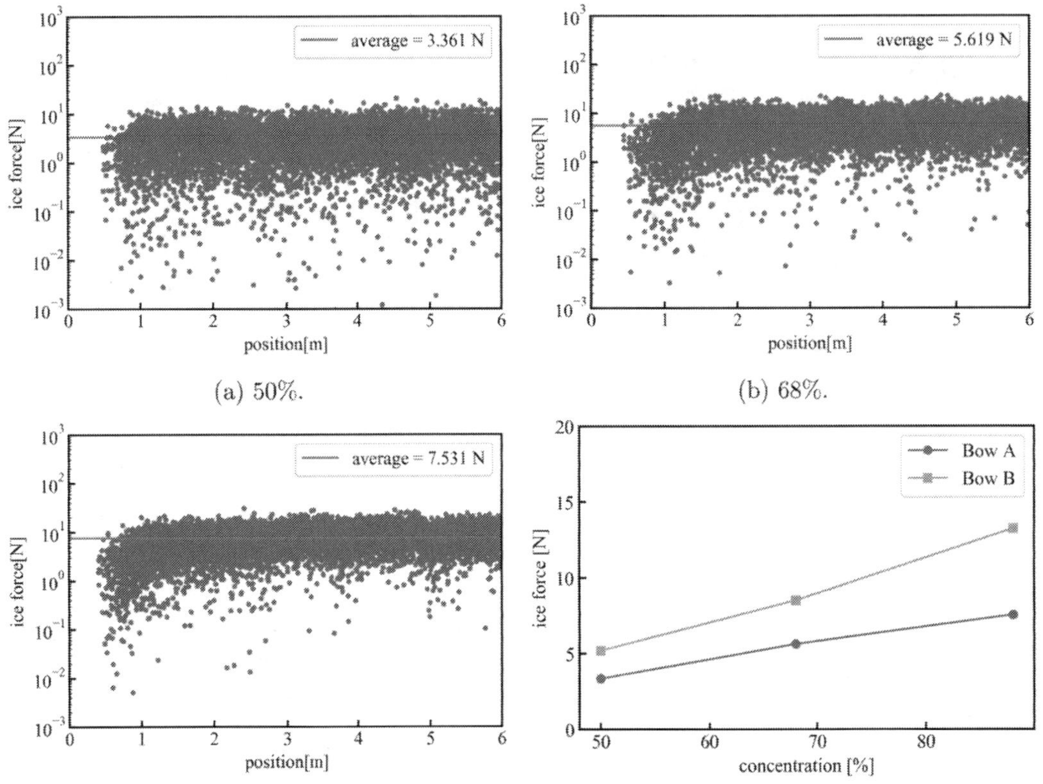

(a) 50%. (b) 68%. (c) 88%. (d) Mean ice force.

Figure 9. Relationship between ice force and concentration. Force signals of bow A are shown.

compared with the ice resistance R_{CH} estimated by the FSICR formula [17]

$$R_{\mathrm{CH}} = C_1 + C_2 + C_3 C_\mu \left(H_{\mathrm{F}} + H_{\mathrm{M}}\right)^2 \left(B + C_\psi H_{\mathrm{F}}\right) + C_4 L_{\mathrm{PAR}} H_{\mathrm{F}}^2 + C_5 \left(\frac{LT}{B^2}\right)^3 \frac{A_{\mathrm{wf}}}{L}, \tag{22}$$

$$C_\mu = 0.15 \cos\phi_2 + \sin\psi \sin\alpha, \tag{23}$$

$$C_\psi = 0.047 \cdot \psi - 2.115, \tag{24}$$

$$H_{\mathrm{F}} = 0.26 + \left(H_{\mathrm{M}} B\right)^{0.5}, \tag{25}$$

$$\psi = \arctan\left(\frac{\tan\phi_2}{\sin\alpha}\right). \tag{26}$$

For ice class 1A, the resistance C_1 and C_2 for breaking the consolidated layer were zero. The resistance coefficients due to friction between the brash ice and the ship were set to $C_3 = 845$ kg/(m^{2}s^2), $C_4 = 42$ kg/(m^{2}s^2), and $C_5 = 825$ kg/s^2. Table 1 shows the dimensions of the ship and some other parameters for the FSICR formula. Table 2 compares the simulation results and the FSICR formula of ice resistance in a class 1A brash ice channel. Previous studies have reported that the FSICR formula tends to estimate ice resistance larger than that of the ice tank experiment [18]. Our simulation results were reasonable because the order of the ice forces was consistent with the FSICR equation, and the simulated ice forces were smaller than the FSICR formula. However, when comparing the ratio of resistance between Bow A

IOP Conf. Series: Materials Science and Engineering 1288 (2023) 012015 doi:10.1088/1757-899X/1288/1/012015

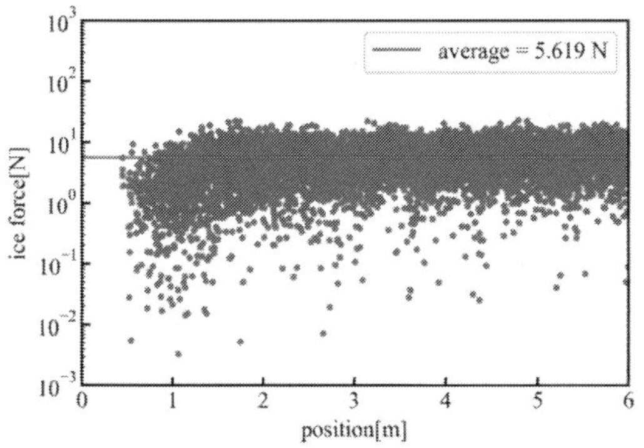

(a) 0.512 m/s (1 knots).

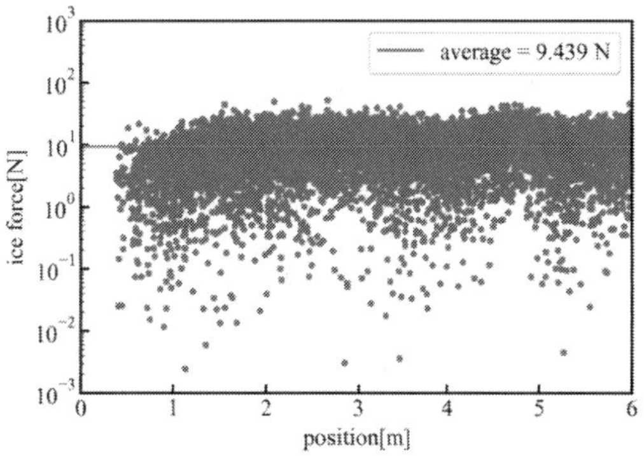

(b) 0.768 m/s (1.5 knots).

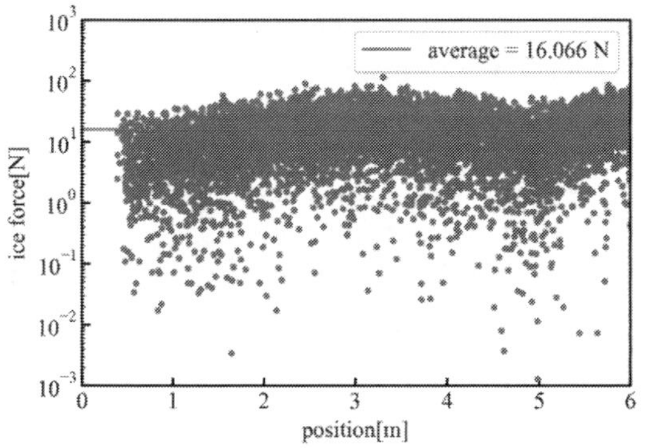

(c) 1.024 m/s (2 knots).

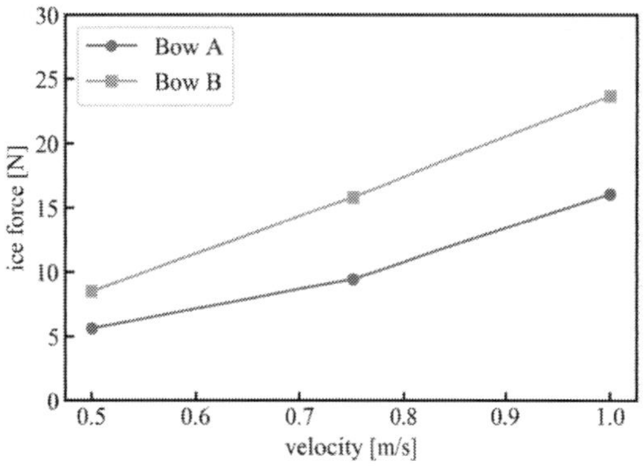

(d) Mean ice force.

Figure 10. Relationship between ice force and ship velocity. Force signals of bow A are shown.

Table 1. Dimensions of model ship and some other parameters for FSICR formula.

Length of the ship between the perpendiculars [m]	L	76.0
Length of the parallel midship body [m]	L_{PAR}	45.7
Maximum breadth of the ship [m]	B	17.5
Actual ice class draughts of the ship [m]	T	11.0
Area of the waterline of the bow [m^2]	A_{wf}	96.25
Angle of the waterline at B/4 [degrees]	α	30
Rake of the stem at the centreline [degrees]	ϕ_1	30 for Bow A, 45 for Bow B
Rake of the bow at B/4 [degrees]	ϕ_2	30 for Bow A, 45 for Bow B
Thickness of the brash ice in mid channel [m]	H_{M}	1

and Bow B cases, the simulation result was 1.76, which was different from 1.21 in the FSICR formula. Although more detailed and accurate validation tests, such as comparisons with tank experiments, are needed, even from the current results, we consider that the proposed ship-ice interaction simulation method helps evaluate ice force in brash ice channels.

Table 2. Comparison of ice resistance between simulation and FSICR equation on real scale.

	Present	FSICR
Bow A	188 kN	293 kN
Bow B	330 kN	355 kN

5. Conclusion

We have developed a numerical method for ship-ice interactions using the LBM and the DEM. The proposed method was applied to model-scale simulations for ship navigation in a brash ice channel to validate the ice piece motions and ice resistance. The simulation reproduced appropriate ice motions, such as the ice pieces sinking under the hull and forming an ice-free channel behind the ship. The ice forces acting on the ship increased with increasing channel ice concentration and ship speed. The simulated ice loads were in the same order of magnitude as the ice forces estimated by the FSICR equation. We conclude that the proposed model was adequate for evaluating ice motions and forces based on the validation results.

Future work includes more detailed and accurate validation compared to ice tank experiments.

Acknowledgements

This research was partly supported by KAKENHI, Grant-in-Aid for Scientific Research (S) JP19H05613, JP22K14181 from Japan Society for the Promotion of Science (JSPS) and 'Joint Usage/Research Center for Interdisciplinary Large-scale Information Infrastructures (JHPCN)' jh210031. The numerical calculations were carried out using the computing resources of the TSUBAME3.0 supercomputer at Tokyo Institute of Technology.

References

[1] Xue Y, Liu R, Li Z and Han D 2020 *Ocean Engineering* **215** 107853
[2] Li F and Huang L 2022 *Journal of Marine Science and Engineering* **10**(2) 165
[3] hwan Kim J, Kim Y, soo Kim H and yeob Jeong S 2019 *Ocean Engineering* **180** 162–174
[4] Akihisa K and Yoshimoto K 2009 *Journal of the Japan Society of Naval Architects and Ocean Engineers* **10** 49–56
[5] Mintu S and Molyneux D 2018 *OTC Arctic Technology Conference* (OnePetro)
[6] Janßen C F, Mierke D and Rung T 2017 **155** 22–32
[7] Watanabe S and Aoki T 2021 *Computer Physics Communications* **264** 107871
[8] Watanabe S, Fujisaki S and Hu C 2021 *Journal of Hydrodynamics* **33**(2) 185–194
[9] Geier M, Schönherr M, Pasquali A and Krafczyk M 2015 *Computers & Mathematics with Applications* **70**(4) 507–547
[10] Geier M, Fakhari A and Lee T 2015 *Physical Review E* **91**(6) 063309
[11] Bell N, Yu Y and Mucha P J 2005 *Proceedings of the 2005 ACM SIGGRAPH/Eurographics Symposium on Computer Animation* SCA '05 (New York, NY, USA: Association for Computing Machinery) p 77–86
[12] Cundall P A and Strack O D 1979 *Geotechnique* **29**(1) 47–65
[13] Tuhkuri J and Polojärvi A 2018 *Philosophical Transactions of the Royal Society A: Mathematical, Physical and Engineering Sciences* **376**(2129) 20170335
[14] Bouzidi M, Firdaouss M and Lallemand P 2001 *Physics of fluids* **13**(11) 3452–3459
[15] Wen B, Zhang C, Tu Y, Wang C and Fang H 2014 *Journal of Computational Physics* **266** 161–170
[16] Aidun C K, Lu Y and Ding E J 1998 *Journal of Fluid Mechanics* **373** 287–311
[17] THE STRUCTURAL DESIGN AND ENGINE OUTPUT REQUIRED OF SHIPS FOR NAVIGATION IN ICE "FINNISH-SWEDISH ICE CLASS RULES" `https://www.sjofartsverket.se/globalassets/isbrytning/pdf-regelverk/finnish-swedish_iceclass_rules.pdf` last accessed 18 April 2023
[18] Hellmann J H, Rupp K H and Kuehnlein W L 2005 *24th International Conference on Offshore Mechanics and Arctic Engineering: Volume 2* pp 923–930

Numerical study of green water loads on a fixed structure with elastic sidewalls by MPS-FEM coupled method

Mengmeng Wu[1], Guang Yang[2] and Weiwen Zhao[1]*

[1] Computational Marine Hydrodynamics Lab (CMHL), School of Naval Architecture, Ocean and Civil Engineering, Shanghai Jiao Tong University, Shanghai, China
[2] Offshore Oil Engineering Co. Ltd., Tianjin, China
*Corresponding author: weiwen.zhao@sjtu.edu.cn

Abstract. In rough seas, large masses of water will exceed the freeboard and cause violent slamming on the deck, known as green water. Due to its high destructive power, many scientists have devoted themselves to the study of green water. Given the complexity, various simplified methods have been used to study the mechanism and stresses of green water. In this paper, the loads and patterns of green water events on a fixed structure with rigid or flexible sidewalls were investigated using the solver MPSFSI-SJTU based on the Moving Particle Semi-Implicit (MPS) method coupled with the Finite Element Method (FEM). The rapid flow-structure interaction is generated by the wet dam-break method. Firstly, the numerical simulation of the water shipping events with the rigid structure was carried out, which was in good agreement with the existing numerical simulation results. Secondly, numerical simulations were carried out for different cases of fixed structures with elastic side wall and elastic upper wall. The influence of elastic modulus on wave loads and waveform was analyzed.

1. Introduction

Marine structures are vulnerable to wave intrusion. In severe sea conditions, these incoming waves are likely to exceed the freeboard, board the deck and spread on the deck. This phenomenon is called green water or water shipping [1]. Green water in rough sea conditions will cause adverse effects such as damage to superstructure and hull structure, accidental overloading of ships, injury to people, damage to equipment, etc. For ships and other marine structures, it is of great practical importance to study the physical mechanism of the wave on deck, to predict the wave on deck and to assess the impact on the safety and performance of ships and offshore platforms.

The phenomenon of green water has been studied by experimental [2], theoretical [3] and numerical methods. The current research mainly focuses on the identification of types of green water events [4], the evolution process of the water on the deck [5], the wave loads [6] and so on. With the development of computer technology, Computational Fluid Dynamics (CFD) methods based on the meshless method and mesh-based method [7-8] have been used to study the green water phenomena. The meshless method needs to introduce special free surface processing methods when dealing with problems such as wave surface rolling, crushing and splashing. Compared with the mesh-based method, the particle-based method does not require additional interface capture or reconstruction algorithms, and can follow free surfaces with arbitrary large deformation. It has natural advantages in simulating large deformation of free surface [9], and scholars have done a lot of research work, for instance, dam-break flow [10-11], water entry problems [10], the liquid sloshing problems [12], wave-ship interaction [13], multiphase flow [14], fluid-structure interaction [15-16] and so on.

Many scientists have used particle methods to study the phenomenon of green water. Shibata et al. [17] simulated the water shipping phenomenon on the bow deck of a three-dimensional fixed ship under the action of solitary waves, using the MPS method to predict the impact pressure on the deck. The wave pattern, wave height and impact pressure on the deck were compared with the experiment, and good consistency is obtained. Kawamura et al. [18] applied the Smoothed Particle Hydrodynamics (SPH) method to predict the motion of fishing vessel under green water conditions produced by steep regular wave and this SPH simulation presented the shipping water events and wave reflection well. Based on the MPS method, Zhang et al. [19] studied the green water phenomenon of solitary waves impacting a plate structure, analyzed the trend of the impact loads on the plate and the wave evolution process under the interaction between solitary waves and structures.

It is worth mentioning that the above simulations simplify the structure as a rigid model. In the recent CFD numerical simulation of the deck wave, some scholars have also carried out numerical simulation of the elastic structure: Qin et al. [20] studied the wave phenomenon on the deck caused by freak waves and the structural response of the deck. A horizontal elastic bare deck with elastic support in the middle of the deck was set in the flume. Qin et al. [20] compared the results of rigid deck, elastic bare deck and elastic deck with intermediate elastic support, and found that although deck elasticity has little effect on the overall evolution of fluid motion, it has a great influence on local fluid pressure. The stiffener (intermediate elastic support) will significantly affect the structural response frequency and deformation. Yu et al. [21] developed a fully coupled nonlinear three-dimensional hydroelastic method to study the vibration response of large ships. The numerical model consists of three-dimensional boundary element method, one-dimensional Euler-Bernoulli beam model and two-dimensional generalized Wagner model. In the experimental research part, the wave load on the deck is considered. The ship model is tested in large amplitude regular waves. The impact pressure and nonlinear vertical bending moment are measured. It is found that the nonlinear model proposed by numerical simulation produces similar results to the experimental model. In addition, the effects of elastic mode and nonlinearity on the numerical results are analyzed. Zhang et al. [22] proposed a coupled MPS and FEM method for the numerical study of the interaction of regular waves with horizontally suspended structures. The fluid domain calculated by the MPS method was discretized into fluid particles and the structural domain solved by the FEM method was discretized into beam elements. The interaction of the regular wave with the rigid structure was analysed and was in good agreement with the experiment. On this basis, the dynamic response of the structure under the action of wave impact was investigated by considering the elasticity of the structure, including the evolution of the free surface, the variation of the wave impact pressure, the velocity distribution and the deformation response of the structure.

Green water events are complex phenomena that occur rapidly. To investigate the features and loads generated by isolated green water events, systematic alternative method has been used. Hernández-Fontes et al. [23] proposed the use of the wet dam-break approach to identify patterns of shipping water on a barge-shaped fixed structure. This method is simple and easily reproduced and can be used to observe the mechanism of the green water phenomenon. In this paper, a numerical simulation study would be carried out based on this approach. The loads and patterns of green water events on a fixed structure with rigid or flexible sidewalls were investigated using the solver MPSFSI-SJTU [22] based on the Moving Particle Semi-Implicit (MPS) method coupled with the Finite Element Method (FEM) in this paper. The rapid flow-structure interaction is generated by the wet dam-break method.

2. Numerical methods

In this paper, the fluid-structure interaction problems were investigated by the MPS-FEM coupled method. The fluid domain was solved by the MPS method, and the deformation and force of the structure were calculated by FEM method. The detailed content of this part of the theory can be found in the paper of Huang et al [24].

2.1. Numerical method for fluid analysis

Our previous papers have introduced the MPS method in detail, including the governing equation, boundary condition, and so on. In this section, these theories are briefly introduced.

The governing equations of the fluid particles for the MPS method are the continuity equation and the momentum equation. And the equations are expressed in Lagrangian form as follows,

$$\nabla \cdot \boldsymbol{V} = 0 \tag{1}$$

$$\frac{D\boldsymbol{V}}{Dt} = -\frac{1}{\rho}\nabla p + \nu\nabla^2\boldsymbol{V} + \boldsymbol{g} \tag{2}$$

As can be seen from the above formulas, due to the Lagrangian nature of the MPS method, there is no convection term as there is in the mesh-based methods.

The interaction between particles is determined by the kernel function in MPS method. Various forms of kernel functions have been proposed. The kernel function used in this paper can be written as

$$W(r) = \begin{cases} \frac{r_e}{0.85r+0.15r_e} - 1 & 0 \leq r < r_e \\ 0 & r_e \leq r \end{cases} \tag{3}$$

where r is the distance between two particles, r_e denotes the influence radius of the target particle. The value of r_e is also different in different particle interaction models. It can be seen that when two particles are quite close together, the value of the kernel function is not too large because the denominator is not close to 0. This avoids the oscillation of the fluid pressure field and makes the calculation more stable.

In order to solve the governing equation, the velocity divergence $\nabla \cdot \boldsymbol{V}$, the pressure gradient ∇p and other terms must be discretised. This is accomplished by the interaction model between particles. The models that are used in this paper can be written as,

$$\langle \nabla \phi \rangle_i = \frac{D}{n^0} \sum_{j \neq i} \frac{\phi_j + \phi_i}{|\boldsymbol{r}_j - \boldsymbol{r}_i|^2} (\boldsymbol{r}_j - \boldsymbol{r}_i) \cdot W(|\boldsymbol{r}_j - \boldsymbol{r}_i|) \tag{4}$$

$$\langle \nabla \cdot \boldsymbol{\Phi} \rangle_i = \frac{D}{n^0} \sum_{j \neq i} \frac{(\boldsymbol{\Phi}_j - \boldsymbol{\Phi}_i) \cdot (\boldsymbol{r}_j - \boldsymbol{r}_i)}{|\boldsymbol{r}_j - \boldsymbol{r}_i|^2} W(|\boldsymbol{r}_j - \boldsymbol{r}_i|) \tag{5}$$

$$\langle \nabla^2 \phi \rangle_i = \frac{2D}{n^0 \lambda} \sum_{j \neq i} (\phi_j - \phi_i) \cdot W(|\boldsymbol{r}_j - \boldsymbol{r}_i|) \tag{6}$$

$$\lambda = \frac{\sum_{j \neq i} W(|\boldsymbol{r}_j - \boldsymbol{r}_i|) |\boldsymbol{r}_j - \boldsymbol{r}_i|^2}{\sum_{j \neq i} W(|\boldsymbol{r}_j - \boldsymbol{r}_i|)} \tag{7}$$

where ϕ and $\boldsymbol{\Phi}$ represent to arbitrary scalars and vectors respectively, D is the number of space dimensions, $\boldsymbol{r}_{i,j}$ is the position vector, λ is a parameter and expressed as equation (7), and n^0 is the initial particle number density. The value of $\boldsymbol{r_e}$ in different particle interaction models is shown in table 1. In the table, l_0 is the initial spacing of particles.

Table 1. The value of $\boldsymbol{r_e}$ in different particle interaction models.

Particle interaction model	r_e
Gradient model	$r_{e\_gra} = 2.1*l_0$
Divergence model	$r_{e\_div} = 2.1*l_0$
Laplacian model	$r_{e\_lap} = 4.01*l_0$

In the MPS method, the pressure of particles is calculated using the Pressure Poisson Equation (PPE). Unlike the explicit method for solving the equation of state in the SPH method, this is a semi-implicit method.

In this paper, a mixed source term method is used to solve the PPE. This solution combines the velocity divergence-free condition and the constant particle number density condition, which can be written as:

$$\langle \nabla^2 p^{k+1} \rangle_i = (1-\gamma)\frac{\rho}{\Delta t}\nabla \cdot V_i^* - \gamma\frac{\rho}{\Delta t^2}\frac{\langle n^k \rangle_i - n^0}{n^0} \tag{8}$$

where P^{k+1}, Δt and V_i^* represent the pressure of the step $k+1$, time step and temporal velocity respectively. γ is the weight parameter. In this paper, the value of γ set to 0.01. n^k is the temporal particle number density and it is defined as the sum of the kernel functions of all particles in the support domain, which can be written as:

$$\langle n^k \rangle_i = \sum_{j \neq i} W(|\boldsymbol{r}_j - \boldsymbol{r}_i|) \tag{9}$$

Dirichlet boundary conditions must be imposed on the free surface particles to solve the PPE. Therefore, accurate detection of free surface particles is very important for accurate calculation of fluid pressure. In this paper, the particle number density is used to judge the free surface particles. If the particle number density satisfies the expression,

$$\left({\langle n \rangle_i}/{n^0} \right) < 0.8 \tag{10}$$

the particle will be determined as a free surface particle. When $\left({\langle n \rangle_i}/{n^0} \right) > 0.97$, the particle is regarded as an internal particle. For the particle with number density which satisfies the expression$0.8 < \left({\langle n \rangle_i}/{n^0} \right) < 0.97$, additional judgment conditions need to be introduced. There is an improved surface particle detection method based on the asymmetry arrangement of neighbouring particles for the surface particles. And a vector function $\langle \boldsymbol{F} \rangle_i$ is introduced for this method which can be written as:

$$\langle \boldsymbol{F} \rangle_i = \frac{D}{n^0} \sum_{j \neq i} \frac{(r_j - r_i)}{|r_j - r_i|} W(|\boldsymbol{r}_j - \boldsymbol{r}_i|) \tag{11}$$

here particles satisfying $\langle \boldsymbol{F} \rangle_i > 0.9|\boldsymbol{F}|^0$ will be regarded as free surface particles, and $|\boldsymbol{F}|^0$ is the initial value of $|\boldsymbol{F}|$ for surface particles.

There are two types of boundary conditions in the MPS method, the free surface boundary conditions and the solid boundary conditions. For the free surface boundary, the pressure of the free surface particles is set to zero. In the MPS method, the solid boundary consists of several layers of particles, including one layer of wall particles and two layers of ghost particles. The wall particles are used to exert repulsive forces on the fluid particles, and the function of the ghost particle is to complete the support domain of the fluid particles near the boundary to avoid the phenomenon of the fluid particle's influence domain being truncated by the boundary. The pressure of the wall particles is calculated by the PPE to be the same as that of the fluid particles. While the pressures of the ghost particles are obtained by interpolation.

2.2. *Numerical method for structure analysis*

The structure domain is calculated using the FEM method. In this paper, the two-dimensional plate element is used to discretise the structure. The governing equation is the dynamic balance equation of discrete elements, which can be written as:

$$\mathbf{M}\ddot{\mathbf{y}} + \mathbf{C}\dot{\mathbf{y}} + \mathbf{K}\mathbf{y} = \boldsymbol{F} \tag{12}$$

$$\mathbf{C} = \beta_1 \mathbf{M} + \beta_2 \mathbf{K} \tag{13}$$

In order to simplify the calculation process, Rayleigh damping matrix is used in this paper, as the equation 13 shows, the matrix C is assumed to be a linear combination of M and K, where β_1 and β_2 are the Rayleigh damping coefficients. According to Newmark, the structural node displacement at $t=t+\Delta t$ can be solved with the help of Taylor's expansions of velocity and displacement:

$$\dot{y}_{t+\Delta t} = \dot{y}_t + (1-\gamma)\ddot{y}_t \Delta t + \gamma \ddot{y}_{t+\Delta t} \Delta t, 0 < \gamma < 1 \tag{14}$$

$$y_{t+\Delta t} = y_t + \dot{y}_t \Delta t + \frac{1-2\beta}{2} \ddot{y}_t \Delta t^2 + \beta \ddot{y}_{t+\Delta t} \Delta t^2, 0 < \beta < 1 \tag{15}$$

where β and γ are parameters in the Newmark- β method and set as $\beta = 0.25$, $\gamma = 0.5$ in this paper. The nodal displacements at $t = t+\Delta t$ can be solved by the following formula:

$$\bar{\mathbf{K}}\mathbf{y}_{t+\Delta t} = \bar{\boldsymbol{F}}_{t+\Delta t} \tag{16}$$

$$\bar{\mathbf{K}} = \mathbf{K} + a_0\mathbf{M} + a_1\mathbf{C} \tag{17}$$

$$\bar{\boldsymbol{F}}_{t+\Delta t} = \boldsymbol{F}_t + \mathbf{M}(a_0 y_t + a_2\dot{y}_t + a_3\ddot{y}_t) + \mathbf{C}(a_1 y_t + a_4\dot{y}_t + a_5\ddot{y}_t) \tag{18}$$

$$a_0 = \frac{1}{\beta\Delta t^2}, a_1 = \frac{\gamma}{\beta\Delta t}, a_2 = \frac{1}{\beta\Delta t}, a_3 = \frac{1}{2\beta} - 1, a_4 = \frac{\gamma}{\beta} - 1, a_5 = \frac{\Delta t}{2}\left(\frac{\gamma}{\beta} - 2\right),$$
$$a_6 = \Delta t(1-\gamma), a_7 = \gamma\Delta t \tag{19}$$

where $\bar{\mathbf{K}}$ and $\bar{\boldsymbol{F}}$ are effective stiffness matrix and effective force vector, respectively.

2.3. Determination of time step

For the fluid domain, the Courant-Friedrichs-Lewy (CFL) condition is used to set the time step Δt_f for MPS method. The CFL condition can be written as:

$$\Delta t_f \le \frac{Cl_0}{u_{max}} \tag{20}$$

where C is the Courant number between 0 to 1.

For the structure domain, the time-step Δt_s for the FEM method should satisfy the central difference method, which can be written as:

$$\Delta t_s \le \frac{L_{min}}{C_s} \tag{21}$$

where C_s can be defined as,

$$C_s = \sqrt{\frac{E_s(1-\mu_s)}{\rho_s(1+\mu_s)(1-2\mu_s)}} \tag{22}$$

where E_s, μ_s and ρ_s denote the Young's modulus, Poisson's ratio, and density of structure, respectively. To ensure numerical stability, the time-step Δt, employed in FSI solver needs to satisfy both the requirement for the fluid and structure domain, which can be written as:

$$\Delta t \le \min\{\Delta t_f, \Delta t_s\} \tag{23}$$

For the flexible structures, the time step of the structure domain is larger than that of the fluid domain. However, for the simulation between the stiff structure and the high velocity fluid, the time step of the structure domain would become smaller than that of the fluid domain. Therefore, when dealing with the fluid-structure interaction problem, the time step for the fluid domain and the structure domain should each be judged according to the specific situation. The correct setting of the time step can improve the efficiency and keep the calculation stable.

2.4. Coupling algorithm between the MPS and FEM method

As shown in Figure 1, the coupling between the MPS method and the FEM method is realised by the CSS method. In Figure 1, **u** represents the velocity of the structure and **p** represents the pressure of the fluid. Once the nodal displacement and velocity of the structure at the nth time step are known, we can get the velocity and pressure from the fluid solver. Then, if we load the pressure on the structure as an external force, we can get the new displacement and velocity from the structure solver. These new structural node displacements and velocities are then transferred to the fluid solver in the next step.

The interface condition of displacement and traction equilibrium needs to be satisfied. Thus, an interface interpolation algorithm is employed, which can be written as:

$$\boldsymbol{u}^F = \boldsymbol{u}^S \tag{24}$$

$$p^F\boldsymbol{n}^F = -p^S\boldsymbol{n}^S \tag{25}$$

where $\boldsymbol{n}^S$ and $\boldsymbol{n}^F$ are normal vectors for interface particles in the structure and fluid domain respectively. The Kernel Function Based Interpolation Technique (KFBI) is then used for data interpolation, including force transformation and displacement transformation. The force is transformed from the fluid domain to the structural domain, and the displacement is transformed from the structural domain to the fluid domain. The schematic diagram of the transforms is shown in Figure 2.

As shown in figure 2, for structure node n, if the distance between the boundary particle and structure node is smaller than the interpolation effective radius r_{e_i} of interpolation, the boundary particle j of the fluid domain will be denoted as a neighbour particle of structure node. In this paper, according to previous research, the value of effective radius r_{e_i} is set as $0.5*r_{e\_gra}$, which is slightly larger than l_0. The weighted value of the fluid force of the neighbour particle $W(|\boldsymbol{r}_j-\boldsymbol{r}_n|)$ is calculated based on (3). Then, the equivalent nodal hydrodynamic pressure p_n^s corresponding to the node n is estimated by the weighted value of pressure components regarding to the neighbour boundary particles and defined as,

$$p_m^S = \frac{\sum_j p_j \cdot \boldsymbol{n}_j^F \cdot W(|\boldsymbol{r}_j - \boldsymbol{r}_m|)}{\sum_j W(|\boldsymbol{r}_j - \boldsymbol{r}_m|)} \tag{26}$$

Figure 2 also shows the deformation transformation. If the distance between the structure node and boundary particle m is smaller than the interpolation effective radius r_{e_i}, the displacement of the boundary particle m is calculated by the neighbouring node displacement. Then displacement of boundary particles $\boldsymbol{w}_m^F$ can be obtained by the interpolation based on the kernel functions $W(|\boldsymbol{r}_i-\boldsymbol{r}_m|)$, and the nodal velocity $\boldsymbol{u}_i$, which can be written as:

$$\boldsymbol{w}_m^F = \frac{\sum_i \delta_i \cdot W(|\boldsymbol{r}_i - \boldsymbol{r}_m|)}{\sum_i W(|\boldsymbol{r}_i - \boldsymbol{r}_m|)} \tag{27}$$

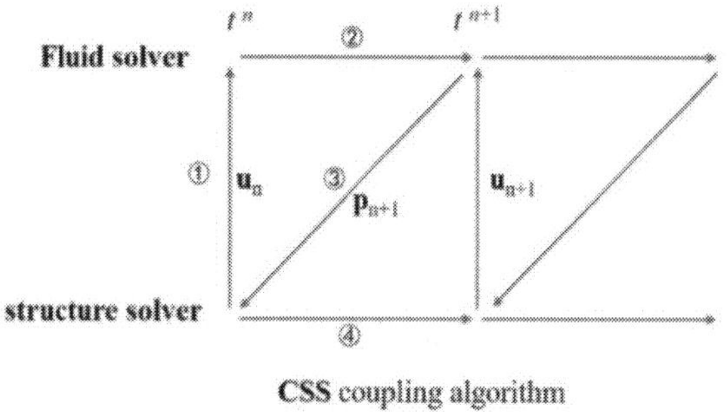

Figure 1. Schematic diagram of partitioned coupling strategy between fluid and structure field.

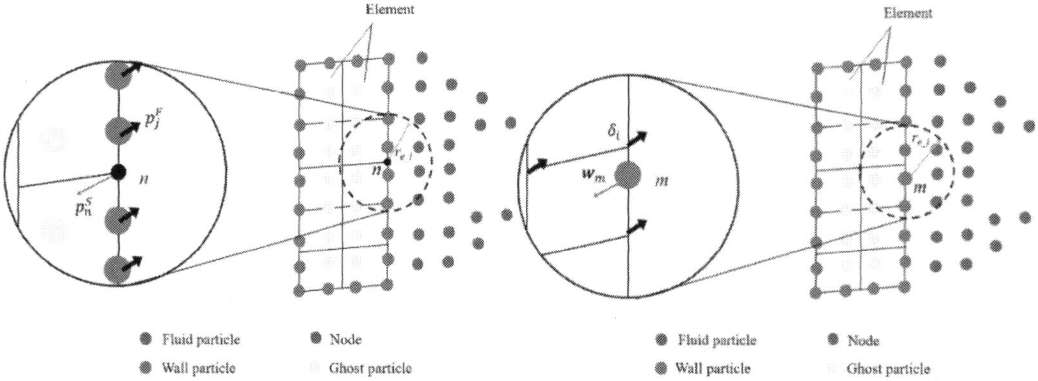

(a) The force interpolation (b) The displacement interpolation

Figure 2. The schematic diagram of the force interpolation and the displacement interpolation method.

3. Numerical Simulations

3.1. Rigid-body case

In this section, the structural dynamic response is ignored. The effectiveness of the MPS method and the rigid body solving part of solver MPSFSI-SJTU was verified by the SPH simulation [25]. SPH simulation used the wet dam-break method to generate the incident wave, and a fixed rectangular structure was placed on the right side of the tank. The right wall of the tank can be regarded as the side wall of the superstructure on the fixed structure.

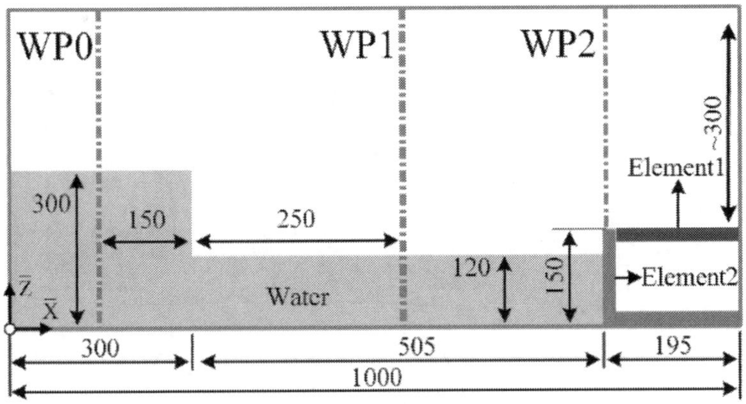

(Dimensions in mm)

Figure 3. The sketch of the water shipping model studied by Areu-Rangel[25], indicating the positions of the three wave probes and two structural elements for the investigation of the hydrodynamic loads.

The simulation model of Areu-Rangel et al [25]. is shown in Figure3. Three wave probes WP0, WP1 and WP2 have been arranged. The upper part of the right side two-dimensional fixed structure is a force measuring element with a length of 0.18 m and a 0.15 m high force measuring element is arranged on the left side of the structure. These settings were to measure water elevations of the incident wave (WP0 and WP1, Figure3), freeboard exceedance (WP2, Figure3), vertical loads on the deck (Element 1, Figure3) and the horizontal loads of left side of structure. The remaining specific dimensions are also shown in Figure3.

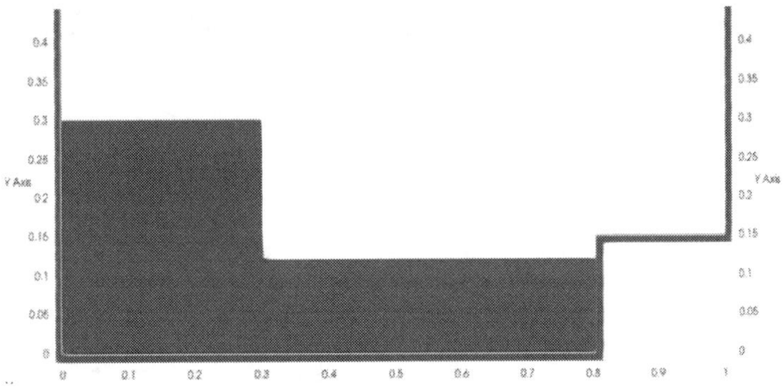

Figure 4. The model used in validation of MPS simulation.

In order to save computing resources, the model showed in Figure4 was used in MPS simulation, which can achieve the same simulation goals as Areu-Rangel et al [25]. Corresponding to the simulation of SPH, the start time of MPS simulation in this paper was also set to 0.5 s. The simulation time interval was 0.0002 s, and the initial particle spacing was 0.001 m in this paper while the time-step algorithm used was the Verlet and the initial particle spacing was 0.0005 m in SPH simulation.

Figure 5 shows the velocity field comparison diagram of SPH simulation and the simulation of the MLParticle-SJTU solver used in this paper. It can be seen that the wave patterns and velocity fields of these two simulations were very similar. After the incident wave reached the left side of the structure, it broke, then the incident wave attacked the upper deck of the structure and climbed up along the right side of the tank wall, then fell back and flowed down from the deck. In this process, a cavity appeared near the left edge of the deck and then disappeared.

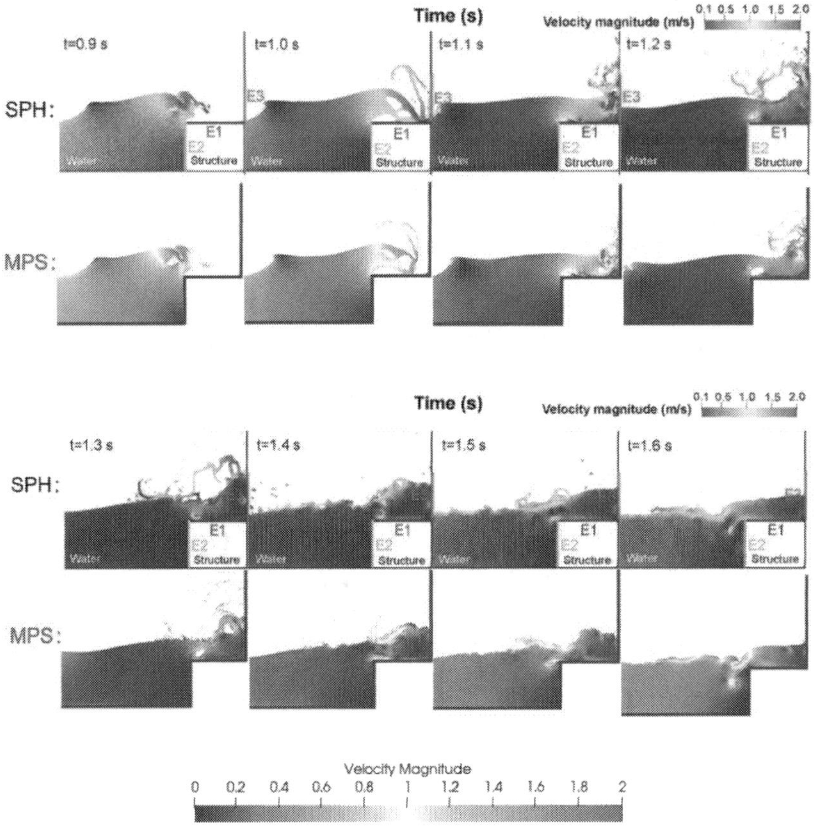

Figure 5. Snapshots of the velocity fields of some relevant stages observed during the simulations using the SPH method[25] and the MPS method.

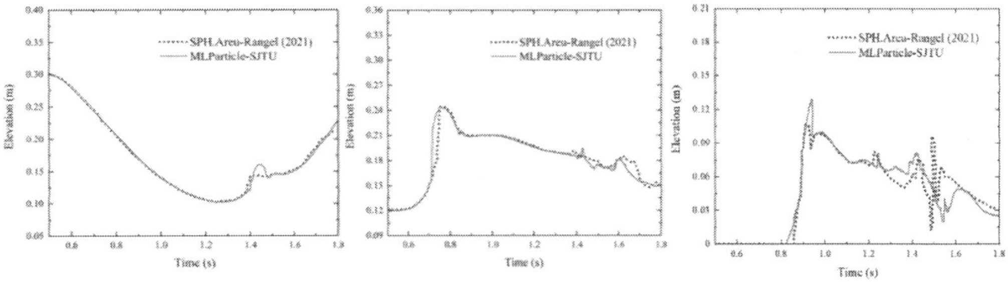

(a) WP0 (b) WP1 (c) WP2

Figure 6. Validation of the numerical approach: comparison of the wave elevations between the SPH results and MPS results. (a) Water elevations at WP0. (b) Water elevations at WP1. (c) Water elevations at WP2.

It can be seen from the water elevations measured at WP0, WP1and WP2 in Fig.6 that the elevations of SPH and MPS simulation were roughly the same, which proved that the wave patterns of MPS simulation and SPH simulation in the left side of the tank were basically the same.

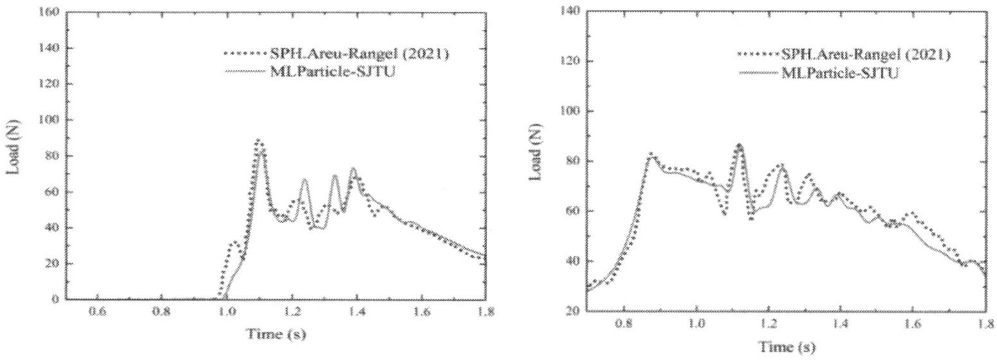

(a) Element1 (b) Element2

Figure 7. Validation of the numerical approach: comparison of the loads of elements between the SPH results and MPS results. (a) Vertical loads of element 1. (b) Horizontal loads of element 2.

Figure7 shows the loads on Element1 and Element2, and the results of SPH and MPS were highly consistent in terms of trends and values. In summary, the numerical simulation results using the MPS method shown in this paper had a high similarity with the results of Areu-Rangel et al. [25], which verified the accuracy of the numerical simulation method of MPS used in this paper.

3.2. The influence of structural elasticity

In this section, the influence of structural elasticity is considered. Element1 and Element2 were taken as elastic sidewalls respectively in this section. The density of the elastic sidewall was 7600 kg/m3, the thickness was 8 mm and the modulus of elasticity was 35, 55 and 75 MPa respectively. The element type used was beam and the effective radius of interpolation r_{e_i} was 0.003 m. Element2 is divided into 151 beam cells and Element1 into 186 beam cells.

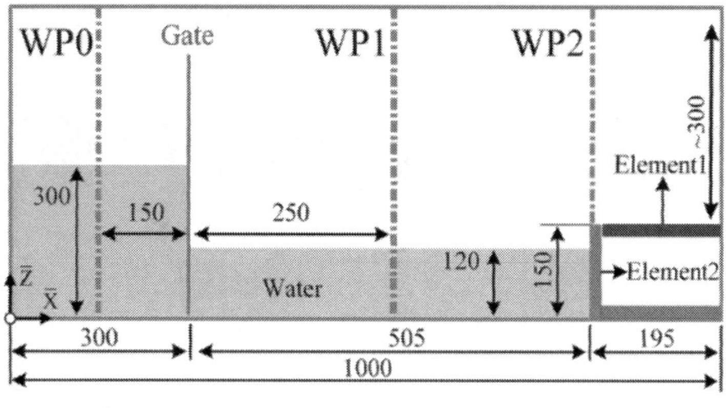

(Dimensions in mm)

Figure 8. The sketch of the water shipping model studied by Hernández-Fontes et al. [23], indicating the positions of the three wave probes and two structural elements for the investigation of the hydrodynamic loads.

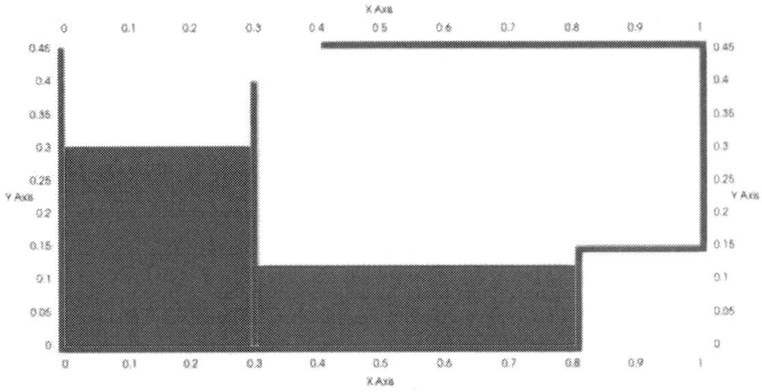

Figure 9. The sketch of the model used in structural elasticity simulation.

The time step of structure and fluid is 0.0002s. For comparison with the experimental results of Hernández-Fontes et al. [23] for a rigid structure, a gate was added to this part of the numerical simulation to simulate the opening of the gate in the experiment, which opened at a uniform speed of 1.6 m/s. Figure 8 shows the scale of the model and the location of the measurement points. Figure 9 shows a schematic of the numerical model.

The simulation was first carried out with the left side of the structure being the elastic side wall. Figure 10 shows the velocity diagram of the flow field and it can be seen that the higher the elastic modulus, the smaller the deformation of the left-hand wall. In the elastic Element 2 case, the edges of the waveform of the wave forming green water events started to become irregular and the velocity field of the wave was less smooth.

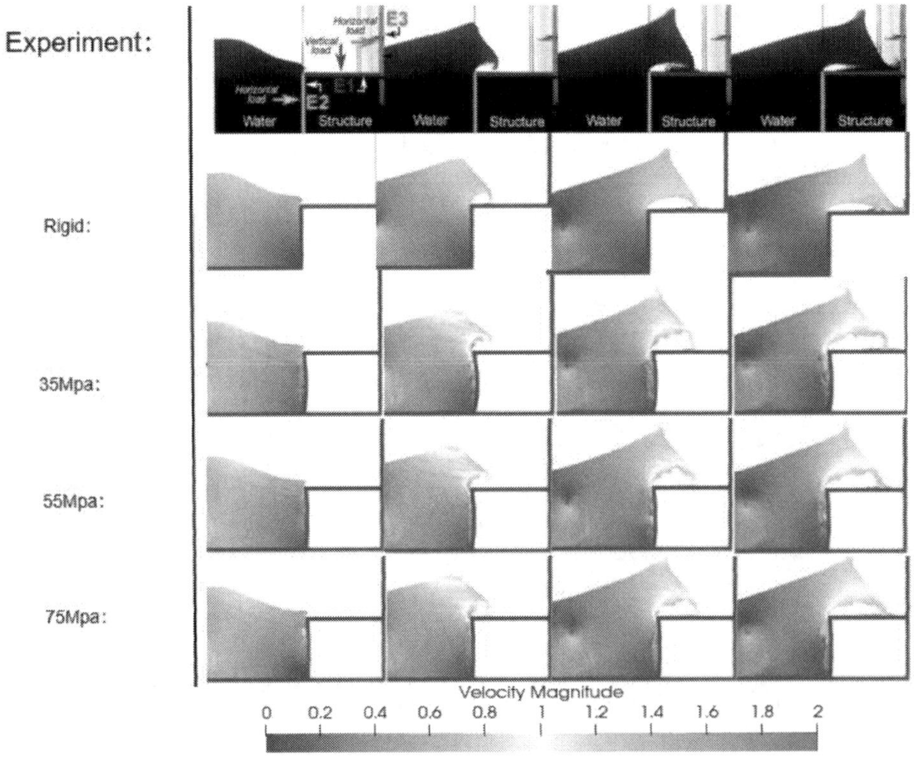

Figure 10. Snapshots found in the experiment of Hernández-Fontes et al. [23] and MPS simulations with different Young's modulus.

The results from the time history curves in Figure 11 show that the simulations in the rigid sidewall condition are in good agreement with the experiments of Hernández-Fontes et al. [23]. The elastic sidewall did not have a significant effect on the waveform of the left-hand side flow field of sidewall. The wave heights calculated for the left side of the structure with or without the elastic sidewall are similar to the experimentally measured wave heights calculated for the rigid object.

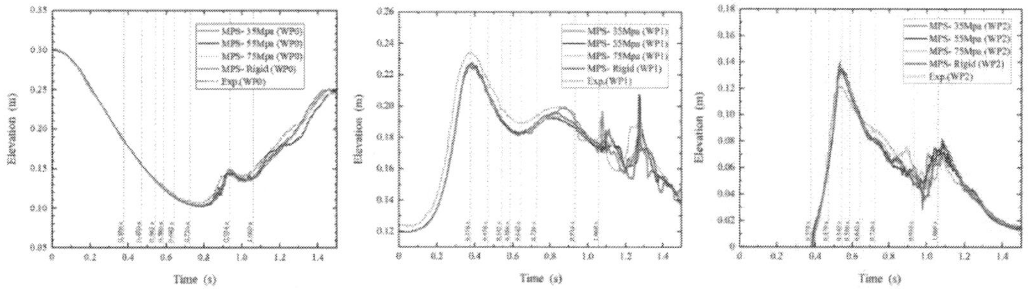

(a) WP0 (b) WP1 (c) WP2

Figure 11. Comparison of the wave elevations in the case of different elastic left walls. (a) Water elevations at WP0. (b) Water elevations at WP1. (c) Water elevations at WP2.

As can be seen from the pressure time history curves in Figure 12, the pressure on the rigid body measured using the MPSFSI-SJTU solver was greater than the experimentally measured pressure, but the trend of the loads was very similar. The peak horizontal load on the Element 2 was greatest when the modulus of elasticity was 75Mpa. The load on the Element1 decreased as the modulus of elasticity decreased and the time of arrival of the peak value was also delayed. When the modulus of elasticity was small, the deformation of the left-hand wall delayed the time at which the vertical load on the Element1 reached its peak.

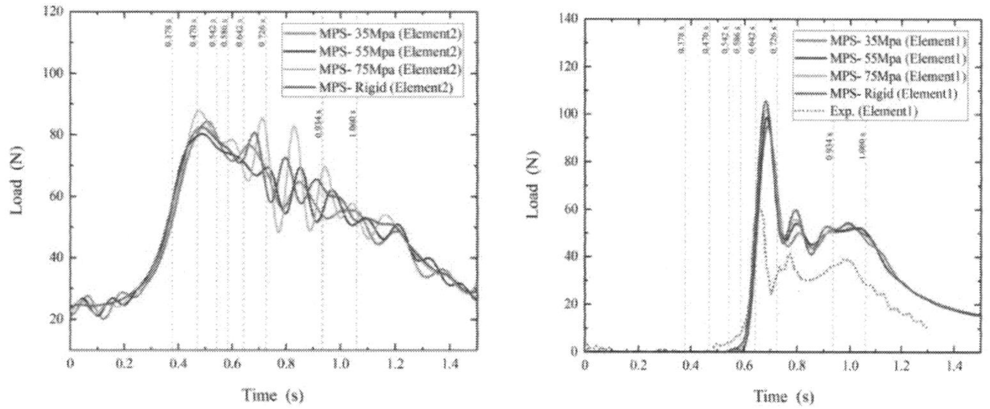

(a) Element1 (b) Element2

Figure 12. Validation of the numerical approach: comparison of the loads of elements between the experimental results and MPS results. (a) Vertical loads of element 1. (b) Horizontal loads of element 2.

Next the simulation of the Element 1 as an elastic sidewall was carried out. Figure 13 shows the velocity diagram of the flow field and it can be seen that the higher the elastic modulus, the smaller the deformation of the Element1.

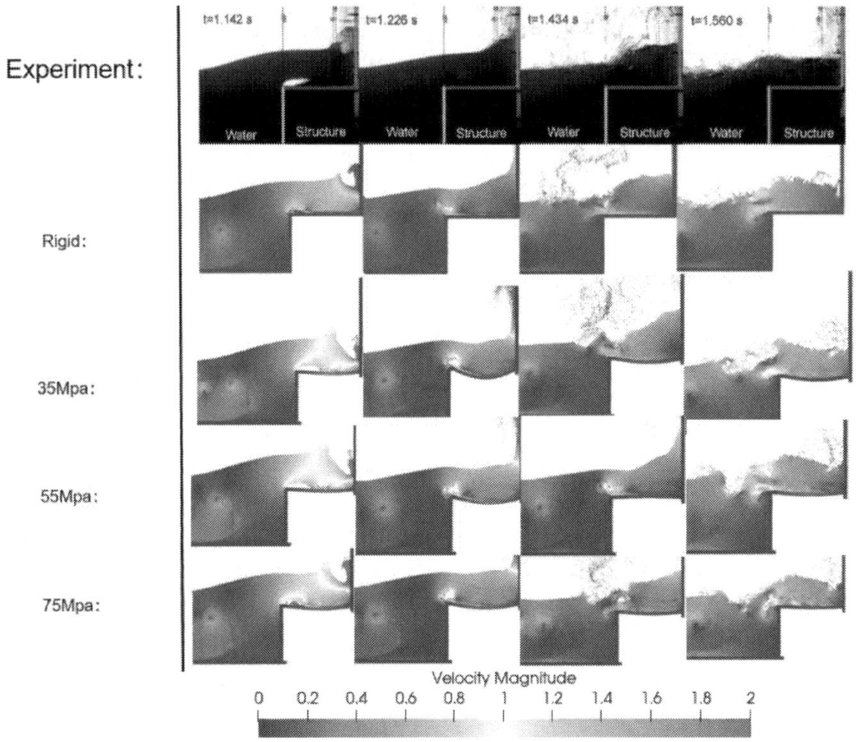

Figure 13. Snapshots found in the experiment of Hernández-Fontes et al. [23] and MPS simulations with different Young's modulus.

The results of the time history curves for the wave probes show that the elastic Element 1 also had less influence on the flow field on the left side of the structure.

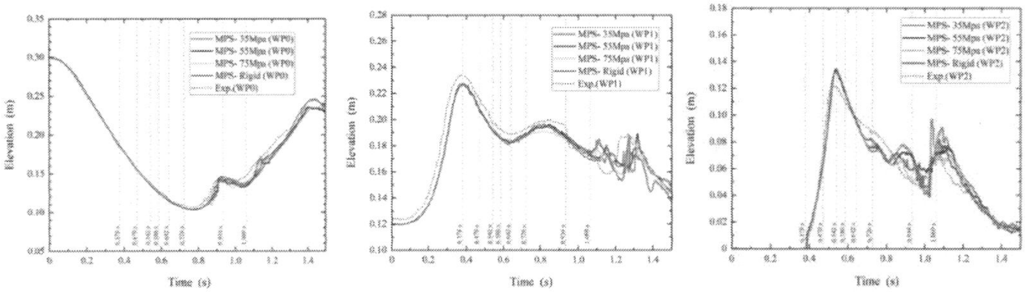

(a) WP0 (b) WP1 (c) WP2

Figure 14. Comparison of the wave elevations in the case of different elastic right walls. (a) Water elevations at WP0. (b) Water elevations at WP1. (c) Water elevations at WP2.

As can be seen from the pressure time history curve in Figure 15, the peak horizontal load of Element2 did not differ for different modulus of elasticity of Element1 because the moment of peak value of Element2 was reached when the wave did not reach Element1, independent of the modulus of elasticity of Element1. As the modulus of elasticity of Element1 becomes smaller, the smaller the peak vertical load applied to Element1 and the more delayed the time to reach the peak load.

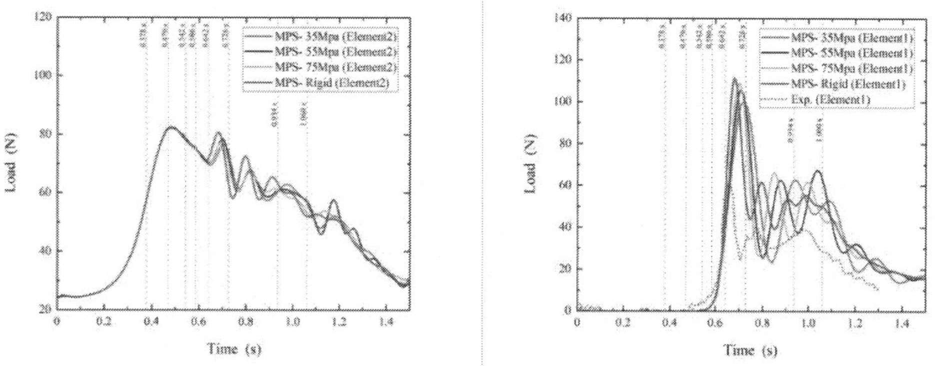

(a) Element1 (b) Element2

Figure 15. Validation of the numerical approach: comparison of the loads of elements between the experimental results and MPS results. (a) Vertical loads of element 1. (b) Horizontal loads of element 2.

4. Conclusions

This paper used the MPSFSI-SJTU solver to numerically simulate green water events with fixed structures with elastic sidewalls as well as rigid sidewalls. The rigid sidewalls were first simulated numerically and the numerical simulation results were in good agreement with those of the SPH numerical simulation. Subsequently, gate was added to the model and numerical simulations of fixed structures with rigid sidewalls as well as elastic sidewalls were carried out and compared with experimental results with rigid sidewalls. From the results we can see that the elastic sidewalls have little effect on the flow field on the left side of the structure. In general, the higher the elastic modulus, the smaller the deformation of the elements. The peak load on the Element1 decreased as the modulus of elasticity decreased and the arrival of the peak value was also delayed.

5. Acknowledgements

This work was supported by the National Natural Science Foundation of China (52131102), and the National Key Research and Development Program of China (2019YFB1704200), to which the authors are most grateful. The authors specially thank to Prof. Decheng Wan for his kind suggestions and supervision.

6. References

[1] GRECO M, LANDRINI M, FALTINSEN O M. 2004 *J. Journal of Fluids and Structures* 19(3): pp251-275.
[2] FONSECA N, SOARES C G. 2005 *J. Journal of Offshore Mechanics and Arctic Engineering-Transactions of The Asme* 127(4): pp322-330..
[3] ZHANG S, YUE DK-P, TANIZAWA, K. 1996 *J. Journal of Fluid Mechanics*, 327: pp221-254.
[4] Greco M, Colicchio G, Faltinsen O M. 2007 *J. Journal of Fluid Mechanics* 581: pp371-399.
[5] LE TOUZÉ D, MARSH A, OGER G, GUILCHER PM. 2010 *J. Journal of Hydrodynamics* 22(5): pp231-236.
[6] Nielsen K B, Mayer S. 2004 *J. Ocean Engineering* 31(3-4): pp363-399.
[7] Rosetti G F, Pinto M L, De Mello P C, et al. 2019 *J. Marine Structures* 65: pp154-180.
[8] Silva D F C, Esperanca P T T, Coutinho A L G A. 2017 *J. Ocean Engineering* 140: pp434-452.
[9] Luo M, Khayyer A, Lin P. 2021 *J. Applied Ocean Research* 114: pp102734.

[10] Tang Z, Wan D, Chen G, et al. 2016 *J. Journal of Ocean Engineering and Marine Energy* 2(3): pp355-364.
[11] Chen X, Wan D. 2019 *J. Ocean Engineering* 171: pp677-694.
[12] Zhang Y, Wan D, Hino T. 2014 *J. Ocean Engineering* 32(4): pp24-32.
[13] Shibata K, Koshizuka S, Sakai M, et al. 2012 *J. Ocean Engineering* 42: pp13-25.
[14] Wen X, Zhao W, Wan D. 2021 *J. Applied Ocean Research* 117: pp102963.
[15] Zhang Y, Tang Z, Wan D. 2016 *J. Chinese Journal of Hydrodynamics* 31(4): pp821-831.
[16] Khayyer A, Gotoh H, Falahaty H, et al. 2017 *J. Ocean Systems Engineering* 7(3): pp299-318.
[17] Shibata K, Koshizuka S. 2007 *J. Ocean Engineering* 34(3-4): pp585-593.
[18] Kawamura K, Hashimoto H, Matsuda A, et al. 2016 *J. Ocean Engineering* 120: pp220-229.
[19] Zhang Y, Tang Z, Wan D. 2016 *J. Chinese Journal of Hydrodynamics* 31(4): pp821-831.
[20] Qin H, Tang W, Hu Z, et al. 2017 *J. Journal of Fluids and Structures* 68: pp322-338.
[21] Yu H, Ren H, Xu Y, et al. 2018 *J. Shock and Vibration.*
[22] Zhang G, Chen X, Wan D, et al. 2019 *J. Journal of Marine Science and Application* 18(4): pp387-399.
[23] Hernández-Fontes JV, Vitola MA, Esperanca PTT, et al. 2020 *J. Ocean Engineering* 197: pp106891.
[24] Huang C, Zhang G, Wan D. 2022 *J. Acta Mechanica Sinica* 38(11).
[25] Areu-Rangel OS, Hernández-Fontes JV, Silva R, et al. 2021 *J. Ocean Engineering* 2(3): 219: pp108392.
[26] Tang Z, Wan D, Chen G, et al. 2016 *J. Journal of Ocean Engineering and Marine Energy* 2(3): pp355-364.

A Numerical Sensitivity Analysis of Fluid-Structure Interaction Simulations on Slamming Loads and Responses

Saeed Hosseinzadeh[1*], Ameen Topa[2], Kristjan Tabri[1]

[1] Department of Civil Engineering and Architecture, Tallinn University of Technology, Tallinn, Estonia
[2] Department of Maritime Technology, Universiti Malaysia Terengganu, Kuala Nerus, Malaysia

*Corresponding author, saeed.hosseinzadeh@taltech.ee

Abstract. This paper presents a parametric study on the Multi Material Arbitrary Lagrangian–Eulerian (MMALE) method for hydroelastic analysis of a non-prismatic stiffened aluminium wedge. An explicit finite element formulation with a penalty-based coupling technique is employed to evaluate the impact-induced loads and responses during free-fall water entry. Based on the penalty factor, damping factor, and number of coupling points, a sensitivity analysis is carried out. It is shown that the penalty coupling method can generate high-frequency oscillations due to the nature of the phenomenon that may affect the predicted results of slamming loads. The computed results on the stiffened and unstiffened plates of the wedge are compared with published experimental data in terms of vertical acceleration, pressure distributions, and strain responses at different locations. A reasonable agreement can be found between the numerical results and the measured values. It is found that a combination of penalty and damping can improve the simulation results and reduce numerical instabilities in hydroelastic slamming simulations.

1. Introduction

Ship slamming is a critical phenomenon that needs to be accurately calculated during ship design. As part of the concept and detailed design process, ship designers need to assess the conditions that lead to slamming and how the structure responds to such events. Numerical simulations have become a powerful tool for predicting and understanding slamming loads, as well as for designing the structure. Simulation-based methods can accurately capture the hydrodynamic and structural interaction between the ship hull and the water surface [1-4]. Moreover, they can provide insight into the physics of the problem and enable designers to identify the areas of the ship that are subjected to the slamming phenomenon. In order to obtain reliable numerical solutions, the physical behaviour of slamming needs to be accurately modelled. This includes the pressure, acceleration and hydrodynamic loads that occur during slamming events. It is also important to consider the hydrodynamic behaviour of the ship in order to properly analyze the structural response. Despite the fact that numerical models are useful tools for predicting impact-induced loads, the simulation results are highly dependent on the solvers used and the setup implemented. Therefore, it is important to conduct a thorough sensitivity analysis to ensure that the results are reliable.

The Arbitrary Lagrangian-Eulerian (ALE) method is commonly used to simulate fluid-structure interaction problems, such as slamming loads on marine structures. Souli et al. [5] presented the ALE formulation for fluid–structure interaction problems. Aquelet et al. [6] developed a coupling algorithm to predict local high pressure loads on a rigid structure using the ALE approach. They pointed out that the penalty coupling algorithm can produce high frequency oscillations due to the nearly incompressible nature of the fluid and the strong coupling between the fluid and the structure [6]. The influence of hydroelastic effects during the water entry impact of an elastic panel was studied by Stenius et al. [7] using an explicit finite element method. They concluded that at high impact velocities, insufficient penalty contact can result in large non-physical penetrations, disrupting the flow field and causing leakage in the Lagrangian structure. An analysis of the penalty factor, mesh density, and contact stiffness on a two-dimensional rigid wedge was conducted by Luo et al. [8]. Souli and Benson [9] highlighted the capability of penalty coupling algorithms by simulating a 2D wedge with 10 deadrise angle and demonstrated that the local peak pressure is sensitive to the penalty factor. Cheon et al. [10] used the ALE algorithm to simulate the water entry problem of a deformable flat stiffened panel. They examined the effects of mesh sensitivity, structural flexibility, and penalty factors on the results. The effect of damping factors on the impact loads was not considered in their study [10]. A study of the numerical uncertainty caused by discretization for the ALE method in predicting impact-induced loads on rigid and elastic flat plates was conducted by Wang et al. [11]. In spite of the extensive research on slamming loads on rigid and elastic structures based on ALE method, there are few papers that study the effect of numerical instabilities on the results of ALE coupling algorithm. The lack of research into numerical instabilities could lead to inaccurate results, and therefore, it is important for further research to be done to ensure the accuracy of the results. Additionally, the study of numerical instabilities could provide insight into how to optimize the ALE coupling algorithm for greater efficiency.

This study presents a numerical sensitivity analysis of impact-induced loads and responses using a multi-material ALE solver in LS-DYNA that allows modelling materials with different densities and viscosities in the same simulation. A two-way coupling technique is applied to model the relative displacement between the fluid and the structure. In order to minimize the high frequency oscillations that occur due to the nature of the water entry problems, a penalty coupling method is used to enforce the boundary conditions at the fluid-structure interface. Both penalty and damping factors are applied in the present numerical simulation. As tuning appropriate contact factors requires an intensive process of trial and error, the numerical results are presented in a way that illustrates how each parameter affects the simulation. The results of vertical acceleration, impact pressure, and strain responses are compared with experimental data.

2. Numerical Setup

The aluminium wedge with non-prismatic characteristics is numerically modelled using an explicit finite element scheme under free-fall impact. To analyze the effect of structural rigidity on the impact induced loads and responses, the bottom of the wedge was designed with two different plates (stiffened and unstiffened) with varying deadrise angles from 20 to 30 degrees. The total mass of the wedge is 55 kg. The details about the material properties and dimensions of the wedge, properties of the longitudinal and transverse stiffeners, and sensors arrangement can be found in Hosseinzadeh et al. [12].

This numerical simulation models the wedge section using the Lagrangian method with shell elements. The material types called Elastic and Rigid are used to define the material of the wedge section and top frame, respectively. The fluid domain including water and air is modelled based on ALE multi-material formulation with solid elements. The penalty-based method is used to model the interaction between the fluid and the structure, allowing for an accurate FSI simulation of the wedge section. This method transports history variables in the direction of the flow, which makes it delicate to prevent leakage through the FSI interface, especially for high pressure flows to the structure [13]. Additionally, a number of integration points must be defined through the shell thickness, and five integration points were selected based on previous work [14]. The Lobatto integration method is adopted as it does not require interpolation for stress and strain values on the top and bottom surfaces of the shell element [15]. The modified advection method is used which employs a looser constrained for the advection process. This method can better preserve the material interface for ALE materials [16]. A summary of the numerical model, mesh study, and domain and elements sizes can be found in Hosseinzadeh et al. [14].

The numerical model setup for simulating the water entry problem of the 3D wedge is shown in Figure 1(a). The mesh regions of the numerical model with water (blue) and air (green) are illustrated in Figure 1(b). A fine mesh is required in the impacts area to minimize numerical issues and achieve accurate results, while coarse meshes are used in areas far from the areas of interest to reduce computational time. Figure 1(c) depicts the stiffened and unstiffened bottom plates, along with the locations of the strain and pressure sensors.

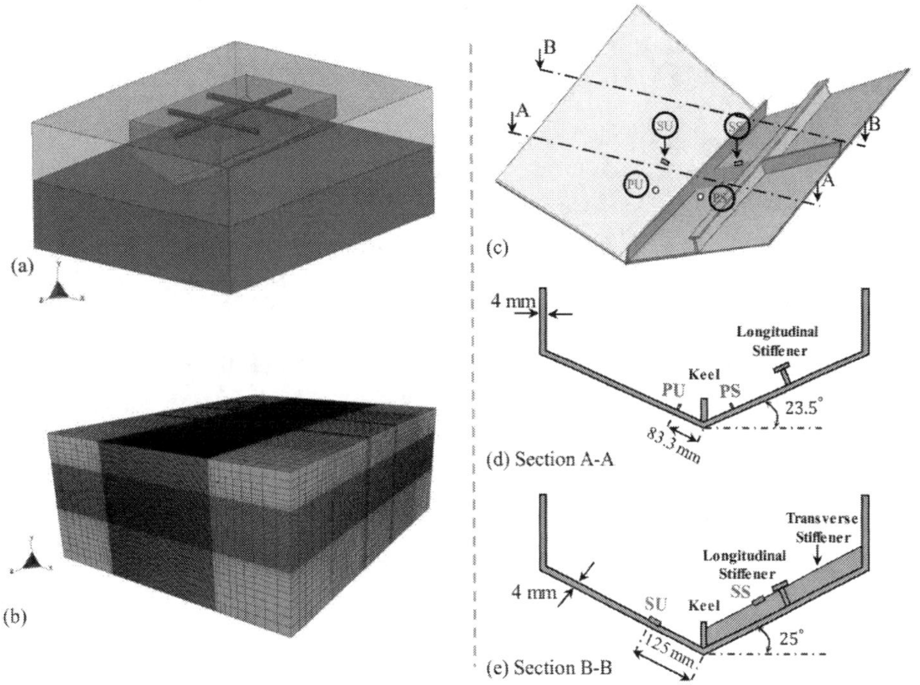

Figure 1. Finite Element setup in LS-DYNA; (a) numerical domain including water and air, (b) mesh regions of fluid domain, (c) bottom stiffened and unstiffened plates and sensor arrangement, (d) location of pressure sensors on section A-A with 23.5° deadrise angle, (e) location of strain gauges on section B-B with 25° deadrise angle.

The pressure sensors are located on 23.5-degree deadrise angle and denoted as PU and PS for unstiffened and stiffened bottom plates, respectively (Figure 1d). In addition, the strain gauges are placed on the 25-degree deadrise angle and indicated as SU and SS for unstiffened and stiffened bottom plates, respectively (Figure 1e). One of the major challenges in modelling fluid-structure interactions with the ALE approach is calibrating the coupling algorithm. The proper coupling parameters are usually determined through a trial-and-error process. Therefore, the effects of the penalty factor and the damping coefficient on the numerical results are fully investigated. Table 1 presents the number of simulations conducted in this study. In order to determine the effect of PFAC (penalty factor) and DAMP (damping factor) on the numerical results, a total of 30 runs were conducted.

Table 1. Number of simulations with different PFAC and DAMP factors

Run No.	PFAC	DAMP				
1-5	0	0	0.05	0.1	0.5	1
5-10	0.004	0	0.05	0.1	0.5	1
10-15	0.02	0	0.05	0.1	0.5	1
15-20	0.1	0	0.05	0.1	0.5	1
20-25	0.5	0	0.05	0.1	0.5	1
25-30	1	0	0.05	0.1	0.5	1

3. Results and Discussion

The effects of Euler-Lagrange coupling parameters are fully examined in this study, including vertical acceleration, total force, pressure, and strain. The numerical results are compared with experimental data [12] both for stiffened and unstiffened plates. An initial impact velocity of 4.00 m/s is applied to all simulations. The numerical simulations with different values of PFAC and DAMP are carried out. The parameter PFAC refers to the penalty factor ($0 < p_f < 1$), a scale factor used for scaling the estimated stiffness of the interacting fluid-structure system. On the other hand, DAMP is referring to the damping factor (ξ) which can be between 0 and 1. This parameter is used to scale down the critical-damping force and a value of 1 is used for a critically damped case [16].

Figure 2 illustrates the numerical results with different values of PFAC and DAMP. It is apparent that a higher value of DAMP reduces the noise in the rigid body acceleration time history. Moreover, in the case of highest damping factor (PFAC = 0.5 and DAMP = 1), there is an error in the finite element calculation which is due to the numerical instability in the simulation. The influence of the damping factor on the rigid body acceleration noise is less at higher values of the penalty factor. That is; with PFAC = 0, there is a notable difference in the maximum acceleration when comparing the cases of DAMP = 0 and DAMP = 1. However, with PFAC = 0.5, the difference is not as much. Therefore, with respect to rigid body acceleration, it can be concluded that the damping factor is more crucial at lower values of PFAC (see Figure 2). The influence of these two parameters is observed on the noise of vertical acceleration but not on its magnitude. In addition, the effects of PFAC and DAMP on the total force acting on the wedge section are illustrated in Figure 3. Unlike the rigid body acceleration, PFAC has a direct influence on the rigid body force. With no damping, the peak force in the case of PFAC = 0 is around 10.5 kN, whereas in the case of PFAC = 0.02 the force is around 9.5 kN. Interestingly, with PFAC = 0.5, the force is around 11.0 kN. As for the damping factor, increasing it has no significant influence on the peak force. The results also demonstrate that a high penalty factor, coupled with a low damping factor (PFAC = 0.5 and DAMP = 0) leads to high frequency oscillations on vertical forces. Generally, it is shown that the maximum vertical force and curve shape do not vary significantly when different coupling configurations are used, except when numerical instabilities are present.

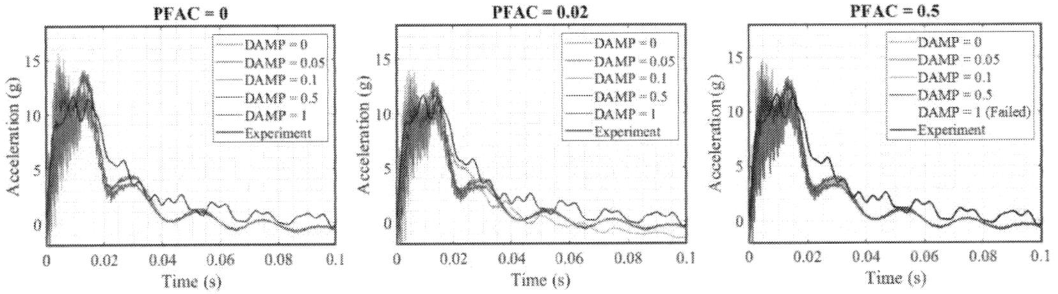

Figure 2. Time histories of vertical acceleration at 25-degree deadrise angle with different PFAC and DAMP factors compared with experimental data.

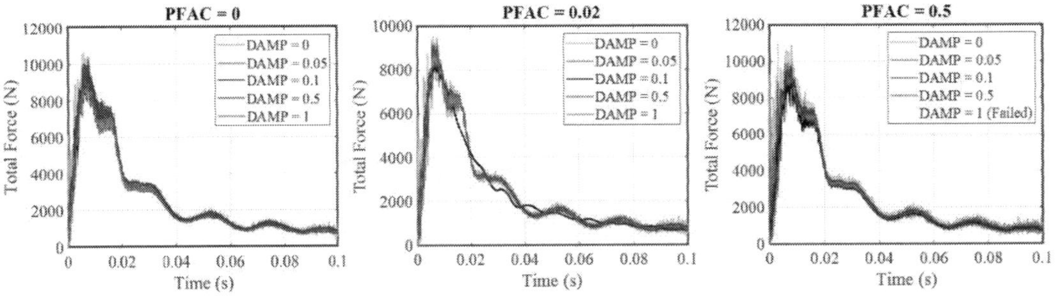

Figure 3. Time histories of computed total force with different penalty and coupling factors.

Figure 4(a) displays the average pressure histories on the unstiffened panel (PU). It is observed that increasing the penalty factor from 0 to 0.5 decreases the average peak pressure from 95 kPa to 79 kPa, in the case of DAMP = 0. Likewise, increasing the damping factor decreases the average peak pressure. For PFAC = 0, the average peak pressures are 95 kPa and 76 kPa for DAMP equal 0 and 1, respectively. The average pressure histories on the stiffened panel (PS) are displayed in Figure 4(b). It is apparent that the influence of PFAC is significant on maximum value of pressure. In comparison to the experimental data, PFAC = 0.02 shows the lowest difference in peak pressure, whereas PFAC = 0 and 0.5 yield higher values. In addition to the peak pressure values, high-frequency fluctuations decrease with increasing damping factor. However, due to the non-physical behavior of the coupling algorithm at high damping factors, the simulation with PFAC = 0.5 and DAMP = 1 crashed (see Figure 4).

The strain responses of the sensors located on the unstiffened (SU) and stiffened (SS) plates with 25-degree deadrise angle are depicted in Figure 5(a) and (b), respectively. Unlike the vertical acceleration, the maximum strain is sensitive to the damping factor. With no damping in the fluid-structure coupling, the numerically calculated strain is larger than the experimental one for the unstiffened panel (SU). On the other hand, DAMP = 0.1 gives the best results for the stiffened plate and larger values will give more conservative results. It is shown that coupling parameters have a more significant effect on the strain response of the unstiffened plate (Figure 5a). This is because the strain sensor of the stiffened plate is located on the stiffener and do not have direct interaction with the water. It is important to note that the behaviour of the coupling system can change by using an excessively high factor that damps out the actual physical oscillations, which leads to inaccurate results.

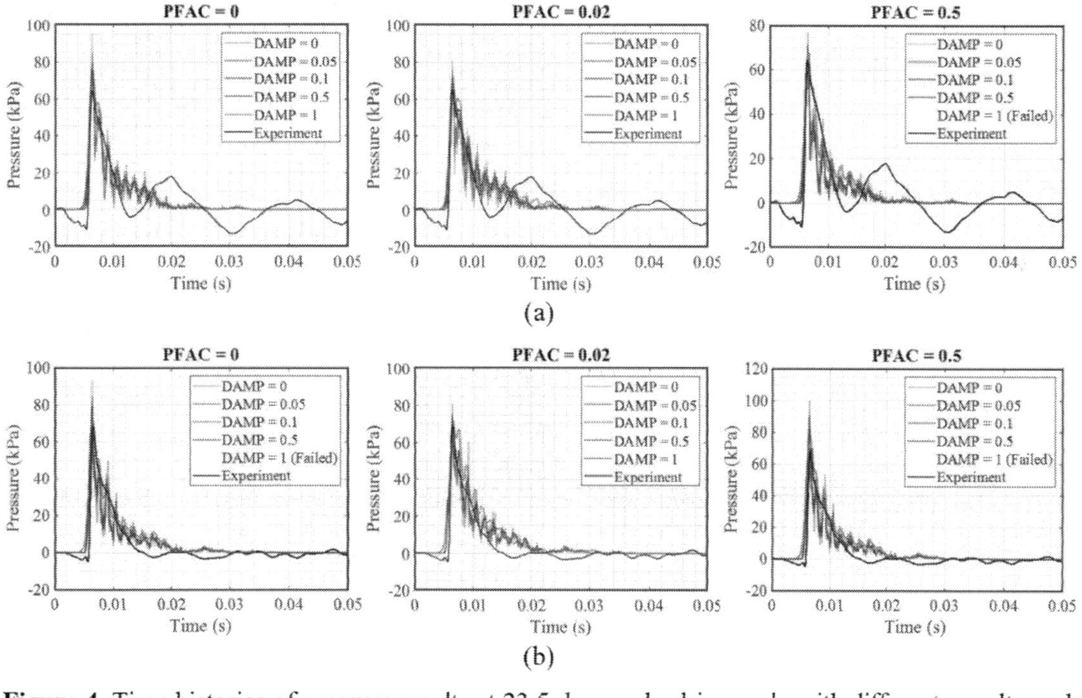

Figure 4. Time histories of pressure results at 23.5-degree deadrise angle with different penalty and damping factors compared with experimental data on: (a) unstiffened plate (PU) (b) stiffened plate (PS).

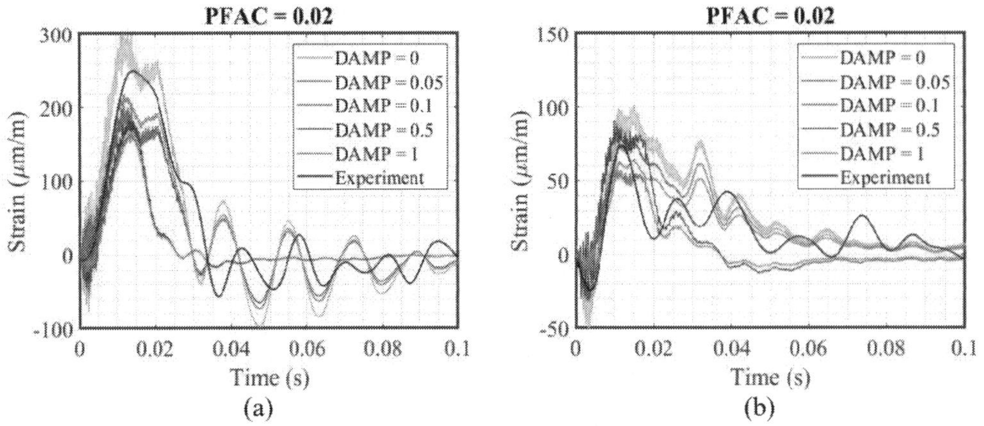

Figure 5. Time histories of strain responses at 25-degree deadrise angle with PFAC = 0.02 and different coupling factor compared with experimental result on: (a) unstiffened plate (SU) (b) stiffened plate (SS).

The obtained results are explained by how the penalty-based method of coupling algorithm works. The coupling algorithm searches for any overlaps or penetration between the structure segments and the fluid mesh. As soon as an interaction is detected, the algorithm applies a coupling force on the penetrating segments pushing them backwards and thus removing the penetration [9]. This force is expressed as follows:

$$F = kd = \left(p_f \frac{KA^2}{V}\right) d \tag{1}$$

where d is the penetration and k is the spring stiffness which depends on p_f is the penalty factor, K is the bulk modulus, A is the average area of the structure element, and V is the volume of the fluid element [16]. With higher values of the penalty factor, the contact stiffness is higher, and penetrations are minimized or eliminated. However, the virtual spring-mass system of the contact becomes of unstable and oscillates more violently. With the introduction of damping, the coupling force becomes:

$$F = kd + C\dot{d} = kd + \left(\xi\sqrt{kM}\right)\dot{d} \tag{2}$$

where ξ is the damping factor and M is the equivalent mass [16]. With higher values of the damping factor, the spring-mass system is more stable. It is important to note that the damping coefficient C depends on the mass as well, which is calculated based on the element size. Models with smaller element size and smaller material stiffness require larger values of damping coefficient.

In order to further analyse the role of coupling parameters in slamming simulations, the maximum values of acceleration, pressure, and strain are compared with the experimental data using the equation $\Delta(\square)_{max} = ((\square)_{\text{max}\_MMALE} - (\square)_{\text{max}\_EFD})/\ (\square)_{\text{max}\_EFD})) \times 100$. Figure 6 summarizes the influence of penalty and damping factors on the maximum acceleration. With PFAC = 0.02, the maximum acceleration is acceptable regardless of the value of DAMP. This reveals that penalty factor is more crucial compared to the damping factor for acceleration results (ΔA_{max} = 3.1%). The most accurate numerical result is obtained when PFAC = 0.02 and DAMP = 0.1. It is worth noticing that these values are not true for all cases of fluid-structure interactions, as they depend on many parameters such as stiffness and thickness of the structure, properties of the fluid material, impact velocity, and the relative element size of the structure with respect to the fluid. However, it can be considered as a benchmark for tuning the most suitable values for coupling parameters in the slamming analysis of aluminium structures. Therefore, it is important to take into account the particularities of the study case in order to correctly setup the numerical model and obtain reliable results. These values can also be used to improve the accuracy of numerical simulations for similar types of problems.

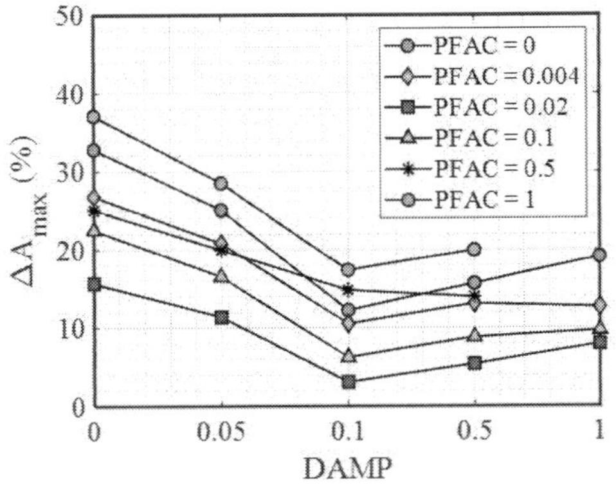

Figure 6. Difference of maximum acceleration using different penalty factor (PFAC) and damping factor (DAMP).

Figure 7 and Figure 8 illustrate the influence of penalty and damping factors on the difference of peak pressure and maximum strain, respectively. The same pattern as in Figure 6 can be observed. Penalty factor of 0.02 and damping factor of 0.1 yields the best numerical results. Higher or lower values in one or both of these factors increase the difference in the maximum pressure and maximum strain responses. Again, it is important that these two values do not hold true for all cases, and they vary from one application to another. It is found that the combination of a low penalty and damping factors leads to inaccurate results. As the penalty factor increases, the high-frequency oscillations increase. In addition, a high damping factor can cause unstable coupling.

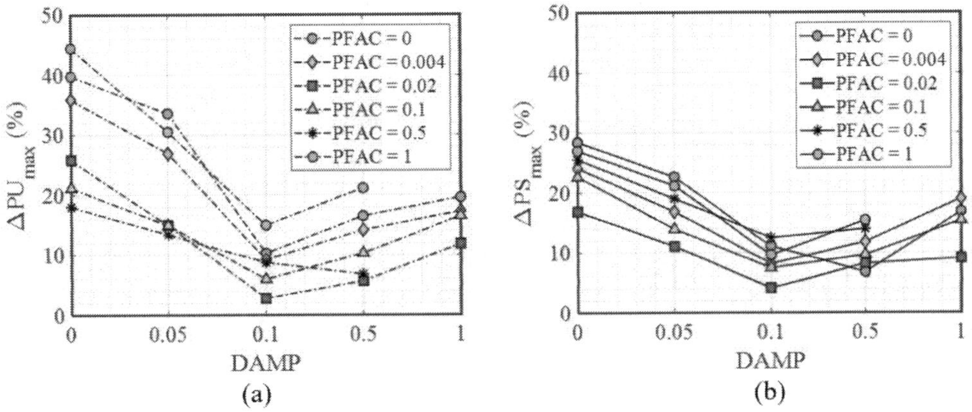

Figure 7. Difference of peak pressure using different penalty factor (PFAC) and damping factor (DAMP) on (a) unstiffened plate (b) stiffened plate.

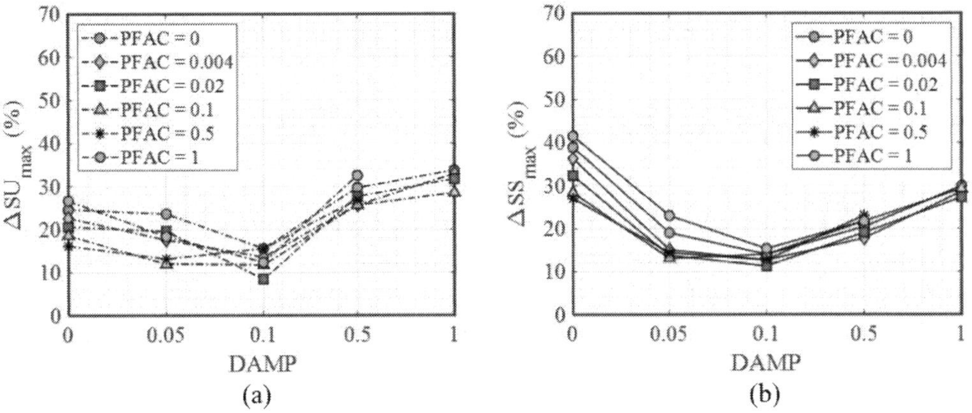

Figure 8. Difference of maximum strain using different penalty factor (PFAC) and damping factor (DAMP) on: (a) unstiffened plate (b) stiffened plate.

4. Conclusions

This paper studied hydroelastic slamming on a non-prismatic aluminium wedge section. A numerical model using an explicit finite element formulation based on the penalty coupling technique was used to simulate the described water entry phenomenon at high impact velocity. The wedge section and fluid domain were discretized with four-nodded shell elements and hexahedral elements, respectively. Understanding the influence of coupling parameters on numerical results is essential to successfully implement the MMALE coupling algorithm. Therefore, this study provides a comprehensive analysis of how the penalty factor and the damping coefficient affect the numerical results. The results of the simulations demonstrated that the presented MMALE method was able to accurately capture the behavior of the water entry phenomenon. The numerical model was validated with experimental data and showed good agreement with the results. It is found that although the coupling parameters have a minimal effect on the total force and its maximum value, they have a considerable influence on hydrodynamic pressure, vertical acceleration, and strain response. The results indicated that numerical instability may occur particularly for simulations with a high penalty factor. In addition, the simulations without a damping coefficient result in high frequency oscillations and make the coupling system unstable. These findings suggest that the coupling parameters should be carefully chosen in order to obtain a stable numerical model. Furthermore, a combination of penalty and damping factors is recommended to guarantee the stability of the FSI coupling system. Based on the benchmark study presented in this paper, the penalty factor and damping coefficient of 0.02 and 0.1 are recommended for hydroelastic slamming problems of aluminium structures.

Acknowledgements

This research work has been financially supported by the Estonian Research Council via the grant PRG1820 (Dynamic response of offshore structures). This help is kindly appreciated.

References

[1] Hosseinzadeh S, Tabri K. Hydroelastic effects of slamming impact loads during free-fall water entry. Ships and Offshore Structures. 2021 Aug 18;16(sup1):68-84.

[2] Yan D, Hosseinzadeh S, Lakshmynarayanana A, Mikkola T, Hirdaris S. Comparative study on numerical hydroelastic analysis of impact-induced loads. InProceedings of the 23rd numerical towing tank symposium. NuTTs 2021 Oct 11 (p. 6).

[3] Hosseinzadeh S, Izadi M, Tabri K. Free fall water entry of a two-dimensional asymmetric wedge in oblique slamming: a numerical study. InInternational Conference on Offshore Mechanics

and Arctic Engineering 2020 Aug 3 (Vol. 84409, p. V008T08A013). American Society of Mechanical Engineers.

[4] Tavakoli S, Babanin AV, Hirdaris S. Hydroelastic Analysis of Hard Chine Sections Entering Water—Observations for Use in Preliminary Design Stage. Journal of Offshore Mechanics and Arctic Engineering. 2023 Oct 1;145(5):051901.

[5] Souli M, Ouahsine A, Lewin L. ALE formulation for fluid–structure interaction problems. Computer methods in applied mechanics and engineering. 2000 Nov 10;190(5-7):659-75.

[6] Aquelet N, Souli M, Olovsson L. Euler–Lagrange coupling with damping effects: Application to slamming problems. Computer methods in applied mechanics and engineering. 2006 Jan 1;195(1-3):110-32.

[7] Stenius I, Rosén A, Kuttenkeuler J. Explicit FE-modelling of fluid–structure interaction in hull–water impacts. International Shipbuilding Progress. 2006 Jan 1;53(2):103-21.

[8] Luo H, Wang S, Soares CG. Numerical prediction of slamming loads on a rigid wedge subjected to water entry using an explicit finite element method. Advances in marine structures. 2011 Mar 14:41-8.

[9] Souli MH, Benson DJ, editors. Arbitrary Lagrangian Eulerian and fluid-structure interaction: numerical simulation. John Wiley & Sons; 2013 Mar 1.

[10] Cheon JS, Jang BS, Yim KH, Lee HD, Koo BY, Ju H. A study on slamming pressure on a flat stiffened plate considering fluid–structure interaction. Journal of Marine Science and Technology. 2016 Jun;21:309-24.

[11] Wang S, Islam H, Soares CG. Uncertainty due to discretization on the ALE algorithm for predicting water slamming loads. Marine Structures. 2021 Nov 1;80:103086.

[12] Hosseinzadeh S, Tabri K, Hirdaris S, Sahk T. Slamming loads and responses on a non-prismatic stiffened aluminium wedge: Part I. Experimental study. Ocean Engineering. 2023 Jul 1;279:114510.

[13] Haufe A, Weimar K, Göhner U. Advanced airbag simulation using fluid-structure-interaction and the eulerian method in LS-DYNA. Proceedings of the LS-DYNA Anwenderforum. 2004 Apr.

[14] Hosseinzadeh S, Tabri K, Topa A, Hirdaris S. Slamming loads and responses on a non-prismatic stiffened aluminium wedge: Part II. Numerical simulations. Ocean Engineering. 2023 Jul 1;279:114309.

[15] Haufe A, Schweizerhof K, DuBois P. Review of Shell Element Formulations in LS-DYNA: Properties, Limits, Advantages, Disadvantages. InProceedings of the LS-DYNA Developer Forum 2013 Sep.

[16] LS-DYNA Theory Manual. Livermore Software Technology (Lst). 2022 An Ansys Company.

IOP Conf. Series: Materials Science and Engineering 1288 (2023) 012018 doi:10.1088/1757-899X/1288/1/012018

Numerical investigations of dam-break flow impacting on an elastic beam using various FSI coupling algorithms

Wenjie Zhang[1], Xiaolong Yang[2] and Jianhua Wang[1]*

[1] Computational Marine Hydrodynamics Lab (CMHL), School of Naval Architecture, Ocean and Civil Engineering, Shanghai Jiao Tong University, Shanghai, China
[2] Offshore Oil Engineering Co. Ltd., Tianjin, China

*Corresponding author: jianhuawang@sjtu.edu.cn

Abstract. The occurrence of green water and slamming in severe sea conditions can often be simplified as a dam-break problem. The flow characteristics of dam-break are similar with those on a ship's deck and can result in structural damage. In this study, the fluid-structure interaction (FSI) of dam-break flow impacting on an elastic beam is investigated based on CFD-FEM method. The FSI coupling is solved in a partitioned manner. The fluid field is solved by OpenFOAM and the structure is calculated by self-developed code using FEM. The two-way coupling algorithm has been implemented to update the mesh of the fluid field. An investigation on mesh independence and the comparison of results with other methods have been conducted to validate the simulation results. After that, three types of FSI coupling algorithms have been utilized to examine the impact on both computational accuracy and efficiency. It is studied that the computational accuracy of all three FSI coupling algorithms is in good agreement. The FSI coupling algorithm utilizing weak iteration exhibits excellent performance in terms of both computational accuracy and efficiency. For the strong coupling algorithm, it is crucial to make a rational selection of the outer and inner iteration numbers in PIMPLE loops.

1. Introduction

In the field of marine engineering, the issue of wave slamming on structures is a prevalent phenomenon that requires specialized attention. In severe sea conditions, vessels frequently encounter the phenomena of green water and slamming. These phenomena can often be theoretically and numerically verified and studied through simplified problems that share similar fundamental flow characteristics, such as the dam-break problem. Greco[1] have concluded that the occurrence of dam-break type (type DB) water-on-deck scenarios in ocean engineering is more prevalent than that of plunging wave type (type PW). Many experiments and numerical simulations have investigated the interactions between dam-break flow and an obstacle , especially in cases where the structures exhibit elasticity.

Liao[2] has conducted experimental studies on dam break with an elastic baffle. Yilmaz[3] devised an experiment to investigate the interaction between a dam-break flow and an elastic sluice gate. Hu[4] has studied the interactions of an elastic vertical wall in periodic waves. The measurements obtained from experimental studies are now widely accepted as standard data for comparing numerical algorithms and developing computational solvers. The impact of a dam-break flow on an elastic beam is a typical case for studying the interaction between nonlinear fluid dynamics and structures. Numerical results of many scholars have been compared with benchmark experiment measurements to validate their solvers and algorithms (Khayyer[5], McLoone[6], Mitsume[7], Ryzhakov[8], Walhorn[9], Wu[10], Zheng[11]). The phenomenon encompasses the dynamics of free surface evolution and breaking, as well as the interaction with structures through climbing and deformation. Accurately predicting impact loads and describing fluid-structure interaction are critical for evaluating structural safety.

In this paper, the problem of dam-break flow impacting on an elastic beam is investigated with various FSI coupling algorithms numerically. The simulation is conducted using the CFD-FEM method with OpenFOAM and a self-developed structural solver to calculate the bending of a cantilever beam. The two-way coupling algorithm has been implemented, and the fluid field mesh is updated using the displacement Laplacian method in OpenFOAM. A study on mesh independence and comparison of results with other methods has been conducted to validate the simulation results. Three types of FSI coupling algorithms have been utilized to examine the impact on both computational accuracy and efficiency.

2. Numerical methods

2.1. *Fluid governing equations*

The continuity and momentum Navier-Stokes equations for incompressible Newtonian fluid are given as follows:

$$\nabla \cdot \boldsymbol{U} = 0 \tag{1}$$

$$\frac{\partial \boldsymbol{U}}{\partial t} = -\frac{1}{\rho}\nabla p + \nu_k \nabla^2 \boldsymbol{U} + \boldsymbol{g} \tag{2}$$

where ρ is the fluid density, p is the pressure, ν_k is the kinetic viscosity, $\boldsymbol{U}$ is the velocity vector of flow fluid ,and $\boldsymbol{g}$ is the gravitational acceleration vector.

2.2. *Structure governing equations*

The structure in this study is simplified as a cantilever beam model with one fixed end and one free end. Based on the theory of Finite Element Method (FEM), structure dynamic governing equations are given as follows:

$$M\ddot{\boldsymbol{x}}(t) + C\dot{\boldsymbol{x}}(t) + K\boldsymbol{x}(t) = \boldsymbol{F}(t) \tag{3}$$

where M, C and K represent mass matrix, damping matrix and stiffness matrix of the cantilever beam respectively. The dimension of each matrix is two times of node numbers of the discrete FE model. The vectors of $\boldsymbol{x}(t)$, $\dot{\boldsymbol{x}}(t)$ and $\ddot{\boldsymbol{x}}(t)$ are displacement vector, velocity vector and acceleration vector at the previous time t, respectively. Vector $\boldsymbol{F}(t)$ is external load acting on the beam and the source term of the structure dynamic governing equation.

The damping term of structural system is very complex, and in the engineering, the format of Rayleigh damping has been widely accepted. Rayleigh damping is a linear combination of mass and stiffness of the structure:

$$C = \alpha_1 M + \alpha_2 K \tag{4}$$

where α_1 and α_2 are the proportional coefficients foe the mass and stiffness, respectively, and given as:

$$\begin{pmatrix}\alpha_1\\ \alpha_2\end{pmatrix} = \frac{2\varsigma}{f_{n1}+f_{n2}}\begin{pmatrix}2\pi f_{n1} f_{n2}\\ \frac{1}{2\pi}\end{pmatrix} \tag{5}$$

where f_{n1} and f_{n2} are the natural frequencies at the first and second modes, respectively; and ς is the damping ratio of the structure.

2.3. *Newmark's method*

The Newmark's method is one of the direct integration methods to solve structure dynamic equations by assuming the relationship between displacement vector, velocity vector and acceleration vector of

FE model[12]. Using the Newmark's method, the velocity and the acceleration at the time t + Δt can be calculated as:

$$\dot{\boldsymbol{x}}(t+\Delta t)=\dot{\boldsymbol{x}}(t)+(1-\gamma)\Delta t\ddot{\boldsymbol{x}}(t)+\gamma\Delta t\ddot{\boldsymbol{x}}(t+\Delta t) \tag{6}$$

$$\boldsymbol{x}(t+\Delta t)=\boldsymbol{x}(t)+\Delta t\dot{\boldsymbol{x}}(t)+\left(\frac{1}{2}-\beta\right)\Delta t^2\ddot{\boldsymbol{x}}(t)+\beta\Delta t^2\ddot{\boldsymbol{x}}(t+\Delta t) \tag{7}$$

where $\boldsymbol{x}(t+\Delta t)$, $\dot{\boldsymbol{x}}(t+\Delta t)$ and $\ddot{\boldsymbol{x}}(t+\Delta t)$ are the displacement, velocity and acceleration at the next time step t + Δt; Δt is the constant time-interval. Newmark factors of γ and β affect the stability and reliability in the integration method, which are usually taken as 0.25 and 0.5, respectively.

2.4. FSI coupling algorithms

The FSI coupling algorithms can be divided as one-way coupling and two-way coupling by judging whether the flow fluid field grid is updated with structure deformation. Actually, one-way coupling is not truly couple between fluid and structure as the structure is still rigid for fluid calculation. Besides, the FSI coupling algorithms can also be classified into partitioned explicit algorithm, partitioned implicit algorithm and monolithic algorithm by judging the FSI solving procedures. The partitioned explicit algorithm is also classified as weak coupling of FSI, because the fluid and structural solutions cannot be satisfied by their respective governing equations at the same time at each time step. With the same judging criterion, partitioned implicit algorithm and monolithic algorithm are classified as strong coupling algorithms of FSI. The details will be described deeply below.

2.4.1. Two-way coupling algorithm. In this study, two-way coupling algorithm between structural deformation and fluid mesh update has been adopted to satisfy the geometrical compatibility and the equilibrium conditions on the interface. The coupling process is shown in Figure 1. It should be emphasized that the developed two-way coupling algorithm in this work is suitable for both 3D and 2D cases.

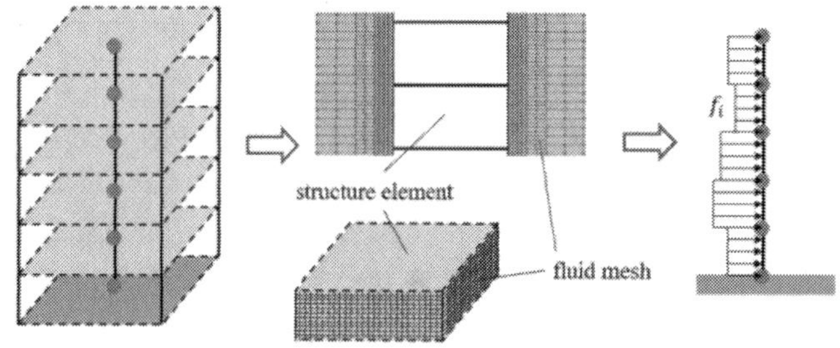

(a) FSI data exchange from fluid to structure.

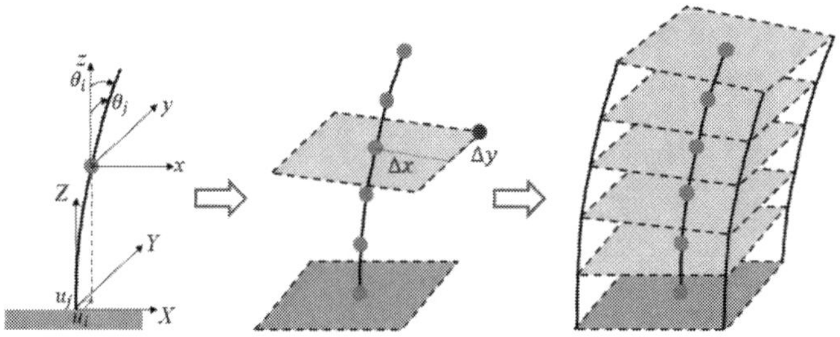

(b) FSI data exchange from structure to fluid boundary mesh.

Figure 1. The schematic diagram of two-way FSI coupling process.

Firstly, the pressure and viscous force are integrated at each structural element to obtain the concentrated force acting on the beam model. Then, the concentrated forces are transformed as distributed loads and the structural deformations at each element node are calculated. After that, a cubic spline interpolation function is adopted to calculate coordinate updates of points of boundary meshes in fluid field. Finally, the update of fluid mesh deformation is executed.

2.4.2. Partitioned algorithm. Figure 2 shows the FSI solution procedures of both partitioned and monolithic algorithms. Partitioned algorithm allows for the application of different solvers for fluid and structure, respectively, in numerical simulations. This approach takes advantage of specialized computing software for both fluid and structural analysis. For this algorithm, data exchange between both fluid and structural solvers needs to be executed in each time step. If FSI calculations do not involve iterations in each time step, the algorithm is categorized as explicit and weak; otherwise, the algorithm is categorized as implicit and strong. The iterations of FSI in each time step ought to have its convergence criteria of structural deformations and fluid forces. In this study, convergence check is set by ensuring that the relative errors between adjacent iterations of structural displacements at each element node and fluid forces acting on each beam element are less than 5×10^{-3}. Besides, in the program, the maximum number for iterations is limited to 50.

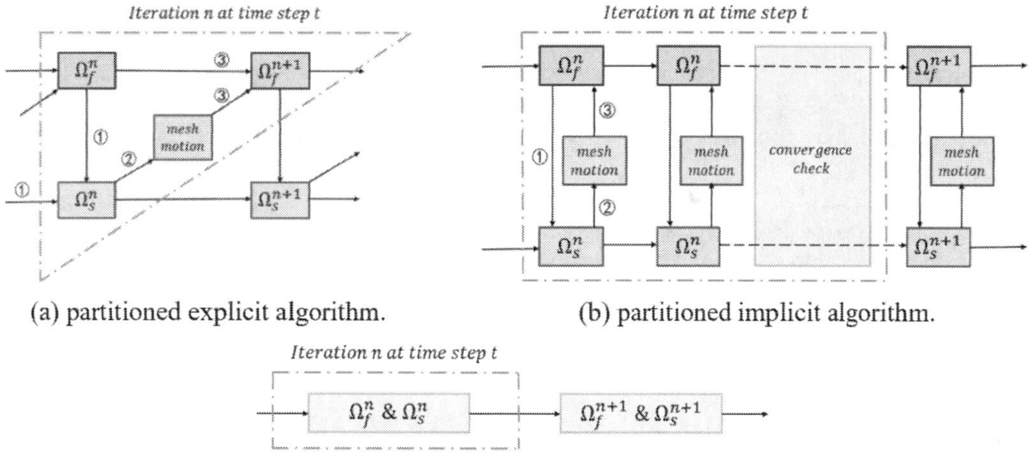

(a) partitioned explicit algorithm. (b) partitioned implicit algorithm.

(c) monolithic algorithm.

Figure 2. The schematic diagram of partitioned and monolithic FSI coupling algorithms.

For the monolithic algorithm, one solution framework ought to be built for fluid and structural calculation. Fluid and structure are discretized with only one method, and their governing equations are solved together in this framework. This algorithm, is absolutely regarded as strong coupling of FSI. However, it's hard to build such solution framework for structures with complex geometric shapes. The studies presented below all pertain to partitioned algorithms.

As the computational resource consumption of partitioned implicit algorithms is significant, some scholars have proposed a weak algorithm as an alternative. However, it still requires FSI iterations at each time step. Before proceeding, it is necessary to introduce the PIMPLE algorithm in OpenFOAM, as depicted in Figure 3. In OpenFOAM, a PIMPLE loop consists of an outer iteration and an inner iteration, with the latter being a PISO loop. Regarding the computation of fluid, it is mandatory to execute at least one outer iteration and one inner iteration. To ensure full solution of the fluid, both outer and inner iterations typically require more than one iteration. Some scholars strive for both accuracy and efficiency in numerical simulations, and have proposed a weak algorithm with iterations[13]. They opt to exchange FSI data after the completion of a PISO loop, rather than waiting for the entire PIMPLE loop to finish. Due to incomplete solution of the fluid field, this algorithm is considered weak. Furthermore, as it necessitates multiple iterations within a single time step, in this paper it is referred as weak iteration algorithm, distinguishing it from the partitioned explicit algorithm without iterations also classified as weak. Moreover, the strong algorithm discussed in this article specifically pertains to the partitioned implicit method. The flowchart for numerical calculating with a strong algorithm differs from that of a weak iterative algorithm, as illustrated in Figure 4.

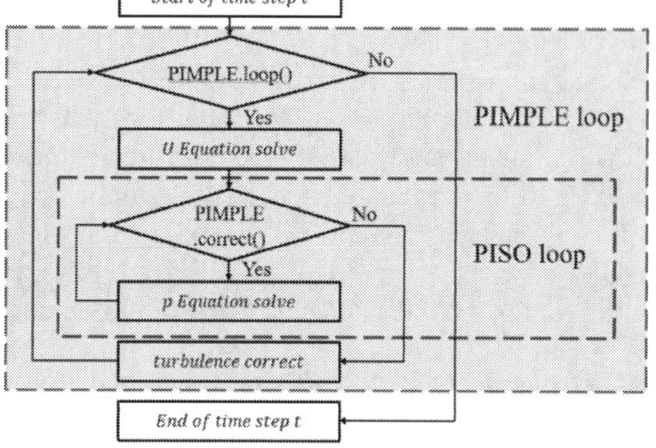

Figure 3. The flowchart of PIMPLE algorithm in OpenFOAM.

In the strong algorithm, FSI data exchange occurs between structural and fluid results after an outer iteration of the PIMPLE algorithm. In contrast, during the weak iteration algorithm, FSI data exchange takes place after an inner iteration of the PIMPLE algorithm. Since each outer iteration encompasses multiple inner iterations, this algorithm provides advantageous in minimizing superfluous PISO calculations. Furthermore, as both algorithms require convergence checks of FSI data, it can be predicted that if the PIMPLE outer iteration number is chosen as one, there will be no differences between the results obtained from either algorithm. However, the strong algorithm involves more unnecessary PISO loops, resulting in a wastage of computational resources.

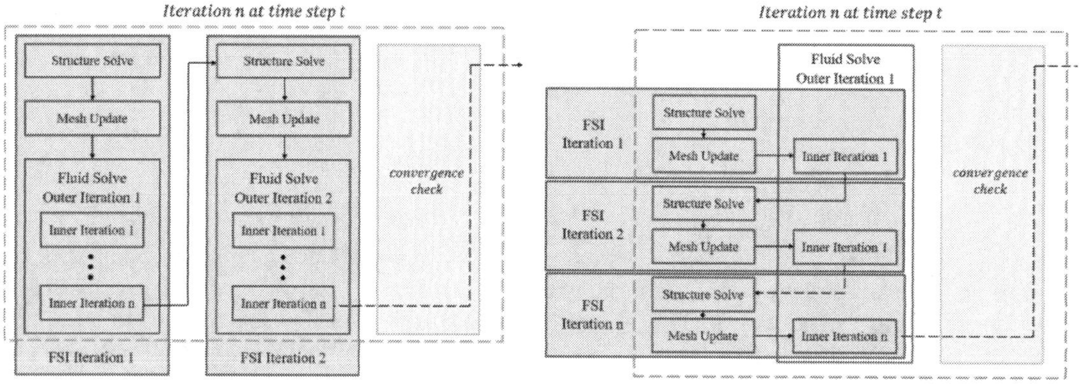

(a) strong algorithm. (b) weak iteration algorithm.

Figure 4. The flowchart of strong algorithm and weak iteration algorithm.

3. Numerical setup

A benchmark numerical simulation is utilized to investigate the interaction between a dam break and an elastic structure. A 2-D numerical domain of length 0.584 m and height 0.40 m is utilized to be consistent with the water tank from the other scholars' studies[7, 8]. The volume of water (0.146 m wide and 0.292 m high) is initially contained. The structure, which is 0.0012 m × 0.08 m, is placed 0.146 m downstream from the end of water volume, as shown in Figure 5. The coordinate system is defined with x and z representing the horizontal and vertical components, respectively. The origin is located at the bottom left-hand corner of the domain. All boundaries are treated as walls with no-slip conditions, except for the top boundary which is considered as an atmosphere boundary with free inlet and outlet flow.

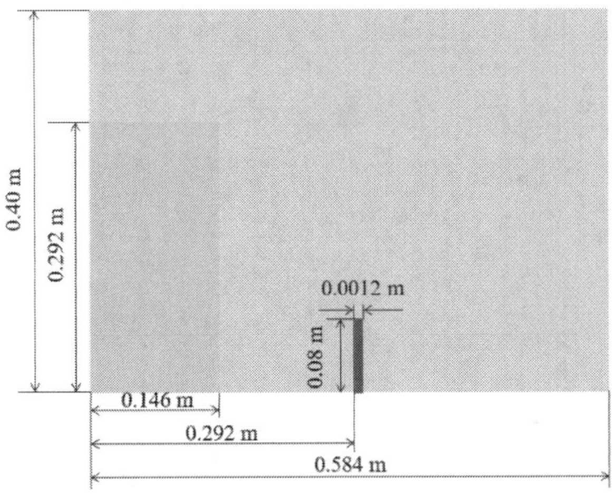

Figure 5. Diagram of the numerical setup for the dam break test case.

The Young's modulus of structure is $E_s = 1.0 \times 10^6\ N/m^2$, density $\rho_s = 2.5 \times 10^3\ kg/m^2$. In this study, the structural model is taken as Euler-Bernoulli beam model. Main parameter settings are shown in Table 1. Based on the results of a mesh independence analysis (section 3.1), the total number of cells in the fluid mesh is 111, 200.

Table 1. Parameter settings of fluid and structure simulations.

Parameters	Value
ρ_s	1.0×10^3 kg/m^2
v_s	1.0×10^{-6} m^2/s
g	9.81 m/s^2
ρ_s	2.5×10^3 kg/m^2
E_s	1.0×10^6 Pa
Δt	0.0001 s

4. Results and discussions

4.1. Mesh independence study

To validate the accuracy of the FSI solution, Table 2 shows the summarized mesh convergence study information and the discretization error resulting from mesh grids. The numerical time history results of horizontal top displacements in the structure are further compared with three meshes, as illustrated in Figure 6.

Table 2. Mesh independence study parameters and structural deformation results.

Mesh	Cell Number	Maximum Top Displacement	Relative Error
Coarse	80, 456	0.04548 m	4.4%
Medium	111, 200	0.04263 m	2.1%
Fine	153, 831	0.04355 m	/

The figure illustrates that all three meshes maintain a favorable time history shape of the beam top displacement, and exhibit relatively small computational errors in peak value. With consideration given to computational cost, it is concluded that a medium-sized mesh is the optimal choice for investigating the influence of different FSI algorithms in this study.

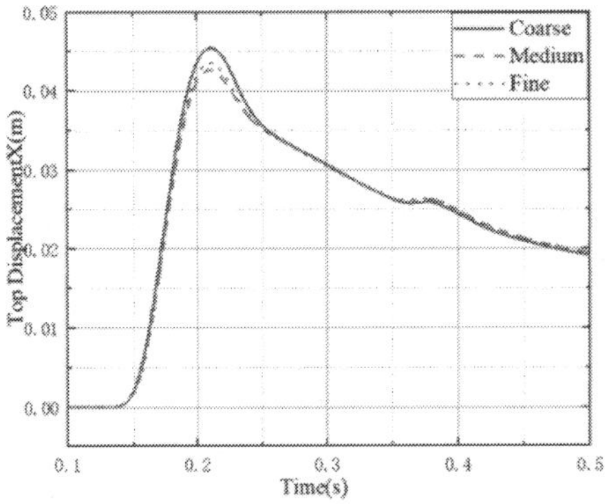

Figure 6. Comparison of the horizontal top displacement of beam with three different meshes.

4.2. Comparison with other methods

Figure 7 shows the horizontal displacement of the uppermost point on the elastic beam versus time. The present results are obtained utilizing a mesh of medium size in this study, and show good agreement with the particle FEM results[8] and MPS-FEM results[7] during the main slamming procedure from 0.15s to 0.2s. The results of this paper are slightly different from those of other methods during t > 0.2s, which could be attributed to the simplification of bending by the Euler-Bernoulli beam model.

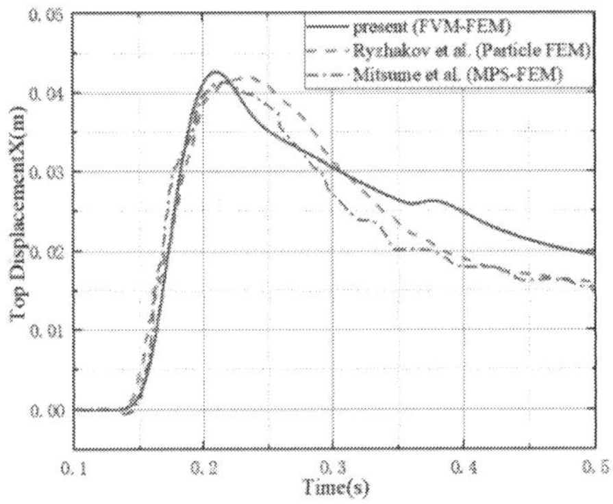

Figure 7. Comparison of the top displacement of beam with different methods.

4.3. Comparison with different FSI coupling algorithm

In this study, the developed program has implemented three types of FSI coupling algorithms: weak algorithm, weak iteration algorithm, and strong algorithm. Four cases are studied with varying PIMPLE outer and inner iteration numbers, as shown in Table 3.

Table 3. Settings of PIMPLE outer and inner iterations with different coupling algorithms.

Coupling Algorithms	PIMPLE Outer Iterations	PIMPLE Inner Iterations
Weak	1	3
Weak Iteration	1	3
Strong (case1)	1	3
Strong (case2)	2	3

As stated in section 1.4.2, selecting the outer iteration number of one renders no discernible difference between the computational results produced by the weak iteration algorithm and strong algorithm. This is due to the fact that both algorithms differ solely in their PISO loop iteration numbers, while the result accuracy is governed by FSI convergence checks. In this scenario, the superfluous PISO loop iterations are proved as redundant in a robust algorithm. The validation of this conclusion can be demonstrated through a comparison of the structural top displacement using three different coupling algorithms, as illustrated in Figure 8 and Table 4.

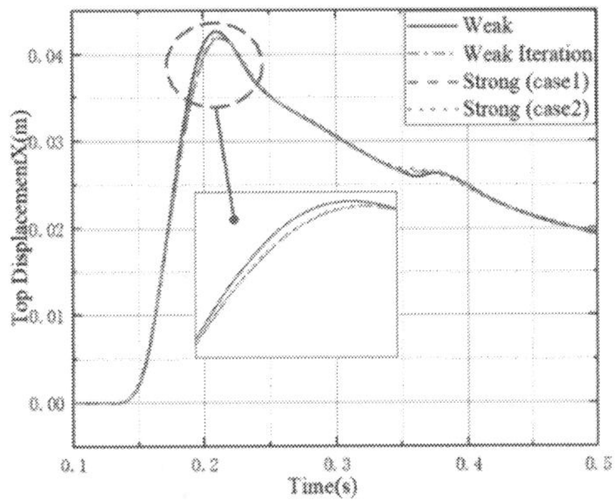

Figure 8. Comparison of the top displacement of beam with different FSI coupling algorithms.

The diagram demonstrates that the outcomes of the weak iteration algorithm are nearly indistinguishable from those of two cases of the strong algorithm. The table shows that weak algorithm and iterative algorithms yield a maximum relative error of 1% in the top displacement results. Moreover, the weak iteration and strong algorithms exhibit identical results in terms of maximum top displacement and time history of structural top displacement, thus confirming the previous conjecture.

Table 4. Comparison of the maximum beam top displacement with different FSI coupling algorithms.

Coupling Algorithms	Maximum Top Displacement	Relative Error
Weak	0.04263 m	1.43%
Weak Iteration	0.04187 m	0.38%
Strong (case1)	0.04187 m	0.38%
Strong (case2)	0.04203 m	/

Comparisons between the results of structural deformation and free surface at typical simulation times have been selected to investigate the influence towards fluid solution of different FSI coupling algorithms, as shown in Figure 9.

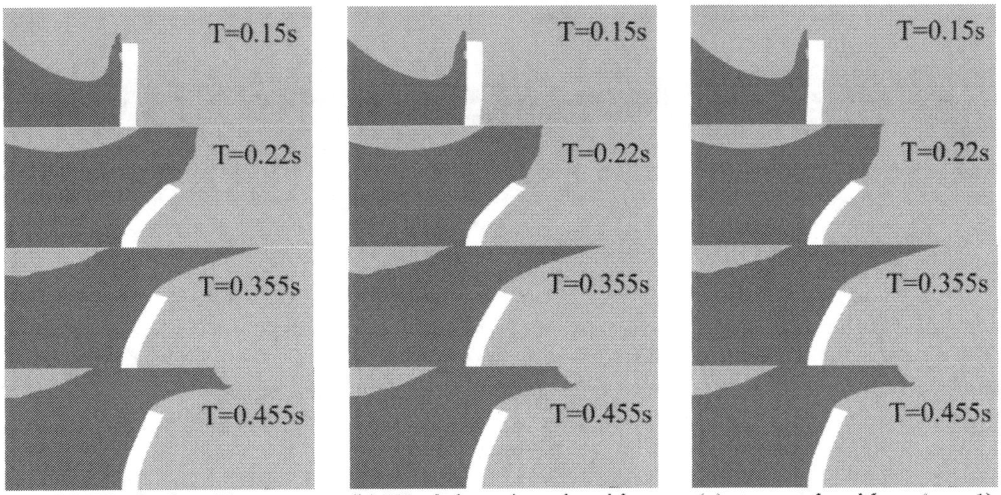

(a) Weak algorithm. (b) Weak iteration algorithm. (c) strong algorithm. (case1)

Figure 9. Snapshots of structural deformation and free surface with different FSI coupling algorithms.

Flow field calculations solved by the three algorithms are indistinguishable in the diagram. The maximum relative error in structural solutions being 1.43% may also imply only slight discrepancies among the results obtained from the three FSI coupling algorithms.

Finally, the computational efficiency has been investigated and compared using three distinct FSI coupling algorithms, as illustrated in Table 5. It is evident that all FSI coupling algorithms have significantly increased the computational time. Weak iteration algorithm requires the least computational time increment among all algorithms and achieves the same computational accuracy as strong algorithm when the outer iteration number is set to one in PIMPLE loops. Besides, it has been observed that when the number of outer iterations doubles in PIMPLE loops for case 1 and case 2 using strong algorithms, there is only a slight increase in computational time instead of doubling. The comparison of FSI iteration numbers between the two cases depicted in Figure 10 can provide an explanation for this phenomenon.

Table 5. Comparison of the computational time with different FSI coupling algorithms.

Coupling Algorithms	Computational Clock Time	Computational Time Increment
Weak	17, 337 s	/
Weak Iteration	58, 714 s	238.7%
Strong (case1)	65, 425 s	277.4%
Strong (case2)	68, 636 s	295.9%

As depicted in Figure 10, case 2 utilizing a strong algorithm requires fewer FSI iterations compared to case 1 employing the same algorithm. This phenomenon implies that while additional outer iterations may result in increased time costs during flow fluid computation directly, it can enhance the convergence rate of FSI iterations and is equivalent to reduce the number of PISO iterations required in the weak iteration algorithms. Therefore, it is crucial to make a rational selection of the outer and inner iteration numbers within PIMPLE loops in FSI strong coupling algorithms.

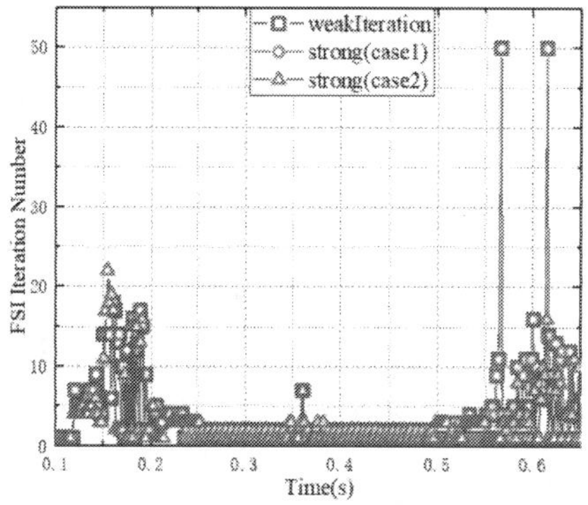

Figure 10. Comparison of the FSI iteration numbers between different coupling algorithms.

5. Conclusions

In this study, a CFD-FEM solver employing various FSI coupling algorithms is developed to simulate the impact of dam-break flow on an elastic beam. The flow fluid field is computed using OpenFOAM, while the structural deformation is solved through a developed FEM code. The FSI algorithms, including the partitioned explicit algorithm and two types of partitioned implicit algorithms, are employed in the simulations to investigate their impact on both computational accuracy and efficiency. A mesh independence study and comparison of results with other methods are conducted to validate the computational findings.

In terms of computational accuracy, all three FSI coupling algorithms exhibit good agreement with respect to free surface evolution and structural deformations with the maximum top displacement relative error being 1.43%. Moreover, when the outer iteration number is set to one in PIMPLE loops, both strong and weak iteration algorithms yield identical solutions for fluid and structure. When the outer iteration number chosen as one, the strong algorithm incurs computational inefficiencies compared to the weak iteration algorithm in terms of computational efficiency. Besides, increasing the number of outer iterations can enhance the convergence of FSI iterations in strong coupling algorithm. Therefore, a higher setting for outer iteration number does not necessarily result in a significant increase in computational time. It is crucial to make a rational selection of the outer and inner iteration numbers within PIMPLE loops in FSI strong coupling algorithms.

6. Acknowledgements
This work was supported by the National Natural Science Foundation of China (52131102), and the National Key Research and Development Program of China (2019YFB1704200), to which the authors are most grateful. The authors specially thank to Prof. Decheng Wan for his kind suggestions and supervision.

7. References
[1] Greco M, Colicchio G and Faltinsen O M 2007 Shipping of water on a two-dimensional structure Part 2 *J. Fluid Mech.* **581** pp 71-99
[2] Liao K, Hu C and Sueyoshi M 2015 Free surface flow impacting on an elastic structure: Experiment versus numerical simulation *Appl. Ocean Res.* **50** pp 192-208
[3] Yilmaz A, Kocaman S and Demirci M 2021 Numerical modeling of the dam-break wave impact on elastic sluice gate: A new benchmark case for hydroelasticity problems *Ocean Eng.* **231** 108870
[4] Hu Z, Huang L and Li Y 2023 Fully-coupled hydroelastic modeling of a deformable wall in waves *Coast. Eng.* **179** 104245
[5] Khayyer A, Gotoh H, Falahaty H and Shimizu Y 2018 An enhanced ISPH–SPH coupled method for simulation of incompressible fluid–elastic structure interactions *Comput. Phys. Commun.* **232** pp 139-64
[6] Mcloone M and Quinlan N J 2022 Coupling of the meshless finite volume particle method and the finite element method for fluid–structure interaction with thin elastic structures *Eur. J. Mech. B-Fluid.* **92** pp 17-31
[7] Mitsume N, Yoshimura S, Murotani K and Yamada T 2014 MPS-FEM Partitioned Coupling Approach for Fluid–Structure Interaction with Free Surface Flow *Int. J. Comp. Meth.* **11** 1350101
[8] Ryzhakov P B, Rossi R, Idelsohn S R and Onate E 2010 A monolithic Lagrangian approach for fluid–structure interaction problems *Comput. Mech.* **46** pp 883-99
[9] Walhorn E, Kölke A, Hübner B and Dinkler D 2005 Fluid–structure coupling within a monolithic model involving free surface flows *Comput. Struct.* **83** pp 2100-11
[10] Wu J, Zhang G, Jiang Y and Yang X 2022 Numerical Simulations on the Flooding into a Damaged Cabin with a Flexible Bulkhead Based on the Mixed-Mode Function-Modified MPS Method *J. Mar. Sci. Eng.* **10** 101582
[11] Zheng Z, Duan G, Mitsume N, Chen S and Yoshimura S 2020 An explicit MPS/FEM coupling algorithm for three-dimensional fluid-structure interaction analysis *Eng. Anal. Bound. Elem.* **121** pp 192-206
[12] NEWMARK N M 1959 A method of computation for structural dynamics *J. Eng. Mech. Div.* **85** pp 67-94
[13] Wei Y, Incecik A and Tezdogan T A Fully 2022 Coupled CFD-DMB Approach on the Ship Hydroelasticity of a Containership in Extreme Wave Conditions *J. Mar. Sci. Eng.* **10** 101778

IOP Conf. Series: Materials Science and Engineering 1288 (2023) 012019 doi:10.1088/1757-899X/1288/1/012019

Numerical Simulation of Ship-Pack Ice-Water Interaction and Motion of Pack Ice Based on CFD-DEM Method

Ming Zou[1, 2], Xiang-Jie Tang[1], Lu Zou[1, 3], Zao-Jian Zou[1, 3*]

[1] School of Naval Architecture, Ocean and Civil Engineering, Shanghai Jiao Tong University, Shanghai 200240, China
[2] Marine Design and Research Institute of China, Shanghai 200011, China
[3] State Key Laboratory of Ocean Engineering, Shanghai Jiao Tong University, Shanghai 200240, China

*Corresponding author. E-mail: zjzou@sjtu.edu.cn

Abstract. When a ship sails in pack ice area, it not only collides with the ice but also interacts with the water, generating ship-generated waves. The role and influence of ship-generated waves on the ship-ice-water interaction have not been thoroughly studied. In this study, a numerical model with ship-generated waves is established using a coupled Computational Fluid Dynamics (CFD) and Discrete Element Method (DEM), and an appropriate contact model is selected for numerical simulations. Meanwhile, a simplified numerical model without ship-generated waves is proposed. By comparing the simulation results under the same simulation conditions with and without ship-generated waves, the effects of ship-generated waves on the phenomena of ship-ice-water interaction, longitudinal and lateral contact forces between the ship and pack ice, and ice resistance are analysed, along with the underlying mechanisms. The results indicate that the ship-generated waves can mitigate and reduce the collision intensity and contact frequency between the ship and pack ice, resulting in a decrease in the contact forces and ultimately achieving a significant reduction in ice resistance. Furthermore, this mitigation effect becomes more pronounced with increasing ship speed.

1. Introduction

As a result of global warming, the melting of sea ice in the Arctic is intensifying, thereby increasing the possibility of ship navigation in the region. Additionally, climate change is leading to a transition in the ice condition from level ice cover to pack ice cover in the Arctic [1]. Studying the interaction between the ship, pack ice, and water, as well as predicting ship resistance in the pack ice area, holds significant importance in enhancing the safety and efficiency of ship navigation in the Arctic.

Early in the study on ship-pack ice interaction, model tests have been conducted. According to test conditions, the materials used for pack ice used can be mainly divided into frozen ice [2-4] and non-frozen ice. The non-frozen ice materials can be further divided into paraffin [5-7] and polypropylene [8]. Although the results of model tests are considered as most reliable, the preparation and implementation of the tests require huge manpower and material resources, and the tests have high requirements on the facilities and conditions; thus, making frequent testing inconvenient. With the rapid advancement of computer techniques, numerical methods that offer promising prediction capabilities have garnered increasing attention.

Currently, the numerical methods for studying ship-ice-water interaction include Finite Element Method (FEM), Smoothed Particle Hydrodynamics (SPH) method, Discrete Element Method (DEM), etc [9-10]. Using these methods, a lot of studies have been conducted to solve the problems of ship-level ice interaction, and some good results have been achieved [11-14]. However, there are relatively few studies conducted specifically on pack ice conditions [15-20]. Among the existing studies on pack ice conditions, FEM and DEM are relatively commonly used [21]. However, in these studies, the influence of flow field was usually simplified, and the influence of ship-generated waves is not only ignored but also cannot be considered, which limits the ability of the established numerical models to accurately reflect ship-ice-water interaction. To consider the influence of flow field and ship-generated waves, some researchers have coupled CFD method with DEM method to investigate ship navigation in pack ice conditions. Vroegrijk [22] utilized the CFD-DEM method to study the ship-pack ice interaction, providing preliminary evidence for the effectiveness of the CFD-DEM method. Furthermore, Mucha [23] presented a general simulation framework for simulating ship advancing through pack ice in Simcenter STAR-CCM+, further validating the capability of the CFD-DEM method. Huang et al. [24-25] developed a pack ice generation method and employed the CFD-DEM method to predict ice resistance for three different ships under varying navigation conditions. By summarizing and analysing the results, a general empirical formula was derived for rapid ice resistance prediction. Additionally, Luo et al. [26] and Zhang et al. [27] investigated ice resistance in restricted pack ice channel using model-scale and full-scale ships, respectively.

Existing study has shown that the ship-generated waves play a crucial role throughout the entire ship navigation process [24]. These waves not only affect the interaction between the ship and pack ice, but also affect the motion state of the ice. The influence of ship-generated waves on the interaction among the ship, ice, and water is an intriguing and significant problem that has not been extensively investigated. This study conducts a comparative analysis by establishing two numerical models: one with ship-generated waves and one without ship-generated waves. The analysis examines the effects of ship-generated waves on the phenomena of ship-ice-water interaction, ship-ice longitudinal contact force, ship-ice lateral contact force, and ice resistance. The mechanism underlying the influence of ship-generated waves is also analysed.

2. Numerical method

2.1. CFD-DEM method

The governing equations of the CFD-DEM method are as follows:

$$\frac{\partial \varepsilon_f}{\partial t}+\nabla\cdot\left(\varepsilon_f \boldsymbol{u}\right)=0 \tag{1}$$

$$\frac{\partial\left(\varepsilon_f \boldsymbol{u}\right)}{\partial t}+\nabla\cdot\left(\varepsilon_f \boldsymbol{u}\cdot\boldsymbol{u}\right)=-\frac{1}{\rho_f}\left[\varepsilon_f \nabla p-\nabla\cdot\left(\varepsilon_f \boldsymbol{\tau}_f\right)+\rho_f \varepsilon_f \boldsymbol{g}-\boldsymbol{F}\right] \tag{2}$$

where ε_f represents the volume fraction of the fluid phase, $\boldsymbol{u}$ and p denote the average velocity and pressure of the fluid phase, respectively. ρ_f represents the mass density and $\boldsymbol{\tau}_f$ represents the stress tensor. The gravitational acceleration is denoted by $\boldsymbol{g}$, and $\boldsymbol{F}$ represents the interaction force between the fluid phase and the discrete phase ice.

In this paper, the turbulence model utilized for the numerical model with ship-generated waves is the k-ω SST (Shear Stress Transport) model [28]. The VOF (Volume of Fluid) method [29] is employed to capture the free surface elevation, while the simulation accuracy of the free surface is enhanced through the implementation of HRIC (High-Resolution Interface Capturing).

For the numerical model without ship-generated waves, this paper limits the solution of CFD within the framework of CFD-DEM, without update of the iterative calculation of CFD in time domain. Therefore, the ship-generated waves are not captured. It should be noted that such a simplified treatment does not account for the variations in the flow field below the free surface.

This paper adopts a one-way coupling mechanism, which means that during the solving process, the fluid phase affects the discrete phase ice, while the discrete phase ice does not affect the fluid phase. This mechanism helps to conserve computational resources [23, 26].

2.2. Contact model

As a key factor in DEM, the contact model directly affects the results, hence it is necessary to select an appropriate contact model. In existing studies, the elastic model consisting of spring and damper, as shown in Figure 1, is commonly used. Among the elastic contact models, the Hertz-Mindlin model [30-31] and the Linear Spring model [32] are the most common. This paper compares the influence of these two contact models on ice resistance.

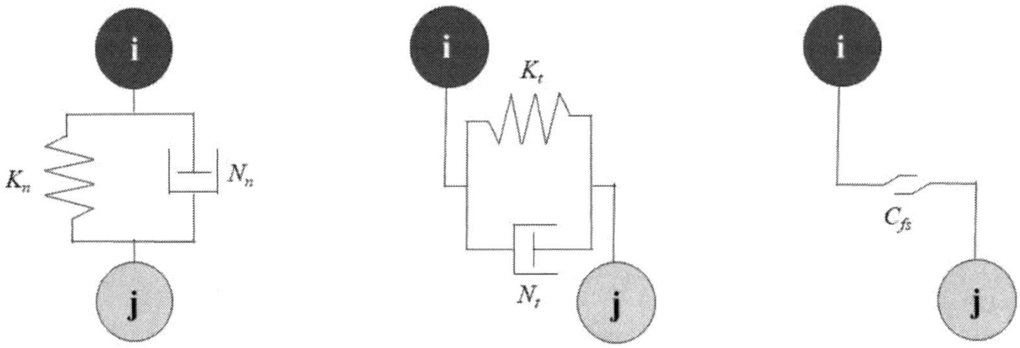

(a) Normal spring and damping (b) Tangential spring and damping (c) Tangential friction

Figure 1. Spring-damper model

The contact force F_c in spring-damper contact model can be expressed as

$$F_c = F_n + F_t \tag{3}$$

where F_n represent the normal contact force and F_t represent the tangential contact force.

The F_n and F_t can be expressed as

$$F_n = -K_n d_n - N_n v_n \tag{4}$$

$$F_t = \begin{cases} -K_t d_t - N_t v_t & |K_t d_t| < |K_n d_n| C_{fs} \\ \dfrac{|K_n d_n| C_{fs} d_t}{|d_t|} & |K_t d_t| > |K_n d_n| C_{fs} \end{cases} \tag{5}$$

where K_n and K_t represent the normal and tangential spring stiffness, d_n and d_t denote the overlap in the normal and tangential directions, N_n and N_t represent the normal and tangential damping, v_n and v_t correspond to the normal and tangential velocity components, and C_{fs} represents the static friction coefficient.

In the Hertz-Mindlin contact model, K_n, K_t, N_n and N_t are defined as

$$K_n = \frac{4}{3} E_{eq} \sqrt{d_n R_{eq}} \tag{6}$$

$$K_t = 8 G_{eq} \sqrt{d_t R_{eq}} \tag{7}$$

$$N_n = N_{nd} \sqrt{5 K_n M_{eq}} \tag{8}$$

$$N_t = N_{td} \sqrt{5 K_t M_{eq}} \tag{9}$$

where E_{eq}, R_{eq}, G_{eq} and M_{eq} represent the equivalent Young's modulus, radius, shear modulus, and particle mass respectively. N_{nd} and N_{td} denote the normal and tangential damping coefficients.

In the Linear Spring contact model, K_n, K_t, N_n and N_t are defined as

$$K_n = \frac{4}{3}\sqrt{\lambda_{\max}} E_{eq} R_{eq} \tag{10}$$

$$K_t = 8\sqrt{\lambda_{\max}} G_{eq} R_{eq} \tag{11}$$

$$N_n = 2N_{nd}\sqrt{K_n M_{eq}} \tag{12}$$

$$N_t = 2N_{td}\sqrt{K_t M_{eq}} \tag{13}$$

where $\lambda_{\max}=\delta_{\max} / R_{eq}$ is the maximum overlap ratio, $\delta_{\max}$ is the maximum overlap.

2.3. The discrete phase motion

The motion of pack ice in the DEM involves both translational and rotational motion. The governing equations for the motion can be expressed as follows:

$$m_i \frac{d\boldsymbol{v}_i}{dt} = \boldsymbol{F}_{c,i} + \boldsymbol{F}_{g,i} + \boldsymbol{F}_{d,i} + \boldsymbol{F}_{p,i} + \boldsymbol{F}_{vm,i} \tag{14}$$

$$\boldsymbol{I}_i \frac{d\boldsymbol{\omega}_i}{dt} = \boldsymbol{M}_{c,i} + \boldsymbol{M}_{f,i} \tag{15}$$

where m_i and $\boldsymbol{v}_i$ represent the mass and velocity of pack ice *i*, respectively. $\boldsymbol{F}_{c,i}$, $\boldsymbol{F}_{g,i}$, $\boldsymbol{F}_{d,i}$, $\boldsymbol{F}_{p,i}$ and $\boldsymbol{F}_{vm,i}$ denote the contact force, gravity, drag force, pressure gradient force, and virtual mass force acting on pack ice i, respectively. $\boldsymbol{\omega}_i$ and $\boldsymbol{I}_i$ represent the angular velocity and moment of inertia of pack ice i, $\boldsymbol{M}_{c,i}$ represents the total contact moment, and $\boldsymbol{M}_{f,i}$ denotes the interaction moment between discrete phase and the fluid phase.

3. Study objects and numerical setup

3.1. Ship model

The KCS container ship model is utilized as the target model, with a scale ratio of 1:52.667, as illustrated in Figure 2. The ship's geometry can be referenced from Kim et al. [33]. The main particulars of the model-scale ship are given in Table 1.

Figure 2. Geometry of KCS hull

Table 1. Main particulars of KCS

Parameter	Unit	Model scale
Length between perpendiculars / L_{PP}	m	4.367
Breadth / *B*	m	0.661
Draught / *d*	m	0.205
Block coefficient / C_B	-	0.651
Wetted surface area / *S*	m^2	3.397

3.2. Pack ice model

Kim et al. [2] compared the influence of using breakable frozen ice and unbreakable non-frozen ice on ship-ice-water interaction and ice resistance, and found that whether the pack ice is breakable or not has no notable influence. Therefore, in order to save computational resources and improve computational efficiency, the pack ice model is constructed in this paper with composite particles that are rigid and do not break. The pack ice model size in this paper is obtained by converting the full-scale pack ice in Luo et al. [7] to model scale. The length, width and thickness of pack ice are shown in Figure 3.

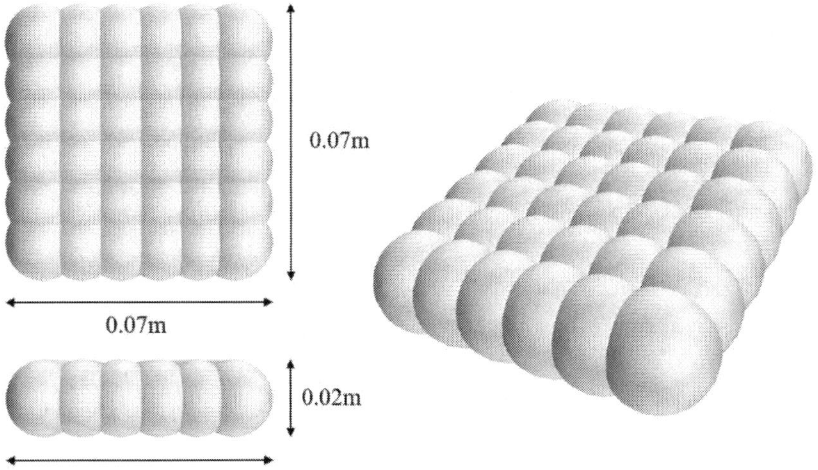

Figure 3. Pack ice model

The pack ice model has a density of 917 kg/m^3 and a Poisson's ratio of 0.30. The Young's modulus of the pack ice model is 170 MPa, which is obtained through scale conversion based on full-scale pack ice [34]. The ship-ice rolling resistance coefficient is 0.035, and the ship-ice static friction coefficient is 0.40 [35]. The detailed parameters are given in Table 2.

Table 2. Parameters of pack ice model

Parameter	Unit	Value
Density / ρ_{ice}	kg/m^3	917
Young's modulus / E	MPa	170
Poisson's ratio / v	-	0.30
Ship-ice static friction coefficient / $C_{fs(\text{ship-ice})}$	-	0.40
Ship-ice rolling resistance coefficient / $\mu_{r(\text{ship-ice})}$	-	0.035
Ice-ice static friction coefficient / $C_{fs(\text{ice-ice})}$	-	0.10
Ice-ice rolling resistance coefficient / $\mu_{r(\text{ice-ice})}$	-	0.03
Normal restitution coefficient / C_{nr}	-	0.50
Tangential restitution coefficient / C_{tr}	-	0.50
Maximum overlap ratio / $\lambda_{\max}$	-	0.10

3.3. Computational domain and boundary conditions

The computational domain used in this study is depicted in Figure 4. The domain extends from –3 L_{pp} to 3 L_{pp} in length, from –2.5 L_{pp} to 2.5 L_{pp} in width, and from –3 L_{pp} to 1.0 L_{pp} in height. The boundary conditions for each side of the computational domain are shown in Figure 4.

The hull is fixed on even keel and moves forward at a constant speed U. Based on the principle of relative motion, it is assumed that the ship is stationary, while there is a flow of water towards the ship with a velocity $-U$. Thus, in the numerical simulations, the ship is fixed while the flow and the pack ice are moving towards the ship at a velocity $-U$.

The pack ice in the computational domain is generated by the part injector, as shown in Figure 4. The ice concentration can be adjusted by changing the number of pack ice generated per injection point per unit time. The ratio of the total area of pack ice generated per unit time to the area of pack ice flow per unit time is defined as ice concentration, represented by C.

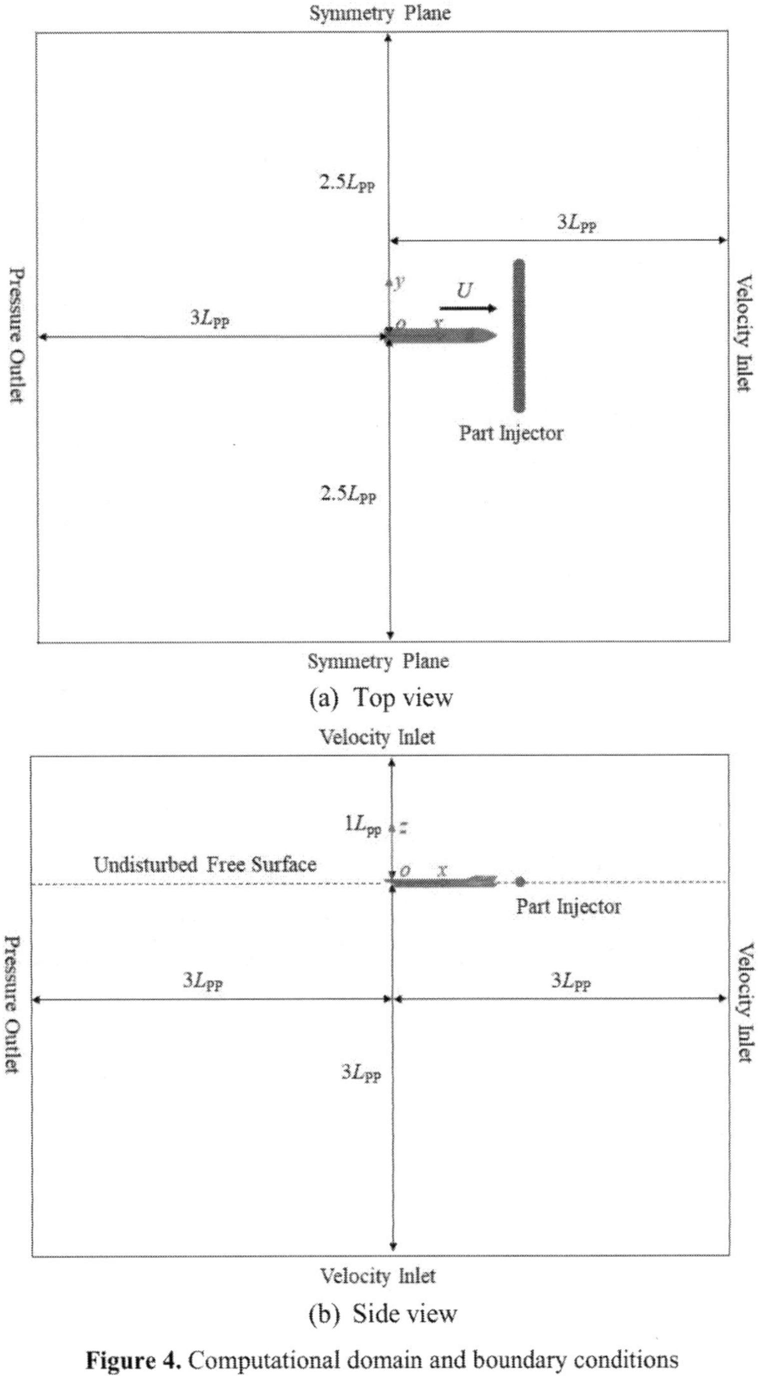

Figure 4. Computational domain and boundary conditions

IOP Conf. Series: Materials Science and Engineering 1288 (2023) 012019 doi:10.1088/1757-899X/1288/1/012019

3.4. Mesh generation

To discretize the computational domain, the Trimmer Cell Mesher is used to generate meshes within the domain. The resulting meshes are presented in Figure 5. To accurately capture the flow details and ship-generated waves, mesh refinement is applied to the free surface as well as the bow and stern regions. For simulations involving near-wall flows, prism layer meshes are generated around the hull surface to capture the flow within the boundary layer.

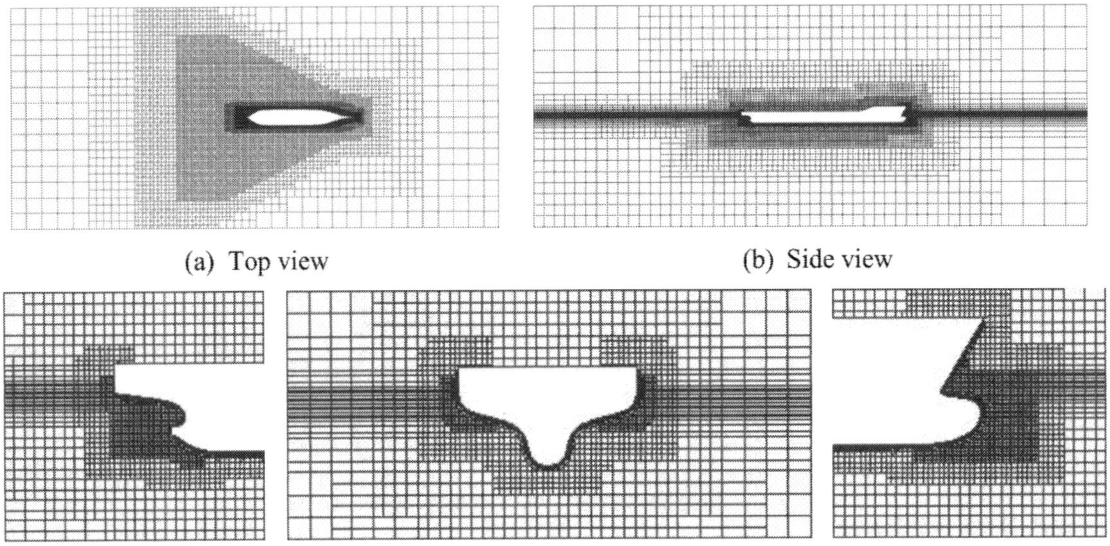

(a) Top view (b) Side view

(c) Stern (d) Midship (e) Bow

Figure 5. Generated meshes

3.5. Numerical simulation conditions

To study the influence of ship-generated waves, the numerical simulation models with and without ship-generated waves are established, and a comparative study on these two simulation models is carried out. The detailed conditions are given in Table 3, including the ship speed U, the corresponding Froude number $Fr = U / \sqrt{gL_{pp}}$, and the ice concentration C, etc.

Table 3. Simulation conditions

Parameter	Unit	Value
Ship speed) / U	m/s	0.4, 0.6, 0.8, 1.0
Froude number / Fr	-	0.06, 0.09, 0.12, 0.15
Ice concentration C	-	80%
Ship-generated waves / -	-	With, Without

4. Verification of the numerical method

4.1. Open-water resistance

The open-water resistance is predicted and verified against the data of model tests [36]. The CFD simulations for open-water resistance are conducted under the same conditions as the model tests, with ship model speeds of Fr = 0.06, 0.09, 0.12, and 0.15. The time step used in CFD computation is 0.02 s.

Figure 6 presents the comparison between the numerical results obtained by CFD simulations and the model test data (EFD). The close agreement between the CFD-computed open-water resistance and the EFD data validates the reliability and accuracy of the CFD method.

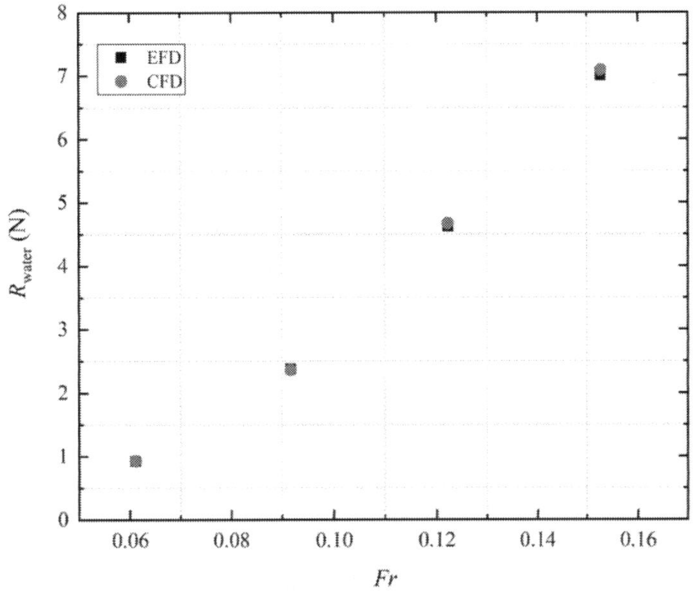

Figure 6. Comparison of numerical and experimental open-water resistance

4.2. *Ice resistance*
Based on the validation of using CFD to predict open-water resistance, this paper further verifies the results of ice resistance prediction obtained by the CFD-DEM method using an empirical formula proposed by Huang et al. [25]. The derivation of this empirical formula utilized the KCS ship as one of the studied objects, which aligns with the ship model used in this study.

$$R_{ice} = A \times \rho_{ice} \times h \times D \times U^2 \times B / L_{pp} \times C^{1.5} \times Fr^{-0.8} \tag{16}$$

where A represents a coefficient with a recommended value of 7.64 for the KCS ship. h denotes the ice thickness, D represents the equivalent diameter of pack ice, determined through conversion and set to 0.079 m.

When DEM and CFD are coupled for solution, a smaller time step is required in the DEM component to obtain more accurate results. Following the time step settings in Huang et al. [24] and Zhang et al. [27], the time step for the coupled CFD-DEM solution is set to 0.005 s in this study.

Figure 7 shows the comparison of ice resistance results obtained using the Hertz-Mindlin contact model, the Linear Spring contact model, and the empirical formula at different ship speeds under the ice concentration $C = 80\%$. Table 4 presents the corresponding error and average error.

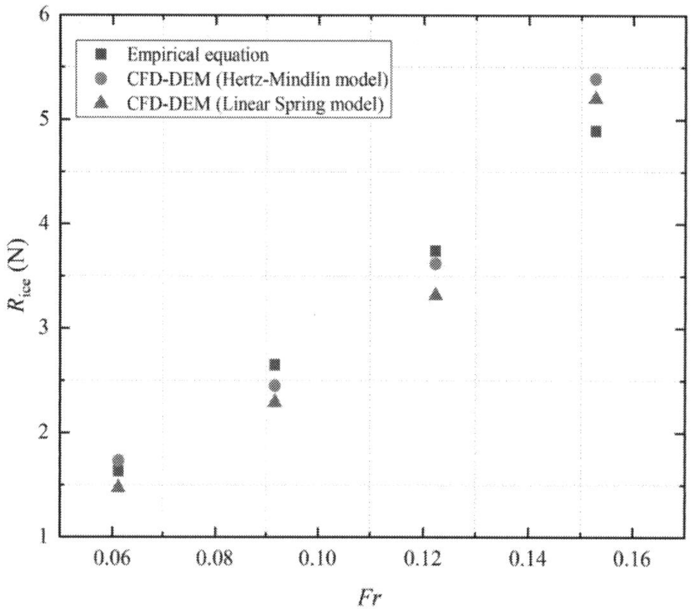

Figure 7. Comparison of numerical and empirical ice resistance (C = 80%)

Table 4. Comparison of ice resistance between two contact models (C = 80%)

Fr	Empirical value (N)	Hertz-Mindlin model (N)	Error (%)	Average error (%)	Linear Spring model (N)	Error (%)	Average error (%)
0.06	1.63	1.73	6.1	6.8	1.47	9.8	10.2
0.09	2.65	2.45	7.5		2.29	13.6	
0.12	3.74	3.62	3.2		3.32	11.2	
0.15	4.89	5.39	10.2		5.2	6.3	

The comparison reveals a close agreement between the ice resistance values obtained from the two contact models and the empirical values. Table 4 demonstrates that the Hertz-Mindlin model yields a minimal error of approximately 3% and a maximal error of around 10%, with an average error of 6.8%. The Linear Spring model exhibits a minimal error of approximately 6% and a maximal error of around 14%, resulting in an average error of 10.2%. Both contact models can effectively predict the ice resistance. Based on the results of comparison with empirical values, the Hertz-Mindlin contact model is used for the subsequent simulations.

4.3. Simulated phenomena

Figure 8 presents the numerical phenomena of ship sailing in open water using the CFD method and ship sailing in pack ice using the CFD-DEM method, both at a ship speed of Fr = 0.15. It can be observed that the ship-generated waves are accurately captured, and the phenomena of ship-ice-water interaction are effectively simulated.

Figure 8. Simulated phenomena ($Fr = 0.15$)

5. Results and discussion

5.1. Numerical phenomena with and without ship-generated waves

Figure 9 shows the numerical phenomena with and without ship-generated waves under the ice concentration $C = 80\%$ and the ship speed $Fr = 0.15$. It can be seen that without the influence of ship-generated waves, the pack ice mainly accumulates around the ship and slides backward along the hull side. In addition, it can be found that the sinking of pack ice is more obvious without wave effect.

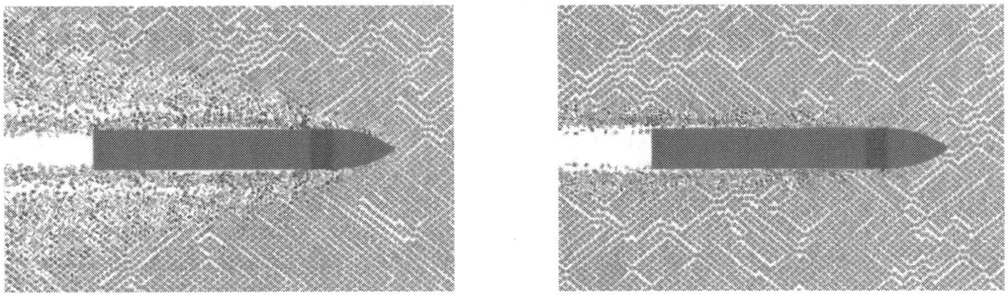

(a) Top view (Left: with waves; Right: without waves)

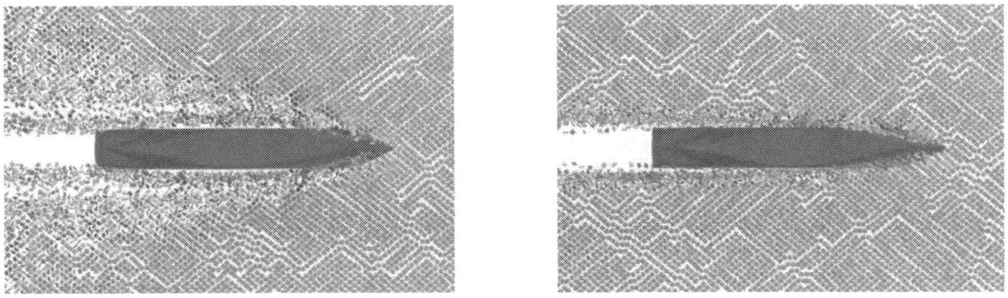

(b) Bottom view (Left: with waves; Right: without waves)

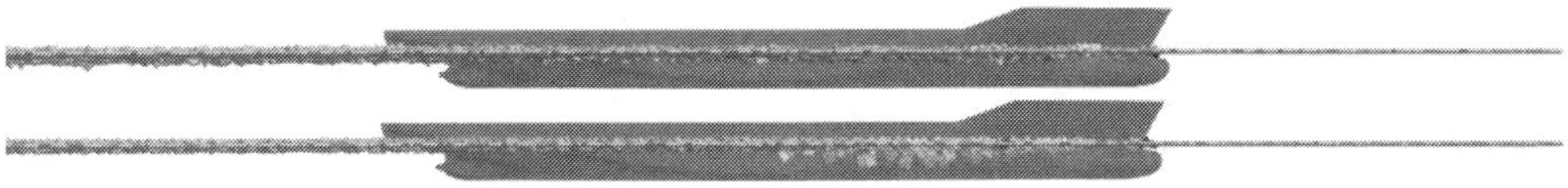

(c) Side view (Above: with waves; Below: without waves)

Figure 9. Phenomena of ship-ice-water interactions ($C = 80\%$, $Fr = 0.15$)

5.2. Motion of pack ice with and without ship-generated waves

Figures 10 and 11 show respectively the velocity distribution and angular velocity distribution of pack ice with and without ship-generated waves. It can be seen that when there is no wave effect, significant changes are observed in the velocity and angular velocity of the pack ice surrounding the hull. Specifically, the velocity of the pack ice decreases at the bow, then increases as it enters into the midship area, ultimately reaching a relatively stable state. The angular velocity of the pack ice is largest at the bow and subsequently stabilizes at the hull sides. When there exists wave effect, the velocity and angular velocity of pack ice in the area of ship-generated waves change.

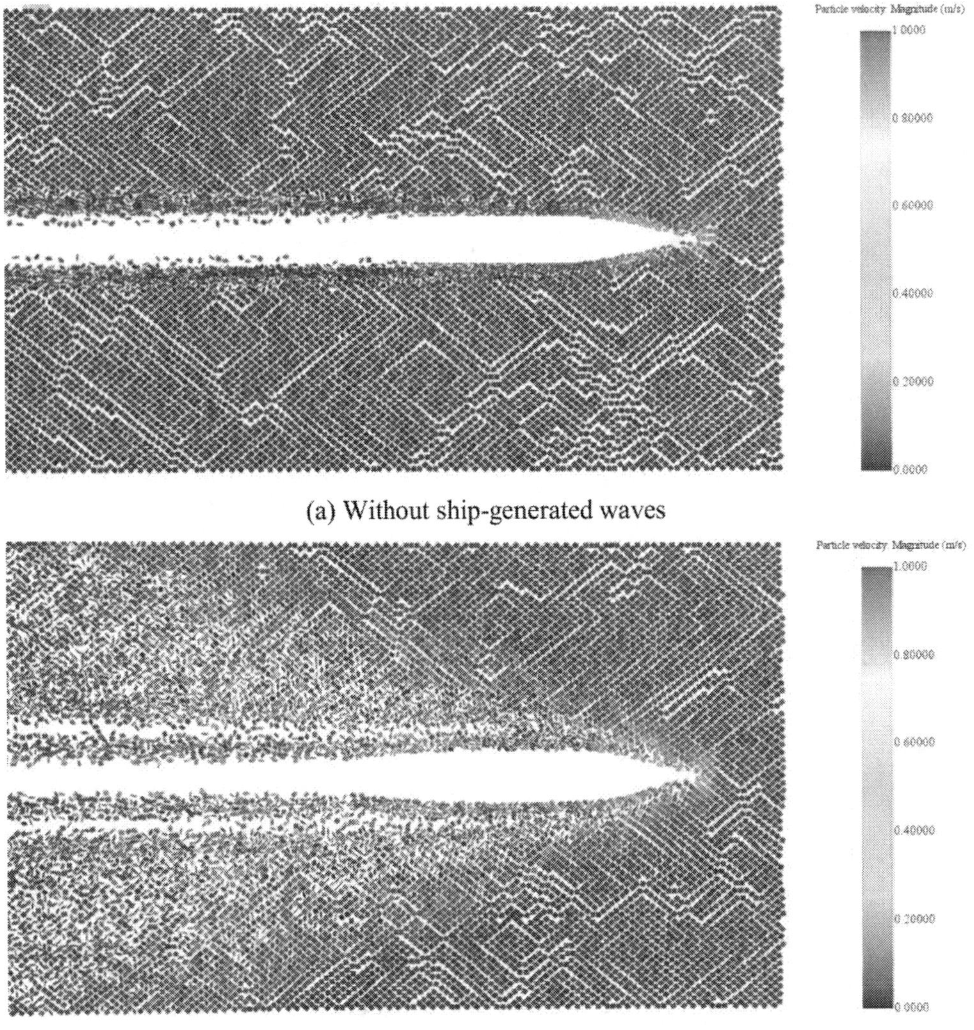

(a) Without ship-generated waves

(b) With ship-generated waves

Figure 10. Velocity distribution of pack ice ($Fr = 0.15$, $C = 80\%$)

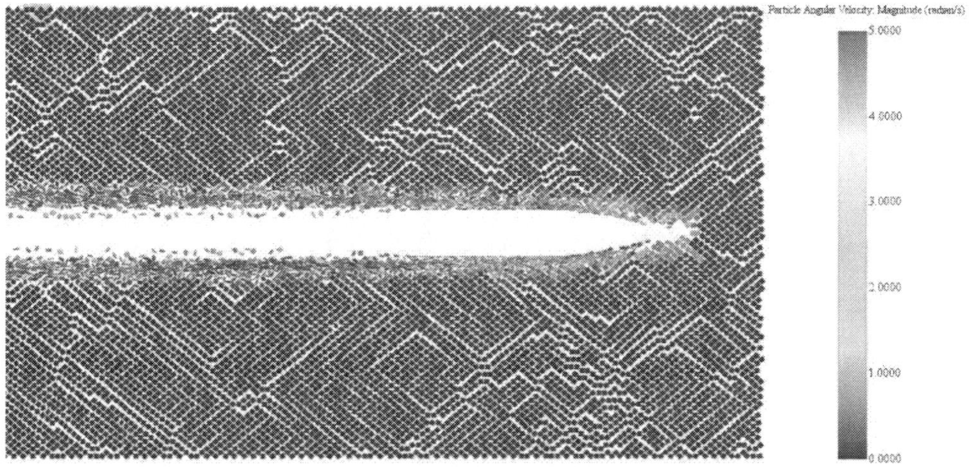

(a) Without ship-generated waves

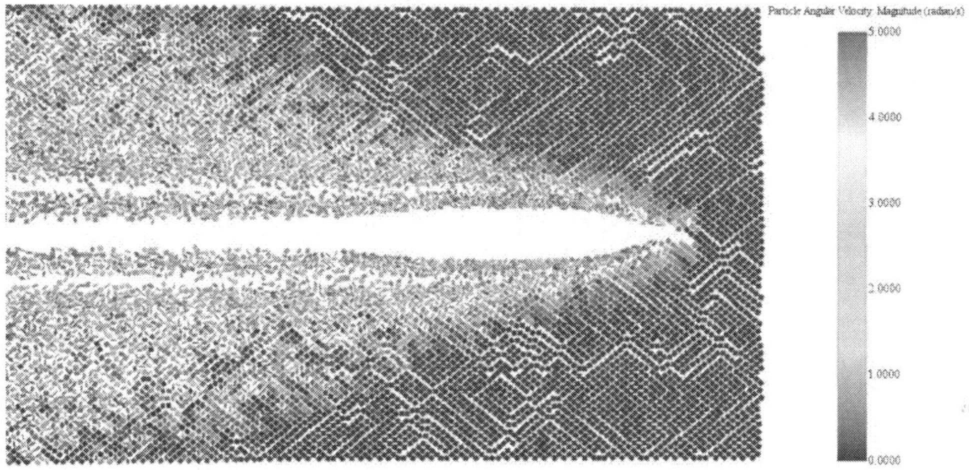

(b) With ship-generated waves

Figure 11. Angular velocity distribution of pack ice ($Fr = 0.15$, $C = 80\%$)

5.3. Contact force with and without ship-generated waves

Figure 12 and Figure 13 present respectively the time history of the longitudinal contact force and the time history of lateral contact force with and without ship-generated waves. It can be seen that when there exists wave effect, the contact frequency and the instantaneous value of longitudinal contact force are obviously smaller than those without wave effect, and the oscillation frequency and amplitude of lateral contact force are significantly smaller than those without wave effect. This is because when there is no wave effect, the pack ice is not dispersed by waves, leading to more frequent contact and more intense collision between the ship and the ice. In addition, it can be observed that when there is no wave effect, the longitudinal contact force basically has no zero value. Combined with the analysis of Figure 9, it can be concluded that when there is no wave effect, the ship and pack ice will be in continual contact, which causes the longitudinal contact force fluctuating without zero value.

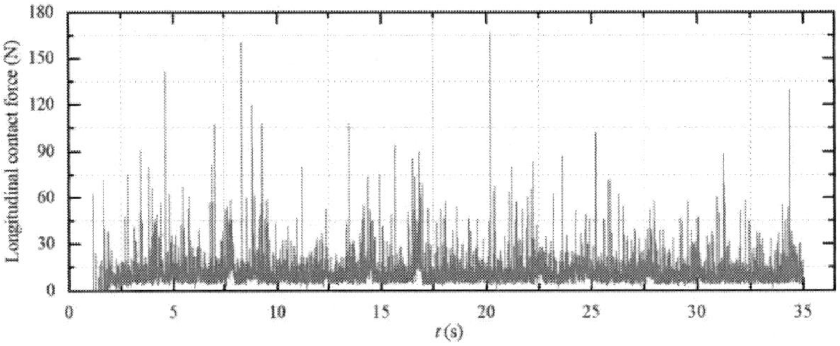

(a) Without ship-generated waves

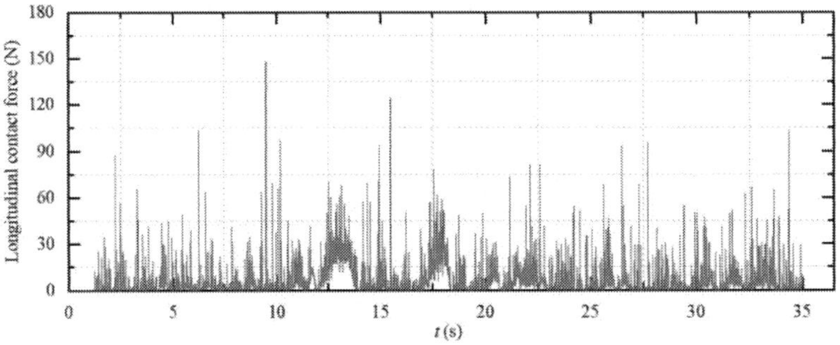

(b) With ship-generated waves

Figure 12. Time history of longitudinal contact force ($Fr = 0.15$, $C = 80\%$)

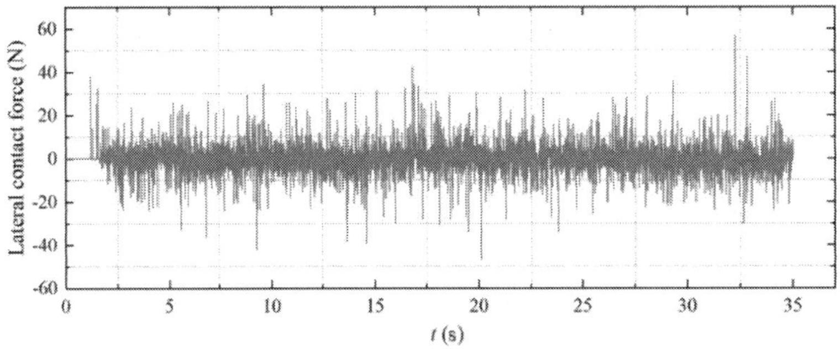

(a) Without ship-generated waves

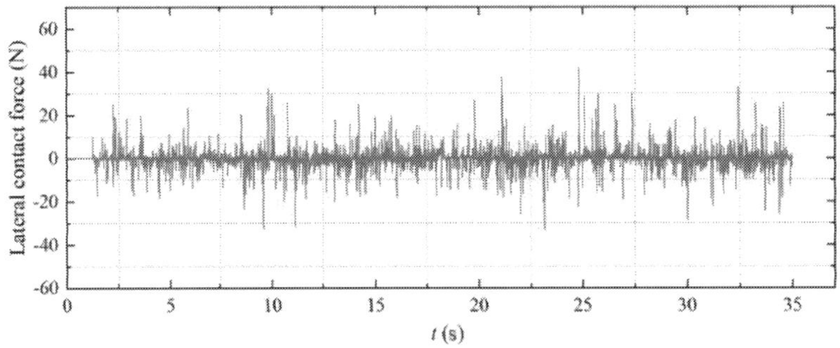

(b) With ship-generated waves

Figure 13. Time history of lateral contact force ($Fr = 0.15$, $C = 80\%$)

In order to further study the influence of ship-generated waves on the lateral contact force, the positive and negative lateral contact forces (positive and negative representing the action direction) are averaged. Figure 14 shows the comparison of the mean values of positive and negative lateral contact forces with and without ship-generated waves. It can be observed that as the ship speed increases, the mean values of positive and negative lateral contact forces also increase. The mean values of positive and negative lateral contact forces increase at a slower rate with wave effect, which is because as the ship speed increases, the wave effect becomes more pronounced, and the mitigation effect of ship-generated waves on the interaction between the ship and pack ice becomes more evident.

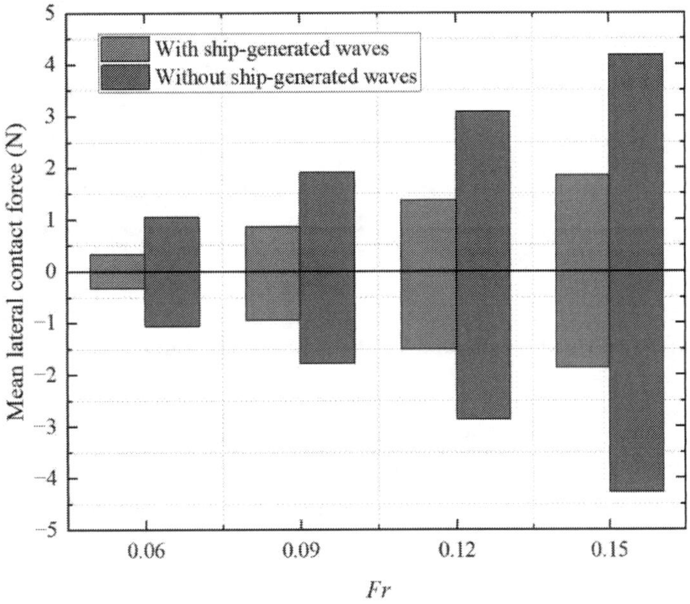

Figure 14. Mean values of positive and negative lateral contanct force with and without ship-generated waves ($C = 80\%$)

5.4. *Ice resistance with and without ship-generated waves*

Figure 15 shows the comparison of ice resistance with and without ship-generated waves. It is evident that ship-generated waves have a notable effect on ice resistance. Both with and without the wave effect, the ice resistance increases as the ship speed increases. However, with the wave effect, the increase in ice resistance is slower as the ship speed increases. This is because as the ship speed increases, the pack ice is more affected by waves, resulting in reduced ship-ice contact and subsequently decreasing ice resistance.

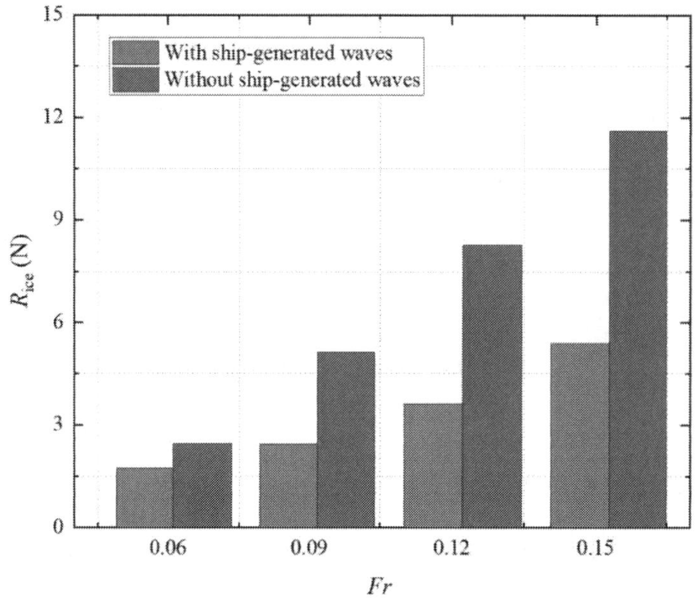

Figure 15. Ice resistance with and without ship-generated waves (C = 80%)

6. Conclusions

In order to investigate the role and influence of ship-generated waves in ship-ice-water interaction, this study establishes a numerical model with ship-generated waves using the CFD-DEM method. Additionally, a simplified numerical model without ship-generated waves is proposed, and a comparative study is conducted between the two models. The study examines the effects of ship-generated waves on the phenomena of ship-ice-water interaction, longitudinal contact forces between the ship and ice, lateral contact forces between the ship and ice, and ice resistance. Some conclusions can be drawn:

1) The CFD-DEM model with ship-generated waves developed in this paper enables accurate prediction of both the open-water resistance and ice resistance, while effectively simulating various typical phenomena of ship-ice-water. Based on the parameter settings in this paper, the ice resistance calculated by Hertz-Mindlin contact model is in better agreement with the empirical value.

2) During the ship-ice interaction, ship-generated waves push aside the pack ice, providing a buffering effect. This helps to mitigate the collision between the ship and pack ice to some extent. The buffering effect of ship-generated waves becomes increasingly significant as the ship speed increases.

3) The influence of ship-generated waves on the motion of pack ice is obvious. Without wave effect, the motion of pack ice changes only around the ship. However, with wave effect, the motion of pack ice changes throughout the entire wave-affected area.

4) Without ship-generated waves, the contact frequency between the ship and pack ice is much higher compared to when waves are present. Additionally, the longitudinal contact force between the ship and pack ice without ship-generated waves shows a nearly non-zero characteristic, which greatly increases the ice resistance.

References

[1] Thomson J, Ackley S, Girard-Ardhuin F, Ardhuin F, Babanin A, Boutin G, et al. 2018. Overview of the Arctic sea state and boundary layer physics program. Journal of Geophysical Research: Oceans, 123(12): 8674-8687.

[2] Kim M C, Lee S K, Lee W J, Wang J Y. 2013. Numerical and experimental investigation of the resistance performance of an icebreaking cargo vessel in pack ice conditions. International Journal of Naval Architecture and Ocean Engineering, 5(1): 116-131.

[3] Jeong S Y, Jang J, Kang K J, Kim H S. 2017. Implementation of ship performance test in brash ice channel. Ocean Engineering, 140: 57-65.
[4] Jeong S Y, Choi K, Kim H S. 2021. Investigation of ship resistance characteristics under pack ice conditions. Ocean Engineering, 219, 108264.
[5] Kim M C, Lee W J, Shin Y J. 2014. Comparative study on the resistance performance of an icebreaking cargo vessel according to the variation of waterline angles in pack ice conditions. International Journal of Naval Architecture and Ocean Engineering, 6(4): 876-893.
[6] Guo C Y, Xie C, Zhang J Z, Wang S, Zhao D G. 2018. Experimental investigation of the resistance performance and heave and pitch motions of ice-going container ship under pack ice conditions. China Ocean Engineering, 32(2): 169-178.
[7] Luo W Z, Guo C Y, Wu T C, Su Y M. 2018. Experimental research on resistance and motion attitude variation of ship-wave-ice interaction in marginal ice zones. Marine Structures, 58: 399-415.
[8] Zong Z, Yang B Y, Sun Z, Zhang G Y. 2020. Experimental study of ship resistance in artificial ice floes. Cold Regions Science and Technology, 176, 103102.
[9] Xue Y, Liu R, Li Z, Han D. 2020. A review for numerical simulation methods of ship-ice interaction. Ocean Engineering, 215, 107853.
[10] Islam M, Mills J, Gash R, Pearson W. 2021. A literature survey of broken ice-structure interaction modelling methods for ships and offshore platforms. Ocean Engineering, 221, 108527.
[11] Warntjen J, Erceg S, Piehl H, et al. The influence of the bow design on structural response due to ice loading. Ships and Offshore Structures, 2018, 13: 302-311.
[12] Jou O, Celigueta M A, Latorre S, Arrufat F, Oñate E. 2019. A bonded discrete element method for modeling ship-ice interactions in broken and unbroken sea ice fields. Computational Particle Mechanics, 6: 739-765.
[13] Zhang N B, Zheng X, Ma Q W, Hu Z H. 2019. A numerical study on ice failure process and ice-ship interactions by Smoothed Particle Hydrodynamics. International Journal of Naval Architecture and Ocean Engineering, 11(2): 796-808.
[14] Li F, Korgesaar M, Kujala P, Goerlandt F. 2020. Finite element based meta-modeling of ship-ice interaction at shoulder and midship areas for ship performance simulation. Marine Structures, 71, 102736.
[15] Guo C Y, Zhang Z T, Tian T P, Li X Y, Zhao D G. 2018. Numerical simulation on the resistance performance of ice-going container ship under brash ice conditions. China Ocean Engineering, 32(5): 546-556.
[16] Wang C, Hu X H, Tian T P, Wang C H. 2020. Numerical simulation of ice loads on a ship in broken ice fields using an elastic ice model. International Journal of Naval Architecture and Ocean Engineering, 12: 414-427.
[17] Løset S. 1994. Discrete element modelling of a broken ice field - Part I: model development. Cold Regions Science and Technology, 22: 339-347.
[18] Hansen E H, Løset S. 1999. Modelling floating offshore units moored in broken ice: model description. Cold Regions Science and Technology, 29(2): 97-106.
[19] Ji S Y, Li Z L, Li C H, Shang J. 2013. Discrete element modeling of ice loads on ship hulls in broken ice fields. Acta Oceanological Sinica - English Edition, 32(11): 50-58.
[20] Xuan S Y, Zhan C S, Liu Z Y, He B F, Zhao Q S, Guo W. 2021. Research on the influences of broken ice parameters on maneuverable forces of ice-going ship oblique sailing. Proceedings of the 40th International Conference on Ocean, Offshore and Arctic Engineering. Virtual, Online.
[21] Li F, Huang L F. 2022. A review of computational simulation methods for a ship advancing in broken ice. Journal of Marine Science and Engineering, 10(2): 165.
[22] Vroegrijk E. 2015. Validation of CFD+DEM against measured data. Proceedings of the 34th International Conference on Ocean, Offshore and Arctic Engineering. Newfoundland, Canada.
[23] Mucha P. 2019. Fully-coupled CFD-DEM for simulations of ships advancing through brash ice.

Proceedings of the SNAME Maritime Convention. Tacoma, WA, USA.
[24] Huang L F, Tuhkuri J, Igrec B, Li M, Thomas G. 2020. Ship resistance when operating in floating ice floes: A combined CFD&DEM approach. Marine Structures, 74, 102817.
[25] Huang L F, Li Z Y, Ryan C, Ringsberg J W, Thomas G. 2021. Ship resistance when operating in floating ice floes: Derivation, validation, and application of an empirical equation. Marine Structures, 79, 103057.
[26] Luo W Z, Jiang D P, Wu T C, Guo C Y, Wang C, Deng R, Dai S S. 2020. Numerical simulation of an ice-strengthened bulk carrier in brash ice channel. Ocean Engineering, 196, 106830.
[27] Zhang J N, Zhang Y, Shang Y C, Jin Q, Zhang L. 2022. CFD-DEM based full-scale ship-ice interaction research under FSICR ice condition in restricted brash ice channel. Cold Regions Science and Technology, 194, 103454.
[28] Menter F R. 1993. Zonal two equation k-ω turbulence models for aerodynamic flows. Proceedings of the 24th Fluid Dynamics Conference. Orlando, Florida, USA.
[29] Hirt C W, Nichols B D. 1981. Volume of fluid (VOF) method for the dynamics of free boundaries. Journal of Computational Physics, 39(1): 201-225.
[30] Johnson K L. 1987. Contact Mechanics. Cambridge University Press. Cambridge.
[31] Renzo A D, Maio F. 2004. Comparison of contact-force models for the simulation of collisions in DEM-based granular flow codes. Chemical Engineering Science, 59(3): 525-541.
[32] Cundall P A, Strack O D L. 1979. A discrete numerical model for granular assemblies. Geotechnique. 29: 47-65.
[33] Kim W J, Van S H, Kim D H. 2001. Measurement of flows around modern commercial ship models. Experiments in Fluids, 31(5): 567-578.
[34] Timco G W, Weeks W F. 2010. A review of the engineering properties of sea ice. Cold Regions Science and Technology, 60(2): 107-129.
[35] Zou M, Tang X J, Zou L, Zou Z J. 2022. Numerical simulation of ship-ice interaction in pack ice area based on CFD-DEM coupling method. Proceedings of the 41st International Conference on Ocean, Offshore and Arctic Engineering. Hamburg, Germany.
[36] Guo C Y, Xie C, Zhang J Z, Wang S, Zhao D G. 2018. Experimental investigation of the resistance performance and heave and pitch motions of ice-going container ship under pack ice conditions. China Ocean Engineering, 32(2): 169-178.

Ship motions and unsteady pressures on ship hull in oblique waves estimated by potential-based solvers

K Suzuki[1*], K Takeda[2*], T Iida[1†] and F Yanagimoto[2]

[1]Department of Naval Architecture and Ocean Engineering, Osaka University, Osaka, Japan
[2]Nippon Kaiji Kyokai (ClassNK), Tokyo, Japan
[*]These authors contributed equally to this work
[†]Corresponding author

E-mail: [†]iida@naoe.eng.osaka-u.ac.jp

Abstract. The seakeeping performances in oblique wave conditions, such as ship motions and unsteady pressures on a ship hull, are getting more important, recently. These physical quantities are used for estimating loads induced by waves and evaluating the safety of a ship. Thanks to the evaluation of various incident wave angles, a rational ship design with sufficient safety and less construction cost can be realized. Therefore, the estimation by numerical schemes in those conditions is demanded for the ship design. However, schemes were usually validated through the comparison with experimental data in only heading wave conditions because few towing tanks can carry out the experiment for a ship advancing in oblique waves. The computed results in oblique waves should be validated well before the industrial usage of schemes. Although Computational Fluid Dynamics is useful to know the seakeeping performance in oblique waves thanks to the development of computer performance in these years, it takes much longer time than the schemes based on a potential theory. For this reason, three solvers based on the potential theory, the Salvesen-Tuck-Faltinsen method, the Green function method, and the Rankine panel method, are validated by comparing ship motions and unsteady pressure on the ship hull obtained from towing tank tests in oblique wave conditions.

1. Introduction

A lot of types of numerical schemes have been developed for predicting seakeeping performances. Especially, Computational Fluid Dynamics (CFD) is well-used nowadays. However, it still requires a very long calculation time compared to schemes based on a potential theory. Therefore, potential-based solvers are suitable when the estimation of ship performance in a lot of conditions is required.

The schemes based on the potential theory can be roughly classified into a two-dimensional method and a three-dimensional method. Thanks to the slender ship theory, a three-dimensional problem for a ship in waves can be regarded as a two-dimensional problem. Salvense-Tuck-Fultinsen (STF) method[1] is one of the most famous theories in two-dimensional schemes. To enhance the modeling accuracy of forward-speed effects and three-dimensional effects, the unified theory was developed[2]. This theory was enhanced more by considering normal vector in x direction[3].

To consider the three-dimensional effects, two famous methods have been studied, i.e. the free surface Green function method[4, 5, 6] and the Rankine panel method (RPM)[7, 8, 9]. Since the Green function method which fully considers the forward-speed terms is sensitive to numerical accuracy, the Green function in which these terms are ignored (zero-speed approximation) has been developed and widely used. In the approximated method, only the Doppler effect expressed as the encounter circular frequency is considered. In the RPM, the free-surface condition is not imposed on the Green function. Because of excluding the free surface condition from the Green function, the double-body flow that expresses the curving flow due to the ship hull can be considered in the RPM.

The computed results have been validated using experimental data, for example, ship motions, wave exciting forces, and added resistance. The accuracy of these estimated results has been generally investigated only in heading waves because few towing tanks enable conducting an experiment for measuring advancing ship performance in oblique waves. From the viewpoint of safety, ship performances in heading waves are important because the maximum vertical bending moment induced by waves occurs in that condition. This might justify such less validation in oblique waves. In recent years, however, more rational ship design with sufficient safety and less construction cost is required. Therefore, the estimation of ship performances in oblique waves is demanded.

For the reliable calculation in the oblique waves, validations by comparison with experimental data should be done more. Thus, experiments in oblique waves were conducted in an Actual Sea Model Basin of National Maritime Research Institute. The 6-DOF ship motions were measured in some oblique wave angles. In addition, pressures on the ship hull were measured because it is an important physical quantity for calculating wave loads. By installing a new type of pressure sensor, FBG pressure sensors[10, 11, 12], pressures were measured at 212 points simultaneously. In addition, 59 points were measured using conventional strain-type pressure sensors. From the viewpoint of the advantage of calculation time in schemes based on potential theory, three potential-based solvers were validated through the comparison with measured data. Used schemes are the STF method, the Green function method with zero-speed approximation[6], and the RPM. The ship speed was in low-speed condition because the STF method and the Green function method do not fully consider the forward speed effects.

2. Towing tank tests

For measuring the ship motions and unsteady pressure on the ship hull in oblique waves, towing tank tests were carried out in an Actual Sea Model Basin of National Maritime Research

Table 1. Principal particulars of container ship in model scale

Item	Value	
Ship length: L_{pp}	4.000	m
Ship breadth: B	0.568	m
Draft: d	0.170	m
Displacement volume: ∇	0.2619	m^3
Block coefficient: C_b	0.676	-
Water-plane area: A_w	1.914	m^2
Horizontal center of gravity: x_G	-0.082	m
Vertical center of gravity: z_G	0.059	m
Vertical center of bouyancy: z_B	-0.078	m
Roll gyrational radius: κ_{xx}/B	0.400	-
Pitch gyrational radius: κ_{yy}/L_{pp}	0.250	-
Transversal metacentric radius: $\overline{BM_T}$	0.176	m
Longitudinal metacentric radius: $\overline{BM_L}$	7.497	m

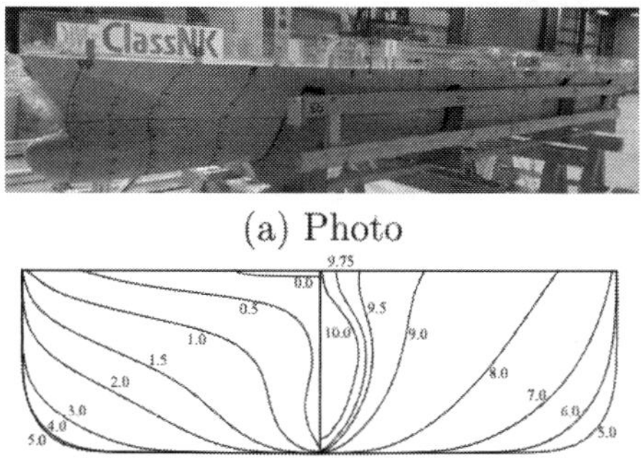

(a) Photo

(b) Body plan. The number along the line denotes Square Station (S.S.).

Figure 1. Photo and body plan of the container ship model

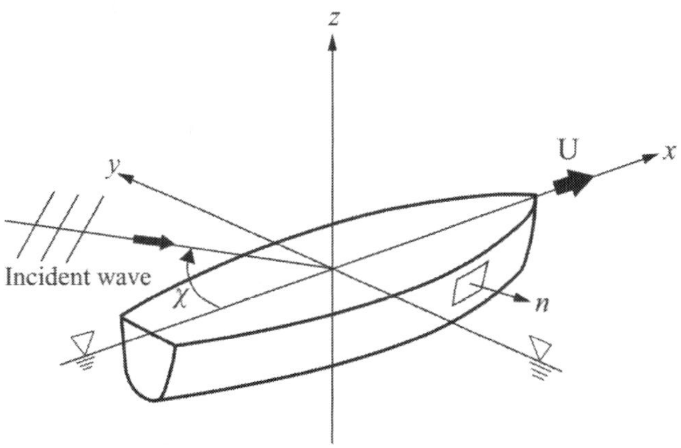

Figure 2. Coordinate system and notations

Institute. A ship model used in the experiment was one eighty-eighth scale of a 14000 TEU container ship. The principal of the model is shown in Table 1. In Fig.1, a photo of the ship model and its body plan are shown. S.S.=0.0 in Fig.1 (b) is a transverse section in the aft perpendicular, and S.S.=10.0 indicates the one in the fore perpendicular.

The definition of the positive direction of coordinate is illustrated in Fig.2. Ship speed is given in the Froude number $F_n = 0.144$. This speed corresponds to 75 percent of the service speed, which is used for strength evaluation and fatigue evaluation. 6-DOF ship motions were measured. In addition, pressure on the ship hull was measured using 212 FBG pressure sensors (version 6)[10, 12] and 59 strain-type pressure sensors. The positions of pressure sensors attached to or installed in sections can be seen in [13]. These physical quantities were measured every 30 deg from χ =0 deg to $\chi = 180$ deg. Here, $\chi = 0$ deg indicates following waves and $\chi = 180$ deg indicates heading waves. The wave conditions were in some regular waves whose ratios between wavelength and ship length λ/L were 0.3, 0.4, 0.6, 0.8, 1.0, 1.5, 2.0, 2.5. Because of the restriction of the paper, some results from all conditions are shown in the later section. Upper waves are defined on the port side.

3. Potential-based solvers

Three different types of potential-based solvers are validated. The first method is the Salvesen-Tuck-Faltinsen (STF) method[1]. This scheme is a two-dimensional scheme constructed under the slender ship theory. In the scheme, terms concerning ship speed U are ignored in free surface conditions. Therefore, the velocity potentials obtained in this theory consider no three-dimensional effects and restricted forward speed effects.

The second potential-based solver is the free-surface Green function method. In the investigated Green function method, terms regarding ship speed U in the free surface condition are ignored[6]. This means the forward-speed effects are considered only in the Doppler effect expressed as the encounter circular frequency. Therefore, although the velocity potentials consider three-dimensional effects, the consideration of forward speed effects is the same as the STF method.

The third method is the Rankine panel method (RPM). In the RPM, the free surface condition is not included in the Green function. Thanks to this separation, the effects of double-body flow that expresses the curving flow due to a ship hull can be taken into account in unsteady velocity potentials. Instead of this advantage, the free surface mesh is also used for satisfying the free

IOP Conf. Series: Materials Science and Engineering 1288 (2023) 012020 doi:10.1088/1757-899X/1288/1/012020

surface condition (for example, see [14]) although only the hull surface mesh is used in the above two methods.

The boundary conditions on the free surface (S_F) and hull surface (S_H) to obtain the velocity potential in the STF method and the three-dimensional Green function method are given in the form

$$[F] \quad \left(-\omega_e^2 + g\frac{\partial}{\partial z}\right)\phi_j = 0 \quad j = 1, 2, \ldots, 7 \quad (\boldsymbol{x} \in S_F) \tag{1}$$

$$[H] \quad \frac{\partial \phi_j}{\partial n} = \begin{cases} n_j + \dfrac{U}{i\omega_e} m_j & j = 1, 2, \ldots, 6 \\ -\dfrac{\partial \phi_0}{\partial n} & j = 7 \end{cases} \quad (\boldsymbol{x} \in S_H) \tag{2}$$

$$\left.\begin{array}{llll} (n_1, n_2, n_3) & = \boldsymbol{n} & (m_1, m_2, m_3) & = (0, 0, 0,) \\ (n_4, n_5, n_6) & = \boldsymbol{x} \times \boldsymbol{n} & (m_4, m_5, m_6) & = (0, n_3, -n_2) \end{array}\right\} \tag{3}$$

where g, ω_e, and $\boldsymbol{n}$ are gravitational acceleration, encounter circular frequency given in the form $\omega_0 - k_0 U \cos\chi$, and normal vector, respectively. Furthermore, ω_0 and $k_0 (= \omega_0^2/g)$ are a circular frequency of the incident wave and a wavenumber of the incident wave, respectively. Note that the normal vector is defined as positive when pointing into the fluid region from the hull surface as shown in Fig.2. ϕ_0 is incident wave velocity potential, and ϕ_7 is scattering velocity potential. $\phi_j (j = 1, 2, \ldots, 6)$ is radiation potential in j-th mode. m_j is so-called m-term proportional to forward speed U.

On the other hand, the free surface condition and hull surface condition in RPM are given as follows:

$$[F] \quad \mathcal{L}[\phi_j] = 0 \quad j = 1, 2, \ldots, 7 \quad (\boldsymbol{x} \in S_F) \tag{4}$$

$$[H] \quad \frac{\partial \phi_j}{\partial n} = \begin{cases} n_j + \dfrac{U}{i\omega_e} m_j & j = 1, 2, \ldots, 6 \\ -\dfrac{\partial \phi_0}{\partial n} & j = 7 \end{cases} \quad (\boldsymbol{x} \in S_H) \tag{5}$$

$$\left.\begin{array}{llll} (n_1, n_2, n_3) & = \boldsymbol{n} & (m_1, m_2, m_3) & = -(\boldsymbol{n} \cdot \nabla)\boldsymbol{V} \\ (n_4, n_5, n_6) & = \boldsymbol{x} \times \boldsymbol{n} & (m_4, m_5, m_6) & = -(\boldsymbol{n} \cdot \nabla)(\boldsymbol{x} \times \boldsymbol{V}) \end{array}\right\} \tag{6}$$

$$\boldsymbol{V} \equiv \nabla\Big(\Phi_{DB}(\boldsymbol{x}) + \varphi_S(\boldsymbol{x})\Big) \tag{7}$$

where $\Phi_{DB}(\boldsymbol{x})$ and $\varphi_S(\boldsymbol{x})$ in (7) denote double-body flow potential and steady scattering potential, respectively. $\Phi_{DB}(\boldsymbol{x})$ is composed of two elements about uniform flow and deformed flow due to a ship hull. It is expressed as $\Phi_{DB}(\boldsymbol{x}) = -x + \phi_{DB}(\boldsymbol{x})$. $\mathcal{L}$ in (4) is the differentiation operator for the free surface as defined below

$$\begin{aligned} \mathcal{L} \equiv & -k_e + 2i\tau\overline{\nabla}\Phi_{DB} \cdot \overline{\nabla} + \frac{1}{K_0}\overline{\nabla}\Phi_{DB} \cdot \overline{\nabla}(\overline{\nabla}\Phi_{DB} \cdot \overline{\nabla}) \\ & + \frac{1}{2K_0}\overline{\nabla}(\overline{\nabla}\Phi_{DB} \cdot \overline{\nabla}\Phi_{DB}) \cdot \overline{\nabla} + \overline{\nabla}^2\Phi_{DB}\left(i\tau + \frac{1}{K_0}\overline{\nabla}\Phi_{DB} \cdot \overline{\nabla}\right) + \frac{\partial}{\partial z} \end{aligned} \tag{8}$$

where $k_e = \omega_e^2/g$, $\tau = U\omega_e/g$, $K_0 = g/U^2$, and $\overline{\nabla}$ illustrates the differentiation operator only in the horizontal plane (x, y).

After solving the boundary-value problem by giving boundary conditions in each theory, velocity potentials can be calculated. Then, hydrodynamic forces are given by integrating pressure on the ship hull expressed by Bernoulli's pressure equation. By giving the hydrodynamic forces to motion equations and solving them, ship motions can be obtained.

Unsteady pressure can be calculated after ship motions are estimated. The complex amplitude of unsteady pressure $p_U(\boldsymbol{x})$ is composed of three components as shown below

$$p_U(\boldsymbol{x}) = p_D(\boldsymbol{x}) + p_R(\boldsymbol{x}) + p_S(\boldsymbol{x}) \tag{9}$$

where $p_R(\boldsymbol{x})$, $p_D(\boldsymbol{x})$, and $p_S(\boldsymbol{x})$ are radiation pressure, diffraction pressure, and static pressure induced by ship oscillation, respectively. $p_S(\boldsymbol{x})$ is given as follows:

$$p_S(\boldsymbol{x}) = -\rho g(X_3 + yX_4 - xX_5) \tag{10}$$

where X_j is the complex amplitude of ship motions in j-th mode. x, y, and z on the ship hull are the time average position of motions. This means a position in calm water. In the STF method and the three-dimensional Green function method, $p_R(\boldsymbol{x})$ and $p_D(\boldsymbol{x})$ are given in the form

$$p_D(\boldsymbol{x}) = \rho g\zeta_a \frac{\omega_e}{\omega_0}\left(1 - \frac{U}{i\omega_e}\frac{\partial}{\partial x}\right)(\phi_0 + \phi_7) \tag{11}$$

$$p_R(\boldsymbol{x}) = -\rho(i\omega_e)^2 \sum_{j=1}^{6} X_j\left(1 - \frac{U}{i\omega_e}\frac{\partial}{\partial x}\right)\phi_j \tag{12}$$

In the case of the RPM, $p_R(\boldsymbol{x})$ and $p_D(\boldsymbol{x})$ are given as follows:

$$p_D(\boldsymbol{x}) = \rho g\zeta_a \frac{\omega_e}{\omega_0}\left(1 + \frac{U}{i\omega_e}\boldsymbol{V}\cdot\nabla\right)(\phi_0 + \phi_7) \tag{13}$$

$$p_R(\boldsymbol{x}) = -\rho(i\omega_e)^2 \sum_{j=1}^{6} X_j\left\{\left(1 + \frac{U}{i\omega_e}\boldsymbol{V}\cdot\nabla\right)\phi_j - \frac{1}{2}\left(\frac{U}{\omega_e}\right)^2(\beta_j\cdot\nabla)(\boldsymbol{V}\cdot\boldsymbol{V})\right\} \tag{14}$$

$$\beta_j = \begin{cases} \boldsymbol{e}_j & (j = 1,2,3) \\ \boldsymbol{e}_{j-3}\times\boldsymbol{x} & (j = 4,5,6) \end{cases} \tag{15}$$

where $\boldsymbol{e}_j = (\delta_{1j}, \delta_{2j}, \delta_{3j})$, and δ_{ij} is Kronecker delta.

4. Results and discussion

Ship motions and unsteady pressures are illustrated in this section. Note that the results of all conditions cannot be shown because of the page restriction. Those results that cannot be shown will be presented at the conference.

4.1. Ship motions

Surge, heave, roll, and pitch motions are illustrated in Figs.3 to 6. A slightly different roll gyration radius from the one written in Table 1 was used in numerical calculations to adjust the roll peak position to experimental results. Note that the roll damping force due to viscosity is given to motion equations in the STF method and Green function method using extinction coefficients. Those coefficients were acquired in the free oscillation test. In the RPM, this damping force is not considered in the results shown below. The results that it is considered in the RPM will be shown in the presentation.

From the figures, the STF method and the Green function method show similar trends and good agreement with experimental data. Although surge motions and pitch motions obtained by the RPM show almost the same results as the other two schemes, heave and roll motions are different. Heave motions over χ = 120 deg show hump and hollow. Roll motions show a sharp peak compared with the other two computational results and different behavior from experimental data even if it is considered that viscous damping force is not included in the motion equations. Unsmooth outputs by RPM can be seen in long wavelengths, for example, in the case of heave motion in $\chi = 150$ or pitch motion in $\chi = 120$. This might be caused by the less mesh region of the free surface. While the RPM calculation has some problems to be improved, three potential-based solvers can generally estimate the ship motions with accuracy in oblique waves in low-speed cases.

4.2. Unsteady pressure

Figures 7 to 14 show unsteady pressure distribution in some transverse sections. The results in the case of the ratio between wavelength and ship length λ/L=0.6 and 1.5 in wave angle χ=0, 60, 120, and 180 deg are shown in each figure. All three methods show good agreement with experimental data from the viewpoint of both amplitude and phase, except for the result of the RPM in Fig.10. In this case, $\chi = 60$ deg and λ/L=1.5, roll amplitude in the RPM is much smaller than the measured amplitude, which can be seen in Fig.5. For this reason, the amplitude of pressure estimated by the RPM in Fig.10 is small compared with the other schemes and experimental results.

5. Conclusions

Ship motions and unsteady pressures are estimated by three different types of potential-based solvers, the STF method, the Green function method, and the RPM. The validation was conducted in the low-speed case. These methods show the tendency of the experimental results well and good agreement in both ship motions and unsteady pressure. However, the roll motions in the RPM need an introduction of the roll damping force and an improvement of estimation accuracy. The motions and unsteady pressures in relatively high-speed cases in oblique waves should be validated for confirming the applicable range of the STF method and the Green function method.

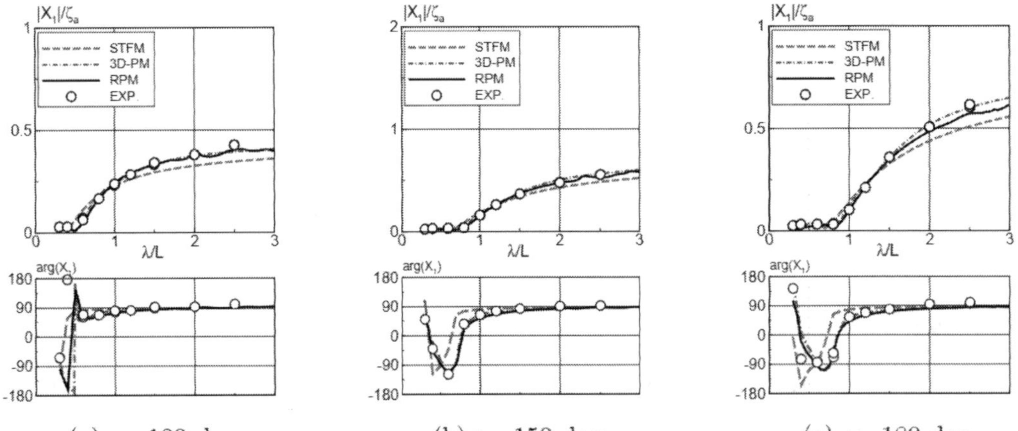

(a) χ=120 deg (b) χ=150 deg (c) χ=180 deg

Figure 3. Surge motions (3D-PM in the caption denotes the results by the Green function method).

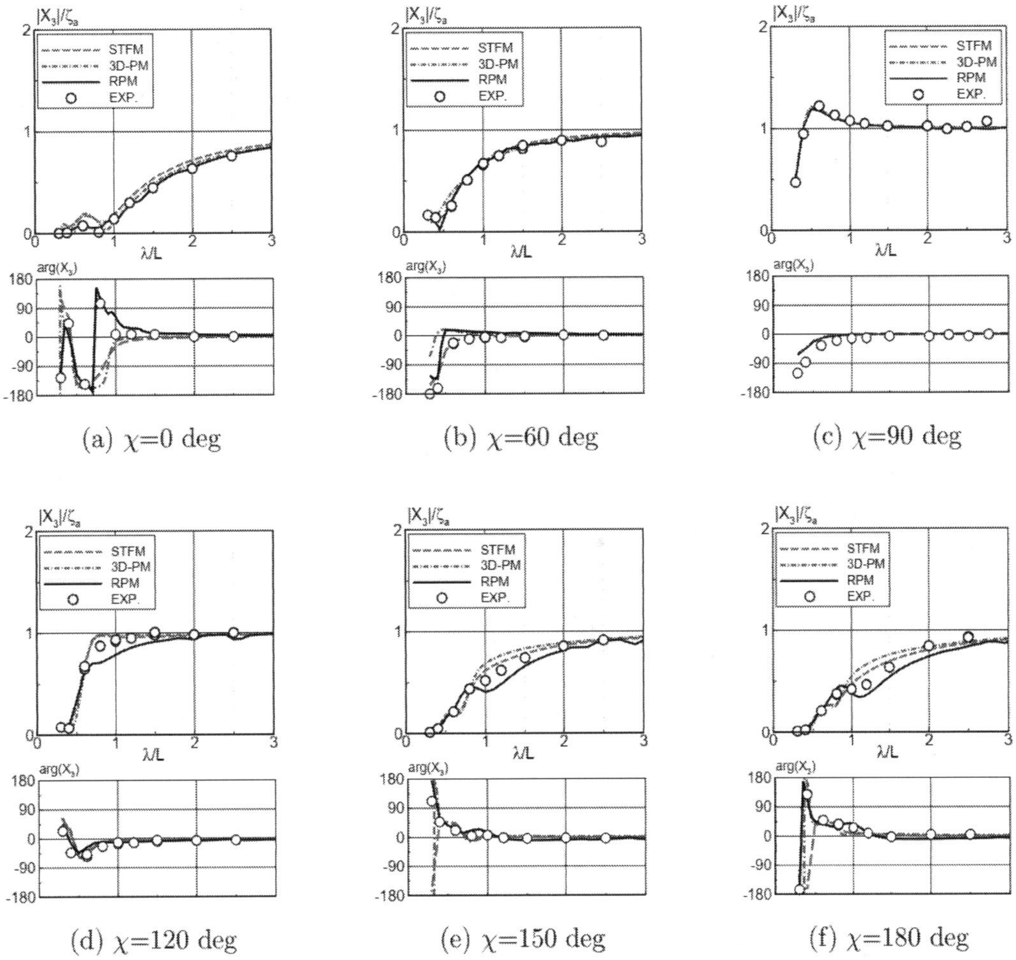

Figure 4. Heave motions (3D-PM in the caption denotes the results by the Green function method).

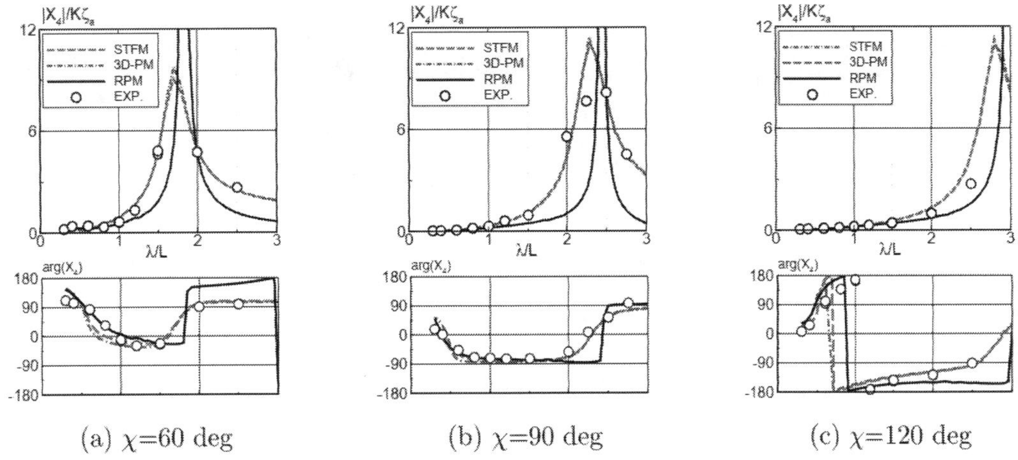

Figure 5. Roll motions (3D-PM in the caption denotes the results by the Green function method).

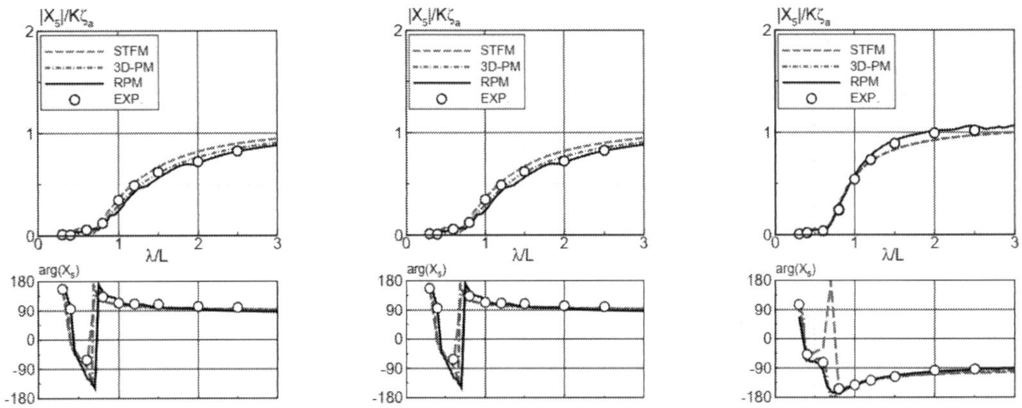

(a) χ=0 deg (b) χ=120 deg (c) χ=180 deg

Figure 6. Pitch motions (3D-PM in the caption denotes the results by the Green function method).

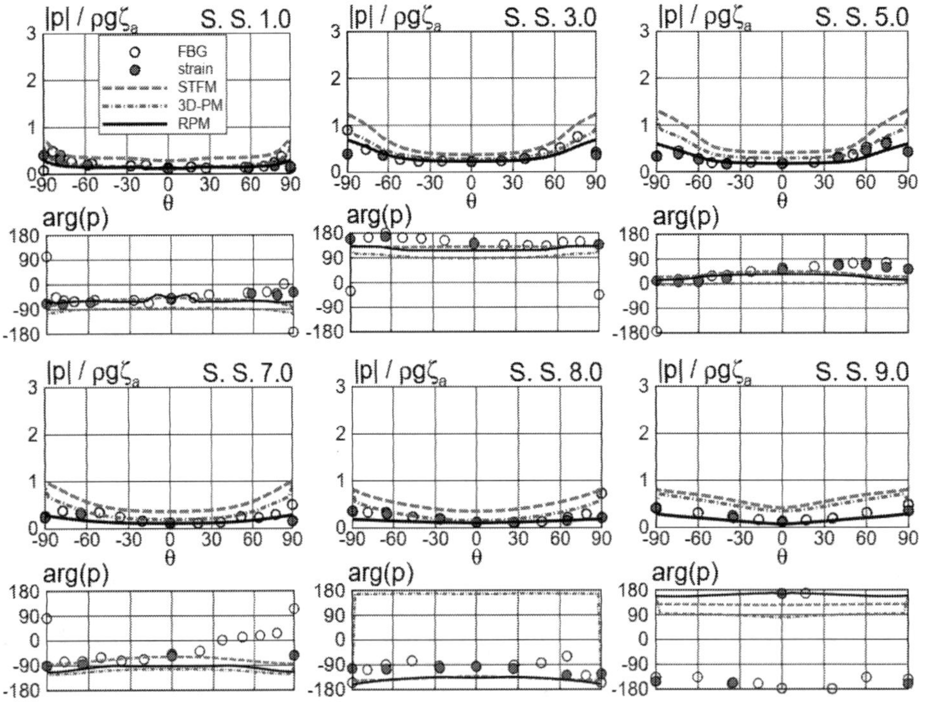

Figure 7. Unsteady pressure distribution in several sections in the case of χ=0 deg, λ/L=0.60 (3D-PM in the caption denotes the results by the Green function method).

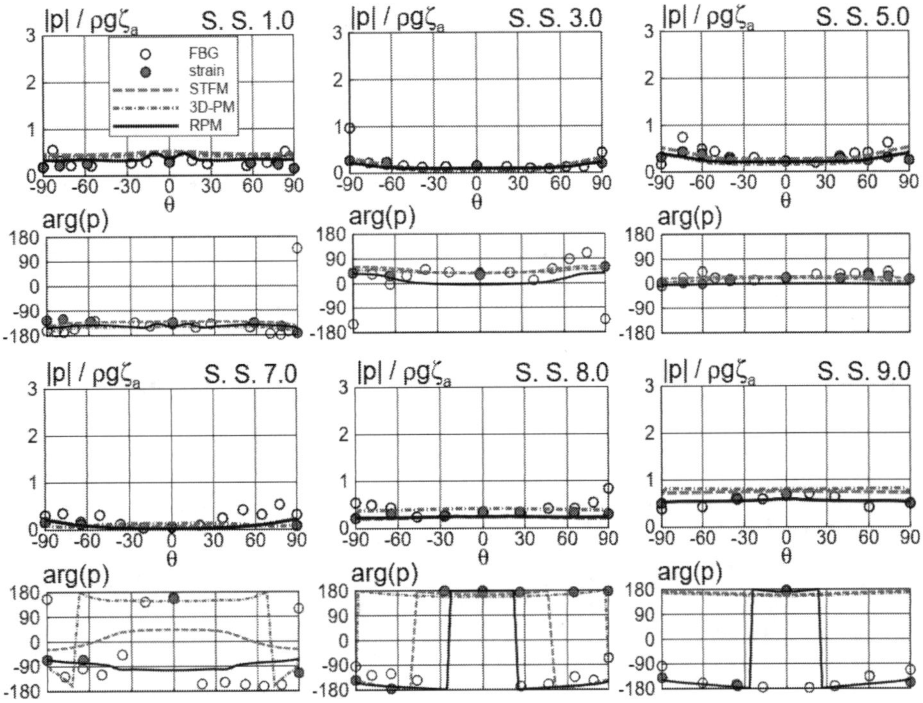

Figure 8. Unsteady pressure distribution in several sections in the case of χ=0 deg, $\lambda/L = 1.50$ (3D-PM in the caption denotes the results by the Green function method).

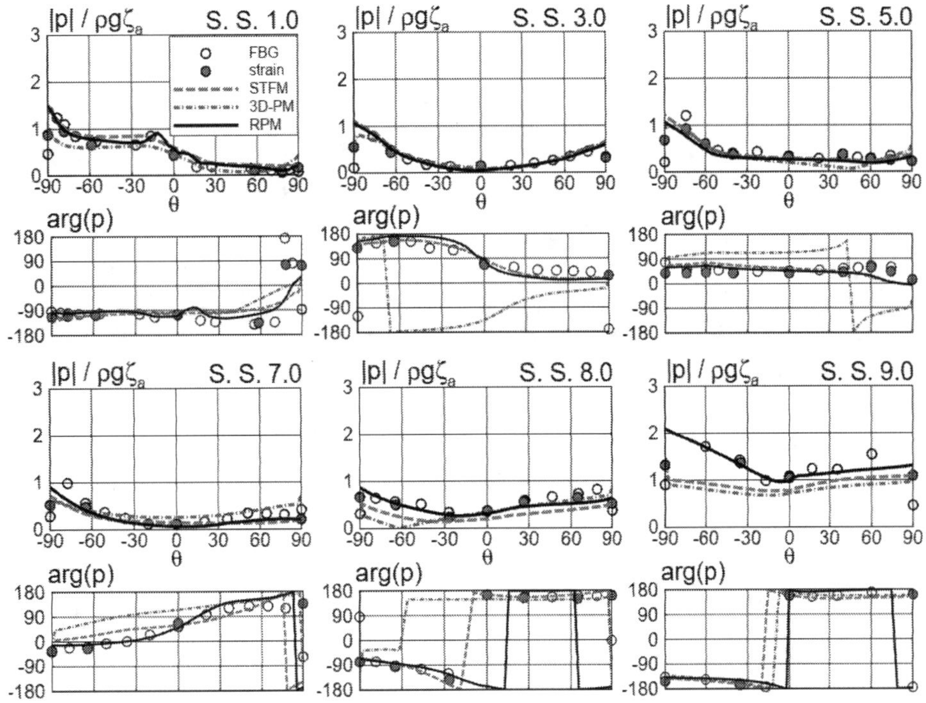

Figure 9. Unsteady pressure distribution in several sections in the case of χ=60 deg, $\lambda/L = 0.60$ (3D-PM in the caption denotes the results by the Green function method).

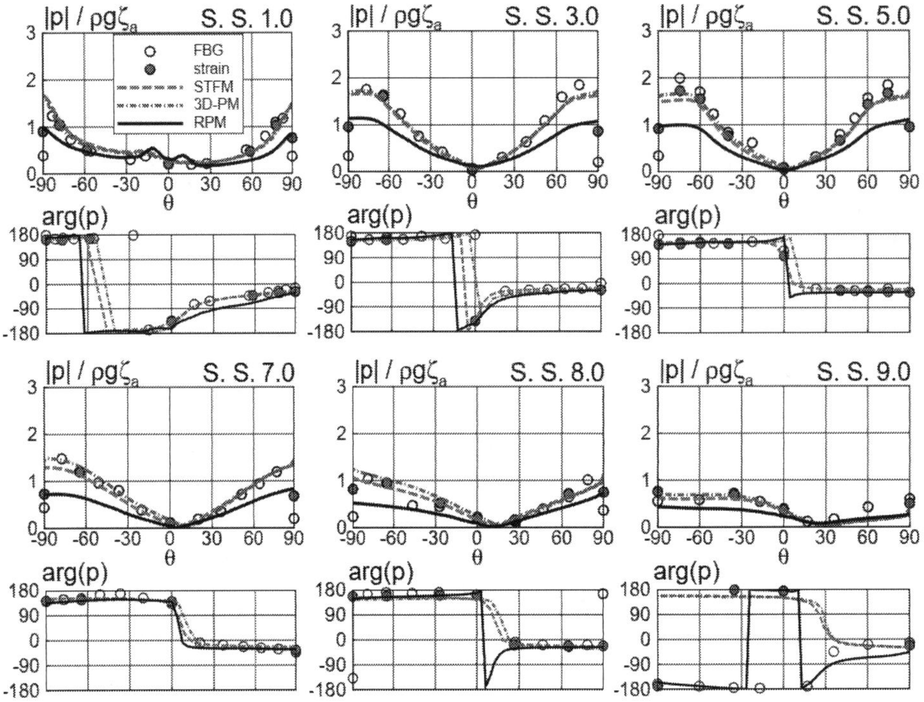

Figure 10. Unsteady pressure distribution in several sections in the case of χ=60 deg, $\lambda/L = 1.50$ (3D-PM in the caption denotes the results by the Green function method).

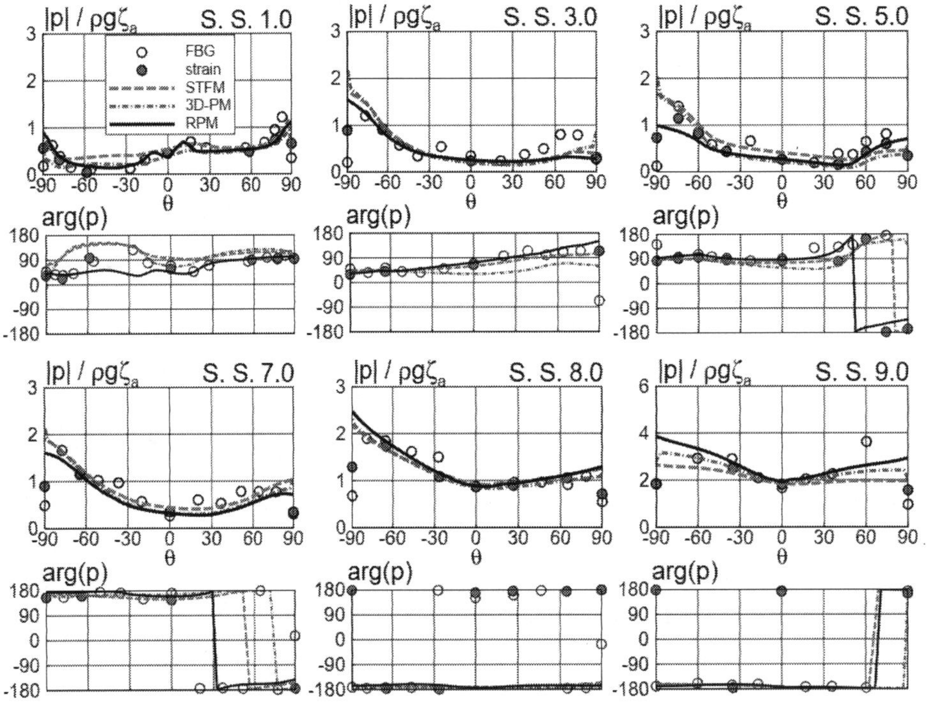

Figure 11. Unsteady pressure distribution in several sections in the case of χ=120 deg, $\lambda/L = 0.60$ (3D-PM in the caption denotes the results by the Green function method).

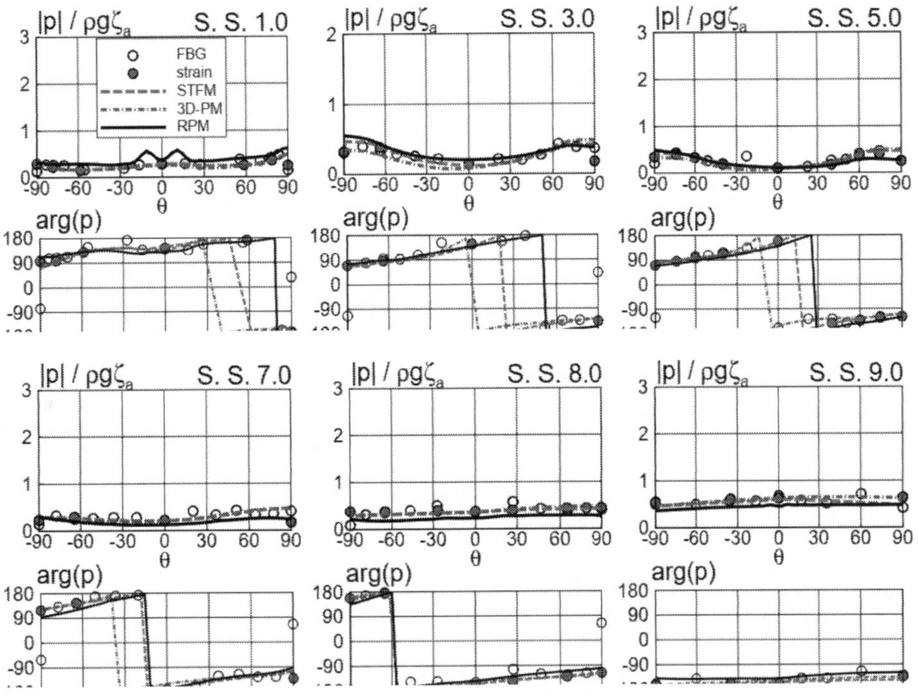

Figure 12. Unsteady pressure distribution in several sections in the case of χ=120 deg, $\lambda/L = 1.50$ (3D-PM in the caption denotes the results by the Green function method).

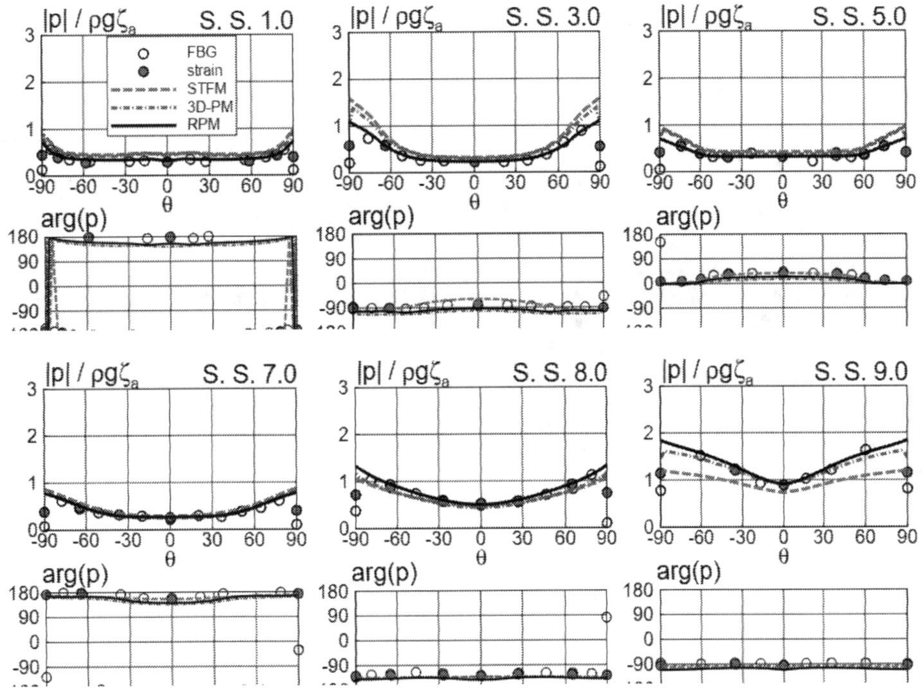

Figure 13. Unsteady pressure distribution in several sections in the case of χ=180 deg, $\lambda/L = 0.60$ (3D-PM in the caption denotes the results by the Green function method).

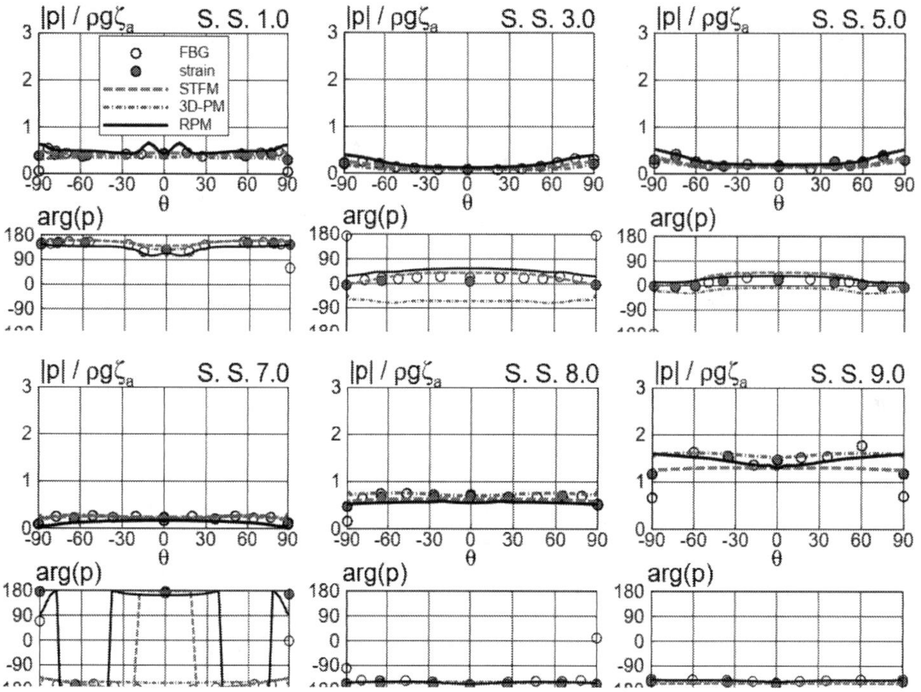

Figure 14. Unsteady pressure distribution in several sections in the case of χ=180 deg, $\lambda/L = 1.50$ (3D-PM in the caption denotes the results by the Green function method).

References

[1] Salvesen N, Tuck E O & Faltinsen O 1970 *Trans. the Society of Naval Architecture and Marine Engineers* **78** 250-287
[2] Newman, J. N. 1978 *Advances of Applied Mechanic* **18** 221-283
[3] Kashiwagi M 1995 *Trans. the West-Japan Society of Naval Architects* **89** 77-89
[4] Chang M 1977 *Proc. the 2nd International conference on numerical ship hydrodynamics* 124-135
[5] Iwashita H & Ohokusu M 1989 *J. the Society of Naval Architects of Japan* (in Japanese) **166** 187-205
[6] Papanikolaou A D & Schellin T E 1992 *J. Ship Technology Research* **39(4)** 147-56
[7] Nokos D E1990 MIT, Massachusetts, USA, Ph.D. Thesis.
[8] Bertram V 1990 *J. Ship Technology Research* **37(4)** 143-152
[9] Kim M & Kim Y 2011 *J. Ocean Engineering* **38** 1357-67
[10] Wakahara M, Nakajima M & Fukasawa T 2008 *J. the Japan Society of Naval Architects and Ocean Engineers* **7** 1-7
[11] Iwashita H & Kashiwagi M 2018 *Proc. 33rd International Workshop on Water Waves and Floating Bodies* 85-88
[12] Suzuki K, Iwashita H, Kashiwagi M, Wakahara M, Iida T & Minoura M 2023 *J. Marine Science and Technology.* (in press)
[13] Takeda K, Fukui T & Zhu T 2021 *Proc. the Japan Society of Naval Architects and Ocean Engineers* **24** 613-16 (in Japanese)
[14] Suzuki K, Iida T, Iwashita H & Minoura M 2022 *Proc. the 32nd International Ocean and Polar Engineering Conference* 2205-12

Optimization of a coupled launching process considering hydrodynamic interaction effects

L Altenbach[1], M Wermbter[1], P Marleaux[1] and M Abdel-Maksoud[1]

[1] Institute for Fluid Dynamics and Ship Theory, Hamburg University of Technology (TUHH), Am Schwarzenberg-Campus 4, 21073 Hamburg, Germany

E-mail: `lukas.altenbach@tuhh.de`

Abstract. Crane-assisted launching and recovering of small boats from a large vessel is common in a wide range of operational scenarios and may be required during research missions or in emergency situations for rescue boats. As the mothership moves in natural irregular seas, the resulting accelerations and motions of the small boat can lead to dangerous situations, such as a collision between the mothership and the small boat or a slamming of the small boat onto the free water surface can occur, posing a great danger to the people involved. In the paper, the development of a coupled simulation method for the motions of the mothership and the small boat during the launching in the time-domain is described. The developed simulation method is based on the impulse response function. The implementation of the suspension on a crane is realised by a holonomic constraint and a Lagrange's equation. The parameter influencing the launching operation include the current sea state, the stationary wave field induced by the motherships forward speed, at which the small boat is launched and the initial position of the small boat relative to the water surface. By varying the above mentioned parameters, the process can be optimised for different sea states in terms of forces and accelerations. In considering the forward speed of the mother boat during the launching process, it can be observed that in many cases the small boat is operated at high Froude numbers of greater than 0.4. This leads to an exceedance of the operational limits of the programmes used in this paper.

1. Introduction

The aim of this paper is to analyse the behaviour of a small craft launched into the water by crane with forward speed from a mothership. Simplified, the launching procedure can be divided into three steps. First, the small craft is suspended freely from the crane without the influence of hydrodynamic forces. In the second step, the boat is lowered and comes into contact with the water surface. Finally, the boat moves independently without the influence of the crane cable next to the mother ship. When touching down on the water surface, the sudden hydrodynamic forces cause high accelerations. These might cause damage to persons and to the hull of the small craft. With regard to human factors for safe operation, limit values for transverse and vertical accelerations have been summarised, as Nordforsk criteria [1]. Many parameters can influence the transverse and vertical accelerations such as the distance between the two ships, the initial trim angle of the small boat, the lowering speed as well as the forward speed of the coupled system of mother ship and boat. Since launching operations with moderate forward motion of the mothership lead to high Froude numbers of the small boat. In this case, the optimal range of applicability of the method applied for the calculation of the ship motion based on the impulse response functions [2, 3], with coefficient calculation by the Rankine Panel Method [4], is exceeded. For example, in such cases programs based on the 2D+t method and explicitly developed for the simulation of fast planing boats can provide reasonable results [5]. Due to the different hull geometries of dinghies, the focus in this paper is not to provide general predictions about the motion behaviour of the boat and the loads on the hull, but to analyse the influence of specific boundary conditions on the parameters mentioned above.

2. Mathematical model

2.1. Impulse response method IMPRES

The simulation program IMPRES [3] is a method for determining ship motions in the time domain. It uses the approach of Cummins [2] to calculate a motion response to a force from the decaying impulses of the past time steps. Due to the calculation with values from transfer functions and the assumption of harmonic motions, only linear motion responses can be realised. The equation of motion to be solved has the following notation:The equation of motion to be solved has the following notation:

$$(\mathbf{M}+\mathbf{A})\ddot{\boldsymbol{\xi}}(t)+\int_0^\infty \mathbf{B}(\tau)\dot{\boldsymbol{\xi}}(t-\tau)d\tau+\mathbf{S}\boldsymbol{\xi}(t)=\sum \mathbf{F}(t). \quad (1)$$

M is the inertia matrix, **A** is the hydrodynamic mass matrix, **S** is the restoring force matrix and **B** is the retardation function matrix, which includes the motion influence from previous time steps. **F** denotes the time-dependent excitation forces. To speed up the calculation, pre-calculated values for hydrodynamic mass and damping and transfer functions of the motion for different wave frequencies are provided in frequency domain. For this purpose, the method *pan*FDS is used, which is explained in 2.2 in more detail. These calculations are performed in the initialisation phase of the IMPRES simulation before the first time step. Subsequently, the acting forces are considered in the equation of motion from equation 1 and the system of equations is solved according to the accelerations in all six degrees of freedom. The motions of the hull can be determined by temporal integration. The generation of the irregular seaway results from the superimposition of individual linear wave components. A Jonswap wave spectrum is used as the distribution basis for the individual wave frequencies.

2.2. Threedimensional panel code panFDS

In order to make time-efficient predictions about motions and forces on ships, calculation methods in the frequency domain are often used. As mentioned above, IMPRES uses such pre-calculated values. These are calculated using the *pan*FDS method. With *pan*FDS, forces on and motions of symmetrical single hull ships in regular waves can be simulated. The fluid around the ship is assumed to be a potential flow. This implies a velocity potential from which pressures and hence forces can be calculated. When travelling with forward speed, a stationary wave system is developed, which moves along with the ship. This is not calculated by *pan*FDS, but read in as input from a Rankine Source method [4, 6]. Three-dimensional panels are used to discretise the hull and the water surface. As output, the program generates files with complex values for hydrodynamic mass and damping, diffraction or drift forces and transfer functions of motion responses to waves of different frequencies.

2.3. Wave interaction between the ships

The wave interaction is mainly caused by the stationary wave system of the mothership. This has an influence on the suspended small craft at higher speeds. At lower speeds, this influence decreases due to the lower wave height and can be considered to be negligible beyond a certain threshold. After launching, a stationary wave field also emanates from the small craft, the influence of which on the large ship can also be neglected. The stationary wave field is calculated in the initialisation process by using the Rankine Panel Method. The wave potential is calculated from the associated source strengths, which are determined iteratively. Two grids are generated for the calculation of the transfer functions. One is the surface grid for the ship hull and the other is the grid for discretising the water surface. The deformation of the water surface is determined from the potential-based velocity and pressure field around the ship. In each iteration, the grids of water surface and body are adapted to the deformation of the free surface and the initial floating position is also adjusted to the prevailing pressure condition due to changes forces and moments. While the grid resolution of the hull grid remains the same, the grid resolution of the free surface changes according to the surface deformation. The shorter the wave length, the smaller the panel size of the free surface grid becomes [4]. This leads to a limitation of the method for both very high and very low forward velocities. For low velocities, the computational effort becomes large due to an increasing number of panels and for large velocities, details of the hull are insufficiently accurately represented due to coarse mesh. On this evidence, the different velocities used in the following have been selected. Figure 1 shows the stationary wave field of the KCS (Kriso Container Ship [7]) at a Froude number of $Fn = 0.211$ in still water conditions. The blue colour stands for a wave crest. Red defines a wave trough.

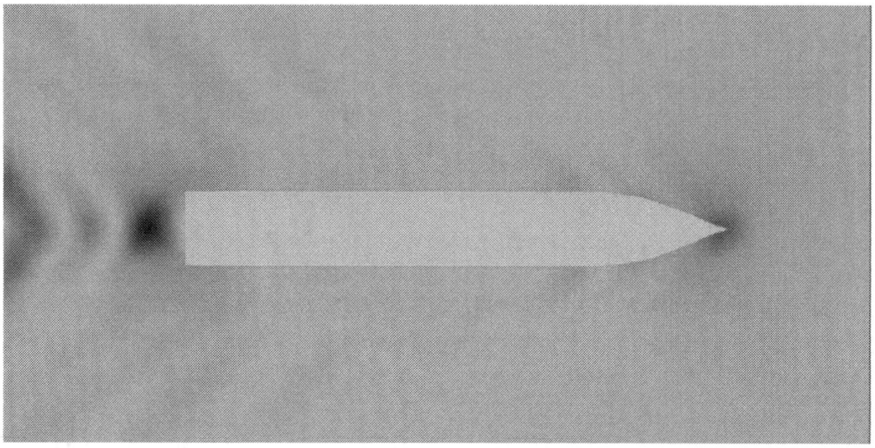

Figure 1. Stationary wave field of the KCS hull at $Fn = 0.211$.

2.4. Modelling of the crane suspension

During the launching process, the small craft is attached to the hook of a crane installed on the mothership. The motions of the mothership cause the tip of the crane and the dinghy suspended from it to move accordingly. By constantly extending the crane rope, the small craft is lowered from its initial position on the crane to the level of the water surface. Until the small craft generates sufficient buoyancy to float on its own, a constraining force acts on the small craft, which in the stationary state acts in the opposite direction to the weight force and is of equal magnitude. The suspension on a crane rope is solved by means of a holonomic constraint and a Lagrangian equation [8]. The holonomic constraint describes a property in the form of a function **g**, which does not change over time. In general,

$$\mathbf{g}(\boldsymbol{\xi}, t) = 0 \tag{2}$$

is the form of the constraint. In the particular case of a rope suspension, the function **g** is defined by

$$\mathbf{g}(\boldsymbol{\xi}, t) = (\boldsymbol{\xi}_{1q} - \boldsymbol{\xi}_{1p})^2 + (\boldsymbol{\xi}_{2q} - \boldsymbol{\xi}_{2p})^2 + (\boldsymbol{\xi}_{3q} - \boldsymbol{\xi}_{3p})^2 - l^2 = 0. \tag{3}$$

Here, the indice p describes the end of the rope at the crane tip and the indice q defines the point on the small boat and l the rope length. With the definition for the vector **Q**

$$\mathbf{Q} = \frac{\partial \mathbf{g}}{\partial \boldsymbol{\xi}} \tag{4}$$

the rope force F_z acting on the boat is given by

$$\mathbf{F}_z = \mathbf{Q}^T \boldsymbol{\lambda}. \tag{5}$$

$\boldsymbol{\lambda}$ is the Lagrangian coefficient resulting from the solution of the extended system of equations in 8. For the right-hand side of the equation system, the vector **G** is needed, which results from the second time derivative of equation 4. Since both temporal and spatial differentiation occur, the total derivative must be formed, which are given in equation 6 and 7.

$$\mathbf{Q}\dot{\boldsymbol{\xi}} = -\frac{\partial \mathbf{g}}{\partial t} \tag{6}$$

$$\mathbf{Q}\ddot{\boldsymbol{\xi}} = -\frac{\partial(\mathbf{Q}\dot{\boldsymbol{\xi}})}{\partial \boldsymbol{\xi}}\dot{\boldsymbol{\xi}} - 2\frac{\partial^2 \mathbf{g}}{\partial \boldsymbol{\xi} \partial t}\dot{\boldsymbol{\xi}} - \frac{\partial^2 \mathbf{g}}{\partial t^2} = \mathbf{G} \tag{7}$$

The original system of equations consisting of six equations for solving the motion is extended by one Lagrange equation for each constraint. In this case, a set of seven equations is created with one constraint.

$$\begin{bmatrix} \mathbf{M}+\mathbf{A} & \mathbf{Q}^T \\ \mathbf{Q} & \mathbf{0} \end{bmatrix} \begin{bmatrix} \ddot{\boldsymbol{\xi}} \\ -\boldsymbol{\lambda} \end{bmatrix} = \begin{bmatrix} \mathbf{F}_H + \mathbf{F}_A + \mathbf{A}\ddot{\boldsymbol{\xi}} - \mathbf{K} \\ \mathbf{G} \end{bmatrix} \tag{8}$$

3. Simulation

For the purpose of optimisation, different simulation setups are investigated and the results are compared. Some settings are kept the same for all simulations. The KCS hull is downscaled by a factor of 4 and used as the mothership and the planing boat's geometry is the prismatic hull shape by Brown and Klosinski [9]. The main dimensions of the two ships are listed in table 1. The initial scenario for the simulation is shown in figure 2.

Table 1. Main dimensions of mothership and small craft.

Parameter	Mothership	Small craft
L_{PP}	57.7 m	12.0 m
B	8.0 m	2.29 m
D	2.7 m	0.348 m

Figure 2. Illustration of the launching process before lowering with the small craft hanging from the crane.

Launching of the small boat from the initial suspended position starts five seconds after the begin of simulation. A wave spectrum is used for modelling the seaway. Two sea states with different zero upcrossing periods T_z are selected from the North Atlantic scatter table [10]. The significant wave height for both sea states is $H_s = 1.5\,\mathrm{m}$ and the periods are $T_{z1} = 7.5s$ and $T_{z2} = 8.5\,\mathrm{s}$, respectively. The seaway is modelled by a Jonswap spectrum and is consists of 30 wave components. For comparability, the phase angles, periods and heights of the individual waves are identical for all simulations. The wave direction is opposite to the ship's direction of travel. The wavelength of the ship's own wave system is approximated by $\lambda_{\text{ship}} = 2\pi\, L_{\text{ship}}\, Fn^2$ and therefore increases quadratically with the speed. The refinement of the free surface grid, which is used in *pan*FDS to calculate the stationary wave field, is proportional to the wavelength. This results in a strongly increased computational effort when the grid of the Rankine source method becomes finer due to the shorter wave. All six degrees of freedom are enabled for the mothership and small craft. The mothership's motions are transmitted to the boat via the suspension system.

Table 2. Froude numbers for different ships and forward speeds.

Parameter	Mothership	Small craft
$Fn\ (v = 3\frac{\mathrm{m}}{\mathrm{s}})$	0.126	0.277
$Fn\ (v = 4\frac{\mathrm{m}}{\mathrm{s}})$	0.168	0.369
$Fn\ (v = 5\frac{\mathrm{m}}{\mathrm{s}})$	0.211	0.461

The lateral gap between the mothership and small boat (1.0 B, 2.0 B and 3.0 B) is considered as the parameter to be varied. B stands for the width of the small craft. Furthermore, the initial trim angle at which the small boat is launched is changed between 0.0 °, 2.0 ° and 4.0 °. A positive trim angle indicates a turn that causes the bow to move upwards. In addition to the release speed of the crane cable ($v_{\mathrm{L}} = 0.4\,\frac{\mathrm{m}}{\mathrm{s}}$ $v_{\mathrm{L}} = 0.6\,\frac{\mathrm{m}}{\mathrm{s}}$ and $v_{\mathrm{L}} = 0.8\,\frac{\mathrm{m}}{\mathrm{s}}$), the forward speed is changed. The Froude numbers matching the vessels are listed in table 2. The forward speed is initially the same for both vessels and is added to the solution from the equation of motion without considering any additional propulsion force acting. This also applies to the small craft in a suspended state. For the smallest speed of $v = 3\frac{\mathrm{m}}{\mathrm{s}}$ the calculation of the stationary wave field is switched off and its influence is neglected as explained in section 2.3. During the launching process, no hydrodynamic forces act on the small craft's hull until the time of first immersion in a wave. Wind and current effects are not considered in the simulations carried out here.

4. Results of parameter study

The results for the root mean square (RMS) values for accelerations and roll angles are presented in figures 3-8. In order to assess the influence of the hydrodynamic forces on the motion during launching procedure, the simulated results are compared with limiting values for lateral and vertical acceleration and roll angle from the Nordforsk criteria [1]. These limits evaluate the general operability of the vessel under consideration. For a small fast boat, the root mean square values of 0.1 g for lateral acceleration, 0.275 g for vertical acceleration and 4.0 ° roll angle must not be exceeded. The values for the accelerations are estimated on the bridge. For this dinghy the position is assumed to be one metre vertically above the centre of gravity. For post-processing, intervals of the calculated time series are examined in which the immersion into the water occurs. The interval length is 10 s. On the x-axis of each figure, the forward speed of the small craft is plotted as the non-dimensional Froude number. The accelerations are provided dimensionless with the gravitational acceleration.

4.1. Motion comparison

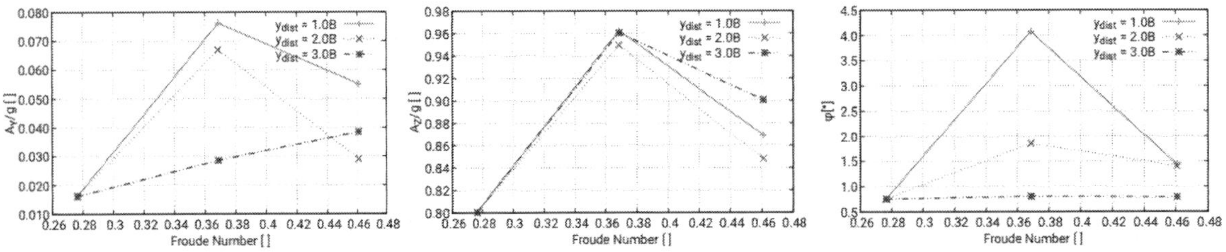

Figure 3. RMS values for lateral acceleration (left), vertical acceleration (middle) and roll angle (right) of the small craft while changing the y-distance between the two ships and $T_z = 7.5\,\mathrm{s}$.

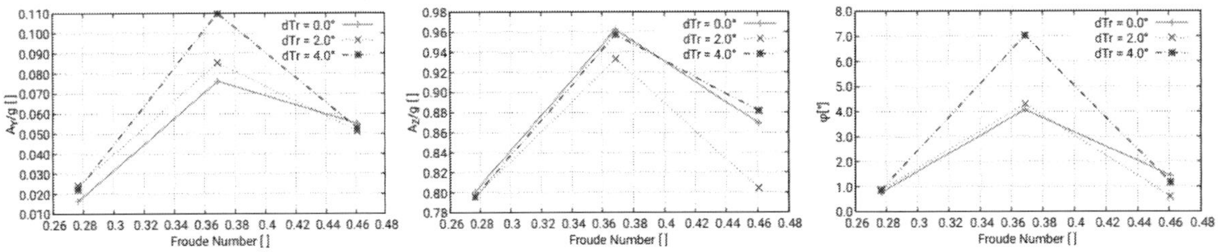

Figure 4. RMS values for lateral acceleration (left), vertical acceleration (middle) and roll angle (right) of the small craft while changing the trim angle of the small craft and d $T_z = 7.5\,\mathrm{s}$.

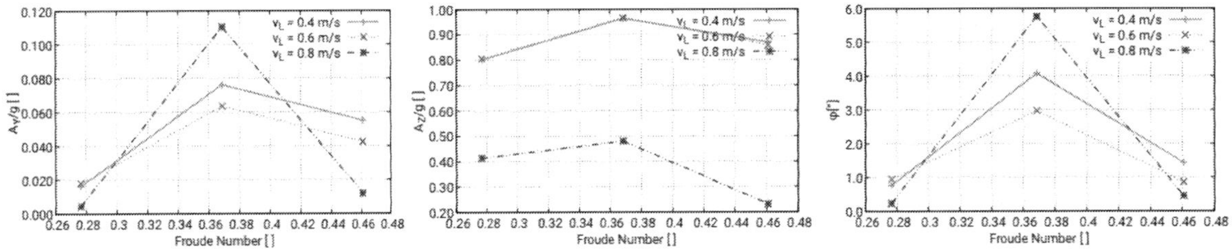

Figure 5. RMS values for lateral acceleration (left), vertical acceleration (middle) and roll angle (right) of the small craft while changing the lowering speed on the crane suspension and d $T_z = 7.5\,\mathrm{s}$.

The results are evaluated for the both sea states. First, the sea state with a smaller period $T_z = 7.5\,\mathrm{s}$ and relatively shorter waves is investigated. To be recognised is that the largest accelerations and roll angles occur with a Froude number of $Fn = 0.369$. The limits of the Nordforsk criteria for vertical accelerations are exceeded in almost all simulations. In some cases, up to 3.5 times the limit value. For the lateral accelerations, the specified limit is only exceeded in two simulations. One with a launch velocity of $v_L = 0.8\,\frac{\mathrm{m}}{\mathrm{s}}$ and one with an initial trim angle of $4.0\,°$ and a Froude number of $Fn = 0.369$. The roll angle exceeds the limit especially close to the mothership (1.0 B), at the highest launch speed and the largest initial trim angle. The reason

IOP Conf. Series: Materials Science and Engineering 1288 (2023) 012021 doi:10.1088/1757-899X/1288/1/012021

for that is the stationary wave system is more pronounced near the mothership. Noticeable is that the launch velocity $v_{\mathrm{L}} = 0.8\,\frac{\mathrm{m}}{\mathrm{s}}$ causes significantly lower vertical accelerations, but the opposite is observed for the roll angle and lateral acceleration. The interaction between lateral acceleration and roll angle is evident. The influence of the individual parameter variations can be read off from the scattering of the values at the same forward speed. The less the scatter, the smaller the influence of this parameter on the considered variable is. The scatter is effected by Froude number especially at $Fn = 0.461$. The trim angle leads to changes in accelerations and roll angle at medium ($Fn = 0.369$) speed. The same applies to the launching speed. For one combination of parameters the Nordforsk criteria limits are not exceeded: $Fn = 0.461$, $v_{\mathrm{L}} = 0.8\,\frac{\mathrm{m}}{\mathrm{s}}$, $y_{\mathrm{dist}} = 1.0\,\mathrm{B}$ and $dTr = 0.0\,°$.

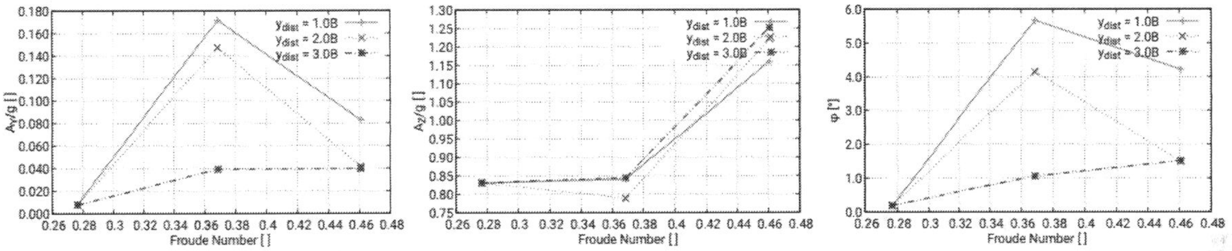

Figure 6. RMS values for lateral acceleration (left), vertical acceleration (middle) and roll angle (right) of the small craft while changing the y-distance between the two ships and $T_{\mathrm{z}} = 8.5\,\mathrm{s}$.

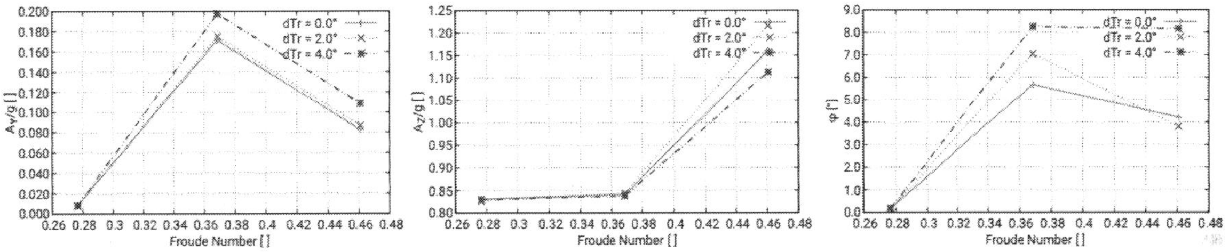

Figure 7. RMS values for lateral acceleration (left), vertical acceleration (middle) and roll angle (right) of the small craft while changing the trim angle of the small craft and $T_{\mathrm{z}} = 8.5\,\mathrm{s}$.

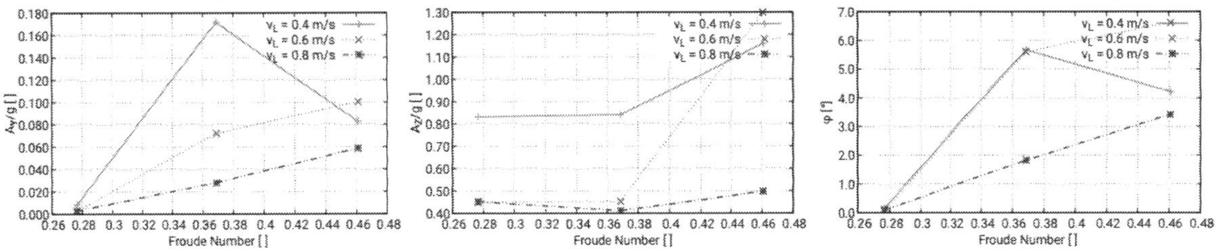

Figure 8. RMS values for lateral acceleration (left), vertical acceleration (middle) and roll angle (right) of the small craft while changing the lowering speed on the crane suspension and $T_{\mathrm{z}} = 8.5\,\mathrm{s}$.

In figures 6-8 the results for the seaway with larger wave period $T_{\mathrm{z}} = 8.5\,\mathrm{s}$ are shown. As with the other seaway, the values for vertical acceleration are greatly exceeding the threshold of Nordforsk criteria. In this case, there is no parameter combination that leads to a values below the limit. Furthermore, the vertical accelerations for a Froude number $Fn = 0.461$ are significantly larger than for the other velocities. Limits for both the lateral accelerations and the roll angle are exceeded for the two larger forward speeds and are greater than for the seaway with a shorter period. The maximum values are twice the limit for lateral acceleration and roll angle, and up to four times higher for vertical acceleration. In comparison, an increase in lateral acceleration is particularly visible. It is noticeable that this time the highest suspension speed causes the lowest accelerations and roll angles over the entire forward speed range. The influence on the lateral acceleration and thus also the roll angle is visible especially with a Froude number $Fn = 0.369$ in combination with the launching speed and the distance between the ships. In case of vertical acceleration, only the forward speed and the launching speed have an significant influence.

4.2. Force comparison

For the force comparison, the vertical hydrodynamic radiation forces on the hull are considered. These are shown for the different parameter variations in figure 9 and 10 for $T_z = 7.5\,\mathrm{s}$ and $T_z = 8.5\,\mathrm{s}$, respectively. The values are plotted over the Froude numbers and depicted dimensionless by using the small craft's weight $\Delta = 5500\,\mathrm{kg}$ and the gravitational acceleration.

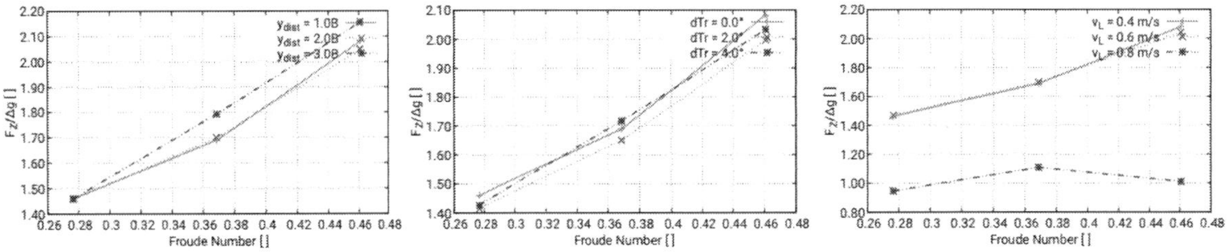

Figure 9. RMS values for the vertical force acting on the small craft for different parameter variances and $T_z = 7.5\,\mathrm{s}$.

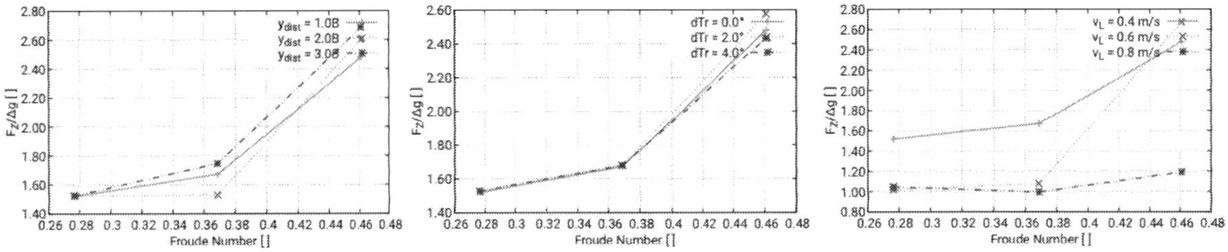

Figure 10. RMS values for the vertical force acting on the small craft for different parameter variances and $T_z = 8.5\,\mathrm{s}$.

Apparently, the forces for both sea states and all parameter variations increase at higher forward speed. Values of $0.9 - 1.75$ times the ship weight are reached for the Froude number $Fn = 0.277$ and up to 2.7 times the ship weight for $Fn = 0.461$. The influence of changes in trim and distance between the ships is small compared to the launching speed. Notably, for a launching velocity of $v_\mathrm{L} = 0.8\,\frac{\mathrm{m}}{\mathrm{s}}$, the forces are only slightly more than half as high in comparison to $v_\mathrm{L} = 0.4\,\frac{\mathrm{m}}{\mathrm{s}}$.

5. Conclusion

In this paper, the influence of different parameters on the accelerations and forces during a launching procedure of a boat with a crane from a mothership was analysed. Besides the transverse distance between small craft and mothership, the initial trim angle of the dinghy, the lowering speed of the crane and different forward speeds with and without the effect of a stationary wave system are investigated. Overall, reasonable results are calculated. The launching process in combination with the selected sea state leads to exceedance of the limit values for the vertical acceleration in almost all simulations and only in some cases for the lateral acceleration and the roll angle. With regard to the accelerations and roll angles, the varied trim angle has less influence compared to the other two parameters. A low Froude number as $Fn = 0.277$ is recommended for the small craft. In terms of forces, it is a higher forward speed that leads to larger forces. In addition, the launch speed has the strongest influence. The effect of changing trim angle and the transverse distance is relatively small. By varying the parameters, a combination is obtained which leads to compliance with all limiting values of accelerations and roll angle. This is applicable for the following launch condition at a sea state with a shorter significant period: a mean transverse distance of 1.0 B, an initial trim angle of 0.0 °, a high launching speed of $v_{\mathrm{L}} = 0.8\,\frac{\mathrm{m}}{\mathrm{s}}$ combined with a forward speed of $v = 5.0\,\frac{\mathrm{m}}{\mathrm{s}}$. Additional reduction of the acceleration values and roll angles can can be achieved by increasing the distance between the vessels and considering a certain initial trim angle of the small craft.

References

[1] Faltinsen O. M. 1990 *Sea Loads on Ships and Offshore Structures* (Cambridge University Press)

[2] Cummins, W. E. and United States. Navy 1962 *The Impulse Response Function and Ship Motions* (Washington DC, United States Department of the Navy, David Taylor Model Basin)

[3] Detlefsen O., Reiner C. and Abdel-Maksoud M. 2022 A Reliable Multilevel Method for Simulation-Based Load Determination in Deck Cell Guides. *Proceedings of the ASME 2022 41st International Conference on Ocean, Offshore and Arctic Engineering, OMAE. ASME, 6*

[4] El Moctar, B. O., Schellin, T. E. and Soeding, H. 2021 *Numerical Methods for Seakeeping Problems* (Springer International Publishing)

[5] Simonis H., Marleaux P. and Abdel-Maksoud M. 2021 Numerical analysis of semi-displacement vessels in head waves *33rd Symposium on Naval Hydrodynamics 2020, published jointly by the U.S. Office of Naval Research and Osaka University*

[6] Soeding H. 1993 A method for accurate force calculations in potential flow, *Ship Technology Research*, volume **40**, pages 176-186

[7] Kim W. J., Van S. H. and Kim D. H. 2001 Measurement of flows around modern commercial ship models, *Experiments in Fluids* volume **31**, pages 567578

[8] Mueller P. C. and Schiehlen, W. O. 1976 *Lineare Schwingungen* (Wiesbaden, Akademische Verlagsgesellschaft)

[9] Brown P. W. and Klosinski W. E. 1995 *Experimental Determination of the Added Inertia and Damping of Planing Boats in Roll* (Hoboken, Davidson Laboratory, Stevens Institute of Technology)

[10] DNV 2010 *Recommended Practice DNV-RP-C205 - Environmental Conditions and Environmental Loads* (Det Norske Veritas)

Asymmetry in sagging and hogging responses of a ship hull

J Matusiak[1], A Karola[1], A Niemelä[2], T Mikkola[1], S Hirdaris[1]
[1] Marine and Arctic Technology Group, Aalto University, Finland
[2] Meyer Turku Oy, P. O Box 666, Telakkakatu 1, FI-20240 Turku, Finland

Abstract. In this paper asymmetry in sagging and hogging of a hull progressing in regular waves is investigated by applying a weakly nonlinear potential flow theory. At first instance ship motions in waves are evaluated using a linear strip-theory yielding transfer functions of ship motions and sectional loads. Then, by assuming time-histories of ship motions the nonlinear part of the Froude-Krylov and restoring forces and moments are evaluated using a panel representation. Linear and weakly nonlinear predictions are compared against available model tests for a roll-on/off (Ro-Pax) ship hull. The results confirm that importance of nonlinearities of restoring and Froude-Krylov part of wave loading on asymmetry of sagging and hogging.

1. Introduction

Linear seakeeping models are well established and provide a fair evaluation of wave loads and responses [1]. For example, frequency domain solvers based on 2D strip-theory [2] and 3D panel methods [3], are fast and straightforward to apply in design assessment for the prediction of short- and long-term wave induced loads and responses. Notwithstanding this, linear models do not disclose important phenomena associated with non-linear hydrodynamic actions. As these should be considered in the design process, there is a need for the development and validation of fast yet accurate nonlinear numerical hydrodynamic methods.

In theory, the rapid development of Computational Fluid Dynamics (CFD) presents an obvious alternative in terms of accounting for the influence of nonlinearities on ship motions and wave induced loads without making usual assumptions and simplifications. However, despite the advent of computational power, CFD solvers remain slow and complex to use at pre- and post-processing stages. Additional arguments against the wide spread use of CFD methods relate with the advanced experience and education required by experienced users as well as the fact that the fluid physics of marine hydrodynamics are not so much governed by viscosity effects, which are the essence of CFD. For these reasons, CFD applications are normally limited to investigation of loads in head regular waves, while seakeeping analysis in waves is often disregarded [4].

This paper is focused on the potential use of weakly nonlinear potential flow theory for the prediction of hydrodynamic actions associated with ship motions and wave loads for use in ship design. Accordingly, pure hydrodynamic wave loads, restoring and inertia loads due to hull motions, that sum-up into wave loads (sheer forces and bending moments and torque) are considered. Hydroelastic effects or extreme hydrodynamic wave loads and ship responses affecting vessel stability and hence safety are not dealt with. These are treated amongst the others in [5].

2. Types of non-linearities of wave structural loads on ships

Nonlinear wave induced structural loads on ships relate to (a) the asymmetry of the first order hydrodynamic actions and responses that depend on the encounter frequency; (b) higher order effects associated with wave impacts and vibratory responses of a ship in waves. In this publication the influence of springing, slamming and whipping induced loads and responses using potential flow or CFD methods are not elaborated and the reader may refer to [9 – 12]. Accordingly, the following sections focuses on issues associated with the modelling of surface waves, sagging and hogging asymmetry.

2.1. Surface waves modelling

Airy waves do not consider the asymmetry of the wave profile that may be pertinent for steep waves and wave breaking is ignored. Airy theory is often used as it makes linear superposition of components regular waves into so called irregular wave pattern possible. This together with an introduction of some non-linear effects (e.g., Wheeler stretching) seems to be sufficient to reproduce numerically the action of the wave environment on a ship [7].

2.2. Sagging and hogging asymmetry

The asymmetry of hull sagging and hogging is a typical non-linear feature of the first order wave loading. Kukkanen [6] and Kukkanen and Matusiak [8] investigated this using model tests and pulsating Green function method [6,8]. Investigation of sagging and hogging asymmetry is usually restricted to the head seas condition. It is somewhat unclear what are the dominant effects causing this asymmetry. For the relatively long waves, such as $\lambda > 0.7L_{pp}$, non-linearities of the restoring and of the Froude-Krylov part of wave excitation are most likely to be the main source of this first order non-linearity. If this is the case, radiation and diffraction can be regarded linear. Moreover, it is assumed that the effect of non-linearities on ship motions.

3. Numerical model

It is often assumed that the biggest contributors to non-linearities of wave loading are the restoring forces and the Froude-Krylov part of wave loading. The interaction of incoming surface waves with the ship-made waves is considered insignificant. Also, diffracted waves are not considered. Radiation and diffraction are regarded as linear ones. In other words, the so-called hybrid approach is used. Within this context it may be suggested that the seakeeping problem could be adequately approximated based on classic strip theory [2]. The linear approximation to sectional wave load can be presented as :

$$\mathbf{F}_{\text{total}}(\text{lin}) = \mathbf{F}_{\text{res}}(\text{lin}) + \mathbf{F}_{\text{F-K}}(\text{lin}) + \mathbf{F}_{\text{diff}}(\text{lin}) + \mathbf{F}_{\text{rad}}(\text{lin}) + \mathbf{F}_{\text{inert}}(\text{lin}), \quad (1)$$

where $\mathbf{F}$ is the vector of sectional forces and moments, subscripts res, F-K, diff, rad and inert stand for restoring, Froude-Krylov, diffraction, radiation, and inertia components respectively. Linear strip theory-based program *sealoads,* based on the theory presented in reference [2], is used to evaluate $\mathbf{F}_{\text{total}}(\text{lin})$.

In a hybrid method we may assume that the contributions to the total sectional load are independent from each other and apply the superposition technique as follows:

$$\mathbf{F}_{\text{total}}(\text{nlin}) = \mathbf{F}_{\text{total}}(\text{lin}) + [\mathbf{F}_{\text{F-K}}(\text{nlin}) - \mathbf{F}_{\text{F-K}}(\text{lin})] + [\mathbf{F}_{\text{res}}(\text{nlin}) - \mathbf{F}_{\text{res}}(\text{lin})], \quad (2)$$

where 'lin' and 'nlin' depict linear and non-linear sectional loads.

4. Numerical implementation of total wave loads evaluation.

Ship motions and linear approximation of the sectional loads were evaluated by applying the linear strip theory software *sealoads* [2]. Time-history of ship linear motion and sectional load approximation, in terms of heave z and pitch θ, was created from the transfer function information for a given wave amplitude A_W and ship speed V_S.

$$x_i(t) = x_{i0}A_w \cos[k(X_G\cos\mu - Y_G \sin\mu) - \omega t - \gamma_i] \text{ for } i = 1,2,3 \quad (3)$$

$$x_i(t) = x_{i0}kA_w \cos[k(X_G\cos\mu - Y_G \sin\mu) - \omega t - \gamma_i] \quad \text{for } i = 4,5,6 \quad (4)$$

In the above equations x_{i0} is the Response-Amplitude-Operator (RAO) representing the modulus of the transfer function of the i-th motion component; $k=\omega^2/g$ is the wave number at an ω wave angular frequency; t is time and γ_i represents the phase angle of the i-th motion component. The position of the ship in the Earth-fixed co-ordinate system is given by (X_G,Y_G). Indices 1, 2 and 3 depict surge, sway and heave while 4, 5 and 6 correspond to angular motion components namely roll, pitch and yaw. For the head seas condition, that is μ=180^0, the ship position is defined as $X_G= -V_S t$, Y_G=0. Linear approximation to the sectional wave loads in terms of time histories is generated from their transfer functions and in the same manner as motions.

The incoming regular surface waves can be idealized according to the expression

$$\zeta(X;t) = A_w \cos(kX - \omega t). \quad (5)$$

The non-linear restoring and Froude-Krylov forces and moments are evaluated using a discretized model of the hull geometry made-up of triangular panels, as presented in Figure 1 below.

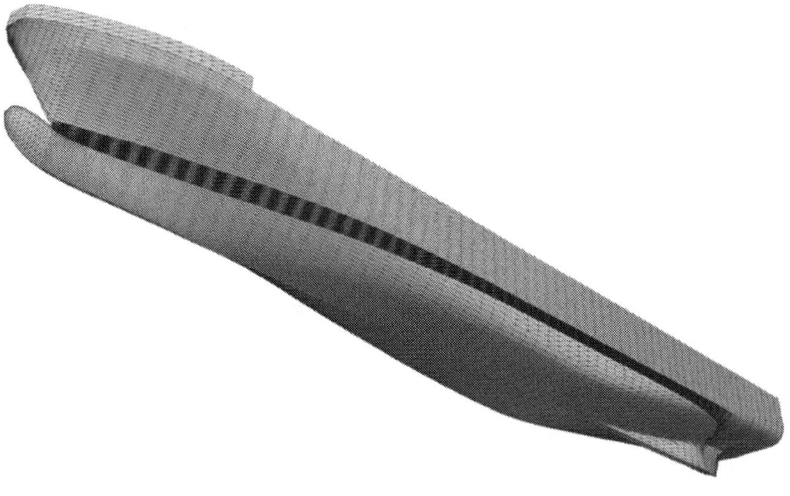

Figure 1. Panel model of the investigated Ro-Pax vessel (20,500 triangular panels).

Forces and moments were evaluated by summing-up the pressure forces at each panel twice for each time step. The hull accounted for ship motions prescribed in Equations (3) and (4). While transforming the panel coordinates from the ship- to the Earth-fixed coordinate system, the vertical distance from the control point of each panel to the wave surface was evaluated. The pressure due to motions and wave surface positioning in relation to the still water surface were consequently evaluated. The details of the algorithm can be found in reference [5]. At first this is done for the panels in contact with water by applying the so-called Wheeler stretching to evaluate the pressure in waves extending to the actual water surface. The same is repeated for the panels extending from the keel to the still waterplane assuming linear waves. The later requires extending pressures to the still water level and then integrating the pressures up to the still water level. Both summations are conducted for the portion of hull extending from the stem to the section in question. The first summation results in the total restoring and Froude-

Krylov forces and moments $\mathbf{F}_{F\text{-}K}$ (nlin) + $\mathbf{F}_{res}$ (nlin), while the second round of pressure integration yields linear approximation $\mathbf{F}_{F\text{-}K}$ (lin) + $\mathbf{F}_{res}$ (lin). Their difference is added to the total linear wave loading, as given by Equation (2).

5. Study case

The results presented in this paper are based on the Ro-Pax vessel namely Seatech-D with principal particulars / weight distributions and hull form depicted in Table 1 and Figure 2 respectively [6]. The wetted surface of the hull displays significant variations; hence it is prone to nonlinear wave loads, particularly in head waves. Model tests were conducted for a stationary ship and for a speed, corresponding to Froude number $Fn=0.25$ in head seas, under regular and irregular wave assumptions. Following the evaluation of global ship motions, vertical shear force and bending moments were respectively measured at frames 6.5 and 4.

Table 1 Main particulars of the vessel Seatech-D

	Full scale	*Model Scale*
	1	39.024
L_{wl} [m]	158.6	4.064
L_{pp} [m]	158	4.049
D_p [m]	4.8	0.123
B [m]	25	0.641
B_{wl} [m]	25	0.641
T_a [m]	6.1	0.156
T_ϕ [m]	6.1	0.156
D [m]	15	0.384
V [m^3]	13766	0.232
M_m [kg]		231.40
S [m^2]	4356	2.860
C_B	0.571	0.571
L_{wl}/B_{wl}	6.344	6.344
B_{wl}/T_{wl}	4.098	4.098
$L_{wl}/\nabla^{(1/3)}$	6.618	6.618
KG [m]	11, 11.5, 12	0.282, 0.295, 0.307

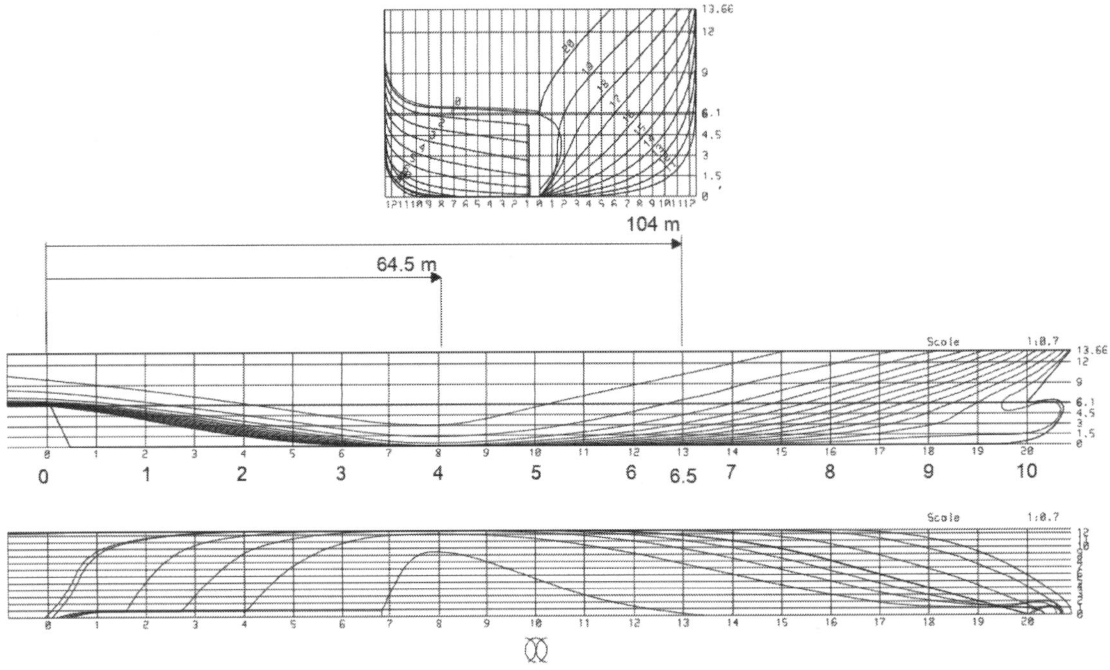

Figure 2. Line drawings of the RoPax. Internal loads were measured and computed for stations #4 (bending moment) and #6.5 (vertical shear force).

6. Results of total wave loads evaluation.

Wave loads were evaluated in regular waves and a head seas condition for the constant amplitude of 2 m and for varying wave lengths while assuming zero speed. The peak values of the vertical shear force F_z and bending moment are depicted as V3 and V5 in Figures 3 - 5. Positive values of V3 correspond to sagging and negative to hogging. The opposite holds for V5. The axis of frequency is presented in non-dimensional form as $\tilde{\omega} = \omega\sqrt{L/g}$, while shear force V_3 and bending moment V_5 are made non-dimensional by dividing them with terms ρgBL and ρgBL^2 respectively, with ρ being water density. The curves labelled 'computed' represent peak values corresponding to the total sectional loading including nonlinear Froude-Krylov and restoring force contributions. A clear asymmetry in wave load predictions is visible. However, the maxima of sagging are bigger than the values measured in model tests. The difference between sagging and hogging values is exaggerated as well. The linear approximation of the sectional loading evaluated by *sealoads* is presented by a pair of curves denoted as 'linear'. The computed hogging values are somewhat closer to the measured values as compared to the linear approximation.

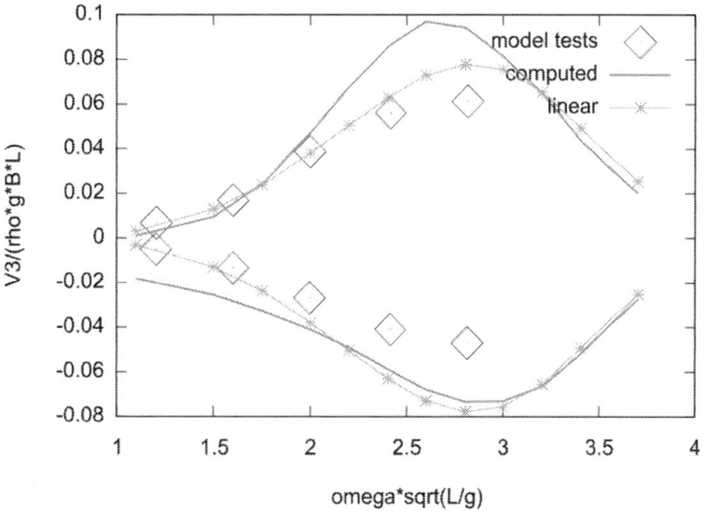

Figure 3. Maximum and minimum peaks of vertical shear force at frame #6.5 at zero speed.

In Figure 5 a time record of bending moment is presented. Time is shown in non-dimensional form as $\tilde{t} = t * \sqrt{\frac{L}{g}}$. Both linear and nonlinear predictions of moment as well as their difference are presented. The nonlinearities, represented by the difference, sharpen and magnify the sagging response thus introducing asymmetry to the time record. From the time record of heave and pitch motion components shown in Figure 6 it is suggested that mainly it is the V-shaped bow immersion (positive pitch) that results to the sudden increase of the sagging bending moment. The same can be seen observing the RAO of heave and pitch presented in Figure 7. Large sagging frequency band corresponds to high RAO values of pitch confirming the importance of this motion component on bending moment at amidships. It is interesting to note that the computation conducted with the hybrid model *LaiDyn* [5] gives values of pitch closer to the measured ones than the linear model of *sealoads* does.

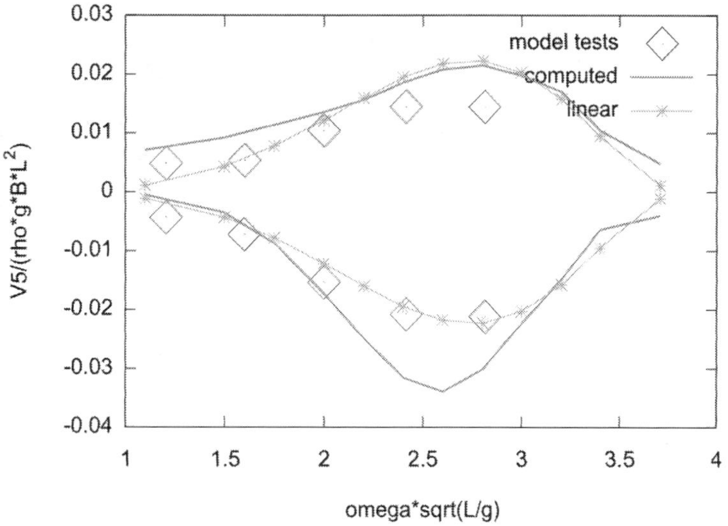

Figure 4. Maximum and minimum peaks of bending moment at frame #4 at zero speed.

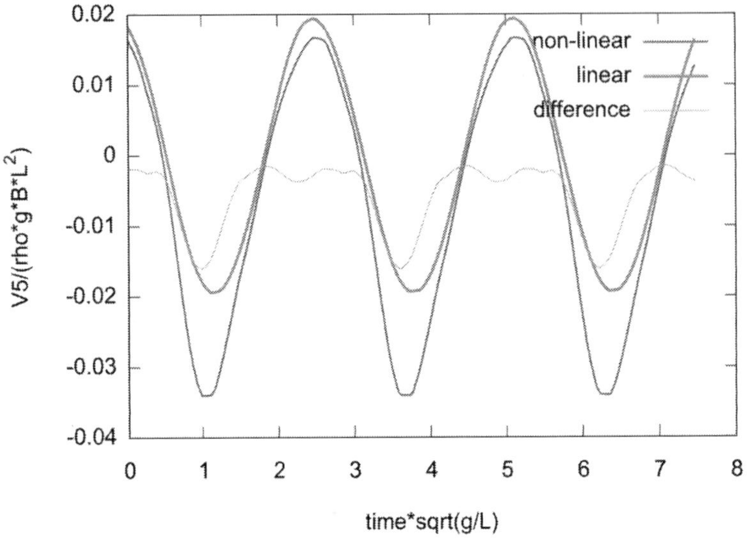

Figure 5. Time history of linear approximation and non-linear bending moment at frame #4 in regular head waves of non-dimensional frequency 2.4 and amplitude 2 m.

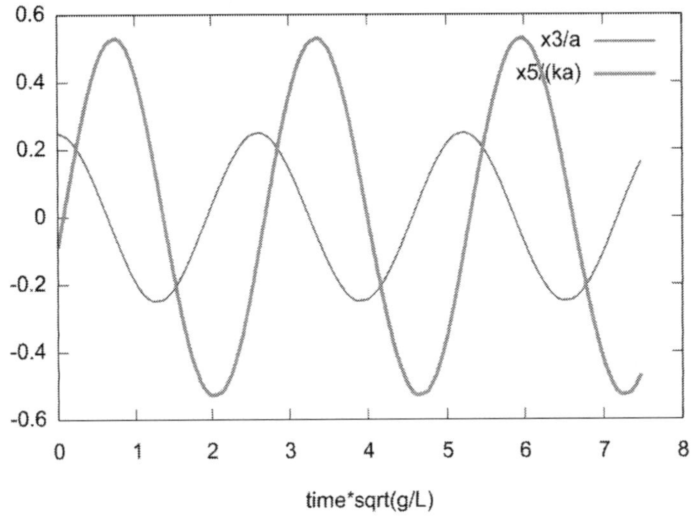

Figure 6. Time history of linear heave and pitch motion in head regular waves of non-dimensional frequency 2.4.

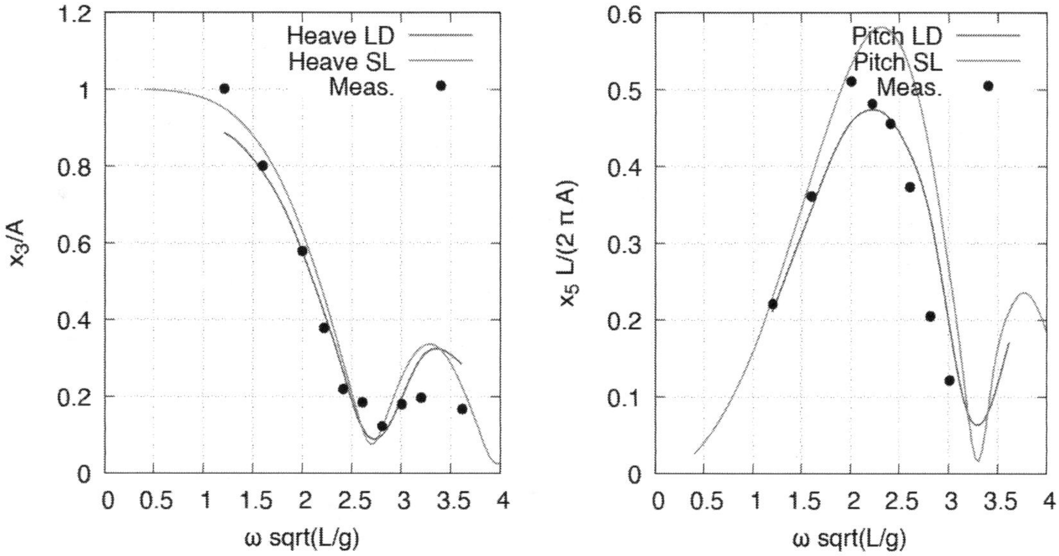

Figure 7. Response -Amplitude-Operator of heave and pitch derived by model tests and computations. Head seas and Fn=0. LD stands for *LaiDyn* while SL corresponds to the results given by *sealoads*. Meas are the experimental results of Kukkanen [6].

7. Conclusions.

The results confirm validity of the main assumption of this study, that is the effect of non-linearities of restoring and Froude-Krylov part of wave loading on asymmetry of sagging and hogging. Exaggerated values of predicted sagging may have several sources. One of them is the assumed linear motion of the ship, that is forced motion, not affected by non-linearities. The method with the consideration of non-linearities of both ship motions and wave loads may clarify this. Another possible reason is the assumption of the hybrid model that is disregarding non-linearities related to diffraction and radiation. An interaction of ship-made waves with the oncoming waves may influence the sectional wave loads. In principle, these effects could be checked by a method like the one found in the references [6, 8].

8. References

[1] Karola A, Mikkola T, Matusiak J and Hirdaris S 2022 *Proceedings of the International Conference on Offshore Mechanics and Arctic Engineering - OMAE* [doi:10.1115/OMAE2022-79702]
[2] Salvesen N, Tuck E O Faltinsen O M 1970 *Ship motions and sea loads* (Transactions of SNAME, vol. 78)
[3] ANSYS *ANSYS 2019 R1 Aqwa Theory Manual*
[4] Hänninen S K, Mikkola T, Matusiak J 2013 *Journal of marine science and technology* pp. 1-13
[5] Matusiak J 2021 *Dynamics of a rigid ship – with applications,* 3rd edition (Espoo: Aalto-University) Pp 173 ISBN: ISBN: 978-952-60-5205-2 (electronic), 978-952-64-0398-4 (printed) https://aaltodoc.aalto.fi/handle/123456789/108000
[6] Kukkanen T 2012 Numerical and experimental studies of nonlinear wave loads of ships *PhD Thesis* Espoo
[7] Matusiak J 2011 *International Journal of Naval Architecture and Ocean Engineering* **3, No 1**
[8] Kukkanen T, Matusiak J 2014 *Ocean Engineering* **75C** pp. 1-14 [doi:10.1016/j.oceaneng.2013.10.008]

[9] Sengupta D, Show TK, Hirdaris S, Datta R 2023 *Proceedings of the Institution of Mechanical Engineers, Part M: Journal of Engineering for the Maritime Environment* **0(0)** [doi:10.1177/14750902231165808]

[10] Hirdaris S, Parunov J, Qui W, Iijima K, Wang X, Wang S, Brizzolara S, Guedes Soares C 2023 *Marine Structures* **89** [https://doi.org/10.1016/j.marstruc.2023.103383]

[11] Show YK, Hirdaris S, Datta R 2022 *Shock and Vibration* Article ID 4564769 pp. 18 [https://doi.org/10.1155/2022/4564769]

[12] Lakshmynarayanana PAK, Hirdaris S 2020 *Ocean Engineering* **203** 107179 [https://doi.org/10.1016/j.oceaneng.2020.107179]

A CFD investigation of the effects of passing ship on a small size ice floe

Z Jiang[1], **M Suominen**[1], **S Tavakoli**[1], **P Kujala**[1] **and S Hirdaris**[1,*]

[1] Marine and Arctic Technology, Department of Mechanical Engineering, Aalto University, Espoo 02150, Finland

[*] Corresponding author, e-mail address: spyros.hirdaris@aalto.fi

Abstract. A RANS (Reynolds Averaged Navier-Stokes) CFD STAR-CCM+ model is used to investigate the hydrodynamic interactions between a ship and a circular ice floe. The overset mesh technique is employed to simulate the advancing movement of the passing ship. The ship has a hull of similar configuration to the KCS MOERI container ship. The ice floe has a diameter of 30% of the ship's length and a thickness of 3 m. The paper primarily focuses on the motions of the ice floe and the hydrodynamic forces induced by the ship progressing in different speeds and calm water conditions. A parametric study on the influence of the ice floe surge and sway motions and associated hydrodynamic loads is presented. It is concluded that ship-generated waves can significantly affect the motions of the ice floe and the influence on sway is greater than that on surge. The ship speed is also found to have a distinct impact on ice floe motions.

1. Introduction

In recent years, the development of shipping corridors in the Arctic has been inevitably interlinked with the adverse effects of climate change leading to the reduction of sea ice cover. An example is the increase of traffic along the Northern Sea Route (NSR), which is recognized as a viable alternative to traditional shipping routes through the Suez and Panama Canals. Arctic shipping raises several unique challenges of relevance to ship safety and efficiency. For example, the collision with sea ice in Arctic waters challenges the structural integrity of ship hulls.

The hydrodynamic interaction between an advancing ship and ice floes may lead to unusual seakeeping responses of the ice floe and ship. For example, the motions of an ice floe induced by the forward ship might lead in a collision with the following ship. In such cases different speeds of the ship and ice floe result in disparate boundary conditions in way of the free surface. If the speeds of an ice floe and a floating structure are zero, it will be easier to describe the boundary condition in way of the free surface boundaries [1]. Some scholars consider the velocity of floaters but disregard the effects of surface waves [2]. Recently, Jiang et al. [3] proposed a potential flow analysis method to tackle the issue of free surface boundary conditions induced by different speeds. This method is based on the linear superposition of velocity potential and the encounter frequency method. Hence, it is constraint to low speeds. Recently, the RANS based CFD method has been extensively applied to study the hydrodynamic interactions between a passing and a moored ship [4].

This paper introduces a RANS based CFD model to simulate the hydrodynamic interactions between a passing ship and a small size ice floe. The overset technique is used to deal with the large amplitude motions of the passing ship. The study examines the surge and sway motions of the ice floe under the influence of three different advancing speeds of the passing ship. The fluid forces in surge and sway acting on the ice floe are presented to investigate dynamics.

2. Theory

2.1. Governing equations

The governing equations of the viscous fluid model are the Navier-Stokes (NS) equations, which consist of the mass conservation equation and the momentum conservation equation. These equations are defined as

$$\frac{\partial \rho}{\partial t} + \frac{\partial \rho u_i}{\partial x_i} = 0 \tag{1}$$

$$\frac{\partial \rho u_i}{\partial t} + \frac{\partial \rho u_i u_j}{\partial x_i} = -\frac{\partial p}{\partial x_i} + \rho \frac{\partial}{\partial x_j}\left[\mu\left(\frac{\partial u_i}{\partial x_j} + \frac{\partial u_j}{\partial x_i}\right)\right] - \frac{\partial}{\partial x_j}\left(-\rho \overline{u_i' u_j'}\right) \tag{2}$$

where t is the time, ρ is the effective fluid density, u_i and u_j are the Cartesian components of velocity vector with $i, j = 1,2,3$, p is the pressure, μ is the effective dynamic viscosity, $\rho \overline{u_i' u_j'}$ is the Reynolds stress.

The Volume of Fluid (VOF) method is adopted for the free surface capture with a volume fraction factor α, which represents the averaging the density ρ and viscosity v in partial elements

$$\rho = \alpha \rho_w + (1-\alpha)\rho_a, v = \alpha v_w + (1-\alpha) v_a \tag{3}$$

where w and a represent the water phase and air phase, respectively.

2.2. Numerical implementations

The Finite Volume Method (FVM) is used to solve NS equations in conjunction with the segregated flow solver that is used to solve the conservation of mass and momentum. The convective flux is computed by the second order upwind scheme. The $SST\ k-\omega$ turbulence model is used to calculate the transport of turbulent shear stress. The all $y+$ wall treatment is used to simulate the boundary layer around the no-slip wall boundaries, which uses a blended wall function to cover all three sublayers [5]. The time step is set in compliance with the Courant-Friedrichs-Lewy condition

$$C_r \geq \frac{U \Delta t}{\Delta x} \tag{4}$$

where C_r is the Courant number, U is the mesh flow speed, Δt is the computational time step, Δx is the cell dimension.

2.3. Dynamic fluid body interaction

The fluid force $\mathbf{F}$ and moment $\mathbf{M_o}$ acting on the structure can be obtained as

$$\mathbf{F} = -\int_S p \cdot \mathbf{n} ds + \int_S \mu \frac{\partial \mathbf{u}_\tau}{\partial \mathbf{n}} ds = \int_S \mathbf{f}^p ds + \int_S \mathbf{f}^s ds \tag{5}$$

$$\mathbf{M_o} = \int_S (\mathbf{x}_f - \mathbf{x}_0) \times (\mathbf{f}^p + \mathbf{f}^s) ds \tag{6}$$

where $\mathbf{u}_\tau$ is the tangent velocity component around the body surface, S is the body surface, $\mathbf{x}_f$ is the centroid of cell face f, $\mathbf{x}_0$ is the specified origin vector.

The motions of ice floe in 6 – Degrees of Freedom (DoF) are calculated by the Dynamic Fluid Body Interaction (DFBI) method. The translation and rotation of the ice floe can be calculated as

$$\mathbf{M} \frac{\partial \mathbf{u}}{\partial t} = \mathbf{F} \tag{7}$$

$$\mathbf{M} \frac{\partial \boldsymbol{\omega}}{\partial t} + \boldsymbol{\omega} \times \mathbf{M} \boldsymbol{\omega} = \mathbf{M_o} \tag{8}$$

where **M** is the inertia matrix, $\boldsymbol{\omega}$ is the angular velocity.

3. Numerical model setup

A Numerical Wave Tank (NWT) is established to conduct the simulation. The boundary conditions and dimensions of the NWT are depicted in Figures 1 and 2. The fluid domain of this NWT is a rectangle with a length of 8.4 times the length of the passing ship (L_p), a breadth of 3 times L_p, and a depth of 2.5 times L_p. The water depth is 1.5 times L_p.

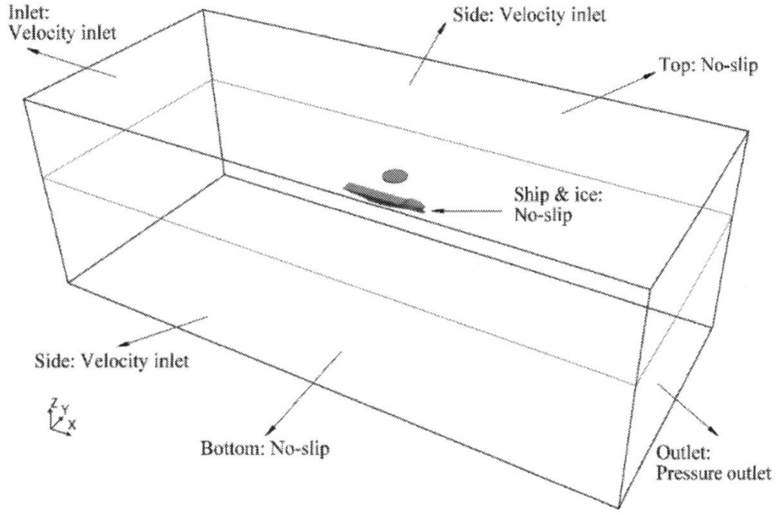

Figure 1. 3D view of Numerical Wave Tank and its boundary conditions.

Figure 2 displays a simplified scenario of a ship corresponding to the MOERI container ship KCS passing by a circular ice floe with a diameter of $0.3L_p$ and a thickness of 3 m at three constant speeds of the order of 6, 8,10 knots (see Figure 1 and Table 1). The transverse distance between the ice floe edge and the port side of the passing ship is $1/3\ L_p$. The simulation starts when the passing ship's bow is $0.5L_p$ from the centre of the ice floe and ends when its stern is $0.5L_p$ from the center of ice floe. The main settings of the numerical simulation are presented in Table 2. The fluid domain is discretized by a trimmed mesh (see Figure 3). An overset mesh technique is used to handle the translation of the passing ship over the long distance at constant speed. The ship is contained in the local region, while the ice stays in the background and idealised by the DFBI method. The mesh around the ship's course and the ice floe is refined to better describe the fluid flow.

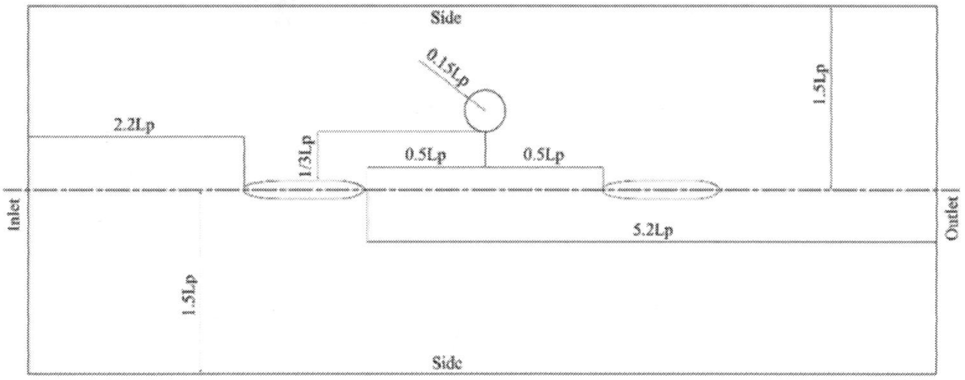

Figure 2. Top view of the Numerical Wave Tank and positions of the ship and ice floe.

IOP Conf. Series: Materials Science and Engineering 1288 (2023) 012023 doi:10.1088/1757-899X/1288/1/012023

Table 1. Main dimensions of KCS.

Item	Symbol	Units	Value
Length between perpendiculars	L_p	m	230
Beam	B	m	32.2
Draft	T	m	10.8
Displacement	Δ	m^3	52030
Vertical center of Gravity	KG	m	7.28
Long. Center of Gravity	LCG	m	111.6
Roll radius of gyration	k_{xx}/B	-	0.4
Pitch & Yaw radius of gyration	k_{yy}/L_p & k_{zz}/L_p	-	0.25

Table 2. Main settings in RANS simulation

Item	Setting
Turbulence model	SST $k-\omega$
Time model	Implicit unsteady
Time step	0.01 s
Solver formulation	Segregated
Velocity-pressure algorithm	SIMPLE
Free surface capture	VOF
DOF of passing ship	Surge
DOF of ice floe	6 DOF

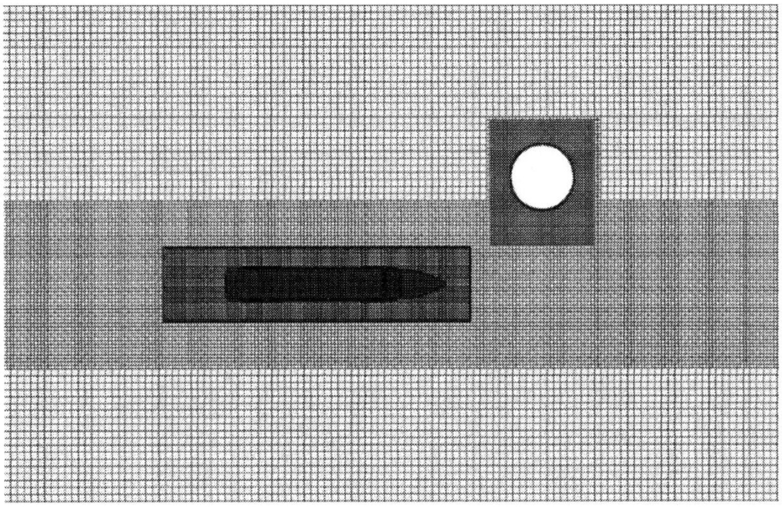

Figure 3. Trimmed mesh in the fluid domain.

4. Results and discussion

When a ship passes by a small size ice floe, the ship generated waves instigate 6-DoF motions. Amongst those, sway and surge may be considered as the most relevant to ship-ice collision scenarios. In this paper, to better analyze the spatial effects, a parameter D_L is introduced to represent the longitudinal distance between the passing ship amidship region and the center of an ice floe. $D_L < 0$ indicates the ship is approaching the ice floe, while $D_L > 0$ indicates the ship is departing.

4.1. Sway motions of the ice floe

Figure 4 illustrates the sway motion of an ice floe with respect to the relative longitudinal position between the ship and ice floe over time at ship speeds 6,8,10 knots. At first instance, the sway of ice may be considered negligible until $D_L/L_p \approx -0.8$. At this point, the curves start to diverge. When the ship speed is 6 knots, sway motion amplitudes become negative. This indicates that the ice floe moves toward the passing ship. In 10 knots forward ship speed, the ice floe initially advances- and then reverses-toward the ship. In 8 knots forward ship speed, the sway motion remains close to zero until $D_L/L_p \approx -0.4$. This phenomenon can be revealed by the sway velocity of the ice floe, as shown in Figure 5. This figure shows that the sway velocity is small when $D_L/L_p < -0.8$. Subsequently, velocity fluctuations become proportional to the ship speed for $D_L/L_p < -0.4$. At this stage, the velocity is negative when the ship speed is 6 knots, and positive when the ship speed is 10 knots. For 8 knots ship speed the velocity curve is approximately evenly positioned above and below 0. Eventually, all three curves indicate a continuous negative pattern.

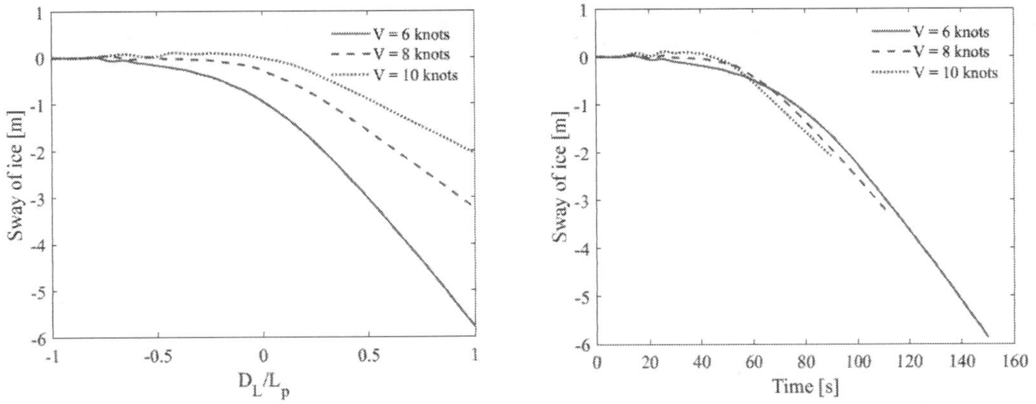

Figure 4. Sway of the ice floe under the waves generated by the passing ship.

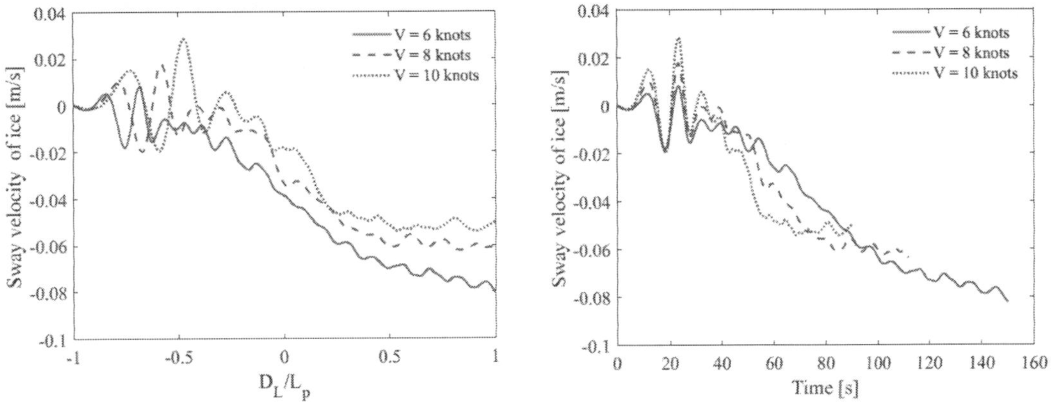

Figure 5. Sway velocity of the ice floe under the waves generated by the passing ship.

IOP Conf. Series: Materials Science and Engineering 1288 (2023) 012023 doi:10.1088/1757-899X/1288/1/012023

Figures 4 and 5 reveal that the ice foe is attracted by the passing ship for all three ship speeds. However, in 10-knots, the ice floe is initially pushed away by the ship waves, and then dragged back when their distance is short enough. Figure 6 shows the field of fluid velocity at the transverse section crossing the centre of the ice floe for 6 and 10 knots forward ship speed. Figure 6 demonstrates that the fluid velocity is higher in way of the part of the ice floe that is close to the passing ship. It complies with the expectation because the fluid velocity is induced by the passing ship. According to the Bernoulli's equation for incompressible fluids, the fluid pressure is inversely proportional to the velocity. Thus, the pressure gradient pushes the ice floe to the passing ship. Figure 7 shows the time history curve of integral fluid force on the ice floe in sway. It is demonstrated that that negative over positive sway forces dominate over time. The change of momentum depends on the product of force with time. Thus, the record of sway force agrees with the record of sway motion of ice. Similar results have been reported in the experiments by Tsarau et al. [2]. The large amplitude of fluctuation of sway forces at the initial stage could be attributed to the divergent bow wave.

Figure 6. Fluid velocity field at the transverse section crossing the center of ice floe.

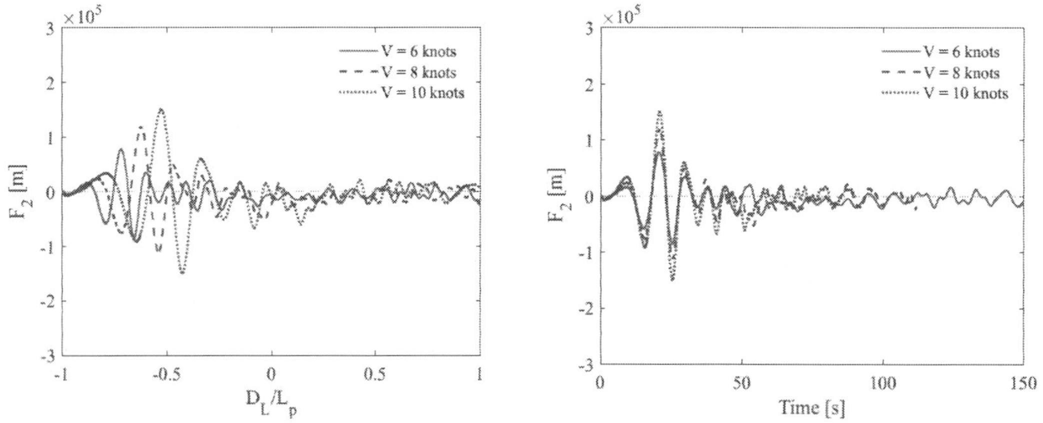

Figure 7. Transient fluid force in sway on the ice floe.

4.2. Surge motions of the ice floe

Figure 8 illustrates the surge motion of the ice floe in waves generated by the passing ship at different speeds. At first instance the amplitude of surge remains low for $D_L/L_p < -0.9$. Then, the ice floe progresses towards the direction of ship speed. When the distance between the ship and the ice floe is marginal, i.e., $-0.5 < D_L/L_p < 0.5$, the ice floe is dragged in the opposite to the ship speed and for $D_L/L_p > 0.5$ it reverses (see Figure 9). The large fluctuation of velocity observed at first may be attributed to the corresponding fluctuation of surge forces (see Figure 10), which could be induced by the divergent bow wave.

The surge and sway displacements of the ice floe are respectively proportional and inversely proportional to ship speed. This could be attributed to the longer acting time of fluid force acting on the ice floe when the ship speed is lower. As an example, Figure 11 illustrates the impulse of sway force during the whole course of simulation. As shown in Figure 11, the impulse of sway force is negative for all three ship speeds, and the absolute values continuously decline with the increase of speed. Correspondingly, the final sway velocity of ice floe is proportional to the ship speed. This can be applied to any time span of sway motion. As a result, the sway motion is largest when the ship passes at 6 knots.

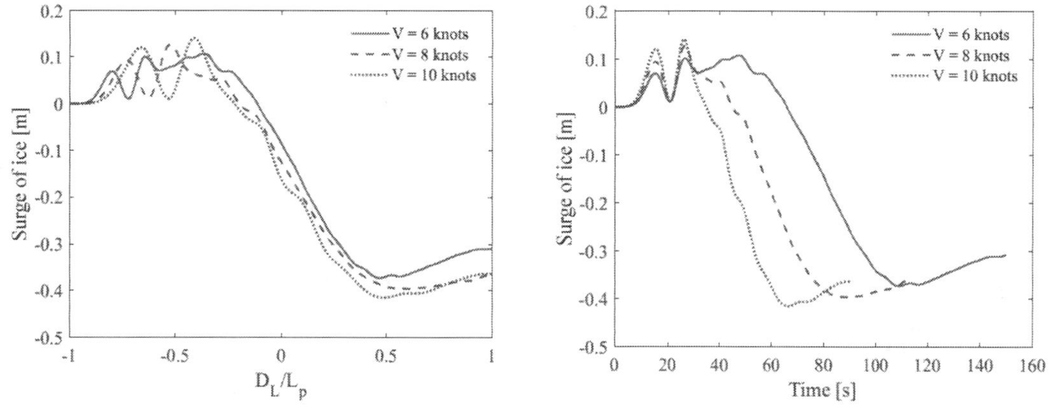

Figure 8. Surge of the ice floe under the waves generated by the passing ship.

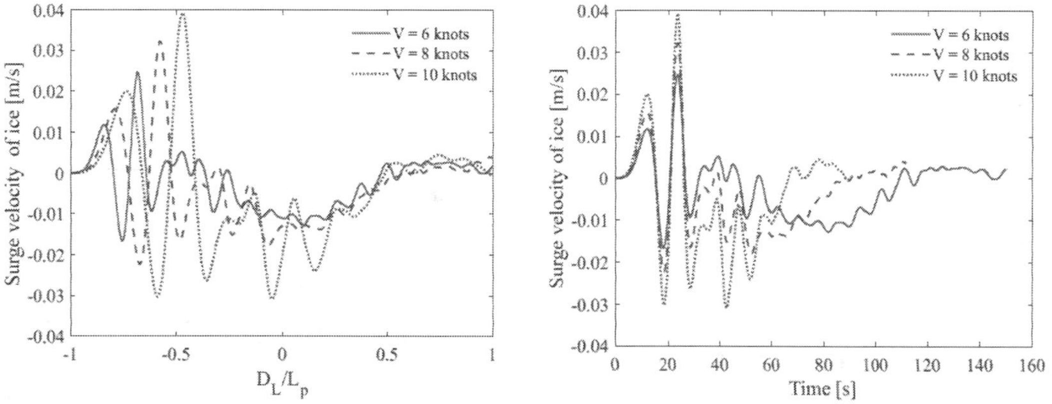

Figure 9. Surge velocity of the ice floe under the waves generated by the passing ship.

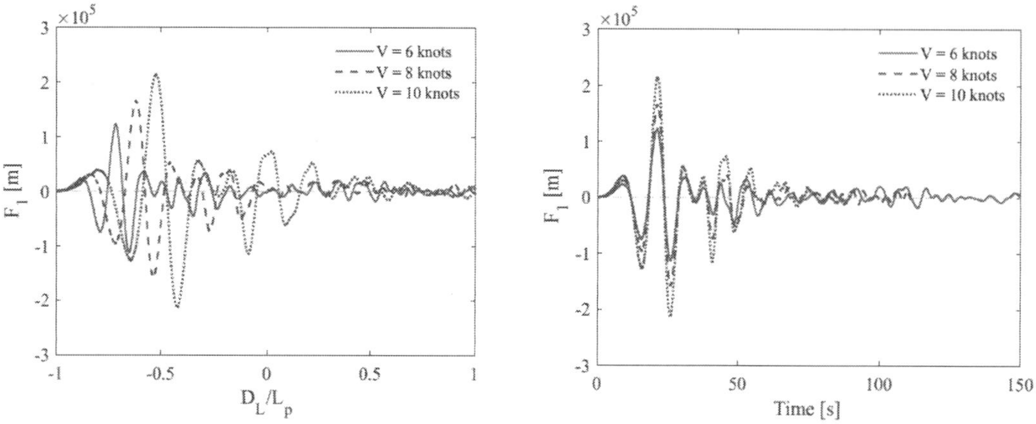

Figure 10. Transient fluid force in surge on the ice floe.

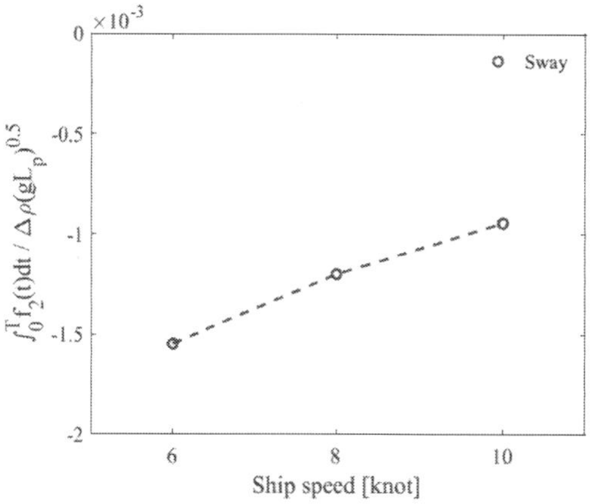

Figure 11. Impulse of sway force.

5. Conclusions

The study presents a RANS based CFD model that examine the hydrodynamic interaction between a passing ship and a small size ice floe. The overset technique is used to deal with the large amplitude motions of the passing ship. The study examines the surge and sway motions of the ice floe under the influence of three different advancing ship speeds. Based on the CFD simulation, the following conclusions can be drawn:

1. The motions of the ice floe can be significantly influenced by the passing ship. Yet, the influence on sway motions is much greater than the influence on surge motions.
2. The ship speed has a distinct effect on the motions of ice floe. The surge displacement of the ice floe is proportional to the ship speed, while the sway is inversely proportional to the ship speed. This is mainly attributed to the longer acting time of the fluid force on the ice floe at lower ship speed.
3. In sway, the ice floe is primarily attracted to the passing ship for all ship speeds. An exception is the scenario of 10 knots where the ice floe is initially marginally pushed away by the ship generated waves.

IOP Conf. Series: Materials Science and Engineering 1288 (2023) 012023 doi:10.1088/1757-899X/1288/1/012023

4. The surge motion of the ice floe manifests in three subsequent stages depicting forth and back alignment to the direction of the ship's speed.

References

[1] Keijdener, C, de Oliveira Barbosa, J. M, and Metrikine, A V 2017 The influence of level ice on the frequency domain response of floaters. *Cold Regions Science and Technology*, **143**, 112-125

[2] Tsarau, A, Lubbad, R, and Løset, S 2014 A numerical model for simulation of the hydrodynamic interactions between a marine floater and fragmented sea ice. *Cold Regions Science and Technology*, **103**, 1-14

[3] Jiang, Z, Li, F, Mikkola, T, Kujala, P, and Hirdaris, S 2023 A Boundary Element Method for the Prediction of Hydrodynamic Ship–Ice–Wave Interactions in Regular Waves. *Journal of Offshore Mechanics and Arctic Engineering*, **145**(6), 061601

[4] Zhou, L, Abdelwahab, H S, and Soares, C G 2021 Experimental and CFD investigation of the effects of a high-speed passing ship on a moored container ship. *Ocean Engineering*, **228**, 108914

[5] Siemens, P L M 2021. *STAR-CCM+ User Guide Version 2021.3*. (Siemens PLM Software Inc: Munich, Germany)

Comparative summary of approaches used with the impulse response method

M. Wermbter[1], L. Altenbach[1], O. Detlefsen[1] and M. Abdel-Maksoud[1]

[1] Institute for Fluid Dynamics and Ship Theory, Hamburg University of Technology (TUHH), Am Schwarzenberg-Campus 4, 21073 Hamburg, Germany

E-mail: malwin.wermbter@tuhh.de

Abstract. This paper presents a comparison of various methodologies to calculate the seakeeping behaviour of ships using the impulse response method. Developments of several authors incorporate non-linearities regarding the radiation, diffraction and wave forces. In contrast to frequency domain methods, advantages are expected if ship motions become non-linear, for example in waves with high steepness ratios. The comparison is structured into analysis for each force component with and without motions in waves. The main contribution is the quantitative analysis for several ships, forward speeds and methodologies. Multiple frequency domain methods are incorporated in the analysis. The focus of the analysis lies on the evaluation of applicability as well as the limits of the method for all states investigated. The results of the impulse response method show the following: The method is applicable to several forward speeds. In long waves, the choice of the parameters calculated in the frequency domain for the computation of the radiation and diffraction forces has a strong influence on the computed motion amplitudes. In short waves, the computation of the added resistance is a considerable challenge. Especially at high forward speeds, the consideration of the stationary wave system is essential for obtaining accurate results with three-dimensional panel codes.

1. Introduction

Motion simulations in seaways are categorised by their level of non-linearity and the required effort of calculation. The impulse response method is a time-domain approach, which allows to incorporate non-linearities in each force component, however the simulations are faster than real-time. Existing publications present a variety of approaches to simulate the ship motion in regular and irregular waves, as well as special topics as green water on deck [1], hydroelastic calculations [35], parametric rolling [18] or dynamic positioning [6]. Although, it is unclear which approach has the most advantages in terms of efficiency and accuracy. In order to categorise various seakeeping methods, Hirdaris [13] developed a six level system. The current state-of-the-art impulse response methods are to be assigned to level 2, Froude-Krylov non-linear methods. Some authors [24, 26] have incorporated body-non-linearity into radiation forces to reach level 3. Other differences between existing publications relate to the coordinate system used for the motion integration and force calculation, the method used for frequency domain coefficients and the calculation of Froude-Krylov, restoring, radiation and diffraction forces.

1.1. Coordinate system

Ship motions are usually described in three coordinate systems. A global, hydrodynamic and ship-fixed system [11]. Newtons 2nd Law is used to calculate the ship motions either in the inertial hydrodynamic or global coordinate system or in the non-inertial ship-fixed coordinate system. If the motions are calculated in the non-inertial frame additional Coriolis forces due to the acceleration of the coordinate system have to be considered. Approaches by Fossen [11], Kristiansen [17], Perez [22], Hirdaris [14],Ballard [3] apply this principle. The reason to use this coordinate frame is either the formulation of the frequency domain coefficients or the application of control forces, which are usually formulated in the ship-fixed frame. Perez [21] compared the developed formulations in both coordinate system, when using the strip theory with and without forward speed. Most publications use the hydrodynamic coordinate system to calculate the ship motion. The publications by Riesner [28, 26, 27, 25] are especially relevant for this paper, since the applied three-dimensional panel code is the same as in this publication. The main advantage of the formulation in the inertial coordinate system is that the Coriolis forces are not required and the frequency domain coefficients can be directly used.

1.2. Frequency domain coefficients

The radiation and diffraction forces in the impulse response method are calculated by using coefficients from frequency domain programs. Frequency domain codes are classified into Green and Rankine type sources and strip or three-dimensional discretisations. Strip theory is mostly based on the formulations by Salvesen [31] or Bertram [4], it is assumed that the change in flow quantities along the ship's length is small and negligible. Three dimensional calculations require higher computational effort, however the stationary wave system and the change in flow quantities along the ship length can be calculated. Green-function methods require only the discretisation of the hull, whereas Rankine-source methods require the discretisation of the hull and the free surface [8, 34]. In principle all methodologies can account for forward speed, however strip theory calculations are only corrected by considering the longitudinal change of the main sectional area, which neglects the influence of the stationary wave system. The coefficients for the hydrodynamic added mass, the hydrodynamic damping are solved by calculating the forces on a moving ship in still water, the so-called radiation problem to determine the added mass and damping. The diffraction coefficients are determined by measuring the forces on a fixed ship and subtracting the Froude-Krylov and restoring forces [8]. The coefficients for the Froude-Krylov and restoring forces may also be used in the impulse-response method, however the computational power has increased to a level, where these forces can be calculated in every time step on a panel grid. When transforming the radiation forces from frequency domain to time domain using the formulation by Cummins [5], the resulting force amplitude and phase in the time-domain is not equal to the frequency domain coefficients. Therefore, Fonseca [10] and Ma [19] introduced the so-called radiation restoring matrix using two different formulations. Riesner [27] developed a radiation restoring coefficient for the three-dimensional panel code used in their calculations.

1.3. Wave and restoring force calculations with panel grids

Froude-Krylov and restoring forces are calculated using panel grids and evaluating the Bernoulli pressure on every panel in each time step. The reason for this methodology is that these forces are dominant and non-linear, if the wave amplitudes are high or the body geometry exhibits large local changes. Hirdaris [14] highlighted the force imbalance which arises from the non-linear and linear mixture of force components. In order to prevent an ever increasing drift motion, soft springs are usually applied to hold the ship in a time-averaged position. The panel grid used in the calculations is usually composed of triangular or quadrilateral panels. Differences in the force calculation arise for the stretching scheme, the order of integration and at the waterline, where panel cutting or panel deformation schemes are used. Stretching schemes are introduced due to the fact, that the Bernoulli pressure is only valid up to the still waterline. Above it, the pressure exponent e^{-kz} is positive and pressures increase exponentially. Although, the pressure at the free-surface is clearly defined as air pressure. In order to correct the inaccuracies of the Airy-wave theory, stretching schemes are introduced which force the pressure to be equal to the air pressure at the free surface. A popular and easy applicable stretching scheme is the so-called Wheeler-Stretching [36]. Alternatively, linear extrapolation is sometimes used, however only accurate for wave amplitudes which are significantly much smaller than the ship dimensions [35]. Du [7] compared mapping methods, Wheeler-Stretching in linear and non-linear waves and showed that the mapping methods are closest to the wave pressures by 5th order Stokes theory, but due to the simplicity of Wheeler-Stretching and acceptable accuracy, this method seems to be the most efficient solution. Riesner [28] deformed the mesh to the current waterline and applied Wheeler stretching for every panel to accurately capture the pressure above the still waterline. The pressure is assumed to be constant above the panel area. Rodrigues [29] developed an adaptive mesh refinement with quadrilateral panels and exact integrations of the Bernoulli pressure by using Green's theorem for integrals. Zhang [37] used the same approach, however formulated the forces directly in the ship-fixed coordinate system. Exact integrations of the Bernoulli pressure is favourable at locations where the panel size is big relative to the pressure gradients. Theoretically, the panels can be made infinitely big, if the geometry is accurately represented, as the computational effort decreases only if the number of panels is reduced. Although, the formulations for Froude-Krylov forces are more accurate than constant pressure calculations, the additional effort for hydrostatic forces is not required, since the hydrostatic pressure is linear and the force always equal to the submergence depth of the center of area multiplied with the area. In relation to the integration of Froude-Krylov pressures, this paper follows a new approach and integrates the pressure exactly along the area of triangular panels, without the need of coordinate transformations.

1.4. Radiation forces

Radiation forces in the impulse response method incorporate the memory effect of ship motions. The memory effect describes the time-dependency of forces and ship motions due to the radiating waves. The calculation of the forces is done with the convolution integral or state-space models. Convolution integrals require higher computational effort than state-space methods. However, the identification of state-space models is challenging and various methods for identifying the parameters exist, see for example Kristiansen [17], Perez [22], Hatecke [12], Roessling [30], Peña-Sanchez [23]. In order to reduce the simulation time, a limited order of states is used, theoretically the state-space formulation is equal to the convolution integral when using infinite states. Input to both formulations is either the hydrodynamic damping or added mass, most publications use the hydrodynamic damping, because the boundary values are clearly defined. Perez [20] discussed the limits of the retardation functions, which are integrated from the frequency domain damping. The proposed boundary value extrapolation is used in many publications and efficiently avoids oscillations in the retardation functions. Recent publications did not further incorporate state-space models, since the simulation time depends mainly on the pressure calculations with panel grids. Although recent developments were made by incorporating body non-linearity into the radiation forces. Rajendran [24] calculated radiation forces for several draughts and for each strip individually. In the time-domain, the convolution integral is reformulated for every time step and saved until the retardation function has decayed. This has a considerable influence on the bending moments and the added resistance. Riesner [26] assumed linear pressure distributions above the still waterline to the current waterline and developed a formulation for the panel radiation pressures, without the need for a new convolution formulation. The influence on the added resistance is significant.

1.5. Diffraction forces

Diffraction forces arise due to the reflection of waves at the ship. It is assumed that the response is linear to the incoming wave in the frequency domain codes. In the time-domain these coefficients are multiplied with the wave elevation at the midship section [6]. Some authors also include impulse response functions for diffraction forces [3, 14, 15, 37]. Differences are only visible if the wave elevation at the midship section becomes non-linear. This effect is only present, if the ship motions or rotations are non-linear, since the Airy-wave theory is always linear.

1.6. Paper structure

The contribution of this paper can be summarised as a comparison of forces for several formulations, extracting knowledge about the formulations and limitations and areas of application. The paper may help other researches and software developers to choose the approach suitable for the field of application. Chapter 2 gives a brief overview of the force calculations as well as the motion integrations. Chapter 3 introduces the simulation parameters and the investigated ship geometries. A post-Panamax container ship DTC [9] and the Kriso container ship [16] are investigated. Chapter 4 shows the results for each force component for fixed DTC and the response amplitude operators (RAO) as well as the added resistance for the DTC and KCS at several forward speeds. In the last chapter 5 the findings are summarised and concluded. The appendix lists additional formulae for the exact pressure integration.

2. Methodologies

2.1. Coordinate systems

Ship motions are described by three coordinate systems. A global coordinate system is fixed to the still water plane with the z-axis pointing downwards. The x and y-axis are orientated arbitrarily, usually in the ship longitudinal and transversal axis. The hydrodynamic coordinate system is also fixed to the still water plane, but moves with the average ship forward speed. The x-axis points in the direction of the time-averaged longitudinal ship axis and y to starboard. The origin of the hydrodynamic system is chosen to be at the main frame, midship and at the still water level. Ship motions are described by the ship moving relative to the hydrodynamic coordinate system, therefore a ship-fixed coordinate system is introduced, which is fixed to the main frame, midship and the initial draft. The ship-fixed coordinate system is participating in all rotations.

The coordinates of a coordinate vector in global coordinates $\vec{\xi_g}$ follow from the hydrodynamic coordinates by adding the longitudinal translation from the average ship speed $\vec{U} = [u, 0, 0]^T$ and the constant yaw angle $\bar{\psi}$

$$\vec{\xi_g} = \vec{U} \cdot t + \boldsymbol{R}_{\mathrm{z},\bar{\psi}} \vec{\xi} . \tag{1}$$

The transformation matrix $\boldsymbol{R}_{\mathrm{z},\bar{\psi}}$ describes a rotation around the vertical z-axis.

A ship-fixed point $\vec{\xi}_{\mathrm{P}}$ in hydrodynamic coordinates is calculated with the Euler angle transformations $\boldsymbol{R}^h_{\mathrm{b}}$ of the ship-fixed position vector $\vec{x}_{\mathrm{P}}$ and the vector of the coordinate origin $\vec{\xi}$ expressed in hydrodynamic coordinates

$$\vec{\xi}_{\mathrm{P}} = \vec{\xi} + \boldsymbol{R}^h_{\mathrm{b}} \cdot \vec{x}_{\mathrm{P}} . \tag{2}$$

The transformation matrix is composed from the yaw-pitch-roll transformation matrices

$$\boldsymbol{R}^h_{\mathrm{b}} = \boldsymbol{R}_{\mathrm{z},\psi} \cdot \boldsymbol{R}_{\mathrm{y},\theta} \cdot \boldsymbol{R}_{\mathrm{x},\phi} , \tag{3}$$

$$= \begin{bmatrix} \cos\psi & -\sin\psi & 0 \\ \sin\psi & \cos\psi & 0 \\ 0 & 0 & 1 \end{bmatrix} \cdot \begin{bmatrix} \cos\theta & 0 & \sin\theta \\ 0 & 1 & 0 \\ -\sin\theta & 0 & \cos\theta \end{bmatrix} \cdot \begin{bmatrix} 1 & 0 & 0 \\ 0 & \cos\phi & -\sin\phi \\ 0 & \sin\phi & \cos\phi \end{bmatrix} . \tag{4}$$

The angular velocities $\vec{\omega}$ obtained from the equation of motions are transformed to Euler angle derivatives by

$$\dot{\vec{\Theta}} = \boldsymbol{T}\vec{\omega} = \begin{bmatrix} 1 & \sin\phi\tan\theta & \cos\phi\tan\theta \\ 0 & \cos\phi & sin\phi \\ 0 & \frac{\sin\phi}{\cos\theta} & \frac{\cos\phi}{\cos\theta} \end{bmatrix} \cdot \vec{\omega}\,, \tag{5}$$

the initial value problem is updated every time step.

2.2. Equation of motion

According to Cummins [5] the equation of motion using impulse response functions includes the hydrodynamic mass matrix at infinite frequency $\boldsymbol{A}(\infty)$, the hydrodynamic damping at infinite frequency $\boldsymbol{B}(\infty)$, the retardation function matrix $\boldsymbol{K}(t)$ and external forces

$$(\boldsymbol{M} + \boldsymbol{A}_\infty)\ddot{\vec{\xi}} + \boldsymbol{B}_\infty\dot{\vec{\xi}} + \int_0^t \boldsymbol{K}(t)\dot{\vec{\xi}}(t-\tau)d\tau + C_\mathrm{R}\vec{\xi} = \vec{F}_\mathrm{D} + \vec{F}_\mathrm{FK+R} + \vec{F}_\mathrm{S} + \vec{F}_\mathrm{ext}\,. \tag{6}$$

The formulation in (6) differs from the original formulation. The restoring matrix $\boldsymbol{C}$ is replaced with non linear restoring and exciting wave forces $\vec{F}_\mathrm{FK+R}$. Additionally, the radiation restoring matrix $\boldsymbol{C}_\mathrm{R}$ is added, which is derived by for example Fonseca [10] and arises due to the differences between the time domain and frequency domain equation. The calculation of these forces and moments, as well as the calculation of the diffraction force $\vec{F}_\mathrm{D}$ and the soft mooring force $\vec{F}_\mathrm{S}$ is explained in the following sections. The vector $\vec{F}_\mathrm{ext}$ may include arbitrary external forces, but is neglected in this paper.

If the Cummins equation is formulated in body-fixed coordinates, additional Coriolis forces arise due to the rotation of the body-fixed frame around the inertial axis of the hydrodynamic coordinate system [11]. An advantage of this formulation is the direct integration of arbitrary ship-fixed forces and a constant right-hand side, which leads to more stable accelerations and allows to use bigger time steps

$$(\boldsymbol{M} + \boldsymbol{A}_\infty)\ddot{\vec{x}} + \boldsymbol{B}_\infty\dot{\vec{x}} + \int_0^t \boldsymbol{K}(t)\delta\dot{\vec{x}}(t-\tau)d\tau + C_\mathrm{R}\vec{x} = \vec{F}_\mathrm{D} + \vec{F}_\mathrm{FK+R} + \vec{F}_\mathrm{S} + \vec{F}_\mathrm{ext} - m\cdot\boldsymbol{L}\dot{\vec{x}}\,. \tag{7}$$

Where $\dot{\vec{x}}$ is the velocity vector expressed in ship-fixed coordinates relative to the global coordinate system. $\delta\dot{\vec{x}}$ is the the velocity expressed in ship-fixed coordinate relative to the hydrodynamic coordinate system, which is obtained by subtracting the constant velocity vector

$$\delta\dot{\vec{x}} = \dot{\vec{x}} - \boldsymbol{R}^T\vec{U}\,. \tag{8}$$

The Coriolis forces in vectorial form are described by the matrix $\boldsymbol{L}$, which simplifies $m\vec{\omega} \times \dot{\vec{x}}$

$$\boldsymbol{L} = \begin{pmatrix} 0 & 0 & 0 & 0 & 0 & 0 \\ 0 & 0 & 0 & 0 & 0 & 1 \\ 0 & 0 & 0 & 0 & -1 & 0 \\ 0 & 0 & 0 & 0 & 0 & 0 \\ 0 & 0 & 0 & 0 & 0 & 0 \\ 0 & 0 & 0 & 0 & 0 & 0 \end{pmatrix}\,. \tag{9}$$

Frequency domain coefficients are determined with the strip theory *PDStrip* and the three-dimensional panel code *panFDS*, which requires the stationary wave system from *GLRankine*. The theory of the programs is extensively described in [8].

2.3. Diffraction forces

Forces arising due to reflection of waves at structures are described as diffraction forces. Their calculation in the impulse response method is based on the coefficients obtained from the frequency domain calculations. Two methods are present in the literature. In one method the frequency domain coefficients are transformed into the time domain by multiplying with the wave elevation at the coordinate origin

$$\vec{F}_{\mathrm{D,linear}}(t) = \Re(\vec{\hat{D}}_{\mu_e,\mathrm{T}}(\omega_e)\hat{\zeta}e^{i(\omega t - kx\cos\mu + ky\sin\mu + \varphi)}) \tag{10}$$

Where $\vec{\hat{D}}_{\mathrm{r},\mu_e,\mathrm{T}}$ is the complex diffraction force coefficient for the current encounter angle μ_e and the current draught T. The wave elevation with the amplitude $\hat{\zeta}$ and the phase φ is evaluated at the position $x(t), y(t)$ of the coordinate origin.
The second method is introduced by Bailey [2] and calculates a diffraction impulse response function $\vec{K}_{\mathrm{D}}$. Hirdaris [14] arguments that the wave elevation at the midship can become non-linear

$$\vec{K}_{\mathrm{D}}(t) = \frac{1}{\pi}\int_0^\infty \vec{D}_{\mathrm{r}}(\omega_e)\cos(\omega_e\tau) - \vec{D}_{\mathrm{i}}(\omega_e)\sin(\omega_e\tau)\ \mathrm{d}\omega_e\,. \tag{11}$$

Where the diffraction force coefficient is split into its real part $\vec{D}_{\mathrm{r}}$ and imaginary part $\vec{D}_{\mathrm{i}}$. The diffraction force is calculated with a convolution integral

$$\vec{F}_{\mathrm{D,conv}} = \int_0^t \vec{K}_{\mathrm{D},\mu_e,\mathrm{T}}(\tau)\zeta(t-\tau)\ \mathrm{d}\tau\,. \tag{12}$$

$\zeta(t)$ is the real part of the wave elevation in (10). The memory consumption of this formulation is significantly higher than the formulation in (10), since all wave elevations of the last n time steps have to be saved, until the diffraction retardation function is fully decayed. The decay time is for the investigated ships around 15-30 seconds. If the body-non-linearity is used, the last n draughts have to be saved additionally to select and interpolate the retardation functions to each draught.

The implementation of the diffraction retardation function is automated with the use of splines. The retardation functions are integrated and afterwards fitted with a C-Spline. The usage of splines allows to evaluate the retardation functions at every time instant. Different draughts than the initial draughts of the frequency domain calculations are interpolated. This requires sufficiently close arranged draughts.

2.4. Froude-Krylov and restoring forces

Froude-Krylov and restoring forces arise from the wave pressures and the submergence of the ship body. The pressure is described by the Bernoulli equation

$$p_{\mathrm{FK+R}} = \rho g z - \rho\dot{\phi} - \rho g\zeta_{\mathrm{st}}\,. \tag{13}$$

With the submergence depth z and the free surface height due to the stationary wave system ζ_{st}. The stationary wave is interpreted as an additional contribution to the hydrostatic pressure. The assumption applied is that the stationary wave system is not changing relative to the ship and the velocities are small. Froude-Krylov forces arise due to the time derivative of the wave potential $\dot{\phi}$ which is described by

$$\dot{\phi}(x,y,z,t) = \hat{\zeta}ge^{-k(z-\zeta)}\cos(\omega t - kx\cos\mu + ky\sin\mu + \varphi)\,. \tag{14}$$

Where the exponent in the classical formulation e^{-kz} [8] is modified to $e^{-k(z-\zeta)}$. This technique is called Wheeler stretching [36], because the pressure for coordinates with $z < 0$ is exponentially growing. The resulting pressure at the free surface is only valid for small wave amplitudes and the stretching forces it to be equal to the atmospheric pressure. The technique is widely used and has been investigated for example by Du [7].

In natural seaways, the pressure is formulated for every wave component individually and summed afterwards. However, the wave elevation in the pressure always refers to the sum of all wave components for every component.

The resulting force and moment vector on the ship can be calculated from the pressure by multiplying it with the normal vector $\vec{n}$ and integrating it over the ship surface

$$\vec{F} = \int_S p_{\mathrm{FK+R}} \vec{n} \, \mathrm{d}S \,, \tag{15}$$

$$\vec{M} = \int_S p_{\mathrm{FK+R}} (\vec{x}_{\mathrm{P}} - \vec{x}_{\mathrm{G}}) \times \vec{n} \, \mathrm{d}S \,. \tag{16}$$

The moment vector is additionally multiplied with the distance of a point $\vec{x}_{\mathrm{P}}$ to the position of the center of gravity x_{G}.

In order to evaluate the pressure on the ship, the ship is meshed with n_{P} triangular panels. Each panel is seen as individual object and the forces of each panel are summed to the total force

$$\vec{F} = \sum_{\mathrm{i}=1}^{n_{\mathrm{P}}} \int_S p_{\mathrm{FK+R,i}} \vec{n}_i \, \mathrm{d}S_{\mathrm{P}} \,, \tag{17}$$

$$\vec{M} = \sum_{\mathrm{i}=1}^{n_{\mathrm{P}}} \int_S p_{\mathrm{FK+R,i}} (x_{\mathrm{P,i}} - x_{\mathrm{G}}) \times \vec{n}_i \, \mathrm{d}S_{\mathrm{P}} \,. \tag{18}$$

The pressure on a panel can be integrated analytically or can be taken as constant and multiplied with the panel area A

$$\vec{F} = \sum_{\mathrm{i}=1}^{n_{\mathrm{P}}} p_{\mathrm{FK+R,i}} \vec{n}_i A_i \,. \tag{19}$$

The point of pressure evaluation $\vec{x}_{\mathrm{P}}$ is taken as the center of the submerged panel area

$$\vec{x}_{\mathrm{P}} = \sum_{\mathrm{i}=1}^{3} \frac{\vec{x}_i}{3} \,. \tag{20}$$

This assumption leads to exact hydrostatic forces, since the hydrostatic pressure is linear and the area above and below the center of area is always equal to the integral of the pressure.

The direct integration by Rodrigues [29] or Zhang [37] is done by using Green's theorem. The following formulae are derived by direct integration of the wave pressures over the panel area. The integration allows to use significantly larger panels for the evaluation of the Froude-Krylov forces. As long as the panel size is not reducing the shape accuracy, the panels can be theoretically infinitely large.

The total force on a panel is the sum of the integrated force for two triangles

$$F_{\mathrm{fk}} = F_{\mathrm{fk,1}} + F_{\mathrm{fk,2}} \,. \tag{21}$$

An example panel is shown in Figure 1. For the integration of the area, the each panel is split into two separate triangular areas. If the panel intersects the free surface, it is additionally split

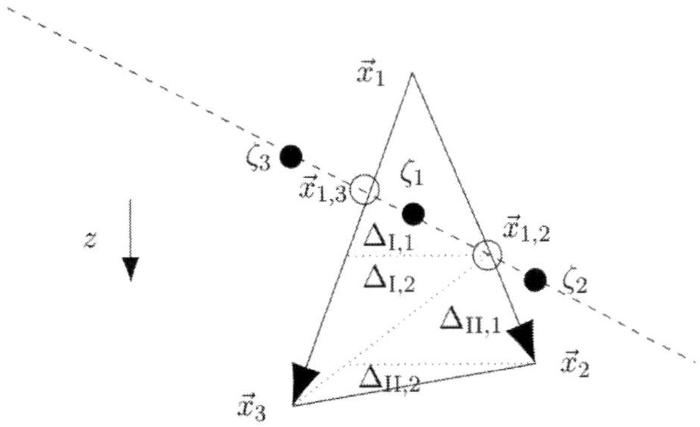

Figure 1: Schematic view of a panel intersecting the free surface

into additional triangles I and II, indicated by the dotted line from $\vec{x}_{1,2}$ to $\vec{x}_3$. The separate panels corner points are sorted by their z-coordinate amplitude. The integration is done for the upper and lower panels separately, the horizontal line from $\vec{x}_{1,2}$ and $\vec{x}_2$ show the triangles.

The derivation of the formulations is shown in the appendix. For non-horizontal panels, inserting the equations (48) into the force at a specific height (52) leads to the forces

$$F_{\mathrm{fk},1} = a_1 \left[\frac{e^{-kz_1} \cdot (k \sin\beta \sin(c_2 L_1 + b_1) + c_2 \cos(c_2 L_1 + b_1)) - e^{-kz_0} (k \sin\beta \sin(b_1) + c_2 \cos(b_1))}{k^2 \sin^2\beta + c_2^2} \right. \tag{22}$$

$$\left. + \frac{e^{-kz_1} \cdot (-k \sin\beta \sin(b_2 L_1 + b_1) - b_2 \cos(b_2 L_1 + b_1)) - e^{-kz_0} (-k \sin\beta \sin(b_1) - b_2 \cos(b_1))}{k^2 \sin^2\beta + b_2^2} \right], \tag{23}$$

$$F_{\mathrm{fk},2} = -a_2 \left[\frac{e^{-kz_1} \cdot (k \sin\beta \sin(c_4 L_2 + b_3) + c_4 \cos(c_4 L_2 + b_3)) - e^{-kz_2} (k \sin\beta \sin(b_3) + c_4 \cos(b_3))}{k^2 \sin^2\beta + c_4^2} \right. \tag{24}$$

$$\left. + \frac{e^{-kz_1} \cdot (-k \sin\beta \sin(b_4 L_2 + b_3) - b_4 \cos(b_4 L_2 + b_3)) - e^{-kz_2} (-k \sin\beta \sin(b_3) - b_4 \cos(b_3))}{k^2 \sin^2\beta + b_4^2} \right], \tag{25}$$

$F_{\mathrm{fk},1}$ is the force for the upper panel and $F_{\mathrm{fk},2}$ for the lower part. The abbreviations are all shown in the appendix.

If a panel is purely horizontal, meaning that $|n_z| = 1$, the sorting of coordinates and the integration is done along the y-direction leading to

$$F_{\mathrm{fk},1} = \begin{cases} a_3(f_{\mathrm{fk},11} + f_{\mathrm{fk},12}) & y_1 - y_0 \neq 0 \\ 0 & y_1 - y_0 = 0 \end{cases} \tag{26}$$

$$F_{\mathrm{fk},2} = \begin{cases} a_3(f_{\mathrm{fk},21} + f_{\mathrm{fk},22}) & y_2 - y_1 \neq 0 \\ 0 & y_2 - y_1 = 0 \end{cases} \tag{27}$$

$$\tag{28}$$

Theoretically, the wave elevation ζ, which appears in the Wheeler stretching, is also a function of the panel length and should be integrated too. However, the resulting formulae include a term $e^{\cos(x)} \cdot \cos(x)$, which is not integratable without series approximations and seems not to be a justifiable effort. Therefore, the amplitude in the Wheeler stretching is assumed as constant over the panel length and is evaluated at the panel center of area.

The center of pressure can be integrated by multiplying the pressure with the position vector $\vec{x}$ and integrating the resulting force per area. This work is ongoing and not yet ready to be published. The moments in this paper are calculated relative to the center of the area.

Emerged and partly submerged are identified by their position relative to the free surface at the corner points. Emerged panels are characterised by the fact that all corner point z-coordinates are lower than the wave elevations at their positions. In Figure 1 a partly submerged panel is shown. The free surface is indicated by the dotted line and the intersections of the panel with the free surface as $\vec{x}_{1,3}$ and $\vec{x}_{1,2}$. The intersection is calculated by

$$\vec{x}_{1,\mathrm{i}} = \vec{x}_1 + \frac{z_1 - \zeta_1}{(\zeta_\mathrm{i} - \zeta_1) - (z_\mathrm{i} - z_1)} \cdot (\vec{x}_\mathrm{i} - \vec{x}_1)\,. \tag{29}$$

In order to integrate the resulting quadrilateral, the area is split into two triangular panels $\vec{x}_{1,3}, \vec{x}_{1,2}, \vec{x}_3$ and $\vec{x}_3, \vec{x}_{1,2}, \vec{x}_2$. If only one corner point is submerged, the triangle is formed by the submerged point and the two intersection points.

2.5. Radiation forces

Radiation forces originate from the ship moving in the waves. They are separated into the acceleration dependent hydrodynamic mass $\boldsymbol{A}$, the radiation restoring matrix $\boldsymbol{C}_\mathrm{R}$ and the velocity dependent convolution integral

$$\vec{F}_\mathrm{rad} = -\boldsymbol{C}_\mathrm{R}\vec{x} - \boldsymbol{A}(\infty)\ddot{\vec{x}} - \boldsymbol{B}(\infty)\dot{\vec{x}} - \int_0^t \boldsymbol{K}(t-\tau)\delta\dot{\vec{x}}(\tau)\,\mathrm{d}\tau \tag{30}$$

$$\boldsymbol{K}(t) = \frac{2}{\pi}\int_0^\infty (\boldsymbol{B}(\omega) - \boldsymbol{B}(\infty))\cos(\omega t)\,\mathrm{d}\omega \tag{31}$$

Perez [20] established formulas for the asymptotic behaviour of the diagonal entries of the hydrodynamic damping $\boldsymbol{B}$. The resulting retardation function $\boldsymbol{K}(t)$ is sensitive to high frequent oscillations, if the damping at high frequencies is not zero. In case of the formulation in (31), the difference at low frequencies should not be lower than zero, leading to an excitation of the motion. The proposed boundary values for the retardation function can also be integrated analytically as done by Riesner [28]. In this publication, the boundary values have been approximated with the formulation by Perez [20], the integration afterwards is done numerically. Since the boundary values at infinity are zero, the radiation forces are integrated without subtracting the damping.

Fonseca [10], Riesner [27], Ma [19] introduced various formulations for radiation restoring coefficients. The resulting forces are required, since the formulation of the Cummins equation is not equal to the frequency domain equation of motion. The formulation by Fonseca [10] and Ma [19] refer to strip theory coefficients. The formulation by Fonseca is used in this paper due to its simplicity of implementation

$$\boldsymbol{C}_{\mathrm{R,2d}} = \omega_\mathrm{e}^2\,(\boldsymbol{A}(\infty) - \boldsymbol{A}(\omega_\mathrm{e})) - \omega_\mathrm{e}\int_0^\infty \boldsymbol{K}(\tau)\sin(\omega_\mathrm{e}\tau)\,\mathrm{d}\tau\,. \tag{32}$$

For the three-dimensional panel code, Riesner [28] subdivided the panel pressure into frequency dependent and frequency independent parts, which leads to the restoring coefficient matrix, when these pressures are integrated over all panels for each radiation problem

$$\boldsymbol{C}_{\mathrm{R,3d}} = \int_A -\left(\vec{U} + \nabla\phi^0\right)(\vec{\alpha} \times \nabla\phi^0)\,. \tag{33}$$

In order to introduce non-linearities into the radiation forces, the convolution integral and the hydrodynamic mass is calculated for several draughts and interpolated in every time step during the simulation

$$\vec{F}_\mathrm{rad} = -\boldsymbol{C}_\mathrm{R}(z)\vec{x} - \boldsymbol{A}(\infty, z)\ddot{\vec{x}} - \int_0^t \boldsymbol{K}(t-\tau, z)\dot{\vec{x}}(\tau)d\tau \tag{34}$$

$$\approx -\boldsymbol{C}_\mathrm{R}(z_n)\vec{x} - \boldsymbol{A}(\infty, z_n)\ddot{\vec{x}}(t_\mathrm{n}) - \frac{\mathrm{dt}}{2}\cdot\left(\boldsymbol{K}(t_n, z_n)\dot{\vec{x}}(t_0) + 2\sum_{\mathrm{i}=1}^{n-1}\boldsymbol{K}(t_{\mathrm{n-i}}, z_{\mathrm{n-i}})\dot{\vec{x}}(t_i) + \boldsymbol{K}(t_0, z_0)\dot{\vec{x}}(t_n)\right). \tag{35}$$

The computational effort for the evaluation of the integral is increased because the retardation functions for a specific draught z_i have to be saved until the function is decayed.

2.6. Soft mooring force

Imbalances due to the different level of non-linearities in the force components and the missing propulsion force can lead to constantly increasing surge, sway and yaw motions [14]. These motions are non-physical and can be prevented by a spring force and moment. The spring forces are calculated based on the position of the ship in the hydrodynamic coordinate frame and are assumed to act in the hydrodynamic coordinate axis. The force is composed of a linear spring force with constant c_s and a damping force linear to the velocities with the damping constant d_s

$$\vec{F}_{\mathrm{sm}} = -\begin{pmatrix} \xi \\ \eta \\ 0 \\ 0 \\ 0 \\ \psi L_{\mathrm{PP}} \end{pmatrix} c_s - \begin{pmatrix} \dot{\xi} \\ \dot{\eta} \\ 0 \\ 0 \\ 0 \\ \dot{\psi} L_{\mathrm{PP}} \end{pmatrix} d_s . \quad (36)$$

The constants are chosen based on the heave eigenfrequency. One sixth of the eigenfrequency is used to calculate the spring constant and the damping constant to ensure that the soft mooring is not influencing the heave, roll or pitch motion

$$c_s = \frac{\Delta \omega_3^2}{36}, \qquad d_s = \frac{\Delta \omega_3}{6} . \quad (37)$$

3. Ship geometries and simulation parameters

The panel grid on the hulls of the Duisburg Test Case (DTC) and the Kriso Container Ship (KCS) are shown in Figure 2a and Figure 2b. The main particulars are listed in Table 1. The investigated speeds orientate at the existing model test data and RANS simulations, which are $Fn = 0.139$ for the DTC and $Fn = 0.26, 0.33, 0.4$ for the KCS.

The DTC is a post-Panamax container ship with a full-scale length of 355 m with a capacity of 14000 TEU [9]. The KCS is smaller with a length of 230 m [33]. Both ships are extensively validated and investigated and validated in the literature.

Table 1: Main particulars of DTC and KCS in model size

	DTC	KCS
λ	59.4	38.59
L_{PP}[m]	5.976	5,96
B_{wl} [m]	0.859	0.83
T_m [m]	0.244	0.279
∇ [m^3]	0.827	0.976
COG_x [m]	2.907	3.06
COG_y [m]	0.0	0,0
COG_z [m]	0.4	0.381
r_{xx} [m]	0.4B	0.4B
r_{yy} [m]	0.25L	0.25L
r_{zz} [m]	0.25L	0.25L

Table 2: Wave length and height combinations for simulations with fixed DTC

λ/L_{PP}	$H[m]$
2.5	0.2
1.8	0.2
1.4	0.2
1.2	0.138
1.09	0.15
1.0	0.158
0.91	0.128
0.8	0.134
0.6	0.1

In the following section, a comparison of the force amplitudes for the DTC in regular waves is presented. These amplitudes are compared with results obtained by Riesner [25] and use the wave length and wave heights shown in Table 2.

(a) DTC (b) KCS

Figure 2: Panel grids used in the time and frequency domain simulation

4. Results

The following section contains results for the fixed DTC container ship, as well as results for the freely moving DTC and KCS. The comparison for the fixed ship is verified with CFD results presented by Riesner [25]. The response amplitude operators for heave and pitch are compared with experimental results by Sigmund [32] (DTC) and Simonsen [33] (KCS). The response amplitude operator for heave is normalised with the wave amplitude ζ

$$Y_3 = \frac{z}{\zeta}\,, \tag{38}$$

and the pitch motion is made non-dimensional with the wave steepness

$$Y_5 = \frac{\theta}{k\zeta}\,. \tag{39}$$

The added resistance in waves is normalised with the ship length and breadth

$$C_{\mathrm{AW}} = \frac{F_{\mathrm{x,avg}} L_{\mathrm{PP}}}{\rho g B^2 \zeta^2}\,. \tag{40}$$

The evaluation of the added resistance is challenging, since the amplitude is small compared to the harmonic amplitudes. The force average of all external forces is subtracted by the average of the longitudinal hydrostatic pressure. This is required, since the stern of the ships is not wetted and therefore not covered with panels. The resulting hydrostatic force in combination with the pressures due to the stationary resistance are not part of the added resistance. It is important, that the equilibrium state of the ship is equal to the desired floating condition, otherwise the added resistance in short waves may become inaccurate.

4.1. Forces on fixed DTC container ship

In Figure 3, the energy of the force calculated from its Fourier transform is compared for different number of panels. The amplitudes are normalised with the amplitude obtained at the highest number of panels, which is assumed to be most accurate. The simulation is done with a wavelength of $\lambda/L_{\mathrm{PP}} = 2.5$ and a steepness of $H/\lambda = 0.01$. The hydrostatic and Froude-Krylov forces are separately calculated and compared for the exact pressure integrations and the constant pressure evaluations.

The differences in between force amplitudes from coarse to fine panel discretisations is greater for the hydrostatic forces. Hydrostatic forces are equal for every discretisation, if the geometry is represented accurately. Therefore, the differences arise due to the inaccurate representation of the surface of the ship with low panel numbers. The difference between exact and constant hydrostatic pressures arises due to the applied splitting scheme. The results for exact panels are subdivided at the waterline, whereas the constant panel calculations approximate the hydrostatic pressure at the waterline by calculating the new center of the submerged area. It is assumed, that the surface is linear between its intersections. This simplification leads to difficulties at the transom stern, which emerges during the simulation and the panels intersecting the free surface may experience highly non-linear intersecting curves.

The Froude-Krylov forces show smaller dependency on the discretisation, the exact pressure integrations are less sensible to the number of panels. The additional effort of integrating the pressure may be beneficial if the grid is refined manually after the generation with *GLRankine*. At the current state, the form accuracy limits the benefit of exact integrations. A typical number of panels for frequency domain calculations is around 2000 panels per ship. This level of discretisation seems to be accurate enough for the time domain simulation and is used in every following simulation.

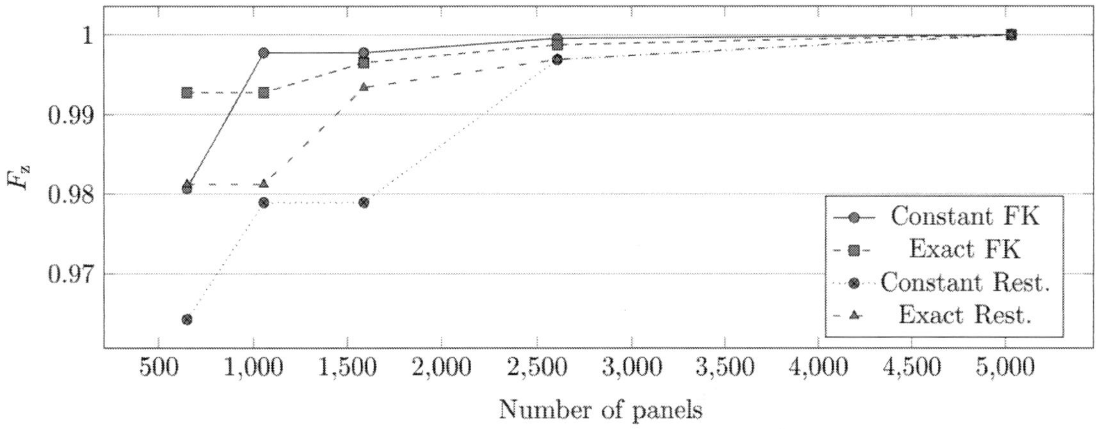

Figure 3: Comparative grid study for hydrostatic and Froude-Krylov forces with exact pressure integrations and constant pressure evaluations in a regular wave of length $\lambda/L_{PP} = 2.5$ and a steepness of $H/\lambda = 0.01$

Figure 4 shows the absolute value of the diffraction force coefficient for heave motion. The frequency coefficients from the strip theory *PDStrip* and the panel code *panFDS* are compared with the force amplitudes from a time domain simulation (TD coefficient/TD convolution) with the coefficient and convolution approach.

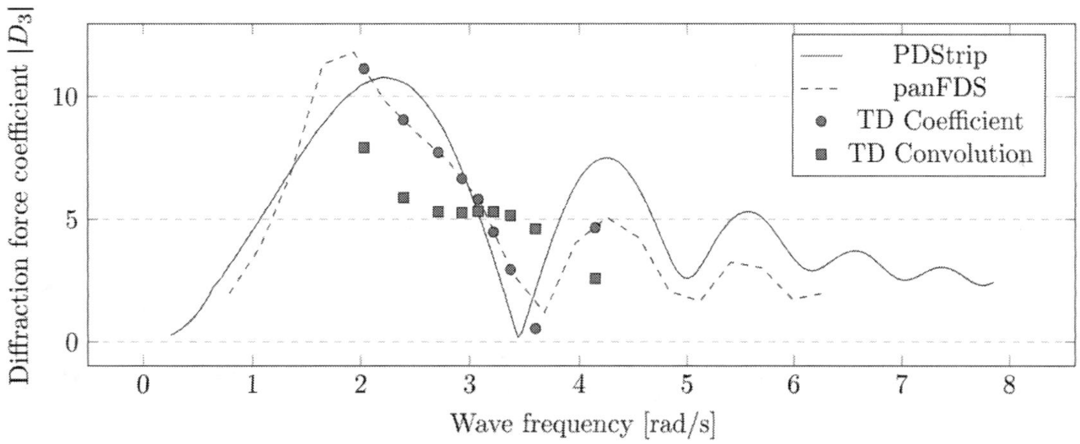

Figure 4: Normalised diffraction force $|D_3| = |F_3/\zeta|$ amplitudes in time and frequency domain

It is visible, that the transformation into time domain and retransformation into frequency domain does not lead to differences between the *panFDS* results and the *TD Coefficient* results. The *PDStrip* coefficients show slightly increased higher order amplitudes, but a lower first order amplitude. The convolution diffraction function significantly affects the force amplitudes. The peaks from the linear calculations are no longer present and the amplitudes are slowly decaying with increasing wave frequency.

In Figure 5 the radiation force amplitudes are compared for a calculation with constant retardation functions and a constant hydrodynamic mass matrix as well as the interpolation approach presented in subsection 2.5. The radiation restoring coefficient is yet neglected.

The comparison of force amplitudes shows that the first order peaks are not influenced by the calculation method, however the second order peak is significantly larger for the variable calculation. The FFT of the force signal calculated with variable radiation forces shows even higher orders of forces,

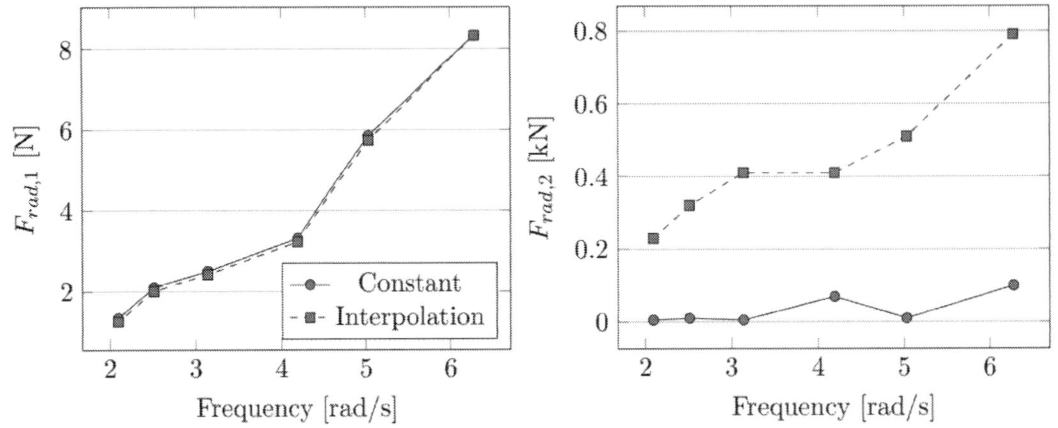

Figure 5: Comparison of vertical radiation force amplitudes for 1st and 2nd order motion frequency due to forced heave motion of the DTC with amplitude of 0.12 m

however their amplitude is negligible small. The amplitudes of the constant calculation show only one peak at the motion frequency and decay afterwards. The higher order force is explainable with the variation of the hydrodynamic mass and the retardation. If it is assumed that the hydrodynamic mass varies linearly with the draught, the force is described by the product of two harmonic functions

$$\boldsymbol{A}(z,t) = (\boldsymbol{A}(\infty) + \boldsymbol{\delta A}(\infty) sin(\omega t))\zeta \sin(\omega t)\,, \tag{41}$$

where the added mass variation $\boldsymbol{\delta A}(\infty)(t)$ oscillates with twice the frequency as the motion frequency ω.

The following figures Figure 6-Figure 8 show comparisons of force calculations with CFD results by Riesner [25]. The DTC container ship is fixed in head waves with wave heights and wavelength according to Table 2. The left figures show the first order force and moment amplitudes and the right figures the second order force and moment amplitudes. Results for the third and fourth order amplitude are also available, but not presented due to negligible amplitudes.

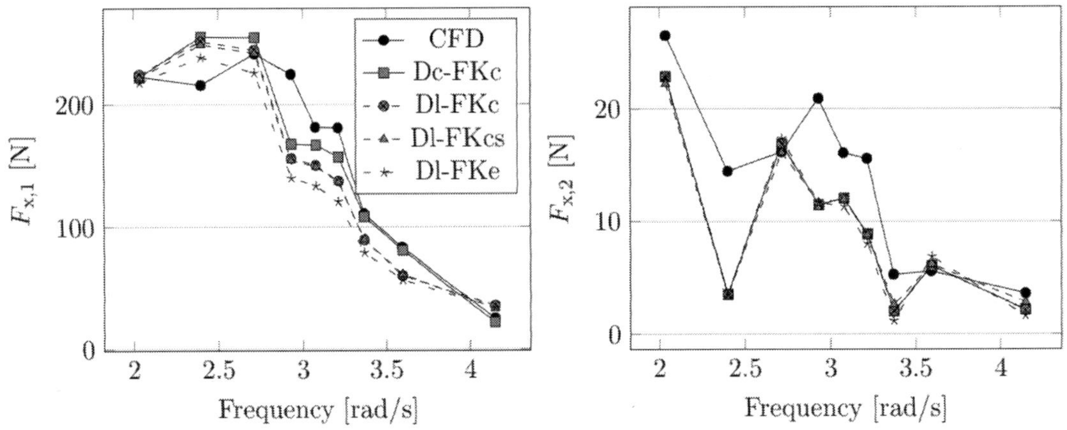

Figure 6: DTC force amplitudes in longitudinal direction of first and second order in head waves with $Fn = 0.139$ for various diffraction (D) and Froude-Krylov (FK) calculations. The diffraction parameters are convolution (Dc), linear (Dl) and the Froude-Krylov parameters stand for constant (FKc), exact (FKe) and constant with stationary surface (FKcs).

Figure 6 shows the longitudinal forces on the DTC ship in head waves at a forward speed corresponding to a Froude number of $Fn = 0.139$. Over all simulation settings the forces are in range of the CFD calculations. The difference for first order forces is large in the range around 3 $\frac{\text{rad}}{\text{s}}$ and is almost equal at the lowest and highest simulated wave frequency. The value of the second order force is around 10% of the first order amplitudes and the impulse response simulations do not differ from each other, however the differences to the CFD results are higher in percentage. The diffraction calculation with a convolution integral (Dc-FKc) leads to higher first order longitudinal forces. The consideration of stationary pressures in the Froude-Krylov calculation (Dl-FKcs) does not influence the calculation of longitudinal forces and results in approximately the same force amplitude as the linear diffraction and Froude-Krylov calculation (Dl-FKc). The exact integration of the panel pressures (Dl-FKe) leads to lower forces across the whole frequency range. In long waves this reduction leads to a better agreement with the CFD results, whereas in shorter waves, the forces are under predicted.

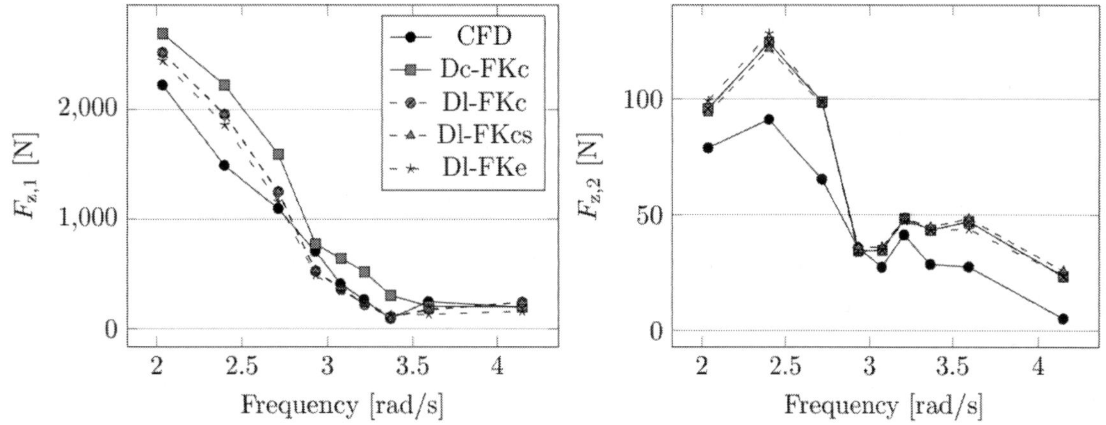

Figure 7: DTC force amplitudes in vertical direction of first and second order in head waves with $Fn = 0.139$ for various diffraction (D) and Froude-Krylov (FK) calculations.

Figure 7 shows the vertical force amplitudes at first and second order of the wave frequency. In comparison to Figure 6 the amplitudes are around 10 times larger and decrease from 2 kN in waves equal to three times the ship length to almost negligible amplitudes in waves shorter than half of the ship length. The second order amplitudes are around 5% of the first order amplitudes. The agreement is better in short waves, whereas in long waves the impulse response method calculations show higher forces than the CFD results. The individual force calculations show a similar trend as for the longitudinal forces, where the diffraction convolution (Dc-FKc) produces higher force amplitudes and the exact pressure integrations (Dl-FKe) lead to smaller force amplitudes than the calculations Dl-FKc and Dl-FKcs, which are closest to the CFD results.

In Figure 8 the pitch moment amplitudes are compared. The calculated values differ in long waves, which is most likely due to the non-linear diffraction forces included in the CFD computations. The reflection of waves is expected to increase at the bow and stern if they are submerged in the wave, which is not covered by the linear frequency domain coefficients. The exact integration leads to significantly higher second order moments, which may originate from the approximation of the point of action at the center of the panels. The determination of the exact point of action is in progress and assumed to reduce the second order force amplitudes.

4.2. Motions and added resistance of DTC in regular waves

The following section shows comparisons of response amplitude operators (RAO) for heave and pitch and the added resistance in head waves at Froude number $Fn = 0.139$. The results are compared with experimental results by Sigmund [32]. The heave and pitch RAO are shown in Figure 9.

The heave motion at the eigenfrequency, indicated by the local peak at 5 $\frac{\text{rad}}{\text{s}}$, is underestimated for the impulse response calculations, except the calculation with convolution diffraction. The diffraction is generally overpredicting the motion amplitudes in heave and pitch. The exact integration of pressures

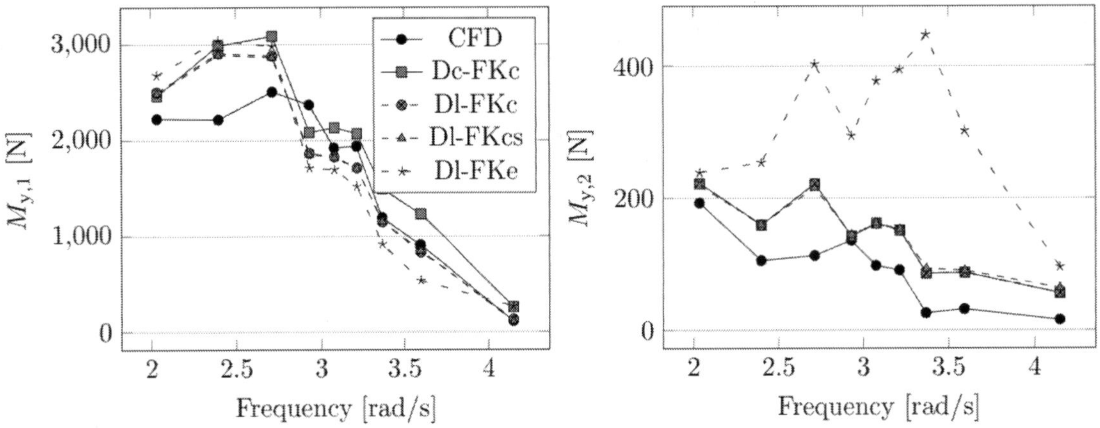

Figure 8: DTC pitch moment amplitudes of first and second order in head waves with $Fn = 0.139$ for various diffraction (D) and Froude-Krylov (FK) calculations.

leads to lower motion amplitudes. The under prediction of the heave and pitch motion may be explained by simplifications related to potential theory, since the frequency domain transfer functions are not representing the peak either.

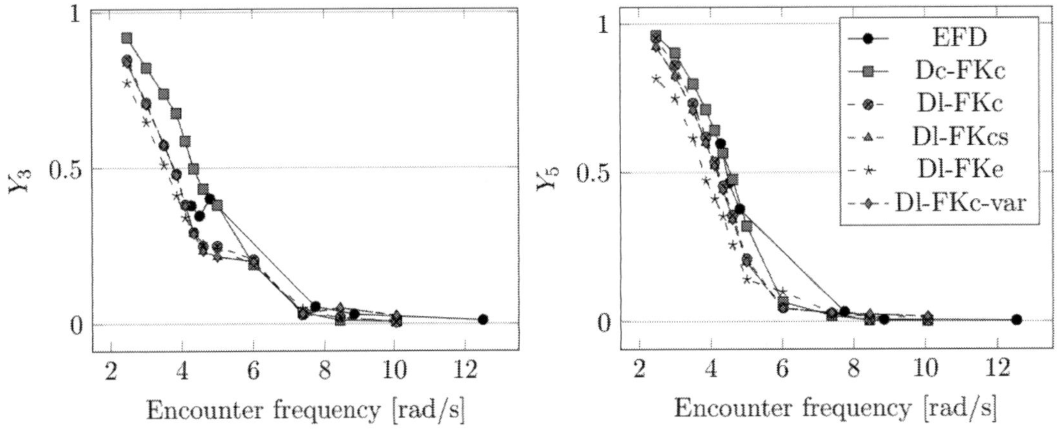

Figure 9: DTC heave (left) and pitch (right) transfer function for $Fn = 0.139$ in head waves

Figure 10 shows the added resistance coefficient C_{AW} plotted over the encounter frequency. The experimental results are taken from Sigmund [32]. The coefficient increases up to the region of the heave and pitch eigenfrequency at approximately 5 $\frac{\mathrm{rad}}{\mathrm{s}}$ and decreases to a constant value at high frequencies, which represents the reflection of incoming waves at the bow. The linear diffraction calculation methods show neglectable differences across the frequency range and a slightly underpredicted peak value, whereas the diffraction convolution method overestimates the peak of the added resistance. The exact integration leads to a slightly negative coefficient at the highest wave frequency. The values at the higher frequencies are highly sensitive to the equilibrium hydrostatic forces. Since the transom is not covered with panels, the longitudinal force is always unequal to zero. If the center of gravity is not exactly above the center of buoyancy, the hydrostatic force will differ at the end of the simulation and cause an error in the calculation of the added resistance.

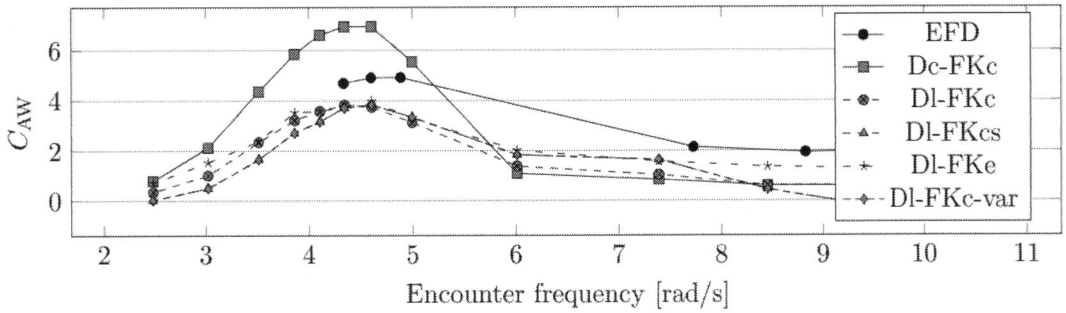

Figure 10: DTC added resistance for $Fn = 0.139$ in head waves

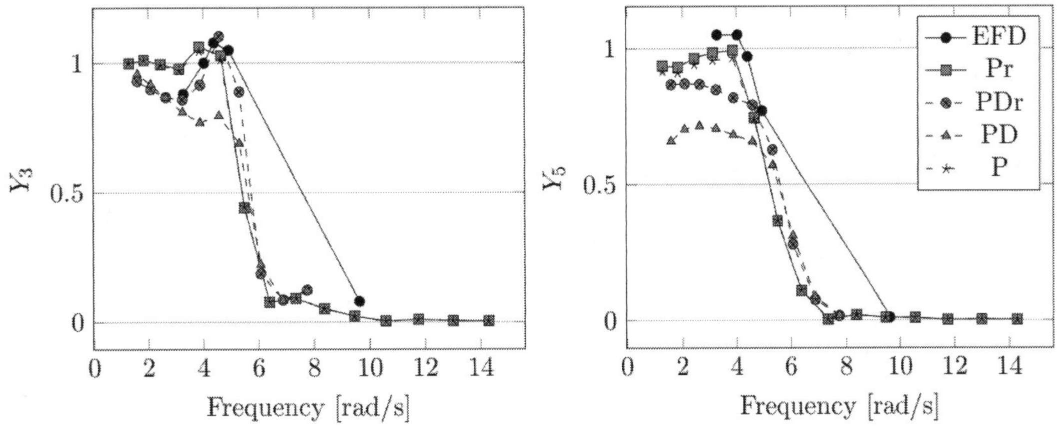

Figure 11: KCS heave (left) and pitch (right) transfer function for $Fn = 0.26$ in head waves

4.3. Motions and added resistance of KCS in regular waves

The simulations for the DTC include a comparison between the force calculation methods. This section contains a comparison for the KCS and the usage of the radiation restoring coefficients as well as strip theory and the three-dimensional panel code *panFDS*. The results are compared against the experimental results by Simonsen [33]. The diffraction force calculation is conducted using the frequency domain coefficients and the panel pressures are calculated with constant pressures along the panel.

Figure 11-Figure 13 show the heave and pitch transfer functions in head waves with a Froude-Number of $Fn = 0.26$, $Fn = 0.33$, $Fn = 0.4$. Different parameter sets are indicated by

- Pr = *panFDS* coefficients and restoring matrix
- P = *panFDS* coefficients
- PDr = *PDStrip* coefficients and restoring matrix
- PD = *PDStrip* coefficients
- EFD = Experimental results by Simonsen [33]

Figure 11 shows the heave and pitch transfer function for $Fn = 0.26$. The heave motion is accurately captured by the simulations, except the strip theory calculations without restoring coefficient. Significant differences are visible for the strip theory coefficient calculations in long waves. Both motions are underestimated, the application of the restoring coefficient leads to an increase in amplitude. As shown by Ma [19], this is a problem when using strip theory coefficients in impulse response formulations with the radiation restoring coefficient by Fonseca [10]. The radiation restoring coefficient by Riesner [25]

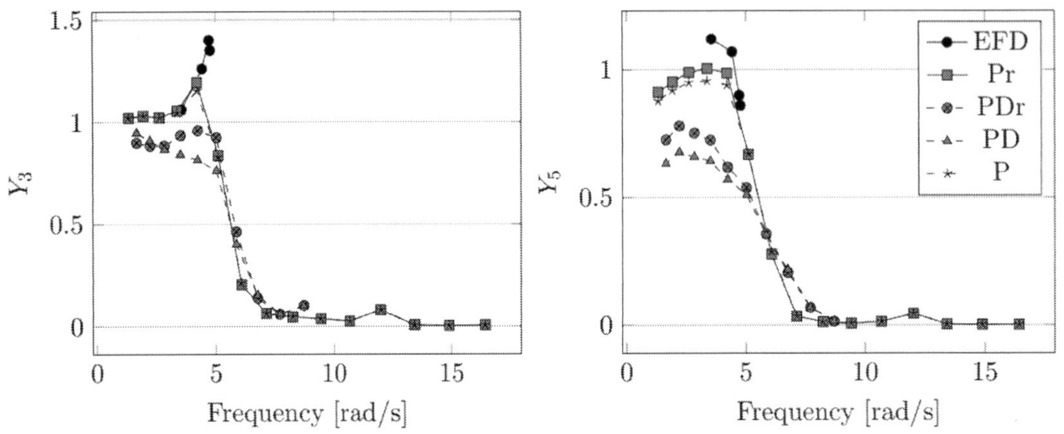

Figure 12: KCS heave (left) and pitch (right) transfer function for $Fn = 0.33$ in head waves

leads to an improvement of the pitch motion amplitudes and should be applied in this speed range, when *panFDS* coefficients are used.

For $Fn = 0.33$ the heave and pitch transfer functions are plotted in Figure 12. The heave and pitch motion are underestimated by the impulse response calculations for the resonance area at approximately $5\,\frac{\text{rad}}{\text{s}}$. The resulting motions are not as close to the experimental results, as the results for the DTC and KCS at lower Froude numbers. The influence of the coupling between the heave and pitch motions is increasing with higher forward speed. The non-diagonal entries of the damping matrix take non-zero values at infinite frequency and the consideration of the boundary values suggested by Perez [20] is no longer straightforward. It is possible to adapt the integration of the retardation functions by subtracting the highest calculated value of the damping, however this often leads to negative damping at low-frequencies which violates the passive system properties [12] and leads to an excitation of the ship motions. The chosen solution procedure presented in subsection 2.5 leads to the most reasonable results for all investigated ship types and velocities.

The results for the highest Froude number $Fn = 0.4$, corresponding to an exceptionally high velocity of 37 kn at full scale, are shown in Figure 13. The simulations results for the heave motion, except for PDr, are close to the experimental results. The pitch motion, however is strictly overpredicted by PD and underpredicted by PDr. Simulations with *panFDS* lead to low pitch motion amplitudes in long waves, but an acceptable agreement in the resonance range. In comparison with results by Riesner [28], the underprediction of the pitch motion is shifted to a higher forward speed and the application of non-inertial motion integration and the restoring coefficient seems to be beneficial for moderate to high forward speeds.

In Figure 14-16 the added resistance of the KCS is plotted over the encounter frequency. The experimental results by Simonsen [33] are denoted by EFD. The accuracy of the measurements peak for $Fn = 0.33$ is questionable and is not further investigated. The calculations and experimental values show a peak of the added resistance coefficient at 4.5-5 $\frac{\text{rad}}{\text{s}}$. The *PDStrip* calculations lead to a higher added resistance than *panFDS* and the experimental results for $Fn = 0.26$ and $Fn = 0.33$. The accuracy at $Fn = 0.4$ is limited due to the overestimated pitch motion, which has significant effect on the added resistance. *panFDS* coefficient calculations differ slightly from experimental results, the calculation with a restoring coefficient is marginally closer to the experiments.

5. Conclusions

In the present paper, different approaches to calculate the hydrodynamic forces by using the impulse response method are presented and compared. The calculations have been conducted with hydrodynamic coefficients obtained from the strip theory code *PDStrip* and the three dimensional rankine source code *panFDS*. The motion behaviours of the Duisburg test case container ship and the Kriso container ship were simulated and the computational results are compared with experimental and computational results.

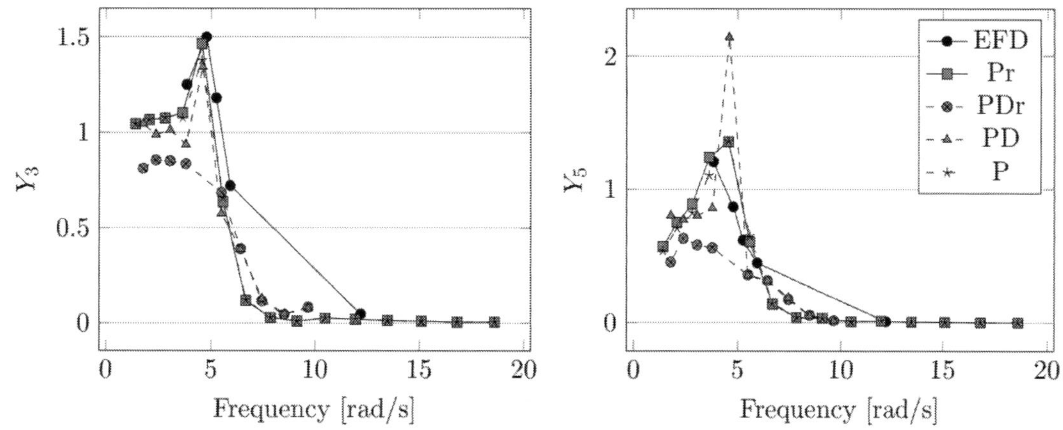

Figure 13: KCS heave (left) and pitch (right) transfer function for $Fn = 0.4$ in head waves

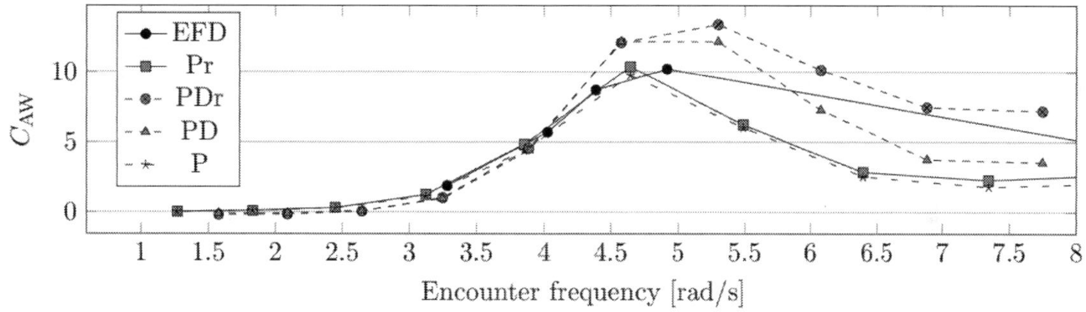

Figure 14: KCS added wave resistance for $Fn = 0.26$ in head waves

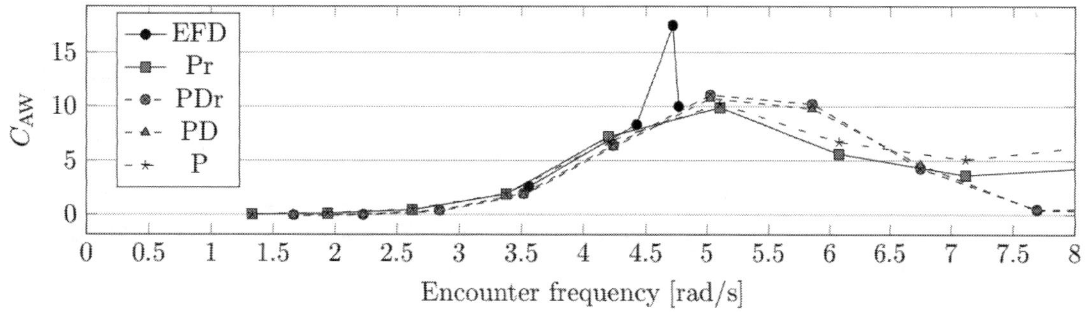

Figure 15: KCS added wave resistance for $Fn = 0.33$ in head waves

Results showed that the diffraction force calculated using a linear coefficient is closer to force amplitudes obtained by RANS simulations than the calculation based on a convolution integral. An exact integration procedure for Froude-Krylov pressures is developed, however numerical accuracy and the simplified determination of the center of pressure limits the accuracy. The application of exact pressure integration should be advantageous in case that the panel dimensions can be adapted to the surface curvature. The current grid generation procedure limits the accuracy, since it is refined uniformly along the ship's body and reduces the shape precision if less panels are used. Another finding is that, the consideration of stationary wave system is beneficial for the derivation of the frequency domain coefficients. This is demonstrated by the fact, that all transfer functions and added resistance curves

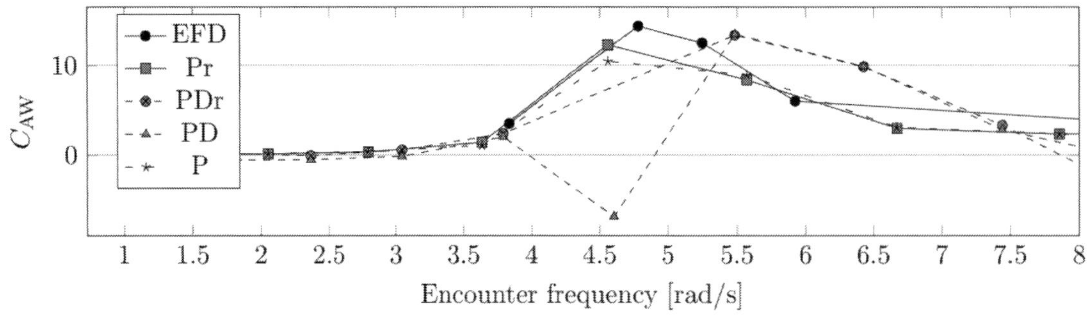

Figure 16: KCS added wave resistance for $Fn = 0.4$ in head waves

with *panFDS* coefficients are closer to experimental results. However, the integration of stationary flow panel pressures in the panel calculations has an negligible effect on heave and pitch forces and motions. The accuracy of the transfer functions with *PDStrip* coefficients is better at lower forward speeds, the applied radiation restoring coefficients increase the accuracy of the calculation for both coefficient sets. The pitch motion at high forward speeds is too small, a limit is identified around $Fn = 0.3$ where the dynamic effects of the flow have a stronger influence on the motion. The added resistance in waves is characterised by the mean value of the additional force in longitudinal direction. It is calculated with high confidence with *panFDS* coefficients, most important finding during the investigation was that the longitudinal center of gravity should be adapted to ensure the desired floating condition.

Future work will focus on implementing the point of pressure integration into the exact Froude-Krylov calculation. Furthermore, an elastic beam theory will be implemented to further investigate whether non linear force calculations are beneficial and if the level of non-linearity of the presented impulse response method can be further increased.

References

[1] K. R. Babu, S. V. K. R. Nelli, A. Bhattacharyya, and R. Datta. Experimental and numerical investigation of green water occurrence for KRISO container ship. *Journal of Ship Research*, 66(01):54–72, dec 2020.

[2] P. Bailey, D. Hudson, W. Price, and P. Temarel. Time simulation of manoeuvring and seakeeping assessments using a unified mathematical model. *Royal Institution of Naval Architects Transactions 2002 Part B*, 144:57–78, 01 2002.

[3] E. J. Ballard, D. A. Hudson, W. G. Price, and P. Temarel. Time domain simulation of symmetric ship motions in waves. *The International Journal of Maritime Engineering*, 145(a2):1, 2003.

[4] V. Bertram, B. Veelo, H. Söding, and K. Graf. Development of a freely available strip method for seakeeping. *Proc. 5th International Conference on Computer and IT Applications in the Maritime Industries*, 01 2006.

[5] W. Cummins. The impulse response function and ship motions. Technical report, David Taylor Model Basin Washington DC, 1962.

[6] O. Detlefsen, L. Theilen, and M. Abdel-Maksoud. Time domain simulation of dynamic positioning manoeuvres based on impulse response functions. In *6th International Conference on Computational Methods in Marine Engineering, MARINE 2015*, page 761, 2015.

[7] S. X. Du, D. A. Hudson, W. G. Price, and P. Temarel. Implicit expressions of static and incident wave pressures over the instantaneous wetted surface of ships. *Proceedings of the Institution of Mechanical Engineers, Part M: Journal of Engineering for the Maritime Environment*, 223(3):239–256, jun 2009.

[8] B. O. el Moctar, T. E. Schellin, and H. Söding. *Numerical Methods for Seakeeping Problems.* Springer International Publishing, 2021.

[9] O. el Moctar, V. Shigunov, and T. Zorn. Duisburg test case: Post-panamax container ship for benchmarking. *Ship Technology Research*, 59(3):50–64, aug 2012.

[10] N. Fonseca and C. G. Soares. Time-domain analysis of large-amplitude vertical ship motions and wave loads. *Journal of Ship Research*, 42(02):139–153, jun 1998.

[11] T. I. Fossen. A nonlinear unified state-space model for ship maneuvering and control in a seaway. *International Journal of Bifurcation and Chaos*, 15(09):2717–2746, sep 2005.

[12] H. Hatecke. The impulse response fitting and ship motions. *Ship Technology Research*, 62(2):97–106, jun 2015.

[13] S. Hirdaris, W. Bai, D. Dessi, A. Ergin, X. Gue, O. Hermundstad, R. Huijsmans, K. Iijima, U. Nielsen, J. Parunov, N. Fonseca, A. Papanikolaou, K. Argyriadis, and A. Incecik. Loads for use in the design of ships and offshore structures. *Ocean Engineering*, 78:131–174, 03 2014.

[14] S. Hirdaris, Y. Lee, G. Mortola, A. Incecik, O. Turan, S. Hong, B. Kim, K. Kim, S. Bennett, S. Miao, and P. Temarel. The influence of nonlinearities on the symmetric hydrodynamic response of a 10,000 TEU container ship. *Ocean Engineering*, 111:166–178, jan 2016.

[15] J. Jiao, Z. Chen, C. Chen, and H. Ren. Time-domain hydroelastic analysis of nonlinear motions and loads on a large bow-flare ship advancing in high irregular seas. *Journal of Marine Science and Technology*, 25(2):426–454, may 2019.

[16] W. Kim, S. Van, and D. Kim. Measurement of flows around modern commercial ship models. *Experiments in fluids*, 31(5):567–578, 2001.

[17] E. Kristiansen, Å. Hjulstad, and O. Egeland. State-space representation of radiation forces in time-domain vessel models. *Ocean Engineering*, 32(17-18):2195–2216, dec 2005.

[18] S. Liu and A. Papanikolaou. Prediction of parametric rolling of ships in single frequency regular and triple frequency group waves. *Ocean Engineering*, 120:274–280, jul 2016.

[19] S. Ma, R. Wang, J. Zhang, W. Y. Duan, R. C. Ertekin, and X. B. Chen. Consistent formulation of ship motions in time-domain simulations by use of the results of the strip theory. *Ship Technology Research*, 63(3):146–158, sep 2016.

[20] T. Perez and T. Fossen. A derivation of high-frequency asymptotic values of 3d added mass and damping based on properties of the cummins equation. *J. Marit. Res*, 5:65–78, 2008.

[21] T. Perez and T. I. Fossen. Kinematic models for manoeuvring and seakeeping of marine vessels. *Modeling, Identification and Control: A Norwegian Research Bulletin*, 28(1):19–30, 2007.

[22] T. Perez and T. I. Fossen. Time- vs. frequency-domain identification of parametric radiation force models for marine structures at zero speed. *Modeling, Identification and Control: A Norwegian Research Bulletin*, 29(1):1–19, 2008.

[23] Y. Peña-Sanchez, N. Faedo, and J. V. Ringwood. A Critical Comparison Between Parametric Approximation Methods for Radiation Forces in Wave Energy Systems. volume All Days of *International Ocean and Polar Engineering Conference*, 06 2019. ISOPE-I-19-521.

[24] S. Rajendran, N. Fonseca, and C. G. Soares. Simplified body nonlinear time domain calculation of vertical ship motions and wave loads in large amplitude waves. *Ocean Engineering*, 107:157–177, oct 2015.

[25] M. Riesner. *Numerical Method to Compute the Wave-Induced Rigid Body and Elastic Response of Ships at Forward Speed.* PhD thesis, University Duisburg-Essen, 2022.

[26] M. Riesner and O. el Moctar. A time domain boundary element method for wave added resistance of ships taking into account viscous effects. *Ocean Engineering*, 162:290–303, aug 2018.

[27] M. Riesner and O. el Moctar. A numerical method to compute global resonant vibrations of ships at forward speed in oblique waves. *Applied Ocean Research*, 108:102520, mar 2021.

[28] M. Riesner, A. von Graefe, V. Shigunov, and O. el Moctar. Prediction of non-linear ship responses in waves considering forward speed effects. *Ship Technology Research*, 63(3):135–145, aug 2016.

[29] J. Rodrigues and C. G. Soares. Froude-krylov forces from exact pressure integrations on adaptive panel meshes in a time domain partially nonlinear model for ship motions. *Ocean Engineering*, 139:169–183, jul 2017.

[30] A. Roessling and J. Ringwood. Finite order approximations to radiation forces for wave energy applications. *Renewable energies offshore*, 359:359–366, 2015.

[31] N. Salvesen, E. Tuck, and O. Faltinsen. Ship motions and sea loads. 1970.

[32] S. Sigmund and O. el Moctar. Numerical and experimental investigation of added resistance of different ship types in short and long waves. *Ocean Engineering*, 147:51–67, jan 2018.

[33] C. D. Simonsen, J. F. Otzen, S. Joncquez, and F. Stern. EFD and CFD for KCS heaving and pitching in regular head waves. *Journal of Marine Science and Technology*, 18(4):435–459, apr 2013.

[34] H. Söding, A. von Graefe, O. el Moctar, and V. Shigunov. Rankine source method for seakeeping predictions. In *Volume 4: Offshore Geotechnics Ronald W. Yeung Honoring Symposium on Offshore and Ship Hydrodynamics*. American Society of Mechanical Engineers, jul 2012.

[35] J. T. Tuitman and Š. Malenica. Fully coupled seakeeping, slamming, and whipping calculations. *Proceedings of the Institution of Mechanical Engineers, Part M: Journal of Engineering for the Maritime Environment*, 223(3):439–456, jun 2009.

[36] J. Wheeler. Method for calculating forces produced by irregular waves. *Journal of Petroleum Technology*, 22(03):359–367, mar 1970.

[37] P. Zhang, T. Zhang, and X. Wang. Hydrodynamic analysis and motions of ship with forward speed via a three-dimensional time-domain panel method. *Journal of Marine Science and Engineering*, 9(1):87, jan 2021.

IOP Conf. Series: Materials Science and Engineering 1288 (2023) 012024 doi:10.1088/1757-899X/1288/1/012024

Appendix A

The derivation of the integration is based on two angles α and β. α describes the angle between the panel longitudinal axis and the transversal axis if seen from above the water surface. It describes, how the y-coordinates at a height z depend on the x-coordinates

$$y(x,z) = y_s(z) + (x(z) - x_s(z)) \cdot \tan(\alpha)\,, \tag{42}$$

$$\alpha = \begin{cases} \arctan(n_x/n_y) & n_y \neq 0 \\ \frac{\pi}{2} & n_y = 0 \end{cases}. \tag{43}$$

The coordinates y_s and x_s describe an outer edge of the panel and the indices x, y of n the entries of the normal vector in the respective directions.
β describes the orientation of the angle between the vertical coordinate axis z and the panel

$$\beta = \arccos(n_z)\,. \tag{44}$$

The triangle is described by three corner points $\vec{x}_0$-$\vec{x}_2$, which are sorted by their z-coordinate. The functions for the x-coordinates of the outer edges are

$$x_{s,1}(z) = x_0 + \frac{z - z_0}{z_1 - z_0} \cdot (x_1 - x_0)\,, \tag{45}$$

$$x_{e,1}(z) = x_0 + \frac{z - z_0}{z_2 - z_0} \cdot (x_2 - x_0)\,, \tag{46}$$

$$x_{s,2}(z) = x_2 + \frac{z - z_2}{z_0 - z_2} \cdot (x_0 - x_2)\,, \tag{47}$$

$$x_{e,2}(z) = x_2 + \frac{z - z_2}{z_1 - z_2} \cdot (x_1 - x_2)\,. \tag{48}$$

At a constant value of z, the Froude-Krylov pressure can be integrated along the breadth $B(z)$ of the panel

$$B(z) = \frac{x_e - x_s}{\cos\alpha}\,. \tag{49}$$

With the force per height defined as the integral of the instationary Bernoulli pressure

$$f(z) = -\rho\hat{\zeta} g e^{-k(z-\zeta)} \int_0^B \cos(\omega t - kx\cos\mu + ky\sin\mu + \varphi)\,\mathrm{d}B\,. \tag{50}$$

If one inserts the relationships from (43) and (49), the resulting force per length is equal to

$$f(z) = \frac{\rho\hat{\zeta} g e^{-k(z-\zeta)}}{k\cos(\alpha+\mu)} \Big[\sin(\omega t - k\frac{x_e - x_s}{\cos\alpha}\cos(\alpha+\mu) - kx_s\cos\mu + ky_s\sin\mu + \varphi) \tag{51}$$

$$- \sin(\omega t - kx_s\cos\mu + ky_s\sin\mu + \varphi) \Big] \tag{52}$$

The integration along the z-direction is done by describing the height L of the two separate areas with the angle β

$$L_1 = \frac{z_1 - z_0}{\sin\beta}\,, \tag{53}$$

$$L_2 = \frac{z_1 - z_2}{\sin\beta}\,. \tag{54}$$

$$F_{fk,1} = a_1 \int_0^{L_{z,1}} e^{-k(L_z\sin\beta + z_0)} (\sin(b_1 + b_2 L_z) - \sin(c_1 + c_2 L_z)) \tag{55}$$

The resulting formulae are shown in subsection 2.4.
The following abbreviations are used:

$$a_1 = \frac{\rho g\hat{\zeta}e^{k\zeta}e^{-kz_0}}{k\cos(\alpha+\mu)} \qquad a_2 = \frac{\rho g\hat{\zeta}e^{k\zeta}e^{-kz_2}}{k\cos(\alpha+\mu)} \qquad a_3 = \frac{\rho g\zeta e^{-k(z_0-\zeta)}}{k\cos\mu}$$

$$b_1 = c_1 = \omega t + \varphi - k(x_0\cos\mu - y_0\sin\mu)$$

$$b_2 = k\sin\beta\Big(-\frac{\cos(\alpha+\mu)}{\cos\alpha}\left(\frac{x_2-x_0}{z_2-z_0}-\frac{x_1-x_0}{z_1-z_0}\right)-\cos\mu\frac{x_1-x_0}{z_1-z_0}+\sin\mu\frac{y_1-y_0}{z_1-z_0}\Big)$$

$$b_3 = c_3 = \omega t + \varphi - k(x_2\cos\mu - y_2\sin\mu)$$

$$b_4 = k\sin\beta\Big(-\frac{\cos(\mu+\alpha)}{\cos\alpha}\left(\frac{x_2-x_0}{z_2-z_0}-\frac{x_2-x_1}{z_2-z_1}\right)-\cos\mu\frac{x_2-x_1}{z_2-z_1}+\sin\mu\frac{y_2-y_1}{z_2-z_1}\Big)$$

$$b_5 = \begin{cases}\omega t+\varphi-kx_0\cos\mu+k\frac{x_2-x_0}{y_2-y_0}y_0\cos\mu & y_2-y_0\neq 0\\ 0 & y_2-y_0=0\end{cases}$$

$$b_6 = \omega t+\varphi-kx_0\cos\mu+k\frac{x_1-x_0}{y_1-y_0}y_0\cos\mu$$

$$b_7 = \begin{cases}-k\frac{x_2-x_0}{y_2-y_0}\cos\mu-k\sin\mu & y_2-y_0\neq 0\\ 0 & y_2-y_0=0\end{cases}$$

$$b_8 = \omega t+\varphi-kx_1\cos\mu+k\frac{x_2-x_1}{y_2-y_1}y_1\cos\mu$$

$$c_2 = k\sin\beta\Big(-\frac{x_1-x_0}{z_1-z_0}\cos\mu+\frac{y_1-y_0}{z_1-z_0}\sin\mu\Big)$$

$$c_4 = k\sin\beta(-\cos\mu\frac{x_2-x_1}{z_2-z_1}+\sin\mu\frac{y_2-y_1}{z_2-z_1})$$

$$c_5 = \begin{cases}-k\frac{x_2-x_0}{y_2-y_0}\cos\mu-k\sin\mu & y_2-y_0\neq 0\\ 0 & y_2-y_0=0\end{cases}$$

$$c_6 = -k\frac{x_1-x_0}{y_1-y_0}\cos\mu-k\sin\mu$$

$$c_7 = \begin{cases}-k\frac{x_2-x_0}{y_2-y_0}\cos\mu-k\sin\mu & y_2-y_0\neq 0\\ 0 & y_2-y_0=0\end{cases}$$

$$c_8 = -k\frac{x_2-x_1}{y_2-y_1}\cos\mu-k\sin\mu$$

$$f_{fk,11} = \begin{cases}\frac{\cos(b_6+c_6y_1)-\cos(b_6+c_6y_0)}{c_6} & c_6\neq 0\\ -\sin(b_6)\cdot(y_1-y_0) & c_6=0\end{cases} \qquad f_{fk,12} = \begin{cases}-\frac{\cos(b_5+c_5y_1)-\cos(b_5+c_5y_0)}{c_5} & c_5\neq 0\\ \sin(b_5)\cdot(y_1-y_0) & c_5=0\end{cases}$$

$$f_{fk,21} = \begin{cases}\frac{\cos(b_8+c_8y_2)-\cos(b_8+c_8y_1)}{c_8} & c_8\neq 0\\ -\sin(b_8)(y_2-y_1) & c_8=0\end{cases} \qquad f_{fk,22} = \begin{cases}-\frac{\cos(b_7+c_7y_2)-\cos(b_7+c_7y_1)}{c_7} & c_7\neq 0\\ \sin(b_7)(y_2-y_1) & c_7=0\end{cases}$$

An example of numerical ice tank based on DEM simulation and physical model testing

Yukui Tian[1,2], Dongbao Yang[2], Xuhao Gang[1], Chaoge Yu[1], Shungying Ji[2]

1. China Ship Scientific Research Center, Wuxi, 214082, China
2. State Key Laboratory of Structural Analysis, Optimization and CAE Software for Industrial Equipment, Dalian University of Technology, Dalian, 116024, China

E-mail address: tianyukui@cssrc.com.cn (Y. Tian); jisy@dlut.edu.cn (S. Ji)

Abstract. The determination of ice loads on polar ships and offshore structures is of great significance for ice-resistant design, safe operation, and structural integrity management in ice-infested waters. The physical model testing carried out in ice tank/basin is usually an important technical approach to evaluate the ice loads, however, the high cost and time consumption make it difficult to perform multiple repetitions or large number of trials for this purpose. Recently, the rapid development of high-performance computation techniques provides a usable alternative where the numerical methods represented by the discrete element method (DEM) have made remarkable contributions to the ice load predictions. On basis of DEM simulation validated by physical model testing, numerical ice tank can be developed as an effective supplement to its counterpart. In this paper, such an example of numerical ice tank adopting GPU computational mechanism and DEM modelling algorithm was established with respect to the small ice model basin of China Ship Scientific Research Center (CSSRC-SIMB). The numerical ice tank was calibrated and further optimized with physical model tests on typical structures of vertical cylinder and inclined flat plate in level ice sheets by making agreements of both globe value and time history of the ice loads. Then it was practiced for modelling the tests of Wass bow advancing in level ice performed in SIMB separately. It is demonstrated by the comparisons of ice failure details and ice loads that the numerical ice tank can precisely simulate the ice-structure interactions and determine the ice loads under the same initial conditions of physical model testing. In the end, the advantages as well as the challenges of the numerical ice tank are discussed.

1. Introduction

The ice load of polar ships and offshore structures in cold regions is influenced by sea environment (temperature, salinity, wind field, ocean current et al.), ice type (floe ice, broken ice, level ice, ridge, ice rubble et al.) and structural motion characteristics (rigid and elastic bodies, fixed and floating) and other factors[1-2]. The most reliable approach to determine the ice load on marine engineering structures is the full-scale measurement. For example, ice load measurement projects for various marine structures (jacket platforms in the Bohai Sea, lighthouses in the Baltic Sea, Confederation Bridge in Canada, caisson platforms, polar icebreakers, etc.) were carried out during the last century[3-5]. However, expensive installation costs and time make it difficult to implement. Accordingly, the implement of physical model testing in an ice tank/basin is an attractive alternative to determine the ice loads on marine structures. The design of the physical model testing in an ice tank/basin usually adopts the Froude and Cauchy scaling laws to ensure the geometry, kinematics and dynamics of the test object (structure) in the field ice environment, which are far more complicated than those in open water. The document of General Guidance and Introduction to Ice Model Testing in the International Towing tank Conference (ITTC)[6-7] presents the information on the nine main overview of grain structure and chemical composition of the model ice in operating facilities. Due to the different of production method, model ice can be divided into fine grained ice and columnar ice. The fine grained ice is represented by Aalto University (Finland) and Aker Arctic (Finland), while columnar ice is represented by the Hamburg Ship Model Basin (Germany) and National Research Council Canada, Ocean, Coastal and Engineering (Canada). Generally, the physical model testing has been widely used to investigate global ice action and structural ice loads. For example, study the ice-structure interaction mode (ice failure mode, ice ride up, ice accumulation, ice blockage, etc.), and provide icebreaker ship design scheme (maneuverability, stability, rapidity, design load, etc.).

Due to the high cost and time of each test run, it is not common to performed multiple repetitions of physical model testing under the same conditions. However, when analyzing test results from ice tank tests, it is very important that the test results be reproducible or repeatable. The repeatability of test results is closely related to the random uncertainty of the test, the higher the test repeatability, the lower the random uncertainty. Although the Procedure for assessing the experimental uncertainty in ship resistance testing in ice document, provided by the ITTC, has described how to independently evaluate the random uncertainty of the results by splitting up the time histories of measured results from a single test, the differences of test conditions in each segment will still affect the assessment of the random uncertainty of the final results. It will be more convenient for us in the interpretation of results from ice-tank tests if we can carry out repeated ice water pool test experiments.

This study proposes an example of numerical ice tank based on the DEM, which is not only limited to the numerical repeatability of physical ice tank tests, but also can be extended to simulation scenarios and calculation parameters on some physical environment. DEM simulations proposed by Cundall and Strack (1979)[8] have been widely used in ice engineering research, such as demonstrate ice failure behaviors, structural ice loads, ice rubble pile up, and sea ice drift. Among them, the element morphology of DEM has also developed from 2D disk to complex 3D shapes such as sphere, polyhedral, and extended polyhedral[9-11]. Compared to polyhedron elements, 3D sphere elements can easily establish high-performance numerical algorithms for numerical ice tank based on the graphics processing units (GPU) technology to achieve rapid simulation. In addition, considering the coupled effect of structural dynamic response or the wave/current and sea ice damage and movement, the coupled method of 3D spherical DEM with FEM or Computational Fluid Dynamics (CFD) has also been developed accordingly[12-13].

2. Numerical ice tank description based on DEM

2.1 Numerical ice tank corresponding to CSSRC-SIMB

The main body of the small ice model basin of China Ship Scientific Research Center (CSSRC-SIMB)[14-15] has a dimension of 8 m in length,2 m in width and 1 m in depth. The ice tank is housed in an insulated room that can be cooled down to an air temperature of -16 ℃, as shown in Figure 1(a). In addition, a small cold room is close to the ice tank, with a size of 3 m (length) × 2.2 m (width) × 2.8 m (height), which is mainly used to test the physical and mechanical properties of model ice. The main carriage on the tank is designed for tractive force or thrust up to 3 kN, and the speed ranging from 0.01m/s to 1m/s. The model ice used is naturally grown columnar ice constructed by sodium chloride solution. Other in-situ test of offshore structure is designed by Froude and Cauchy scaling laws to carried to physical model testing. According to the dimensions and ice modeling methodology, the physical and mechanical parameters of ice in both full and model scales are listed in Table 1. According to the Froude-Cauchy similarity, the scale ratio in both full and model scales can be controlled between 10 and 20 in the CSSRC-SIMB.

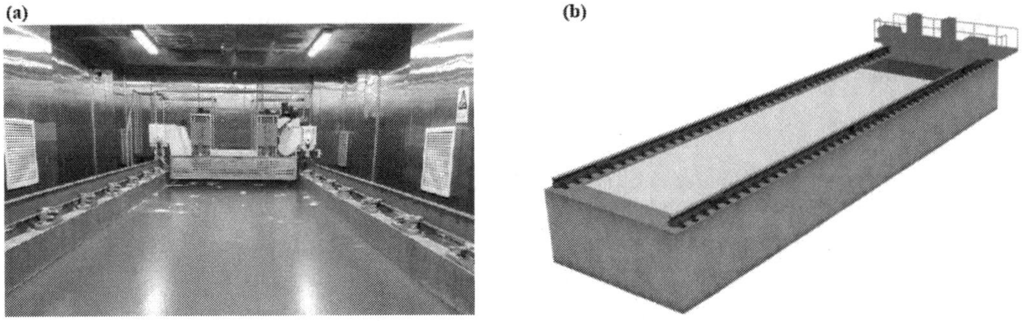

Figure 1. Physical and numerical ice tank based on CSSRC-SIMB(a) physical ice tank (b) numerical ice tank

Table 1. Physical and mechanical parameters of ice in both full and model scales

Parameter	One-year columnar sea ice	Model ice
Ice thickness (m)	0.5~2.0	0.03~0.05
Ice density (g/cm^3)	0.91	0.9~0.92
Young's Modulus (GPa)	2~5	0.06~0.2
Compression strength (MPa)	0.5~12	0.05~0.2
Flexure strength (MPa)	0.5~2	0.03~0.1
Tensile strength (MPa)	0.2~0.8	0.01~0.03

Using the physical ice tank of the CSSRC-SIMB as a template, this paper describes a numerical ice tank based on the DEM to virtual offline mode tests of various offshore structure, as shown in Figure 1(b). The geometric model and dimensions of the numerical ice tank are the same as those of the physical ice tank, and the physical and mechanical properties of the model ice and offshore structure in numerical and model tests are also the same, that is, both use the Froude-Cauchy similarity in the scheme design. The model ice simulation in numerical ice tank employs an 3D spherical DEM code of the Dalian University of Technology [16-18]. The DEM code is developed with a GPU-based parallel algorithm to achieve high efficiency simulation analysis.

2.2 Ice model based on 3D spherical DEM in Numerical ice tank

The model ice in numerical ice tank is simplified to a particle system of 3D spherical elements with same size and mass, which are regularly arranged in space according Hexagonal Close Packing (HCP) arrangement, as shown in Figure 2. In essence, the mechanical behavior of the numerical model ice is described by the motion of element and the forces and moments acting between elements. The motion is performed explicit calculation by Newton's second law, and the contact force between two elements is calculated with an elastic-viscous contact model based on the spring and dashpot. In addition, considering the macroscopic continuous characteristics of the model ice, all 3D spherical elements are bonded together using the parallel bonding model. In the parallel bonding model, the parallel-bond disk with circular cross-section is set to transfer through the forces and moments between two adjacent elements. The maximum normal and shear stresses acting on the section of parallel-bond disk are determined with beam theory, which can be written as follows [8,13]:

$$\sigma_{\max} = \frac{-F_n}{A} + \frac{|M_s|}{I}R, \tau_{\max} = \frac{|F_s|}{A} + \frac{|M_n|}{J}R \qquad (1)$$

where, F_n and F_s are the normal and shear forces, respectively; M_n and M_s are the normal and shear moment, respectively; R is the radius of spherical element; $A = \pi R^2, I = \frac{1}{4}\pi R^4$ and $J = \frac{1}{2}\pi R^4$ are the area, moment of inertia, and polar moment of inertia of the parallel-bond disk, respectively.

(a)

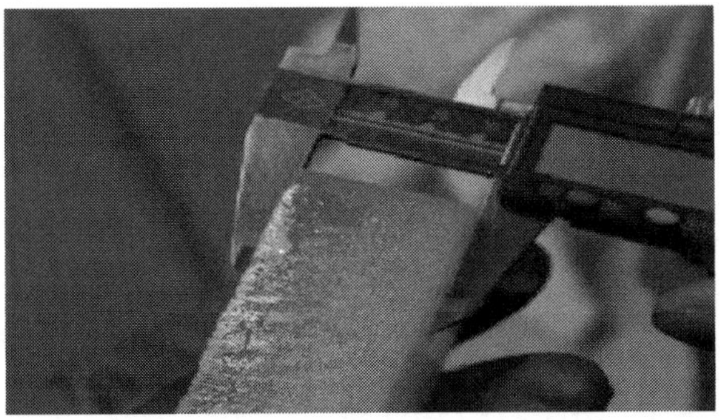

(b)

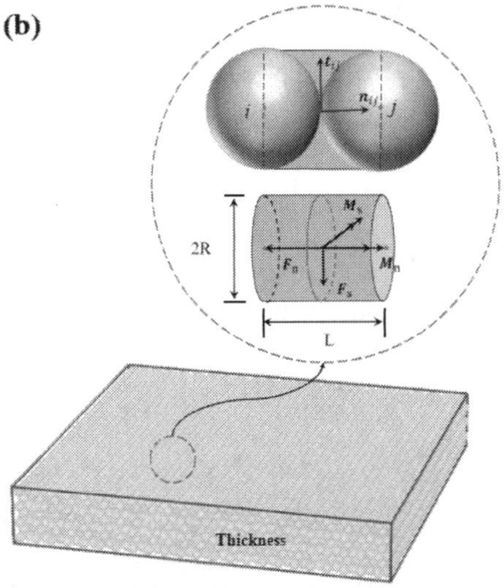

Figure 2. Physical and numerical models of model ice in CSSRC-SIMB (a) model ice (b) construction of model ice with bonded spheres

The macroscopic failure of model ice is closely related to the generation and propagation of internal cracks. In the DEM of model ice, the failure of the bonding disk. between elements simulates the generation of internal cracks in the material. During the failure process, the bond failure modes between elements can be divided into tensile failure and shear failure based on the normal and tangential stress of parallel-bond disk. When the maximum tensile stress ($\sigma_{\max}$) exceeds the tensile strength (σ^{t}) between adjacent elements, $\sigma_{\max} > \sigma^{\mathrm{t}}$, the parallel-bond disk occurs tensile failure; When the maximum shear stress ($\tau_{\max}$) exceeds the shear strength (τ^{s}) between adjacent elements, $\tau_{\max} > \tau^{\mathrm{s}}$, the parallel-bond disk occurs shear failure, as shown in Figure 3. According to the Mohr-Coulomb shear friction law, the tensile strength σ^{t} and shear strength τ^{s} can be determined as follows:

$$\sigma^{\mathrm{t}} = \sigma_{\mathrm{b}}^{\mathrm{n}} \tag{2}$$

$$\tau^{\mathrm{s}} = \sigma_{\mathrm{b}}^{\mathrm{s}} + \mu_{\mathrm{b}} \cdot \max(0, \sigma_{\max}) \tag{3}$$

where μ_{b} is the friction coefficient between the bonding elements; $\sigma_{\mathrm{b}}^{\mathrm{n}}$ and $\sigma_{\mathrm{b}}^{\mathrm{s}}$ are the normal and shear bonding strengths of the parallel-bond disk, respectively.

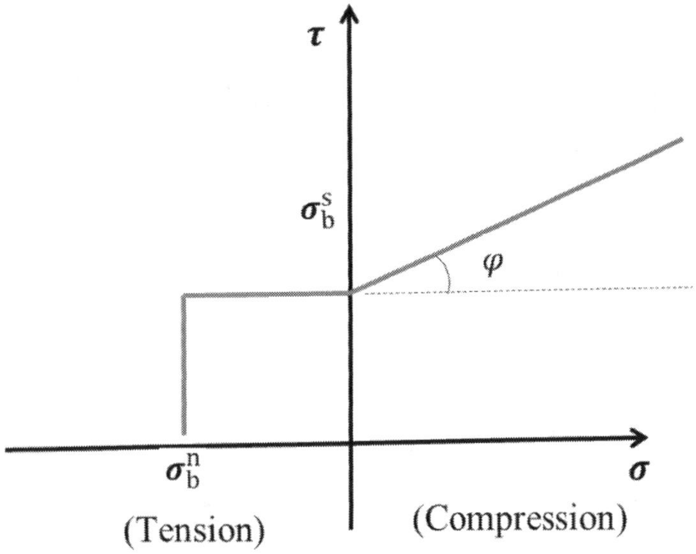

Figure 3. Failure criterion of the parallel-bond disk

3. Verification of physical model tests in the numerical ice tank

3.1 Description of physical model tests

The model tests of typical offshore structures represented by vertical cylinder and inclined plate were performed at the CSSRC-SIMB as part of the exploration of the model tests method. The similarity between model ice and sea ice is emphasized in the design of relevant tests to ensure that the physical and mechanical parameters of sea ice are within a certain range, while the size and carriage speed of structure do not need to strictly meet the similarity law of model tests. Considering the test capacity of the CSSRC-SIMB, the model tests of the vertical cylinder and inclined plate structures are selected as test models, as shown in Figure 4. The relevant test information of each condition, including sea ice strength, ice density, speed, structural geometry and so on, is listed in Table 2 and 3. The former is used to analyze the brittle crushing of the ice, and the latter is used to analyze the bending failure of the ice.

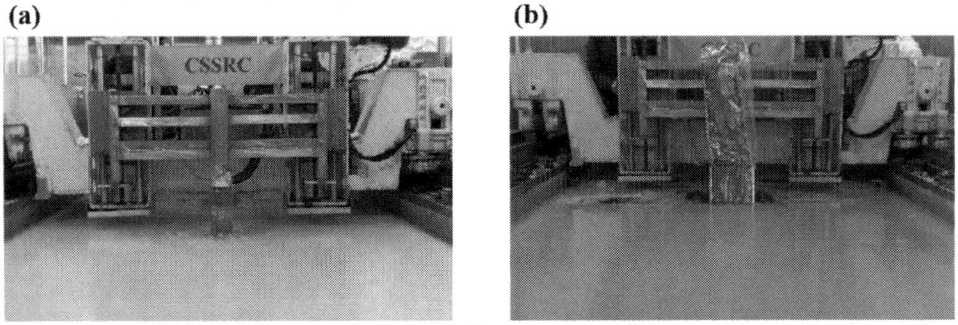

Figure 4. Physical model tests of interaction between the offshore structures and level ice (a) vertical cylinder (b) inclined plate

Table 2. Test matrix and ice conditions of interaction between the vertical cylinder and level ice

Test No.	Compressive strength (kPa)	Ice density (kg/m^3)	Ice thickness (mm)	Diameter of cylinder (mm)	Speed (mm/s)
#101					50
#102	51.6	901	36	100	100
#103					150
#201					50
#202	55.3	902	37	150	100
#203					150
#301					50
#302	57.3	915	37	200	100
#303					150

Table 3. Test matrix and ice conditions of interaction between the inclined plate and level ice

Test No.	Flexural strength (kPa)	Ice density (kg/m^3)	Ice thickness (mm)	Angle of inclined plater (°)	Speed (mm/s)
#401					50
#402	33.4	924	37	120	100
#403					150

3.2 Comparison of the results

A visual result comparison is presented for test cases #202 and #402, as shown in Figure 5. The computational parameters of numerical simulation in numerical ice tanks are listed in Table 4. Figure 5 shows the interaction process of ice-structure including crushing failure, bending failure and channel ice accumulation, which occurred in the physical and numerical ice tank. In the process of the interaction between ice sheet and vertical cylinder, the ice sheet is subjected to crushing brittle failure, and the broken ice fragments are extruded from the upper and lower surfaces of the ice sheet to form a temporary accumulation body. During the interaction between ice and inclined plane, the ice sheets have periodic bending failure, and the length of each failure of the ice sheets is basically the same. In addition, the phenomenon of sea ice climbing and accumulation during the action is very obvious.

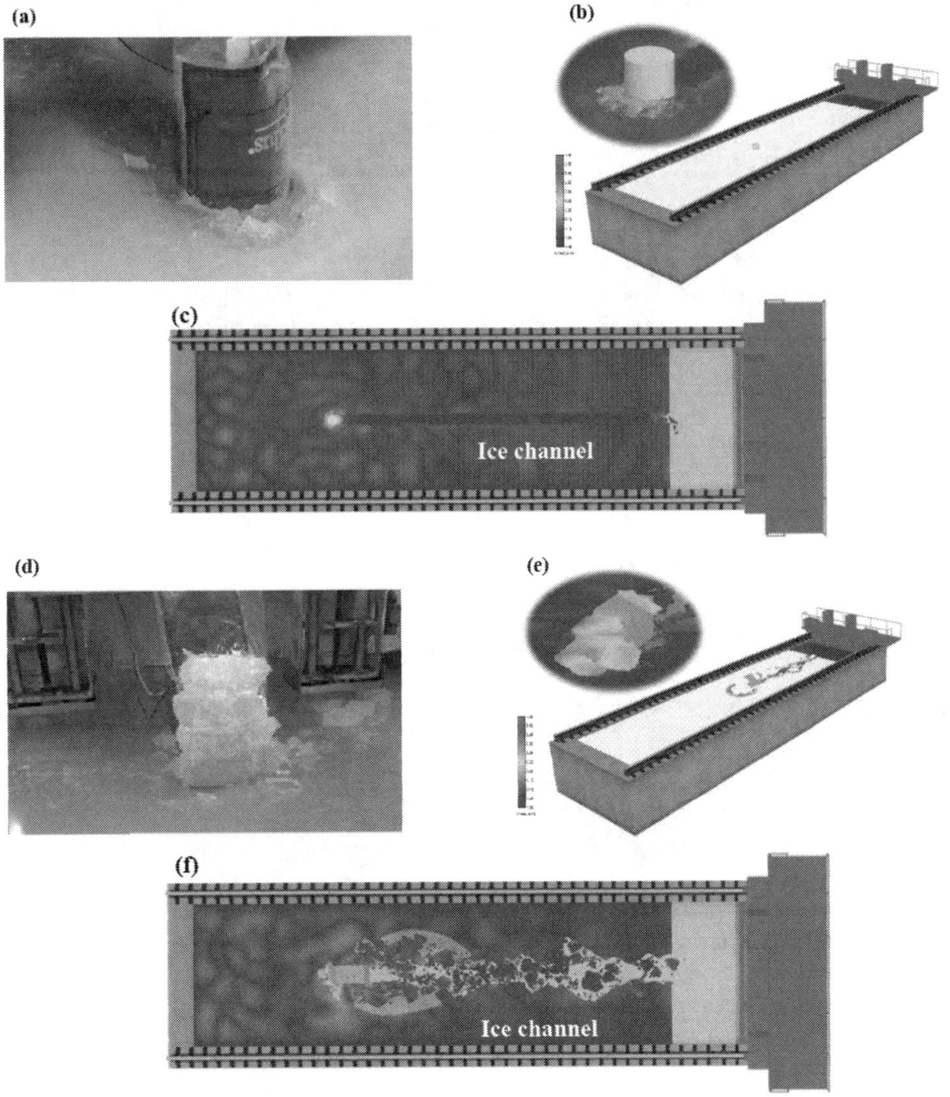

Figure 5. Visual comparison of the interaction between structures and model ice in test #202 and #402 (a) physical ice tank tests of vertical cylinder (b) numerical ice tank tests of vertical cylinder (c) ice channel in numerical ice tank tests of vertical cylinder (d) physical ice tank tests of inclined plate (e) numerical ice tank tests of inclined plate (f) ice channel in numerical ice tank tests of inclined plate

Table 4. Computational parameters of numerical simulation in numerical ice tanks

Definition	Symbol	Value
Young's modulus of elements	E_p	7.5 MPa
Friction coefficient of elements	μ_p	0.2
Normal and shear bonding strengths	$\sigma_\mathrm{b}^\mathrm{n}$, $\sigma_\mathrm{b}^\mathrm{s}$	62kPa
Ice-structure friction coefficient	μ_s	0.15
Number of elements	N	24W
Element size	R	0.014 m

Figure 6 shows the numerical comparison of ice loads from vertical cylindrical structures with different diameters. Both in the ice-tank test results and in the numerical ice tank results, the ice load of the structure presents continuous brittle extrusion and crushing, and the load presents loading and unloading forms, but there is no periodicity. Meanwhile, Figure7 shows the numerical comparison of ice loads from inclined plate structures with different speeds. The dynamic load on the inclined plate is ascribed to the continuous bending failure occurring around the front of the structure. In each period of loading and uploading, the ice load gradually increases over time and then rapidly drops once it reaches a certain level.

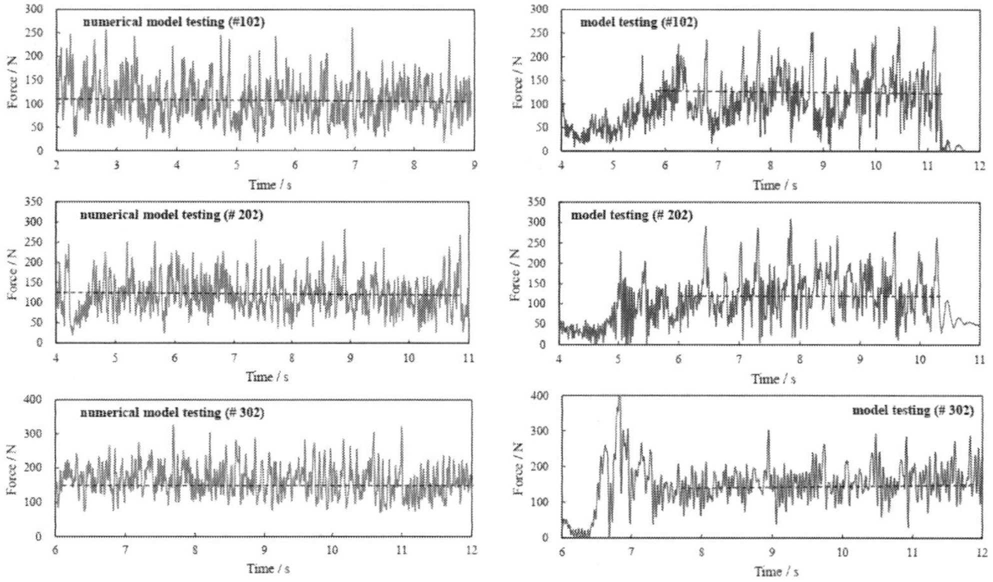

Figure 6. Ice load histories of vertical cylinder structure in physical and numerical ice tank

The mean load and maximum load are of comparable magnitude. The mean and maximum values of the ice loads of the vertical cylindrical structure in the simulation and the test are compared and analyzed. Where, the mean value and maximum value are the time history of ice load in the stable period, as shown by the dotted line in Figure 6. Meanwhile, the mean peak load and maximum load of the plate structure are also analyzed. The peak load represents the maximum value of structure within a single loading and unloading cycle. Table 5 summarizes the statistical characteristics and relative errors of ice loads in 12 working conditions. Although the relative error between the test and the simulation is relatively large in some cases (#103, #301), the mean load (mean peak load) and the maximum load error of the numerical ice tanks are still within the acceptable range. These error sources are not only the unreasonable selection of numerical simulation parameters, but also the random uncertainty of some experiments.

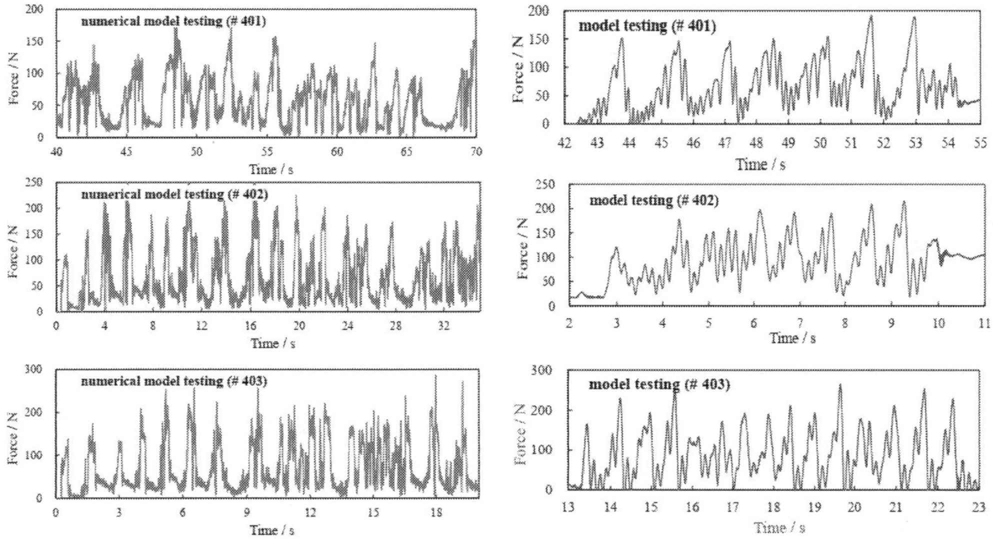

Figure 7. Ice load histories of inclined plate in physical and numerical ice tank

4. Application of numerical ice tank

The Wass bow is named from Dr. Heinrich Wass of Germany. The main characteristics of the simplified Thyssen Wass bow include small stem angle, large waterline angle and large outward drift angle, which can increase the proportion of bending failure components of the ice sheet and greatly reduce the ice breaking resistance when the ship is sailing through the ice region. In this section, a new Wass bow test model is designed with the simplified Wass bow [19] as the mother ship, as show in Figure 8 (a) and (b). The numerical ice tank model above is used to evaluate the ice breaking resistance of the bow under different ice thickness conditions, as shown in Figure 8 (c). At full-scale test, the ice breaking conditions for the new Wass bow are as follows: ice flexural strength is 600kPa, ice thickness is 0.6~1.5 m, and ship speed is 2kn. According to the ship's bow model and the size of the CSSRC-SIMB, the scale of the model test set as 30. The selection of discrete element parameters in the numerical ice tank refer to the parameters in Table 4, and other parameters of the relevant model tests are listed in Table 6.

Table 5. Comparison of the ice loads between the simulation and test with different condition

Structure	Test No.	Mean load (Mean peak load) / N			Maximum load / N		
		Simulation	test	Relative error	Simulation	test	Relative error
vertical cylinder	#101	90.6	92.3	1.9%	233.2	246.8	5.5%
	#102	105.6	110.4	4.4%	247.4	253.3	2.3%
	#103	145.6	116.7	24.7%	299.8	220.9	35.7%
	#201	105.4	119.7	12.0%	260.6	260.5	0.1%
	#202	117.6	128.8	8.7%	283.0	308.8	8.4%
	#203	157.2	153.2	2.6%	326.0	302.0	7.9%
	#301	110.8	151.1	26.7%	286.9	334.0	14.1%
	#302	165.4	178.3	7.2%	297.6	311.0	4.3%
	#303	183.0	204.8	10.6%	342.9	357.8	4.2%
	Mean error	-	-	11.0%	-	-	9.2%
inclined plate	#401	115.2	152.7	24.6%	172.2	187.1	8.0%
	#402	194.0	176.1	10.2%	225.3	206.1	9.3%
	#403	204.2	187.5	8.9%	286.8	259.7	10.4%
	Mean error	-	-	14.5%	-	-	9.2%

A series of model tests of the Wass bow breaking the level ice were conducted in the numerical ice tank. Figure 9 shows the simulation snapshot of interaction between ice sheet and Wass bow in ice thickness 0.03m. The ship bow interacts with the ice sheet at a constant carriage speed, resulting in crushing and bending failure of the ice sheet, as show in Figure 9 (a). The bending failure of ice plate is the most obvious, and the development of circumferential and radial cracks can be observed during the failure process. In addition, Figure 9 (b) compares the circumferential and radial cracks of this paper with other simulation results [20], showing a good consistency. In this paper, simplified Wass bow is also adopted for the icebreaker, which is the same as the design ship bow selected in this paper. Figure 9 (c) also shows the shape of the ice channel that was created behind the ship bow.

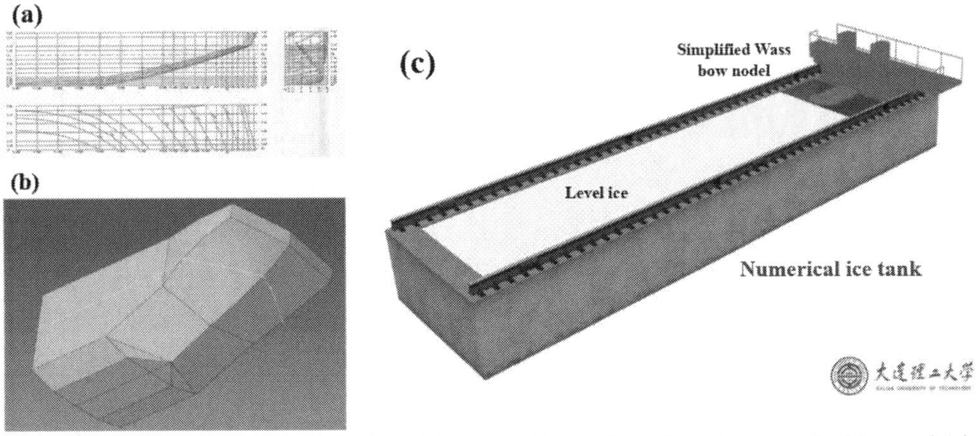

Figure 8. Numerical model of level ice and of new Wass bow in numerical ice tank(a) line drawing of simplified Wass bow (b) model of new Wass bow (c) numerical model of level ice and of new Wass bow

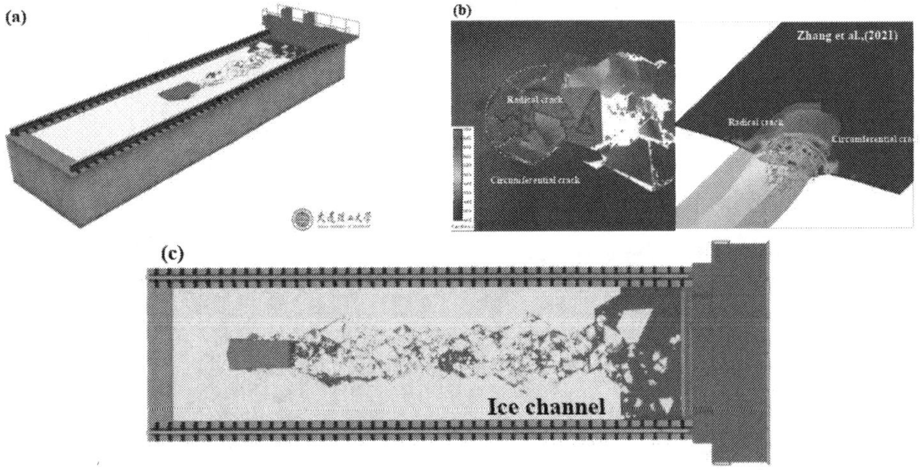

Figure 9. Simulation snapshot of the interaction between ice sheet and Wass bow in different (a) snapshot of ship-ice action at t=20s (b) comparison snapshot radical and circumferential crack (c) ice channel in numerical ice tank tests

The time histories of ice force on the ship bow in three directions is shown in Figure 10. With the bow continuously entering the ice area, its ice force also gradually increases. About 5s later, the bow completely enters the ice area, and the variation trend of its ice force tends to be stable. Since then, it can be seen that the ice force of ship bow has periodic load characteristics due to the continuous bending failure of ice sheet. The ice force in the x direction represents the breaking resistance of the bow, which consists of two parts: bending resistance and crushing resistance. The mean value of breaking resistance in this state is analysed and compared, which is the red dotted line in Figure 10 (a).

The validity of the numerical ice tank results can be verified by empirical formulas, thus improving the ability of the numerical ice tank to comprehensively describe the ice load characteristics. In this paper, Lindqvist (1989) [21] ice resistance formula is selected to check the rationality of the numerical ice tank results. The ice breaking resistance of ship bow in Lindqvist's formula can be divided into bending resistance and crushing resistance, which can be expressed as:

$$R_{\mathrm{ice}} = (R_{\mathrm{c}} + R_{\mathrm{b}})\left(1 + \frac{1.4v_{\mathrm{ship}}}{\sqrt{gh_{\mathrm{ice}}}}\right) \tag{4}$$

$$R_{\mathrm{c}} = 0.5\sigma_{\mathrm{ice}}h_{\mathrm{ice}}^{2}\frac{\tan\varphi + \mu\cos\varphi/\cos\psi}{1 - \mu\sin\varphi/\cos\psi} \tag{5}$$

$$R_{\mathrm{b}} = \frac{27}{64}\sigma_{\mathrm{ice}}B_{\mathrm{ship}}\frac{{h_{\mathrm{ice}}}^{1.5}}{\sqrt{\frac{E}{12(1-v^2)g\rho_{\mathrm{w}}}}}\frac{\tan\psi + \mu\cos\varphi}{\cos\psi\sin\alpha}\left(1 + \frac{1}{\cos\psi}\right) \tag{6}$$

where, R_{ice}, R_{c}, and R_{b} is the breaking resistance, crushing resistance and bending resistance, respectively. ψ is normal angle, $\psi = \arctan(\tan\varphi/\sin\alpha)$. E and v is Young's module and Poisson's ratio of ice. ρ_{w} the density of water, $\rho_{\mathrm{w}} = 1035$ kg/m3. g is the gravitational acceleration.

Table 6. model test parameters of new Wass bow in numerical tank

Definition		Symbol	Value
Wass bow model	Length	L_{ship}	0.96 m
	Front / Back breadth	B_{ship}	0.40 / 0.46 m
	Depth	D_{ship}	0.30 m
	Draft	T_{ship}	0.18 m
	Stem angle	φ	24 deg
	Waterline angle	α	84 deg
	Drift angle	β	52 deg
	Ice-structure friction coefficient	μ	0.1
	Speed	v_{ship}	0.188 m/s
Model ice	Thickness	h_{ice}	0.02~0.05 m
	Flexural strength	σ_{ice}	20.0 kPa
	Density	ρ_{ice}	920 kg/m^3
	Poisson's ratio	v	0.3
	Young's modulus of elements	E_{p}	4.5 MPa
	Normal and shear bonding strengths	$\sigma_{\mathrm{b}}^{\mathrm{n}}$, $\sigma_{\mathrm{b}}^{\mathrm{s}}$	37 kPa
	Element size	R	0.014 m
Scale		λ	30

The mean breaking resistance results of numerical ice tank on the Wass ship bow under different ice thickness are shown in Figure 11. It can be observed that the trend of the breaking resistance is the same as that obtained with the Lindqvist's formula. Overall, the result of numerical ice tank is slightly smaller than that calculated by the Lindqvist's formula, Overall, the result of the ice pool is slightly smaller than that calculated by the formula, but the error is within an acceptable range. In conclusion, the numerical ice tank model proposed in this paper has certain advantages in carrying out virtual ice-breaking resistance evaluation tests.

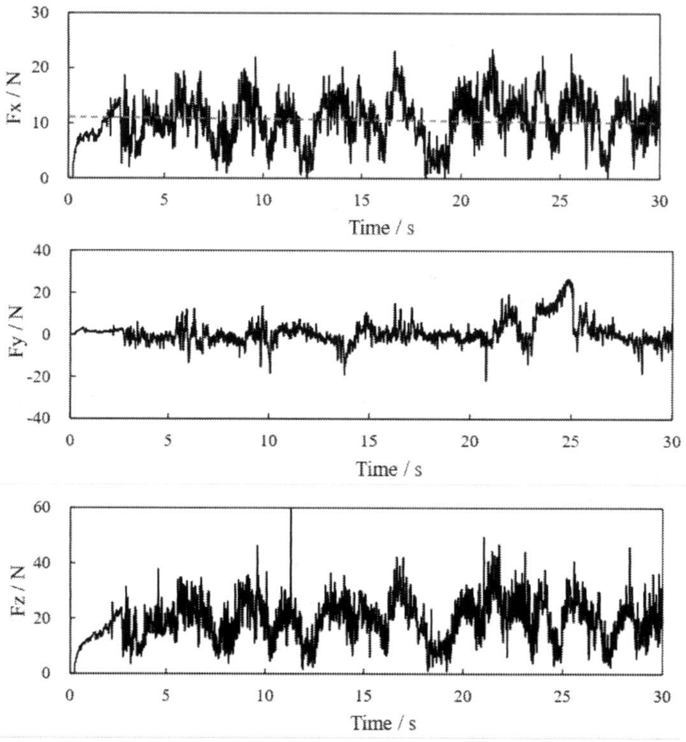

Figure 10. Time histories of ice force on the ship bow (h_{ice} = 0.03m)

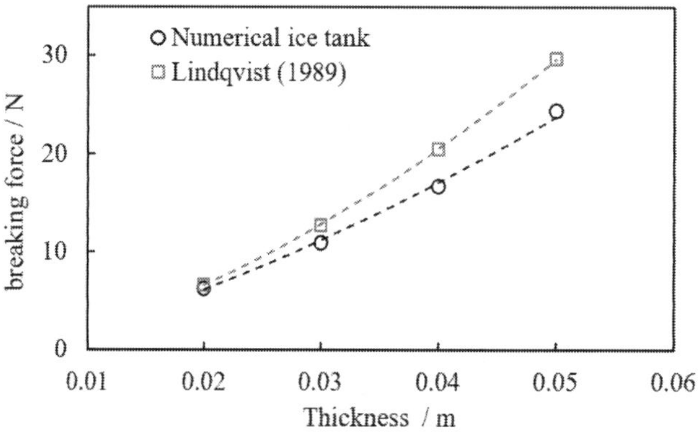

Figure 11. breaking resistance on the Wass ship bow under different ice thickness

5. Conclusion

In this study, the numerical ice tank model corresponding to CSSRC-SIMB was proposed using the 3D spherical DEM to investigate the interaction processes between offshore structure and ice. A series of ice-tank test cases including three structure types (vertical cylinder, inclined flat, and new Wass bow) were analysed. The numerical reproduction of the model tests in the numerical ice tank provides convenient information on offshore structure-ice interaction processes, ice failure models and time histories of ice force on the structure. The results of this study support the following conclusions: The numerical ice tank model is a reasonable method to performing repetitions of the model test in physical ice tank/ basin, which can reproduce ice failure performance including curding, bending, accumulation and dynamic ice load on offshore structure; The visual and numerical comparison of the simulation results of 12 ice-tank tests demonstrates the rationality of the numerical ice tank. The mean load (mean peak load) and the maximum load error of the numerical ice tanks are still within the acceptable range; The numerical ice tank model was used to evaluate the breaking resistance of the new Wass bow. In tests with various ice thicknesses, the numerical ice tank produced results that were found to be consistent with those obtained by Lindqvist's formula.

Acknowledgements

This work was supported by the Fundamental Research and Development Project of China [Grant No. JCKY2020206B073], the National Natural Science Foundation of China [Grant Nos. 52192690, 52192694], and the Hi-Tech Ship Project of the Ministry of Industry and Information Technology [Grant No. [2021]-342]

References

[1] Islam M, Mills J, Gash R, et al. A literature survey of broken ice-structure interaction modelling metho ds for ships and offshore platforms[J]. Ocean Engineering, 2021, 221: 108527.
[2] Xue Y, Liu R, Li Z, et al. A review for numerical simulation methods of ship–ice interaction[J]. Ocean Engineering, 2020, 215: 107853.
[3] Timco G, Johnston M. Ice loads on the caisson structures in the Canadian Beaufort Sea[J]. Cold Regio ns Science and Technology, 2004, 38(2): 185-209.
[4] Noorma S, Thomas GB. 20 years of monitoring of ice action on the Confederation Bridge piers[J]. Col d Regions Science and Technology, 2018, 151: 208-236.
[5] Yue Q, Guo F, Kärnä T. Dynamic ice forces of slender vertical structures due to ice crushing[J]. Cold Regions Science and Technology, 2009, 56: 77-83.
[6] ITTC. ITTC – Recommended procedures and guidelines: experimental at uncertainty analysis for ship resistance in ice tank testing. Quality Systems Group of the 28th ITTC,2017.
[7] ITTC. ITTC – Recommended procedures and guidelines: Resistance in ice tank testing. Specialist Com mittee on Ice of the 28th ITTC,2017.
[8] Cundall P.A, Strack O. A discrete numerical model for granular assemblies[J], 1979, 29(1): 47-65.
[9] Liu L, Ji S. Dilated-polyhedron-based DEM analysis of the ice resistance on ship hulls in escort operat ions in level ice[J]. Marine Structures, 2021, 80: 103092.
[10] Sun S, Shen HH. Simulation of pancake ice load on a circular cylinder in a wave and current field[J]. Cold Regions Science and Technology, 2012, 78: 31-39.
[11] Berg M., Lubbad R, Løset S. Repeatability of ice-tank tests with broken ice[J], 2020, 74.
[12] Huang L, Tuhkuri J, Igrec B, et al. Ship resistance when operating in floating ice floes-A combined CF D-DEM approach[J]. Marine Structures, 2020, 74: 102817.
[13] Yang D, Liu L, Ji S. Numerical analysis of interaction between sea ice and propeller based on coupled DEM-FEM model[J]. Ocean Engineering, 2023, 268: 113469.
[14] Tian Y, Wang Y, Ji S, et al. Design and Realization of CSSRC Small Ice Model Basin for Ice-related F undamental Researches[C]. 25th International Conference on Port and Ocean Engineering under Arcti c Conditions (POAC 2019). Delft, Netherlands,2019.
[15] Gang X, Tian Y, Ji S, et al. Investigation on Ice-Breaking Resistance of a Ship Bow Advancing in Lev el Ice[C]. The 32nd International Ocean and Polar Engineering Conference, Shanghai, China, 2022.
[16] Ji S, Yang D. Ice loads and ice-induced vibrations of offshore wind turbine based on coupled DEM-FE M simulations[J]. Ocean Engineering, 2022, 243: 110197.
[17] Liu L, Ji S. Comparison of sphere-based and dilated-polyhedron-based discrete element methods for th e analysis of ship–ice interactions in level ice[J]. Ocean Engineering, 2022, 244: 110364.
[18] Long X, Liu S, Ji S. Discrete element modelling of relationship between ice breaking length and ice lo ad on conical structure[J]. Ocean Engineering, 2020, 201: 107152.
[19] Fernando MP. Experimental and numerical research on the interaction between ice flows and a ship's hull during icebreaking[D]. Delft University of Technology, 2003.

[20] Zhang Y, Tao L, Wang C, et al. Numerical study of icebreaking process with two different bow shapes based on developed particle method in parallel scheme[J]. Applied Ocean Research, 2021, 114: 102777.

[21] Lindqvist, G. A straightforward method for calculation of ice resistance of ships. Proceedings of the 10th International Conference on Port and Ocean Engineering Under Arctic Conditions. 722–735, 1989

IOP Conf. Series: Materials Science and Engineering 1288 (2023) 012026 doi:10.1088/1757-899X/1288/1/012026

An attempt to predict planing hull motions using machine learning methods

Ahmed Shehata[1*] **and Abbas Dashtimanesh**[2]

[1] Estonian Maritime Academy, Tallinn University of Technology, 11712, Tallinn, Estonia
[2] Center of Naval Architecture, Department of Engineering Mechanics, School of Engineering Science, KTH Royal Institute of Technology, 11428, Stockholm, Sweden

ahmed.shehata@taltech.ee

Abstract Designing a high-speed craft for better seakeeping in waves can contribute significantly to higher safety and human comfort. Early in the design process, mathematical models such as the 2D+T method are commonly used, while high-fidelity computational fluid dynamics (CFD) and experimental models are used later in the process. Some of the limitations of such models are that they are not fast enough to be used in the ship's system for real-time monitoring or to develop a digital twin. Recently, machine learning methods have demonstrated great promise in building surrogate models from data. These methods include deep learning and recurrent neural network (RNN). In this paper, a systematic investigation of the network architectures and the used optimizers to train the network is presented. Adam, Adagrad, RMSprob and SGD are investigated in training the network. To train the model almost 35000 data points were collected for Fridsma hull operating in 18 regular waves using a 2D+T model. The result showed that gated recurrent unit (GRU) outperformed long short-term memory (LSTM) and RNN in predicting the heave motion. Also, one hidden layer with 5 neurons was enough to achieve mean absolute error of 0.000298 and to predict unseen waves when trained with more than 24000 data points.

1. Introduction

Planing high-speed boats designed to be able to operate at high speeds and use planing to skim over the surface of the water [1]. Designing the boats for better seakeeping can lead to better safety and human comfort. In calm water, planing hull is known as a good performer. Whether it's plowing through water at low speeds or planing over water at high speeds. Mathematical models in this area are based on the works of Von Karman [1] which describes the problem of a two-dimensional wedge impacting on the water surface and Wagner [2] on the impact of the floats landing seaplanes on the water surface. It is known as the 2D+T theory. Martin [3] predicted the linearized response of the seakeeping of a high-speed boat with a constant deadrise angle. It was shown in this work that the non-linear effect is more significant in higher speeds-to-length ratios. One of the earliest models was introduced in Zarnick [4]. The accuracy of the model was good in moderate operating conditions but in severe sea states the heave acceleration was one-half of the experimental values. This model was later extended by Keuning [5] to include hulls with variable deadrise angles and include wider speed ranges. Payne [6] implemented some modifications to coefficients employed in the "added mass" equations.

Recently, a mathematical model to predict heave, pitch, and roll motions in regular waves was developed by Ghadimi et al. [7]. The model was an extension of the one developed by Zarnick [4] and based on the framework developed by Sebastian et al. [8]. The model considered the unsymmetrical water impact of the section to estimate roll, heave, and pitch motions Moreover, momentum and added mass theories were applied for the evaluation of forces acting on the 2-dimensional sections. The model assumed the wavelength to be longer than the boat length. As a result, the accuracy of the model for longer waves was much better and in a good agreement with experimental results. This model was later extended and validated with experimental and CFD results in Tavakoli et al. [9]. To extend the model for head sea and try to generalize it in the computation of 3D forces, Ghadimi et al. [10] solved the potential flow around the wedge. They concluded that the accuracy of the model was acceptable for practical applications in the initial stages of the design process. The 2D+T method provides a faster alternative to CFD. Yet, it is not fast enough to be used in the ship system for real-time monitoring or to develop a digital twin. Furthermore, its implementation differs from one case to another. Therefore, the model that works with a planing craft with a fixed deadrise angle doesn't always work with planing craft with a variable deadrise or the model that works with double stepped planing hull doesn't always work with other types of planing hulls. Other limitations were reported in Dashtimanesh et al. [11]. For instance, it cannot solve the problem of a two-phase flow, consider the environmental effect, or give an accurate estimation of the hull geometry. In the meanwhile, surrogate machine learning models can predict future output in a few seconds and can be generalized, if trained with sufficient data.

In fact, machine learning methods have recently shown promises in building surrogate models from data. Machine learning is a subset of artificial intelligence. Artificial intelligence (AI) refers to the ability of any computer to perform a human task. While machine learning deploys the field in which a computer learns to perform the task by identifying patterns in stored data. Furthermore, a surrogate model is an approximation model that is fitting an output to an input. If the model is fed with an input similar in behavior to the one that was used to train the model, the model will be able to predict the output, accurately. It can imitate the behavior of the system that was used to generate this data. These surrogate models could be faster, cheaper and in some cases, more general than the original model or surpass it. These models could be generated using white, grey or black box techniques. In white box techniques, the structure of the model is known. An example of a white box model is linear or higher-order regression. While in black box techniques, the structure of the system is totally unknown. Some examples of the black box techniques such as support vector regression (Suykens and Vandewalle [12]), Feed forward neural network, recurrent neural network (RNN) (Rumelhart et al. [13]) or long short-term memory (LSTM) (Hochreiter and Schmidhuber [14]). In grey box techniques, part of the model structure is known while the other part is unknown. An examples of the grey box techniques is the physics-informed neural network (Raissi et al. [15]). In a physics informed neural network, the linear term is developed using a white box model while the non-linear term is developed using a black box technique.

The use of machine learning to develop seakeeping models has become a subject of interest as it can develop fast models. LS-SVM with kernel and experimental data was used in [16] to predict roll motion. LS-SVM has an advantage over neural networks in training time. Yet, after training both models neural network is faster in making prediction than the support vector machine. del Águila Ferrandis et al. [17] developed a model for heave, pitch and roll motions for catamaran vessel and DTMB battleship for 5th order regular waves and stochastic waves that represent real life sea-states. To summarize the workflow conducted in this research, one of the contributions of this work was trying two types of RNN (Gating recurrent unit (GRU) and LSTM. The results of the two networks were quite acceptable for 1 hidden layer and 20 neurons. Furthermore, LSTM outperformed GRU. As reported in this work, LSTM can predict the heave and pitch motions given a sufficient amount of data with high accuracy, however, when roll motion was included in the analysis the roll prediction accuracy was lower. As reported, the reason was inaccuracy in the roll data generation from the unsteady Reynolds averaged Navier-Stokes (URANS). The model was trained in half an hour when trained on GPU and to obtain a result, it takes a fraction of a second which was a great achievement in comparison to the CFD models. Some purely data-driven approach was developed by Xu et al. [18]. The framework was similar to the previous work.

The proposed structure was more complicated with 6 LSTM hidden layers which is quite a very big network. Recently, Marlantes and Maki [19] developed a hybrid method for the prediction of planing craft dynamics in waves. In this hybrid method, the equation of motion is used for predicting the linear terms while the non-linear terms were predicted using the data. This approach could be a good solution if the model cannot be trained with purely data-driven models due to the size of the data needed to train the model. The hybrid method could be used in the future to override the computational complexity in training a complex seakeeping model. The objective of the current paper is to act as a guideline for researchers interested in using data-driven models in developing a seakeeping model. A systematic investigation of the network architectures and the used optimizers to train the network is presented. Adam, Adagrad, RMSprob and SGD are investigated in training the network.

2. Methods

The framework to build the model is described in figure 1. The aim is to use neural networks to build a surrogate model from data. The work will investigate the use of RNN, LSTM and GRU in developing a seakeeping model. Data-driven models are based on using data that describes the model behavior to train the current model. The first step is to generate the data for the mentioned hull in regular waves. To generate the data, a mathematical model developed by Tavakoli et al. [9] is used. The mathematical model is based on the 2D+T method. The model can predict heave and pitch motions in regular waves as well as the accelerations at the bow and the center of gravity (COG). The model is valid for a planing hull with a constant deadrise angle. The case study is the Fridsma planing hull. This data is then divided, one portion is for training the network, another is for validating it while training and the last part is unseen test data to evaluate the performance of the model. This way of dividing the data into three portions is to avoid overfitting (Ying [20]). Overfitting is building a model that violates the principle of parsimony (Hawkins [21]). The principle of parsimony indicates that a model needs to include all the necessary elements. Overfitting could be building a non-linear model to describe a linear behavior or building a model that works fine with the training data but cannot achieve the same results with unseen data. To evaluate the model, the mean absolute error (MAE) is used in this study. The MAE is the mean of the difference between the actual values and estimated values and can be calculated by equation (1).

$$MAE = \frac{1}{n}\sum_{i=1}^{n} Y_i - \hat{Y}_i \tag{1}$$

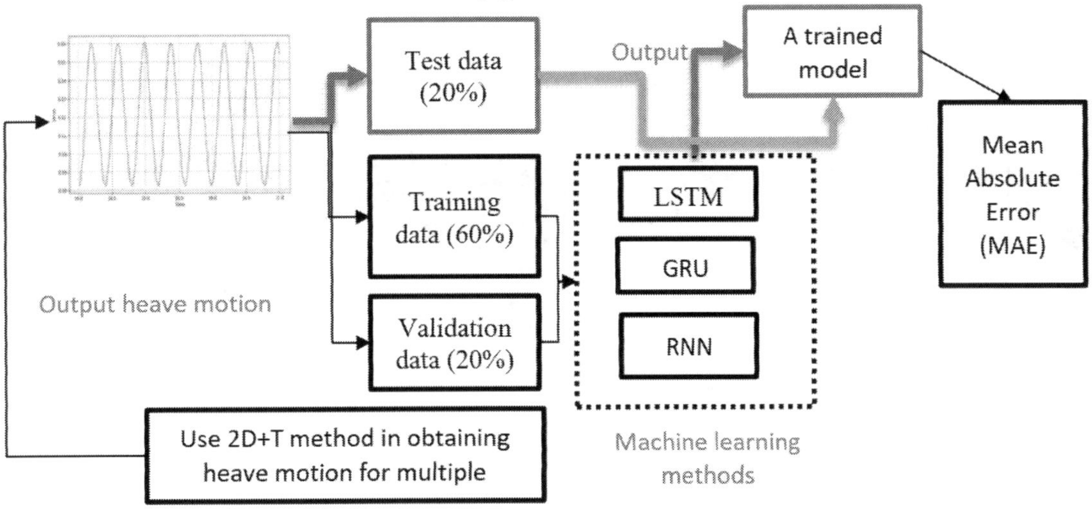

Figure 1. The framework to build the machine learning model. First the data is obtained for series of waves using the 2D+T mathematical model. Then, this data is divided into training, validation, and testing data. The training and validation data are used to train three different models LSTM, RNN and GRU. The output models are evaluated using the test data with mean absolute error.

2.1. Recurrent neural networks and supervised learning

The structure of the artificial neural network was inspired by the structure of the neurons in the human brain. A deep neural network is an implementation of a neural network where the network consists of several layers of a single-layer neural network. Each layer can be described mathematically using equation (2),

$$y = f(z) \qquad where \quad z = w_o + wx \tag{2}$$

The number of neurons in the first layer is based on the number of variables in the model. The number of neurons in the output layer is based on the number of outputs or classes in the model. The hidden layer/s is the reason why neural networks can capture nonlinear complex behavior. Each neuron in a hidden layer acts as a linear classifier. The goal of the algorithm is to find the weights (w) when given an input vector x to optimize a loss function. The loss is defined as the deviation between the prediction and the real value for the training data. Training a neural network is an optimization problem. The goal of the algorithm is to minimize the loss defined in equation (1). The loss is the difference between the prediction of the network and the real label or value. Stochastic gradient descent (SGD) optimization algorithm is used to train the network and the learning rate is the most important hyperparameter when training the network. The learning rate is the parameter that is concerned with updating the weights in equation (2) when using stochastic gradient descent (SGD) to optimize the loss function. A very high learning rate will allow the model to learn the features in the data rapidly but, it might not converge to a global minimum or generalize the model. Moreover, using a very small learning rate means the model will learn the features very slowly and as a result, the model could require huge computational resources and could not be able to take big steps to converge into a global minimum. Furthermore, the most popular learning rate optimization method is the adaptive learning rate methods. The advantages of these methods are that it can adapt to the model without extensive work in advance to define the learning rate. Examples of such optimizers are adaptive moment estimation (Adam) (Kingma and Ba [22]),Adaptive Subgradient Methods (Adagrad) (Duchi et al. [23]) or RMSprob (Tieleman and Hinton [24]). Ruder [25] provides a great explanation of the advantages and disadvantages of every optimizer. RNN is the type of artificial neural network that includes one or more recurrent layers. In the recurrent layers, as shown in figure 2, the output is fed again to the hidden layers. These recurrent layers allow the network to store data from previous time steps. Unlike feed-forward neural networks which just predict the output based on the current input [26].

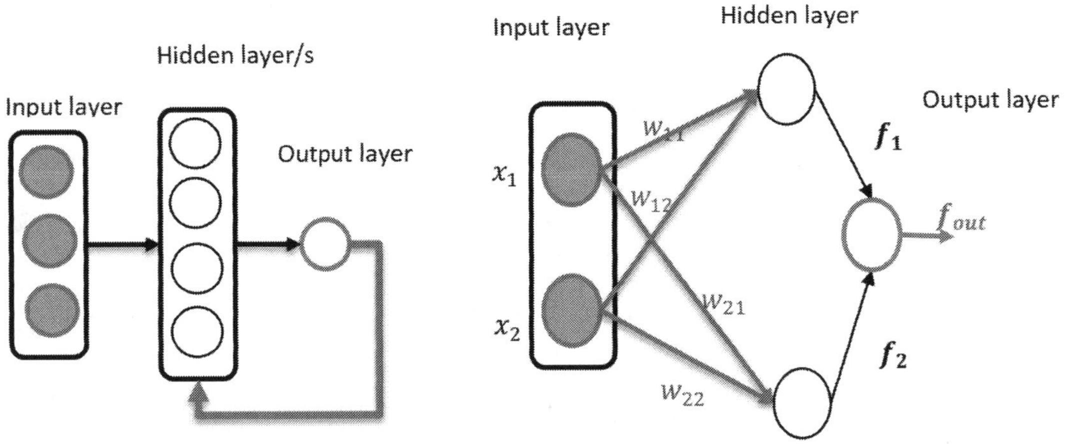

Figure 2. Recurrent neural network on the left vs feed forward neural network on the right.

In Hochreiter and Schmidhuber [13], the gating principle was introduced into the recurrent neural network. This structure is what is known now as long short-term memory (LSTM). The gating principle

IOP Conf. Series: Materials Science and Engineering 1288 (2023) 012026 doi:10.1088/1757-899X/1288/1/012026

is a way to decide which information to keep in the system and which one to forget or override. Equations 3 to 7 explain how a gated network works.

$$f_t = sigmoid(W^{f,s}h_{t-1} + W^{f,x}x_t) \quad (3)$$

$$i_t = sigmoid(W^{i,s}h_{t-1} + W^{i,x}x_t) \quad (4)$$

$$o_t = sigmoid(W^{o,h}h_{t-1} + W^{o,x}x_t) \quad (5)$$

$$c_t = f_t \odot c_{t-1} + i_t \odot tanh(W^{c,h}h_{t-1} + W^{c,x}x_t) \quad (6)$$

$$h_t(u, v, r, \delta_R) = o_t \odot tanh(c_t) \quad (7)$$

In a gating network, there is a weight assigned to every gate which determines the effect of the current time step or the previous information in the current output. Every network has a forget gate, $\mathrm{f_t}$, input gate, $\mathrm{i_t}$, hidden state, $\mathrm{h_t}$ and cell state, $\mathrm{c_t}$. GRU is another structure which requires less computational time to train. The main difference from LSTM is that it doesn't have an output gate. In fact, LSTM doesn't expose all the information stored in the network but just shows the important information while GRU lacks this feature [27]. A recurrent neural network with a very simple structure can have a very high number of parameters. For example, 3 layers with 5 neurons and 3 input variables have led to 51 parameters for the simple RNN, 156 for GRU and 186 for the LSTM. Thus, for a long sequence or a high number of variables, it can be non-trainable, computationally. Consequently, attention mechanisms were introduced in [28,29]. The attention mechanisms work by adding another layer of classification before the classification occurs. Vaswani et al. [29] also introduced what is known as Transformers.

In supervised learning, the goal is to predict an output (label) from some input/s. Using neural networks in supervised learning has a specific advantage. It can detect the features without defining the features by the developer. For instance, a convolutional neural network can detect images where ever it located in the photos. The developer of the model doesn't need to specify the features. The model will be able to detect it by itself. To use recurrent neural networks in predicting time series data, the data need to be rearranged to have inputs and labels. The goal is to predict the variable x at any point from previous time steps n. These time steps will be defined as a window. The size of the window is based on how much the future prediction is affected by the history of the data. Therefore, it is assumed that the time series data has just one variable x consisting of m data points. The input data is rearranged as a tensor [30]. A tensor is a multi-dimension array or an array consisting of embedding arrays or vectors. Each embedding vector has a size i which refers to the number of variables in our data. The number of columns of the main array is n which is the chosen window size. The full input size will be $m - n - 1$ data points as described in equations (8) and (9). Mathematically speaking, the training data will be a tensor of size $(m - 1, n, i)$. The label is the following data point in the target variable (x_1). Therefore, for the first-row data of size n, the target will be point $n + 1$.

$$x = \begin{bmatrix} [x_{11} \dots x_{1i}] & [x_{21} \dots x_{2i}] & [x_{n1} \dots x_{ni}] \\ [x_{21} \dots x_{2i}] & [x_{31} \dots x_{3i}] & [x_{(n+1)1} \dots x_{(n+1)i}] \\ [x_{31} \dots x_{3i}] & [x_{21} \dots x_{2i}] & [x_{(n+2)1} \dots x_{(n+2)i}] \\ & \vdots & \\ [x_{(m-n-1)1} \dots x_{(m-n-1)i}] & [x_{(m-n)1} \dots x_{(m-n)i}] & [x_{(m-1)1} \dots x_{(m-1)i}] \end{bmatrix} \quad (8)$$

$$y = \begin{bmatrix} x_{n+1} \\ x_{n+2} \\ \vdots \\ \vdots \\ x_{m-1} \\ x_m \end{bmatrix} \quad (9)$$

3. Model investigation

Since the objective of this work is to act as a guideline for researchers interested in using data-driven models in developing a seakeeping model, a more systematic description of the network architectures and the used optimizers was presented in the previous section. In this section, we investigate which parameters affect the structure of a neural network. These parameters include the number of neurons and hidden layers. It also includes the different types of optimizers and the used network. To investigate different neural network structures and the effect of using different approaches to defining learning rate instead of just optimizing the hyperparameter by trial and error, we will use the heave motion as a case study. We have gathered data for different waves' amplitudes and lengths. We have 18 waves included in this study described in Table 1.

Table 1. Regular waves amplitudes and lengths.

No	Wave amplitude in (m)	wave length λ in (m)
1	0.0126873	1.143
2	0.0190881	1.143
3	0.0253746	1.143
4	0.0064008	2.286
5	0.0126873	2.286
6	0.0190881	2.286
7	0.0190881	2.286
8	0.0253746	2.286
9	0.0381762	2.286
10	0.0381762	2.286
11	0.0126873	3.429
12	0.0253746	3.429
13	0.0381762	3.429
14	0.0126873	4.572
15	0.0253746	4.572
16	0.0381762	4.572
17	0.0126873	5.715
18	0.0381762	5.715

In this paper, the heave motion will be used as a case study. Pitch and heave motions in regular waves both are simple harmonic motions. The same model can be used for both but we are just showing the result of the heave motion. To evaluate the models, the paper will use the mean absolute error described in equation (1). The planing hull used in this study is the Fridsma hull (Fridsma [31]). The hull parameters are defined in Table 2. The chosen speed is $4.6\ m/s$ which will lead to a F_r of 3.072 which lie within the planing range. For the sake of evaluating the network structure and learning rate, we will start with the simple RNN. Furthermore, the training, testing and validation data will include samples from all 18 waves. As mentioned before, to avoid overfitting, the test data was totally unseen by the model before. The simple structure that was used to test the network consists of 4 layers. The input layer is fed with two inputs which are the number of variables and a window size. In our case, we have 3 variables heave (the targeted variable), wavelength and wave amplitude. Then, a simple recurrent hidden layer with 100 neurons. A dense layer with 8 neurons and a 'Relu' activation function. Finally, an output dense layer with 1 neuron and a linear activation function. The reason for using the third layer is that RNN, as we can see in figure 2, has fully connected layers. That means every neuron of the hidden layer is connected to the neurons from the output layer or the next layers. The 100-neuron RNN layer would be fully connected with the output layer with 1 neuron. Every neuron act as a linear classifier. It is like connecting 100 lines with 1 line which is not feasible. As a result, this structure is affecting the ability of the network to build a model that describes the data, accurately. Therefore, including one more layer with 8 neurons make the training process smoother. Due to that structure, the number of the trainable parameters in the surrogate model for this network is 10400 parameters which is quite a high number. The reason why this number of parameters is high is that there is no physical system that requires a 100th order differential equation to present it. Moreover, the paper will investigate the usage of RNN, LSTM and GRU in building a network that can predict heave and pitch motions. It will also study the effect of the size of the training data on the accuracy of the model and the accuracy of the prediction for unseen regular waves. In summary, this paper will investigate the possibilities and limitations of using data-driven models in building an accurate coupled heave, pitch and roll model.

Table 2. The Fridsma hull characteristics.

Parameter	Value
L_{OA} (length overall in m)	1.9
B (Beam in m)	0.424
Gr (Radii of gyration in m)	0.128
β (Deadrise angle in degree)	20°
LCG (Longitudinal position of the centre of gravity from transom in m)	0.667
V (Speed in m/sec)	4.6

4. Results

To evaluate learning rates and different optimizers, this paper uses an Adaptive learning rate with Adam, Adagrad, RMSprob and plain SGD. The common way to define a learning rate is just by using different values and see how this learning rate affects the loss. The learning rate common values range from $[0{,}01, 10^{-6}]$. Another way is by scheduling a learning rate range from some high value to a very low value and check how the learning rate affects the training process. This can be done by scheduling a learning rate in Keras. Using Adam, the optimal learning rate is lower than $8 * 10^{-2}$. The safer option is to take 10^{-3}. The same process was repeated with SGD, Adagrad and RMSprob. Measuring the performance was using the mean absolute error after 50 epochs.

To build an RNN model, a window size n must be specified as described in equation 8. In this work, the window size for the past results was 5. Increasing the window size means more past data is included in the calculation of the current moment while decreasing it makes the model more relying on the near past data. Increasing the window size come with some computational cost for sure. In this model, 5 samples for window size was satisfactory to achieve a good accuracy. Table 3 and figure 3 show the results of using different optimizers with its optimal learning rate in training the mentioned network.

Adam was the best optimizer. Adam was faster in training than the other optimizers. Adam stopped improving near the 20th epoch which means it didn't even need the 50 epochs to obtain a better mean absolute error than the other optimizers. The mean absolute errors for all optimizers were good and that is expected due to the size of the training data and because of the nature of the heave motion in regular waves. Regular waves consist of a single frequency and because of that, the heave motion behavior is the same as a simple harmonic motion. Therefore, moving forward Adam will be used in the evaluation of the other network structure. The number of trainable parameters was high which can lead to a very high computational time while training the network. For such an application, it might be possible to obtain good accuracy with a simple structure such as using 5 neurons instead of 100 neurons for the hidden layer. To develop a more systematic way of building a neural network, the simplest structure is to have one input layer, one hidden layer and one output layer. The output layer consists of 1 neuron and one hidden layer is just simple RNN with 5 neurons.

Table 3. The performance of different optimizers. Adam achieved the best results.

Optimizer	Learning rate	Mean absolute error after 50 epochs	Time taken
SGD	$4 * 10^{-2}$	$1.37 * 10^{-3}$	8 min 51.0 sec
Adam	10^{-3}	$8.92 * 10^{-4}$	5 min 51.3 sec
Adagrad	10^{-2}	$3.45 * 10^{-3}$	6 min 8.7 sec
RMSprob	10^{-4}	$2.07 * 10^{-3}$	6 min 54.6

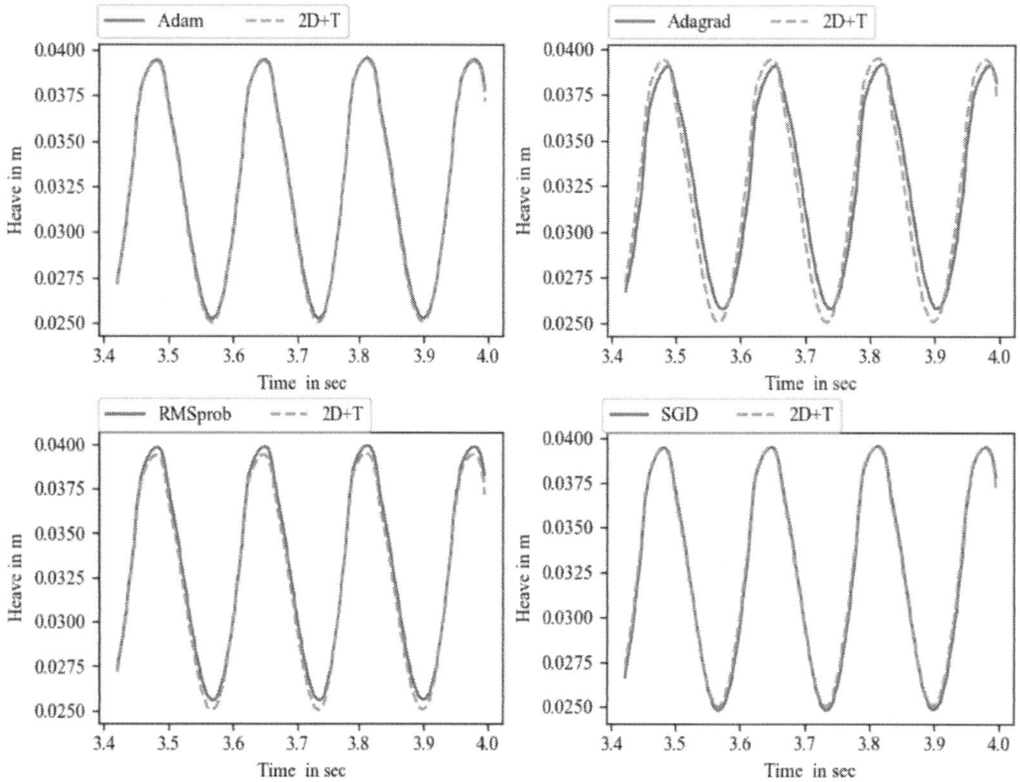

Figure 3. The accuracy of the training with 4 different optimizers after 50 epochs. Adam outperformed the other optimizers in training time and accuracy.

RNN with 1 hidden layer and 5 neurons were faster in comparison to all other networks, but it was also the lowest in mean absolute error. Furthermore, GRU outperformed both LSTM and RNN. GRU was faster than LSTM and has better mean absolute error. To try to draw a conclusion from this finding, for the heave motion in regular waves, a hidden GRU layer with 5 neurons was enough to obtain very good accuracy. del Águila Ferrandis et al. [17] suggested using 1 hidden LSTM layer with 20 neurons while Xu et al. [18] suggested using a structure consists of 5 LSTM hidden layers, each consists of 50 neurons. Both structures will defenitely achieve the task. As we explained earlier, every neuron in the hidden layer is a linear classifier and every hidden layer acts as a linear connection between these linear classifiers. More layers in the network or more neurons will lead the model to be able to capture more nonlinearity in the data. The cost in this case as we explained earlier is the computational time. Moreover, for this simple models, it doesn't matter as the difference is in 1 minute, but for the case of training more complex and general models, that difference is significant in the training process. Consequently, the parameter investigation was concerned with the size of the network, the structure of the hidden layers and the optimizer used in training the network. The next part of the study will be concerned with the size of the data used to train the network and the ability of the network to predict newly encountered waves.

Table 4. RNN (5 and 100 neurons), LSTM (5neurons) and GRU (5 neurons).

Network	Number of neurons	Mean absolute error after 50 epochs	Time taken
RNN	100	$8.922 * 10^{-4}$	5 min 51.3 sec
RNN	5	$1.139 * 10^{-3}$	4 min 43 sec
LSTM	5	$8.31 * 10^{-4}$	5 min 44.15 sec
GRU	5	$6.1155 * 10^{-4}$	5 min

To investigate the impact of the size of the data on the accuracy of the prediction (figure 4), 25% of the training data will be used. Reducing the size of the training data didn't affect the model accuracy significantly. The model obtained a mean absolute error of 0.001 and was trained in 3 minutes and 5.68 seconds. This was almost 40% decrease from the 5 minutes taken by the same GRU model that was trained with the whole data.

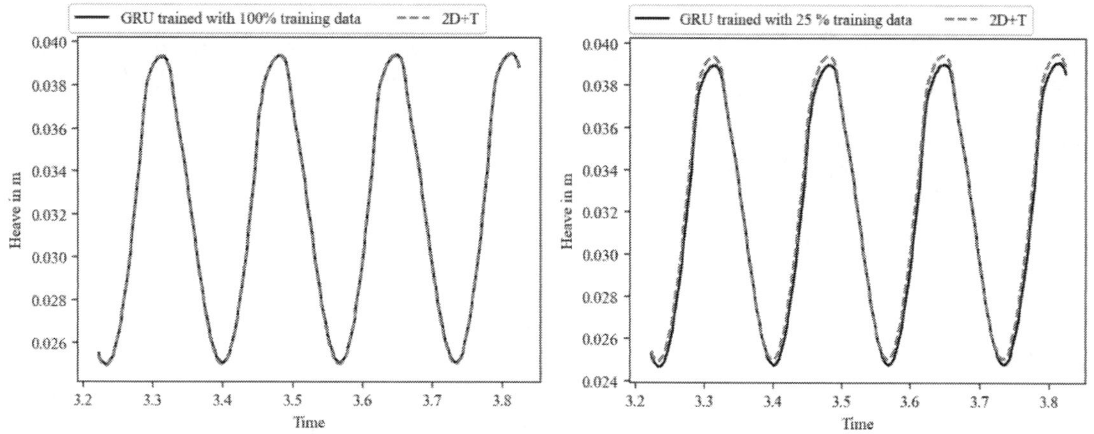

Figure 4. Using 25% of the training data decreased the model quality. It took for the model in the right 3 minutes and 5.68 seconds while the model in the left was trained in 5 minutes.

Another important challenge is the ability of the model to predict the motion in new regular waves. New wave is a wave that was not used to train the model. Let's say the model was trained using some sort of waves with a specific amplitude and heights, will it be able to predict unseen regular waves? To answer this question, this paper will exclude two waves from the training data and use them as test data. Wave number 4 and 18 will be used as testing data. The remaining 16 waves will be used for training and validating the model. The reason why these two waves were chosen is because wave number 4 has a unique wave amplitude in comparison to the other waves while wave number 18 is the highest in amplitude and length in comparison to the other 16 waves.

Figure 5 shows the accuracy of the prediction for the unseen regular waves. The prediction was very satisfactory. The mean absolute error was 0.000298 which was an improvement from all the previous models. The explanation for that is due to the size of the training data. The number of data points included in this model was bigger than the one that was included in the previous models.

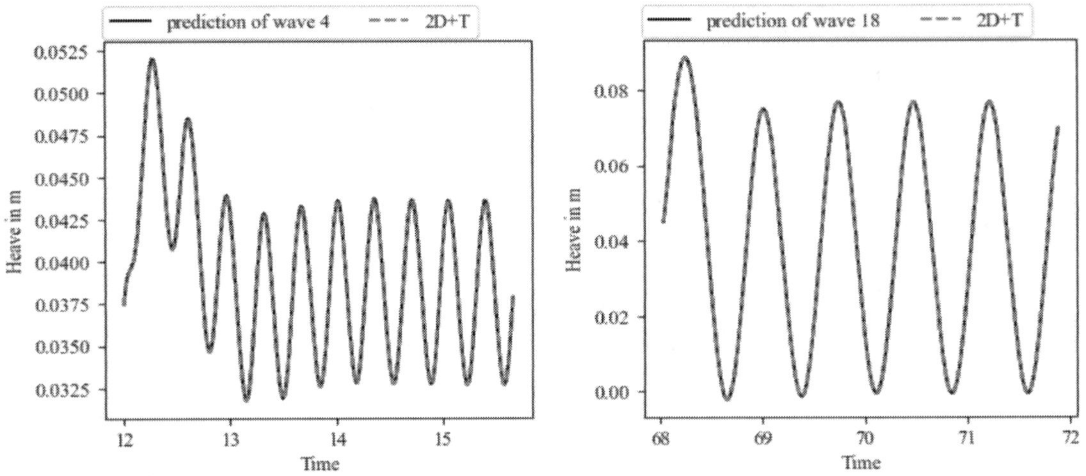

Figure 5. The converging of the model due to using the different optimizers. Adam was able to obtain this accuracy in almost 20 epochs.

5. Discussion and future outlook

This paper was an attempt by the authors to investigate the possibilities and the limitations of the data-driven models in building a seakeeping model in regular waves. Four different optimizers and 3 different network structures were investigated. Moreover, the paper studied the number of hidden layers and the neurons in each hidden layer. Furthermore, the effect of the size of the training data was explored. And finally, the possibility of predicting the heave motion in a regular wave that was not used to train the model was also explored. In this work, the heave motion was the case study, but the same method can be used for all 6 degrees of freedom.

All optimizers are based on SGD. SGD's main characteristic is that it guarantees convergence. The question here is converging to local or global minimum and how long it will take to converge. For instance, in Figure 5, it might converge to the first two peaks in the movement and predict the whole wave like these two but in our case, it was able to predict the right behavior. Adam optimizer was able to update the network weights to achieve good accuracy in a shorter time and epochs than the others. Adam outperformed the other optimizers. The paper doesn't say Adam is the best optimizer. All optimizers with the right parameter definition are guaranteed to train the network. For instance, using SGD with 0.9 momentum would achieve the task but would take more time and effort to achieve it. Adam is the smartest optimizer of all the optimizers, so it rarely stuck at a local minimum. However, Adagrad and RMSprob would outperform the other optimizers if the data were sparse or incomplete.

When investigating the network structure, RNN, LSTM and GRU with 5 neurons were enough to achieve mean absolute error of 0.001139, 0.000831 and 0.000611, respectively. del Águila Ferrandis et al. [17] proposed using 1 hidden layer with 20 neurons to achieve a good accuracy for regular wave model. This paper agrees with the finding that 1 hidden layer was sufficient for a high number of training samples to achieve a very good accuracy. Xu et al. [18] proposed a structure with 5 hidden LSTM layers for the neural network. This model was for the motion in irregular waves which is much more complex in nature in comparison to the motion in regular wave that was considered in this study. Furthermore, they both proposed LSTM was the best network in terms of structure. This may be due to that LSTM doesn't expose all the information stored in the network but just show the important information while GRU lacks this feature. This feature allowed LSTM to learn irregular waves much better. In this paper, the model GRU was faster and had lower parameters and mean absolute error than LSTM. However, this paper could say with confidence any RNN would achieve the task perfectly. Some would need more neurons such as RNN or more epochs but in the end, the simple RNN is able to capture the non-linearity in the model. To put that into context, LSTM and GRU biggest advantage is that they contain some gates that can help the structure to remember or choose to forget certain information. For the heave motion in regular waves, the future prediction is just relying on the near past information which is a simple task that could be achieved by any RNN structure. For irregular waves, and due to the existence of several frequencies in single wave, RNN might not be able to develop a model but indeed, LSTM and GRU would be able to achieve it. Maybe, the 5 neurons structure cannot achieve it or the learning rate would need adjusting but using the same methodology that was used in this paper the model is achievable.

One of the major finding of this paper is that the model actually was able to predict two waves that was not used to train the model with high accuracy in agreement with the work done in [17] and [18]. The model was able to predict unseen data with high accuracy. The major factor here which could affect the prediction accuracy is the size and the generalization of the training data. The model included different wavelengths which resulted in different heave motion magnitudes as shown in Table 1. Thus, it was able to build an assumption on how this model will react to some new waves amplitudes or heights. Therefore, the model avoided overfitting as it was able to predict new information. Furthermore, simplifying the model further will affect the accuracy. Using 3 neurons for instance or replacing GRU with RNN will affect the model accuracy. So, the model avoided underfitting, as well.

This model is limited to regular waves and Fridsma planing hull. One way to improve this model is to include irregular waves by using some data combining regular and irregular waves. Furthermore, to improve the generalization of the model, the data can include motions for 8 or 10 planing hulls that are within the feasible hull characteristics. Moreover, one of the approaches can be to use geometry modification software and investigate the design space and create data that consider the hull characteristics as variables. One problem that can arise from this procedure is that due to the size of the training data may not be possible to find a pattern in this data or for the model to understand the behavior correctly. One way to move around such a challenge is by using the attention models [29] that were used in improving the speech recognition problem. Another approach is to use the hybrid surrogate models that were used in predicting only the non-linear terms in the equation of motion in [19]. This approach is one of the methods that can help in reducing the computational complexity, removing the inconsistency in the captured data and might be able to improve the prediction if the captured data is not enough. At the end, the good thing about using surrogate models is that it is full of potential and can be implemented soon. Once a model is trained, it is ready to predict the future with very good accuracy.

6. References

[1] Von Karman T H 1929 The impact on seaplane floats during l anding
[2] Wagner H 1932 Über Stoß- und Gleitvorgänge an der Oberfläche von Flüssigkeiten *ZAMM - J. Appl. Math. Mech. Z. Für Angew. Math. Mech.* **12** 193–215
[3] Martin M 1978 THEORETICAL PREDICTION OF MOTIONS OF HIGH-SPEED PLANING BOATS IN WAVES *J. Ship Res.* **22**
[4] Zarnick E E 1978 A nonlinear mathematical model of motions of a planing boat in regular waves
[5] Keuning J A 1994 Nonlinear Behaviour of Fast Monohulls in Head Waves
[6] Payne P R 1995 Contributions to planing theory *Ocean Eng.* **22** 699–729
[7] Ghadimi P, Dashtimanesh A, Djeddi S R and Maghrebi Y F 2013 Development of a mathematical model for simultaneous heave, pitch and roll motions of planing vessel in regular waves *Int. J. Adv. Sci. Res.* **1** 44–56
[8] Sebastiani L, Bruzzone D, Gualeni P, Rambaldi G, Ruscelli D and Viviani M 2008 A Practical Method for the Prediction of Planing Craft Motions in Regular and Irregular Waves *Volume 4: Ocean Engineering; Offshore Renewable Energy* ASME 2008 27th International Conference on Offshore Mechanics and Arctic Engineering (Estoril, Portugal: ASMEDC) pp 687–96
[9] Tavakoli S, Niazm and Bil andi R, Mancini S, De Luca F and Dashtimanesh A 2020 Dynamic of a planing hull in regular waves: Comparison of experimental, numerical and mathematical methods *Ocean Eng.* **217** 107959
[10] Ghadimi P, Tavakoli S and Dashtimanesh A 2016 Coupled heave and pitch motions of planing hulls at non-zero heel angle *Appl. Ocean Res.* **59** 286–303

[11] Dashtimanesh A, Ghaemi M H, Wang Y, Karczewski A, Bil andi R N and Hirdaris S 2022 Digitalization of High Speed Craft Design and Operation Challenges and Opportunities *Procedia Comput. Sci.* **200** 566–76

[12] Suykens J A K and V andewalle J 1999 Least Squares Support Vector Machine Classifiers *Neural Process. Lett.* **9** 293–300

[13] Rumelhart D E, Hinton G E and Williams R J 1986 Learning representations by back-propagating errors *Nature* **323** 533–6

[14] Hochreiter S and Schmidhuber J 1997 Long Short-Term Memory *Neural Comput.* **9** 1735–80

[15] Raissi M, Perdikaris P and Karniadakis G E 2019 Physics-informed neural networks: A deep learning framework for solving forward and inverse problems involving nonlinear partial differential equations *J. Comput. Phys.* **378** 686–707

[16] Sclavounos P D and Ma Y 2018 Artificial Intelligence Machine Learning in Marine Hydrodynamics ASME 2018 37th International Conference on Ocean, Offshore and Arctic Engineering (American Society of Mechanical Engineers Digital Collection)

[17] del Águila Ferr andis J, Triantafyllou M S, Chryssostomidis C and Karniadakis G E 2021 Learning functionals via LSTM neural networks for predicting vessel dynamics in extreme sea states *Proc. Math. Phys. Eng. Sci.* **477** 20190897

[18] Xu W, Maki K J and Silva K M 2021 A data-driven model for nonlinear marine dynamics *Ocean Eng.* **236** 109469

[19] Marlantes K E and Maki K J 2022 A neural-corrector method for prediction of the vertical motions of a high-speed craft *Ocean Eng.* **262** 112300

[20] Ying X 2019 An Overview of Overfitting and its Solutions *J. Phys. Conf. Ser.* **1168** 022022

[21] Hawkins D M 2004 The Problem of Overfitting *J. Chem. Inf. Comput. Sci.* **44** 1–12

[22] Kingma D P and Ba J 2017 Adam: A Method for Stochastic Optimization

[23] Duchi J, Hazan E and Singer Y 2011 Adaptive Subgradient Methods for Online Learning and Stochastic Optimization *J. Mach. Learn. Res.* **12** 2121–59

[24] Tieleman T and Hinton G 2012 Lecture 6.5-rmsprop: Divide the gradient by a running average of its recent magnitude *COURSERA Neural Netw. Mach. Learn.* **4** 26–31

[25] Ruder S 2017 An overview of gradient descent optimization algorithms

[26] DiPietro R and Hager G D 2020 Chapter 21 - Deep learning: RNNs and LSTM *Handbook of Medical Image Computing and Computer Assisted Intervention* The Elsevier and MICCAI Society Book Series ed S K Zhou, D Rueckert and G Fichtinger (Academic Press) pp 503–19

[27] Chung J, Gulcehre C, Cho K and Bengio Y 2014 Empirical Evaluation of Gated Recurrent Neural Networks on Sequence Modeling

[28] Parikh A, Täckström O, Das D and Uszkoreit J 2016 A Decomposable Attention Model for Natural Language Inference *Proceedings of the 2016 Conference on Empirical Methods in Natural Language Processing* EMNLP 2016 (Austin, Texas: Association for Computational Linguistics) pp 2249–55

[29] Vaswani A, Shazeer N, Parmar N, Uszkoreit J, Jones L, Gomez A N, Kaiser Ł and Polosukhin I 2017 Attention is All you Need *Advances in Neural Information Processing Systems* vol 30 (Curran Associates, Inc.)

[30] Vasilescu M A O 2009 *A Multilinear (Tensor) Algebraic Framework for Computer Graphics, Computer Vision and Machine Learning* Thesis

[31] Fridsma G 1971 A systematic study of the rough water performance of planing boats *Stevens Inst. Technol. Davidson Lab. Castle Point Stn. Hoboken New Yersey USA Rep. No 1275*

IOP Conf. Series: Materials Science and Engineering 1288 (2023) 012027 doi:10.1088/1757-899X/1288/1/012027

Numerical study on the influence of nozzle divergence half angle on thrust oscillation characteristic of underwater jet power

Pengfei Si[1,*], Ningfei Wang[1], Shitong Xia[1], Lin Shi[1], Shipeng Li[1] and Jianing Tang[2]

[1] School of Aerospace Engineering, Beijing Institute of Technology,
5 Zhongguancun South Street, Beijing 100081, China
[2] Institute of Unmanned Autonomous System, Yunnan Minzu University, 134 Yieryi Street, Kunming 650031, China
*Corresponding author, Email: sipengfei2014@163.com

Abstract. Underwater rocket propulsion is increasingly used in advanced marine vehicles that may be subject to supersonic gaseous jets while submerged. This work investigates the influence of supersonic nozzle divergence on the multi-flow field evolution process and the thrust oscillation characteristics by multiple numerical simulations using a VOF multiphase flow model. Four nozzle divergence half angles are taken into consideration. The analysis indicates that the thrust rises sharply at nozzle closure opens, then rapidly decreases and follows another rise. The thrust oscillates at high frequencies. As a result, increase of the nozzle divergence half angle leads to the size decrease of flow bubbles but does not affect the changing trends of the thrust. Increase of the nozzle divergence half angle leads to a reduction of the frequency of the thrust oscillations.

1. Introduction

Using a solid rocket motor (SRM) as an underwater propulsion unit allows marine vehicles to achieve higher speeds. Gaseous jets injected into water are typically found in underwater rocket propulsion, and the high water-to-gas density ratio can result in oscillatory thrust. Hence it remains a challenging issue to stabilize the thrust of underwater rocket motors. At present, underwater rocket motors often use small expansion ratio nozzles making the thrust more stable [1].

Studies of submerged gaseous jets have been carried out during past decades. Shuqun Yi et al. [2] used the stagnant isobaric bubble model and MAC method to solve the one-dimensional model of underwater gaseous jets. And the gas bubble evolution process of underwater gaseous jets was calculated successfully. Jiayi Cao et al. [1] performed simulations of underwater supersonic gaseous jets generated by SRM using the Mixture multiphase flow model and found that there is no fixed oscillation frequency of the pressure of the underwater gas flow field. Haipeng Wei et al. [3] studied the structure of underwater supersonic gaseous jets in under-expanded conditions using the Mixture multiphase flow model and found that there is a Mach disk structure in the core area inside the gas bubble. Jianing Tang [4-5] found that expansion, necking, and back-attack are fundamental phenomena in underwater gaseous jets. She also used a finite-time Lyapunov exponent (FTLE) field to identify the structures of multiphase flows and captured the interface vortex region. By taking phase transitions and buoyancy effects into consideration, Yunlong Tang et al. [6] performed numerical simulations of gaseous jets for underwater rocket propulsion systems and found that buoyancy effects can result in the momentum of force and the thrust shows different orders of oscillatory characteristics. Haiyu Xu et al. [7] studied the unsteady characteristics of underwater supersonic gaseous jets and found that the pressure oscillations of the flow field have certain periodic characteristics. In addition, the mass flow rate of the throat of nozzle oscillates greatly in over-expanded conditions. Xiaoyuan Zhang et al. [8] adopted both experimental and computational methods in studying the flow patterns of underwater gaseous jets. Ge Jin et al. [9] studied the unsteady characteristic of underwater gaseous jets and thrust stability.

Previous studies of underwater supersonic gaseous jets focus mainly on the influence of the nozzle expansion ratio, chamber pressure, water depth, and so on. It is necessary to study the influence mechanism of nozzle divergence half angle on the gas bubble evolution and the thrust oscillation characteristics. RANS method and VOF multiphase flow model were adopted in simulating the flow field of underwater solid rocket motor ignition at a depth of 100 m. In detail, four nozzle divergence half angles were taken into consideration to investigate the change of gas bubble structure and thrust. The structure of this paper is as follows: Section 2 describes the numerical methods and computational domain; detailed results and discussion are presented in Section 3; and the conclusions are shown in Section 4.

2. Numerical methods and computational domain

2.1. Numerical methods

For underwater supersonic gaseous jets, all of the simulations were performed using the VOF multiphase flow model which tracks the change of gas-liquid interface. The VOF multiphase flow model is widely used in the prediction of jet break-up, the motion of large bubbles in liquid, and dam-break problems in engineering. In the CFD (Computational Fluid Dynamics) simulation, the injected gas is regarded as ideal compressible air, and water is treated as an incompressible liquid. Both two phases have no additional generated sources, and no mass transfer occurs between them. The continuity, momentum, energy, and VOF model equations are listed as follows.

Continuous conservation equation:

$$\frac{\partial \rho_{\mathrm{m}}}{\partial t}+\frac{\partial}{\partial x_i}\left(\rho_{\mathrm{m}} u_i\right)=0 \tag{1}$$

Momentum conservation equation:

$$\rho_{\mathrm{m}} \frac{\mathrm{D} u_i}{\mathrm{D} t}=-\frac{\partial p}{\partial x_i}+\frac{\partial}{\partial x_j}\left[\mu_{\mathrm{m}}\left(\frac{\partial u_i}{\partial x_j}+\frac{\partial u_j}{\partial x_i}\right)\right] \tag{2}$$

Energy conservation equation:

$$\frac{\partial}{\partial t}\left(\rho_{\mathrm{m}} E\right)+\frac{\partial}{\partial x_i}\left[u_i\left(\rho_{\mathrm{m}} E+p\right)\right]=\frac{\partial}{\partial x_i}\left(k_{\mathrm{eff}} \frac{\partial T}{\partial x_i}\right) \tag{3}$$

The equation for conservation of phase volume fraction:

$$\frac{\partial}{\partial t}\left(\alpha_{\mathrm{n}} \rho_{\mathrm{n}}\right)+\frac{\partial}{\partial x_i}\left(\alpha_{\mathrm{n}} \rho_{\mathrm{n}} u_i\right)=0 \tag{4}$$

Where, t is the time; x_i , x_j are coordinate components; u_i, u_j are velocity components; ρ_{m} is the density coefficient of the mixed phase; μ_{m} is the viscosity coefficient of the mixed phase; p is the pressure, T is the temperature, E is the energy; k_{eff} is the effective thermal conductivity; α_{n} is the volume fraction of gas or liquid phase and ρ_{n} is the density of gas or liquid phase.

The shear-stress transport (SST) k-ω two-equation model was adopted to describe the turbulent flow, and SIMPLE pressure-velocity coupling algorithm was used to resolve the momentum equation. The finite volume method was adopted to solve the governing equations. The Body Force Weighted scheme was used in the pressure term. The density, momentum, and energy term followed a first-order upwind scheme. The time discretization term was based on the first order implicit scheme.

2.2. Numerical method validation
Experiments were performed by Chun Zhang et al. [10] to investigate the characteristics of underwater SRM ignition. To validate the numerical method in Section 2.1, a two-dimensional axisymmetric model was established according to the size of the underwater SRM used in the experiment [7]. Pressure inlet and other boundary conditions were selected according to the experiment [7] cases. Experimental and numerical simulation results of underwater gaseous jets are shown in Figure 1. The upper part of the figure is experimental results, and the lower part is numerical simulation results. It can be seen from Figure 1 that the size of simulated gas bubbles is almost identical to that of experimental gas bubbles. Hence the numerical method can effectively simulate underwater supersonic gaseous jets.

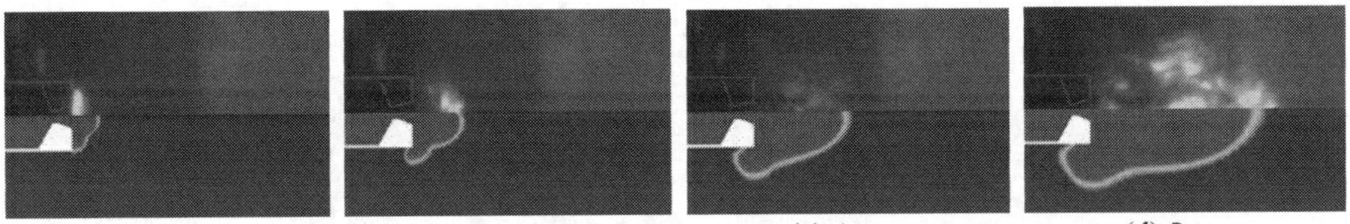

(a) 1 ms (b) 2 ms (c) 4 ms (d) 8 ms

Figure 1. Comparison of experimental [7] and numerical simulation results.

2.3. Computational grid and boundary conditions
Figure 2 shows the structural schematic diagram of the SRM used in this study. As shown in Figure 2, the nozzle expansion ratio is the ratio of the nozzle exit area to the nozzle throat area. In Figure 2, the angle between the extension of the nozzle expansion part and the central axis is α which is the nozzle divergence half angle.

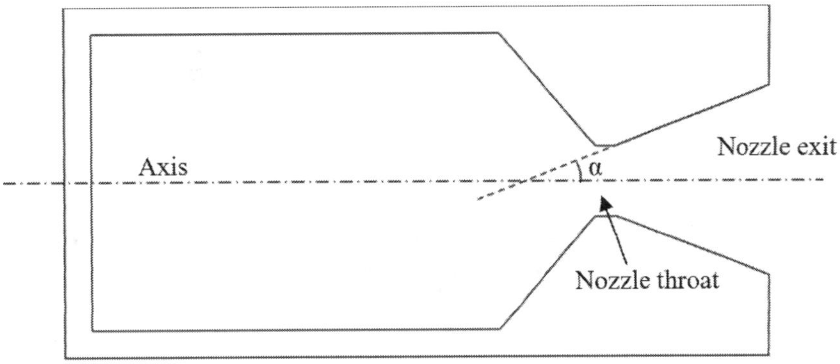

Figure 2. Structural schematic diagram of the SRM.

The computational mesh is a two-dimensional axisymmetric model. The computational domain includes internal and external flow fields of the SRM nozzle as shown in Figure 3. The nozzle throat diameter D_t and nozzle exit diameter D_e of the SRM are 7 mm and 14 mm respectively. To overcome the influence of the boundary conditions on the gaseous jets core area, the total length of the computational domain is set to 160 times of the nozzle throat diameter, and the width is set to 60 times of the diameter of the nozzle throat. The mesh is refined in areas where the flow field parameters change drastically, such as the nozzle wall and the tail wall near the nozzle exit.

The nozzle inlet is set to pressure inlet boundary condition, and the total pressure is 5 MPa. To simplify the calculation process, the gas parameters are set to constants, where the total temperature of the gas is set to 3050 K and the specific heat capacity is set to 1750 J/kg·K. The boundary condition of the outlet is set to pressure outlet, and the static pressure is set to 1.1 MPa. The temperature of the water is set to 300 K. Walls of the domain are subjected to a non-slip wall boundary condition. The domain is initialized with the outlet value. To meet the requirement of CFL (Courant-Friedrichs-Lewy) calculation convergence conditions, the computational time step is selected as 1×10^{-6} s.

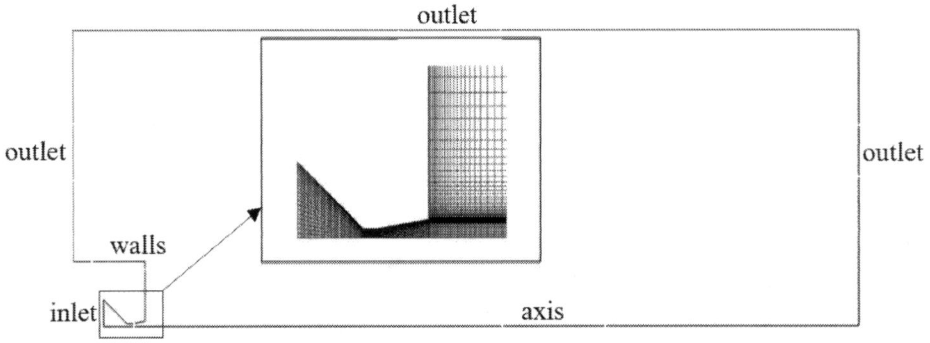

Figure 3. Computational mesh and computational domain.

The grid independent study is shown in Figure 4. Three different mesh densities are performed in calculating the force on the SRM tail wall. The total number of different meshes is about 50 000, 90 000, and 140 000, respectively. As shown in Figure 4, the number of computational mesh basically has no effect on the simulation results, which shows the numerical results are independent with mesh. As a result, the total mesh number of 90 000 is chosen under a comprehensive evaluation of computing accuracy and economy in resources.

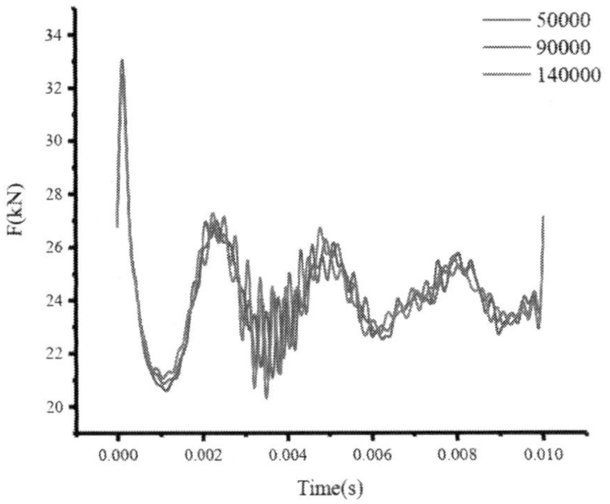

Figure 4. Force on the SRM tail wall.

3. Results and discussion

3.1. Gas bubbles evolution process underwater

The moment when the nozzle closure opens is regarded as 0 s. Table 1 shows the development of the underwater gaseous jet structure under four different nozzle divergence half angle conditions within 4.2 ms. It can be seen from Table 1 that changing the nozzle divergence half angle has no effect on the

IOP Conf. Series: Materials Science and Engineering 1288 (2023) 012027 doi:10.1088/1757-899X/1288/1/012027

basic development of bubbles formed by high-temperature supersonic gaseous jets. However, increase of the nozzle divergence half angle causes slight differences in the shape of gas bubbles. As indicated by the black arrow in Table 1, a surface depression occurs in the centre of the gas bubble formed under larger nozzle divergence half angle conditions. As shown in the table, the size of the gas bubble decreases with the increase of nozzle divergence half angle.

Table 1. Gaseous jets development under four different nozzle divergence half angle conditions.

	0.2 ms	0.7 ms	1.2 ms	2.2 ms	3.2 ms	4.2 ms
9.9°	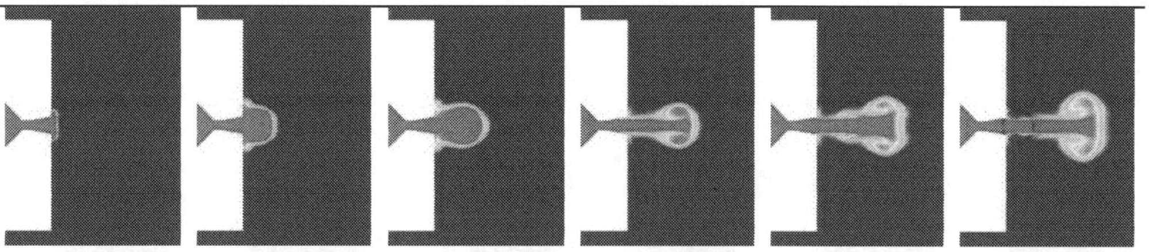					
15°	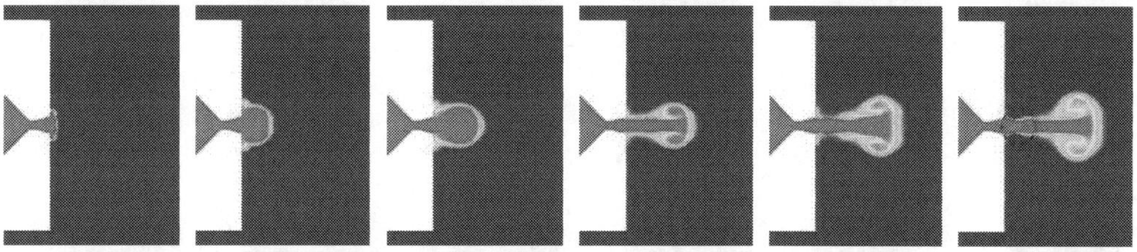					
20°	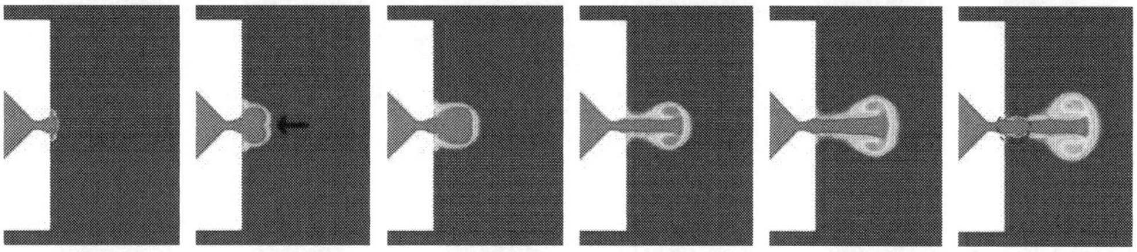					
25°	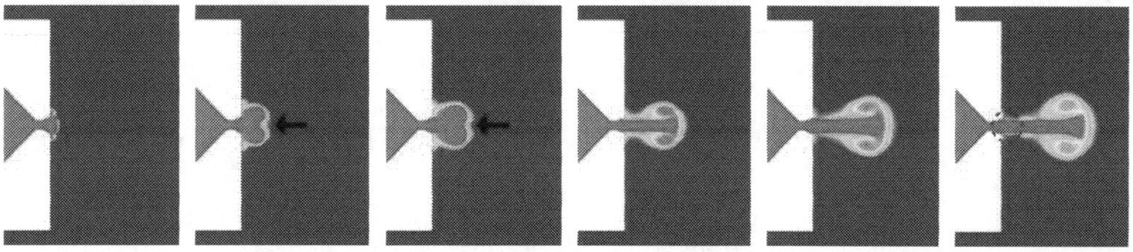					

For the development of underwater gaseous jets, the flow field is extremely unstable and shows expansion-necking oscillation characteristics. When the nozzle closure opens at 0 ms, the gaseous jets enter the water rapidly. Due to the pressure difference of 3.9 MPa between gas and water, the high-temperature gas in the nozzle expands quickly and pushes water aside, forming an initial gas bubble, as shown in Table 1. At the moment of 0.2 ms, the speed of the jets in the initial gas bubble is subsonic. Hence the initial gas bubble expands along the SRM tail wall rather than the axis, and the shape of the initial gas bubble loos like a "hat" covering the nozzle exit. Then the gas gradually accelerates in the nozzle to form a supersonic jet, and drives the gas bubble to expand in the axial direction. At 0.7 ms, the shape of the gas bubble is hemispherical, as shown in the table. At 1.2 ms, it can be seen from Table 1 that the diameter of the gaseous jet near the nozzle exit becomes narrow under the high pressure of the surrounding water, which is the so-called necking phenomenon. Meanwhile, the gas bubble keeps expanding and spreading away from the nozzle exit. After 1.2 ms, the gas bubbles undergo expanding and necking again, as indicated by the black circle in Table 1. However, the size of the bulge after initial expanding only equivalent to the diameter of the nozzle exit. As a result, the overall shape of the gas bubble formed by underwater supersonic gaseous jets is a slender cone.

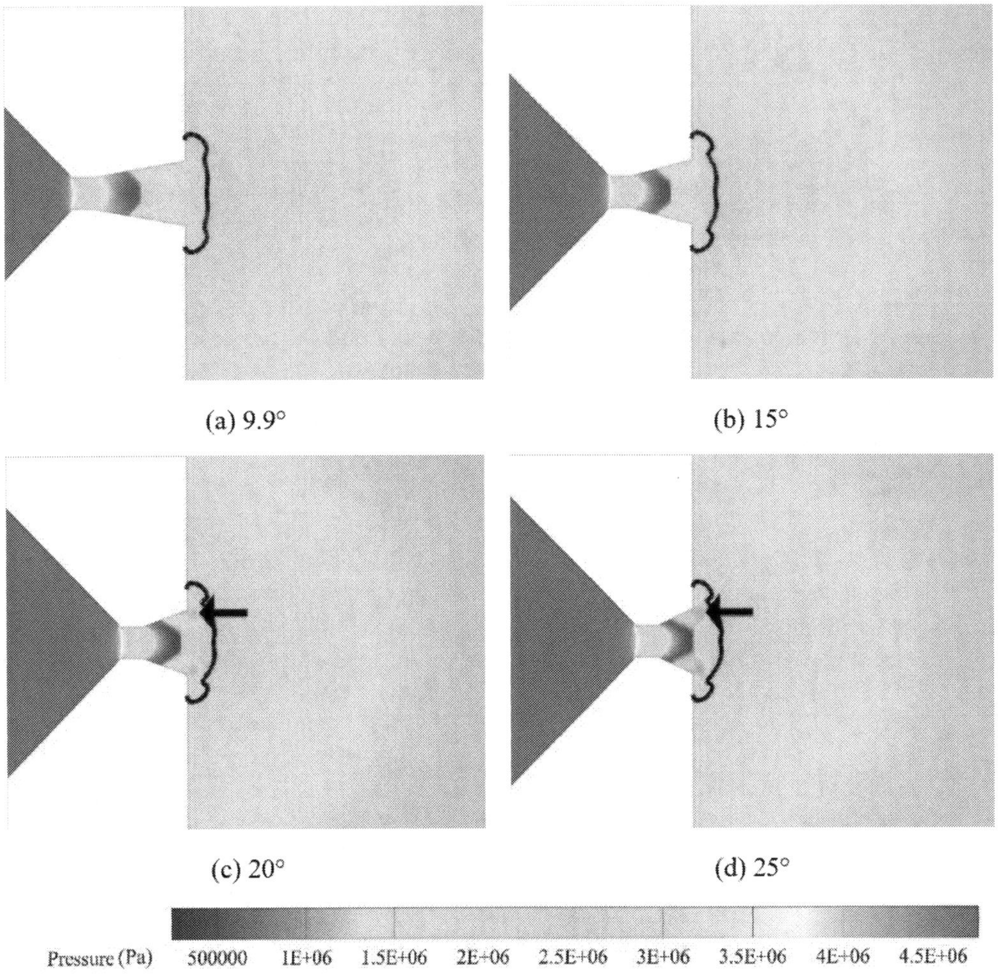

Figure 5. Pressure contours at 0.2 ms.

As the nozzle divergence half angle increases, the central bulge of the hat-shaped initial gas bubble becomes more prominent, as shown in Table 1. Figure 5 shows the pressure contour of the flow field under four nozzle divergence half angle conditions. The interface of the two phases is plotted by iso-volume of gas fraction of 0.6, which is indicated by the black solid line. It can be seen from the figure that a strong shock wave appears inside the nozzle at this moment. Because for underwater gaseous jets, in the initial process of ignition, the nozzle is likely to be working under over-expanded conditions due to the high ambient pressure of surrounding water.

As shown in Figure 5, the shock wave moves closer to the nozzle exit with the increase of the nozzle divergence half angle. As a result, there is a lower-pressure region, as marked by the black arrow in the figure, appearing at the edge of the nozzle exit under larger nozzle divergence half angle conditions. This phenomenon subsequently causes lower pressure in the edge of the gas bubble and higher pressure in the centre of the gas bubble. And as a result, the central bulge of the initial gas bubble becomes more prominent with the increase of the nozzle divergence half angle.

In addition, at the moments of 0.7 ms and 1.2 ms, as shown in Figure 1, there is a surface depression appearing in the centre of the gas bubble under nozzle divergence half angle of 25° condition. The reason for this phenomenon is that increasing nozzle divergence half angle reduces the ability of nozzle to concentrate gas.

3.2. Process of establishing supersonic gaseous jets inside the nozzle

To maintain a constant nozzle expansion ratio, different nozzle divergence half angles can be obtained by changing the length of the nozzle divergence part. In this section, the development of gaseous jets was compared by selecting nozzle divergence half angles of 7°, 9.9°, and 20°.

Table 2 shows the pressure and Mach number contours over a certain period of time (0.2-3.2 ms) of gaseous jets under three nozzle divergence half angle conditions. In the table, the black solid line represents the interface between the two phases. In addition, the upper half of each image in Table 2 shows the pressure of the flow field, and the lower part shows the Mach number.

Table 2. Contours of flow field pressure and Mach number.

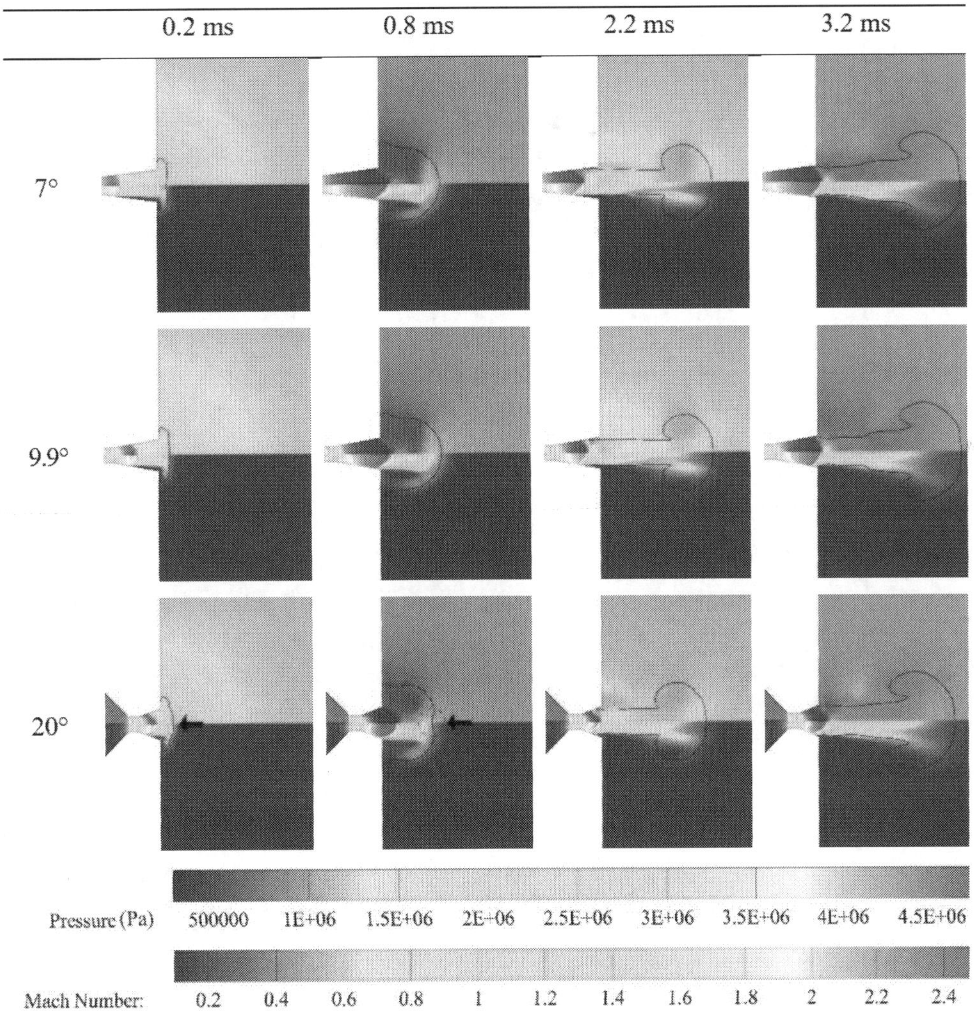

At 0.2 ms, according to the Mach number contour in Table 2, the speed of gaseous jets inside the nozzle is transonic. Then the Mach number of the gaseous jets rises to 2.45 at 0.8 ms. It can be seen

from the table that the supersonic region of the gaseous jets keeps oscillating between 0.2 ms to 3.2 ms. In addition, the great ambient pressure even causes the flow separation phenomenon in the nozzle at 2.2 ms, which will lead to an extremely unstable SRM working state. And this phenomenon occurs with changing the nozzle divergence half angle. The core area of the supersonic gaseous jets is compressed hard when the nozzle divergence half angle is large enough, such as 20°, and the shock wave is very close to the nozzle throat, in which condition the SRM may stop. As a result, in the process of establishing supersonic gaseous jets inside the nozzle, there is a shock wave moving back and forth between the nozzle throat and the nozzle exit.

3.3. Thrust oscillation

As shown in Figure 6, the pressure on the internal and external surfaces of the SRM is unbalanced. Therefore, an axial unbalanced force is generated, which is the thrust of SRM. Based on this definition, the total thrust of underwater SRM is the integration of pressure on all internal and external surfaces of the SRM.

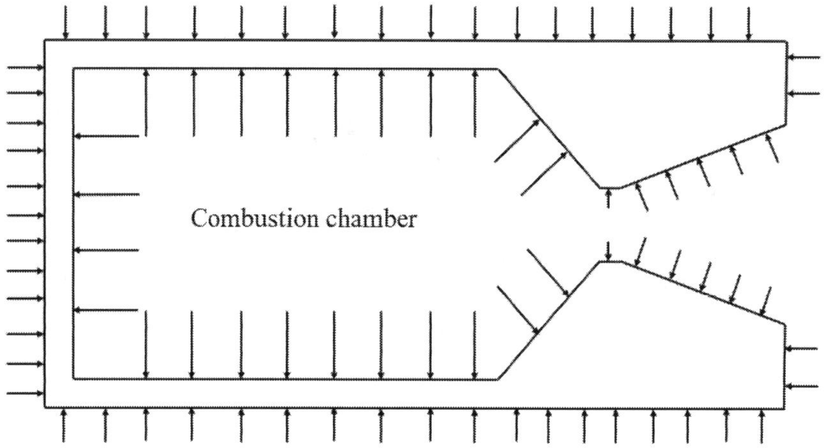

Figure 6. Pressure distribution on the internal and external surfaces of the SRM.

Figure 7 shows the thrust of underwater SRM under three nozzle divergence half angle conditions. As shown in Figure 7, the thrust of underwater SRM has obvious oscillation characteristics within 70 ms after ignition. There is a peak value of the SRM thrust at the moment when the nozzle closure opens. Then the thrust decreases sharply followed by another rise.

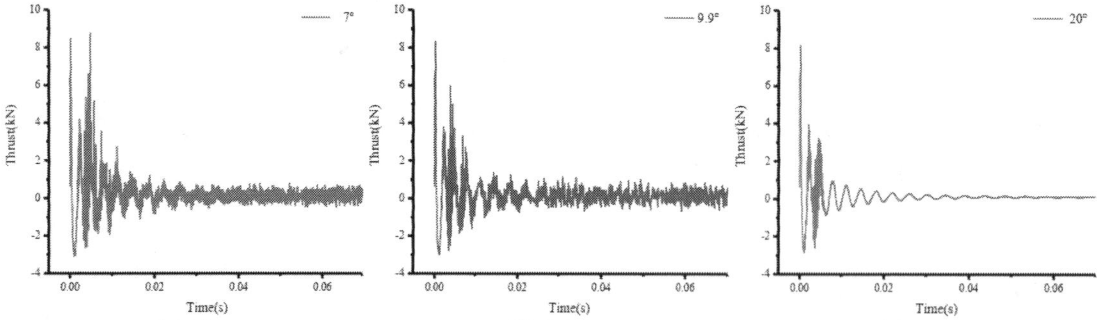

(a) 7° (b) 9.9° (c) 20°

Figure 7. The thrust of underwater SRM.

It can be seen from Figure 7 that increasing nozzle divergence half angle can reduce the frequency of SRM thrust oscillation. According to Figure 7 (c), the thrust of SRM with a nozzle divergence half angle of 20° nearly stops oscillating after 50 ms, which explains that a larger nozzle divergence half angle makes the thrust of underwater SRM more stable. While the thrust of underwater SRM with a smaller nozzle divergence half angle keeps oscillating at high frequencies. The thrust of the SRM with nozzle divergence half angle of 20° finally remains at about 130 N, while the thrust of the other two SRMs is going to be negative from time to time.

To investigate the influence of nozzle divergence half angle on the changing trend of the thrust, filtered thrust curves of underwater SRM with different nozzle divergence half angles are illustrated in Figure 8. As shown in the figure, three filtered SRM thrust curves are almost coincided with each other, which illustrates that changing nozzle divergence half angle has no obvious effect on the SRM thrust value and oscillation trend. As a result, the thrust of underwater SRM with a larger nozzle divergence half angle is more stable.

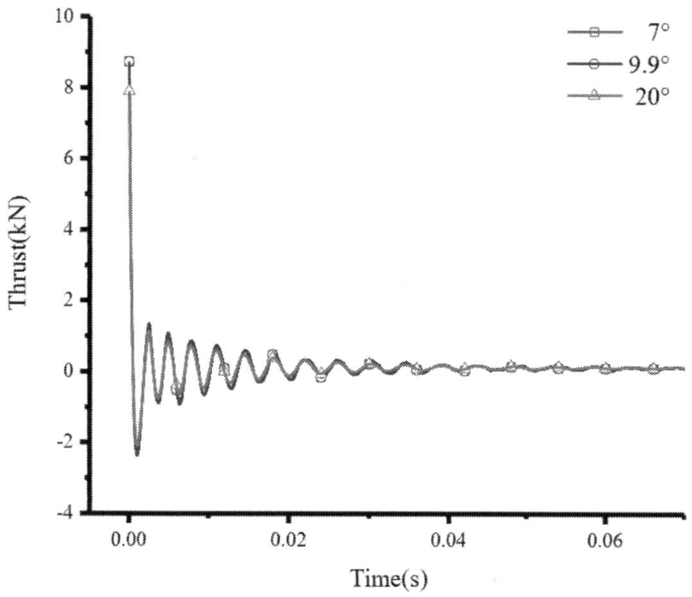

Figure 8. Comparison of three filtered thrust curves.

4. Conclusion

In the present work, the RANS method and VOF multiphase flow model are adopted to study the influence of nozzle divergence half angle on the thrust oscillation characteristics of underwater solid rocket motors.

In the early stage of igniting the underwater SRM, the initial gas bubble has typical underwater gaseous jet morphological characteristics such as expansion and necking. While the size of the bulge after the initial expansion is only equivalent to the diameter of the nozzle exit. Hence, the overall shape of the gas bubble formed by underwater supersonic gaseous jets is a slender cone. As a result, the size of the gas bubble decreases with the increase of the nozzle divergence half angle.

Due to the high ambient pressure of surrounding water, a shock wave appears inside the nozzle and moves back and forth in it, causing the oscillation of the supersonic region of gaseous jets.

There is a peak value of the SRM thrust at the moment when the nozzle closure opens. Then the thrust decreases sharply followed by another rise and oscillates above and below the average thrust. Changing the nozzle divergence half angle has no obvious effect on the SRM thrust value and oscillation trend. To a certain extent, a larger nozzle divergence half angle reduces the frequency of underwater SRM oscillation. As a result, the thrust of underwater SRM with a larger nozzle divergence half angle is more stable.

References

[1] Cao J, Lu C, Li J and Chen X 2009 Study on the dynamic characteristics of underwater supersonic gas jets *Chinese Journal of Hydrodynamics* 2009 24(05):575-582

[2] Yi S, Xu C, Yan K and Chen J 2002 Study of underwater high-speed jet gas flow field *Chinese Journal of Hydrodynamics* 2002(04):448-453

[3] Wei H, Guo F and Quan X 2009 Numerical study of underwater gas jets *Missiles and Space Vehicles* 2009(02):37-39+47

[4] Tang J 2011 Study on the Characteristics of the underwater solid rocket engines *Beijing Institute of Technology* 2011

[5] Tang J, Wang N and Shyy W 2011 Flow structures of underwater supersonic gas jets *Acta Mechanica Sinica* 2011 27(4):461-472

[6] Tang Y, Li S, Liu Z, Sui X and Wang N 2015 Study on the initial characteristics of horizontal jets of underwater rockets *Acta Physica Sinica* 2015 64(23):193-205

[7] Xu H, Luo K and Liu R 2019 Study on the unsteady characteristics of underwater supersonic air flow field *Journal of Propulsion Technology* 2019 40(11):2618-2625

[8] Zhan X, Li S, Yang B and Wang N 2020 Flow structure of over-expanded supersonic gaseous jets for deep-water propulsion *Ocean Engineering* 2020 213:107611

[9] Jin G, Wang D and Li S 2022 Study on the influence of incoming flow velocity on thrust stability of underwater solid rocket motor *AIAA Scitech Forum* 2022

[10] Zhan C, Yu W and Wang B 2018 Study on the initial flow field characteristics of underwater supersonic gas jets *Acta Armamentarii* 2018 39(05):961-968

Numerical study of second-order wave forces using a Rankine panel method

Guohua Dong[1,2], Jianghao Huang[1,2], Chaobang Yao[1,2,*], Dakui Feng[1,2], Fanliang Meng[3]

[1]School of Naval Architecture and Ocean Engineering, Huazhong University of Science and Technology, Wuhan, China, 430074;
[2]Hubei Key Laboratory of Naval Architecture and Ocean Engineering Hydrodynamics (HUST), Wuhan, China, 430074;
[3]China Special Vehicle Research Institute, Jingmen, China, 448035;
*yaochaobang@hust.edu.cn

Abstract. Numerical analysis of the wave forces acting on ships can be complex primarily because of the interactions between incident waves and structures. In this paper, the three-dimensional frequency domain Rankine panel method is used to study the seakeeping and second-order wave drift forces of a ship in waves. The model considers second-order wave forces acting on the ship hull and therefore accounts for slow drift ship motions and the added resistance. The second-order wave forces are evaluated using the pressure integral method, also known as the near-field approach. The unsteady velocity potential is divided into three parts namely (a) incident, (b) diffraction and (c) radiation to satisfy the Laplace equation. To avoid wave reflection the Rayleigh's artificial friction is used in way of the free surface boundary. This method is equivalent to introducing a numerical damping beach for energy dissipation. The numerical results are compared with those of experiments for Wigley-III hull in head waves and the SCb-84 hull in oblique waves. Good agreement is obtained.

1. Introduction

With increasing maritime trade over the years, more and more ships are sailing the world's waters. However, maritime transportation faces safety challenges, particularly regarding ship motion in waves. It is important to study the motion characteristics of ships in waves. This paper aims to explore the motion characteristics of ships in waves and provide some insights.

In term of seakeeping issue, the three-dimensional panel method is widely used for ships and offshore structures to solve velocity potentials by discretizing the body surface into triangles or quadrilateral elements and using Green's function. The three-dimensional panel method was initially used for zero-speed offshore floating structures and was applied in engineering software such as WAMIT, Maxsurf and HydroStar. In order to deal with the problem with the forward speed, Haskind[1] proposed a Three-Dimensional Translating-pulsating Source Green Function (3DTP), which satisfied the free surface condition and the radiation condition with forward speed and could be expressed in various forms. Lv et al.[2][3] calculated the motions and mean second-order wave-induced forces of Wigley and S-175 ships based on the frequency-domain Rankine source boundary element method. Both the uniform flow and double-body flow models were employed to linearize the free surface conditions, and the effects of these two methods on the results were compared. Riesner and Moctar[4] applied a partial nonlinear time-domain boundary element method to investigate the added resistance of DTC container and KVLCC2 tanker ship models with sailing speed. Chen et al.[5] used the time-domain Taylor expansion boundary element method to calculate the motions and mean second-order wave forces of the SCb-84 ship.

When sailing in waves, ships are significantly affected by the low-frequency wave drift forces, known as the second-order wave forces. These forces can cause increased ship resistance and drifting motions. Nonaka[6] has proposed that the high-frequency wave-induced motions and low-frequency motions of the ship can be separated, and only the low-frequency wave drift forces will affect the ship's maneuverability. The mean drift forces method is helpful for the rapid prediction of ship maneuverability in waves as it only considers the low-frequency second-order wave drift forces and ignores the high-frequency first-order wave forces. Fournarakis et al.[7] predicted the turning maneuvers of DTC container ship model in waves by combining the 4-DOF maneuvering equations and the mean second-order wave forces obtained from the three-dimensional panel method. Chillcce and Moctar[8] combined the MMG model and the second-order drift forces calculated by the three-dimensional Rankine panel method to predict the turning circles of DTC in waves. The hydrodynamic coefficients for MMG were obtained by a RANS-based CFD solver.

In this study, the proposed three-dimensional frequency-domain Rankine panel method accounts for the N-K linear free surface condition modified with Rayleigh's artificial friction. Moreover, a time-averaging method based on the near-field pressure integration is derived to facilitate the calculation of mean second-order wave drift forces. The accuracy and effectiveness of the present method are demonstrated through validation using experimental data from the Wigley and SCb-84 ship models.

2. Materials and methods

To describe the motions of a ship in waves, three different coordinate systems are established: the inertial coordinate system O_0-$X_0Y_0Z_0$, the body coordinate system *O-XYZ*, and the reference coordinate system *o-xyz*. The O_0-$X_0Y_0Z_0$ is fixed to the earth. The origin *O* of the body-fixed frame *O-XYZ* is located at the projected center of gravity on the waterline plane, with the *X*-axis positive towards the bow, the *Y*-axis positive towards the port side, and the *Z*-axis positive upward. The *o-xyz* moves with the ship at velocity U, and is parallel to the O_0-$X_0Y_0Z_0$. The difference between the body coordinate system *O-XYZ* and the reference coordinate system *o-xyz* is employed to describe the 6-DOF motions of the ship in waves. The generalized displacement of the ship is defined as $\boldsymbol{\eta}$, the generalized velocity is $\dot{\boldsymbol{\eta}}$, and the generalized acceleration is $\ddot{\boldsymbol{\eta}}$.

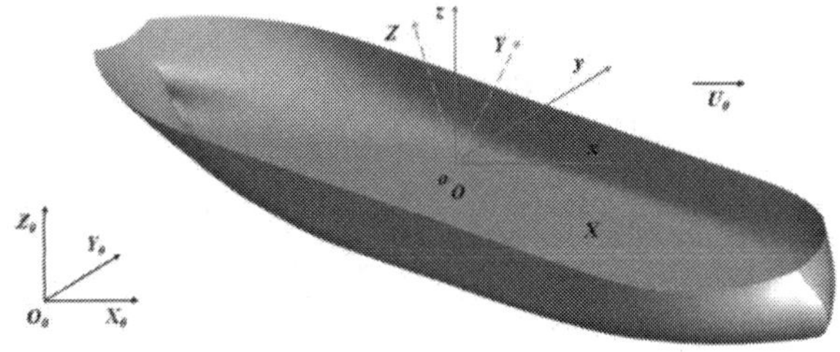

Figure 1 Coordinate System Definition.

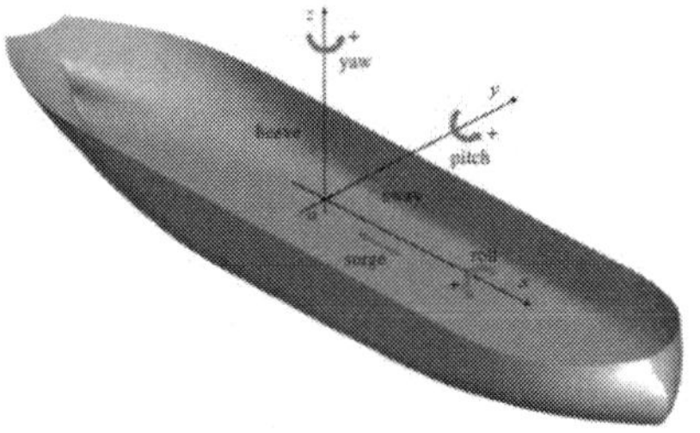

Figure 2 Description of 6-DOF motion.

The water in the flow field is assumed to be inviscid and irrotational. The total velocity potential is defined as ψ, which satisfies the Laplace equation,

$$\nabla^2\Psi(x,y,z,t)=\frac{\partial^2\Psi}{\partial x^2}+\frac{\partial^2\Psi}{\partial y^2}+\frac{\partial^2\Psi}{\partial z^2}=0 \tag{1}$$

The total velocity potential can be divided into uniform flow velocity potential, steady velocity potential and unsteady velocity potential. Unsteady velocity potential includes incident velocity potential, diffraction velocity potential and radiation velocity potential. In the frequency domain, the total velocity potential is expressed as

$$\Psi=-U_0x+\Phi+\phi=-U_0x+\Phi+(\phi_{0I}e^{-i\omega_e t}+\phi_{0D}e^{-i\omega_e t}+\phi_{0R}e^{-i\omega_e t}) \tag{2}$$

In present study, the uniform incoming flow assumption is adopted, the effect of the steady velocity potential is ignored, and only the unsteady velocity potential is solved[9][10]. The incident velocity potential in the reference coordinate system with infinite water depth can be expressed as

$$\phi_I = -\frac{Ag}{i\omega_0} e^{kz} e^{i(kx\cos\beta + ky\sin\beta)} e^{-i\omega_e t} = \phi_{0I} e^{-i\omega_e t} \tag{3}$$

The relationship between encounter frequency ω_e and wave circular frequency ω_0 is

$$\omega_e = \omega_0 - kU_0 \cos\beta \tag{4}$$

Where β is the wave direction, which satisfies β=180° when sailing in the head sea and β=0° when sailing in the following sea.

The incident potential and diffraction potential satisfy the body surface boundary condition

$$\left.\frac{\partial \phi_D}{\partial n}\right|_{S_0} = -\left.\frac{\partial \phi_I}{\partial n}\right|_{S_0} \tag{5}$$

Due to the 6-DOF motions of the ship, the radiation velocity potential satisfies the body surface boundary conditions

$$\begin{aligned} &\left.\frac{\partial \phi_R}{\partial n}\right|_S = \sum_{j=1}^{6} \dot{\eta}_j n_j + \sum_{j=1}^{6} \eta_j m_j, j = 1,2,3,4,5,6 \\ &(n_4, n_5, n_6) = (x, y, z) \times (n_1, n_2, n_3) \\ &(m_1, m_2, m_3) = (0,0,0) \\ &(m_4, m_5, m_6) = (0, U_0 n_3, -U_0 n_2) \end{aligned} \tag{6}$$

In present study, the Neumann-Kelvin (N-K) condition is used to linearize the free surface, and the Rayleigh's artificial friction is incorporated to modify the far-field radiation condition. The free surface condition with damping is

$$-(\omega_e + i\nu)^2 \phi_{0j} + 2i(\omega_e + i\nu)U_0 \frac{\partial \phi_{0j}}{\partial x} + U_0^2 \frac{\partial^2 \phi_{0j}}{\partial x^2} + g\frac{\partial \phi_{0j}}{\partial z} = 0 \tag{7}$$

The damping zone is set up by referring to Yasuda et al.[11]. The damping factor on the free surface is

$$\nu = \begin{cases} 0 & ,0 \le R \le 1 \\ \alpha[1 - e^{-\gamma(R-1)}] & ,1 < R \end{cases} \tag{8}$$

Where $R = \sqrt{x^2 + y^2} / L$, α=0.5, γ=1.5.

To obtain the source strength distribution on the body surface boundary, the boundary integral equation using the distributed source method is employed based on the third Green's theorem.

$$\frac{1}{2}\sigma(p) + \frac{1}{4\pi}\iint_S \sigma(q) \frac{\partial G(p,q)}{\partial n_q} dS_q = \frac{\partial \phi(p)}{\partial n_q} \tag{9}$$

The right term of the equation involves the velocity potential gradient, which is derived from the body surface boundary conditions. In the left term of the equation, we employed the Green's function $G(p,q) = 1/\sqrt{(x-\xi)^2 + (y-\eta)^2 + (z-\zeta)^2}$, which represented the reciprocal of the distance between the field point $p(x,y,z)$ and the source point $q(\xi,\eta,\zeta)$ and is a specific solution for the Laplace equation.

The 6-DOF motion equation of a ship is expressed in the frequency domain as

$$\begin{aligned} &(\boldsymbol{M} + \boldsymbol{A})\ddot{\boldsymbol{\eta}} + \boldsymbol{B}\dot{\boldsymbol{\eta}} + \boldsymbol{C}\boldsymbol{\eta} = \boldsymbol{F}_W \\ &[-\omega^2(\boldsymbol{M} + \boldsymbol{A}) - i\omega\boldsymbol{B} + \boldsymbol{C}]\boldsymbol{\eta} = \boldsymbol{F}_W \end{aligned} \tag{10}$$

Where $\boldsymbol{A}$ is the added mass matrix, $\boldsymbol{B}$ is the damping coefficient matrix, $\boldsymbol{C}$ is the restoring force coefficient matrix, $\boldsymbol{M}$ is the inertia matrix of the ship, and $\boldsymbol{\eta}$ is the generalized displacement. $\boldsymbol{F_W}$

represents the wave forces acting on the ship, which is calculated by the Bernoulli equation.

In present study, the near-field method is used to evaluate the instantaneous pressure on the wet surface S_B of the hull. The second-order term of the pressure integral is taken to obtain the second-order forces F and moments M of the ship. The angle of hull flare at water plane between the hull and still water is defined as α. The second-order forces and moments on the ship are

$$\begin{aligned} F^{(2)} &= \iint_{S\bar{B}} [p^{(0)} \cdot \boldsymbol{n}^{(2)} + p^{(1)} \cdot \boldsymbol{n}^{(1)} + p^{(2)} \cdot \boldsymbol{n}^{(0)}] dS + \frac{1}{2} \rho g \int_{WL} [\zeta^{(1)}]^2 \frac{\boldsymbol{n}^{(0)}}{\sin\alpha} dl \\ M^{(2)} &= \iint_{S\bar{B}} [p^{(0)} \cdot \boldsymbol{m}^{(2)} + p^{(1)} \cdot \boldsymbol{m}^{(1)} + p^{(2)} \cdot \boldsymbol{m}^{(0)}] dS + \frac{1}{2} \rho g \int_{WL} [\zeta^{(1)}]^2 \frac{\boldsymbol{m}^{(0)}}{\sin\alpha} dl \end{aligned} \tag{11}$$

Where $\zeta^{(1)} = \zeta - (\eta_3 + Y\eta_4 - X\eta_5)$, $p^{(0)} = -\rho g Z$, $p^{(1)} = \rho(\phi_t - U_0\phi_x) - (\xi_T + \xi_R \times X)\nabla(\rho g Z)$, $p^{(2)} = \rho(\frac{1}{2}\nabla\phi\cdot\nabla\phi) + (\xi_T + \xi_R \times X)\nabla[\rho(\phi_t - U_0\phi_x)] - HX\nabla(\rho g Z)$, $\boldsymbol{n}^{(0)} = N$, $\boldsymbol{n}^{(1)} = \boldsymbol{\xi_R} \times N$, $\boldsymbol{n}^{(2)} = HN$, $\boldsymbol{m}^{(0)} = \boldsymbol{X} \times N$, $\boldsymbol{m}^{(1)} = \xi_T \times N + \boldsymbol{\xi_R} \times (\boldsymbol{X} \times N)$, $\boldsymbol{m}^{(2)} = H(\boldsymbol{X} \times N) + \xi_T \times (\boldsymbol{\xi_R} \times N)$, $\boldsymbol{H} = \frac{1}{2}\begin{bmatrix} -(\eta_5^2+\eta_6^2) & 0 & 0 \\ 2\eta_4\eta_5 & -(\eta_4^2+\eta_6^2) & 0 \\ 2\eta_4\eta_6 & 2\eta_5\eta_6 & -(\eta_4^2+\eta_5^2) \end{bmatrix}$, linear displacement $\boldsymbol{\xi_T} = (\eta_1, \eta_2, \eta_3)^T$, angular displacement $\boldsymbol{\xi_R} = (\eta_4, \eta_5, \eta_6)^T$.

It can be seen from the expression for the second-order forces that most terms are the multiplication of the small-amplitude ship motions and velocity potential partial derivatives in the frequency domain except several basic terms that are real numbers. In this paper, the frequency-domain boundary element method is employed, which could separate the space and time terms of the second-order forces. The steady-state mean value of the second-order wave forces can be obtained directly by considering the real part of the space term. Compared with the time domain method which took several stable second-order wave force periods, this method is more effective to solve the mean second-order wave drift forces.

Taking the generalized displacement as an example, the basic process of time-averaged mean second-order wave drift forces in the complex frequency domain is illustrated. By summing and averaging with its conjugate, the real part of the imaginary number could be obtained.

$$\eta_j = \mathrm{Re}[\eta_{0j} e^{-i\omega_e t}] = (\eta_{0j} e^{-i\omega_e t} + \eta_{0j}^* e^{i\omega_e t})/2, \quad (j = 1,2,3,4,5,6) \tag{12}$$

The multiplication of two generalized displacements can be expressed as the space term and whose conjugate to eliminate the time-oscillating term. The-time averaged process can then be derived as

$$\begin{aligned} \overline{\eta_i \cdot \eta_j} &= \overline{\mathrm{Re}[\eta_{0i} e^{-i\omega_e t}] \cdot \mathrm{Re}[\eta_{0j} e^{-i\omega_e t}]} \\ &= \overline{(\eta_{0i}\eta_{0j} e^{-2i\omega_e t} + \eta_{0i}\eta_{0j}^* + \eta_{0i}^*\eta_{0j} + \eta_{0i}^*\eta_{0j}^* e^{2i\omega_e t})/4} \\ &= (\eta_{0i}\eta_{0j}^* + \eta_{0i}^*\eta_{0j})/4, \quad (i, j = 1,2,3,4,5,6) \end{aligned} \tag{13}$$

The first term of the second-order wave forces obtained by integrating the wet surface is

$$\begin{aligned} & p^{(0)} \cdot \boldsymbol{n}^{(2)} \\ &= -\rho g Z \cdot \boldsymbol{HN} \\ &= -\frac{\rho g Z}{2} \begin{bmatrix} -(\eta_5^2+\eta_6^2) & 0 & 0 \\ 2\eta_4\eta_5 & -(\eta_4^2+\eta_6^2) & 0 \\ 2\eta_4\eta_6 & 2\eta_5\eta_6 & -(\eta_4^2+\eta_5^2) \end{bmatrix} \begin{bmatrix} n_1 \\ n_2 \\ n_3 \end{bmatrix} \end{aligned} \tag{14}$$

Where $\eta_4\eta_4 = (\eta_{04}\eta_{04}^* + \eta_{04}^*\eta_{04})/4$, $\eta_5\eta_5 = (\eta_{05}\eta_{05}^* + \eta_{05}^*\eta_{05})/4$, $\eta_6\eta_6 = (\eta_{06}\eta_{06}^* + \eta_{06}^*\eta_{06})/4$, $\eta_4\eta_5 = (\eta_{04}\eta_{05}^* + \eta_{04}^*\eta_{05})/4$, $\eta_4\eta_6 = (\eta_{04}\eta_{06}^* + \eta_{04}^*\eta_{06})/4$, $\eta_5\eta_6 = (\eta_{05}\eta_{06}^* + \eta_{05}^*\eta_{06})/4$.

3. Results and discussion

3.1. *Numerical calculation of Wigley Ship*

Journee et al.[12] have conducted extensive seakeeping experiments on the Wigley ships. The Wigley I and Wigley III ships are simulated, and their main parameters are shown in Table 1. Quadrilateral meshes with 3000 and 2000 panel elements are employed to calculate the Wigley I and Wigley III, and the front and three-dimensional views of the two ships' mesh are shown in Figure 3.

Table 1 Main parameters of Wigley Ship.

Ship type number	I	III
Midship section coefficient C_m	0.9090	0.6667
Aspect ratio L/B	10	10
Length L(m)	3.0000	3.0000
Beam B(m)	0.3000	0.3000
Draft T(m)	0.1875	0.1875
Volume ∇(m^3)	0.0946	0.0780
CG above keel KG(m)	0.1700	0.1700
Pitch radius of gyration r_{yy}(m)	0.7500	0.7500

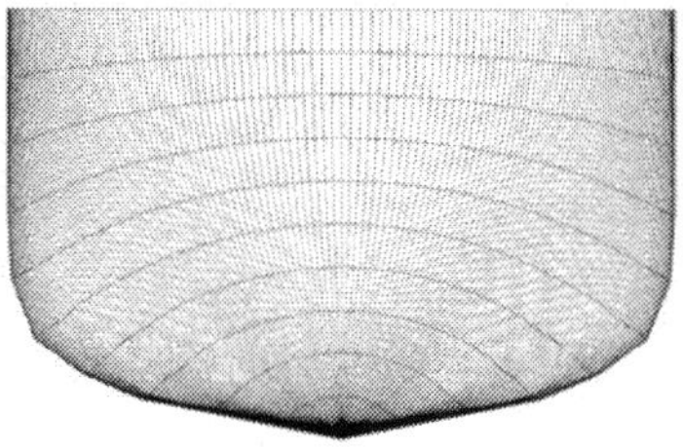
(a) The front view of Wigley-I.

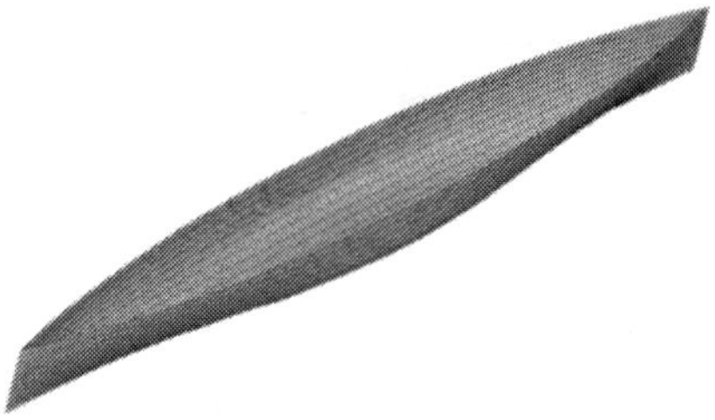
(b) The 3D view of Wigley-I.

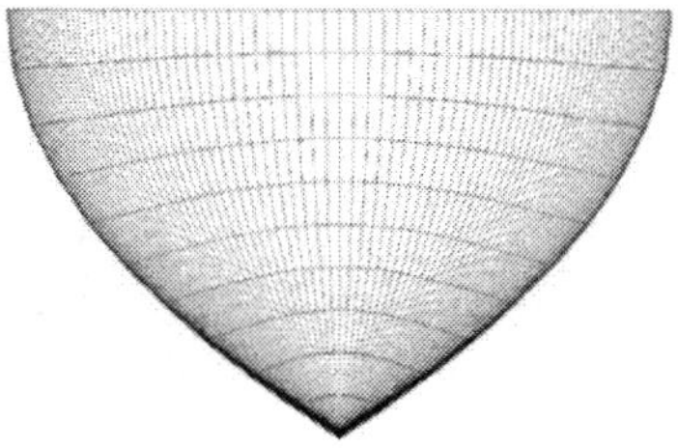
(c) The front view of Wigley-III.

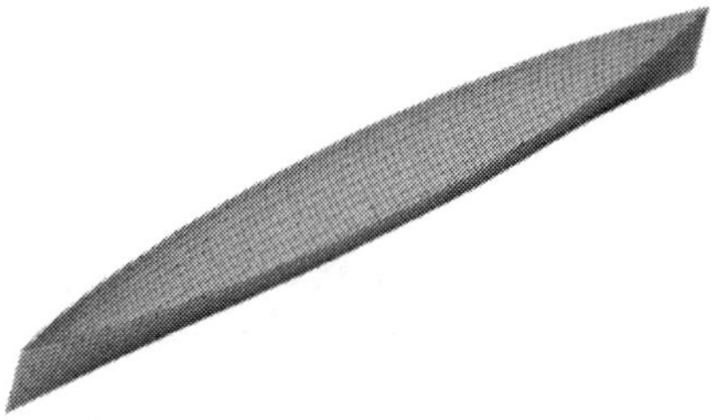
(d) The 3D view of Wigley-III.

Figure 3 Views of Wigley ships' mesh.

In this section, only the heave and pitch motions of Wigley ships are considered. The added mass A_{ij}, damping coefficient B_{ij}, first-order heave wave force F_{W3}, first-order pitch wave force F_{W5}, heave motion amplitude η_3, pitch motion amplitude η_5 and added resistance R_{AW} of Wigley-I and Wigley-III are extracted. The grid convergence study of Wigley-I is conducted under three different hull meshes at Fr=0.2 (Number of meshes NB=2000, 3000, 4000). The heave and pitch hydrodynamic coefficients of the Wigley-I ship are shown in Figure 4 and Figure 5. The potential flow calculation results are compared with the experimental data of Journee et al.[12]. The x-axis represents the encounter frequency of ships, while the y-axis represents the added mass and damping coefficient.

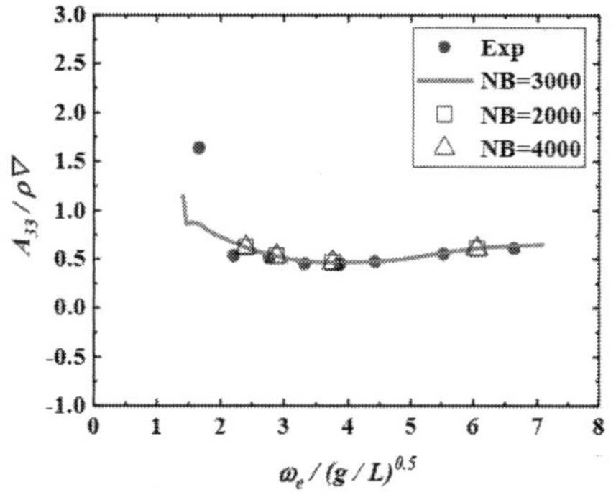

(a) Added mass A33.

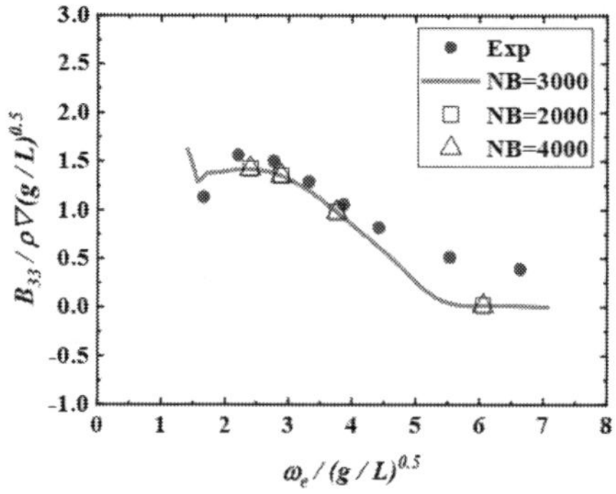

(b) Damping coefficient B33.

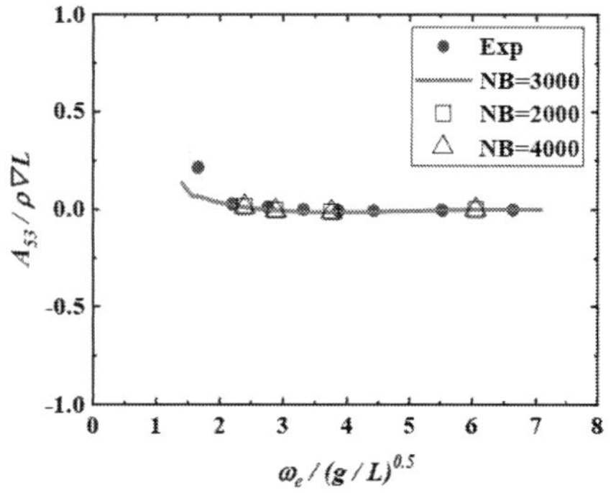

(c) Added mass A53.

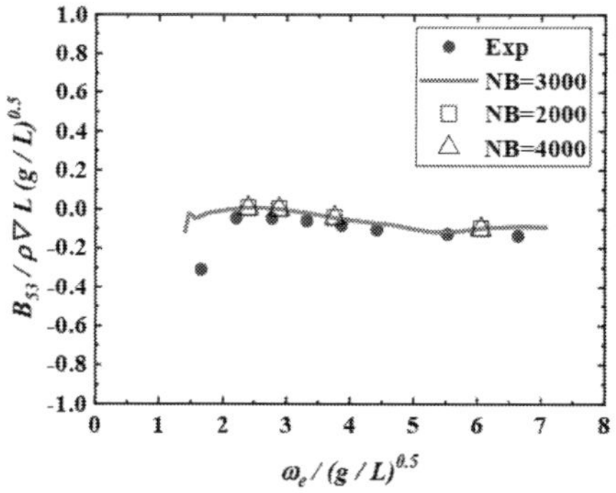

(d) Damping coefficient B53.

Figure 4 Heave hydrodynamic coefficient of Wigley-I (Fr = 0.2).

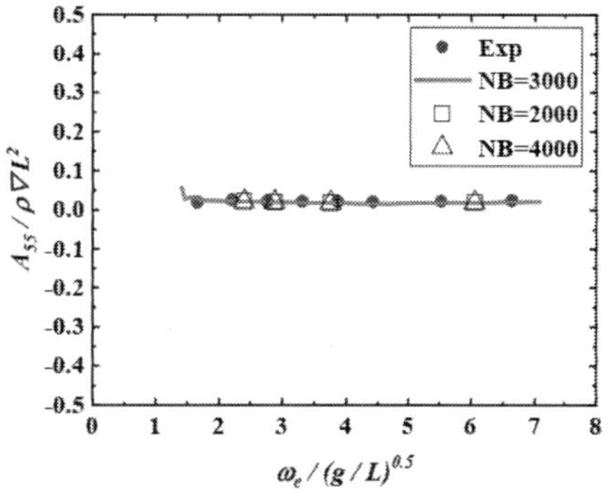

(a) Added mass A55.

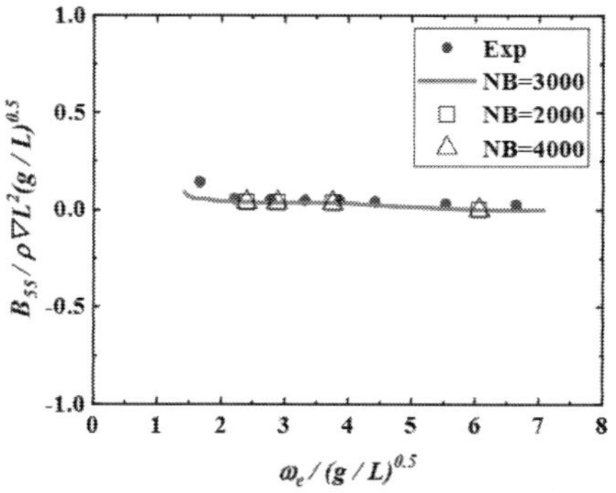

(b) Damping coefficient B55.

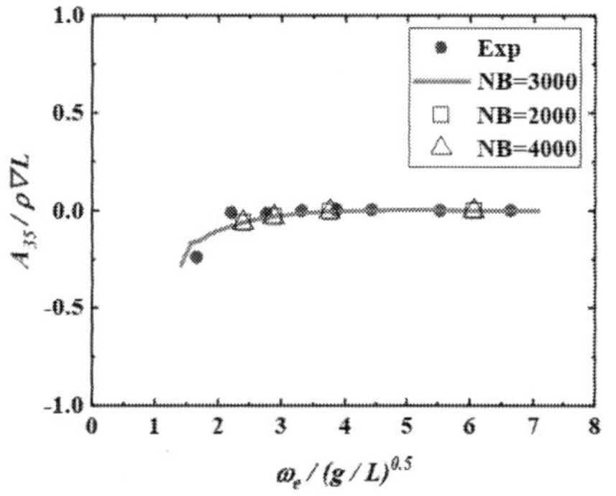

(c) Added mass A35.

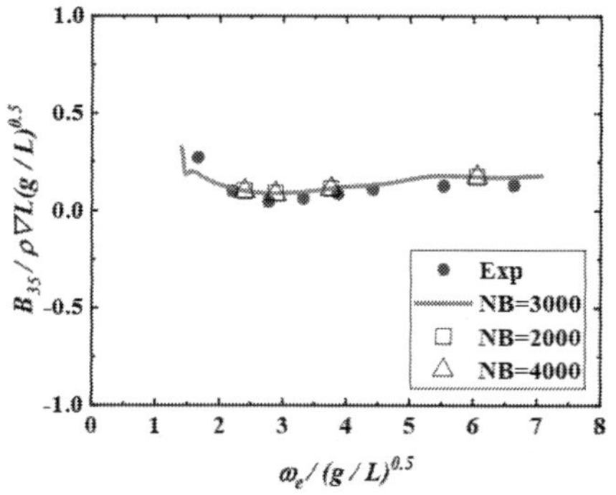

(d) Damping coefficient B35.

Figure 5 Pitch hydrodynamic coefficient of Wigley-I (Fr = 0.2).

The calculation results are not sensitive to the mesh because there is no significant change in the results with increasing mesh size. The numerical results exhibit significant differences between low and high encounter frequencies, which may be due to the relatively low accuracy of the calculation of the radiation velocity potential and its partial derivatives using the Rankine panel method adopted in this study. Overall, the numerical results agree well with the experimental data and follow consistent trends. The accuracy of this method for calculating ship hydrodynamic coefficients has been verified.

The wave-induced forces and motions of the Wigley-I ship at Fr=0.2 are shown in Figure 6. The motions of the Wigley-III ship at Fr=0.2 are shown in Figure 7. The numerical results of the first-order wave forces are in good agreement with the experimental data. Overall, the heave and pitch motions agree well with the experimental data, except for resonance frequency, which is overestimated by the N-K free surface linear condition.

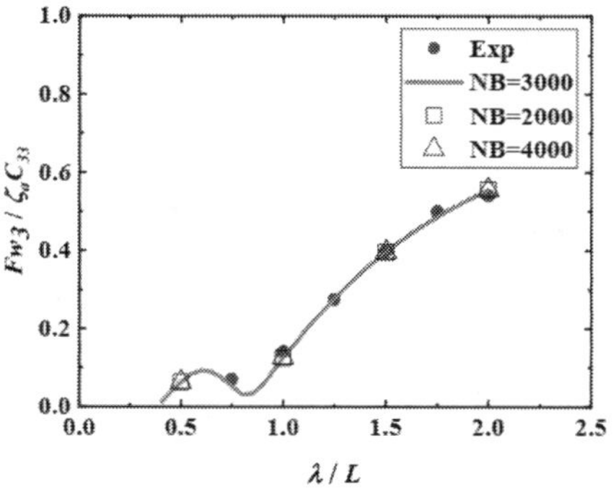

(a) Heave wave forces. (b) Pitch wave forces.

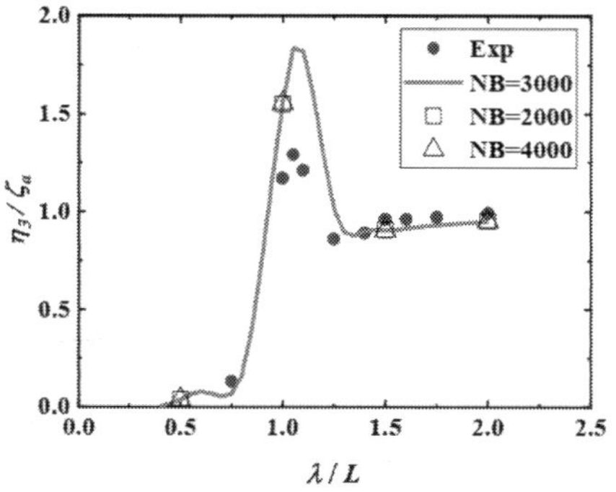

(c) Heave motions. (d) Pitch motions.

Figure 6 First-order wave forces and motions of Wigley-I ship (Fr=0.2).

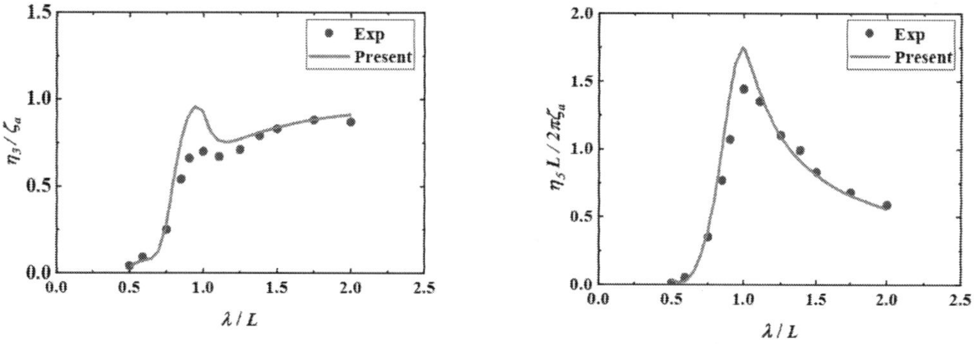

(a) Heave motions. (b) Pitch motions.

Figure 7 Motions of Wigley-III (Fr=0.2).

Figure 8 compares the numerical and experimental results of added resistance of two Wigley ships in head waves. By using the time-averaged method, the second-order wave forces are basically consistent with the experimental data, and the added resistance in the head wave is effectively obtained. The effectiveness of the second-order wave forces time-averaged method is demonstrated.

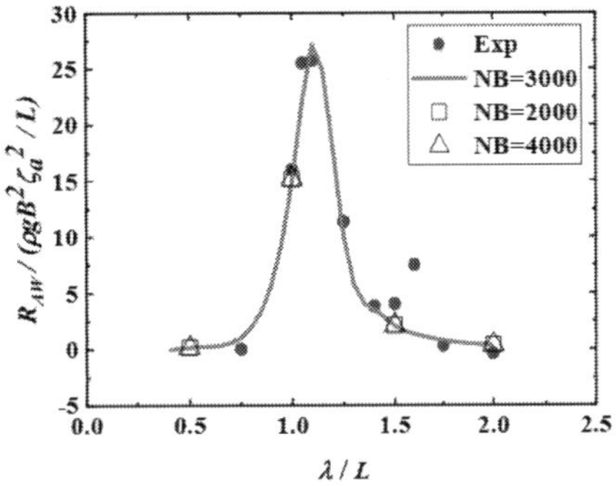

(a) Wigley-I.

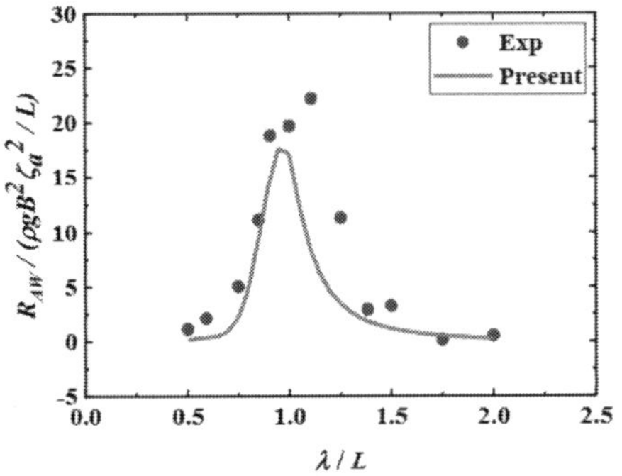

(b) Wigley-III.

Figure 8 The added resistance in head wave (Fr=0.2).

3.2. Numerical calculation of SCb-84 Ship

The bulk carrier SCb-84 is a fully formed ship modified from KVLCC2, with a block coefficient of C_B=0.84. The main parameters are shown in Table 2. The underwater part of the SCb-84 ship is discretized into 3748 quadrilateral meshes. The meshes are shown in Figure 9.

Table 2 Main parameters of SCb-84 Ship.

	Real scale	Model scale
Length *L* (m)	178.00	2.909
Beam *B* (m)	32.26	0.527
Depth *D* (m)	14.46	0.236
Draft *T* (m)	11.57	0.189
Volume ∇ (m^3)	55810	0.224
Radius of gyration r_{xx}, r_{yy}, r_{zz} (m)	10.0, 44.5, 44.5	0.163, 0.727, 0.727
Center of gravity to waterline z_g (m)	-4.51	-0.074
Block coefficient C_B	0.84	0.84
Metacentric height GM (m)	6.61	0.108
Natural rolling period T (s)	9	1.15

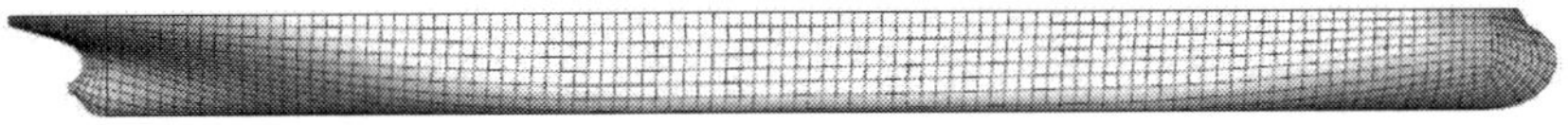

(a) Side view.

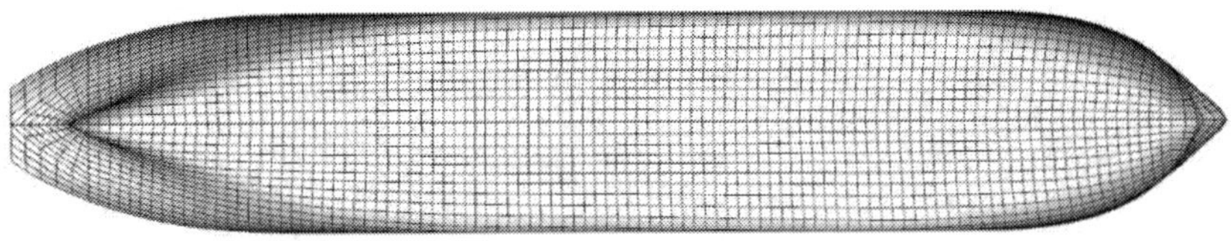

(b) Vertical view.

Figure 9 SCb-84 ship meshes.

According to the ship model test carried out by Yasukawa et al.[13], the calculation conditions of the SCb-84 ship are shown in Table 3.

Table 3 Calculation conditions of SCb-84 ship.

Conditions	Value
Real-scale forward speed U(kn)	0, 4, 8
Wavelength/ship length λ/L	0.4, 0.6, 0.7, 0.8, 1.0, 1.2, 1.5, 2.0
Wave direction angle β(°)	180, 150, 90, 30

In this section, the surge, heave, and pitch motions and the second-order wave forces (including the added resistance R_{AW}, mean lateral forces Y_W, and mean yaw moments N_W) of the SCb-84 ship at different wave angles and sailing speeds are compared with experimental data carried out by Yasukawa et al.[13] as shown in Figure 10~Figure 14.

Under the head seas or bow seas with the wavelength greater than the length of the ship, SCb-84 experiences severe surge motions, which are stronger at low speeds. Under beam sea conditions, the surge motions are significant only at short wavelengths. Moreover, the amplitudes of surge motions with the quartering seas are higher at high speeds. The numerical results are consistent with the experimental trends, and the numerical values are also in good agreement.

Under head sea conditions, the numerical results of the heave motions at high speed and long wavelength are higher than the experimental data. This is mainly because the significant ship motions result in pronounced fluid viscosity and nonlinearity. Under beam sea conditions, the heave motions of the ship with the same wavelength at different speeds are basically the same, and the maximum amplitude is nearly 1.5 times the incident amplitude in the resonance region.

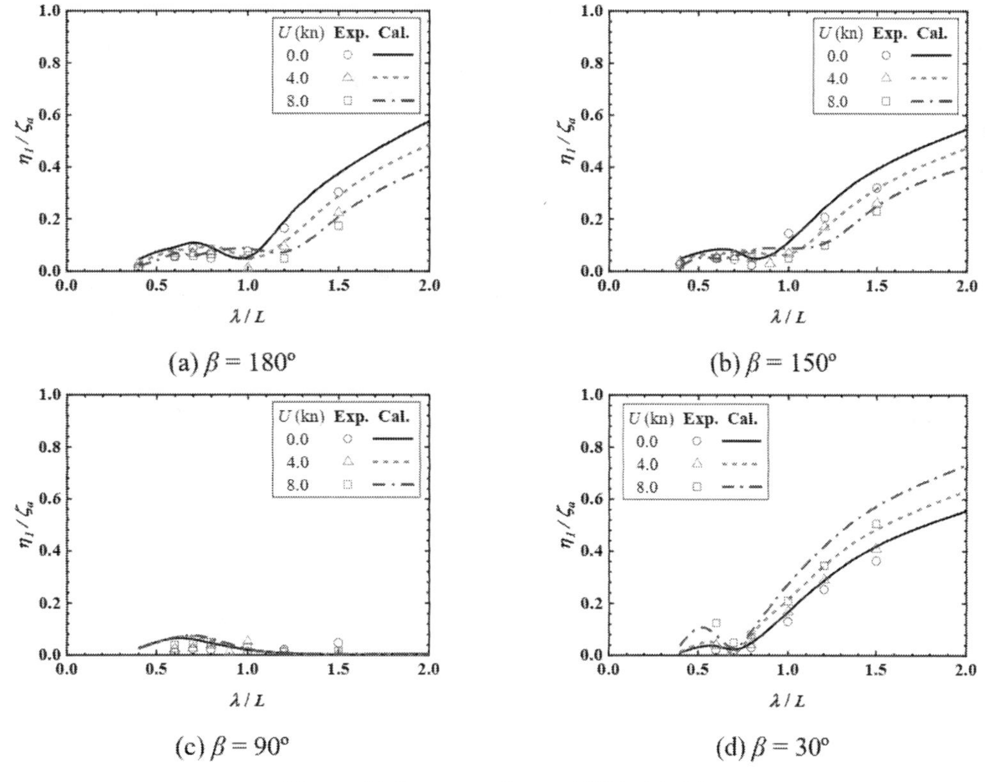

(a) $\beta = 180°$ (b) $\beta = 150°$

(c) $\beta = 90°$ (d) $\beta = 30°$

Figure 10 Surge motions of SCb-84 ship.

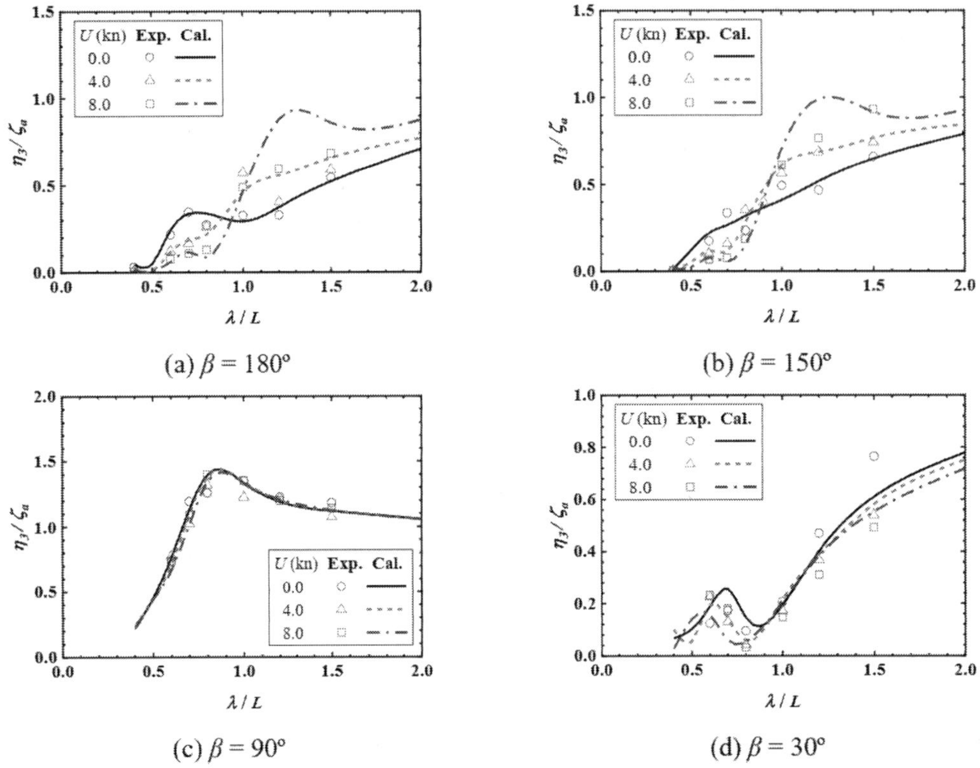

(a) $\beta = 180°$ (b) $\beta = 150°$

(c) $\beta = 90°$ (d) $\beta = 30°$

Figure 11 Heave motions of SCb-84 ship.

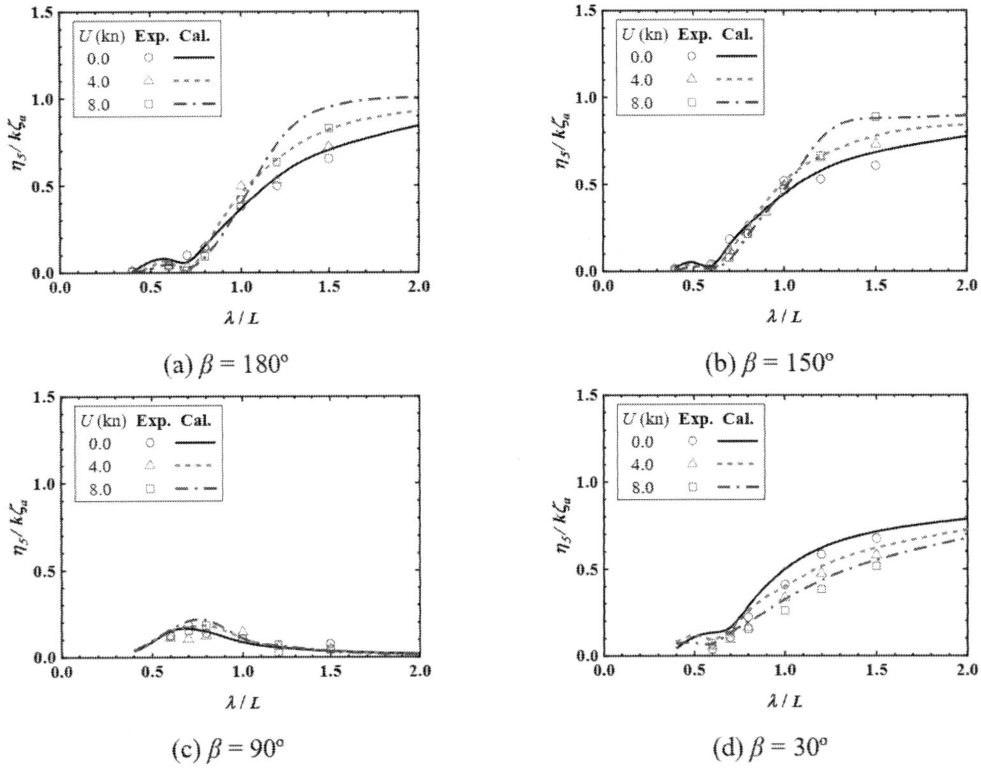

(a) $\beta = 180°$ (b) $\beta = 150°$

(c) $\beta = 90°$ (d) $\beta = 30°$

Figure 12 Pitch motions of SCb-84 ship.

Sailing in the long wavelength of the head seas and bow seas, the non-dimensional results of the pitch motions for the ship at high speeds are gradually stable with the increase of wavelength. The motion results of the ship at different speeds and different wave directions calculated in present study are generally in good agreement with the experimental data.

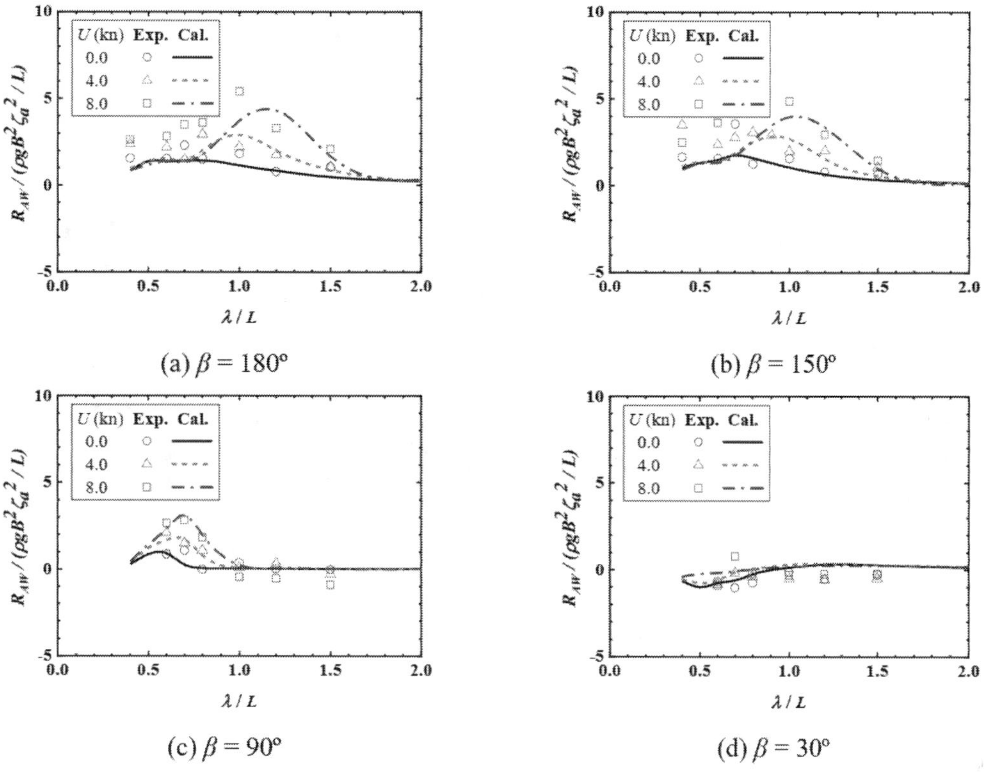

Figure 13 The added resistance of SCb-84 ship.

The experimental data and numerical results of the mean second-order drift forces are compared as shown in Figure 13 and Figure 14. There is a significant deviation between the numerical and experimental results in the short wavelengths under the head sea conditions, though the differences among different speeds are relatively small. This is due to the difficulties in the numerical simulations and experimental measurements of the added resistance in short wavelengths. Specifically, the short waves are easier to attenuate during the experiment, resulting in the small-amplitude ship motions which causes a large uncertainty in measuring the added resistance. For numerical simulations, more refined meshes are required to obtain accurate results. Under the beam sea conditions, the numerical results agree well with the experimental data, and the trend of added resistance is consistent with the experiment.

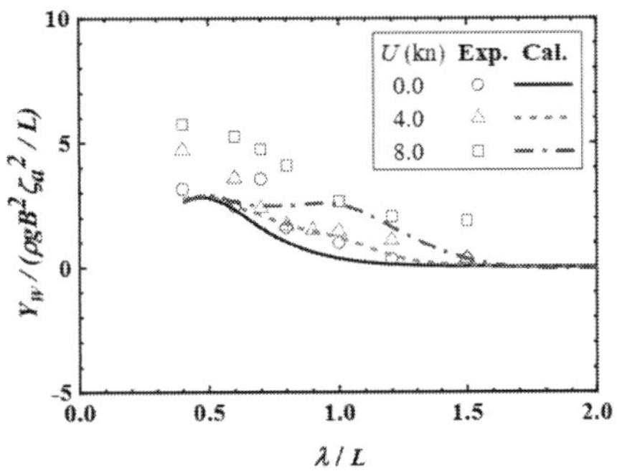

(a) Mean second-order lateral forces ($\beta = 150°$).

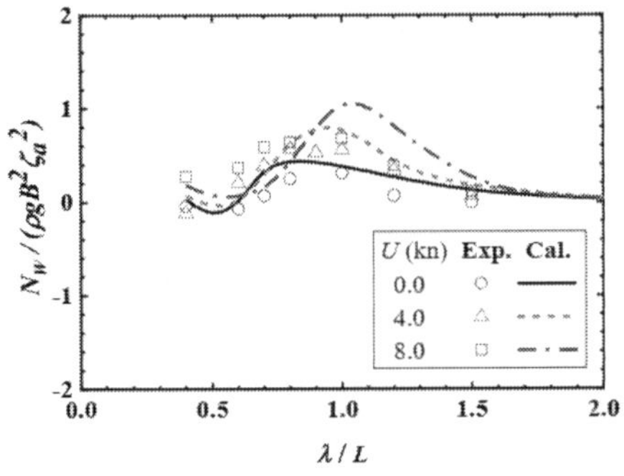

(b) Mean second-order yaw moments ($\beta = 150°$).

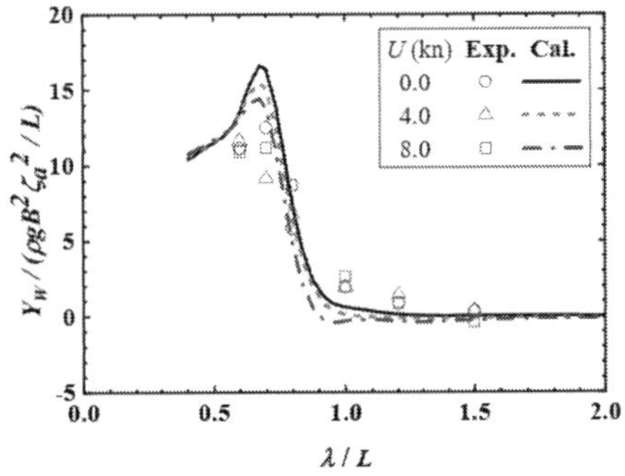

(c) Mean second-order lateral forces ($\beta = 90°$).

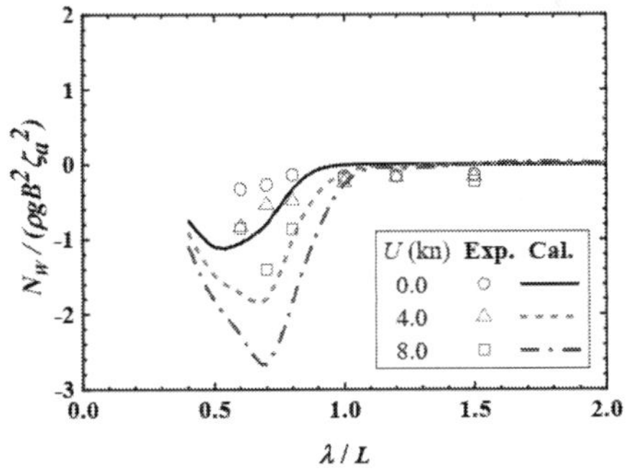

(d) Mean second-order yaw moments ($\beta = 90°$).

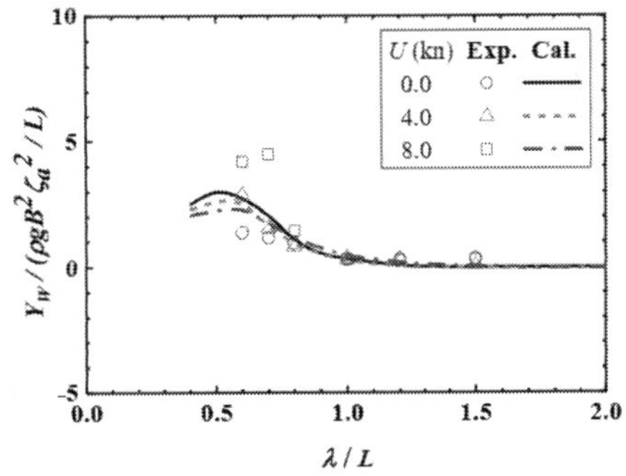

(e) Mean second-order lateral forces ($\beta = 30°$).

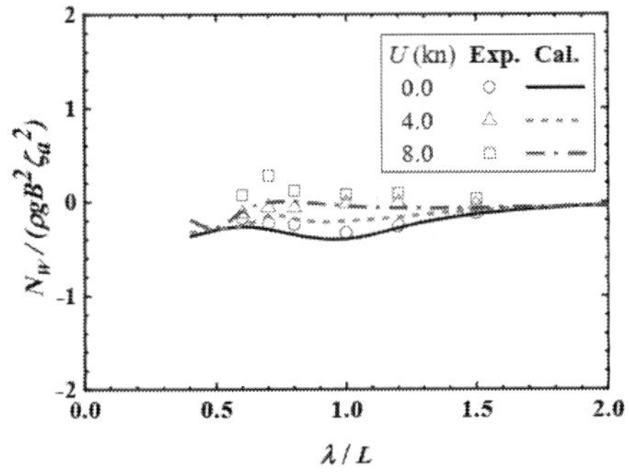

(f) Mean second-order yaw moments ($\beta = 30°$).

Figure 14 Second-order wave forces of SCb-84 ship.

4. Conclusion

Based on the three-dimensional frequency domain Rankine panel method, the seakeeping and second-order wave drift forces of Wigley ship hulls and SCb-84 ships in waves are investigated. The motions and second-order wave forces are compared with the experimental data.

The numerical results of added mass, damping coefficients and first-order wave forces of Wigley-I ship at Fr=0.2 are compared with those of experiments to verify the effectiveness of the present Rankine panel method in calculating radiation forces and first-order wave forces. The motions and added resistance of Wigley-I and Wigley-III ships under the head seas are also compared, and the numerical results are in good agreement with the experimental data.

Meanwhile, the motions and the second-order wave forces of the SCb-84 ship at different speeds under head and oblique waves are compared with the experimental data. The results are generally consistent, which verifies the effectiveness of the frequency domain boundary element method with different wave directions and speeds.

References

[1] Haskind M D 1946 The hydrodynamic theory of ship oscillations in rolling and pitching *Prikl. Mat. Mekh* **10** 33-66
[2] Lv X Q, Zhang X S, Zhang Z H, Song X Y 2021 Calculation of ship seakeeping and second-order wave force based on Rankine source in frequency domain. In: *16th National Hydrodynamics Conference and 32nd National Hydrodynamics Symposium Proceedings* (in Chinese)
[3] Lv X Q, Zhang Z H, Zhang X S, Song X Y 2023 Seakeeping computation based on Rankine panel method in frequency domain *The Ocean Engineering* **41** 68-8132 (in Chinese)
[4] Riesner M and El Moctar O 2018 A time domain boundary element method for wave added resistance of ships taking into account viscous effects *OCEAN ENG* **162** 290-303
[5] Chen J K, Duan W Y, Ma S and Liao K P 2021 Time-domain TEBEM method for mean drift force and moment of ships with forward speed under the oblique seas *J MAR SCI TECH-JAPAN* **26** 1001-13
[6] Nonaka K 1990 On the Manoeuvring Motion of a Ship in Waves *Transaction of the West-Japan Society of Naval Architects* **80** 73-86
[7] Fournarakis N, Chroni D, Plessas T and Papanikolaou A 2016 Estimation of the maneuvering characteristics of the DTC containership using URANS based simulations. In: *3rd International Conference on Maritime Technology and Engineering*
[8] Chillcce G and Moctar O el 2018 A numerical method for manoeuvring simulation in regular waves *OCEAN ENG* **170** 434-444
[9] Das S and Cheung K F 2012 Scattered waves and motions of marine vessels advancing in a seaway *WAVE MOTION* **49** 181-977
[10] Das S and Cheung K F 2012 Hydroelasticity of marine vessels advancing in a seaway *J FLUID STRUCT* **34** 271-90
[11] Yasuda E, Iwashita H and Kashiwagi M 2016 Improvement of Rankine Panel Method for Seakeeping Prediction of a Ship in Low Frequency Region. In: *Asme International Conference on Ocean*
[12] Journee J M J 1992 Experiments and calculations on Four Wigley Hullforms *TUDelft, Faculty of Marine Technology, Ship Hydromechanics Laboratory*
[13] Yasukawa H, Hirata N, Matsumoto A, Kuroiwa R and Mizokami S 2018 Evaluations of wave-induced steady forces and turning motion of a full hull ship in waves *J MAR SCI TECH-JAPAN* **24** 1-15

Acknowledgments

The present work was supported by the National Natural Science Foundation of China (Grant no.52071148, 51509256), the Aeronautical Science Foundation of China (Grant no. 20200002307900l), Special Study on Civil aircraft of Ministry of Industry and Information Technology (MJZ5-3N21-1).

IOP Conf. Series: Materials Science and Engineering 1288 (2023) 012029 doi:10.1088/1757-899X/1288/1/012029

Numerical Evaluation of Hydrodynamic Coefficients in Roll Motion

E C Hwang and K K Yang*

Department of Autonomous Vehicle System Engineering, Chungnam National University, Daejeon, Republic of Korea

*Corresponding author, E-mail: kkyang@cnu.ac.kr

Abstract. The determination of ship motions in a seaway heavily depends on hydrodynamic coefficients. However, the coefficients related to roll-motion exhibit nonlinear behavior due to viscous effects and relatively small wave damping, which pose significant challenges. To overcome these challenges, several studies have employed computational fluid dynamics (CFD) to analyze hydrodynamic performance, leveraging the advancements in computational performance. Nevertheless, limited research has investigated the error and uncertainty of these numerical solutions. In this study, we conducted a forced oscillation test on a ship-like section that induces lateral vortex shedding using the OpenFOAM with overset grid method. We validated the numerical results of the roll hydrodynamic coefficients with experiments and the results of potential theory. We used the procedure proposed by the ITTC to perform uncertainty analysis on the grid size and time step size, which presented the overall uncertainty of the numerical analysis results. Our findings indicate that the grid uncertainty of the added moment of inertia was significant in the low frequency region, while the time step uncertainty was significant in the high frequency region.

1. Introduction

The precise determination of a ship's hydrodynamic coefficient poses a considerable challenge, particularly concerning roll motion, which results from nonlinearity arising from viscosity and vortex shedding. This challenge is further intensified by the growing significance of ship stability. Historically, hydrodynamic coefficients have been acquired through experiments, potential-based methods, and empirical formulas. For example, Vugts [1] conducted experiments on various two-dimensional sections to investigate hydrodynamic coefficients such as added mass and damping across different frequencies and amplitudes. While potential-based methods and experimental results display similar tendencies for heave and sway, significant discrepancies have been observed in the case of roll motion. The Ikeda model [2] is a widely used empirical formula for obtaining roll damping, as it considers viscosity effects. However, recent research by Yildiz et al. [3] suggests that it may not be effective in several situations such as shallow draft conditions.

The coefficient obtained through the potential or empirical formula in roll motion can result in significant errors, as previously described. As computational performance has advanced significantly over the past few decades, there has been increased interest in methods for predicting the hydrodynamic coefficients of roll motion through computational fluid dynamics (CFD). In a study conducted by Yu et al. [4], forced roll motion was performed on rectangular floating production storage and offloading (FPSO) sections with and without bilge keels, using an incompressible Navier-Stokes solver. This study observed that the added mass and damping increase according to the width of the bilge keel, and there is a difference in damping between horizontal and vertical bilge keels. Kinnas [5] also performed forced roll motion on rectangular FPSO sections, using the incompressible Navier-Stokes solver to predict the flow around the hull. The study revealed a strong dependence on nonlinearity due to viscosity around the hull. Wilson [6] carried out roll decay and forced motion on an entire ship (combatant), using the Reynolds-Average-Navier-Stokes equation (RANS). Their research showed good correlation with experiments. More recently, Irkal et al. [7] investigated the nonlinearity through the free roll decay test by examining the flow around the model using CFD and particle image velocimetry (PIV) methods. They reported nonlinearity in the fully appended and linearity in the bare hull. A study conducted by Lavrov et al. [8] involved the comparison of numerical results to experiments performed on a rectangular section in Vugts [1] for the purpose of validating the method. The researchers conducted a forced oscillation test through a morphing grid in open-source shapes that hardly induced vortex shedding. The results of the study revealed that shapes causing a large vortex shedding, characterized by a pointed shape, resulted in significant errors in both added mass and damping.

Despite the considerable research on various techniques and commercial codes for the roll motion, there is a lack of studies that present the error and uncertainty of numerical results of hydrodynamic coefficients. To address this gap, the current study performs a forced oscillation test using an overset grid in OpenFOAM v2106 on a triangular shape that induces a large vortex, as previously studied by Vugts [1]. The validation of the results involves comparing them with the experiment and potential-based results, and employing the procedure proposed by ITTC [9] to verify the uncertainty of the added mass and damping coefficient predicted through the forced oscillation test. The variables used in the uncertainty analysis include grid size and time step. The study performs an uncertainty analysis for these variables and overall uncertainty to provide information on the uncertainty of numerical analysis results.

2. Mathematical background

2.1. Governing Equations for Fluid Flows

The calculation of fluid flow is performed using the unsteady Reynolds-averaged-Navier-Stokes (URANS) equations. These equations can be expressed for two-phase flows as follows:

$$\frac{\partial \bar{u}_j}{\partial x_j} = 0 \tag{1}$$

$$\frac{\partial(\rho \bar{u}_i)}{\partial t} + \frac{\partial(\rho \bar{u}_i \bar{u}_j)}{\partial x_j} = -\frac{\partial(\bar{p}')}{\partial x_i} + 2\mu \frac{\partial \bar{S}_{ij}}{\partial x_j} + \frac{\partial(R_{ij})}{\partial x_j} \tag{2}$$

where ρ is the fluid density, $\bar{u}_i$ represents the time-averaged velocity in the i-th direction, R_{ij} is the Reynolds stress, and μ is dynamic viscosity coefficient. The time-averaged hydrodynamic pressure is expressed as $\bar{p}' = \bar{p} + \rho g z$ where $\bar{p}$ is the time-averaged pressure, z is the vertical coordinate, and g is the gravitational acceleration. The present study employs the k-ω shear stress transport (SST) turbulence models.

To capture the water-air interface, we employed the volume-of-fluid (VOF) method, from which the density and viscosity may be derived as follows:

$$\rho = \alpha \rho_{water} + (1-\alpha)\rho_{air} \tag{3}$$

$$\mu = \alpha \mu_{water} + (1-\alpha)\mu_{air} \tag{4}$$

where α is the volume fraction of water phase in each cell, which is governed by:

$$\frac{\partial \alpha}{\partial t} + \bar{u}_j \frac{\partial \alpha}{\partial x_j} = 0. \tag{5}$$

The *overInterDyMFoam* solver of OpenFOAM version 2106 was used for solving the governing equations and the PIMPLE algorithm, which is combined method of the PISO and SIMPLE algorithms, was used for the pressure-velocity coupling.

2.2. Computation of Hydrodynamic Coefficient

A sectional body was subjected to harmonic roll motion, and the resulting roll moment and sway force were analyzed in order to obtain the linearized hydrodynamic coefficients. The forced roll motion can be expressed as follows:

$$\varphi = \varphi_0 \sin \omega t \tag{6}$$

where φ_0 is the amplitude of the roll motion, ω is the circular frequency, and t indicates time. The present investigation pertains to the study of forced oscillation with a single degree of freedom in still water, where the resulting roll moment can be expressed as follows:

$$M_{44}(t) = -A_{44}\ddot{\varphi} - B_{44}\dot{\varphi} - C_{44}\varphi \tag{7}$$

where A_{44} is the added roll moment of inertia and B_{44} is the linearized roll damping coefficient, C_{44} is the restoring coefficient and $M_{44}(t)$ is the roll moment due to the forced roll motion. Additionally, the measured roll moment can be written as the Fourier series and the first harmonic component can then be calculated as follows:

$$M_{44}(t) = M_{44}^{(s)}(t)\sin \omega t + M_{44}^{(c)}(t)\cos \omega t + \cdots \tag{8}$$

$$M_{44}^{(s)} = \frac{2}{T}\int_0^T \{M_{44}(t)\sin \omega t\}\, dt \tag{9}$$

$$M_{44}^{(c)} = \frac{2}{T}\int_0^T \{M_{44}(t)\cos \omega t\}\, dt \tag{10}$$

where T represents the oscillation period. By substituting equation (6) into (7) and comparing it with equation (8), the added roll moment of inertia and roll damping coefficients can be obtained as follows:

$$A_{44} = \frac{M_{44}^{(s)} + C_{44}\varphi_0}{\varphi_0 \omega^2} \tag{11}$$

$$B_{44} = -\frac{M_{44}^{(c)}}{\varphi_0 \omega^2} \tag{12}$$

Similarly, the coupling coefficients between sway and roll motions can be calculated as follows:

$$A_{24} = \frac{F_{24}^{(s)}}{\varphi_0 \omega^2} \tag{13}$$

$$B_{24} = -\frac{F_{24}^{(c)}}{\varphi_0 \omega^2} \tag{14}$$

where F_{24} is the sway force due to the forced roll motion and the symbols "(s)" and "(c)" in superscript denote the sine and cosine components, respectively. A_{24} and B_{24} represent the added mass and damping coefficients in the sway direction, respectively, due to roll motion. The resulting hydrodynamic coefficients are non-dimensionalized as follow:

$$A_{44}^{'} = A_{44} / \left(\rho A_S B^2\right) \tag{15}$$

$$B_{44}^{'} = B_{44}\sqrt{\frac{B}{2g}} / \left(\rho A_S B^2\right) \tag{16}$$

$$A_{24}^{'} = A_{24} / \left(\rho A_S B\right) \tag{17}$$

$$B_{24}^{'} = B_{24}\sqrt{\frac{B}{2g}} / \left(\rho A_S B\right) \tag{18}$$

$$\omega^{'} = \omega\sqrt{\frac{B}{2g}} \tag{19}$$

where A_S is the area of the submerged section, B is the breadth of a section at the waterline, and g is the gravitational acceleration.

2.3. Procedure of Uncertainty Analysis

An uncertainty analysis was conducted in accordance with the procedure outlined by ITTC [9], wherein the uncertainties associated with grid size and time step size were estimated in this study. The input variables for the uncertainty analysis procedure encompass grid size δ_G, time step size δ_T, and other variables δ_P. The resultant total errors and uncertainties are expressed as follows:

$$\delta_{SN} = \delta_G + \delta_T + \delta_P \tag{20}$$

$$U_{SN}^2 = U_G^2 + U_T^2 + U_P^2. \tag{21}$$

In order to determine the convergence state, it is necessary to employ three different grids or time steps. The solutions obtained through the use of these systems – namely fine, medium, and coarse – are denoted as $S_{i,1}$, $S_{i,2}$, and $S_{i,3}$. The subscript "i" represents G for the grid size, and T for the time step. Then, the convergence ratio, R_i can be calculated as follows:

$$\varepsilon_{i,21} = S_{i,2} - S_{i,1} \tag{22}$$

$$\varepsilon_{i,32} = S_{i,3} - S_{i,2} \tag{23}$$

$$R_i = \varepsilon_{i,21} / \varepsilon_{i,32} \tag{24}$$

Three convergence conditions can be distinguished based on the value of the convergence ratio. If the ratio is less than zero, it is oscillatory convergent; if it exceeds one, it is monotonic divergent. Otherwise, it is monotonic convergent. In the case of monotonic convergence, the uncertainty can be estimated using Richardson extrapolation. If the refinement ratio between two different grid or time-step systems is equal to r_i, the order of accuracy P_i can be expressed as:

$$P_i = \frac{\ln\left(\varepsilon_{i,32} / \varepsilon_{i,21}\right)}{\ln\left(r_i\right)} \tag{25}$$

Finally, the estimated error and uncertainty can be obtained as follows:

$$C = \frac{r_i^{P_i} - 1}{r_i^{P_{i,est}} - 1} \tag{26}$$

$$\delta_i^* = C\left(\frac{\varepsilon_{i,21}}{r_i^{P_{i,est}} - 1}\right) \tag{27}$$

$$U_i = \left|C\left(\frac{\varepsilon_{i,21}}{r_i^{P_{i,est}} - 1}\right)\right| + \left|(1-C)\left(\frac{\varepsilon_{i,21}}{r_i^{P_{i,est}} - 1}\right)\right| \tag{28}$$

IOP Conf. Series: Materials Science and Engineering 1288 (2023) 012029 doi:10.1088/1757-899X/1288/1/012029

In this study, $P_{i,est}$ is equal to four and the uncertainty for the oscillatory convergence case is simply estimated as follows:

$$U_i = \frac{1}{2}\left(S_U - S_L\right) \tag{29}$$

where S_U and S_L are maximum and minimum oscillation values, respectively.

3. Numerical Setup

To validate the current computational setup, a triangular section was considered. The geometric dimensions of both the triangular section and the computational domain are depicted in Figures 1 and 2. The characteristic breadth B of the triangular section is 0.3464m, and the draft d is 0.3m. The inertial coordinate system is located at the center of gravity and aligned with the still water level, with the z-direction pointing upwards and the y-direction pointing towards the right. The x-axis extends from the back to the front.

In this study, the overset method was utilized, requiring two distinct meshes, one for the body and the other for the background. The background computational domain must have sufficient depth in the vertical downward direction to prevent the shallow water effect. For the present cases, the horizontal domain size was set to 3.5λ and the vertical depth was $10B$, which is always greater than $\lambda/2$. To reduce computational time, the cell size was densely set near the water plane and the body, while relaxation zones were placed on both sides using the *waves2foam* tool [10] to exclude the effects of reflected waves. The *BlockMesh* tool in OpenFOAM was used to generate the background mesh.

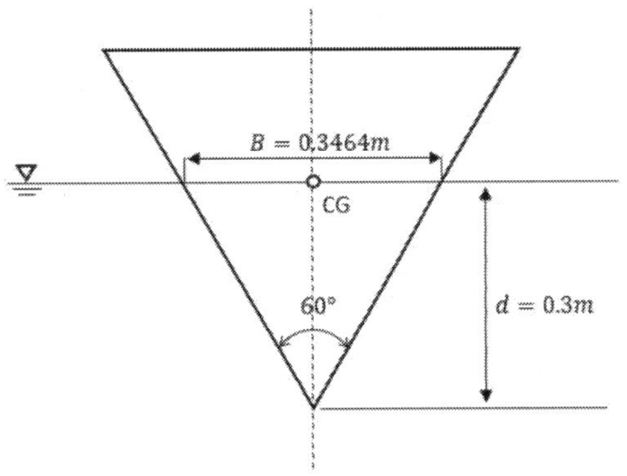

Figure 1. Geometric dimensions of the triangle section

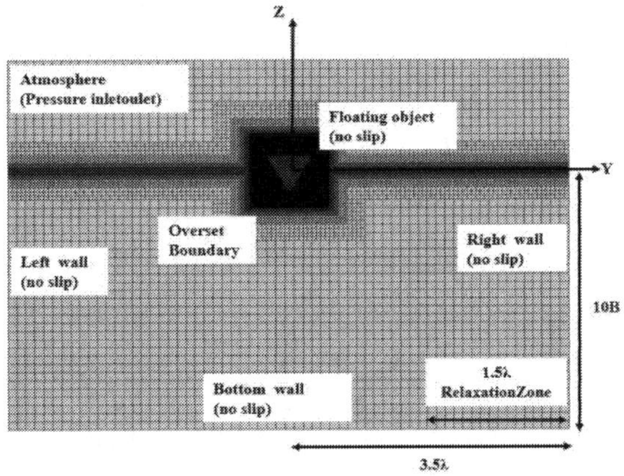

Figure 2. View of domain and boundary condition

The mesh around the body was generated using the *snappyHexMesh* tool within the OpenFOAM software. Initially, a uniform mesh was created in the y- and z-directions in a square region, and the triangular geometry was subsequently embedded within this region. The *snappyHexMesh* tool then eliminated grid points that were entirely inside the geometry and deformed the intersection grid between the geometry boundary and uniform mesh to match the body boundary. Finally, a boundary layer mesh was added near the body surface. To create a two-dimensional mesh, the *ExtractMesh* and *Createpatch* tools were employed, using only one cell in the x-axis.

The time step size was chosen in accordance with the ITTC CFD guidelines [11] and the CFL number. For periodic phenomena, such as the forced oscillation test conducted in this study, ITTC recommends using at least 100 time steps per each period. In order to ensure stable calculations, the period of forced oscillation was divided into approximately 200 time steps. Additionally, the time step for each frequency was adjusted to match the same CFL number, using the following equation:

$$\frac{\varphi_0 \cdot \omega \cdot d \cdot \Delta t}{\Delta s} = \text{constant} \tag{30}$$

where Δt is the time step, and Δs is the length of one axis of the cell.

The no-slip boundary condition was enforced on the surface of the body as well as the left, right, and bottom walls of the background mesh. To satisfy the atmospheric pressure condition, the *totalPressure* boundary condition with zero value was imposed on the top boundary. For the time integration of the simulation, the first-order backward Euler method was applied, while the van Leer limiter was employed for the discretization of divergence terms. Moreover, the overset interpolation scheme was carried out using the cell volume weighting method.

4. Results and Discussions

4.1. Grid size and time step uncertainties

In order to address the uncertainty arising from grid size, three distinct grids were generated, namely coarse, medium, and fine, as shown in Figure 3. The number of grids was increased by 1.5 times in each direction while maintaining the same background grid and the boundary layer of the floating body. The time histories of the roll moment for each grid are compared in Figure 4. It can be observed that there is a tendency for the roll moment to converge as the grid increases. Figure 5 presents the added roll moment of inertia with respect to the time window size used for the Fourier analysis. The sensitivity to the size of the time window can be considered negligible. In this study, Fourier analysis was performed using the time series of the last three cycles.

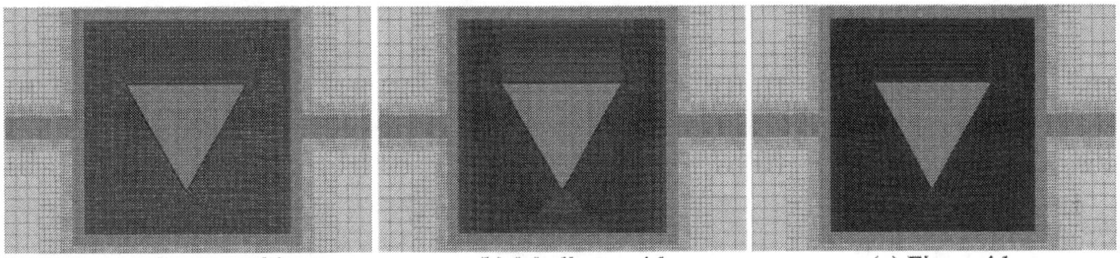

(a) Coarse grid (b) Medium grid (c) Fine grid

Figure 3. Enlarged view of three grids for uncertainty analysis

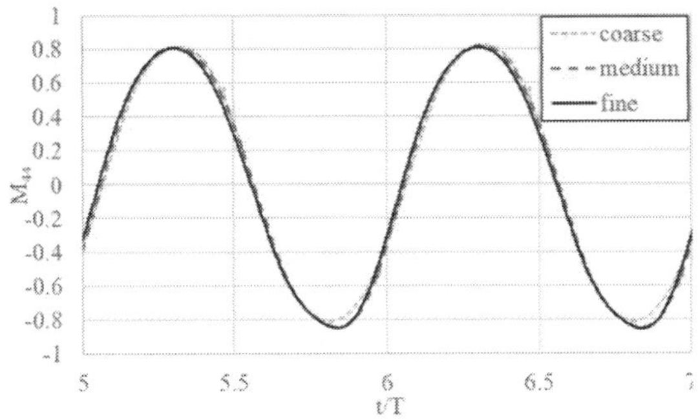

Figure 4. Time histories of roll moment with different grids

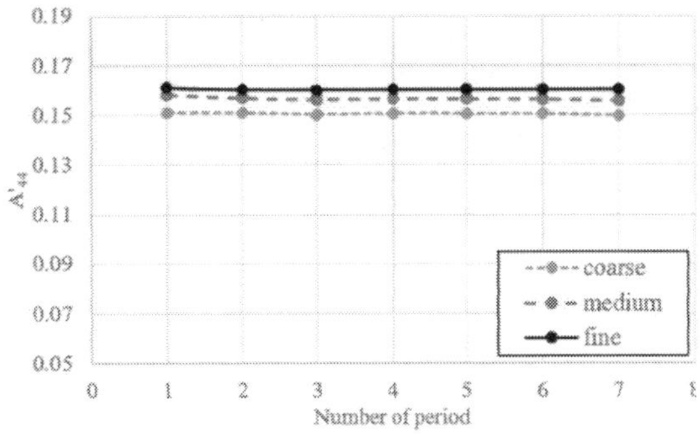

Figure 5. Convergence of added moment of inertia in roll with respect to the size of time windows in Fourier analysis

The results of the uncertainty analysis for the grid spacing and time step are presented in Tables 1 and 2, respectively. Five different excitation frequencies, each with the same excitation roll amplitude, were considered. The added moment of inertia shows a higher degree of uncertainty for the grid spacing than for the time interval. Furthermore, the uncertainty increases as the excitation frequency decreases. Conversely, the damping coefficient exhibits large uncertainty regarding the time step, and the uncertainty increases as the excitation frequency increases. The damping coefficient shows oscillatory convergence in relation to the grid spacing, while it exhibits monotonic convergence with respect to the time step. Additionally, the added moment of inertia shows monotonic convergence with respect to both the time step and grid spacing.

Table 1. Grid size uncertainty of various ω' with $\varphi_0 = 0.05rad$

	$\omega' = 0.531$	$\omega' = 0.797$	$\omega' = 1.063$	$\omega' = 1.329$	$\omega' = 1.594$
$R_G(A'_{44})$	0.875	0.777	0.785	0.450	0.593
$U_G(\%, S_{G,1}, A'_{44})$	12.82	7.58	7.16	1.21	2.87
$R_G(B'_{44})$	-1.468	-0.981	-1.150	-1.253	-1.268
$U_G(\%, S_{G,1}, B'_{44})$	0.90	2.94	3.10	2.73	3.46
$R_G(A'_{24})$	0.883	0.811	0.794	0.277	0.459
$U_G(\%, S_{G,1}, A'_{24})$	10.86	11.53	7.10	1.38	1.22
$R_G(B'_{24})$	-0.472	-0.329	-0.512	-0.851	-1.170
$U_G(\%, S_{G,1}, B'_{24})$	1.33	1.80	1.43	1.44	2.15

Table 2. Time step size uncertainty of various ω' with $\varphi_0 = 0.05rad$

	$\omega' = 0.531$	$\omega' = 0.797$	$\omega' = 1.063$	$\omega' = 1.329$	$\omega' = 1.594$
$R_T(A'_{44})$	-1.479	0.366	0.384	0.062	0.579
$U_T(\%, S_{T,1}, A'_{44})$	0.45	0.60	0.45	0.13	0.79
$R_T(B'_{44})$	0.528	0.535	0.334	0.168	0.653
$U_T(\%, S_{T1}, B'_{44})$	1.04	2.12	1.58	1.34	4.76
$R_T(A'_{24})$	0.458	-0.286	0.111	-0.555	0.366
$U_T(\%, S_{T,1}, A'_{24})$	0.21	0.48	0.10	0.41	0.23
$R_T(B'_{24})$	0.486	-0.155	-0.506	0.326	0.723
$U_T(\%, S_{T,1}, B'_{24})$	0.55	0.60	0.78	2.05	7.74

IOP Conf. Series: Materials Science and Engineering 1288 (2023) 012029 doi:10.1088/1757-899X/1288/1/012029

4.2. Hydrodynamic coefficients and local flows

The added moment of inertia and damping coefficients for each frequency were compared to experimental and potential theory analysis results and presented in Figure 6. The horizontal axis represents the dimensionless excitation frequency, and the vertical axis represents the dimensionless hydrodynamic coefficients. The solid line in the figure represents the potential theory result, while the symbols represent the results of the experiment and the present study. The results of this study were presented with error bars representing the uncertainty of the grid and time intervals. The added moment of inertia and added mass estimated through numerical analysis show slight differences compared to the experimental results but are similar to the potential theory results. As added mass is a physical quantity dominated by inertial forces, the reliability of the potential theory result is high. Therefore, considering this aspect, the accuracy of the numerical analysis results can be considered reasonable.

The shape of the object studied in this research is a triangle, and vortices occur at the vertex of the triangle. Hence, the accuracy of the damping coefficient estimation through the potential theory is low. On the other hand, the damping coefficients of the present study, estimated using CFD, show very similar results to the experimental results, validating the numerical analysis technique used in this study for estimating damping coefficients.

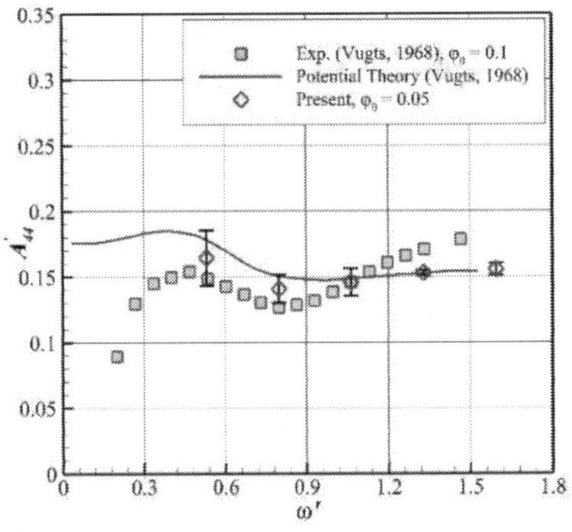

(a) Added moment of inertia in roll

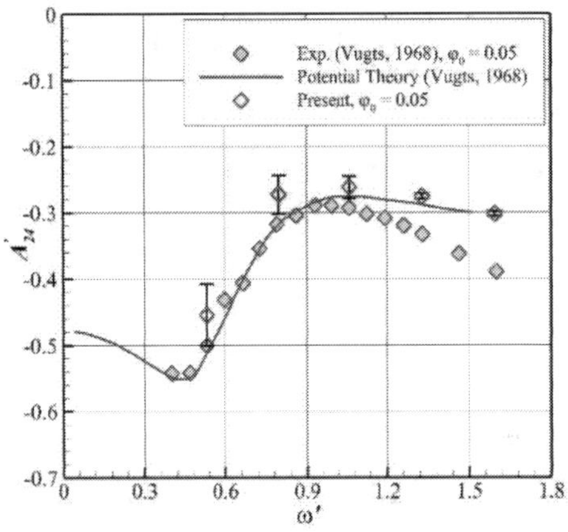

(b) Added mass in roll into sway

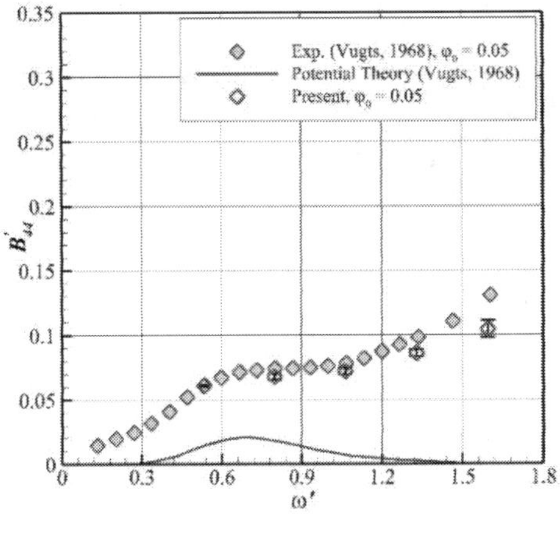

(c) Damping coefficient in roll

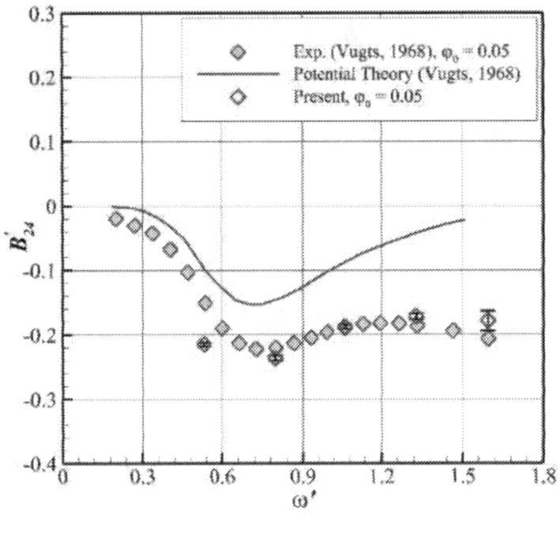

(d) Damping coefficient in roll into sway

Figure 6. Comparison of added moment of inertia and damping coefficients in roll and roll into sway

Figure 7 illustrates the pressure and velocity vector fields around the triangle at each grid system at the two specified times. The dimensionless excitation frequency is 1.594 rad/sec, and the pressure is non-dimensionalized as $\rho g\varphi_0 d$. When the triangle rotates counterclockwise, negative pressure is generated on the opposite side, leading to flow acceleration. Consequently, vortex shedding occurs in pairs under the conditions studied here. However, the second opposite vortex in the medium grid case appears somewhat fainter than in the cases of coarse and fine grids, leading to the conclusion that this effect causes the damping coefficient to oscillatory converge. Therefore, it is necessary to employ a sufficient number of grids in the vortex shedding region to achieve an accurate damping coefficient estimation.

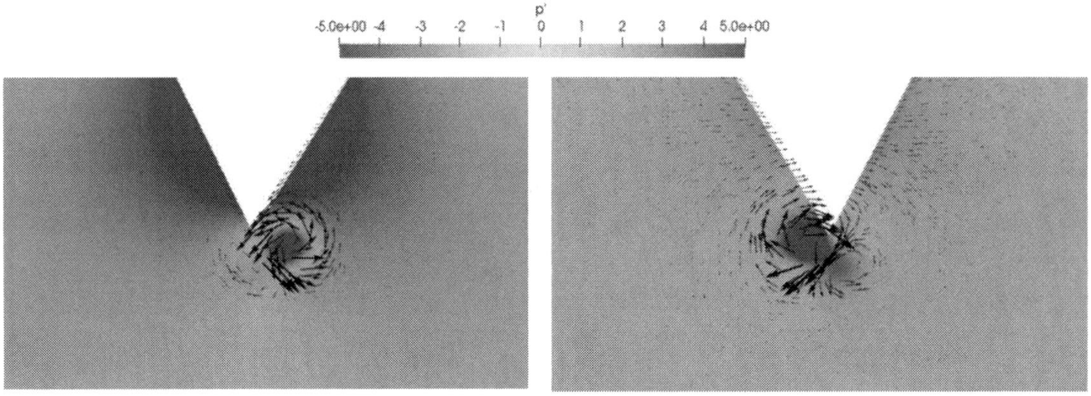

(a) coarse grid at $t = t_0$ (b) Coarse grid at $t = t_0 + 2T/5$

(c) Medium grid at $t = t_0$ (d) Medium grid at $t = t_0 + 2T/5$

(e) Fine grid at $t = t_0$ (f) Fine grid at $t = t_0 + 2T/5$

Figure 7. Dynamic pressure and velocity distribution around the rolling triangle cylinder : $\omega' = 1.594$, $\varphi_0 = 0.05rad$

5. Conclusion

Forced oscillation tests were conducted on a two-dimensional triangular section using the overset grid technique in OpenFOAM at different excitation frequencies. The numerical results indicates that the present grid system converged to a reasonable level of uncertainty, thus increasing the reliability of the numerical results. Hydrodynamic coefficients were compared with results from previously conducted experiment and potential theory. Both the damping coefficients of the roll-into-roll and the roll-into-sway agreed well with the experimental results. On the other hand, in the case of added mass or moment of inertia, it was more in line with the potential theory results rather than the experimental results. The grid size contributed the most uncertainty in the added mass, whereas the time step size had a relatively low level of uncertainty. In terms of the damping coefficient, it was established that a sufficiently large number of grids should be used to consider the influence of smaller vortices.

Acknowledgments

This work was supported by the Industrial Strategic Technology Development Program (20014578, Development of assessment software for IMO second generation intact stability criteria) funded by the Ministry of Trade, Industry & Energy(MOTIE, Korea). This research was also supported by the research fund of Chungnam National University.

References

[1] Vugts J H 1968 The hydrodynamic coefficients for swaying, heaving and rolling cylinders in a free surface *Shipbuilding Laboratory, Technical University of Delft* Report No. 112

[2] Ikeda Y, Himeno Y and Tanaka N 1978 A prediction method for ship roll damping *University of Osaka Prefecture* Report No. 00405

[3] Yildiz B, Cakici F, Katayama T and Yilmaz H 2016 URANS prediction of roll damping for a ship hull section at shallow draft *J. Mar. Sci. Technol* **21** pp 48-56

[4] Yu Y H, Kinnas S A and Vinayan V 2005 Prediction of flow around FPSO hull section subject to roll motions: effect of the separated flow around bilge keels *Proceedings in the Fifteenth International Offshore Polar Engineering Conference 2005* **8** pp 163-170

[5] Kinnas S, Yu Y and Vinayan V 2006 Prediction of flows around FPSO hull sections in roll wsing an unsteady Navier-Stokes solver *Proceedings in the sixteenth International Offshore Polar Enginering conference* **4** pp 384-393

[6] Wilson R V, Carrica P M and Stern 2006 Unsteady RANS method for ship motions with application to roll for a surface combatant *Comput. Fluids* **35** pp 501-524

[7] Irkal M, Nallayarasu S and Bhattacharyya S 2016 CFD approach to roll damping of ship with bilge keel with experimental validation *Appl. Ocean Res* **55** pp 1-17

[8] Lavrov A, Rodrigues J M, Gadelho J F M and Guedes Soares C 2017 Calculation of hydrodynamic coefficients of ship sections in roll motion using Nacier-stokes equations. *Occean Eng.* **133** pp 36-46

[9] ITTC 7.5-03-01-01 2021 Recommended Procedures and Guidelines: Uncertainty Analysis in CFD Verification and Validation, Methodology and Procedures

[10] Jacobsen N, Fuhrman D and Fredsoe J 2012 A wave generation toolbox for the opensource CFD library:OpenFoam *Int. J. Numer. Methods Fluid* **70** pp 1073-1088

[11] ITTC 7.5-03-02-03 2011 Recommended Procedures and Guidelines: Practical guideline for ship CFD

IOP Conf. Series: Materials Science and Engineering 1288 (2023) 012030 doi:10.1088/1757-899X/1288/1/012030

Comparative analysis of the long-term extreme VWBM calculated according to IACS Rec. No 34 rev 1 and rev 2

A Mikulić[1], M Ćorak[2] and J Parunov[1]

[1] Department for Naval Architecture and Ocean Engineering, Faculty of Mechanical Engineering and Naval Architecture, University of Zagreb, Ivana Lučića 5, 10000 Zagreb, Croatia

[2] Maritime Department, University of Dubrovnik, Ćira Carića 2, 20000 Dubrovnik, Croatia

josko.parunov@fsb.hr; antonio.mikulic@fsb.hr; maro.corak@unidu.hr

Abstract. After facing some criticism regarding assumptions contained in IACS Rec. No. 34 Rev 1 (2001), efforts have been put in by IACS members resulting in a new revision of this document, which provides a procedure for computation of the long-term extreme wave loads on ocean-going ships. The main changes include a new scatter diagram, modification of the wave spectral model and energy spreading function, as well as probability level of design load. Preliminary, the reduction of extreme loads from 10% to 30%, depending on the vessel and response type, is estimated. As this reduction could have serious implications for ship structural design, the present research aims to further investigate consequences on the most important load component, i.e., vertical wave bending moment (VWBM) at midship. Transfer functions of VWBM are calculated using closed-form expressions, which are convenient for comparative studies where many computations are required. Ships of different types and sizes are subjected to the long-term VWBM analysis comparing extreme values based on the procedures proposed in Rev 1 and Rev 2. Although this study is not intended to provide extreme VWBM in absolute terms, the comparison with IACS UR 11 is nevertheless presented as a reference. It is found that Rev 2 long-term VWBMs never exceed corresponding IACS Rule values, which can't be stated for Rev 1. Results also indicate that differences in scatter diagrams between Rev 1 and Rev 2 could explain a large decrease in the extreme VWBM.

1. Introduction

The vertical wave bending moment (VWBM) is arguably the most important global load component in ship structural design. The design value of VWBM for ocean-going ships without operational restriction is given by simple formula from Unified Requirement (UR) S11 of the International Association of Classification Societies, corresponding to 10^{-8} exceeding probability, i.e., a return period of about 25 years for navigation in North Atlantic (NA) [1]. The rule VWBM is obtained by incorporating computations of long-term VWBM, full-scale measurement campaigns, and experienced knowledge of classification societies. UR S11 is recently revised for container ships only, providing different formula [2].

For innovative or non-standard ships outside the scope of validity of UR S11, the long-term extreme VWBM prediction based on seakeeping analysis is mandatory instead of using the Rule formula [3]. Additionally, the rule formula cannot be used in cases when the combined influence of hull flexibility, slenderness, and environmental actions on global dynamic response is important [4]. However, depending on the seakeeping tool employed and assumptions used in the long-term analysis, discrepancies in predicted extreme VWBM had been observed, even for standard ship types covered by UR S11 [5]. The most important sources of uncertainties in the long-term predictions include the seakeeping method used, the choice of the wave scatter diagram, the shape of the wave energy spectrum, wave energy spreading, and the ship operational profile, i.e., probability of wave heading angles and ship speeds. To standardize long-term predictions, IACS issued Recommendation No. 34 rev 1 (2001), detailing how to perform long-term computations of VWBM [6]. Thus, IACS proposed a standard wave scatter diagram, two-parameter Pierson-Moskowitz spectrum, energy spreading function proportional to cos^2, zero speed, and equal probability of all headings between ship and waves. The seakeeping method is not proposed, but it is understood that hydrodynamic software based on potential flow theory, approved by major classification societies, is to be used.

Though Recommendation No. 34 rev 1 (2001) has extensively been used in the design of innovative ships [3] and calibration of partial safety factors in Common Structural Rules [7], some issues have persisted. Namely, VWBM from the long-term predictions systematically overestimated the rule formula from UR S11 [3]. This was unexpected, as a direct long-term analysis requires more detailed and elaborated input data, while the rule formula represents an envelope that should normally include an additional safety margin to compensate for the simplicity of the input data. This was so even some unconservative assumptions have been employed in Recommendation No. 34 rev 1 (2001), e.g., zero speed, as even in extreme wave conditions, ships could maintain minimum cruising speed higher than zero, and higher speed normally tends to increase VWBM. Another unconservative assumption was related to the short-crested waves, as it is known that this assumption can reduce extreme VWBM by about 10% compared to the long-crested waves. Finally, the assumption that heading angles between ship and waves were uniformly distributed was also conservative, as ship master in extreme seas tends to turn a ship in head waves to avoid excessive rolling, while it is well known that the largest VWBM is achieved in head seas.

From these considerations, it appears that the choice of the wave scatter diagram in Recommendation No. 34 rev 1 (2001) was largely conservative and the most important among uncertainty sources [8]. This could explain the overestimation of Rule VWBM by the long-term prediction. The wave scatter diagram in Recommendation No. 34 rev 1 (2001) was developed for the North Atlantic based on visual observations published in the wave atlas Global Wave Statistics (GWS) [9]. Wave data in GWS are collected by observations from merchant ships along frequent shipping routes and, as such, are relevant for ocean-going ships. On the other hand, data from GWS suffer from the inaccuracies of visual observations, especially wave periods. It was found that wave observation underestimated wave periods, i.e., overestimated wave steepness, so the original scatter diagram from GWS had to be "smoothed" before practical application [10]. Although data from GWS inherently include weather routing, it is not clear if and how they consider heavy weather avoidance of individual ships [11]. Another drawback of data collected in the past century is that they do not include the effect of climate changes that became relevant in the past decade [8].

As a response to these technical drawbacks and inaccuracies of Recommendation No. 34 rev 1 (2001) [6], IACS came out with the revised document Recommendation No. 34 rev 2 (2022) [12]. In the second revision, all assumptions are revised, and some modifications are introduced. The most important change is the introduction of the new IACS NA scatter diagram, based on the hindcast wave database, filtered for the actual shipping route using Automatic Identification System (AIS), a satellite ship positioning system available for all merchant ships [13]. Other modifications compared to the first revision are to propose the JONSWAP wave energy spectrum instead of the Pierson-Moskowitz spectrum, to use an energy spreading function proportional to $\cos^3$ instead of $\cos^2$, and to use a minimum ship speed of 5 knots instead of zero speed. Furthermore, 75% of the design speed is recommended for the evaluation of design wave loads for fatigue assessment. IACS decided to maintain uniform distribution of heading angles between ship and wave. Another difference is that instead of extracting design VWBM for probability level 10^{-8}, a return period of 25 years is proposed, meaning that the mean response period and number of responses in 25 years are to be calculated. Several studies have been published in the past years showing the motivation and theoretical background of the development of Recommendation No. 34 rev 2 (2022). Thus, GWS is revised in [14] and [15], wave parametrization in [16], while the consequence assessment, indicating that the extreme wave loads are reduced from 10% to 30%, is described in [13].

The aim of the present study is to perform an independent assessment of design VWBMs, calculated by the application of Recommendation No. 34 rev 2 (2022), in comparison to Recommendation No. 34 rev 2 (2001). The analysis is performed on four oil tankers and four container ships of different sizes. Scantlings of oil tankers are chosen as representative of different classes [17], while container ships are taken from previous studies [18] and [19]. Transfer functions required for long-term analysis are determined according to closed-form expression formulated by Jensen et al. [20]. Although not so suitable for providing accurate values of VWBM in absolute terms, these expressions are conveniently used in conceptual studies like the present one, where only relative values between two approaches are required [21]. Having this in mind, relative values of calculated extreme VWBM are determined with respect to linear IACS VWBM from UR S11 and UR S11A for oil tankers and container ships, respectively, without the intention to get conclusions about the adequacy of the rule formulation. In addition, the comparative analysis of scatter diagrams from two revisions is performed, being the most important reason for large differences in calculated extreme VWBMs.

The paper is structured into five sections. The methodology used for the calculation of the long-term distribution of VWBM is presented in Section 2. Section 3 provides descriptions of ships used in the study. Results of the comparative long-term analyses of VWBM according to two revisions of IACS Rec No. 34 are presented in Section 4. Discussion is provided in Section 5 comparing scatter diagrams from Rev 1 and Rev 2 in terms of the most probable wave amplitudes and environmental contours. The paper ends with the Conclusions given in Section 6.

2. Method for long-term distribution of VWBM

The key wave load component, the VWBM, is a stochastic process influenced by the randomness of the ocean wave environment. Such a stochastic process can be analyzed by statistical methods where spectral moments of ship response are one of the most important parameters. Spectral moments are determined for the short-term sea states, while the long-term distribution combines short-term Rayleigh-distributed ship responses. Due to the assumption of linearity, ship response in irregular seas can be obtained as a product of the square of the transfer function and spectral wave density in the wave frequency domain:

$$S_R(\omega|H_S, T_Z, v, \beta, C) = |H(\omega|v, \beta, C)|^2 S_\eta(\omega|H_S, T_Z) \tag{1}$$

Transfer functions of VWBM are calculated using the semi-analytical approach proposed and validated by Jensen and Mansour [21]:

$$\frac{\Phi_M}{\rho g B L^2} = \kappa \frac{1-kT}{(k_e L)^2}\left[1 - cos\left(\frac{k_e L}{2}\right) - \frac{k_e L}{4} sin\left(\frac{k_e L}{2}\right)\right] F_V(Fn) F_C(Cb) \sqrt[3]{|cos\beta|}. \tag{2}$$

A ship hull is approximated by a pontoon of equal length L, equivalent breadth B, and draught T, while block coefficient C_b (3) and speed (Froude number, Fn) (4) are accounted for through correction factors, $F_C(C_b)$ and $F_V(Fn)$, respectively.

$$\begin{gathered} F_C(C_b) = [(1-\vartheta)^2 + 0.6\alpha(2-\vartheta)] \\ \vartheta = 2.5(1 - C_b); \quad C_b = \max(0.6, C_b) \end{gathered} \tag{3}$$

$$F_V(Fn) = 1 + 3Fn^2, for\ Fn < 0.3 \tag{4}$$

The Smith correction factor κ from (2) is approximated by (5), where k_e is the effective wave number.

$$\kappa = \exp(-k_e T)$$
$$k_e = |k\cos\beta| \tag{5}$$

Considering the forward speed of the ship, the n-th spectral moment of the short-term ship response can be derived as:

$$m_n = \int_0^\infty \left|\omega + \frac{\omega^2}{g} v\cos\beta\right|^n S_R(\omega|H_S, T_Z, v, \beta, C)d\omega \tag{6}$$

The short-term VWBM corresponds to a random sea state of several hours in duration. Since considered stationary, a random sea state follows the Gaussian process as well as the corresponding response. By assuming the narrow-banded Gaussian process, the short-term probability distribution of response amplitudes $F_S(M_W)$ can be described by the Rayleigh distribution as:

$$F_S(M_W|H_S, T_Z, v, \beta, C) = 1 - e^{\left(-\frac{M_W{}^2}{2\sigma_R{}^2}\right)} \tag{7}$$

where the response variance of VWBM is obtained as zero spectral moment, i.e., the area below the response spectrum:

$$\sigma_R^2 = m_0 = \int_0^\infty S_R\left(\omega|H_S, T_Z, v, \beta, C\right)d\omega \tag{8}$$

The long-term formulation of VWBM response amplitudes can be structured differently, but essentially, it is defined as a series of Rayleigh-distributed short-term responses fully determined by a process variance. The response variance of the short-term sea state can be considered as a random variable f_R, with the probability density function given as [22]:

$$f_R\left(H_S, T_Z, v, \beta, C\right) = f_0\left(\beta|H_S\right) f_D\left(\beta|H_S\right) f_M\left(\beta|H_S\right) f_M\left(v|H_S\right) f_{H_S,T_Z}(h,t) f_C(C) \tag{9}$$

where $f_0(\beta|H_S)$ is the probability density function of heading angles, $f_D(\beta|H_S)$ describes the directional spreading of waves, $f_M(\beta|H_S)$ and $f_M(v|H_S)$ are the maneuvering effects that have an influence on the ship speed and heading angle, $f_{H_S T_Z}(h,t)$ is joint probability density function of wave heights and zero crossing periods, while $f_C(C)$ is the probability density function of the loading condition.

Under the assumption that all heading angles have an equal probability of occurrence, omitting maneuvering effects, and considering only one loading condition, the long-term distribution of VWBM amplitudes $F_L(M_w)$ may be obtained as:

$$F_L\left(M_w\right) = \sum_i^{n_\beta} \frac{\Delta\beta}{2\cdot\pi} \cdot \left(\sum_{j,k}^{n_H, n_T} F_{i,j,k}\left(M_w \middle| H_{S_j}, T_{Z_k}, \beta_i\right) \cdot r_{w_{k,i}}\left(T_{Z_k}, \beta_i\right) \cdot p_{j,k}\left(H_{S_j}, T_{Z_k}\right)\right) \tag{10}$$

where $p_{j,k}\left(H_{S_j}, T_{Z_k}\right)$ is the probability of occurrence of short-term sea state and $r_{w_{k,i}}(T_{Z_k}, \beta_i)$ is the relative number of response cycles in each short-term sea state derived by equation (11):

$$r_w\left(T_{Z_k},\beta_i\right)=\frac{\overline{T_Z^*}}{T_{Z_{k,i}}^*} \tag{11}$$

The mean zero up-crossing period of the response of a single short-term sea state $T_{Z_{k,i}}^*$, is calculated from:

$$T_Z^* = 2\pi\sqrt{\frac{m_0}{m_2}} \tag{12}$$

while $\overline{T_Z^*}$ and $\overline{v_w}$, the mean response zero up-crossing period and mean zero up-crossing frequency of all sea states, respectively, are determined as follows.

$$\overline{v_w}=\sum_i^{n_\beta}\frac{\Delta\beta}{2\cdot\pi}\left(\sum_k^{n_T}\frac{1}{T_{Z_{k,i}}^*\left(T_{Z_k},\beta_i\right)}p_k\left(T_{Z_k}\right)\right)=\frac{1}{\overline{T_Z^*}} \tag{13}$$

A Weibull 2-parameter model is usually used to approximate the discrete long-term probability distribution of vertical wave bending moments computed by the above-described procedure:

$$F\left(M_W\right)=1-e^{\left(-\left(\frac{M_W}{\vartheta}\right)^{\lambda}\right)} \tag{14}$$

where ϑ and λ are the Weibull scale parameter and the shape parameter, respectively.

The effect of wave directional spreading is embedded into the calculation of spectral moments (15) using spreading function $f_s(\theta)$ defined by (16), where n_s is cosine wave spreading power, θ_0 is the mean wave direction, and θ is the wave direction at which the spectrum is evaluated.

$$m_n=\int_0^{\infty}\sum_{\theta_0-90°}^{\theta_0+90°} f_s(\theta)\,\omega_e^n\cdot S_R(\omega|H_S,T_Z,\theta,v)d\omega \tag{15}$$

$$f_s(\theta)=k\,cos^{n_s}(\theta-\theta_0) \tag{16}$$

The normalization factor k is selected such that:

$$\sum_{\theta_0-90°}^{\theta_0+90°} f_s(\theta)=1 \tag{17}$$

The procedure for both revisions follow previously described methodology with already described modifications summarized in Table 1.

Table 1. Changes between Rev 1 and 2 [13].

Ship class	Rev 1	Rev 2
Source of wave data	Visual observations	Hindcast
Wave spectrum	Pierson-Moskowitz	JONSWAP, gamma=1.5
Cosine wave spreading power	2	3
Return period for extreme loads	At least 20 years (10^{-8})	25 years
Vessel speed for strength assessments	0 kn	5 kn

3. Studied ships

In the present study, the previously described analysis is performed for four oil tankers, representing full-form ships with large block coefficients, and four containerships, representing fine-form ships with low block coefficients.

The actual range of sizes of oil tankers operating worldwide, namely Panamax, Aframax, Suezmax, and VLCC, is used with the average dimensions of typical ships representing each class as given by Michel and Osborne [17], and presented in Table 2.

Table 2. Main dimensions of typical oil tankers.

Ship class	Length	Breadth	Draught	Cb
Panamax	174.4	31.4	11.3	0.8
Aframax	229.7	41.9	13.1	0.8
Suezmax	260.8	45.8	15.9	0.8
VLCC	318.6	58.4	21.1	0.8

Analyzed containerships exhibited in Table 3 represent actual ship structures previously studied in [18] and [19], also aiming to cover ships of different sizes.

Table 3. Main dimensions of examined containerships.

Ship class	Length	Breadth	Draught	Cb
S175	175	25.4	9.5	0.572
4400TEU	261	37.1	12.6	0.610
Flokstra	270	32.2	10.85	0.598
9400TEU	333	42.8	13.17	0.668

4. Results

Results of the analysis are presented as long-term distributions of linear VWBM, presented in figures 1 and 2 for oil tankers and container ships, respectively. The ordinate in both figures represents the ratio of VWBM from direct analysis and IACS Rule VWBM, taken from UR S11 and UR S11A, for oil tankers and container ships, respectively. It should be clarified that only a linear part of IACS UR S11A VWBM is used, neglecting thus non-linear aspects of global loads calculation for container ships. One may notice that long-term VWBM obtained using Rev 2 (full lines) are always lower compared to Rev 1 (dashed lines). Also, interestingly, Rev 1 leads to long-term distributions represented with almost straight lines, indicating near-exponential long-term distribution. For Rev 2, however, long-term distributions are curves with increasing slopes resulting in decreased probability levels.

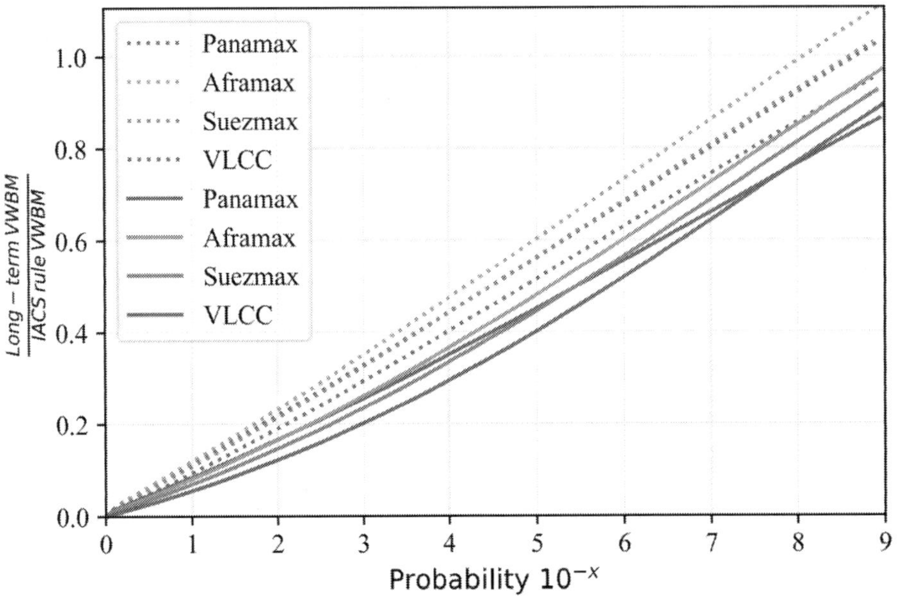

Figure 1. The long-term distributions of linear VWBM for oil tankers. Dashed lines represent Rev 1, and full lines Rev 2 results.

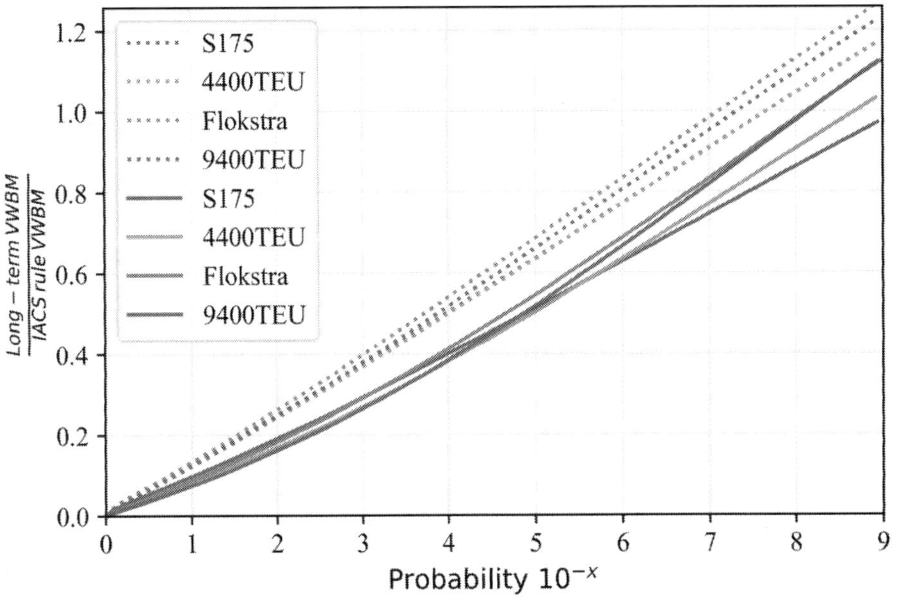

Figure 2. The long-term distributions of linear VWBM for container ships. Dashed lines represent Rev 1, and full lines Rev 2 results.

IOP Conf. Series: Materials Science and Engineering 1288 (2023) 012030 doi:10.1088/1757-899X/1288/1/012030

Differences in design VWBM caused by using Rev 2 instead of Rev 1 may be observed in Figure 3a) and 3b) for tankers and container ships, respectively. The ratio between linear VWBM obtained by direct analysis and those from IACS rules is plotted on the ordinate. Design values in Figures 3a) and 3b) are obtained for the probability level of 10^{-8} for Rev 1 and the return period of 25 years for Rev 2. In the latter case, the probability level is determined using an average response period. It may be seen that the design VWBM determined by Rev 1 is reduced by 13-19% if Rev 2 is employed, where a higher reduction is obtained for smaller ships. For containerships, these reductions read 16–23%, also increasing with decreasing ship size. These figures, as well as their trends, correspond to those reported by IACS [13].

Furthermore, it is interesting to compare the design VWBM by long-term analysis to the IACS rule VWBM. For oil tankers, both Rev 1 and Rev 2 lead to the design VWBMs lower than IACS UR S11. For containerships, however, the long-term analysis leads to larger VWBMs compared to the IACS UR S11A for all analyzed ships. For Rev 2, however, long-term design VWBM is lower than the Rule value. It should be emphasized, however, that long-term analysis is performed based on simplified transfer functions and that computed long-term VWBM is not considered an accurate estimate in the absolute term.

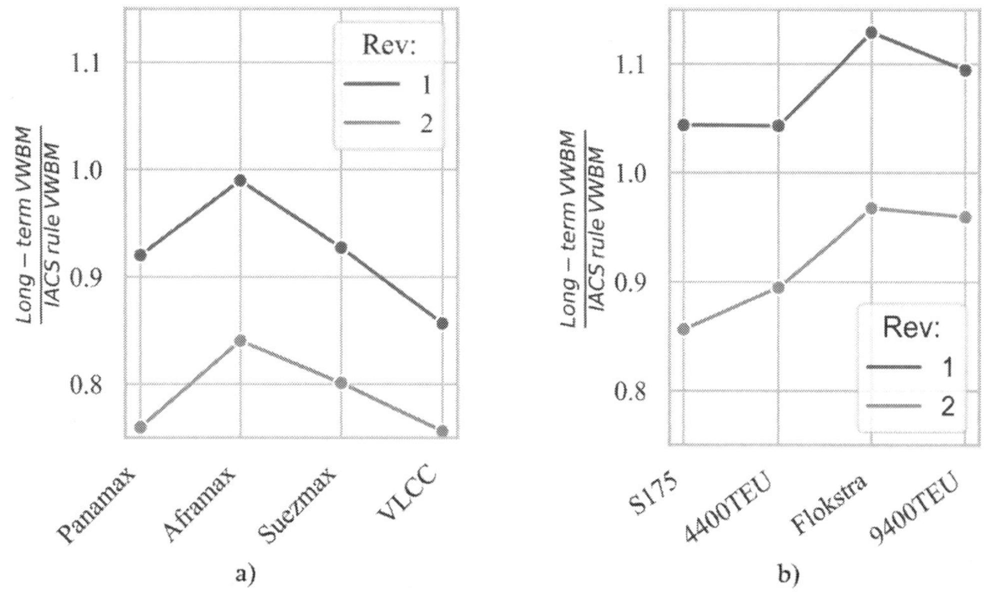

Figure 3. The design VWBM ratio comparison between Rev 1 and 2 for different ships: a) oil tankers; b) container ships.

5. Discussion

The choice of an appropriate wave scatter diagram is the most important parameter in the long-term prediction of extreme wave loads. Therefore, a more detailed analysis of scatter diagrams from Rev 1 and Rev 2 is conducted to understand better differences in design VWBM obtained in Section 4.

The scatter diagram in Rev 1 describes the wave data of the North Atlantic, covering areas 8, 9, 15, and 16, as defined in GWS. The original wave data from GWS are modified to describe more accurately the average wave steepness of extreme sea states. The IACS diagram includes significant wave heights (H_S) up to 16.5m and contains 100000 observations. It should be noted that the sea states in the IACS diagram with a low probability of occurrence are defined by numerical simulation, namely because, in the written version of GWS, the probabilities are rounded to four decimal places.

Thus, the sea states with a very low probability of occurrence are removed from the scatter diagram. The simulation is performed based on theoretical distribution fitted to the empirical data set.

The scatter diagram in Rev 2 is obtained by combining numerical data from the IOWEGA hindcast wave database and those from AIS for the geographical area of the North Atlantic, very similar to the area from Rev 1 ([13],[14]). Significant wave heights up to 18.5 m are contained in the scatter diagram in Rev 2, although these sea states have a rather low occurrence probability. The total number of observations is the same as in Rev 1, 100000. One difference between the two scatter diagrams is that the mean wave period (T_{0m1}) is used in Rev 2, while the average zero up-crossing period (T_Z) is used in Rev 1.

The severity of the scatter diagram may be quantified by determining the most probable long-term wave amplitude. This may be achieved by analyzing long-term distributions of wave amplitudes contained in the scatter diagram. The procedure for computing the long-term distribution of wave amplitudes is the same as for wave loads, given in Section 2. The resulting long-term distributions of the wave amplitudes are presented in Figure 4, where one may notice that extreme amplitude for 10^{-8} probability level reads 16.2 m and 14 m for Rev 1 and Rev 2, respectively. A similar analysis is performed in ref. [3], where the same value is obtained for Rev 1, while 13.7 m extreme wave amplitude is obtained for the DNV World-Wide scatter diagram used for fatigue analysis. Similarly, as for VWBM, the long-term distribution of wave amplitudes is an almost straight line for Rev 1, while it is curved for Rev 2.

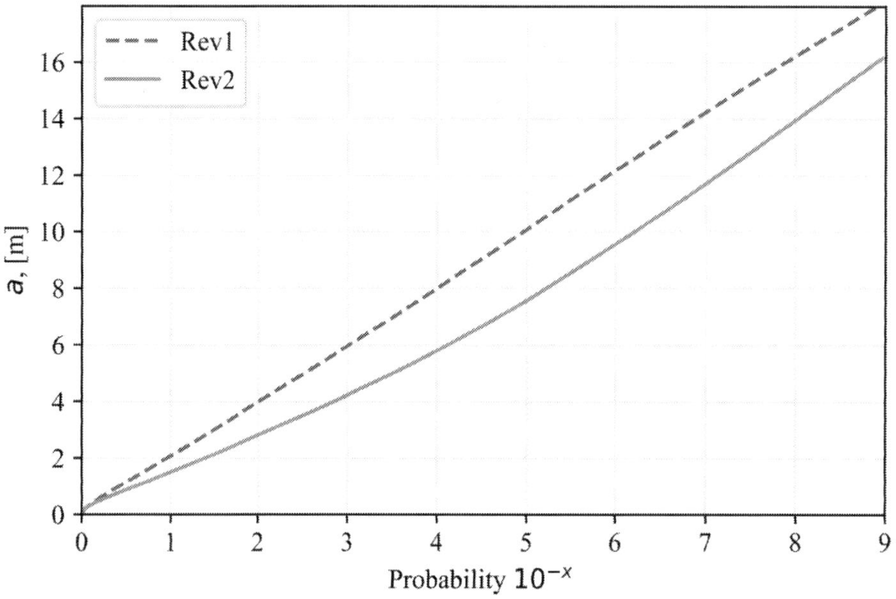

Figure 4. The long-term distributions of most probable wave amplitude for container ships. Dashed lines represent Rev 1, and full lines Rev 2 results).

Another approach to visually estimate the severity of wave scatter diagrams is to present environmental contours, which is a relatively novel approach for defining and presenting extreme sea states [23]. Contours are created employing the IFORM method assuming a 3h sea state duration. The zero-upcrossing wave periods T_z from Rev1 are converted to the mean wave periods T_{0m1} for the purpose of the presentation on the same diagram. Environmental contours for the 25 years return period are presented in Figure 5 for scatter diagrams from Rev 1 and Rev 2. It may be observed that Rev 1 greatly overestimates significant wave heights encapsulating even Rev 2 100 years contour, therefore, confirming the greatest differences depends on the selection of the scatter diagram. Large shift of the contour toward lower wave periods, i.e., higher wave steepness, may be observed for Rev 1 compared to Rev 2. Similar trend is observed for the environmental contours for Zone 14, presented in [14].

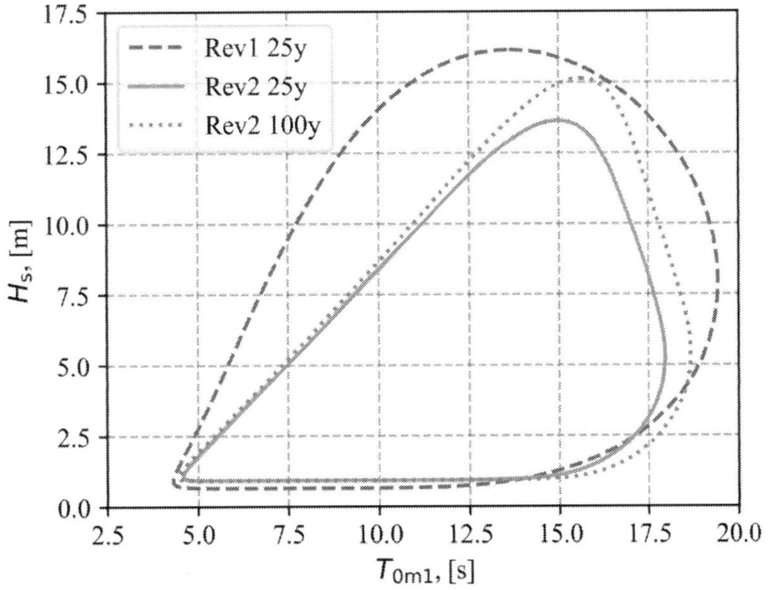

Figure 5. F The comparison of the IFORM contour based on joint models representing scatter diagrams form Rev 1 and 2.

6. Conclusion

Comparison of two revisions of IACS Rec. No. 34 procedure for direct calculation of design VWBM is presented in the study. Transfer functions of VWBM are defined using semi-analytical closed-form expressions, conveniently used in conceptual studies. Thus, design VWBMs are calculated for four tankers and four container ships of different sizes. Design linear VWBM from the direct analysis are compared to IACS Rule values from UR S11 and UR S11A for tankers and container ships, respectively. As discrepancies in scatter diagrams between two revisions of Rec No. 34 are the most important reason for differences in long-term VWBM, long-term distributions of wave amplitudes for two scatter diagrams are analyzed, and IFORM environmental contours are compared. The following conclusions are drawn from the analysis:

1. Design values of VWBM calculated by Rev 2 are reduced by 13-19% for tankers and 16-23% for container ships, where a larger reduction is obtained for smaller ships.
2. Although quantitative results should be taken with cautiousness because of the simplistic approach of calculation of transfer functions, it is noted that IACS URS11A Rule values are exceeded for Rev 1 by the results of direct analysis for all container ships. For Rev 2, however, direct analysis results are lower compared to IACS Rule values through all cases.
3. The extreme wave amplitude for exceeding probability level 10^{-8} is 2.2 m higher for Rev 1 compared to Rev 2. This could explain large differences in long-term VWBM between the two revisions.
4. Environmental contours indicate significant conservatism of the Rev 1 scatter diagram. Rev 1 25-year contour substantially overestimates significant wave heights compared to Rev 2,

encapsulating both 25- and 100-year contours. It also appears that the wave steepness contained in Rev 1 is much higher than in Rev 2.

It is expected that IACS Rec. No. 34 Rev 2 could unlock great, still unused potential of direct wave load analysis in s reliability-based structural design and optimization of ocean-going ships. Namely, ship designers can obtain reduced Rule wave loads and, consequently, material saving if they perform direct long-term analysis, which was not the case with Rev 1.

Acknowledgment
This work has been fully supported by the Croatian Science Foundation under the project IP-2019-04-2085.

References

[1] International Association of Classification Societies (IACS) 2010 *UR S11 longitudinal strength standard. Reb 6* (London: IACS)
[2] International Association of Classification Societies (IACS) 2015 *UR S11A longitudinal strength standard for container ships* (London: IACS)
[3] Parunov J, Senjanović I and Pavičević M 2004 Use of vertical wave bending moment from hydrostatic analysis in design of oil tankers *Transactions of the Royal Institution of Naval Architects Part A: International Journal of Maritime Engineering* **146** a4
[4] Tilander J, Patey M and Hirdaris S 2020 Springing Analysis of a Passenger Ship in Waves *J. Mar. Sci. Eng.* **8** 7 p 492
[5] Parunov J, Guedes Soares C, Hirdaris S and Wang X 2022 Uncertainties in modelling the low-frequency wave-induced global loads in ships *Mar. Struc.* **86** 103307
[6] International Association of Classification Societies (IACS) 2001 *Recommendation No. 34: standard wave data. Rev 1* (London: IACS)
[7] International Maritime Organization (IMO) 2006 *Maritime Safety Committee MSC 81/INF.6, Goal-based new ship construction standards - linkage between FSA and GBS*
[8] Bitner-Gregersen E, Waseda T, Parunov J, Yim S, Hirdaris S, Ma N and Guedes Soares C 2022 Uncertainties in long-term wave modelling *Mar. Struc.* **84** 103217
[9] British Marine Technology (Primary contributors Hogben N, Da Cunha L F and Oliver H N) 1986 *Global Wave Statistics* (London: Unwin Brothers Limited)
[10] Bitner-Gregersen E M, Cramer E H and Løseth R 1995 Uncertainties of Load Characteristics and Fatigue Damage of Ship Structures *Mar. Struc.* **8** pp 97-117
[11] Moan T, Shu Z, Drummen I and Almashi H 2006 Comparative reliability analysis of ships considering different ship types and the effect of ship operations on loads *Trans. SNAME* **114** pp 16–54
[12] International Association of Classification Societies (IACS) 2022 *Recommendation No. 34: standard wave data. Rev* 2 (London: IACS)
[13] Austefjord H N, de Hauteclocque G, Johnson M C and Zhu T 2023 Update of wave statistics standards for classification rules *Advances in the Analysis and Design of Marine Structures* ed J W Ringsberg and C Guedes Soares (Oxon: CRC Press Balkema) pp 95-205
[14] de Hauteclocque G, Maretic N V and Derbanne Q 2023 Hindcast based global wave statistics. *Appl. Ocean Res.* **130** 103438.
[15] de Hauteclocque G, Zhu T, Johnson M, Austefjord H and Bitner-Gregersen E 2020 Assessment of Global Wave Datasets for Long Term Response of Ships *OMAE2020, Volume 2A: Structures, Safety, and Reliability* OMAE2020-18874
[16] de Hauteclocque G and Lasbleis M 2022 Extreme Seastate Parametrization and Its Consequences on Ship Responses *Proceedings of PRADS2022* (Dubrovnik) pp 187-202
[17] Michel R K and Osborne M 2004 Oil Tankers *Ship Design and Construction* ed T Lamb (Jersey City: SNAME) pp 29.1–29.41

[18] Parunov J and Ćorak M 2010 Influence of environmental and operational uncertainties on wave bending moments of containerships *RINA William Froude Conference: Advances in Theoretical and Applied Hydrodynamics – Past and Future*

[19] Ćorak M, Parunov J and Guedes Soares C 2018 Structural Reliability Analysis of Container Ships under Combined Wave and Whipping Loads *J. Ship Res.* **62** 3 pp 115-33

[20] Jensen J J, Mansour A E and Olsen A S 2004 Estimation of ship motions using closed-form expressions *Ocean Eng.* **31** pp 61–85.

[21] Jensen J J, Mansour A E 2002 Estimation of Ship Long-Term Wave-Induced Bending Moment Using Closed-Form Expressions *RINA Trans.* **144** pp 41–55

[22] Guedes Soares C 1998 Stochastic Modelling of Waves and Wave Induced Loads *Risk and Reliability in Marine Technology* (Rotterdam: Balkema) pp 197-211

[23] Det Norske Veritas (DNV) 2019 *Recommended Practice DNV-RP-C205: Environmental Conditions and Environmental Loads* (Oslo: DNV)

IOP Conf. Series: Materials Science and Engineering 1288 (2023) 012031 doi:10.1088/1757-899X/1288/1/012031

Impact of direct seakeeping computation on ship Underwater Keel Clearance prediction

Filippo Gallucci and Giuliano Vernengo

Department of Naval Architecture, Electric, Electronic and Telecommunication Engineering (DITEN), Polytechnic School, University of Genoa, Via Montallegro 1, 16145, Genoa, Italy

giuliano.vernengo@unige.it

Abstract. The continuous increase in worldwide ship traffic poses relevant challenges on the management of the onboard safety during navigation and specific operations. Among these issues, navigation in relatively shallow water areas, being either channels or port areas, is of major interest especially when rough weather conditions occur due to the increased risk of grounding. Such a problem, which is deeply related to the seakeeping performance of a ship, can be tackled by different perspectives. In the present paper, the approach proposed by the PIANC recommendations, typically used to design port and channels areas and infrastructures, are analysed from the perspective of ship design hence dealing with its operability. In particular, the dynamic Underwater Keel Clearance (UKC) of a ship is studied based on direct seakeeping computations to assess the effect of the operability conditions in terms of sea states and ship speed at different wave encounter angles.

1. Introduction

The analysis of the Underwater Keel Clearance (UKC) of ships is a topic that is typically very related to operations in real environments. In these situations, the significant parameters which contribute to the realization of the actual UKC are mainly measured by a set of sensors on board, so that the trend of the ship related to such a specific performance criterion might directly be observed [1,2,3,4]. On the other hand, the same performance metric is used by regulatory institutions to provide a reference for the design of coastal infrastructures such as for example, channels and other structures [5]. This metric has also been used for example in the light of ship manoeuvring prediction in shallow waters [6,7].

In the PIANC recommendations some approximate seakeeping performance prediction approaches are provided, based on different assumptions. In general, since the overall idea is to design a relatively large infrastructure, such as a channel, the seakeeping of the possible ships that might be involved in the use of that structure is only a part of the process. Moreover, typically the shape of the ships is not available, so approaches that can be easily used based on a relatively small set of data for the ships are needed. In this light, the proposed approximate solutions will provide results that intrinsically include a safety margin, which means that the seakeeping performance of a real ship might be overestimated. This might be a problem if seen from the perspective of the actual ship operations. Indeed, if the UKC would be included into an operability analysis, overestimation of the ship performance can result in possible failures of the criteria. In this study, the UKC performance is chosen as a criterion for short-term operability assessment of a reference ship. Such a metric could indeed be used at least in two perspectives. First it might be adopted as an operability criterion to be included every time the operations of a ship will involve the transit in confined waterways. Second, this assessment might be used at a very early design stage to compare different design solutions relying on the maximum operability with respect to this unusual quantity of interest.

The direct seakeeping prediction, made according to a Strip-Theory based approach, is compared to the formulations proposed by the PIANC recommendations to assess the differences between the models. Then a preliminary sensitivity analysis of the results computed by the numerical prediction with respect to operating and environmental parameters is carried out.

2. Selected test case: the DTC hull

The analysis of the UKC has been carried out considering the well-known DTC hull [10], shown in Figure 1, as a reference. The main parameters of such a Post-Panamax container ship hull are listed in Table 1. Since the prediction of the UKC must be referred to real operational scenarios, the full-scale hull size has been considered.

Table 1. Main particulars of the DTC hull.

L_{PP} [m]	B_{WL} [m]	T [m]	Volume [m^3]	CB [-]
355.0	51.0	14.5	173467 .0	0.66

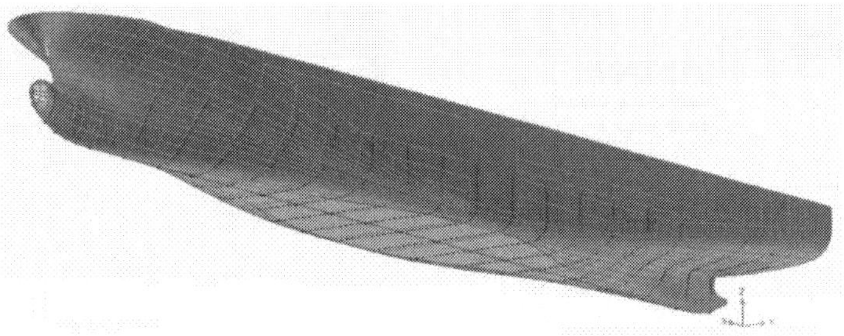

Figure 1. Perspective view of the DTC hull.

3. Approaches to seakeeping evaluation for the UKC

The UKC of the DTC hull has been predicted by using four different approaches, namely the Trigonometric Method, the Japanese Method, the Spanish ROM method, and direct seakeeping numerical predictions. The first three approaches are proposed by the PIANC, in which the direct assessment is also mentioned, without reference to any specific numerical method.

The trigonometric method is based on the rough assumption that heave, pitch and roll motions would be in phase to each other and the wave lengths to be considered are twice the ship length for the pitch induced motion and twice the ship breadth for the roll induced vertical motion, respectively. The heave motion is equal to the geometric centre of a half-sine wave of given amplitude (or height), $\theta_5 = 1\ deg$ and $\theta_4 = 5\ deg$ are considered for pitch and roll induced vertical motions, respectively. The maximum vertical motion Z_{Max}^{Trigo} is then computed according to Eq. (1).

IOP Conf. Series: Materials Science and Engineering 1288 (2023) 012031 doi:10.1088/1757-899X/1288/1/012031

$$Z_{Max}^{Trigo} = \frac{1}{2} \cdot \left(\frac{\pi}{8} H_{\frac{1}{3}} + B \cdot \sin\theta_4 + L_{BP} \cdot \sin\theta_5 \right) \quad (1)$$

The Japanese approach considers two possible situations, namely the vertical motion at the bow, caused by a combination of heave and pitch, and the vertical displacement at the bilge, due to combined heave and roll motions, respectively. The first contribution, Z_2, is directly provided in terms of RAOs of the bow vertical motion of a representative ship with CB=0.7 at Fn=0.1, at various wave encounter angles. These RAOs are shown in Figure 2. RAOs of the bow vertical motion used in the Japanese approach, referred to a ship with CB =0.7 and Fn=0.1.. The second term, Z_3, is computed according to Eq. (2), being θ_4^* the maximum roll angle based on the wave slope and on the encounter wave angle. Then for each encounter angle $Z_{Max}^{Jap} = \max(Z_2, Z_3)$.

$$Z_3 = \frac{1}{2} \cdot \left(0.7 \cdot H_{\frac{1}{3}} + B \cdot \sin\theta_4^* \right) \quad (2)$$

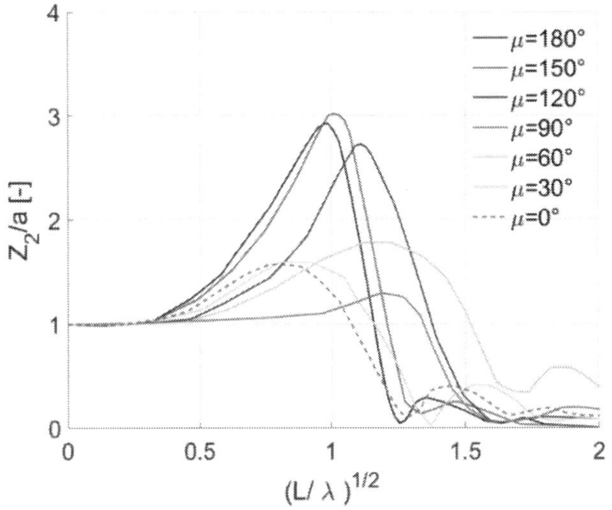

Figure 2. RAOs of the bow vertical motion used in the Japanese approach, referred to a ship with CB =0.7 and Fn=0.1.

The third method proposed is the so-called Spanish ROM approach, a parametric, semi-probabilistic method based on the definition of specific coefficients as shown in Eq. (3). The coefficients are used to represent specific effects contributing to the ship response. The first and the second coefficients are related to the maximum wave height and to the ship vertical displacement, respectively. The other coefficients are corrections accounting for the ship loading condition, C3, the ship speed, C4, the depth of the bottom, C5, and the encounter angle, C6. They are provided in tabular form, with the exception of $C_1 = 0.707 \cdot \sqrt{\ln\left(N_w / \ln\left(1/(1-P_m)\right)\right)}$, being NW the expected number of waves encountered by the ship in the area of interest during each occurrence over its design life (NW=200 is a suggested value) and $P_m = 1 - (1 - P_{DL})^{1/N_C}$ the exceeding probability for each occurrence, based on the acceptable failure probability during the design life, $P_{DL} = [0.05; 0.50]$, and the consistent number of critical cases over the same time range, NC.

$$Z_3 = H_{\frac{1}{3}} \prod_{i=1}^{6} C_i \tag{3}$$

In the framework of the PIANC approach, the direct seakeeping assessment is described as a further, more refined, possibility which can be used in case the geometric description of the ships under investigation will be available. In this study, the seakeeping performance of the DTC hull have been computed by using a strip-theory based approach [8], available through the Ship Motion Program (SMP) [9]. Such a method relies on the classic strip-theory approach, including some non-linear corrections for lateral motions, in particular, for the roll motion prediction. Appendages that provide a damping non-linear (viscous) correction term, such as the stern skeg, the rudders and the bilge keels have been included in the numerical prediction too. Once the Response Amplitude Operators (RAOs) of the motions of the ship (at full scale) have been computed, the RAO of the vertical motion at specific locations has been derived and the statistics of the response, exactly in terms of vertical displacements, have been computed according to the Rayleigh hypothesis on the distributions of the maxima. Three locations have been selected to assess the UKC of the ship. He vertical displacement have been evaluated at the keel of the hull, both at bow and at stern, and at the bilge at midship. These locations are the most probable points to be interested by the possibility of collision with the bottom of the sea.

4. Results of the analysis of different seakeeping methods on the UKC prediction

In this preliminary analysis of the effect of different seakeeping prediction methods on the UKC, the performance assessment has been carried out considering a set of operating conditions. The range of wave heights, wave periods and ship speeds are shown in Table 2. **Operating** conditions for the seakeeping analysis.. The UKC performance has been investigated in the perspective of a short-term operability analysis. Hence, all the encounter angles from stern to head seas have been included, uniformly spaced each 15 deg. Long-crested (mono-directional) sea and Bretshneider sea spectrum formulation have been considered.

Table 2. Operating conditions for the seakeeping analysis.

$H_{1/3}$ [m]	T_W [sec]	V_S [knots]
[0.5; 1.0; 2.0; 4.0]	[6.5; 13.5]	[0; 10]

First, the adoption of the different seakeeping performance prediction methods has been analysed. It is worth noting that all the other components of the UKC, namely the heel due to turning and wind, the squat, the clearances for tides and salinity of the water and the sea bottom, have been kept equals among the proposed seakeeping solutions to make the comparison consistent.

The values of the UKC predicted by using all the seakeeping methods over the entire range of encounter angles for V=4knots, To=11.5sec, H1/3=1m and H1/3=4m are displayed in Figure 3 and Figure 4, respectively, for the location at the bow. The trigonometric approach (light blue curve), which is the simplest among the selected ones, provides larger values compared to all the other methods. In particular, the UKC resulting from the application of such an approach is more than twice the values predicted on average by the others. Moreover, these values does not change too much with respect to the wave height.

At the lower wave height, H1/3=1m, the other three methods, namely the Japanese, the ROM and the direct assessment, are all below the limit of the criteria, arbitrarily imposed at 3m. However, the direct assessment and the Japanese methods present quite similar results both in terms of values and shape of the UKC over the encounter angles while the ROM approach, despite similar for bow and stern waves, shows very lower values of the response over the rest of the range of angles. This might be due to the inclusion of the default RAOs for the bow vertical motion in the Japanese method (see Figure 2). Moving to the higher wave height there is a much larger influence of this parameter on the direct assessment and on the ROM approach rather than that on the Japanese method. The Japanese method is subject to a greater effect of the wave height close to beam seas. This is due to the dependency of Z3, Eq. (2), by the H1/3, which is smoothed in the Z2 term in which the wave height is include in the RAO which in turns depends on the wave frequency, so the sea period.

These trends change if the location at midship is considered. The same conditions have been analysed in Figure 5 and Figure 6 for the midship location. At H1/3=1m, overall, the application of the methods produces the same result in terms of final operability, i.e. the trigonometric approach shows values of the UKC much higher than the limit, while the other three methods are well below this frontier. However, the ROM approach and the direct computation seem to have a better fit to each other, while the Japanese method provides slightly higher results. It is worth noting that the Japanese and the ROM approaches remain in the same relative order, so the first shows higher values of the UKC rather than the latter. This provides a sort of consistency verification for the methods proposed by the PIANC. Considering the higher wave height, both the Japanese and the ROM methods present a similar trend compared to the previous case of the response at the bow location, while the direct assessment provides much lower results, with larger values in the range of angles closer to beam sea. These values of the UKC are still below the imposed limit of operability. This is due by the implicit inclusion of the phases of the heave, roll and pitch RAOs in the computation of the vertical motion at midship location which is not considered at all in the other approaches. If seen in the light of an operability analysis this might significantly change the final evaluation of the ship under investigation.

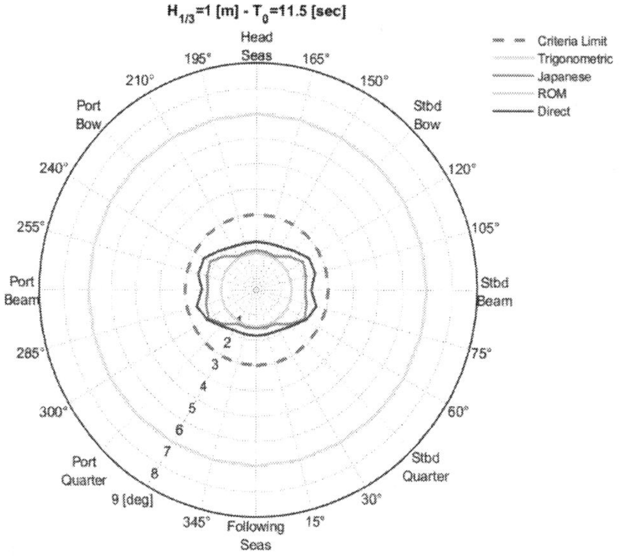

Figure 3. UKC predicted by using the four proposed seakeeping methods. Dashed red line: criteria limit at 3m. V=4knots. To=11.5sec. H1/3=1m. Bow.

IOP Conf. Series: Materials Science and Engineering 1288 (2023) 012031 doi:10.1088/1757-899X/1288/1/012031

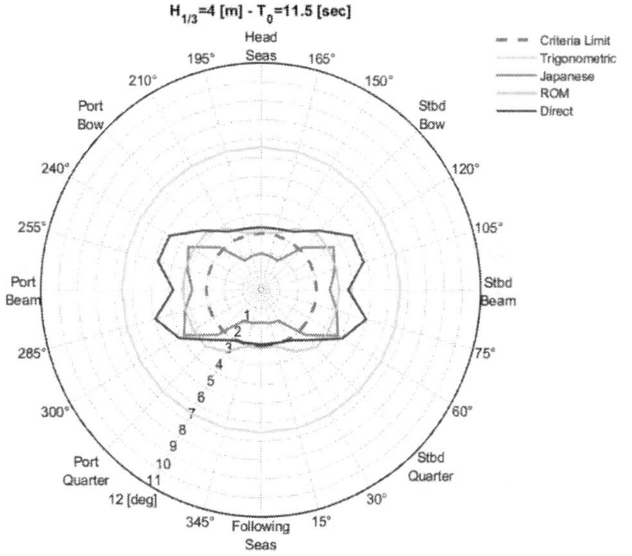

Figure 4. UKC predicted by using the four proposed seakeeping methods. Dashed red line: criteria limit at 3m. V=4knots. To=11.5sec. H1/3=4m. Bow.

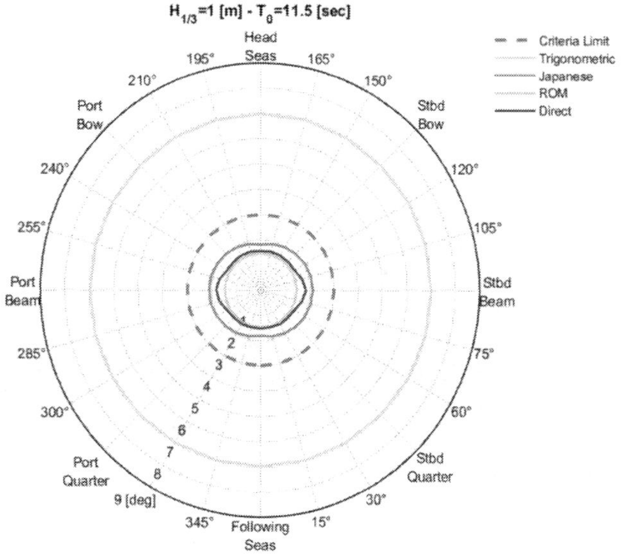

Figure 5. UKC predicted by using the four proposed seakeeping methods. Dashed red line: criteria limit at 3m. V=4knots. To=11.5sec. H1/3=1m. Mid.

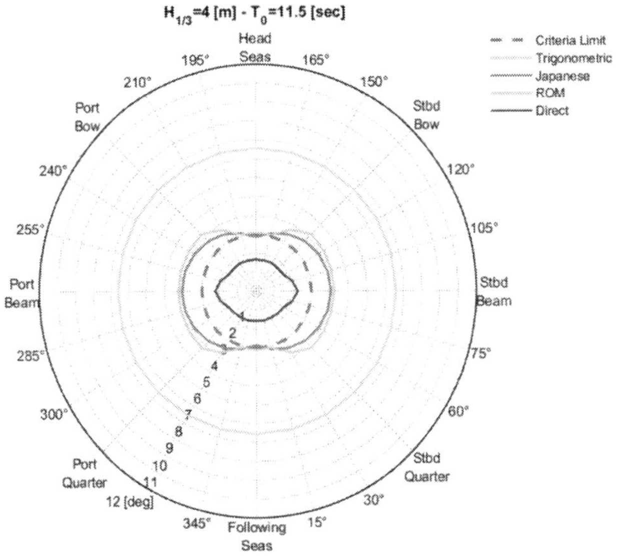

Figure 6. UKC predicted by using the four proposed seakeeping methods. Dashed red line: criteria limit at 3m. V=4knots. To=11.5sec. H1/3=4m. Mid.

Considering the direct seakeeping computation, the sensitivity of the UKC with respect to the ship speed, an operating parameter, and to the sea modal period, an environmental parameter, has been studied. Results obtained on the UKC for significant wave height H1/3=2m, modal periods in the range To=[6.5,13.5] sec, for ship speeds in the range Vs=[0,10] knots, are shown in Figure 7 to Figure 10, for the bow location.

The speed variation does not have a vary strong impact on the results. This might be partially due the strip-theory based speed correction of the hydrodynamic terms of the equation of motion combined to the relatively slow speeds considered. At lower sea modal periods the speed shows a larger effect for stern waves while it becomes more effective for bow waves shifting towards higher sea modal periods. This is mainly due to the combination of the sea spectrum peak position related to the changes in the encounter frequency, which depends on both the ship speed and the angle of encounter. A greater effect is seen by considering the sea modal period. Indeed, the response is overall amplified for larger modal periods showing a stronger dependency closer to bow waves. This trend is reasonable considering the main particulars of the ship with respect to the sea period. For example, accounting for the length of the DTC hull, which is LBP=355m, a modal period of To=13.5sec corresponds to a 'apparent' wave length with λ/L~0.8, which is closer to an expected peak of the vertical motion RAOs for head waves.

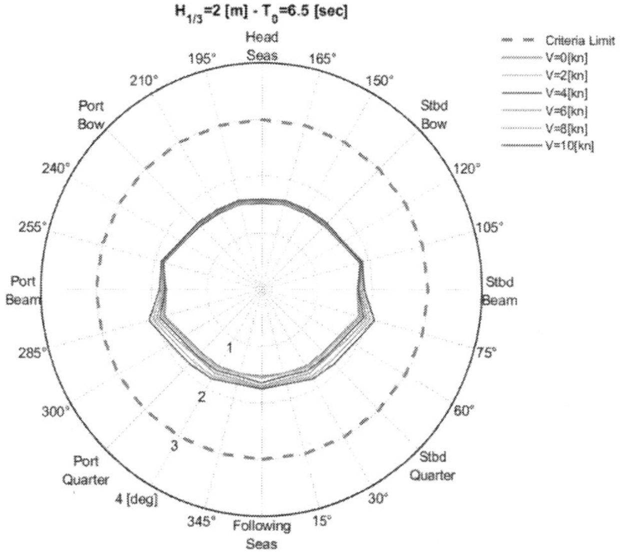

Figure 7. UKC predicted by using the direct seakeeping assessment. Dashed red line: criteria limit at 3m. V=[0;10]knots. To=6.5sec. H1/3=2m. Bow.

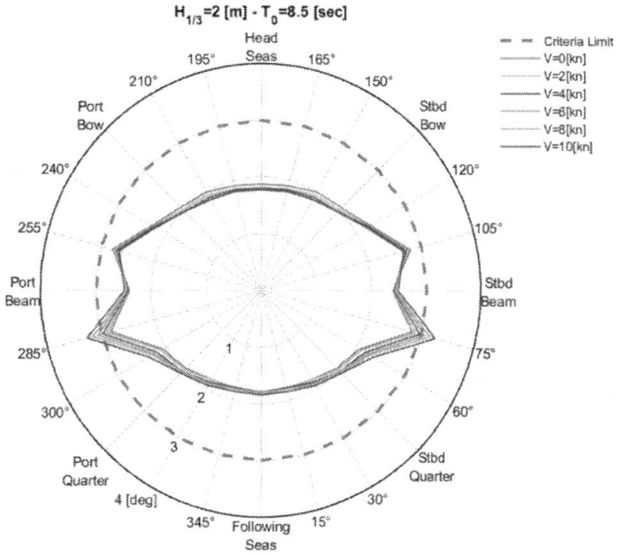

Figure 8. UKC predicted by using the direct seakeeping assessment. Dashed red line: criteria limit at 3m. V=[0;10]knots. To=8.5sec. H1/3=2m. Bow.

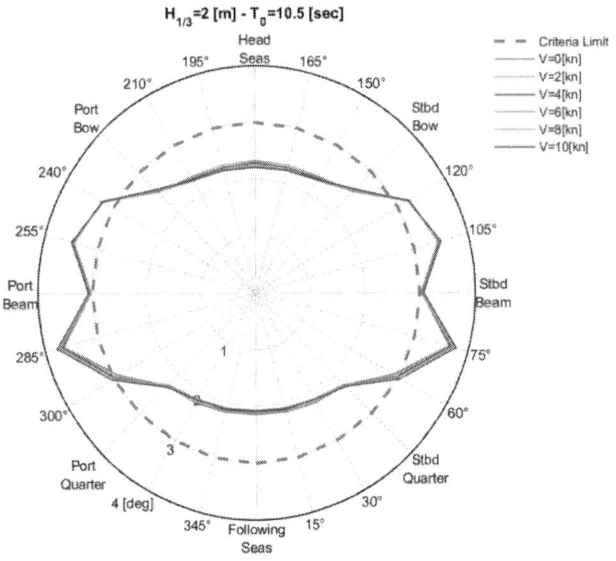

Figure 9. UKC predicted by using the direct seakeeping assessment. Dashed red line: criteria limit at 3m. V=[0;10]knots. To=10.5sec. H1/3=2m. Bow.

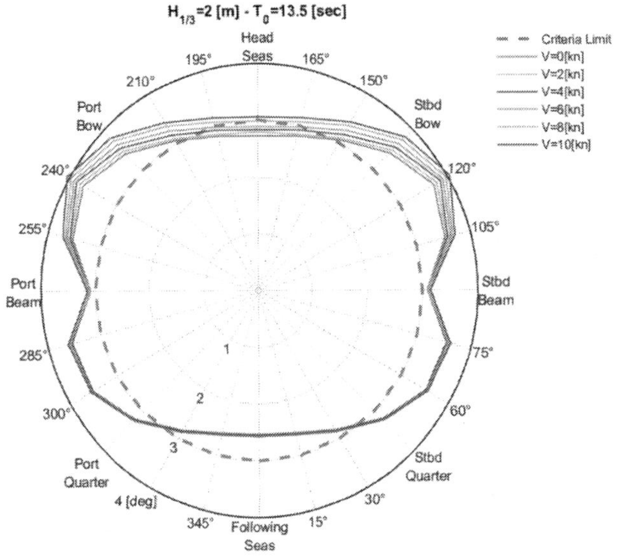

Figure 10. UKC predicted by using the direct seakeeping assessment. Dashed red line: criteria limit at 3m. V=[0;10]knots. To=13.5sec. H1/3=2m. Bow.

5. Conclusions

The Underwater Keel Clearance (UKC) of the DTC hull has been studied. In particular, since this performance metric is typically used to design coastal infrastructures according to the PIANC recommendations, the methods proposed in that rules have been analysed and compared to a direct assessment of the seakeeping performance of the ship. The seakeeping direct computations have been carried out by using a Strip-Theory based approach with non-linear corrections for lateral motions. Results have been first analysed comparing the obtained values of the UKC from the applications of the different methods and then a sensitivity analysis of the direct assessment of the UKC with respect to wave period and ship speed has been conducted.

Overall, two of the approximate methods proposed in the PIANC provides quite accurate results for low values of the significant wave height, especially if the bow location, which is typically the most problematic with respect to this topic, is considered. However, when other locations are investigated, such as the midship location at the bilge, the approximate methods seem to overestimate the UKC values by a large extent compared to the direct assessment. This is primarily due to the phase motion effect in the evaluation of the vertical motion at a location over the hull where heave, roll and pitch motions needs to be combined. The UKC does not shows a very relevant dependency on the ship speed, especially in the range of the low speeds considered. Such a dependency is then mixed with the effect of the sea period, which is mainly related to the interaction with the main dimensions of the ship itself, and with that of the encounter frequency that changes according to the encounter angle too.

In conclusion, the PIANC recommendations provide reliable results if seen from the perspective of the design of port and coastal infrastructures considering that a safety margin will always be wanted and accepted. However, if the UKC would be included into a ship operability analysis, a direct assessment of the seakeeping performance will certainly be needed to provide reliable predictions which won't be too harsh for the performance of the ship.

6. Reference

[1] Gourlay, Tim P., and Wallace G. Cray. "Ship under-keel clearance monitoring using RTK GPS." In Coasts and Ports. 2009.
[2] Ha, J. H., Gourlay, T., & Nadarajah, N. (2016). Measured ship motions in Port of Geraldton approach channel. In 4th MASHCON-International Conference on Ship Manoeuvring in Shallow and Confined Water with Special Focus on Ship Bottom Interaction (pp. 236-250).
[3] Giorgi, G., Teunissen, P. J., & Gourlay, T. P. (2012). Instantaneous global navigation satellite system (GNSS)-based attitude determination for maritime applications. IEEE Journal of oceanic engineering, 37(3), 348-362.
[4] Ruggeri, F., Watai, R., Tannuri, E. A., Rodrigues, C. P. C., Souza Filho, C. A., Tavares, B. R., & Geraldes, V. N. (2016). On the development of a dynamic underkeel clearance computational code for Santos Port (SP, Brazil). PIANC-COPEDEC IX.
[5] MarCom, W.G. Report n. 121 - Harbour Approach Channels Design Guidelines; The World Association For Waterborne Transport 117 Infrastructure, PIANC, 2014.
[6] Galor, W. (2007, July). Optimization of under-keel clearance during ship's manoeuvring in port water areas. In European Safe and Reliability Conference (ESREL 2007) (Vol. 1, No. 6, pp. 2681-2688).
[7] Vantorre, M., Delefortrie, G., Laforce, E., De Vlieger, H., & Claeys, S. (2003). Ship manoeuvring at very small and negative under keel clearance. IFAC Proceedings Volumes, 36(21), 37-42.
[8] Salvesen, N., Tuck, E. O., & Faltinsen, O. (1970). Ship motions and sea loads. Transactions SNAME, 1970.
[9] Meyers, W. G., Applebee, T. R., & Baitis, A. E. (1981). User's Manual for the Standard Ship Motion Program, SMP (No. DTNSRDC/SPD-0936-01).
[10] Moctar, O. E., Shigunov, V., & Zorn, T. (2012). Duisburg Test Case: Post-panamax container ship for benchmarking. Ship Technology Research, 59(3), 50-64.

Dynamic response of offshore triceratops under regular waves with tether postulated failure

Srinivasan Chandrasekaran[1] and Ganta Shanmukha Rao[2]

[1]Professor and Corresponding author, Dept of Ocean Engg, IIT Madras, email: drsekaran@iitm.ac.in
[2]Senior Research Scholar, Dept of Ocean Engg, IIT Madras, email: oe19d008@smail.iitm.ac.in

Abstract. Triceratops are innovative offshore platforms that consist of a deck, three buoyant legs, and ball joints. Their deck is supported by buoyant legs, which, in turn, are restrained on the sea bed using taut-moored tubular tethers. Ball joints partially isolate the deck from the buoyant legs by restraining the transfer of rotations from the legs to the deck. Triceratops platforms exhibit a monolithic structural action in the vertical plane and a hybrid dynamic behavior resembling tension leg platforms and spar buoys. The present study investigates their dynamic response in regular waves. Response amplitude operators (RAOs) are estimated for intact and postulated failure conditions of the tethers. Based on detailed numerical investigations, it is seen that the pitch response of the deck increases significantly under the postulated failure of tethers, which are otherwise absent due to the presence of ball joints. Marginal change in surge response, even under tether failure, confirms the suitability of triceratops for ultra-deepwater installations. The fatigue life of tethers before and after postulated failure conditions is compared using the S-N curve approach and noticed that the adjacent tethers' fatigue life reduces significantly under postulated failure conditions of tethers.

1. Introduction

Over the last several decades, as demand for gas and oil has grown, the industry has moved deeper and ultra-deep-waters to look for oil and gas. Because the environmental loads acting on the structures are severe in greater water depths, the geometric form of offshore platforms proposed for deep and ultra-deep waters needs special attention [1]. The offshore triceratops is cost-effective; the proposed geometric shape shall also have better motion characteristics under the encountered forces arising from the harsh environment [2,3]. Introducing ball joints between the deck and buoyant legs makes the triceratops a more novel geometric configuration from other newly generated offshore compliant platforms. Triceratops consists of three buoyant legs, a deck, and three ball joints. Buoyant legs are positioned and restrained by tethers, which are connected to the seabed. Ball joints transfer translational displacements to the deck from the legs and isolate the rotational displacements to the deck [4,5]. The different motion characteristics of structural members, such as buoyant legs and deck, uniqueness its structural behaviour under lateral loads [6]. Offshore triceratops resembles both spar platform and TLP due to their deep draft and positively buoyant nature [7]. The offshore platform's buoyancy is very high compared to the platform's self-weight, and the excess buoyancy is taken care of by pretension tethers. The fatigue damage studies are very much useful to analyze the service life of the compliant taut moored platforms and floating structures [8–14]. The stability of positively buoyant structures is more tactful to tethers, so the postulated failure is a fundamental analysis scenario that must be considered during the compliant taut moored platforms design process [15,16]. As seen from the critical literature survey, it is very clearly shown that the scarcity of the response behavior of offshore triceratops is under the postulated failure condition of the tethers. In the present study, the detailed numerical analysis was conducted under the action of regular waves with a zero-degree heading angle for tethers intact and postulated failure conditions.

2. Numerical model

The geometric details of the platform proposed by [4] are considered for the present study. The legs are modeled as tubular elements (Morison elements, D/L < 0.2) and their self-weight is computed by program control. The ballast load is applied as a point mass to lower the vertical CG of the legs by -112.74m and the forces acting on the legs are calculated by using the Morison equation:

$$q_n = \frac{1}{2}\rho C_d dA(v_n - \dot{x}_n)|v_n - \dot{x}_n| + \rho dV a_n + (C_m - 1)\rho dV(a_n - \ddot{x}_n) \quad (1)$$

where C_d, C_m represents the drag and inertia coefficients; v_n and a_n are water particle velocity and acceleration; $\ddot{x}_n$, $\dot{x}_n$ are acceleration and velocity of the structure; ρ is the density of the seawater; dA and dV are exposed area and displaced volume per unit length, respectively. The deck weight and payload are applied as a point mass at its CG and tubular tethers are connected to the legs for position-restraining. The convolution integration technique solves the following equation of motion in AQWA-NAUT (ANSYS AQWA PACKAGE) to obtain the response history in all degrees of freedom.

$$[M + M_a]\ddot{x}(t) + [C]\dot{x}(t) + [K]x(t) = F(t) \quad (2)$$

where, $\ddot{x}(t)$, $\dot{x}(t)$, $x(t)$ and $F(t)$ is acceleration, velocity, displacement, and force vectors; $[M]$ and $[M_a]$ are the structural mass and added mass matrices; $[C]$ is the damping matrix; $[K]$ is the stiffness matrix. The geometric details of the platform and proposed tubular tethers are shown in Table 1 and Table 2, respectively. Figure 1 and Figure 2 show the numerical model and plan view of the platform. The tether systems used for taut-moored compliant platforms like TLPs and Triceratops have various forms, including tubular, solid rods or bar shapes, and stranded construction such as parallel or helical wire ropes. Steel tubular has been the preferred material for the tethers because the bore may be considered for use in routing umbilical between the seafloor and surface or as a fluid flow conduit [17]. The environmental loading conditions of very high (H =10m) and phenomenal sea states (H =15m) on the Douglas Sea scale are used in the present study.

Figure 1. Numerical model

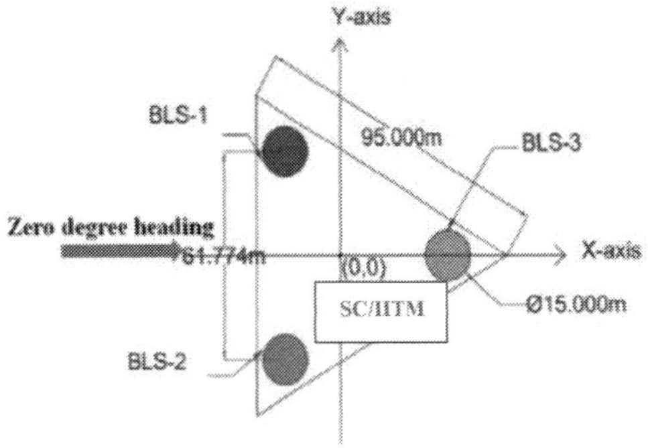

Figure 2. Plan view of the platform

Table 1. Geometric details

Geometric details	Value
Centre to Centre distance of the Legs	61.77 m
Diameter of Leg	15
Freeboard	20.24 m
Draft	154 m
Tether length	2246 m
Water depth	2400 M
Self-weight + payload	562 MN
Buoyancy force	821 MN
Total pretension	259 MN

Table 2. Tether details

Total number of tethers	12
Leg-1, leg-2, and leg-3	Tethers 1-2-3-4, tethers 5-6-7-8 and tethers 9-10-11-12
Tether outer dia	1.29 m
Tether thickness	0.043 m
Total elements	250
Stiffness of the tether	15.8 MN/m
Length of each element	8.89 m
Grade of steel	X65
Yield strength	450 N/mm^2
Tensile strength	535 N/mm^2
Postulated failure tethers	Tether 9,11 under leg-3

3. Numerical model validation

Free oscillation tests are conducted numerically for getting natural periods and damping ratios. An external disturbing force is given in all respective degrees of freedom to measure the free vibration responses. The damping is calculated by using the logarithmic decrement method. The obtained natural periods and damping ratios of tethered triceratops are compared with [4]. The natural periods of flexible degrees of freedom are slightly more and stiff degrees of freedom are less due to the alignment of the tether configuration of the present study. The results are shown in Table 3.

Table 3. Natural periods and damping ratios

Active-degrees-of-freedom	Intact condition (Present study)		Chandrasekaran et al. (2018)	
	Natural periods (Sec)	Damping ratio (%)	Natural periods (Sec)	Damping ratio (%)
Surge	221.4	5.38	215	5.14
Sway	222.3	5.16	215.4	5.03
Heave	3.45	0.58	4.3	0.51
Roll	4.1	6.4	6.2	6.23
Pitch	4.3	6.28	6.1	6.08
Yaw	219.7	5.9	215.9	5.68

4. Results and discussions

The numerical analysis is carried out in the time domain by using the AQWA-NAUT module under regular waves by varying the incident wave period. The simulations are carried out for up to 2000 seconds until obtaining a steady-state response. The response amplitude operators (The plot between the ratio of output response of the system to unit wave amplitude versus wave period or wave frequency) of both the deck and buoyant legs are presented under very high and phenomenal sea state conditions with intact and postulated failure tethers. The natural periods in all degrees of freedom under intact and postulated failure conditions are shown in Table 4 including damping ratio values.

Table 4. Natural periods and damping ratios under intact and postulated failure conditions

Degree-of-freedom	Intact condition		Postulated failure condition	
	Natural periods (Sec)	Damping ratio (%)	Natural periods (Sec)	Damping ratio (%)
Surge	221.4	5.38	228.6	5.14
Sway	222.3	5.16	229.0	5.03
Heave	3.45	0.58	3.7	0.51
Roll	4.1	6.4	4.89	6.23
Pitch	4.3	6.28	5.51	6.08
Yaw	219.7	5.9	227.4	5.68

4.1 Responses of the deck and buoyant legs

Figure 3 shows the response amplitude operators (RAOs) of the deck under very high and phenomenal sea state conditions for active degrees of freedom in intact and postulated failure conditions.

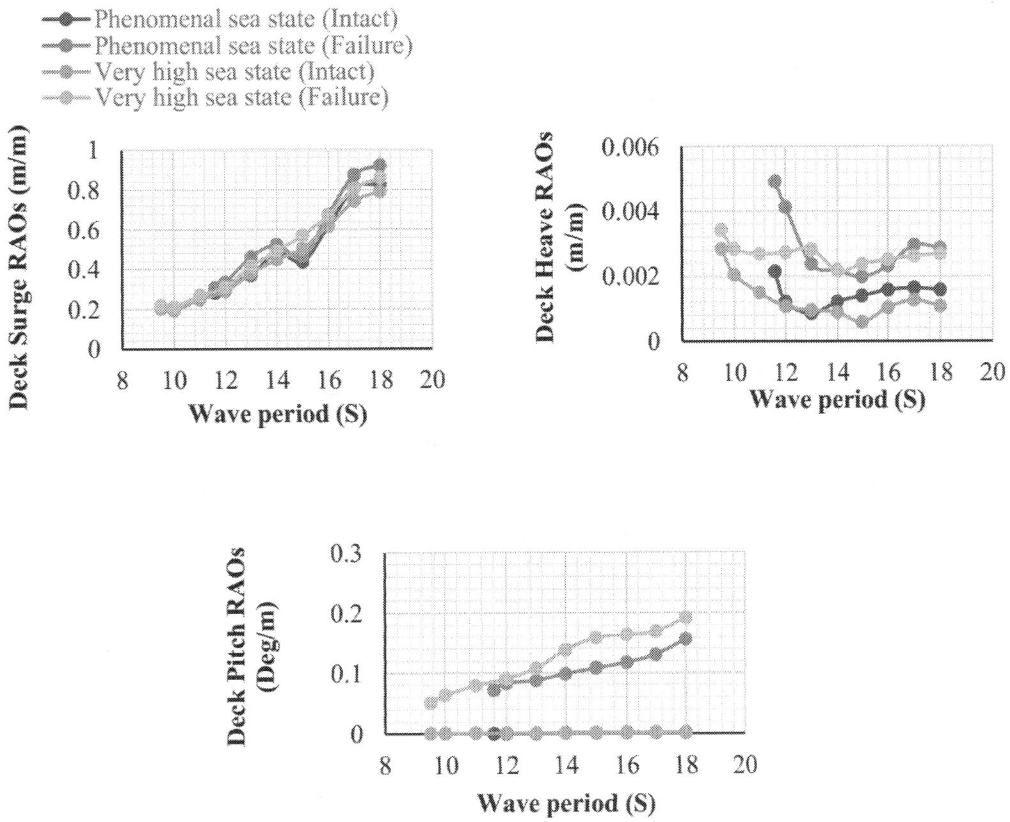

Figure 3. RAOs under regular waves for intact and postulated failure conditions

It is noticed that the surge response is not altered in both the intact and postulated failure conditions of tethers in the two loading cases due to no change in the horizontal tension component of tethers in both conditions. This indicates that under the postulated failure of tethers 9 and 11, the excessive tether tension is taken care of by adjacent tethers 10 and 12. The surge response has a highly coupling effect with heave and pitch responses; however, the deck pitch response is insignificant in the intact condition of tethers due to the presence of ball joints between the legs and the deck.

Table 5. Legs responses under intact condition for phenomenal sea state

Wave period (Sec)	Intact condition								
	Surge (m)			Heave (m)			Pitch (Deg)		
	L1	L2	L3	L1	L2	L3	L1	L2	L3
11.6	4.39	4.41	4.53	0.03	0.03	0.031	1.7	1.7	1.8
12	5.50	5.54	5.98	0.02	0.02	0.021	1.56	1.56	1.5
13	5.41	5.44	5.83	0.02	0.02	0.023	1.43	1.43	1.31
14	7.08	7.12	7.54	0.02	0.02	0.02	1.67	1.67	1.67
15	8.22	8.26	8.80	0.02	0.02	0.021	1.82	1.82	1.7
16	11.65	11.72	12.99	0.03	0.03	0.04	2.06	2.06	2.03
17	12.43	12.51	13.56	0.04	0.04	0.041	2.11	2.11	2.15
18	12.79	12.82	13.70	0.04	0.04	0.042	2.2	2.2	2.54

Table 6. Legs responses under postulated failure condition for phenomenal sea state

Wave period (Sec)	Postulated failure condition								
	Surge (m)			Heave (m)			Pitch (Deg)		
	L1	L2	L3	L1	L2	L3	L1	L2	L3
11.6	5.17	5.17	4.82	0.068	0.072	0.08	1.91	1.91	3.16
12	5.54	5.54	4.75	0.05	0.05	0.05	1.65	1.65	2.69
13	6.61	6.61	6.44	0.04	0.04	0.06	1.57	1.57	2.52
14	8.20	8.20	7.24	0.03	0.03	0.043	2.1	2.1	3.25
15	9.30	9.30	8.04	0.03	0.03	0.046	1.98	1.98	3.12
16	11.14	11.14	10.80	0.03	0.03	0.051	2.32	2.32	3.8
17	14.43	14.43	13.61	0.04	0.04	0.06	2.28	2.28	3.8
18	13.67	13.67	13.80	0.03	0.03	0.052	3.12	3.12	4.2

Under failure conditions of tethers, the significant differential heave motion generates imbalanced forces in the legs, creating more pitch response in the deck. The pitch response increases with an increase in wave period in both sea state conditions. As the intensity of the sea state increases, the pitch response decreases by about 23.1% at the 18-second wave period. The heave response of the deck is maximum at 11.7 seconds for phenomenal sea state conditions under postulated failure, which gradually decreases with an increase in wave period. Table 5 and Table 6 indicates the leg's responses under phenomenal sea state conditions for intact and postulated failure condition of tethers respectively. It is demonstrated that the surge response increases as the wave period increases. It is also noticed that the responses of legs 1, 2, and 3 in the surge degree of freedom are about 12.8m and 13.7m, respectively, in the intact condition due to the platform's geometry. The differential heave motion of leg 2 and leg 3 is about 5% and 53%, respectively, for a wave period of 15 seconds under intact and postulated failure conditions. This indicates that leg 3 is more critical compared to leg 1 and 2.

4.2 Response comparison of deck and leg-3

Figure 4 illustrates the deck and critical leg-3 surge and pitch responses. The response amplitude operators of the deck and leg 3 are almost the same, which indicates that the deck and legs are monolithic in the vertical plane, highly desirable for ultra-deep waters. In intact conditions, the pitch response of the deck is negligible, and the leg-3 response is 0.15 deg/m, indicating the importance of ball joints in triceratops. Nevertheless, under failure conditions, the pitch RAO is significant for the deck, which is lower than the leg-3 RAO.

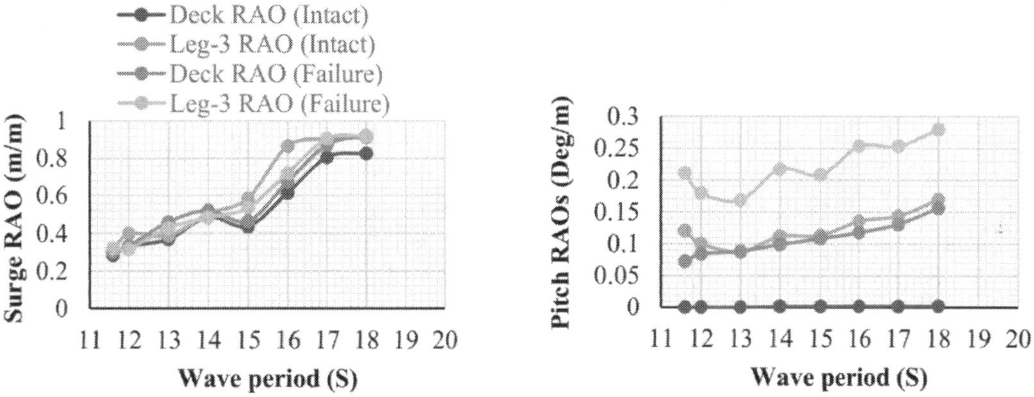

Figure 4. Response comparison of Deck and Leg-3 under phenomenal sea state

4.3 Fatigue life assessment

The arrangement of tethers to the legs is shown in Figure 5. T9 and T11 are considered postulated failure tethers under Leg-3 due to the subjecting of more tension force by its geometry. Figure 6 indicates the dynamic tether tension variation of tethers for legs 2 and 3. Under intact conditions, the mean pretension is about 21.5 MN and 22.1 MN for tethers 5 and 10 under legs 2 and 3, respectively, while tether 10's mean dynamic tension variation shifted to 43.2 MN under postulated failure tethers, which indicates that the failed tether tension is taken care of by adjacent tethers.

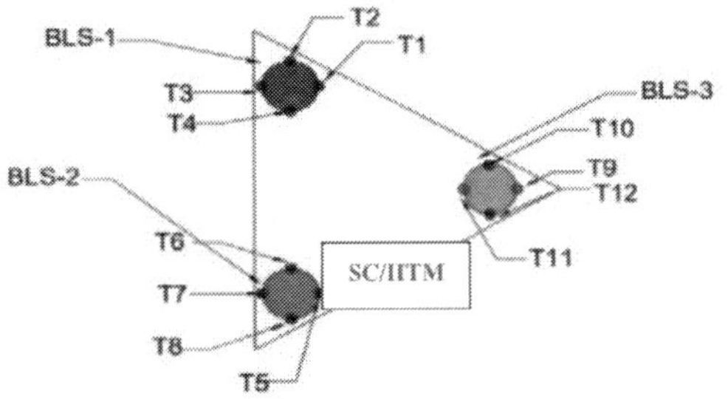

Figure 5. Tether connections to each leg

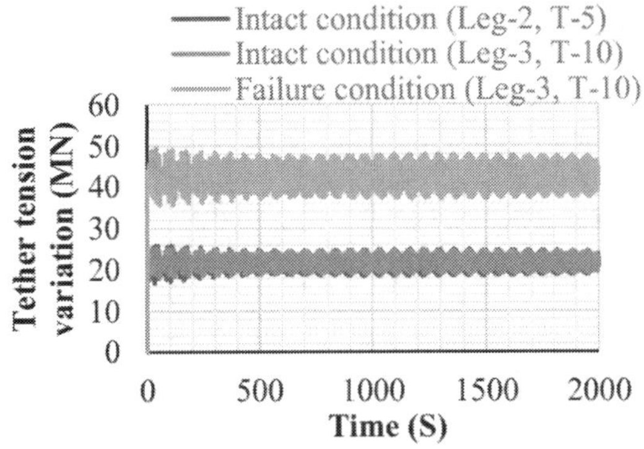

Figure 6. Tether tension variation under intact and postulated failure conditions for phenomenal sea state

The taut moored offshore compliant platforms like offshore triceratops tethers are more susceptible to fatigue failure due to the following reasons: a) platform drift under the action of wind, waves, and current, b) variable submergence of the platform, c) vibration induced by vortex shedding. In the present study, the fatigue life of the tethers under postulated failure conditions of tethers (tethers 9, and 11 of leg-3) is presented. The fatigue life is calculated by using S-N curves. Palmgrem-Minar's is applied as a correction factor to account for environmental loading conditions. The fatigue damage is estimated by following the formula.

$$FD = \sum_{i=1}^{m} \frac{n_i}{N_i}$$

Where n_i is the number of cycles per year for a given tension range interval and N_i is the cycles to failure under tension range 'i' according to the S-N curve. The dynamic variation of pretension in the tethers makes the stress response as a non-zero mean process that could not use directly in S-N curves. Therefore, after obtaining the cycle ranges and cycle averages, each cycle is converted into an equivalent stress range of zero-mean using Goodman Diagram. The rain flow counting method which is suggested by ASTM E 1049-85 (2005) is used to find the stress ranges corresponding to the simulation time. The parameters for the S-N curve are chosen according to [18]. Table 7 indicates the Fatigue life of tethers under intact and postulated failure conditions. The fatigue life of tethers under leg-3 is lesser than that of legs 1, and 2 for both the sea state conditions under intact and postulated failure tethers. The minimum fatigue life is about 8.39 years and 0.024 years for tethers under leg-3 in both the conditions for phenomenal sea state condition.

Table 7. Fatigue life of tethers under intact and postulated failure conditions

Sea state	Tethers under legs	Fatigue life in intact Condition (Years)	Fatigue life in failure Condition (Years)
Very high sea state	Legs -1,2	12.46	12.29
	Leg-3	11.23	0.79
Phenomenal sea state	Legs -1,2	9.75	9.48
	Leg-3	8.39	0.024

5. Conclusions

The detailed numerical analysis of offshore triceratops is carried out for two loading conditions with intact and postulated failure tethers. The following findings are listed below:

- The same surge responses of the deck and legs show that the platform is monolithic in the vertical plane, which is highly desirable for ultra-deep waters
- The significant differential heave motion is observed under postulated failure tethers
- The pitch response of the deck is significantly lower than that of buoyant legs in intact and postulated failure conditions of tethers, indicating the added advantage of ball joints
- Under postulated failure conditions, the fatigue life of adjacent tethers reduces remarkably.

6. References

[1] Shaver CB, Capanoglu CC and Serrahn CS 2001 Buoyant leg structure preliminary design, constructed cost and model test results *Proc. Int. Offshore Polar Eng. Conf. (Norway)* **1**: 432-439

[2] White CN, Copple RW and Capanoglu C 2005 Triceratops: An effective platform for developing oil and gas fields in deep and ultra deep water *Proc. Int. Offshore Polar Eng. Conf. (Korea)* **2005**:133-139

[3] Chandrasekaran S and Madhuri S 2015 Dynamic response of offshore triceratops: Numerical and experimental investigations *Ocean Eng.* **109**: 401-409

[4] Chandrasekaran S and Nagavinothini R 2018 Dynamic analyses and preliminary design of offshore triceratops in ultra-deep waters *Innovative Infrastructure Solutions*. **3**: 1-13

[5] Nagavinothini R and Chandrasekaran S 2019 Dynamic analyses of offshore triceratops in ultra-deep waters under wind, wave, and current *Structures*. **20**: 279-289

[6] Chandrasekaran S, Madhuri S and Jain AK 2013 Aerodynamic response of offshore triceratops *Ships Offshore Struct.* **8**: 123-140

[7] Chandrasekaran S and Nannaware M 2014 Response analyses of offshore triceratops to seismic activities *Ships Offshore Struct.* **9**: 633-642

[8] Wirsching PH and Chen YN 1987 Fatigue Design Criteria for TLP Tendons *J. Struct. Eng.* **113**: 1398-414

[9] Siddiqui NA and Ahmad S 2001 Fatigue and fracture reliability of TLP tethers under random loading *Marine Structures*. **14**: 331-352

[10] Bachynski EE and Moan T 2012 Design considerations for tension leg platform wind turbines. *Marine Structures*. **29**: 89-114

[11] Hirdaris SE et al. 2014 Loads for use in the design of ships and offshore structures. *Ocean Eng.* **78**: 131-174

[12] Chandrasekaran S and Rao MH 2019 Numerical Analysis on triceratops restraining system: failure conditions of tethers. *International Journal of Environmental and Ecological Engineering*. **13**: 588-592

[13] Chandrasekaran S and Nagavinothini R 2020 Parametric studies on the impact response of offshore triceratops in ultra-deep waters. *Struct. Infrastruct. Eng.* **16**: 1002-1018

[14] Chandrasekaran S, Shah B and Chauhan YJ 2022 Fatigue Assessment of Offshore Triceratops Restraining System Under Hurricane-driven Metocean Conditions. *Int. J. Steel Struct.* **23**: 208-224

[15] Tabeshpour MR, Ahmadi A and Malayjerdi E 2018 Investigation of TLP behavior under tendon damage *Ocean Eng.* **156**: 580-95

[16] Ren Y, Venugopal V and Shi W 2022 Dynamic analysis of a multi-column TLP floating offshore wind turbine with tendon failure scenarios. *Ocean Eng.* **245**: 110472

[17] American Petroleum Institute 2010 *Planning, Designing and Constructing Tension Leg Platforms*

[18] DNV G (2014) Recommended Practice DNVGL-RP-0005: 2014-06 RPC203: fatigue design of

offshore steel structures

IOP Conf. Series: Materials Science and Engineering 1288 (2023) 012033 doi:10.1088/1757-899X/1288/1/012033

Numerical simulations of towing a jacket foundation with triple buckets

Lei Tan[1,2], Changhong Hu[1,*] and Yingyi Liu[1]

[1] Research Institute for Applied Mechanics, Kyushu University, 6-1 Kasuga-koen, Kasuga, Fukuoka 816-8580, Japan
[2] Department of Oceanic Architecture and Engineering, Nihon University, 7-24-1 Narashinodai, Funabashi, Chiba 274-8501, Japan

hu@riam.kyushu-u.ac.jp

Abstract. Offshore wind energy is developing fast as a promising renewable energy resource that can be helpful in achieving the global net-zero emissions goal. In recent years, offshore wind turbine projects have gradually moved towards far-offshore sea areas, where massive jacket foundations with suction buckets are suitable options. To reduce transportation costs, the wet-towing method is usually favourable since the required buoyancy of the foundation can be provided by the air cushions in the suction buckets. This engineering background brings us a research topic that involves towing and air cushions in waves. This study considers the towing problem of a jacket foundation with triple suction buckets. First, numerical analysis in the frequency domain is conducted using WAMIT®. In particular, the NEWMODES subroutine is extended to consider three separate motion modes in the buckets. Numerical predictions of the natural period show satisfactory agreement with the available experimental data in the literature. Moreover, a time-domain numerical model is developed to simulate towing of the jacket foundation in regular waves based on the potential flow theory. The time-domain model numerically solves the towing force in regular waves, the body motions and the air pressures in the buckets. The simulation results generally agree well with the published experimental data. Besides, the present simulation results suggest that the transit towing force and the forward speed of towed triple-bucket jacket may be considerable in the initial towing stage.

1. Introduction

To achieve the global net-zero emissions goal, offshore wind energy as a promising source of renewable energy has been attracting extensive attentions and continuous investments. As shown in Figure 1, the global offshore wind capacity in operation has been increasing and approached 57.6×10^3 MW by 2022 [1]. The offshore wind developments at the early stage were mainly located in the shallow waters near the coastline. Over the past years, offshore wind turbines have been gradually moving towards far-offshore sea areas, which raises new challenges in technical and economic feasibility. Various fixed and floating types of foundations have been used to support wind turbines, including, gravity, monopile, jacket, spar, TLP, semisubmersible and so on. Among these types, for the intermediate water depths of 25~50 meters, the jacket foundation is one of the suitable options [2]. A traditional jacket wind turbine is usually founded on several vertical piles which are driven deep into the soil at sea bottom. In this way the piles can transfer the environmental loads acting on the jacket into the seabed. Relevant studies on jacket-type wind turbine performance under environmental loading can be found in [3-7].

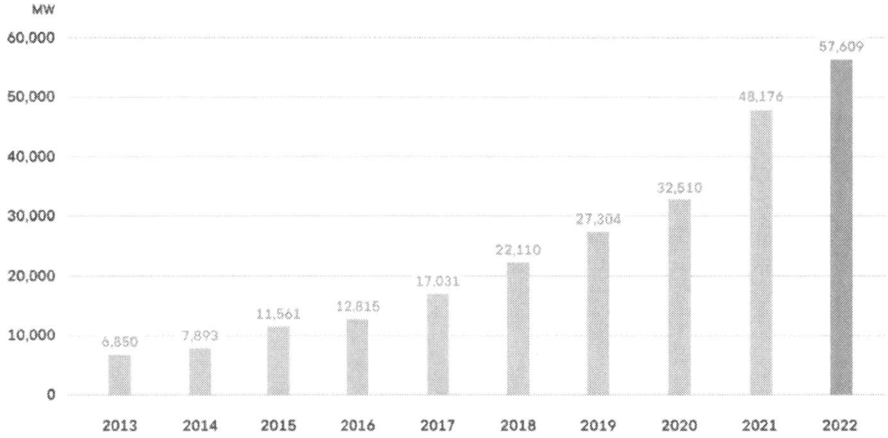

Figure 1. Global offshore wind capacity in operation. (source: global offshore wind report 2022)

Apart from pile-type foundations, suction bucket is another option for offshore jacket wind turbines. The suction bucket foundation is not a new concept, which has been used for an offshore natural gas platform in 1995 (see figure 2(a)). Over the recent years, the suction bucket foundations have been proposed to support offshore wind turbines, as shown in figures 2(b) and 2(c). The relevant design and safety analysis of suction-bucket wind turbines can be found in [8-11].

(a) Troll A platform (Shell Norway) (b) Suction bucket jacket (Ramboll) (c) Sleipner T Statoil platform (Ørsted)

Figure 2. Offshore structures with suction bucket foundations [27-29].

To reduce the costs of offshore wind projects, an economical transportation manner is to utilize wet towing rather than using transportation vessels since the latter manner often requires a relatively high ship renting expanse. Regarding suction buckets, the required buoyancy can be naturally provided by the air cushions inside the buckets. Figure 3 shows an overview of the manufacture, transportation and installment of a jacket offshore wind turbine with suction-type foundation. The interactions between water waves and air-cushion supported structures have been widely studied in the past decades [12-16]. Generally speaking, the problem of floating structures with air cushions in waves involve not only hydrodynamics but also aerodynamics, which brings difficulties to the traditional hydrodynamic analysis approaches. From physical viewpoint, the restoring stiffness of air-cushion supported floaters in vertical direction results from air and water restoring forces. This characteristic has a significant influence on the natural periods of suction bucket foundations.

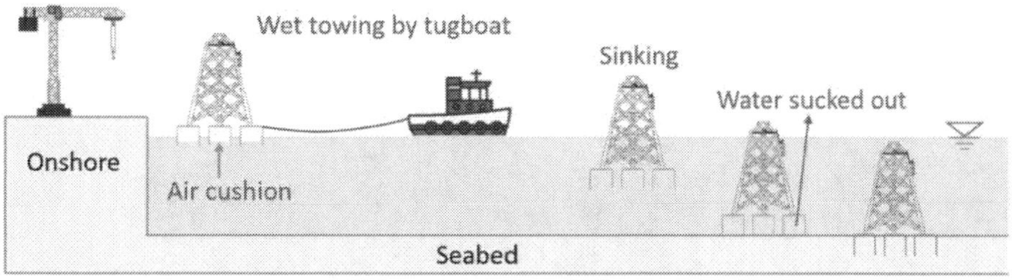

Figure 3. The manufacturing, transporting and installing processes of a jacket-type offshore wind turbine with suction bucket foundation.

Recently, wet towing of a floating structure with suction buckets in waves has been studied due to its engineering significance [17-18]. Guo et al. (2021) [19] analyzed the wet towing process of a jacket foundation with four suction buckets by model tests and numerical simulations. Ding et al. (2020; 2022) [20-21] studied a composite bucket and found that the compartment pressure changes are small during towing process. Zhang et al. (2021) [22] studied the wet towing resistance of a composite bucket foundation. Liu et al. (2022) [23] carried out an experimental study of the motion responses of buckets with air cushions. Zhang et al. (2022; 2023) [24-25] investigated the sinking process of a triple-bucket foundation via time domain simulations based on the hydrodynamic software Moses®. To authors' knowledge, apart from hydrodynamic software, the studies that reported the development of numerical models of simulating the wet towing of suction buckets are rare so far.

In this study, a numerical investigation of the wet towing of a jacket wind turbine with suction buckets is carried out. The main purpose of this work is to develop a reliable and efficient in-house program code to simulate the towing process of the floater with multiple air cushions in waves. Hence, a numerical model which can solve 9-DoF motion equations in time domain is established, where the hydrodynamic coefficients as inputs are computed by the WAMIT® code (with extensions to the subroutine NEWMODES). Then, numerical validations of the developed model are carried out against available experimental data [26] under various towing conditions. On this basis, numerical analysis is further carried out to shed insights into the towing problem of air-cushion supported floating structures and provide guidance to engineering practice.

2. Numerical models

The wet towing of a triple-bucket supported jacket foundation with air cushions was numerically investigated by the hydrodynamic software WAMIT® in frequency domain and an in-house program code in time domain. By assuming the effect of slow towing speed on hydrodynamic coefficients is negligible, WAMIT (Version 7.4) was used to compute the hydrodynamic coefficients and wave exciting forces to provide input data for time-domain simulations. In addition, we also calculated the natural frequencies of the bucket structure with air cushions by WAMIT. Regarding air cushion problems, since the original version of WAMIT cannot consider each motion mode separately in the three buckets, the subroutine NEWMODES code was modified according to the user manual.

Figure 4 shows the definition of coordinate system and the motion modes in the numerical models. The original of the coordinate was positioned at the center of gravity (CoG). Figure 5 shows a sketch of the considered air cushion problem of this study. Note that there are 9 degrees of freedom (DoFs) were considered in the numerical computations, because the vertical piston mode water motions in the three buckets were also included (i.e., 7th, 8th and 9th DoF, respectively) in addition to the traditional 6-DoF motions of a rigid body.

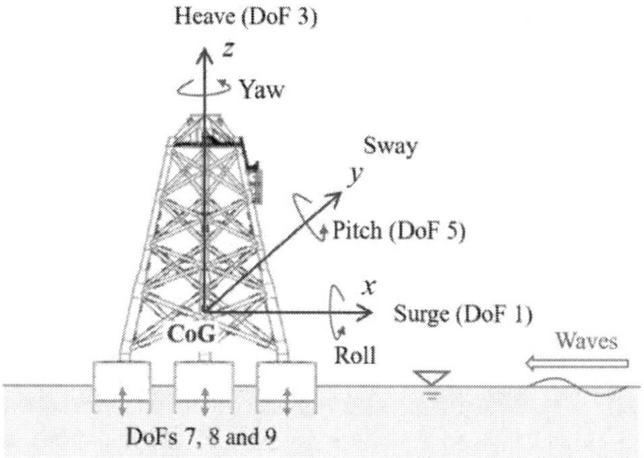

Figure 4. The coordinate system and motion modes considered in numerical computations.

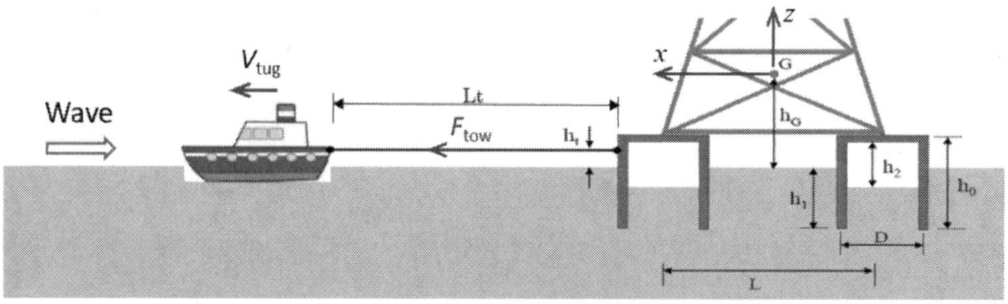

Figure 5. The symbols defining the parameters in this study.

2.1. *Modification of NEWMODES code*

To conduct the hydrodynamic analysis of the suction bucket jacket, it is necessary to consider the motion modes in the three separate air cushions. The reason is that the difference of the water motions in different buckets have an influence on the natural period of pitch. If one considers heave motions only, the water motions in the triple buckets may be regarded as the same motion mode in the motion equations. In this manner, there will be 7 motion modes in total including the piston mode water motions. To take the water motion differences into account in the simulations, the subroutine NEWMODES code of WAMIT was modified. The three air-water interfaces in the buckets were identified by the different indexes in the code. The newly compiled newmodes.dll file was made for WAMIT computations.

2.2. *Time-domain simulation model*

A time-domain numerical model has been developed for the wet towing of a floating structure with multiple air cushions. Within the framework of linear potential flow theory, the motion equation of the floating structure in time domain is written as

$$Q\left[t, \{X(t)\}, \{\dot{X}(t)\}\right] = [\boldsymbol{M} + \boldsymbol{M_a}(\infty)]^{-1}\left\{F_e(t) - \int_0^t [\boldsymbol{K}(t-\tau)]\{\dot{\xi}(\tau)\}\,\mathrm{d}\tau - [\boldsymbol{C}]\{\xi(t)\}\right\} \tag{2.1}$$

where Q is the acceleration, $X(t)$ and $\dot{X}(t)$ are the motion displacement and velocity, respectively, M is the mass matrix (the matrix is 9×9 hereafter), $M_a(\infty)$ is the mass matrix at infinite frequency, K is retardation function, C is restoring matrix, and

$$F_e(t) = F_{wave}(t) + F_{tow}(t) + F_{vis}(t) + F_{aw}(t) \tag{2.2}$$

The towing force reads

$$F_{tow}(t) = \begin{cases} k\Delta L = k\big(V_{tug}t - X(t) - L_0\big) & (\Delta L \geq 0) \\ 0 & (\Delta L < 0) \end{cases} \tag{2.3}$$

where k is the spring constant of towing line, V_{tug} is the forward speed of tug ship, L_0 is the initial length of the slack towing line minus the initial distance between tugboat and bucket model.

Since the frictional damping is usually negligible, the viscous drag force is estimated by

$$F_{vis}(t) = -0.5\rho C_d S V_m(t)|V_m(t)| \tag{2.4}$$

where ρ is water density, C_d is drag coefficient, S is projected area, $V_m(t)$ is the forward speed of bucket model. According to Higo et al. (1988) [33], the wave added resistance can be written as

$$F_{aw}(t) = -\frac{1}{12}\rho g H^2 D \bar{R}_{aw} \tag{2.5}$$

where g is gravitational acceleration, H is wave height, D is bucket diameter, $\bar{R}_{aw}$ is the mean added resistance coefficient, which can be roughly estimated by the wavelength and Froude number based on the experimental results of [33]. In this study, to explore the largest towing resistance, only head sea is considered as the wave direction, which is illustrated in figure 6.

In equation (2.1), the required hydrodynamic coefficients and wave excitation forces were calculated by WAMIT with the modified NEWMODES. The mass and restoring matrices can be determined according to the method of [12]. The motion equation is numerically solved by a developed program code based on fourth-order Runge-Kutta method (Liu et al., 2014; 2021) [30-31]. This in-house code was written using Fortran language.

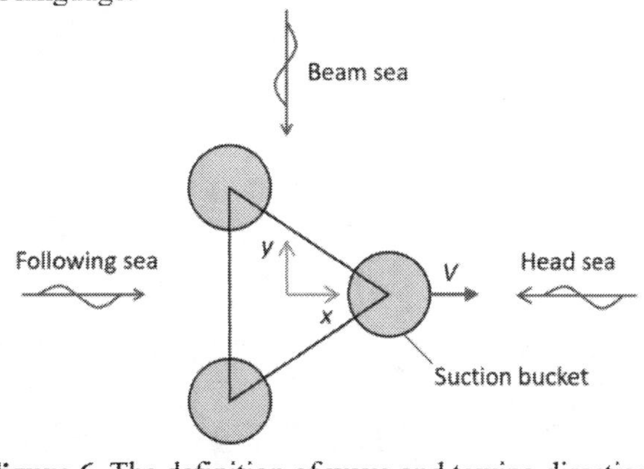

Figure 6. The definition of wave and towing directions.

3. Results and discussion

To validate the developed time-domain model and the WAMIT solutions with the modified NEWMODES, comparisons were made with the available experimental data in the literature. Although the relevant published data is not many so far, fortunately, we found that Yan et al. (2021) [26] recently reported their experimental data which was therefore chosen in the numerical validations. In addition to the data in [26], more experimental data of their research work (obtained from the master dissertation [32]) was also employed in this work, which includes not only the stable towing stage but also the initial towing process.

In this section, we first briefly introduce the experiments of [26] and then show the comparisons with the numerical solutions of the present time-domain model and the modified version of WAMIT. Following the numerical validations, more details of wet towing of suction bucket jacket, which may be to some extent challenging for direct experimental measurements, will be shown via time-domain simulations.

3.1. Introduction of the experiments of Yan et al. (2021)

It was reported in [26] that a series of laboratory model tests on wet towing were carried out in the wave tank at Tianjin University. Figure 7 was taken from their paper, which shows the main experimental setup and the suction bucket model. Their experiment considered a scaling ratio of 1:40 in geometry when using a Froude number similarity. The similarity theory reads

$$\text{Froude number} = \frac{V_s}{\sqrt{gL_s}} = \frac{V_m}{\sqrt{gL_m}} \tag{3.1}$$

where V and L are the velocity and wavelength, respectively; the subscripts s and m denote prototype and model, respectively. Both still water and regular wave conditions were considered in the towing tests, while the wind and current effects were ignored. The main parameters of the experiment are shown in table 1. Regarding wave direction, in this study only head sea condition was considered. More details of the experiment can be found in [26] and the relevant dissertation [32].

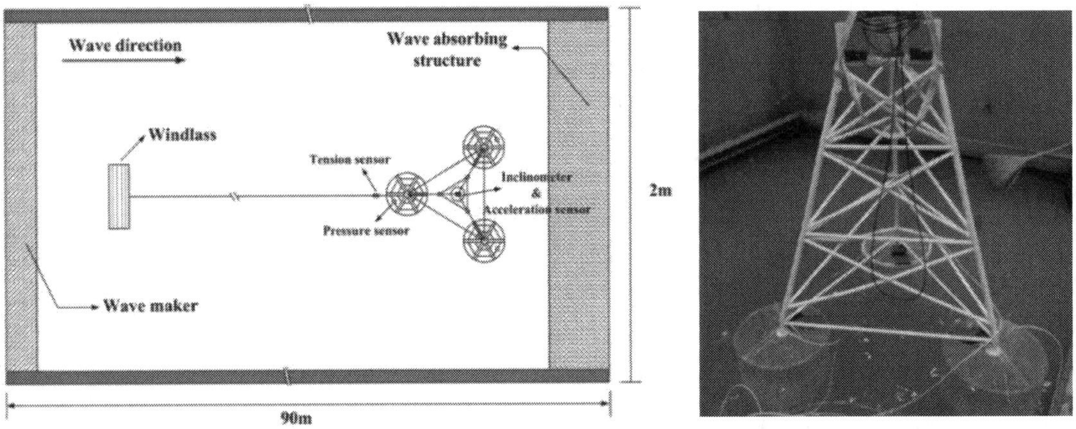

Figure 7. The experimental setup and suction bucket jacket model. Taken from [26].

Table 1. The considered prototype and experimental condition in [26].

	Prototype	Model	Unit
Bucket diameter D	20.0	0.5	m
Bucket height	20.0	0.5	m
Buckets' distance	44	1.1	m
CoG height	29.74	0.7426	M
Self-weight	3.18×10^6	48.43	kg
Bucket wall thickness	10.00	0.25	cm
Draft depth d	10.00	0.25	m
Air ratio in bucket	66.3%	66.3%	-
Water ratio in bucket	33.7%	33.7%	-
Towing-point height h_t	2.00	0.05	m
Towing speed V_{tug}	2.59	0.41	knots
Wave period	8.47	1.34	sec
Wave height H	1.00	0.025	m

3.2. Validation of natural periods

The first attention was paid to the predictions and validations of the natural periods of the suction bucket jacket model. Only heave and pitch motions (corresponding to DoFs 3 and 5) are shown here since the motions in the other DoFs were fairly small for towing in head sea, which are therefore of little engineering significance. The computational panels generated by Rhino® (version 6.0) are shown in figure 8(a), which were used as input data for WAMIT computations. In particular, figures 8(b) and 8(c) show in detail the wet surfaces that are covered by the panels. As shown in the figure, the thin walls of the buckets were considered by two-layer panels in close proximity. Although WAMIT includes the function of dipole panels for then submerged elements, in this study the normal constant (low-order) panels were used. It is worth noting that, one should make sure that the generated normal vectors of the panels (particularly for the air-water surfaces) by Rhino® is consistent with the WAMIT requirements.

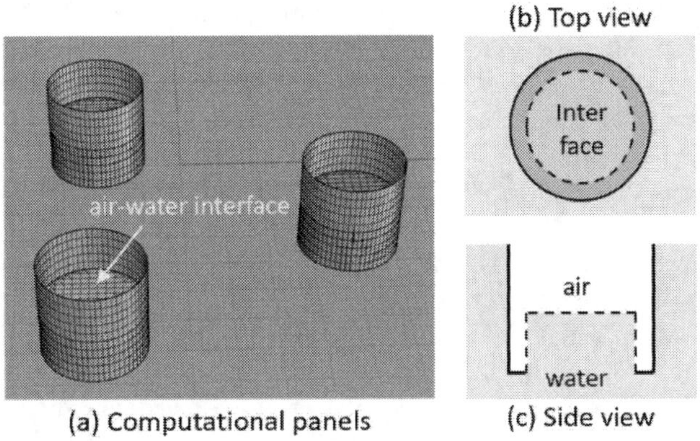

Figure 8. Panels for WAMIT computations. (a) generated panels of triple buckets and interfaces; (b) an illustration showing wet surfaces of each bucket: top view; (c) same to (b) but for side view.

Figure 9 shows the numerical solutions of the wave-induced motion responses in heave and pitch of the 3-bucket jacket model, where RAO is the response amplitude operator. Within the showed period range, only one peak was observed in each period-amplitude response curve, that is, the natural period. The values of calculated natural periods are shown in table 1, which were found to be approximately 1.19 seconds in heave and 1.67 in pitch seconds, respectively. Regarding pitch motion, the radius of gyration of the model was calibrated against the experimental measurements since this parameter was not found in the references [26,32].

Besides, the experimental data of Yan et al. (2021) [26] and the present time-domain solutions were also included in table 2 for comparisons, where the predicted natural periods were obtained via FFT (Fast Fourier Transformation) with the simulated time histories. The experimental results were determined by FFT as well based on the free-decay tests [26]. The reasonable agreements shown in table 2 indicate that the present time-domain and WAMIT solutions are capable of accurately estimating the heave and pitch resonance frequencies of a jacket model with suction buckets.

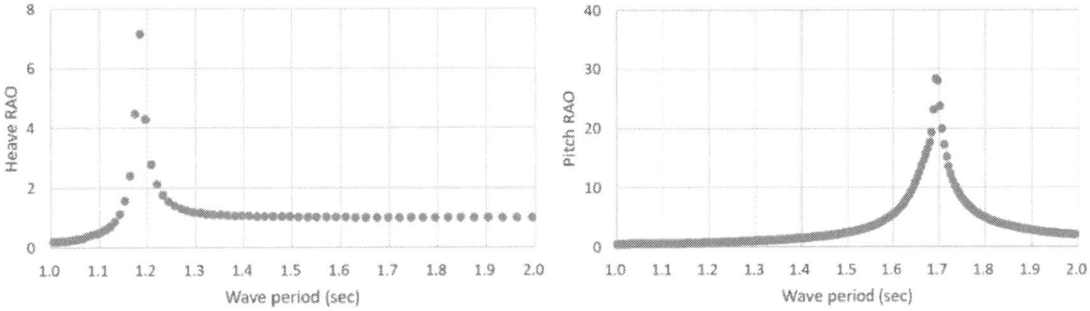

Figure 9. Calculated heave (left) and pitch (right) wave-induced responses by WAMIT.

Table 2. The comparisons of measured and calculated natural periods (sec).

	Experiment [26]	Present WAMIT	Present Time-domain
Heave	1.18	1.19	1.18
Pitch	1.67	1.69	1.64

3.3. Validation of towing in still water

Further attention was put on the towing of suction bucket model in still water. Figures 10(a) and 10(b) show the experimentally measured and numerically simulated towing forces, respectively. The experimental data in plot (a) was taken from the dissertation [32] (the pink highlight and dashed lines were added as marks), since the experimental time series in [26] are incomplete (the initial towing stage are not shown). The present time-domain solution corresponds to the draft d = 0.25m. The drag coefficient C_d= 1.2 was used in the present simulations. Additionally, the stiffness of the mooring wire between the tugboat and the bucket model was calibrated against the measured towing forces in initial stage by trial and error since it was not available in the reference [26,32].

From the comparison of experimental and numerical results, it can be seen that the simulated towing force is basically satisfactory, not only in the stable stage (figure 10(b), after t = 10 sec) but also during the transit stage (before t = 10 sec). As the towing operation started, the measured maximum transit towing force reached approximately 34 N, and the counterpart of simulation result was 30N. Then, the minimum towing force was observed in the model test and the simulations, both of which were roughly 3N. After a few rounds of oscillations in the transit stage, the towing force gradually yielded to a constant value of 10N approximately. Our numerical tests suggest that the stiffness of the towing line has a strong influence on the transit towing force, but has little effect on the final stable towing force. This finding is consistent with our physical understanding of towing in still water. Finally, apart from the good agreement in force amplitudes, it can be also seen that the transit stage took different time periods in figures 10(a) and 10(b). Although the main reason for this discrepancy is not clear, it may be caused by certain nonlinear effects of the towing line in the experiments. One evidence is the high-frequency oscillations within t = 30 ~ 35 sec in figure 10(a).

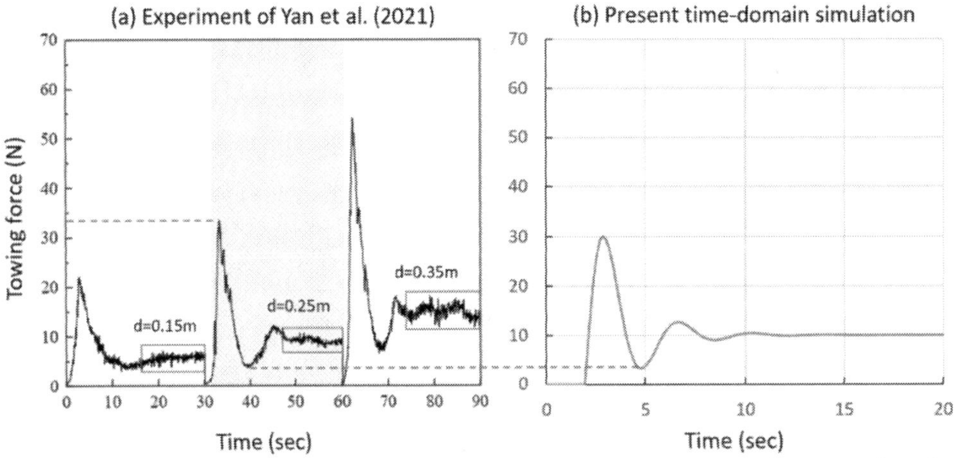

Figure 10. Comparison of towing force results. Measured data [32] versus time-domain solution.

3.4. Validation of towing in regular waves

In this subsection, we consider the towing of the suction bucket jacket in regular waves. The numerical solutions were validated against the experimental data of [32]. Moreover, the numerical model was used to provide more detailed results of towing compared with the available experimentally measured results.

3.4.1. Towing force. Figure 11 shows the comparison of measured and calculated results of towing force. Again, the draft d = 0.25m was used as the simulation condition. It can be seen that, the predicted mean value of towing force in the steady state was slightly smaller than the measured result of [32]. Since the original experimental data is not available (we have figures only), we can hardly estimate the exact value of relative error. From the comparison of the two figures, the error seems to be not over 20% by roughly estimation, suggesting that the simulation accuracy is generally acceptable. Moreover, the observed discrepancy can be improved by artificially tuning the wave added resistance and viscous coefficients in the simulations. On the other hand, the predicted maximum towing force in transit stage was about 30N, whereas the measured data of its counterpart was roughly 50N, larger than the time-domain solution by 40%. As explained before, this difference may be caused by the nonlinear effects of the towing line. Despite the non-negligible discrepancy, the prediction of the maximum transit towing force by the present time-domain model is still of engineering significance since it can provide guidance to the towing operations in practice.

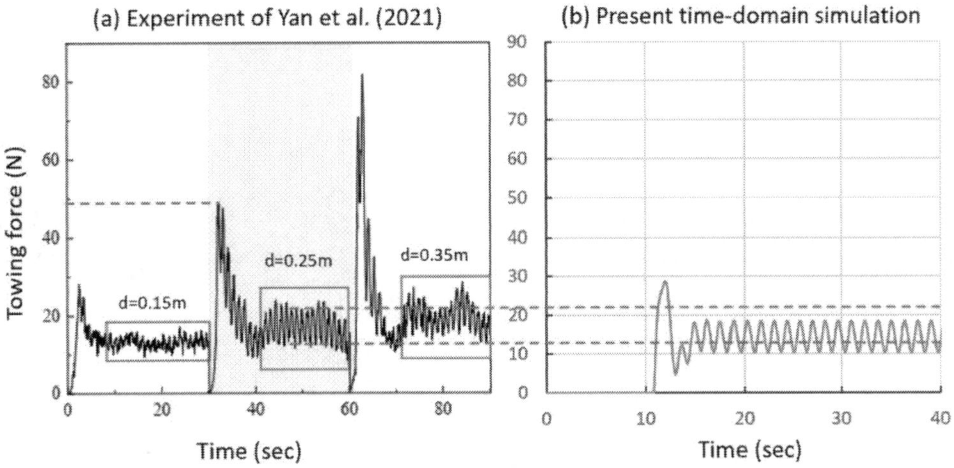

Figure 11. Comparison of towing force results. Measured data [32] versus time-domain solution.

3.4.2. Pressure in buckets. Our attention was then paid towards the air pressures in the suction buckets when towing in waves. Figure 12 shows the comparisons between the measured data and simulation results. The bucket 1 was on the up-wave side, while the bucket 2 (and bucket 3) was on the down-wave side. From figure 12, regarding the bucket 1, the averaged pressure was slightly over-estimated, while the oscillating amplitude was roughly captured by the numerical model. For the bucket 2, the mean pressure was estimated satisfactorily, but the fluctuating amplitude was to some extent over-predicted. Generally speaking, the numerical predictions of the pressures in the air cushions are acceptable. Particularly, one important characteristic of the experimental data has been well captured by numerical simulations, that is, the mean pressure in the bucket 2 larger than that of the bucket 1.

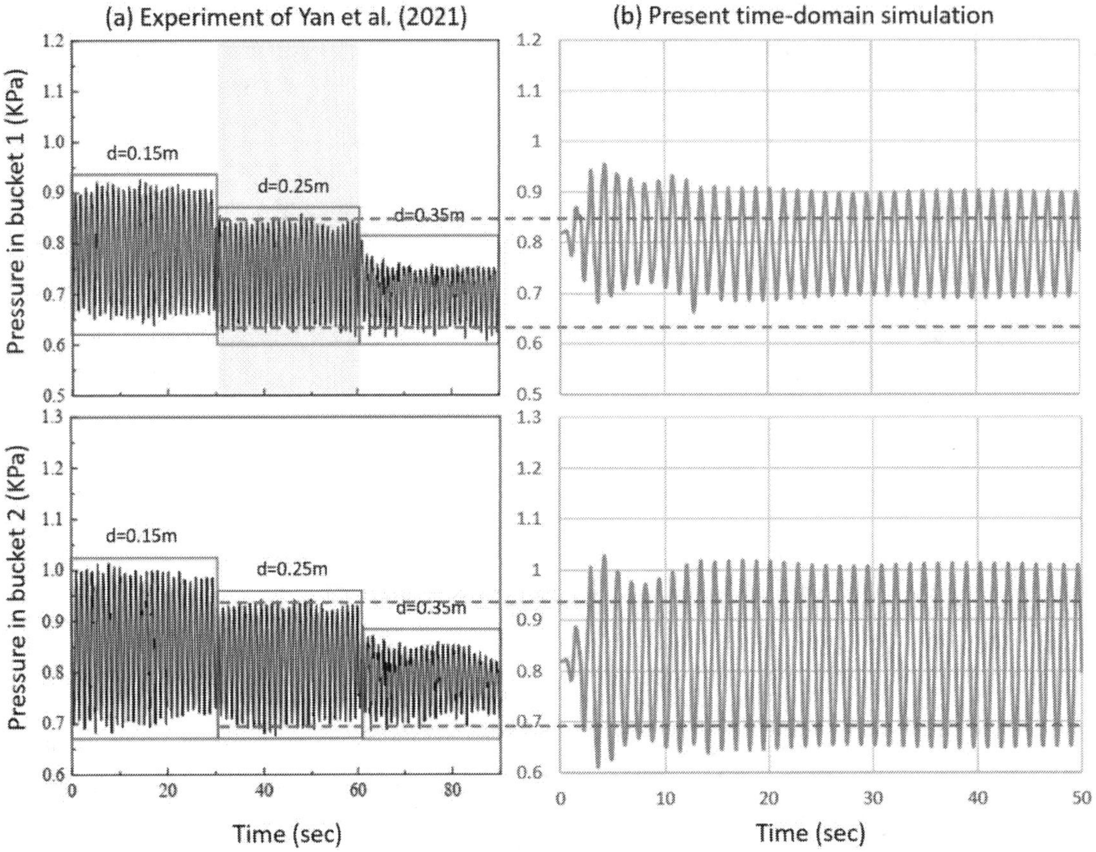

Figure 12. Comparison of pressures in suction buckets (bucket 1 on up-wave side, bucket 2 on down-wave side). Measured data [32] versus time-domain solution.

3.5. *Predicted displacement and velocity*

Compared with experimental measurements, numerical simulations can easily provide more detailed results of the towing process. After the numerical validations of towing force and air pressure, attention was paid on the body motions in waves and the associated forward speed. Figure 13 shows the surge and pitch motions of the triple bucket model, where the tugboat starts to move forward at t = 10 sec. From *figure* 13(a), it can be seen that there were only wave-induced motions before t = 10 sec; then it took approximately 1~2 seconds after t = 10 sec for the slack towing line to become tight. After that, the bucket model was driven to start to move forward with a nearly (not exactly, see *figure* 13(a)) constant speed. Figure 13(b) shows that the averaged pitch angle slightly varied as the towing started, but the pitch motions were basically rather small when towing in waves. The simulation result suggests that the pitch motion seems more sensitive to incident waves rather than the towing behaviour. Generally speaking, *figures* 12 and 13 suggest that the heave motion seems to play a more important role in the towing problems compared with pitch motion.

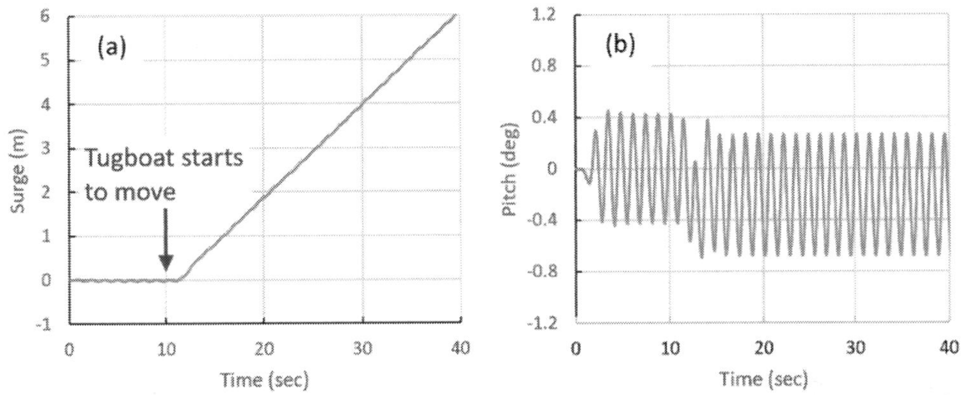

Figure 13. Predicted surge and pitch motions of the 3-bucket jacket by time-domain simulations.

Figure 14 shows the calculated forward speed of the towed suction bucket model. Basically speaking, the towing behaviour was found to change only the mean velocity in surge, whereas the amplitude of wave-induced oscillating velocity in the steady state was hardly influenced by towing. In the transit stage, the maximum motion velocity in surge approached 0.34 m/s, which was about 1.5 times of the averaged speed in the steady state. The observed considerable moving speed of the bucket model during the initial towing stage should be considered with care in engineering practice. We believe this issue deserves further investigations in future works.

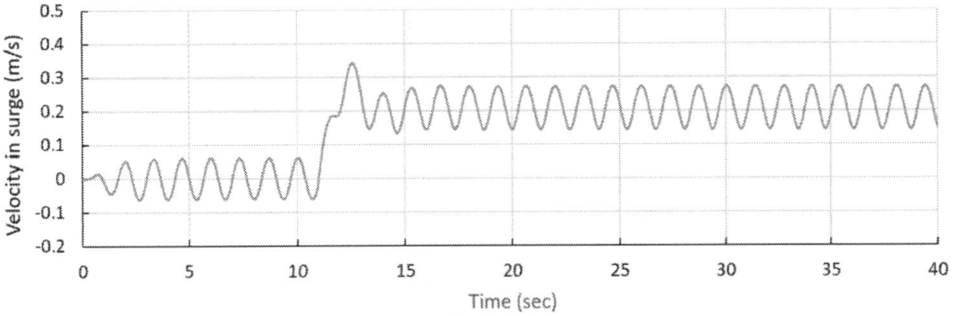

Figure 14. Predicted velocity in surge of the 3-bucket jacket by time-domain simulations.

4. Conclusions

The wet towing of a jacket foundation with triple suction buckets has been studied. A numerical model in time domain was developed to simulate the towing forces and responses in waves. The subroutine NEWMODES code of WAMIT was modified to carry out the hydrodynamic analysis of triple buckets with air cushions in waves. The time-domain and frequency-domain predictions were validated against the available experimental data found in the literature. Generally, reasonable agreements were obtained between the measurements and predictions of towing force, natural periods and air pressures in the buckets. The performance of the developed time-domain model was found to be satisfactory not only for the predictions in the steady state of towing but also for the initial transit stage.

The present numerical results suggest that the transit towing force and the motion speed of towed 3-bucket jacket in the initial towing stage may be considerably large, which should be considered with care in engineering practice. The towing behavior was found to have an influence in the averaged pitch angle of the 3-bucket floater but have little effect on the pitch amplitude. Overall, the pitch motion of the triple-bucket jacket was found to be fairly small when towing under the considered wave conditions. The heave motion was found to play a relatively more important role in the wet towing of jacket foundation with triple buckets.

Acknowledgements
This research was funded by a joint research project between Kyushu University and Obayashi Corporation in FY2022. We would like to take this opportunity to express our deepest gratitude. Lei Tan would like to thank Dr. Lee Chang-Ho of WAMIT company for his valuable comments and suggestions on the modifications of the subroutine NEWMODES. Lei Tan also would like thank Mr. Yan Chao-Jun of Tianjin University for the discussions on their experiments which have been used in the numerical validations of this study.

References
[1] Global offshore wind report 2022, World Forum Offshore Wind (WFO). https://wfo-global.org/wp-content/uploads/2023/03/WFO_Global-Offshore-Wind-Report-2022.pdf
[2] Bailey H, Brookes K L and Thompson P M 2014 *Aquat. Biosyst.* **10** 1-13
[3] Shi W, Park H C, Chung C W, Shin H K, Kim S H, Lee S S and Kim C W 2015 *INT J PR ENG MAN-GT* **2** 139-148
[4] Seidel M 2007 *Conference Proceedings European Offshore Wind* (Berlin)
[5] Abhinav K A and Saha N 2015 *Soil Dyn. Earthq. Eng* 2015, 73, 66-79
[6] Liu F, Li H, Li W and Wang B 2014 *Renew. Energ.* **72** 174-181
[7] Wei K, Arwade S R and Myers A T 2016 *Renew. Energ.* **97** 709-721
[8] Wang X, Zeng X, Li J 2019 *Ocean Eng.* **180** 40-48
[9] Liu B, Zhang Y, Ma Z, Andersen K H, Jostad H P, Liu D and Pei A 2020 *Appl. Ocean Res.* **104** 102358
[10] Kourkoulis R S, Lekkakis P C, Gelagoti F M and Kaynia A M 2014 *Géotechnique* **64** 171-185
[11] Zhang J, Cheng W and Cheng X 2021 *Ocean Eng.* **232** 109159
[12] Lee C H and Newman J N 2000 *Mar. Struct.* **13** 315-330
[13] Pinkster J A and Meevers Scholte E J A 2001 *Mar. Struct.* **14** 163-179
[14] Ikoma T, Masuda K, Maeda H and Rheem C K 2002 *21st International Conference on Offshore Mechanics and Arctic Engineering (Oslo)* OMAE2002-28090 pp 745-752
[15] Ikoma T, Masuda K, Rheem C K and Maeda H 2006 *25th International Conference on Offshore Mechanics and Arctic Engineering (Hamburg)* OMAE2006-92072 pp 371-377
[16] Thiagarajan K P and Morris-Thomas M T 2006 *Ocean Eng.* **33** 1143-1160
[17] Zhang P, Ding H, Le C and Huang X 2013 *J. Renew. Sustain. Energy* **5** 053117
[18] Zhang P, Ding H and Le C 2013 *J. Renew. Sustain. Energy* **5** 063126
[19] Y Guo, Y Zhao, S Zhao, Wang H and Lian J 2023 *Ships Offshore Struct.* **18** 205-216
[20] Ding H, Feng Z, Zhang P, Le C and Guo Y 2020 *Energies* **13** 882
[21] Ding H, Feng Z, Zhang P and Le C 2022 *China Ocean Eng.* **36** 133-143
[22] Zhang P, Zhao X, Ding H and Le C 2021 *Mar. Struct.* **80** 103089
[23] Liu X, Le C, Zhao M, Ding H, Zhang P, Lv N and Luo S 2022 *China Ocean Eng.* **36** 258–267
[24] Zhang P, Li Y, Ding H and Le C 2022 *Renew. Energ.* **185** 564-584
[25] Zhang P, Qi X, Wei Y, Zhang S, Le C and Ding H 2023 *China Ocean Eng.* **37** 73-84
[26] Yan C, Wang H, Guo Y, Wang Z and Liu X 2021 *J. Mar. Sci. Eng.* **9** 1152
[27] Saito T, Yoshida Y, Itoh M and Masui N 2006 Skirt suction foundation-application to strait crossings. https://www.pwri.go.jp/eng/ujnr/tc/g/pdf/22/22-6-4saito.pdf
[28] Dührkop J, Borstel T, Pucker T and Nielsen M B 2016 *Stahlbau* **85** 612-619
[29] Ørsted company homepage. Our experience with suction bucket jacket foundations
[30] Liu Y, Hu C, Sueyoshi M, Yoshida S, Iwashita H and Kashiwagi M 2021 *Ocean Eng.* **232** 109078
[31] Liu Y, Hu C, Sueyoshi M, Yoshida S, Honda Y, Ohya Y 2014 *24th International Ocean and Polar Engineering Conference (Busan)* ISOPE-I-14-005
[32] Liu X 2021 Research on the foundation of offshore wind power three-tube jacket and the floating transportation of the whole machine (Master dissertation: Tianjin University) (in Chinese)
[33] Higo Y and Sato K 1988 *J. Jpn. Soc. Nav. Archit. Ocean Eng.* **163** 205-213 (in Japanese)

IOP Conf. Series: Materials Science and Engineering 1288 (2023) 012034 doi:10.1088/1757-899X/1288/1/012034

Effects of variable porosity net on flow field inside deep-sea aquaculture cage

Xiaonan Yuan, Chenyi Wang, Quanming Miao* and Changfang Zou

Jiangsu Ocean University, China
*Corresponding Author: qmmiao98@163.com

Abstract. This paper focuses on the effect of the porosity of nets on the flow field inside the deep-sea aquaculture cage. Based on CFD method, the 2D flow fields inside and surrounding the deep-sea aquaculture cage from different views are simulated. The effects of certain porosity net and variable porosity net with different configurations on the fishability area, aquaculture efficiency and benefit of the cage are investigated. The results showed that the variable porosity net with suitable configurations in vertical direction can increase the fishability area and change the flow field inside the cage. This can be conducive to improving the efficiency of fish aquaculture. In horizontal direction, increasing the porosity of the nets parallel to the incoming flow direction can increase the fishability area and improve the water exchange capacity. This can be beneficial to improve fish aquaculture benefit. The results of this paper can guide for the selection of nets for deep-sea aquaculture cages.

1. Introduction

As the world's population grows and people's living standards improve, the demand for fish products and the pursuit of quality are increasing by the day [1]. At present, aquatic products are mainly obtained through freshwater aquaculture and coastal aquaculture. Deep-sea cage farming can be carried out in water depths exceeding 25 meters, which expands the aquaculture area and can well relieve the pressure of freshwater and coastal aquaculture. As shown in Figure 1-1 is Jinghai 001, which is the largest deep-sea aquaculture cage in Asia. This cage encloses about 70,000 m^3 of water and can raise 1,000 tons of fish in a single run. Daily aquaculture requires only a few people to manage. Because of its characteristics of high yield and high quality of aquatic products, deep-sea cage aquaculture is gradually coming into the limelight [2].

Figure 1-1 Deep-sea aquaculture cage - Jinghai 001

During fish aquaculture the fish are confined to the aquaculture cages and the efficiency and benefit of fish aquaculture is not only closely related to the composition of the feed fed, but also to the flow field inside the cages. Excessively high flow velocity can easily cause fish muscle fatigue, which affects fish growth and lead to lower aquaculture efficiency. Excessively low flow velocity can affect the water exchange capacity of the cage, which affects the quality of fish and lead to lower aquaculture benefit.
It is therefore essential to research the fishability velocity of different aquaculture fish. Fishes exhibit different swimming behaviors at different flow velocities. According to the length of time, swimming behaviors can mainly be divided into continuous swimming performance and transient swimming performance [3,4]. Continuous swimming performance is a swimming behavior that fish can maintain for a long time without muscle fatigue. Transient swimming performance is an explosive swimming behavior of fish, which does not exceed 15-20 s generally [5]. Obviously, cultured fish should not always be in the transient swimming state. Continuous swimming performance can be further divided into steady swimming and unsteady swimming [6,7]. For the selection of velocity inside the aquaculture cage, many scholars consider that the maximum sustained swimming speed when fish are swimming steadily and the maximum domed swimming speed when they are not swimming steadily can be referred to as the maximum velocity of the aquaculture area for fish [6,8]. Different fish have different swimming abilities, for example, the American red drum (Sciaenops ocellatus) is a fish with strong swimming ability. After many studies by scholars, we can learn that the maximum sustained swimming speed of the American red drum is 0.75 m/s and the maximum dome swimming speed is 0.75 m/s [6,8]. Therefore, the velocity inside the aquaculture cage should be less than 0.75 m/s. The flow velocity inside the cage should not be too low, excessive low flow velocity affects the water exchange capacity of the cage [9-11]. This leads to the reduction of dissolved oxygen content, which threatens the basic survival conditions of cultured fish. Therefore, the fishability velocity should not only ensure sufficient water exchange capacity but also not be higher than the maximum sustained swimming speed or maximum domed swimming speed of the aquaculture fish.

For the velocity distribution inside the cage, it should consider the habit of the cultured fish and the settlement path of fish feed and feces at the same time. The fish feed should be fully diffuse inside the cage to improve the efficiency of fish feeding. The residual feed and fish feces should be separated from the cage in time, or it will pollute the environment. This can lead to eutrophication of the aquaculture sea [12-15] and substrate environment [16], threaten the health of cultured fish [17]. Ultimately affect the quality of fish and reduce the aquaculture benefit.

In summary, the regional proportion of fishability area and velocity distribution inside the aquaculture cage have a strong effect on the fish aquaculture efficiency and benefit, but few scholars have studied these contents. Within this paper, the 2D flow field inside and surrounding the deep-sea aquaculture cage is simulated by CFD from different views. The effects of certain porosity net and variable porosity net with different configurations on the fishability, aquaculture efficiency and benefit of the cage are investigated. The results of this paper can guide for the selection of netting for deep-sea aquaculture cages.

2. Mathematical Model

2.1. Governing equations

The flow field inside and surrounding the deep-sea aquaculture cages can be described by the incompressible Navier-Stokes and RANS equations, with the following control equations:

(1) Navier-Stokes equation:

$$\frac{\partial \rho}{\partial t}+\frac{\partial}{\partial x_i}\left(\rho u_i\right)=0 \tag{1}$$

(2) RANS equation:

$$\rho\frac{\partial \overline{u_i}}{\partial t}+\rho\overline{u_j}\frac{\partial \overline{u_i}}{\partial x_j}=-\frac{\partial \overline{p}}{\partial x_i}+\frac{\partial}{\partial x_j}\left(\mu\frac{\partial \overline{u_i}}{\partial x_j}-\rho\overline{u_i' u_j'}\right)+S_i \tag{2}$$

where u_i is the velocity in direction i, $\overline{u_i}$ is the average velocity. u_i' is the undulating velocity. $\overline{p}$ is the total pressure on the fluid element. μ is the dynamic viscosity and ρ is the density. S_i is the source term of the momentum equation. S_i is 0 in the fluid region outside the porous medium boundary. For the region inside the porous medium, S_i can be evaluated as follows:

$$S_i=-\left(\sum_{j=1}^{2} D\mu v_j+\sum_{j=1}^{2}\frac{1}{2}C\rho|v|v_j\right) \tag{3}$$

$$D=\begin{pmatrix} D_n & 0 \\ 0 & D_t \end{pmatrix} \tag{4}$$

$$C=\begin{pmatrix} C_n & 0 \\ 0 & C_t \end{pmatrix} \tag{5}$$

where D and C are the coefficient matrix for porous media. D_n is the normal viscous resistance coefficient, D_t is the tangential viscous resistance coefficient, C_n is the normal inertial resistance coefficient and C_t is the tangential inertial resistance coefficient. Due to the flow-blocking effect of the net, the pressure gradient approaches to the net is large and the fluid moves more intensely in the vicinity. Given that the Realizable k-ε model can simulate this situation more accurately, it is applied in this paper. The transport equations of the Realizable k-ε model are as follows:

$$\frac{\partial}{\partial t}(\rho k)+\frac{\partial}{\partial x_j}(\rho k u_j)=\frac{\partial}{\partial x_j}\left[\left(\mu+\frac{\mu_t}{\sigma_k}\right)\frac{\partial k}{\partial x_j}\right]+G_k+G_b-\rho\varepsilon-Y_M+S_k \tag{6}$$

$$\frac{\partial}{\partial t}(\rho\varepsilon)+\frac{\partial}{\partial x_j}(\rho\varepsilon u_j)=\frac{\partial}{\partial x_j}\left[\left(\mu+\frac{\mu_t}{\sigma_\varepsilon}\right)\frac{\partial \varepsilon}{\partial x_j}\right]+\rho C_1 S\varepsilon-\rho C_2\frac{\varepsilon^2}{k+\sqrt{v\varepsilon}}+C_{1\varepsilon}\frac{\varepsilon}{k}C_{3\varepsilon}G_b+S_\varepsilon \tag{7}$$

$$C_1=\max\left[0.43,\frac{\eta}{\eta+5}\right],\eta=S\frac{k}{\varepsilon},S=\sqrt{2S_{ij}S_{ij}} \tag{8}$$

where G_k denotes the turbulent kinetic energy due to the mean velocity gradient, G_b denotes the turbulent kinetic energy due to buoyancy, Y_M denotes the contribution of fluctuating expansion in compressible turbulence to the total dissipation rate, and C_2 and $C_{1\varepsilon}$ are constants. σ_k and σ_k are the turbulent Prandtl numbers for k and ε, respectively. S_K and S_ε are user-defined source terms.

2.2. *CLSVOF method for free surface tracking*

Due to the wave action, the free surface has violent normal movement. In order to simulate the flow field inside the cage more accurately, CLSVOF is used to capture the free surface. Its control equation is as follows:

$$\frac{\partial \varphi}{\partial t}+u_i\cdot\frac{\partial \varphi}{\partial x_i}=0 \tag{9}$$

where u_i is the fluid velocity field, φ is the continuous signed distance function in the Level Set function. The CLSVOF method has the advantage of more accurate estimation of the interface normal vector with the following interface curvature:

$$\kappa(\varphi)=-\frac{\partial}{\partial x_i}\left(\frac{\partial \varphi}{\partial x_i}/\sqrt{\frac{\partial \varphi}{\partial x_j}\frac{\partial \varphi}{\partial x_k}}\right) \tag{10}$$

The physical properties of the discontinuity near the interface are smoothed by applying the smoothing Heaviside function, as follows:

$$H(\varphi)=\begin{cases}0 & ,\varphi<-1.5l\\ \dfrac{1}{2}+\dfrac{\varphi}{3l}+\dfrac{1}{2\pi}\left(\dfrac{2\pi\varphi}{3l}\right) & ,|\varphi|\le 1.5l\\ 1 & ,\varphi>1.5l\end{cases} \tag{11}$$

where l is the grid spacing. When the cell contains two phases, the equivalent density ρ' and equivalent viscosity μ' are calculated by $H(\varphi)$ as follows:

$$\rho' = H(\varphi)\rho_w + [1 - H(\varphi)]\rho_a \tag{12}$$

$$\mu' = H(\varphi)\mu_w + [1 - H(\varphi)]\mu_a \tag{13}$$

Where ρ is the density, μ is the viscosity, and w and a are water and air respectively.

2.3. *Resistance coefficient of porous media*

The resistance effect of net can be expressed as follows:

$$F_i = S_i \lambda A \tag{14}$$

where λ is the thickness of the net, A is the area of the net and F_i is the hydrodynamic force in the direction i. Due to the anisotropic inertial resistance of the porous medium is strong and the viscous resistance can be ignored in this paper, the resistance effect of net on the flow field can be simulated by alternative formulas. The alternative formula is shown as follows:

$$S_i = \frac{1}{2}\rho C_i |v_i| v_i \tag{15}$$

Substituting equation (15) into equation (14) can obtain the drag force (F_d) and lift force (F_l) formulas for the net:

$$F_d = \left(C_n \frac{1}{2}\rho|v|v\right)\lambda A \tag{16}$$

$$F_l = \left(C_t \frac{1}{2}\rho|v|v\right)\lambda A \tag{17}$$

The drag and the lift force on net also can be calculated by the Morison equation:

$$F_d = \frac{1}{2}\rho C_d A v^2 \tag{18}$$

$$F_l = \frac{1}{2}\rho C_l A v^2 \tag{19}$$

where the resistance coefficients C_d and C_l can be calculated by the empirical formula proposed by Aarsnes et al. [18]:

$$C_d = 0.04 + \left(-0.04 + S_n + 1.24S_n^{\ 2} + 13.7S_n^{\ 3}\right)\cos\theta \tag{20}$$

$$C_l = \left(0.57S_n - 3.54S_n^{\ 2} + 10.1S_n^{\ 3}\right)\sin 2\theta \tag{21}$$

$$\theta = 90° - \theta' \tag{22}$$

where S_n is the solidity of the net and S_n is smaller than 0.35 in this paper. θ' is the angle between the net and the flow. When θ'=90 °, the tangential inertial resistance coefficient (C_l) is 0. The normal inertial resistance coefficient (C_n) and the tangential inertial resistance coefficient (C_t) can be obtained by fitting the curve of force and velocity by the least squares method.

3. Numerical Models and Boundary Conditions

3.1. Numerical models and parameters

Figure 3.1-1 shows the schematic diagram of the deep-sea aquaculture environment. The area enclosed by the black lines is the aquaculture cage with the geometric shape of a quadratic prism. The cage consists of nets except for the top and thickness of the nets is 0.05 m. In this paper, two representative views of the flow fields inside and surrounding the cage are selected and studied by means of numerical simulations. The symmetry plane in the width direction of the cage is selected as the first view, which is marked in red in Figure 3.1-1. The green area, as shown in Figure 3.1-1, is the second view, chosen the plane of symmetry in the height direction of the cage. Figure 3.1-2 shows the cross section in the first view. The flow field computational domain is 56×33 m (L×D) and the cage is 16×17 m (l×h). The cage is exposed to the water surface for 1 m, and the distance between the bottom of the cage and the bottom of the calculation domain is 16 m. Figure 3.1-3 shows the cross section in the second view. The calculation domain of the flow field is 56 × 56 m (L × W) and the cage is 16 × 16 m (l × w).

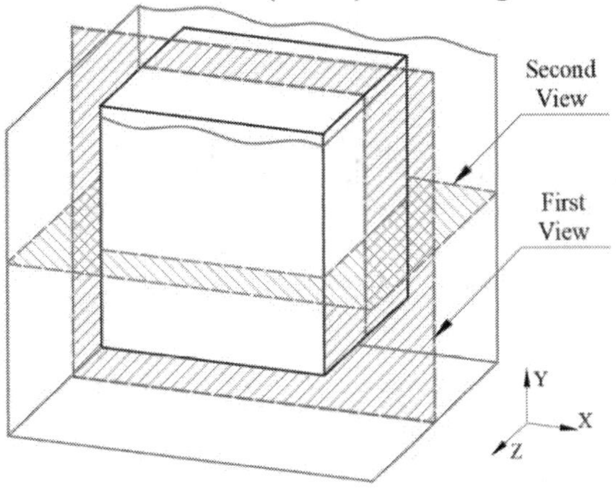

Figure 3.1-1 Schematic diagram of deep-sea aquaculture environment.

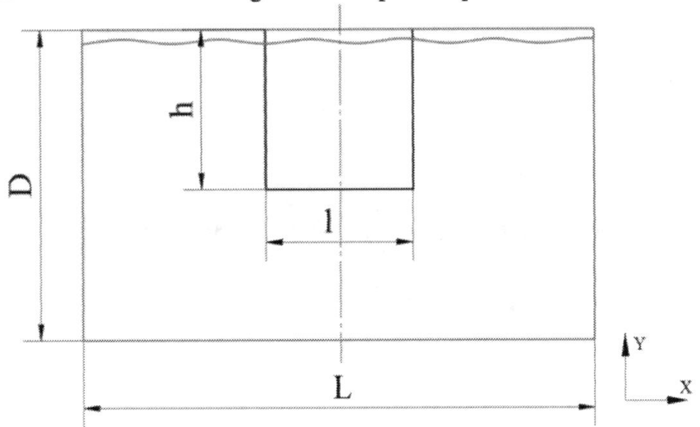

Figure 3.1-2 Numerical model in the first view.

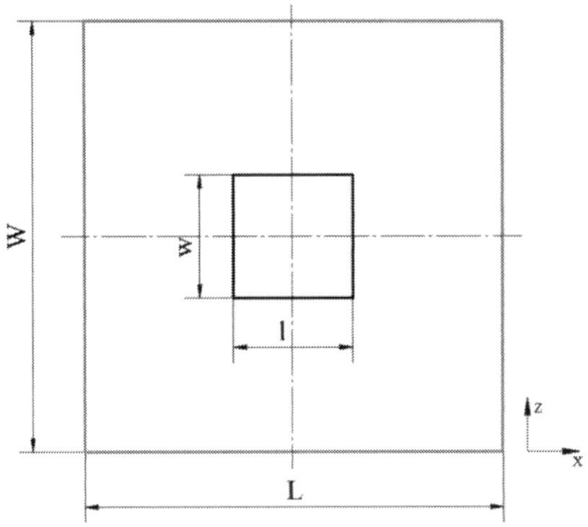

Figure 3.1-3 Numerical model in the second view.

3.2. Boundary condition

As shown in Figure 3.1-2 and Figure 3.1-3, the boundary conditions of the numerical models are as follows:

(1) Velocity inlet boundaries:

The left sides of the computational domain in Figure 3.1-2 and Figure 3.1-3 are the velocity inlet. The direction of flow velocity is along with X direction. The velocity distribution along Y direction in Figure 3.1-2 is defined as follows:

$$v = 0.63 \cdot (1.349 - 0.01 d_w) \tag{23}$$

where d_w is the depth. The velocity in Figure 3.1-3 is 0.8 m/s.

(2) Pressure outlet boundaries:

The right side and top of the computational domain in Figure 3.1-2 and the right side of the flow field in Figure 3.1-3 are the pressure outlet, all of them are set as 0 Pa.

(3) Wall boundaries:

The bottom of the calculation domain in Figure 3.1-2 and the top and bottom of the calculation domain in Figure 3.1-3 are walls, all of them are set as free-slip boundaries.

3.3. Numerical stability criterion

Due to the large scale of the cage, the effects of surface tension and capillary force can be ignored. Therefore, the homogeneous spatial resolution is $\delta x = \delta y = 2.5 \times 10^{-2}$ m, which has been carried out through independent verification and sensitivity analysis. For the time step, there are two stability criteria often used to make the scheme stable. As follows, it is related to the convection process and diffusion process respectively:

$$\delta t \le \min \left\{ \frac{Cr \times \delta x}{u_{\max}}, \frac{Cr \times \delta y}{v_{\max}}, \frac{Cr \times \delta z}{w_{\max}} \right\} \tag{24}$$

$$\delta t \le \min \left\{ \frac{\rho \times \delta x^2}{2\Gamma}, \frac{\rho \times \delta y^2}{2\Gamma}, \frac{\rho \times \delta z^2}{2\Gamma} \right\} \tag{25}$$

where δt is the time step, u, v, and w are the velocity in X, Y, and Z directions, respectively. Cr is the Courant number. The value of Cr in this paper is chosen to be 0.5.ρ is the density of the fluid. Γ is the diffusion coefficient. δt can be calculated by equation (24) to obtain a value of 1×10^{-2} s, which has been adopted after the sensitivity analysis.

4. Results and Discussion

4.1. The resistance effect of nets with different porosities

The resistance effect of nets with different porosities on flow field is different, this is mainly reflected on the change in the velocity of the fluid after flowing through net. Since the fish aquaculture area is circled by the nets and the flow filed inside the aquaculture cage is related to the efficiency of fish aquaculture, it is necessary to figure out the change in the velocity of the fluid after flowing through different porosity nets before determining the porosity of deep-sea aquaculture net. In this paper, the velocity change of the fluid is defined by the deceleration ratio:

$$R_n = \frac{u_c - u_n}{u_c} \times 100\% \tag{26}$$

where R_n is the deceleration ratio for different positions after the net, u_c is the flow velocity at velocity inlet, and u_n is the average flow velocity for different positions after the net.

Referring to the second view in Figure 3.1-1, a 2D numerical model to study the deceleration ratio is given in Figure 4.1-1. The black area is the net. The computational domain of the flow field is 32.05 m×48 m (L'×W'). The length l of the net is 16 m, the thickness of the net is 0.05 m. The angle between the net and the flow is 90 °. The length of d_0 is 8m. Average flow velocities for different positions behind the net are monitored. The length of the velocity sensors is same as the net. After the net are sensor 1 to n in that order. The distance d_1 between the Sensor 1 and the net is 2 m, and the distance d between adjacent sensors is 2 m. The normal inertial resistance coefficient (*Cn*) and the tangential inertial resistance coefficient (*Ct*) of the different porosity nets are estimates by the method in Section 2.3, as shown in Table 4.1-1.

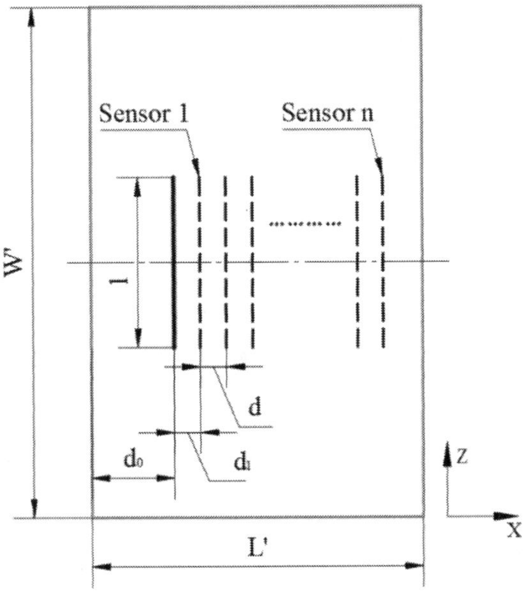

Figure 4.1-1 2D numerical model for analysis of deceleration ratio.

Table 4.1-1 Inertial resistance coefficient of the nets with different porosity.

Porosity	C_n (m^{-1})	C_t (m^{-1})
0.66	14.702	
0.7	11.166	
0.75	7.731	0
0.8	5.200	
0.85	3.367	
0.9	2.026	

Figure 4.1-2 shows the velocity distribution in the symmetric region around the net with different porosities. The left boundary is the velocity inlet boundary with a velocity of 0.75 m/s. As can be seen from Figure 4.1-2, the velocity decreases after fluid flowing through the net in the normal direction of the net. The smaller the porosity, the lower the flow velocity after the net. In the tangential direction of the net, bypassing occurs at the edge of the net, and the velocity increases after fluid flowing through the edges of the net. It can also be found that the velocity change close to net is more pronounced than that further away. Due to the large size of the net, the flow field behind the net does not show significant vortices due to the large Reynolds number values. The spatial and temporal distributions of the flow field are relatively stable.

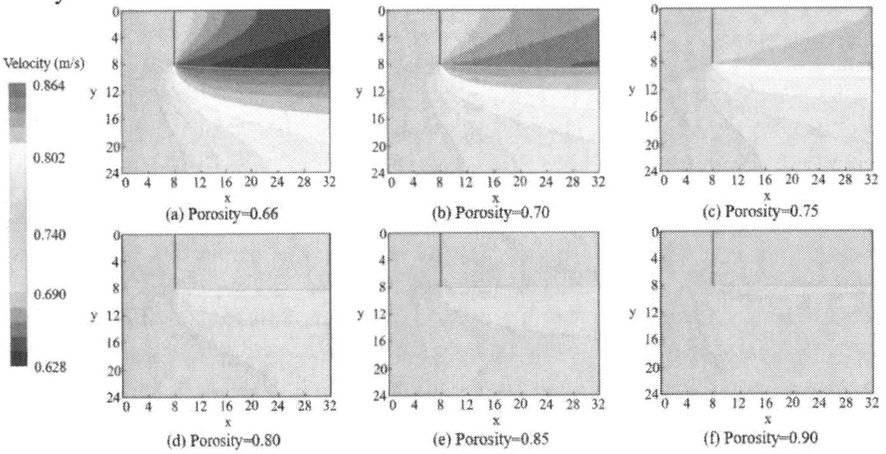

Figure 4.1-2 Flow field of the vicinity of the net with different porosities.

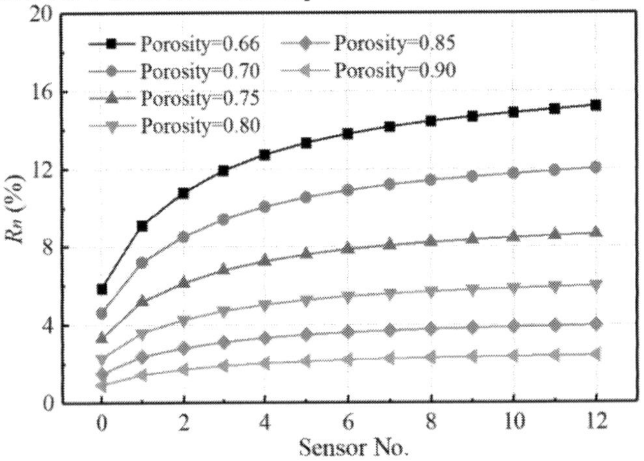

Figure 4.1-3 Deceleration ratio for different porosities at specified sensors.

Figure 4.1-3 shows the deceleration ratios for different distances after the net. Sensor 0 indicates that the position of velocity sensor coincides with that of net. In Figure 4.1-3, the deceleration ratio of 0.75 porosity is about 3.57 times of that of 0.90 porosity and the deceleration ratio of 0.66 porosity is about 6.28 times of that of 0.90 porosity at Sensor 0 position. The deceleration ratio increases with the porosity decrease. In addition, for 0.66 porosity, the deceleration ratio of Sensor 1 is about 1.54 times that of Sensor 0, while the deceleration ratio of Sensor 11 is about 1.01 times that of Sensor 12. The deceleration ratio increases with the increase of distance and the rate of ratio increase decreases.

In this paper, with reference to the maximum sustained swimming speed of the American red drum (aquaculture species) and the velocity inside the aquaculture area, the proportion of the flow velocity area suitable for the growth of the American red drum in the cage is studied by numerical simulation. The maximum sustained swimming speed of the American red drum is 0.75 m/s. The velocity inside the aquaculture cage should not be higher than 0.75 m/s. Combining the maximum flow velocity of 0.75 m/s with the deceleration ratio of different porosity nets, the flow velocity V_0 suitable for American red drum culture when using different porosity nets can be evaluated.

Considering the flow velocities at different positions behind the net are variable and the variation of flow velocities closer to the net are more pronounced, R_1, R_2, R_3 and R_4 in the cage are applied to evaluate the suitable flow velocity for different porosities. Table 4.1-1 shows the flow velocities for the deceleration ratios at different positions and the mean values. When the flow field of the aquacultural environment is certain, the suitable porosity for efficient fish farming also can be determined by Table 4.1-1.

Table 4.1-2 Suitable flow velocity V_0 for cages with different porosities.

Porosity	V_0 (m/s)				
	R_1	R_2	R_3	R_4	Mean
0.66	0.825	0.841	0.852	0.859	0.844
0.70	0.808	0.820	0.828	0.834	0.823
0.75	0.791	0.799	0.805	0.809	0.801
0.80	0.778	0.784	0.787	0.790	0.785
0.85	0.768	0.772	0.774	0.776	0.773
0.90	0.7611	0.763	0.765	0.766	0.764

4.2. Flow field inside the cage with certain porosity

Deep-sea aquaculture cages are generally arranged in waters deeper than 25 m. The flow velocity in these areas always decreases with the increase of the water depth. The relationship between flow velocity and water depth in this paper is given by equation (23). The flow velocity is 0.85 m/s at the top of the cage and 0.75 m/s at the bottom. The wave height is 0.5 m and the period is 2 s. In order to improve the efficiency and benefit of the aquaculture of fish, the flow velocity inside the cage should be within a suitable range. Excessive flow velocities tend to cause muscle fatigue in fish, this can affect growth and lead to a reduction in aquaculture efficiency. Excessively low flow velocity tends to affect the water exchange capacity in the cage, this can affect the quality of fish and leads to a reduction in aquaculture benefit. In this paper, $0.80V_s$-$1.05V_s$ is selected as the fishable flow velocity, the area with this velocity is the fishable area. For the American red drum, V_s is 0.75 m/s and the range of suitable velocity is 0.600-0.788 m/s.

Considering the fishability of the flow field, in this section, the flow velocity (0.8 m/s) at the depth where the symmetry plane (Second view) locates is used as a criterion for the selection of net. To enable more areas to meet fishable conditions, the net with a porosity of 0.75 is chosen according to Table 4.1-1. Figures 4.2-1 and 4.2-2 show the results of the numerical calculations in the first and second views respectively.

In the first view (Figure 4.2-1), the flow velocity inside the cage decreases with the increase of water depth, this is consistent with the distribution of flow velocity near velocity inlet. However, the flow velocity varies intensely inside the cage and in the downstream direction of the cage due to the resistance of the net. The flow velocity inside the cage decreases rapidly in the direction of water depth, at the same water depth, this makes the flow velocity inside the cage lower than the flow velocity near velocity inlet. Although the flow velocity inside the cage basically meets the fishability, the high flow velocity in the upper part of the cage and the low flow velocity in the lower part of the cage are not suitable for fish. The larger flow velocity in the upper part of the cage does not allow sufficient residence time for effective diffusion of the feed. The smaller flow velocity in the lower part of the cage makes it easy for the feed and fish feces to remain. This leads to scaling or parasites attached to the bottom of the cage. Additionally, the presence of vortices makes the velocity distribution at the bottom of the cage uneven. The flow velocity in most areas near the bottom is close to 0.7 m/s but the flow velocity exceeds $1.05V_s$ in local areas and the velocity gradient is large. This not only causes fish muscle fatigue but also causes irritation to the fish and reduces the aquaculture efficiency.

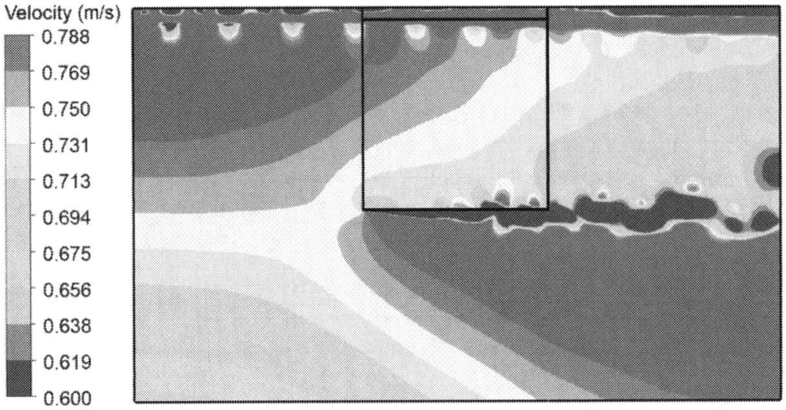

Figure 4.2-1 Velocity contour of computational domain in the first view.

In the second view (Figure 4.2-2), the velocity decayed after fluid flowing through the nets, the flow velocity inside the cage decreased gradually along the direction of the flow. The whole flow field inside the cage almost satisfied the fishability condition. Since the net is a porous medium with highly anisotropic inertial resistance, only a small portion of the fluid enters the cage when flowing through the net parallel to the direction of the incoming flow. This results in the effect of the net parallel to the direction of the incoming flow on the flow being almost equivalent to that of a solid wall. Influenced by the bypass flow and boundary layer effect, the flow velocity near the wall outside the net vertical to the direction of inflow is low, the flow field inside the cage is largely dominated by the fluid in the direction of inflow.

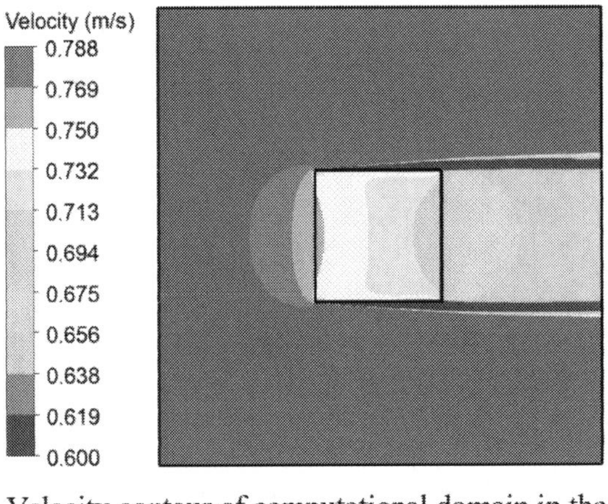

Figure 4.2-2 Velocity contour of computational domain in the second view.

4.3. Flow field inside the cage with variable porosity in the vertical direction

In order to increase the fishability of the flow field inside the cage and improve the efficiency and benefit of aquaculture, this section focuses on the effect of variable porosity net on the flow field inside the cage in the first view. The porosity of the whole net of the side wall is divided into 4 equal parts along the vertical direction in the first view. The porosity of the whole net of the side wall is variable. The porosity of each part is determined by the flow velocity corresponding to the average depth of each part. With the increase of water depth, the porosities of the net can be determined by Table 4.1-1 and they are in the following order: 0.66, 0.70, 0.80, 0.90.

Figure 4.3-1 shows the flow field inside the variable porosity net cage in the first view. Affected by the variable porosity net, the relationship between flow velocity and water depth has changed. The flow velocity does not decrease exactly with the increase in water depth. The low flow velocity in the upper part of the cage and the high flow velocity in the lower part is benefit to spread the feed and inhibit the scaling in the bottom of the cage. In addition, the area with a larger velocity gradient at the bottom of the cage was reduced. However, the flow velocity in the lower part of the cage was higher than $1.05V_s$. The fishable area in such variable porosity net cage is reduced by about half compared to the fishable area in certain porosity net cage. This drastically reduces the efficiency of aquaculture inside the cage. It can be seen that it is not reasonable to only consider the velocity distribution of incoming flow along the water depth when selecting configuration of porosity.

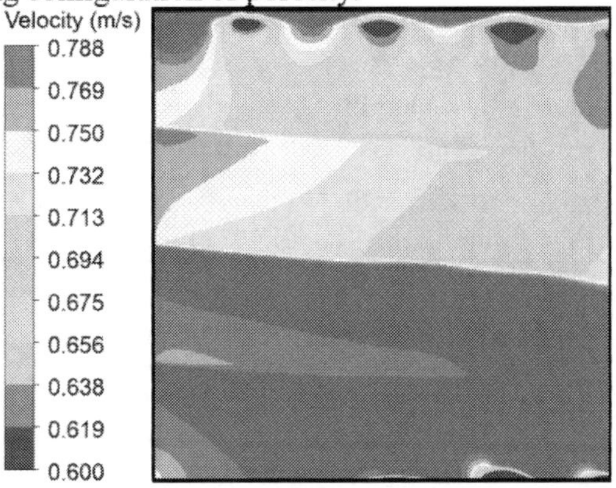

Figure 4.3-1 Velocity contour of the domain inside the cage with variable porosity in the first view.

The flow field inside the modified variable porosity net cage is shown in Figure 4.3-2. The modified configuration is 0.66, 0.66, 0.80, and 0.75 in order of increasing water depth. It can be seen from Figure 4.3-2 that the area satisfying fishability is not significantly reduced inside the cage with modified variable porosity. In the upper part of the cage, the velocity distribution in the water depth direction is more uniform and reasonable. Due to the abrupt change of porosity, the flow velocity in the middle of the cage has increased more obviously, but most of the area in the middle and lower parts of the cage still meet the fishability conditions.

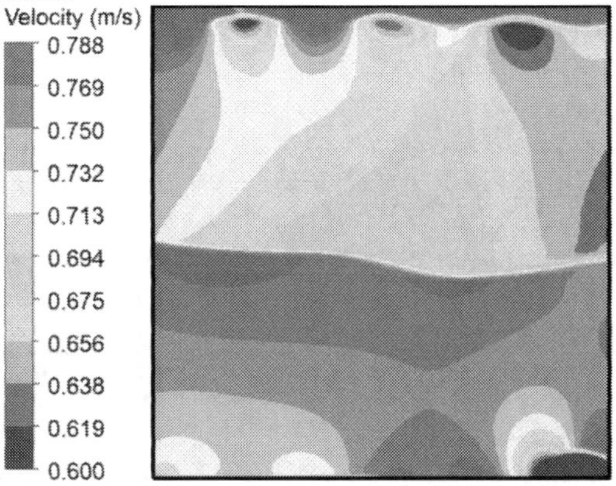

Figure 4.3-2 Velocity contour of the domain inside the cage with modified variable porosity in the first view.

In order to further analyse the impact of porosity configuration on the flow field inside the cage and aquaculture efficiency, the regional proportion corresponding to different flow velocity ranges inside the cage are calculated. Table 4.3-1 presents the ranges of flow velocity and their regional proportion inside the cage for different porosity configurations. The flow velocity increases from Range 1 to Range 6 and the flow velocity of Range 2-5 meets the fishability condition. For the certain porosity, the area meets the fishability condition accounted for about 89.92% of the total area inside the cage. The flow velocity in most of the area is about 0.69-0.79 m/s. For the unmodified porosity configuration, the area meets the fishability conditions accounted for about 60.63% of the total area inside the cage. The flow velocity in most of the area is greater than 0.69 m/s. More than one-third of the area did not meet the fishability conditions. For the modified variable porosity configuration method, the area meets fishability conditions accounted for about 91.28% of the total area inside the cage. The flow velocity in most of the area was about 0.65-0.79 m/s. Comparing with the certain porosity, the modified variable porosity configuration increased the percentage of areas meets fishability conditions inside the cage by 1.36%. The flow velocity in some areas of the flow field is further decreased without affecting the water exchange capacity. It can be seen that the variable porosity configuration can increase the area inside the cage that satisfies the fishability condition, but the configuration of porosity is crucial. A suitable configuration can not only improve the fishability of the cage, but also change the velocity distribution inside the cage to improve the aquaculture efficiency from various aspects.

Table 4.3-1 Regional proportion of flow velocity ranges inside the cages for different configurations of porosity in the first view.

No.	Velocity range（m/s）	Regional proportion (%)		
		Certain porosity	Variable porosity	Modified variable porosity
1	0-0.60	0.723	0.607	1.327
2	0.60-0.65	0.245	7.543	8.009
3	0.65-0.69	1.070	9.645	19.313
4	0.69-0.74	45.970	25.387	24.278
5	0.74-0.79	42.632	18.053	39.685
6	0.79-1.82	9.360	38.764	7.389

4.4. Flow field inside the cage with variable porosity in the horizontal direction

From section 4.2, it is clear that the effect of the net parallel to the incoming flow direction on the flow field is almost equal to that of the solid wall. Therefore, this section increases the porosity of the net parallel to the incoming flow direction. The porosity of the net increases from 0.75 to 0.85 parallel to the direction of flow, and the flow field inside the cage is studied in the second view.

Figure 4.4-1 shows the flow field inside the variable porosity net cage in the second view. Figure 4.4-1 shows the flow field inside the cage is still dominated by the water in the direction of the incoming flow after increasing the porosity. The area inside the cage almost meets the fishability condition completely and the velocity distribution is uniform. Table 4.4-1 presents the flow velocity ranges and the corresponding regional proportion inside the cage for different porosity configurations in the second view. As can be seen from Table 4.4-1, the regional proportion of fishability area inside the cage is increased by 0.66%. However, the flow velocity inside the cage does not increase. On the contrary, the flow velocity in some areas inside the cage further decreases. It can be seen that an appropriate increase in the porosity of the net parallel to the direction of flow velocity can improve the water exchange capacity, aquaculture efficiency and benefit.

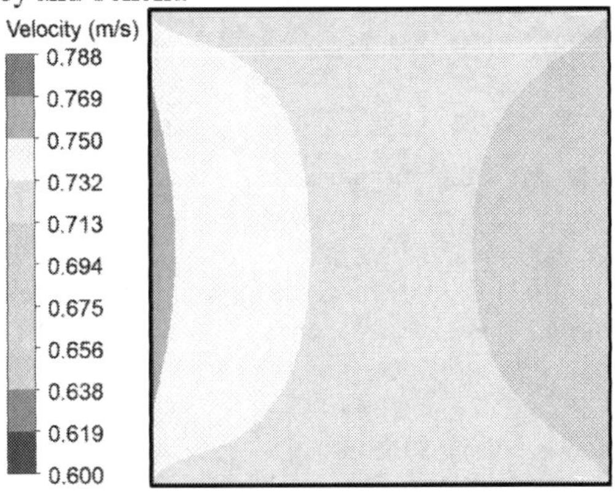

Figure 4.4-1 Velocity contour of the domain inside the net cage with modified variable porosity in the second view.

Table 4.4-1 Regional proportion of flow velocity ranges inside the cages for different configurations of porosity in the second view.

No.	Velocity range (m/s)	Regional proportion (%)	
		Certain porosity	Variable porosity
1	0-0.60	1.200	0.541
2	0.60-0.65	0.266	0.316
3	0.65-0.69	0.259	0.125
4	0.69-0.74	77.333	80.818
5	0.74-0.79	20.943	18.200
6	0.79-1.82	0	0

5. Conclusions

The flow field inside the deep-sea aquaculture cage are closely related to the efficiency and benefit of fish aquaculture. There should be as many fishability areas inside the cage as possible. The flow field inside the cage should be conducive to the diffusion of feed and the flee of fish feces. In this paper, the 2D flow field inside and surrounding the deep-sea aquaculture cage is simulated by CFD from different views. The effects of certain porosity net and variable porosity net with different configurations on the fishability, aquaculture efficiency and benefit of the cage are investigated. The main findings are as follows:

(1) In a larger-scale aquaculture cage, the net near the upstream has higher flow velocity and a larger velocity gradient. Behind this net, both flow velocity and velocity gradient decrease with increasing distance.

(2) Inside the cage with a certain porous net, the velocity distribution in the direction of water depth is the same as that near the velocity inlet. When the flow velocity decreases with increasing water depth, the certain porosity net is not benefit for fish aquaculture.

(3) Variable porosity net with suitable configuration can not only change the velocity distribution in the direction of water depth but also adjust the flow velocity inside the cage. However, it is not reasonable to select the porosity configuration completely according to the velocity distribution near the velocity inlet.

(4) Due to the anisotropic characteristics of the resistance effect of the net, the effect of the net parallel to the incoming flow direction on the flow field is almost equivalent to that of a solid wall. The flow field near these regions inside the cage is basically dominated by the fluid in the incoming flow direction.

(5) After increasing the porosity of the net parallel to the incoming flow direction, the effect of flow field outside the cage on the flow field inside the cage has no significant change. The regional proportion of fishability inside the cage increases and the flow velocity in some areas inside the cage further decreased. It not only increases the water exchange capacity of the cage, but also improves the aquaculture efficiency and benefit.

6. References

[1] Weitzman J Far-field and near-field effects of marine aquaculture 2019 *World Seas: an Environmental Evaluation (Second Edition)* 197-220 https://doi.org/10.1016/B978-0-12-805052-1.00011-5

[2] Zhao Y, Zhang J, Liu Y, Sun K, Zhang C, Wu W and Fei T Numerical assessment of the environmental impacts of deep sea cage culture in the yellow sea, china 2020 *Sci. Total. Environ.* **706** 135752.1-135752.10 https://doi.org/10.1016/j.scitotenv.2019.135752

[3] Plaut I Critical swimming speed: Its ecological relevance 2001 *Comp. Biochem. Phys.* **131(1) A** 41-50 https://doi.org/10.1016/S1095-6433(01)00462-7

[4] Hammer C Fatigue and exercise tests with fish 1995 *Comp. Biochem. Phys.* **112(1) A** 1-20 https://doi.org/10.1016/0300-9629(95)00060-K

[5] Bailey D M, Bagley P M, Jamieson A J, Collins M A and Priede I G In situ investigation of burst

swimming and muscle performance in the deep-sea fish antimora rostrata 2003 *J. Exp. Mar. Biol. Ecol.* 285 295-311 https://doi.org/10.1016/s0022-0981(02)00534-8

[6] Gui F, Wang P and Wu C Evaluation approaches of fish swimming performance 2014 *Ag. Sci.* **05(02)** 106-13 https://doi.org/10.4236/as.2014.52014

[7] Yu X, Sbma B, Han A, App A, Jahbc D, TQT C, MAMG A, HK A and HJM A Quantitative trait loci controlling swimming performance and their effect on growth in Nile tilapia (Oreochromis niloticus) 2022 *Aquaculture* **560** 738522 https://doi.org/10.1016/j.aquaculture.2022.738522

[8] Chai R, Lou Y, Huo R, Yin H, Huang L, Wang H and Wang P Effects of constant flow velocity on endurance swimming and fatigue metabolism in red drum and blackhead seabream 2023 *Comp. Biochem. Phys. A* **275** 111331 https://doi.org/10.1016/j.cbpa.2022.111331

[9] Diego-Mcglone M, Azanza RV, Villanoy CL and Jacinto GS Eutrophic waters, algal bloom and fish kill in fish farming areas in bolinao, pangasinan, philippines 2008 *Mar. Pollut. Bull.* **57(6-12)** 295-301 https://doi.org/10.1016/j.marpolbul.2008.03.028

[10] Klebert P, Lader P, Gansel L and Oppedal F Hydrodynamic interactions on net panel and aquaculture fish cages: a review 2013 *Ocean Eng.* **58** 260 – 74 https://doi.org/10.1016/j.oceaneng.2012.11.006

[11] Weitzman J Far-field and near-field effects of marine aquaculture 2019 *World Seas: an Environmental Evaluation (Second Edition)* 197-220 https://doi.org/10.1016/B978-0-12-805052-1.00011-5

[12] Kelly L A Dissolved reactive phosphorus release from sediments beneath a freshwater cage aquaculture development in West Scotland 1992 *Hydrobiologia* **235(1)** 569-74 https://doi.org/10.1007/BF00026244

[13] Phillips MJ, Beveridge MCM and Ross L G The environmental impact of salmonid cage culture on inland fisheries: present status and future trends 1985 *J. Fish Biol.* **27** 123-37 https://doi.org/10.1111/j.1095-8649.1985.tb03236.x

[14] Wu RSS The environmental impact of marine fish culture: Towards a sustainable future 1995 *Mar. Pollut. Bull.* **31** 4-12 https://doi.org/10.1016/0025-326X(95)00100-2

[15] Grant CJ A numerical model of flow modification induced by suspended aquaculture in a Chinese bay 2001 *Can. J. Fish Aquat. Sci.* **58(5)** 1003-11 https://doi.org/10.1139/cjfas-58-5-1003

[16] Gao QF, Cheung KL, Cheung SG and Shin PKS Effects of nutrient enrichment derived from fish farming activities on macroinvertebrate assemblages in a subtropical region of hong kong 2005 *Mar. Pollut. Bull.* **51(8/12)** 994-1002 https://doi.org/10.1016/j.marpolbul.2005.01.009

[17] Porrello S, Tomassetti P, Manzueto L, Finoia MG, Persia E, Mercatali I and Stipa P The influence of marine cages on the sediment chemistry in the Western Mediterranean Sea 2005 *Aquaculture* **249(1-4)** 145-58 https://doi.org/10.1016/j.aquaculture.2005.02.042

[18] Aarsnes JV, Rudi H and Loland G Current forces on cage, net deflection 1990 *Engineering for Offshore Fish Farming* Thomas Telford London 137–52

IOP Conf. Series: Materials Science and Engineering 1288 (2023) 012035 doi:10.1088/1757-899X/1288/1/012035

Hydrodynamic performance of manta rays under different motion parameter

Pengcheng Gao[1,2,3], Qiaogao Huang[1,2,3*], Guang Pan[1,2,3], Jing Liu[1,2,3], Yao Shi[1,2,3], Xing He[1,2,3], and Xushun Tian[1,2,3]

[1]School of Marine Science and Technology, Northwestern Polytechnical University, Xi'an 710072, People's Republic of China
[2]Unmanned Vehicle Innovation Center, Ningbo Institute of NPU, Ningbo 315048
[3]Key Laboratory of Unmanned Underwater Vehicle, Northwestern Polytechnical University, Xi'an, 710072, PR China

*Corresponding author, E-mail: huangqiaogao@nwpu.edu.cn

Abstract. This paper presents a numerical simulation of the steady propulsion state of manta rays and investigates the influence of single motion parameters and the addition of perturbation signals on the hydrodynamic characteristics and vortex evolution of manta rays. A numerical model and the motion equations of the manta ray were established by observing the living organisms, and then a computational method combining the immersed boundary method (IBM) and the Sphere function-based Gas Kinetic Scheme (SGKS) was used to simulate the active propulsion state of the manta ray. The results show that in a single motion parameter, as the motion frequency increases, the thrust force increases subsequently, but the propulsion efficiency decreases; with the increase of motion amplitude, the thrust also increases, and the propulsion efficiency reaches the maximum at the dimensionless amplitude of 0.35; as the wavenumber increases, the thrust reaches its maximum at wavenumber of 0.4, and the propulsion efficiency increases subsequently. When the same amplitude low-frequency sinusoidal perturbation is added, both thrust and efficiency decrease when the perturbation frequency is less than or equal to 0.4, and increase when the perturbation frequency is greater than 0.6. This work provides a new perspective to study the influence of manta ray motion parameters and perturbation parameters on its hydrodynamic characteristics.

1. Introduction

Underwater vehicles, as a kind of intelligent small equipment that can navigate autonomously underwater for a long time, play an increasingly important role in marine resource extraction and marine ecological protection [1-4]. Conventional underwater vehicles mostly use propulsion and operating systems with propeller systems and thus have limitations in terms of speed, efficiency, maneuverability, and stealth. The development of novel underwater vehicles using bio-inspired designs can be useful in challenging marine environments and can enrich the operational capabilities of underwater drones [5].

The long natural evolution has endowed fish with extraordinary locomotor and survival capabilities: low energy consumption for long-distance swimming, high thrust for efficient propulsion, powerful explosive and maneuverable performance, sensitive perception and localization, excellent self-protection, and precision attack capabilities. Due to the above advantages, bionic underwater vehicles/robots borrowed from aquatic animal concepts have received a lot of interest in recent decades [6]. Manta rays belong to the ray family, and their pectoral fins have a large span chord ratio, which can meet the demand of a large lift-to-drag ratio and also have a large load space [7]. Moreover, the cruising speed of manta rays is suitable for long-distance navigation of underwater vehicles, so their morphological characteristics are very suitable as bionic objects for bionic vehicles. Manta rays use MPF (Median and/or Paired Fin) mode as a steady-state cruise [8], with a cruise speed of 0.25-0.47 m/s [9] and an efficiency of up to 89% [10], which is higher than that of aquatic organisms such as eels and trevally, so their efficient propulsion performance also determines their great bionic value.

Recently, several scholars have designed and developed prototype bionic vehicles based on the MPF model. Lamas [11] established a flexible waving propulsion device and combined the propulsion device with a catamaran to develop a bionic propulsion catamaran. Zhou [12] and Wang [13] developed a bionic underwater robotic fish with dual waving fins, which can flexibly perform direct and maneuvering movements under the propulsion of the dual waving fins. Li [14] used dielectric elastomer artificial muscle drive to develop a kind of flexible body robotic fish imitating lionfish, which successfully dived into the Marianas Trench at a depth of 10, 900 m.

A deep understanding of the fish swimming mechanism will contribute to the development of high-performance bionic vehicles. Bottom Ii [15] numerically simulated self-propelled stingrays and showed that a horseshoe-shaped vortex formed at the leading edge of the swimmer enhances thrust generation. Fish [16] found that the fin motion of rays is asymmetric, producing a complex three-dimensional wake structure. Zhang [17] used the IB-SGKS method to study the vortex formation and related hydrodynamic effects caused by chordal deformation and semi-amplitude asymmetry relative to the longitudinal axis of the body of manta rays during forward swimming, providing a new perspective to study the role of vortices in swimming and force enhancement mechanisms. Luo [18] conducted a numerical study of a manta ray-shaped swimmer with asymmetric left and right pectoral fin movements in tethered mode to reveal the effect of fin kinematics on torque production.

As mentioned above, current studies on the hydrodynamic properties of manta rays have focused on single motion parameters and left-right asymmetric motions, while there is almost no research on the effect of perturbation parameters on the hydrodynamic performance of manta rays. At present, studies on perturbation parameters are still limited to two-dimensional flapping foils, and no studies have been conducted for three-dimensional real-life organisms. Lehn [19] first added sinusoidal perturbation to the flapping foil to show that applying perturbation can improve thrust and propulsion efficiency, and Gao [20-22] conducted numerical studies on the effects of perturbation parameters on the hydrodynamics of heaving foils and obtained the effects of phase difference, perturbation frequency on flap thrust, efficiency, and wake structure.

In this work, we first established a numerical computational model and motion equations for manta rays by observing living organisms and then simulated the active propulsion state of manta rays using a combination of the immersed boundary method (IBM) and the Sphere function-based Gas Kinetic Scheme (SGKS). The most significant difference from previous studies is that we have not only studied a single kinematic parameter, but we have also investigated the effect of perturbation parameters on the hydrodynamic properties of manta rays, filling a gap in the current research.

The subsequent sections are organized as follows. Section 2 shows the manta structure model and kinematic model, and gives the definition of perturbation parameters; Section 3 describes and verifies the computational methods used in this paper; Section 4 gives the research results mainly containing hydrodynamic performance and wake structure; Section 5 concludes the whole paper and gives the recommendations on the setting of motion parameters for the manta-like navigator.

2. Manta ray model and problem description

In our previous study [23], we photographed real manta rays in an oceanarium and established a manta ray coordinate system with the midpoint of the manta ray head as the origin, the body length direction as the X-axis, the span direction as the Y-axis, and the vertical direction as the Z-axis. Subsequently, we measured the 3D coordinates of the manta ray-key points, established shape curves and shape surfaces, and obtained a manta ray model, as shown in Figure 1(a), and the reliability of this model was also confirmed in our previous study [17,24,25]. In Figures 1(a) and (b), we define the geometric and kinematic quantities, where the body length and span length are denoted by *BL* and *SL*, respectively, and the z-directional half-amplitude values are denoted by A_u and A_d, respectively.

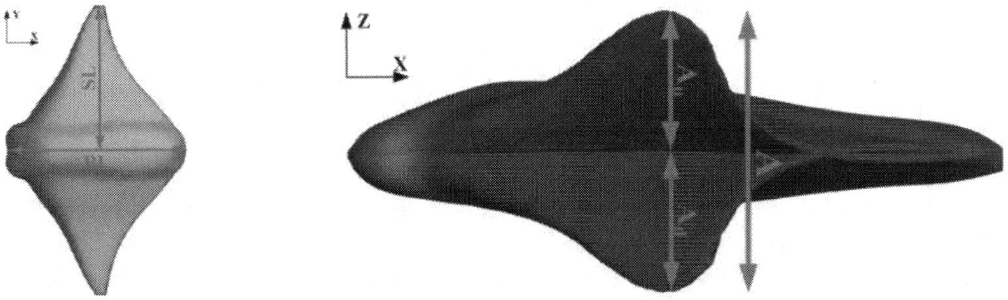

(a) Geometric model (b) Kinematic model

Figure 1. The geometric and kinematic model of the manta ray.

Previously [17], we used a high-resolution digital camera to record the free-swimming process of manta rays and DLT software to track the head, tail, and fin tip of manta rays, and finally summarized the kinematic equations with the help of statistical knowledge as follows.

When $y_f \geq 0$:

$$\begin{aligned} x(x_f, y_f, t) &= x_f \\ y(x_f, y_f, t) &= y_f(1-(1-k)\left|\theta(t,x_f)\right|y_f / SL)\cos[\theta_{\max} y_f / SL \cdot \theta(t,x_f)] \\ z(x_f, y_f, t) &= z_f + y_f(1-(1-k)\left|\theta(t,x_f)\right|y_f / SL)\sin[\theta_{\max} y_f / SL \cdot \theta(t,x_f)] \\ \theta(x_f, t) &= \sin(\omega t - 2\pi W x_f / BL) \end{aligned} \tag{1}$$

When $y_f \leq 0$:

$$\begin{aligned} x(x_f, y_f, t) &= x_f \\ y(x_f, y_f, t) &= -(-y_f)(1-(1-k)\left|\theta(t,x_f)\right|(-y_f) / SL)\cos[\theta_{\max}(-y_f) / SL \cdot \theta(t,x_f)] \\ z(x_f, y_f, t) &= z_f + (-y_f)(1-(1-k)\left|\theta(t,x_f)\right|(-y_f) / SL)\sin[\theta_{\max}(-y_f) / SL \cdot \theta(t,x_f)] \\ \theta(x_f, t) &= \sin(\omega t - 2\pi W x_f / BL) \end{aligned} \tag{2}$$

$$k = \sin\theta_{\max} / \theta_{\max} \tag{3}$$

$$SL(1-(1-k)SL / SL\sin(\theta_{\max} SL / SL) = A_u \tag{4}$$

Where x_f, y_f, z_f are the pectoral fin coordinate points in the initial state; $\theta(x_f, t)$ is the chordwise deformation equation; W is the dimensionless wave number, defined as $W = BL / \lambda$, where λ is the chordwise wavelength(The computational procedure and detailed definition of the motion equation can be found in our previous study [17] and won't be repeated here).

The above description is the motion equation for a single motion parameter, but in this study, the effect of the perturbation parameter on the hydrodynamic performance is also discussed. The perturbation parameter in the equations of motion is mainly reflected in the chordwise deformation equation. The sinusoidal perturbation with varying frequency is introduced into the equation of chord deformation. The new chordwise deformation equation is as follow:

$$\theta(x_f,t)_{new} = \sin(\omega t - 2\pi W x_f / BL) + \sin(\omega_{new} t - 2\pi W x_f / BL) \tag{5}$$

3. Numerical method and Validations

During the active propulsion of manta rays, the pectoral fin flapping will produce large deformation. To effectively simulate the fluid flow at the boundary of pectoral fin motion, a sphere function-based gas kinetic scheme solver (SGKS) based on the immersion boundary method is used in this paper. Detail of this method can be found in the works of Yang, Shu & Wu [26]; Yang et al. [27,28]. In previous studies, the solver was applied to the calculation of flapping propulsion, and its accuracy was well verified [29-31].

As shown in Fig. 2, The computational domain in the present study is a rectangular computational domain of 16BL×16BL×16BL, and the computational model is located at the center of the computational domain. The computational domain is discretized by an orthogonal Cartesian grid, and there is a rectangular inner domain of 1.62BL×1.95BL×1.62BL around the model. A high-resolution uniformly encrypted grid is used in the inner domain, and the grid size in the outer domain gradually becomes uniformly sparse from the inner domain to the far field. To study the hydrodynamic performance of the manta ray in the uniform incoming flow, the calculations are carried out using the constrained model method, i.e., the model system stays in place, and the motion of the model in the water is simulated by a given velocity boundary condition at the entrance, the velocity outlet boundary condition at the outlet, and the rest of the boundary conditions are wall boundary conditions. The manta ray model is divided by a triangular mesh.

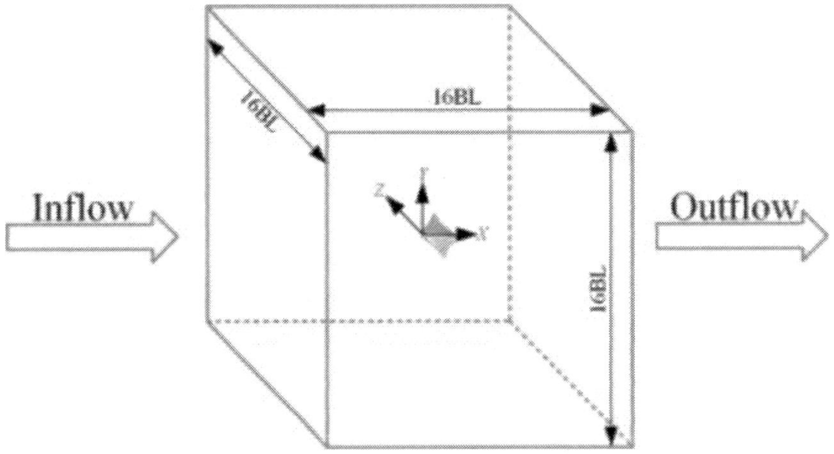

Figure 2. The computational domain (not to scale).

To perform the grid-independent study, the model motion at frequency f = 1Hz, amplitude A_u=A_d=0.35BL, and wave number W= 0.4 was calculated. Specifically, three grid schemes, Coarse, Nominal, and Finer, were selected, in which the grid sizes in the encrypted region were 0.016BL, 0.014BL, and 0.012BL, corresponding to 6.85 million, 8.93 million, and 11.77 million grids, respectively, and the time step Δt= 5.848×10^{-4}s (CFL=0.5). For time-step-independence verification, CFL=0.5 and 0.3 were chosen for the calculation. We plot the computational results with different grid schemes and time steps in Fig. 3, from which we can see that the computational results of the Nominal grid have reached the grid-independent requirement, and the time step at CFL=0.5 has reached the time-step-independent requirement, to save computational resources and improve the computational efficiency we choose the Nominal grid scheme with the time step set to 5.848×10^{-4}s in the subsequent computation.

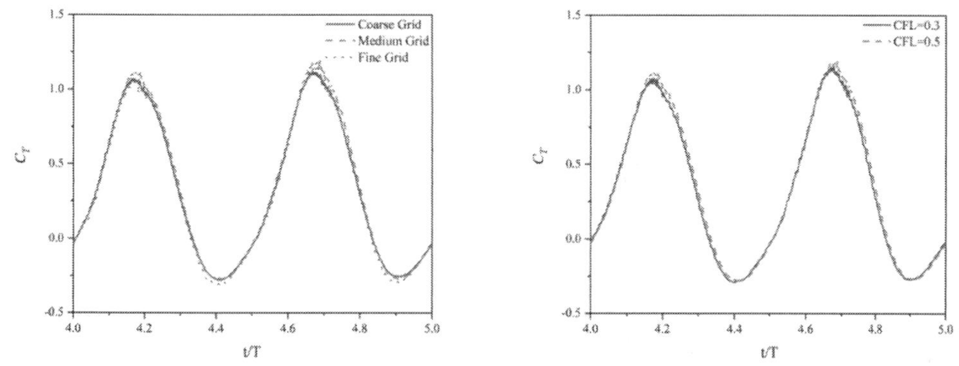

(a) Grid independence validation (b) Time-step independence validation

Figure 3. Grid and time-step independence validations.

4. Results

In this section, we present the results and discussion of C_T, C_L, η, and wake structure. The interrelationships between C_T, η, and wake structure are discussed. First, we outline the variation of instantaneous/time-averaged thrust coefficients and efficiency with motion parameters, and finally, explain the thrust and efficiency performance in the connection with the wake structure. First, we give the defining equations of the hydrodynamic parameters as follows.

$$C_{L,T} = \frac{F_{L,T}}{0.5\rho U^2 BL^2} \tag{6}$$

$$C_P = \frac{P}{0.5\rho U^3 BL^2}, P = \int_{\Omega} F_{Lagrangian} \cdot U_{body} ds \tag{7}$$

$$\eta = \frac{\overline{C_T} + C_D}{\overline{C_P}} \tag{8}$$

Where ρ is density, U is the inlet velocity, BL is the body length, P is the input power, Ω denotes the model surface, $F_{Lagrangian}$ is the Lagrangian force, U_{body} is the deformation velocity of the fin at each boundary element ds, C_D is the towed drag coefficient, which is taken as 0.1044 in this model [17].

4.1. Thrust, lift, and efficiency

First, we explore the single motion parameters on the hydrodynamic performance of manta rays, specifically we simulate the motion of manta rays at different amplitudes, frequencies, and wave numbers.

The effect of motion amplitude on the hydrodynamic performance of the manta ray was the first consideration. We fixed the frequency at 1 Hz and the wave number at 0.4, selected four dimensionless amplitudes (A/BL = 0.25, 0.3, 0.35, 0.4), and plotted the instantaneous thrust, lift, average thrust, and efficiency at these four amplitudes in Figure 4. From Figure 4(a), it can be seen that there are two peaks of thrust which correspond to the upper and lower strokes in one motion cycle, and the time of peak thrust is almost the same regardless of the change in amplitude, but the thrust increases with the increase in amplitude. The instantaneous lift curve has only one wave peak in one motion cycle, and the peak also increases with the increase of motion amplitude, as shown in Figure 4(b). With the increase of the motion amplitude, the average thrust shows a nearly linear increase, and the propulsive efficiency first increases and then decreases, reaching the maximum at A/BL=0.35.

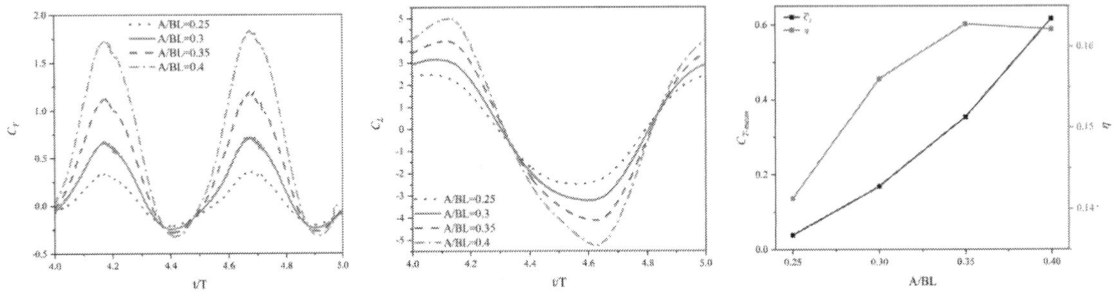

(a) C_T (b) C_L (c) $C_{T\text{-}mean}$ and η

Figure 4. Hydrodynamic performance at different amplitudes.

To investigate the effect of wave number on the kinematic performance of manta rays, we fixed the frequency at 1 Hz and the dimensionless amplitude at 0.35. Four wave numbers (W = 0.2, 0.4, 0.6, 0.8) were selected for calculation, and the variation laws of instantaneous thrust, lift, average thrust, and efficiency with wave number were plotted in Figure 5. As can be seen in Fig. 5(a), two wave peaks of the instantaneous thrust still appear in one motion cycle, but with the increase of the wave number, there is a lag in the time of wave peak appearance. The peak instantaneous thrust also does not increase monotonically with the increase of wave number but reaches the maximum instantaneous thrust at W = 0.4. In one motion cycle, a wave peak appears in the instantaneous lift, and with the increase of the wave number, the peak time appears lagged and the peak size decreases, as shown in Figure 5(b). The wave valley value of thrust basically does not change with the wave number, so the average thrust value also reaches the maximum value at W=0.4, and then the average thrust value keeps decreasing as the wave number continues to increase, which also explains the real manta ray to keep the wave number in the range of 0.34-0.46 during the fast swimming process [32]. From the efficiency curve, we can see that although the manta ray can reach the maximum average thrust at W=0.4, it consumes more energy, so a large thrust is not equal to high efficiency.

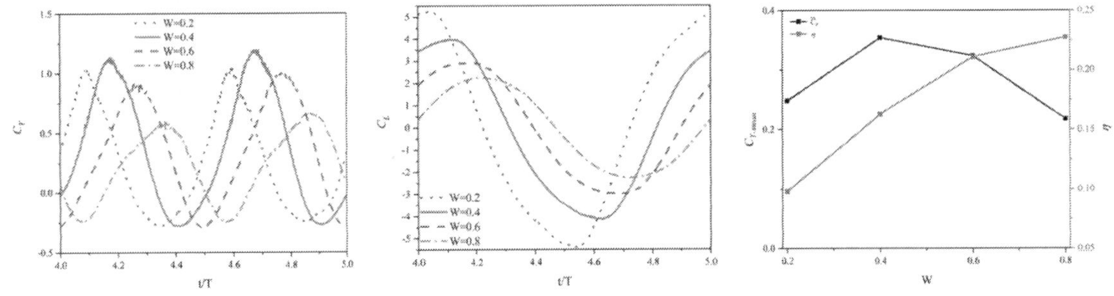

(a) C_T (b) C_L (c) $C_{T\text{-}mean}$ and η

Figure 5. Hydrodynamic performance at different wave numbers.

We fixed the dimensionless amplitude at 0.35 and the wave number at 0.4 and carried out the calculations at five frequencies, and the average thrust, lift, and efficiency were tallied in Table I. From the data in the table, it can be seen that the manta ray cannot produce positive thrust when the frequency is less than 0.6 Hz. As the frequency increases, the average thrust is increased, but the propulsion efficiency decreases. This is because a higher motion frequency will inevitably lead to a significant increase in the input work, so the thrust performance is improved, but the propulsion efficiency is bound to decrease.

Table 1 Hydrodynamic performance at different frequencies

Motion frequencies	$\overline{C_T}$	$\overline{C_L}$	η
0.2 Hz	-0.115349779	-0.004598776	\
0.4 Hz	-0.065350911	-0.009985247	\
0.6 Hz	0.021059417	-0.016342623	0.208602431
0.8 Hz	0.158029896	-0.024925409	0.191740223
1 Hz	0.354024689	-0.033919732	0.162781767

Subsequently, we explored the effect of adding sinusoidal perturbations to the chordal wave signals of manta rays on their hydrodynamic performance, specifically simulating the motion of manta rays under variable frequency perturbations.

Furthermore, we added the same amplitude and phase perturbation at variable frequencies to the original chordwise deformation equation, selected five frequencies (0.2-1Hz), defined the unperturbed case as $F_p=0$ as before, and plotted the hydrodynamic performance for all cases in Fig. 6. It can be seen from Fig. 6 (a) that the instantaneous thrust still shows two peaks in one cycle, but with the addition of the variable frequency perturbation, the peaks appear slightly earlier. Apart from this, we also find that when a frequency perturbation is added, the two peak thrust magnitudes in one cycle are no longer similar, and the first peak is slightly lower than the second peak. Moreover, the first peak will be larger than unperturbed only when $F_p > 0.6$ Hz and the second peak will be larger than unperturbed when $F_p > 0.4$ Hz. With the increase of the perturbation frequency, the instantaneous lift increases slightly, and the peak occurrence time is the same. From Fig. 6(c), it can be seen that not all the variable frequency perturbations are beneficial to the propulsion performance, and both the average thrust and the propulsion efficiency decrease significantly when $F_p \leq 0.4$ Hz. With the increase of the perturbation frequency, the average thrust appears to be continuously enhanced, and when $F_p=1$Hz, the thrust is raised to 139% of the original. The propulsion efficiency still has an optimal value, which occurs at $F_p=0.8$Hz, with an efficiency increase of about 1.53%.

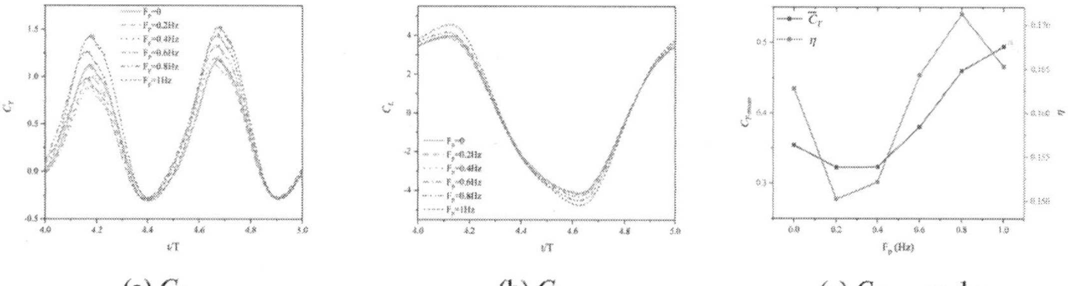

(a) C_T (b) C_L (c) $C_{T\text{-}mean}$ and η

Figure 6. Hydrodynamic performance under different frequency perturbations.

4.2. Wake structure

In this section, the wake evolution of manta rays under active propulsion is analyzed, and the effects of frequency, amplitude, wave number, and the introduction of perturbation signals on the hydrodynamic performance of manta rays are explained in terms of wake structure. During the swimming process, the vortex structure of manta rays is divided into three main categories: leading edge vortex, tail edge vortex, and tip vortex.

We plotted the 3D wake structure of the manta ray for various amplitudes at the moment t/T=4.2 in Figure 7, where the 3D wake structure is visualized by the isosurface of the Q-criterion, taking Q=90. We find that when the motion amplitude increases, the vortex intensity in the manta ray wake field is greater. Due to the viscous effect of the fluid, the vortex will slowly dissipate as it propagates in the flow direction, and the dissipation time is longer for a strong vortex structure, so we can see from Figure 7 that the T1 vortex outline becomes clearer as the amplitude increases.

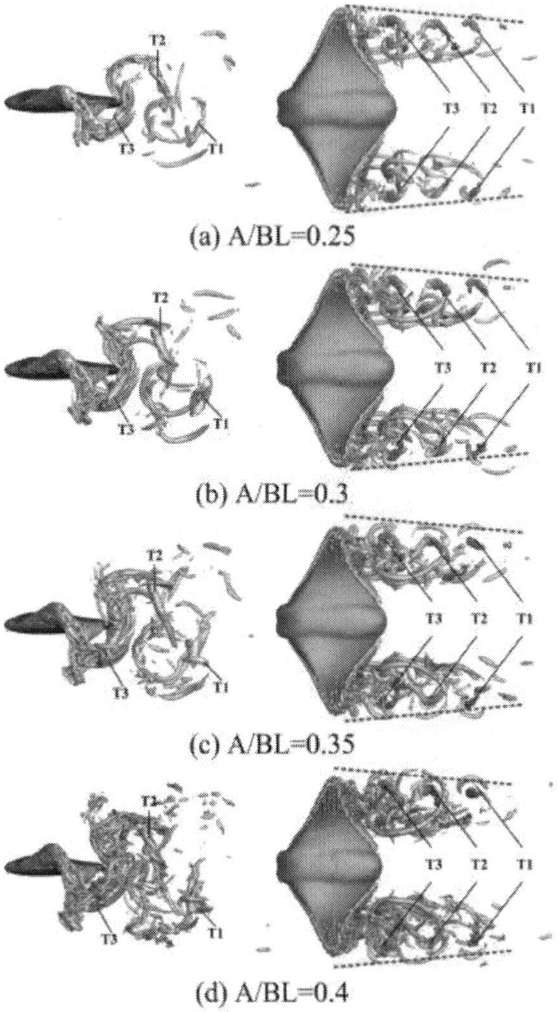

(a) A/BL=0.25

(b) A/BL=0.3

(c) A/BL=0.35

(d) A/BL=0.4

Figure 7. The 3D wake structure at different amplitudes is (t/T=4.2).

As mentioned above, the increase in motion amplitude brings an increase in vortex intensity, which inevitably leads to an increase in thrust, but a large thrust does not mean high propulsion efficiency. We can see from Fig. 14 that when A/BL=0.4, the vortex breakup in the flow field is quite large, and the mutual interference between the tip vortex and tail edge vortex is significant, which inevitably leads to energy waste, so the propulsion efficiency decreases at this amplitude, which is consistent with the hydrodynamic law in Fig. 4.

To show more clearly the variation of vortex structure under different motion amplitudes, the variation of leading and tailing edge vortices as well as tip vortices are explained below by chordwise and spanwise vortex cloud plots, respectively. From Fig. 8, we can see that the vortex intensity increases significantly with the increase of motion amplitude, but when A/BL=0.4, it is found that the trailing edge vortex interferes with the fin tip vortex, and a large number of broken vortices appear in the flow field, which is the same as the previous three-dimensional vortex structure conclusion, which further explains the reason that with the increase of amplitude, the thrust gradually rises but the propulsion efficiency will first rise and then fall.

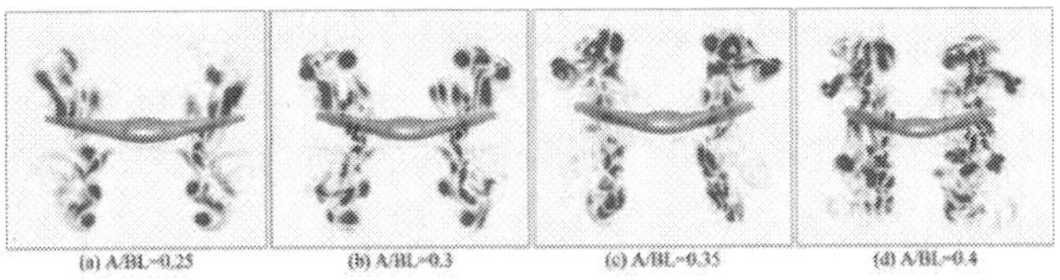

(a) Chordwise vortex plot at different amplitudes, x/BL=1.15

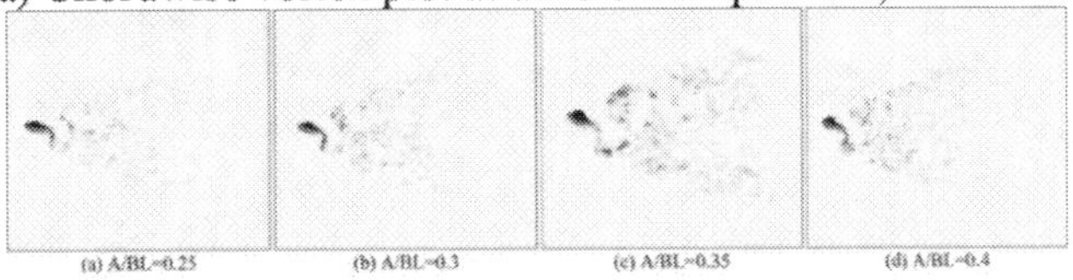

(b) Spanwise vortex plot at different amplitudes, y/SL=0.6

Figure 8. The 2D wake structure at different amplitudes(t/T=4.2).

As before, we plot in Fig. 10 the 3D wake structure of the manta ray at different wave numbers at the moment t/ T = 4.2, where Q = 90. As the wave number increases, the chordwise wave guidance is stronger, resulting in a more regular vortex structure in the wake field, less interference between tip vortices and other vortices, and less broken vortices and inter-vortex interference leading to less energy consumption. This is the reason why the efficiency increases as the wave number increases.

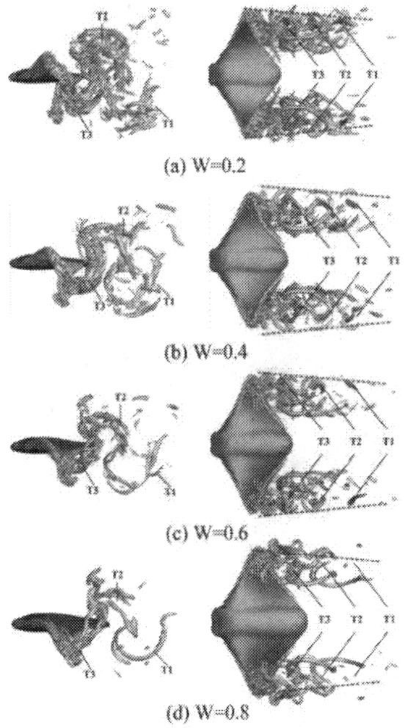

Figure 9. The 3D wake structure at different wave numbers (t/T=4.2).

Then we analyze the effect of wave number change on thrust in Figure 9. We find that as the wave number increases, the wake field changes from numerous broken vortices to vortices with regular structure, and the inter-vortex interference gradually decreases, so the thrust first shows a trend of increasing, but as the wave number further increases, the vortex intensity decreases and the dissipation

IOP Conf. Series: Materials Science and Engineering 1288 (2023) 012035 doi:10.1088/1757-899X/1288/1/012035

speed accelerates, and when W=0.8, the T1 vortex is almost not observed, so the thrust decreases, which is consistent with Fig. 5.

As can be seen from the spanwise and chordwise vortex cloud plots in Fig. 10, with the increase of wave number, the broken vortex and inter-vortex interference decrease, which inevitably reduces the energy loss and brings the efficiency improvement, but the vortex intensity shows a trend of increasing and then decreasing, so the thrust generated by the manta ray also increases and then decreases.

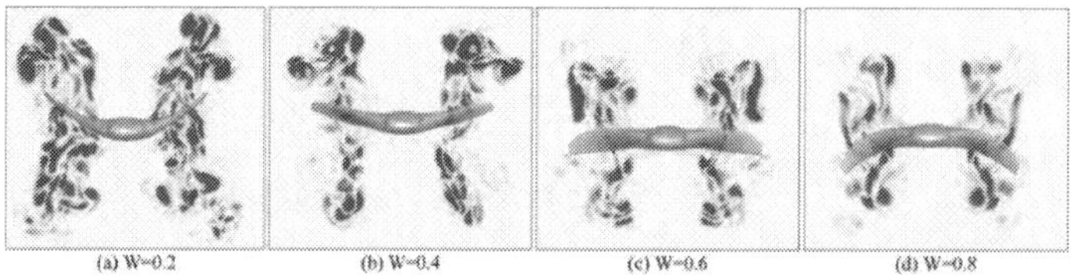

(a) Chordwise vortex plot at different amplitudes, x/BL=1.15

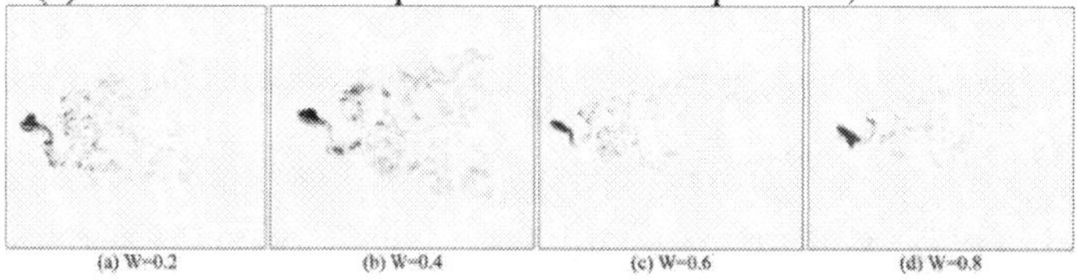

(b) Spanwise vortex plot at different amplitudes, y/SL=0.6

Figure 10. The 2D wake structure at different wave numbers(t/T=4.2).

In Fig. 11, we plot the 3D wake structure of the manta ray at different frequencies at t/ T = 4.2, where Q = 90. From the figure, we can see that when f = 0.2 Hz, the manta ray only produces leading-edge vortices attached to the leading edge, and there is no vortex shedding phenomenon; when f = 0.4 Hz, in addition to leading-edge vortices, tip vortices are also generated, accompanied by vortex shedding phenomenon; when the frequency is further increased, the vortex generation is more obvious and the vortex intensity is further increased, and the vortex dissipation speed is reduced. Therefore, in Table I, we find that when $f \geq$ 0.6Hz, the thrust starts to be generated, and the thrust gradually increases with the increase of frequency.

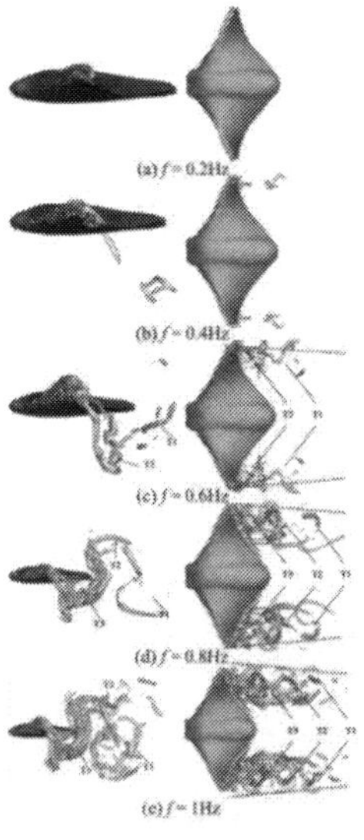

Figure 11. The 3D wake structure at different frequencies0 (t/T=4.2).

We only combine the 2D wake structure to analyze the propulsion efficiency at $f \geq 0.6$ Hz, we can see that when the frequency increases, the original regular wake structure is destroyed by broken vortex, and the inter-vortex interference also becomes larger, so the efficiency decreases with the increase of frequency. We can also see from the spanwise and chordwise vortex cloud plots in Fig. 12 that when the frequency increases, the vortex intensity increases, the dissipation rate slows down, and it propagates further along the flow direction, which is beneficial to boost the thrust. However, as the frequency increases, the shedding of the tailing edge vortex interferes with the tip vortex, causing a large number of broken vortices in the flow field, which leads to a decrease in propulsion efficiency.

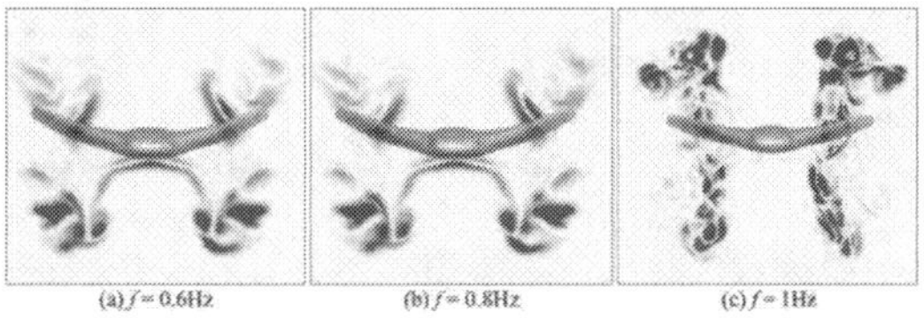

(a) Chordwise vortex plot at different amplitudes, x/BL=1.14

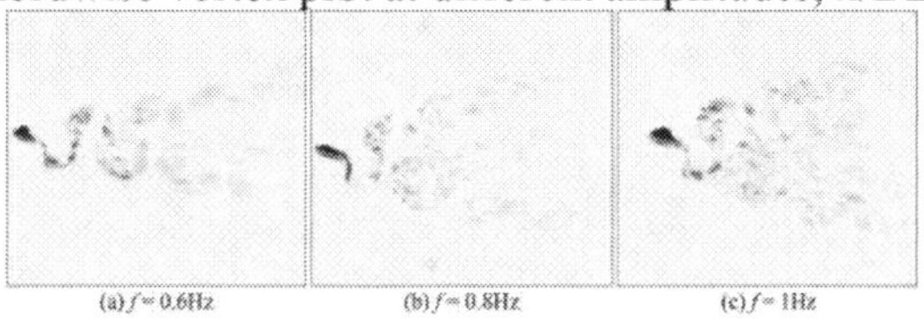

(b) Spanwise vortex plot at different amplitudes, y/SL=0.62

Figure 12. The 2D wake structure at different frequencies(t/T=4.2).

From Fig. 6(c), we can see that the introduction of different frequencies of sinusoidal perturbations in the manta ray chordal wave perturbation has little effect on its hydrodynamic performance, so we can see by comparing Fig. 13 and Fig. 9(b) that the 3D wake structure of the manta ray is similar before and after the addition of the perturbation with only a small amount of detail difference.

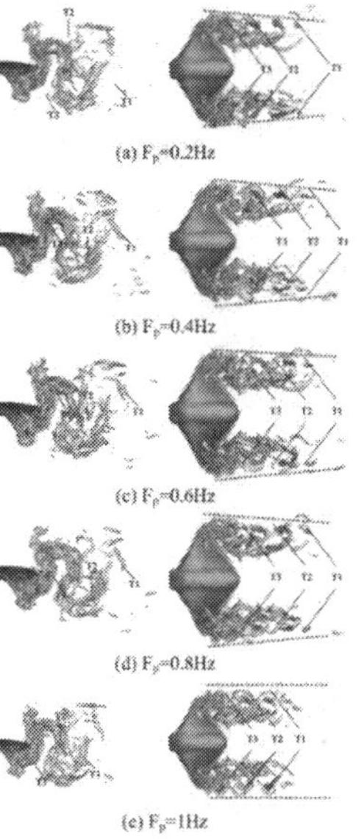

Figure 13. The 3D wake structure under different frequencies perturbations (t/T=4.2).

From Figs. 13 and 14, we find that when Fp≤0.6Hz, the vortex intensity decreases slightly and the number of broken vortices in the flow field is higher, so both the thrust and the propulsion efficiency are not improved or even decreased compared with that when no perturbation is applied. However, when F_P≥0.8Hz, the vortex strength is improved, the vortex distribution distance is reduced, and the number of broken vortices in the flow field is also reduced, so the thrust and propulsion efficiency are improved.

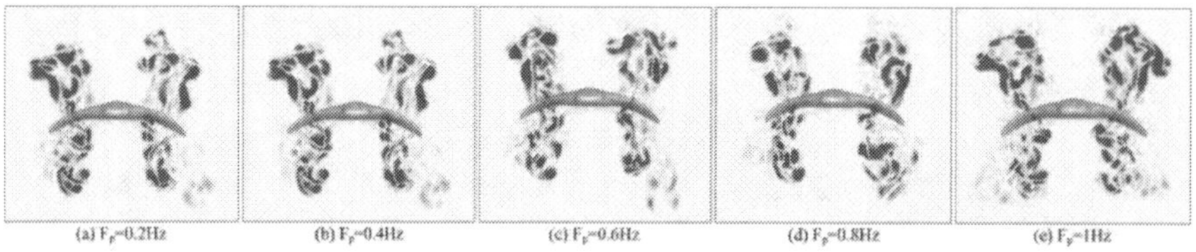

(a) Chordwise vortex plot under different amplitude perturbations, x/BL=1.14

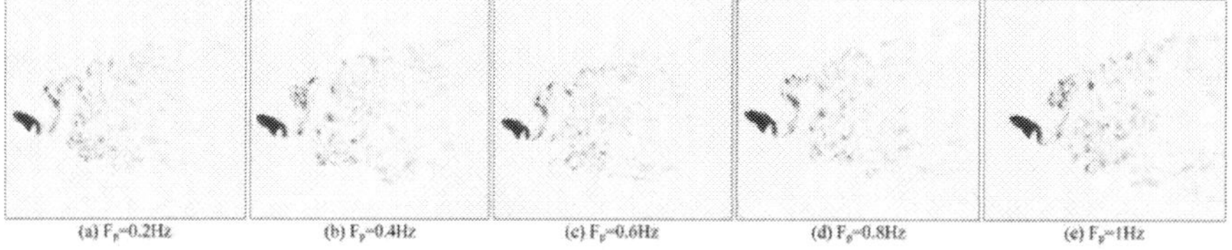

(b) Spanwise vortex plot under different amplitude perturbations, y/SL=0.62

Figure 14. The 2D wake structure under different amplitude perturbations(t/T=4.2).

5. Conclusions

In this paper, the IB-SGKS method is used to simulate the active propulsion of manta rays under a single motion parameter and applied perturbation parameter, and the variation curves of lift, thrust, and propulsion efficiency with motion parameter are obtained and analyzed with the wake structure. At a single motion parameter, a large frequency or large magnitude can bring an increase in thrust, but it does not mean that the propulsive efficiency increases as well. We found that the maximum propulsive efficiency was achieved at an amplitude of 0.35BL when the frequency was fixed at 1 Hz. When the amplitude is fixed at 0.35BL and the frequency is fixed at 0.6Hz, the propulsive efficiency increases monotonically with the increase of wave number, but the thrust reaches the maximum value when W = 0.4. When sinusoidal perturbations of different frequencies are applied to the chordwise wave equation, both thrust and propulsive efficiency do not vary monotonically with increasing frequency. Applying small frequency ($F_P \leq 0.4$ Hz) perturbations has a detracting effect on the manta ray propulsive performance, and applying large frequency perturbations can slightly improve the propulsive performance.

Declaration of Competing Interest

The authors declare that they have no known competing financial interests or personal relationships that could have appeared to influence the work reported in this paper.

Acknowledgments

This work was supported by the National Key Research and Development Program of China (Grant No. 2020YFB1313200), the National Natural Science Foundation of China (Grant No. 12002282 and 52001260), and Innovation Foundation for Doctor Dissertation of Northwestern Polytechnical University (CX2022025).

Reference

[1] Hodges, B. A., & Fratantoni, D. M. (2014). AUV observations of the diurnal surface layer in the North Atlantic salinity maximum. Journal of Physical Oceanography, 44(6), 1595-1604.
[2] Kim, M. G., Kang, H., Lee, M. J., Cho, G. R., Li, J. H., & Kim, C. (2021). UUV Platform Optimal Design for Overcoming Strong Current. Journal of Ocean Engineering and Technology, 35(6), 434-445.
[3] Hallawa, A., Iacca, G., Sariman, C., Rahman, T., Cochez, M., & Ascheid, G. (2020). Morphological evolution for pipe inspection using Robot Operating System (ROS). Materials and Manufacturing Processes, 35(6), 714-724.
[4] Gorma, W., Post, M. A., White, J., Gardner, J., Luo, Y., Kim, J., ... & Xiao, Q. (2021). Development of modular bio-inspired autonomous underwater vehicle for close subsea asset inspection. Applied Sciences, 11(12), 5401.
[5] Fish, F. E. (2020). Advantages of aquatic animals as models for bio-inspired drones over present AUV technology. Bioinspiration & biomimetics, 15(2), 025001.
[6] Low, K. H. (2011, August). Current and future trends of biologically inspired underwater vehicles. In 2011 Defense Science Research Conference and Expo (DSR) (pp. 1-8). IEEE.
[7] Clark, R., & Smits, A. (2006). Visualizations of the Unsteady Wake of Manta Ray Model. In 44th AIAA Aerospace Sciences Meeting and Exhibit (p. 352).
[8] Tangorra, J. L., Davidson, S. N., Hunter, I. W., Madden, P. G., Lauder, G. V., Dong, H., ... & Mittal, R. (2007). The development of a biologically inspired propulsor for unmanned underwater vehicles. IEEE Journal of Oceanic Engineering, 32(3), 533-550.
[9] Rosenberger, L. J. (2001). Pectoral fin locomotion in batoid fishes: undulation versus oscillation. Journal of Experimental Biology, 204(2), 379-394.
[10] Graham, R. T., Witt, M. J., Castellanos, D. W., Remolina, F., Maxwell, S., Godley, B. J., & Hawkes, L. A. (2012). Satellite tracking of manta rays highlights challenges to their conservation. PloS one, 7(5), e36834.
[11] Lamas, M., Rodríguez, J., Rodríguez, C., & González, P. (2011). Three-dimensional CFD analysis to study the thrust and efficiency of a biologically-inspired marine propulsor. Polish Maritime Research, 18(1), 10-16.
[12] Zhou, H., Hu, T., Xie, H., Zhang, D., & Shen, L. (2010). Computational and experimental study on dynamic behavior of underwater robots propelled by bionic undulating fins. Science China Technological Sciences, 53, 2966-2971.
[13] Wang, Y., Wang, R., Wang, S., Tan, M., & Yu, J. (2019). Underwater bioinspired propulsion: From inspection to manipulation. IEEE Transactions on Industrial Electronics, 67(9), 7629-7638.
[14] Li, T., Li, G., Liang, Y., Cheng, T., Dai, J., Yang, X., ... & Yang, W. (2017). Fast-moving soft electronic fish. Science advances, 3(4), e1602045.
[15] Bottom Ii, R. G., Borazjani, I., Blevins, E. L., & Lauder, G. V. (2016). Hydrodynamics of swimming in stingrays: numerical simulations and the role of the leading-edge vortex. Journal of Fluid Mechanics, 788, 407-443.
[16] Fish, F. E., Schreiber, C. M., Moored, K. W., Liu, G., Dong, H., & Bart-Smith, H. (2016). Hydrodynamic performance of aquatic flapping: efficiency of underwater flight in the manta. Aerospace, 3(3), 20.
[17] Zhang, D., Huang, Q. G., Pan, G., Yang, L. M., & Huang, W. X. (2022). Vortex dynamics and hydrodynamic performance enhancement mechanism in batoid fish oscillatory swimming. Journal of Fluid Mechanics, 930, A28.
[18] Luo, Y., Xu, T., Huang, Q., Hou, Z., & Pan, G. (2022). A numerical investigation on thrust and torque production of a batoid fish with asymmetric pectoral fins flapping. Ocean Engineering, 263, 112342.
[19] Lehn, A. M., Thornycroft, P. J., Lauder, G. V., & Leftwich, M. C. (2017). Effect of input perturbation on the performance and wake dynamics of aquatic propulsion in heaving flexible foils. Physical Review Fluids, 2(2), 023101.

[20] Gao, P., Huang, Q., & Pan, G. (2021). Propulsion performance and wake dynamics of heaving foils under different waveform input perturbations. Journal of Marine Science and Engineering, 9(11), 1271.
[21] Gao, P., Huang, Q., Pan, G., & Ma, Y. (2022). Effect of perturbations with different phases on the propulsive performance of rigid heaving foils. Ocean Engineering, 252, 111264.
[22] Gao, P., Liu, G., Huang, Q., Song, D., Pan, G., & Ma, Y. (2022). Heaving foil propulsion performance under combined base and perturbation signal inputs. Ocean Engineering, 266, 113065.
[23] Huang, Q., Zhang, D., & Pan, G. (2020). Computational model construction and analysis of the hydrodynamics of a Rhinoptera Javanica. IEEE Access, 8, 30410-30420.
[24] Gao, P., Huang, Q., Pan, G., Ma, Y., & Song, D. (2022). Research on the hydrodynamic performance of double manta ray gliding in groups with variable attack angles. Physics of Fluids, 34(11), 111908.
[25] Ma, Y., Huang, Q., Pan, G., & Gao, P. (2022). Investigation of the Hydrodynamic Characteristics of Two Manta Rays Tandem Gliding. Journal of Marine Science and Engineering, 10(9), 1186.
[26] Yang, L. M., Shu, C., & Wu, J. (2015). A three-dimensional explicit sphere function-based gas-kinetic flux solver for simulation of inviscid compressible flows. Journal of Computational Physics, 295, 322-339.
[27] Yang, L. M., Shu, C., Wang, Y., & Sun, Y. (2016). Development of discrete gas kinetic scheme for simulation of 3D viscous incompressible and compressible flows. Journal of Computational Physics, 319, 129-144.
[28] Yang, L. M., Shu, C., Yang, W. M., Wang, Y., & Wu, J. (2017). An immersed boundary-simplified sphere function-based gas kinetic scheme for simulation of 3D incompressible flows. Physics of Fluids, 29(8), 083605.
[29] Lin, X., Wu, J., Zhang, T., & Yang, L. (2019). Phase difference effect on collective locomotion of two tandem autopropelled flapping foils. Physical Review Fluids, 4(5), 054101.
[30] Lin, X., Wu, J., Zhang, T., & Yang, L. (2020). Self-organization of multiple self-propelling flapping foils: energy saving and increased speed. Journal of Fluid Mechanics, 884, R1.
[31] Lin, X., Wu, J., & Zhang, T. (2021). Self-directed propulsion of an unconstrained flapping swimmer at low Reynolds number: hydrodynamic behaviour and scaling laws. Journal of Fluid Mechanics, 907, R3.
[32] Rosenberger, L. J. (2001). Pectoral fin locomotion in batoid fishes: undulation versus oscillation. Journal of Experimental Biology, 204(2), 379-394.

Numerical Investigation on Drag and Vortex Induced Vibration Characteristics of Multiple Cylinders

M Araki

Ocean Engineering Department, National Maritime Research Institute, Japan, 6-38-1 Shinkawa, Mitaka, Tokyo, Japan

araki@m.mpat.go.jp

Abstract. The drag and vortex induced vibration (VIV) characteristics of a multiple cylinder are investigated using the detached eddy simulations. The multiple cylinders consist of a single main cylinder and twin auxiliary cylinders for a deep sea mining riser. Tandem and side-by-side arrangements are tested and compared with the single cylinder. The drag characteristics are investigated with the stationary case in the range of $0.5\sim2.0\times10^5$ in Reynolds number (Re). The drag of the tandem arrangement shows smaller drag than those of the single cylinder while it's consisted with the three cylinders. The stationary vortexes are observed at the clearance between the main and auxiliary cylinder which make the tandem arrangement become airfoil-like section as a total which induce the smallest drag than the other cases. The VIV characteristics are studied with the cases of forced oscillation in the cross-flow direction. The damping and added mass coefficients in cross-flow direction are estimated from the time series of the lift forces. Flow field analysis reveal the difference in VIV characteristics due to the variety of the arrangements and the single pipe.

1. Introduction

Over the past decade, several deep sea mining projects are showing progress and the conceptual production design have been proposed. Some of these systems consist of ore storage vessel, seabed mining / harvesting machines and risers hanged from the vessels that connect the vessels to the harvesting machines [1]-[4]. The harvester crawls around the seabed to correct ore, and the vessel tracks the movement of the harvester. During the tracking, the uniform flow, which is not assumed in conventional oil and gas production riser or drilling riser, will act on the entire riser. In such situation, due to the drag force, there could be excessive bending moment at the top end of the riser. Furthermore, in the aspect of the fatigue damage, the vortex-induced vibration (VIV) of the riser in the uniform flow could be more severe than that of general tidal current. Moreover, the deep sea mining riser is usually composed with multiple pipes (e.g. ore-lift pipe, drain and/or water injection pipes) which induces more complex hydrodynamic phenomena than single pipe. The conventional drilling risers and some of the production risers in offshore oil and gas field are also composed with multiple pipes. Therefore, in this paper, drag and VIV characteristics of the multiple cylinders are numerically investigated.

Number of the experimental and numerical studies have been published on drag and VIV characteristics of single cylinder [5], [6]. And several experimental and numerical research have been done on twin cylinders in tandem, side-by-side, and staggered arrangements [7], [8]. However, due to too many combinations of parameters of diameter and distance between the cylinders, a few studies have been published on both drag and VIV characteristics on multiple cylinders [9], [10]. From an engineering point of view, as mentioned above, many productions are composed with multiple pipes, and it is time and cost consuming to investigate the drag and VIV characteristics using the experimental method. Therefore, it is necessary to develop and validate the numerical method to predict these characteristics while the computational cost is still high but has decreased exponentially in recent decades. Therefore, as a general numerical method, the viscous computational fluid dynamics (CFD) method is applied on this paper.

In this paper, multiple pipes are composed with three pipes, single main pipe and twin auxiliary pipes, with tandem and side-by-side arrangement. Firstly, the drag characteristics are investigated and validated on single pipe and multiple pipes composed with single main pipe and twin auxiliary pipes in steady flow. Secondary, the VIV characteristics are examined with forced oscillation simulation in cross-flow direction.

2. Dimensions and Arrangements of Subject Cylinders

Recently, a few deep sea mining pilot test have been conducted on reduced commercial scale [2], [3]. However, the only deep sea mining riser examples manufactured on a commercial scale are GMC's seafloor mining riser for the Solwara 1 project developed by Nuatilus Minerals as shown in figure 1 [11]. The riser was composed with three pipes, single ore-lift pipe and twin water injection pipes. In this paper, subject multiple pipes are designed based on the GMC's seafloor mining riser. The arrangement and the dimensions of the subject pipes are shown in Fig. 2 and table 1. The dimensions are roughly estimated based on Fig. 1. In this paper, two arrangements are examined, the tandem arrangement aligned in in-line direction, and the side-by-side arrangement aligned in cross-flow direction as shown in figure 2, since completely different hydrodynamic characteristics can be expected between these two arrangements.

Figure 1. GMC's mining riser joints [11]

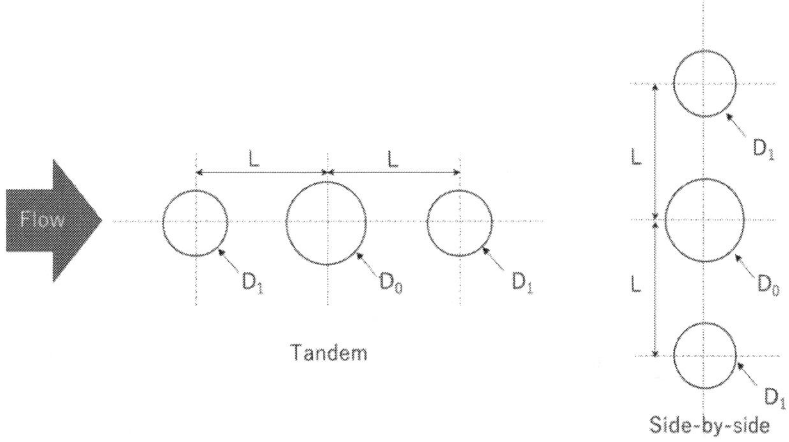

Figure 2. Tandem and side-by-side arrangement of the subject cylinders

Table 1. Dimensions of the subject cylinders.

Item		Value
Diameter	Main cylinder: D_0	250mm
	Auxiliary cylinder: D_1	200mm
Distance between main and aux. cylinders: L_{01}		400mm

3. Computational Modelling

This section describes the details of the computational modelling including the CFD solver, computational grids, and simulation conditions.

3.1 CFD solver

In-house CFD solver NAGISA [12] is applied for the numerical simulation. The governing equation of NAGISA is the incompressible three-dimensional Reynolds-averaged Navier-Stokes (RANS) equation with the artificial compressibility approach for the coupling of pressure and velocities. The dual time stepping method is used for the transient computation. The spatial discretization is based on a finite volume method with structured grid of cell-centred layout. The convective flux is evaluated by MUSCL approach with the third order upwind scheme based on the flux difference splitting scheme of Roe. The viscous fluxes are the second order centred differencing. For the fast computational convergence, full multi gird method is applied. The details of the CFD solver are described in the reference [12].

It is well known that unsteady RANS (URANS) simulation for highly separating flow, such as this subject, shows relatively poor performance compared to large eddy simulations (LES). However, the computational cost of LES is still too expensive for practical Reynolds number (Re) application. Therefore, detached eddy simulation (DES), a hybrid method with URANS behaviour near the wall and LES further away, is applied in this study. The DES are based on the k-ω SST method [13]. The switching limiter between the DES and URANS area of k-ω SST using second blending function F_2 is show in equation (1)-(3). Here, Δ is computed from the grid spacing of (x, y, z) direction as $\Delta = (\Delta_x^2 + \Delta_y^2 + \Delta_z^2)^{1/2}$. C_{DES} is DES model parameter usually 0.61 is applied. L_t is turbulent length scale computed from the turbulent kinetic energy k, the turbulent specific dissipation rate ω, and the model constant $\beta^*=0.09$. y_w is distance from the nearest wall. The detail of the DES based on k-ω SST model is shown in reference [13].

IOP Conf. Series: Materials Science and Engineering 1288 (2023) 012036 doi:10.1088/1757-899X/1288/1/012036

$$F_{DES} = \max\left(\frac{L_t}{C_{DES}\Delta}(1 - F_{SST});\ 1\right) \tag{1}$$

$$L_t = \frac{\sqrt{k}}{\beta^* \omega} \tag{2}$$

$$F_{SST} = F_2 = \tanh(\arg_2^2), \arg_2 = max\left(\frac{2\sqrt{k}}{\beta^* \omega y_w}; \frac{500\nu}{y_w^2 \omega}\right) \tag{3}$$

3.2 Computational grids

In this study, multi-block structured grids generated with Pointwise® are applied. Figure 3 shows horizontal grid arrangements of single, tandem, and side-by-side cylinders. The span-wise length L_s is set to be $L_s=\pi D_0$ and grid spacing of span-wise direction Δ_z is set to be $\Delta_z= \pi D_0/64 \approx 0.049\ D_0$. The minimum spacing of nondimensional distance from the wall y^+ is controlled to be $y^+ \approx 1$. The summary of the cell number of the grids are shown in table 2. For the grid convergence study, single cylinder arrangement of three different resolutions are generated, i.e. 'coarse', 'base line', and 'fine' grids. The refinement ratio between 'coarse', 'fine', and 'base line' grids are design to be around $2^{\pm 1/2}$ on the grid spacing for the each IJK-directions [14]. Based on the 'base line' single cylinder arrangement, the tandem and side-by-side 'base line' grids are generated.

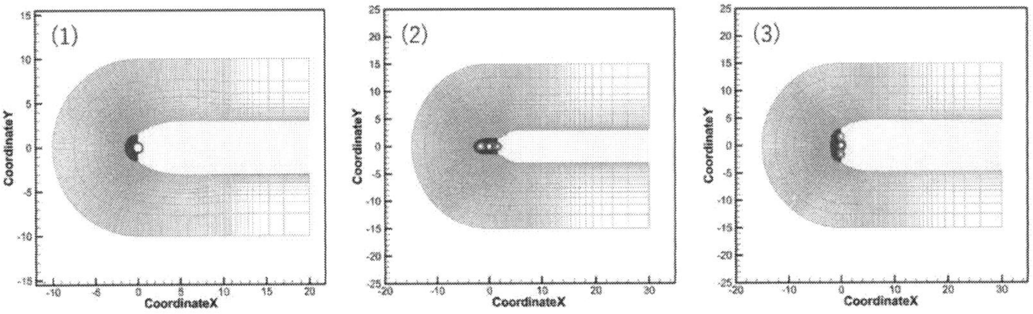

Figure 3. Horizontal arrangements of the multi-block structured grids ((1): single; (2): tandem; (3): side-by-side).

Table 2. Cell number of the computational grids.

Item	Cell number
Single	1,204,862 (coarse), 3,407,872 (base line), 9,638,918(fine)
Tandem	3,604,480 (base line)
Side-by-side	3,899,392 (base line)

The boundary conditions of the grid, e.g. tandem arrangement, are shown in figure 4. The viscous wall conditions are defined on the cylinder surfaces. The inflow conditions are set on the front and side faces, and the rear face is the outflow condition. The periodic/cyclic boundary condition is set at the end faces of the spanwise direction.

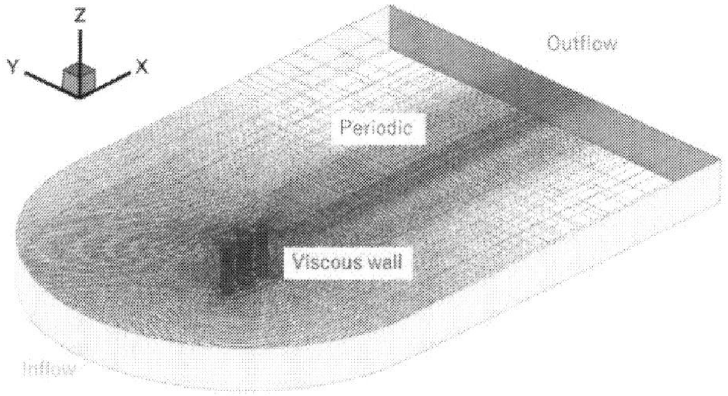

Figure 4. Boundary conditions of the computational grid in case of tandem arrangement.

3.3 Simulation conditions

In this study, two different simulations are conducted, i.e. towing with/without oscillation in the cross-flow direction. Firstly, the numerical simulations on stationary cylinders are performed with three different Re to estimate the drag, RMS (root mean square) lift coefficients. Secondary, the oscillating cylinders in cross-flow direction is carried out to estimate the VIV characteristics, in particular the added mass and the damping coefficients in the cross-flow direction. Note that the main cylinder and the auxiliary pipes are oscillate with the same amplitude and phase. Table 3 summarises the simulation conditions. Hereafter, all nondimensionalization is calculated based on the diameter of the main cylinder D_0. The Re of the simulation is around the critical Re, since in full scale riser, the high current condition could reach to the critical Re. The surface condition of the cylinders is assumed to be smooth. The time interval Δt is set to be Δt=0.02 D_0/U=5/250 D_0/U because it is well known that the Strouhal number (St) of smooth cylinder is around 0.2 (=1/5) in range of 1,000 < Re < 2.00×10^5.

Table 3. Simulation conditions.

Towing simulation	Item	Value
Stationary cylinder	Re ($\times10^5$)	0.50, 1.00, 2.00
Oscillating cylinder	Re ($\times10^5$)	2.00
	Oscillation amplitude ($\times D_0$)	0.50
	Oscillation frequency ($\times U/D_0$)	0.10, 0.15, 0.20

4. Results and discussion

The section presents and discusses the results of the stationary and oscillating simulations. The results and discussion of the simulation with/without oscillation are described in each subsection.

4.1 Stationary cylinder

Figure 5 shows the time series of drag coefficient C_D and lift coefficient C_L of the main cylinders of the single, tandem, side-by-side arrangement in Re = 2.0×10^5. The main cylinder in the single arrangement shows the largest drag, followed by the side-by-side and tandem arrangements. The main cylinder of the side-by-side arrangement show the largest fluctuation in drag and lift forces while the main cylinder of the tandem arrangement shows almost no fluctuation.

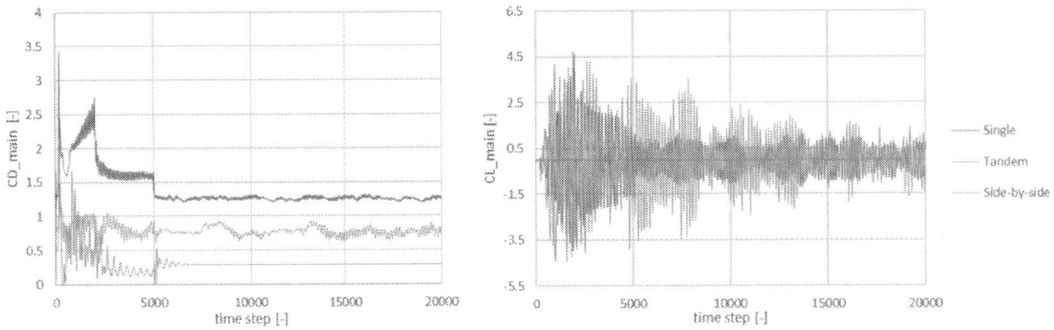

Figure 5. Time series of drag and lift coefficients in Re=2.0×10^5 in stationary cases (Left: drag coefficient; Right: lift coefficient).

Table 4 shows the summary of the drag coefficient C_D and the root-mean-square (RMS) lift coefficient C_{L_RMS} of the stationary cylinders. The C_D and C_{L_RMS} are computed by averaging the last 2000 steps of the simulations and nondimensionalized with equation (4). Here F_X is in-flow force, F_{Y_RMS} is RMS of cross-flow force, and ρ is fluid density (=1000kg/m^3).

$$C_{D,L\_RMS} = \frac{F_{X,Y\_RMS}}{\frac{1}{2}\rho U^2 D_0 L_S} \tag{4}$$

The C_D and C_{L_RMS} of main and auxiliary pipes in Re = 2.0×10^5 are shown in figure 6. In the tandem arrangement, the front pipe shows the smallest and the rear pipe shows the largest drag. The fluctuations of the lift coefficients are nearly negligible in the tandem arrangement. On the other hands, in the side-by-side arrangement, the fluctuations of the lift forces are larger than those of the single pipe. Especially, the top and bottom pipe of the side-by-side arrangement shows very large fluctuation of lift force.

Table 4. Drag and RMS lift coefficient in stationary cases.

Cylinder		Re =0.5×10^5		Re =1.0×10^5		Re =2.0×10^5	
		C_D	C_{L_RMS}	C_D	C_{L_RMS}	C_D	C_{L_RMS}
Single		1.2760	0.4695	1.2128	0.4781	1.2474	0.4974
Tandem	Main	-0.0686	0.0022	0.1824	0.0101	0.2878	0.0084
	Front	0.9034	0.1450	0.5597	0.0467	0.5085	0.0006
	Rear	0.1939	0.0076	0.2007	0.0207	0.3101	0.0569
	Total	1.0287	0.1504	0.9429	0.0775	1.1064	0.0479
Side-by-side	Main	0.7348	0.9651	0.9135	0.8707	0.7718	0.6786
	Top	0.8097	1.2703	0.8534	1.3761	0.8268	1.3033
	Bottom	0.8470	1.5860	0.6415	0.5874	0.8774	1.8536
	Total	2.3915	1.1205	2.4084	0.9878	2.4760	0.9892

The simulation results of the total C_D and C_{L_RMS} are compared with the experimental data of single pipe [15], [16] in figure 7. The simulation results of the single pipe show good agreement with those of the experimental results. The tandem arrangement shows smaller total C_D and C_{L_RMS} than those of the single pipe while the tandem arrangement includes main and two auxiliary pipes. On the other hands, the total C_D and C_{L_RMS} of the side-by-side arrangement shows almost twice of those of the single pipe.

The spectra of the total drag and lift forces in Re=2.0×10^5 are show in figure 8. In the spectrum of the drag force, the single pipe has large peak around nondimensional frequency f' of 0.2 which is the well-known St. The spectrum of the side-by-side arrangement shows larger peak at f'=0.2 than that of the single pipe and another peak around f'=0.1 can be found. No significant peak can be found in the tandem arrangement. In lift force spectra, the single and side-by-side arrangements show similar peak around f'=0.1 while the tandem shows very small peak around f'=0.35.

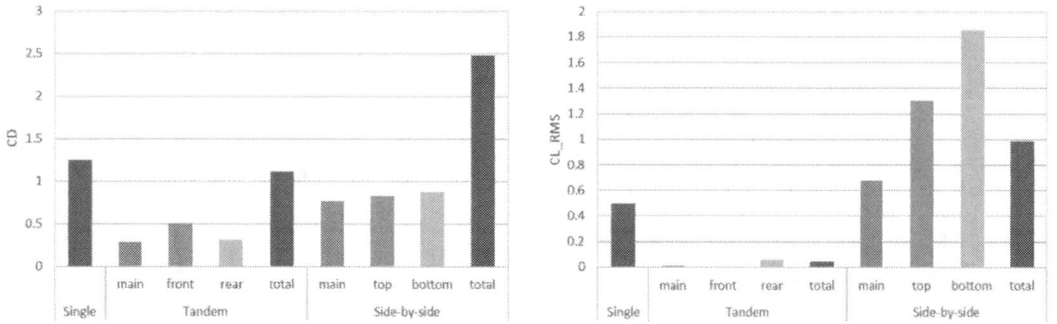

Figure 6. Drag and lift RMS coefficients of the stationary cylinders in Re =2.0×10^5.

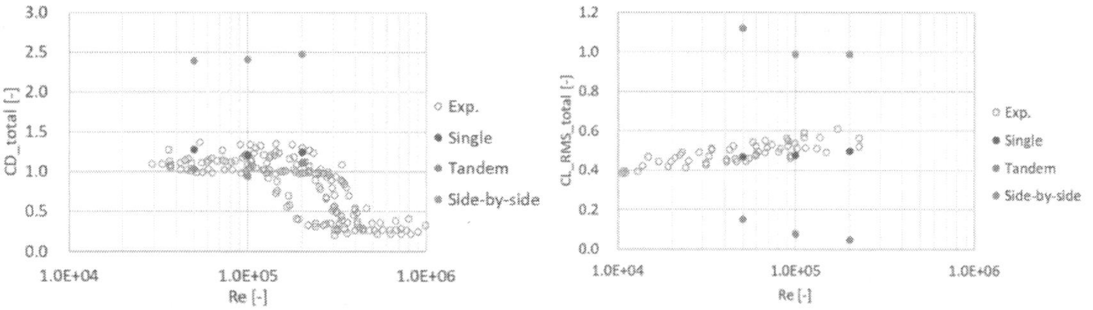

Figure 7. Drag and lift RMS coefficients of total pipes in stationary cases (Left: drag coefficient; Right: lift RMS coefficient).

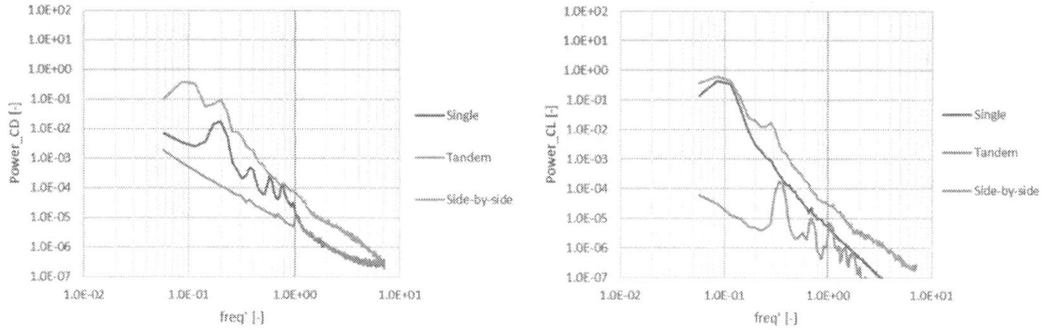

Figure 8. Power spectra of total pipe drag and lift coefficients in stationary cases, Re=2.0×10^5 (Left: drag coefficient; Right: lift coefficient).

Figure 9 shows instantaneous iso-surface of the second invariant of gradient of velocity tensor Q=1.0 which shows the vortical structure in Re=2.0×10^5. The vortical structure of the side-by-side arrangement is much more complicated than that of single pipe because of the interaction between the vortexes generated by each pipe. On the other hands, the tandem arrangement shows small, simple, and two-dimensional vortical structure which cause the stable drag and lift force.

The instantaneous vorticity and streamlines are shown in figure 10. The vorticities and streamlines are on the Z=L_S/2 plane in spanwise direction. In the tandem arrangement, As shown in figure 10, stationary vortexes reside between the main and auxiliary pipes while unsteady vortexes periodicity shed in the single and side-by-side arrangements. In the tandem arrangement, the stationary vortexes are generated due to a so-called "flatback" airfoil effect [17]. Moreover, due to the stationary vortexes between the pipes, the tandem arrangement become airfoil-like section as a total.

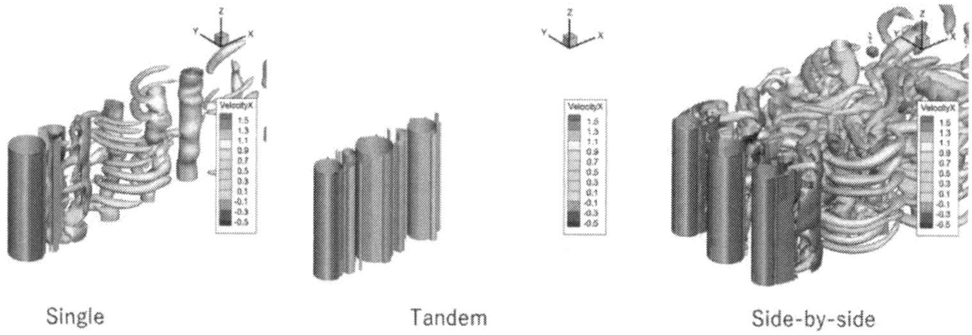

Figure 9. Instantaneous wake configuration. Vortical structure presented by iso-contours of Q=1.0 coloured by in-flow velocity in Re=2.0×10^5, stationary cases.

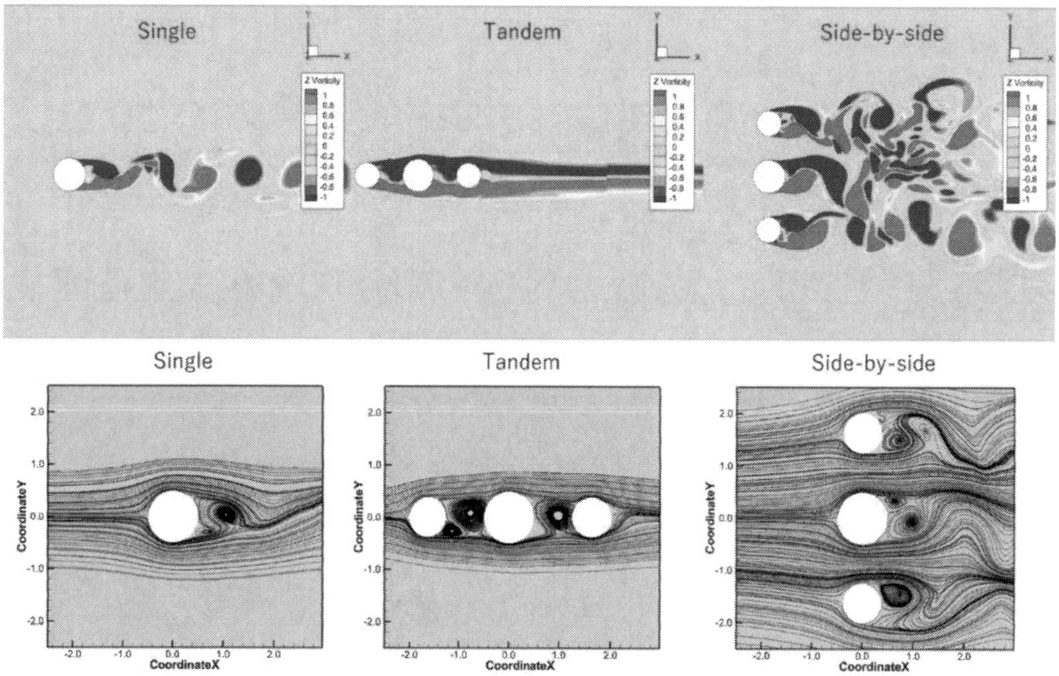

Figure 10. Instantaneous wake configuration. Vortical structure presented by vorticity and streamline on Z=L_s/2 plane in Re=2.0×10^5, stationary cases.

4.2 Oscillating cylinder in cross-flow direction

Figure 11 shows time series for three cycles of the total lift forces in the oscillating cases. The tandem arrangement shows the largest total lift forces in all oscillation cases since the projected area in cross flow direction is the largest than the other arrangements. On the other hand, the nonlinearity of the signal is large in the following arrangement: side-by-side, single, and tandem which shows a similar trend to the time series of the drag. Furthermore, the smaller the nondimensional oscillation frequency f', the greater the nonlinearity.

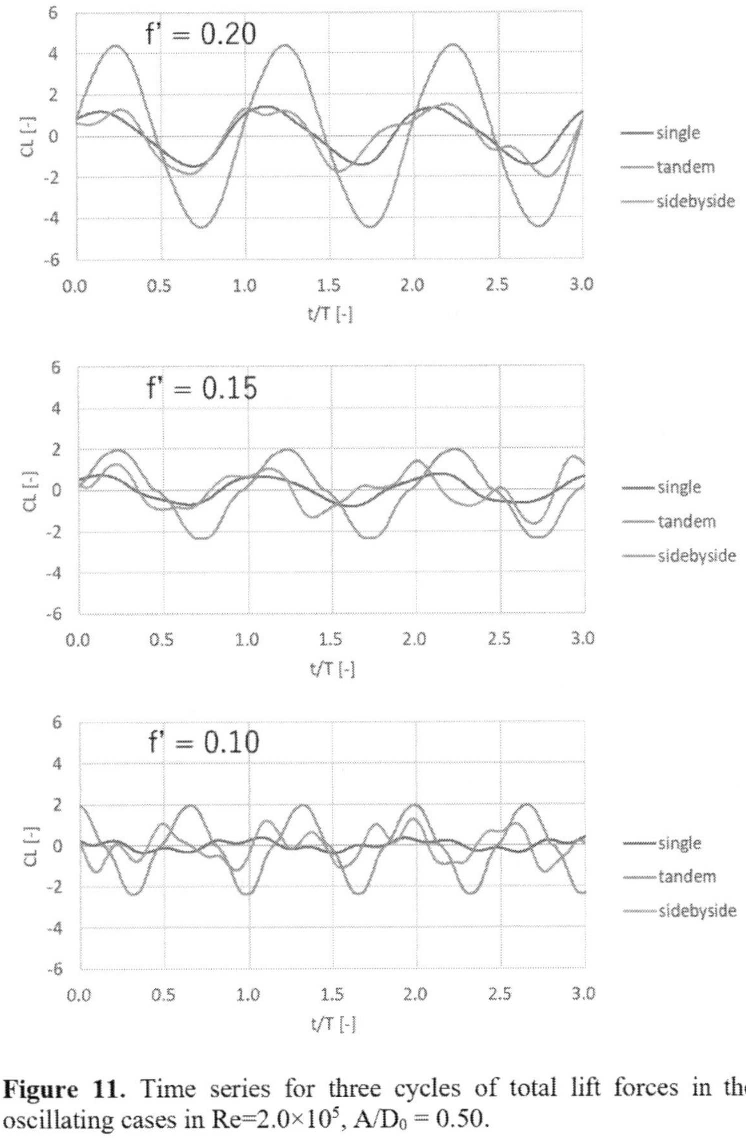

Figure 11. Time series for three cycles of total lift forces in the oscillating cases in Re=2.0×10^5, $A/D_0 = 0.50$.

To estimate the VIV characteristics of the multiple pipes, the damping and added mass coefficients in cross-flow direction is estimated from the oscillation test.

The nondimensional coefficient C_{L_V0} which define the magnitude of the power transfer, is the lift coefficient in phase with oscillation velocity are estimated by equation (5) following the analytical method of [5].

$$C_{L\_V_0} = \frac{L_0 \sin(\phi_0)}{\frac{1}{2}\rho D_0 L_S U^2} \tag{5}$$

Where L_0 is the amplitude of the total lift force $L(t) = L_0\sin(2\pi ft+\phi_0)$. Here ϕ_0 is lift force phase angle.

The added mass coefficient is defined by equation (6). Here A is oscillation amplitude, V is volume of the total volume of the main and auxiliary pipes. The V in single arrangement is $D_0^2L_s/4$, and in tandem and side-by-side is $(D_1^2/2 + D_0^2/4)\ L_s$.

$$C_{M0} = \frac{L_0(-\cos(\phi_0))}{\rho VA(2\pi f_0)^2} \tag{6}$$

The estimated C_{L_V0} and C_{M0} coefficients are compared with the single pipe experimental data [5] in figure 12. Note that the conditions of the experimental data is $Re=1.0\times10^4$, $A/D_0 = 0.30$ which is different from the present simulation conditions ($Re=2.0\times10^5$, $A/D_0 = 0.50$).

The C_{L_V0} results of the single arrangement, shows good agreement with the experimental results in f' = 0.10 and 0.15. However, in f'=0.20, the simulation C_{L_V0} shows positive while the experimental result show negative value. Regarding to added mass, the simulation single C_{M0} show agreement with the experimental results in f' = 0.10 and 0.20. However, in the single experimental results, the C_{M0} rise rapidly around f'=0.16~0.17 while the simulation value seems to start rising around f'=0.15. These discrepancies could be caused by the numerical error of the simulation or the mismatch of the experiment and simulation condition in Re and A/D.

In the side-by-side arrangement, the trend of the C_{L_V0} and C_{M0} with respect to the oscillation frequency seem to be consistent while the magnitude of the C_{L_V0} is larger and the C_{M0} is smaller than those of the single pipe. The projection area in the oscillation direction of the single and side-by-side arrangements are same but the V: total volume of the cylinders in equation (6) of the side-by-side is much larger than the single cylinder which could lead the C_{M0} of the side-by-side arrangement to be smaller than those of the single pipe.

In tandem arrangement, the C_{M0} shows larger value than those of the single and side-by-side arrangements, because of larger project area in the oscillation direction than those of the single and side-by-side arrangements. While the trend of C_{M0} with respect to the oscillation frequency match to the single and side-by-side pipes. Whereas the trend of C_{L_V0} with respect to the oscillation frequency in tandem arrangement show different trend with other arrangements: large positive value at f'=0.10 and small negative values at f' = 0.15 and 0.20.

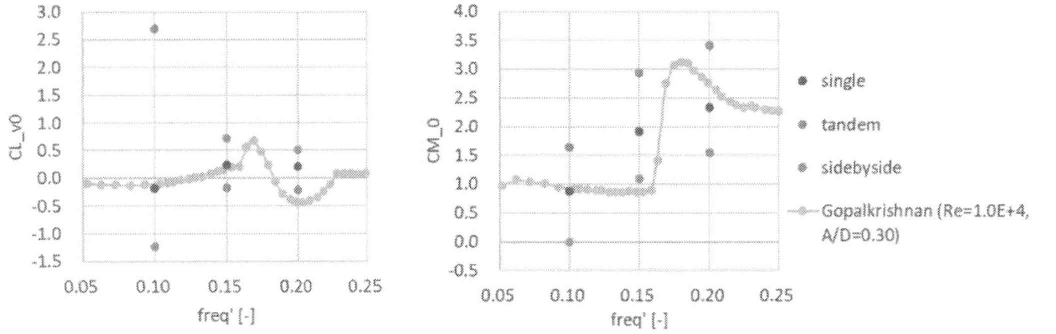

Figure 12. Damping and added mass coefficients of cross-flow direction with respect to oscillation frequency in $Re=2.0\times10^5$, $A/D_0=0.50$.

To investigate C_{L_V0} trend, The pressure distributions on the cylinder are investigated. The polar coordinate around the pipe is defined in figure 13. Instantaneous pressure distribution on the main cylinder surface on $Z=L_s/2$ is shown in figure 14. The oscillation phase in figure 14 is t/T=1/2 where displacement velocity is maximum and acceleration velocity is zero so that the damping effect is predominate. The pressure in figure 14 nondimensionalized with $0.5\rho U^2$. The stagnation point of the main pipe is shifted to the pressure side of the cylinder: negative θ of figure 13 in all case. Since the pipes are move from positive to negative in the Y direction. In f'=0.10 case, the pressure distribution in the pressure side (negative θ) is resemblance to each other while the suction side shows different degrees of the pressure drop and recovery. Especially the pressure drop in the suction side of the tandem main pipe has smaller pressure drop and shifts towards the stagnation point compared to the other cases. In f'=0.20 case, the pressure distribution of the side-by-side arrangements are resemblance to that of the single pipe while large discrepancy can be found with the tandem arrangement. In the tandem arrangement, the peaks of the pressure point and pressure drop are shifted and smaller compared to the other cases.

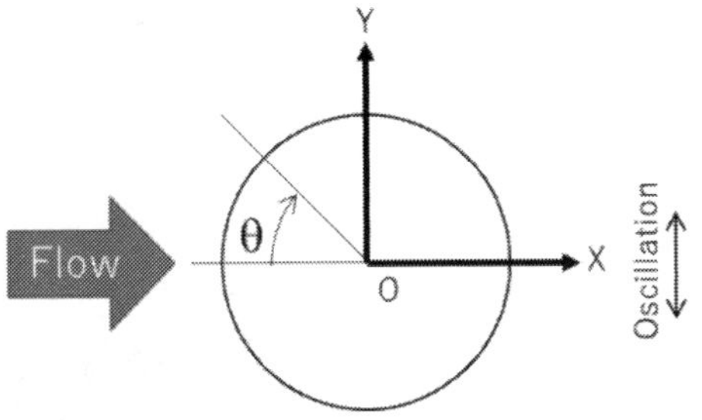

Figure 13. Polar coordinate around the main pipe.

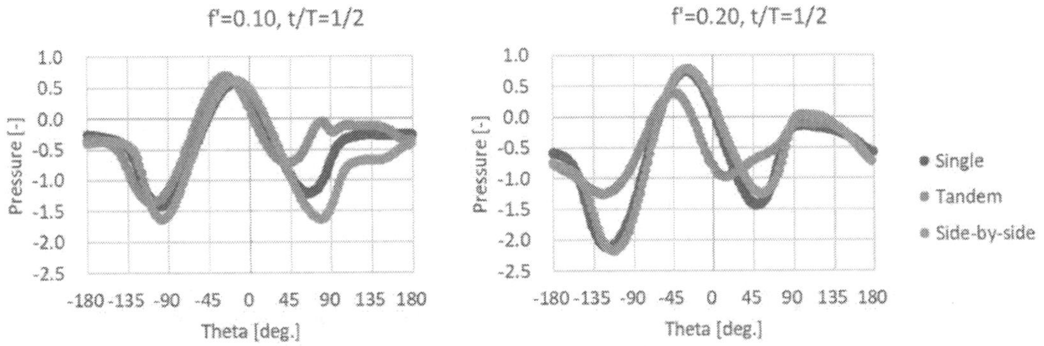

Figure 14. Instantaneous pressure distribution on the main pipe surface at $Z=L_s/2$ in $Re=2.0\times10^5$, $A/D_0=0.50$, t/T=0.5.

Figure 15 show the flow velocity magnitude field around the pipes on $Z=L_s/2$ plane. The velocity magnitude in figure 15 is nondimensionalized by U. In both f'=0.10 and 0.20 cases, the flow acceleration on the suction side of the tandem main pipe is smaller than the other cases because of the existence of the wake from the front pipe which might cause the smaller pressure drop than the other cases. In the side-by-side arrangement, the velocity magnitude between the main and side pipes are relatively large. The flow between main and auxiliary pipes would be accelerated by the contraction effect. Since the aperture ratio of the clearance between main and auxiliary pipes is the largest at $|\theta|$ = 0, 180deg. and the smallest at $|\theta|$ = 90deg. This flow acceleration would enlarge the peaks of the pressure drop and could increase the magnitude of the damping coefficient. In f'=0.20 case of the tandem arrangement, the front part of the main pipe is almost covered by the wake from the front pipe, which would reduce the magnitude of the peaks of pressure point and pressure drop and could reduce the magnitude of the damping coefficient.

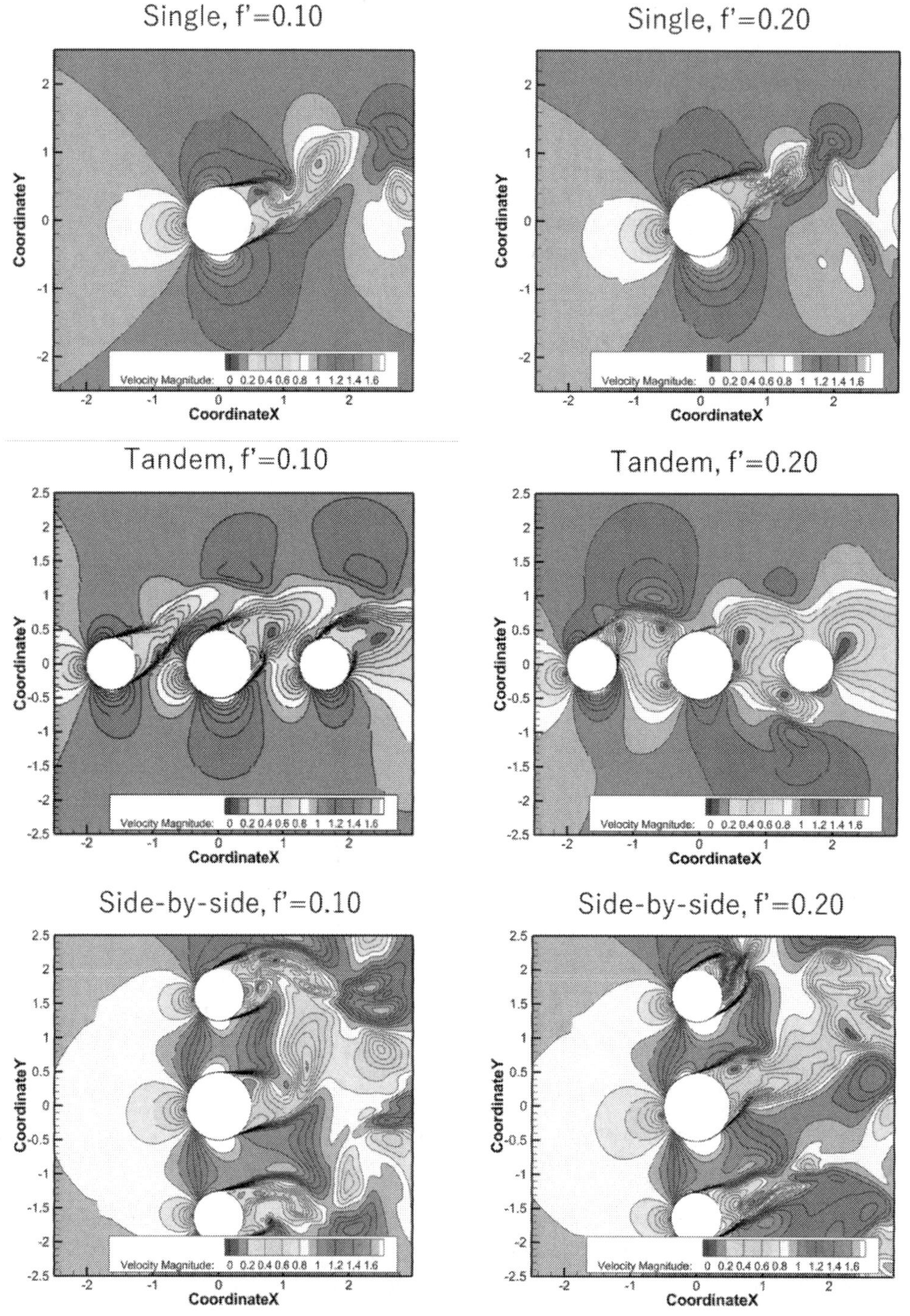

Figure 15. Instantaneous velocity magnitude field around the pipes on $Z=L_s/2$ plane in $Re=2.0\times10^5$, $A/D_0=0.50$, $t/T=0.50$.

5. Conclusions

The drag and VIV characteristics of the multiple pipes consist with single main pipe and twin auxiliary pipes are investigated by the stationary and oscillating simulations using DES of in-house CFD solver NAGISA. The single, tandem, and side-by-side arrangements are the subject of the analysis. The numerical results of the single pipe are validated with the previous experimental results.
In the stationary case, characteristics of the drag and lift forces are investigated. The tandem arrangement shows smaller drag than that of the single pipe despite it is composed with three pipes. In the tandem arrangement case, stationary vortexes are generated at the clearance between the main and auxiliary pipes which make the tandem arrangement become airfoil-like section as a total which induce the smallest drag and the most stable lift force than the other cases.

In the oscillation case, the VIV characteristics: added mass and damping coefficients are studied. In side-by-side arrangement, the trend of the added mass and damping coefficients with respect to the oscillation frequency are consistent with the single pipe while the magnitude of the damping is larger, and the added mass is smaller than those of the single pipe. In the tandem arrangement, the trend of the added mass is consistent with the single and side-by-side arrangements while the magnitude is larger than the other cases. Since, the project area of the tandem arrangement in oscillation direction is larger than the other cases. However, in the tandem arrangement, the trend of the damping coefficient is different with the other cases. To investigate the damping trend, pressure distribution around the main pipe and the velocity magnitude field around the pipes are compared in all arrangements. Those comparison shows, in tandem arrangement, the wake caused by the front pipe has significant effects on the following pipes which could reduce the peak of the pressure point and/or pressure drop. In side-by-side arrangement, the contraction effect around the clearance between the main and side pipes would accelerate the flow around the main pipe and could increase the magnitude of the damping.

References

[1] Blue Nodule, https://blue-nodules.eu/.

[2] JOGMEC, 2018, Comprehensive assessment report on the development plan for subsea hydrothermal deposit, https://www.jogmec.go.jp/news/release/news_10_000279.html, (in Japanese).

[3] The Metal Company, https://metals.co/.

[4] Benbia A, Ngim B, Espinasse P, Parenteau T, 2013, Subsea mining field development concept using subsea crushing and feeding unit, *Proc. of OTC2013*, OTC-23953-MS.

[5] Gopalkrishnan R, 1993, Vortex-Induced Forces on Oscillating Bluff Cylinders, Ph.D. Thesis of Massachusetts Institute of Technology, Dept. of Ocean Engineering.

[6] Dong S, Karniadakis G, 2005, DNS of Flow Past a Stationary and Oscillating Cylinder at Re = 10000, *Journal of Fluids and Structures*, Vol. **20**, pp. 519-531.

[7] Summer D, 2010, Two Circular Cylinders in Cross-flow: A Review, *Journal of Fluids and Structures*, Vol. **26**, pp. 849-899.

[8] Zhao M, Cheng L, Teng B, Dong G, 2007, Hydrodynamic Forces on Dual Cylinder of Different Diameters in Steady Currents, *Journal of Fluids and Structures*, Vol. **23**, pp. 59-83.

[9] Lai LSH., 2017, Cross Sectional Flow Modelling of a Drilling Riser System for Hydrodynamic Drag Consideration, *Proc. OTC 2017*, OTC-27631-MS.

[10] Igarashi T, 1993, Aerodynamic forces acting on three circular cylinders having different diameters closely arranged in line, *Journal of Wind Engineering and Industrial Aerodynamics*, Vol. **49**, Issues 1-3, pp. 369-378.

[11] GMC, https://gmcdeepwater.com/gmcs-seafloor-mining-risers/.

[12] Ohashi K, Hino T, Kobayashi H, Onodera N, Sakamoto N, 2019, Development of a Structured Overset Navier-Stokes Solver including a Moving Grid with a Full Multigrid Method, *Journal of Marine Science and Technology*, Vol. **24**, No. 3, pp. 884-901.

[13] Strelets M, 2001, Detached Eddy Simulation of Massively Separated Flows, *Proc. of 39th AIAA Fluid Dynamics Conference and Exhibit*, 2001-0879.
[14] ITTC, 2011, ITTC-Recommended Procedure and Guidelines- Uncertainty Analysis in CFD Verification and Validation Methodology and Procedure, 7.5-03-01-01.
[15] Lehmkuhl O, Rodriguez I, Borrell R, Chiva J, Oliva A, 2014, Unsteady Forces on Circular Cylinder at Critical Reynolds Numbers, *Physics of Fluid*, Vol. **26**, 125110.
[16] Norberg C, 2001, Flow Around a Circular Cylinder: Aspects of Fluctuating Lift, *Journal of Fluids and Structure*, Vol. **15**, Issues 3-4, pp. 459-469.
[17] Winnenmoller T, Van Dam C.P, 2007, Design and Numerical Optimization of Thick Airfoils Including Blunt Trailing Edges, *Journal of Aircraft*, Vol. **44**, No. 1, pp. 232-240.

Acknowledgment
This work was supported by JSPS KAKENHI Grant Number JP22K04572.

IOP Conf. Series: Materials Science and Engineering 1288 (2023) 012037 doi:10.1088/1757-899X/1288/1/012037

Uncertainty Quantification of Turbulence Model Applied to a Cavitating Hydrofoil

Simone Romani, Mitja Morgut, Lucia Parussini

Department of Engineering and Architecture, University of Trieste, 34127 Tieste, Italy

E-mail: simone.romani@phd.units.it, mmorgut@units.it, lparussini@units.it

Abstract. This paper presents the Global Sensitivity Analysis of the coefficients of the standard k-ε turbulence model used in RANS (Reynolds Averaged Navier-Stokes) simulations aimed to predict the flow around a bi-dimensional hydrofoil operating at non-cavitating and cavitating flow regimes. The sensitivity analysis is treated by the Sobol Decomposition, where the Sobol Indices are computed through the Polynomial Chaos Expansion of the 2-nd order with a Point-Collocation Non-Intrusive approach. From the current results, it seems that the considered cavitating flow regime is less sensitive to the variability of the input parameters, at least for the prediction of lift and drag.

1. Introduction

Owing to the continuous improvements of the CFD (Computational Fluid Dynamics) technologies and continuous increase of the computer performances, numerical simulations are today extensively used for design purposes, allowing the experimental test to be performed only at the final stages of the project. But, despite the progress of the computational technologies, the uncertainties related to the operative conditions, the geometry (due to manufacture tolerances) as well as the physical model can still affect the reliability of a CFD-based design procedure.

In this work, we focus on the uncertainties related to a physical model. In particular, we investigate, using a stochastic approach, based on the Polynomial Chaos Expansion, the possible influence of the propagation of the uncertainties imposed on the coefficients of the standard k-ε turbulence model on the prediction of the lift and drag forces of a NACA66(MOD) hydrofoil operating at a given angle of attack [1].

The study is carried out for a selected cavitating flow regime as well as for the fully-wetted flow condition. The flow is treated as a homogeneous mixture of water and vapour, and the mass transfer rate is modelled using the Zwart et al. model [2].

At this stage, it is important to point out that uncertainties can be classified into two families: *aleatory uncertainties*, which describe the natural/intrinsic variability of a certain quantity; *epistemic uncertainties*, which depend closely on the set of knowledge available [3].

Moreover, the Sensitivity Analysis which allows to quantify the contribution of the uncertainties related to given model inputs (as turbulence model coefficients considered in this study, for instance) to the uncertainty of the model output (as lift and drag evaluated in this study, for instance) can be preformed using two different approaches: *Local Sensitivity Analysis*, which analyses the effects of local changes of a parameter in a system; *Global Sensitivity Analysis*, which evaluates the entire parameter space and the contribution of input variables to the outputs.

Here, a Global Sensitivity Analysis has been carried out, relying on the Variance Based Method and Sobol Decomposition.

Using such an approach, the importance of a given input parameter x_i to the uncertainty of the specific response y is measured using *variance* as a yardstick. Then, Sobol indices are used to rank the influence of input parameters, as well as its combination, on the predictions of selected output parameters. [3, 4, 5, 6]

In this study, as previously mentioned, the uncertainty propagation and the Sobol Indices for the sensitivity analysis have been determined exploiting a method based on response surfaces. Specifically, the Polynomial Chaos Expansion based on a Point-Collocation Non-Intrusive Approach [7, 8].

The selected approach has been practically implemented using *Dakota*, an optimization and uncertainty quantification framework [9], in combination with *Ansys-CFX*, a general purpose CFD (Computational Fluid Dynamics) solver [10]. In particular, Dakota has been used to evaluate the random values of the input coefficients as well as the values of the Sobol Indices, while Ansys-CFX has performed the required simulations.

From the overall results, it seems that the uncertainty propagates differently for the two selected flow regimes. As a matter of fact, the variance of the global values (lift and drag) observed in the cavitating flow regime is less than that obtained for the fully-wetted flow conditions. Moreover, from the comparison of the Sobol indices it seems that the ranking of the contributions of the single input parameters to the uncertainty is also different, as well as the ranking of the contributions of mutual parameter interactions.

In the following, the approach for the uncertainty quantification and sensitivity analysis is described. Then, for the sake of clarity the standard $k-\varepsilon$ turbulence model is presented along with the mathematical model for the cavitating flow. Next, the computational strategy is presented. Finally the results are discussed and concluding remarks are provided.

2. Uncertainty Quantification and Sensitivity Analysis

As introduced in the above section, the importance of the turbulence model coefficients has been quantified by Sobol Indices. Thus, for the sake of completeness, a brief description in accordance with [6, 11] is given in the following.

Since the Sobol Indices are composed by total and partial variances, we start by defining the total variance, D, named also Global Sensitivity as:

$$D = \text{Var}\left\{f\left(\underline{x}\right)\right\} = \int_{K^n} f^2\left(\underline{x}\right) d\underline{x} - f_0^2, \tag{1}$$

where:

$$K^n = \left\{\underline{x} : 0 \leq x_i \leq 1\,,\, i = 1, \ldots, n\right\} : n - \text{dimensional unit cube} \tag{2}$$

and generalized response:

$$f\left(\underline{x}\right) = f\left(x_1, \ldots, x_n\right) = f_0 + \sum_{i=1}^{n} f_i\left(x_i\right) + \sum_{1\leq i<j\leq n}^{n-1} \left[f_{ij}\left(x_i, x_j\right) + \cdots + f_{1,\ldots,n}\left(x_1, \ldots, x_n\right)\right]. \tag{3}$$

Now, the total variance can be decomposed as:

$$D = \sum_{i=1}^{n} D_i + \sum_{1\leq i<j\leq n}^{n-1} \left[D_{ij} + \cdots + D_{1,\ldots,n}\right], \tag{4}$$

where, the partial variances in (4) are specified in the below equation:

IOP Conf. Series: Materials Science and Engineering 1288 (2023) 012037 doi:10.1088/1757-899X/1288/1/012037

$$D_{i_1,\dots,i_s} = \int_{K^s} f^2_{i_1,\dots,i_s}(x_{i_1},\dots,x_{i_s})\, dx_{i_1},\dots,dx_{i_s} \quad \text{for } 1 \le i_1 < \dots < i_s \le n \text{ and } s = 1,\dots,n. \tag{5}$$

Sobol's Indices are defined as follows:

$$S_{i_1,\dots,i_s} = \frac{D_{i_1,\dots,i_s}}{D} \tag{6}$$

and by the total variance decomposition, the indices satisfy the property:

$$\sum_{i=1}^{n} S_i + \sum_{1\le i<j\le n}^{n-1} [S_{ij},\dots,S_{1,\dots,n}] = 1. \tag{7}$$

At this stage, it is important to point that, $S_{i_1},\dots,S_{i_s}$ are useful for studying uncertainty because they classify which amount of the total variance is influenced by the uncertainties in the set of input parameters $\{i_1,\dots,i_s\}$.
Instead, $\sum_{i=1}^{n} S_i$ are first order terms which measure the influence of each parameter taken alone without taking into account mutual interactions.
On the other hand higher order indices evaluate the possible mixed influence of various parameters.

2.1. Polynomial Chaos Expansion

The Polynomial Chaos Expansion is a non-sampling-method capable of determining the evolution of uncertainty in a dynamic system, where random functions or random variables are expressed in deterministic and aleatory components. It is based on the spectral representation of uncertainties.

In practice an output response of a model can be approximated/expressed in truncated polynomial form [8]. The goal is to evaluate the polynomial coefficients because they give a measure of the uncertainties and allow us to perceive which input variables weight more on the response function [12]. The coefficients can be determined by an intrusive or non-intrusive approach on the original equations .

This work is based on the Point-Collocation Non-Intrusive Polynomial Chaos approach [7]. In the case of the PCE the Sobol indices can be evalutated as:

$$D = \sigma^2\left[Y\left(t,\underline{\xi}\right)\right] = \sum_{i=1}^{M} \alpha_i^2(t) = \sum_{i=1}^{M}\sum_{k=1}^{K} Y^2(t,\xi_k)\,\varphi_i^2(\xi_k)\,w_k^2, \tag{8}$$

where:

- $\underline{\xi} = (\xi_1,\dots,\xi_K)$: set of K random variables
- $\varphi_i(\xi_k)$: i-th basis function with respect to the k-th random variable
- $Y(t,\xi_k)$: stochastic process associated with the k-th random variable
- $\alpha_i(t)$: i-th Polynomial Chaos coefficient
- w_k: k-th weight.

In this work, we have considered a continuous uniform distribution function so that basis functions are Legendre polynomials [13] and all weights, w_k, are equal to 1. Then, Polynomial Chaos coefficients are directly evaluated solving the system:

$$\begin{Bmatrix} Y(t,\xi_1) \\ \vdots \\ Y(t,\xi_K) \end{Bmatrix} = \begin{bmatrix} \varphi_0(\xi_1) & \cdots & \varphi_M(\xi_1) \\ \vdots & \ddots & \vdots \\ \varphi_0(\xi_K) & \cdots & \varphi_M(\xi_K) \end{bmatrix} \begin{Bmatrix} \alpha_0(t) \\ \vdots \\ \alpha_M(t) \end{Bmatrix}. \tag{9}$$

3. Mathematical model for cavitating flow

The cavitating flow has been modelled as a homogeneous mixture flow (liquid-vapour mixture) governed by the following continuity, momentum and volume fraction transport equations:

$$\nabla \cdot \underline{v} = \dot{m}\left(\frac{1}{\rho_l} - \frac{1}{\rho_v}\right) \tag{10}$$

$$\frac{\partial}{\partial t}(\rho \underline{v}) + \nabla \cdot (\rho \underline{v}\underline{v}) = -\nabla p + \nabla \cdot \left\{(\mu + \mu_t)\left[\nabla \underline{v} + (\nabla \underline{v})^T\right]\right\} \tag{11}$$

$$\frac{\partial \gamma}{\partial t} + \nabla \cdot (\gamma v) = \frac{\dot{m}}{\rho_l} \tag{12}$$

with: $\underline{v}$, time averaged mixture velocity, p time averaged pressure, $\dot{m}$ interphase mass transfer rate due to cavitation, ρ_v vapour density, ρ_l liquid density, μ_t turbulent viscosity. The vapour α and water volume fractions are defined as:

$$\gamma = \frac{V_{liquid}}{V_{total}} \qquad \alpha = \frac{V_{vapour}}{V_{total}}$$

which relates as:

$$\gamma + \alpha = 1. \tag{13}$$

and the mixuture density, ρ, and mixture viscosity, ν are defined as:

$$\rho = \gamma\rho_l + (1-\gamma)\rho_v \tag{14}$$

$$\mu = \gamma\mu_l + (1-\gamma)\mu_v. \tag{15}$$

Furthermore, in order to close the system of the governing equation, the standard k-ε turbulence model and the native CFX mass transfer model, i.e. the Zwart et all model, are employed to evaluate the turbulent viscosity, μ_T, and the mass transfer rate, $\dot{m}$, respectivelly.

3.1. Zwart et al. model

The Zwart model is based on the simplified Rayleigh-Plesset equation for bubble dynamics:

$$\dot{m} = \begin{cases} -F_e \dfrac{3r_{nuc}(1-\alpha)\rho_v}{R_B}\sqrt{\dfrac{2}{3}\left(\dfrac{p_v - p}{\rho_l}\right)}, & \text{if } p < p_v \\ \\ F_c \dfrac{3\alpha\rho_v}{R_B}\sqrt{\dfrac{2}{3}\left(\dfrac{p - p_v}{\rho_l}\right)} & \text{if } p > p_v, \end{cases} \tag{16}$$

where p_v is vapour pressure, $r_{nuc} = 5.0 * 10^{-4}$ nucleation site volume fraction, $R_B = 1.0 * 10^{-6}$ radius of a nucleation site. $F_e = 300$ and $F_c = 0.03$ empirical calibration coefficient set according to [2].

3.2. Standard k-ε turbulence model
As pointed out previously, in order to evaluate the turbulent viscosity:

$$\mu_t = \rho C_\mu \frac{k^2}{\varepsilon} \tag{17}$$

the standard k-ε model has been employed [14].

In the following the governing equations of the turbulence model along with the standard model constants, subsequently modified by the sensitivity analysis, are presented.

$$\frac{\partial}{\partial t}(\rho k) + \nabla \cdot (\rho \underline{v} k) = \nabla \cdot (\mu_{eff,k} \nabla k) + P_k - \rho\varepsilon = \nabla \cdot (\mu_{eff,k} \nabla k) + S^k \tag{18}$$

$$\frac{\partial}{\partial t}(\rho \varepsilon) + \nabla \cdot (\rho \underline{v} \varepsilon) = \nabla \cdot (\mu_{eff,\varepsilon} \nabla \varepsilon) + C_{\varepsilon 1} \frac{\varepsilon}{k} P_k - C_{\varepsilon 2} \frac{\varepsilon^2}{k} \rho = \nabla \cdot (\mu_{eff,\varepsilon} \nabla \varepsilon) + S^\varepsilon, \tag{19}$$

The above equations are the turbulent kinetic energy and turbulent energy dissipation rate transport equations, and where

$$\mu_{eff,k} = \mu + \frac{\mu_t}{\sigma_k} \quad , \quad \mu_{eff,\varepsilon} = \mu + \frac{\mu}{\sigma_\varepsilon}$$

.

P_K is the production of turbulent energy term. In Table 1 the native values of the model coefficients are provided.

Table 1. Calibration constants of the Standard k-ε Model

Constant	Native value
$C_{\varepsilon 1}$	1.440
$C_{\varepsilon 2}$	1.920
C_μ	0.090
σ_k	1.000
σ_ε	1.300

4. Computational strategy
In this section the logic employed to perform the uncertainty quantification and sensistivity analysis is first described. Then, the numerical setup used to performed the numerical simulations, required by the PCE method, is presented.

4.1. Logic
The PCE method has been applied taking advantage of Dakota. In this case, the study has been practically performed in three separate stages.

Stage 1: Using Dakota, the collocation points (samplings) for the deterministic CFD simulations were generated using Latin Hypercube Sampling (LHS). In this study, the number of the collocation points, N_s, was set following [9] and for the second order of the polynomial chaos expansion.

$$N_s = n_p \left[\frac{(n+p)!}{n!p!} \right] \tag{20}$$

In the above equation p is the order of the polynomial expansion, n_p is the oversampling ratio, n is the number of the uncertain variables. Here: $p = 2$, $n_p = 2$, $n = 5$, thus $N_s = 42$.

Stage 2: The overall CFD simulations were carried out for the given set of collocation points. The related results were saved in proper text files.

Stage 3: Using again Dakota, in combination with a proper and ad-hoc implemented interface, the polynomial chaos coefficients were determined based on the results obtained at stage 2. The related probabilistic values as well as sensitivity indices were calculated.

Fo the sake of clarity, it is important to point out that the overall computational strategy was subdivided in three separate stages in order to perform the CFD simulation independently from Dakota which was conveniently used at the first and third stages.

4.2. CFD setup

As pointed out previously, the current study has been performed considering the NACA66MOD hydorofoil at angle of attack of 4 degrees. The hydrofoil had camber ratio $f/c = 0.020$, mean line $a = 0.8$, thickness ratio t/c=0.9, and chord c=0.150. In the above fractions f is the maximum thickness, t maximum camber.

In order to simulate the turbulent bi-dimensional flow the computational domain depicted in Figure 1 has been used.

Figure 1. Geometry of the domain and boundary conditions

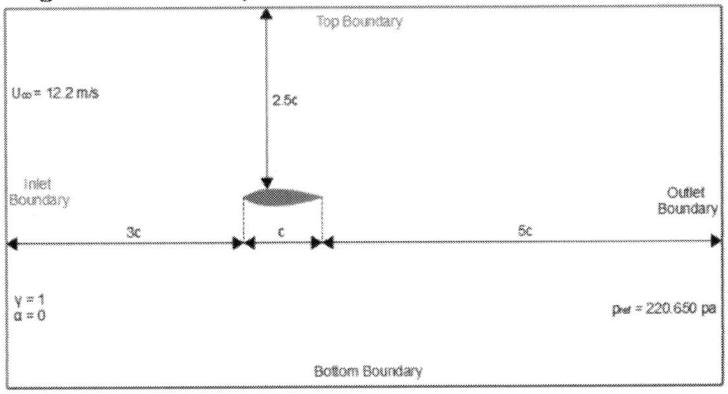

For all the simulations the free-stream velocity imposed at inlet boundary has been kept equal to U=12.2 m/s, as well as the values of the turbulent kinetic energy $k = 0.0223\ m^2/s^2$ and turbulent dissipation rate $\varepsilon = 0.1837\ m^2/s^3$. The pressure on outlet boundary, P_{out} has also been kept constant, in this case equal to 202650 Pa. For this reason in order to simulate, respectively, the fully wetted-flow and and the selected cavitating flow regime - in this case $\sigma = (P_{out} - P_v)/(0.5\rho_L U^2) = 0.91$ - the value of the vapour pressure, P_v, has been properly modified. The water density has been set equal to $\rho_L = 998\ kg/m^3$ and maximum water-vapour density ratio has been limited to $\rho_L/\rho_V = 1000$ in order to ensure solver stability.

The simulations have been carried out using a steady state approach, on a hexa-structured mesh with about 58.7k nodes which proved to ensure mesh independent results [2]. The average value of y^+ on hydrofoil surface, evaluated considering the fully wetted flow regime, was equal to 28.

5. Results

The computational strategy, described in the former section, has been practically used considering the variability of the model coefficients within own sensitivity intervals. As first point of reference, the ranges described in [15] and the Pope equation for σ_ε have been taken into account (see Table 2 and eq. (21))

$$\sigma_\varepsilon = \frac{k^2}{C_\mu^{\frac{1}{2}} \left(C_{\varepsilon_2} - C_{\varepsilon_1}\right)} \tag{21}$$

Table 2. First Intervals of k-ε coefficients

Coefficient	L. Bound	U. Bound	Interval
C_μ	0.0500	0.150	0.100
σ_k	0.800	1.400	0.600
C_{ε_1}	1.000	1.500	0.500
C_{ε_2}	1.500	3.000	1.500
σ_ε	0.290	1.500	1.210

It is important to point out that in the preliminary study, carried out considering the intervals reported in Table 2, several simulations run to overflow. For this reason, the sensitivity intervals have been properly reduced as shown in Table 3.

Table 3. Final Intervals of k-ε coefficients

Coefficient	L. Bound	U. Bound	Interval
C_μ	0.0634	0.146	0.0826
σ_k	0.840	1.365	0.525
C_{ε_1}	1.064	1.464	0.400
C_{ε_2}	1.561	3.000	1.439
σ_ε	1.300	1.300	0.000

5.1. Hydrodynamics Characteristics

As for hydrodynamic coefficients, from Table 4 it is possible to note that the standard deviation, σ_{std}, is very low regardless of the flow regime. The predicted mean values, m, of lift and drag are, in this case, in line with the results obtained with the native values of the turbulence model coefficients [2]. However, the predictions of lift and drag for the cavitating flow regime seems to be less influenced by the variability (uncertainty) of the input parameters.

Table 4. Mean value and standard deviation of C_D and C_L

H. coefficient	Condition	m	σ_{std}
C_D	no cavitation	$2.266 \cdot 10^{-2}$	$5.781 \cdot 10^{-3}$
C_D	cavitation	$1.764 \cdot 10^{-2}$	$1.804 \cdot 10^{-6}$
C_L	no cavitation	$6.499 \cdot 10^{-1}$	$1.137 \cdot 10^{-2}$
C_L	cavitation	$6.533 \cdot 10^{-1}$	$3.512 \cdot 10^{-5}$

5.2. Sobol indices

Here, the Sobol indices are first presented separately for the two different flow regimes and next, some consideration regarding the comparison between the obtained results is given.

Non cavitating flow

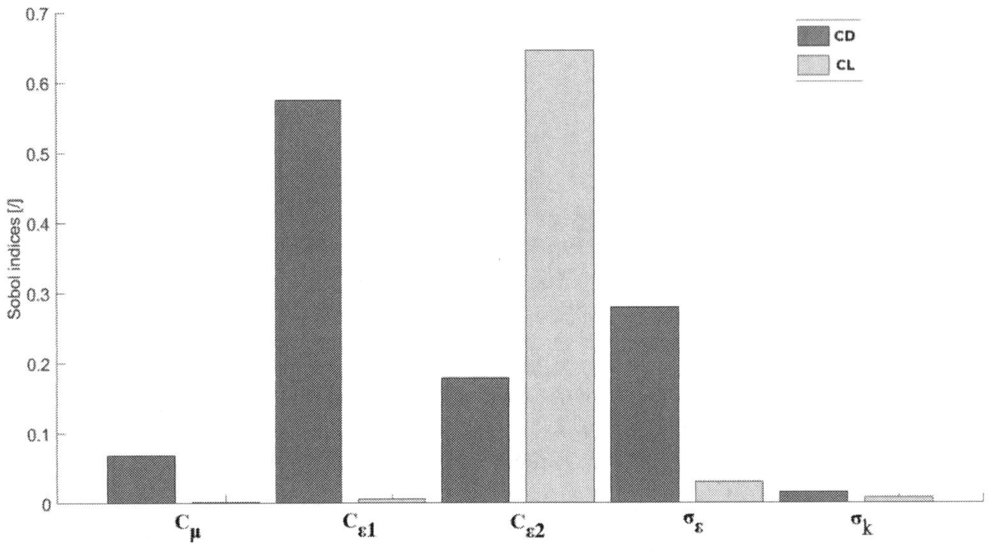

Figure 2. Comparison between the mean values of Sobol indices for C_D and C_L in non cavitating case, for a Polynomial Chaos of the second order

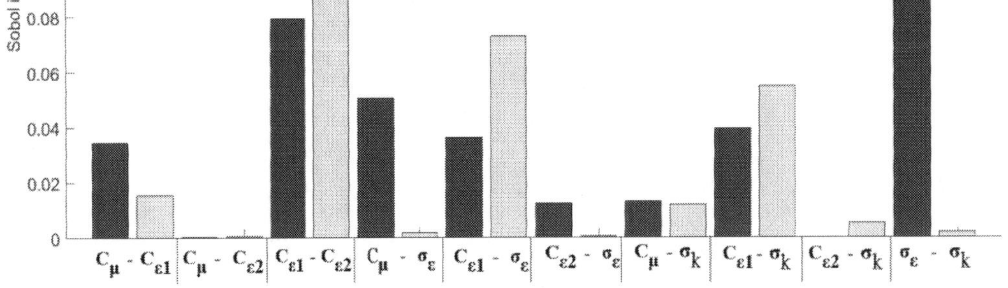

Figure 3. Comparison between interactions (pairs) of Sobol indices for C_D and C_L in non cavitating case, for a Polynomial Chaos of the second order

From Figure 2 it is possible to note that drag is most influenced by C_{ε_1}, while lift is most influenced by C_{ε_2}. Figure 3 shows that $\sigma_\varepsilon - \sigma_k$ is the pair that most influences the prediction of drag and $C_{\varepsilon_1} - C_{\varepsilon_2}$ is the pair that most influences the lift.

Cavitating flow

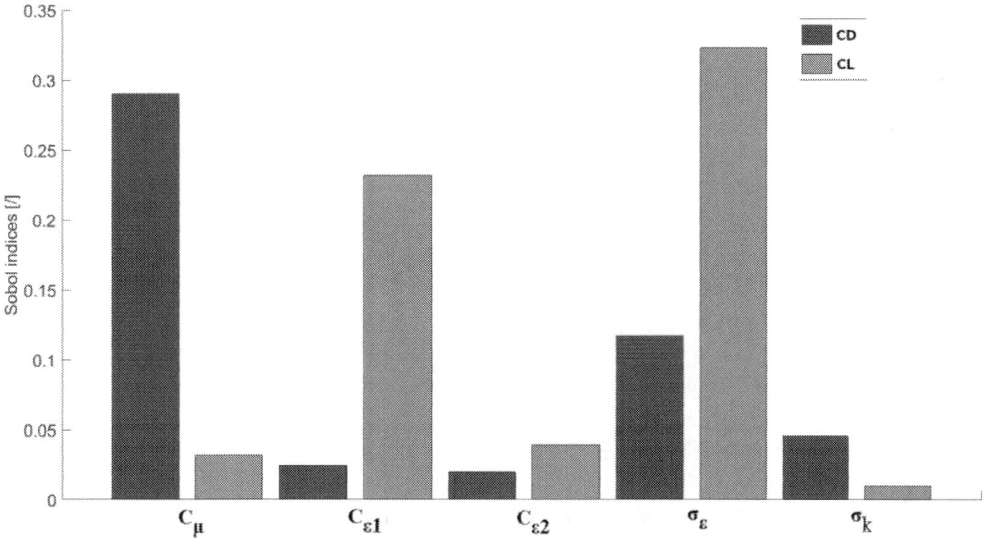

Figure 4. Comparison between the mean values of Sobol indices for C_D and C_L in cavitating case, for a Polynomial Chaos of the second order

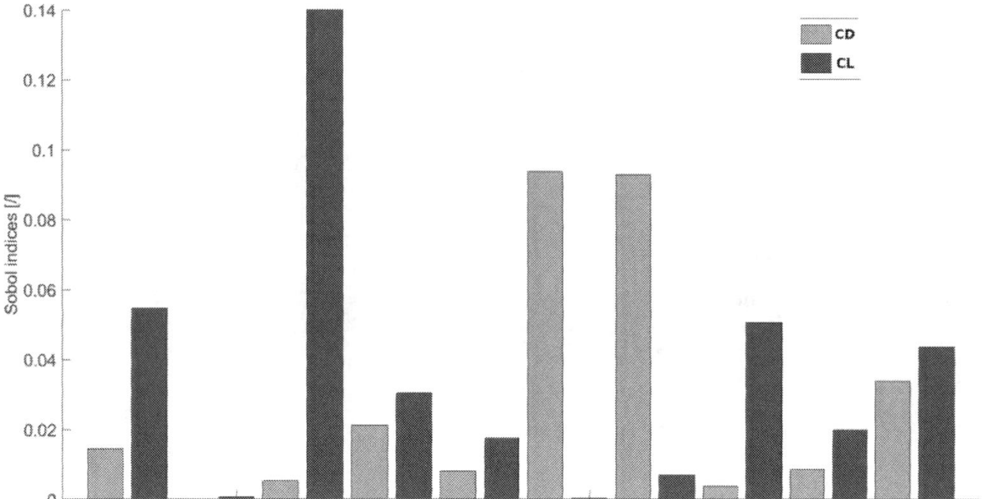

Figure 5. Comparison between interactions (pairs) of Sobol indices for C_D and C_L in cavitating case, for a Polynomial Chaos of the second order

From Figure 4 it is possible to note that drag is most influenced by C_μ, while lift is most influenced by C_{ε_1} and σ_ε. Figure 5 shows that $c_{\varepsilon_2} - \sigma_\varepsilon$ and $C_\mu - \sigma_k$ are the pair that most influence the prediction of drag and $C_{\varepsilon_1} - C_{\varepsilon_2}$ is the pair that most influences the lift.

Moreover considering Figures 2 - 4 it is possible to observe that for lift as well as drag the most important coefficients are different for the two different flow regimes. Eventually comparing Figures 3 and 5 the same behaviour can be observed for mutual interactions.

6. Conclusions

In this study, the Global Sensitivity Analysis has been applied to the case of the RANS simulations of the flow around the NACA66(MOD) hydrofoil operating at a given angle of attack. In particular, for non cavitating and cavitating flow regimes the possible effect of the uncertainties of the coefficient of the standard $k - \varepsilon$ turbulence model on the predicted lift and drag coefficients has been evaluated.

The uncertainty propagation and the Sobol Indices for the sensitivity analysis have been determined using the Polynomial Chaos Expansion, of the second order, based on a Point-Collocation Non-Intrusive Approach.

The solution strategy has been based on Dakota, which has been employed to evaluate the random values of the input coefficients as well as the values of the Sobol Indices, while Ansys-CFX has been used to perform the required simulations.

From the obtained results it seems that, for the considered flow conditions and with the considered ranges for uncertainties, the hydrodynamic characteristics are less sensitive to the variability of the turbulence model coefficients for the cavitating flow condition than for the non-cavitating flow regime. Moreover, considering the ranking based on Sobol indices, it seems that the most important input parameters are related to the given output (lift or drag) as well as to the considered flow regime. The same trend has been observed for parameter interactions.

7. References

[1] Shen Y T and Dimotakis P E 1989 The influence of surface cavitation on hydrodynamic forces *Proc. 22nd ATTC, St. Johns, Canada* pp 44–53
[2] Morgut M, Nobile E and Biluš I 2011 *International Journal of Multiphase Flow* **37** 620–626
[3] Zhang J, Yin J and Wang R 2020 *Mathematical Problems in Engineering* **2020**
[4] Eck V G, Donders W P, Sturdy J, Feinberg J, Delhaas T, Hellevik L and Hunberts W 2015 *International Journal for Numerical Methods in Biomedical Engineering* **32** 1–32
[5] Pathmanathan P, Cordeiro J and Gray R 2019 *Frontiers in Physiology* **10**
[6] Sudret B 2008 *Reliability Engineering and System Safety* **93** 964 – 979
[7] Hosder S and Walters R *Computattional Uncertainty in Military Vehicle Design*
[8] Kaintura A, Dhaene T and Spina D 2018 *Electronics* **7**
[9] Adams B M, Bohnhoff W J, Dalbey K R, Ebeida M S, Eddy J P, Eldred M S, Hooper R W, Hough P D, Hu K T, Jakeman J D, Khalil M, Maupin K A, Monschke J A, Ridgway E M, Rushdi A A, Seidl D T, Stephens J A, Swiler L P and Winokur J G 2020 *Dakota, A Multilevel Parallel Object-Oriented Framework for Design Optimization, Parameter Estimation, Uncertainty Quantification, and Sensitivity Analysis: Version 6.13 User's manual* Sandia National Laboratories
[10] URL `www.ansys.com`
[11] Katsuno E, Lidtke A, Düz B, Rijpkema D, Dantas J and Vaz G *Computers and Fluids* **230** 16
[12] Mourousias N, García-Gutiérrez A, Malim A, Fernández D, Marinus B and Runacres M 2022 *Aerospace Science and Technology* **133** 16
[13] Doma B 2015 *The classical Orthogonal Polynomial* (Singapore)
[14] Moukalled F, Mangani L and Darwish M 2015 *The Finite Volume Method in Computational Fluid Dynamics* vol 113
[15] Shirzadi M, Mirzaei P and Naghashzadegan M 2017 *Journal of Wind Engineering and Industrial Aerodynamics* **171** 366 – 379

IOP Conf. Series: Materials Science and Engineering 1288 (2023) 012038 doi:10.1088/1757-899X/1288/1/012038

Developing methods of decomposing the physical resistance components acting on a hydrofoil

Lev Chernyshev[1]*, Natalia Kabaliuk[1], Mark Jermy[1], Simon Corkery[2], Daniel Bernasconi[2]

Department of Mechanical Engineering, The University of Canterbury, Christchurch, NZ
Emirates Team New Zealand,

*E-mail: lev.chernyshev@pg.canterbury.ac.nz

Abstract. The America's Cup is a sailing competition contested by teams fielding bespoke yachts designed and built to the AC75 class regulations. These yachts make use of hydrofoils, which permit them to lift their hulls above the water surface. When in this "foiling" regime, the drag forces on the hull are drastically lower, permitting greater sailing speed. Predicting hydrodynamic forces produced from hydrofoils on race yachts remains challenging, due to a wide operating regime and complex hydrodynamical phenomena. In this study, existing wake survey methods of decomposing the total drag force of a hydrofoil into physical subcomponents were developed through examining their sensitivities and accuracy. Computational fluid dynamics (CFD) simulations of a rectangular NACA4412 hydrofoil were performed to generate data for the wake surveys. The hydrofoil was simulated at a submergence depth of one chord. Viscous flow was simulated with the $k-\omega$ SST turbulence model in steady state; and the volume-of-fluid (VOF) model was used to capture free surface effects. Chord Froude numbers of 0.5, 1, 2 and 4 were investigated, corresponding to Reynolds numbers of 2.57×10^5, 5.15×10^5, 1.03×10^6 and 2.06×10^6. Force decomposition was performed using two different wake survey methods – one each for the wave and viscous drag – as well as Trefftz plane analysis for the induced drag. Wave drag forces could be reliably isolated at all tested Froude numbers, while reliable predictions of the viscous drag were dependent on the Froude number and method of isolating the hydrofoil's viscous wake region. Accurately isolating the induced drag component proved challenging due to free surface interactions. Careful selection of the wake survey location was also necessary for accurate predictions of any of the drag components.

1. Introduction

Hydrofoil-equipped craft have become popular in many fields of yachting and boating. In yacht racing, their adoption became widespread after Emirates Team New Zealand introduced them as a design element for their 2013 America's Cup campaign. There are a myriad of other race classes that also incorporate hydrofoils (henceforth "foils") such as the IMOCA60, Waszp, and Olympic Nacra 17. In such cases, foil design is dictated by the class rules - a set of technical regulations that impose geometrical, structural and/or hydrodynamic restrictions on design. Aside from racing, foiling craft find uses in recreational and general-use boating and sailing: on superyachts, sport boats, multihulls, passenger ferries, and others. Here their design is less constrained and has more scope. The use of foils on watercraft adds substantial design complexity and challenge that requires significant effort from designers to overcome.

Foils used on race yachts face a large spectrum of hydrodynamic phenomena, since the boat speed varies greatly depending on wind speed, direction, and whether the boat is foiling or not. The operating regime encompasses lift coefficients from zero up to stall. Furthermore, the lift and drag forces are affected by the presence of the free surface. Typical Froude numbers for contemporary America's Cup yachts range from <1 to >10 depending on their speed. This range encompasses a shift in wave kinematics from subcritical to supercritical, as well as the possible onset of cavitation and natural ventilation at higher speeds[1, 2]. In parallel, the Reynolds numbers range from about 5×10^5 up to over 3×10^6 - hence appendage flows are often in the transitional regime. In addition, foil forces are highly sensitive to foil orientation, immersion, and interactions with other hull appendages [3, 4]. Such a wide operating regime makes it challenging for foil designers to carry out holistic hydrodynamic analyses of candidate foil designs. This kind of analysis is especially important for America's Cup teams due to the current generation of technical regulations, which restrict the degree of physical testing allowed for built foil prototypes [5].

Yacht designers have many modelling tools to characterise the hydrodynamics of foils and appendages at their disposal. These can be broadly split into two classes, based on the desired exactness of solution ("fidelity") sought by the designer. The first are methods based on computational fluid dynamics (CFD). These methods predominantly use Reynolds-Averaged Navier–Stokes (RANS) formulations for viscous flow, and have found extensive applications for America's Cup yacht design [4, 6, 7]. These days, RANS models such as the $k - \omega$ SST turbulence closure [8] are popular for their ability to accurately capture complex 3D flows at a wide range of Reynolds numbers and over complicated geometries for a feasible computational cost. However, in the fast-paced design cycles typical of the America's Cup, where potentially hundreds of candidate designs need to be evaluated in a wide range of operating conditions, RANS models are still too inefficient.

This is where the second class of hydrodynamic tools available to designers stands out: potential flow models. Unlike RANS, these cannot capture effects due to turbulence since they rely on the assumption of irrotational flow [9, 10]. There are many types of potential flow models. On the hydrodynamic side, some popular tools are lifting-line models, vortex lattice methods (VLM), and various forms of panel methods. Additionally, tools such as XFOIL [11] are popular for rapid evaluation of lift-drag polars and foil section design subject to a desired pressure distribution. Potential flow codes can be adapted to capture free surface effects and hence study the impacts of immersion and wave kinematics on appendage hydrodynamics [12]. On top of this, some of them can be run in real-time as part of a velocity prediction program (VPP) or dynamics simulator [3]. This makes them ideally suited for rapid design evaluation, at the cost of reduced fidelity compared to RANS codes.

The often fine performance margins between competing yachts necessitates an optimisation-heavy design cycle by sailing teams to extract maximum race prowess from their boats. In the case of appendages, often the focus is on optimising things such as their hydrodynamic efficiency, which is a function of lift and drag. Hence more accurate and detailed force predictions would be desirable, since they would constitute a more precise objective function with respect to optimisation goals for appendages. One way to achieve this is by developing a framework for drag decomposition. This is the breaking up of total drag forces into subcomponents with different physical meanings - for example, a hydrofoil has drag forces arising from interactions with the free surface (wave drag, spray drag); ones due to shed trailing vortices (induced drag) and ones from viscous effects such as profile and skin friction drag [13]. Obtaining such a breakdown would offer greater insight into the performance of a hydrodynamic appendage, since a designer could look to refine the design with a deeper understanding of how different drag mechanisms will be affected.

A well-developed framework for decomposing drag forces would be based on a thorough assessment of the sensitivities underlying the methods of drag decomposition, as well as their accuracy. The aim of this investigation is to assess these things for existing methods of extracting the wave, viscous and induced components of drag. These were implemented on RANS simulations of a fully submerged hydrofoil with free surface effects, as this would incorporate many of the hydrodynamic flows relevant to appendage design evaluation.

2. CFD methodology

2.1. Geometry, mesh and setup

A rectangular NACA4412 foil model with an aspect ratio of 6 and chord of $c = 0.3$ m was used for simulations. No fillet radii were used at the wing tips. A geometric angle of attack of 5° was prescribed for all runs, with the foil's trailing edge submerged one chord length beneath the undisturbed free surface. The global coordinate system had its origin on the undisturbed free surface and vertically above the foil's leading edge, with x pointing in the direction of motion, z directed vertically upwards against gravity, and y perpendicular to the first two. A rectangular prism was used as the computational domain, with a plane of symmetry on the x-z plane (i.e. only half of the foil was modelled), top, bottom and side walls. The foil walls were modelled as no-slip boundaries. Figure 1 shows the domain dimensions and inlet/outlet boundary condition specifications. An ANSYS polyhedral unstructured mesh was generated using ANSYS Fluent

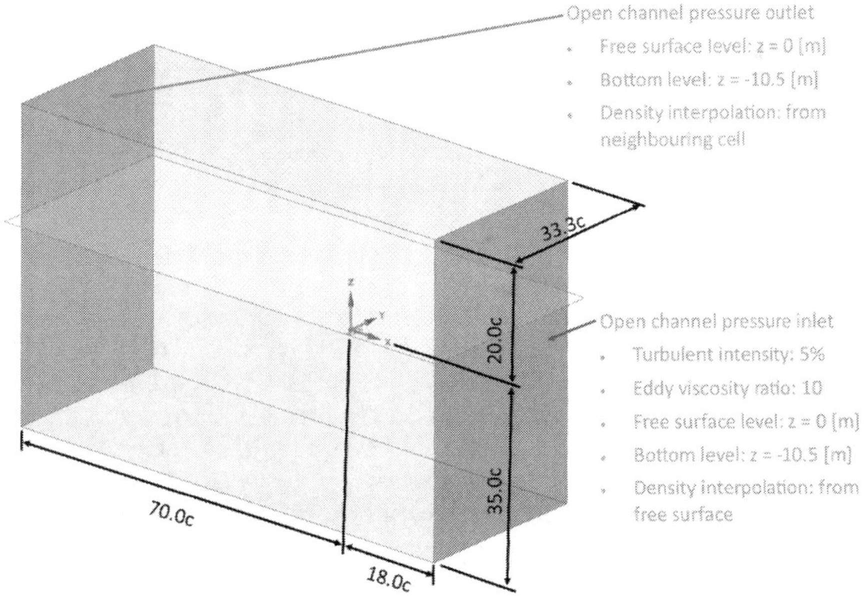

Figure 1. Computational domain used for simulations, with indicative dimensions and the inlet/outlet boundary conditions labelled. Detailed descriptions of the VOF settings can be found in [14].

Meshing 2020 R2. Bodies of influence in the foil's wake and around the nominal free surface level were used to resolve the wake and sharp surface interface. Fifteen inflation layers were used on the foil surface, with a first layer height of 1.5×10^{-5} m, to achieve a y_+ of maximum

1 on all walls. The final mesh had around 6 million cells before adaption - a number deemed suitable from prior mesh convergence checks.

Flows were analysed at chord-based Froude numbers F_c of 0.5, 1, 2 and 4, which corresponded to inlet velocities of 0.858 ms^{-1}, 1.716 ms^{-1}, 3.431 ms^{-1} and 6.862 ms^{-1} respectively. The corresponding Reynolds numbers were 2.57×10^5, 5.15×10^5, 1.03×10^6 and 2.06×10^6. The density of water was taken as $\rho = 1000$ kgm^{-3} and kinematic viscosity $\nu = 1.00 \times 10^{-6}$ m^2s^{-1} at 18°C for freshwater. To ensure a margin to cavitation, pressure distributions were computed at the same Reynolds numbers in XFOIL [11] and checked against the water's vapour pressure. At the highest Froude number, a cavitation number of 2.67 was maintained.

The $k-\omega$ SST turbulence model [8] was used to simulate viscous effects, and ANSYS Fluent's open channel volume-of-fluid (VOF) [14] model was used to capture the free surface. The sharp, implicit formulation was chosen, along with implicit body force. Air was the primary phase and water the secondary. Air properties were taken as $\rho_{\text{air}} = 1.225$ kgm^{-3} and $\nu_{\text{air}} = 1.46 \times 10^{-5}$ m^2s^{-1}.

2.2. *Solution scheme*

ANSYS Fluent 2020 R2 was used for simulations. The couple pressure-velocity solver was used, with second order upwind schemes for turbulence and momentum, the compressive scheme for VOF, body-force-weighted scheme for pressure and least squares cell based scheme for gradients. Simulations were initialised using "flat" open channel initialisation, where the free surface was set uniformly at $z = 0$ and other variables were taken from the inlet. Afterwards, the simulation was run in steady state with pseudo-transient enabled for 500 iterations. Automatic time-stepping and a user-specified length scale equal to the foil chord were used.

Next, adaptive mesh refinement was turned on subject to a refinement/coarsening criterion based on the globally-scaled gradient of volume fraction which captures the free surface location. With these settings, the solution was evolved for about 2000 iterations with mesh adaption occurring every 100 iterations.

3. Drag force decomposition methodology

The drag force, or resistance, experienced by foils moving through water has numerous physical sources. The most common way of extracting the drag force from CFD simulations is by integrating the normal and tangential stresses on the foil projected in the direction of motion [15]. This way, components usually termed "pressure drag" and "skin friction drag" are obtained. Figure 2 depicts how further decompositions may be performed, along with the methods of calculating each component. The following sections explain the methods pertaining to the wave, induced and viscous drag components and how each of them was implemented for this study. Due to the absence of any surface-piercing geometry in the present study, spray drag was not considered.

A MATLAB script was written to process CFD data and calculate the drag components per the methods explained below. Figure 3 shows a flowchart describing the code architecture. Details of each process are described in the following sections. Common to all methods was the Voronoi tessellation of the wake plane, which was used since cell face areas on the wake plane could not be directly obtained from Fluent. This was performed with the `VoronoiLimit` function in MATLAB, with the boundary being defined by the convex hull surrounding all of the nodes in the wake plane.

3.1. *Wave drag*

A foil moving under water near the free surface generates waves which propagate away from it through the action of a gravitational field [16]. A certain energy expenditure is required to generate such waves, which results in wave resistance component of the total drag. Isolating this

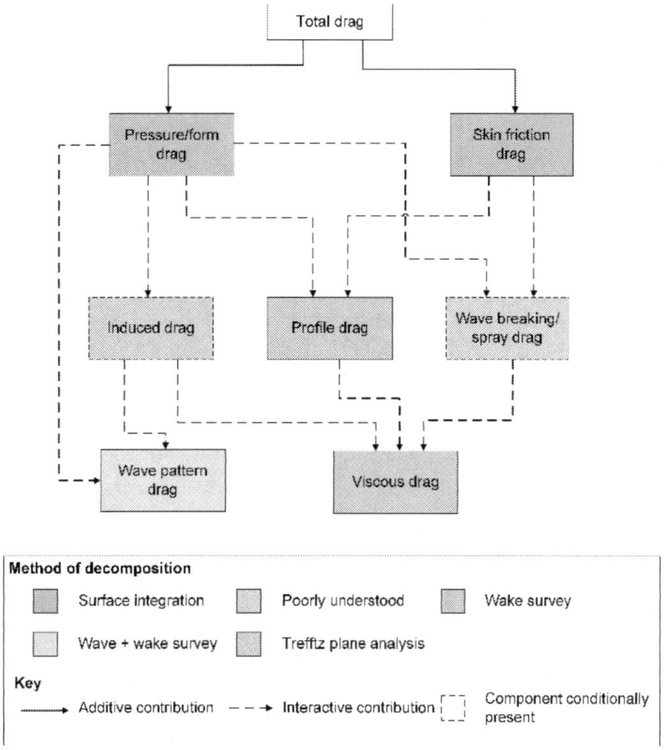

Figure 2. Hierarchical decomposition of drag forces on sailing yacht foils and appendages.

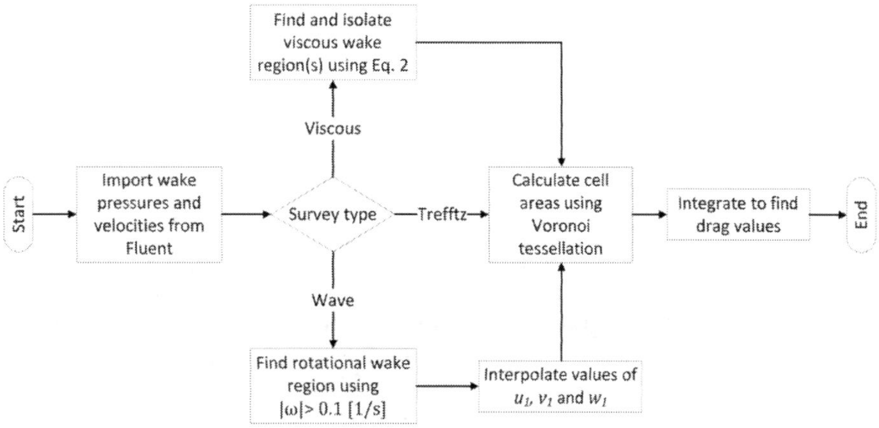

Figure 3. Flowchart showing the post-processing implemented in MATLAB to calculate various drag forces from wake survey data extracted from CFD.

part of drag is usually done by way of wave pattern analysis, descriptions of which may be found in [17]. However, these methods were tailored more for experimental measurement when they were first derived, and are somewhat unwieldy to implement. Instead, a more straightforward method of calculating the wave drag is described in [18]. This one requires only a single planar wake survey at a wake plane A', as in Figure 4. The wave drag can then be expressed as

$$D_w = \frac{1}{2}\iint\limits_{A'} \rho(v_1^2 + w_1^2 - (U-u_1)^2)dA + \frac{1}{2}\rho g \int_{-y_{\max}}^{y_{\max}} \zeta^2(x_{\text{survey}}, y)dy, \quad (1)$$

where D_w is the total wave drag force; ρ is the water density (here included inside the integral because the density in cells close to the surface deviates from the bulk value in the VOF formulation); u_1, v_1 and w_1 are the x, y and z components of the irrotational velocity field; U is the freestream velocity; g is the acceleration due to gravity; and $\zeta(x_{\text{survey}}, y)$ is the wave height distribution at the survey plane. Per the method, the velocity field in the rotational wake region (with non-zero vorticity) must be replaced with an irrotational one. This may be done by interpolating into the rotational flow region [19] using the surrounding irrotational velocity field. To do this, the rotational wake was first identified using the criterion on the vorticity magnitude $|\omega| > 0.1\text{s}^{-1}$ (small value chosen to filter out cells with non-zero vorticity due to numerical diffusion). Then, the wake velocity field was interpolated in that region using MATLAB's `scatteredInterpolant` function. The integral in Eq. 1 could then be evaluated using the computed Voronoi cell areas over the whole wake plane.

3.2. Viscous drag

Viscous drag is the component of resistance that arises from a momentum deficit between the freestream and wake behind a foil [20]. The momentum deficit in the wake may be integrated over a wake plane that is perpendicular to the foil's direction of motion to yield a value of viscous resistance - an idea first put forth in [21]. Many different formulations have since been developed and used in naval and yachting contexts [19, 22–24]. A summary of formulas is given in [25]. Integration must be performed only over part of the wake where viscous effects are appreciable. A small error in the chosen area can lead to large errors in the computed viscous drag [19]. Referring to the rectangular control volume enclosing a foil in motion beneath a free surface in Figure 4, we are hence interested in a criterion that can isolate the viscous wake region labelled "G". Usually, criteria based on vorticity magnitude or total pressure deficit have been employed [18, 24, 26]. However, for this investigation, a criterion based on the integrand of Eq. 3 was used instead (which in essence is a total pressure deficit too), where the viscous wake region was defined by the inequality

$$I > KI_{\max} \quad (2)$$

where I is the integrand of Eq. 3 evaluated in each cell on the wake plane A' and $I_{\max}$ is the maximum value of I. K is a percentage factor used to isolate the viscous wake region. For the present study, K was taken as 5% except for the sensitivity analysis presented in Figure 7. This value was deemed to adequately filter out small non-zero values of I in the outer computational domain arising from numerical diffusion at all Froude numbers.

Prior comparisons between formulations [25] showed no appreciable difference in the computed values. Therefore, Eq. 7 from that reference was chosen since it was the most precise formulation. The equation is

$$D_v = \iint\limits_{w} p_\infty - p + \frac{\rho}{2}((U-u_1)^2 - v_1^2 - w_1^2)dA, \quad (3)$$

where w denotes integration only over the foil's viscous wake region; p_∞ is the freestream static pressure; p is the static pressure at the wake survey plane A'; U is the freestream velocity; and u_1, v_1 and w_1 are the x-, y- and z-velocities at the wake survey plane, defined as explained previously.

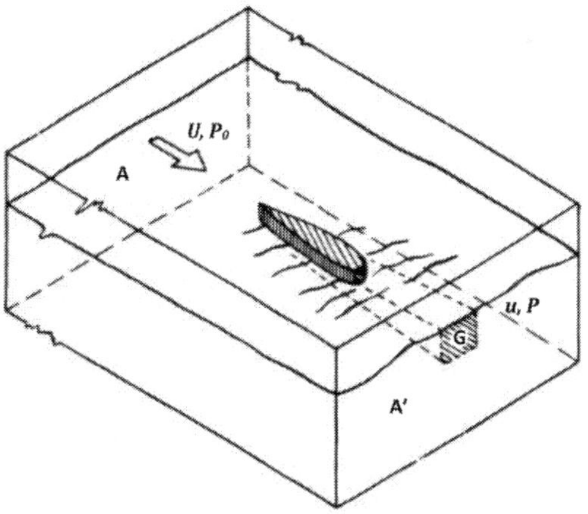

Figure 4. Illustration of a control volume used to compute the momentum deficit to quantify viscous drag of a foil, taken and modified from [27]. Plane A is in the freestream and plane A' is in the wake. Symbols U and P_0 denote freestream velocity and total pressure; likewise u and P are corresponding values in the wake. Region G is the viscous wake.

3.3. Induced drag

Induced drag is the component of drag associated with the generation of lift and the trailing vortex system left by a finite-aspect-ratio foil. It can be isolated by performing a Trefftz plane analysis [28], whereby the induced cross-flow velocities are integrated over a wake plane (in a similar manner to the wake surveys described previously). The corresponding equation is

$$D_i = \frac{1}{2} \iint_{A'} \rho(v^2 + w^2) dA, \tag{4}$$

where D_i is the induced drag and v, w are the y, z components of velocity at the wake plane A'.

4. Results

Drag force decompositions were computed using the methods outlined previously, and are presented in §4.2 as functions of both the Froude number and the wake plane offset (WPO) – that is, the distance between the foil's leading edge and the downstream survey plane. All drag forces are in coefficient form, computed by dividing a given drag force component by $q_\infty S$ where q_∞ is the freestream dynamic pressure and S is the foil's planform area. Some comparisons with reference data were made in §4.1 alongside an examination of the Froude decomposition to check the accuracy of our computed drag forces.

4.1. Assessment of accuracy

The accuracy of our decompositions was examined by comparing our computed drag forces with available theoretical values. Theories of wave and induced drag described [13] were used for this purpose. The induced wave drag coefficient is given as

$$C_{Di,\text{theory}} = \frac{C_L^2}{\pi\lambda}\left(1 - \frac{1}{8}\ln\left(\frac{1.05\lambda^2k^2 + 4}{0.000625\lambda^2k^2 + 4}\right)\right), \tag{5}$$

where C_L is the foil's lift coefficient; λ is the foil's aspect ratio; and $k = \frac{c}{h}$ is the ratio of the foil's chord length to the submergence depth (which in our simulation case equals unity). For the wave drag coefficient, the theoretical expression is

$$C_{Dw,\text{theory}} = \frac{C_L^2}{2F_c^2} \exp\left(\frac{-2}{F_h^2}\right), \tag{6}$$

where F_c is the chord-based Froude number and F_h is the submergence-depth-based Froude number. Since in the present study $h = c$, both of these variables are equivalent.

Figure 5a shows the computed wave and induced drag at $F_c = 1$ normalised by the corresponding theoretical values derived using Eq. 5 and 6. There was no clear dependence on the WPO of either the wave or induced drag coefficient. At all WPOs except $x/c = 2$, the calculated wave drag coefficient exceeded the theoretical value – at most by around 36% at $x/c = 12$. The calculated induced drag coefficient showcased greater disagreement with its theoretical value, exceeding it by about 100% at $x/c = 12$. Good agreement for both sets of values was seen at $x/c = 8$, though the absence of a clear trend in neighbouring points meant this could not be reliably verified as a suitable WPO for carrying out the drag decompositions in general.

The discrepancies could be attributed to the simplifications underlying the theoretical predictions. In deriving Eq. 6 the foil was modelled as a single infinite vortex filament translating beneath the free surface, whence finite-span effects were neglected. Furthermore, the model did not account for the foil's thickness either. Calculating the induced drag coefficient using Trefftz plane analysis, while a proven approach for flows in an infinite fluid, has not been demonstrated to yield reliable, accurate predictions of the induced drag for flow with a free surface. Nevertheless, it has been assumed for this investigation that the method is suitable for use with our flow conditions.

Another appraisal of accuracy can be made by considering the Froude decomposition of the total drag – i.e. that the total drag force is the sum of viscous and wave drags. In Figure 5b, the sum of computed viscous and wave drag coefficients is plotted normalised by the coefficient of total drag calculated using surface integration, as a function of WPO and the Froude number. At $F_c = 4$, very good agreement was seen through the whole range of WPOs. As the Froude number decreased, the sum of wave and viscous drag coefficients deviated more from $C_{D,\text{tot}}$. The greatest discrepancy was seen at $F_c = 0.5$, where the drag component sum was, depending on the WPO, almost double or half the actual total drag coefficient. Since this Froude number corresponded to the slowest flow condition, numerical diffusion likely impacted the accuracy of flow variables at the wake surveys more than at higher Froude numbers, which may have contributed to those discrepancies. The fact that the drag components overshot the total drag force more as the WPO increased supports this hypothesis. With reference to Figure 6b, it also appears that the over-prediction of the viscous drag component is the primary contributor to this overshoot, since the corresponding wave drag coefficients remained relatively level as the WPO increased in Figure 6a.

4.2. Effect of wake survey plane offset and wake criterion on drag component predictions

Figure 6 shows the computed drag component coefficients as a function of the WPO and Froude number. The wave drag coefficient in Figure 6a showed little to no dependence on the WPO, with the curve of each Froude number remaining relatively flat as the WPO increased. Predictions of the wave drag coefficient as the WPO decreased through $x/c = 2$ markedly increased, suggesting a strong influence from the more powerful wake in that region compared to stations further downstream. In this area as well, some of the wave drag coefficients overlapped between different Froude numbers, which demonstrated that predictions made using wake surveys in those areas could be unreliable at isolating Froude number effects. Considering a wake survey where no such overlap occurred, such as at $x/c = 4$, it was seen that the wave drag coefficient exhibited

IOP Conf. Series: Materials Science and Engineering 1288 (2023) 012038 doi:10.1088/1757-899X/1288/1/012038

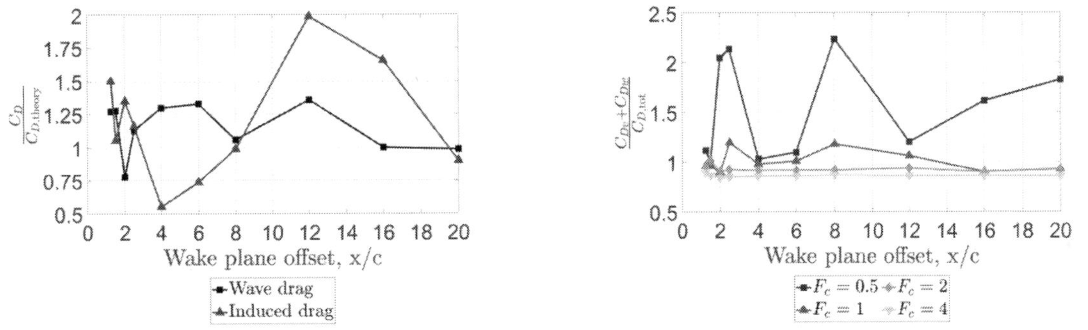

(a) Comparing computed values of the induced and wave drag coefficients with theoretical ones, at $F_c = 1$.

(b) Assessing the accuracy of viscous and wave drag coefficient predictions by comparing their sum with the total obtained from CFD.

Figure 5. Examination of the accuracy of methods for decomposing the foil's total drag into wave, viscous and induced drag components.

the classic "hump" around a Froude number of unity. A higher resolution of Froude numbers in this area would be desired to resolve this pattern and the hump better.

Looking to the coefficient of viscous drag in Figure 6b, it was observed that as the Froude number decreased, the dependence of C_{Dv} on the WPO became stronger. This was revealed by C_{Dv} significantly increasing as the WPO increased when $F_c = 0.5$; while a weaker increase was observed at $F_c = 1$. At higher Froude numbers there was no apparent dependence of the viscous drag coefficient on the WPO, with the predictions remaining relatively constant through the whole range of WPOs examined.

The trends relating to the induced drag coefficient in Figure 6c were less clear. The predictions seemed to be centred around approximately the same mean value of 0.018, to which they converged at a WPO of $x/c = 8$ and $x/c = 20$ but otherwise deviated from significantly, by up to about 100% in the case of $F_c = 1$. The induced drag depends on both the lift and the interaction of the trailing vortices with the free surface, the latter of which was suspected to be the main reason for the highly variable predictions seen here. While the observations from Figure 5a suggest that the Trefftz plane analysis was able to capture the theoretically-predicted C_{Di} at some of the WPOs, it was not able to do so reliably. Because of this, it was difficult to judge the accuracy of this method in the simulated flow conditions. Perhaps using a more sophisticated method, such as Maskell's integral [29] with appropriate boundary conditions for the Laplace equation, would assist in dealing with the problem of interaction between the wake and free surface.

The effect of changing the factor K on the predicted coefficients of viscous drag is shown in Figure 7, where the sum of viscous and wave drag coefficients normalised by the total coefficient extracted from CFD has been plotted as a function of WPO and K for two different Froude numbers. With the previously observed reliability of the wave drag predictions, this plot permits us to judge the accuracy of the C_{Dv} predictions as a function of K. The chosen Froude numbers were selected to showcase the stark difference in the resulting values of C_{Dv}, which were seen to depend on both F_c and K. In the case of $F_c = 1$, $K = 5\%$ seemed to yield more accurate values of C_{Dv} than the smaller K values, deviating by at most 20% from the total drag coefficient across all WPOs. It was hypothesized that this was because effects of numerical diffusion were more pronounced at the lower Froude numbers, so that a greater K would be needed to adequately isolate the foil's viscous wake region for computing the viscous drag with Eq. 3.

In contrast, at $F_c = 4$ in Figure 7b, the opposite effect was observed. The lowest value of K gave the most accurate prediction of the viscous drag coefficient across the whole range of WPOs. As the value of K increased, the predictions became less accurate, with a maximum discrepancy of 16% seen when $K = 5\%$. In contrast to the case of $F_c = 1$, this was likely because the effects of numerical diffusion were weaker, hence permitting the use of a smaller K value to isolate the viscous wake. Since an inversion of dependence on K was seen between $F_c = 1$ and $F_c = 4$, it would be desirable to examine a finer resolution of Froude numbers in this region to accurately determine the location of this inversion.

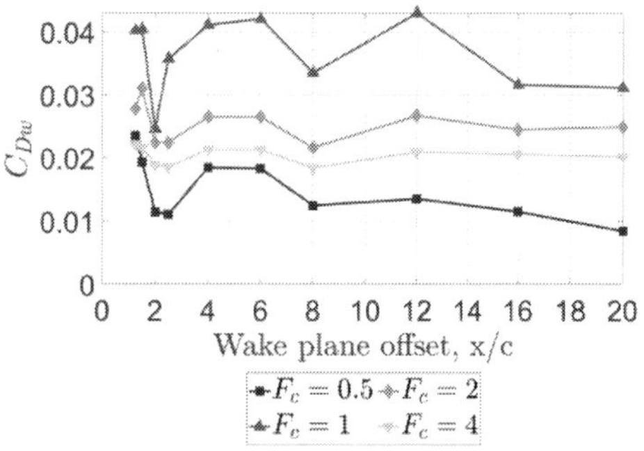

(a) Wave drag

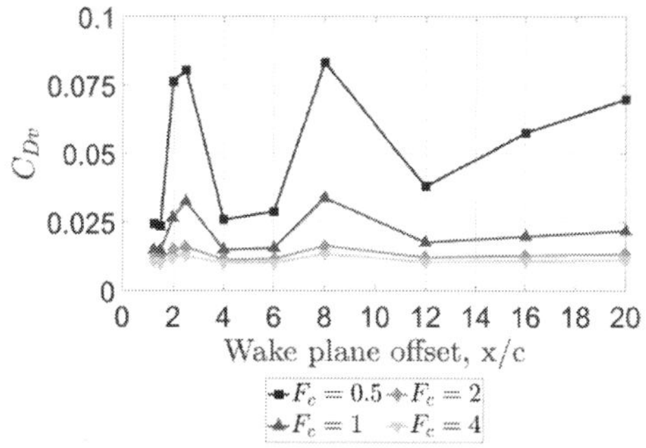

(b) Viscous drag

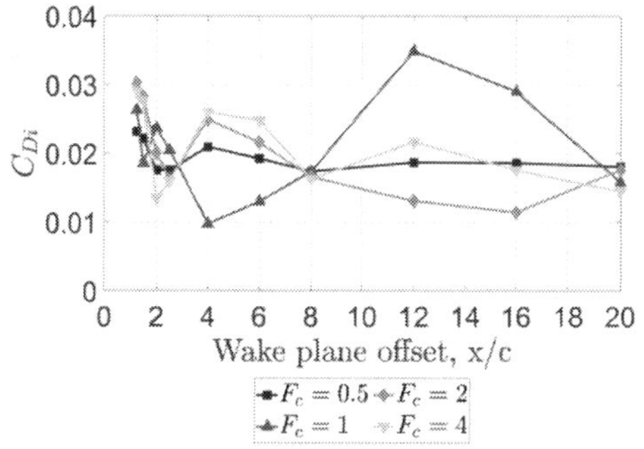

(c) Induced drag

Figure 6. Dependence of the decomposed coefficients of drag on the downstream location of the wake survey plane, as a function of the Froude number.

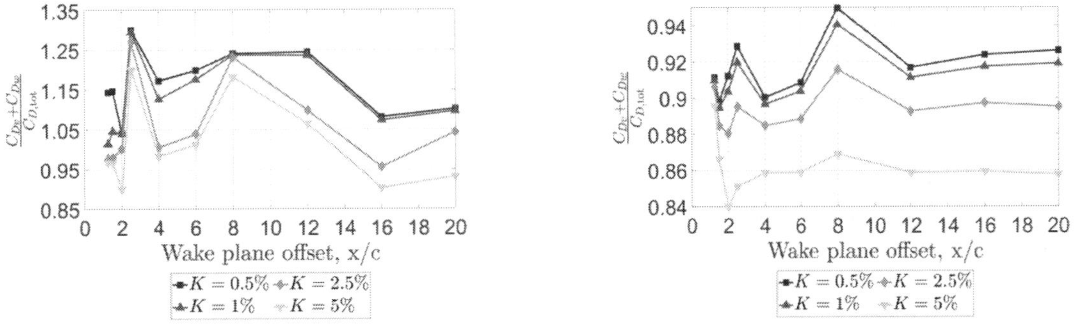

(a) $F_c = 1$ (b) $F_c = 4$

Figure 7. Effect of the factor K on the viscous drag coefficient, shown by normalising the Froude decomposition of the total drag coefficient with the total obtained directly from CFD.

5. Conclusion

The drag force acting on a rectangular NACA4412 hydrofoil moving at a constant Froude number was decomposed using various methods based on wake surveys. The wave, viscous and induced drag components were extracted. Sensitivity of the methods to the offset of the wake survey plane was examined at all simulated Froude numbers too. It was observed that there was weak dependence of the predicted wave drag on this offset, while it was significantly stronger in the case of viscous drag, where the error increased as the survey plane was placed further downstream. No concrete trends were observed in the case of induced drag, where it was concluded that free surface interactions hampered a reliable prediction of the induced drag; and that a more sophisticated method that is capable of taking these interactions into account might be preferable for isolating this component.

The sum of viscous and wave drags was in good agreement with the total extracted directly from simulations across all examined wake plane offsets, but only at Froude numbers greater than unity. This evidence supported the Froude decomposition of hydrodynamic drag. Decent agreement between the computed wave drag and corresponding theoretical values was seen, while with induced drag this agreement was not as good and varied greatly depending on the location of the wake survey plane.

The influence of a wake criterion cut-off factor, used to identify the foil's viscous wake region, on predictions of viscous drag was examined. At Froude numbers around one and less, a higher cut-off was needed to adequately filter out numerical diffusion on the wake survey plane and adequately isolate the wake region. The opposite was true at a Froude number of four, where a smaller cut-off factor yielded more accurate computations of the viscous drag.

In evolving this study, it would be desirable to examine a larger range of Froude numbers with finer resolution to ascertain some of the observed trends. The next natural step would also be to examine surface piercing geometries while taking care to account for additional effects brought by this addition, such as spray drag and ventilation.

Acknowledgements

We gratefully acknowledge Emirates Team New Zealand for their discussions and insights into the hydrodynamics of hydrofoils.

References

[1] Acosta A 1973 *Annual Rev. Fluid Mechanics* **5** 161–184 ISSN 0066-4189
[2] Faltinsen O M 2005 *Hydrodynamics of high-speed marine vehicles* (Cambridge university press) ISBN 0521845688
[3] Oliver C and Gauvain E 2022 Practical performance prediction of foil-configured monohull yachts *SNAME 24th Chesapeake Sailing Yacht Symp.* (OnePetro)
[4] Turnock S R and Holroyd N J 2001 Appendage design for the america's cup using cfd *European Congr. on Comput. Methods in Appl. Sci. and Eng.*
[5] Royal New Zealand Yacht Squadron and Royal Yacht Squadron 2022 AC technical regulations
[6] Graf K, Boehm C and Renzsch H 2009 Cfd-and vpp-challenges in the design of the new ac90 americas cup yacht *SNAME 19th Chesapeake Sailing Yacht Symp.* (OnePetro)
[7] Jones P and Korpus R 2001 International america's cup class yacht design using viscous flow cfd *SNAME Chesapeake Sailing Yacht Symp.*
[8] Menter F 1993 Zonal two equation k-w turbulence models for aerodynamic flows *23rd Fluid Dynamics, Plasmadynamics, and Lasers Conf.*
[9] Hess J L 1990 *Annual Rev. Fluid Mechanics* **22** 255–274 ISSN 0066-4189
[10] Smith A 1990 *Appl. Comput. Aerodynamics* 3–17
[11] Drela M 1989 Xfoil: An analysis and design system for low reynolds number airfoils *Low Reynolds Number Aerodynamics* ed Mueller T J (Berlin, Heidelberg: Springer Berlin Heidelberg) pp 1–12 ISBN 978-3-642-84010-4
[12] Rosen B S and Laiosa J P 1997 Splash nonlinear and unsteady free-surface analysis code for grand prix yacht design *SNAME 13th Chesapeake Sailing Yacht Symp.* (OnePetro)
[13] Vladimirov A 1955 Approximate hydrodynamic design of a finite span hydrofoil Report Central Aero-Hydrodynamical Institute
[14] ANSYS Inc 2020 Ansys fluent theory guide Report ANSYS Inc.
[15] Hoerner S 1965 *Fluid-dynamic Drag: Practical Information on Aerodynamic Drag and Hydrodynamic Resistance* (Hoerner Fluid Dynamics)
[16] Lamb H 1924 *Hydrodynamics* (University Press)
[17] Eggers K 1967 *Sname Trans.* **75** 112–157
[18] Wehausen J V 1973 *Advances Appl. Mechanics* **13** 93–245 ISSN 0065-2156
[19] Tzou K T and Landweber L 1968 *J. Ship Res.* **12** 105–115 ISSN 0022-4502
[20] Anderson J 2017 *Fundamentals of Aerodynamics* (McGraw Hill) ISBN 0077147669
[21] Betz A 1925 *Zeitschrift Für Flugtechnik Und Motorluftschiffahrt*
[22] Werner S, Larsson L, Regnström B and Chalmers University of Technology 2007 *Ship Technol. Res.* **54** 28–42 ISSN 0937-7255
[23] Baba E 1969 *J. Soc. Naval Architects Japan* 9–22 ISSN 0514-8499
[24] Landweber L and Wu J 1963 *J. Ship Res.* **7** 1–6
[25] Shin-hyoung K and Beom-soo H 1982 *Bulletin Soc. Naval Architects Korea* **19** 19–25 ISSN 1225-1178
[26] Wu J 1962 *J. Ship Res.* **6**
[27] Tulin M P 1951 The separation of viscous drag and wave drag by means of the wake survey Report David Taylor Model Basin
[28] Kroo I 2001 *Annual Rev. Fluid Mechanics* **33** 587 ISSN 0066-4189
[29] Kusunose K and Crowder J 2001 Physical properties of maskell's induced drag integral *39th Aerospace Sci. Meeting and Exhibit* p 421

IOP Conf. Series: Materials Science and Engineering 1288 (2023) 012039 doi:10.1088/1757-899X/1288/1/012039

Drag reduction and power optimization due to an innovative, toroidal hull form of an AUV

Lakshmi Madhavan Miller[1,6], **Taylor Njaka**[2,3], **Stefano Brizzolara**[4], **Thomas Curtin**[5]

[1] Graduate Research Assistant, Aerospace and Ocean Engineering, Virginia Polytechnic Institute and State University, Blacksburg, VA, USA
[2] Research Associate, Aerospace and Ocean Engineering, Virginia Polytechnic Institute and State University, Blacksburg, VA, USA
[4] Professor of Ship Hydromechanics, Aerospace and Ocean Engineering, Virginia Polytechnic Institute and State University, Blacksburg, VA, USA
[5] Senior Scientist, Applied Physics Laboratory, University of Washington, WA, USA

E-mail: lakshmimiller@vt.edu

Abstract. The use of unconventional design has become a norm for mission-specific underwater autonomous craft. Geometrical characteristics and drag optimized shapes of single-hulled underwater vehicles are known since 1970s and a few significant improvements have been possible since then. These are mostly related to the efficiency of a wake adapted propeller, positioned at the stern in wake of the hull. In a novel effort, when the propeller is integrated into the craft at center-line, and the hull in its entirety is designed to act as the duct, the drag of the hull is further reduced. This craft uses a toroidal hull form that distributes payload axi-symmetrically around an internal ducted thruster. The craft is designed to operate at an economic cruise speed of 2kn. The power consumption is optimized to achieve an endurance of over 500 miles through this implementation of a ducted thruster. The hydrodynamic characteristics of the new toroidal hull are compared to those of a reference design which is slender and conventionally shaped. This conventional design is intended for the same volume and dimensions of payload. A reduction of 13% in drag from the first to final design iteration is achieved. This reduction in drag is due to a higheer effective length to thickness ratio of the ducted AUV, quantified in terms of a new parameter defined as I_h. The reduction of the wetted surface area also contributes to the drag optimization. The article shows that the shape of the leading edge of the toroidal hull becomes important to achieve low drag. The process of design and optimization using CFD, with the most appropriate turbulence models is detailed. The beneficial effect of the acceleration induced in the flow at the leading edge and trailing edge of the hull by the integrated thruster is studied and explained with the help of CFD results.

1. Introduction

Drag-optimized, unconventional AUVs (Autonomous underwater vehicles) have applications in areas such as underwater surveillance and environmental monitoring. For example, AUVs can be used to detect oil spills or monitor the health of coral reefs. The improved maneuverability

[6] Present address: Research Assistant Professor, Virginia Tech- NSI, Blacksburg, VA, USA
[3] Present address: Senior Mechanical Engineer, Applied Materials, Santa Clara, CA, USA

and energy efficiency of these AUVs make them ideal for conducting long-term, high-resolution surveys of underwater environments.

Mission requirements for AUVs include two important parameters: endurance and payload capacity. These two parameters are affected by the propulsion and control systems apart from design characteristics of the hull form. Such requirements to create a long-range AUV with a specific payload capacity has led to the design of the craft 'Seppia' which is detailed in this article. The Seppia maintains the same payload and battery capacity as that of the Pencil carrier. The endurance of this AUV is achieved through reduction in drag by using a tunnel thruster at the center of the craft as its sole propulsion system. This craft is required to be stable in a linear forward course with very less maneuverability. There are multiple ways to improve the endurance of an AUV of which some are surveyed below:

1.1. Energy reduction using control system design

The work of [1] targets control system design for reduction in energy consumption. The power consumption of servo motors used in control surfaces is studied in [1] resulting in an energy efficient controller for the 690 craft. The study by [2] uses a control uses an adaptive dynamic control allocation strategy. This control strategy considers the moment caused by the asymmetry of the hull in vertical motion. Moreover, a mathematical model of energy consumption of motion actuators is established for this long-range AUV.

1.2. Energy reduction using buoyancy

The study by [3] performs a study of the effects of buoyancy on the range (and hence endurance) on a flight-class AUV that is propelled by an open propeller. In this study, a buoyancy compensated range model is used to asses various concepts. This study [3] shows that a design with wings is required to overcome the positive buoyancy of the craft to remain energy efficient with a pre-determined power consumption. The wings generate an extra thrust when the hull is at an inflow angle due to buoyancy. The study of [3] deals with a faster AUV or a glider whose mission requirement is a higher speed than that of the current study.

1.3. Energy reduction by drag optimization: using delay in onset of transition from laminar to turbulent flow

The turbulence of the flow around an axisymmetric body can be used to reduce drag. The delay in transition of laminar to turbulent flow results in lower drag for a body of revolution. The delay in onset of transition on surfaces can be established using micro-structures, hydrophobic coatings and micro-bubbles. This study has not been attempted on underwater craft so far. Raised surface structures such as those used in the work of [4] also use the principle of delay in the onset of transition. the work of [4] uses biomimetic structures on a hydrodynamic skin to resemble shark denticles. This research study surveys the use of multiple such structures to result in the conclusion that the synergistic effect of multiple drag reduction technologies will break through the upper limit of drag reduction rate using just one method. In the work of [5], air jets or microbubbles are used to reduce the frictional component of the drag experienced by an underwater craft named Afterbody1. The study of [5] uses the k-ω-SST turbulence model to determine the multi-phase flow and resultant drag effects due to air jets. These air jets, comprised of micro-bubbles lowers the viscosity of the fluid along the length of the craft, thereby delaying onset of turbulence and hence reducing the drag. But these types of improvements are cost-intensive in terms of material, maintenence and implementation.

1.4. Energy reduction by drag optimization: using shape of the hull

Another way to increase the endurance of AUVs is to design a drag-optimized hull form. The work by [6] derives a relationship between friction drag, form drag and the slenderness ratio of the craft. The direct effect of hull profile on reduction in drag is evident from this research work, for axisymmetric bodies at zero angle of incidence. Traditional shapes of AUVs were systematically studied by [7]. This study determines detailed theoretical expressions for different axisymmetric hull profiles. It also defines a method of prediction of friction drag using a mathematically defined hull profile, a value of free stream turbulence level, and the point of onset of transition. In a subsequent study, [8] through theoretical calculation of drag, in agreement with [6], shows that a slenderness ratio of 5.5 is optimal for low drag. Slenderness (fineness) ratio is defined as the ratio of length of the craft to its maximum diameter. In addition, in [8] it is found that opportunely rounded noses and tapered tails have a beneficial effect on drag. This has led to many submerged craft having a slender, drop-shaped hull. Such designs are used commercially as recently as Virginia Tech's 690 craft, [9].

A drag optimized hull has been developed for the DIVE-LD craft as shown in [10]. The hull form is drag optimized for a specific forward speed of 4kn, with determination of an optimal propeller position determined for the same. While this hull form does not have a cylindrical mid-body, it still remains true to the type of hull form considered in [8]. Another example of a drag optimized hull is shown in the study of [11]. This study investigates seven different combinations of the conventional shape with a slender midbody and tapered nose and tail. The most drag efficient form is chosen based on CFD (Computational fluid dynamics) assessments.

1.5. The proposed innovative method of drag reduction using a toroidal hull design

The use of tunnel thrusters is widely understood for large surface craft as done in [12]. Ducted propulsion is advantageous because a well designed duct results in a positive thrust apart from the thrust of the propeller. The tip vortices generated in an open propeller are reduced in ducted propulsion upon appropriate choice of tolerance in duct diameter. Smaller crafts such as AUVs use ducted thrusters such as [13]. The research by [13] optimizes the drag of an AUV with a ducted propeller using a commercial optimization code. The work of [13] not only optimizes the hull-form, but also optimizes the duct shape to ensure reduction in drag at a speed range of 1-3kn. The reference [14] is a detailed survey of the advances in ducted thrusters, starting from the first experimental evidence of shrouded propellers in ([15]). But yet another application of such a propulsion concept is used in this research study.

The baseline design used in this article, is the 'Pencil carrier', which is a conventional, torpedo shaped AUV with a slender and cylindrical mid-body. This hull has a step-decrease in its diameter as shown in figure 1. This study uses the principles of drag reduction through the radical optimization of the hull form of an AUV by using a tunnel through its entire length. This tunnel houses the sole propulsion system of this AUV. There is reduction in drag due to such a design and is compared with the Pencil carrier. The resultant drag-optimized design, Seppia, has a 13.09% lower drag in the presence of the thruster but has a 11.8% higher drag in the absence of propulsion. The new toroidal design is quantified using a new parameter, I_h. This parameter serves as a surrogate definition for slenderness of such a craft. The new toroidal design is progressively refined until a desired minimum is reached in drag of the craft. This drag-optimized design reached at iteration 11 of the design process, is termed 'Seppia'. The baseline and final are shown in figure 1 (a) and (b).

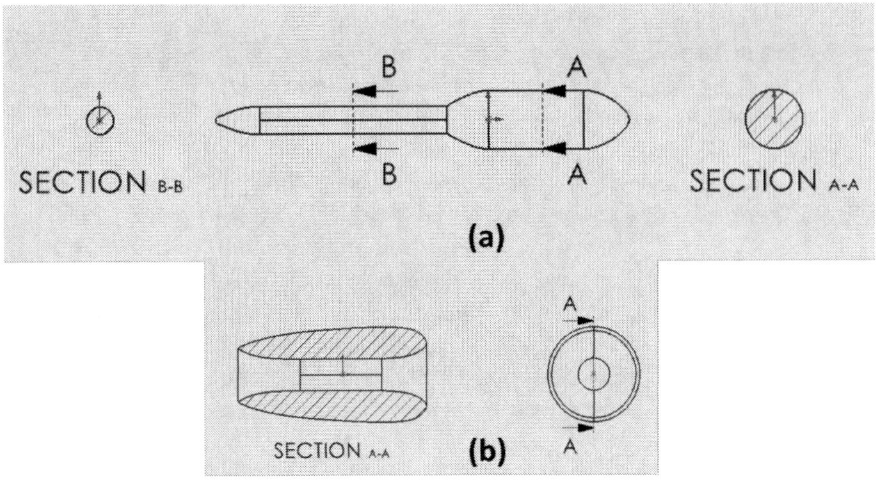

Figure 1. (*a*) The baseline is a conventional design with an open propeller at the stern (*b*) the final design is a toroidal design with ducted propeller at the center of the craft.

2. CFD informed design study

The pencil carrier with its slender form is a fairly conventional design for a craft aforementioned requirements. Therefore, instead of further optimizing the slender form, a radically different method of using a toroidal hull form was considered for the same payload. The following sections outline the CFD model set-up, different approaches to design modifications on the toroidal design for drag optimization, and a detailed comparison of significant designs.

2.1. CFD model set-up

The two crafts were simulated using the commercial RANSE (Reynolds averaged Navier Stokes) solver, documented in [16]. In each simulation, the vehicle is fixed with an inflow-velocity induced on the surrounding fluid (water). For the coarse mesh, computational domain extends approximately 15 meters ahead and 20 meters behind the vehicle, containing approximately 3M cells. To model the boundary layer, the mesh contains 12 prism layers, where each element connected to the hull is approximately 0.2% of the vehicle length. A propeller is simulated using the "virtual disc" method, where a disk-shaped region of space is selected to add propeller-like momentum to any encapsulated fluid. A propulsive thrust is then set equal to the drag on the hull, and the virtual disk is placed accordingly. Figure 2 shows the CFD mesh structure and resolution, as well as the placement of the virtual propulsor discs, for the baseline model and a toroidal designs. This coarse mesh is used only for baseline and toroidal designs upto iteration 10.

Upon reaching iteration 10, a mesh resolution study is performed on iteration 10 and the baseline designs, shown in figure 3. The final value of drag for baseline and best toroidal design is using this fine mesh of 26M and 19M cells respectively. Two types of simulations are used for the baseline and final designs. The first type involved the craft propelled using a momentum source in the place of a thruster. This is achieved using a steady state solver with the 2-layer realizable $K-\epsilon$ model with a wall y+ of 65 at the surface of the craft. The fine mesh for both crafts are shown in figure 4. The thrust used in the momentum source is the same as the value of the converged total drag of the craft. The use of a momentum source in place of the actual model of the thruster is recognized as shown in [17]. Since this study involves a comparison between hull forms, using an identical momentum source in place of the full CAD (Computer aided design) model is justified. The second type of simulation is the simple estimation of drag with the same CFD setup but without the momentum source for propulsion.

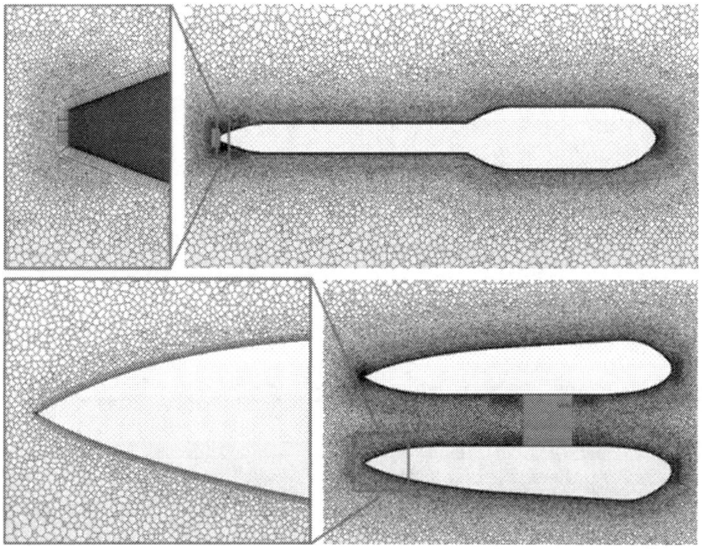

Figure 2. A coarse mesh of 3.5 M cells to reach a significant point in the design modification process. This mesh was chosen based on the dimensions of the crafts under consideration. The momentum source is modelled as a virtual disc and highlighted. Both the baseline and iteration 10 of the toroidal design are shown in this mesh.

2.2. Approach to design modifications

Multiple designs were attempted before resulting in the final hull form with the lowest drag. These hull forms contained a tunnel through the length of the craft with the thruster within. But the modifications remained in the outer hull profile and the profile of the curve leading to the duct where the thruster is placed. An arbitrary tunnel is chosen for now, to be refined as part of future work.

2.2.1. Initial toroidal design variants The first toroidal design approach consisted of four designs with jet-like ducts. Drag was measured and compared at 1m/s such that different qualities could be better quantified. Figure 5 compares the results of the first four design iterations at 1m/s. In this case, it is noted that the hull form is more open and less curved towards incoming flow. This led to high pressure drag at the bow section. The details of this design are shown in figure 5.

The first and second place designs have the best characteristics for drag, showing that the bow needs curvature to lead the flow into the duct with less pressure drag at the bow. The slight increase in outer and inner diameter is observed in the third and fourth designs in comparison to first and second. This results in worse flow characteristics at the bow and poorer drag.

The second approach incorporated a general reclining outer surface and utilized re-expansion of the duct outlet to make the outlet velocity close to the surrounding water velocity for better efficiency. This design modification continues to have a curvature leading to the duct which

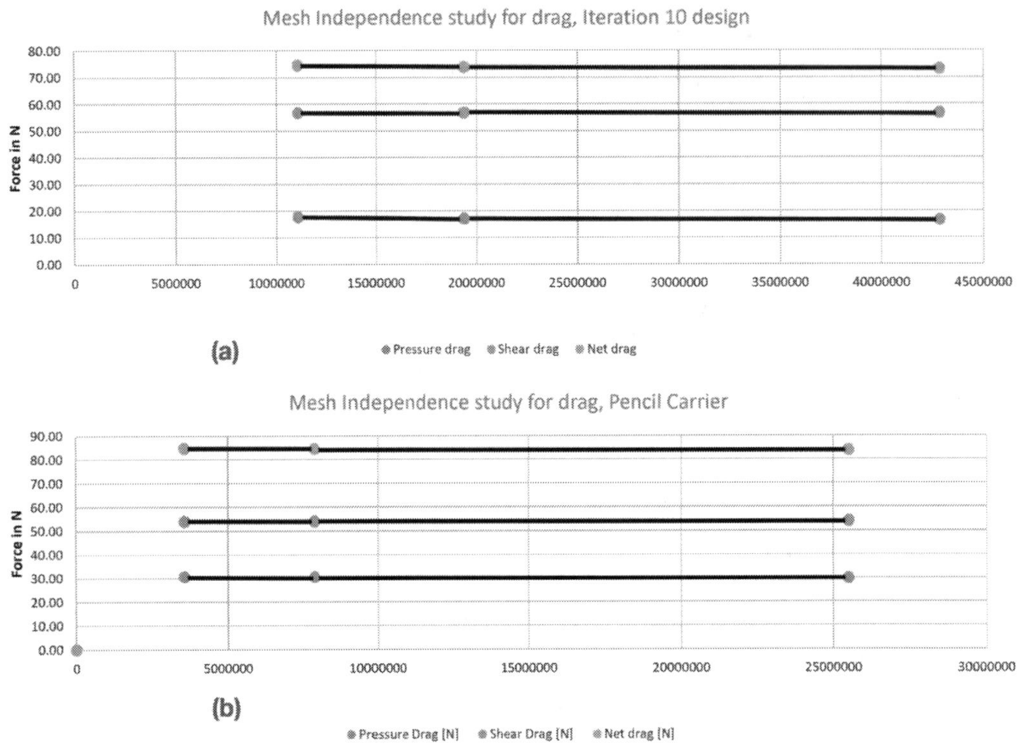

Figure 3. A fine mesh of (a) 26M cells for the baseline and (b) 19M cells for the iteration 10 of the toroidal design.

is more pronounced that the curvature of the outermost hull surface. This type of design was intended to have a larger area of cross-section leading to more flow into the duct. Since the inner duct velocity is now relatively low, control surfaces can be placed at the duct outlet for leverage at low speed and minimal drag.

2.2.2. Further modification including payload reversal The third approach involves taking a closer look at maximizing efficiency of drag through payload arrangement. It was observed that reversal of the orientation of the payload facilitates the reduction in the thickness of the hull at the bow. This orientation of the payload also results in the reduction of the cross section at the tail. Therefore the drag is reduced further from iteration 5. The third approach of payload arrangement also indicates that a more tapering external hull profile, if done well, reduces pressure drag at the bow and stern sections. It was observed that most of the hydrodynamic drag in iteration 7 was due to skin friction. The fourth approach attempts to improve from design iteration 7 by increasing the duct radius while maintaining the same outer hull profile. By increasing the duct radius, the corresponding reduction of flow velocity could potentially reduce the overall drag. Instead, the overall increase in both volume and front cross-section causes led to more wetted surface area and an increased drag for iteration 8. Therefore, iteration 7 was modified at the bow section instead of changing the duct. This modification brought the point of inflection of the bow profile closer to the duct. The shape of the duct inlet is guided using the pressure coefficient. By moving the nose tip in order to get as close as possible to high pressure bands, the vehicle gains further drag reduction. Figure 6 compares design iterations 9 and 10 with iteration 7, showing the improvement in drag at 2 knots. For comprehensive comparison

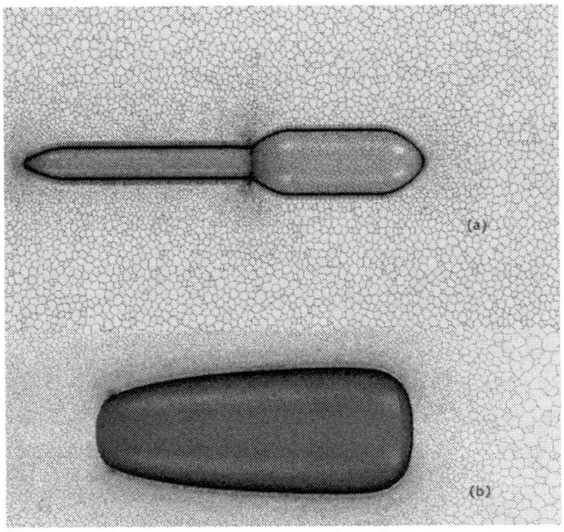

Figure 4. A fine mesh of (a) 26M cells for the baseline and (b) 19M cells for the final design. The two meshes are not shown in scale with respect to each other so that the designs and mesh details are legible.The surface of the Seppia needed to be refined more to capture the continuous curvature of the hull profile.

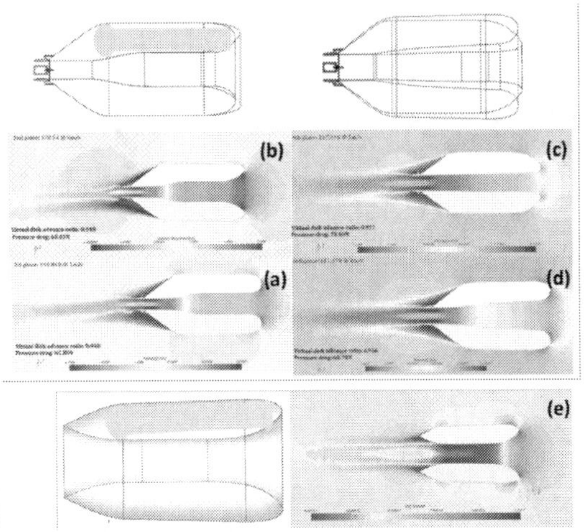

Figure 5. Four designs (a) to (d) attempted under this approach are shown in clockwise order. The contour indicates velocity characteristics of each design at 1m/s of forward velocity. (e) indicates an iteration with more significant change in hull form. Payload is greyed out for confidentiality.

at this significant point in the design process, iterations 7, 9, and 10 are compared at various low speeds, to ensure that any efficiency improvements are maintained over a range of cruise speeds, in this case, 1-4 knots.

IOP Conf. Series: Materials Science and Engineering 1288 (2023) 012039 doi:10.1088/1757-899X/1288/1/012039

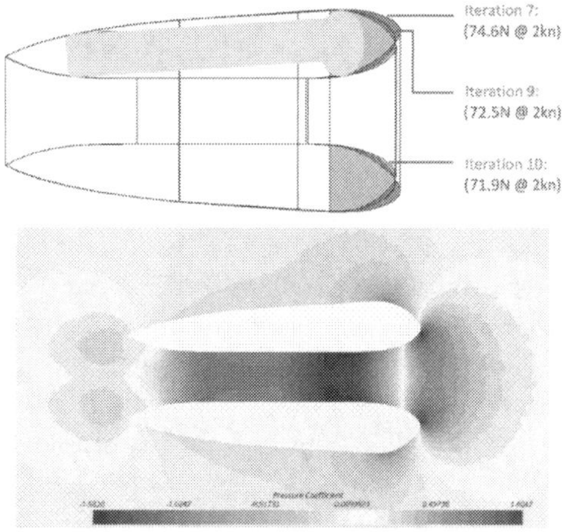

Figure 6. Iteration 7, 9 and 10, indicating differences. These are the last few designs with coarse meshing. Payload is greyed out for confidentiality.

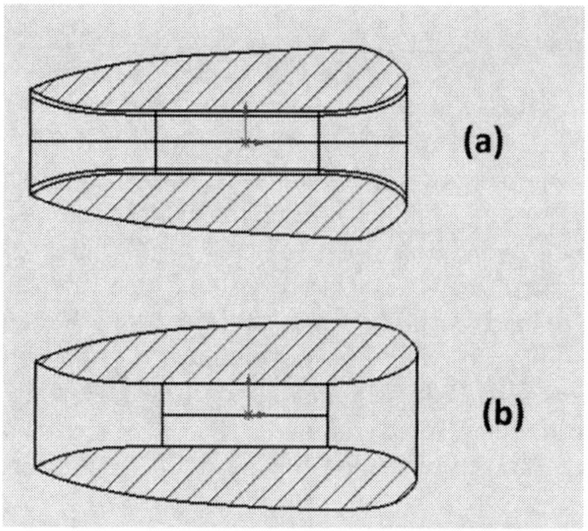

Figure 7. (*a*) Iteration 10 sectional view along the length. (*b*) Iteration 11 (Seppia) sectional view along the length. The minor volume change strategically changes the curvature to result in lower drag at the bow section.

2.2.3. Final design variants resulting in Seppia While iteration 10 already shows lower drag than the baseline while propelled, yet another design modification is made to the spline of the curve at the bow. This is established by removing the extra volume near the payload and adding length at the bow section for a more gradually tapered hull profile, shown in figure 7. Since this change is relatively small in magnitude with 2.08% volume loss, a mesh resolution study is performed as mentioned in section 2.1. Using the finer mesh, it is determined that iteration 11 has a lower drag than iteration 10 as shown in table 1.

2.3. Definition of a new design parameter for toroidal shapes, I_h

The slenderness or fineness ratio used by [6] and [7] are suited for the conventional hydrofoil-shaped designs. But a toroidal design such as figure 7 needs a parameter that uses the internal tunnel length for a more honest understanding of the design. Therefore a new design parameter shown in equation 1 is defined.

$$I_h = \frac{l}{t} \tag{1}$$

Design	Pressure	Shear	Net drag	Volume	Drag loss % w.r.t Baseline	Volume loss % w.r.t Iteration 10
Iteration 10	17.70	56.73	74.44	6.26	11	
Iteration 11/Seppia	16.88	56.37	73.25	6.13	12.6	2.08
Pencil Carrier	29.82	53.98	83.81		NA	

Table 1. Drag results of the final design and baseline design in comparison to iteration 10 using a fine mesh for each. The final design, (iteration 11 or) Seppia is highlighted. The payload volume is the same across all designs. The net volume or the total volume of the design varies.

where I_h is the design parameter for annular designs, l is the length of the craft and t is the toroidal thickness obtained by the difference of the maximum outer diameter and minimum inner thruster diameter. Design parameter for annular designs is I_h.

This parameter is used as a surrogate for the slenderness ratio and is calculated in table 2 for all the toroidal designs. The slenderness ratio is determined for crafts based on the maximum width in [6] and [7]. By using maximum thickness rather than width, I_h can quantify the true slenderness of a toroidal design. While the length could have been modified to include the length of the thruster, such inclusion does not account for the flow interaction within a duct rather than free flow. Thus, instead of using a modified slenderness ratio, the new I_h values for some significant designs are shown in table 2. This new parameter has not been studied yet for compliance with the recommended slenderness ratio of 5.5 for drag optimized hull forms that comes recommended by [8].

Design Iteration	Outer	Inner	Length	I_h
5	2.05	0.71	3.68	2.75
6	1.82	0.58	3.59	2.90
7	1.86	0.60	3.59	2.86
8	2.06	0.80	3.59	2.86
9	1.85	0.60	3.59	2.88
10	1.84	0.60	3.59	2.90
11	1.79	0.6	3.69	3.10

Table 2. The parameter I_h calculated for design iterations 5 through 11. This parameter represents toroidal designs better than the slenderness ratio which would indicate that the toroidal design is less drag efficient than the pencil carrier.

2.4. Comparison of flow characteristics among the last two designs and the baseline

The three important designs of iteration 10, Seppia and Pencil carrier are compared for various characteristics to ensure that the advantages in drag are well justified.

2.4.1. Comparison of the friction and pressure drag The figure 8 shows a comparison of C_f (skin friction coefficient) for the final two designs. The design change has caused a reduction of C_f at the bow due to the change in hull profile. The change creates a more gradual curvature in Seppia than that of iteration 10, while adding a length of 0.1m to the craft. The red region of the C_f contour in Seppia is reduced but the yellow region is extended past the bow to part of the mid-body, seen in figure 8 (c). However this increase does not compensate for larger red region in figure 8 (a), hence making the Seppia have better drag characteristics than iteration 10. The comparison of stern sections in figure 8 (b) and (d) does not yield much difference due to the hull shape remaining almost identical. The figure 9 (a) indicates the C_p (pressure coefficient) in contours on iteration 10 of the toroidal design. Comparing figure 9 (a) and (c), it is observed that there is a larger band of negative C_p on iteration 10 than Seppia. This agrees with the observation on the C_f plots, making Seppia have lower shear and pressure drag than iteration 10.

The figure 10 indicates the bow and stern views of the pencil carrier. Figures 10 (a) and (b) show that the C_f does not reach a contour value as high as the Seppia. However the wetted surface area shown in table 3 indicates that the increase in surface area in the pencil carrier with the extended green contour results in a higher friction drag in the presence of propulsion.

Design	Length	Dmax	Dmin	Vol	WSA	Dp	Ds	Dtot	Dr
	[m]	[m]	[m]	[m3]	[m2]	[N]	[N]	[N]	[N]
Seppia	3.69	1.79	0.60	6.26	37.1	17.7	56.7	74.4	104
Pencil Carrier	11.14	1.07	0	9.56	34.8	29.8	54.0	83.8	112
Change %						-40.6	5.1	-11.2	-7.1

Table 3. Design parameters characterizing each design, along with drag values. Roughness is added based on an arbitrary value for 3D printed hull construction material. Change is shown in the last row of data for Seppia with respect to the Pencil carrier. In this article, Dmax is the maximum diameter, Dmin is the internal minimum diameter, Vol is volume, WSA is the wetted surface area, Dp is pressure drag, Ds is the shear drag, and Dr is the drag with added roughness

The rest of the craft shows a mid-negative pressure drag. the low C_p regions indicate a high velocity near the surface, resulting in a high friction drag.

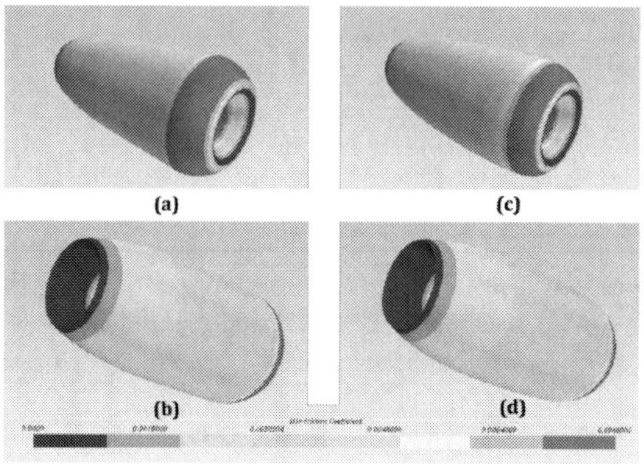

Figure 8. Coefficient of skin friction compared for design iterations 10 and 11 (Seppia): (*a*) Front perspective view of iteration 10, (*c*) Front perspective view of iteration 11, (*b*) Back perspective view of iteration 10, and (*d*) Back perspective view of iteration 11.

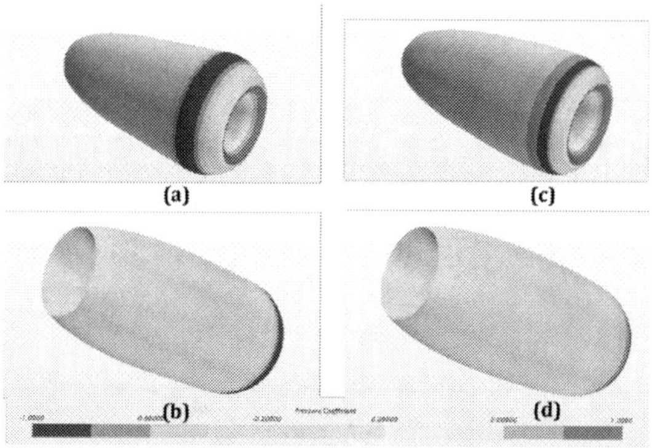

Figure 9. Coefficient of pressure compared for design iterations 10 and 11 (Seppia): (*a*) Front perspective view of iteration 10, (*c*) Front perspective view of iteration 11, (*b*) Back perspective view of iteration 10, and (*d*) Back perspective view of iteration 11.

2.4.2. Velocity characteristics The figure 11 indicates the velocity and wake of the water when the craft is moving at a constant forward velocity of 1m/s. It is observed that there is a detachment and partial re-attachment of the boundary layer, facilitated by the step in the hull profile. This step is gradual and reduces the hull into a smaller diameter since the payload is carried at the bow. The open propulsion concept used in this craft shows relatively higher acceleration since it is placed at stagnating flow at the tail. This is one of the main reasons that the Seppia has an advantageous design over the Pencil carrier.

The Seppia craft is shown in figure 12 in forward constant motion at 1m/s. This figure indicates that the acceleration of this craft is lower in comparison to the Pencil carrier, indicating that the propulsion is more efficient. This is due to the duct accelerating the flow towards the internal momentum source. This also explains that the drag in absence of propulsion does not indicate the true comparison between these two crafts, as outlined in the following section. the

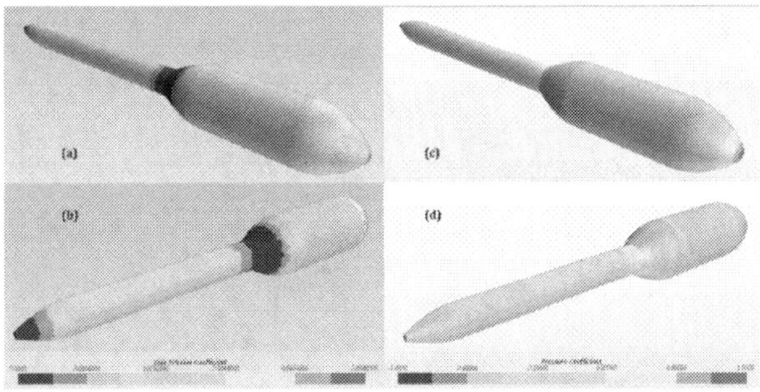

Figure 10. Coefficient of skin friction and pressure compared for the pencil carrier: (*a*) Front perspective view of C_f (*b*) Back perspective view of C_f (*c*) Front perspective view of C_p (*d*) Front perspective view of C_p.

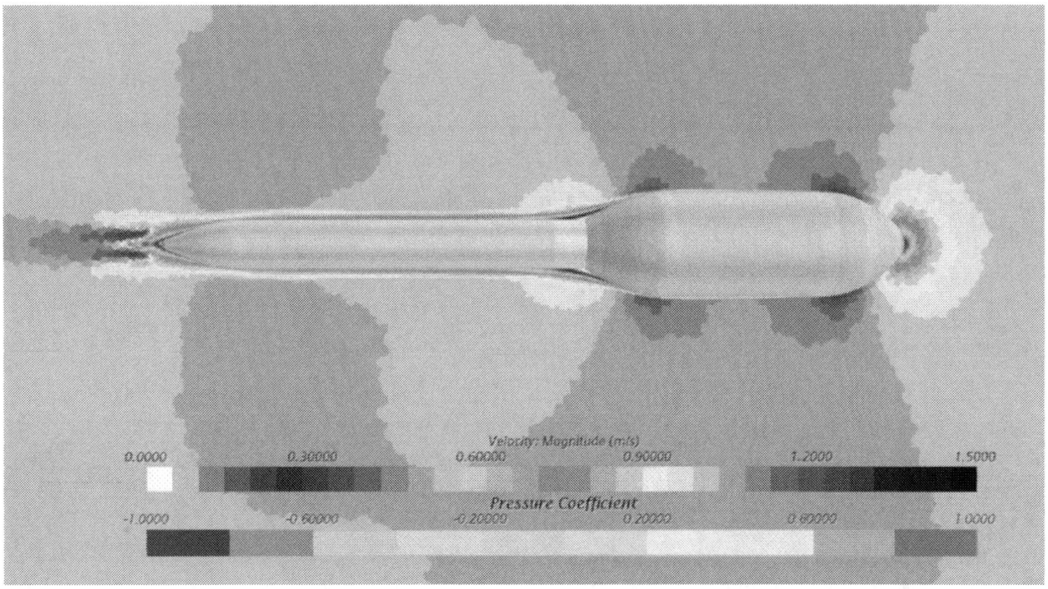

Figure 11. Velocity in the plane of motion for the pencil carrier is indicated. The surface of the craft is contoured with C_p values.

absence of a propeller would cause the duct to accelerate the flow, increasing the velocity in the internal duct and hence the C_f. The craft also shows almost no boundary layer separation, with C_p being low at the stern. There is a high region for the C_p pictured in figure 9 (*c*), at the inlet of the duct. This becomes a disadvantage during the absence of propulsion, making the slender Pencil carrier perform better. This region of highly negative C_p causes a localized region of high velocity and hence a high friction drag. However the ducted propulsion more than compensates for this issue.

2.4.3. On the presence and absence of propulsion The pencil carrier is a slender hull form. This design is in compliance with the recommended slenderness or fineness ratio of 5.5. The resultant drag in the absence of a propulsion system is 76N. However, in the presence of propulsion, it is 85N for a suitable size of disc used as the momentum source (0.5m diameter). Most of the drag

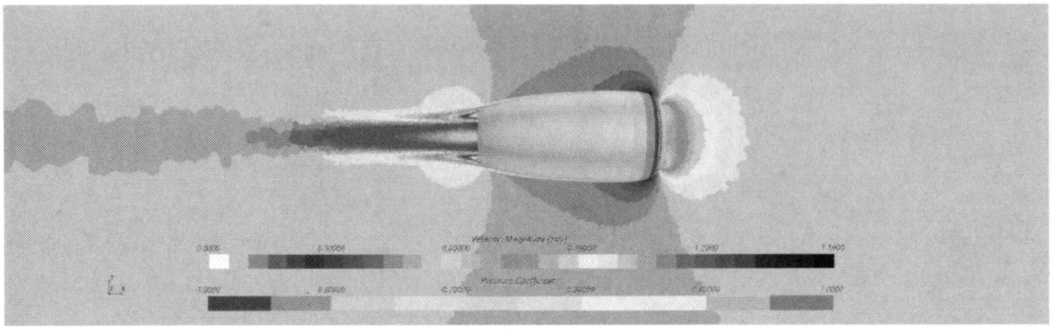

Figure 12. Velocity in the plane of motion for the Seppia is indicated. The surface of the craft is contoured with C_p values.

Design	PT	η_p	PB	PH	Ptot [W]	Battpack #	Alto fl. #	Speed
	[W]		[W]	[W]	[W]			[m/s]
Seppia	107	0.5	214	100	314	24	18	1.03
Pencil Carrier	115	0.5	230	100	330	24	18	1.03
Change %	-7.1	0.0	-7.1	0.0	-4.9			

Table 4. Parameters important for design decisions such as power consumption and range. % Change is shown in the last row of data for Seppia with respect to the Pencil carrier. In this article, PB is battery power, PH is hotel power, P_{Tot} is total power, η_p is assumed propulsive efficiency, Battpack is the no. of battery packs, Alt.fl. is the number of Alto floats.

optimization is from the slender tail section of the craft. The payload is distributed in the bow of the craft.

Pressure drag is most prevalent at the bow of the Pencil carrier since the nose is not sharp. The friction drag is minimal, with the flow reattaching past the reduction in diameter at the mid section of the craft. But the wetted surface area of the longer hull form by 7 m leads to a drag that can be reduced if the wetted surface area could be decreased.

The final toroidal hull design, named Seppia, is just 3.69 m long with a outer diameter of 1.79m and inner diameter of 0.6m. This design is not quite slender enough to be compliant with the recommendations in [7] with a slenderness ratio of 2.06. However, the thruster placed in the tunnel through the craft accelerates flow in such a way that the pressure drag at the bow is minimized and the skin friction within the tunnel is minimized. Moreover, the design of the tunnel can be further optimized with the guidelines in [18]. This design has a drag of 84N in the absence of a propeller but just 73N in the presence of the propeller. This shows that this hull form proves advantageous for the efficiency of propulsion. Moreover, accelerated flow past the propeller leads to the moving fin of the craft experiencing a high flow rate and thus being more effective for a relatively small angle of deflection. This is why the toroidal design advantageous. It must be noted that the design comparisons discussed here are at a 0° angle of incidence.The table 4 shows the change in power consumption between the two crafts. The table 5 shows the increase in endurance and range of the Seppia in comparison to the pencil carrier.

Thus, this research study shows that the slender and traditional design of AUVs can be further drag optimized by the use of unconventional hull-form. The drag improvement is 13.09% in the presence of a thruster in both designs. The pencil carrier drag is lowers than that of the 'Seppia' in the absence of a propeller by 11.8% in the absence of a propeller.

Design	En.	En_Tot	En_av.	Endur.	Endur.	Range	Range [nm]
	[Wh/pp]	[kWh]	[kWh]	[h]	[days]	[km]	[nm]
Seppia	6300	151	144	458	19.1	1695	915
Pencil Carrier	6300	151	144	435	18.1	1611	870
Change %	5.2	5.2	5.2	5.2			

Table 5. Parameters important for design decisions such as power consumption and range. % Change is shown in the last row of data for Seppia with respect to the Pencil carrier. En is the energy consumed by the craft, Battpack is the no. of battery packs, En.Tot is the total energy consumed by the craft, En.Av is the average energy consumed by the craft, Endur is the endurance of the craft, and Alt.fl. is the number of Alto floats.

3. Conclusions and Further Studies

The paper summarizes the hydrodynamic design principles for a new class of underwater vehicle based on the toroidal hull form. As demonstrated in the paper, this unconventional type of hull form has the potential of significant drag reduction with respect to equivalent conventional, single, monolithic hull forms of the same payload capacity and trans-critical Reynolds numbers.

- The drag reduction achieved by the toroidal hull in comparison to a conventional hull form is due to a higher effective length to thickness ratio for the ducted AUV, quantified in terms of a new parameter defined as I_h.
- The action of the internal mid-ship propulsor, on the local flow field at the nose and tail of the hull is discussed by comparing the results of CFD simulations of the hull advancing at cruise speed in towed or self-propelled conditions, at a $0°$ angle of incidence.
- It is shown that the bare hull drag, measured in towing condition, drops of 11.8% when the propulsor in activated. The toroidal hull does in fact act as a long duct around the inner thruster, so that its effect on the leading and trailing edge flow, as well known, is significant.

Thus, by following a practical design study aimed at an optimal long range, large payload craft, the paper shows how a reduction of more than 13% in drag is achieved.

Current studies are considering the design of the inner propeller and its effect on the hull advancing in self-propulsion. By incorporating the action of a real multi-bladed impeller spinning into the hull inside the CFD simulation model, the overall propulsion power and the overall propulsion efficiency of the toroidal hull configuration will be estimated and compared with those of the conventional hull with a single open propeller at stern. Finally, the effect of drag reduction devices known to be effective at transitional Reynolds numbers will be considered and evaluated.

4. References

[1] Kleiber J, Stilwell D, Miller L M and Brizzolara S 2022 Preliminary Analysis of Control Design for Long-Endurance AUVs *Conference: 14th IFAC Conference on Control Applications in Marine Systems, Robotics and Vehicles* (Kongens Lyngby)

[2] Xu H, Zhang G C, Sun Y S and Pang S 2020 Energy-saving control of long-range autonomous underwater vehicle vertical plane based on human simulating intelligent control method *International Journal of Advanced Robotic Systems* **17** ISSN 17298814

[3] Furlong M E, McPhail S D and Stevenson P 2007 A concept design for an ultra-long-range survey class AUV *OCEANS 2007 - Europe* (IEEE Computer Society) ISBN 1424406358

[4] Tian G, Zang Y, Feng X and Hu Y 2022 Focus on Bioinspired Textured Surfaces toward Fluid Drag Reduction: Recent Progresses and Challenges *Advanced Engineering materials* **24**

[5] Shereena S G, Vengadesan S, Idichandyxs V G and Bhattacharyya S K 2013 CFD study of drag reduction in axisymmetric underwater vehicles using air jets *Engineering Applications of Computational Fluid Mechanics* **7** 193–209 ISSN 1997003X

[6] Young A D 1939 The calculation of the total and skin friction drags of bodies of revolution at zero incidence Tech. rep. Director of Scientific Research, Air ministry

[7] Myring D F 1976 Theoretical Study of Body Drag in Subcritical Axisymmetric Flow. *Aeronautical Quarterly* **27** 186–194 ISSN 0001-9259

[8] Myring D F 1981 A Theoretical Study of the Effects of Body Shape and Mach number on the Drag of Bodies of Revolution in Subcritical Axisymmetric Flow Tech. Rep. Technical Report 81005 Royal Aircraft Establishment

[9] McCarter B, Portner S, Neu W L, Stilwell D J, Malley D and Minis J 2015 Design elements of a small AUV for bathymetric surveys *2014 IEEE/OES Autonomous Underwater Vehicles, AUV 2014* (Institute of Electrical and Electronics Engineers Inc.) ISBN 9781479943449

[10] Miller L and Brizzolara S 2019 Optimum propeller positioning and sizing for underwater vehicles *SNAME Maritime Convention 2019, SMC 2019*

[11] Uppal H, Isaac M T and M S B 2019 Numerical investigation on the drag characteristics of AUV hulls Tech. rep. URL `https://www.researchgate.net/publication/365488252`

[12] Saari A 2014 *Hydrodynamic study on a ducted propeller in a large vessel by time-accurate self-propulsion simulation with Reynolds-Averaged Navier- Stokes -equations* Ph.D. thesis Aalto University

[13] Joung T H, Sammut K, He F and Lee S K 2012 Shape optimization of an autonomous underwater vehicle with a ducted propeller using computational fluid dynamics analysis *International Journal of Naval Architecture and Ocean Engineering* **4** 44–56 ISSN 20926790

[14] Sacks A H and Burnell J A 1962 Ducted propellers-a critical review of the state of the art *Progress in Aeronautical Sciences* ISSN 03760421

[15] Stipa L 1931 On the use of propellers of various types *L'Aeronautica* **3**

[16] Siemens Digital Industries 2022 Simcenter Star-CCM+ v2022

[17] Miller L M and Brizzolara S 2022 Qualitative and quantitative study about hull-propeller interaction for an X-fin AUV by different propeller models *OCEANS 2022, Hampton Roads* (Hampton roads, VA)

[18] Brizzolara S and Brizzolara E B 2017 Long Tunnel Configurations for High Efficiency Thrusters *Proceedings of the Fifth International Symposium on Marine Propulsors* vol 3 (Espoo)

IOP Conf. Series: Materials Science and Engineering 1288 (2023) 012040 doi:10.1088/1757-899X/1288/1/012040

Numerical simulations of the flows past two parallel square columns by MPS method

Congyi Huang[1], Xiaolong Yang[2] and Decheng Wan[1]*

[1] Computational Marine Hydrodynamics Lab (CMHL), School of Naval Architecture, Ocean and Civil Engineering, Shanghai Jiao Tong University, Shanghai, China
[2] Offshore Oil Engineering Co. Ltd., Tianjin, China
*Corresponding Author: dcwan@sjtu.edu.cn

Abstract. The hydrodynamic interactions between an offshore platform and surrounding water mainly occurs in way of the pile legs, and the flow past the pile legs can be simplified to the flow past the columns. This is the reason why it is of great significance to simulate the flow around multiple columns with accuracy. In this paper, based on a mesh-less method, Moving Particle Semi-implicit (MPS) method, the MLParticle-SJTU solver is used to simulate the flow past two parallel square columns. Firstly, the reliability and accuracy of the solver is verified by simulating the Poiseuille flow, in which the no-slip boundary condition is applied. Then, the flow past two parallel square columns with different spacing between the two columns is simulated when the Reynold number is 40. The influence of spacing on flow pattern, force on columns and the vortex shedding frequency is analyzed. It is found that when the spacing is relatively small, small vortices will be attached to the surface of the columns. As the spacing ratio increases, periodic shedding of vortices will occur.

1. Introduction

In the field of naval architecture and ocean engineering, simulating the motion of offshore platforms exactly is a very important problem. The interaction between the platforms and the fluid mainly happens at platform's pile legs, which can be simplified as the flow past columns. The flow around a cylinder is a very classic and representative problem, and many researchers have conducted relevant research and discussion on this issue. Although the calculation model of this problem is very simple, there are many kinds of complex flow phenomena in this process, such as flow separation, vortex generation and shedding. Therefore, it is of great significance to simulate this process accurately. When simulating the flow around the columns of an offshore platform, each pile leg cannot be simply regarded as an independent column. The interaction among the columns is very important and should be considered. Therefore, the study of the flow around two columns has great engineering significance and academic value.

In recent years, with the rapid development of computer, CFD method has been used in simulating engineering problems more and more widely. And many scholars have studied the flow around a cylinder through numerical simulations. According to whether the computational domain needs to be divided into meshes, CFD methods can be divided into mesh-based methods and meshless particle methods. Different from traditional mesh-based method, the meshless particle methods use discrete particles to represent the computational domain. There is no fixed topological relationship between each particle, and each carries physical quantities such as mass, pressure and velocity. The traditional mesh methods often encounter the problem of mesh fragmentation when dealing with the problems with large deformation characteristics of free surface. However, meshless particle methods naturally overcome this problem due to their Lagrangian property. Therefore, the meshless methods have received widespread attention and application in recent years. In 2009, Yildiz et al. [1] proposed a multi-layer boundary particle approach to deal with boundaries, which solved the problem that the particle method could not deal with curved boundaries correctly. For the first time, the SPH method was used to simulate the flow around a cylinder with Reynolds number 50. In 2015, Shibata et al. [2] simulated the flow around a cylinder with Reynolds number of 100 and 200 using MPS method, and successfully simulated the Karman vortex street behind the cylinder and square cylinder. Zhou et al. [3] used direct numerical simulation methods to analyze the changes in turbulence intensity and energy of the flow field around parallel columns. Jamshed et al. [4] conducted a numerical simulation on the flow around two parallel porous square columns. Based on the analysis of the influence of Reynolds number and spacing ratio, the influence of permeability on the flow is further analyzed. Sahu et al. [5] investigated the turbulent flow properties experimentally in the vicinity of two side-by-side circular cylinders, along with the influence of the third cylinder of the same dimension placed in the upstream and successively in the downstream forming an equilateral triangle.

In this paper, the methods and boundary conditions used in this article is briefly introduced firstly. After that, the accuracy of the method and the stability of the solver is verified by simulating the Poiseuille flow. Finally, the flow past two parallel columns is simulated and the influence of spacing on the flow pattern is analyzed. The present work contributes to better optimization of platform design in engineering application.

2. Numerical method

The MPS method is a meshless particle method based on the Lagrange representation. The computational domain is represented by discrete particles. These particles are not connected by grids or nodes, but carry physical quantities such as mass, velocity and acceleration separately. The flow field is controlled by establishing the governing equation.

2.1. Governing equations

The governing equations include the continuity equation and the momentum equation. The governing equation for viscous incompressible fluid can be written as:

$$\frac{1}{\rho}\frac{D\rho}{Dt} = -\nabla \cdot \boldsymbol{V} = 0 \tag{1}$$

$$\frac{D\boldsymbol{V}}{Dt} = -\frac{1}{\rho}\nabla P + \nu\nabla^2 \boldsymbol{V} + \boldsymbol{g} \tag{2}$$

where ρ is fluid density, $\boldsymbol{V}$ is velocity vector, P presents pressure, ν is kinematic viscosity, $\boldsymbol{g}$ is gravitational acceleration vector, t indicates time.

2.2. Discretization of the governing equations

In the MPS method, the computational domain is composed of discrete particles. Therefore, the governing equations need to be discretized.

2.2.1. Kernel function In the MPS method, the interaction between particles is realized by the kernel function, which can be written as:

$$W(r) = \begin{cases} \dfrac{r_e}{0.85r + 0.15r_e} - 1 & 0 \le r < r_e \\ 0 & r_e \le r \end{cases} \tag{3}$$

where $r = |r_j - r_i|$ represents the distance between particle i and j, r_e is the influence radius.

2.2.2. Density of the particle number. The particle number density is the sum of kernel functions of all the particles within the influence radius, which can be written as:

$$\langle n \rangle_i = \sum_{j \neq i} W\left(\left|\boldsymbol{r}_j - \boldsymbol{r}_i\right|\right) \tag{4}$$

for incompressible fluid, the particle number density remains constant.

2.2.3. Gradient model. The gradient model is used to discretize the pressure gradient in the governing equation. The expression is:

$$\langle \nabla P \rangle_i = \frac{D}{n^0} \sum_{j \neq i} \frac{P_j + P_i}{|\boldsymbol{r}_j - \boldsymbol{r}_i|^2} (\boldsymbol{r}_j - \boldsymbol{r}_i) W(|\boldsymbol{r}_j - \boldsymbol{r}_i|) \tag{5}$$

where D represents the dimension and represents the initial particle number density.

2.2.4. Divergence model. Similar to the gradient model, the divergence model is used to discretize the velocity divergence in the governing equation. The expression is:

$$\langle \nabla \cdot \boldsymbol{V} \rangle_i = \frac{D}{n^0} \sum_{j \neq i} \frac{(\boldsymbol{V}_j - \boldsymbol{V}_i) \cdot (\boldsymbol{r}_j - \boldsymbol{r}_i)}{|\boldsymbol{r}_j - \boldsymbol{r}_i|^2} W(|\boldsymbol{r}_j - \boldsymbol{r}_i|) \tag{6}$$

2.2.5. Laplacian model. Laplacian model is used to discretize the second derivative in the governing equation, which can be expressed as:

$$\langle \nabla^2 \phi \rangle_i = \frac{2D}{n^0 \lambda} \sum_{j \neq i} (\phi_j - \phi_i) W(|\boldsymbol{r}_j - \boldsymbol{r}_i|) \tag{7}$$

where λ represents the correction of the error introduced by the kernel function, and it can be written as:

$$\lambda = \frac{\sum_{j \neq i} W\left(\left|\boldsymbol{r}_j - \boldsymbol{r}_i\right|\right) \left|\boldsymbol{r}_j - \boldsymbol{r}_i\right|^2}{\sum_{j \neq i} W\left(\left|\boldsymbol{r}_j - \boldsymbol{r}_i\right|\right)} \tag{8}$$

2.3. Pressure Poisson equation

In the MPS method, the Poisson equation is used to solve the particle pressure. The incompressibility of fluid is determined by divergence-free condition and constant particle number density condition. The Poisson equation adopted in this paper is as follows:

$$\langle \nabla^2 P^{k+1} \rangle_i = (1-\gamma) \frac{\rho}{\Delta t} \nabla \cdot \boldsymbol{V}_i^* - \gamma \frac{\rho}{\Delta t^2} \frac{< n^k >_i - n^0}{n^0} \tag{9}$$

where superscripts k and k+1 represent k and k+1 time steps. is a variable parameter, representing the proportion of particle number density in the source term of Poisson equation. In the numerical simulation in this paper, takes 0.01. (Tanaka and Masunaga[6], Lee et al, [7])

2.4. Detection of free surface particles

When solving the Poisson's pressure equation, it is very important to determine whether a particle is located on a free surface. The number density of particles can be used to determine whether a particle is on a free surface in MPS method. When $\langle n \rangle_i < 0.8n^0$, the particle is considered to be on a free surface. When $\langle n \rangle_i > 0.97n^0$, the particles are thought to be inside the fluid. For particles with particle number density between 0.8 and 0.97, it's difficult to tell whether the particle is on the free surface or inside the fluid. Khayyer and Gotoh [8] first proposed the criterion which simply based on the fact that for a free-surface particle, the distribution of neighboring particles is asymmetric. In this paper, the vector function $\boldsymbol{F}$ presented by Zhang et al. [9] is introduced, as follows:

$$< F >_i = \frac{D}{n^0} \sum_{j \neq i} \frac{(\boldsymbol{r}_i - \boldsymbol{r}_j)}{|\boldsymbol{r}_i - \boldsymbol{r}_j|} W(|\boldsymbol{r}_i - \boldsymbol{r}_j|) \tag{10}$$

If $\langle|\boldsymbol{F}|\rangle_i > 0.9|\boldsymbol{F}|^0$, the particle i will be considered to be on a free surface. $|\boldsymbol{F}|^0$ stands for $|\boldsymbol{F}|$ at the initial time of the free surface particle.

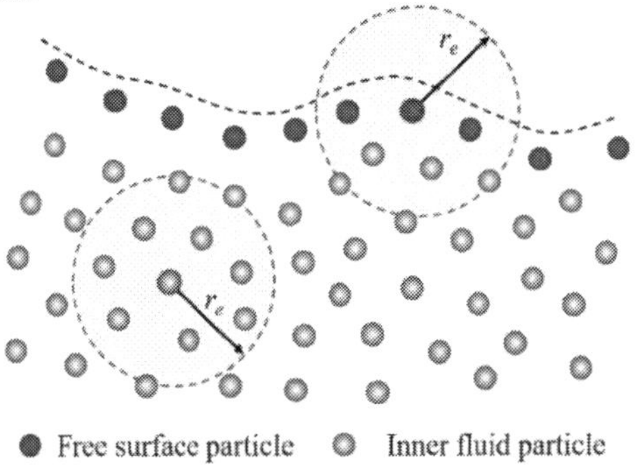

Figure 1. Schematic diagram of free surface particle judgment.

3. Numerical simulation

3.1. Reliability verification

In this section, the reliability of the MLParticle-SJTU solver used in this article is verified by simulating the Poiseuille flow. The theoretical velocity distribution of Poiseuille flow is given in the following equation.

$$u = \frac{6Q}{h^3}(y^2 - hy) \tag{11}$$

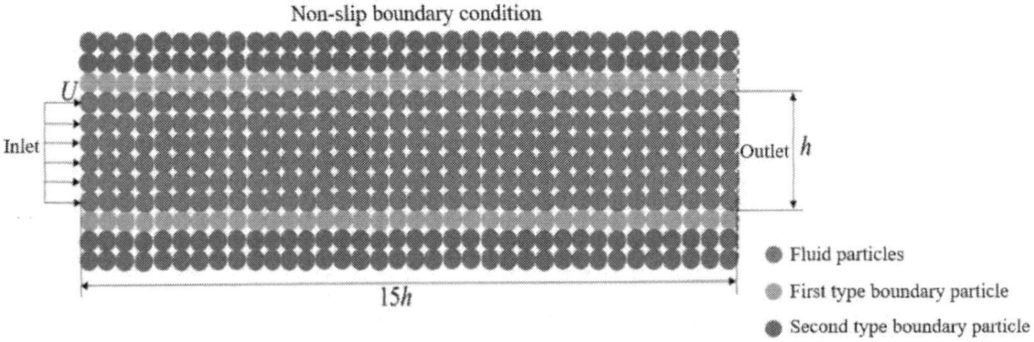

Figure 2. Schematic diagram of computational domain.

The computational domain is shown in Figure 2. It can be seen that the boundary conditions used in this paper is composed of multiple layers of particles, one layer of first type boundary particles and two layers of second type boundary particles. The first type of boundary particles is used to provide repulsive force to fluid particles, while the second type of boundary particles is used to ensure the integrity of the fluid particle support domain. The first type boundary particles participate in solving the pressure Poisson equation, while the pressure of the second type boundary particles is obtained by interpolation between fluid particles and the first type boundary particles. In addition, non-slip boundary conditions need to be applied when simulating Poiseuille flow. In this paper, the method of moving virtual mirror particles proposed by Lee et al. [7] is used, as shown in Figure 3.

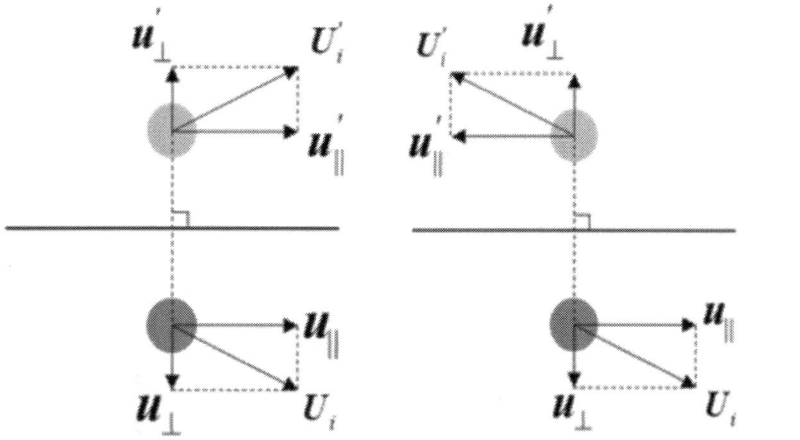

Figure 3. Schematic diagram of slip and non-slip boundary conditions.

In figure 3, U_i, $u_\perp$ and $u_\parallel$ represent the velocity, vertical and parallel components of fluid particles respectively. U_i', $u_\perp'$ and $u_\parallel'$ represent the velocity and components of virtual mirror particles corresponding to fluid particles. When calculating the viscous force at each time step, virtual mirror particles are generated on the outer side of the boundary. The virtual mirror particles and fluid particles are axisymmetric about the boundary. When the boundary is stationary, under the non-slip boundary condition, the vertical and parallel velocity components of virtual mirror particles are opposite to those of fluid particles. Under slip boundary condition, the vertical velocity component of virtual mirror particles is opposite to that of fluid particles, while the parallel velocity component is the same as that of fluid particles.

The calculation model is shown as figure 2, the diameter *h* of the pipeline is 0.2m and the length of the pipeline is 15*h*, which is an enough distance for the flow to fully develop. The Poiseuille flow is simulated at Reynolds number is 4, and the velocity distribution is shown in Figure 4. It can be seen that the flow velocity in the pipe distributs evenly, showing the characteristics of high velocity near the central line and low velocity near the pipe wall, which conforms to physical characteristics and proves the validity and rationality of the boundary conditions used in this article.

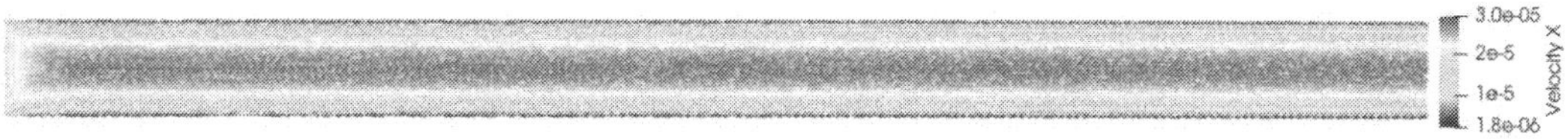

Figure 4. Velocity distribution of the Poiseuille flow.

Next, the velocity profile at x=10h cross-section at different times is drawn as shown in figure 5. It can be seen that when Ut/h=0.5, the flow has developed to a stable state, and the velocity profile almost coincides with the analytical solution, which can verify the correctness of the proposed method and the accuracy and stability of the MLParticle-SJTU solver.

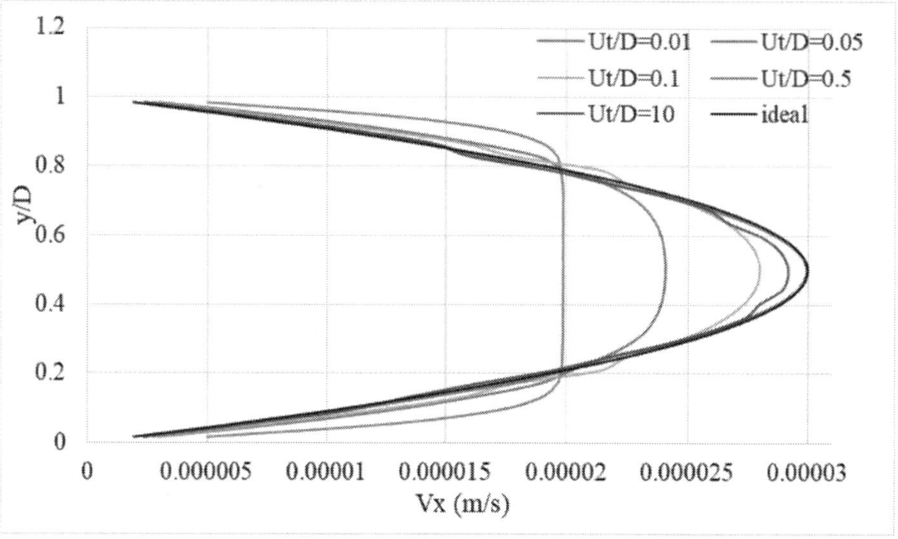

Figure 5. Velocity profiles at different times at x=10h.

3.2. Numerical Simulation of the flow past two parallel square columns
Since the accuracy and validity of the numerical method used in this paper is confirmed in the previous subsections, the flow past two parallel square columns will be simulated in this part. The Reynolds number is set as 40, and the influence of the spacing between square cylinders on the vortex shedding frequency, lift drag on the columns is analysed.

3.2.1. Calculation domain selection and boundary condition setting
The schematic diagram of the computational domain is shown in Figure 6. The side length D of the square column is 0.1 m, H is the distance between the two square columns, and the spacing ratio is set as H/D=1.2-5.0. The initial particle spacing is set to D/Δx=20. Schematic diagrams of the velocity distribution obtained from the simulation with different spacing ratios at Ut/D=200 are shown in Figure 7. In the following parts, the fluid is assumed to be viscous and incompressible, and the friction between the fluid and the square column is not considered

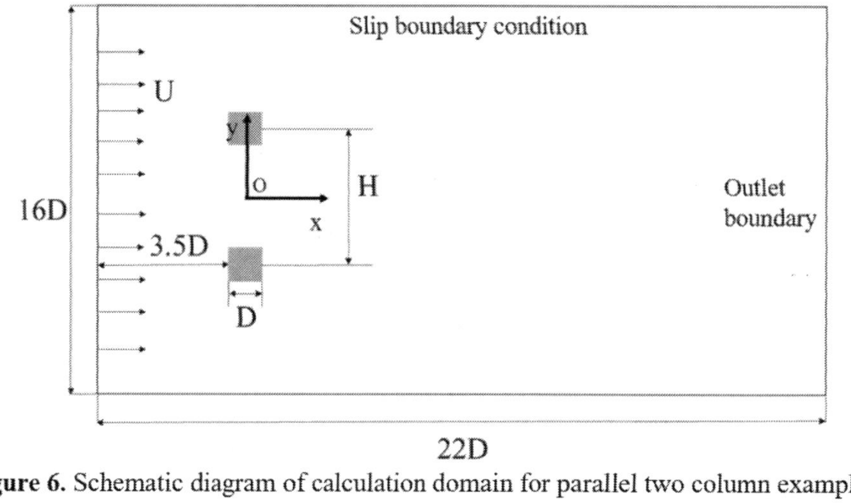

Figure 6. Schematic diagram of calculation domain for parallel two column examples.

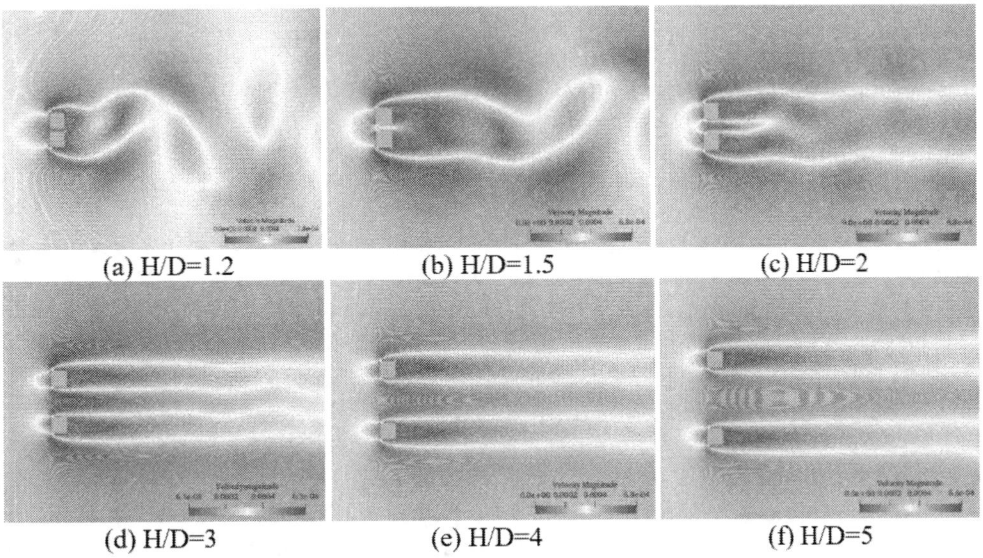

Figure 7. Velocity distribution at different spacing ratios at Ut/D=200.

From Figure 7, it can be seen that the velocity distribution obtained by MPS method is uniform and very stable. When H/D=1.2 and 1.5, the wake behind the two square columns is connected to form a large wake region and oscillates periodically, but the length of the wake is significantly shorter under H/D=1.2 than H/D=1.5. When H/D=2, the wake behind the square columns is separated near the trailing edge of the square column, but at a certain distance from the square column, the two wakes merge. When H/D=3, the two wakes have completely separated, but as can be seen from the distant wakes, mutual interference still exists. When H/D=4 and 5, the wake in the computational domain is almost straight and there is almost no obvious mutual interference.

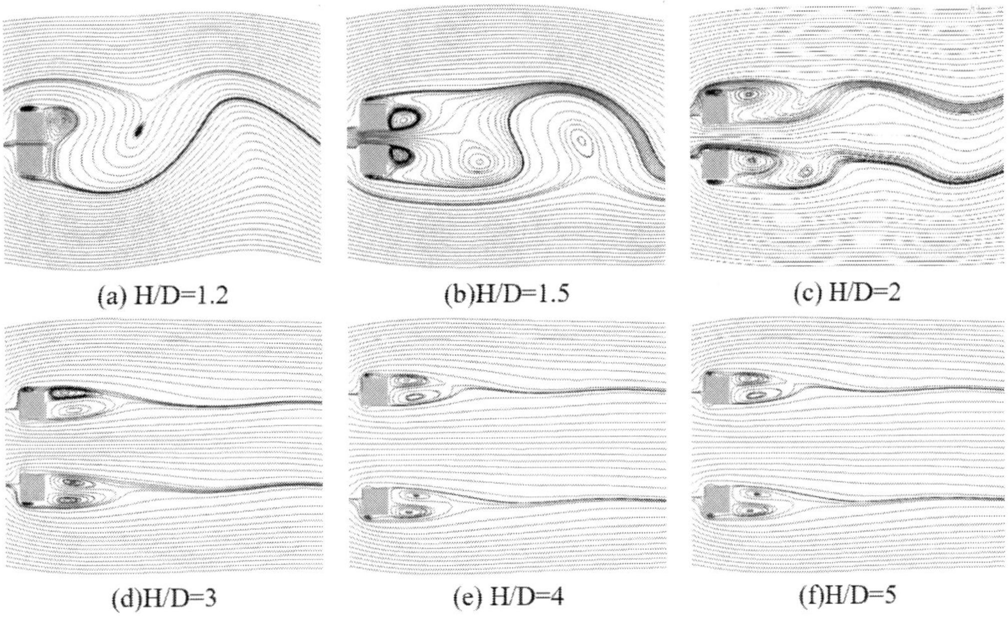

Figure 8. Streamlines around two parallel square columns at different H/D with Re=40.

To further investigate the formation and shedding of the vortex near the square column, the streamline diagram is drawn, as shown in Figure 8. It can be seen from that when the spacing ratio is 1.2, 1.5, and 2, vortices detach behind the square column, and small vortices are generated near the sharp corners of the column. As the spacing ratio increases, the mutual influence between the square columns weakens. Under the conditions of H/D=3, 4, and 5, the vortex shedding phenomenon behind the square column disappears and returns to a steady flow state. However, the wake behind the square column exhibits obvious asymmetric characteristics.

Figure 9 shows the average velocity profile curves at different positions under different spacing ratios. Under the condition of H/D=1.2, the x/D=0 curve exhibits negative values near y/D=±1.1. Based on the analysis in Figure 8 (a), this may be due to the formation of small eddies here. At the same time, it can be seen that the velocity between the two columns is very small, so the flow past the parallel two columns can be approximated as the flow around a single square column under this spacing ratio. The velocity profiles of each section behind the columns also exhibit a "V" shape, and tend to flatten along the flow direction, which is consistent with the results shown in Figures 7 and 8.

Under the condition of H/D=1.5, there are still small vortices on the outer surfaces of the two square columns, so the x/D=0 curve shows a negative value at y/D=± 1.25. The flow velocity between the two square columns is much higher than that under the H/D=1.2 condition. The x/D=2 curve shows a region below zero behind both columns, possibly because that two stable vortices is formed and attached to each column at this time. At x/D=5, the velocity curve has only a minimum value, which is less than zero at y=0. The reason for this phenomenon may be that at the position where y=0, periodically shed vortices form a reflux zone.

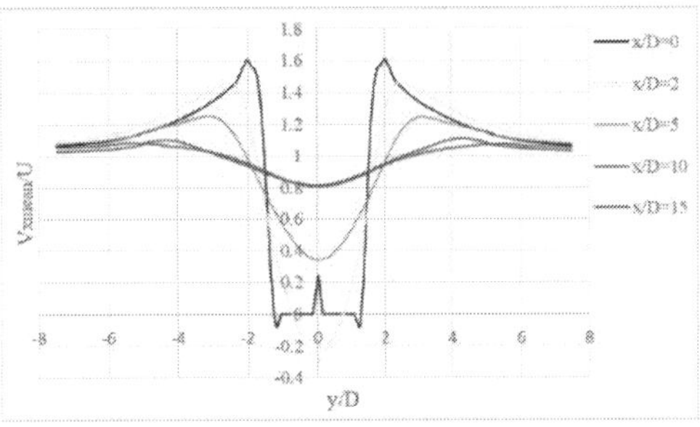

(a) H/D=1.2

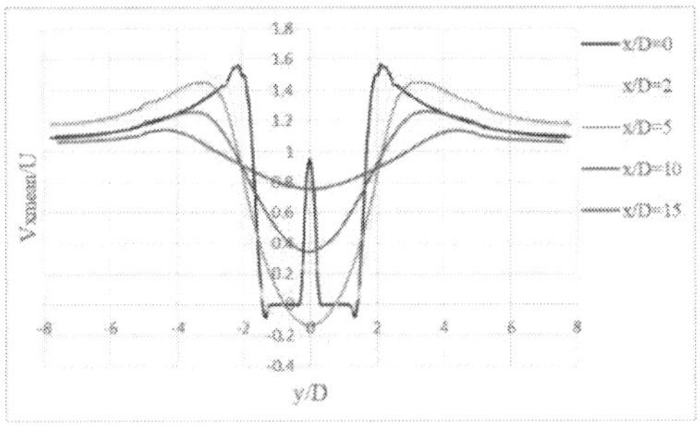

(b) H/D=1.5

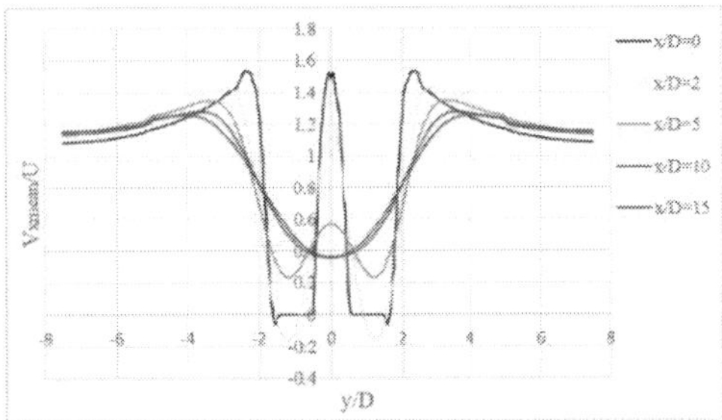

(c) H/D=2

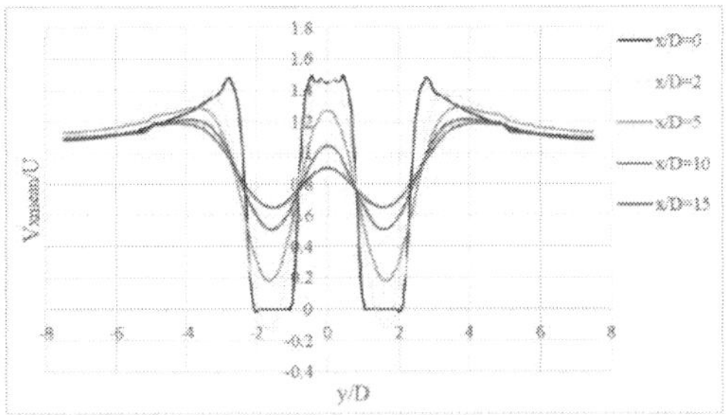

(d) H/D=3

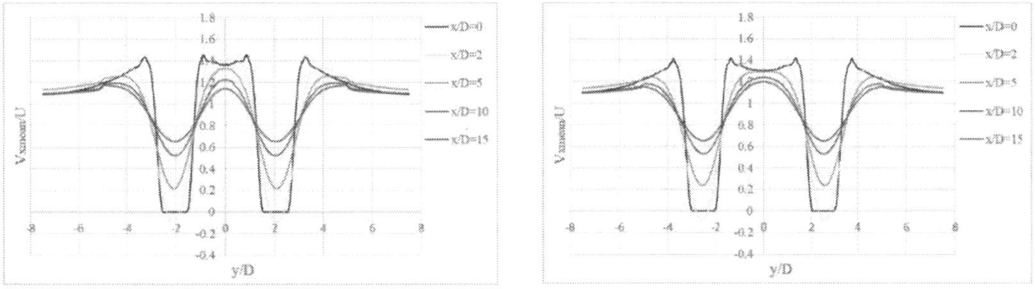

(e) H/D=4 (f) H/D=5

Figure 9. Average velocity profile curves at different H/D with Re=40.

When H/D=2, it can be seen from the x/D=0 curve that small vortices still exist on the columns, and the flow velocity between the two square columns is higher than the results mentioned earlier. The x/D=2 curve still has two minimum values less than zero like under H/D=1.5 condition, but the formation reasons for these two cases may be different. At this time, there is no stable attached vortex behind the column, and the less than zero average velocity may be formed by the periodic shedding vortices after the two square columns. The x/D=5 curve still has two minimum values, but at this point, the minimum value is higher than zero. At x/D=10, the wake merges and only one minimum value appears at the central line between the two columns.

Under H/D=3, 4, and 5 conditions, there are two minimum values for sections with x/D⩾2, because the wake behind the two columns does not merge. However, due to the mutual influence between the square columns, it can be seen that the minimum values are not located on the central line of the square column, but slightly off central line.

Table1. Peak average velocity at x/D=0.

H/D	1.2	1.5	2	3	4	5
$V_{meanmax}/U$	1.604	1.562	1.535	1.480	1.429	1.411

Table 1 shows the peak value of the average velocity curve of the flow around the two columns at x=0 under different spacing ratios. It can be seen that with the increase of spacing ratio H/D, the velocity peak decrease. As the Reynolds number is 40, it is close to the critical Reynolds number when the flow changes from steady to unsteady state. The fluid flows past the column, leading to an increase velocity on both sides, which may be one of the reasons for the periodic vortex shedding phenomenon when the spacing ratio H/D⩽2.

4. Conclusions

The flow past two parallel square columns is simulated based on MPS method in this particle. Firstly, the stability of the MLParticle-SJTU solver used in this paper is verified by simulating Poiseuille flow at Reynolds number 4. Next, the flow around two parallel columns is simulated. By analyzing the velocity distribution under different spacing ratios at Ut/D=200, it is found that when the spacing ratio H/D<2, the interaction between the two square columns is very obvious. The wake behind the two columns is very similar to that of a single square column, presenting a large oscillating wake. However, as the spacing ratio increases, the wake behind the square column becomes longer. After H/D>3, the wake behind the two square columns are already independent of each other, but mutual interference can also be observed.

Then, the vortices formation and shedding in the flow around the square column are further analyzed by streamlines and average velocity profiles at different sections. The results indicate that when H/D<2, small vortices generates at the sharp corners of the two columns, so the average velocity there is less than 0. At x/D=2, except for only one minimum value under the H/D=1.2 condition, there are two minimum values less than 0 in the other five conditions. It may be due to the formation of stable vortices attached to the square column here, or it may be due to the formation of stable shedding vortices here. For H/D=3, 4, and 5, the wakes of the two square columns are independent of each other, but there is still mutual influence. The streamline diagrams of the two square columns show asymmetric display, and the minimum values of the velocity profile do not appear on the central line of the square column, but appear slightly off the central line.

In this paper the two-dimensional square column winding process is simulated. In the future, the work will be extended to three dimensions, and the flow past multi-columns will be simulated.

Acknowledgements
This work was supported by the National Natural Science Foundation of China (52131102), and the National Key Research and Development Program of China (2019YFB1704200), to which the authors are most grateful.

References
[1] Yildiz M, Rook R A and Suleman A 2009 SPH with the multiple boundary tangent method *Int. J. Numer. Methods Eng.* **77(10)** 1416-38.
[2] Shibata K, Koshizuka S, Murotani K, Sakai M and Masaie I 2015 Boundary conditions for simulating Karman vortices using the MPS Method *J. Adv. Simul. Sci. Eng.* **2(2)** 235-54.
[3] Zhou Y, Nagata K, Sakai Y. Watanabe T and Hayase T 2020 Energy transfer in turbulent flows behind two side-by-side square cylinders *J. Fluid Mech.* **903(A4)** 1-31.
[4] Jamshed S, Dhiman A 2021 Channel-Confined Wake Structure Interactions Between Two Permeable Side-by-Side Bars of a Square Cross-Section *J. Fluid Mech.* **143** 091301.
[5] Sahu C, Eldho T I, Mazumder B S 2023 Experimental Study of Flow Hydrodynamics Around Circular Cylinder Arrangements Using Particle Image Velocimetry *J. Fluids Eng.-Trans. ASME* **145(1)** 011302.
[6] Tanaka M and Masunaga T 2010 Stabilization and Smoothing of Pressure in MPS Method by Quasi-Compressibility *J. Comput. Phys.* **229(11)** 4279–90.
[7] Lee B H, Park J C, Kim M H and Hwang S C 2011 Step-by-step improvement of mps method in simulating violent free-surface motions and impact-loads. *Comput. Meth. Appl. Mech. Eng.* **200(9-12)** 1113-25.
[8] Khayyer A, Gotoh H and Shao S 2009 Enhanced predictions of wave impact pressure by improved incompressible SPH methods, *Appl. Ocean Res.* **31(2)** 111-31.
[9] Zhang Y, Wan D 2012 Apply MPS method to simulate liquid sloshing in LNG tank. *Proc. of the 22nd Int. Offshore and Polar Eng. Conf.* (Rhodes, Greece) 381-91.
[10] Koshizuka S, Nobe A and Oka Y 1998 Numerical Analysis of Breaking Waves Using the Moving Particle Semi-implicit Method *Int. J. for Numer. Methods Fluids* **26(7)** 751-69.

IOP Conf. Series: Materials Science and Engineering 1288 (2023) 012041 doi:10.1088/1757-899X/1288/1/012041

Simulation of a floating solar farm in waves with a novel sun-tracking system

Yujia Wei[1], Binjian Ou[1], Junxian Wang[1], Liang Yang[1], Zhenhua Luo[1], Sagar Jain[1], Wolter Hetharia[2], Soegeng Riyadi[3], IKAP Utama[3] and Luofeng Huang[1]*

[1]School of Water, Energy and Environment, Cranfield University, Building 52, Cranfield, 45 College Rd, Cranfield, MK43 0AL UK

[2]Department of Naval Architecture, University of Pattimura, Ambon, 97223, Indonesia

[3]Department of Naval Architecture, Institut Teknologi Sepuluh Nopember, Surabaya, 60111, Indonesia

*luofeng.huang@cranfield.ac.uk

Abstract. The awareness of the energy and climate crisis has accelerated the development of renewable energy sources. Photovoltaic (PV) solar power plants harvest clean solar energy and convert it to electricity, which will be one of the most promising alternatives to the power industry in the context of a low-carbon society. Due to its low power density, the traditional deployment of PV systems on land or inland rivers requires much space. Therefore, industries are increasingly interested in expanding offshore Floating PhotoVoltaics (FPV) to oceans, where FPV has less influence on the marine environment and does not occupy precious space for land resources and human activities. This study performs a hydrodynamics-based structural response analysis for a novel FPV system in OpenFOAM. The wave-proof FPV platform is newly designed for this work, which integrated breakwater technologies to sustain the system's survivability in harsh ocean-wave environments. Firstly, the rational mooring types for FPVs installed close to the island are studied considering seabed effects. Subsequently, extensive parametric studies have been conducted to determine a rational design strategy for the mitigation of wave impact. Several potential effects of the proposed platforms on the hydrodynamics in a coastal sea are evaluated for the first time.

1. Introduction

The Floating PhotoVoltaic system (FPV) has garnered significant interest in the industry due to its unique combination of extendable PV size and floating structure benefits. By utilizing a floating structure, the system can absorb solar radiation energy, which reduces temperature and ultimately improves efficiency. The efficiency of floating solar plant is 11% higher than the conventional solar power plant and reduces the water evaporation by 70%, however, the investment of such power plant is 1.2% higher [1]. The development of FPV is particularly important in areas that lack land and lake space. Some review papers have addressed for FPV development, such as, FPV installation guidelines [2], general FPV technologies [3], Hybrid FPV system that combins with other renewable energy sources [4] and the economic aspects of FPV systems [5], to name a few. Several design regulations for the FPV system have been given by DNV GL [6].

Despite previous efforts, the development of FPV systems has remained limited to calm or near-shore regions, with offshore development still in the trial and design stage. The unpredictable ocean environment restrains the development of FPV systems in deep-sea regions. Because PV panels are relatively fragile, the platform design should support them and prevent wave attacks. Some pioneering industrial designs, such as those from Ocean Sun [7] and Oceans of Energy, have shown promising field test results. As reported by Oceans of Energy [8], the FPV prototype test in the North Sea Two (NS2) [9] project successfully operated in offshore regions for 11 months. During this time, it withstood storms such as "Ciara," "Dennis," and "Bella," with maximum wave heights of 13.5m.

Parallel with the experimental investigations numerical simulations were developed to predict the dynamics structure motions of the FPV system in waves. Choi, Park [10] conducted a series of simulations to investigate the motions and stress distributions of a floating PV system based on a one-way coupled CFD and FEA simulation with various inlet angles of wind and wave loads. The authors found that the first or last row of solar panels in a solar farm was exposed to the highest wind loads and stress. The drawback of this method is the pressure distributions on the solar panel were measured in the steady state conditions. As a recommendation for future work, the transient analysis of pressure distributions on solar panel should be studied in transient state. Song, Kim [11] presented a global performance analysis for a moored FPV system based on a combination of software. The selected FPV system consists of 100 floating modules and the whole system is tested in an inner harbour, and the wave height was set to 1.5m as the severe condition. The large scale effects of the FPV system were studied and the results showed that the largest motions and shear forces are detected at the corner modules. The reinforcements of the frame at the corner should be applied.

Based on the lessons learned from previous studies, a novel solar farm concept is proposed in this study, takeing two primary advantages: 1) A hexagonal shape breakwater is designed for the solar farm to avoid direct contact from all-directional waves. 2) A installation of a sun-tracking system that rotates the solar farm to maximise the overall energy production.

The tracking system in floating is easier to implement than that in ground . This is because PV panels float on water hence it requires little energy to rotate the whole structure. The tracking system in FPV consists of a rotary structure to track the azimuth angle and a tilt variable structure to track the tilt angle. The pinoneering prototype from K-water (Korea Water Resources Corporation) has installed the world's first 100 kW tracking-type floating photovoltaic generation system in Korea for commercial power generation [12]. The tracking system has evidence that it will cost 1% of the total energy output and give a 10% of the energy feedback.

In this context, the present study fills in the gap that no research has investigated on the dynamic behaviours of a rotary solar farm in waves based on a fully nonlinear CFD method. The computational results, including the breakwater performance and the dynamic behaviors of the FPV system with/without mooring lines, are presented.

This paper is structured as follows: in Section 2, the problem definition on the solar farm is discussed. In Section 3, the numerical setting ups and methodology used are presented. In section 4, the simulation results of the solar farm with/without a mooring system are illustrated accordingly. Section 5 summarises this work with its main conclusions.

2. Problem definition

The prototype design of the novel offshore solar farm consists of three parts: an octagonal shape of breakwater, a rotary PV supporting structure and a mooring system, as shown in Figure 1. The breakwater is designed to protect the platform from all-directional waves and currents. The rotary PV supporting structure is where the photovoltaic panels are installed, and it rotates relatively to the breakwater with the rotational speeds controlled by the tracking system. The mooring system keeps the platform anchored in place.

This innovative design allows for scalability and flexibility in energy output. The designed energy output of a single moduluar is a 1MW with a 100m in diameter, and it has the capacity to extend to 4MW by connecting multiple modules together using a hard or soft connection, as shown in Figure 1(b). The middle region of the four modules can be designed to house an energy conversion and storage system. This system can include energy storage options such as batteries, along with a conversion system that can efficiently transfer energy to the grid.

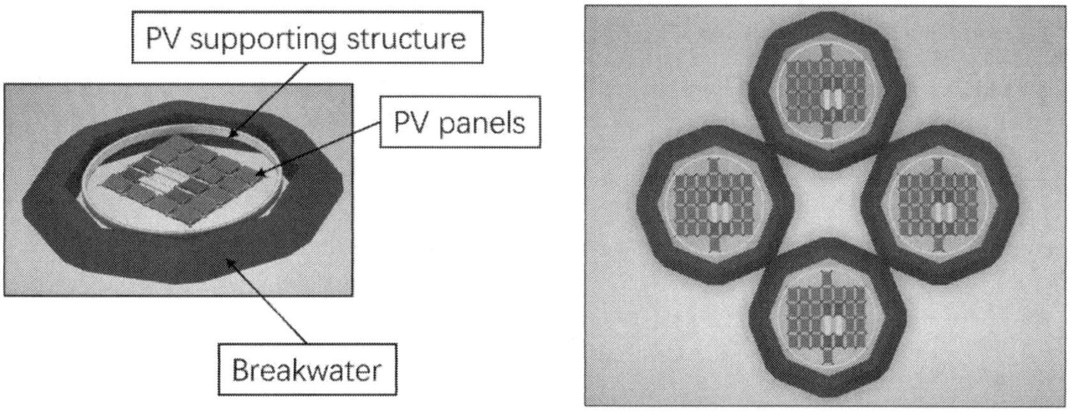

(a) (b)

Figure 1. The configuration of the FPV solar farm: (a) Single modular design, (b) multiple modular design.

The FPV platform is equipped with a tracking system that ensures a constant rotational speed. The active control system is currently in place to ensure that the platform functions smoothly, as shown in Figure 2(a-c). However, as technology advances and evolves, a passive control system (Figure 2d) can be implemented in the future. This would allow for a more efficient and streamlined approach to the management of the platform.

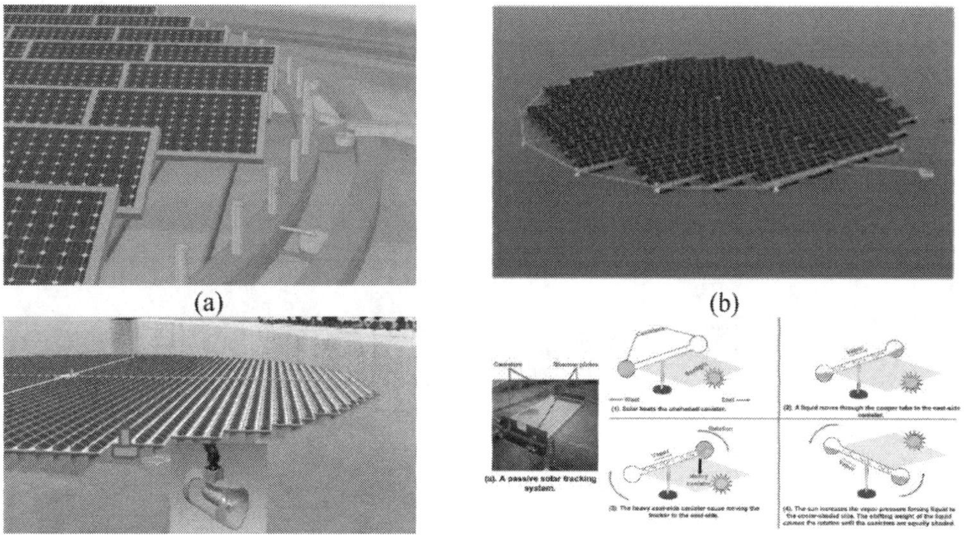

(a) (b) (c) (d)

Figure 2. Tracking system used on solar farms: (a) Tracking with a confining structure [13], (b) rope system [13], (c) bow thrusters [13], (d) passive tracking [14].

3. Computational approach

In this study, a numerical framework is presented to investigate the dynamic motions of a rotary solar farm with a tracking system in waves based on the open-source toolbox OpenFOAM. The flow field is calculated by solving the Navier-Stokes equations, while the structure motions are calculated by rigid body assumptions. The relative rotation between the PV supporting structure and breakwater is accounted for using the nested sliding mesh technique in OpenFOAM. The catenary shape mooring lines and their effects on the structural motions are evaluated by a quasi-static method. The detailed numerical principles of each are described in the following sub-sections.

3.1. Governing equations

The governing equations for a transient, incompressible, and viscous fluid can be written as

$$\nabla \cdot U = 0 \tag{1}$$

$$\frac{\partial \rho U}{\partial t} + \nabla \cdot (\rho UU) - \nabla \cdot \tau = -\nabla P_d + \rho g + F \tag{2}$$

where U refers to the velocity of flow field, ρ is the mixed density of water and air, g is the gravity acceleration, P_d refers to the dynamic pressure, τ is the dynamic viscosity, F is the surface tension. The VOF technique was adopted to simulate the free surface using an additional transport equation to solve for the volume fraction a. The volume fraction a is assigned a value of 0 when the cells are filled with air, and a =1 the cell is filled with water.

3.2. Computational domain and mesh generation

The finite volume mesh is generated by the mesh generation tool "SnappyHexMesh" based on a cell splitting and body fitting technique in OpenFOAM. The model scale of solar farm geometry is used in this study, with a scale of 1:100 applied. The purpose of using a model scale is to validate the experimental data in future work. The wave domain extends in three directions, i.e., $-3.0L < x < 4.0L$, $-2.0L < y < 2.0L$ and $-4.0L < z < 1.5L$, where L refers to the characteristic length of the solar farm (1.2m), as shown in Figure 3(a). The mesh region at the free surface is progressively refined several times, however, the mesh uncertainty tests are not conducted in this study. The mesh region close to the structure requires a more refined mesh in the area immediately around the solar farm, as shown in Figure 3(b).

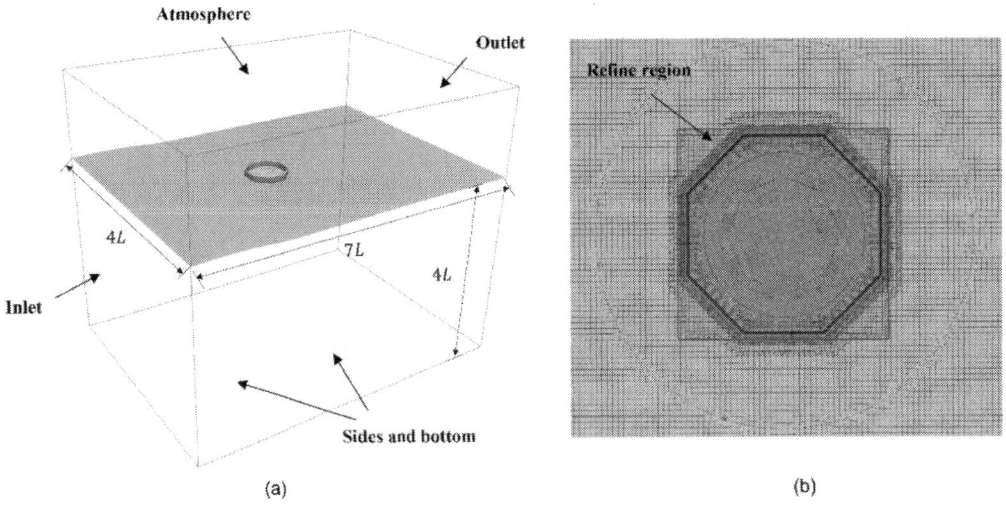

Figure 3. The layout of (a) the overall computational domain and (b) detailed mesh regions.

3.3. Boundary conditions

The boundary conditions for the wave-structure interaction problem are as follows: at the inlet on the left boundary, the velocity and alpha phases are prescribed as the incident waves. An in-house wave absorption boundary condition is used on the outlet side of the wave domain. The top domain is set as atmosphere, and the bottom of the domain is set as a symmetry plane, representing a deep-water condition. The moving wall boundary condition with zero pressure gradient is defined on the geometrical surface of the solar farm.

3.4. Wave generation and absorption

The wave generation library "waves2Foam" used in this study to generate the regular heading waves inside the wave domain [15]. The second order Stokes waves theory was used to generate the regular waves throughout all the cases with the following equations:

$$\eta = \frac{H}{2}\cos(\theta) + k\frac{H^2}{4}\frac{3-\sigma^2}{4\sigma^3}\cos(2\theta) \tag{3}$$

$$u = \frac{H}{2}\omega\frac{\cosh(kz)}{\sinh(kz)}\cos(\theta) + \frac{3}{4}\frac{H^2\omega k\cosh(2kz)}{4\sinh^4(kh)}\cos(2\theta) \tag{4}$$

$$w = \frac{H}{2}\omega\frac{\sinh(kz)}{\sinh(kz)}\sin(\theta) + \frac{3}{4}\frac{H^2\omega k\sinh(2kz)}{4\sinh^4(kh)}\sin(2\theta) \tag{5}$$

where, H is the wave height, wave angle $\theta = kx - \omega t + \psi$ with k is the wave number, ω is the angular frequency and ψ is the wave phase and $\sigma = \tanh(kh)$.

The relaxation zone technique is used to improve wave quality near the inlet boundary and eliminate spurious reflections at the outlet boundary. The primary function of the relaxation zones is specified by the following equations.

$$a_R(\chi_R) = 1 - \frac{\exp(\chi_R^{3.5})-1}{\exp(1)-1} \tag{6}$$

$$\phi_R = \omega_R\phi_R^{computed} + (1-\omega_R)\,\phi_R^{target} \tag{7}$$

where ϕ_R refers to either the velocity or volume fraction of water a. The definition of χ_R is that the weighting function a_R is always 1 at the interface between the non-relaxed computational domain and the relaxation zones, and χ_R is a value between 0 and 1.

3.5. Model description

3.5.1. Rigid body motion. In OpenFOAM, the global motion of a solar farm can be calculated based on the rigid body motion equations, which are written as follows:

$$H(q)\ddot{q} + C(q,\dot{q}) = \tau \tag{8}$$

where $q, \dot{q}, \ddot{q}$ denote as the position, velocity and acceleration of the structures, H is the inertia matrix, C is the force which produces zero acceleration, which accounts for forces like gravity, Coriolis and centrifugal forces, τ is the total force acting on the bodies. In this equation, multi-values can be calculated based on what has been provided. Therefore, the forward dynamics method is applied to calculate the acceleration based on the initial position, velocities and forces with the equations shown below:

$$\ddot{q} = FD(model, q, \dot{q}, \ddot{q}) \tag{9}$$

This calculated acceleration will be integrated for the first time step to get the new position and velocity, which will be used as input for the next time step. The forces at bodies are updated simultaneously. Boundary conditions can be applied to a rigid body by applying restraints (forces acting on the body) and constrained (the degree of freedom is restricted).

3.5.2. The nested-sliding mesh technique. The tracking system raise complexity in numerical modellings. To account for the relative rotation between the PV supporting structure and breakwater, the nested sliding mesh technique is applied in OpenFOAM. Two pairs of AMI (Arbitrary Mesh Interface) surfaces with three cell zones are defined accordingly, as shown in Figure 4. The outer cell zone completely covers the structure, allowing the solar farm to move globally in 6 DOF. The inner cell zone covers the PV supporting structure, allowing for relative rotational movement to the breakwater. AMI surfaces are sandwiched between three cell zones for data transmission and exchange. It is worth noting that the usage of nested sliding mesh should avoid any overlap or gap between each pair of AMI surface. Overall, the nested sliding mesh allows for the modelling of the solar farm global responses with free surface, self-rotations of PV supporting structure and their coupled motions. The rotary speed is controlled constant in this study.

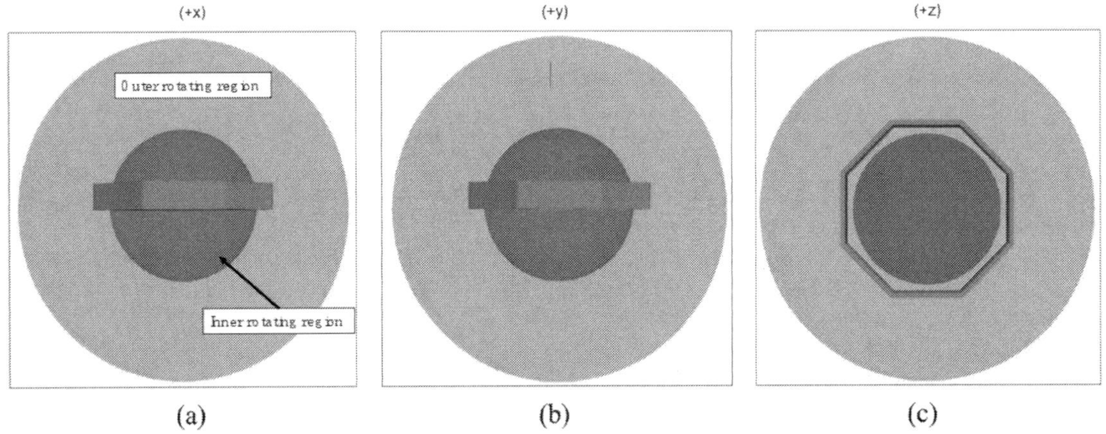

Figure 4. The layout of the nested sliding mesh and cell zone regions: (a) +x, (b) +y, (c) +z.

3.6. Mooring system

The mooring tension is calculated using the catenary equation, derived from [16]. Figure 5(a) shows the profile of a two-dimensional mooring line. The initial pretension is T_f at the top point of the line. The water depth, or the depth of the anchor point, is h. Quasi-static methods assume that the shape and tension of a mooring line at any given instant are static while ignoring the dynamic loadings from fluids.

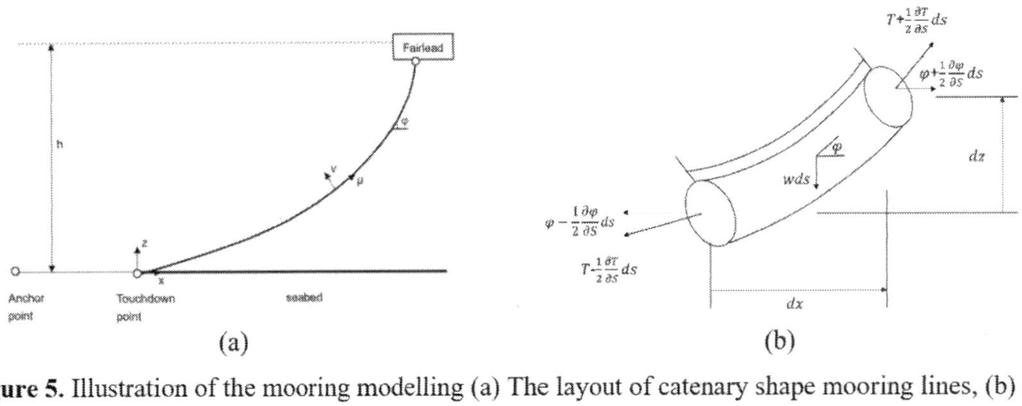

Figure 5. Illustration of the mooring modelling (a) The layout of catenary shape mooring lines, (b) An example of small segment of catenary mooring line.

An infinitely small element of mooring line is shown in Figure 5(b). In the absence of fluid forces, it is possible to resolve the forces acting normally and tangentially to the element with the following equations:

$$\frac{d\varphi}{ds} = \frac{w}{T}cos\varphi \tag{10}$$

$$\frac{dT}{ds} = wsin\varphi \tag{11}$$

where T is element's tension, w is submerged weight per unit length, φ is the angle between the axial and horizontal directions.

$$H = T_f - wh \tag{12}$$

$$L = h\sqrt{\frac{2T_f}{wh} - 1} \tag{13}$$

$$X = \frac{H}{w}cosh^{-1}(\frac{wh}{H} + 1) \tag{14}$$

$$V = wL \tag{15}$$

where L is the length of the line in its elongation under the given pretension T_f, X is the horizontal distance from touchdown point to fairlead, and V is the vertical force component at fairlead. The mooring line seabed interaction is modeled by a simplified spring damper model. When a node touches the seabed, a vertical reaction force is applied to that node. To obtain the tension force and coordinates of each node of a mooring line, the above equations are solved in a piecewise manner.

4. Results

In order to evaluate the performance of the solar farm, a series of numerical simulations were conducted in a regular heading wave with a frequency $f_w = 0.91Hz$ and a wave height $H_w = 0.10m$. The time step is selected as 0.005s through all simulations to ensure the model progress smoothly. The numerical results are presented in this section, including the analysis of the breakwater performance, tracking system evaluation and the dynamic motions of the solar farm with and without mooring lines.

4.1. Breakwater performance

The octagonal shape breakwater installed surrounding the PV supporting structures to absorb the wave energy and avoid direct contact between the structure and waves. Two wave gauges were placed: one in the wave propagation area and one between the breakwater and supporting structure. It was found that approximately 40% of wave heights are blocked due to the breakwater's design. However, breakwater performance cannot be accurately determined based on a simulation at a single frequency. It is worth repeating numerical models with different wave elevations and frequencies to properly evaluate breakwater performance.

4.2. Tracking motion visualization

An active control tracking system is installed for the solar farm. It allows for relative rotation between the breakwater and the PV supporting structure. The rotational speed for this design is modelled as 1 rad/s, but in reality it should be calibrated with the angle of sunlight radiation. Figure 6 shows screenshots of five rotating angles, from 45 to 135 degrees, taken at 5-second intervals over 25 seconds. Due to the relatively slow rotational speeds, the rotation of the structure may not have an effect on the surrounding flow field. However, the rotation of the upper PV panel will change the wind field, which is worth further investigation.

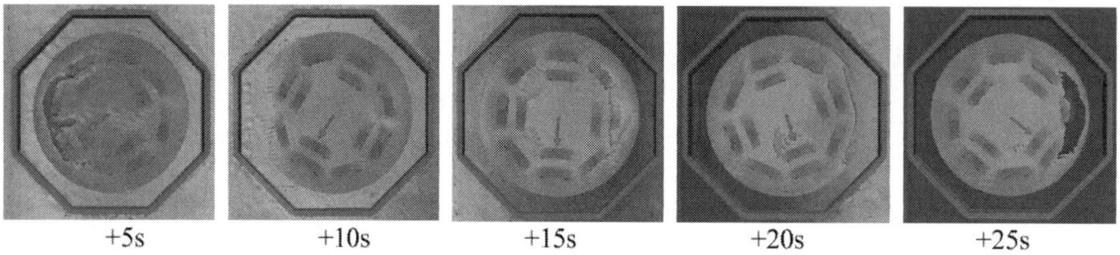

Figure 6. Virtual observation of the active control system.

4.3. Dynamic motion of solar farm

The dynamic response of the solar farm without a mooring system is evaluated at wave frequency $f_w = 0.91Hz$ and wave height $H_w = 0.10m$. The time-series heave and pitch motions are plotted in Figure 7. The signals have a main frequency component coincident with the wave frequency. However, there are also noticed nonlinearity components that may come from the implementation of breakwater. The three-dimensional virtual observation of the wave-structure interaction is shown in Figure 8.

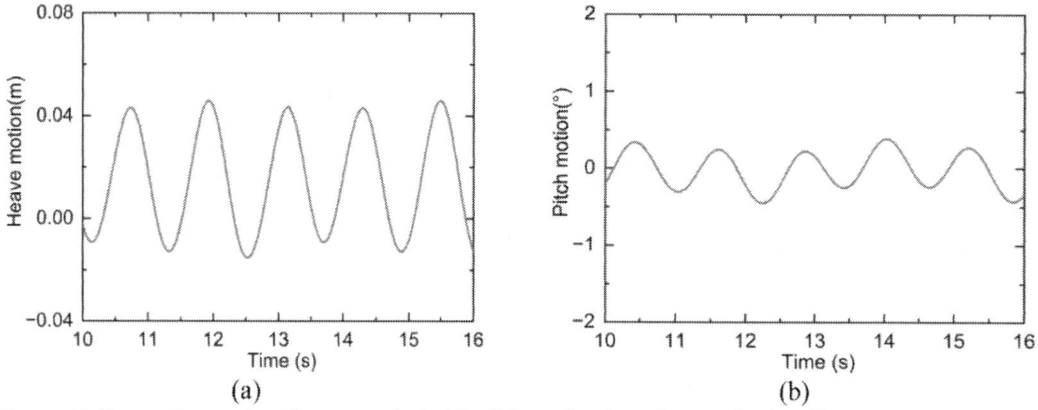

Figure 7. Dynamic motion (heave and pitch) of the solar farm in regular heading waves: (a) Heave, (b) Pitch.

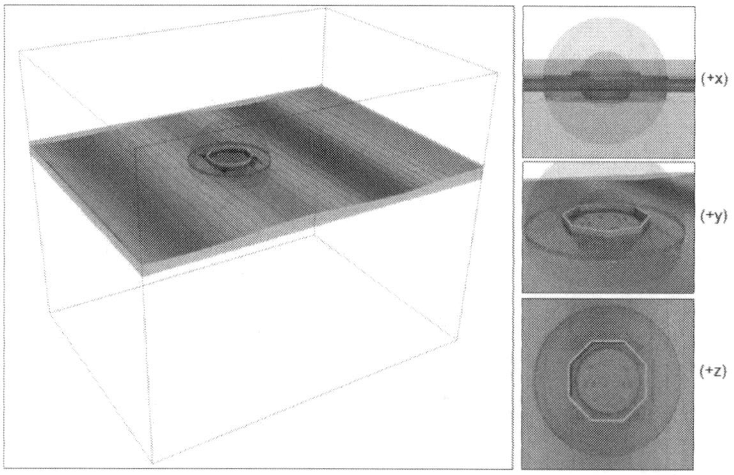

Figure 8. Virtual observation of the solar wave farm in waves.

4.4. FPV with catenary mooring system

The mooring system is essential in solar farm design, as it provides restoring forces and moments to balance environmental loadings and help maintain the farm's position at sea. In this study, the catenary type of mooring lines was implemented, which is suitable for floating structures operating in medium-deep water conditions. It consists of a chain or wire rope that hangs between the fairlead and anchor points, forming a curve known as the catenary. Eight catenary mooring lines, each with their fairleads attached to the vertex of the breakwater, distributed at a 45°angle, as shown in Figure 9. The overall length of the mooring lines is 7.8m, with 1.56m (8 segments) lying down on the seabed and 6.24m hung in water. The main mooring characteristics are summarized in Table 1.

Table 1: Properties of the catenary type mooring lines.

Properties	Value
Arrangement angle	45°
Mooring segment	40
Mooring length	7.8m
Mooring line diameter	3.656mm
Mooring line mass per unit length	0.0607
Mooring line stiffness	$2.78e^6$
Seabed stiffessnesss	$1e^6$

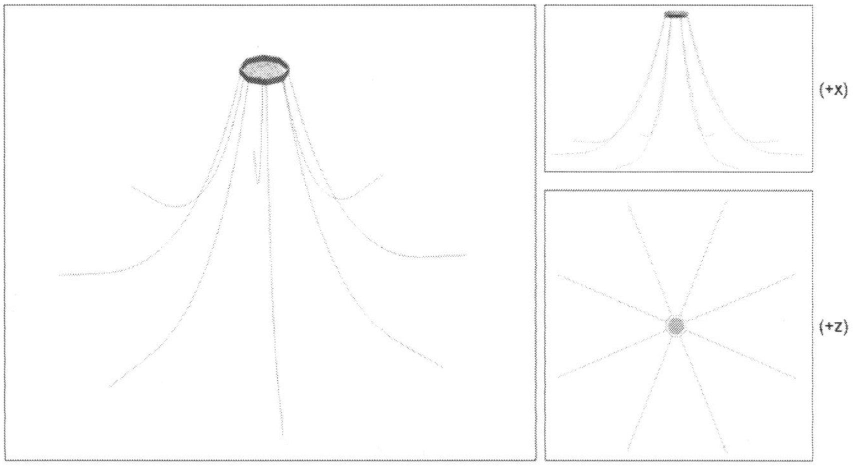

Figure 9. Catenary type of mooring system configuration of the proposed FPV platform.

Figure 10 compares the effects of attaching a mooring system on the structural motions of the solar farm. The stabilization provided by the mooring lines is particularly noticeable in pitch motions, where the peak magnitude is reduced by approximately 12.5%. However, heave motions are not significantly affected by the mooring lines, as catenary mooring lines restrain the structure through gravity. To achieve optimal performance of the mooring lines, an optimization procedure is required to determine their characteristics. This process involves a detailed analysis of the environmental conditions, including wind, waves, and currents, to ensure that the mooring system can withstand the expected loads.

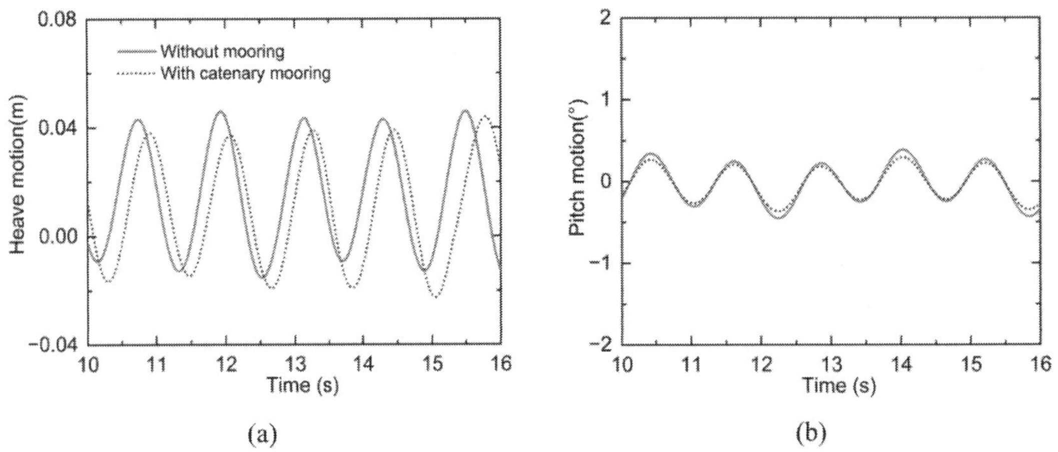

Figure 10. Comparison on the vertical motions of the solar farm with and without mooring system: (a) Heave, (b) Pitch.

5. Conclusions

The development of offshore solar farms is a crucial step towards sustainable energy production. This paper presented a novel design for a solar farm that includes a breakwater, PV supporting structure, mooring system, and tracking system. This design has the potential to shed light on the industry for future prototype designs.

To study the dynamic motions of such a system in waves, high fidelity CFD methods are applied. Some important factors, including breakwater performance, dynamic motions with/without a mooring system, are studied. The main conclusions are outlined below:

- Designing breakwaters to avoid wave attacks is important for the FPV section, as it reduces the risk of damage due to rough seas.
- The tracking system should be modelled so that the PV supporting structure rotates along with the sun, improving conversion efficiency. This would ensure that the solar panels are always positioned to receive the maximum amount of sunlight.
- The mooring system should be carefully designed, as the catenary mooring lines tested in this paper, can effectively reduce the global pitch motion. This would minimize the amount of movement experienced by the solar farm so as to improve its stability.
- Experimental studies will be conducted in the future to validate the present CFD framework, using the wave tank facility at Cranfield University.

Acknowledgements

This work has been supported by funding from the Innovate UK "Solar2Wave" project (grant No. 10048187), The authors would like to acknowledge the usage of the Delta 2 High-Performance Computing based at Cranfield University.

Referencence

[1] Sahu A, Yadav N, and Sudhakar K, Floating photovoltaic power plant: A review. *Renewable and Sustainable Energy Reviews*, 2016. **66**: p. 815-824.

[2] Trapani K and Redon M, A review of floating photovoltaic installations: 2007-2013. *Progress in Photovoltaics*, 2015. **23**(4): p. 524-532.

[3] Essak, L and Ghosh A, Floating Photovoltaics: A Review. *Clean Technologies*, 2022. **4**(3): p. 752-69.

[4] Solomin E, Soritkin E, Cuce E, and Selvanathan S P, Hybrid Floating Solar Plant Designs: A

Review. *Energies*, 2021. **14**(10).
[5] Cazzaniga R and Rosa-Clot M, The booming of floating PV. *Solar Energy*, 2021. **219**: p. 3-10.
[6] DNVGL_AS, *Recommended Practice DNVGL-RP-0584: Design, Development and Operation of Floating Solar Photovoltaic Systems*. 2021: Technical Report.
[7] OceanSun. *https://oceansun.no/*. [cited 2023 10/04].
[8] Oceans_of_Energy. *https://oceansofenergy.blue.com*. 2020 [cited 2023 19/03].
[9] Safety4Sea, *Offshore floating solar farm installed at the Dutch North Sea*. 2020.
[10] Choi S M, Park C D, Cho S H and Lim B J, Effects of various inlet angle of wind and wave loads on floating photovoltaic system considering stress distributions. *Journal of Cleaner Production*, 2023. **387**.
[11] Song J, Kim J, Chung W C, Jung D, Kang Y J and Kim S, Wave-induced structural response analysis of the supporting frames for multiconnected offshore floating photovoltaic units installed in the inner harbor. *Ocean Engineering*, 2023. **271**.
[12] Choi Y K and Lee Y G, A Study on Development of Rotary Structure for Tracking-Type Floating Photovoltaic System. *International Journal of Precision Engineering and Manufacturing*, 2014. **15**(11): p. 2453-60.
[13] Cazzaniga R, Cicu M, Rosa-Clot M, Rosa-clot P, Tina G M and Ventura C, Floating photovoltaic plants: Performance analysis and design solutions. *Renewable and Sustainable Energy Reviews*, 2018. **81**: p. 1730-1741.
[14] El Hammoumi A, Chtita S, Motahhir S and El Ghzizal A. Solar PV energy: From material to use, and the most commonly used techniques to maximize the power output of PV systems: A focus on solar trackers and floating solar panels. *Energy Reports*, 2022. **8**: p. 11992-2010.
[15] Jacobsen N G, Fuhrman D R and Fredsøe J, A wave generation toolbox for the open-source CFD library: OpenFoam®. *International Journal for numerical methods in fluids*, 2012. **70**(9): p. 1073-1088.
[16] Ong P P and Pellegrino S. Modelling of seabed interaction in frequency domain analysis of mooring cables. in *International Conference on Offshore Mechanics and Arctic Engineering*. 2003.

Experimental study on the hydrodynamic characteristics of a Bragg breakwater with triple vertical plates

Jing-Ping Wu[1, *], Hong-Ye Luo[1], Zao-Jian Zou[2], Peng Chen[3], Jing-Jing Qi[4], Xing-Yu Xu[4]

[1] School of Naval Architecture, Ocean and Energy Power Engineering, Wuhan University of Technology, Wuhan 430063, China

[2] School of Naval Architecture, Ocean and Civil Engineering, Shanghai Jiao Tong University, Shanghai 200240, China

[3] Zoomlion Heavy Industry Science and Technology Company Limited, Changsha 410013, China

[4] Shengli Oilfield Technology Inspection Center, SINOPEC, Dongying 257000, China

[*] Corresponding author *E-mail address*: wujp@whut.edu.cn

Abstract. The hydrodynamic characteristics of a Bragg breakwater with triple vertical plates in monochromatic regular waves are experimentally studied. The two spacings between the every two adjacent plates are denoted by S_1 and S_2 respectively which are both wide enough for the necessary of Bragg reflection, and S_2 is equal or unequal to S_1. A series of experiments are carried out in the wave flume equipped with a piston-type wave generator, and the hydrodynamic performance are analysed with the variation of the spacing S_2, the plate draft d_1 and freeboard d_2, the wavelength L and the wave steepness H/L (H is the wave height). Special attention is paid to the effects of nonuniform spacings on the hydrodynamic performance and the characteristics of Bragg reflection. In addition, the comparisons between the fixed and the unrestrained positioning are also presented. The results show that the transmission coefficients of the two unequal spacings mostly are smaller than the two equal spacings, and the intensity and bandwidth of Bragg reflection are increased by the non-uniform spacings. Under the unrestrained positioning, the movements have no serious effects on the wave attenuation performances of the breakwater. Moreover, the trough of transmission coefficient occurs at the left of the primary Bragg reflection condition due to both of Bragg resonance and wave energy dissipation not just Bragg resonance.

1. Introduction

Breakwaters are commonly used to protect ports, coasts, docks from destroyed by severe waves. They may be also used to provide a tranquil sheltered area for marine operation. A traditional breakwater is bottom-founded and strongly limited by water depth. Hence, another kind of alternative structures, known as free surface breakwater (or floating breakwater), were studied by many researchers as it is more economic, environmentally friendly, etc.

There is a large number of free surface breakwaters which can be catalogued into different types: box-type, pontoon-type, mat-type, frame-type, plate-type, flexible-membrane-type and so on [1-4]. The mechanism of wave attenuation mainly relies on the wave reflection and various kinds of wave energy dissipation. While encountering the relatively longer period waves, it is necessary to enlarge the sizes of the free surface breakwaters. However, the sizes may be so large that it is impossible for the design and construction to put it into practical use. Hence, the measures to improve the ability of free surface breakwaters to mitigate long period waves keep continuously being developed.

These improving measures tend to be classified into three main kinds: (1) Adding appendages. The box-type breakwaters are mostly investigated, Mani [5] fixed a row of pipes beneath a trapezoidal float; Kee [6] hanged triple vertical tensioned porous membranes beneath a floating pontoon; Huang et al. [7] attached three rows of slotted barriers to the bottom of a floating box-type breakwater; Wu et al. [8] attached double arc-shaped flexible membranes to a box-type breakwater; He et al. [9] attached five rows of vertical plates with fabricated slots to a rectangular floating breakwater, etc. The purpose of adding appendages is mainly to improve wave energy dissipation. (2) Assembling multiple structures. Usually, two or more individual breakwaters were connected together to form a new single breakwater, Williams et al. [10] presented a pair of floating pontoon breakwaters of rectangular section; Yang et al. [11] connected two rectangular pontoons with two horizontal plates; Cho et al. [12] presented a dual vertical-flexible-membrane wave barrier; Panduranga and Koley [13] connected dual porous cylinders with a rectangular porous box; Zhou et al. [14] investigated a pair of parallel parabolic arc pontoons. This kind of measure can reinforce the efficacy of wave attenuation compared to only one monomer. (3) Bragg breakwaters. A Bragg breakwater is consisted of an array of breakwaters that are completely independent of each other, and the spacings of the adjacent breakwaters are wide enough relative to the draft of the breakwaters. Because the wave reflections are strengthened as Bragg reflection occurs when the spacings of the breakwaters are half of the incoming wavelength, the wave transmissions correspondingly are substantially decreased.

Bragg reflection used earliest to block water waves was in the investigations of interaction between regular waves and periodic sinusoidal sand dams at sea bottom in 1980s (Davies, [15]; Heathershaw, [16]; Mei, [17]). Nearly 30 years later, it was adopted to free surface breakwaters. In 2012, Karmakar et al. [18] presented a Bragg breakwater consisted of multiple vertically moored surface-piercing tensioned porous membranes, and analyzed its interactions with surface gravity waves based on the linear theory of water waves. Ding et al. [19] also investigated a group of flexible membranes where the multiple floating tensioned flexible membranes were laid out horizontally and over submerged periodical rectangular bars on the flat seabed. The effects of various parameters including the number, the tension, the spacing, and the length of the flexible membranes on Bragg reflection were numerically investigated. Beside the flexible membranes, the box- and pontoon-type breakwaters have been mostly studied. Ouyang et al. [20] studied a train of fixed floating pontoon breakwaters, and the characteristics of Bragg reflections with the dependence on the spacing, the number, the draft and width of the breakwater, and the wavelength were investigated numerically. Ding et al. [21] studied three kinds of surface-piercing fixed structures with rectangular, cosinoidal and triangular sections respectively, and the influences of the number, the draft, the width, the gap of the structures on the characteristics of Bragg reflection were investigated numerically. Ji et al. [22] experimentally studied double-row floating breakwaters with a dual rectangular mooring pontoon, and found that the double-row floating breakwaters significantly reduced the transmission coefficients, but the reason was attributed to dissipation of eddies and moon-pool effect, not wave reflection. Tseng et al. [23] numerically investigated the Bragg reflections of oblique water waves by various periodic surface-piercing structures over periodic parabolic bottoms based on the linear theory of water waves. Chen and Wu [24] numerically studied the wave attenuation characteristics of two surface-piercing fixed box structures by the desingularized boundary integral equation method. Chen et al. [25] investigated a pair of floating pontoon breakwaters of rectangular section partially restrained by linear moorings. Besides, plate-type structures are also often used as breakwaters, such as Bragg breakwaters. Xu et al. [26] experimentally studied ten rigid plates which were fixed vertically, penetrating the free surface. It was found that the curve of transmission coefficient appeared a trough when the wavelength was about twice the plate spacing. Shin and Cho [27] studied theoretically and experimentally two vertical thin plates on the free surface, and found that the wave diffractions depended on the relationship between the wavelength and the spacing of the two adjacent plates, and the peaks of reflection and transmission coefficients became sharper and narrower at the resonance regions. Ding et al. [28] investigated numerically and experimentally the multiple vertical thin plates by varying the number, the immerged depth and the spacing of the plates, and analyzed the characteristics of Bragg reflection involving the occurring points, the primary reflection coefficients and the effective bandwidths.

Using Bragg breakwater to mitigate long period waves should have good application prospective. Theoretically, the wavelengths of waves which a Bragg breakwater could effectively attenuate should be about twice the Bragg breakwater spacings. However, a Bragg breakwater has a narrow wave reflection resonance band. This should be improved because the incoming waves in sea circumstances commonly have a wide wave period range. To overcome this defect, Zhang et al. [29] numerically studied doubly- and multiply-sinusoidal bottom undulations consisting of both slowly-varying and rapidly-varying components, and found that the bandwidth of high-performance region was increased due to many sub-harmonic and higher-harmonic Bragg resonances. Hsu et al. [30] numerically investigated the Bragg scattering of multiply composite artificial bars by varying the relative bar height and the number of bars with different intervals, and showed that the bandwidth of high-performance region was increased. Tsai et al. [31] investigated theoretically and experimentally the Bragg scattering of water waves by multiple composite artificial bars with different shapes, spacings, bar heights, bar footprint, and the number of bars, and the increased bandwidth was observed. Recently, in the respect of free surface breakwater, Ding et al. [32] numerically studied the interaction of water waves with multiple composite flexible membranes by changing the relative group spacing, the end condition, the number, the tension, the draft and the protrusion of the membranes, and found that the bandwidth of Bragg reflection was increased compared to the multiple uniform-spacing flexible membranes. However, few relevant experimental studies can be found in literature.

From the studies mentioned above, it can be found that the plate-type structures were frequently used. In this paper, we use triple vertical plates as Bragg breakwaters with non-uniform wide spacings. The hydrodynamic performance of the breakwaters and the characteristics of Bragg reflection are experimentally studied in a wave flume under monochromatic regular waves. The rest of the paper is structured as follows: Section 2 describes the experiments in the wave flume. Section 3 is devoted to the hydrodynamic coefficients. In Section 4, the experimental method is validated by comparing the experimental results with the numerical results. In Section 5, the effects of the spacings, freeboard and draft of plates are discussed. In addition, the comparisons between two positioning modes, fixed and unrestrained, are presented. In Section 6, the conclusions drawn from this study are summarized.

2. Experiments

2.1. Experimental facilities

The wave flume in the Laboratory of Fluid Mechanics at Wuhan University of Technology has the sizes of 18 m × 0.6 m × 0.8 m, as shown in Figure 1. Its piston-type wave generator controlled by an AC servo is installed to generate monochromatic regular waves in this experimental study. At the wave-absorbing section, a group of triangular sun shields with vertical through holes is fixed at another end of the flume.

Figure 2 shows the sketch of the experimental setup. The triple vertical plates are positioned at the experimental section of the wave flume, and four capacitance-type digital wave probes (see Figure 1) with an accuracy of ±1mm are fixed in front of and behind the plates respectively.

The wave elevations are acquired by the four wave probes. There are two wave probes, P1 and P2, fixed in front of the plates, and two other probes, P3 and P4, fixed behind the plates. The distances between the wave probes are shown in Figure 2. The wave probes P1 and P2 are used to decompose the reflected waves; while P2 is in actual use, P1 is in reserve in case of P2 failed. Both P3 and P4 measure the transmitted waves directly. Similarly, P3 is in actual use while P4 is in reserve.

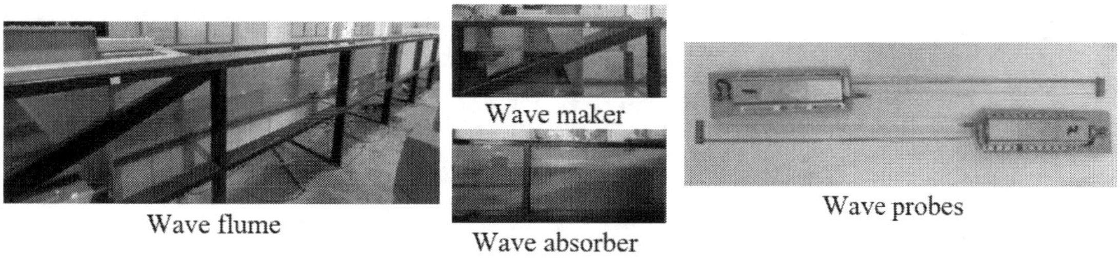

Figure 1. The wave flume and wave probes.

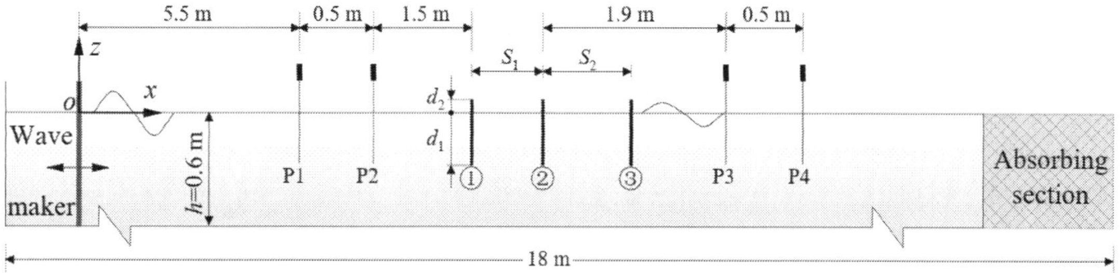

Figure 2. The sketch of the experimental setup.

2.2. Physical models

Herein, the Bragg breakwater consists of three vertical plates made of transparent perspex sheet. Each plate is 5 mm in thickness, 0.595 m in length, and there are three plate widths: 0.12 m, 0.15 m, and 0.18 m (see Figure 3), to investigate the effect of the draft and freeboard of the plates on the hydrodynamic characteristics.

The plates are positioned in two modes: (1) Fixed. The plates are fixed to the flume by aluminum rods, the breakwater model installed are shown in Figure 3. In fixed cases, the spacing S_1 is 0.40 m, the spacing S_2 is changed as 0.40 m, 0.48 m, 0.50 m, 0.52 m, 0.55 m, 0.60 m. The spacing S_2 =0.40 m stands for the uniform spacing breakwater, while others stand for the non-uniform spacing breakwaters. The freeboard d_2 is changed as 0 and 0.03 m; correspondingly, the draft d_1 is determined by subtracting the freeboards from the plate widths. The S_1, S_2, d_1 and d_2 are defined in Figure 2. (2) Unrestrained. There are four tube floats to supply buoyancy for the plates. The floats and the floating breakwater model are shown in Figure 4. The floats are made of PVC flexible membrane; each of them is a cylindrical tube of 5 cm diameter, and atmospheric air is sealed in. The lengths of the floats are equal to the spacing where they are placed. In the unrestrained cases, the spacing S_1 is still 0.40 m, while the spacing S_2 is only 0.48 m; the freeboard d_2 is just 0, the draft d_1 is correspondingly equal to the plate widths of 0.12 m, 0.15 m, 0.18 m. Table 1 summarizes the parameters of the experimental models.

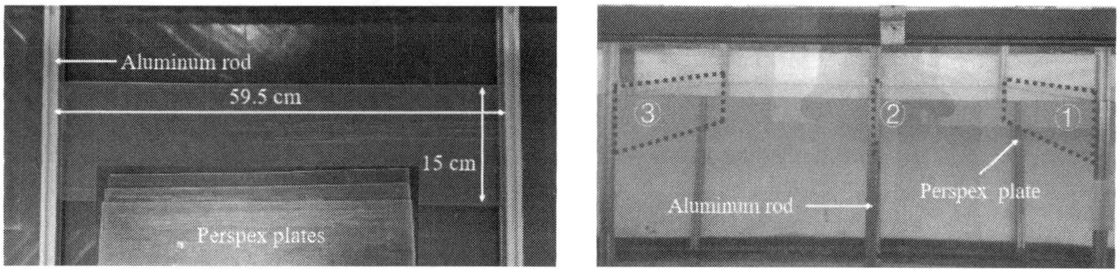

Figure 3. The photos of the fixed plates.

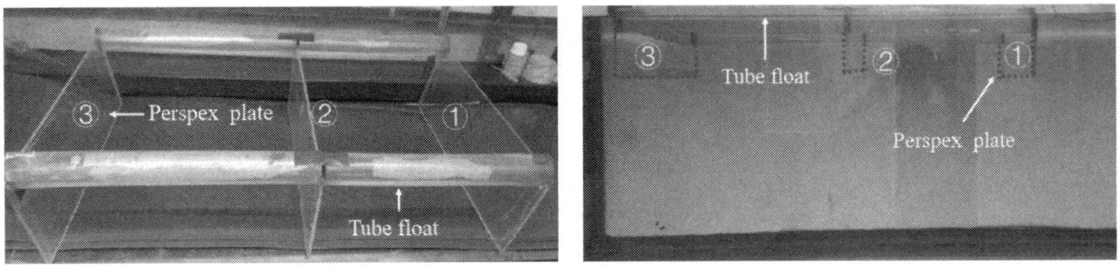

Figure 4. The photos of the unrestrained plates.

Table 1. The parameters of the experimental models.

Positioning	Plate width (m)	S_1 (m)	S_2 (m)	d_1 (m)	d_2 (m)
Fixed	0.12, 0.15, 0.18	0.40	0.40, 0.48, 0.50, 0.52, 0.55, 0.60	0.12, 0.15, 0.18; 0.09, 0.12, 0.15	0, 0.03
Unrestrained	0.12, 0.15, 0.18	0.40	0.48	0.12, 0.15, 0.18	0

2.3. *Wave parameters*

In this study, the regular waves are generated at a constant water depth $h = 0.6$ m. Two wave steepness, H/L=0.02 and 0.05, are selected, where H is the wave height, L is the wavelength; the wave period T ranges from 0.6 s to 1.2 s, and the corresponding wavelength range is 0.561 - 2.121 m. The wave parameters are given in Table 2. The actual experimental wave heights are slightly smaller than the wave heights H in Table 2 due to the wave generation principle, viscous dissipation, and the slits between the wave maker and the flume walls.

Table 2. The wave parameters.

T (s)	L (m)	H/L	H (mm)
0.6-1.2	0.561-2.121	0.02, 0.05	11.2- 42.4; 28.1-106.1

3. Hydrodynamic coefficients

In this paper, the hydrodynamic performance of the breakwater is expressed in transmission coefficient K_t, reflection coefficient K_r, and energy dissipation coefficient K_l.

The transmission coefficient K_t is defined as the ratio of the transmitted wave height H_t to the incident wave height H_i, that is,

$$K_t = \frac{H_t}{H_i} \tag{1}$$

The incident wave and the transmitted wave are recorded by the wave probe P3 (see Figure 2) without and with the breakwater model placed in the flume respectively. The results are shown in Figure 5, where the wave heights H_i and H_t are averaged by the three wave heights in the black box respectively.

The reflection coefficient K_r is defined as the ratio of the reflected wave height H_r to the incident wave height H_i:

$$K_r = \frac{H_r}{H_i} \tag{2}$$

For the decomposition of reflected wave, the fixed single-point method proposed in Wu et al. [33] is adopted, where the reflected wave is obtained by subtracting the incident wave from the combined wave. The incident wave and the combined wave are recorded by the wave probe P2 (see Figure 2) without and with the breakwater model placed in the flume respectively.

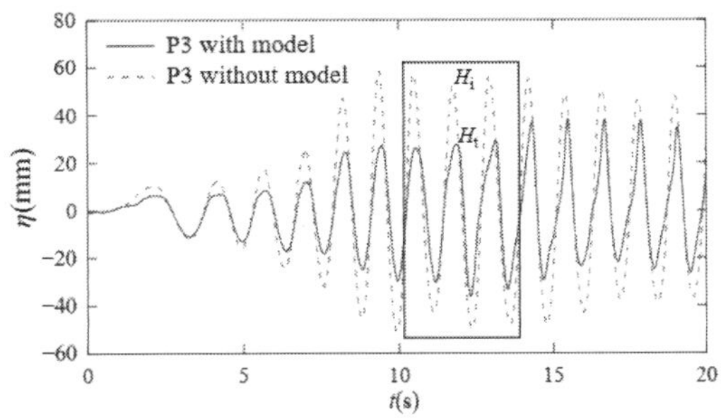

Figure 5. The transmission coefficient K_t (h=0.6 m; T=1.2 s, H/L=0.05; S_1=0.40 m, S_2=0.48 m; d_1=0.15 m, d_2=0; H_i =101.62 mm, H_t =61.68 mm, K_t=0.607).

In Figure 6, the incident wave, the combined wave, and the reflected wave are shown and compared with the numerical results (Num) of CFD (Computational Fluid Dynamics), where the wave heights H_i and H_r are averaged by the three wave heights in the black box respectively. On this basis, the reflection coefficient K_r is obtained.

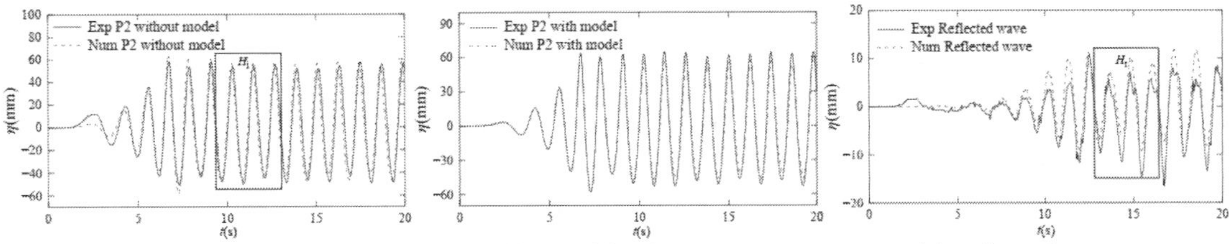

(a) Incident waves (b) combined waves (c) reflected waves

Figure 6. The reflection coefficient K_r (h=0.6 m; T=1.2 s, H/L=0.05; S_1=0.40 m, S_2=0.48 m; d_1=0.15 m, d_2=0; Exp: H_i=102.26 mm, H_r=18.1 mm, K_r=0.177; Num: H_i=102.48 mm, H_r=16.38 mm, K_r=0.16).

It can be seen from Figure 6 that for both the incident wave and the combined wave, the experimental wave trains are in good agreement with the numerical results. However, the experimental reflection coefficient K_r is 0.177, while the numerical one is 0.16, giving an error of about 10%. The reason for this error may be that comparing the contours of the reflected waves, the wave crests and wave troughs of the experimental results are not so smooth as those of the numerical results. In experimental operations, the run-up times of the P2 without and with the breakwater model synchronized with the startup time of the wave generator respectively. There might be a quite small time difference in the transmission of signals, which can lead to a tiny phase difference between the two wave trains of P2. Overall, it is demonstrated that the experimental results are reliable.

In addition, the following relationship exists between the energy dissipation coefficient K_l and the other two coefficients:

$$K_r^2 + K_t^2 + K_l^2 = 1 \tag{3}$$

Therefore, the energy dissipation coefficient K_l can be obtained by the following formula:

$$K_l = \sqrt{1 - K_r^2 - K_t^2} \tag{4}$$

4. Validation of experimental results by numerical method

4.1. Numerical method

In order to demonstrate the effectiveness of the physical experiments, a numerical wave flume is set up by using ANSYS Fluent. The numerical wave flume is similar to the physical wave flume. In the numerical flume, the air and water are viscous and incompressible, and subject to the force of gravity.

The RANS equations are discretized and solved by finite volume method under physical boundary conditions and initial conditions. The SST k-ω model is adopted as the turbulence model, the Volume of Fluid (VOF) is adopted to capture the free surface. The PISO method is used to solve the pressure-velocity coupling problem. The convection term and diffusion term are discretized by the second order upwind scheme. The temporal terms are discretized using a first-order backward Euler scheme. The UDF (User Defined Function) and dynamic mesh technique are used to realize the movement of the piston-type wave generator.

The mesh resolutions of the horizontal and vertical directions are about $L/80$ and $H/10$ respectively. The meshes are refined near the free surface and the plates. A gradually sparse grid technique is used in the wave-absorbing section starting 4 m away from the flume end. The time step is about $T/1000$, and the convergence precision is 1.0×10^{-6}.

4.2. Comparisons of experimental results with numerical results

Figure 7 shows the comparisons of the hydrodynamic coefficients between the experimental results and the numerical results under the conditions of S_1=0.40 m, S_2=0.48 m, d_2=0, H/L=0.05, and d_1 =0.12 m, 0.15 m and 0.18 m.

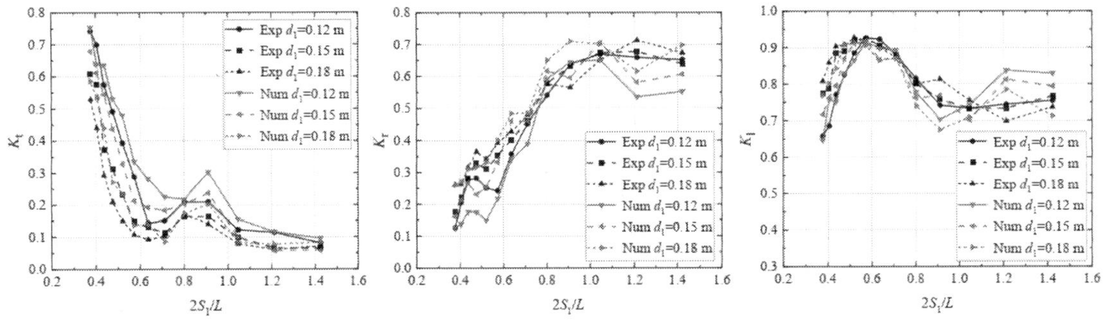

Figure 7. The comparisons between the experimental and numerical results.

In Figure 7, it can be seen that the trends of the experimental and numerical hydrodynamic coefficients are fairly consistent. The transmission coefficients K_t are in non-monotonic descent with the increase of $2S_1/L$, and appear troughs in the range of $2S_1/L$=0.6-0.8. The reflection coefficients K_r roughly increase with the increase of $2S_1/L$, and a strong reflection of Bragg resonance appears in the vicinity of $2S_1/L$=0.9-1.0, which is more obvious in the numerical results. The energy dissipation coefficients K_l are non-monotonic too, and a peak appears in the vicinity of $2S_1/L$=0.5-0.6. Besides, the effects of d_1 of the experimental and numerical results are generally identical. A larger d_1 gives a smaller transmission coefficient, and the reflection coefficient is the converse. From these comparisons, it is validated that the experimental tests are reliable.

On the other hand, there are some deviations between the experimental and numerical results. With respect to the transmission coefficient K_t, the experimental data are smaller than the numerical results, and it is converse for the energy dissipation coefficients K_l. The reason for the deviations may be that there is a systematic error of the experimental wave probes, and a smaller energy dissipation of the numerical simulation due to the ignored plate thickness.

5. Results and discussions

5.1. The effect of spacing S_2

Figure 8 shows the variations of the hydrodynamic coefficients with dimensionless plate spacing $2S_1/L$ under the conditions of S_2=0.40-0.60 m with S_1=0.40 m, d_1=0.15 m, d_2=0 m, and H/L=0.05.

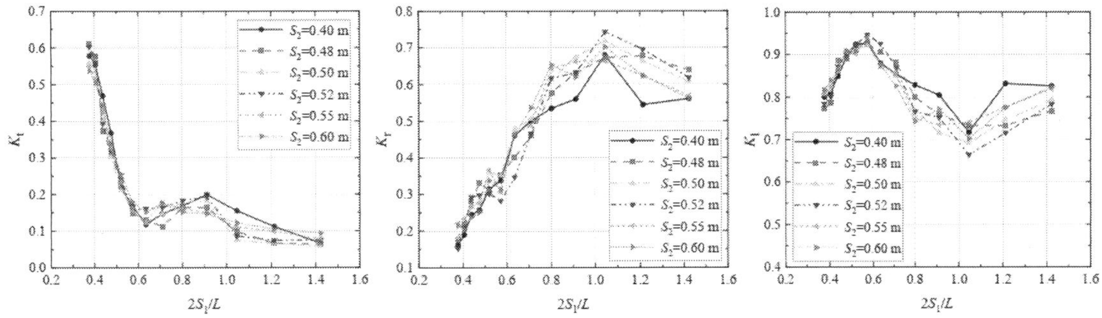

Figure 8. The effect of spacing S_2 (S_1=0.40 m, d_1=0.15 m, d_2=0, H/L=0.05).

In Figure 8, a steep decrease of the transmission coefficients K_t until $2S_1/L$ approaching to 0.65 is first observed; after a short increase, they decrease again. The reflection coefficients K_r generally increase with the increase of $2S_1/L$, and reach peaks in the vicinity of $2S_1/L$=1.05, indicating the occurrence of the primary Bragg resonance. The variation tendency of the energy dissipation coefficients K_l is non-monotonic, and a peak value appears in the vicinity of $2S_1/L$=0.55. According to the reflection coefficients and the energy dissipation coefficients, the mechanism that the transmission coefficients K_t appear troughs in the vicinity of $2S_1/L$=0.65 is due to the combination of Bragg resonance and wave energy dissipation. In addition, it is noted that while the stronger Bragg resonance occurs in the vicinity of $2S_1/L$=1.05, the wave energy dissipation gets weaker.

As for the effect of S_2 on the transmission coefficients K_t, generally the transmission coefficients of S_2=0.40 m are slightly larger than the others. Since S_2=0.40 m stands for the uniform spacing breakwater, it implies that the non-uniform spacing breakwaters are more conducive than the uniform spacing breakwater. Among the non-uniform spacing breakwaters, the transmission coefficients of S_2=0.48 m are slightly smaller than the others, where S_2=0.48 m is closest to the uniform spacing. The reflection coefficients of the non-uniform spacing breakwaters are visibly above those of the uniform spacing breakwater in the resonance range of $2S_1/L$=0.8-1.2, and both the reflection strength and bandwidth are increased. These should be attributed to the primary Bragg reflection. The energy dissipation coefficients of S_2=0.40 m, 0.48 m are roughly larger than others, it may be attributed to that the closer plates could lead to a more vortex energy dissipation.

5.2. *The effect of freeboard d_2*

Figure 9 shows the comparisons of the hydrodynamic coefficients between d_2=0 and d_2=0.03 m in the case of d_1=0.15 m, S_1=0.40 m, S_2=0.48 m, H/L=0.05.

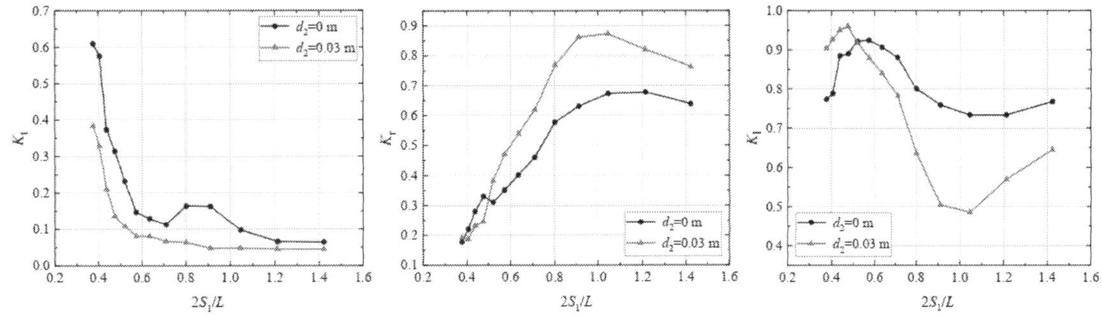

Figure 9. The effect of freeboard d_2 (d_1=0.15 m, S_1=0.40 m, S_2=0.48 m, H/L=0.05).

As can be seen from Figure 9, the transmission coefficients K_t and the energy dissipation coefficients K_l of d_2=0.03 m are generally smaller than those of d_2=0, and the reflection coefficients K_r of d_2=0.03 m are larger than those of d_2=0, except at several smaller $2S_1/L$. It indicates that the freeboard d_2 can effectively decrease the wave transmission, and the wave-attenuation mechanism of freeboard d_2 is due to more wave reflection than energy dissipation. The freeboard d_2=0.03 increases the reflection strength and bandwidth in the vicinity of $2S_1/L$=1.0, and brings about the occurrence condition of Bragg reflection shifting to left.

5.3. *The effect of draft d_1*

Figure 10 shows the effect of draft d_1 on the hydrodynamic coefficients at d_2=0, 0.03 m under S_1=0.40 m, S_2=0.48 m and H/L=0.05. The three plate widths are 0.12 m, 0.15 m, 0.18 m respectively.

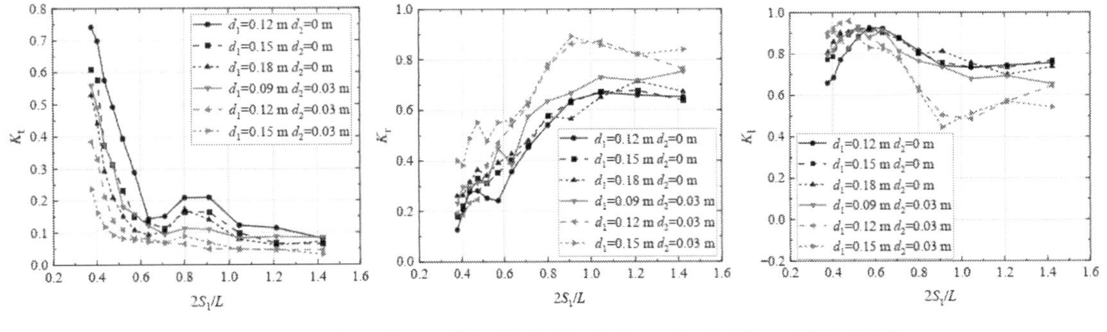

Figure 10. The effect of draft d_1 (S_1=0.4 m, S_2=0.48 m, H/L=0.05).

From Figure 10, it can be seen that the effects of draft d_1 on the hydrodynamic coefficients are generally identical for d_2=0 and d_2=0.03 m. First, while the d_1 is smallest, all the transmission coefficients are largest, and the reflection coefficients are generally smallest. Second, while $2S_1/L>0.6$, the three coefficients of the two larger d_1 almost coincide with each other. It indicates that there exists a limit of the draft, and further increase in draft will not improve the wave attenuation performance. While $2S_1/L<0.6$, the larger the draft d_1 is, the smaller the transmission coefficients are, and the larger the reflection coefficients are. Third, in the case of d_2=0 m, the reflection coefficients of three plates are generally close to each other; in the case of d_2=0.03 m, increasing the draft d_1 results in the increase of the reflection strength and bandwidth of Bragg reflection.

5.4. *The effect of wave steepness H/L*

Figure 11 presents the effect of wave steepness H/L on the hydrodynamic coefficients under H/L=0.02, 0.05 in the cases of S_1=0.40 m, S_2=0.48 m, and d_2=0, 0.03 m. The plate width is 0.15 m.

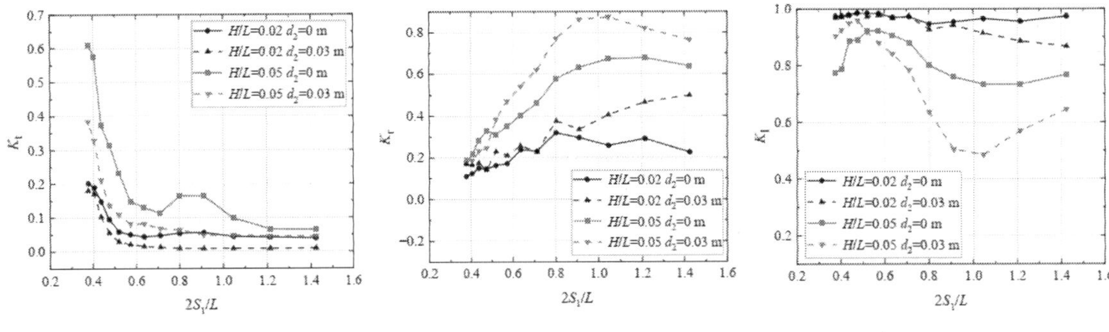

Figure 11. The effect of steepness H/L (S_1=0.40 m, S_2=0.48 m).

From Figure 11, it can be seen that for d_2=0 and d_2=0.03 m, the trends of effect of the wave steepness are consistent. The larger the wave steepness H/L is, the larger the transmission coefficients and the reflection coefficients are, and the smaller the wave dissipation coefficients are. Under larger wave steepness, the decline of wave dissipation coefficients is more than the increase in reflection coefficients. The effects of the two wave steepnesses are quite obvious.

5.5. Comparisons between the fixed and unrestrained positionings

Figure 12 displays the comparisons between the fixed and unrestrained positionings with d_1=0.12 m, 0.15 m, 0.18 m under the condition of S_1=0.40 m, S_2=0.48 m, d_2=0, H/L=0.05.

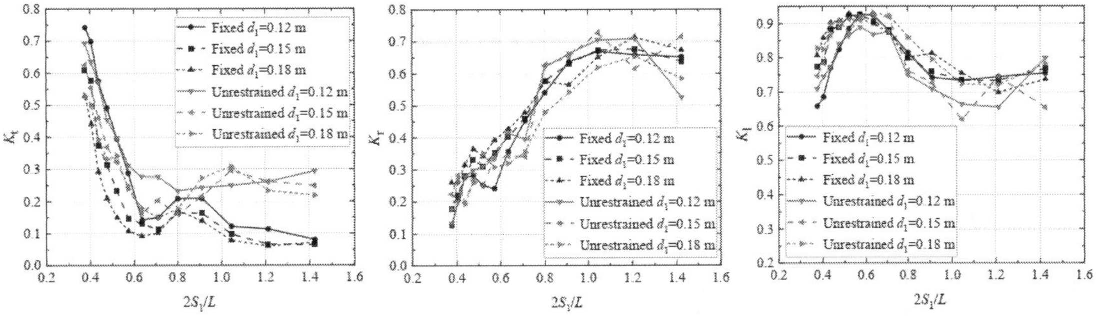

Figure 12. Comparisons of two modes of positionings (d_2=0, S_1=0.40 m, S_2=0.48 m, H/L=0.05).

As can be seen from Figure 12, while $2S_1/L$>0.6, the transmission coefficients of the unrestrained breakwater are obviously larger than those of the fixed one. It is because that the wave dissipations of the unrestrained breakwater are smaller than those of the fixed one. While $2S_1/L$<0.6, they are close to each other. Roughly, for both the two modes of positioning, the smaller the draft d_1 is, the larger the transmission coefficients are. For the unrestrained breakwater, the reflection coefficients of d_1=0.15 m, 0.18 m are unexpectedly smaller than those of d_1=0.12 m; the reason is probably due to the larger sway and heave motions of the breakwater with larger d_1. Moreover, the reflection strength and bandwidth of Bragg reflection of d_1=0.18 m are smaller. Nevertheless, the wave dissipation coefficients of d_1=0.18 m is still larger than those of d_1=0.12 m.

In addition, while $2S_1/L$<0.6, all the transmission coefficients, the reflection coefficients and the wave dissipation coefficients of the fixed and unrestrained breakwaters are almost overlapped. It implies that the movements of the Bragg breakwaters with multiple vertical plates have no obvious effects on the wave attenuation performance. This conclusion shows that these plate-type breakwaters do not need to be fastened with the mooring system to keep their wave attenuation performance, and the mooring system could be lightened.

6. Conclusion

In this paper, a kind of Bragg breakwater consisting of triple vertical plates with non-uniform spacings is investigated. A series of preliminary tests are conducted in a wave flume. The effects of the spacing, the freeboard and the draft of the plates, the wave steepness, and the positioning mode on the hydrodynamic performance and the characteristics of Bragg reflection are discussed through the transmission coefficients, the reflection coefficients and the wave dissipation coefficients. From this study, the following conclusions can be drawn:

1) Compared with the uniform spacing, the non-uniform spacings can improve the wave attenuation performance through both Bragg reflection and wave energy dissipation, and can increase the intensity and bandwidth of Bragg reflection.

2) Increasing the draft and freeboard of the plates can improve the wave attenuation performance and strengthen the Bragg reflection. However, there exists a limit of the draft. Besides, the wave attenuation performance of the breakwater becomes weaker in larger wave steepness.

3) Compared with the fixed breakwater, the wave attenuation performance of the unrestrained breakwater is slightly decreased. Nevertheless, the movements of the breakwater with multiple vertical plates have no serious effects on the wave attenuation performance.

In this study, the cases of the spacing S_2 larger than the spacing S_1 are investigated. In the future study, the reversed cases, that is, the spacing S_1 larger than the spacing S_2, is worthy of investigation. Besides, the experimental waves in this paper are regular; in the further study, the irregular waves need to be considered. Moreover, the draft of the plates should be investigated in more detail to find out the limit rule, and the moored breakwater needs to be investigated with respect to not only the hydrodynamic coefficients but also the mooring force and the breakwater motion.

Acknowledgements

The study was supported by Shengli Oilfield Science and Technology Research Project of Sinopec under Grant Nos. YKD2303 and Shandong Provincial Natural Science Foundation of China under Grant Nos. ZR2022QD074. The authors would like to thank the reviewers for his constructive guides which helped to improve the quality of this work significantly.

References

[1] McCartney B L. 1985. Floating breakwater design. *Journal of Waterway, Port, Coastal, and Ocean Engineering* **111(2)** 304-318.

[2] Teh H M. 2013. Hydraulic performance of free surface breakwaters: A review. *Sains Malaysiana* **42(9)** 1301-1310.

[3] Dai J, Wang C M, Utsunomiya T and Duan W. 2018. Review of recent research and developments on floating breakwaters. *Ocean Engineering* **158** 132-151.

[4] Zhao X L, Ning D Z, Zou Q P, Qiao D S and Cai S Q. 2019. Hybrid floating breakwater-WEC system: A review. *Ocean Engineering* **186** 106126.

[5] Mani J S. 1991. Design of Y-frame floating breakwater. *Journal of Waterway, Port, Coastal, and Ocean Engineering* **117(2)** 105-119.

[6] Kee S T. 2004. Floating pontoon-membrane breakwater. *Journal of Ocean Science and Technology* **1(1)** 49-57.

[7] Huang Z H, He F and Zhang W B. 2014. A floating box-type breakwater with slotted barriers. *Journal of Hydraulic Research* **52(5)** 720-727.

[8] Wu J P, Mei T L and Zou Z J. 2022. Experimental study on wave attenuation performance of a new type of free surface breakwater. *Ocean Engineering* **244** 110447.

[9] He F, Li J D, Pan J P and Yuan Z M. 2023. An experimental study of a rectangular floating breakwater with vertical plates as wave-dissipating components. *Applied Ocean Research* **133** 103497.

[10] Williams A N, Lee H S and Huang Z. 2000. Floating pontoon breakwaters. *Ocean Engineering* **27(3)** 221-240.

[11] Yang B, Chen Z J, Wang G Y and Wang Y X. 2014. Experimental investigation of twin pontoon-twin horizontal plate floating breakwater. *Chinese Journal of Hydrodynamics* **29(1)** 40-49.

[12] Cho I H, Kee S T and Kim M H. 1998. Performance of dual flexible membrane wave barriers in oblique waves. *Journal of Waterway, Port, Coastal, and Ocean Engineering* **124(1)** 21-30.

[13] Panduranga K and Koley S. 2021. Water waves scattering by cylindrical dual porous floating breakwaters connected with a rectangular porous box and floating over an undulated seabed. *Meccanica* **56(12)** 3001-3024.

[14] Zhou B Z, Zheng Z, Jin P, Wang L and Zang J. 2022. Wave attenuation and focusing performance of parallel twin parabolic arc floating breakwaters. *Energy* **260** 125164.

[15] Davies A G. 1982. On the interaction between surface waves and undulations on the seabed.

Journal of Marine Research **40(2)** 331-368.

[16] Heathershaw A D. 1982. Seabed-wave resonance and sand bar growth. *Nature* **296** 343-345.

[17] Mei C C. 1985. Resonant refection of surface water waves by periodic sandbars. *Journal of Fluid Mechanics* **152** 315-335.

[18] Karmakar D, Bhattacharjee J and Guedes Soares C. 2013. Scattering of gravity waves by multiple surface-piercing floating membrane. *Applied Ocean Research* **39** 40-52.

[19] Ding W W, Zou Z J and Wu J P. 2019. Bragg reflection of water waves by multiple floating horizontal flexible membranes with submerged rectangular bars on the seabed. *Applied Ocean Research* **83** 103-111.

[20] Ouyang H T, Chen K H and Tsai C M. 2015. Investigation on Bragg reflection of surface water waves induced by a train of fixed floating pontoon breakwaters. *International Journal of Naval Architecture and Ocean Engineering* **7(6)** 951-963.

[21] Ding W W, Zou Z J, Wu J P and Huang B G. 2019. Investigation of surface-piercing fixed structures with different shapes for Bragg reflection of water waves. *International Journal of Naval Architecture and Ocean Engineering* **11(2)** 819-827.

[22] Ji C Y, Deng X K and Cheng Y. 2019. An experimental study of double-row floating breakwaters. *Journal of Marine Science and Technology* **24(2)** 359-371.

[23] Tseng I F, You C S and Tsai C C. 2020. Bragg reflections of oblique water waves by periodic surface-piercing and submerged breakwaters. *Journal of Marine Science and Engineering* **8(7)** 522.

[24] Chen C Z and Wu J P. 2020. Study on wave attenuation characteristics of two box structures. *Proceedings of the 30th International Ocean and Polar Engineering Conference, Virtual, Online*. 2020-October, 2164-2170.

[25] Chen Y K, Liu Y, Meringolo D D and Hu J M. 2023. Study on the hydrodynamics of a twin floating breakwater by using SPH method. *Coastal Engineering* **179** 104230.

[26] Xu L, Wu J P, Ding W W, Zou Z J and Li T Q. 2016. Experimental study on the attenuating waves performance of multiple vertical plates based on Bragg resonance. *Chinese Journal of Hydrodynamics* **31(4)** 449-453.

[27] Shin D M and Cho Y. 2016. Diffraction of waves past two vertical thin plates on the free surface: A comparison of theory and experiment. *Ocean Engineering* **124** 274-286.

[28] Ding W W, Yue W Z, Sheng S W, Wu J P and Zou Z J. 2022. Numerical and experimental study on the Bragg reflection of water waves by multiple vertical thin plates. *Journal of Marine Science and Engineering* **10(10)** 1464.

[29] Zhang L, Kim M H, Zhang J and Edge B L. 1999. Hybrid model for Bragg scattering of water waves by steep doubly-sinusoidal bars. *Journal of Coastal Research* **15(2)** 486-495.

[30] Hsu T W, Tsai L H and Huang Y T. 2003. Bragg scattering of water waves by multiply composite artificial bars. *Coastal Engineering Journal* **45(2)** 235-253.

[31] Tsai L H, Kuo Y S, Lan Y J, Hsu T W and Chen W J. 2011. Investigation of multiply composite artificial bars for Bragg scattering of water waves. *Coastal Engineering Journal* **53(4)** 521-548.

[32] Ding W W, Zou Z J and Wu J P. 2018. Bragg reflection of water waves by multiple composite flexible membranes. *International Journal of Offshore and Polar Engineering* **28(2)** 135-142.

[33] Wu J P, Bo H, Zhang M, Pan J. 2017. A reflection wave decomposition method based on the linear superposition principle of water waves. Wuhan University of Technology, Wuhan, China. Patent No. CN201710605897.4.

IOP Conf. Series: Materials Science and Engineering 1288 (2023) 012043 doi:10.1088/1757-899X/1288/1/012043

Adaptive grid deformation method for CFD application to hull optimization

Shuguang Wang[1], **Yonghwan Kim**[1,2]

[1] Seoul National University, Department of Naval Architecture and Ocean Engineering,1 Gwanak-ro, Gwanak-gu, Seoul 08826, Republic of Korea
[2] Corresponding author (Prof. Yonghwan Kim), yhwankim@snu.ac.kr

Abstract. This work introduces a new grid deformation method for an efficient CFD-based optimization of ship hull forms. The method uses a two-level point transformation technique to manipulate grid points with a small number of design points. At the first level, generic B-spline is employed to transfer grid points based on movements of control points sampled inside a control box, ensuring accuracy and smoothness of surface modification. At the second level, Radial Basis Function with Wendland's C2 continuity is adopted to interpolate movements of control points based on relatively few design points. It is shown to be effective in preserving good mesh quality and efficiency. The method is applied to the deformation of hull surfaces for ship models KVLCC2 and KCS, and to investigate the effects of bulbous bow on calm-water resistance with fixed Lpp. A regression model is proposed for ship length, location of buoyancy, wet surface area, and displacement. Numerical results show that the present method is well-suited for CFD-based hull form optimization with better efficiency.

1. Introduction

Computational Fluid Dynamics (CFD) provides engineers with a potent tool to test the hydrodynamic performance of ships and offshore structures. However, its industrial-level applications are mainly restricted by the uncertainty of results and simulation efficiency, particularly for hull optimization problems where hundreds of hull forms, often derived from an initial parent hull, need testing to obtain hydrodynamic coefficients such as the resistance of each hull form. Therefore, an effective simulation strategy to accommodate modified geometry is crucial for minimizing computational expenses in CFD-based hull optimization.

Re-meshing the entire computation domain and deforming initial grids are two common approaches. The first method is a natural choice linked to a geometry modification, commonly using shifting methods [1], B-spline-based techniques like Non-Uniform Rational B-spline (NURBS) adopted in many commercial CAD packages [2], Free-Form Deformation method (FFD) [3], or Radial Basis Function (RBF) interpolation [4]. The details of this technique and its applications to hull optimization can be referred in the review of [5]. FFD-based applications later than this review can be found in [6-8]. This is so far the most popular method and has shown its applicability in ship design. However, it is still time-consuming when used with high-fidelity CFD solver computing unsteady flows. As the re-meshing of a complex grid is a nontrivial task, it may inevitably increase the grid uncertainty. Therefore, a reliable and efficient grid deformation technique is necessary to make it feasible to use a high-fidelity CFD solver in the main optimization loop.

Point-by-point schemes are preferred to deform unstructured grids as each grid point is moved individually based on its location in space while preserving mesh topology, including connectivity between different points to form edges, faces, and cells unaltered. This makes it particularly suitable for CFD solvers, and it facilitates parallel computation. Grid deformation methods of this nature mainly include RBF interpolation and generic B-spline [9]. RBF-interpolation-based mesh deformation is generally carried out by a two-step procedure. The surface points on the original hull are first modified by the aforementioned methods. Then, the displacements of internal grid points are interpolated with respect to the displacements of modified surface points via RBF interpolation. It has been applied in hull optimizations [10,11] and shown to provide smooth distortion. However, one possible issue may be critical when using this method with a large number of surface points (RBF nodes). One is the evident deformation time [12], and the other is the large demand for memory to solve RBF interpolation coefficients, which increases quadratically with the number of surface points. This makes it impractical when the number of RBF nodes is more than a thousand. Directly deforming mesh using RBF interpolation by moving RBF nodes is also possible and has been applied to some problems with simple geometries like aerofoil optimization [13,14].

The generic B-spline method shares a similar idea of FFD but directly deforms grids without any coupling with geometry modification. It also uses a control box and defines the maps between grid points and control points using B-spline basis. It offers important advantages including similar accuracy, flexibility, smoothness of surface modification with NURBS, and local deformation of grids. An additional computational burden is inevitable because of the computations of parametric coordinates of grid points. As each movable control point can be used as design variables, it is usually used with a dimensional-free optimization method like adjoint methods [15]. The generic B-spline method has been applied to aerodynamic shape design [9] and car optimization [16], while its applications to hull optimization is hard to find.

This paper introduces an adaptive grid deformation method for efficient CFD computations with the following objectives:

- Develop a universal point-by-point scheme for both 2D (surface) and 3D grid deformation based on generic B-spline. The point cloud (surface points and grid points) can be directly adapted to the desired positions relative to the given movements of sampled control points. The method should enable local deformation and provide similar accuracy, smoothness, and flexibility of surface modification as NURBS, which is originally used to produce hull forms.
- Reduce the number of design variables by interpolating the movements of control points with respect to design points whose displacements in space are treated as design variables. RBF interpolation is used to provide at least Wendland's C2 continuous interpolation and enable arbitrary settings of design points inside the control box.

In this paper, the mathematical background of the proposed method is explained first. Grid adaptation behavior is then presented and measured by two mesh quality metrics, including non-orthogonality and skewness. The uncertainty and efficiency of CFD computations using the developed method are also discussed to demonstrate its ability to accelerate CFD simulations. Finally, two applications to hull optimization problems, including a regression model of hull geometric parameters and CFD-based calm-water resistance predictions, are illustrated.

2. Methodology

2.1. Generic B-spline

The 3D generic B-spline, also known as volumetric B-spline [9], can be expressed as:

$$B(u,v,w) = \sum_{i=1}^{I} \sum_{j=1}^{J} \sum_{k=1}^{K} U_{i,pu}(u) V_{j,pv}(v) W_{k,pw}(w) C_{i,j,k} \tag{1}$$

Here, $B(u,v,w)$ represents the Cartesian coordinates of the grid point, (u,v,w) is the grid point parametric coordinates. *I*, *J* and *K* denote the number of control points in x-, y- and z- directions, respectively. *U*, *V* and *W* are the B-spline basis functions, and *pu*, *pv* and *pw* their respective degrees. $C_{i,j,k}$ refers to the cartesian coordinates of the *i,j,k*-th control point.

Taking *U* as an example. the B-spline basis can be obtained using a recursive expression:

$$\mathrm{U}_{\mathrm{i},1}(\mathrm{u}) = \begin{cases} 1 & \mathrm{u_i} \le u \le \mathrm{u_{i+1}} \\ 0 & otherwise \end{cases} \tag{2}$$

$$\mathrm{U}_{\mathrm{i,k}}(\mathrm{u}) = \frac{(\mathrm{u}-\mathrm{u_i})\mathrm{U}_{\mathrm{i,k-1}}}{\mathrm{u_{i+k-1}}-\mathrm{u_i}} + \frac{(\mathrm{u_{i+k}}-\mathrm{u})\mathrm{U}_{\mathrm{i+1,k-1}}}{\mathrm{u_{i+k}}-\mathrm{u_{i+1}}} \tag{3}$$

Given the points of control points and the B-spline basis, the parametric coordinates $(\mathrm{u}, \mathrm{v}, \mathrm{w})$ of a gird point with Cartesian coordinates $X = (x_g, y_g, z_g)$ can be computed by solving the following root-finding problem by Newton-Raphson method:

$$\begin{cases} B_x(u, v, w) - x_g = 0 \\ B_y(u, v, w) - x_g = 0 \\ B_z(u, v, w) - x_g = 0 \end{cases} \tag{4}$$

This process to compute parametric coordinates only has to be conducted once at the beginning. Moreover, the process of each point is independent from any other grid point, so that it can run efficiently in parallel. Once all parametric coordinates are available, the Cartesian coordinates of each grid point inside the control box can be calculated using Equation (1) at a low cost. This makes the 3D generic B-spline a powerful tool to adapt 2D surface and 3D grid points instantly.

By using unit weight for all control points, Equation (1) can be easily extended in a rational form having better accuracy, smoothness and continuity similar to the NURBS.

$$B(u, v, w) = \frac{\sum_{i=1}^{I}\sum_{j=1}^{J}\sum_{k=1}^{K} U_{i,pu}(u)V_{j,pv}(v)W_{k,pw}(w)C_{i,j,k}}{\sum_{i=1}^{I}\sum_{j=1}^{J}\sum_{k=1}^{K} U_{i,pu}(u)V_{j,pv}(v)W_{k,pw}(w)} \tag{5}$$

2.2. RBF interpolation

Radial Basis Function (RBF) interpolation is a method of approximating a function by constructing an interpolating surface that passes through scattered data points in n-dimensional space. In contrast to polynomial interpolation methods, RBF interpolation uses a set of radial basis functions to construct the interpolating surface, which can be more effective when dealing with high-dimensional or noisy data. It can expressed as:

$$d(X) = \sum_{i=1}^{N} \alpha_i \, \phi(r) + \beta p(X) \tag{6}$$

Here, d can be the movement of each control point, N denotes the number of RBF nodes. $r = \|X - X_i\|$ is the radial distance and X_i represents coordinates of i-th RBF node. The radial basis function is defined as $\phi(r) = (1-r)^4(4r+1)$in this work to ensure Wendland's C2 continuity. $p(X)$ is any polynomial function, α_i and β are interpolation coefficients. Further details on the process of building a RBF interpolation can be found in [4]. Here, only the procedure for using RBF interpolation to obtain movements of control points used in 3D generic B-splines with respect to design variations is explained.

As depicted in Figure 1 and 2, the boundary control points (red circle markers) are usually fixed to avoid discontinuous changes to the outside part, while the interior control points (black circle markers) are moveable. To use RBF interpolation, RBF nodes include all the boundary control points, along with some design points (red X marker, here only one example is shown) that can be set arbitrarily inside the control box for optimization purposes. The displacements of design points are handled as design variables. Once design variations are generated by an optimization algorithm, the movements of moveable control points are computed using Equation (6) The new locations of grid points inside the control box are then calculated using either Equation (1) or (5). Using this two-level point transformation, grid points are finally adapted based on design variations using a few design points.

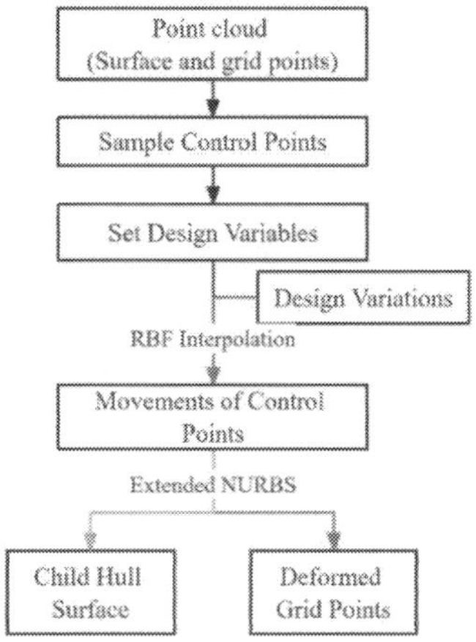

Figure 1. The flowchart of adaptive grid deformation

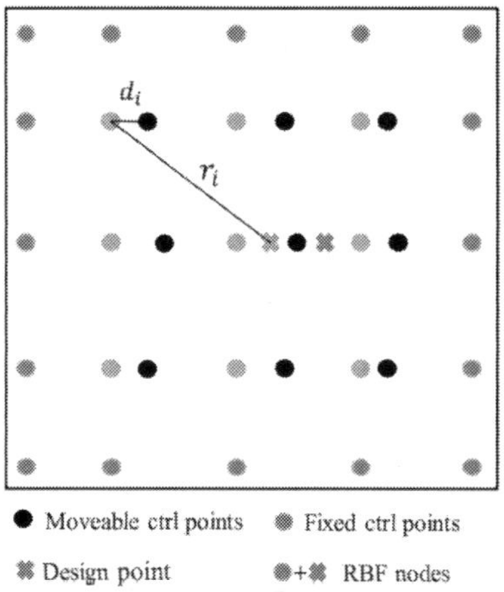

Figure 2. Illustration of RBF interpolation

2.3. CFD solver

CFD solver snuMHLFoam, developed on OpenFOAM by the Marine Hydrodynamics Lab at Seoul National University, is utilized in this work to compute ship performance. It solves the unsteady Reynolds-Averaged Navier-Stokes (URANS) with a k-omega SST turbulence model. Details of this model can be found in [17]. The Volume of Fluid (VoF) method [18] was used for the multiphase modelling.

Governing equations were discretized applying the finite volume method [19]. First-order accurate Euler implicit scheme was used for temporal terms to seek quasi steady-state solutions. Second-order accurate linear schemes were utilized for the discretization of convection and diffusion terms. Velocity, pressure, and volume fraction were coupled by the PIMPLE algorithm, allowing the computation to advance with large timesteps. In addition, a redesigned ship motion module based on rigid body dynamics algorithms [20] is coupled with the flow solver. Further details regarding this CFD solver and its applications can be found in [21-23].

3. Applications to calm-water cases

This section discusses the applications of the developed method to CFD computations on ship flows in calm water. The studied problems are first concisely stated, followed by a presentation of grid adaption and mesh quality changes. Subsequently, an analysis on the uncertainty and efficiency of a single case is addressed to show the feasibility of developed method. Finally, this method is applied to study the variations of calm-water resistance with respect to the changes of bulbous bow.

Table 1. Main particulars of ship model KVLCC2.

Item	Symbol	Value
Scale	-	58
Froude number	*Fr*	0.142
Length between perpendiculars	*Lpp* (m)	320
Maximum beam of waterline	*B* (m)	58
Draft	*T* (m)	20.8
Displacement	Δ (m^3)	312622
Moment of Inertia	K_{xx}/B (m)	0.4
Moment of Inertia	K_{yy}/Lpp, K_{zz}/Lpp (m)	0.25

IOP Conf. Series: Materials Science and Engineering 1288 (2023) 012043 doi:10.1088/1757-899X/1288/1/012043

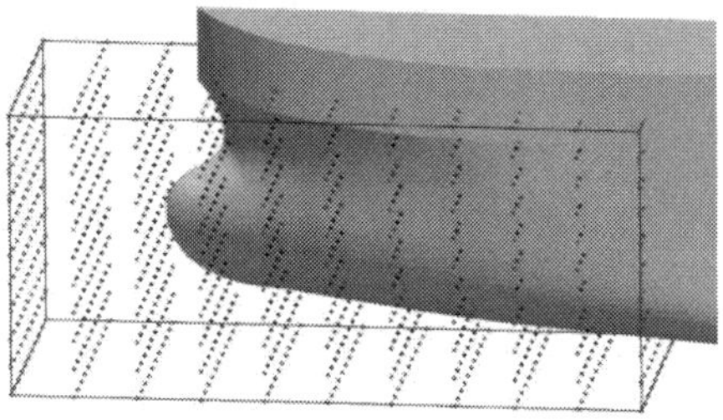

a. Control points

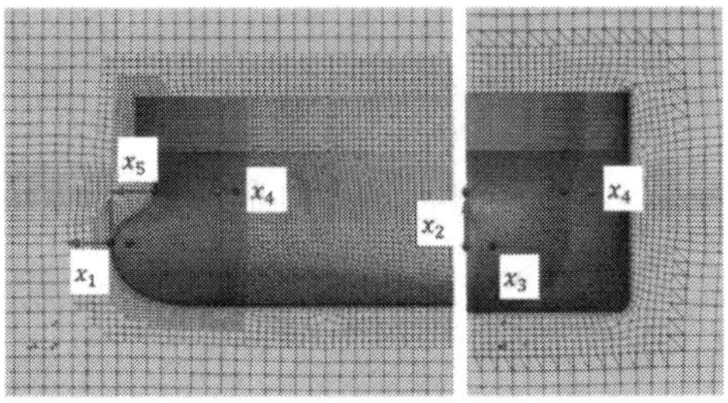

b. Design points

Figure 3. Setting of control points and design points

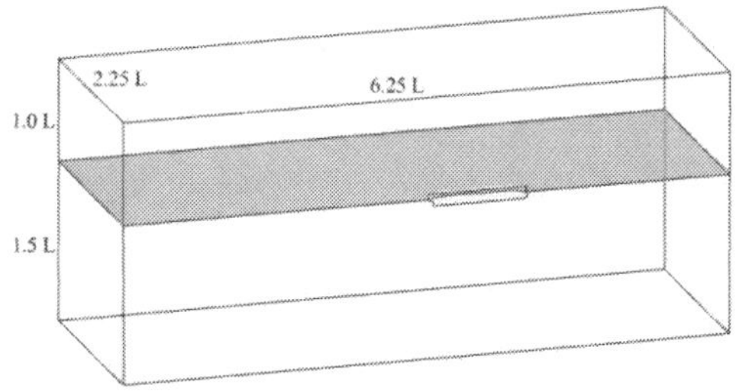

a. Computational domain

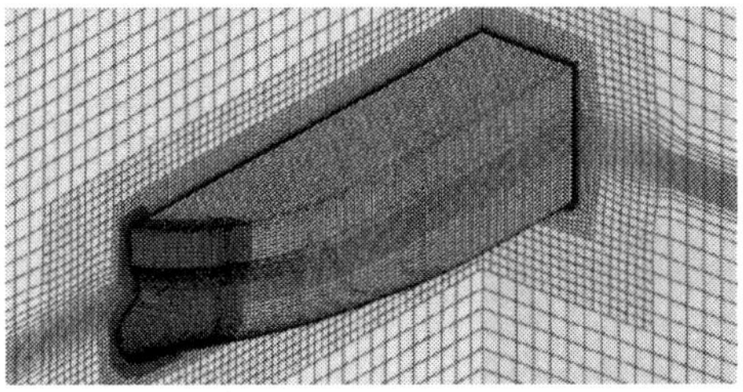

b. Near-ship mesh

Figure 4. The computational domain and mesh

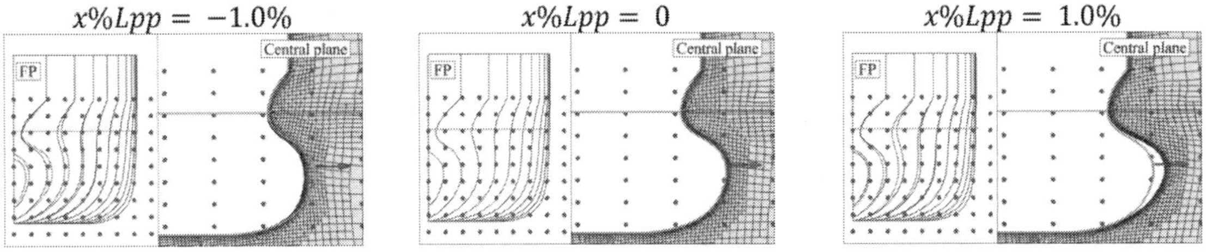

a. x1

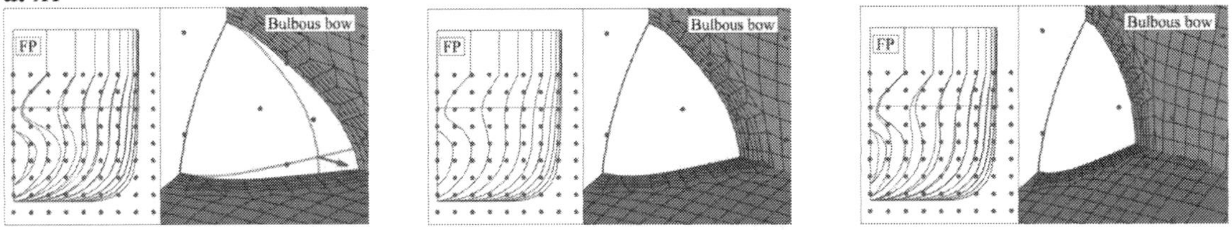

b. x3

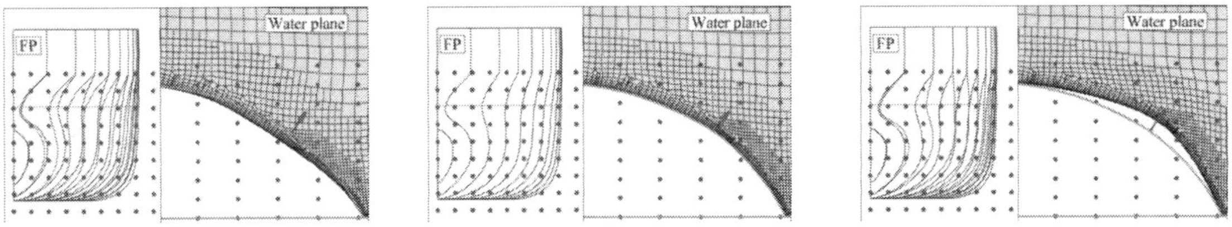

c. x4

Figure 5. The computational domain and mesh

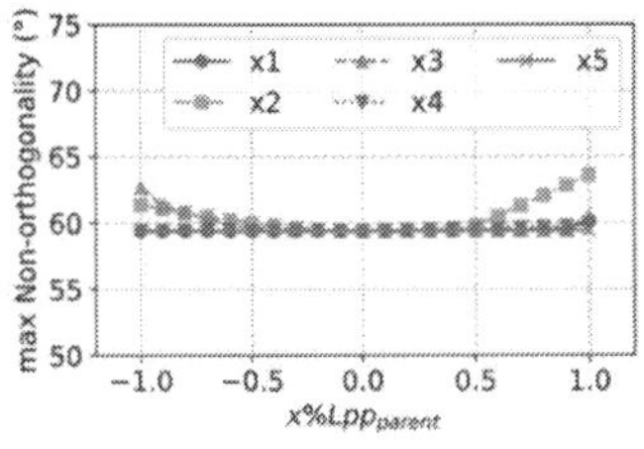

a. Max non-orthogonality

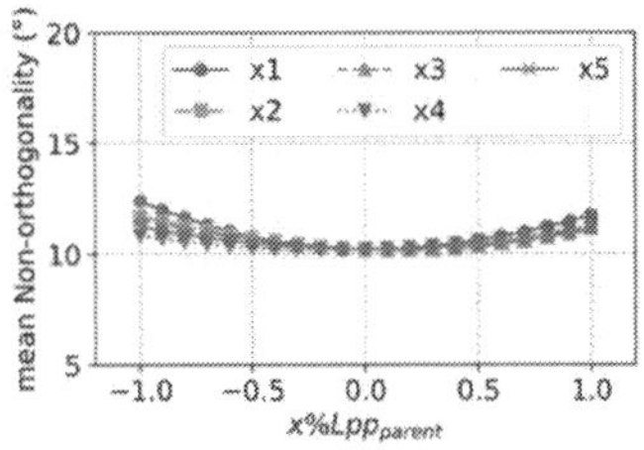

b. Mean non-orthogonality

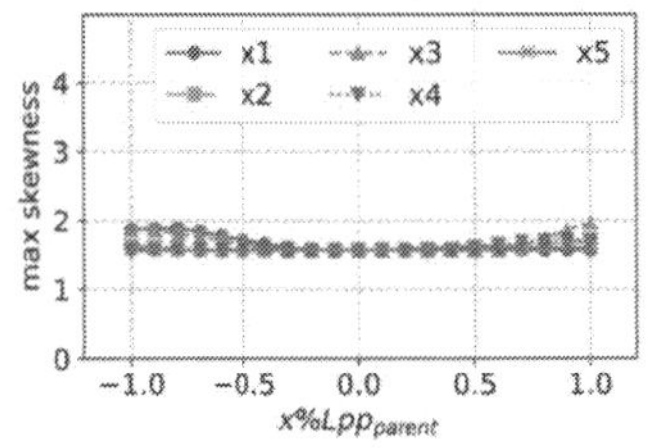

c. Max skewness

Figure 6. Mesh quality metrics

3.1. Problem statements

Ship model KVLCC2 was used, and main particulars are summarized in Table 1. More than a thousand (11x11x11) control points shown in Figure 3.a were set near the ship bow. The boundary points (red markers) were also employed as fixed RBF nodes to ensure that only gird points inside the control box can be adapted by the moveable control points (black markers).

Four design points are shown in Figure 3.b. The first design point was set ahead of the bulbous bow, moving longitudinally (x1) and vertically (x2) to adjust the length and height of bulbous bow, respectively. The second was located on the side of bulbous bow, moving transversely (x3) to alter the width. The third design point was set on the waterplane, moving normal (x4) to it to change the waterplane's profile, which can significantly modify the displacement. The last was placed at the free surface, moving in the longitudinal direction (x5) to change the L_{pp}.

Figure 4.a shows the computational domain of CFD simulations. Figure 4.b shows grids near the hull. Only half domain is built, and the total number of grids is 2.05M.

3.2. Grid adaption and quality metrics

3.2.1. Gird adaption behaviour Figure 5 displays the adapted mesh close to the ship, in response to variations of x1, x3, and x4, respectively. The changes caused by x2 and x5 were similar. It is evident that both the hull surface and grids can be both smoothly adapted according to the variations of design variables. The structure of prism layer mesh was also kept well. The usage of RBF interpolation enables the settings of continuous movements of all moveable control points via a single design variable alone. Moreover, the usage of extended NURBS inherently preserves the accuracy of surface modification as the hull form is originally defined based on NURBS.

3.2.2. Mesh quality metrics. Non-orthogonality and skewness are utilized to present the quality of adapted girds in a quantitative manner. These metrics are crucial for most CFD solvers, ensuring stable computations. A bad mesh with larger non-orthogonality and skewness values can lead to numerical diffusion error based on the finite volume discretization. The precise definitions of these two parameters may vary slightly, and OpenFOAM-based definitions [19] are used in this work.

The non-orthogonality used in OpenFOAM is defined as the angle, made by the vector between the two adjacent cell centers across the common face and the unit face normal. It is considered a type of face non-orthogonality and ranges from 0, indication orthogonality, to 90 degrees, and mainly affects the accurate computation of gradient terms. With the non-orthogonal correction, OpenFOAM-based computations generally require a non-orthogonality angle less than 80 degrees. The skewness describes the normalized distance from the face center to the intersection point on the common face by the velocity vector between the cell centers of two adjacent cells. The skewness of a cell is defined as the maximum skewness value of all cell faces. It affects the accuracy of flux calculation. An acceptable value of max skewness for CFD grids should be less than 4.0.

Mesh quality of girds inside the control box was checked and results including the max & mean non-orthogonality and max skewness, are presented in Figure 6. A total of 77909 grid points were set inside the control box. It is observed that the mesh quality was preserved very well with different design variations, as long as the initial mesh quality satisfied the requirements. The max non-orthogonality varied less than 5 degrees and mean varied less than 3 degrees. The max skewness changed even less than 1.0.

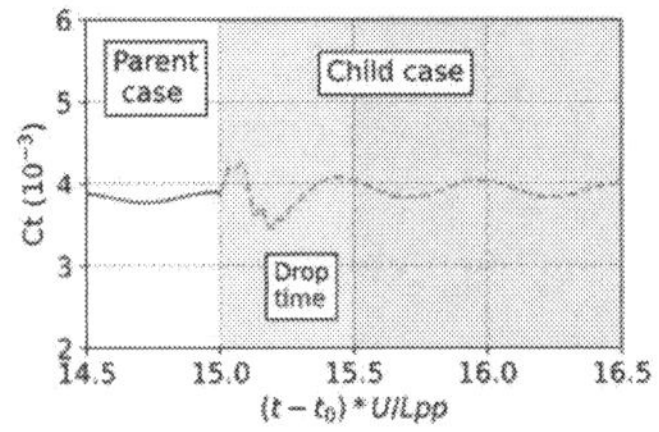

Figure 7. An example of time signal

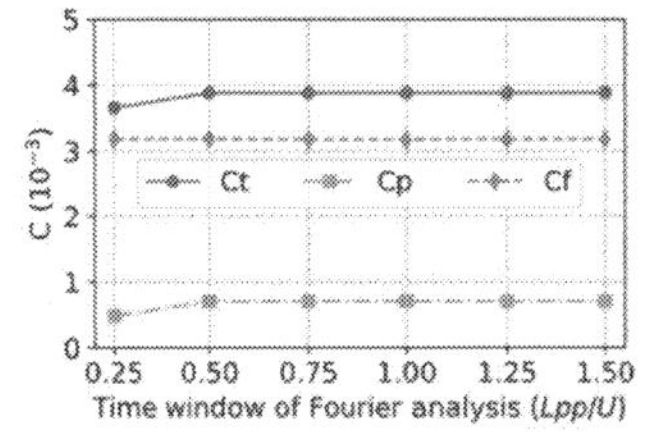

Figure 8. Uncertainty of time window

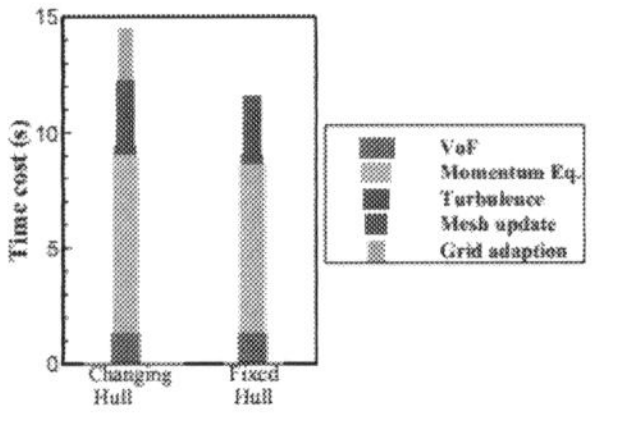

Figure 9. Time cost inside one timestep

3.3. Uncertainty and efficiency analysis of CFD computations

Through the use of developed adaptive grid deformation method, the grids for parent case can be utilized directly for hundreds of child cases with local adaptation regrading design variables. As previously discussed, the mesh quality remained largely unchanged after adaptation and the grid resolution was unchanged as well. For these reasons, it is only necessary to examine the grid uncertainty of mesh used for the parent case. Grid uncertainty analysis for calm-water cases has been done in previous works [21-23], following the method and procedure in [24].

To conduct the efficiency analysis, the simulation time (time window) of each child case must be first determined before. As illustrated in Figure 7, each child case starts at the end of the parent case, thereby using the quasi-steady flow as initial conditions. This significantly reduces the simulation time since there is no longer a need for the flow to attain full development. Points of hull surface and near-ship grids are altered smoothly to the desired locations in response to the variations of design variables at the beginning. Then the simulation continues and eventually converges to a new quasi-steady state. To set the simulation time of each child case, the uncertainty analysis of the time window for Fourier analysis to obtain means values of calm-water resistance is shown in Figure 8. Time signals are cut-off from the beginning based on a given value of 'drop time' (0.5 Lpp/U) to exclude inadequately developed results due to hull changing. It is observed that results of calm-water resistance, sinkage, and trim converged rapidly and a time window of 1.0 Lpp/U is sufficient for post-processing. Hence, the simulation time of each child case was ultimately set to be 1.5 Lpp/U.

Table 2. Efficiency of CFD computations.

Item	Parent case	Child case
Grid number	2.05 M	2.05 M
Simulation time (*Lpp/U*)	15	1.5
# of CPUs	24	24
Clock time (hour)	22.41	2.79

The efficiency analysis of the parent and child cases is summarized in Table 2. The presented clock time may vary depending on the computational resources, though the computational time was significantly reduced due to the much shorter simulation time. This reduction was achieved through the use of developed adaptive grid deformation method which made it possible to use one mesh system for all child cases in a hull optimization problem, as well as providing better initial conditions for the child case. It is also observed that the reduction in clock time was not proportional to the decrease in simulation time. This is because additional burden was introduced by computing the movements of control points and new locations of grid points at the beginning of each child case when the hull is changing. Figure 9 shows the components of time cost within a single timestep. The time cost marked with 'Mesh update' indicates the time required to deform grids in relation to ship motions.

It is observed that this additional burden of adapting grid points accounted for 15.36% of the total time cost when the hull form is changing. However, as hull-changing time is very short (6.25% of the total simulation time), it does not significantly affect the overall efficiency.

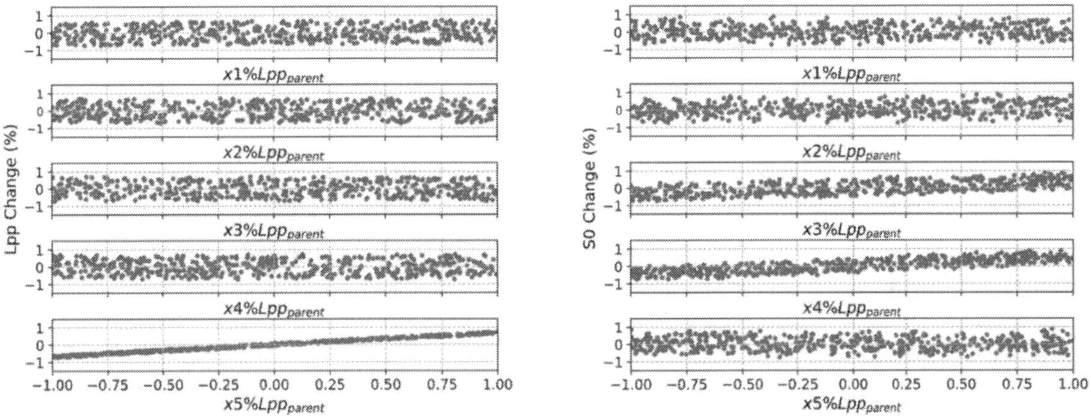

a. Lpp change **b.** S0 change

Figure 10. Changes in geometric parameters with respect to design variations

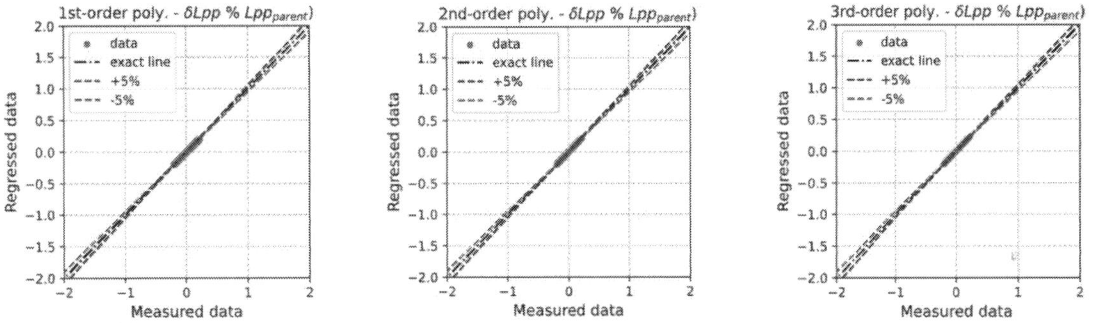

a. Lpp change

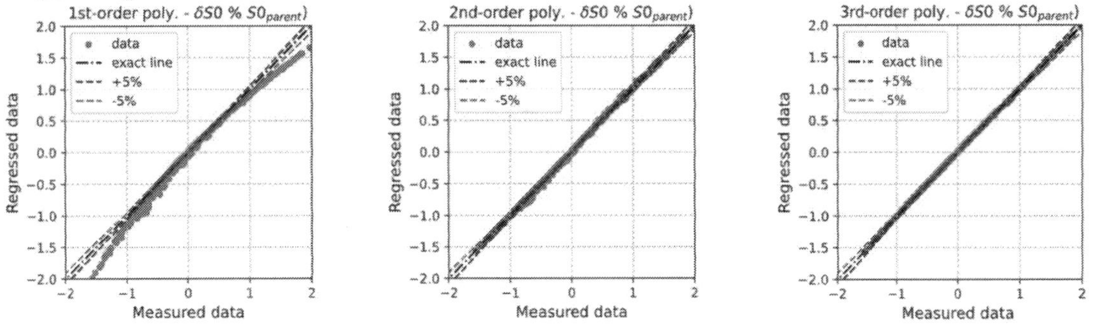

b. S0 change

Figure 11. Cross-validation of regression models

3.4. Application 1 -Regression model of hull geometric parameters

The developed method enables the production of numerous hull forms with minimal effort, allowing for the examination of changes in ship geometric parameters with respect to the design variables. A total of 500 hull forms were randomly created based on the control box shown in Figure 3 with $|x_1, x_2, x_3, x_4, x_5|\%Lpp_{parent} \leq 1.0\%$. The changes in Lpp, wet surface area (S0) are shown in Figure 10 as an example. Evident linear correlations were observed for all the four geometric parameters. The variation of Lpp showed a visible linear relationship to changes of x5, while the change of wet surface area demonstrated a strong correlation to variations of x4.

$$f_1 = \sum_{i=1}^{N} \alpha_i x_i \tag{7}$$

$$f_2 = f_1 + \sum_{i=1}^{N} \sum_{j=1}^{N} \beta_{i,j} x_i x_j \tag{8}$$

$$f_3 = f_2 + \sum_{i=1}^{N} \sum_{j=1}^{N} \sum_{k=1}^{N} \gamma_{i,j,k} x_i x_j x_k \tag{9}$$

Based on these observed tendencies, a regression model was tested with different polynomials defined as Equations (7)-(9). Here, f represents any change in ship geometric parameters such as $\delta Lpp\%Lpp_{parent}$, $\delta LCB\%Lpp_{parent}$, $\delta S0\%S0_{parent}$, among others. The coefficients of α_i, $\beta_{i,j}$, and $\gamma_{i,j,k}$ were solved through the least-square method. An additional 500 hull forms were randomly created to validate the regressed models. As displayed in Figure 11, the pure linear regression model (Equation (7)) proved effective for length changes including Lpp but demonstrated apparent errors for the changes of wet surface area. The model with a form of second- and third-order polynomial fitted data well. By comparing contributions from different orders, shown in Table 3, it can be concluded that the contributions from third-order terms are negligible, amounting only 1.0% of the total change. Consequently, the second-order polynomial Equation (8) was finally used to express the changes of ship geometric parameters with respect to design variations. Its matrix form can be written as $f_2 = AX + X^T BX$.

Table 3. Efficiency of CFD computations.

Item	$f = \sum_{i=1}^{N} \alpha_i x_i + \sum_{i=1}^{N}\sum_{j=1}^{N} \beta_{i,j} x_i x_j + \sum_{i=1}^{N}\sum_{j=1}^{N}\sum_{k=1}^{N} \gamma_{i,j,k}\, x_i x_j x_k$			
	$f = \delta Lpp$	$f = \delta LCB$	$f = \delta S_0$	$f = \delta\Delta$
$\lvert \alpha_{i_{max}} \rvert$	0.0995	0.3368	0.5094	0.8271
$\lvert \beta_{i,j_{max}} \rvert$	0.0044	0.0139	0.0375	0.0387
$\lvert \gamma_{i,j,k_{max}} \rvert$	0.0010	0.0020	0.0052	0.0055
$\lvert \beta_{i,j_{max}} \rvert \% \lvert \alpha_{i_{max}} \rvert$	4.42%	4.13%	7.36%	4.68%
$\lvert \gamma_{i,j,k_{max}} \rvert \% \lvert \beta_{i,j_{max}} \rvert$	22.73%	14.39%	13.87%	14.21%

3.5. Application 2 - CFD-based calm-water resistance prediction

The present section discusses the applications to CFD computations in calm water. To investigate the effects of the shape of bulbous bow on the calm-water resistance, three design variations (x1 - x3) were evenly distributed (9x9x9) in the range of [-1.0%, 1.0%]Lpp. To show a more realistic and practical application, two geometric constraints were introduced: fixed Lpp and limited variance of displacement.

$$\delta Lpp = 0 \tag{10}$$

$$\left|\delta\Delta\%\Delta_{parent}\right| \leq 1.0\% \tag{11}$$

$$\delta Lpp\%Lpp_{parent}(X) = A_1 X + X^T B_1 X \tag{12}$$

$$\delta\Delta\%\Delta_{parent}(X) = A_2 X + X^T B_2 X \tag{13}$$

With provided values of x1 - x3, two unknowns, x4 and x5, can be directly solved with the two regression models analytically or iteratively by Newton-Raphson method. A total of 76 cores were

IOP Conf. Series: Materials Science and Engineering 1288 (2023) 012043 doi:10.1088/1757-899X/1288/1/012043

used to compute these 729 cases. By running four cases simultaneously, it took approximately 28.85 days to complete all the cases. The efficiency is acceptable for further practical applications regarding hull optimization and can be improved by using more cores.

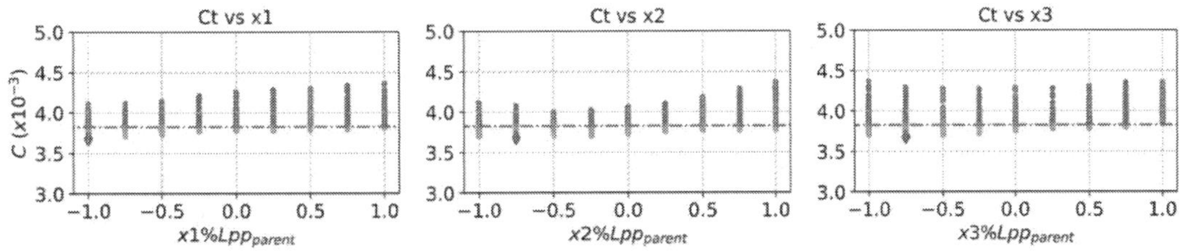

Figure 12. Changes in calm-water resistance (Ct) with respect to design variations

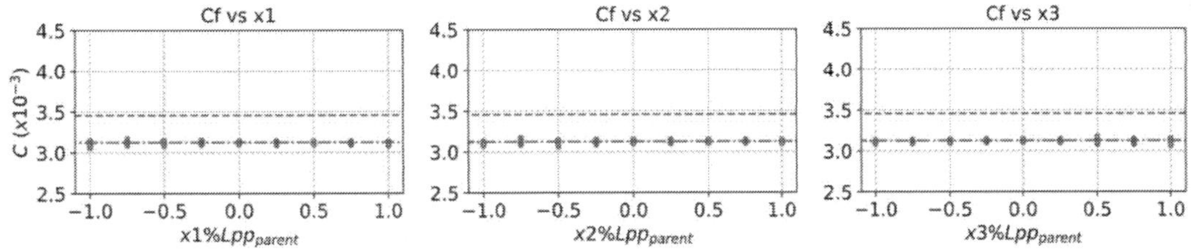

Figure 13. Changes in frictional component (Cf) with respect to design variations

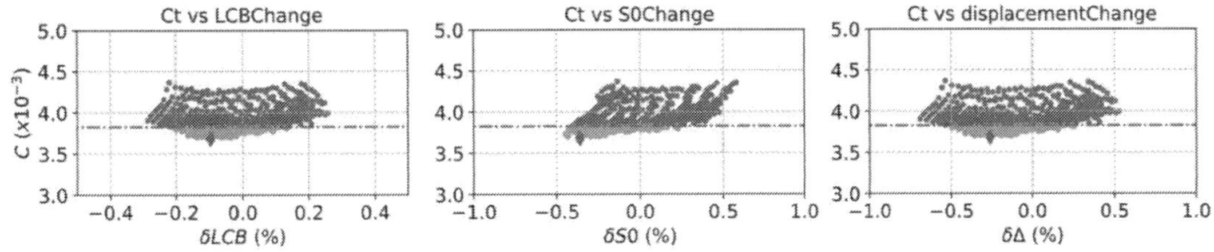

Figure 14. Changes in calm-water resistance (Ct) with respect to geometric parameters

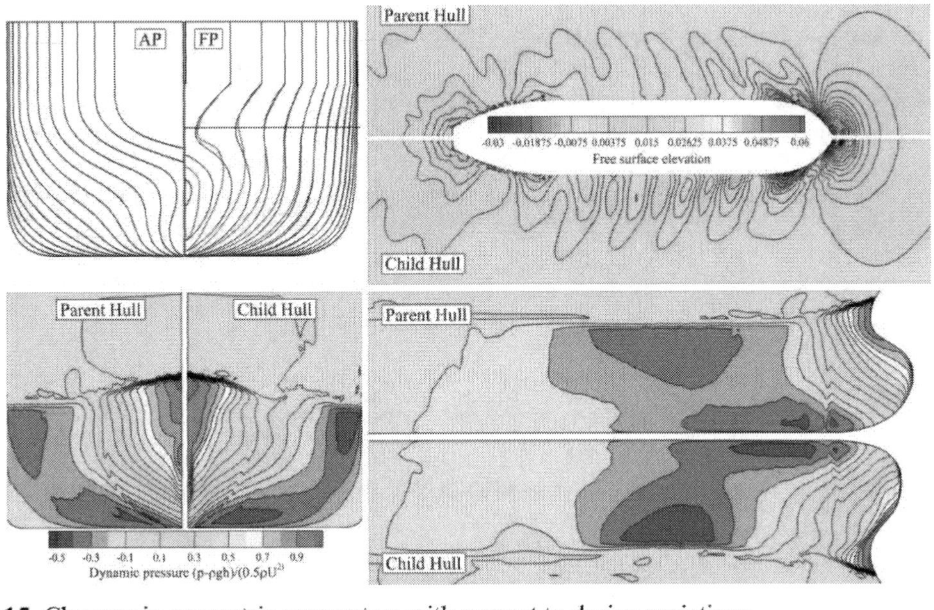

Figure 15. Changes in geometric parameters with respect to design variations

Figures 12 and 13 present the summary on the changes of calm-water resistance and its frictional component regarding design variables (x1 - x3), respectively. Within the 3D design space formed by x1 to x3, cases with smaller Ct predominantly located at the left-bottom corner (negative x1 - x3), where the bulbous bow is reduced in length, height, and width. As for the frictional resistance coefficient, there was little change due to variations of three design variables. This is because the primary dimensions of hull remained the same. The frictional component of resistance in calm water relies largely on the Reynolds number which was consistent for all cases ($Re = 4.55E06$) with the constraint of fixed Lpp. This further indicates that variations in total resistance mainly stem from the changes of pressure component.

Figure 14 directly shows the trends in changes of total resistance coefficient (Ct) with respect to the changes of LCB, wet surface area, and the displacement, respectively. It is observed that Ct showed similar non-linear tendencies with changes of LCB and displacement, decreasing to the minimum as these two geometric parameters decreased, and then increasing with smaller LCB and displacement. As for the wet surface area, Ct did not exhibit an evident trend among cases having larger Ct than the parent hull, but kept increasing among cases better than the parent case. The best case among (green marker) all the cases showed a 0.10% reduction of LCB, 0.4% of S0, and 0.28% of displacement.

Figure 15 displays the body plan, free surface elevation, and dynamic pressure of the best case among all the computed cases. As mentioned earlier, the bulbous bow was evidently reduced, leading to a smaller high-pressure region near the ship bow in the longitudinal direction. In addition, it is apparent from the free-surface elevation that the induced wave was notably reduced near the ship bow, indicating a smaller wave-making resistance.

4. Conclusions

An adaptive grid deformation method has been shown to be capable of efficiently deforming grids. By sampling control points inside a volumetric box, the generic B-spline defines a map between displacements of grid points and movements of control points. It can preserve the accuracy, continuity, smoothness, and efficiency nature of traditional NURBS and provides good deformation behaviours for both surface and mesh with a high degree of mesh quality preservation. The RBF interpolation is further employed to interpolate C2 continuous movements of control points based on variations of design points. By this two-level point transformation, it is possible to control lots of grid points by a few design points which is typically set for optimization purposes.

CFD computations with developed adaptive grid deformation showed a significantly improved efficiency for child cases. As the computation of parametric coordinates of grid points used in B-splines and RBF interpolation coefficients is only solved once at the beginning, there are no expensive calculations during an unsteady CFD simulation. The additional burden has been shown to be minimal in each timestep, and child cases can converge quickly since the developed method enables the directly use the quasi-steady flow of parent ship as initial conditions.

Two applications have been presented to further demonstrate the feasibility of adaptive grid deformation in hull optimization problems. The first application proposed a regression model to express the changes of ship geometric parameters with respect to design variations and showed a good fit with test data. In the second application, the usage of develop method in this work allows for comprehensive CFD simulations within the design space. The changes of calm-water resistance with respect to design variables and optimization algorithms will be further studied in the future.

Acknowledgments

This study was supported by the Lloyd's Register Foundation (LRF)-Funded Research Center at Seoul National University. Their support is greatly appreciated.

References

[1] Lackenby H 1950. On the Systematic Geometrical Variation of Ship Forms. *Transaction of Royal Institute of Naval Architects*, vol. 92.

[2] Piegl L and Tiller W 1996. The NURBS book. *Springer Science & Business Media.*

[3] Sederberg T W and Parry S R 1986. Free-form deformation of solid geometric models. *In Proceedings of the 13th annual conference on Computer graphics and interactive techniques* (pp. 151-160).

IOP Conf. Series: Materials Science and Engineering 1288 (2023) 012043 doi:10.1088/1757-899X/1288/1/012043

[4] Buhmann M D 2003. Radial basis function. *Cambridge University Press*, 2003.
[5] Chi Yang and Huang F 2016. An overview of simulation-based hydrodynamic design of ship hull forms. *Journal of Hydrodynamics*, Ser. B, 28(6), 947-960.
[6] Diez M, Campana E F and Stern F 2018. Stochastic optimization methods for ship resistance and operational efficiency via CFD. *Structural and Multidisciplinary Optimization*, 57, 735-758.
[7] Miao A and Wan D 2020. Hull form optimization based on an NM+ CFD integrated method for KCS. *International Journal of Computational Methods*, 17(10), 2050008.
[8] Nazemian A and Ghadimi P 2021. CFD-based optimization of a displacement trimaran hull for improving its calm water and wavy condition resistance. *Applied Ocean Research*, 113, 102729.
[9] Martin M J, Andres E, Lozano C and Valero E 2014. Volumetric b-splines shape parametrization for aerodynamic shape design. *Aerospace Science and Technology*, 37, 26-36.
[10] Demo N, Tezzele M, Mola A and Rozza G 2021. Hull shape design optimization with parameter space and model reductions, and self-learning mesh morphing. *Journal of Marine Science and Engineering*, 9(2), 185.
[11] Nazemian A and Ghadimi P 2022. Shape optimization of trimaran ship hull using CFD-based simulation and adjoint solver. *Ships and Offshore Structures*, 17(2), 359-373.
[12] Jeong K L and Jeong S M 2020. A mesh deformation method for CFD-based hull form optimization. *Journal of Marine Science and Engineering*, 8(6), 473.
[13] De Boer A, Van der Schoot M S and Bijl H 2007. Mesh deformation based on radial basis function interpolation. *Computers & structures*, 85(11-14), 784-795.
[14] Morris A M, Allen C B and Rendall T C S 2008. CFD - based optimization of aerofoils using radial basis functions for domain element parameterization and mesh deformation. *International journal for numerical methods in fluids*, 58(8), 827-860.
[15] Papoutsis-Kiachagias E M and Giannakoglou K C 2016. Continuous adjoint methods for turbulent flows, applied to shape and topology optimization: industrial applications. *Archives of Computational Methods in Engineering*, 23(2), 255-299.
[16] Papoutsis-Kiachagias E M, Magoulas N, Mueller J, Othmer C and Giannakoglou K C 2015. Noise reduction in car aerodynamics using a surrogate objective function and the continuous adjoint method with wall functions. *Computers & Fluids*, 122, 223-232.
[17] Menter F R, Kuntz M and Langtry R 2003. Ten years of industrial experience with the SST turbulence model. *Turbulence, heat and mass transfer*, 4(1), 625-632.
[18] Weller H G 2008. A new approach to VOF-based interface capturing methods for incompressible and compressible flow. *OpenCFD Ltd.*, Report TR/HGW, 4, 35.
[19] Jasak H 1996. Error analysis and estimation for the finite volume method with applications to fluid flows.
[20] Featherstone R 2014. Rigid body dynamics algorithms. *Springer*.
[21] Kim, B S, Park D M and Kim Y 2022. Study on nonlinear heave and pitch motions of conventional and tumblehome hulls in head seas. *Ocean Engineering*, 247, 110671.
[22] Zhu Z, Kim B S, Wang S and Kim Y 2022. Study on numerical PMM test and its application to KCS hull. *Applied Ocean Research*, 127, 103327.
[23] Wang S, Kim B S, Zhu Z and Kim Y 2023. Numerical course-keeping tests of ONR tumblehome in waves with different rudder control strategies. *Ocean Engineering*, 269, 113584.
[24] ITTC Specialist Committee 2017. Recommended procedures and guidelines-uncertainty analysis in CFD verification and validation methodology and procedures. *In 28th International Towing Tank Conference*.

IOP Conf. Series: Materials Science and Engineering 1288 (2023) 012044 doi:10.1088/1757-899X/1288/1/012044

The use of Ctrl+Z in ship design: removing a bulbous bow

F P Arribas[1], S Oyuela[2], A D Otero[3,4], R Sosa[2,3], H R Diaz-Ojeda[5]

[1] ETSIN-UPM, Av. de la Memoria, Madrid 28040, SPN
[2] Universidad de Buenos Aires, Facultad de Ingeniería - CEAN – INTECIN, Av. Paseo Colon, Ciudad de Buenos Aires C1063, ARG
[3] Consejo Nacional de Investigaciones Científicas y Técnicas, Godoy Cruz, Ciudad de Buenos Aires C1425FQC, ARG
[4] Universidad de Buenos Aires, Av. Paseo Colon, Ciudad de Buenos Aires C1063, ARG
[5] Universidad de Las Palmas de Gran Canaria, Instituto Universitario de Sistemas Inteligentes y Aplicaciones Numéricas en Ingeniería, Las Palmas de Gran Canaria 35017, SPN
Production Editor, *Journal of Physics: Conference Series*, IOP Publishing, Dirac House, Temple Back, Bristol BS1 6BE, UK

E-mail: hectorruben.diaz@ulpgc.es

Abstract. A bulbous bow is a widely employed feature to enhance the hydrodynamic resistance of ships. More than 95% of the merchant fleet incorporates various types of bulbous bows to reduce wave-making resistance and wave breaking. Bulbous bow is also utilized in smaller vessels like fishing vessels. In certain cases, the initial ship design includes the addition of a bulbous bow by default. In this study, which focuses on a typical Argentinian trawler fishing vessel, we investigate the impact of the bulbous bow on resistance. Taking a reverse approach, we modify the fishing vessel by removing the bulbous bow and we evaluate its hydrodynamics. This research aims to assess the reduction in ship resistance achieved by the bulbous bow under different load conditions and speeds, comparing the vessel with and without the bulbous bow. The numerical analysis uses OpenFOAM, and the results are validated through towing tank experiments. The research demonstrates that the performance of the bulbous bow varies across different conditions, indicating that not always the vessel with a bulbous bow is better.

1. Introduction

When ordering a new ship, one of the critical requirements that must be obtained during its operation is the contract speed. This refers to the specified speed at which the ship must navigate while maintaining a specific power consumption. Significant deviations from the contract speed upon delivery of the ship can lead to penalties and conflicts for the designers. Such variations in powering can have a profound impact on various aspects of ship operations, including stability, profitability, and overall performance. Therefore, the accurate determination of the speed-resistance curve in ship design is of utmost importance.

In the early stages of the ship design, when the hull form is nearly finalized, statistical and regression methods are commonly used for initial estimations. The Holtrop and Mennen regression method [1], which is based on extensive sea trials and towing tank experiments involving more than 200 ship models, has emerged as the most widely employed approach for estimating ship resistance. However, this method tends to yield higher uncertainties when compared to experimental or numerical results. For instance, a study by Bilec and Obreja [2] founds discrepancies of up to 30% in resistance when comparing the Holtrop-Mennen results with experimental tests for fishing vessels.

As ship design progresses, more accurate and reliable methods are required to account for various shape, viscous, and pressure effects that cannot be adequately addressed through statistical approaches alone. Experimental Fluid Dynamics (EFD) involving towing tank tests has traditionally been the most reliable method in such cases. However, with the rapid advance of computational resources, Computational Fluid Dynamics (CFD) has gained prominence due to its enhanced post-processing capabilities, becoming a reliable alternative [3].

The International Towing Tank Conference (ITTC) regulates the use of these methods through its "Recommended Procedures." These procedures establish guidelines to ensure a consistent methodology for obtaining accurate results, offering rules for both numerical and experimental approaches.

The fishing industry in Argentina has experienced significant transformations since the 1960s. The hake fishery began between 1961 and 1969, followed by the incorporation of fresh fish vessels from 1970 to 1975 to improve catch quality and freshness. Frozen fish processing vessels were introduced between 1976 and 1981, while the 1980s witnessed the development of coastal fisheries and the expansion of the shrimp fishery [4]. In March 2019, a task force was established in Argentina to implement measures aimed to reduce bycatch in fisheries. This task force brings together scientists and stakeholders from both the public and private sectors. Furthermore, Argentina has initiated a fleet modernization process to enhance crew safety and adopt newer, more environmentally sustainable technologies. It is worth noting that more than 50% of the Argentine fleet currently exceeds the recommended age limits set by international standards [5].

In general ship hull design, the inclusion of a bulbous bow below the waterline is a common feature due to its effectiveness in reducing total resistance and installed power, thereby improving the economic efficiency of the design through reducing fuel consumption. A well-designed bulbous bow can lead to favorable interference between the bow waves generated by the bulb and those generated by the hull, resulting in resistance reduction ranging from 12% to 15% [[6], [7]], assuming the vessel maintains a consistent speed. Consequently, this principle guides the design of typical Argentine fishing vessels.

However, an incorrect design can result in a negative interference between these two wave systems, leading to a significant increase in ship resistance. Given the complex interactions between the hull and bulbous bow waves, designs are not universally efficient for all ship speeds and operating conditions, considering the varying load conditions and speeds at which fishing vessels operate. For instance, trawler fishing vessels experience different displacements when traveling to the fishing grounds without load and when returning to port with fish on board. Speeds can also vary depending on external factors such as weather conditions and competition with other fishing vessels trying to reach the same fishing grounds.

The design of a bulbous bow has always been a complex and controversial task. Although model tests involving a wide range of hull forms and bulb shapes have been conducted to select an optimal bulbous bow design with minimal resistance [8], it is not always feasible due to the considerable time and cost involved. In such cases, numerical methods come to the aid of the design process. One of the most renowned experimental studies was carried out by Kracht [9], who tested numerous bulbous bow models and employed statistical analysis of the tests to develop design charts commonly used for bulbous bow design. However, these charts are not applicable to smaller fishing vessels, and if designers use them, the resulting bulbous bows may not be optimal due to the differences between the designed bulb and those used in the tests (which were intended for larger vessels).

Existing studies often rely on model testing based on Kracht charts [10] or [11]. Some studies also employ CFD analysis to optimize design variables such as length, frontal area, breadth, and depth of a bulbous bow. The number of references involving CFD is higher compared to experimental studies, as presented in works by Peri et al. [12], Sharma and Shenoy [13], Watle and Faltinsen [14].

This work aims to evaluate the efficiency of a bulbous bow for a fishing trawler. To achieve this, authors decided to remove the bulbous bow from the original design, representing a unique and novel approach in ship design. There are no previous references to this inverse procedure, hence the title, a "Ctrl + Z" in the design process. By employing CFD, the study investigates the hydrodynamic behavior of the hull without the bulbous bow. The results provide valuable insights into the range of application of the original bulbous bow design and lay the groundwork for future bulbous bow designs, leveraging the knowledge gained from these studies.

2. Experimental Setup

As mentioned earlier, accurately quantifying ship resistance is crucial in ship design. Towing tank experiments are the conventional method for obtaining the ship resistance-velocity curve. The total resistance is measured at various Froude numbers (Fn) ranging from 0.10 to 0.45.

2.1. Model Ship

For this study, a typical Argentine fishing vessel was selected. A 1:20 scale model of the ship was built using Glass-Fiber Reinforced Plastic (GFRP) for the towing tank experiments. The model was manufactured in accordance with the ITTC recommended procedures [15], paying close attention to model manufacturing tolerances, surface finish, and the size and positioning of turbulence stimulation elements. A trip wire, located at 5% of LPP (Length Between Perpendiculars) aft from the fore perpendicular, was used for turbulence stimulation. Additionally, a wire was employed to promote turbulence at the bulbous bow, situated 1/3 of the bulb length from its fore end. Both wires had a diameter of 1.0 mm. Only the bulbous bow ship model was built and tested to optimize costs.

The main characteristics of the full-scale ship and the ship model with the bulbous bow are shown in Table 1. The characteristics for the case without the bulbous bow are presented in Table 2. To distinguish between the models, the letter "b" is used to represent the case with the model that includes the bulbous bow. A graphical comparison of the ship with and without the bulbous bow can be seen in Figure 1, where the Sectional Area Curves of the two models are compared. The study analyzes three load conditions: ballast condition (ship draught of 3.3 m), full load condition (ship draught of 3.9 m), and an intermediate load condition (ship draught of 3.6 m).

2.2. Towing Tank

The towing tank experiments were conducted in the towing tank at the University of Buenos Aires (UBA), Argentina. The tank, known as the Canal de Experiencias de Arquitectura Naval (CEAN), has dimensions of 72 m in length, 3.6 m in width, and a variable depth of up to 2 m. For this study, the water depth was set at 1.80 m.

The towing tank is equipped with a running carriage capable of reaching a maximum speed of 4.0 m/s. The ship model, properly ballasted to achieve the required displacement and waterline, was attached to the resistance dynamometer. A one-component force transducer of type R 47 manufactured by Kempf & Remmers was used in the dynamometer for resistance tests. Coupling the model to the carriage via the resistance dynamometer prevented surge, yaw, and heel, allowing the model to be free in trim and sinkage. The resistance transducer is designed to handle a full-scale load of ±100 N and has a sensitivity of approximately ±1 mV/V of supply voltage.

Table 1. Main parameters of full-scale ship and ship model

Main Particulars	Symbol	Unit	Ship	Model 1b	Model 2b	Model 3b
Model scale	λ	[-]	-	20	20	20
Length on waterline	L_{WL}	[m]	32.68	1.634	1.662	1.641
Length over wetted surface	L_{OWS}	[m]	34.795	1.670	1.740	1.740
Breadth	B	[m]	9.28	0.464	0.464	0.464
Draught	D	[m]	3.30	0.165	0.180	0.195
Displacement volume	∇	[m^3]	599.40	0.075	0.085	0.095
Wetted surface area	S	[m^2]	392.67	0.982	1.058	1.124
Block Coefficient	C_B	[m]	0.60	0.60	0.61	0.64
Midship section area coeff.	C_X	[m]	0.86	0.86	0.87	0.89

Table 2. Main parameters of full-scale ship and ship model without bulbous bow

Main Particulars	Symbol	Unit	Model 1	Model 2	Model 3
Model scale	λ	-	20	20	20
Length on waterline	L_{WL}	[m]	1.515	1.565	1.575
Breadth	B	[m]	0.464	0.464	0.464
Draught	D	[m]	0.165	0.180	0.195
Displacement volume	∇	[m^3]	0.068	0.077	0.087
Wetted surface area	S	[m^2]	0.969	1.049	1.122
Block Coefficient	C_B	[m]	0.58	0.59	0.61
Midship section area coeff.	C_X	[m]	0.86	0.87	0.89

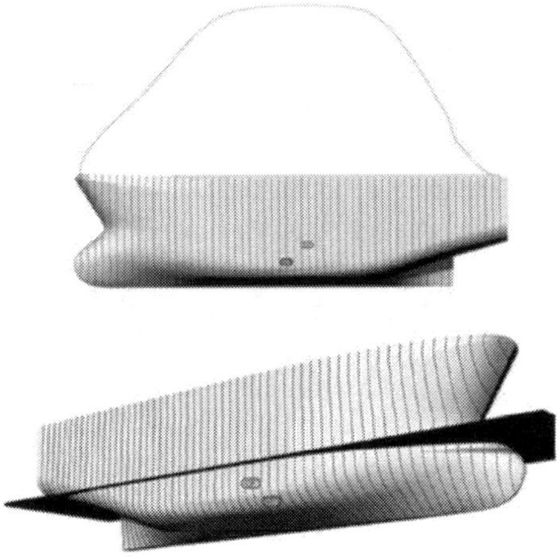

(a)

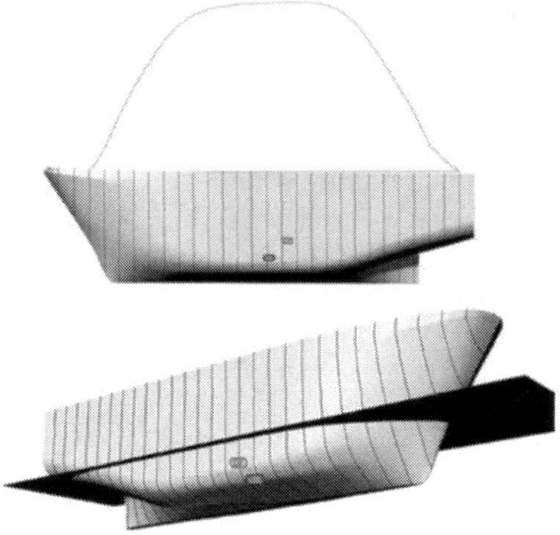

(b)

Figure 1. Comparison in light condition. a: original model, b: without bulbous bow

The mid-sectional area of the vessel is 1%, 1.1%, and 1.2% of the towing tank section area in the ballast, intermediate, and full load conditions, respectively. To account for the blockage effect caused by the presence of the model in the towing tank, Schuster's blockage correction method is applied following the ITTC (International Towing Tank Conference) procedure [16].

The analysis of uncertainty for the resistance measurements in the towing tank experiments considers several significant uncertainty components. These components include the hull geometry, towing speed, water temperature, calibration of the dynamometer, and repeat tests. The standard uncertainty associated with each component is estimated and then combined using the Root-Sum-Square (RSS) method, as recommended by the ITTC (International Towing Tank Conference) [17].

The uncertainty components related to the hull geometry include the wetted surface area and the representative length of the model, which are estimated in terms of the standard uncertainty of the displacement mass. The uncertainty associated with the hull geometry is quantified by measuring the model ballasting. The resulting uncertainties for different speeds (Fn = 0.14, Fn = 0.26, Fn = 0.37) are presented in Table 3.

Table 3. Combination of uncertainty in measurement for resistance at Fn = 0.14, Fn = 0.26 and Fn = 0.37. Repeat test N = 4

Uncertainty Components	Fn = 0.14	Fn = 0.26	Fn = 0.37
Hull geometry	0.05	0.05	0.05
Speed	0.067	0.067	0.067
Water temp.	0.03	0.03	0.03
Dynamometer	4.73	1.04	0.35
Repeat test, Deviation[a]	5.00	3.50	1.50
Combined for single test	6.88	3.65	1.54
Repeat test, Deviation of mean	2.50	1.75	0.75
Combined for repeat mean	5.35	2.04	0.83
Expanded for repeat mean	10.70	4.08	1.66

[a] Repeat test, Deviation = (Repeat test, Deviation of mean)*$N^{1/2}$

The uncertainty of towing speed affects the resistance measurement through the dynamic pressure and the Reynolds (Re) number. The relative uncertainty of the towing speed, which is limited by the bias of the towing carriage, is considered. The resulting uncertainties in resistance due to velocity are also presented in Table 3.

Water temperature variation has a significant effect on water viscosity, Re number, and frictional drag. The deviation of water temperature during the tests showed a variation of less than 0.5°C. The corresponding uncertainties in resistance due to temperature variation are included in Table 3.

The resistance dynamometer is calibrated before and after each test, and the standard deviation of the linear regression is adopted as the standard uncertainty of calibration. The uncertainties related to the calibration of the dynamometer are presented in Table 3. The uncertainty associated with the weights used for ballasting is considered negligible.

The resistance measurements are obtained using a R47 dynamometer with a sampling rate of 200 Hz. The measurements are averaged over a time interval of at least 10 seconds, and the standard uncertainty of the average of the sampling history is determined. The uncertainty of one reading from the Data Acquisition System (DAS) is considered negligible.

Repetition tests are performed, and the mean of the resistance measurements is adopted as the best estimate. The standard uncertainty of the mean from N repeated tests, with N = 4 in this case, is estimated according to the ITTC procedure [17]. The standard uncertainties for repeat tests are shown in Table 3.

Finally, the significant components of uncertainties are combined using the RSS method, and the results are presented in Table 3. The dynamometer and the precision of measurement are identified as the major sources of uncertainty in the repeated tests. The uncertainty of resistance measurements decreases with increasing speed, as the contribution from the dynamometer decreases. The expanded uncertainties in Table 3 correspond to a confidence level of 95%. The uncertainty of resistance measurements for Fn = 0.14, Fn = 0.26, and Fn = 0.37 are 5.35%, 2.04%, and 0.83%, respectively.

3. CFD Model

3.1. Numerical Procedure

The use of Computational Fluid Dynamics (CFD) for ship resistance is widespread. OpenFOAM is one of the most commonly used numerical codes in the community. It is a free and open-source tool that implements the well-known Navier-Stokes equations, along with various numerical schemes and procedures used to solve fluid flow problems. Based on the finite volume method, this code considers incompressible Newtonian flow in the present work. Consequently, the equations of interest are the Navier-Stokes mass and continuity equations, represented by Equation 1 and Equation 2.

Due to the presence of turbulence in ship resistance, a turbulence model is necessary. The Reynolds-Averaged Navier-Stokes (RANS) method is employed to model turbulent flow and perform a time-average of flow quantities. This method offers different models, such as SST k-ω, which includes two eddy-viscosity equations. In this model, k represents the turbulent kinetic energy, while ω denotes the specific dissipation rate.

Two different types of simulations were conducted in this study. Firstly, a simulation involving a single fluid in a double hull configuration, and secondly, a simulation with two phases present, as depicted in Figure 2. In the numerical analysis, the previous equation descriptions are utilized when considering only one fluid. However, when both water and air are present as two phases, an equation capable of handling the presence of both fluids is required. The Volume of Fluid (VOF) method, as shown in Equation 3, is employed for this purpose. This method utilizes a scalar volume fraction α ranging between 0 and 1 to indicate the proportion of each phase contained within each fluid cell. Specifically, α takes a value of 0 for air and 1 for water.

$$\nabla \cdot \overline{U_f} = 0 \tag{1}$$

$$\frac{\partial(\rho_f \overline{U_f})}{\partial t} + \nabla(\rho_f \overline{U_f U_f}) = \rho_f g - \nabla p + \mu_f \nabla^2 \overline{U_f} \tag{2}$$

$$\frac{\partial(\alpha)}{\partial t} + \overline{U_f} \cdot \nabla \alpha = 0 \tag{3}$$

3.2. Numerical Domain and Boundary Conditions

The computational domain utilized in the simulations is depicted in Figure 2. To prevent any blockage effects, the distances within the domain have been chosen sufficiently far from the ship model. The boundary conditions applied in the simulations are summarized in Table 4.

3.3. Mesh Convergence

An analysis of mesh convergence was performed for a draft of 0.165 meters, light condition, and a Froude number of 0.45. The results obtained from different meshes were compared to assess convergence. The results are summarized in Table 5. Once a consistent trend in cell number was observed, Mesh 4 was selected as the final mesh for the remaining cases. Figure 3 illustrates the final mesh, Mesh 4, where refined cells are employed to capture accurately the flow around the ship, while a thin refinement is applied to the entire domain to capture the air-water interface.

IOP Conf. Series: Materials Science and Engineering 1288 (2023) 012044 doi:10.1088/1757-899X/1288/1/012044

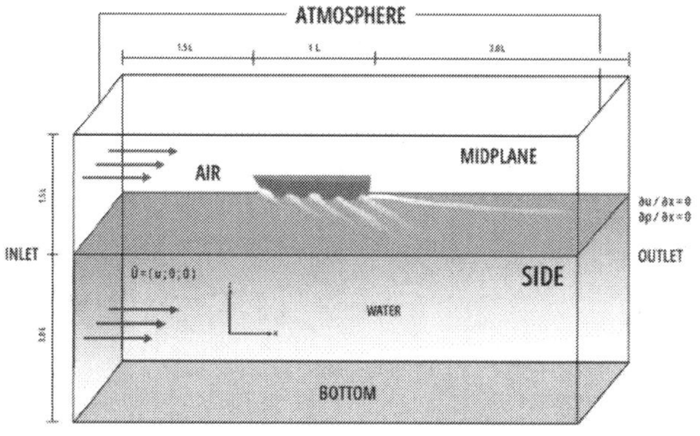

Figure 2. Problem set up and main parameters

Table 4. Boundary conditions

Patch	U	p	k	ω	μ turbulent
Inlet	uniform	fixed flux	fixed value	fixed value	fixed value
Outlet	zero gradient	zero gradient	zero gradient	zero gradient	zero gradient
Atmosphere	zero gradient	zero gradient	zero gradient	zero gradient	zero gradient
Bottom	symmetry	symmetry	symmetry	symmetry	symmetry
Midplane/Side	symmetry	symmetry	symmetry	symmetry	symmetry
Ship	no slip	fixed flux	wall function	wall function	wall function

Table 5. Mesh convergence analysis

Case	Mesh cells	Fn	F [N]	Error (%)
EFD	-	0.45	40.27	-
Mesh 1	633.938	0.45	40.26	0.02
Mesh 2	903.744	0.45	41.76	3.70
Mesh 3	1.213.419	0.45	40.34	0.17
Mesh 4	2.485166	0.45	41.16	2.21

4. CFD Validation

Before analysing the effect produced by removing the bulbous bow in the trawler vessel, an exercise of validation is necessary. Figure 4 shows the non-dimensional total resistance curve (C_T), numerical and experimental, for Froude numbers between 0.1 and 0.45, considering different draughts. A good agreement in C_T between EFD and CFD results is observed for the three studied draughts and the different Froude numbers.

Normally, fishing vessels operate in different conditions and velocities, i.e. when the fishing vessel is leaving port, when it is fishing or when it comes back to port with the fish. Normally, the design criteria are based on the speed to the fishing area and the coming back to port with the capture. For the present case the ship is designed for sailing at $0.25 < \text{Fn} < 0.35$, being the defined operation speed Fn= 0.35. The curves show the typical behaviour for this kind of vessel.

For higher Fn numbers, the slope of the total resistance curve increases continuously, indicating that C_T is increasing at a rate higher than model ship velocity squared. Wave

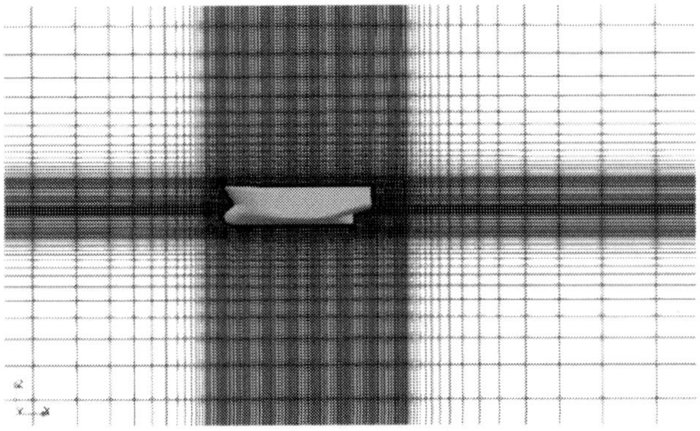

Figure 3. Side view of Mesh 4

resistance is negligible at low speeds, but with increasing velocity, wave resistance grows faster than viscous one. In fact, for most fishing vessels above Fn = 0.30 the wave resistance becomes the largest component of its total resistance, abruptly increasing the ship power and therefore making the vessels inefficient from a consumption point of view.

In ship hydrodynamics it is usual, and useful, to split the wave resistance into two components: wave pattern resistance and wave breaking resistance. If the Fn is large, the waves may be steep enough to break down into eddies and foam. The energy thus removed from the wave system is found in the wake of the ship and the corresponding resistance component is called wave breaking resistance.

The remaining wave energy is radiated away from the ship through the wave system and gives rise to the wave pattern resistance.

The breaking phenomena can be observed in the EFD results for Fn > 0.3, as shown in Figure 5d (Fn = 0.37) and Figure 5f (Fn = 0.45). The CFD results seem to capture the wave breaking phenomena only for the higher Froude number, as seen in Figure 5e (Fn = 0.45). A more detailed description of these phenomena using the CFD model would require a finer mesh in the interface region. However, it is important to note that the wave breaking resistance is considerably smaller than the wave resistance for this type of fishing vessel. From these visualizations, it can be observed that the hull's shoulders play an important role in the distribution of waves along the hull. In this context, it is noticeable that for higher Froude numbers, a maximum hollow wave appears in the ship hull section 14.5 and moves back towards section 11.5 in the case of Fn=0.45.

5. Discussion

Once the numerical procedure is validated, a comparison of force coefficients between the two cases is presented in Figure 6. The calculations for both ships have been conducted for the same draft instead of maintaining a constant displacement value. It should be noted that the lightship weight of the modified ship without the bulbous bow is unknown unless structural calculations are performed. Therefore, in this paper, the draft has been kept the same, resulting in different displacements but maintaining the frontal area of both hulls.

Figure 6 illustrates the comparison for three draughts: a) D1 (0.165 m), b) D2 (0.180 m), and c) D3 (0.195 m), and speeds ranging from Fn = 0.1 to Fn = 0.35.

From Figure 6, it can be observed that for D1 (the lightest draught), the bulbous bow performs better for Fn > 0.25, about 9 knots. This observation seems reasonable since the ship travels to the fishing grounds at higher speeds. However, if the ship operates within the range of Fn [0.2 – 0.25] [7 – 9 knots], the design without a bulbous bow exhibits lower resistance. This

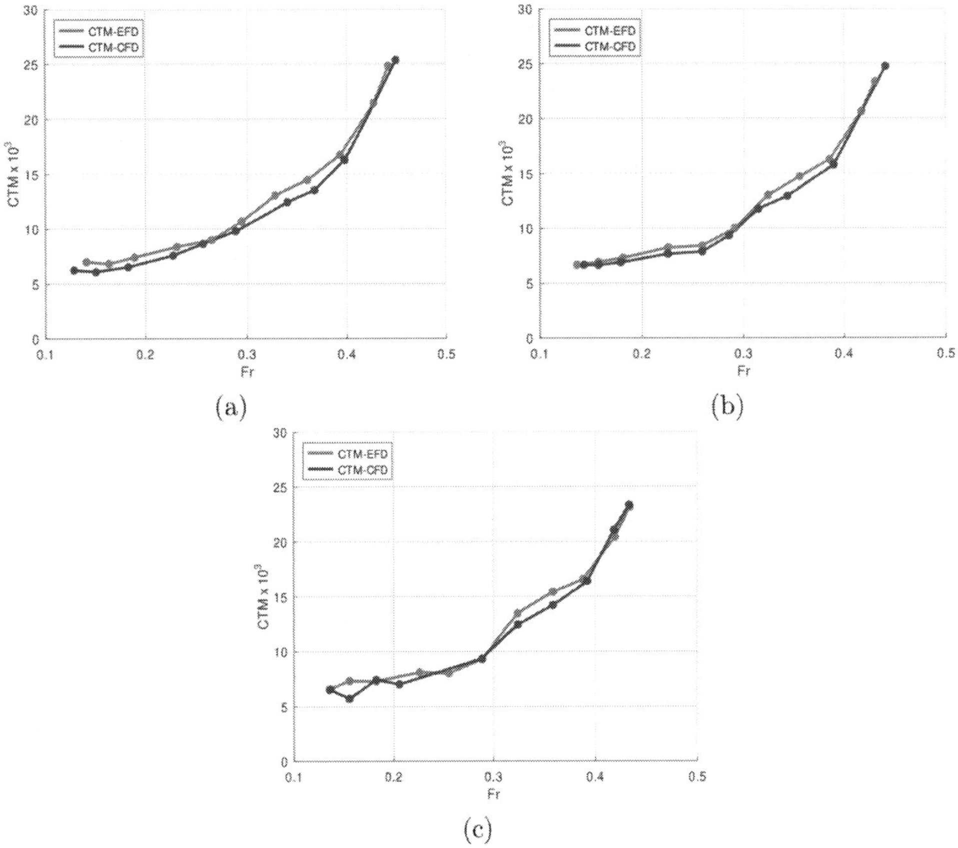

Figure 4. C_T vs Fn for different draughts. a: D1 (0.165 m), b: D2 (0.180 m), c: D3 (0.195 m).

could be advantageous when weather conditions may reduce the ship's speed.

For the intermediate condition D2, the bulbous bow performs worse for Fn $<$ 0.25, which corresponds to speeds lower than 9 knots. Conversely, the bulbous bow improves resistance beyond this range. This observation is particularly interesting as it represents the sailing condition between fishing points. It becomes evident that the bulbous bow works well unless the ship exceeds 9 knots.

For the heavy condition D3, the bulbous bow performs better for Fn $>$ 0.25. However, for lower speeds, it is unclear which hull shape would be better.

Based on the findings from Figure 6, it can be inferred that Fn = 0.25 is a critical velocity. Beyond this point, if the speed increases, the bulbous bow yields better performance for the various draughts investigated. At lower speeds, there is uncertainty regarding the better hull shape, as the hull without a bulbous bow occasionally exhibits superior resistance characteristics.

In summary, when incorporating a bulbous bow into a design, careful considerations must be given to the operational speeds. If an economical speed is maintained for a significant portion of the operational time, the bulbous bow may not provide a distinct advantage that becomes apparent at higher speeds. The different loading conditions encountered during operations need to be thoroughly examined, as demonstrated in the intermediate condition D2. Environmental factors, such as sailing in waves, should also be taken into account.

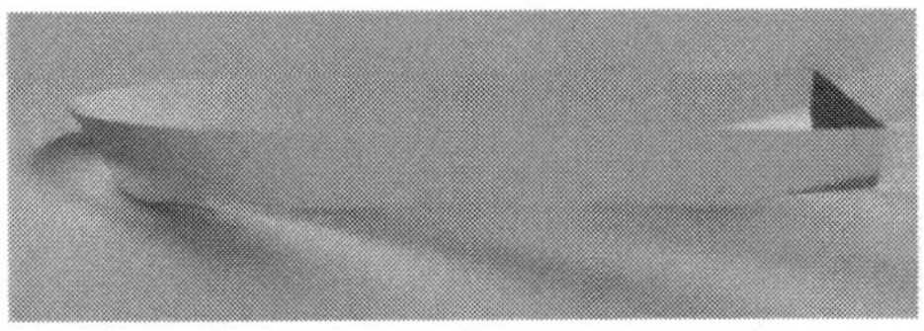
(a)

(b)

(c)

(d)

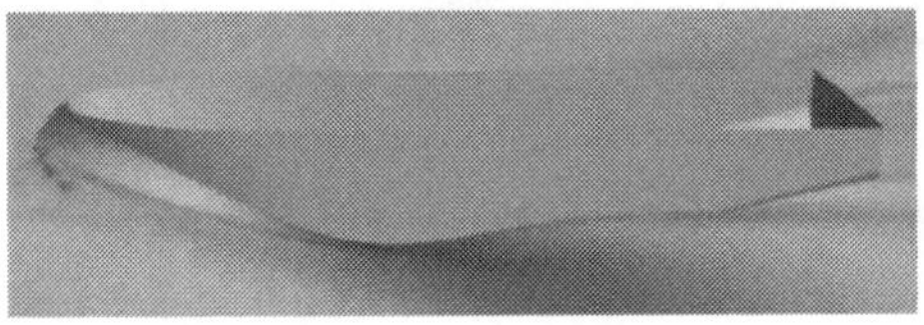
(e)

(f)

Figure 5. Wave pattern for different Fn and draught = D3. Left column: 3D CFD model. Right column: EFD model. First row: Fn = 0.26, second row: Fn = 0.37, third row: Fn = 0.45.

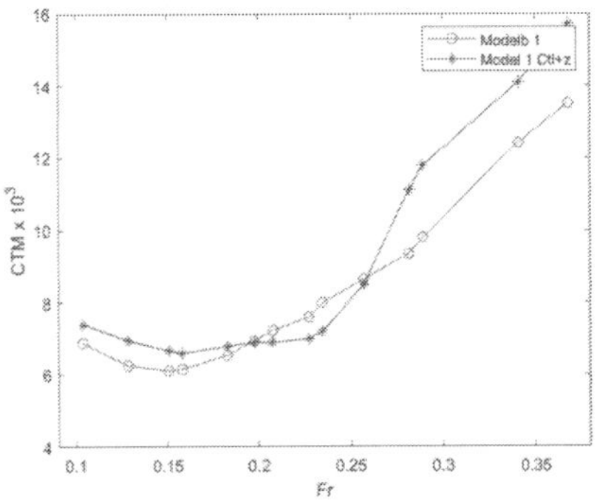

(a)

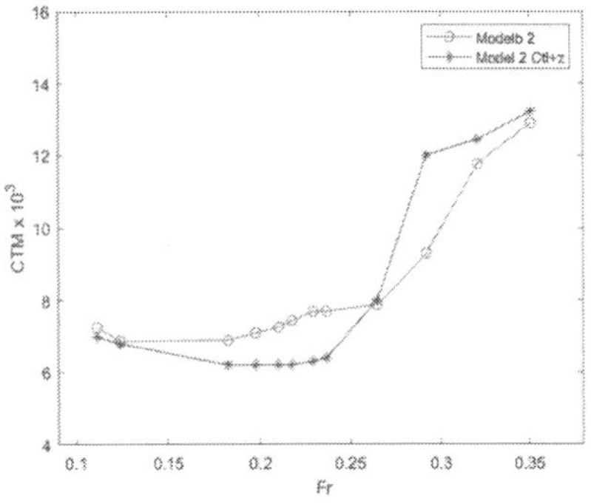

(b)

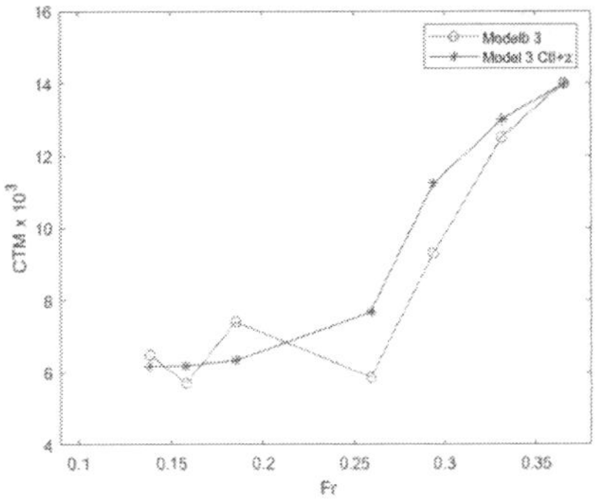

(c)

Figure 6. Total drag coefficient comparison for the bulbous bow hull and the case without bulbous bow for three different draughts: a: D1 (0.165 m), b: D2 (0.180 m), c: D3 (0.195 m).

6. Conclusions

Based on the preceding findings, it is evident that a critical point exists at Fn = 0.25. This Froude number represents a critical velocity, as it is uncertain whether the bulbous bow provides superior performance compared to the hull without one at lower velocities for any draught. Conversely, higher speeds clearly demonstrate that the hull with a bulbous bow experiences lower total resistance. This observation may be attributed to wave interference.

Based on the results, it can be deducted that the inclusion of a bulbous bow ideally enhances the ship's performance. However, under adverse sailing conditions or other factors that result in reduced speed, the absence of a bulbous bow becomes a better alternative. Further investigations are required to quantify the performance of different hull shapes under these non-ideal conditions.

References

[1] Holtrop J 1984 *International shipbuilding progress* **31** 272–276
[2] Bilec M and Obreja C D 2020 *IOP Conference Series: Materials Science and Engineering* **916** 012011 ISSN 1757-899X publisher: IOP Publishing
[3] Larsson L, Raven H and Paulling J 2010 *Ship Resistance and Flow* Principles of naval architecture (Society of Naval Architects and Marine Engineers)
[4] Bertolotti M, Verazay G, Pagani A, Errazti E and Buono J 2001 *El mar argentino y sus recursos pesqueros* ISSN 987-20245-0-2
[5] OECD 2021 . Fisheries and Aquaculture in Argentina
[6] Ferreiro L D 2011 *Technology and culture* **52** 335–359
[7] Barrass B 2004 *Ship design and performance for masters and mates* (Elsevier)
[8] Chakraborty S 2017 . What's Importance of Bulbous Bow of Ships, Marine Insight
[9] Kracht A 1978 *Transactions of the Society of Naval Architects and Marine Engineers*
[10] Hoyle J W, Cheng B H and Hays B 1986 *Society of Naval Architects and Marine Engineers-Transactions* **94**
[11] Hagen G e a 1983 . A Guide for Integrating Bow Bulb Selection and Design into the U.S. Navy's Surface Ship Hull Form Development Process, Naval Sea Systems Command Technical Note No. 885-55W-TN0001.
[12] Peri D, Rossetti M and Campana E F 2001 *Journal of ship research* **45** 140–149
[13] Sharma R and Sha O P 2005 *Naval Engineers Journal* **117** 57–76
[14] Watle A 2017 *Flexible Bulbous Bow Design-A Hydrodynamic Study* Master's thesis NTNU
[15] Specialist Committee: Procedures for Resistance P and of 23rd ITTC 2002 P O W T 2002 *ITTC – Recommended Procedures and Guidelines: Model Manufacture, Ship Models* (ITTC)
[16] 26th ITTC Resistance Committee 2011 *ITTC – Recommended Procedures and Guidelines - Resistance Test* (ITTC)
[17] 28th ITTC Quality Systems Group 2017 *ITTC - General Guidelines for Uncertainty Analisys in Resistance Tests* (ITTC)

Acknowledgments

This work was supported by the Ministry of Science, Technology and Innovation of the Argentine Republic (MINCyT - Proyecto de Investigación y Desarrollo Tecnológico de la Iniciativa Pampa Azul - B2) and the Universidad de Buenos Aires, Facultad de Ingeniería (Grant Peruilh). We thank the Argentine shipyard TecnoPesca Argentina for providing the fishing vessel lines.

IOP Conf. Series: Materials Science and Engineering 1288 (2023) 012045 doi:10.1088/1757-899X/1288/1/012045

Experimental and numerical Investigation on drag reduction using polymer coatings

Xinru Du[1], **Hidemi Mutsuda**[1], **Yuya Wasada**[1], **Takuji Nakashima**[1]

[1]Graduate School of Advanced Science and Engineering, Hiroshima University, Japan

mutsuda@hiroshima-u.ac.jp

Abstract. The phenomenon of polymer drag reduction (DR) has primarily been investigated in the context of internal flows featuring mixed solutions. However, research on non-uniform polymer DR in external flows, such as polymer coatings on the surfaces of marine applications, remains scarce. To address this gap, experimental and numerical investigations were conducted in this research to find out the DR and turbulence performance. Firstly, the experimental results are shown to demonstrate the effectiveness of polymer DR. Specifically, the drag reduction performance of a silicon polymer coating is tested under different Reynolds numbers. Then, the mesoscopic dissipative particle dynamics (DPD) method is coupled with DNS and LES methods, and the finitely extensible nonlinear elastic approach is used to validate the results obtained using this DNS/LES-DPD method. Additionally, the model is used to investigate the influence of the height of the polymer region on non-uniform polymer DR and turbulence behaviour is also investigated. The model also reveals that the polymer DR efficiency is significant even when the polymer region does not extend across the entire fluid area such as the application of polymer coating.

1. Introduction

Ships play a crucial role in global trade by transporting more than 80% of the world's goods and commodities [1]. During their voyages, however, ships emit a substantial volume of greenhouse gases, contributing to climate change. Over the course of a decade, from January 2012 to April 2022, there was a concerning increase in CO2 emissions, which rose by 23.83% [1]. Recent studies in numerical research have centered on exploring ways to reduce drag in marine applications. Ahmadzadehtalatapeh et al. have categorized the various techniques used for drag reduction (DR) in marine vessels into three distinct categories: hull form optimization, air lubrication, and fouling and coating. Among them, fouling and coating, which belongs to passive control, is one of the most convenient methods which makes a significant reduction in vessel operating costs [2]. In addition, there are several passive control methods. Mohamed et al. [3] analysed recent advancements in biomimetic superhydrophobic materials and their potential applications in reducing drag of engineering in marine environments. Mai et al. evaluated how different types of riblets on submarine surfaces impact drag force, which had a direct impact on the operating efficiency of the submarine [4]. Several studies have suggested that incorporating compliant material surfaces can result in a significant reduction of drag, estimated to be in the range of 10-12% compared to a rigid surface [5]. After a comprehensive comparison, it can be inferred that DR through coatings holds greater potential for development in various fields, as it does not require any alteration to the structure of the object undergoing DR.

Recently, Rowin et al. [6] developed a new polymer coating which can attach to any substrate. And it shows a good performance in reducing drag. In fact, DR using polymer solution has been investigated for many years. Ptasinski et al. [7] investigated three different polymer solution concentrations and found that DR can reach up to 60%-70% at the maximum concentration. Then, White et al. [8] provided a comprehensive overview of numerical simulations on viscoelastic turbulent flows, as well as extensive turbulence measurements in dilute polymer solution flows using advanced laser-based optical techniques. When it comes to vehicles, injecting a polymer solution into external flows is not a practical option, whereas applying a polymer coating is a more viable alternative.

However, the numerical method mainly focused on the polymer solution rather than polymer coating. In recent research, a common concept for modeling polymer molecules is to approximate them as elastic dumbbells that experience a spring force [9][10][11][12]. The most widely adopted model for this purpose is finitely extensible nonlinear elastic (FENE), which utilizes a viscoelastic constitutive equation to capture the polymer stress contribution to momentum balance [9][10]. Currently, there is a lack of comprehensive research on DR methods specifically pertaining to polymer coatings.

In this research, we conducted the experiments firstly, where the DR performance of a silicon polymer coating is tested under different Reynolds numbers. Then the DNS/LES-DPD method is employed, and the FENE approach is used to validate the results. The result of the numerical model is also compared with experimental ones to show the similar DR performance. Moreover, the model investigates the impact of the height of the polymer region on non-uniform polymer DR, and turbulence behavior is also examined. The research findings provide a foundation for further studies on local polymer DR, polymer coatings on marine applications, and the behavior of heterogeneous polymer solutions.

2. Experiment work

2.1. Overview of the plate with polymer painting

The aim of this study was to determine the DR of a silicone-based lubricant coating in practical applications. Two plates were used in the experiment, as shown in Figure 1: one without a polymer coating (Figure 1a), and the other with a polymer coating (Figure 1b). The plates had dimensions of 0.2m in length and 0.16m in width, with and without the polymer coating, respectively. The upper area of the plate was used to fix and measure the drag force. The flat plate without a polymer coating was made smooth to avoid any frictional effects that may have been caused by the roughness of the plate surface.

(a) The plate without polymer (b) The plate with polymer

Figure 1. Plates used for measurements in experiments.

2.2. Experimental setup

The experiment was conducted in the circulating water tank with the observation window in Hiroshima University that was 1.4m in length, 0.3m in width, and 0.1m in height as shown in Figure 2. The flat plates were placed in the center of the circulating tank with the surface parallel to the streamwise direction. The plates' center was located at 0.7m from the front and 0.15m from the side of the circulating tank and all the measurements taken while the plates were submerged in water. Then the Friction Drag

(FD) in the three directions (streamwise, wall normal and spanwise) was measured using the load cell (KYOWA, LSM-B-100NSA1) placed at the connection between the plate and the fixed frame. Through the A/D converter (KEYENCE, NR-600), the raw data was transmitted to the computer for filtering, and the FD under still water was subtracted to obtain the final data. The coefficient of friction $\mathbf{C}_f, \mathbf{C}_{f,N}$ can be calculated based on Equation 1.

$$\mathbf{C}_f, \mathbf{C}_{f,N} = \frac{\mathbf{D}_f}{\frac{1}{2}\rho u^2 \mathbf{S}_{plate} \times 2} \quad [1]$$

Here, u is the uniform velocity, $\mathbf{C}_f$ is the coefficient of friction of the polymer solution, $\mathbf{C}_f, \mathbf{C}_{f,N}$ is that of the Newtonian fluid, $\mathbf{D}_f$ is the average FD during the measurement period, 30 s, ρ is the fluid density and $\mathbf{S}_{plate}$ is the area of the flat plate. Considering the flat plate has two faces in this experiment, the plate area in the denominator of Equation 1 is multiplied by two.

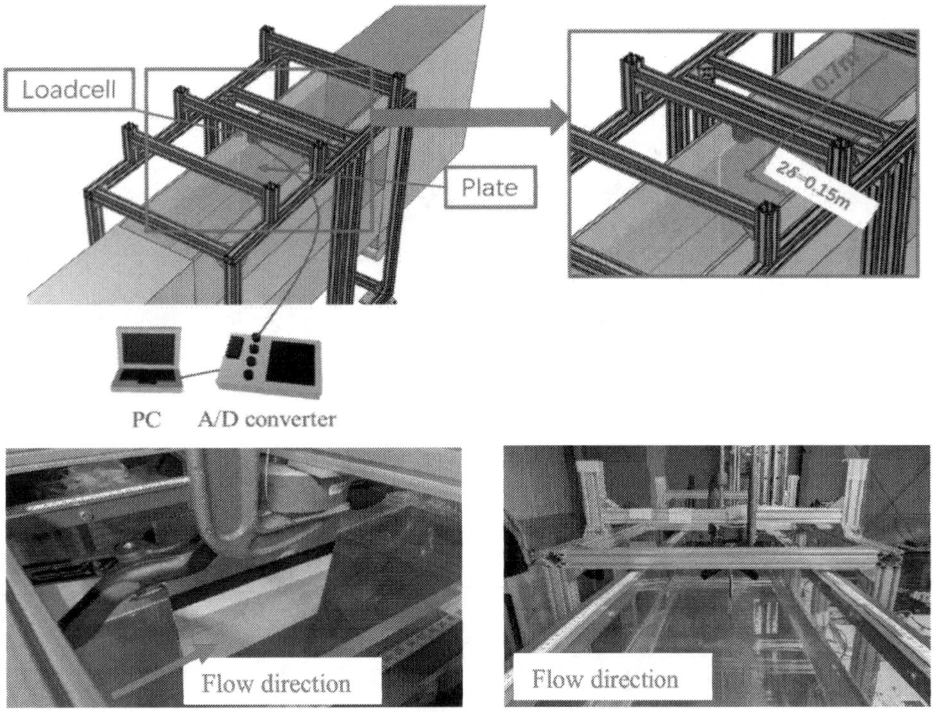

Figure 2. Experimental setup in the circulating water tank and the location of the plate

2.3. Experimental Conditions

In the stationarity test, the uniform flow and still water level were confirmed as the stable state with less than 1% error. Table 1 shows the experimental conditions. Five cases would be conducted and the setting velocity, u, is from 0.138 to 0.637m/s. So the corresponding bulk Reynolds number, Re_b, is from 13032 to 60015. The density for all the cases is 999.965kg/m3 and the viscosity of water is $1.59\times10^{(-3)}$pa•s.

Table 1. Experiment conditions (u is the mean bulk velocity in the streamwise direction, Re_b is the bulk Reynolds number based on the mean bulk velocity)

Case	u(m/s)	Re_b
i	0.138	13032
ii	0.295	27767
iii	0.432	40703
iv	0.551	51920
v	0.637	60015

3. Numerical method

3.1. Numerical model

3.1.1. Governing Equations of the CFD part. This study utilizes a numerical model consisting of two components: the DNS/LES component and the DPD component. The DNS/LES component is utilized to simulate Newtonian fluids, while the DPD component is used for polymer systems. The DNS method is applied for low Reynolds numbers, and the LES method is used for high Reynolds numbers. For greater precision in simulation results, the DNS method is recommended, while the LES method can be utilized to improve computational efficiency.

Equation 2 and Equation 3 are the Navier-Stokes equations for incompressible flows, where equation 3 describes the continuity of the incompressible fluid.

$$\frac{\partial \mathbf{u}}{\partial \mathrm{t}} + \nabla \cdot (\mathbf{uu}) = -\frac{1}{\rho}\nabla \mathbf{P} + \nu\Delta\mathbf{u} + \mathbf{f} \qquad [2]$$

$$\nabla \cdot \mathbf{u} = 0 \qquad [3]$$

Here, $\mathbf{u}$ is the flow velocity, t is time step, ∇ is the divergence operator, ρ is the fluid density, $\mathbf{P}$ is the pressure, ν is the kinematic viscosity, Δ is the Laplacian operator, and $\mathbf{f}$ is an external body force.

The governing equations for filtered turbulent flow used in the LES method are presented in Equation 4, while the Poisson's equation utilized in the LES method is the same as that in the DNS method (Equation 3)

$$\frac{\partial \bar{\mathbf{u}}}{\partial \mathrm{t}} + \nabla \cdot (\bar{\mathbf{u}}\bar{\mathbf{u}}) = -\frac{1}{\rho}\nabla \bar{\mathbf{P}} + \nu\Delta\bar{\mathbf{u}} + \nabla \cdot \tau^{SGS} + \mathbf{f} \qquad [4]$$

Here, *SGS* represents sub-grid-scale. The sub-grid-scale stress tensor, τ^{SGS}, represents the influence of sub-grid-scale flow phenomena on grid-scale flow.

$$\tau^{SGS} = \bar{\mathbf{u}}\bar{\mathbf{u}} - \overline{\mathbf{uu}} \qquad [5]$$

3.1.2. Governing Equations of the DPD part. This research employs DPD model to capture the interaction between particles in close proximity. Equation 6 is the governing equation of DPD method.

$$m\frac{d\mathbf{v}_i}{dt} = \sum_{j\neq i}\mathbf{F}_{ij}^{C} + \sum_{j\neq i}\mathbf{F}_{ij}^{D} + \sum_{j\neq i}\mathbf{F}_{ij}^{R} \qquad [6]$$

Here, m denotes the mass of DPD particles. Subscripts i and ij are introduced to distinguish between individual particles and the effects that one has on the other. The velocity of each particle is represented as $\mathbf{v}_i$. Additionally, the equation incorporates three key forces: $\mathbf{F}_{ij}^{C}$, a conservative force that acts as a soft repulsive force between particles, $\mathbf{F}_{ij}^{D}$, a dissipative force and $\mathbf{F}_{ij}^{R}$, a random force. More introduction is described in [13].

3.1.3. Re-transformation between CFD and DPD parts. The local shear stress τ_d is the result of multiple factors. It includes the momentum of the particle and the forces of particle interaction, which is presented in reference [13]. Specifically, these forces are comprised of the viscosity of the dissipative force, ν_{FD}, and the conservative force, v_{FC}. v_{mm} is calculated by Equation 8. So, the local shear stress can be expressed as:

$$\tau_d = v_{mm} + v_{FD} + v_{FC} \quad [7]$$

$$v_{mm} = \mathbf{v}_y \times (\mathbf{v}_x + r_y \cdot \dot{\gamma}_{DPD}) \quad [8]$$

Here, $\mathbf{v}_x$ and $\mathbf{v}_y$ represent the velocity of DPD particles in the streamwise and wall-normal direction, respectively. r_y and $\dot{\gamma}_{DPD}$ represents the y position and shear velocity of each DPD particle.

$$u_d = \sqrt{\frac{\tau_d}{\rho}} \quad [9]$$

$$\mathrm{Re}_{DPD} = \frac{u_d \delta}{v} \quad [10]$$

Equation 10 introduces the method for calculating the Reynolds number of individual DPD particles, denoted as Re_{DPD}, of each DPD particle. The value of δ is equal to half the hydraulic diameter, and u_d is determined using Equation 9.

Next, Figure 3 shows the schematic of the relationship between the Reynolds number in the DPD and DNS/LES. In the figure, Re_{NF} is the Reynolds number in the DNS/LES part, x is the streamwise direction, and y is the wall-normal direction. DPD particles appear only in the light blue area. H_{DPD} is the height of the DPD particle region. Y_{NF} is the y position of the DNS/LES grid. The superscript of Re_{NF} and Y_{NF}, j is the sequence number of the grid in the y direction, k is the label of the kth DPD particle.

C_0 is the length of a single DPD cell. L^k denotes the distance between the position of the y grid being calculated, Y_{NF}^j, and the kth DPD point, r_y^k. The DNS/LES Reynolds number is determined using Equation 11 and Equation 12.

$$Re_{NF}^j = (1 - \Phi) \times \mathrm{coff}_{Re} \times \sum_{L^k < C0} Re_{DPD}^k \times \frac{L^k}{\mathrm{C}_0} + \Phi \times Re_{ini} \quad [11]$$

$$L^k = |r_y^k - Y_{NF}^k| \quad [12]$$

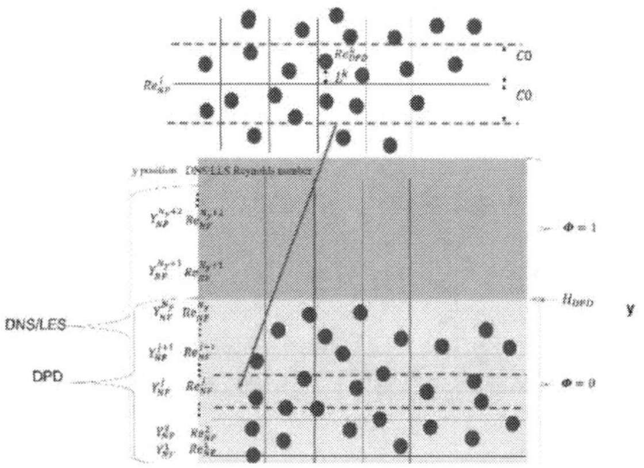

Figure 3. Schematic of the transformation from the DPD Reynolds number to the DNS/LES value

The value Φ is the discriminant value based on the relationship between r_y^k and H_{DPD}. In the case where r_y^k is smaller than H_{DPD}, Φ equals to 0, which is visualized by the light blue area associated with the DNS/LES-DPD method. Conversely, when r_y^k is larger than H_{DPD}, Φ equals to 1, which is depicted by the dark blue area representing solely the DNS/LES method. The coefficient of Re conversion, coff_{Re}, and the initial Reynolds number, Re_{ini} are also utilized in this context.

To calculate the shear velocity of the DPD component, the DNS/LES shear velocity, $\dot{\gamma}_{NF}$, is first determined using Equation 13.

$$\dot{\gamma}_{NF} = \frac{\partial u}{\partial y} \quad [13]$$

$$\dot{\gamma}_{DPD} = \frac{coff_{\dot{\gamma}}}{(Y_{NF}^{j+1} - Y_{NF}^{j})} \times [Y_{NF}^{j+1} - r_y \quad r_y - Y_{NF}^{j}] \cdot \begin{bmatrix} \dot{\gamma}_{NF}^{j} \\ \dot{\gamma}_{NF}^{j+1} \end{bmatrix} (Y_{NF}^{j} \leq r_y \leq Y_{NF}^{j+1},\ j+1 \leq N_y) \quad [14]$$

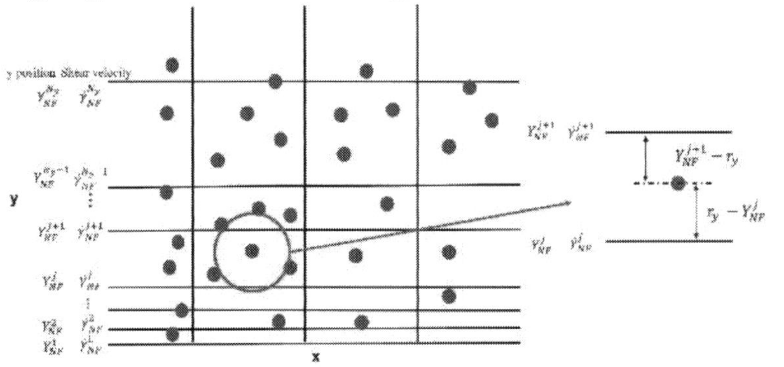

Figure 4. Schematic of the transformation from the DNS/LES shear velocity to the DPD one

Equation 14 is used to convert the DPD shear velocity to the DNS/LES shear velocity. The coefficient of shear velocity conversion, denoted as $coff_{\dot{\gamma}}$, is also included in the equation. The y position of the DNS grid is represented by Y_{NF}^{j} and corresponds to the y position of the DPD particle, denoted as r_y.

In Figure 4, the black lines depict the DNS/LES grid, while the green dots represent the DPD particles. To calculate the DPD shear velocity, a sample point is selected, as indicated by the red circle, using the Newtonian shear velocities, $\dot{\gamma}_{NF}^{j+1}$ and $\dot{\gamma}_{NF}^{j}$, with Equation 14.

3.2. Numerical conditions

Figure 5 illustrates the numerical domain of the model with the x-direction representing the streamwise direction, the y-direction representing the wall-normal direction, and the z-direction representing the spanwise direction, with corresponding velocities u, v, and w.

The red dots are the polymer systems represented by DPD particles and H_{DPD} represented the height of the DPD particle region, which is as same as that in Figure 3. In other words, the red line in Figure 5 indicates the boundary between regions with and without DPD particles, where the upper region represents a Newtonian fluid simulated using DNS or LES methods, and the lower region represents a polymer solution simulated using DNS-DPD or LES-DPD methods. It should be noted that the CFD methods employed in the upper and lower regions are consistent. Table 2 summarizes the operational parameters for nine distinct cases in the lower regions.

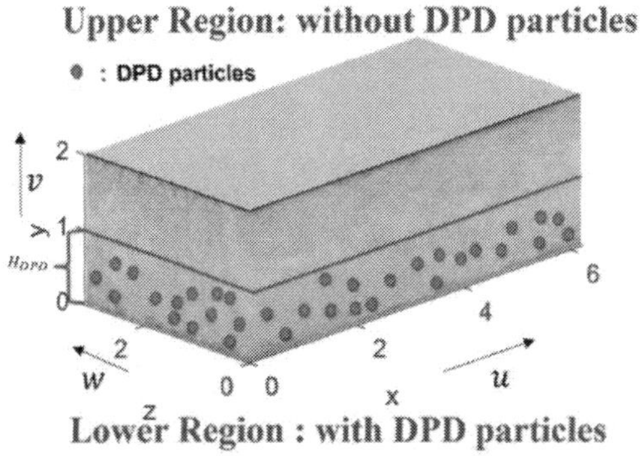

Figure 5. Schematic of the numerical domain

Table 2. Experiment conditions of lower region

Case	Method	H_{DPD}	Re_b	Re_τ	Den
1	LES-DPD	δ	3204	160	21.5
2	LES-DPD	δ	3204	154	43.0
3	LES-DPD	δ	3204	145	64.5
4	DNS-DPD	$\frac{1}{3}\delta$	3204	169	387
5	DNS-DPD	$\frac{1}{2}\delta$	3204	143	387
6	DNS-DPD	δ	3204	135	387

Table 3. Parameters for the cases using FENE method from reference

Case	Method	Re_τ	We_c	$We_{\tau 0}$	DR(%)
A	FENE	160	2	15.4	14.2
B	FENE	154	3	30.8	33.8
N2	DNS	145	0	0	36.6

3.3. Preliminary validation

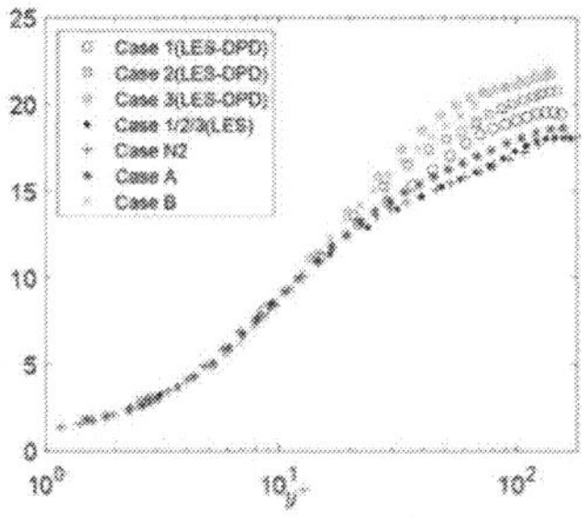

(a) Comparison between Cases 1, 2 and 3's profiles of the streamwise mean velocity.

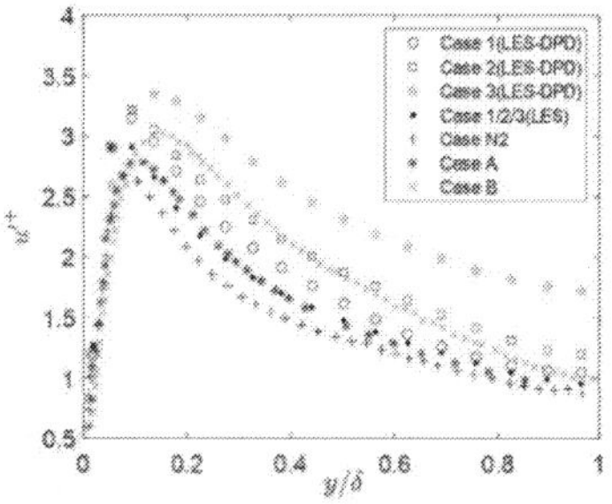

(b) Streamwise RMS velocity

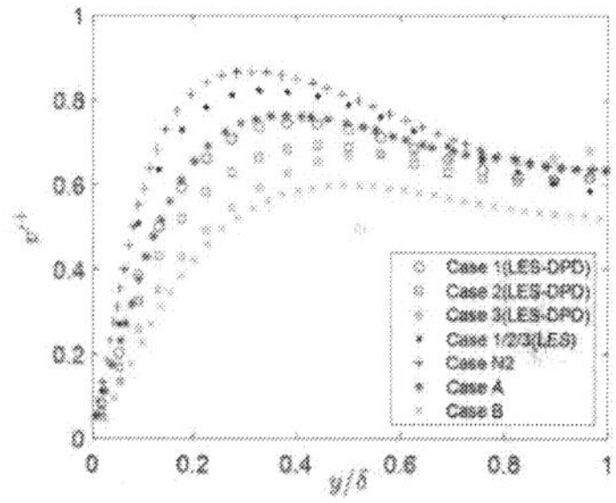

(c) Wall-normal RMS velocity

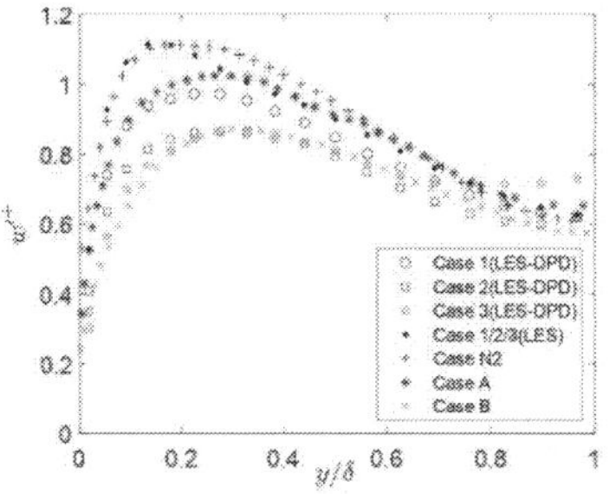

(d) Spanwise RMS velocity

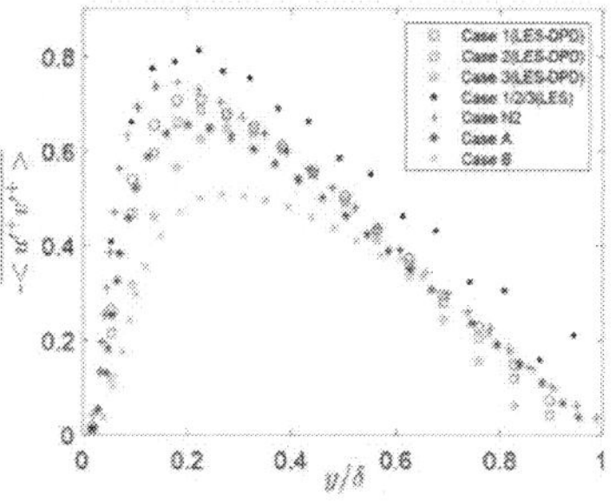

(e) Reynolds shear stress

Figure 6. Schematic of the numerical domain

In this section, we investigate the correlation between Den and DR (%) performance using the FENE method. Table 3 summarizes the characteristics of the FENE method and several cases studied by Dallas et al. [14]. The Weissenberg number, We_c, and the friction Weissenberg number, $We_{\tau 0}$, are related to the polymer relaxation time, and further details can be found in [14]. Figure 6 shows the results for the Newtonian fluid obtained using the LES method for Cases 1, 2, and 3 in Table 2 as a black dot. These results are consistent with those of Case N2. We then focus on the lower region of Cases 1, 2, and 3, which represent the polymer solution using the LES-DPD method. It is evident that as the particle energy density increases, Re_τ drops, DR (%) increases, and the U+ away from the wall shows a monotonic upward trend. Importantly, the shift distance of Case A, which has the same DR and Re_τ as Case 1, corresponds to that of Case 1 (the blue circle and asterisk in Figure 6). Similarly, the profiles of Case B and Case 3 shift up to the same and maximum distance under the same DR and Re_τ (the green star and multiplication sign in Figure 6.

4. Results and discussion

4.1. Comparison between computational results with experimental ones

Based on the experimental conditions described in the Table 2, Figure 7 indicates the time histories of the FD during the measuring time. It is evident that during the stable 20-second period, the FD on the flat plate increases as the Re_b increases. Furthermore, a comparison of plates coated with and without polymer indicates that the FDs of the former are a little bit smaller than those of the latter for five different Re_b. This comparison confirms that the polymer coating has a positive impact on DR.

Then, numerical models with parameters identical to those of the experimental conditions (Table 1) were developed. Figure 8(a) displays the relationship between FD and Re_b, where the experimental data indicates the average value in three times to produce the repeatability during the measurement time. All of the results show an increase with increasing Re_b. Upon comparison of the flat plate with polymer, it can be observed that the FD of the plate with polymer is reduced in both experiments and simulations, which demonstrating that effective DR has been achieved.

Then, the relationship between DR and Re_b are presented in Figure 8(b), where black circles indicate the experimental DR results and black particles represent the computational ones. It is worth noting that although the boundary conditions of the experiment and the numerical simulation are not the same, both show similar trends in the results. It can be inferred that the polymer also changes the laminar viscosity to achieve drag reduction, and reduces the turbulence generated by the corners of the plates under the experimental conditions.

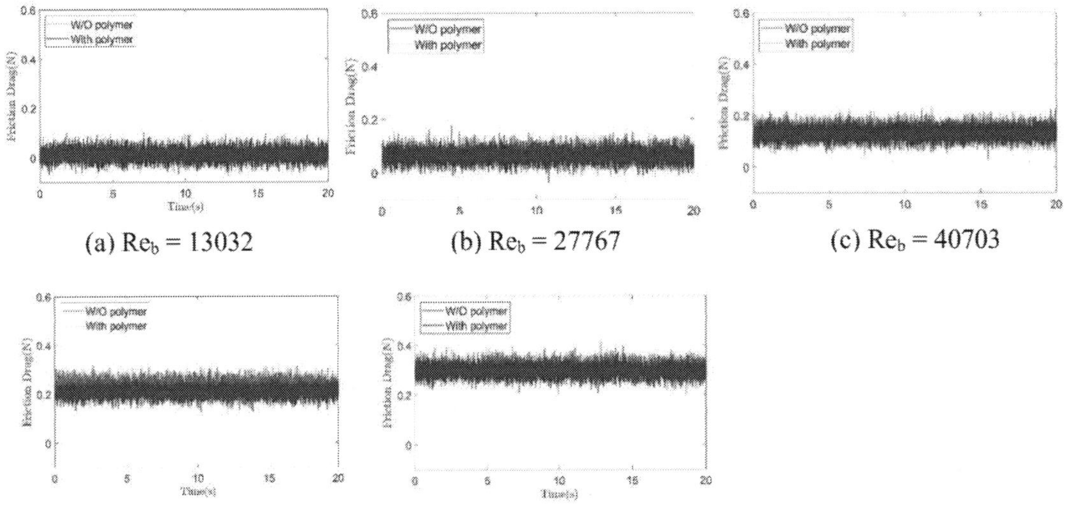

(a) Re_b = 13032 (b) Re_b = 27767 (c) Re_b = 40703

(d) Re_b = 51920 (e) Re_b = 60015

Figure 7. Comparison of time histories of FD under the different Re_b in the experiment

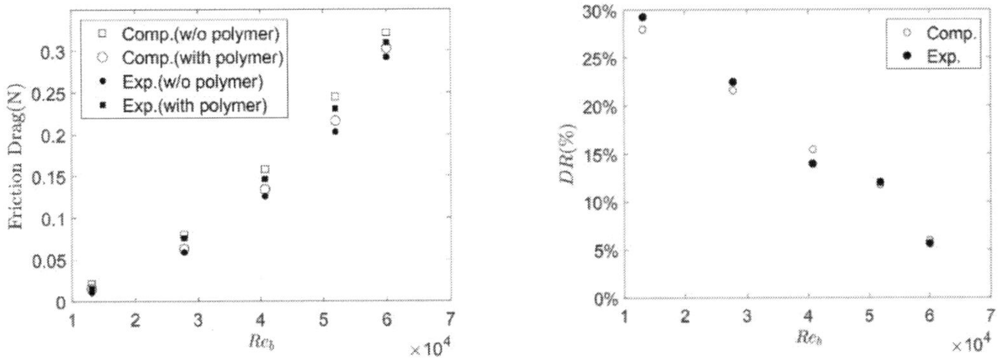

(a) Average FD (b) DR

Figure 8. Comparison between the experimental results and the computational ones

4.2. *Visualization of Flow Field Effects of Polymer Coatings*

Subsequently, this section presents the visualization of the effects of polymer coatings on the flow field. Three different cases (Cases 4, 5, and 6) are considered for the purpose of examining the impact of the DPD particle region's height. The height of the DPD particle region (HDPD) set to $\frac{1}{3}\boldsymbol{\delta}$, $\frac{1}{2}\boldsymbol{\delta}$, and $\boldsymbol{\delta}$, respectively, while the density of the DPD particles (Den) remains constant across all three cases.

Figure 9 depicts the vortex of the Q criterion as obtained through the DNS-DPD method. The presented results are based on instantaneous observations of the same simulation domain at the final time step. Top and side views of all three cases are shown in the Figure.

It can be observed that the vortex becomes smaller and more localized as the HDPD increases, as evidenced in Figures 9(a), 9(b), and 9(c). In the side view of Case 4 (Figure 9(d)), the vortex is seen to grow nearly from the wall, and the reduction in the size of the vortex is not prominent. However, in the side view of Case 5 (Figure 9(e)) and Case 6 (Figure 9(f)), the vortex near the wall completely disappears.

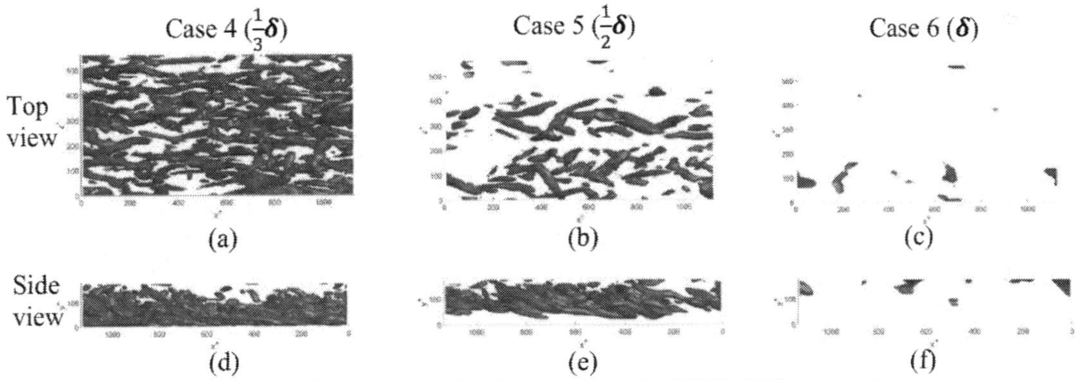

Figure 9. Vortex of the Q criterion using the DNS-DPD method

5. Conclusions

In this research study, we conducted an experiment on polymer coating and proposed a numerical method, DNS/LES-DPD, to simulate the distribution of polymer in any fluid region. Experiments and numerical simulations were conducted. The model involving the DNS/LES method or the DNS/LES-DPD method was found to be in excellent agreement with the results of previous reference studies using FENE, thus indicating its reliability and potential for DR prediction. The study demonstrated that an increase in the Den leads to an improvement in DR. The mean velocity, RMS velocity fluctuations, and Reynolds shear stress also exhibit a monotonic trend in the same direction. Experimental results provided strong evidence of notable DR resulting from polymer coating. The observed changes in DR across various Reynolds numbers were consistent with the simulation model outcomes, further validating the numerical approach's reliability and broad applicability. Our findings suggest that releasing polymer particles near the wall alone can achieve DR, and the DR efficiency of releasing polymer particles in half of the region is remarkably high. The new vortex grows above the polymer region once again, indicating the potential for further DR. Overall, the results of this study contribute to our understanding of turbulent flow and its relationship to DR, as well as highlighting the effectiveness of the DNS/LES-DPD method for simulating polymer distribution in fluid regions.

References

[1] United Nations Conference on Trade and Development 2022 *Review of maritime transport*
[2] Ahmadzadehtalatapeh M and Mousavi M 2015 *Int. J. Marit. Technol.* **4** 51–64
[3] Samaha M A, Tafreshi H V and Gad-el Hak M 2012 *CR Mecanique* **340** 18–34
[4] Mai T D, Kim S, Chang K, Lee S W, Kim M and Ryu J 2022 *Ocean Eng.* **266** 113068
[5] Jayakumar P et al. 2005 *Def. Sci. J.* **55** 37
[6] Rowin W A, Asha A B, Narain R and Ghaemi S 2021 *Ocean Eng.* **240** 109895
[7] Ptasinski P, Nieuwstadt F, Van Den Brule B and Hulsen M 2001 *Flow Turbul. Combust.* **66** 159–182
[8] White C M and Mungal M G 2008 *Annu. Rev. Fluid Mech.* **40** 235–256
[9] Sureshkumar R, Beris A N and Handler R A 1997 *Phys. Fluids* **9** 743–755
[10] Bird R B, Curtiss C F, Armstrong R C and Hassager O 1987 *Dynamics of polymeric liquids, volume 2: Kinetic theory* (Wiley)
[11] Yu B and Kawaguchi Y 2004 *J. Non-Newton. Fluid Mech.* **116** 431–466
[12] Tamano S, Itoh M, Hoshizaki K and Yokota K 2007 *Phys. Fluids* **19** 075106
[13] Satoh A 2004 *Kyoritsusyuppan (Tokyo)* 55
[14] Dallas V, Vassilicos J C and Hewitt G F 2010 *Phys. Rev. E* **82** 066303

Shallow waters resistance of an accelerating ship

Y H Liu[1], M Terziev[1]

[1]Department of Naval Architecture, Ocean and Marine Engineering, Henry Dyer Building, University of Strathclyde, 100 Montrose Street, Glasgow, UK.

yihan.liu.100@strath.ac.uk

momchil.terziev@strath.ac.uk

Abstract. This paper investigates the acceleration resistance of a vessel in shallow waters using potential flow and CFD methods. Results indicated a pronounced resistance peak near the critical depth Froude number. The peak's location and magnitude were sensitive to acceleration intensity and water depth. Excellent agreement between potential flow and CFD methods was found in low and high depth Froude number ranges, suggesting their effectiveness and cost-efficiency. However, non-linear effects affected the resistance peak's magnitude and position, leading to slight disagreements between the methods. The solvers' variation was observed to be sensitive to the investigated parameters.

1. Introduction

Ship resistance prediction has traditionally focused on steady conditions, determining a single resistance value for each combination of speed and water depth. However, recent research has shed light on the unsteady nature of ship resistance. Studies by Havelock [1], Lunde [2][3], and [4] have explored transient wave resistance, considering the effects of acceleration and proposing mathematical theories. Calisal [5] extended these theories to incorporate wave pattern development in unrestricted water. Furthermore, Day et al. [6] investigated the influence of finite depth and width effects in a towing tank.

The investigation of unsteady resistance has been motivated by the performance considerations of unconventional vessels, such as hovercraft [7][10]. In these studies, the vessels are often simplified as two-dimensional pressure distributions on the water surface [11][12]. Researchers have observed oscillations in the time-history signal of towed craft, presenting challenges in accurately measuring steady resistance values [13]. While various theories on unsteady wave drag have been proposed, there has been limited exploration of the scenario involving a ship accelerating past a critical depth Froude number.

In the field of aerodynamics, exceeding the speed of sound results in the occurrence of a sonic boom. Similarly, phenomena analogous to a sonic boom have been observed when a vessel encounters variations in water depth [4]. A change in water depth can cause the wave speed to decrease below the vessel's speed, leading to the generation of significant amplitude waves [4]. As the vessel passes over this change, the depth Froude number undergoes a transition from a subcritical value (F_h <1) to a supercritical value (F_h >1). Unlike in aerodynamics, where trans-critical phenomena are primarily observed through changes in speed, hydrodynamics allows for manipulation of water depth and/or vessel speed to achieve similar effects.

2. Methodology

In this study, the potential flow solver MHydro [20] is employed, which utilizes a 3D boundary element method (BEM) based on a Rankine type Green function. This solver is specifically chosen to investigate unsteady free surface effects. MHydro has been successfully utilized in previous research on ship hydrodynamics in both deep and shallow water by Yuan et al. [21].

To obtain the solution, it is crucial to establish correlations between all results at each time step, allowing for the preservation of unsteady terms in the free-surface boundary conditions. This capability enables the simulation of unsteady effects on the free surface. The methodology employed in this study will be described in detail in the subsequent sections.

2.1. Description of the problem

The problem is described using two right-handed coordinate systems, as illustrated in Figure 1. The first coordinate system, denoted as *O-XYZ*, is fixed to the Earth. The second coordinate system, denoted as o_0-$x_0y_0z_0$, is fixed to the ship hull and moves along with the ship. In both coordinate systems, the positive x-axis points towards the bow of the ship, the positive y-axis points towards the port side of the ship, and the positive z-axis points upward. The undisturbed calm free surface is situated at z=0.

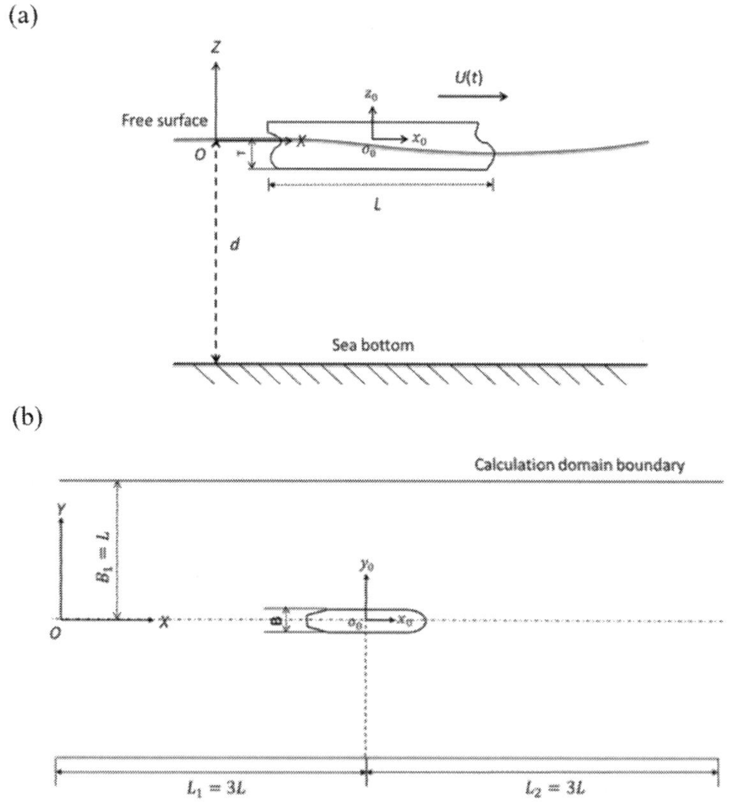

Figure 1. Sketch of the problem. In (a), L is the length of the ship and $U(t)$ is the ship velocity which is changing with time. T represents the draft of the ship while the water depth is denoted as d. (b) shows a top view of the problem, where it can be seen that the distance between the front and rear boundaries of the computational domain and the midship point are both $3L$. On the other hand, both the port and starboard boundaries are L away from the centreline of the ship. Finally, B in this figure represents the ship breadth.

2.2. Boundary value problem

There are only two boundary conditions in this problem: the body-surface condition and the free-surface condition. The body-surface condition is that the surface is impenetrable and the normal velocity of all cells on the body is equal to the speed U of the ship:

$$\frac{\partial \varphi}{\partial \boldsymbol{n}} = Un_1 \tag{1}$$

where $n = (n_1, n_2, n_3)$ is the unit normal vector on body-surface.

The free-surface condition in this method can be separated into two components: the dynamic free-surface condition and the kinematic free-surface condition. Since the method utilized in this study is based on potential flow, the fluid is assumed to be ideal, which implies that it is inviscid, irrotational, and incompressible. As a result, the free-surface conditions can be expressed using the velocity potential $\varphi(x, y, z, t)$ and the wave elevation $\zeta(x, y, t)$ as follows:

$$\varphi_t - U\varphi_x + g\zeta + \frac{1}{2}({\varphi_x}^2 + {\varphi_y}^2 + {\varphi_z}^2) + \frac{P}{\rho} = 0, on\ z = \zeta \tag{2}$$

$$\zeta_t - U\zeta_x + \varphi_x\zeta_x + \varphi_y\zeta_y - \varphi_z = 0, \quad on\ z = \zeta \tag{3}$$

where g is the gravitational acceleration, ρ is the fluid density and P is the pressure on free surface, and $\varphi_t$and ζ_t are the respective time derivatives. Similarly, φ_x, φ_y and φ_z are directional derivatives.

To simplify the solution process, the nonlinear terms are disregarded while retaining the unsteady terms, as discussed earlier. Consequently, the simplified linear unsteady free-surface condition can be expressed as follows:

$$\varphi_t - U\varphi_x + g\zeta + \frac{1}{2}({\varphi_x}^2 + {\varphi_y}^2 + {\varphi_z}^2) = 0, on\ z = \zeta \tag{4}$$

$$\zeta_t - U\zeta_x - \varphi_z = 0, \quad on\ z = \zeta \tag{5}$$

With the boundary conditions established, the subsequent step involves solving the resulting equations. To accomplish this, a three-level scheme is employed to discretize the free surface conditions:

$$(\varphi_t)_{i,j}^{m+1} = \frac{1}{\Delta t}\left(\frac{3}{2}\varphi_{i,j}^{m+1} - 2\varphi_{i,j}^{m} + \frac{1}{2}\varphi_{i,j}^{m-1}\right) \tag{6}$$

$$(\zeta_t)_{i,j}^{m+1} = \frac{1}{\Delta t}\left(\frac{3}{2}\zeta_{i,j}^{m+1} - 2\zeta_{i,j}^{m} + \frac{1}{2}\zeta_{i,j}^{m-1}\right) \tag{7}$$

where m represents the mth time step, i and j indicate location of the cell on the free surface. Substituting the Eq. (6) and Eq. (7) into the kinetic free-surface condition, allows one to obtain the value of $(\zeta)_{i,j}^{m+1}$:

$$\frac{1}{\Delta t}\left(\frac{3}{2}\zeta_{i,j}^{m+1} - 2\zeta_{i,j}^{m} + \frac{1}{2}\zeta_{i,j}^{m-1}\right) - U\frac{1}{\Delta x}\left(\frac{3}{2}\zeta_{i,j}^{m+1} - 2\zeta_{i,j+1}^{m+1} + \frac{1}{2}\zeta_{i,j+2}^{m+1}\right) - \frac{\partial \varphi_{i,j}^{m+1}}{\partial z} = 0 \tag{8a}$$

$$\left(\frac{1}{\Delta t} - \frac{3}{2}\frac{U}{\Delta x}\right)\zeta_{i,j}^{m+1} + 2\frac{U}{\Delta x}\zeta_{i,j+1}^{m+1} - \frac{1}{2}\frac{U}{\Delta x}\zeta_{i,j+2}^{m+1} = \frac{2}{\Delta t}\zeta_{i,j}^{m} - \frac{1}{2}\frac{1}{\Delta t}\zeta_{i,j}^{m-1} + \frac{\partial \varphi_{i,j}^{m+1}}{\partial z} \tag{8b}$$

Then, all variables in the dynamic free-surface condition can be updated with the results obtained:

$$\frac{1}{\Delta t}\left(\frac{3}{2}\varphi_{i,j}^{m+1} - 2\varphi_{i,j}^{m} + \frac{1}{2}\varphi_{i,j}^{m-1}\right) - U\frac{\partial \varphi_{i,j}^{m+1}}{\partial x} + g\zeta_{i,j}^{m+1} = 0 \tag{9a}$$

$$\frac{3}{2}\frac{1}{\Delta t}\varphi_{i,j}^{m+1} - U\frac{\partial \varphi_{i,j}^{m+1}}{\partial x} = \frac{2}{\Delta t}\varphi_{i,j}^{m} - \frac{1}{2}\frac{1}{\Delta t}\varphi_{i,j}^{m-1} - g\zeta_{i,j}^{m+1} \tag{9b}$$

In the above process, as the number of iterations k increases, all variables will be continuously updated until both $\left|\varphi_{i,j}^{m+1,k} - \varphi_{i,j}^{m+1,k-1}\right| < \varepsilon$ and $\left|\zeta_{i,j}^{m+1,k} - \zeta_{i,j}^{m+1,k-1}\right| < \varepsilon$ are satisfied.

In order to solve the above-mentioned boundary value problem, we developed a programme, namely MHydro, based on the Rankine source panel method. Once the potential is obtained, it is possible to calculate the pressure distribution and the forces/moments acting on the hull using Bernoulli's equation, which can be expressed as follows:

$$P = -\rho\left[\varphi_t - U\varphi_x + \frac{1}{2}\left(\varphi_x^2 + \varphi_y^2 + \varphi_z^2\right)\right] \tag{10}$$

$$F_i = \iint_S P n_i ds, i = 1, 2, \ldots, 6 \tag{11}$$

where i represents the degree of freedom, and:

$$n_i = \begin{cases} \boldsymbol{n}, & i = 1, 2, 3 \\ \boldsymbol{x} \times \boldsymbol{n}, & i = 4, 5, 6 \end{cases} \tag{12}$$

3. Case study selection

For the case studies in this research, the Wigley hull has been selected, and its specifications are provided in Table 1. Although there are several experimental results available for the Wigley hull, there is currently no data specifically replicating the trans-critical behaviour of an accelerating ship. By choosing the Wigley hull, we aim to maximize the potential for other researchers to utilize the results presented here in their future studies. This is due to the hull's ease of mathematical modelling and its common usage in potential flow-based investigations [13], [22]-[24].

Table 1. Wigley hull principal dimensions.

Parameter	Symbol	Value	Units
Length	L	3	m
Draft	T	0.1875	m
Beam	B	0.3	m

The depth Froude number serves as the basis for varying the parameters in the test matrix, namely the ship speed U and the water depth d. Following the approach of Day et al. [6] and Li et al. [13], multiples of the gravitational acceleration g are utilized to modify the ship speed. Previous studies focused on rapidly accelerating the hull to a target speed and maintaining that speed, using values ranging from $0.08g$ to $0.02g$. However, the objective of the current study is to observe transient phenomena during acceleration. Therefore, the maximum acceleration intensity chosen for this study is $0.02g$, which corresponds to the lowest acceleration intensity used by Day et al. [6] and Li et al. [13]. Additionally, acceleration intensities of $0.01g$, $0.005g$, and $0.002g$ are also employed.

To assess the impact of water depth, the test matrix in Table 2 combines acceleration intensities with d/T ratios of 1.2 and 1.5. The cases progress up to a maximum water depth Froude number of 2. In the CFD simulations, steady and constant velocity cases are used to estimate the deviation in resistance between fixed speed and accelerating conditions. Table 3 provides the water depth Froude numbers corresponding to different water depths for constant velocities.

IOP Conf. Series: Materials Science and Engineering 1288 (2023) 012046 doi:10.1088/1757-899X/1288/1/012046

Table 2. Case studies modelling the acceleration resistance of the Wigley hull.

Case number	Acceleration	d/T
1	0.02g	1.2
2	0.01g	
3	0.005g	
4	0.002g	
5	0.02g	1.5
6	0.01g	
7	0.005g	
8	0.002g	

Table 3. Constant velocity cases modelled using CFD.

Case number	Depth Froude number	d/T
9	0.60	1.2
10	0.80	
11	0.89	
12	0.95	
13	1.00	
14	1.05	
15	1.20	
16	1.40	
17	1.70	
18	0.60	1.5
19	0.80	
20	0.89	
21	0.95	
22	1.00	
23	1.05	
24	1.20	
25	1.40	
26	1.70	

4. Results and discussion

The potential flow theory provides results for wave-making resistance, whereas the CFD solution yields three sets of resistance results: total resistance (R_T), frictional resistance (R_F), and pressure resistance (R_P). The pressure resistance is further divided into normal and tangential components, with R_F encompassing the effects of surface curvature. On the other hand, R_P represents a combination of viscous pressure resistance and wave resistance [25]. However, the viscous pressure resistance typically constitutes a small portion of the total resistance across all speeds [26]. Since wave resistance predominates over pressure resistance, it is possible to compare the results without the need to separate the pressure resistance into its individual components.

4.1. Pressure and wave resistance

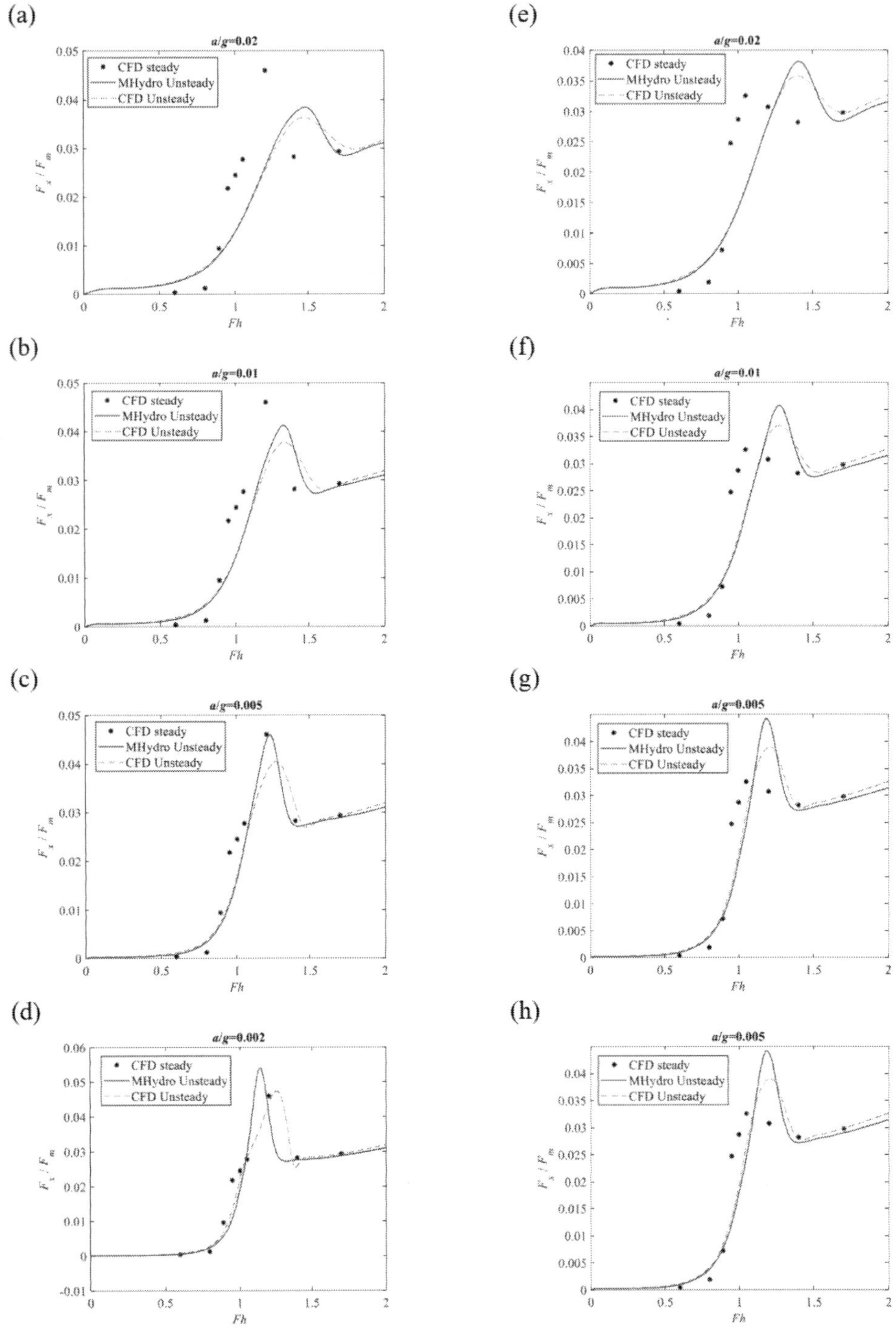

Figure 2. Changes of resistance during ship acceleration. The tiles on the left show d/T =1.2 while the tiles on the right show d/T =1.5. F_x is the wave resistance obtained by MHydro and pressure resistance obtained through the URANS method described previously. It also should be noted that the results of CFD here are pressure resistance while MHydro results are only wave making resistance.

This section compares the wave and pressure resistance values obtained as the ship accelerates for the two modeled depth-to-draft ratios. Figure 2 illustrates the comparison between the CFD method and MHydro for all cases listed in the test matrix (Table 2), including the constant velocity cases modeled using CFD (Table 3). All forces are made dimensionless using the ship mass force F_m, as described in [6]. A pronounced peak can be observed in all cases near F_h =1. In the steady problem (i.e., constant speed with no acceleration), such a peak is generated by the mutual interference of the fore and aft wave systems produced by the ship. The resistance curves in Figure 2 indicate that this type of interference reaches its maximum between F_h =1.05 and F_h =1.2, depending on the value of the dimensionless acceleration (a/g).

It should be noted that the peak values shown in Figure 2 for the constant speed case are based on a limited number of selected points, and they may not precisely indicate the exact peak value. The curve only provides an approximate range within which the peak value is expected to appear. However, it is evident that the increase in resistance is significantly influenced by the water depth, as demonstrated by the higher F_x values when d/T =1.2 compared to d/T =1.5. This phenomenon undergoes changes when acceleration is introduced. In other words, the presence of acceleration generates an unsteady wave that has a notable impact on the resistance.

In terms of the discrepancy in the position of the two peaks, both solvers provide remarkably similar results for high accelerations. Interestingly, the difference between the predictions increases monotonically from 0.75% and 0.68% to 10.13% and 5.54% for d/T =1.2 and d/T =1.5, respectively. Further investigations are necessary to understand why the two sets of predictions performed better for the lowest acceleration when the water depth is slightly larger. This effect could potentially be attributed to the presence of nonlinear effects, which are known to become more significant with decreasing water depth. However, drawing definitive conclusions would require more detailed investigations, preferably incorporating experimental data to eliminate any bias in numerical or potential solutions.

When comparing the results obtained from MHydro and CFD, it can be observed that regardless of whether d/T =1.2 or d/T =1.5, the peak in the MHydro results occurs earlier than the peak predicted by CFD, and the resistance value obtained from MHydro is consistently higher. This difference becomes particularly pronounced when the acceleration is small. One possible explanation for this disparity is the inclusion of viscous and nonlinear terms in CFD. MHydro, on the other hand, provides a linear and unsteady result by neglecting nonlinear terms.

As the acceleration increases, the relative importance of the unsteady effect grows, overshadowing the nonlinear effects. Consequently, the unsteady effect becomes dominant when the acceleration is sufficiently large, which explains why there is less agreement between the results obtained from MHydro and CFD when the acceleration is small. Conversely, the difference between the results obtained from the two prediction methods is smallest when the acceleration is large. This discussion aligns with observations made in other studies concerning linear potential flow solutions near the critical depth Froude number [18], [20], [28].

4.2. Acceleration effects

As mentioned earlier, two key factors influencing the problem under investigation are the magnitude of acceleration and the water depth. This paper will discuss each of these factors in detail. Upon introducing acceleration, Figure 3 demonstrates that both the peak resistance value and its corresponding position undergo changes. Considering the satisfactory agreement between CFD and MHydro, additional case studies are explored in Figure 3 using only the potential flow solver. This choice is made due to its faster turnaround time and lower computational requirements.

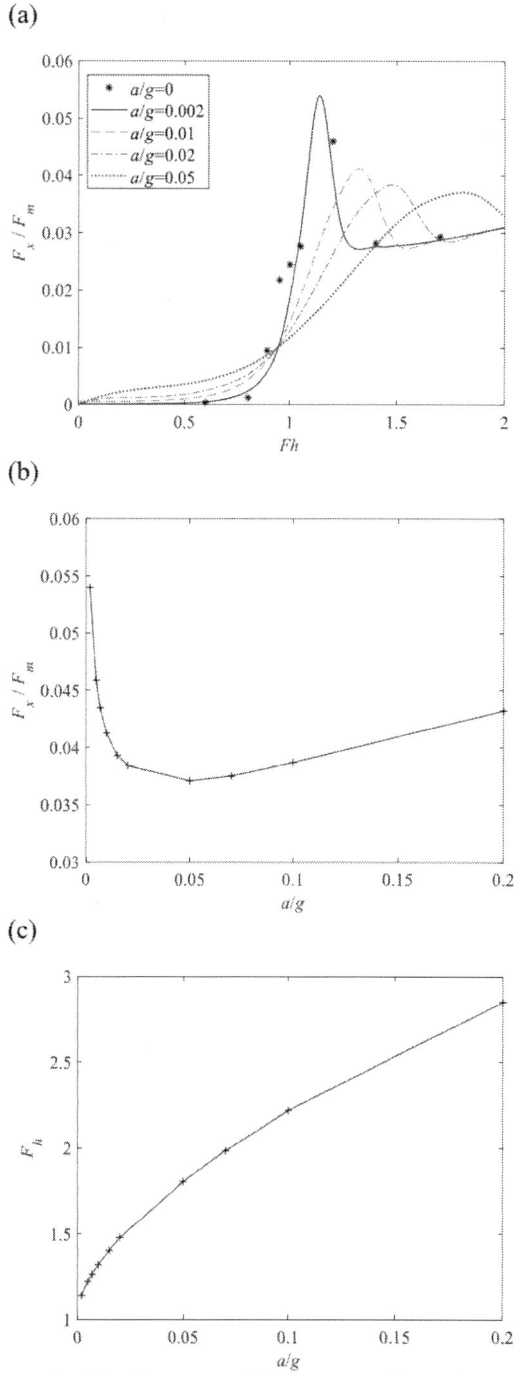

Figure 3. (a) Changes of fully unsteady resistance with varying acceleration; (b) Resistance peak value for different accelerations; (c) Peak location at different acceleration; Cases depicted correspond to $d/T = 1.2$.

Figure 3 illustrates that as the acceleration increases, the resistance peak becomes smaller and shifts towards higher values of the examined depth Froude number. This indicates that the unsteady effect generated by the acceleration significantly influences the results. The shifting position of the resistance peak can be attributed to the shock wave formed at the bow due to the acceleration. During the initial stages when the ship speed is low, a shock wave is emitted forward, resulting in additional wave/pressure resistance. As the ship speed matches the speed of the wave, the wave-making resistance reaches its maximum, which is depicted as the peak in Figure 2.

In simpler terms, as the acceleration increases, the velocity of the shock wave generated by the ship also increases. This means that the ship needs to reach a higher speed to catch up with the shock wave. Consequently, the position of the resistance peak is delayed at higher accelerations. As the ship continues to accelerate, it eventually overtakes the shock wave, resulting in a sharp drop in resistance for a narrow range of F_h values, followed by a linear increase.

Another notable observation from the presented results is that irrespective of the acceleration, the time-history of resistance converges to a single point near F_h =1. Beyond this point, the resistance exhibits approximately linear growth until it reaches the resistance peak. Figure 4 (a) and Figure 4 (b) illustrate these findings for d/T =1.2 and d/T =1.5, respectively.

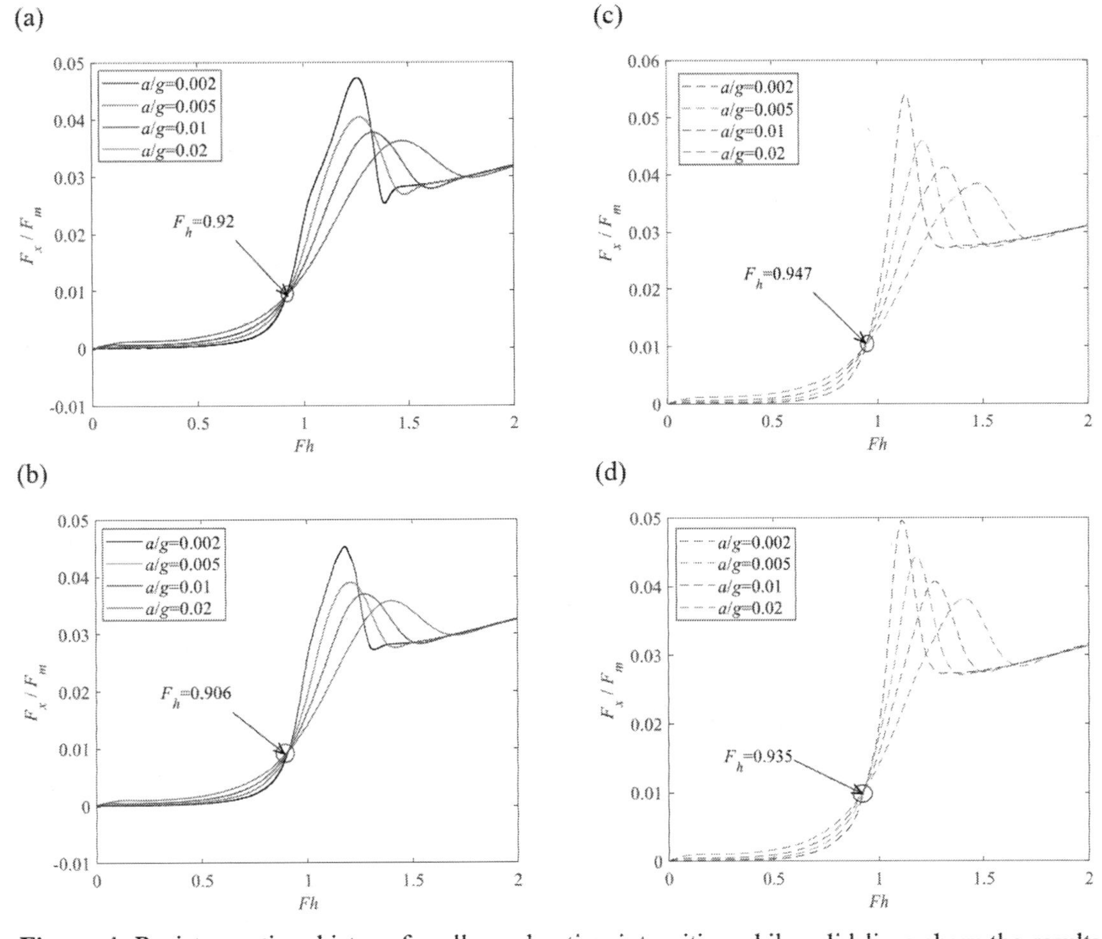

Figure 4. Resistance time-history for all acceleration intensities while solid lines show the results obtained using CFD and dashed lines show the results obtained using MHydro. (a) Resistance time-history at d/T =1.2 for CFD; (b) Resistance time-history at d/T =1.5 for CFD; (c) Resistance time-history at d/T =1.2 for MHydro; (d) Resistance time-history at d/T =1.5 for MHydro.

For the water depth corresponding to d/T =1.5, the CFD results indicate that the resistance time-histories coincide at approximately F_h =0.906, whereas the potential flow results predict this point at approximately F_h =0.935. Similarly, for d/T =1.2, CFD predicts a shift in this point to approximately

F_h =0.92, while MHydro places it at approximately F_h =0.947. Before and after the point of intersection in the resistance time-histories, the resistance follows distinct paths. To the best of the authors' knowledge, such phenomena have not been documented in the existing literature.

To investigate the cause of this phenomenon, Figure 5 presents wave fields at different accelerations near the intersection point. As F_h or the acceleration varies, wave elevations at the bow and stern also change correspondingly. The wave at the bow is primarily influenced by the ship's velocity, which is relatively straightforward to comprehend. However, the situation is different for waves at the stern.

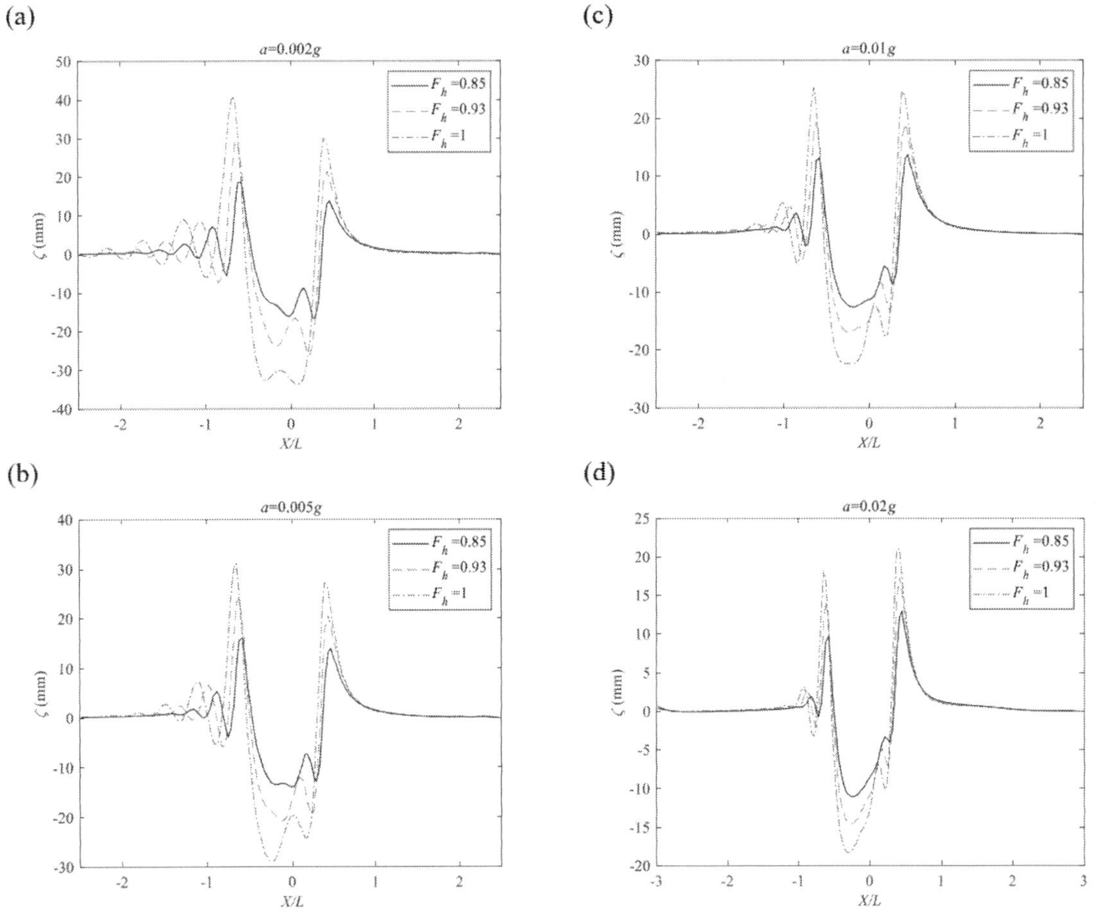

Figure 5. Wave elevations on the port side of ship with different accelerations (y/L=0.0167). These results were computed using MHydro for d/T=1.2.

Figure 6 (a) illustrates that the wave elevation at the ship stern is inversely proportional to the acceleration. A higher acceleration leads to more pronounced unsteady effects, which, when combined with the original wave field at the ship stern, result in smaller wave peaks and troughs. Additionally, it should be noted that waves at the ship stern exhibit peaks before the critical speed and troughs after it. This indicates that when F_h <1, a smaller acceleration brings the wave elevation at the bow and stern closer together, resulting in lower wave resistance. However, after this point, as the acceleration decreases, the wave trough at the stern significantly increases. According to Bernoulli's equation, this leads to a decrease in the pressure field at the stern of the ship, resulting in greater wave resistance. This explains the observed intersection of wave resistance in Figure 4 and why the peak resistance value decreases with an increase in acceleration.

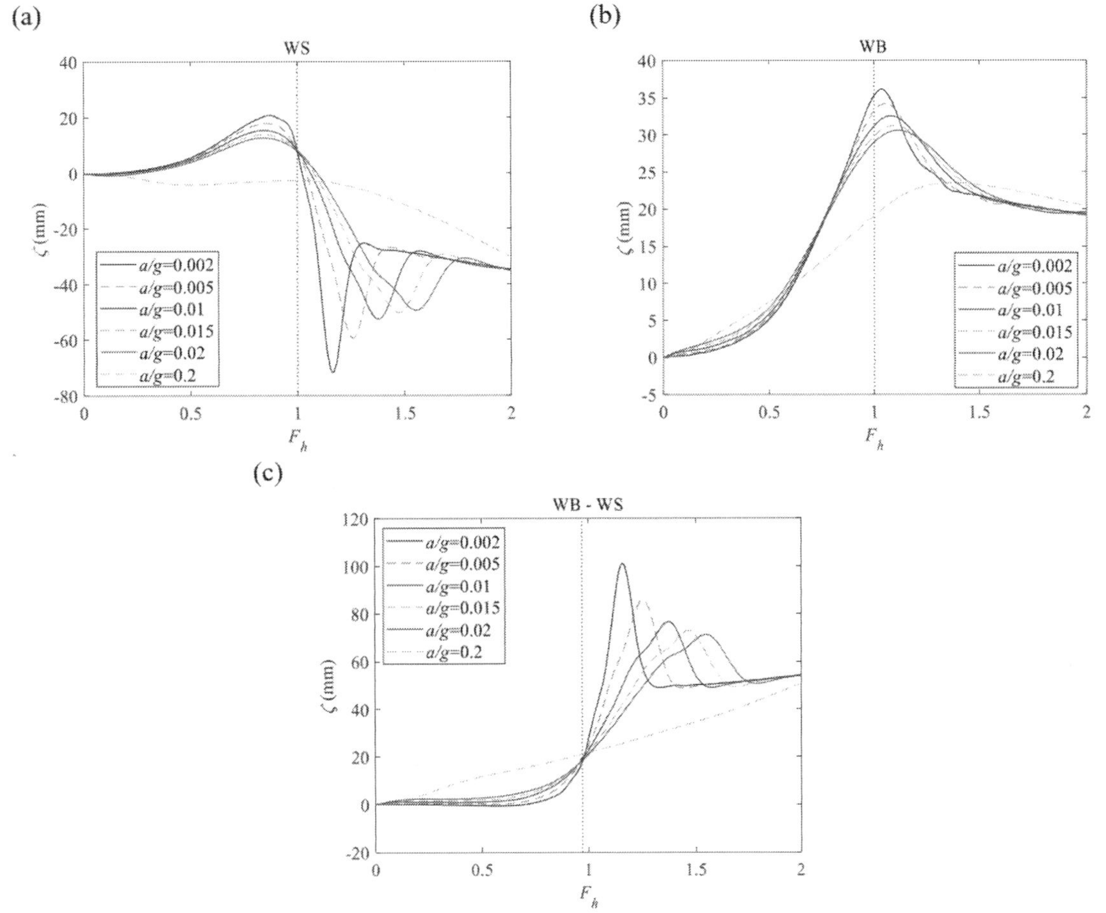

Figure 6. Wave elevations around the ship stern and bow with different accelerations. (a) Stern wave (WS); (b) Bow wave (WB); (c) Wave difference between ship bow and stern. These results were computed using MHydro for d/T=1.2.

4.3. Finite depth effects

The agreement between the constant speed and accelerating results depicted in Figure 2(d) demonstrates that the effect of acceleration at a =0.002g and d/T =1.2 is minimal. However, there is a significant disparity between the corresponding values at d/T=1.5. This observation, combined with the shifting of the resistance peak along the F_h range due to varying acceleration, suggests that the water depth Froude number may not be the most suitable parameter to represent the time-varying velocity accurately.

Figure 7 shows a series of example cases computed with MHydro, with a fixed acceleration of a/g =0.02 for different depth-to-draft ratios (d/T =1.2-4). Within this framework, larger depth-to-draft ratios shift the resistance peak towards higher F_h values and diminish its magnitude. This effect is particularly noticeable between d/T =2 and d/T =4. The resistance penalty resulting from the shallow water effect at higher F_h values diminishes rapidly with increasing water depth. If the water depth were to exceed d/T=4, the distinctive resistance peak observed in shallow waters would disappear.

Another observation from Figure 7 is that shallow water conditions may provide an advantage in terms of reduced resistance when accelerating a vessel. Specifically, the values illustrated for F_h =1 in Figure 7 (d) indicate that higher water depths correspond to higher resistance. For the given acceleration, the increase in resistance between very shallow water (d/T =1.2) and moderately shallow water (d/T =4) exceeds 100%. This increase is a result of the delay in the occurrence of the resistance peak to higher depth Froude numbers, as shown in Figure 7 (c).

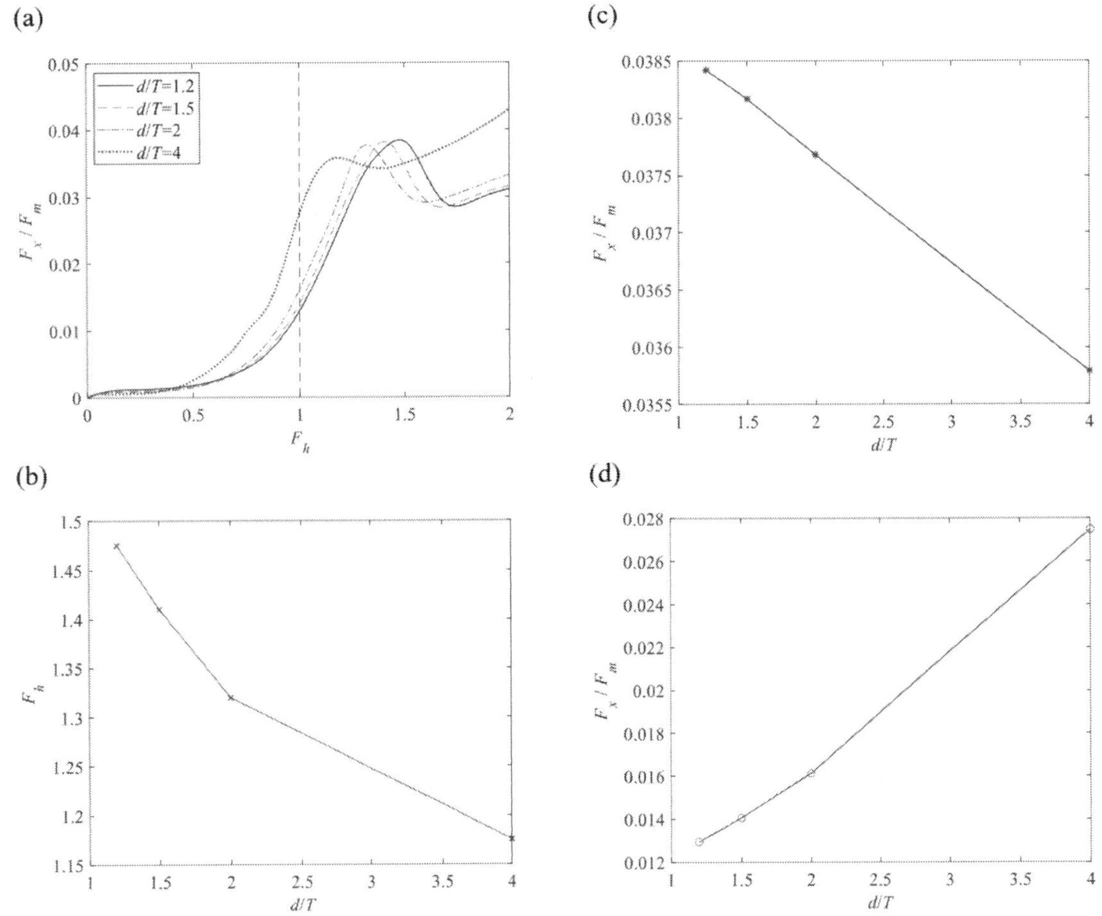

Figure 7. Wave resistance and associated peaks at different water depths, computed using MHydro. (a) shows the time-history of the resistance, (b) shows the magnitude of the peak for varying d/T ratios, (c) show the position of the peak as a function of d/T while (d) shows the resistance value at F_h=1.

4.4. *Unsteady free surface effects*

The results presented in the preceding sections, obtained through the solution of the fully unsteady boundary value problem (BVP) defined in Eq. (1), (4), and (5), provide valuable insights. To further investigate the effect of unsteadiness on the free surface boundary condition, a quasi-steady solution is proposed. In this approach, the time-dependent terms in Eq. (4) and (5) are ignored at each time step, while the body surface boundary condition in Eq. (1) remains unchanged. For both the steady and quasi-steady methods, the unsteady Bernoulli's equation in Eq. (12) is utilized to calculate the hydrodynamic force, specifically the wave-making resistance in the present study.

Figure 8 provides a direct comparison between the results obtained from the quasi-steady and unsteady free-surface solutions mentioned earlier. The observed discrepancies highlight the significance of considering the unsteady free-surface condition in the context of the acceleration problem investigated. Notably, the free-surface effects exert a predominant influence in the vicinity of the critical speed region. Given the unsteady nature of the acceleration problem, it is crucial to take into account the unsteady terms in the free-surface condition.

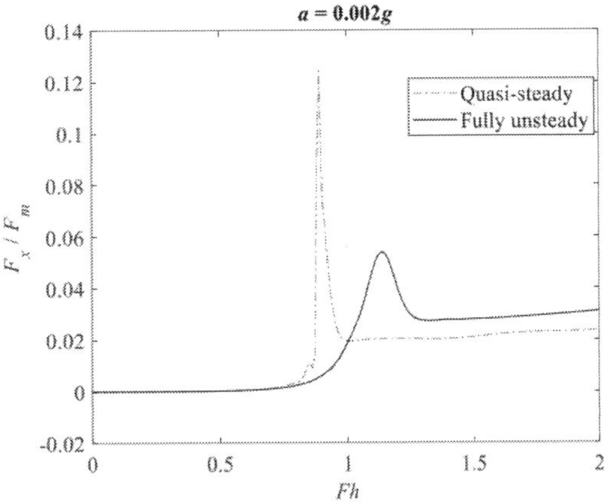

Figure 8 Wave-making resistance obtained by MHydro with and without considering the time-dependent terms on free-surface boundary condition. The water depth used here is d/T=1.2.

5. Conclusions and recommendations for future work

When a watercraft accelerates beyond the speed of the fundamental wave in a given medium, it experiences fluctuations in resistance. While such effects are well-known in aerodynamics, they had not been previously documented in the field of marine hydrodynamics, to the best of the authors' knowledge. In shallow water, the wave speed remains constant and depends only on gravitational acceleration and water depth. This provides an opportunity to investigate hydrodynamic equivalents to the sonic boom. The aim of the present study was to fill this gap in the literature by combining URANS and potential flow solutions.

The research focused on modelling four acceleration intensities and two water depths using the Wigley parabolic hull. In all cases, a pronounced resistance peak was observed near the critical depth Froude number. The results showed a maximum disagreement in predicting the location of the resistance peak along the examined depth Froude number, ranging from 5.54% to 12.44% for d/T=1.2 and 6.23% to 11.75% for d/T=1.5.

The findings indicated excellent agreement in the low and high-speed range beyond the critical transition, suggesting that resistance in that range is linear and primarily influenced by unsteady rather than nonlinear terms. Some deviations were observed between the two solvers near the critical speed, suggesting that nonlinear effects influence the solution in that range. An intersection point was identified near, but less than, F_h =1 for both modelled water depths, where all acceleration intensities intersect. This point's existence was attributed to the proximity of the trans-critical boundary and the fact that no steady flow is possible when the depth Froude number is unity.

While the results presented in this study demonstrate good agreement between the two solvers, it is important to note that experimental data on trans-critically accelerating ship resistance would be invaluable for validating the accuracy of the predictions. Furthermore, models of trans-critical acceleration in fully confined water are of particular interest, as the blockage ratio would also play a significant role. Such cases are intriguing because the trans-critical range can occupy a substantial portion of the depth Froude number, depending on the blockage [29].

References

[1] HAVELOCK TH 1949 The resistance of a submerged cylinder in accelerated motion *The Q. J. Mech. Appl. Mathematics* **2** 419–427

[2] Lunde J K 1957 The linearized theory of wave resistance and its application to ship shaped bodies in motion on the surface of a deep, previously undisturbed fluid (Technical University of Trondheim)

[3] Lunde J K 1951 On the linearized theory of wave resistance for deplacement ships in steady and accelerated motion (Technical University of Trondheim) pp 25-85

[4] Wehauden, J V 1961 *Effect of the initial acceleration upon the wave resistance of ship models* (Berkeley: California Univeristy Berkeley)

[5] Calisal S 1977 Effect of initial acceleration on ship wave pattern and wake survey methods *J. Ship Res.* **21** 239-247

[6] Day A H, Clelland D and Doctors L J 2009 Unsteady finite-depth effects during resistance tests on a ship model in a towing tank *J. of Mar. Sci. Technol.* **14** 387-397

[7] Doctors L J and Sharma S D 1972 The wave resistance of an air-cushion vehicle in steady and accelerated motion *J. Ship Res.* **16** 248-260

[8] Yeung R W 1975 Surface waves due to a maneuvering air-cushion vehicle *J. Ship Res.* **19** 224-242

[9] Haussling H J and Van Eseltine R T 1978 Waves and wave resistance for air-cushion vehicles with time-dependent cushion pressures *J. Ship Res.* **22** 170-177

[10] Barratt M J 1965 The wave drag of a hovercraft *J. Fluid Mech.* **22** 39-47

[11] Doctors L J 1975 The experimental wave resistance of an accelerating two-dimensional pressure distribution *J. Fluid Mech.* **72** 513-527

[12] Doctors L J 1993 On the use of pressure distributions to model the hydrodynamics of air-cushion vehicles and surface-effect ships *Naval Eng. J.* **105** 69-89

[13] Li M X, Yuan Z M and Yeung R W 2020 *Int. Conf. on Offshore Mechanics and Arctic Engineering* Vol 84379 (American)

[14] Jiang T, Henn R and Sharma S D 2002 *24th Symposium on Naval Hydrodynamics* (Japan: Fukuoka) pp 441-457

[15] Grue J 2017 Ship generated mini-tsunamis *J. Fluid Mech.* **816** 142-166

[16] Tuck E O 1966 Shallow-water flows past slender bodies *J. Fluid Mech.* **26** 81-95

[17] EO T 1967 Sinkage and trim in shallow water of finite width *Schiffstechnik* **14** 92-94

[18] Lea G K and Feldman J P 1972 Transcritical flow past slender ships *9th Symposium on Naval Hydrodynamics* (France: Paris) pp 1527–1542

[19] Gourlay T P and Tuck E O 2001 The maximum sinkage of a ship *J. Ship Res.* **45** 50-58

[20] Yuan Z M 2019 Ship hydrodynamics in confined waterways *J. Ship Res.* **63** 16-29

[21] Yuan Z M, Li L and Yeung R W 2019 Free-surface effects on interaction of multiple ships moving at different speeds *J. Ship Res.* **63** 251-267

[22] Yuan Z M, Incecik A and Jia L B 2014 A new radiation condition for ships travelling with very low forward speed *Ocean Eng.* **88** 298-309

[23] Andrun M, Blagojević B and Bašić J 2019 The influence of numerical parameters in the finite-volume method on the Wigley hull resistance *J. Eng. Marit. Environ.* **233** 1123-1132

[24] Bašić J, Blagojević B and Andrun M 2020 Improved estimation of ship wave-making resistance *Ocean Eng.* **200** 107079

[25] Molland A, Turnock S and Hudson D 2017 *Components of Hull Resistance In Ship Resistance and Propulsion: Practical Estimation of Ship Propulsive Power* (Cambridge: Cambridge University Press) pp 12-69

[26] Terziev M, Tezdogan T, Demirel Y K, Villa D, Mizzi S and Incecik A 2021 Exploring the effects of speed and scale on a ship's form factor using CFD *Int. J. Naval Architecture Ocean Eng.* **13** 147-162

[27] Tuck E O 1978 Hydrodynamic problems of ships in restricted waters *Annu. Rev. Fluid Mech.* **10** 33-46
[28] Tuck E O 1970 Shallow wave problems in ship hydrodynamics *8th Symposium Naval Hydrodynamics* (Pasadena)
[29] Lataire E, Vantorre M and Delefortrie G 2012 A prediction method for squat in restricted and unrestricted rectangular fairways *Ocean Eng.* **55** 71-80

An experimental and numerical hydrodynamic study on the Argentinian fishing vessels

S Oyuela[1], R Sosa[1,2], A D Otero[2,3], F P Arribas[4], H R Diaz-Ojeda[5]

[1] Universidad de Buenos Aires, Facultad de Ingeniería - CEAN – INTECIN, Av. Paseo Colon, Ciudad de Buenos Aires C1063, ARG

[2] Consejo Nacional de Investigaciones Científicas y Técnicas, Godoy Cruz, Ciudad de Buenos Aires C1425FQC, ARG

[3] Universidad de Buenos Aires, Av. Paseo Colon, Ciudad de Buenos Aires C1063, ARG

[4] ETSIN-UPM, Av. de la Memoria, Madrid 28040, SPN

[5] Universidad de Las Palmas de Gran Canaria, Instituto Universitario de Sistemas Inteligentes y Aplicaciones Numéricas en Ingeniería, Las Palmas de Gran Canaria 35017, SPN

Production Editor, *Journal of Physics: Conference Series*, IOP Publishing, Dirac House, Temple Back, Bristol BS1 6BE, UK

E-mail: `hectorruben.diaz@ulpgc.es`

Abstract. The purpose of this work is the validation of the hydrodynamic resistance for a small fishing vessel providing more details about the ship's hydrodynamics. The selected ship is a traditional vessel that has been sailing in Argentina for more than 20 years without a previous numerical resistance evaluation. Because of the absence of previous numerical optimization, the ship might be a non-efficient vessel from a consumption, operating and polluting point of view. For this reason, in order to bring a more detailed evaluation of the flow around this hull, numerical studies are carried out with the open source OpenFOAM code and validated with experimental results obtained at the University of Buenos Aires towing tank. Reynolds Averaged Navier Stokes (RANS) method together with volume of fluid (VOF) are used for the numerical procedure. Once the resistance results are validated, the total resistance coefficient is split into viscous and wave coefficients in order to provide insight into ship hydrodynamics. Pressure distributions on the hull and wave patterns will be also presented for different ship loading conditions, focusing on the flow around the hull that might give information for a future hull optimization.

1. Introduction

When a new ship is ordered, one of the strictest specifications that must be fulfilled during the ship operation is the contract speed. This refers to the speed at which the ship must sail with a specific power consumption. If this velocity differs considerably from the contracted speed, the contract might be cancelled, creating a conflict for the designers. This is due to the fact that a difference in powering might alter, for instance, the entire ship's operation, stability, and profitability. Therefore, the determination of the speed-resistance curve in ship design is one of the most important aspects to study.

In the early design stages, with the hull form almost defined, statistical and regression methods are the common procedures for an initial approach. Holtrop and Mennen [1] developed a regression method based on sea trials and more than 200 models tested in towing tanks. This method has become the most commonly used for ship resistance estimation. However, this method presents higher uncertainty results when compared against experimental or numerical results. For instance, Bilec and Obreja [2] showed that Holtrop-Mennen provided differences up to 30% in resistance when compared against experimental tests, particularly for fishing vessels.

Once the ship design advances, reliable and more precise methods need to be used to quantify shape, viscous, or pressure effects that cannot be quantified with statistical methods. For instance, different bow designs, appendages, etc. Within this context, Experimental Fluid Dynamics (EFD) tests in towing tanks have always been the most reliable procedure. However, with the fast development of computational resources, Computational Fluid Dynamics (CFD) has been increasingly used, showing higher capabilities regarding post-processing and reliability [3]. As identified by the Combined CFD/EFD Methods ITTC Specialist Committee, the combination of EFD and CFD could be a feasible solution to increase the accuracy of power predictions instead of choosing EFD or CFD exclusively at the moment [4]. If a part of the model testing or extrapolation procedure causes higher uncertainty than than those caused by the CFD model, accuracy will be increased [4].

Argentina's fishing industry has undergone significant changes since the 1960s. The hake fishery had its beginnings between 1961 and 1969. In the following decade, from 1970 to 1975, the industry incorporated fresh fish vessels to improve the quality and freshness of the catch. This was followed by the introduction of frozen fish processing vessels from 1976 to 1981. The 1980s saw the development of coastal fisheries and the expansion of the shrimp fishery [5]. In March 2019, a task force on measures to reduce bycatch in the fisheries was launched, bringing together scientists and stakeholders from the public and private sectors. Additionally, Argentina has embarked on a process of fleet modernization with the aim of improving crew safety and integrating newer and more environmentally sustainable technologies. More than 50% of the Argentine fleet is older than recommended by relevant international standards [6].

The objective of this work is the validation of the CFD results obtained using the free open-source OpenFOAM code, with the experimental results obtained in the towing tank. Once the resistance results are validated, the total resistance coefficient is divided into viscous and wave coefficients to provide insight into hydrodynamic resistance. The force coefficient decomposition involves determining the form factor k. In this work, the values of k were obtained numerically by the double-hull CFD method and experimentally by Prohaska's method. It is interesting to note that in the 1978 ITTC Performance Prediction method, the form factor has been identified by the CFD/EFD Methods ITTC Specialist Committee as one of the major causes of uncertainties, specifically due to the Prohaska method [4, 7]. Although it is beyond the scope of this work, the obtained values of k will serve as the basis for future research that seeks to answer the question of whether CFD-based form factors can be an alternative or supplement to the Prohaska method.

Finally, pressure distributions on the hull and wave patterns will be presented for different ship loading conditions, focusing on the flow around the hull, that might provide information for future hull optimization. The conclusion and future research directions are discussed.

2. Experimental Setup

As mentioned in the introduction, the quantification of ship resistance is a fundamental aspect in ship design. The common procedure involves conducting towing tank experiments to obtain the ship resistance-velocity curve. The total resistance was measured at Froude numbers (Fn) varying between 0.10 and 0.45.

2.1. Model Ship

For this work, a typical Argentinian fishing vessel was selected. Towing tank experiments were conducted using a ship model at a 1:20 scale, which was built in Glass-Fiber Reinforced Plastic (GFRP). The model was manufactured according to the recommended procedure by the International Towing Tank Conference (ITTC) [8], with particular attention to model manufacturing tolerances, surface finishing, and the size and positioning of turbulence stimulation. Turbulence stimulation was achieved using a trip wire located at 5% of LPP (Length between Perpendiculars) aft from the fore perpendicular. Additionally, a wire was used to promote turbulence at the bulbous bow, positioned at 1/3 of the bulb length from its fore end. In both cases, a wire with a diameter of 1.0 mm was employed.

Three load conditions were analyzed in this work: the ballast condition (ship draught 3.3 m), the full load condition (ship draught 3.9 m), and an intermediate load condition (ship draught 3.6 m). The main characteristics of the full-scale ship in the ballast condition and those of the ship model for the different draughts are shown in Table 1. The hull body lines are depicted in Figure 1.

Table 1. Main parameters of full-scale ship and ship model

Main Particulars	Symbol	Unit	Ship	Model 1	Model 2	Model 3
Model scale	λ	[-]	-	20	20	20
Length on waterline	L_{WL}	[m]	32.68	1.634	1.662	1.641
Length over wetted surface	L_{OWS}	[m]	34.795	1.670	1.740	1.740
Breadth	B	[m]	9.28	0.464	0.464	0.464
Draught	D	[m]	3.30	0.165	0.180	0.195
Displacement volume	∇	[m^3]	599.40	0.075	0.085	0.095
Wetted surface area	S	[m^2]	392.67	0.982	1.058	1.124
Block Coefficient	C_B	[m]	0.60	0.60	0.61	0.64
Midship section area coeff.	C_X	[m]	0.86	0.86	0.87	0.89

2.2. Towing Tank

The experimental tests were conducted in the towing tank of the University of Buenos Aires (UBA), Argentina. The Channel of Naval Architecture Experiences (Canal de Experiencias de Arquitectura Naval, CEAN) has dimensions of 72 m in length, 3.6 m in width, and a variable depth of up to 2 m. For the present tests, the water depth was 1.80 m.

The towing tank is equipped with a running carriage that can achieve a maximum speed of 4.0 m/s. The model was ballasted to the required displacement and waterline, and attached to the resistance dynamometer. A one-component force transducer manufactured by Kempf & Remmers (type R 47) was used in the dynamometer for the resistance tests. By coupling the model to the carriage through the resistance dynamometer, surge, yaw, and heel are prevented, while allowing the model to be free in trim and sinkage. The resistance transducer is designed to handle a full-scale load of ±100 N, with a sensitivity of approximately ±1 mV/V of the supply voltage.

The mid-sectional area of the vessel corresponds to 1%, 1.1%, and 1.2% of the towing tank section area in the ballast, intermediate, and full load conditions, respectively. Therefore, Schuster's blockage correction method is applied following the procedure outlined by the International Towing Tank Conference (ITTC) [9].

2.3. Uncertainty Analysis of Resistance Tests

The analysis of uncertainty for resistance measurements must consider various significant components related to the total resistance. These components include uncertainties associated with hull geometry, towing speed, water temperature, dynamometer calibration, and repeated tests. Each of these standard uncertainty components will be estimated and then combined using the root-sum-squares (RSS) method, as recommended by the International Towing Tank Conference (ITTC) [10].

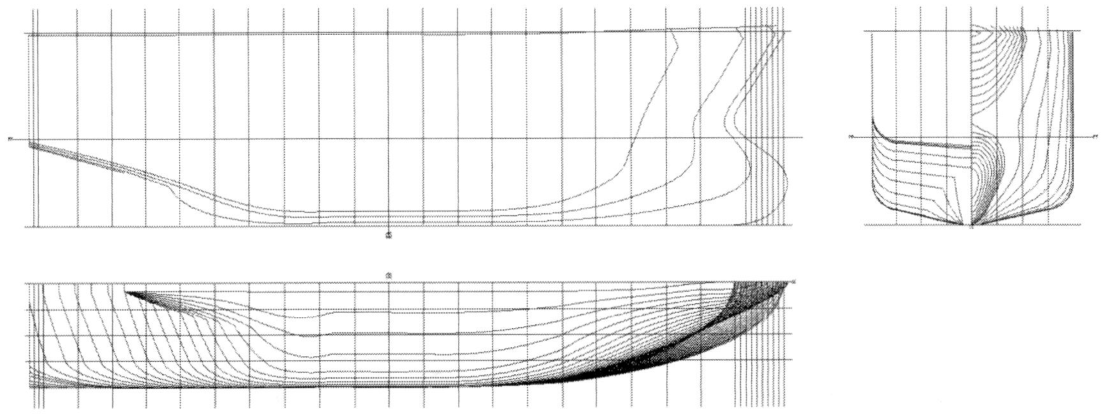

Figure 1. Body lines of the case study ship.

The total resistance of a hull model at a specific Froude number (Fn) depends on the hull's wetted surface area and Reynolds number (Re). It is generally assumed that the geometry uncertainty of a ship model hull has no effect on the residuary resistance coefficient. The uncertainties associated with the wetted surface area and representative length of the model (i.e., Re number) can be estimated in terms of the standard uncertainty of the displacement. Additionally, the uncertainty related to the hull geometry is quantified by measuring the model ballasting. The model and the weights used for ballasting were weighed using a portable beam scale. The resulting uncertainties in hull geometry are presented in Table 2 for speeds corresponding to Fn = 0.14, Fn = 0.26, and Fn = 0.37.

The uncertainty in the towing speed propagates into the resistance through the dynamic pressure and the Re number. The relative uncertainty of the towing speed is considered based on the bias limit of the towing carriage. The resulting uncertainties in resistance due to velocity are presented in Table 2.

Variations in water temperature have a relatively significant effect on water viscosity, and subsequently, on the Re number and frictional drag of the hull model. The water temperature during the tests exhibited a variation of less than 0.5°C. The corresponding uncertainties in resistance due to temperature variation are presented in Table 2.

The resistance dynamometer was calibrated before each test and checked immediately after. Linear curve fitting was employed, and the standard deviation of this linear regression was adopted as the standard uncertainty of calibration. The accuracy of the weights used is so high that its uncertainty is considered negligible. The corresponding uncertainties in dynamometer calibration are presented in Table 2.

The R47 dynamometer was used for measuring the resistance with a sampling rate of 200 Hz. The measurement at each speed was obtained by averaging the time history of the signal from the Data Acquisition System (DAS) over a time interval of at least 10 seconds. The signal from the load cell was low-filtered at 4 Hz before reaching the DAS. The standard uncertainty of the average of the sampling history ranged between 0.0007% and 0.003%, while the average of all repetitions was 0.0012%. Therefore, the uncertainty associated with a single reading from the DAS is considered negligible.

The mean of repeated resistance measurements is considered as the best estimate. The standard uncertainty component of the mean from N repeated tests is estimated following the ITTC procedure [10]. The standard uncertainties for repeated tests, with N = 4, are shown in Table 2.

Table 2. Combination of uncertainty in measurement for resistance at Fn = 0.14, Fn = 0.26 and Fn = 0.37. Repeat test N = 4

Uncertainty Components	Fn = 0.14	Fn = 0.26	Fn = 0.37
Hull geometry	0.05	0.05	0.05
Speed	0.067	0.067	0.067
Water temp.	0.03	0.03	0.03
Dynamometer	4.73	1.04	0.35
Repeat test, Deviation[a]	5.00	3.50	1.50
Combined for single test	6.88	3.65	1.54
Repeat test, Deviation of mean	2.50	1.75	0.75
Combined for repeat mean	5.35	2.04	0.83
Expanded for repeat mean	10.70	4.08	1.66

[a] Repeat test, Deviation = (Repeat test, Deviation of mean)*$N^{1/2}$

Finally, the significant uncertainty components are combined using the Root-Sum-Square (RSS) method, and the results are presented in Table 2.

As observed in Table 2, the dynamometer and the measurement precision are the major sources of uncertainty originating from the repeated tests. The uncertainty of resistance measurements for Fn = 0.14, Fn = 0.26, and Fn = 0.37 are 5.35%, 2.04%, and 0.83%, respectively. It can be seen that the uncertainty decreases with increasing speed because the contribution of the dynamometer decreases at higher speeds. The expanded uncertainties in Table 2 correspond to a confidence level of 95%.

2.4. Experimental Form Factor Determination

The ITTC procedure for splitting the total resistance coefficient adopts Hughes's hypothesis and introduces the form factor k to augment the flat plate frictional resistance coefficient C_{F0}, accounting for the Reynolds number-dependent viscous pressure resistance and the form effects on the boundary layer itself [9].

$$C_T = (1+k)C_{F0} + C_W \tag{1}$$

Once a form factor k is known, the wave resistance coefficient C_W is found by solving:

$$C_W = C_T - (1+k)C_{F0} \tag{2}$$

where C_{F0} is the ITTC model–ship correlation line coefficient for the Reynolds number of the model test.

Prohaska [9] proposed a method to determine the form factor from model tests and it is considered the most reliable procedure available. It assumes that for Fn< 0.2, C_W can be approximated by $\alpha\ Fn^4$ so the Equation 1 may be rewritten as

$$\frac{C_T}{C_{F0}} = (1+k) + \alpha\frac{Fn^4}{C_{F0}} \tag{3}$$

From the experimental results, we plot the left and right-hand sides of Equation 3 as data points (C_T/C_{F0} , Fn^4/C_{F0}). It is expected that these data points align in a straight line for model speeds with Fn < 0.2. ITTC recommends using only data points within the range of Fn in [0.1, 0.2]. The form factor can then be determined through linear regression. In this work, the regression lines were obtained using the method explained by York [11].

As an example, Figure 2 displays the (C_T/C_{F0} , Fn^4/C_{F0}) points for Fn in [0.1, 0.2] and D1: 0.165 m (ballast condition), along with the resulting regression line using York's method. This method takes into account the experimental uncertainties in the regression process and predicts the uncertainties in the form factor as well. The error bar at $Fn^4/C_{F0} = 0$ illustrates the uncertainty in the form factor, which is calculated as 0.163, corresponding to 13% of $(1 + k)$ for the 95% confidence interval.

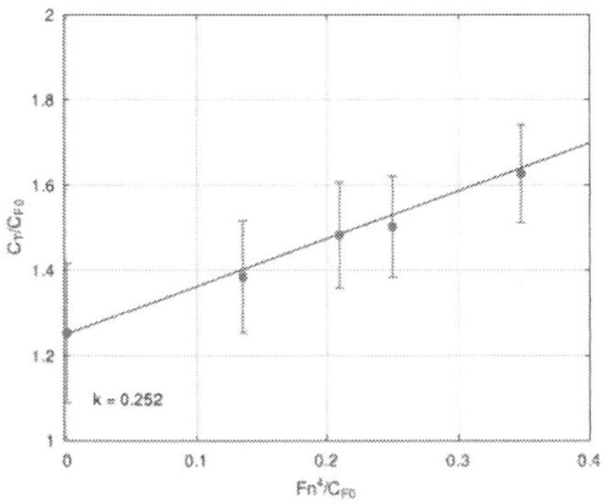

Figure 2. Experimental form factor determination by Prohaska's method with a regression line obtained by York's method (D1: 0.165 m).

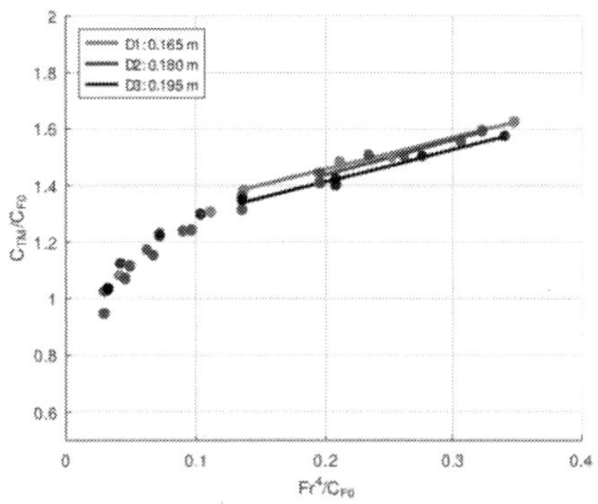

Figure 3. Experimental form factor determination by Prohaska's method applied to the three different draughts.

The model test results of the resistance curves for $0.1 < \text{Fn} < 0.2$ are plotted as shown in Equation 3 for the three different draughts and presented in Figure 3.

The form factors obtained for D1, D2, and D3 were 0.252, 0.205, and 0.194, respectively. It is important to note that, as can be seen from Figure 3, at very low Fn values, the measured points deviate from a straight line, indicating possible laminar flow effects. Therefore, a suitable data set has been selected to determine the form factor.

Additionally, it is important to mention that although the Prohaska method is considered the most reliable procedure available, there is still a high degree of uncertainty in the results. Furthermore, apart from the uncertainty issues associated with the low model speed, the vessel considered in this work has a bulbous bow and transom, which could affect the linearity of the Prohaska data points [12].

3. CFD Model

3.1. Numerical procedure

The use of Computational Fluid Dynamics (CFD) for ship resistance analysis is widely employed. Various different codes are used within the community, with OpenFOAM being one of the most commonly utilized. OpenFOAM is a freely available open-source numerical tool that implements the well-known Navier-Stokes equations along with different numerical schemes and procedures used for solving fluid flow problems. This code is based on the finite volume method, and in the present work, it considers incompressible Newtonian flow. Thus, the equations to be considered are the Navier-Stokes mass and continuity equations (Equations 4 and 5). As ship resistance typically involves turbulent flow, turbulence modeling is necessary. The Reynolds-Averaged Navier-Stokes (RANS) method is employed to model turbulent flow through time-averaging flow quantity decomposition. This method encompasses different turbulence models, such as the SST k-ω model, which includes two eddy-viscosity equations. In this model, k represents turbulent kinetic energy, and ω denotes the specific dissipation rate.

In this work, two types of simulations were conducted. Firstly, simulations were performed when only one fluid phase is present in a double hull setup. Secondly, simulations were carried out when two phases are present, namely water and air, as depicted in Figure 4. From a numerical perspective, when considering only one fluid phase, the aforementioned equations are used. However, when dealing with two phases, an equation that accounts for the presence of the two fluids is required. The Volume of Fluid (VOF) method is employed, as described in Equation 6. This method is based on the scalar volume fraction, α, which takes values between 0 and 1, indicating the proportion of each phase contained in each fluid cell. Specifically, α is set to 0 for air and 1 for water.

$$\nabla \cdot \overline{U_f} = 0 \tag{4}$$

$$\frac{\partial(\rho_f \overline{U_f})}{\partial t} + \nabla(\rho_f \overline{U_f U_f}) = \rho_f g - \nabla p + \mu_f \nabla^2 \overline{U_f} \tag{5}$$

$$\frac{\partial(\alpha)}{\partial t} + \overline{U_f} \cdot \nabla \alpha = 0 \tag{6}$$

3.2. Numerical domain and boundary conditions
The computational domain to be used is presented in Figure 4. The distances in the domain have been selected to be far enough from the ship model in order to avoid any blockage effects. The boundary conditions used are presented in Table 3.

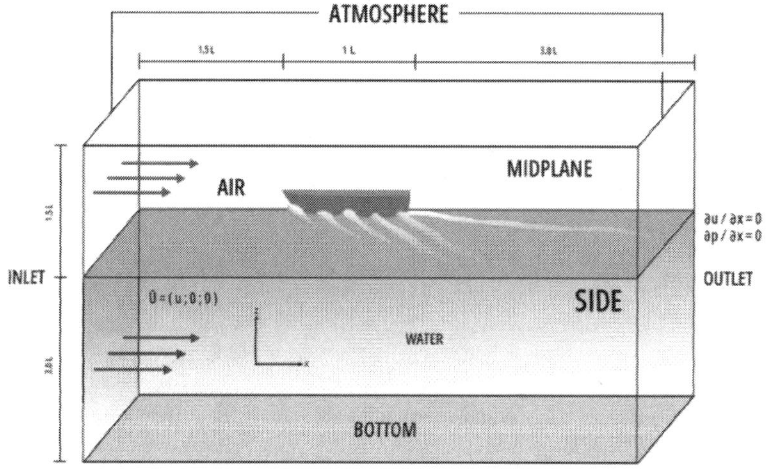

Figure 4. Problem set up and main parameters.

3.3. Mesh convergence
The analysis for mesh convergence was conducted for a draught of 0.165 meters and a Froude number of 0.45. The results of the analysis were compared with different meshes to determine convergence. The results are presented in Table 4. Once an agreement in cell number was reached, Mesh 4 was selected as the preferred mesh for the remaining cases.

Final mesh, Mesh 4, is shown in Figure 5. It can be observed that refined cells are positioned around the ship to accurately capture the flow dynamics, and a thin refinement is applied along the entire domain to capture the air-water interface.

Table 3. Boundary conditions.

Patch	U	p	k	ω	μ turbulent
Inlet	uniform	fixed flux	fixed value	fixed value	fixed value
Outlet	zero gradient	zero gradient	zero gradient	zero gradient	zero gradient
Atmosphere	zero gradient	zero gradient	zero gradient	zero gradient	zero gradient
Bottom	symmetry	symmetry	symmetry	symmetry	symmetry
Midplane/Side	symmetry	symmetry	symmetry	symmetry	symmetry
Ship	no slip	fixed flux	wall function	wall function	wall function

Table 4. Mesh convergence analysis.

Case	Mesh cells	Fn	F [N]	Error (%)
EFD	-	0.45	40.27	-
Mesh 1	633.938	0.45	40.26	0.02
Mesh 2	903.744	0.45	41.76	3.70
Mesh 3	1.213.419	0.45	40.34	0.17
Mesh 4	2.485166	0.45	41.16	2.21

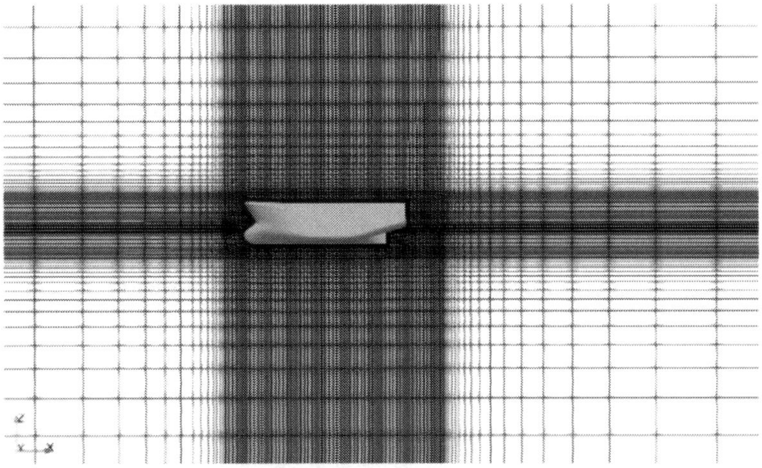

Figure 5. Side view of Mesh 4.

3.4. Numerical form factor determination

The method used in this study for determining the form factor based on CFD follows the same assumptions as Hughes [13] and is derived using the following relation:

$$(1+k) = \frac{C_F + C_{PV}}{C_{F0}} = \frac{C_V}{C_{F0}} \tag{7}$$

where the frictional resistance coefficient C_F and viscous pressure coefficient C_{PV} are obtained from the double body CFD simulation. C_{F0} in the denominator of Equation 7 represents the equivalent flat plate resistance obtained at the same Reynolds number as the computations, using the ITTC-57 Friction Line. Table 5 presents the form factors obtained numerically using the double hull CFD method and experimentally using Prohaska's method (see section 2.4) for different draughts.

Table 5. Form factor at different draughts, calculated by Prohaska's method and double hull CFD method.

Method	D1: 0.165 m	D2: 0.180 m	D3: 0.195 m
EFD (Prohaska)	0.252	0.205	0.194
CFD (Double Hull) Fn 0.10	0.280	0.258	0.265
Difference	11%	26%	37%

4. EFD and CFD results

In naval hydrodynamics, the total resistance can be divided into viscous resistance and wave resistance. This division will be used in this section to analyze the results.

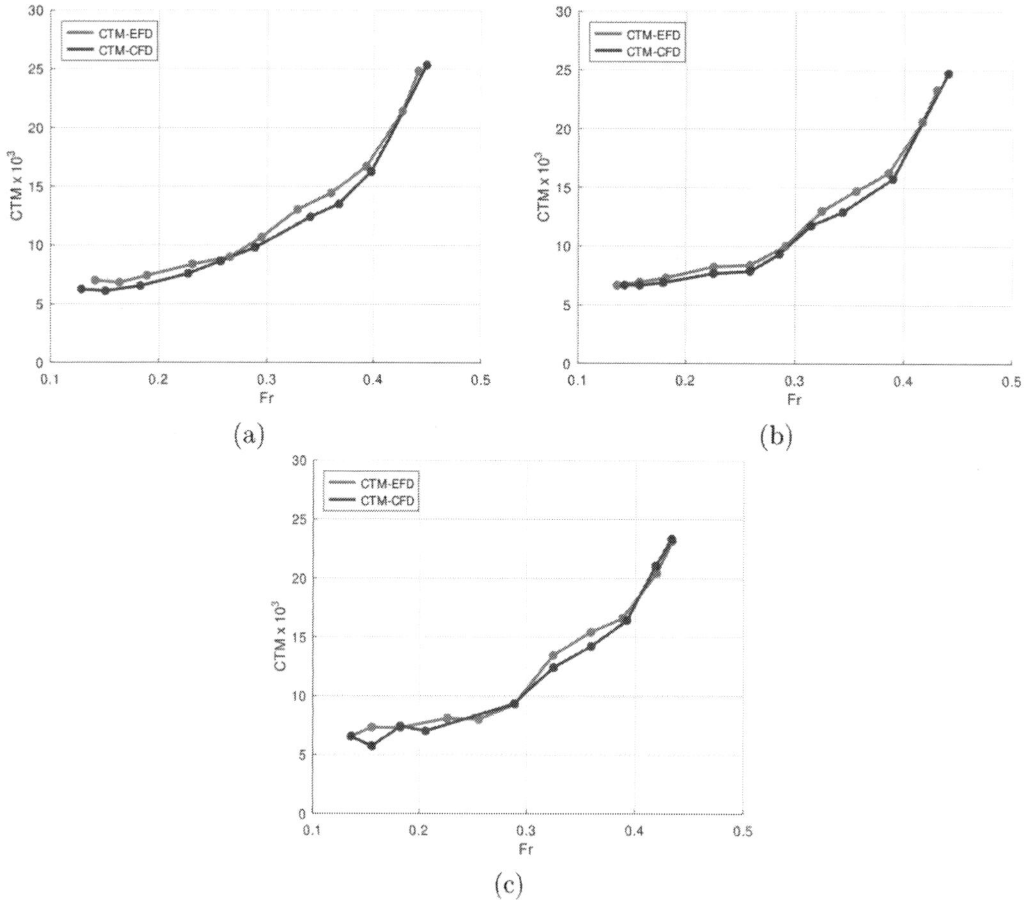

Figure 6. C_T vs Fn for different draughts. a: D1 (0.165 m), b: D2 (0.180 m), c: D3 (0.195 m).

The viscous resistance includes the frictional resistance caused by shear stress between the fluid and the hull, as well as the resistance caused by pressure changes over the hull due to viscous flow effects. The wave resistance is associated with the generation of the ship's wake system.

Figure 6 presents the comparison of the non-dimensional total resistance curve (C_T) obtained numerically and experimentally for Froude numbers ranging from 0.1 to 0.45, considering different draughts. The results show a good agreement in C_T between the EFD and CFD results for all the three draughts and across different Froude numbers.

Normally, fishing vessels operate under different conditions, velocities, and draughts. These conditions can vary when the vessel is leaving port to fish, during fishing operations and when returning to port with the catch. The design criteria for such vessels typically consider the speed required to reach the fishing area and return to port with the catch. In the present case, the ship is designed to operate within the range of $0.25 < \mathrm{Fn} < 0.45$, with a specified operational speed of $\mathrm{Fn} = 0.35$. Therefore, the vessel is classified as a displacement ship according to [14]. The resistance curves exhibit the typical behavior observed in this type of vessel.

In Figure 6, it can be observed that at low speeds ($\mathrm{Fn} < 0.3$), the total resistance is nearly equal to the viscous resistance and the C_T value increases almost linearly, but at a slightly lower rate than the square of the model ship velocity. As the Fn number increases, the slope of the total resistance curve rises, indicating that the C_T value increases at a rate higher than the square of the model ship velocity. At low speeds, wave resistance is negligible, but as the velocity increases, wave resistance grows more rapidly than viscous resistance. In fact, for most fishing vessels, above $\mathrm{Fn} = 0.30$, wave resistance becomes the dominant component of total resistance, causing a significant increase in the overall resistance of the ship. This can make the vessel inefficient in terms of fuel consumption.

Figure 7 displays separate C_T vs Fn curves for the EFD and CFD results, highlighting the influence of draught on C_T. The left graph presents the EFD resistance results for different draughts, while the right graph shows the CFD model results.

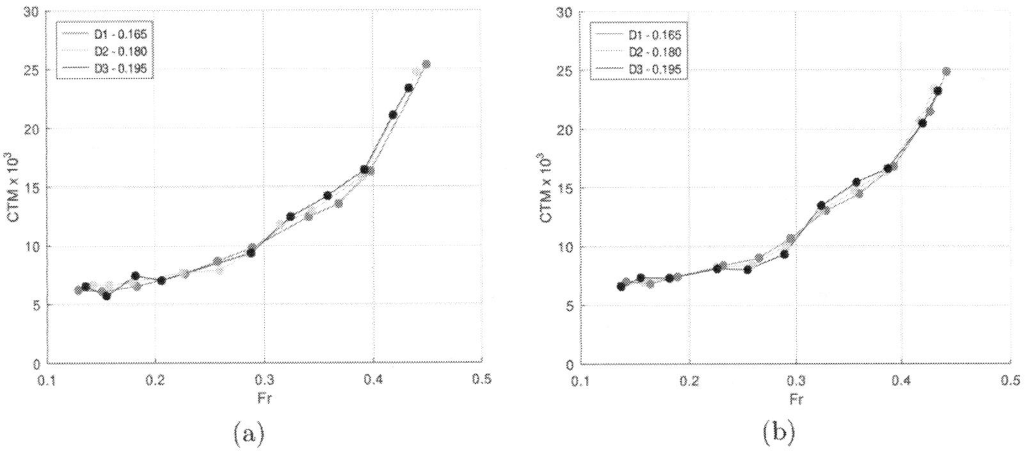

Figure 7. C_T vs Fn. a: all draughts on EFD, b: all draughts on CFD.

As anticipated, the EFD results demonstrate that for $\mathrm{Fn} < 0.3$, the modification in draught does not have a significant impact on the total resistance of the ship. At these low Fn values, the resistance increases with an increase in the hull's wetted surface area. However, due to the very low resistance values in this Fn range, the changes in C_T are not clearly visible in Figure 7. On the other hand, the draught effects become notable for $0.3 < \mathrm{Fn} < 0.4$. In this range, although wave resistance becomes the largest component, viscous resistance still contributes significantly to the total resistance. Thus, an increase in draught leads to an increase in C_T.

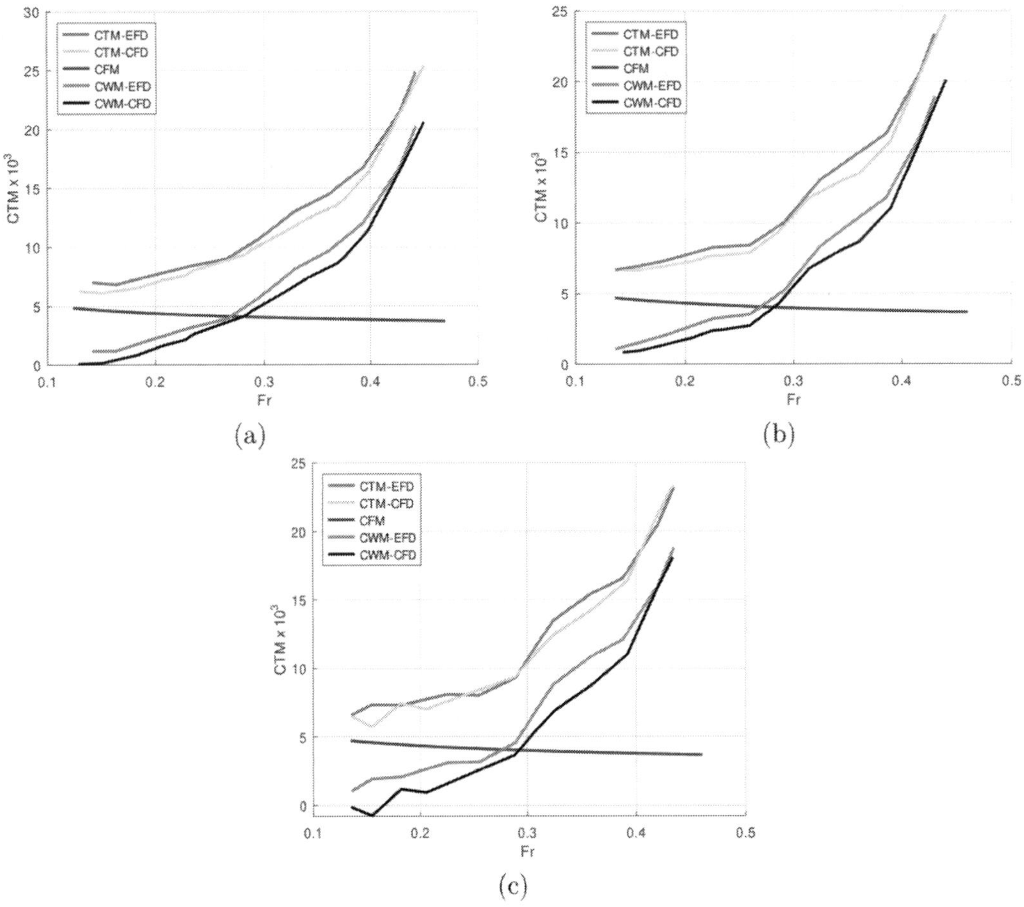

Figure 8. C_T C_V C_W vs Fn. for different draughts. a: D1 (0.165 m), b: D2 (0.180 m), c: D3 (0.195 m).

Finally, the three curves tend to converge to a similar value due to the substantial increase in wave resistance, which masks the effect of draught and viscous resistance for Fn > 0.4.

To gain a better understanding of how the viscous and wave resistance components contribute to the overall resistance, the total resistance coefficient is separated into viscous and wave coefficients using the obtained form factor.

Figure 8 illustrates the numerical and experimental curves of C_T, $C_V = (1 + k)C_{F0}$, and C_W (calculated as C_T - C_V) plotted against Fn for different draughts. The form factor k used in the EFD curves is obtained from experimental data (see section 2.4), while the k used in the CFD curves is obtained numerically (see section 3.4). In both cases, C_{F0} is determined using the ITTC-57 friction line.

Figure 8 demonstrates the ability of the CFD model to decompose the total resistance of the model into wave and viscous resistance using the double body approach to obtain the form factor. Overall, there is a good agreement between the EFD and CFD results for the three studied draughts and various Froude numbers. From Figure 8, it is observed that for Fn < 0.2, the viscous resistance is significantly larger. As Fn exceeds 0.3, the wave component, initially comparable to the viscous component, increases proportionally to Fn^4 and quickly becomes the dominant resistance component. For example, at Fn = 0.30, Fn = 0.37, and Fn = 0.40, the wave resistance represents approximately 47%, 63%, and 70% of the total resistance, respectively.

Finally, for Fn > 0.4, the C_W curves tend to approach the C_T curves, with wave resistance accounting for over 80% of the total resistance.

Based on the above observations, it is evident that wave resistance plays a significant role in the performance of this type of fishing vessel. Therefore, understanding the wave pattern along the ship's hull is crucial for designing and operating efficient and effective vessels. Figure 9 provides a comparison of the wave pattern distribution along the ship hull obtained from the CFD model and EFD for draught D3 and different Fn values. The three cases presented demonstrate the CFD model's capability to accurately reproduce the wave pattern generated by the ship at various Fn numbers.

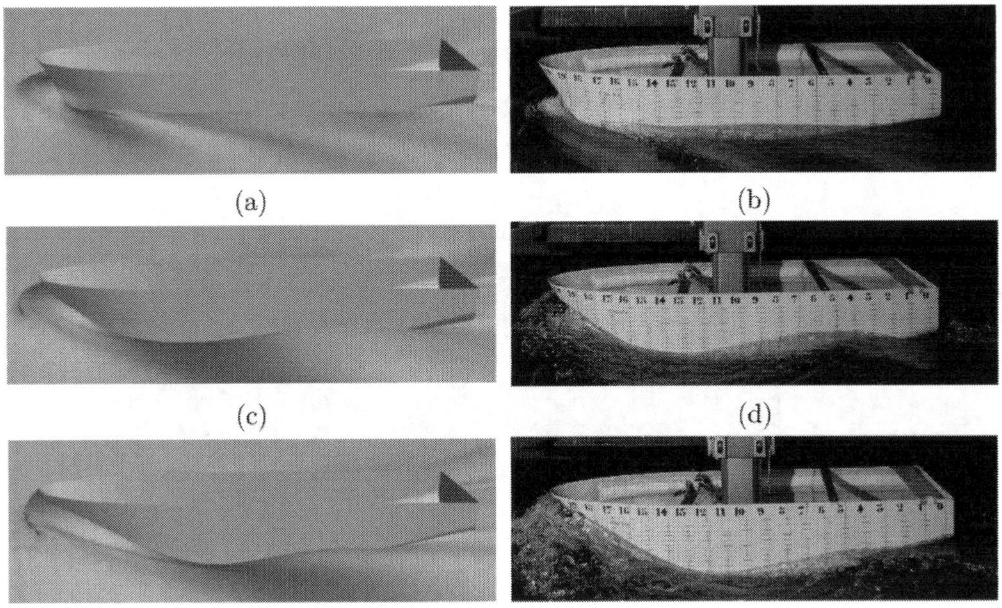

(a) (b) (c) (d) (e) (f)

Figure 9. Wave pattern for different Fn and draught D3. Left column: 3D CFD model. Right column: EFD model. First row: Fn = 0.26, second row: Fn = 0.37, third row: Fn = 0.45.

In ship hydrodynamics, it is common to split wave resistance into two components: wave pattern resistance and wave breaking resistance. When the Fn value is sufficiently high, the waves can become steep enough to break into eddies and foam. The energy removed from the wave system is then found in the ship's wake, resulting in wave breaking resistance. The remaining wave energy is radiated away from the ship through the wave system, leading to wave pattern resistance. The breaking phenomena can be observed in the EFD results for Fn > 0.3, as shown in Figure 9d (Fn = 0.37) and Figure 9f (Fn = 0.45). The CFD results seem to capture the wave breaking phenomena only for the higher Froude number, as seen in Figure 9e (Fn = 0.45). Obtaining a more detailed description of this phenomenon with the CFD model would require a finer mesh in the interface region. However, it is important to note that the wave breaking resistance is considerably smaller than the wave pattern resistance for this type of fishing vessel. The visualization also reveals that the ship's shoulders play a significant role in wave generation and distribution along the hull. For higher Froude numbers, a maximum hollow wave appears in ship hull section 14.5 and moves back until approximately section 11.5 in the case of Fn = 0.45.

The validation of the CFD model enables the determination of variables that would be challenging or costly to obtain using EFD techniques. One variable is the pressure distribution

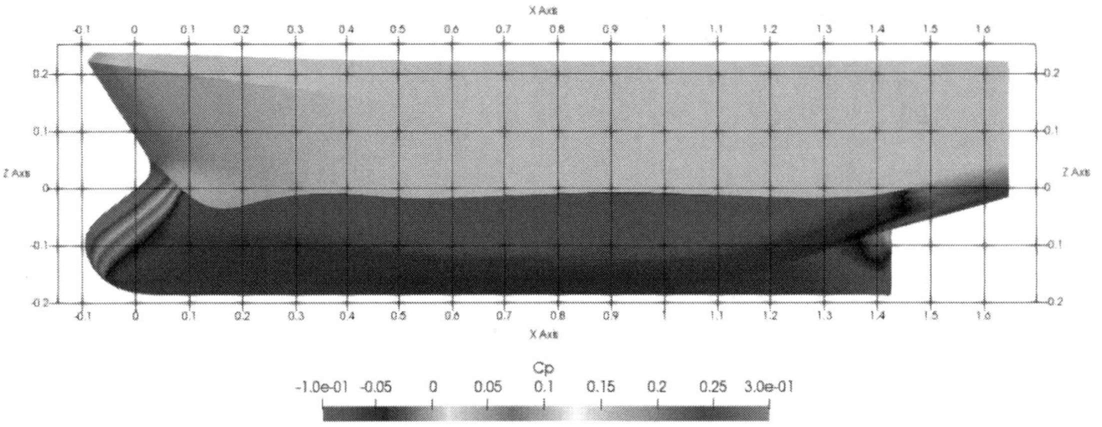

(a)

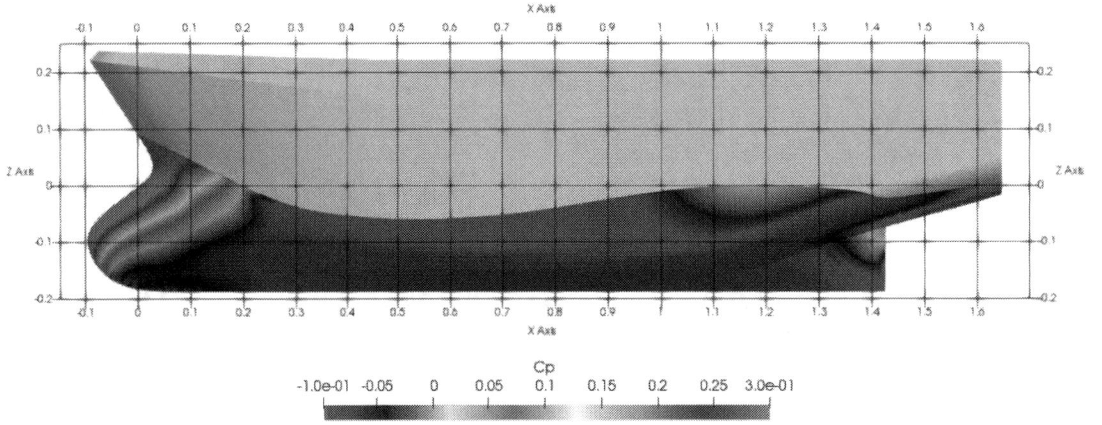

(b)

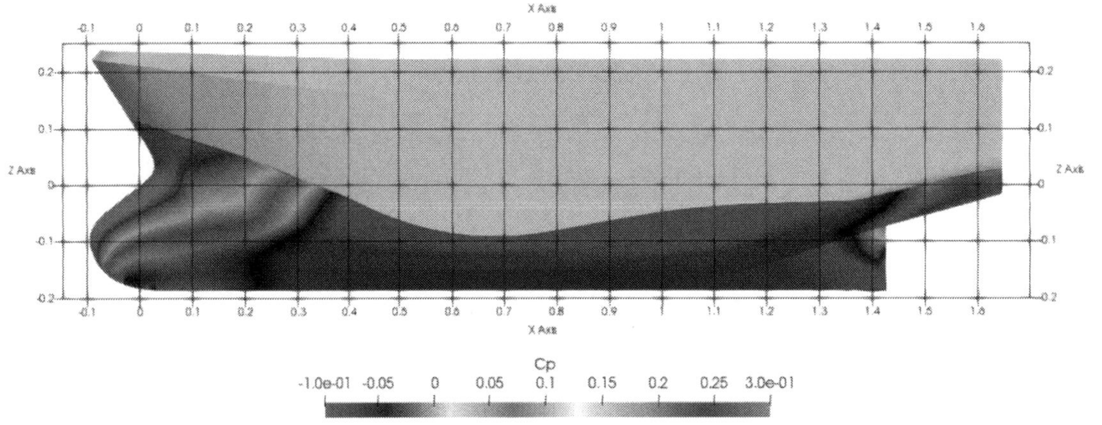

(c)

Figure 10. Cp distribution for different Fn and draught D2. a: Fn = 0.26, b: Fn = 0.37, c: Fn = 0.45.

on the hull. Figure 10 displays the nondimensional pressure distribution (C_P) on the hull for different Fn values and draught D2.

$$Cp = \frac{p}{\frac{1}{2}\,\rho\,U^2\,S} \tag{8}$$

The pressure (p) distribution on the hull is critical in determining its wave generation and resistance characteristics, as pressure disturbances propagate outward from the hull, shaping and sizing the waves. As shown in Figure 10, high and low pressure zones align with wave crests and troughs.

In general, a smooth pressure distribution over the hull results in more efficient and less resistant motion through the water. This is because a balanced pressure distribution creates fewer disturbances, leading to a reduced energy loss in wave generation, and lower drag of the ship. Conversely, an uneven pressure distribution can cause significant disturbances, resulting in increased wave generation and higher resistance. This can occur when the hull shape is not optimized for the vessel's speed and operating conditions. Therefore, understanding and managing the pressure distribution on the ship's hull is crucial for designing and operating efficient and effective vessels.

5. Conclusions

In conclusion, this study successfully obtained the ship resistance for a small fishing vessel through numerical and experimental approaches. The results demonstrates good agreement between numerical and experimental results, with less than 8% of error. The form factors were obtained both numerically and experimentally, allowing for a comparison of the different resistance components based on Hughes's hypothesis. A general agreement was observed between the EFD and CFD results for the viscous and wave resistance coefficients across the three studied draughts and various Fn numbers.

The numerical analysis also revealed that the wave distribution around the model hull closely matched the image obtained from the experimental towing tank test.

Considering that wave resistance is a crucial factor in the performance of this type of fishing vessel, the CFD model was used to determine the pressure distribution on the hull. Understanding the relationship between wave patterns and pressure distribution on the hull is essential for designing and operating efficient and effective vessels.

This work serves as a foundation for future studies on wave resistance and hydrodynamic optimization of fishing vessels of this type. The obtained values of the form factor k, both numerically and experimentally, can be used as a basis for further research exploring whether CFD-based form factors can serve as an alternative or supplement to the Prohaska method for this type of ships.

References

[1] Holtrop J 1984 *International shipbuilding progress* **31** 272–276
[2] Bilec M and Obreja C D 2020 *IOP Conference Series: Materials Science and Engineering* **916** 012011 ISSN 1757-899X publisher: IOP Publishing
[3] Larsson L, Raven H and Paulling J 2010 *Ship Resistance and Flow* Principles of naval architecture (Society of Naval Architects and Marine Engineers)
[4] 29th ITTC Resistance Committee 2021 *ITTC - The Specialist Committee on CFD and EFD Combined Methods Final Report and Recommendations* (ITTC)
[5] Bertolotti M, Verazay G, Pagani A, Errazti E and Buono J 2001 *El mar argentino y sus recursos pesqueros* ISSN 987-20245-0-2
[6] OECD 2021 . Fisheries and Aquaculture in Argentina
[7] Korkmaz K B, Werner S, Sakamoto N, Queutey P, Deng G, Yuling G, Guoxiang D, Maki K, Ye H, Akinturk A, Sayeed T, Hino T, Zhao F, Tezdogan T, Demirel Y K and Bensow R 2021 *Ocean Engineering* **220** 108451

[8] Specialist Committee: Procedures for Resistance P and of 23rd ITTC 2002 P O W T 2002 *ITTC – Recommended Procedures and Guidelines: Model Manufacture, Ship Models* (ITTC)
[9] 26th ITTC Resistance Committee 2011 *ITTC – Recommended Procedures and Guidelines - Resistance Test* (ITTC)
[10] 28th ITTC Quality Systems Group 2017 *ITTC - General Guidelines for Uncertainty Analisys in Resistance Tests* (ITTC)
[11] York D, Evensen N, Ló M, Nez M, De J and Delgado B 2004 *American Journal of Physics - AMER J PHYS* **72**
[12] 14th ITTC Performance Committee 1975 *Form factor - Proc. of the 14th ITTC* (ITTC)
[13] Hughes G 1954 *National Physical Laboratory, NPL, Ship Division, Presented at the Institution of Naval Architects, Paper No. 7, London, April, RINA Transactions 1954-16*
[14] Blount D 2014 *Performance by Design: Hydrodynamics for High-speed Vessels* (Donald L. Blount)

Acknowledgments

This work was supported by the Ministry of Science, Technology, and Innovation of the Argentine Republic (MINCyT - Proyecto de Investigación y Desarrollo Tecnológico de la Iniciativa Pampa Azul - B2) and the Universidad de Buenos Aires, Facultad de Ingeniería (Grant Peruilh). The fishing vessel lines were provided by the Argentine shipyard TecnoPesca Argentina.

Numerical study on wave-induced drift forces and motions of a DTC ship in head and oblique waves using functional decomposition model

Xianjin Fang[3,4], Jiawei Yu[1,2,*], Chaobang Yao[1,2], Zhiguo Zhang[1,2], Xianzhou Wang[1,2]

[1] School of Naval Architecture and Ocean Engineering, Huazhong University of Science and Technology, Wuhan, Hubei, 430074, China
[2] Key Laboratory of Ship and Ocean Hydrodynamics of Hubei Province, Wuhan, Hubei, 430074, China
[3] Harbin Engineering University, Harbin, 150001, China
[4] Guangzhou Marine Engineering Corporation, Guangzhou, 510250, China

*Corresponding author, E-mail: yu_jiaw@hust.edu.cn

Abstract. It is imperative to investigate the adequacy of propulsion power and steering mechanisms in order to maintain the manoeuvrability of ships in adverse wave conditions, as the drift forces caused by wave can result in deviation from the desired course. The primary objective of this research is to develop a hybrid approach, combining potential and viscous flows, known as SWENSE (Spectral Wave Explicit Navier-Stokes Equations), to accurately assess the wave-induced drift forces and motions experienced by ships in waves. SWENSE is a hybrid approach based on the functional decomposition model, in which the total field is composed of the incident field and complementary field. In this study, the incident filed waves are calculated by the Airy wave theory and the complementary field is solved by the SWENSE model. The wave conditions are selected according to the benchmark cases of the European Union research project SHOPERA (Energy Efficient Safe Ship Operation). The wave-induced loads and motions of DTC in head and oblique waves are calculated. Results of CFD (Computational Fluid Dynamics) are compared with those of EFD (Experimental Fluid Dynamics) and other numerical methods. In addition, the wave-induced drift forces of the fixed/3DOF (Degrees of Freedom) ships are compared with each other (3DOF includes heave, roll and pitch). SWENSE model is proven to be reliable for the predictions of the wave-induced drift forces and motions of ships in waves.

1. Introduction

The ships travelling in the seas are inevitably affected due to wave-induced forces and motions, which are closely related to the operation and safety of ships. At present, the head wave conditions are mainly focused on in the research on ship seakeeping performance. However, the ships rarely travel in head waves [1] and always encounter the waves at variable heading angles. Moreover, one approach to meet the Energy Efficiency Design Index (EEDI) requirements, proposed by the International Maritime Organization (IMO), involves reducing engine power [2-3]. However, lowering the velocity can lead to diminished manoeuvrability of ships. Therefore, it becomes indispensable to thoroughly evaluate the sufficiency of propulsion power and steering mechanisms, particularly in adverse wave conditions, where wave-induced drift forces can cause deviation from the intended course. Consequently, the comprehensive assessment of wave-induced drift forces and motions in head and oblique waves holds significant importance for ship operation and safety. Typically, the prediction of these performance parameters employs both experimental and computational fluid dynamics (EFD and CFD). In this context, the SHOPERA project was initiated by the European Union [4-5], aiming to investigate ship drift forces and manoeuvrability. The project involved the selection of the 6DOF Duisburg Test Case (DTC) and KVLCC ship models, which underwent a series of wave tests, encompassing the measurement of ship motions, added resistance, drift forces, and moments. The valuable data obtained from these experimental investigations served as crucial validation material for CFD simulations.

The potential theory is widely used and considered the most efficient method for seakeeping analyses in typical scenarios. To explore the variations and underlying causes of added resistance in oblique sea conditions, Park et al. implemented the strip method and the 3D Rankine panel method [6]. Zhang et al., on the other hand, employed the time-domain Rankine panel method to assess wave-induced motions and steady wave drift forces for the ship model S175 in waves [7-8]. However, the potential theory has limitations, as it cannot account for strongly nonlinear phenomena. To address this, the viscous CFD is applied. Jin et al. investigated wave-induced loads and motions of two zero-speed ship models using the boundary wave-generation approach coupled with a fifth-order Stokes wave model [9]. Nevertheless, it should be noted that the viscous computational fluid dynamics (CFD) method is still not considered cost-effective for routine seakeeping analyses [10].

In recent years, there has been a growing interest among researchers in the hybrid methods that combine potential and viscous flows for seakeeping performance prediction [11]. These hybrid approaches offer the advantages of both potential flow and viscous flow methods. Gong et al. [12] employed the zonal hybrid solver, called QaleFOAM, which is based on the domain decomposition model, to predict the ship motions and added resistance of a trimaran at different headings. Vukčević et al. [13] developed the SWENSE approach, which utilizes the functional decomposition model proposed by Ferrant et al. [14]. This method incorporated the two-phase level set model and the ghost fluid method for seakeeping studies on the KCS ship model in regular head and oblique waves.

The primary objective of this study is to develop a functional-decomposition hybrid solver HUST-SWENSE [11] that combines potential and viscous flows to accurately and quickly predict the wave-induced drift forces and motions experienced by ships in waves. In this study, the total field of HUST-SWENSE is composed of the incident field, calculated by the Airy wave theory, and complementary field, solved by the SWENSE model. The research object DTC ship model and wave conditions are selected according to the benchmark cases of the European Union research project SHOPERA [3-5]. The wave-induced drift forces and motions of DTC in head and oblique waves are calculated. Results obtained from HUST-SWENSE are compared with those of EFD and other numerical methods. In addition, the wave-induced drift forces of the fixed/3DOF (including heave, pitch and roll) ships are compared with each other.

2. Numerical method

2.1. Governing equations

In the SWENSE method [14], the total field (T) consists of the velocity field (U_{Ti}), pressure (P_T), and distance function (φ_T) in the flow field. It is decomposed into the incident field (I) with components (U_{Ii}, P_I, ξ_I) and the complementary field (C) with components (U_{Ci}, P_C, φ_C), as shown in figure 1. Thus, the total field can be written as:

$$\begin{cases} U_{Ti} = U_{Ii} + U_{Ci} \\ P_T = P_I + P_C \\ \varphi_T = \xi_I + \varphi_C \end{cases} \tag{1}$$

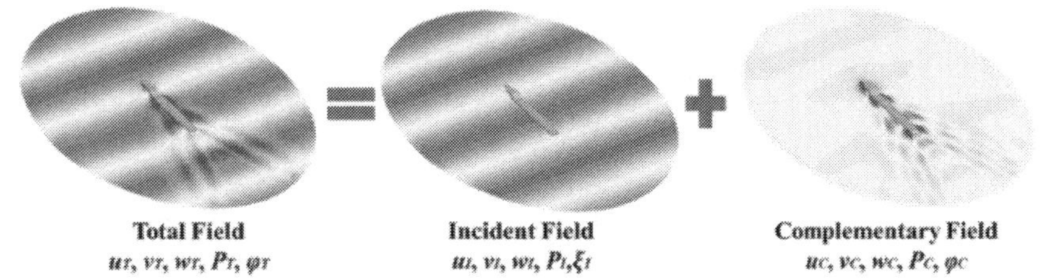

Figure 1. the schematic diagram of the functional decomposition model [11].

2.2. The incompressible RANS equations and single-phase level-set model

The total physics field is governed by the incompressible RANS equations and the level-set function. The incompressible RANS equations, based on the Boussinesq assumption, are written as follows:

$$\frac{\partial U_{Ti}}{\partial x_i} = 0 \tag{2}$$

$$\frac{\partial U_{Ti}}{\partial t} + \left(U_{Tj} - \frac{\partial v_t}{\partial x_j}\right)\frac{\partial U_{Ti}}{\partial x_j} = -\frac{1}{\rho}\frac{\partial \hat{P}_T}{\partial x_i} + (v + v_t)\frac{\partial^2 U_{Ti}}{\partial {x_j}^2} + \frac{\partial v_t}{\partial x_j}\frac{\partial U_{Tj}}{\partial x_i} \tag{3}$$

where the turbulent viscosity v_t and the turbulent kinetic energy k are solved using the SST $k-\omega$ model [15].

The distance function $\varphi_T(x, y, z, t)$ represents the distance from any point in flow field to the free surface, where the value of the distance function at the interface is 0 [11]. The single-phase level-set model [16] is shown as follows:

$$\frac{\partial \varphi_T}{\partial t} + U_{Tj}\frac{\partial \varphi_T}{\partial x_j} = 0 \tag{4}$$

2.3. The incident wave model

The variables of the incident field are obtained by the Airy wave theory and satisfy the incompressible Euler equations:

$$\frac{\partial U_{Ii}}{\partial x_i} = 0 \tag{5}$$

$$\frac{\partial U_{Ii}}{\partial t} + U_{Ij}\frac{\partial U_{Ii}}{\partial x_j} + \frac{1}{\rho}\frac{\partial \hat{P}_I}{\partial x_i} = 0 \tag{6}$$

The physical field in the conventional linear wave model is only solved in the water zone $x_3 \leq 0$, thus, the incident field in the air zone will be treated by Wheeler stretching method [17].

$$z^* = \begin{cases} x_3 & x_3 \leq 0 \\ x_3 - \xi & x_3 > 0 \end{cases} \tag{7}$$

The transient free surface position ξ for the incident wave can be written as follows [11]:

$$\xi_I(x_1, x_2, t) = A\cos(k(x_1 \cos\beta - x_2 \sin\beta) - \omega t + \phi) \tag{8}$$

Furthermore, the velocities and pressure can be easily given as follows [11]:

$$U_{I1}(x_1, x_2, z^*, t) = \omega A e^{kz^*} \cos(k(x_1 \cos\beta - x_2 \sin\beta) - \omega t + \phi)\cos\beta \tag{9}$$

$$U_{I2}(x_1, x_2, z^*, t) = -\omega A e^{kz^*} \cos(k(x_1 \cos\beta - x_2 \sin\beta) - \omega t + \phi)\sin\beta \tag{10}$$

$$U_{I3}(x_1, x_2, z^*, t) = \omega A e^{kz^*} \sin(k(x_1 \cos\beta - x_2 \sin\beta) - \omega t + \phi) \tag{11}$$

$$P_I(x_1, x_2, z^*, t) = \rho g A e^{kz^*} \cos(k(x_1 \cos\beta - x_2 \sin\beta) - \omega t + \phi) \tag{12}$$

The equations for the wave properties in the incident field are described as follows: A represents the wave amplitude, ϕ is the wave phase, β is the wave encounter angle, k is the wave number calculated as 2π divided by the wavelength (λ), and ω is the circular wave frequency determined by the square root of the product of the gravitational acceleration (g) and the wave number (k) under the assumption of infinite water depth. To obtain the velocity derivatives, an analytical approach is utilized.

2.4. SWENSE model

The SWENSE model is derived from the governing equations of the total field and the incident field, which is used to solve the complementary field.

$$\frac{\partial U_{Ci}}{\partial x_i} = 0 \tag{13}$$

$$\frac{\partial U_{Ci}}{\partial t} + \left(U_{Tj} - \frac{\partial v_t}{\partial x_j}\right)\frac{\partial U_{Ci}}{\partial x_j} - (v + v_t)\frac{\partial^2 U_{Ci}}{\partial {x_j}^2} = -\frac{1}{\rho}\frac{\partial \hat{P}_C}{\partial x_i} + \frac{\partial v_t}{\partial x_j}\frac{\partial U_{Cj}}{\partial x_i} - U_{Cj}\frac{\partial U_{Ii}}{\partial x_j} + 2\frac{\partial v_t}{\partial x_j}\frac{\partial U_{Ii}}{\partial x_j} \tag{14}$$

The functional-decomposition level-set equation can be written as follows [11]:

$$\frac{\partial \varphi_C}{\partial t} + U_{Tj}\frac{\partial \varphi_C}{\partial x_j} = -\frac{\partial \xi_I}{\partial t} - U_{Tj}\frac{\partial \xi_I}{\partial x_j} \tag{15}$$

The adapted RANS equations are discretized using the finite difference method and the PISO algorithm is employed for the solution of the velocity-pressure coupling. The HUST-SWENSE code [11], which is developed based on the URANS code HUST-Ship and the SWENSE model, is utilized for the computations. Structured grids with overset technology are employed in the simulations.

The adapted RANS equations are discretized by the finite difference method and the PISO algorithm is used for the velocity-pressure coupling solution [11]. Computations have been made using structured grids with overset technology.

3. Geometries and grids

3.1. Geometry of ship model and simulation conditions

The research object DTC ship model and wave conditions are selected according to the benchmark cases of the European Union research project SHOPERA [3-5], where the experimental data of forces and motions in waves is provided. Figure 2 shows the geometry of the DTC ship model and the principal particulars of the container ship model DTC of scale ratios of 1:63.65 are presented in Table 1. The DTC ship model is appended with a twisted rudder with a Costa bulb and a NACA 0018 base profile.

Figure 2. Geometry of DTC ship model.

Table 1. Principal particulars of DTC ship model.

Main particulars		
Length between perpendiculars	$L_{pp}\ (m)$	5.577
Maximum beam of waterline	$B_{wl}\ (m)$	0.800
Draft	$T\ (m)$	0.230
Displacement volume	$\Delta\ (m^3)$	0.673
Wetted surface area	$S_w\ (m^2)$	5.437
Block coefficient (C_B)	$\nabla/L_{pp}B_{wl}T$	0.661
Longitudinal Center of Gravity	$LCG\ (m)$	2.743
Vertical Center of Gravity	$KG\ (m)$	0.390
Moment of Inertia	k_{xx}/L_{pp}	0.057
	k_{yy}/L_{pp} , k_{zz}/L_{pp}	0.246

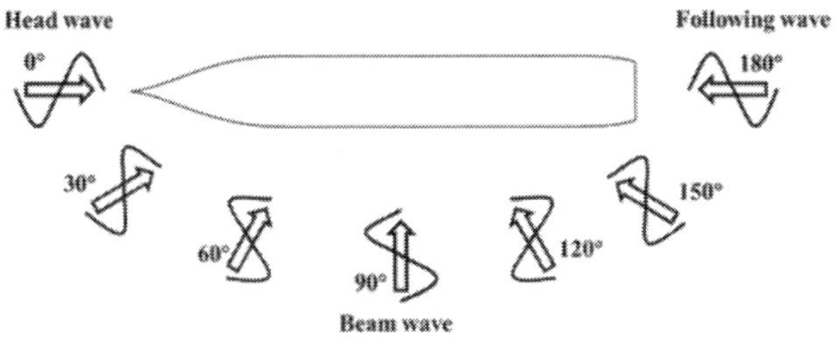

Figure 3. Relationship between ship heading and wave encounter angle.

According to the benchmark experimental cases of the European Union research project SHOPERA [3-5], the CFD simulation cases of wave-induced forces and moments for DTC in head and oblique waves are listed in table 2 with zero velocity, wavelength $\lambda/L_{pp} = 0.356$, wave height $H/\lambda = 0.028$ and wave encounter angle $\chi = 0^{\circ}, 30^{\circ}, 60^{\circ}, 90^{\circ}, 120^{\circ}, 150^{\circ}, 180^{\circ}$. The relationship between ship heading and wave encounter angle is described in figure 3. Wave encounter angles $\chi = 0^{\circ}, 90^{\circ}$ and 180° correspond to the head, beam and following waves from the port, respectively. In addition, the effects of different DOFs (fixed and 3DOF) on the wave-induced drift forces are studied in head and oblique waves, and the simulation conditions are presented in table 2 (3DOF includes heave, roll and pitch).

Table 2. The CFD simulation cases of wave-induced forces and moments for DTC.

Fr	Wave Length (λ/L_{pp})	Wave Height (H/λ)	Wave Direction ($\chi(°)$)	DOF
0	0.356	0.028	0 30 60 90 120 150 180	Fixed/3DOF

3.2. Computational domain and boundary conditions

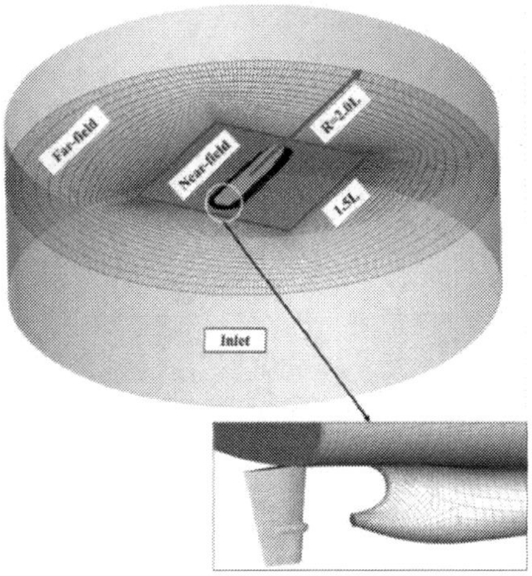

igure 4. Computational domain and boundary conditions for oblique-wave simulations using HUST-SWENSE.

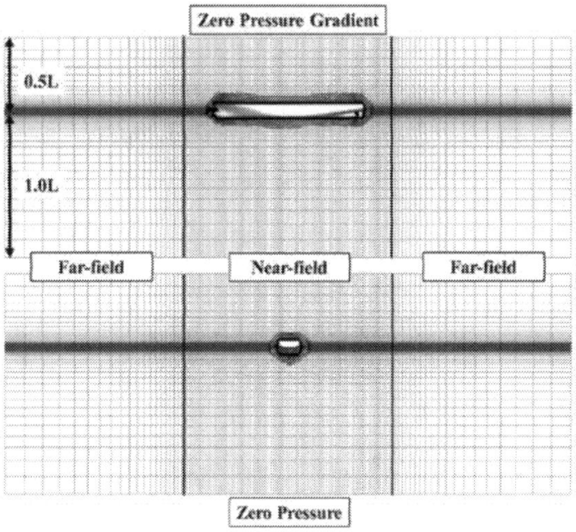

Figure 5. structured overset grid for oblique-wave simulations using HUST-SWENSE.

IOP Conf. Series: Materials Science and Engineering 1288 (2023) 012048 doi:10.1088/1757-899X/1288/1/012048

The cylindrical computational domain of the radius $R/L = 2.0$ is used for the simulations of head and oblique waves using HUST-SWENSE. As seen in figures 4 and 5, the grids are refined in both x and y directions using the uniform grid distributions in the near-field (the square domain of 1.5L×1.5L near the hull) and the coarse grids near the boundary in the far field combined with the moving domain method are used to avoid the wave reflection, with reference to the previous studies on the accuracy of HUST-SWENSE wave generation [11]. Moreover, 100 time steps per wave encounter period are used. The boundary condition on the side is specified as "Inlet". The top boundary is set as "Zero Pressure Gradient" and the bottom boundary as "Zero Pressure". For all the simulations, an in-house overset-grid code called HUST-Overset is employed, which is based on the structured grid and grid-point-overset approach [20]. For the oblique-wave simulations using HUST-SWENSE, a total of 3.0 million grids are used. Out of these, 1.0 million grids are allocated for the numerical tank, while 2.0 million grids are used for the hull and rudder.

4. Result and discussion

The present investigation utilizes time-averaged forces and moments to examine the wave-induced drift forces, while transfer functions (TF) are computed to provide quantitative analysis of the ship motions when subjected to waves. The nondimensional wave-induced drift forces, including surge force F_x, sway force F_y and yaw moment M_z, are defined as $C_{fx} = F_x/(\rho g {B_{wl}}^2 A^2/L_{pp})$, $C_{fy} = F_y/(\rho g {B_{wl}}^2 A^2/L_{pp})$ and $C_{mz} = M_z/(\rho g {B_{wl}}^2 A^2)$. The nondimensional motion responses of heave, roll and pitch motions are defined as $TF_3 = x_3/A$, $TF_4 = x_4/kA$ and $TF_5 = x_5/kA$, where x_3, x_4 and x_5 correspond to the first Fourier series harmonic amplitudes of heave, roll and pitch time histories.

4.1. Wave-induced motions of ship model

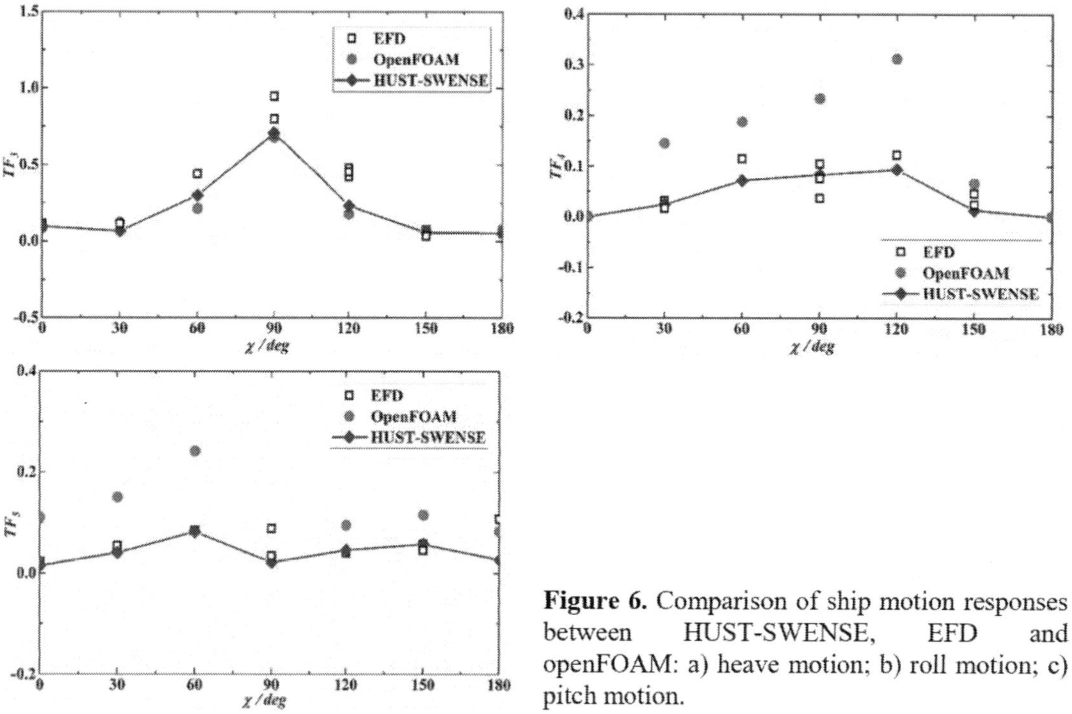

Figure 6. Comparison of ship motion responses between HUST-SWENSE, EFD and openFOAM: a) heave motion; b) roll motion; c) pitch motion.

The CFD simulation cases of wave-induced forces and moments for DTC in head and oblique waves are carried out with zero velocity, wavelength $\lambda/L_{pp} = 0.356$, wave height $H/\lambda = 0.028$ and wave encounter angle $\chi = 0^\circ, 30^\circ, 60^\circ, 90^\circ, 120^\circ, 150^\circ, 180^\circ$. The ship motion responses of HUST-SWENSE, EFD [3] and OpenFOAM [19] are shown in figure 6. Due to $B/\lambda = 0.4$ in the beam wave, which doesn't belong to the very long wave condition, the ship can't simply move up and down parallel with the wave, thus, the pitch motion transfer function is close to zero and the heave motion transfer function $TF_3 = 0.70$ which is less than 1.0. The roll motion responses increase with the wave encounter angle approaching beam wave conditions and the maximum occurs in $\chi = 90^{\circ}$. The largest pitch motion response occurs at the wave encounter angles of 60°. In addition, as shown in figure 6, the amplitudes of heave motion calculated by HUST-SWENSE are close to those of EFD and OpenFOAM. The results of roll and pitch motion responses have good agreement with the experimental data. Thus, it is reliable for HUST-SWENSE to predict ship motions in head and oblique waves.

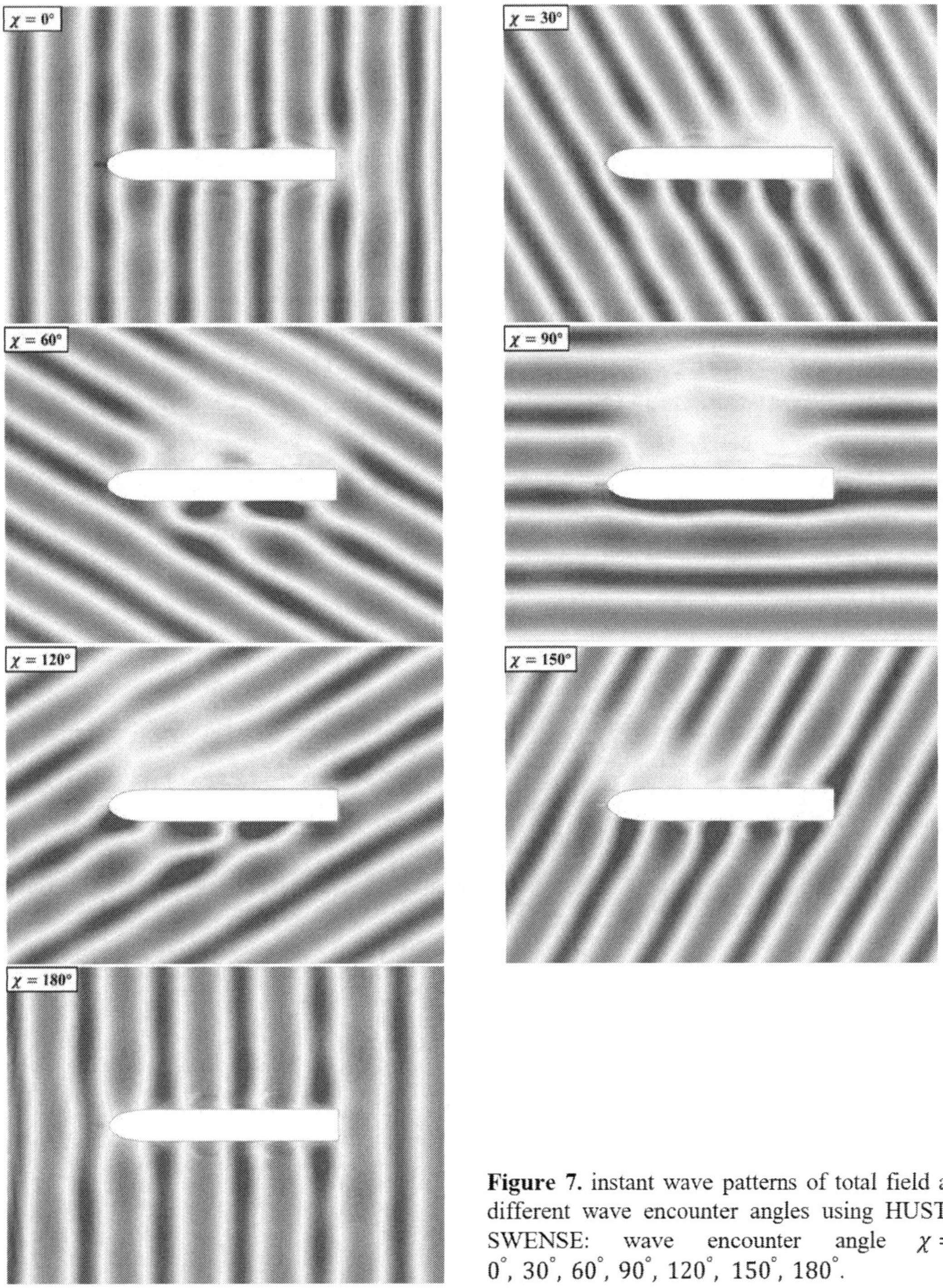

Figure 7. instant wave patterns of total field at different wave encounter angles using HUST-SWENSE: wave encounter angle $\chi = 0^{\circ}, 30^{\circ}, 60^{\circ}, 90^{\circ}, 120^{\circ}, 150^{\circ}, 180^{\circ}$.

IOP Conf. Series: Materials Science and Engineering 1288 (2023) 012048 doi:10.1088/1757-899X/1288/1/012048

Figure 7 shows instant wave patterns of the total field at different wave encounter angles using HUST-SWENSE. As seen in figure 7, under oblique wave conditions, wave peaks and troughs with large amplitudes occur on the port side, especially for the wave encounter angle $\chi = 60^\circ$ and 120°. The 60° wave direction of the incident wave is nearly perpendicular to the forward shoulder of DTC, thus, the interaction between the wave and ship is more violent which leads to the largest wave-induced surge force. Meanwhile, the wave elevations are quite small on the starboard side. The wave elevations on the two sides of the hull are asymmetric and very different, resulting in the wave drift forces and the dramatic change of the instantaneous wetted surface.

4.2. Time-average wave-induced forces and moments

Figure 8 shows the comparison of wave drift forces between HUST-SWENSE, EFD [3] and other numerical methods. Liu and Papanikolaou [18] used the potential flow theory and Liu et al. [19] employed OpenFOAM to predict the wave-induced forces and moments for variable headings. The differences of surge forces for $\chi = 60^\circ$, 150°, sway forces for $\chi = 60^\circ$, 90°, 120°, as well as yaw moments for $\chi = 120^\circ, 150^\circ$ between EFD and CFD are relatively larger, especially for sway force. The reason is that 6DOF motions are free for the DTC ship model in tests and the surge, sway and yaw motions have an effect on the wave-induced forces and moments due to the decrease of the corresponding radiation effects, however, these non-restoring motions aren't included in simulations.

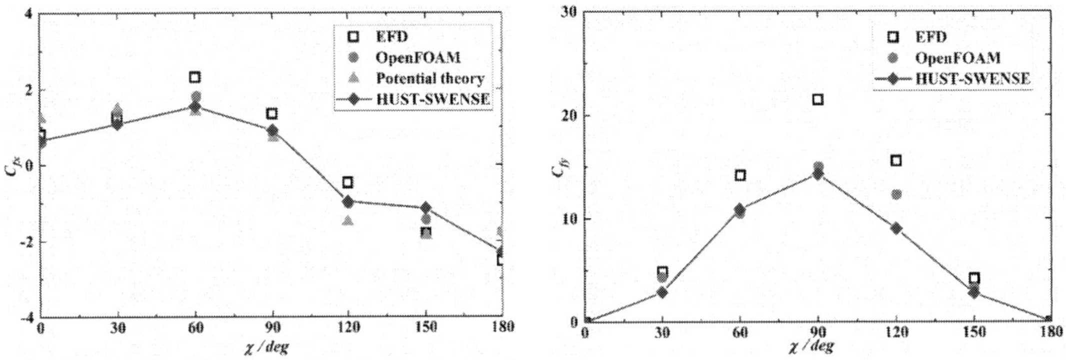

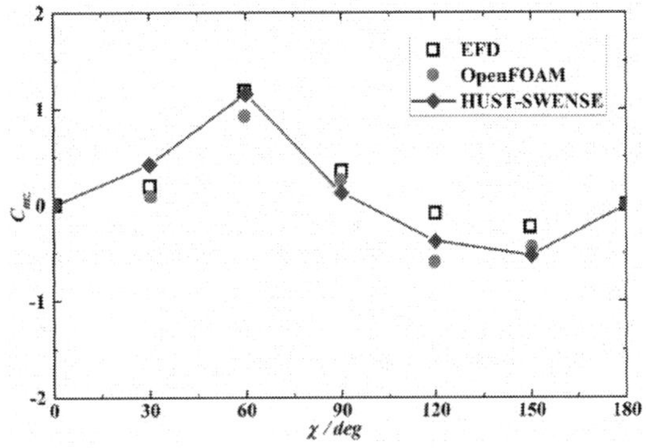

Figure 8. Comparison of wave drift forces between HUST-SWENSE, EFD and other numerical methods: a) surge force coefficient; b) sway force coefficient; c) yaw moment coefficient.

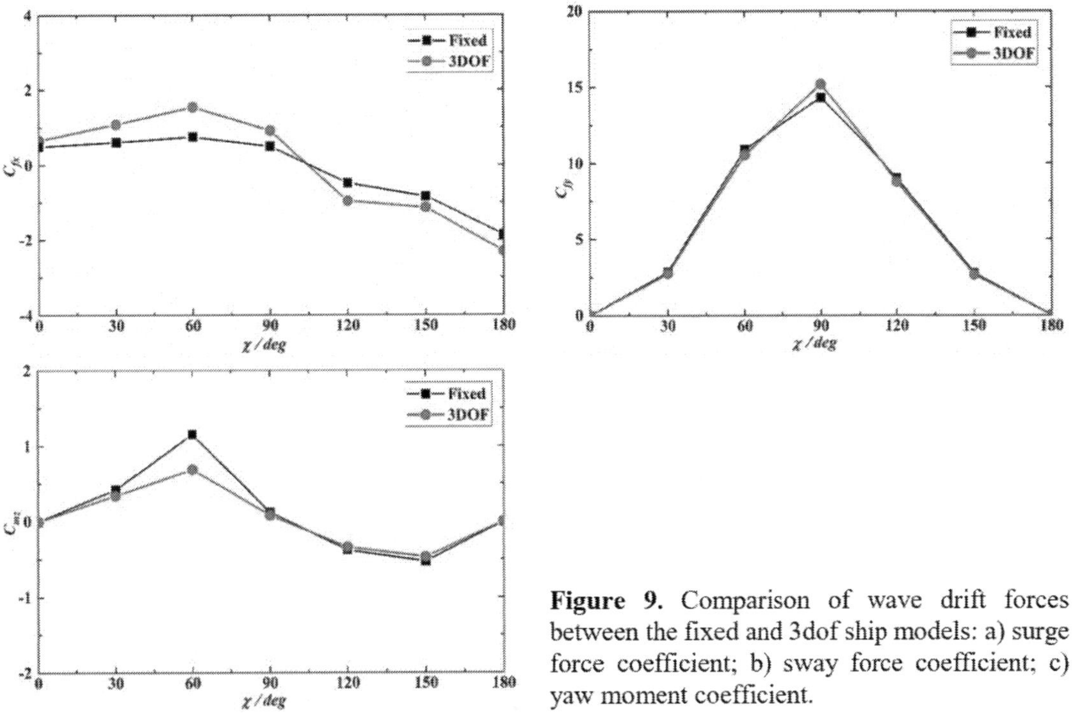

Figure 9. Comparison of wave drift forces between the fixed and 3dof ship models: a) surge force coefficient; b) sway force coefficient; c) yaw moment coefficient.

The results predicted by HUST-SWENSE have good agreement with others' numerical results. The results of wave drift forces obtained by all the numerical methods have the same trend as those of EFD. Thus, it is reliable for HUST-SWENSE to predict wave drift forces in head and oblique waves. As seen in figure 8, the surge forces in astern waves are larger than those of bow waves. Same to the roll motions, the sway forces increase with the wave encounter angle approaching beam wave conditions and the maximum occurs in $\chi = 90^{\circ}$. The largest yaw moment takes place at the wave encounter angles of 60°.

In addition, a comparison of wave drift forces between the fixed and 3DOF ship models is carried out in figure 9. The obvious differences in surge forces can be observed. The surge forces of the fixed ship are evidently smaller than those of the ship with 3DOF since only the diffraction forces are generated for the fixed ship conditions while both the diffraction and radiation forces are included for the 3DOF conditions. The sway forces and yaw moments between the two have almost no difference, except for the conditions of the maximum values (sway force for $\chi = 90^{\circ}$ and yaw moment for $\chi = 60^{\circ}$).

5. Conclusions

In the present study, a functional-decomposition solver HUST-SWENSE is developed to evaluate the wave-induced drift forces and motions of a container ship DTC sailing in head and oblique waves. Simulation conditions are set according to the benchmark case of the European Union research project SHOPERA, and the results of CFD were compared with those of EFD and other numerical methods. Results of the present study indicate that the functional-decomposition solver HUST-SWENSE is a reliable tool for evaluating the wave-induced drift forces and motions of a container ship DTC sailing in both head and oblique waves. Moreover, a comparison of the wave-induced drift forces of the fixed/3DOF (including heave, pitch and roll) ships reveals that while the surge forces of fixed ships are evidently smaller due to the absence of radiation forces, the sway forces and yaw moments exhibited little to no variation, except under the conditions of the maximum values.

References

[1] Islam H, Rahaman M and Akimoto H 2019 Added Resistance Prediction of KVLCC2 in Oblique Waves *American Journal of Fluid Dynamics* **9(1)** 13-26
[2] International Maritime Organization (IMO) 2009 Interim Guidelines on the Method of Calculation

of the Energy Efficiency Design Index MEPC.1/Circ.682

[3] Shigunov V, Moctar O, Papanikolaou A, Potthoff R and Liu S 2018 International benchmark study on numerical simulation methods for prediction of manoeuvrability of ships in waves *Ocean Engineering* **165** 365-385

[4] Sprenger F, Maron A, Delefortrie G, Zwijnsvoorde V, Cura-Hochbaum A, Lengwinat A and Papanikolaou A 2017 Experimental Studies on Seakeeping and Maneuverability of Ships in Adverse Weather Conditions *Journal of Ship Research* **61(3)** 131-152

[5] Liu S, Sprenger F, Papanikolaou A, Dafermos G and Zaraphonitis G 2021 Experimental and numerical studies on linear and nonlinear seakeeping phenomena of the DTC ship in regular waves *Ship Technology Research* **68(1)** 41-61

[6] Park D, Lee J, Jung Y, Lee J, Kim Y and Gerhardt F 2019 Experimental and numerical studies on added resistance of ship in oblique sea conditions *Ocean Engineering* **186** 106070

[7] Zhang W, Moctar O and Schellin T 2020 Numerical simulations of a ship obliquely advancing in calm water and in regular waves *Applied Ocean Research* **103** 102330

[8] Zhang W, Moctar O and Schellin T 2020 Numerical study on wave-induced motions and steady wave drift forces for ships in oblique waves *Ocean Engineering* **196** 106806

[9] Jin Y, Chai S, Duffy J, Chin C and Bose N 2017 URANS predictions of wave induced loads and motions on ships in regular head and oblique waves at zero forward speed *Journal of Fluids and Structures* **74** 178-204

[10] International Towing Tank Conference (ITTC) 2021 The Seakeeping Committee Final Report and Recommendations to the 29th ITTC Committee Reports

[11] Yu J, Yao C, Liu L, Dong G and Zhang Z 2022 Numerical study on wave-structure interaction based on functional decomposition method *Ocean Engineering* **266(4)** 113067

[12] Gong J, Li Y, Cui M, Fu Z and Hong Z 2021 The effect of side-hull position on the seakeeping performance of a trimaran at various headings *Ocean Engineering* **239** 109897

[13] Vukčević V, Jasak H, Gatin I and Malenica Sime 2016 Seakeeping Sensitivity Studies Using the Decomposition CFD Model Based on the Ghost Fluid Method *In 31th Symposium on Naval Hydrodynamics* Monterey California US

[14] Ferrant P, Gentaz L and Touzé D 2002 A new RANSE/Potential approach for water wave diffraction *In Proceedings of Numerical Towing Tank Symposium* Pornichet France

[15] Menter F 1994 Two-equation eddy-viscosity turbulence models for engineering applications *AIAA Journal* **32(8)** 1598-605

[16] Carrica P, Wilson R and Stern F 2007 An unsteady single-phase level set method for viscous free surface flows *International Journal for Numerical Methods in Fluids* **53** 229-256

[17] Wheeler J 1970 Method for calculating forces produced by irregular waves *Journal of Petroleum Technology* **249** 359–367

[18] Liu S and Papanikolaou A 2016 Prediction of the added resistance of ships in oblique seas *The 21st International Journal of Offshore and Polar Engineering* Rhodes Greece

[19] Liu C, Wang J and Wan D 2018 CFD Computation of Wave Forces and Motions of DTC Ship in Oblique Waves *International Journal of Offshore and Polar Engineering* **28** 154-163

[20] Yu J, Yao C, Liu L, Feng D and Wang X 2023 Application of virtual disk propulsion model for self-propelled surface ship in regular head wave *Journal of Marine Science and Technology* **28** 471-495

Acknowledgments

The present work was supported by the National Natural Science Foundation of China (Grant No.52071148), the YEQISUN Joint Funds of the National Science Foundation of China (Grant No. U2141228), and Special Study on Civil aircraft of Ministry of Industry and Information Technology (MJZ5-3N21-1). The essential supports are greatly acknowledged.

IOP Conf. Series: Materials Science and Engineering 1288 (2023) 012049 doi:10.1088/1757-899X/1288/1/012049

Study of Influencing Factors on CFD Method and Separated Flow by the Virtual Static Captive Test

C Yang[1,2], K Zeng[1,2], J L Chu[1,2], Y Han[1], M Gu[1,2]

[1] China Ship Scientific Research Center, National Key Hydrodynamics, Taihu
[2] Laboratory of Deepsea Technological Science, 214082, Wuxi, China
E-mail:yangchen@cssrc.com.cn

Abstract. In this paper, Computational Fluid Dynamic (CFD) methods are adopted to predict the manoeuvring behaviour of the KCS benchmark model. At first instance, the model is calibrated on the basis of three factors namely (a) turbulence model, (b) grid size and (c) time step. Secondly, a Static Oblique Towing Test (OTT) that utilises the SST κ-ω turbulence model with a grid number of 1.64 million and a time step of 0.01 s is conducted. The maximum difference of the lateral force between the port and starboard ship sides is observed at x/L_{pp} = 0.055. Finally, application of Reynold Averaged Navier Stokes (RANS) and Dettached Eddy Simulation (DES) CFD models indicates that the latter may be preferable in terms of capturing the oblique flow field.

1. Introduction

Taking both viscous and rotational effects of fluid into consideration, Computational Fluid Dynamics (CFD) methods have their strength in providing hydrodynamic performance and specific flow details compared with simplified potential theory [1]. With the rapid development of computing power, direct CFD simulation [2] has demonstrated its capability in reproduction of flow characteristics. After decades of development, researchers have realized simulating model tests for manoeuvrability of ship by solving Reynolds-averaged Navier-Stokes (RANS) equations [3], target ships involve KRISO Container Ship (KCS) [4], KVLCC2 [5], S175 [6], DTMB 5415 [7] and so on.

The acquisition of hydrodynamic derivatives by means of CFD methods has attracted great attention from academic and industrial fields. Researchers conducted CFD simulation on captive tests such as OTT [8], circular motion test (CMT) [9] and planar motion mechanism (PMM) [10], and made comparison with experiments for verification. Various computation settings could lead to diverse prediction results. Take static drift test as an instance, the need for calculating acting forces and moment or capturing flow features are essentially different. Through the application of two linear isotropic and non-linear anisotropic RANS models [11], results of forces and moment could satisfy the requirement for accuracy. When it comes to local flow characteristics, large unsteady scales model has authenticated its unsubstituted role in reproducing onset and development of longitudinal vortices. In order to analyze the separation mechanism at the sonar dome, Visonneau et al. [12] conducted a numerical simulation of OTT on DTMB 5415 at drift angles of 10° and 20°. Three different turbulence models were selected, with a comparison of hydrodynamic forces/moment and detailed flow field. In contrast to IIHR TPIV experiments, results indicated that hybrid RANS/LES DDES-SST model was the only model capable of catching complex flow physics linked with unsteady vortex shedding, which appeared at the leeward side and trailing edge.

From published literatures, few work has been devoted to systematic analysis of computation settings in OTT. This paper applied an orthogonal experimental method to discuss effects of three influence

IOP Conf. Series: Materials Science and Engineering 1288 (2023) 012049 doi:10.1088/1757-899X/1288/1/012049

factors at three levels. Application of turbulence model, selection of grid size and setting of time step were taken into account, and an uncertainty analysis concerned the quantifiable influence factor was performed when calculating hydrodynamic forces and moment. On this basis, virtual static captive static model tests at drift angles in the range of 0° ~12° were simulated. The characteristics of flow field are given in details, including elevation of wave surface and contour of velocity field, to find out the relationship between oblique flow field and non-uniform force. For a deep insight into the asymmetric flow field, vortical structures obtained by means of different turbulence model are presented to find out the physical mechanism in hydrodynamic performance.

2. Numerical modeling

2.1. Governing equations

The commercial software STAR-CCM+ is applied for implementation of numerical simulation. For the sake of saving calculating resource and computing time, the method of solving *NS* equations directly is not of vital necessity. With the application of RANS equations, some terms characterizing turbulence effect can be approximated, which becomes a substitute for direct solving method. In computational simulation of this work, results are obtained by solving unsteady RANS equations. The incompressible unsteady RANS equations for two-phase flow are presented as follows:

$$\frac{\partial(\varphi_m \rho_m)}{\partial t} + \frac{\partial}{\partial x_i}\left(\varphi_m \rho_m \overline{u_i}\right) = 0 \tag{1}$$

$$\frac{\partial \rho_m \overline{u_i}}{\partial t} + \frac{\partial}{\partial x_j}\left(\rho_m \overline{u_i u_j}\right) = -\frac{\partial \overline{P}}{\partial x_i} + \frac{\partial}{\partial x_j}\left[\mu_m\left(\frac{\partial \overline{u_i}}{\partial x_j} + \frac{\partial \overline{u_j}}{\partial x_i}\right)\right] + \frac{\partial}{\partial x_j}\left(-\rho_m \overline{u_i' u_j'}\right) + \rho_m \overline{f_i} \tag{2}$$

where i and j represent coordinate components; m refers to the phase id, including water and air in this work; u is velocity; ρ and μ denote density and viscosity of the fluid, respectively; P stands for pressure; f means body force, which is gravity in this study; $\rho_m \overline{u_i' u_j'}$ is the Reynolds stress term, explanation of which can be found in Bhushan's work [8], which will not be introduced in this work.

2.2. Ship geometry

The hull used in the simulation was the well-known benchmarking KCS with a scale factor λ = 75.5, and a non-rotating rudder was installed at the stern of the nearly three-meter-long-ship model. Main parameters of the ship model and particulars involved in the simulation are listed in **Table 1**, which is obtained from SIMMAN (2014) [13]. Experimental results for validation analysis were provided from NMRI (National Maritime Research Institute).

Table 1. Introduction of main parameters and particulars about KCS.

Main particulars	Symbol	Full scale	Model scale
Scale factor	λ	1	75.5
Length between perpendiculars	L_{pp} (m)	230.0	3.0464
Beam at waterline	B_{wl} (m)	32.2	0.4265
Draft	d (m)	10.8	0.1430
Displacement	Δ (m^3)	52030	0.1209
Wetted Surface area	S (m^2)	9530	1.6877
Block coefficient	C_B	0.651	0.651
Metacentric height	GM (m)	0.60	0.097
Scale factor	λ	1	75.5

Figure 1. Geometry of KCS.

2.3. Computational domain and grid generation

As depicted in **Figure 2**, the computational domain is positioned 2.2 times the length between perpendiculars (L_{pp}) in front of the hull and extends up to 3.5 L_{pp} behind the model. For the sake of asymmetry in oblique incoming flow, the width of the computational domain is approximately 5.7 times the L_{pp}, with a distance of 2.5 L_{pp} on the portside of the hull. The height is 1.5 times the L_{pp} with 0.5 L_{pp} below the barycenter position of the model, which is constructed according to the depth of the tank in NMRI. The direction of coordinates is also given in **Figure 2**. The inlet, left and right sides, top and bottom of the virtual tank are set as velocity inlet, while the outlet of the flow field domain is defined as pressure outlet. Meanwhile, hull surfaces are defined as non-slip walls. It is worth mentioning that a wave damping zone is placed along the outlet to eliminate the effect of boundaries.

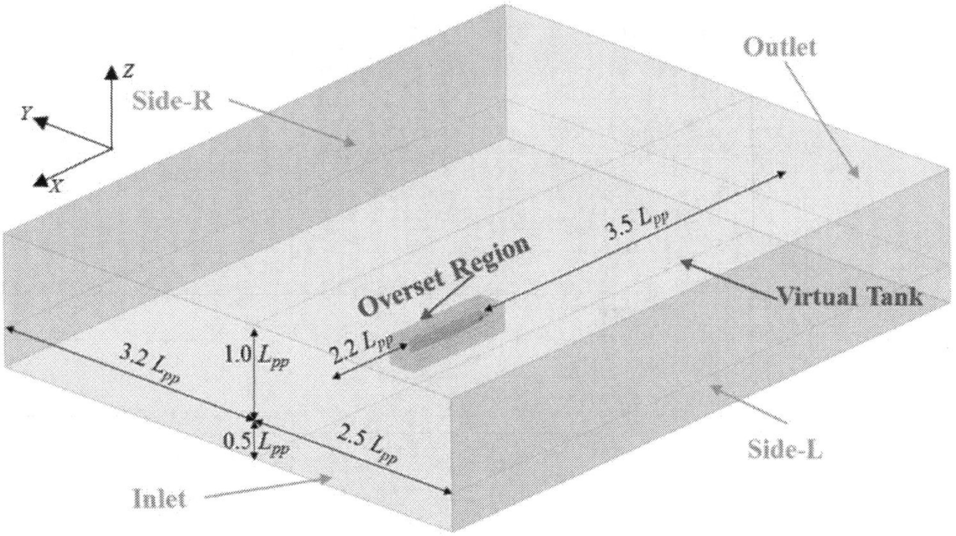

Figure 2. Computational domain for simulation.

As illustrated in **Figure 3**, the trimmed volume method and surface remesher are employed to generate volume meshes and surface meshes, respectively. When it comes to boundary layer, a prism mesh model is selected to generate six layers of orthogonal prismatic cells close to surfaces. The total thickness is constrained to ensure that wall *Y+* falls within the range of 30 to 100. Multiple refinement levels are set up to achieve a reasonable mesh distribution, where finer grids are positioned closer to the hull. For a better capture of Kelvin waves, grids are refined at the interface of the air-water multiphase flow, with the adoption of the Volume of Fluid (VOF) method. Apart from the free-surface region, the trimmer mesh is automatically generated with three refinement regions, including the rudder part, the bow part and the stern part, taking specific geometric structures into consideration. With regard to solution of ship's motion, the Dynamic Fluid Body Interaction (DFBI) method is applied, which allows heave, pitch and roll during the simulation.

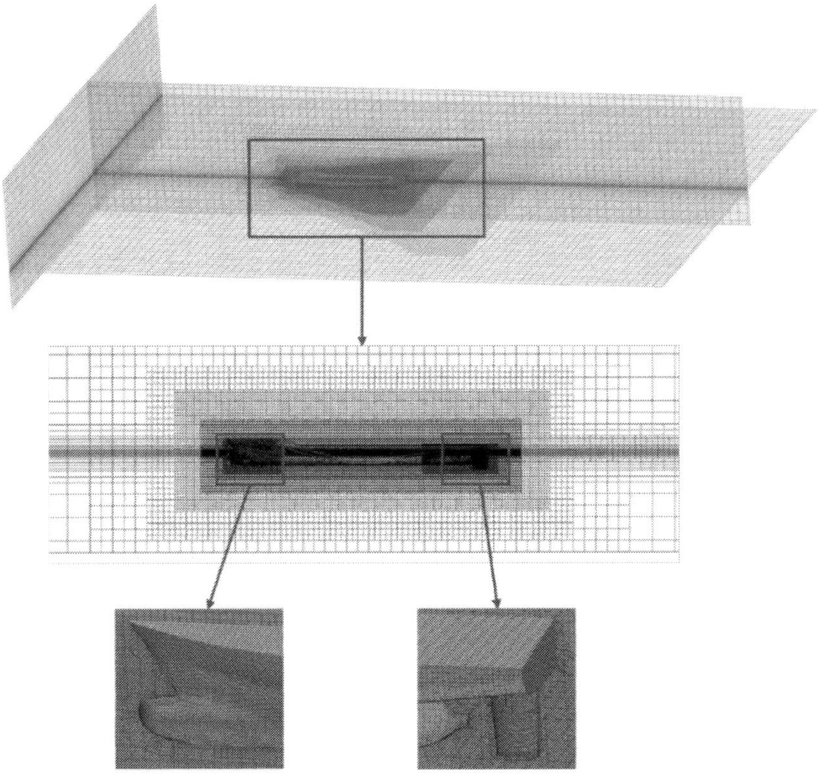

Figure 3. Computational domain for simulation.

2.4. Static drift test conditions

Virtual static captive model tests simulated in this work involve drift angles of 0°, 4°, 8° and 12° with a velocity of 1.1 m/s. For simplification, non-dimensional parameters X', Y', N' can be obtained by using the following relations

$$\begin{cases} X' = \dfrac{X}{\frac{1}{2}\rho V_m^{\ 2} L_{pp} d} \\ Y' = \dfrac{Y}{\frac{1}{2}\rho V_m^{\ 2} L_{pp} d} \\ N' = \dfrac{N}{\frac{1}{2}\rho V_m^{\ 2} L_{pp}^{\ 2} d} \end{cases} \tag{3}$$

where V_m indicates velocity of ship model; L_{pp} stands for length between perpendiculars; d is defined as draft of the ship model.

2.5. Orthogonal uncertainty analysis

Estimating numerical errors and uncertainties is an essential process during usage of numerical tools. Regarded as a problem involving multi-factors, uncertainty analysis can provide a quantitative metric for verification. At least three sets of numerical results are required in calculation. Aimed at improvement of analyzing efficiency, the orthogonal experimental method is adopted in the research, which can incorporate interaction of factors in the meantime.

In the context of the orthogonal experimental method, average values at different levels K_{ij} are crucial for intuitive analysis and variance analysis. These values can be calculated as follows:

$$K_{E,ij} = \frac{q}{n}\sum_{a=1}^{n/q} Y_{E,ij}^{(a)} \tag{4}$$

$$K_{S,ij} = \frac{q}{n}\sum_{a=1}^{n/q} Y_{S,ij}^{(a)} \tag{5}$$

where i and j stand for number of factor and level, respectively. $Y_{ij}^{(a)}$ represents value of No.(a) experiment under influence factor i and level j. Subscripts S means numerical results, while E is relative error between numerical and experimental data

Take relative error $Y_{E,ij}^{(a)}$ as an indicator, analysis of variance will be carried out with the condition of intuitive analysis. The influence factor with the minimum extreme deviation can be chosen as the baseline for comparison. This baseline R_j is defined as:

$$R_j = \max\left(K_{ij}\right) - \min\left(K_{ij}\right) \tag{6}$$

Sum and average of each indicator can be calculated as follows

$$T = \sum Y_{ij}^{(a)} \tag{7}$$

$$\overline{T} = \frac{1}{n}\sum Y_{ij}^{(a)} = \frac{T}{n} \tag{8}$$

where T and $\overline{T}$ denote sum and average of test data, respectively.

Sum of standard deviation concerned each influence factor S_i is defined as

$$S_i = \frac{n}{q}\sum_{j=1}^{q}\left(K_{ij} - \overline{T}\right)^2 \tag{9}$$

$$S_T = \sum\left(Y_{ij}^{(a)} - \overline{T}\right)^2 \tag{10}$$

where S_T is total sum of standard deviation including all indicators.

Mean square MS_i is given as follows

$$MS_i = S_i/f_i = \frac{n}{q}\sum_{i=1}^{q}\left(K_{ij} - \overline{T}\right)^2 \Big/ (p-1) \tag{11}$$

Test statistic F_i is used to discriminate whether influence factor i is significant, which can be written as

$$F_i = S_i/S_e \tag{12}$$

By means of variance analysis, the impact of influence factors can be quantified. A smaller value indicates that the influence factor has less significance. Those influence factors with F_i less than reference value will be excluded in the subsequent uncertainty analysis, which aims to identify quantifiable influence factors. Relevant procedures can be found in predecessors' work [14].

3. Results and discussion

3.1. *Influence of computation settings*

During the simulation of static captive model test, the motion of the ship is confined to the horizontal plane, including surge, sway and yaw. With particular attention to lateral force, the non-dimensional parameter Y' is calculated, and the relative error is obtained for orthogonal uncertainty analysis. To comprehensively analyze the effects of various settings on numerical results, an orthogonal experimental method is applied during the virtual static captive model test. The turbulence model, computational grid and time step are selected as influence factors, denoted by symbols A, B, and C, respectively. Each factor has three different levels, as presented in Table 2. Besides the widely adopted two-equation models, Reynolds Stress Models (RSM) can also be used to address complex flow phenomena, particularly detailed flow separation in manoeuvring tasks.

Table 2. Factors and level of orthogonal experimental method.

Number of levels	Influence factor		
	Turbulence model	Grid size/mm	Time step/s
1	SST κ-ω	0.05	0.005
2	Realizable κ-ε	0.07	0.010
3	RST	0.10	0.020

Table 3 presents the orthogonal array, which determines the parameter configuration of the numerical simulation.

Table 3. Orthogonal array of $L_9(3^3)$.

Number	Factor *A*	Factor *B*	Factor *C*	State of conditions
1	1	1	1	$A_1B_1C_1$
2	1	2	2	$A_1B_2C_2$
3	1	3	3	$A_1B_3C_3$
4	2	1	2	$A_2B_1C_2$
5	2	2	3	$A_2B_2C_3$
6	2	3	1	$A_2B_3C_1$
7	3	1	3	$A_3B_1C_3$
8	3	2	1	$A_3B_2C_1$
9	3	3	2	$A_3B_3C_2$

Figure 4 displays the associated relative error, where the horizontal axis represents the condition numbers and the vertical axis represents the relative error of non-dimensional lateral force Y' between the numerical values and experimental results. In general, the data obtained using the CFD method employed in this study are smaller than those obtained from experimental measurements. The maximum relative error reaches -12.5%, while the minimum value is around -1.8%.

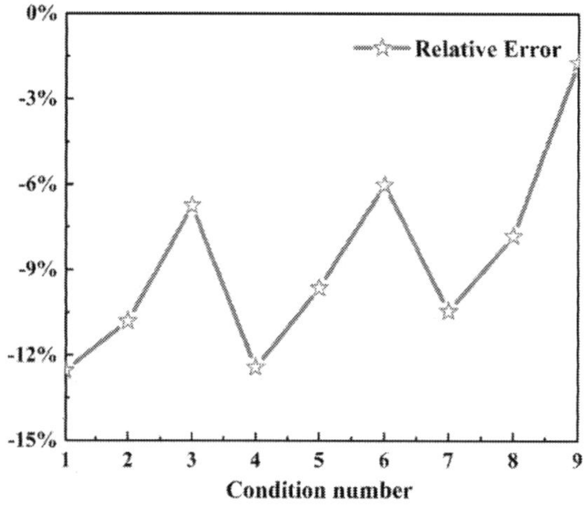

Figure 4. Results of relative error concerned non-dimensionless lateral force Y' under different conditions.

According to the results of intuitive analysis shown in **Table 4**, distinction exists between extreme deviations of each influence factor, which vary in the same order of magnitude. In such qualitative analysis, the impacts exerted by the time step is negligible compared with other two influence factors *A* and *B*. The extreme deviation is only 0.62%, when the time step varies from 0.01 s to 0.02 s.

Table 4. Intuitive analysis results of non-dimensionless lateral force Y' concerned influence factors.

	Factor A	Factor B	Factor C
$K_{E,i1}$	-10.03%	-11.81%	-8.79%
$K_{E,i2}$	-9.36%	-9.42%	-8.32%
$K_{E,i3}$	-6.67%	-4.83%	-8.94%
R_i	3.36%	6.98%	0.62%

Analysis of variance is implemented as an indispensable quantitative analysis, the results are listed in **Table 5**. Corresponding test statistic of factor C is less than 1, indicating that changing the time step from 0.005 s to 0.02 s in the virtual static captive test has a negligible effect. By comparison, the effect of grid size on numerical results cannot be ignored.

Table 5. Variance analysis results of non-dimensionless lateral force Y' concerned influence factors.

Source of error	Sum of standard deviation S_i	Degree of freedom F_i	Mean square MS_i	Test statistic F_i
Factor A	1.905×10^{-3}	2	0.952×10^{-3}	8.154
Factor B	7.550×10^{-3}	2	3.775×10^{-3}	32.32
Factor C	0.064×10^{-3}	2	0.032×10^{-3}	0.272
Sum	9.752×10^{-3}	8	$F_{0.95}(2,8)=19$	$F_{0.99}(2,8)=99$

For an intuitive comparison of results relative to varisized grid, distribution of forces along x-axis is delineated in **Figure 5**, including pressure component and shear component at longitudinal direction and transverse direction. The velocity of model is 1.1 m/s and the drift angle is 12° in the virtual test. The horizontal axis indicates the variation of dimensionless parameter x/L_{pp}, and the vertical axis stands for the value of force. For the convenience of comparison, negative value of longitudinal force is plotted in **Figure 5**. From the magnitude of ordinate, component induced by pressure gives a significant contribution to force. A decrease in grid size leads to an improvement in the level of precision, as evidenced by the gentle variation of the curve with grid refinement. However, the improvement in computational accuracy becomes negligible after a certain number of grids. In order to reach a balance between calculating precision and computing time, the grid convergence index (*GCI*) is often adopted for estimating the uncertainty caused by the grid.

An uncertainty analysis based on the orthogonal experimental test is conducted as shown in **Table 6**. It has to be stressed that average values at different levels $K_{S,ij}$ are used instead of numerical results S_{ij}. Number 1, 2, 3 correspond to finer grid, fine grid and coarse grid in sequence, with corresponding grid numbers of 4.15×10^6, 1.64×10^6 and 0.66×10^6. The time step is set as 0.01 s when performing the grid convergence study, using the SST κ-ω turbulence model.

Table 6. Parameters in grid convergence study.

Parameter	X'	Y'	N'
$K_{S,1}$	-2.0246×10^{-2}	6.0802×10^{-2}	2.7925×10^{-2}
$K_{S,2}$	-2.0260×10^{-2}	6.2447×10^{-2}	2.8157×10^{-2}
$K_{S,3}$	-2.0658×10^{-2}	6.5614×10^{-2}	2.8621×10^{-2}
ε_{21}	-1.420×10^{-5}	1.6453×10^{-3}	2.319×10^{-4}
ε_{32}	-3.978×10^{-4}	3.1664×10^{-3}	4.635×10^{-4}
R	0.0357	0.5196	0.5003
p_{GCI}	3.7033	0.7524	0.7928
e_a^{21}	0.070%	2.706%	0.830%
GCI_{21}	0.003%	3.332%	0.948%
e_a^{32}	1.963%	5.071%	1.646%
GCI_{32}	0.091%	6.537%	1.970%

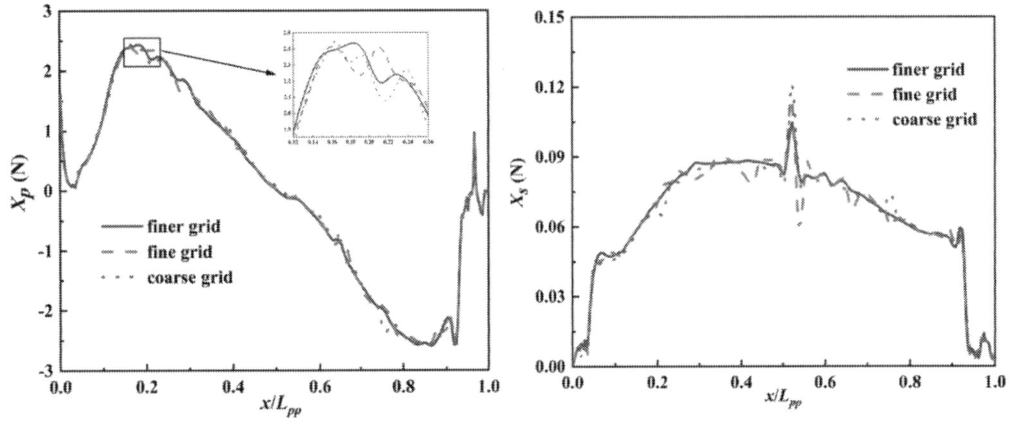

(a) Pressure component of longitudinal force (b) Shear component of longitudinal force

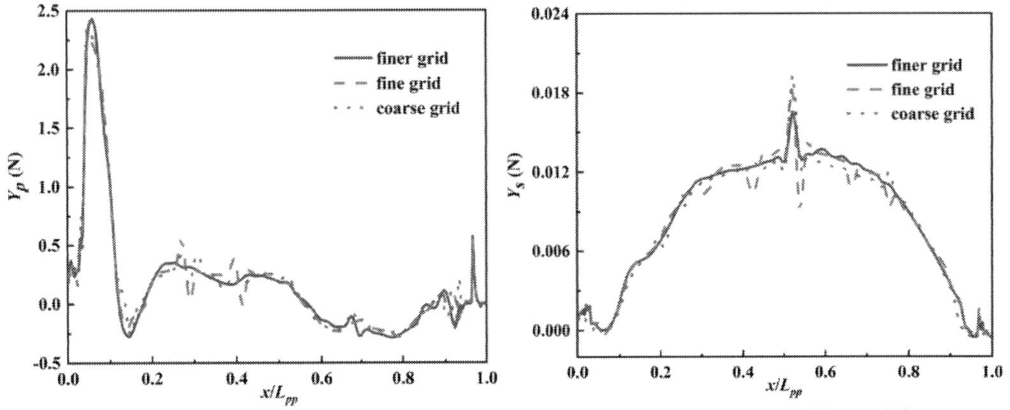

(c) Pressure component of lateral force (d) Shear component of lateral force

Figure 5. Longitudinal distribution of forces under varisized grids (V_m = 1.1 m/s, β = 12°)

Table 6 only considers the effects of grid size on hydrodynamic forces and moment. When calculating dimensionless longitudinal force X', lateral force Y' and yaw moment N', all refinement ratios are in the range of 0 to 1, indicating a monotonic convergence condition. Both the approximate relative error e_a^{21} and GCI_{21} index of adjacent convergence model are below the acceptable upper limit 5%, consequently, the intermediate set of cases achieves the required level of calculation accuracy, with a grid size of 0.07 m.

When it comes to the effect of turbulence model on virtual static captive model test, the longitudinal distribution of lateral force integrated by sections is calculated, as displayed in **Figure 6**. In the study, the 12° drift test at model velocity V_m = 1.1 m/s is considered. Three different turbulence models are selected for numerical simulation: SST κ-ω model (grey line), Realizable κ-ε model (red line), and RST model (blue line). Lateral force acted on the portside and starboard side of the hull is plotted as **Figure 6**(a) and (b). The discrepancy between curves in different colors is small, especially in comparison with that of **Figure 5**. The conclusion is in conformity with the values of the test statistics in **Table 5**.

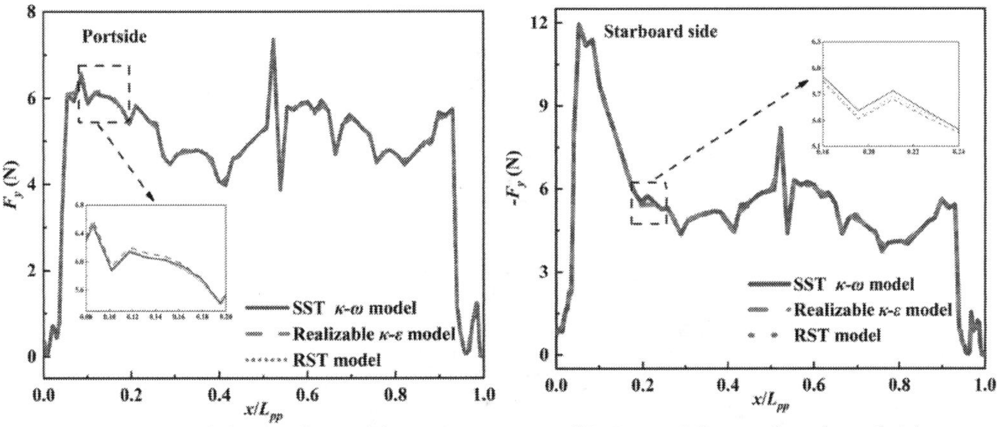

(a) Lateral force of portside (b) Lateral force of starboard side

Figure 6. Longitudinal distribution of lateral force with application of different turbulence models (V_m = 1.1 m/s, β = 12°)

3.2. Simulation results at different drift angles

Through qualitative and quantitative analysis about effects of influence factors at different levels, suitable computation settings for virtual static captive model test can be received. For drift angle β = 12°, minimum relative error to experimental result is obtained, when selecting RST turbulence model, with a grid size of 0.07 m and a time step of 0.01 s. Considering stability of solution and complexity of convergence over the computational domain, the SST κ-ω turbulence model is utilized in conditions at various drift angles. Solving results are in comparison with experimental data provided by NMRI, in **Table 7**, numerical results of dimensionless hydraulic forces are smaller than that achieved by experiments, while dimensionless yaw moment N' calculated by method of CFD is larger. The average relative error is about 10%, when the maximum value of 15.02% is achieved for N' at a drift angle of 12°. In general, adopted computation settings can satisfy accuracy requirement for oblique towing test.

Table 7. Relative error between numerical result and experimental data from NMRI.

Parameter		X'	Y'	N'
	Experimental data	-1.938×10^{-2}	2.338×10^{-2}	1.271×10^{-2}
$\beta = 6°$	Numerical result	-1.916×10^{-2}	2.015×10^{-2}	1.340×10^{-2}
	Relative error	-1.13%	-13.82%	5.47%
	Experimental data	-2.211×10^{-2}	6.894×10^{-2}	2.455×10^{-2}
$\beta = 12°$	Numerical result	-2.066×10^{-2}	6.149×10^{-2}	2.823×10^{-2}
	Relative error	-6.58%	-10.81%	15.02%

Figure 7 demonstrates volume fraction of water at bow region. Under zero drift angle case, the distribution of water at portside of the hull is exactly the same as that of starboard side. In oblique towing tests, the symmetry of waterline observed at β = 0° vanishes as depicted in **Figure 7**. With the increase of drift angle, the characteristic of asymmetry becomes more conspicuous in flow field around bow region. The larger the drift angle is, the higher the wave crest at the starboard side will be. In addition, the opposite situation can be found at the portside.

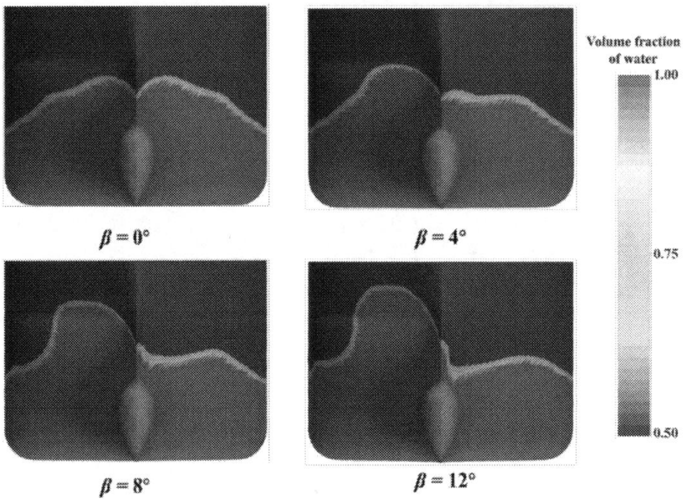

Figure 7. Volume fraction of water at bow under different drift angles (V_m = 1.1 m/s)

The top view of free surface is given in **Figure 8** as supplementary, presenting varying situations under an interval of 4° drift angle. The direction of incoming fluid contributes to the position of fluid in the vertical direcition. Accompanied by the increase of drift angle, disperse waves at the portside vanish gradually. Transformation of wave pattern with incremental drift angle reveals the extent of acting force between hull and fluid. **Figure 8** also displays the longitudinal distribution of wave height at the section of y/L_{pp} = ±0.75 under different drift angles, the grey line represents data of the portside while red the starboard side. Diversity mostly lies in the region of x/L_{pp} = 0~1 along the longitudinal direction, especially near x/L_{pp} = 0.15 and x/L_{pp} = 0.90, in consistent with regions of fore body and aft body. In case of 4° drift test, difference in elevation of wave surface at both two sides around the mid body is obscure, which extends to a distinct contrast.

As an indicator of flow field, distribution of longitudinal velocity component is also asymmetrical. **Figure 9** illustrates specific conditions at sections of x/L_{pp} = 0.12 and x/L_{pp} = 0.88, respectively, which is close to bow and stern regions of the ship. The dimensionless parameter u' is chosen for intuitive analysis. There is trade-off between velocity component in the x-axis and y-axis, the former experiences continuous reduction with increase of drift angle, as demonstrated in **Figure 9**(a) and (b). From the perspective of stern, the right half part of figures corresponds to the starboard side. The asymmetrical distribution can be attributed to lateral flow, which leads up to leaking vortex around the bulbous bow. An interaction between leaking vortex and boundary layer at the starboard side increases the deviation extent of fluid. At the section of x/L_{pp} = 0.12, regions of fluid with higher longitudinal velocity speed can be observed at the starboard side, especially under large drift angle conditions. The opposite appearance can be viewed in **Figure 9**(b).

There is no denying that hydrodynamic forces and moment acting on the hull is closely bound up with flow field. An included angle exists between the hull and incoming fluid during the oblique towing test. In numerical simulation process, to realize alteration of drift angle, inflow direction of in-coming fluid is adjusted while orientation of the hull is maintained. The asymmetry of fluid velocity at two sides of a symmetrical geometry gives birth to the inhomogeneous force. Consequently, lateral force Y is generated, as illustrated in **Figure 10**, where solid line indicates condition of portside and dashed line corresponds to that of starboard side. Similar to **Figures. 7-9**, presence of symmetry properties can also be detected in zero drift angle test. When drift angle changes from 0° to 12°, diversity between force in the transverse direction becomes noticeable by comparison with both two sides. It is worth mentioning that maximum difference exists at the location of x/L_{pp} = 0.055. Under the condition of β = 12°, extreme point of F_y reaches 11.94 N at the starboard side, which increases about 33% compared with the portside.

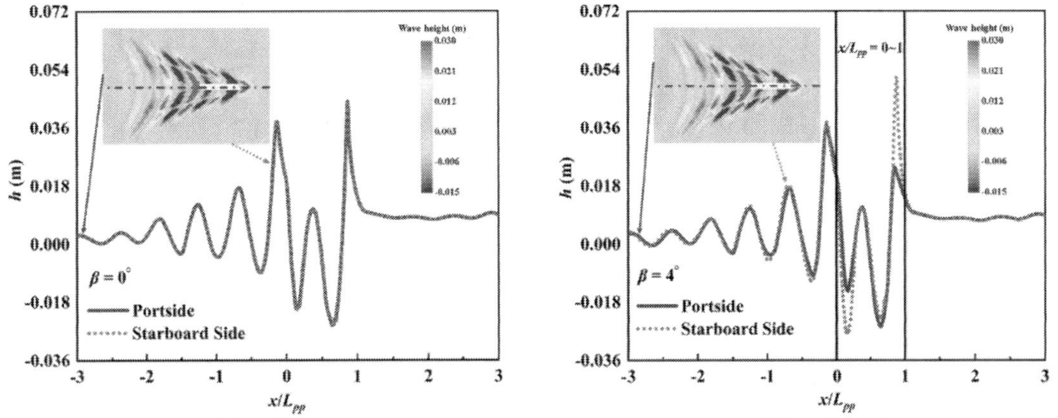

(a) $\beta = 0°$ (b) $\beta = 4°$

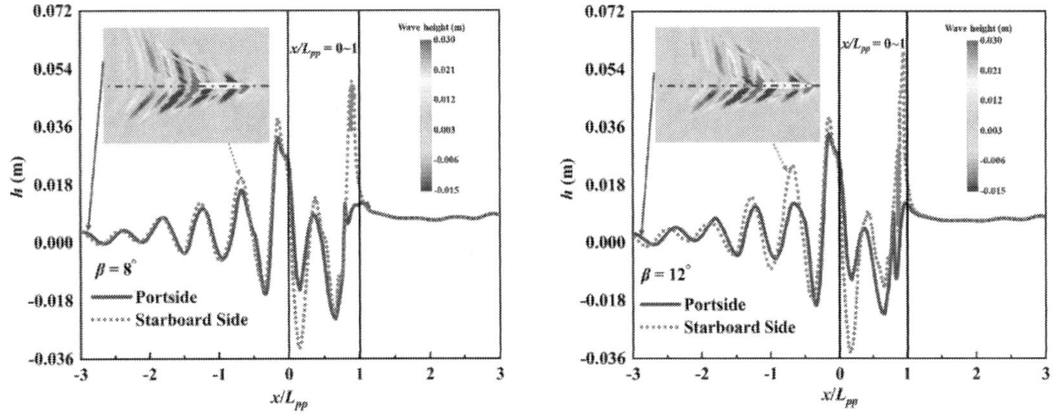

(c) $\beta = 8°$ (d) $\beta = 12°$

Figure 8. Longitudinal distribution of wave height under different drift angles

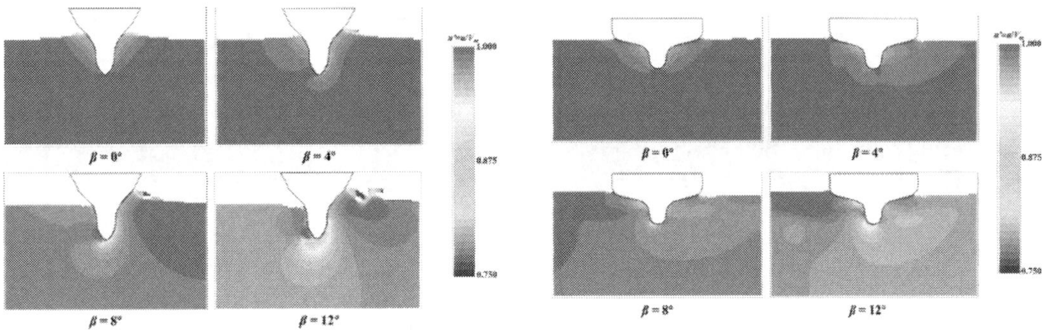

(a) $x/L_{pp} = 0.12$ (b) $x/L_{pp} = 0.88$

Figure 9. Distribution of $u'=u/V_m$ under different drift angles ($V_m = 1.1$ m/s)

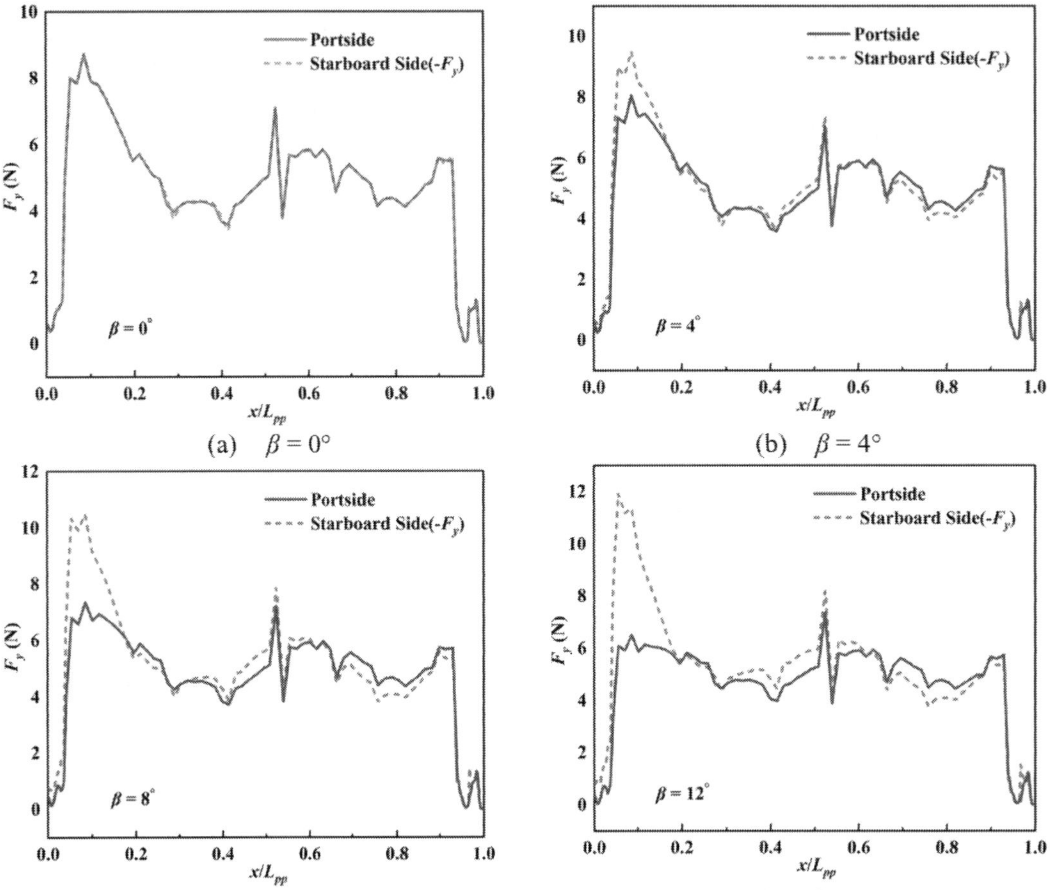

(a) $\beta = 0°$ (b) $\beta = 4°$

(c) $\beta = 8°$ (d) $\beta = 12°$

Figure 10. Longitudinal distribution of horizontal force under different drift angles

3.3. Vortical structures

The inequality in the velocity filed, derived from oblique flow direction, contributes to the disparity of hydraulic forces and moment. In a gesture to obtain a deep insight into the asymmetrical flow field, investigation about vortex structures is conducted with the method of Detached Eddy Simulation (DES) [15] model and Reynolds Averaged Navier Stokes (RANS) model, which is applied in section 3.2. **Figure 11** provide a global view of the main time-averaged vortical structures created around the KCS at 12° static drifts. For vortex identification, Q-criterion [16] is proposed with a threshold of 50, in the form of vorticity magnitude for display. In addition, position of fluid in the z-axis is also presented. In the procedure of grid generation, identical parts are created with refinement of grid for depicting kelvin waves around the hull. In **Figure 11** (b), smooth transition can be observed between regions with and without waves. To put it another way, utilization of DES model has a better solution over the whole computation domain.

Details of vorticities are demonstrated in **Figure 12**, with iso-surfaces of Q-criterion colored by normalized levels same as **Figure 11**. The existence of bulbous bow realizes an effectively reduction in fore body keel vortex. As concerned with aft body keel vortex, more vortex structures can be captured with the deployment of DES method. Simultaneously, more tip vortex can be perceived surrounding the hull in **Figure 12**(b). Admittedly, DES method has signature strength in capturing flow details, which is accord with conclusion from published literatures [17]. For the sake of investigation on flow separation and vortex shedding, DES turbulence model is preferable to analyze the effect of vortical structures on acting force and moment.

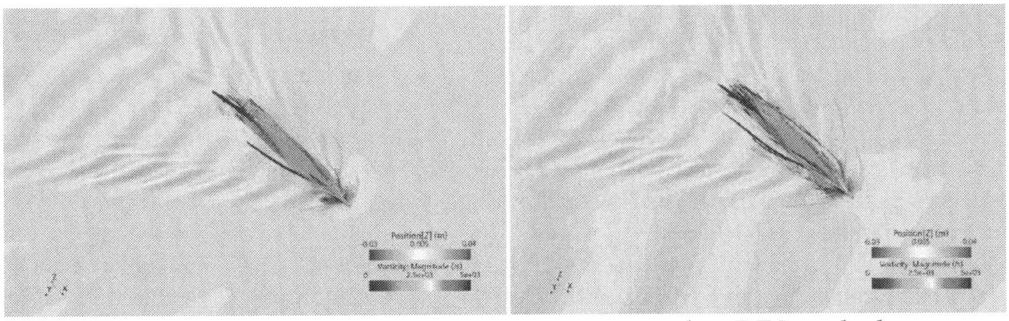

(a) RANS method (b) DES method

Figure 11. Flow field by means of different simulation methods (V_m = 1.1 m/s, β = 12°)

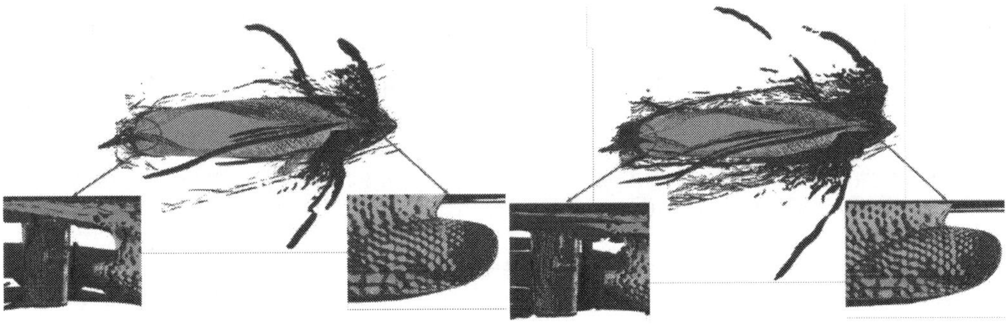

(a) RANS method (b) DES method

Figure 12. Vortical structures related to different simulation methods (V_m = 1.1 m/s, β = 12°)

4. Conclusions

In the virtual static captive static model tests on the KCS ship, systematic analysis about effect of computation settings is conducted through numerical computations. After verification of numerical results in comparison with experimental data, qualitative and quantitative analysis of oblique flow field is discussed in detail and some conclusions can be drawn as follows:

Under fixed grid size, there is not much difference between results from SST κ-ω turbulence model, Realizable κ-ε turbulence model and RST turbulence model. Based on orthogonal uncertainty analysis, relative error of yaw moment is more significant compared with hydrodynamic forces.

With increase of drift angle, the diversity between force in the transverse direction at starboard side and portside becomes more noticeable especially in the bow region. Simultaneously, disperse waves at the portside vanish gradually.

By contrast, DES method has irreplaceable merits in capturing flow characteristics in detail. For better explanation of physical mechanisms in hydrodynamic behaviours, future work will be done by means of DES turbulence model.

References

[1] Kim D, Song S and Tezdogan T 2021 *Ocean Eng.* Free running CFD simulations to investigate ship manoeuvrability in waves **236** 109567.

[2] Wang J H and Wan D C 2018 *J. Mar. Sci. Appl.* CFD investigations of ship maneuvering in waves using naoe-FOAM-SJTU solver **17** 443-458.

[3] Cura-Hochbaum A 2006 *Proc. of 26th ONR Symposium on Naval Hydrodynamics* Virtual PMM Tests for Manoeuvring Prediction (Rome).

[4] Simonsen C D and Stern F 2005 *Comput. Fluids* Flow pattern around an appended tanker hull form in simple maneuvering conditions **34** 169–198.

[5] Xing T, Bhushan S and Stern F 2012 *Ocean Eng.* Vortical and turbulent structures for KVLCC2 at drift angle 0, 12, and 30 degrees **55** 23–43.
[6] Shenoi R R, Krishnankutty P and Selvam R P 2016 *Ships Offshore Struc.* Study of Computational fluid dynamics prediction of hydrodynamic forces on a manoeuvring ship including effects of dynamic sinkage and trim ship by static and dynamic simulations using a RANSE-based solver **11** 316–334.
[7] Sakamoto N, Carrica P M and Stern F 2012 *J. Mar. Sci. Tech.* URANS simulations of static and dynamic maneuvering for surface combatant: Part 1. Verification and validation for forces, moment, and hydrodynamic derivatives **17** 422–445.
[8] Bhushan S, Yoon H and Stern F 2016 *Int. J. Comput. Fluid D.* Large grid simulations of surface combatant flow at straight-ahead and static drift conditions **30** 356-362.
[9] Liu Y, Zou L and Zou Z J 2019 *P. I. Mech. Eng. M-J. Eng.* Computational fluid dynamics prediction of hydrodynamic forces on a manoeuvring ship including effects of dynamic sinkage and trim *233* 251-266.
[10] Yao J X, Jin W and Song Y 2016 *Ocean Eng.* RANS simulation of the flow around a tanker in forced motion **127** 236–245.
[11] Visonneau M, Guilmineau E and Rubino G 2018 *Proc. of 32nd Symposium on Naval Hydrodynamics* (Germany).
[12] Visonneau M, Guilmineau E and Rubino G 2020 *Proc. of 33rd Symposium on Naval Hydrodynamics* (Osaka).
[13] SIMMAN, 2014. Workshop on verification and validation of ship manoeuvring simulation methods. https://simman2014.dk.
[14] Stern F, Wilson R and Shao J 2006 *Int. J. Numer. Method. H.* Quantitative V&V of CFD simulations and certification of CFD codes **50** 1335-1355.
[15] Durbin PA 2018 *Annu. Rev. Fluid Mech.* Some Recent Developments in Turbulence Closure Modeling **50** 77-103.
[16] Zhang Y N, Wang X Y, Zhang Y N and Liu C Q 2019 *J. Hydrodyn.* Comparisons and analyses of vortex identification between Omega method and Q criterion **31** 224-230.
[17] Liu Y, Zou L and Zou Z J 2019 *P. I. Mech. Eng. M-J. Eng.* Computational fluid dynamics prediction of hydrodynamic forces on a manoeuvring ship including effects of dynamic sinkage and trim *233* 251-266

IOP Conf. Series: Materials Science and Engineering 1288 (2023) 012050 doi:10.1088/1757-899X/1288/1/012050

A CFD-based evaluation of ship speed at EEXI draft

Shuai Yuan[1,2,*], Lei Shi[1,2], Fan Hu[1,2], Hai Yu[1,2] and Xiaoqing Xiong[1,2]

[1] Marine Design and Research Institute of China, Shanghai, China
[2] Shanghai Key Laboratory of Ship Engineering, Shanghai, China

*Corresponding author, Email:yuanshuai@maric.com.cn

Abstract. This paper presents a simulation of ship speed for an existing containership at EEXI draft using Computational Fluid Dynamics (CFD). The simulation is divided into two steps namely (a) the numerical prediction of ship performance at model scale, and (b) the extrapolation of results to full scale. Model scale numerical simulations of ship resistance, open-water propeller and ship-propeller-rudder interactions at demonstrated. The method is at first validated by comparing against available experiments and at second stage full-scale ship performance is assessed by an extrapolation method. Comparison of the ship speed predicted by CFD against experiments demonstrates a difference of 0.05 knots when the same extrapolation coefficients are applied.

1. Introduction

The initial IMO (International Maritime Organization) strategy on reduction of GHG emissions from ships was adopted in April 2018.[1] The ambition of this strategy is to reduce the carbon intensity of international shipping by at least 40% by 2030 and to reduce the total annual GHG emissions by at least 50% by 2050, compared to 2008 levels, while also pursuing efforts towards phasing out GHG as soon as possible in this century.

Energy Efficiency Design Index (EEDI) is a metric to predict the energy efficiency of newly built ships. IMO determined the purpose of EEDI as an equitable base for benchmarking and to promote innovations related to energy efficiency on marine vessels. It has determined the base level of efficiency for new ships on the bases of ship size and type. EEDI has been a mandatory application since 2013.[2] Lately, following up on its plans to reduce carbon emissions from shipping by half by 2050, and to achieve full decarbonization by the end of the century, the IMO has introduced two new indices, Energy Efficiency Existing Ship Index (EEXI) and Carbon Intensity Indicator (CII) which will be used to enforce requirements on the reduction of GHG emissions from both new-built and existing ships.[3] The requirements set in MARPOL Annex VI by IMO have been entered into force on November 1, 2022. EEXI and CII calculation and reporting have been mandatory on January 1, 2023.

The EEXI is a design index that represents an estimate of the theoretical energy efficiency of an existing ship, regardless of how it is operated. It is defined to be substantially equivalent to the EEDI approved in 2018 by the MEPC.[4] It is very important to investigate how existing ships will meet the novel EEXI requirements set by the IMO.[5] Shipowners will be required to provide a certified calculation and compliance with the EEXI of their ship by the end of 2023. At this time, ships that will have passed the EEXI test will be allowed to keep sailing, while the others will not. Compliance with the EEXI is obtained by comparing the EEXI achieved by the ship to a reference, which is a function of the ship type, propulsion principle, and of its deadweight: if the EEXI is lower than the reference, the ship is compliant. Otherwise, it has to undergo a retrofit that allows for achieving the required EEXI level. After classification societies have accepted the mitigation actions (if required), the ship will undergo a statutory survey that, if successful, will be concluded with the emission of an International Energy Efficiency Certificate IEEC, that the ship will need to carry on board and show to port authorities if asked to.[3]

The calculation formula for the EEXI consists of two main parts: the first one is the expected carbon dioxide emissions from the ship when its engines are operated at 75% of their maximum continuous rating; the second one is the expected transport work that the ship can produce when sailing in reference conditions, i.e., the speed achievable when the engines are running at 75% of their maximum continuous rating.[6] Reference speed (V_{ref}) is the vessel speed at the summer load line and at rated engine power. When the findings of the tank test or the speed trial are unavailable, the estimated ship speed including the margin V_{ref} must be determined using the reference formula. If an energy-saving application is implemented to comply with the requirements, V_{ref} can be obtained from curves of model tests/CFD (Computational Fluid Dynamics) simulations, admiralty equations, and the formula provided in MEPC.333(76).[6] As mentioned above, to comply with the EEXI requirements, mitigation measures for the reduction of CO2 emissions can be applied. An important short-term operational measure for lowing CO2 emissions is speed reduction (or slow steaming).[7] For existing containerships, which operate at a broad profile, such operational measure is of great concern to shipowners. Thus, to be consistent with the latest policy, speed evaluation at EEXI draft for containerships seems to be essential at the time being.

In the present work, rather than focusing on the complete assessment of EEXI, the simulation of ship speed for an existing containership at EEXI draft is mainly studied. Instead of using the reference formula, a fully CFD-based numerical evaluation is carried out. The simulation is divided into two steps namely (a) the numerical prediction of ship performance at model scale and (b) the extrapolation of results to full scale. Numerical simulations of ship resistance, open-water propeller and ship-propeller-rudder interaction at model scale are performed in sequence. The method is first validated by comparing against available experiments and at second stage full-scale ship performance is assessed by an extrapolation method. According to the working condition of the main engine, the ship speed to be expected for sea trial is finally predicted. To further validate the predicted ship speed based on this fully CFD-based numerical evaluation, it is also compared against the ship speed predicted by experiments.

2. Study object and cases in concern

An 1100 TEU containership appended with a bow thruster, bilge keels and a rudder is taken as the study object. In the current study, speed evaluation of the ship is carried out for the EEXI draft in a traditional way, which is the extrapolation from the model-scaled results. As model test and sea trial data are available for this ship, a scaled ship model of the same scale factor as in the model test is used for CFD calculations. Only the bow thruster tunnel and the rudder are considered as in the model tests. The contributions of the bilge keels and the ship's superstructures to the ship performances are only considered in the extrapolation through an empirical way (see Appendix A). The CAD model for the target ship with the bow-thruster tunnel, propeller and rudder is illustrated in figure 1.

Figure 1. CAD model.

Main particulars of the ship model at EEXI draft are listed in table 1, where L_{PP} is the length between perpendiculars; B is the moulded breadth; T is the ship model's EEXI draft; ∇ is the moulded volumetric displacement; S_W and $S_{B.K.}$ are the wetted surface areas of hull and bilge keel, respectively;

A_T is the frontal projected area of the hull above waterline; LCB is the longitudinal center of buoyancy, positive forward.

Table 1. Main particulars of the ship model.

Parameters	L_{PP} (m)	B (m)	T (m)	∇ (m³)	S_W (m²)	$S_{B.K.}$ (m²)	A_T (m²)	LCB
Values	5.7774	0.9788	0.2972	1.0784	6.7036	0.0923	1.0147	-0.448%L_{PP}

The main particulars of the propeller model are listed in table 2, where D_M is the diameter of propeller model; $P_{0.70R}$ and $C_{0.70R}$ are the pitch and chord at $0.70R$ (R represents the radius of propeller), respectively; A_E/A_O is the blade area ratio; Z is the number of blades; d is the diameter of hub.

Table 2. Main particulars of the propeller model.

Parameters	D_M (m)	$P_{0.7R}/D_M$	$C_{0.7R}/D_M$	A_E/A_O	Z	d/D_M	Sense of Rotation
Values	0.2252	1.0042	0.3380	0.701	5	0.185	Right-handed

Parameters of the main engine at different settings are listed in table 3. With a shaft transmission efficiency of 0.985, the delivered power P_D at 75%MCR is calculated as: P_D = 8730kW × 0.75 × 0.985 = 6449kW. The corresponding revolution rate N_S for the main engine is estimated as: N_S = 127r/min × $0.75^{1/3}$ = 115.4r/min. Here, N_S is only an approximation with an assumption of engine power related to revolutions to power three. The actual value of N_S at 75%MCR, which represents the characteristics of the main engine, together with the actual revolution rate of the propeller at this working condition for the target ship, can be used to evaluate the Light Running Margin (LRM) for the target propeller. However, as the main purpose is to evaluate ship speed, the prediction of delivered power is the priority of current work.

Table 3. Parameters of the main engine.

Setting	Shaft Power (kW)	Revolution Rate (r/min)
MCR	8730	127
CSR	7857	122.6

The evaluation of the ship speed by CFD analysis is divided into two steps namely (a) the numerical prediction of ship performance at model scale and (b) the extrapolation of results to full scale. Based on the numerical simulations of viscous flow around the ship-propeller-rudder at model scale by CFD methods, numerical predictions of ship resistance in calm water, propeller characteristics in open water and the ship self-propulsion factors in calm water at model scale are first carried out. Then, the results at model scale are extrapolated to the full scale through a two-dimensional extrapolation method. Corrections for scale effects are made by following the recommended procedures of the ITTC1978 method. Correction factors for predictions of full-scale ship performance, including the roughness allowance and correction factors for delivered power and revolution rate of the propeller, are reasonably chosen based on experience from model tests at MARIC and sea trials. Based on the obtained results, the ship speed at EEXI draft to be expected on full-scale sea trials in deep unrestricted sea water of 15 degrees Celsius with no effects of wind and waves at 75%MCR with an estimated shaft transmission efficiency of 0.985 is predicted at last. Based on the ship speed on sea trials for a ship of similar main particulars, a roughly estimated ship speed of 18kn for the target ship is first considered. To reduce the computational cost, at least three ship speeds close to this estimated value are chosen for CFD analysis. Meanwhile, the estimated ship speed is within the range of the selected ship speeds so that the curves of delivered power versus ship speed to be analyzed hereinafter are obtained through interpolation of CFD results at different ship speeds. The considered ship speeds for analysis are V_S = 17.0kn, 18.0kn, and 19.0kn.

3. Methodology

Under the guidance of ITTC recommended procedures, CFD-based numerical methods for viscous flow around ship-propeller-rudder system at model scale are established. All the numerical simulations are carried out at the model scale in deep unrestricted fresh water of 15 degrees Celsius with no effects of wind and waves. Numerical simulations are carried out as the physical model tests in the towing tank. In other words, a ship is modeled as it is towed at a certain forward speed, and it is free to heave and pitch but fixed in all the other directions. The open-water propeller is modeled in a similar configuration to that of the propeller model tests in the towing tank. In other words, the inflow reaches the dummy hub first. But, in the numerical simulations, the propeller shaft is extended long enough in the downstream direction. Hereinafter, elaboration is given to the numerical method and modeling, including the numerical solver, algorithm and models, numerical modeling of motion, computational domain and boundary conditions, mesh generation, and numerical configurations of the solver.

3.1. Numerical solver, algorithm and models

To carry out the numerical simulations of viscous flow fields around the ship and the propeller, a steady RANS method is adopted for open-water propeller simulations while an unsteady RANS method is adopted for ship-propeller-rudder simulations. The CFD package Simcenter STAR-CCM+ (version 16.06.008) is used as the RANS solver. SST k-ω turbulence model is adopted for the closure of the governing equations, which is widely used and proves to be suitable for modeling viscous flows around container ships and propellers through validation of benchmark. The finite volume method is used to discretize the governing equations. A segregated solver based on the SIMPLE algorithm is applied to solve the pressure-velocity coupling equations. A second-order upwind scheme is used to discretize the convection terms, and a central-differencing scheme is used to discretize the diffusion terms. A first-order Euler implicit scheme is used for temporal discretization. The Volume of Fluid (VOF) multiphase model is used to capture the free surface, and the High-Resolution Interface Capturing (HRIC) scheme is used for the convection terms to capture a sharp interface.

3.2. Modelling and solving of motion

The principle of relative velocity is adopted to account for the effect of the ship forward speed and propeller advance speed. To be specific, an additional velocity equal to the negative ship (or propeller) forward speed is added to the velocity field of fluids. For solving ship motions, an equilibrium method in STAR-CCM+ software is adopted to quickly achieve a quasi-steady-state equilibrium position of the ship that is subject to fluid forces. By this method, the equation of motion is not solved, so there is no true physical motion of the ship. Instead, the ship is positioned and oriented in the flow in a stepwise manner so that the forces and moments acting on it are balanced in the chosen directions according to a purely numerical iterative procedure. The body mass only plays a role for the gravity force to be balanced to zero in the equilibrium state.

For the modeling of ship motion, deforming mesh technique based on the dynamic mesh method is adopted. Gravity force is considered for the ship-propeller-rudder system as a free surface exists, but not for the open-water propeller. To account for the unsteady interaction between the propeller and the hull, a discretized propeller is modeled by the sliding mesh technique based on the dynamic mesh method. A rotating sub-domain surrounding the propeller is established and embedded into the static domain of the ship. A pair of internal interfaces join the two sub-domains for the exchange of the information of flow fields within the same physical continuum. The DFBI morphing motion is assigned to the static ship domain, while a rotation motion is assigned to the rotating sub-domain. As the ship is not fixed, the rotation motion of the propeller has to be defined under the reference of a ship coordinate system with its origin fixed at the ship's center of gravity and orientation the same as the ship. For open-water simulations, a discretized propeller is also modeled with the same method.

However, the rotation motion assigned to the rotating sub-domain is modeled by the MRF (Multiple Reference Frame) method in a steady manner. Besides, to be computationally efficient, a single-channel method is used. By this method, only the flow around one single propeller blade is numerically simulated. For open-water propeller simulations, the rotation rate is fixed, while the advance velocity is changed as in propeller model tests in the towing tank.

3.3. Computational domain and boundary conditions

Figure 2 shows the computational domain.

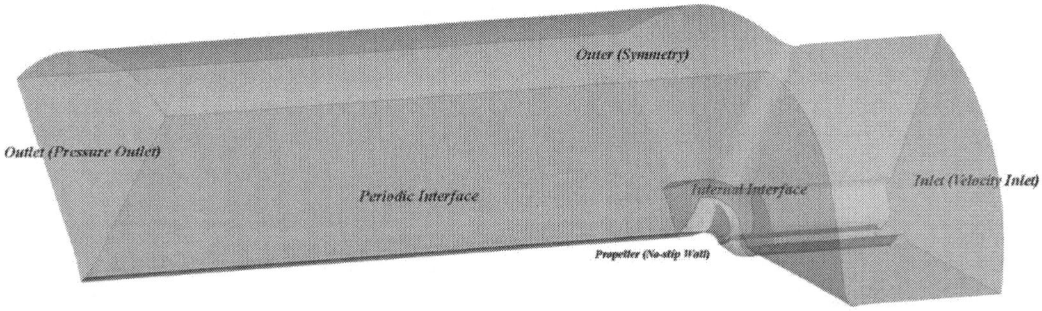

(a)

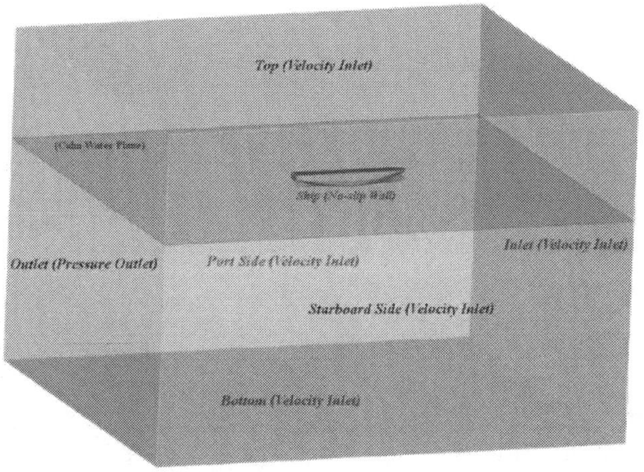

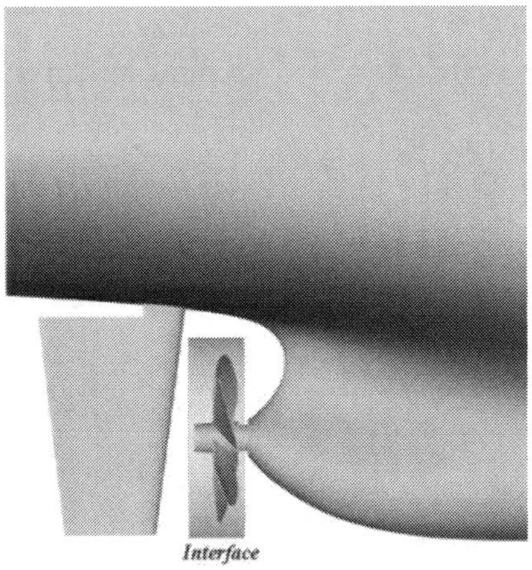

(b) (c)

Figure 2. Computational domain: (a) open-water propeller; (b) ship-propeller-rudder system (static sub-domain); (c) ship-propeller-rudder system (rotating sub-domain).

For simulations of the ship-propeller-rudder system, a rectangular domain is used as the static sub-domain, and a cylinder domain is used as the rotating sub-domain. For the static sub-domain, the upstream (inlet), the top, the bottom and the side boundaries are modeled as velocity inlets, where the volume fraction and velocity of water and air are specified; the downstream (outlet) boundary is modeled as pressure outlet, where hydrostatic pressure and volume fraction of water and air for calm water are specified; a no-slip wall boundary condition is set to the surface of the ship (including the hull, the rudder and the propeller). For simulations of the ship-propeller-rudder system, the whole ship is modeled, while for those of ship resistance, only half the ship is modeled given a symmetric flow field around the hull. For the latter, a symmetric boundary condition is applied to the longitudinal centerplane of the domain. For the pair of interfaces that join the two sub-domains, an internal interface boundary condition is specified. The size of the domain is large enough to ensure full development of flow fields near the ship. For the static sub-domain, the top and the bottom boundaries are $1.0L_{PP}$ above and $2.0L_{PP}$ below the undisturbed free surface, respectively; the side boundaries have a lateral distance of $2.0L_{PP}$ from the ship's longitudinal centerplane; the inlet and the outlet boundaries are $1.5L_{PP}$ ahead of and $2.0L_{PP}$ behind the ship, respectively. To reduce wave oscillation that may occur at the boundaries, wave damping based on a sponge layer boundary technique is applied to the vicinity of the inlet, the outlet and the side boundaries within a zone of $1.25L_{PP}$ in width. For the rotating sub-domain, the diameter of the cylinder is about $1.20D_M$. For simulations of the open-water propeller, cylinder domains are used for both the static and rotating sub-domains. As only the flows around one single blade are numerically simulated, a neutral surface between two adjacent blades is established and used to split both sub-domains into five equally divided portions of each. For the static sub-domain, the upstream (inlet) boundary is modeled as a velocity inlet boundary, where velocities of water at different advance ratios are specified; the downstream (outlet) boundary is modeled as a pressure outlet, where hydrostatic pressure equal to zero is specified; the outer boundary is a symmetry boundary; The two side boundaries that coincide with the neutral surfaces are set as periodic interfaces. For the pair of interfaces that join the static and rotating sub-domains, an internal interface boundary condition is specified. The wall surfaces of the propeller, including the blades, hub and shaft are no-slip walls. The size of the static sub-domain is about $2.5D_M$ upstream and $10D_M$ downstream from the propeller plane respectively, and $6.0D_M$ in diameter. That of the rotating sub-domain is about $0.5D_M$ upstream and $0.5D_M$ downstream from the propeller plane respectively and $1.5D_M$ in diameter.

3.4. Mesh generation

When the finite volume method is used to discretize the governing equations, the computational domain has to be discretized spatially. Figure 3 and figure 4 illustrate the generated meshes.

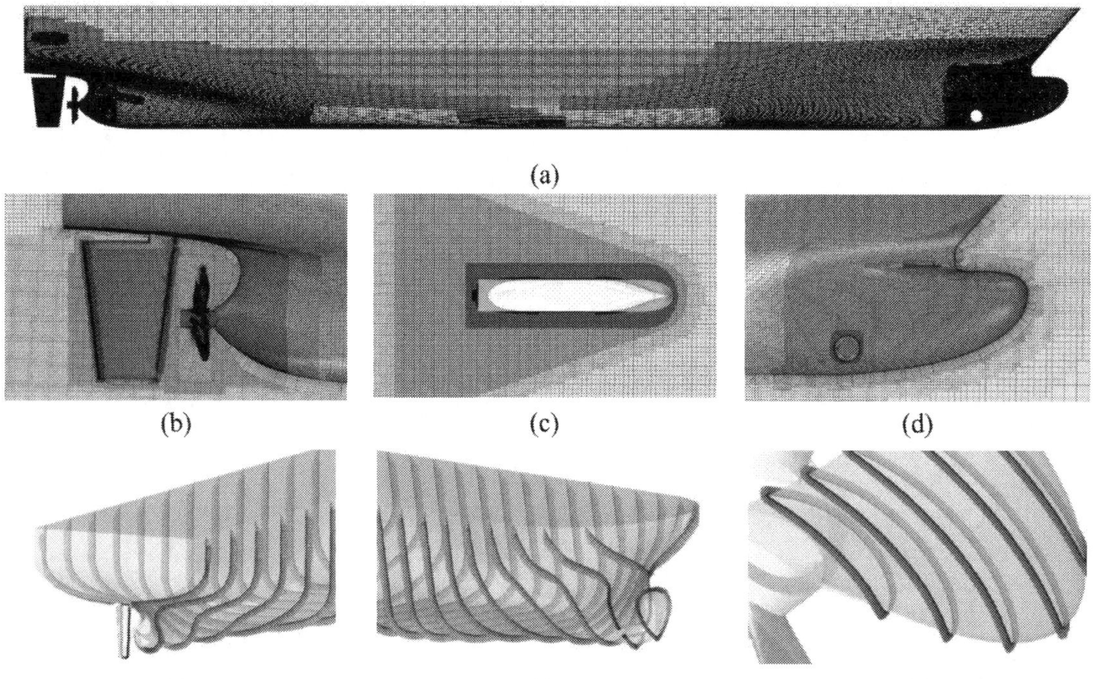

Figure 3. Computational mesh for ship-propeller-rudder system: (a) hull; (b) stern refinement; (c) free surface refinement; (d) bow refinement; (e) boundary layer mesh (stern); (f) boundary layer mesh (bow); (g) boundary layer mesh (blade).

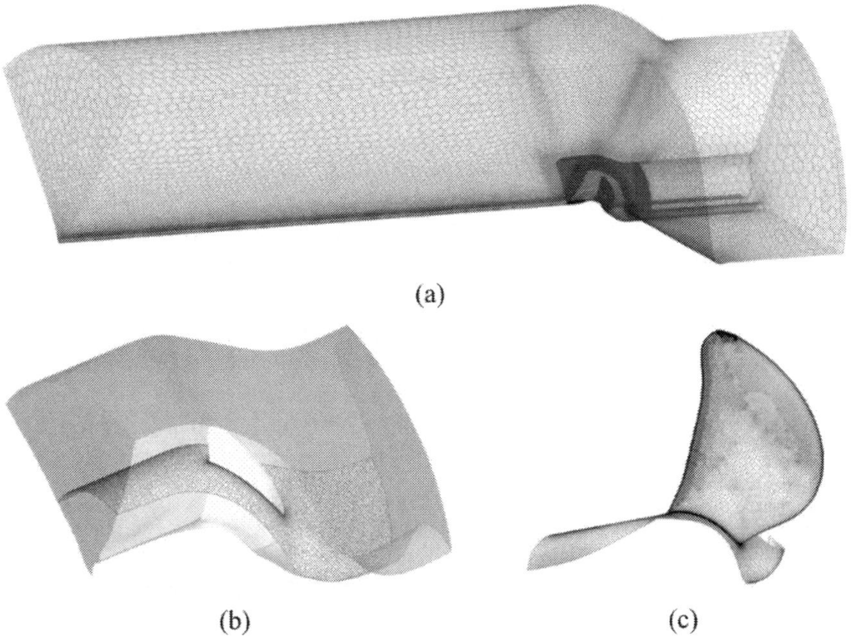

Figure 4. Computational mesh for open-water propeller: (a) static sub-domain; (b) rotating sub-domain; (c) blade.

Unstructured hexahedral cells with polyhedral trimmed cells adjacent to the hull surface are generated for the static ship domain. Surface mesh for the bow and stern areas is refined, and that for the bulbous and bow-thruster tunnel is further refined by an extra level. Progressive volumetric mesh refinements are applied in the vicinity of the hull, near the expected free surface, and in the wake field of ship and propeller. To accurately capture the ship-generated wave patterns at a certain forward speed, approximately 20 cells are used in the vertical direction where the wave profile along the hull is expected and a minimum of 40 cells per wavelength are used to resolve the ship-generated waves, and the number of cells per wavelength is doubled in the vicinity of the hull. Orthogonal prismatic cells next to wall surfaces are generated for solving boundary layer flows for the hull and the rudder. The size of the first layer adjacent to the hull and the rudder gives a non-dimensional value of wall distance around an average value of 45, making sure that the wall functions for the log-law layer are valid. For the rotating sub-domain, a quad-dominant surface mesh is generated for the propeller blades. Additional edge refinements are applied to the leading and trailing edges, and surface mesh refinements are applied to the blade tips and blade surface at r/R>0.70 of the propeller, where r represents the radial coordinate. Polyhedral cells of high quality for complex geometric forms are then generated for the domain. Orthogonal prismatic cells next to wall surfaces are also generated for solving boundary layer flows for the propeller blades. The size of the first layer adjacent to the blades gives a non-dimensional value of wall distance of around unity so that the boundary layer flow can be resolved all the way to the wall cells with a sufficiently fine-layered mesh.

3.5. Numerical configuration of solver

Numerical simulations of ship resistance and ship-propeller-rudder are carried out in an unsteady manner. The time-step for ship resistance simulations is set to be $0.005L_{PP}\cdot V_M^{-1}$, which is smaller than the ITTC recommended value ($0.01L_{PP}\cdot V_M^{-1}$), where V_M is the model speed. Based on this time-step, the global convective Courant number and the convective Courant number on the free surface may not be strictly below unity, especially where the mesh adjacent to the hull is refined by several levels. Moreover, due to the nature of unstructured cells, a Courant number above unity is always expected where the boundary layer cells intersect the free surface. However, as an implicit algorithm is used, no strict requirement of the global convective Courant number for stability has to be followed. On the other hand, to maintain a sharp interface as time advances, a pure HRIC convection scheme is adopted so that the method does not fall back to the first-order upwind scheme to sustain stability which causes an irreversibly smeared interface where the local Courant number exceeds the limitation. For the ship-propeller-rudder simulations, to reduce the CPU time required for full convergence of flow fields, the rotation of the propeller is first modeled in a steady manner by MRF method at a time-step of $1/(Z \cdot N_M)$ (N_M is the revolution rate in $r \cdot s^{-1}$) to obtain a rough estimate of the converged flow fields and then switched to an unsteady manner based on the sliding mesh method with a time-step of $1/(180N_M)$ as the time advances once the propeller rotates two degrees, which is within the ITTC recommended range. As this time-step is much smaller than the convection time scale for the ship ($L_{PP} \cdot V_M^{-1}$), relatively small under-relaxation factors of the solver are set to improve the numerical stability. The specific values for velocity, pressure, VOF, and turbulence solvers are 0.6, 0.2, 0.7 and 0.6, respectively. 15 inner iterations for each unsteady time step are used, which is found to be sufficient for the convergence of concerned variables within one time step, including the ship resistance, propeller thrust, and torque. A maximum of 50 inner time steps between two sequent updates of the ship's position and orientation is specified for the use of the equilibrium method for motion. The convergence tolerance of forces and moments for equilibrium is set to be 0.01. A maximum physical time of $50L_{PP} \cdot V_M^{-1}$ is specified, which is long enough for the flows to reach a converged state.

4. Results and discussions

Figure 5 shows the distribution of non-dimensional wall distance (y^+) on the hull and propeller for V_S=17.0kn at an instant for ship-propeller-rudder simulations. It is clear that y^+ is almost within the range of 30-100 for the hull, while y^+ is around unity for the propeller. The y^+ distributions for all the other cases are identical.

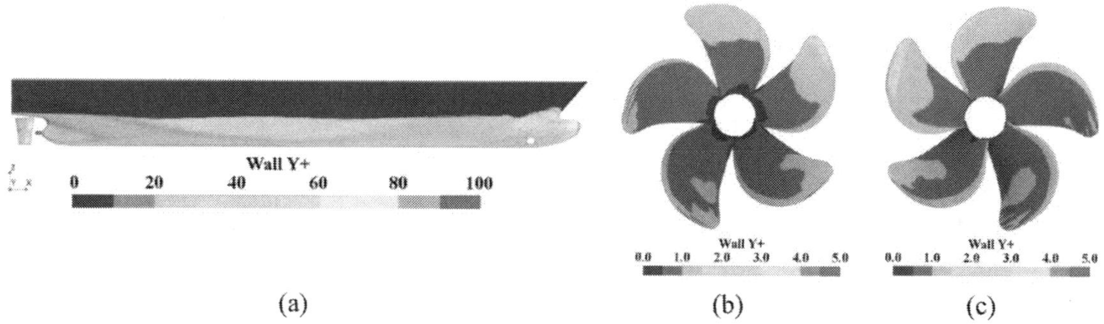

Figure 5. Distribution of non-dimensional wall distance: (a) hull and rudder; (b) propeller face; (c) propeller back.

4.1. Ship resistance

Numerical predictions of model resistance and full-scale extrapolated results are listed in table 4. To be consistent with the model test, the frictional resistance coefficient (C_{FM} for model scale and C_{FS} for full scale) is calculated by use of the ITTC 1957 formula instead of the numerically calculated one, and the results of C_R in table 4 are obtained by subtracting C_{FM} from the numerical results of C_{TM} at different ship speeds. Added resistance due to bilge keels and air resistance above the waterline is included in the total resistance at full scale, as in Appendix A.

To validate the numerical method, the numerical results are compared with the experimental results. As the water temperature in the model test is different from that of CFD calculations, the comparisons are made only for C_R and P_E. As listed in table 5, where EFD (Experimental Fluid Dynamics) stands for experimental results, the relative errors are quite small. It has to be mentioned that the experimental uncertainty of model resistance, though not evaluated for this particular ship but for similar containerships, is less than 1%, which indicates the reliability of EFD data. As for the numerical uncertainty mainly due to discretization errors and iteration errors, though not completely carried out for this particular ship, the spatial and temporal resolutions for discretization applied in our work are based on convergence studies for mesh size and time-step. On the other hand, the numerical method used in the current work has been already validated for benchmark ships (such as KCS, KVLCC2, etc.) and similar ships studied in our institute. Based on previous experience, the numerical uncertainty of model resistance is expected to be less than 1%.

Table 4. Numerical and extrapolated results of ship resistance.

V_S (kn)	Fn (-)	V_M (m·s^{-1})	R_{TM} (N)	C_{TM} (×10^{-3})	C_{FM} (×10^{-3})	C_R (×10^{-3})	ΔC_F (×10^{-3})	C_{FS} (×10^{-3})	C_{AA} (×10^{-3})	C_{TS} (×10^{-3})	P_E (kW)
17.0	0.241	1.820	43.954	3.963	3.039	0.924	0.20	1.533	0.151	2.833	3472.5
18.0	0.255	1.927	50.431	4.056	3.009	1.047	0.20	1.522	0.151	2.944	4285.0
19.0	0.269	2.034	60.928	4.398	2.980	1.417	0.20	1.512	0.151	3.304	5655.4

Table 5. Errors of ship resistance.

V_S (kn)	V_M (m·s^{-1})	C_R (×10^{-3}) CFD	EFD	Absolute Error	P_E (kW) CFD	EFD	Relative Error
17.0	1.820	0.924	0.930	-0.006	3472.5	3481.0	-0.24%
18.0	1.927	1.047	1.066	-0.019	4285.0	4313.6	-0.66%
19.0	2.034	1.417	1.365	0.022	5655.4	5619.2	0.64%

4.2. Open-water propeller characteristics

Numerical predictions of propeller open-water characteristics and full-scale extrapolated results are listed in table 6. Corrections for scale effects are made by following the recommended procedures of the ITTC1978 method, as in Appendix B. To validate the numerical method, the numerical results are compared with the experimental results. As blind calculations are carried out, the revolution rate of the propeller for CFD calculations is set to ensure that the Reynolds number of the propeller (evaluated at 0.70R) is larger than the critical Reynolds number. As the difference in the revolution rate of the propeller between CFD calculations and model tests is small, scale effects can be neglected. As listed in table 6, the largest relative error is found to be less than 4% for the coefficient of thrust, while the relative errors for the coefficients of torque and open-water efficiency are smaller than 3%.

Table 6. Results of open-water propeller characteristics and errors.

J	K_{TM} CFD	EFD	E%	$10K_{QM}$ CFD	EFD	E%	η_{0M} CFD	EFD	E%	K_{TS}	$10K_{QS}$	η_{0S}	K_{TS}/J^2
0.20	0.439	0.432	1.53%	0.619	0.604	2.37%	0.226	0.228	-0.86%	0.439	0.613	0.228	10.979
0.30	0.387	0.379	1.87%	0.551	0.539	2.19%	0.335	0.336	-0.37%	0.387	0.546	0.339	4.302
0.40	0.333	0.326	2.23%	0.483	0.474	1.87%	0.439	0.437	0.30%	0.333	0.477	0.445	2.084
0.50	0.279	0.271	2.86%	0.415	0.409	1.37%	0.535	0.528	1.41%	0.280	0.409	0.544	1.119
0.55	0.252	0.245	3.05%	0.380	0.377	0.74%	0.580	0.568	2.24%	0.253	0.375	0.591	0.836
0.60	0.225	0.219	2.64%	0.346	0.346	0.05%	0.619	0.604	2.54%	0.225	0.341	0.631	0.626
0.65	0.196	0.192	2.58%	0.311	0.312	-0.32%	0.653	0.635	2.86%	0.197	0.306	0.667	0.467
0.70	0.169	0.162	3.96%	0.277	0.275	0.71%	0.678	0.657	3.18%	0.169	0.272	0.694	0.346

4.3. Ship delivered power

Numerical predictions of propeller load at the ship's self-propulsion point, ship self-propulsion factors and full-scale extrapolated results are listed in table 7, table 8, and table 9, respectively. The correction factor for delivered power C_P and the correction factor for propeller rate of revolution C_N are both 1.0.

According to table 7, the numerically predicted propeller load at the ship's self-propulsion point at model scale is in quite good agreement with that of the model test. The relative errors of revolution rate and propeller thrust at model scale are below 1%, and that of propeller torque is no larger than 2%. According to table 8, the model-scale results show that the hull efficiency is underpredicted by about 3%; the propeller relative rotative efficiency is underpredicted by less than 1%; the propeller open-water efficiency at ship self-propulsion point is overpredicted by less than 5%. Although the individual efficiency is predicted not as accurately as ship resistance, the overall propulsive efficiency has a relative error of less than 2%. From the perspective of the ship delivered power, which is directly relevant to the ship effective power and overall propulsive efficiency, the results are in good agreement with experimental results. As can be seen in table 9, the relative error of the CFD-predicted ship delivered power is only 2%, and that of the propeller revolution rate is less than 0.5%. In general, the largest error is expected for propeller open-water efficiency at the ship's self-propulsion point at model scale, which can be explained by the overprediction of propeller efficiency in open-water conditions at relatively large advance ratios. From this point, for ship-propeller-rudder simulations, the most essential and difficult part of the numerical accuracy of ship delivered power is likely to be propeller-related. As the numerical accuracy of propeller open-water characteristics can be a guidance, validation of the numerical method for simulating viscous flow around the propeller in open-water conditions is quite essential before carrying out the numerical simulations of the ship-propeller-rudder interaction.

Table 7. Numerical results of propeller load at ship self-propulsion point and errors.

V_S (kn)	V_M (m·s^{-1})	F_D (N)	T_M (N) CFD	T_M (N) EFD	T_M (N) E%	Q_M (N·cm) CFD	Q_M (N·cm) EFD	Q_M (N·cm) E%	N_M (r·s^{-1}) CFD	N_M (r·s^{-1}) EFD	N_M (r·s^{-1}) E%
17.0	1.820	12.539	42.52	42.90	-0.88%	145.5	148.0	-1.67%	8.61	8.60	0.16%
18.0	1.927	13.820	49.35	49.50	-0.30%	169.1	171.0	-1.10%	9.23	9.20	0.37%
19.0	2.034	15.151	61.31	61.50	-0.32%	208.9	212.5	-1.69%	10.09	10.04	0.50%

Table 8. Numerical results of ship model self-propulsion factors and errors.

V_S (kn)	V_M (m·s^{-1})	t_M	w_M	η_{HM} CFD	η_{HM} EFD	η_{HM} E%	η_{RM} CFD	η_{RM} EFD	η_{RM} E%	η_{OM} CFD	η_{OM} EFD	η_{OM} E%	η_{DM} CFD	η_{DM} EFD	η_{DM} E%
17.0	1.820	0.261	0.357	1.150	1.175	-2.14%	1.016	1.025	-0.92%	0.622	0.595	4.46%	0.726	0.717	1.23%
18.0	1.927	0.258	0.354	1.148	1.188	-3.38%	1.012	1.022	-0.94%	0.619	0.592	4.50%	0.719	0.719	0.00%
19.0	2.034	0.253	0.349	1.148	1.179	-2.65%	1.011	1.012	-0.14%	0.606	0.578	4.86%	0.703	0.690	1.89%

Table 9. Extrapolated results of ship self-propulsion factors and errors.

V_S (kn)	t_S	w_S	η_{HS}	η_{RS}	η_{OS}	η_{DS}	P_{DT} (kW) CFD	P_{DT} (kW) EFD	P_{DT} (kW) E%	N_T (r·min^{-1}) CFD	N_T (r·min^{-1}) EFD	N_T (r·min^{-1}) E%
17.0	0.261	0.330	1.102	1.016	0.646	0.723	4804.8	4916.8	-2.28%	109.5	109.7	-0.22%
18.0	0.258	0.326	1.101	1.012	0.643	0.717	5980.3	6095.3	-1.89%	117.3	117.6	-0.28%
19.0	0.253	0.322	1.101	1.011	0.630	0.701	8063.9	8235.1	-2.08%	128.0	128.0	0.03%

Figure 6 and figure 7 present the non-dimensional free surface elevation and the distribution of non-dimensional hydrodynamic pressure on the ship model for different ship speeds. The waves and pressure distributions on different sides of the ship near the stern area are no longer symmetrical about the ship's longitudinal centerplane due to the rotation of the propeller.

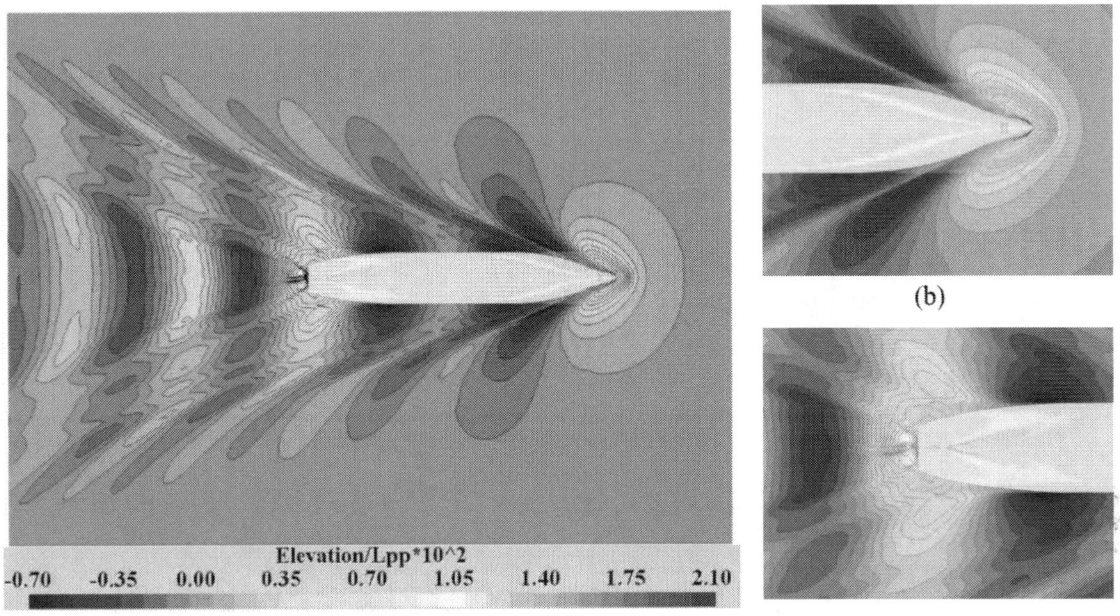

Figure 6. Non-dimensional free surface elevation (V_S=18.0kn): (a) general view; (b) bow; (c) stern.

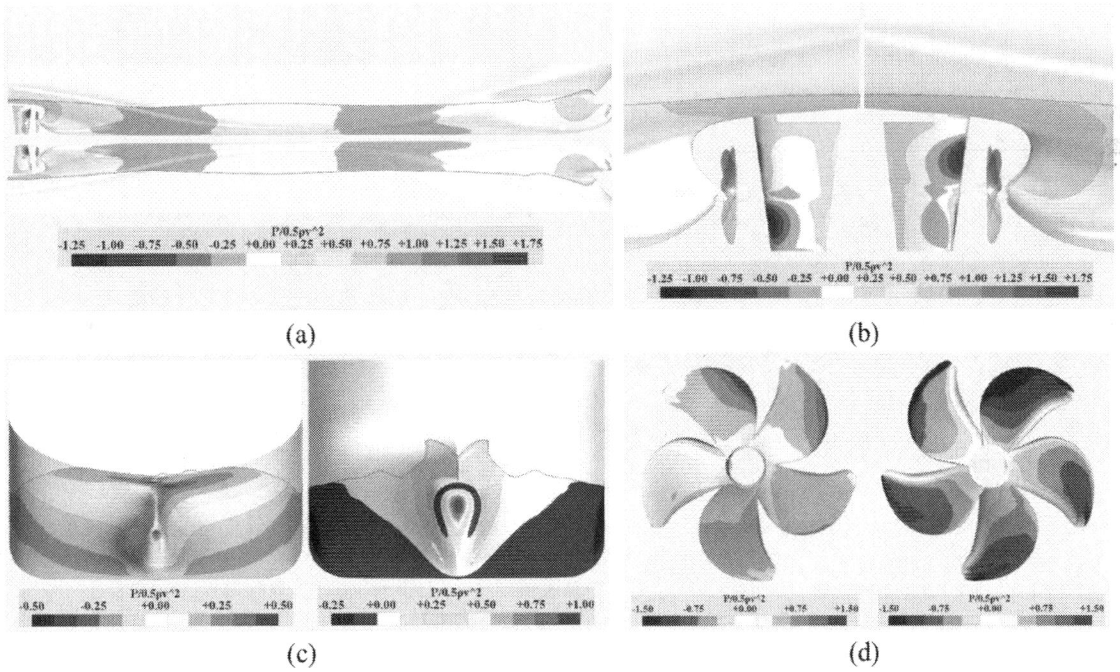

Figure 7. Distribution of non-dimensional hydrodynamic pressure (V_S=18.0kn): (a) general view; (b) rudder; (c) bow and stern; (d) propeller.

4.4. Evaluation of ship speed

Based on the extrapolated results hereinbefore, the ship speed at EEXI draft to be expected on full-scale sea trials in deep unrestricted water of 15 degrees Celsius with no effects of wind and waves at 75%MCR with an estimated shaft transmission efficiency of 0.985 is evaluated. The curves of ship delivered power and propeller revolution rate versus ship speed are presented in figure 8. The results at V_S=17.5kn and 18.5kn in these figures are obtained by interpolation through those at V_S=17.0kn, 18.0kn and 19.0kn. The prediction by model test results is also shown in figure 8 for comparison. The same extrapolation coefficients are applied for predictions based on CFD and EFD results.

From figure 8, with a delivered power of 6449kW, the ship speed is predicted to be 18.27kn by the current CFD method, and the corresponding revolution rate of the propeller is 119.9r·min^{-1}, while the model test predicts a ship speed of 18.22kn, and the corresponding revolution rate of the propeller is 119.5r·min^{-1}. The absolute difference in the predicted ship speed is only 0.05kn, which shows the high accuracy of the CFD method in the present work.

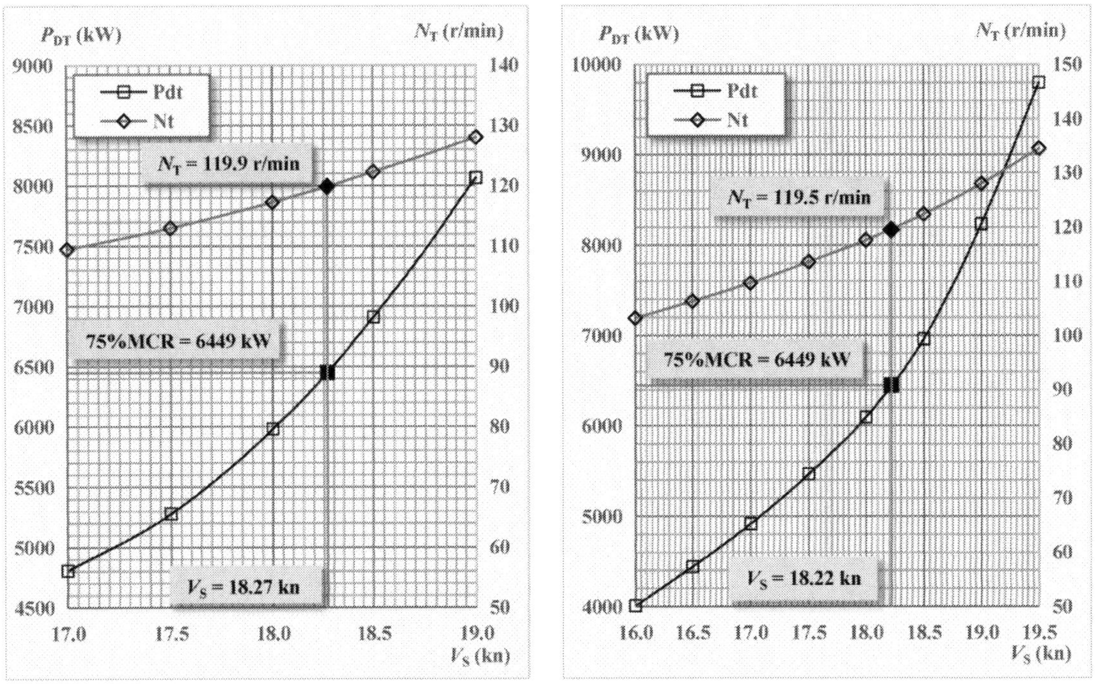

(a) (b)

Figure 8. Curves of delivered power and propeller's revolution rate versus ship speed: (a) CFD predictions; (b) model test predictions.

5. Conclusions

A fully CFD-based numerical evaluation of ship speed for an existing containership at EEXI draft is carried out. The evaluation is divided into two steps namely (a) the numerical prediction of ship performance at model scale, and (b) the extrapolation of results to full scale.

In the first step, numerical predictions of ship resistance in calm water, propeller characteristics in open water and ship self-propulsion factors in calm water at model scale are first carried out. The corresponding numerical results are compared with the model test results to validate the numerical method. For ship resistance, the relative error for ship effective power and the absolute error for the coefficient of residual resistance are both small. For open-water propeller characteristics, the largest relative error is found to be less than 4% for the coefficient of thrust, while the relative errors for the coefficients of torque and open-water efficiency are smaller than 3%. For the propeller load at the ship's self-propulsion point at model scale, relative errors of revolution rate and propeller thrust are below 1%, and that of propeller torque is no larger than 2%. For ship self-propulsion factors at model scale, the hull efficiency is underpredicted by about 3%; the propeller relative rotative efficiency is underpredicted by less than 1%; the propeller open-water efficiency at ship self-propulsion point is overpredicted by less than 5%, but the overall propulsive efficiency has a relative error of less than 2%. For ship delivered power, the relative error is only 2%, and that of the propeller revolution rate is less than 0.5%. From the perspective of ship delivered power, the results are in good agreement with experimental results. In general, the largest error is expected for the propeller open-water efficiency at the ship's self-propulsion point at model scale, which can be explained by the overprediction of the propeller efficiency in open-water conditions at relatively large advance ratios.

Based on the numerical and extrapolated results, the ship speed at EEXI draft to be expected on full-scale sea trials in deep unrestricted water of 15 degrees Celsius with no effects of wind and waves at 75%MCR with an estimated shaft transmission efficiency of 0.985 is evaluated. When the same extrapolation coefficients are applied, the absolute difference in predicted ship speed based on CFD and EFD results is only 0.05kn, which shows the high accuracy of the CFD method in the present work.

Appendix A

Correlation method for ship resistance is described as below.

The numerical model results are extrapolated to full scale ship at 15 degrees Celsius of sea water by means of the two-dimensional method recommended by 15th ITTC. The frictional resistance coefficient is calculated by use of ITTC 1957 formula, as in equation (A.1), where Rn is the Reynolds number.

$$C_F = 0.075\left(\lg Rn\text{-}2\right)^{-2} \tag{A.1}$$

The total resistance coefficient of ship is calculated as in equation (A.2). Roughness allowance (ΔC_F) for two-dimensional extrapolation is based on experience from model tests and sea trials.

$$C_{TS} = \frac{S_W + S_{B.K.}}{S_W}\left(C_{FS} + \Delta C_F\right) + C_R + C_{AA} \tag{A.2}$$

Air resistance coefficient C_{AA} is calculated as in equation (A.3).

$$C_{AA} = 0.001 A_T S_W^{-1} \tag{A.3}$$

Ship effective power is calculated as in equation (A.4).

$$P_E = 0.5 \rho S_W V^3 C_{TS} \tag{A.4}$$

Appendix B

Correlation method for open-water propeller is described as below, where subscripts "M" and "S" stand for ship model and full-scale ship, respectively.

Thrust (T_M), torque (Q_M) and revolution rate (N_M) of propeller model are calculated in open water. The advance coefficient (J), thrust coefficient (K_{TM}), torque coefficient (K_{QM}) and open-water efficiency (η_{0M}) are expressed as in equation (B.1) to equation (B.4).

$$J = V_M\left(N_M \cdot D_M\right)^{-1} \tag{B.1}$$

$$K_{TM} = T_M\left(\rho_M \cdot N_M^2 \cdot D_M^4\right)^{-1} \tag{B.2}$$

$$K_{QM} = Q_M\left(\rho_M \cdot N_M^2 \cdot D_M^5\right)^{-1} \tag{B.3}$$

$$\eta_{0M} = \frac{J}{2\pi} \cdot \frac{K_{TM}}{K_{QM}} \tag{B.4}$$

The open water characteristics of full-scale propeller are calculated from the model characteristics as in equation (B.5) to equation (B.11), where the values of Rn are taken at $0.70R$ of the blade.

$$K_{TS} = K_{TM} - \Delta K_T \tag{B.5}$$

$$K_{QS} = K_{QM} - \Delta K_Q \tag{B.6}$$

$$\Delta K_T = -0.3\Delta C_D \left(\frac{P_{0.70R}}{D}\right)\left(\frac{C_{0.70R}}{D}\right) Z \tag{B.7}$$

$$\Delta K_Q = 0.25\Delta C_D \left(\frac{C_{0.70R}}{D}\right) Z \tag{B.8}$$

$$\Delta C_D = C_{DM} - C_{DS} \tag{B.9}$$

$$C_{DM} = 2\left(1 + 2\frac{t_{0.70R}}{C_{0.70R}}\right)\left(\frac{0.044}{Rn^{1/6}} - \frac{5}{Rn^{2/3}}\right) \tag{B.10}$$

$$C_{DS} = 2\left(1 + 2\frac{t_{0.70R}}{C_{0.70R}}\right)\left(1.89 + 1.62\lg\frac{C_{0.70R}}{k_P}\right), \quad k_P = 30.0\,\mu m \tag{B.11}$$

Appendix C

Correlation method for ship self-propulsion is described as below, where subscripts “M” and “S” stand for ship model and full-scale ship, respectively.

The model self-propulsion factors are obtained by the thrust identity method, i.e. assuming that the thrust coefficient of propeller model behind ship and in open water are the same.

The load of the full-scale propeller is obtained from equation (C.1).

$$\frac{K_T}{J^2} = \frac{C_{TS} \cdot S_W}{2D^2(1\text{-}t_S)(1\text{-}w_S)^2} \tag{C.1}$$

The wake fraction of full-scale ship (w_S) is calculated from the model wake fraction as in equation (C.2).

$$w_S = (t_M + 0.04) + (w_M - t_M - 0.04) \cdot \frac{(1+k)C_{FS} + \Delta C_F}{(1+k)C_{FM}}, \quad 1+k = 1.0 \tag{C.2}$$

The thrust deduction (t_S) and relative rotative efficiency (η_R) of full-scale ship are considered to be the same as those of model ship.

The rate of revolution (N_S), delivered power (P_D), effective power for trial prediction (P_{ET}), delivered power for trial prediction (P_{DT}), and revolution rate for trial prediction (N_T) are calculated as in equation (C.3) to equation (C.7), where C_P is the power correction factor considering the model-ship correlation.

$$N_S = V_S(1\text{-}w_S)J_S^{-1}D^{-1} \tag{C.3}$$

$$P_D = 2\pi N_S Q_S = 2\pi\rho_S N_S^3 D_S^5 K_{QOS}\eta_R^{-1} \tag{C.4}$$

$$P_{ET} = P_E \cdot C_P \tag{C.5}$$

$$P_{DT} = P_{D} \cdot C_{P} \tag{C.6}$$

$$N_{T} = N_{S} \cdot {C_{P}}^{1/3} \tag{C.7}$$

References

[1] IMO 2018 Initial IMO strategy on reduction of GHG emissions from ships *Resolution MEPC* **304** p 1
[2] Bayraktar M and Yuksel O 2023 *Ocean Engineering* **278** 114295
[3] Baldi F and Coraddu A 2022 *Sustainable Energy Systems on Ships* ed F Baldi *et al* (Amsterdam: Elsevier) p 513-17
[4] IMO 2021 Guidelines on the method of calculation of the attained Energy Efficiency Design Index (EEDI) for new ships *Resolution MEPC* **308** p 1
[5] Farkas A, Degiuli N, Martić I and Grlj C G 2022 *Journal of Cleaner Production* **374** 133915
[6] IMO 2021 Guidelines on the method of calculation of the attained Energy Efficiency Existing Ship Index (EEXI) *Resolution MEPC* **333** p 1
[7] Tezdogan T, Demirel Y K, Kellett P, Khorasanchi M, Incecik A and Turan O 2015 *Ocean Engineering* **97** 186-206

IOP Conf. Series: Materials Science and Engineering 1288 (2023) 012051 doi:10.1088/1757-899X/1288/1/012051

Numerical simulations of water-air-bubble mixed flows around hydrofoil and propeller

Tianyi Mao, Xinyi Li and Decheng Wan*

Computational Marine Hydrodynamics Lab (CMHL), School of Naval Architecture, Ocean and Civil Engineering, Shanghai Jiao Tong University, Shanghai, China
*Corresponding author: dcwan@sjtu.edu.cn

Abstract. Water-air-bubble mixed flow is a complex multiphase flow usually generated due to the intense interaction between the sailing ship and the free surface[1].A large number of bubbles scour down along the ship which gather around the propulsion system, making a significant effect on the hydrodynamic performance of propeller and hydrofoil. In this paper, the hydrodynamic performance of two-dimensional hydrofoil and three-dimensional propeller in uniformly mixed water-air-bubble incoming flow is studied by using the Computational Fluid Dynamics (CFD) method. Euler-Euler two-fluid model is used to simulate the uniformly mixed water-air-bubble incoming flow with the open-source CFD software OpenFOAM. The feasibility of numerical simulation is verified by comparing the numerical simulation results with experimental data. On this basis, the changes of physical fields around two-dimensional hydrofoil and three-dimensional propeller caused by water-air-bubble flow under multiple air fraction conditions are discussed. In addition, the differences in dimensionless coefficients are compared between single-phase flow conditions and two-phase flow. Furthermore, the Population Balance Model (PBM) is used in the simulation of two-dimensional hydrofoil to observe the coalescence and break of bubbles.

1. Introduction

Water-air-bubble mixed flow is a complex flow which will be generated when ships or marine engineering structures interact with the surrounding fluid. It usually involves air, water and a large number of bubbles with different sizes. In the water-air-bubble mixed flow, lots of bubbles will sweep down from bow to the stern. Also, the whirlpool turbulence in the stern of the ship will suck in a large amount of gas, which also produce water-air-bubble mixed flow. The bubbles cluster near the propulsion system, which leads to the unreliability of hydrodynamic performance, resistance performance and propulsion performance forecast. Under the condition that the similarity rule between the actual scale structures and models is still unclear, it is difficult to predict the hydrodynamic performance of marine structures or ships by experiments. Therefore, it is particularly important to adopt highly reliable numerical simulation methods to study the force situation of water-air-bubble mixed flow around the ships and analyze the effect.

Nowadays, the numerical simulation methods have been adopted to the water-air-bubble mixed flow. The numerical simulation methods can be divided into two types—Interface capture model and Non-interface capture model. Non-interface capture model mainly includes Euler-Lagrange method, and Euler-Euler method. And the Interface capture model mainly includes Volume of Fluid Method (VOF)[2], Level-Set method. Euler-Lagrange method defines bubbles as tiny particles in the flow[3].So it can track the bubbles through the Lagrange method and solve the two phase based on Euler method. The Euler-Lagrange method simulates collisions between bubbles. However, this method requires large amount of calculation which costs lots of time and computing resources. Euler-Euler method solves the two-phase flow problems by N-S equation based on Euler mesh. It views all the discrete bubbles in the flow as a continuous fluid and it can't analyse the movement of each bubble. Since the discrete phase is solved in Euler mesh, the velocity of different particles in the mesh is considered to be the same. So the Euler-Euler method only needs small amount of calculation, and gives the results quickly. In the traditional Euler-Euler method, we can't consider the coalescence and fragmentation of bubbles. So the PBM is introduced into the method to consider the coalescence and fragmentation of bubbles. Some studies apply PBM to investigate the skin-friction drag reduction on a flat plate with bubble injection[4]. It is difficult to use interface capture method or Euler-Lagrange method. This paper uses the Euler-Euler method to model the water-air-bubble mixed flow.

2. Simulation background

2.1. Governing Equations

In the Eulerian-Eulerian two-fluid solution framework, the governing equations including the mass conservation equation and the momentum conservation equation are solved separately for each phase. When simulating, the liquid phase is considered as the continuum phase and the gas phase is taken as the dispersed phase. In this paper, the two phases are both regarded as incompressible flows which are immiscible, and the heat transfer between phases is neglected.

The mass conservation equation and the momentum conservation equation are as follows :

$$\frac{\partial}{\partial t}(\alpha_i \rho_i) + \nabla \cdot (\alpha_i \rho_i \boldsymbol{u}_i) = 0 \tag{1}$$

$$\frac{\partial}{\partial t}(\alpha_i \rho_i \boldsymbol{u}_i) = -\nabla \cdot (\alpha_i \rho_i \boldsymbol{u}_i \boldsymbol{u}_i) - \alpha_i \nabla p - \nabla \cdot \left[\alpha_i \mu_i (\nabla \boldsymbol{u}_i + \nabla \boldsymbol{u}_i^T)\right] + \alpha_i \rho_i g + \boldsymbol{F}_{ij} \tag{2}$$

where the subscript i represent the phase, in which i = l or g means the water phase or air phase respectively. **u** is velocity, p is pressure, g is the gravity acceleration ρ is density, and α represents the volume fraction.

$\mathbf{F_{ij}}$ means the interphase force between gas and liquid phases, mainly used to represent the momentum transfer between phases, which is combined with various forces as follows:

$$\boldsymbol{F}_{ij} = \boldsymbol{F}_D + \boldsymbol{F}_L + \boldsymbol{F}_{WL} + \boldsymbol{F}_{TD} + \boldsymbol{F}_{VM} \tag{3}$$

$\mathbf{F_D}$ stands for drag force, which is caused by the relative motion between the gas and liquid phases. $\mathbf{F_L}$ stands for lift force, the lateral force caused by the pressure difference perpendicular to the direction of the bubble's motion. $\mathbf{F_{WL}}$ stands for wall lubrication force, which is generated by the slip velocity near the wall. $\mathbf{F_{TD}}$ is the turbulent diffusion force, and represents the traction force caused by liquid turbulence on the bubble. $\mathbf{F_{VM}}$ stands for virtual mass force is the one generated by bubbles' acceleration in the water. Here we only take the drag force and virtual force into account when simulating hydrofoil and all interphase forces are considered for propeller.

In this paper, F_D is calculated as:

$$\boldsymbol{F}_D = -\frac{1}{2} C_D \rho_l |\boldsymbol{u}_r| \boldsymbol{u}_r A_d \tag{4}$$

where C_D is the drag coefficient, ρ_l is liquid density, u_r is the relative velocity between two phases which can be calculated as $\boldsymbol{u}_g - \boldsymbol{u}_l$, and A_d is the projected area of the bubble.

Nowadays, drag models such as Schiller-Naumann model, Grace model, Ishii-Zuber model and Tomiyama model have been widely used in numerical simulation of water-air-bubble mixed flow, and have been implanted in commercial CFD software for users to choose. In this paper, we choose the drag force model proposed by Schiller[5]for the two dimensional hydrofoil and the drag force model proposed by Ishii[6] for the three dimensional propeller.

The virtual mass force F_{VM} can be calculated as:

$$\boldsymbol{F_{VM}} = C_{VM}\alpha_g\rho_l \cdot \left(\frac{D\boldsymbol{u_g}}{Dt} - \frac{D\boldsymbol{u_l}}{Dt}\right) \tag{5}$$

where the total derivate is defined as:

$$\frac{D}{Dt} = \frac{\partial}{\partial t} + \boldsymbol{u} \cdot \boldsymbol{\nabla} \tag{6}$$

C_{VM} is the coefficient of virtual mass force which is always be regarded as a constant in most related literatures. So in this paper, bubbles are considered as spheres and C_{VM} is taken as 0.5.

The lift force $\boldsymbol{F_L}$ is calculated as:

$$\boldsymbol{F_L} = -\boldsymbol{C_L\rho_l\alpha_g u_r} \times (\boldsymbol{\nabla} \times \boldsymbol{u_l}) \tag{7}$$

where C_L is the lift coefficient and similar to drag models, different lift models which mainly focuses on the lift coefficient can be choosen in CFD softwares. The lift model proposed by Tomiyama[7] is adopted here.

The wall lubrication force $\boldsymbol{F_{WL}}$ can be described as:

$$\boldsymbol{F_{WL}} = C_{WL}\frac{\rho_l\alpha_g}{d_b}\boldsymbol{u_{rw}^2 n_w} \tag{8}$$

where C_{WL} represents the wall lubrication force coefficient, u_{rw} represents the tangential component of relative velocity at the wall and $\boldsymbol{n_w}$ represents the external unit normal vector on the wall. The Antal wall lubrication force model[8] is used here.

For the turbulent diffusion force, current turbulent diffusion force model can be divided into two categories:Lopez de Bertodano model and Favre-averaged-drag model. The former supposes that the bubble motion approximates the thermal diffusion of air molecules in atmosphere and is applied here as follows:

$$\boldsymbol{F_{TD}} = -C_{TD}\rho_l k_l \nabla\alpha \tag{9}$$

where k_l represents the turbulent kinetic energy of the liquid phase. The turbulent diffusion coefficient C_{TD} usually ranges from 0.1 to 1.0, and it is a constant 1.0 here.

2.2. Population balance model

In the real environment, the merging and breaking of bubbles are very common, which leads to the multi-scale distribution of bubble size. Also because bubble size distribution will change with the interphase transfer phenomenon in the polyphase system, using the Population-Balance-Model to describe the distribution of bubbles of different sizes in the liquid phase has become a hot topic in the field of gas-liquid two-phase flow.

The transport equation considering bubble collapse and coalescence can be described as:

$$\frac{\partial}{\partial t}\big(n(V)\big) + \nabla \cdot \big(\boldsymbol{u}n(V)\big) + \nabla \cdot \big(G_v n(V)\big) = B_A + D_A + B_B + D_B \tag{10}$$

where G_v represents the growth rate of bubbles, B_A and D_A are the production and disappearance terms related to bubble merging.

$$B_A = \frac{1}{2}\int_0^V \Gamma_c(V - V', V')n(V - V')n(V')dV' \tag{11}$$

$$D_A = \int_0^\infty \Gamma_c(V, V')n(V)n(V')dV' \tag{12}$$

B_B and D_B are the production and disappearance terms related to bubble breaking, V represents the volume of sub-bubbles after crushing and n(V) is the bubble number density function of volume V.

$$B_B = \int_{\Omega_V} p\, g_b(V')\beta_b(V|V')n(V')dV' \tag{13}$$

$$D_B = g_b(V)n(V) \tag{14}$$

where V' represents the bubble volume before breakin, $g_b(V')$ is breaking frequency, $\beta_b(V|V')$ is the probability density function of bubble breakage, $\Gamma_c(V, V')$ is bubble coalescence rate.

The coalescing models generally require collision frequency and coalescing-efficiency-model to describe the coalescing frequency, and the crushing models are generally expressed as models of bubble-breaking frequency and sub-bubble size distribution.

Various coalescing-and-crushing models have been created to simulate the merging and breaking of bubbles. In this paper, the coalescing-and-crushing model proposed by Luo[9] is applied to the PBM.

3. Numerical simulation of hydrofoil

3.1. Calculation conditions

A two-dimensional NACA0012 hydrofoil profile is taken as the research object in this paper. The setting of velocity and angle of attack of the NACA0012 hydrofoil profile is based on the experiment of the same type of hydrofoil profile in single-phase incoming water flow conditions[10]. The geometric model of the hydrofoil profile is set as Figure 1 and the parameters are set as Table 1.

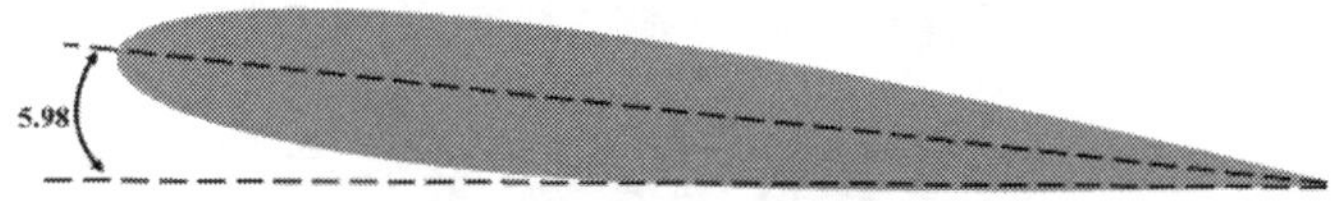

Figure 1. Geometric model of the hydrofoil profile.

Table 1. Parameters of calculation conditions.

Parameters	Unit	Value
Attack angle of hydrofoil	degree	5.98
Chord length of hydrofoil L	m	1
Velocity of incoming flow	m/s	2.24
Kinematic viscosity of water	m2/s	1.12×10^{-6}
Kinematic viscosity of air	m2/s	1.46×10^{-5}
Re		200 000

In the first model, the bubble is assumed to have a constant diameter. The diameter is set as 0.5 mm. The volume fraction of air is set as 0%, 5%, and 10%.

IOP Conf. Series: Materials Science and Engineering 1288 (2023) 012051 doi:10.1088/1757-899X/1288/1/012051

3.2. *Computational domain*

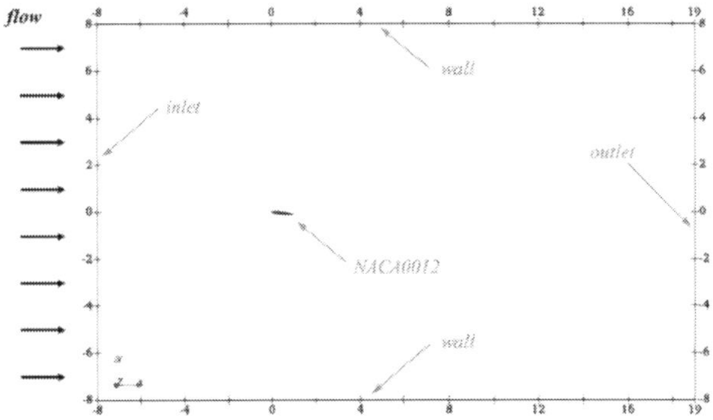

Figure 2. General layout of computational domain.

The general layout of the computational domain is set in Figure 2. This paper applies the oxy system. L is the chord length of the hydrofoil. The length of the domain is set from -8L to 19L and the direction is set along the x-axis. The inlet boundary is located at x=-8L. The outlet boundary is located at x=19L. The width of the domain is set from -8L to 8L and the direction is set along the y-axis. So the distance between the hydrofoil profile and the fixed boundary can be considered large enough to avoid the far-field effect.

The water-air-bubble mixed flow is generated at the inlet boundary and propagates along the x-positive direction.

3.3. *Mesh distributions*

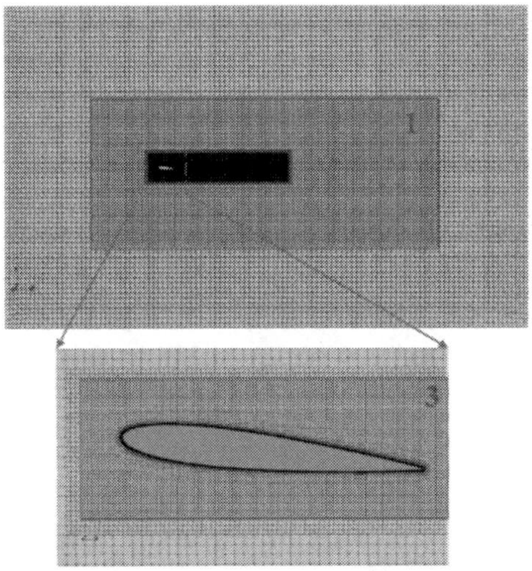

Figure 3. The mesh of computational domain.

This paper applies the mesh by the software Hexpress. The mesh distribution is set in Figure 3. The mesh division is to apply block densification near the hydrofoil profile based on the background grid. Two boxes are applied, as shown in the Figure 3, to densify the background mesh at level 1 and level 3 in the specific location.

The layer thickness of the boundary layer of the hydrofoil profile is limited to $y^+<10$, and the layer thickness of the first boundary layer is 0.00012 m. The total number of meshes is 72885.

The timestep is calculated through Courant number (Co) value as follows:

$$Co = \frac{\Delta t|\boldsymbol{U}|}{\Delta x} < 1 \tag{15}$$

where Δt is the timestep size, $|\mathbf{U}|$ is the modulus of velocity, and Δx is the length of the cell along the velocity direction.

3.4. Boundary conditions

The incoming flow is set at the inlet boundary. The velocity boundary condition is set as $\boldsymbol{u}_l = \boldsymbol{u}_g = \boldsymbol{u}_\infty$. The fixed value boundary condition is applied for α_g and α_l. The pressure condition is set according to the fixedFluxPressure condition in OpenFoam. The turbulent kinetic energy k, turbulence specific dissipation rate ω and the turbulence viscosity v_t is calculated as in equation (16-18).

$$k = \frac{3}{2}(I|\boldsymbol{u}|)^2 \tag{16}$$

$$\omega = \frac{k^{0.5}}{C_\mu^{0.25} l} \tag{17}$$

$$v_t = \frac{a_1 k}{max(a_1\omega, SF_2)} \tag{18}$$

where I is the turbulence intensity, C_μ and a_1 are empirical constants. ($C_\mu = 0.09$, $a_1 = 0.31$), and l is the turbulent reference length scale.

At the outlet boundary, the velocity boundary applies the zero normal gradient condition. The pressure condition is set at fixed value. The inlet and Outlet boundary condition is applied for α_g, α_l,k, and ω. The turbulence viscosity v_t is calculated as in equation (18).

At the hydrofoil boundary, the velocity boundary applies no-slip boundary condition. And the pressure condition applies the zero normal gradient condition. It also goes for volume fraction and ω. For k and v_t.

3.5. Model validation

This paper validates the numerical model in single-phase incoming water flow conditions. The numerical results of lift and drag coefficients are compared with experimental results. The lift and drag coefficients are calculated through the following formulas.

$$C_L = \frac{F_{lift}}{\frac{1}{2}\rho_l \boldsymbol{u}_\infty{}^2 L} \tag{19}$$

$$C_D = \frac{F_{drag}}{\frac{1}{2}\rho_l u_\infty{}^2 L} \tag{20}$$

where F_{lift} and F_{drag} are lift and drag forces of the hydrofoil profile.

Table 2. Comparison of C_L and C_D between experiment and numerical model.

Condition	CL	Cd
Experimental results	0.6084	0.0134
Pure water environment	0.6338	0.0130
Error	2.54%	2.99%

In Table 2, the lift and drag coefficients calculated in this paper are compared with the experimental results. The error is less than 3%. Based on the table, the numerical model in this paper conforms to the experimental results. So the model is validated.

4. Numerical simulation of propeller

4.1. Geometric Model and Computational Domain

The geometry of the propeller is the KP505 propeller model. The propeller diameter D is 0.25m, the rotate speed of the propeller is 9.5 r/s. and other geometric features are shown in the Figure 4.

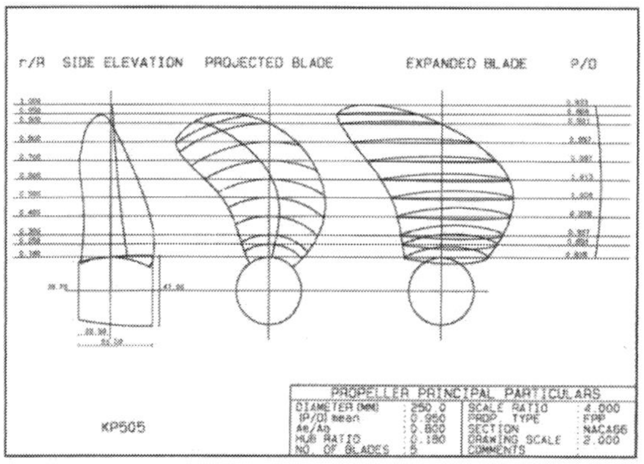

Figure 4. The Geometric model of propeller.

The general layout of the computational domain is shown in Figure 5. To simulate an open deep water environment, this paper takes the computational domain as a box with a cross-section of a square whose side length is equal to 4D, where D means the propeller diameter. The length of this rectangular body is 8D, and the distance from the inlet and outlet to the propeller are both 4D which is far enough to prevent unexpected disturbance from the boundary. In this field, the Cartesian coordinate o-xyz system is adopted and the origin is set at the center of the propeller. The x-axis is along the length of propeller, while the y and the z-axes both are along the radial direction of the propeller.

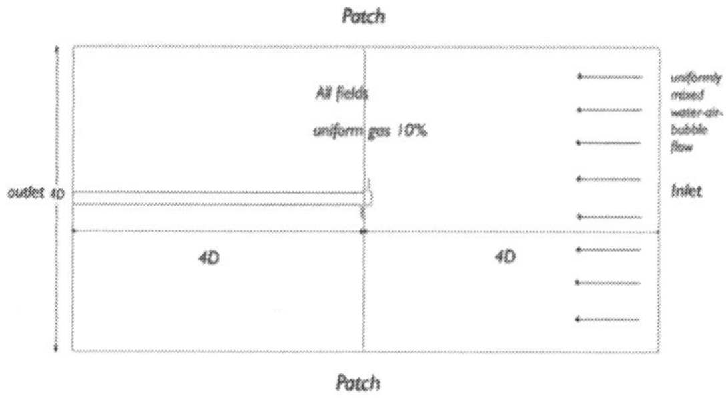

Figure 5. General layout of computational domain.

In this paper, the bubble in water-air-mixed flow is assumed to be spherical and has a constant diameter. $d = 0.1\text{mm}$. The coalescence and break-up of bubbles are not considered in this paper. The volume fraction of air is set as 10% and the volume fraction of water is set as 90% at the inlet and in the computational domain. At the initial moment, the water-air-mixed-flow was assumed that it has already filled the entire computational domain which means the propeller starts rotating in a uniformly mixed water-air-flow to improve the stability of the simulation and to reduce time costs. And the uniformly mixed water-air-bubble incoming flow also is generated at the inlet where x=-4D and flows forward along the x-positive direction.

4.2. Mesh Distributions and time step

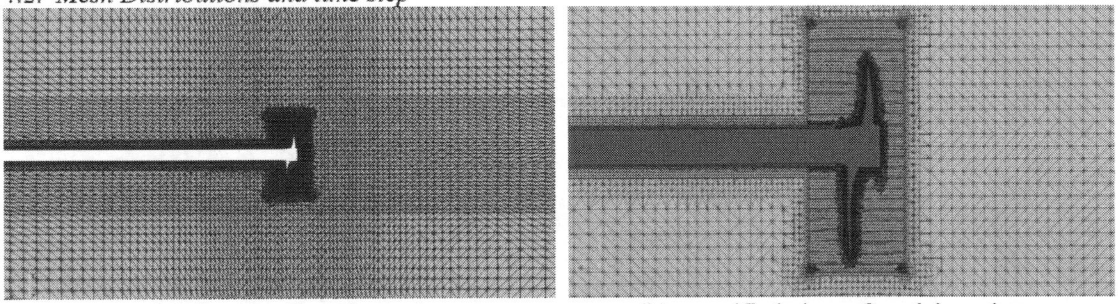

(a) the overall meshes. (b) magnified view of partial meshes.

Figure 6. The mesh of computational domain.

In this paper, the mesh is generated by the preprocessing code snappyHexMesh in OpenFOAM, which is in the form of unstructured mesh. The overall mesh distribution are shown in Figure 6(a) and partial meshes around the KP505 propeller is shown in Figure 6(b).

The background mesh is a gradient mesh with a gradual change ratio of 4 along x-axis and 2 along the y and z-axis. The mesh around the propeller and in the rotating area is much finer than that elsewhere not only for capturing the details of the propeller motion but also to save computational resources. The first layer thickness of the boundary layer near the propeller is about 0.0001m and the stretching ratio is 1.4. The y^+ ranges from 30 to 70. The total number of mesh is 982171.

The timestep size is determined based on a prescribed Courant number (Co) value as follows:

$$Co = \frac{\boldsymbol{u}\Delta t}{\Delta x} \leq 1.5 \tag{21}$$

where Δt is the timestep size, u is the normal velocity, and Δx is the distance between the cell centre and the centre of its neighbour cell.

4.3. Boundary conditions

At the inlet, $\boldsymbol{u}_g$ and $\boldsymbol{u}_w$ use the fixed value condition which equals 1.665m/s, α_g and α_l also take the fixed value condition which is 0.1 and 0.9 respectively. For the pressure, the fixedFluxPressure condition is adopted. The initial value of the turbulent kinetic energy k and turbulence-specific dissipation rate ω are calculated as follows:

$$k = 1.5(I|\boldsymbol{u}|)^2 \tag{22}$$

$$\omega = \frac{k^{0.5}}{C_\mu^{0.25} l} \tag{23}$$

where I means turbulence intensity, C_μ is a constant equals to 0.09 and l is the turbulent reference length scale.

At the outlet, the fixed value which equals 0 is used for the pressure boundary condition, and the zerogradient condition is applied to velocity and volume fraction. The inletOutlet boundary condition is applied to α_g, α_l, k and ω.

On the propeller boundary, moving wall velocity condition is used for $\boldsymbol{u_l}$ and $\boldsymbol{u_g}$ to adapt the rotation of the propeller awe, and no-slip boundary condition is used for the propeller shaft. The zero normal gradient condition is adopted for pressure, volume fraction and the wall function method is for k and ω .

On the patch boundary, fixed value boundary condition, zero gradient condition is used for all scalars and vectors.

In this paper, the rotation of the propeller is achieved through sliding mesh called AMI, the boundary for the AMI is cyclicAMI applied to all scalars and vectors.

4.4. *Validation and verification*

In this research, simulation of the propeller is conveyed under the condition that advance ratio J=0.7 and is validated in single-phase incoming water flow conditions. The numerical results about thrust coefficient K_T, moment coefficient K_Q and open water efficiency $\eta_0$are compared with experimental data.

Advance ratio J, thrust coefficient K_T, moment coefficient $\mathrm{K_Q}$ and openwater efficiency η_0 are shown as follows:

$$J = \frac{V_A}{Dn} \tag{24}$$

$$K_T = \frac{T}{\rho n^2 D^4} \tag{25}$$

$$K_Q = \frac{Q}{\rho n^2 D^5} \tag{26}$$

$$\eta = \frac{J}{2\pi} \frac{K_T}{K_Q} \tag{27}$$

where n is propeller rotating speed, T is thrust, Q is moment, D is the propeller diameter, V_A is incoming flow velocity and ρ is the density of water.

Table 3. Comparison between Numerical results and Experimental results.

Results	K_T	$10K_Q$	η_0
Numerical results	0.1897	0.3008	0.703
Experimental results	0.1850	0.3110	0.665
Error between Numerical results and Experimental results	2.54%	3.28%	5.65%

From the Table 3, the error between numerical results and experimental results are all below 6%, which is acceptable for propeller numerical simulation. On the basis of the comparison, the numerical model established in this paper agrees with the experimental results.

5. Results and discussion

5.1. *Results and discussion of hydrofoil*

5.1.1. *Comparison of flow fields*

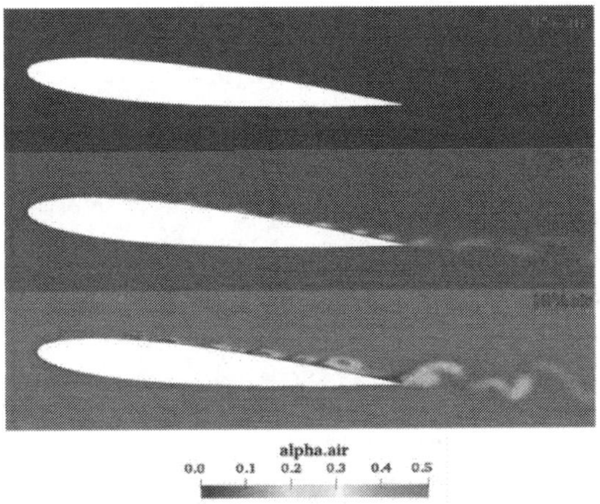

Figure 7. Air volume fraction distribution and LIC streamline distribution near the hydrofoil.

Figure 7 shows the air volume fraction, near the hydrofoil in water-air-bubble mixed flow conditions at different α_g. The LIC is also adopted to visualize the streamline. From the fig, the distribution of air volume is different on the suction surface and the pressure surface. The suction surface of the hydrofoil has higher air volume fraction and the change is more distinct than the pressure surface.

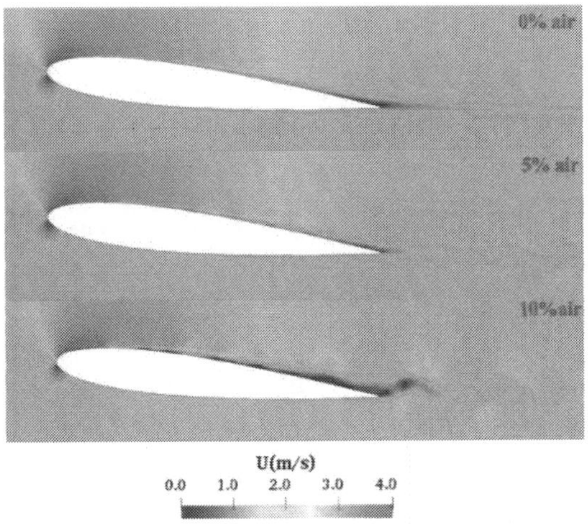

Figure 8. The velocity field near the hydrofoil.

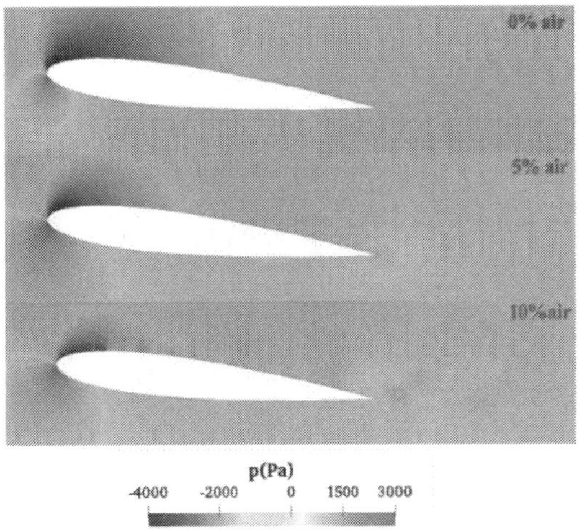

Figure 9. The pressure field near the hydrofoil.

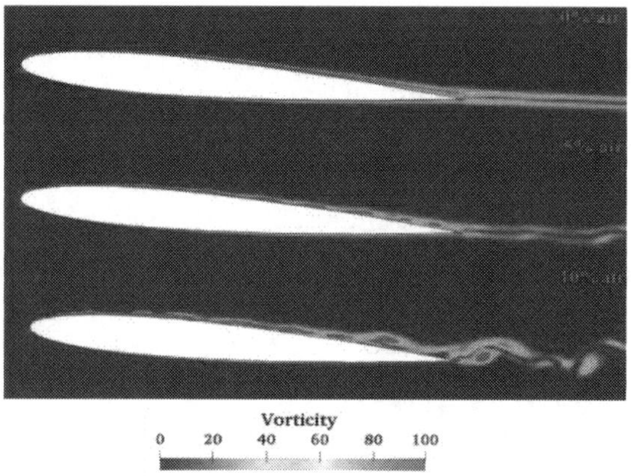

Figure 10. The vortex structure near the hydrofoil.

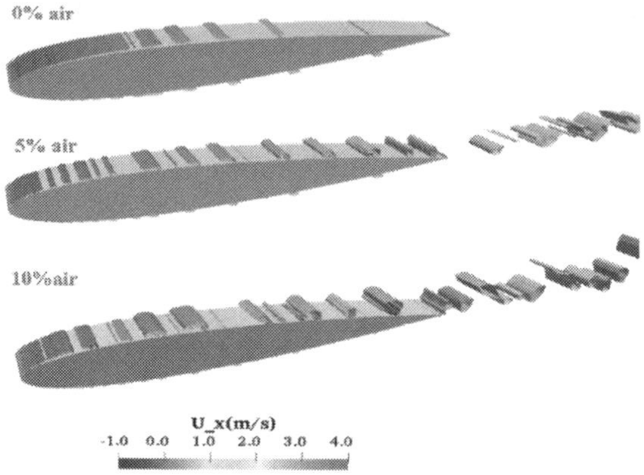

Figure 11. The vorticity near the hydrofoil.

Figure 11 shows the vorticity near the hydrofoil. Figure 10 shows the vortex structure near hydrofoil extracted from the third generation vortex recognition method Liutex The vortexes are generated and propagated along with higher air volume fraction. Also, the phenomenon of vortexes is more obvious in the condition of water-air-bubble mixed flow.

5.1.2. Comparison of drag and lift coefficients

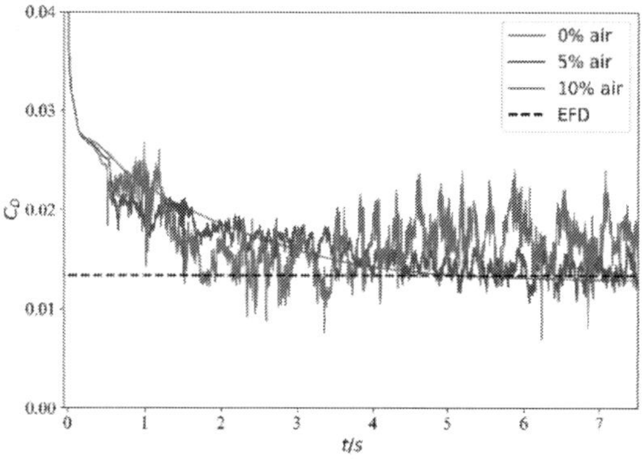

Figure 12. Comparison of time history of lift coefficient C_D.

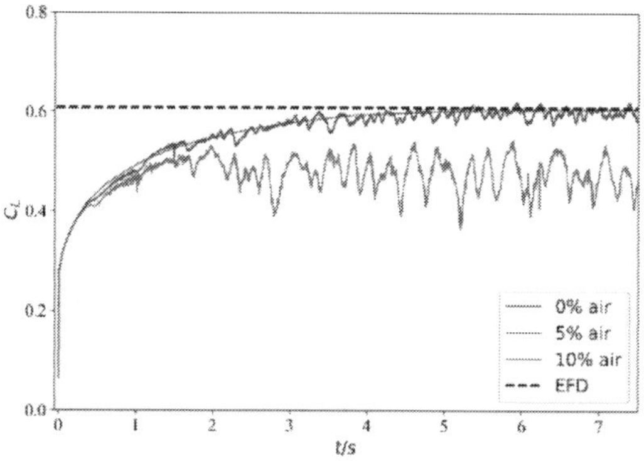

Figure 13. Comparison of time history of lift coefficient C_L.

Table 4. Comparison of average C_L and C_D.

Condition	C_L	C_D
0% air in incoming flow	0.6338	0.0130
5% air in incoming flow	0.5884	0.0142
10% air in incoming flow	0.4744	0.01733

This paper calculates the average C_L and C_D from t=5s to t=8s. Table 4 shows the results. One can see that the lift coefficient decreases and the drag coefficient increases with the increasing of α_g.

5.1.3. Population balance model

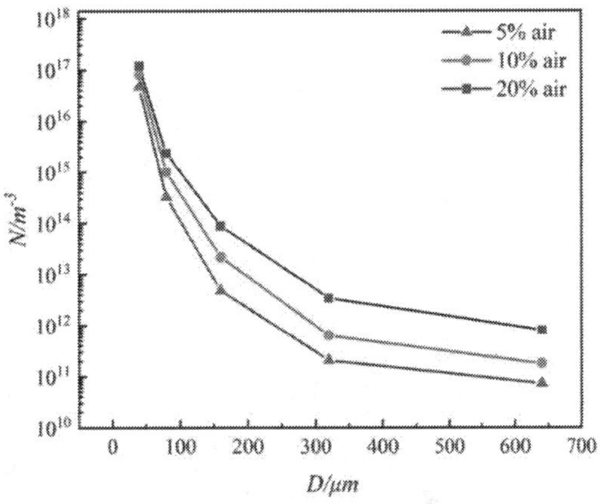

Figure 14. The bubble number density distribution

In this study, 5 bubble diameter groups were selected: $D_0 = 640\mu\text{m}$、$D_1 = 320\mu\text{m}$、$D_2 = 160\mu\text{m}$、$D_3 = 80\mu\text{m}$、$D_4 = 40\mu\text{m}$. Relevant studies show that: if the logarithm of the bubble radius in micrometers is taken as the horizontal mark and the logarithm of the number of bubbles in a certain radius is taken as the ordinate of the water body per cubic meter, then all bubbles are located in the linear strip area with a slope of -10/3. So the relationship between the initial volume fraction rate of the 5 groups of bubbles can be deduced: $f_0 = 0.1218, f_1 = 0.1523, f_2 = 0.1904, f_3 = 0.2380, f_4 = 0.2975$.

Figure 9 shows the bubble number density distribution in the flow field at t=7s. We can observe that, for different values of α_g, the tendency of bubble aggregation and fragmentation is the same, and that higher values of α_g result in the production of more large-diameter bubbles.

5.2. Results and discussion of propeller

5.2.1. Comparison of propeller performance

To take the error into consideration, two-phase results are compared with single-phase results instead of the standard open water experimental results.

Table 5. Comparison between single phase results and two-phase results

Results	K_T	$10K_Q$	η_0
Single Phase results	0.1897	0.3008	0.703
Two Phase results	0.1663	0.2739	0.677
Decrease between Single Phase results and two-phase results	12.33%	8.96%	3.70%

From the Table 5, it can be found that thrust coefficient K_T, moment coefficient K_Q and openwater efficiency η_0 have decreased by 12.33%, 8.96% and 3.70% respectively in the water-air mixed flow of 0.1 α_g. The decrease in thrust coefficient exceeds 10% and is too huge to be neglected. It reveals the fact that the existence of air affects thrust more than moment and that propeller performance will decrease obviously if it is put in the water-air-mixed flow of high air volume fraction like 10%.

5.2.2. *Propeller Static Pressure*

(a) propeller pressure surface ($\alpha_g = 0.1$).

(b) propeller suction surface ($\alpha_g = 0.1$).

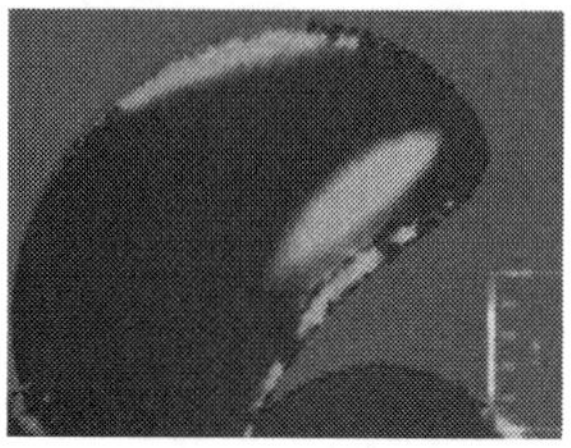

(c) propeller pressure surface ($\alpha_g = 0.1$).

(d) propeller pressure surface ($\alpha_g = 0$).

(e) propeller suction surface ($\alpha_g = 0$).

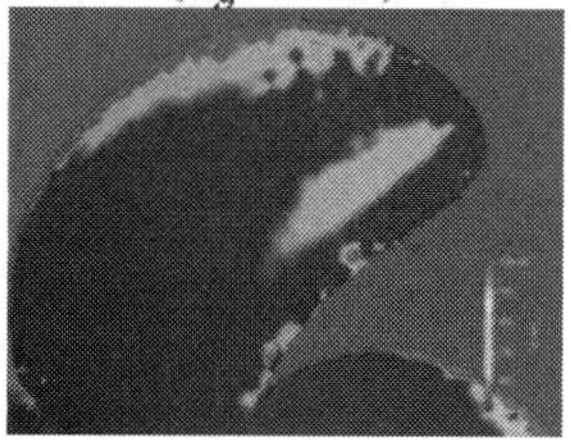

(f) propeller pressure surface ($\alpha_g = 0$).

Figure 15. The pressure distribution of propeller.

In Figure 15(a) and Figure 15(d) shows that on the pressure surface, there is little significant difference in pressure distribution between the condition of $\alpha_g = 0$ and $\alpha_g = 0.1$. But at the tip of the blade, it can be found that high static pressure ranges wider in pure water than in water-air-mixed flow. Also, this means the pressure gradient under $\alpha_g = 0.1$ is huger than that under $\alpha_g = 0$, thus leading to the decrease of η_0 and K_Q. According to the research results of Kawakita[11], there is also some similar conclusion. On the suction surface of the propeller, the pressure distribution changes much more obvious. When $\alpha_g = 0.1$, the absolute value of static pressure is lower than that under $\alpha_g = 0$. This leads to the thrust deduction because it is the difference of static pressure between two surfaces that generate the thrust.

5.2.3. *Comparison of Flow Fields*

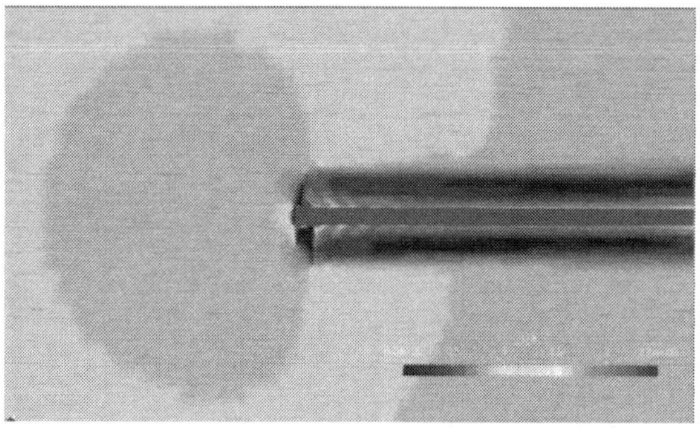

(a) water-air-bubble mixed flow condition.

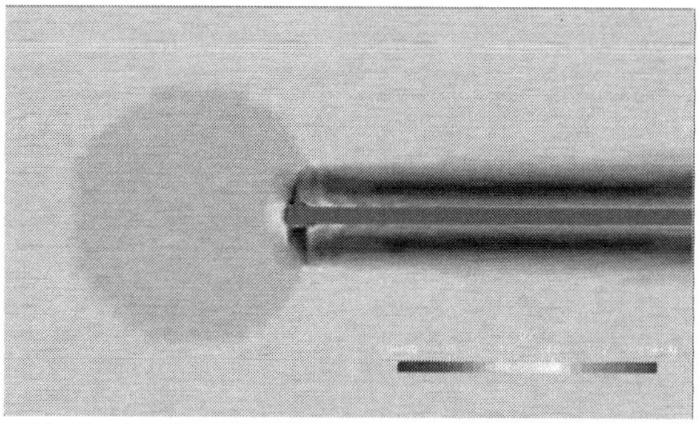

(b) single-phase incoming water flow condition.

Figure 16. The velocity field near the hydrofoil propeller

Figure 16 shows the comparison of instantaneous velocity field, comparing with single-phase incoming water flow, the existence of a bubble intensifies the asymmetry of the flow field. Furthermore, there is a velocity augment in the propeller wake when $\alpha_g = 0.1$.

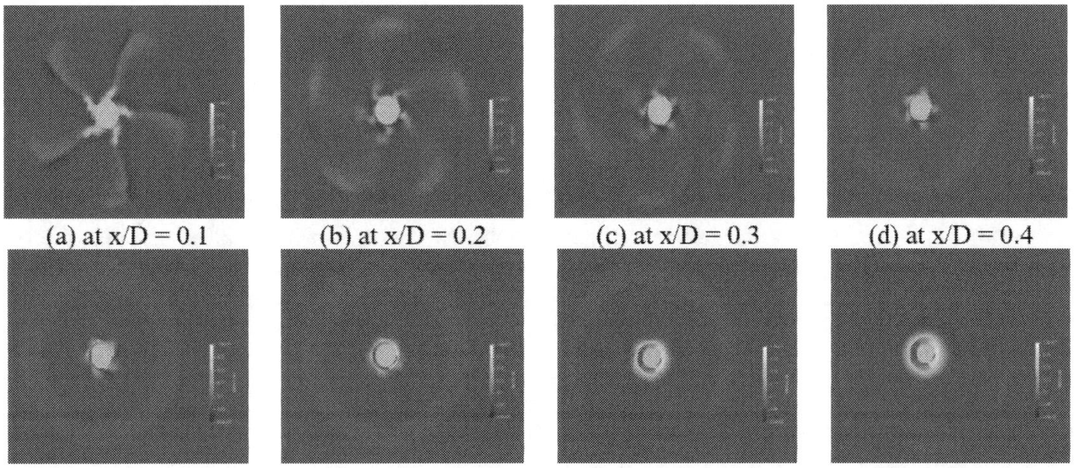

(a) at x/D = 0.1 (b) at x/D = 0.2 (c) at x/D = 0.3 (d) at x/D = 0.4

(e) at x/D = 0.5 (f) at x/D = 0.7 (g) at x/D = 1.3 (h) at x/D = 1.5

Figure 17. Instantaneous air volume fraction distribution at t=1s

The Figure 17 reflects the evolution process of bubbles in the propeller wake flow field. From the above figures, it can be seen that at a distance of 0.1D from the propeller (a), the gas gathering in the water presents a clear blade shape, and the bubbles mostly accumulate at the blade root and blade trailing edge, while the gas content decreases at the blade guide edge. This phenomenon corresponds to the pressure distribution of the propeller blades, indicating that the gas accumulation phenomenon in the propeller wake flow field is caused by the pressure distribution on the propeller. The gas is compressed from high-pressure region to low-pressure region, and then is influenced by the propeller induced-velocity, rotating and diffusing towards the surrounding area. As the distance from the propeller increases, the shape of the coalesced gas gradually changes from the blade (a b c) to a circle (e f). Finally, the coalesced gas disappears (g h), and the air volume fraction in this area returns to uniformly mixed 0.1.

6. Conclusions

In this paper, the effects of uniformly mixed water-air-bubble incoming flow on hydrodynamic performance and fields characteristic of hydrofoil and propeller has been studied. Eulerian-Eulerian two-fluid solution framework is adopted to simulate the uniformly-mixed water-air-bubble multiphase incoming flow. Validation of simulation is conducted by comparing previous experimental and numerical results in single-phase incoming flow with numerical results in this paper, and come to a conclusion that numerical model established here is precise enough to imitate the motion of hydrofoil and propeller.

The differences caused by the mixed water-air-bubble multiphase flow on the flow field, lift and drag coefficients are discussed in hydrofoil simulation. The number density of bubble modeled by PBM is also discussed. And the variations of the static pressure distribution, flow field and performance coefficients are studied in propeller simulation.

Main conclusions about hydrofoil are summarized as follows:

- Air volume on suction surface of the hydrofoil is higher and the change is more distinct than the pressure surface.
- The existence of bubble causes fluctuation in flow field of hydrofoil, and with the increase of α_g, the fluctuation and instability are more obvious.
- The phenomenon of vortexes is more obvious at the condition of water-air-bubble mixed flow.

- With the increasing of air volume fraction, the oscillation of lift and drag coefficient tends to be more obvious and the lift coefficient decreases and the drag coefficient increases with the increasing of α_g.
- The increasing of α_g leads to the generation of more large diameter bubbles.

Main conclusions about propeller are summarized as follows:

- The existence of air causes decreases in hydrodynamic performance of propeller.
- High static pressure of pressure surface ranges wider in pure water than in water-air-mixed flow, the pressure gradient is huger in water-air-mixed flow than in pure water, the absolute value of static pressure on suction surface when bubble exist is lower than that in pure water.
- The existence of a bubble intensifies the asymmetry of the flow field.
- The static pressure on propeller influences the bubble distribution in propeller wake field.

Acknowledgements
This work was supported by the National Natural Science Foundation of China (52131102), and the National Key Research and Development Program of China (2019YFB1704200), to which the authors are most grateful.

References

[1] Li, Z., Zhang, X. S., and Wan, D. 2022 Research progress on the hydrodynamic performance of water-air-bubble mixed flows around a ship *J. Hydrodyn.* 34(2) pp 171-188
[2] Hirt, C.W., and Nichols, B.D. 1981 Volume of Fluid (VOF) Method for the Dynamics of Free Boundaries *J. Comput. Phys.* 39 pp 201–225
[3] Zhang, X., Wang, J., and Wan, D. 2020 An Improved Multi-Scale Two Phase Method for Bubbly Flows *Int. J. Multiphas. Flow* 133 p 103460
[4] Qin, S., Chu, N., Yao, Y., Liu, J., Huang, B., and Wu, D. 2017 Stream-wise distribution of skin-friction drag reduction on a flat plate with bubble injection *Phys. Fluids* 29(3) p 037103
[5] Schiller, L. and Naumann, A. 1935 A Drag Coefficient Correlation *Zeit Ver. Deutsch. Ing.* 77 pp 318-320
[6] Ishii, M., and Zuber, N. 1979 Drag coefficient and relative velocity in bubbly, droplet or particulate flows *Aiche J.* 25(5) pp 843-855
[7] Tomiyama, A. , Kataoka, I. , Zun, I. , and Sakaguchi, T. 2008 Drag coefficients of single bubbles under normal and micro gravity conditions *Jsme Int. J.* 41(2) p 472
[8] Antal, S., Lahey, R., and Flaherty, J. 1991 Analysis of phase distribution in fully developed laminar bubbly two-phase flow *Int. J. Multiphas. Flow* 17(5) pp 635-652
[9] Luo, H., and Svendsen, H. 1996 Theoretical model for drop and bubble breakup in turbulent dispersions *Aiche J.* 42(5) pp 1225-1233
[10] Shang, Y., and Horrillo, J.J., 2021. Numerical Simulation and Hydrodynamic Performance Predicting of 2 Two-Dimensional Hydrofoils in Tandem Configuration *J. Mar Sci. Eng.* 9 p 462
[11] Kawakita, C. 2013 Study on marine propeller running in bubbly flow *In Proceedings of the Third International Symposium on Marine Propulsors* 13 pp 405-411

Propeller-Ice Contact modelling with Peridynamics

Kang Han[1,2]*, Chao Wang[1], Zeyu Yang[2], Chunyu Guo[2], Chunhui Wang[1]

[1] Harbin Engineering University, College of Shipbuilding Engineering, Harbin 150001, China
[2] Qingdao Innovation and Development Centre of Harbin Engineering University, Qingdao 266000, Shandong, China

*Corresponding author: hankang@hrbeu.edu.cn

Abstract. Ice loads pose a significant threat to ice-class propellers in frozen waters, making it crucial to analyse the magnitude of ice loads. However, the complex mechanical properties of ice and its failure process make it challenging to investigate the correlation between ice-propeller contact phenomena and ice loads. This paper uses a PC3 class propeller blade to study the propeller-ice contact process. This study involves developing ice failure and propeller blade models based on ordinary and non-ordinary state-based peridynamics theory and validating them using experimental. Convergence analysis was performed before blade ice loads evaluation. Additionally, we investigate the effect of the pitch angle on the ice-milling process.

1. Introduction

Arctic shipping routes have become the preferred choice for numerous shipowners due to the shorter travel distances they offer between different economic poles, resulting in reduced costs and emissions [1-3]. In addition to commercial and ecological benefits, navigating in polar areas challenges the structural strength of ice-going ships. For example, exposing the propeller to a massive ice load during the ice-propeller interaction process can cause damage when the vessel crosses the ice sheet [4][5]. Therefore, it is crucial to have a reliable method to evaluate the propeller's strength under ice contact conditions for designing and safely operating ice-class ships. However, predicting the ice load and simulating the propeller's structural response is challenging due to the complex mechanical properties, failure mode of ice, and complex interaction scenarios [6-8].

With the development of computational technologies, numerical methods have gained widespread application for analysing the ice-propeller interaction due to their excellent balance between efficiency, economy, and accuracy compared with other techniques. Veitch et al. established the original numerical model for predicting the hydrodynamic and ice contact loads simultaneously using the panel method [9-11]. More recently, Wang et al. developed a cohesive model in LS-DYNA software to calculate the ice load, with which they analysed the influence of operating conditions [12][13]. Xiong et al. introduced a new ice-propeller interaction model with bond-based peridynamics (PD), which significantly progressed in simulating ice failure under a blade stroke [14][15]. However, the simulation of ice property is limited due to Poisson's ratio, and their model neglects the structural response of the blade.

Simulating the structural response of a blade under ice milling conditions is closely tied to the evaluation of ice load. In 2007, the International Association of Classification Societies introduced a set of unified requirements (URI3) for ice-class propellers based on previous studies and rules. Since then, several scholars have used the ice load specified by IACS as the input condition and analysed the structural response of the blade using a cantilever beam and finite element methods (FEM) [16][17]. Recent studies have used coupling methods to investigate the structural response, including combining FEM and PD and FEM and DEM (Discrete Element Methods) [18]. This work is to develop and use a propeller-ice contacting numerical model to predict the ice loads applied on the blade and to evaluate the structural response of the blade.

2. Method and Verification

This section describes the numerical method and corresponding verification examples.

2.1. Calculation method

Peridynamics is a method where the continuum is discretised into finite material particles. By integrating the interactions between neighbouring particles and using Newton's second law, the momentum and motion information of the particles can be obtained [19]. As shown in Figure.1, the horizon δ is the critical distance, and particle interactions occur when the distance between particles is less than the horizon.

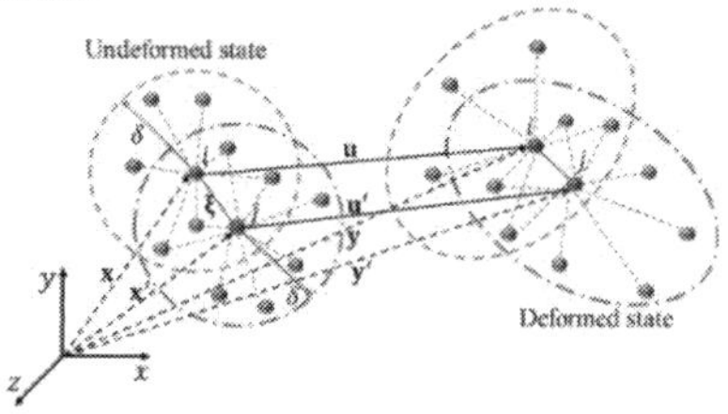

Figure 1. Material point interacts with its neighbours through bonds.

The PD governing equation is derived as follows [20]:

$$\rho(\mathbf{x})\ddot{\mathbf{u}}(\mathbf{x},t) = \int_{H_x} \{\underline{\mathbf{T}}[\mathbf{x},t]\langle\mathbf{x}'-\mathbf{x}\rangle - \underline{\mathbf{T}}[\mathbf{x}',t]\langle\mathbf{x}-\mathbf{x}'\rangle\} dV_{x'} + \mathbf{b}(\mathbf{x},t) \quad (1)$$

where ρ is the density of the material, $\ddot{\mathbf{u}}$ is the acceleration, $V_{x'}$ is the volume occupied by the particle j, and $\mathbf{b}$ is the density of the body force; the vectors $\mathbf{x}$ and $\mathbf{x}'$ represents the spatial locations of the material point i and its neighbour j and vectors $\mathbf{y}$ and $\mathbf{y}'$ are their new positions after experiencing displacements $\mathbf{u}$ and $\mathbf{u}'$, respectively; the force vector state $\underline{\mathbf{T}}(\mathbf{x},t)\langle\mathbf{x}'-\mathbf{x}\rangle$ is the force exerted by particle j on particle i.

In ordinary state-based peridynamics (OSB-PD), considering that materials have force vector states acting in the direction of the deformation vector state, they can be described as [21]:

$$\underline{\mathbf{T}} = \underline{t}\underline{\mathbf{M}} \quad (2)$$

where $\underline{\mathbf{M}}$ is the deformed unit direction vector state that can be estimated by $\underline{\mathbf{M}} = (\mathbf{y}'-\mathbf{y})/|\mathbf{y}'-\mathbf{y}|$. $\underline{t}$ is the dynamic force scalar state and can be defined as [22]:

$$\underline{t} = \frac{3K\theta}{m}\underline{\omega}\underline{x} + \frac{15\mu}{m}\underline{\omega}\underline{e}^d \quad (3)$$

where m is the weighted volume, θ is a measure of dilatation, K and μ are the bulk modulus and shear modulus, respectively. $\underline{x}$ and $\underline{e}^d$ are the scalar reference state and scalar deviatoric extension state, respectively, and $\underline{\omega}$ is the influence function.

In non-ordinary state-based peridynamics (NOSB-PD), force vector states can be developed using an arbitrary stress tensor from the classical theory that can be identified as follows [21]:

$$\underline{\mathbf{T}} = \underline{\omega}\boldsymbol{\sigma}\mathbf{K}^{-1}\boldsymbol{\xi} \quad (4)$$

where $\boldsymbol{\sigma}$ is the Piola stress tensor related to the strain energy density through the approximate deformation gradient tensor $\bar{\mathbf{F}}$, $\sigma_{ij}(\bar{\mathbf{F}}) = \partial\Omega(\partial\bar{F}_{ij})^{-1}$. $\Omega(\mathbf{F})$ is the given strain energy density function in classical theory, $\mathbf{F}$ is the actual deformation gradient tensor, and $\bar{\mathbf{F}}$ is its approximate value in PD and can be represented as:

$$\bar{\mathbf{F}} = (\underline{\mathbf{Y}} * \underline{\mathbf{X}})\mathbf{K}^{-1} \tag{5}$$

where $\mathbf{K}$ is the shape tensor, and can be estimated by $\mathbf{K} = \underline{\mathbf{X}} * \underline{\mathbf{X}}$. The operator $*$ denotes a dyadic product. $\underline{\mathbf{X}}$ and $\underline{\mathbf{Y}}$ are reference vector state and deformation vector state, respectively.

Failure criteria are required in PD theory to model damage initiation and propagation. Here, an energy-based damage criterion which compares the micro-bond potential with specific critical values proposed by Willberg et al. is adopted [23]. If the micro potential is more significant than these critical values, an initial failure occurs and the bond between particles with be removed.

With the assumption of the distinction of crack modes, the following criterion is used in this paper to take mixed-mode fracture into account [23]:

$$w_{\text{cmp}}^2 + w_{\text{ten}}^2 + w_{\text{shr}}^2 > w_{C,\text{cmp}}^2 + w_{C,\text{ten}}^2 + w_{C,\text{shr}}^2 \tag{6}$$

where w_{cmp}, w_{ten}, and w_{shr} are the compression, tension, and shear part of the bond potential. They can be calculated using the following equations [23]:

$$\begin{cases} w_{\text{cmp}} = w_{\text{ten}} = \dfrac{1}{4}\underline{\mu}\left(\dfrac{K[\mathbf{x}]}{m[\mathbf{x}]}\theta^2[\mathbf{x},t] + \dfrac{K[\mathbf{x}']}{m[\mathbf{x}']}\theta^2[\mathbf{x}',t]\right)\underline{x}^2 \\ w_{\text{shr}} = \dfrac{1}{4}\underline{\mu}\left(B[\mathbf{x}]\left(\underline{e}\langle\boldsymbol{\xi}\rangle - \dfrac{1}{3}\theta[\mathbf{x},t]\underline{x}\right)^2 + B[\mathbf{x}']\left(\dfrac{1}{3}\theta[\mathbf{x}',t]\underline{x} - \underline{e}\langle\boldsymbol{\xi}\rangle\right)^2\right) \end{cases} \tag{7}$$

where the coefficient $B = 15G(m)^{-1}$, and $\underline{\mu}$ is the history-dependent value used to indicate whether the bond is removed.

A scalar value, local damage φ, is introduced to quantify the failure of the material and is noted as [19]:

$$\varphi(\mathbf{x},t) = 1 - \int_H \underline{\mu}\langle\boldsymbol{\xi}\rangle dV' \left(\int_H dV'\right)^{-1} \tag{8}$$

Here, φ varies from 0 to 1, and the closer to 1 the greater the degree of the material damage.

2.2. *Validation of the numerical method in modelling ice failure*

Since ice often exhibits brittle elastic failure under propeller milling conditions, which involves high strain rates, this study considers spherical ice-impacting tests as a method for algorithm verification due to the similarity of ice failure modes and satisfactory test results. Tippmann et al. conducted a spherical ice impact experiment using a gas gun apparatus in California, as shown in Fig. 2 [24]. They recorded the dynamic force history and high-speed videos of ice failure. Similar experiments were conducted by Xiong et al. in 2018 [25].

Figure 2. Gas gun apparatus to project spherical ice [24].

The PD simulation's calculation scenarios are consistent with the experiments, where the spherical ice's diameter is 61 mm, and its impacting velocity is 61.8 m/s. The Young's modulus of ice is 1.8 GPa, the bulk modulus is 3.75 GPa, density is 900 kg/m3. The Poisson's ratio of the ice material is chosen using values obtained from [26], who suggest that an effective Poisson's ratio value of 0.42 would be more suitable for short-term calculations than the dynamic value of 0.33.

Figures 3 and 4 compare the test and simulation results regarding the time-history curve of the impact force and the dynamic impact phenomenon, respectively [27]. Based on the peak value of the total impact force and its change trend, the numerical simulation results shown in Figure. 3 agree with the test results, although the simulated value is slightly larger than the experimental one [27]. The main reason for this error is that the target is modelled as an ideal rigid body during the simulation to prevent the plate from deforming and absorbing elastic energy. Additionally, factors such as temperature changes, model ice production, and micro-cracks can affect sea ice breaking and lead to errors.

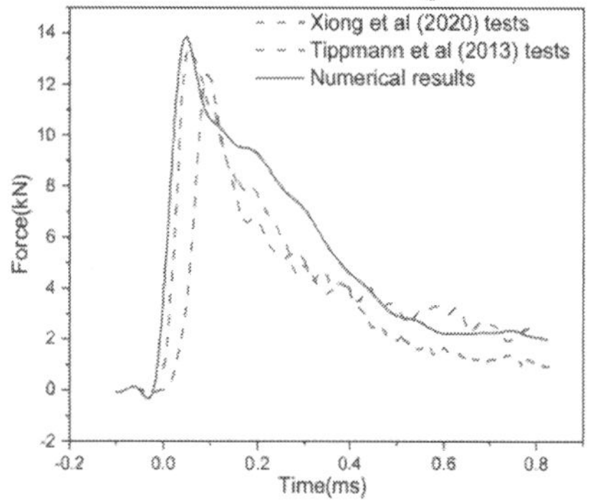

Figure 3. Comparison of the impact loads between experiments results and simulations result [27].

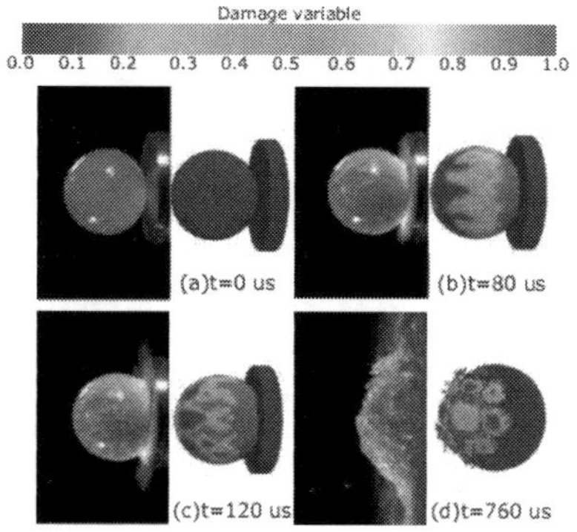

Figure 4. Comparisons of the failure progress of the dynamic impact of the spherical ice between test photographs (left) and the PD simulation snapshots (right) [27].

Figure 4 displays a series of snapshots of the dynamic impact process of the spherical ice taken from experimental photographs and the present numerical simulation at different times. The test and numerical results show the initial contact and crack evolution process, including the propagation of longitudinal cracks oriented along the ejection direction. During the initial stage of the event, the projectile maintains an overall spherical shape except for the contact area. By 760 μs, the spherical ice has broken into small fragments, with forces less than 10% of the peak being exerted.

3. Propeller-ice contact modelling

3.1. Convergence analysis and verification

This study uses the ice paddle contact test carried out by Wang Jungyong in the IOT ice water pool as a verification [28]. As shown in Figure 5, Wang used a motor-driven pod to provide the power for the propeller to mill sea ice and obtained the blade load and propeller shaft load during the contact process [28]. Similar to the type of model ice in the VTT ice pool, the IOT ice pool uses EG/AD/S model ice, combined with the existing model ice mechanical parameters, to summarise the mechanical properties of the model ice used for numerical simulation: In this paper, the model ice is simplified as Homogeneous material, density $\rho = 930$ kg/m^3, Poisson's ratio $\nu = 0.42$, elastic modulus E=0.2 GPa, bulk modulus k=0.046 GPa.

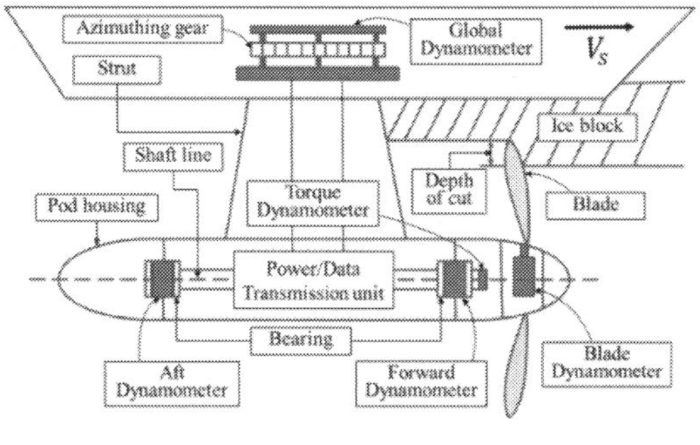

Figure 5. Sketch for the model podded propulsor system with measurement devices [28].

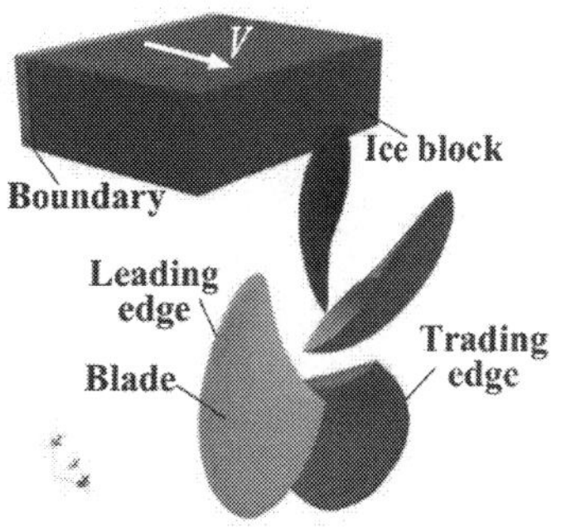

Figure 6. Sketch for the model podded propulsor system with measurement devices.

According to the measurement diagram of the ice loads in Figure 5, the ice-propeller contacting model is simplified, as shown in Figure 6. The central position of the propeller is fixed and rotates around the -x axis. The ice cube moves along the x-axis at a uniform speed and comes into contact with the propeller.

This study uses numerical research methods, so it is necessary to analyse the convergence and reliability of numerical calculation results. Generally speaking, there are three aspects to the convergence analysis in PD simulations: time step size, particle distance and near-field range. According to the research in the literature [27], particle distance convergence is the most important among the three kinds of convergence, so the present study will only explain particle distance convergence.

According to the discussion of the influence of carriage speed on ice loads, the vanishing point of the shadowing coefficient of the Rclass propeller is about J=0.4 [28]. As a result, scenarios of J=0.2 and J=0.4 have been considered verification cases for the convergence analysis. Ye et al. used a particle spacing of 15 mm when studying the interaction between ice and paddles, and Xiong et al. believed that the particle spacing of 7.5 mm has good convergence [25]. Considering that this study improves the damage criterion of sea ice and is more sensitive to sea ice damage, the convergence analysis is performed on a more detailed discrete model. The specific information of the three sets of discrete particle models used in the analysis process and the calculation results of the ice load are shown in Table 1.

Table 1. Mesh convergence analysis in terms of ice load in X-axis direction.

J	Grid name	dx	Particle number	K_T(Ave)	K_T(Max)
	Coarse	4 mm	46500	0.092	0.296
0.2	Medium	2.82 mm	100000	0.120	0.333
	Fine	2 mm	258000	0.132	0.348
	Coarse	4 mm	46500	0.197	0.539
0.4	Medium	2.82 mm	100000	0.311	0.681
	Fine	2 mm	258000	0.325	0.695

It can be seen from Table 1 that when the distance between particles is reduced from 2.82 mm to 2 mm, the number of particles increases by 158%, and the corresponding load peaks change by 4.34% and 2.02% at the two advancing speeds, respectively. Given the stochastic effect of ice damage, it can be considered that the calculation results converged when the particle distance was 2.82 mm.

Table 2 compares the ice load calculated in this paper with the loads measured by the Wang test. Overall, the numerical and experimental results have an error of about 10%, and the simulation effect is good. Among them, when J=0.4, there is a significant difference between the maximum value of the numerical load and the experimental value. There are mainly three reasons for the error: a) Influence of calculating time; b) Influence of the difference between experimental conditions and numerical simulation; c) The numerical means represented by the "constitutive" model of sea ice are insufficient.

Table 2. Comparison of numerical and experimental results.

J	K_T	Numerical	Experimental	Error
0.2	Ave.	0.12	0.108	11.49%
	Max	0.333	0.306	8.54%
0.4	Ave.	0.311	0.299	4.13%
	Max	0.681	0.802	-15.1%

In addition, Fig.7 compares breakage of sea ice, and it can be seen that the test results are in good agreement with the test phenomena.

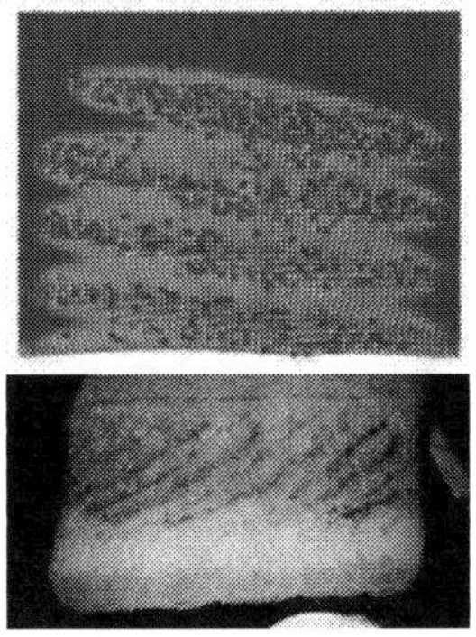

Figure 7. Comparison of the breakage of sea ice after blade milling.

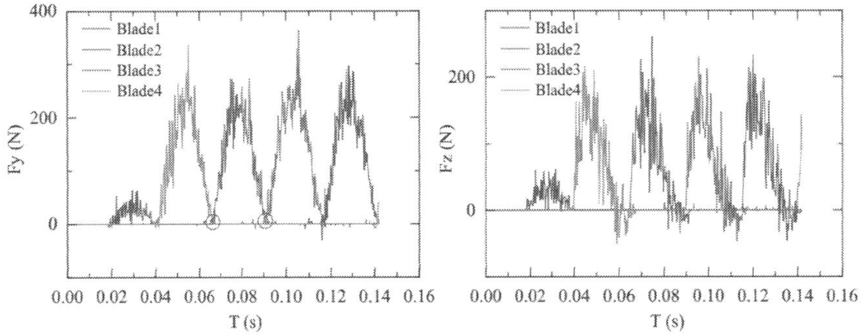

Figure 8. Ice loads in Y-direction and Z-direction.

Figure 8 shows the ice load in the Y-direction and Z-direction (the load direction is opposite to the direction of the coordinate system). Compared with load in Table 2, the axial load is significantly higher than the load in the other two directions, which is related to the projected area of the blade in different directions. In addition, the ice load in the Z direction in Figure. 8 has different characteristics from those in other directions. The load's direction changes in the late contact period, which is related to the rotation direction of the blade. At the initial contact stage, the blade cuts into the ice block upwards, and the load is along the -Z axis direction. In the later contact stage, the blade cuts down the ice block and the load direction reverse.

To intuitively illustrate the dynamic characteristics of the milling process, Figure 9 shows the snapshots of blade four at ordinary moments. It can be seen that the load peaks at the position where the milling depth is the largest, indicating that the ice load is very sensitive to changes in the milling depth (contact area). Figure 9 also shows the formation process of scratches on the ice. The scratches extend as the blade rotates. From the last snapshot, the later blade started to cut into the ice before the current blade cut out the ice. The process of overlapping two consecutive blades is reflected in Figure 8, which is the overlapping part of the load (the circled portion in the figure). Figure 10 shows the stress distribution of the blade at typical moments. It can be seen that the stress concentration area of the blade gradually expands from the tip to the root. However, due to the complex motion state changes involved in the blade rotation process and the influence of the zero-energy mode, the stress calculation results are not uniform, and the numerical calculation method still needs to be further improved.

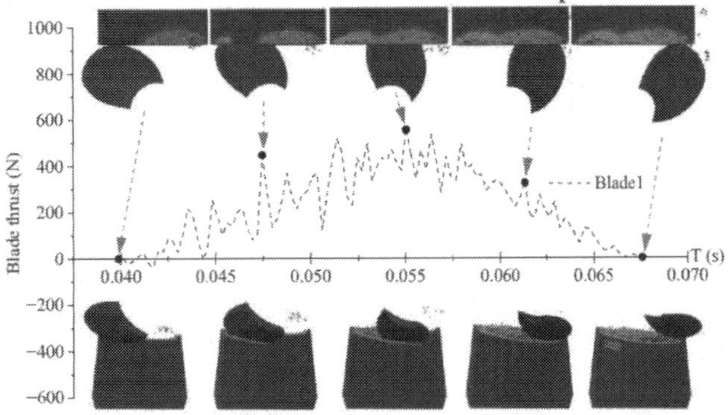

Figure 9. Time-history load and corresponding snapshots

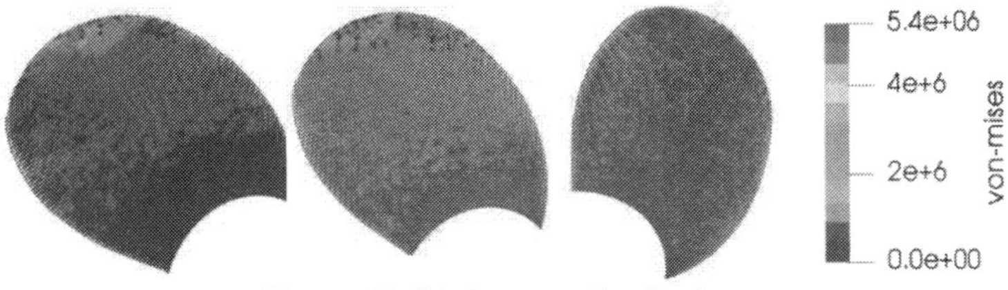

Figure 10. Blade stress distribution

3.2. Influence of pitch distribution on propeller-ice contact characteristics

The pitch is the core parameter reflecting the hydrodynamic characteristics of the propeller. Therefore, it is not of great reference value to analyse the influence of pitch change (increase or decrease) on milling characteristics. Once the propeller design task is determined, the pitch change range is limited. The average pitch is determined by the hydrodynamic performance rather than the ice loads or strength characteristics. Based on the above design considerations, the first prerequisite for analysing the influence of pitch on the milling characteristics of propellers is to ensure that the average pitch remains unchanged, and the average pitch is calculated by Equation (9):

$$\bar{p} = \int_{x=x_h}^{1.0} pxdx\left(\int_{x=x_h}^{1.0} xdx\right)^{-1} \tag{9}$$

The pitch distribution of propellers in ice regions has three primary forms, fixed pitch, increased outer radius pitch and decreased outer radius pitch. According to the existing design experience, set the pitch distribution, as shown in Figure 11. The average pitch of the three propellers is the same.

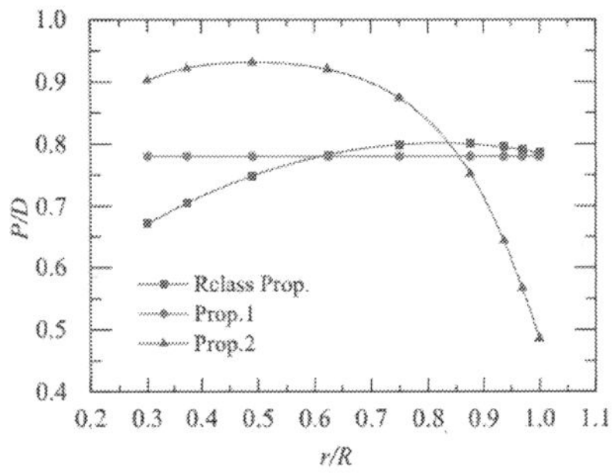

Figure 11. Sketch for the model podded propulsor system with measurement devices.

The ice-propeller contact PD model established in this study is used to analyse the dynamic milling characteristics of these propellers. The blade loads are shown in Table 3.

Table 3. Comparison of ice loads in different directions of various propellers with different pitch distribution

Prop.	Fx(Ave)	Fx(Max)	Fy(Ave)	Fy(Max)	Fz(Ave)	Fz(Max)
Rclass	237.6	551.5	126.1	318.3	64.2	234.9
Prop. 1	216.8	514.5	118.0	279.0	62.5	282.3
Prop. 2	215.9	562.1	121.2	284.1	101.1	268.4

There is no consistent variability in the mean and extreme load in Table 3. The propeller experiencing a smaller average load may suffer a more significant extreme load due to the randomness of the sea ice breakage and ice-propeller contact. And the mean load is more meaningful in the analysis process. To further discuss the influence of the pitch distribution change on the ice propeller's contact load, the degree of deviation between the load and the mean level in different directions is plotted, as shown in Figure. 12.

It can be seen from Figure 16. that under the premise of keeping the average pitch of the blade unchanged, adjusting the pitch distribution will not cause a significant change in the load level of the blade. For the load amplitude, the deviation of the load of the three propellers is 5%. For the load in the X direction that has an essential impact on the blade's strength, the pitch distribution change will not cause a drastic change in the load.

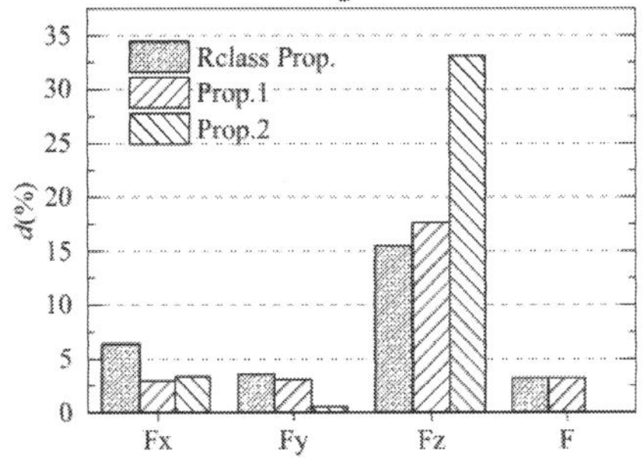

Figure 12. Time-history load and corresponding snapshots.

However, the loads in the Z direction show significant deviations in the three propellers. It can be seen from Table 3 that the load applied on propeller 2 in the Z direction is 60% higher than that of the parent propeller and propeller 1.

To further investigate the load in the Z direction, a load time history curve and the corresponding frequency spectrum of the variable parameter propeller in the Z direction were depicted in Figure 13 and Figure 14, respectively. From the figures, it is evident that variable parameter propeller 2 exhibits distinct dynamic load characteristics compared to the other propellers. Propeller 2 does not exhibit alternating load directions, resulting in an average load value that is 60% higher than the other two propellers. However, upon observing the frequency spectrum, it can be noted that the load fluctuation amplitudes of the three propellers at the blade frequency, twice the blade frequency, and thrice the blade frequency are similar. Additionally, the amplitude of variable parameter propeller 2 at the blade frequency is slightly smaller than that of the parent propeller.

The phenomenon of alternating load direction is linked to changes in the projected area of the blade profile in the Z-axis direction. The alteration in pitch at the blade tip leads to changes in the projected area of the blade section in the Z-axis direction. Figure 11 illustrates a significant change in pitch for variable parameter propeller 2 in the milling section (0.75R-1.0R) and noticeable blade torsion. Consequently, the projected area of the blade surface (pressure surface) in the Z-axis direction increases (represented by the blue area in the figure). The proportion of the projected area increases from 6% to 27%, and the normal direction of the blade surface aligns with the positive direction of the Z-axis, as depicted in Figure 15. During the final 1/4 cycle of a single blade making contact with sea ice, the twisted blade surface continues to impede the leakage of crushed ice, thereby maintaining the load direction without any alterations. As a result, the average load on the blade increases.

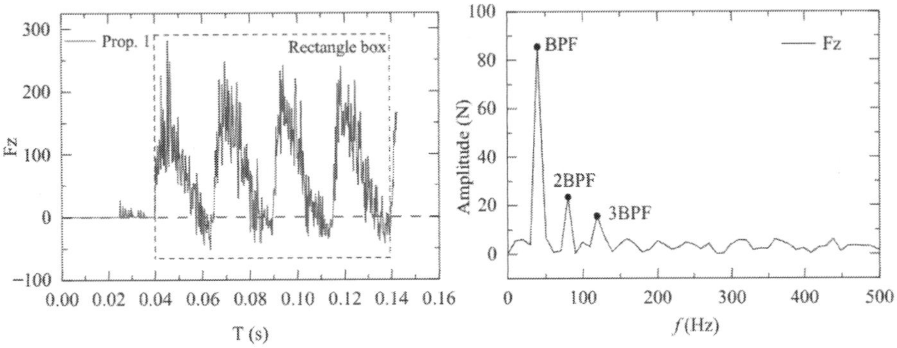

Figure 13. Time-history load and corresponding frequency spectrum of Prop.1.

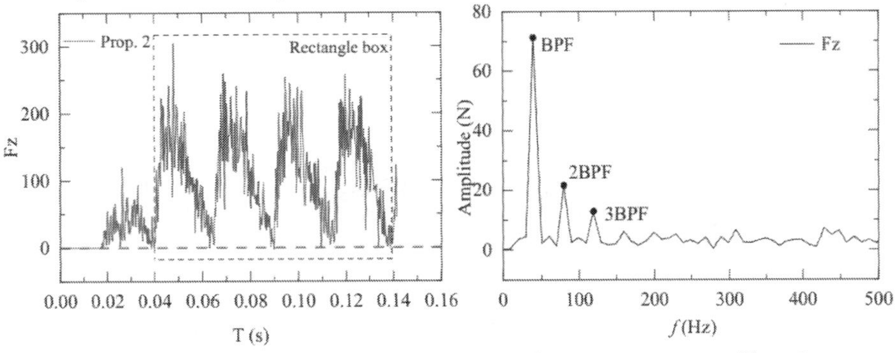

Figure 14. Time-history load and corresponding frequency spectrum of Prop.2.

Figure 15 displays the relatively stable pitch of the Rclass propeller and the tip of variable parameter propeller 1, with similar projected areas. And the normal direction of the blade surface deviates from the positive direction of the Z-axis. Consequently, crushed ice impacts the blade surface, leading to a change in the load direction.

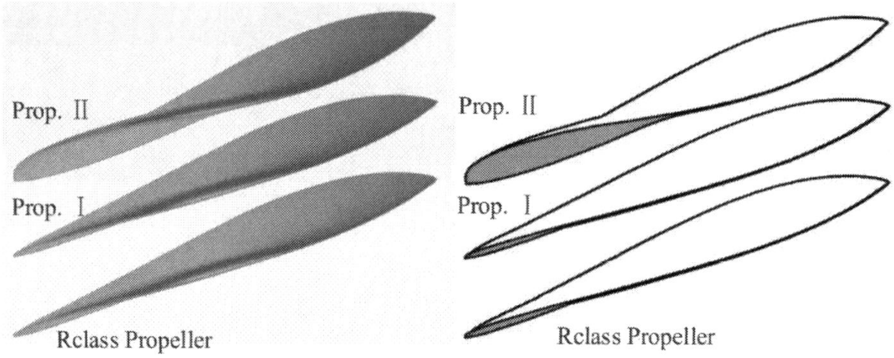

Figure 15. Blade tip area of three propellers.

The drastic change of the load in the Z direction will affect the cantilever effect brought by the propeller, increase the load on the stern bearing of the shafting system, and induce the vibration of the shafting system and the hull. Therefore, from the perspective of hull vibration, the pitch distribution form used by propeller 2 is unfavourable. However, podded propellers unaffected by shaft vibration can adopt this form with a sharply reduced pitch distribution at the tip to improve ice-breaking abilities.

4. Conclusions

This paper focuses on developing a propeller-ice contact model based on the state-based peridynamics theory to simulate the ice failure process and predict ice loads. As part of the study, the presented ice failure model realised by the OSB-PD is verified by spherical ice-impacting results. Then the propeller-ice contact model was established and verified using Wang's tests and applied to investigate ice milling characteristics.

The main conclusions are as follows: (a) The presented propeller-ice contact model can simulate the dynamic characteristics of the blade during the milling process, which means that peridynamics has a broad application prospect in the field of ice-structure interaction. (b) When milling, the ice load in the axial direction is higher than the ice load in the other two directions, and the ice load along Z-axis will change direction. (c) Changes in the blade pitch distribution do not cause significant changes in the axial ice load but significantly affect the load in the Z-axis direction.

References

[1] Lee, S.-W., Song, J.-M., 2014. Economic possibilities of shipping though Northern Sea route. The Asian Journal of Shipping and Logistics 30 (3), 415-430.
[2] Melia, N., Haines, K., Hawkins, E., 2016. Sea ice decline and 21st century trans-Arctic shipping routes. Geophysical Research Letters 43 (18).
[3] Smith, L.C., Stephenson, S.R., 2013. New Trans-Arctic shipping routes navigable by midcentury. Proc Natl Acad Sci U S A 110 (13), E1191-1195.
[4] Bach, C., Myland, D., 2017. An experimental method for model propeller-ice interaction in air: Concept and first results, International Conference on Offshore Mechanics and Arctic Engineering. American Society of Mechanical Engineers, p. V008T007A018.
[5] Kotras, T., Humphreys, D., Baird, A., Morris, G., Morley, G., 1987. Determination of propeller-ice milling loads.
[6] Guo, C., Han, K., Wang, C., Ye, L., Wang, C., 2021. Evaluation and analysis of the static strength of ice-class propellers. Ocean engineering 235, 109336.
[7] Petrovic, J., 2003. Review mechanical properties of ice and snow. Journal of materials science 38 (1), 1-6.
[8] Timco, G., Weeks, W., 2010. A review of the engineering properties of sea ice. Cold Regions Science and Technology 60 (2), 107-129.
[9] Veitch, B., Bose, N., Meade, C., Liu, P., 1997. Predictions of hydrodynamic and ice contact loads on ice-class screw propellers, Proceedings of the international

conference on offshore mechanics and arctic engineering. American society of mechanical engineers, pp. 119-126.
[10] Doucet, J., Liu, P., Bose, N., Veitch, B., 1998. Numerical Prediction of Ice-Induced Loads on Ice-Class Screw Propellers Using a Synthesised Contact/Hydrodynamic Code. Ocean Engineering Research Centre Report No. OERC-1998–004.
[11] Liu, P., 2000. Numerical prediction of ice-induced hydrodynamic loads on propellers due to blockage. Ocean Eng Int 4 (1), 31-38.
[12] Wang, F., Zou, Z.-J., Li, Z., Wang, Y., Yu, H., Zhang, H., 2019. Numerical simulation of ice milling loads on propeller blade with cohesive element method. Brodogradnja: Teorija i praksa brodogradnje i pomorske tehnike 70 (1), 109-128.
[13] Zhou, L., Wang, F., Diao, F., Ding, S., Yu, H., Zhou, Y., 2019. Simulation of ice-propeller collision with cohesive element method. Journal of Marine Science and Engineering 7 (10), 349.
[14] Wang, C., Xiong, W., Chang, X., Ye, L., Li, X., 2018. Analysis of variable working conditions for propeller-ice interaction. Ocean engineering 156, 277-293.
[15] Xiong, W.P., Wang, C., Ye, L.Y., Xu, P., Wang, C.H., 2022. Analysis of the mechanism for the three-dimensional hydrofoil and propeller contacting the ice body. Applied Ocean Research 122, 103107.
[16] Chang, X., Li, P., Wang, C., Ye, L., 2018. Effect of ice-load and trim angle on the strength of propeller. Chinese Journal of Ship Research 13 (4), 7-15.
[17] Lee, S.-K., 2008. Ice Controllable Pitch Propeller Strength Check Based on IACS Polar Class Rule, Originally Presented at the Ice Tech Conference Held July, pp. 20-23.
[18] Dongbao, Y., Junsong, G., Jianping, L., Chu, S., Shunying, J., 2021. Analysis of ice-inducted structure vibration of offshore wind turbines based on DEM-FEM coupled method. Chinese Journal of Theoretical and Applied Mechanics 53 (3), 682-692
[19] Silling, S.A., Askari, E., 2005. A meshfree method based on the peridynamic model of solid mechanics. Computers & structures 83 (17-18), 1526-1535.
[20] Silling, S.A., 2000. Reformulation of elasticity theory for discontinuities and long-range forces. Journal of the Mechanics and Physics of Solids 48 (1), 175-209.
[21] Silling, S.A., Epton, M., Weckner, O., Xu, J., Askari, E., 2007. Peridynamic states and constitutive modeling. Journal of elasticity 88 (2), 151-184.
[22] Foster, J.T., 2016. Constitutive modeling in peridynamics, Handbook of Peridynamic Modeling. Chapman and Hall/CRC, pp. 181-216.
[23] Willberg, C., Wiedemann, L., Rädel, M.J.J.o.M.o.M. and Structures, 2019. "A mode-dependent energy-based damage model for peridynamics and its implementation". 14(2): 193-217.
[24] Tippmann, J.D., K`im, H., Rhymer, J.D., 2013. Experimentally validated strain rate dependent material model for spherical ice impact simulation. International Journal of Impact Engineering 57, 43-54.
[25] Xiong, W., Wang, C., Wang, C., Ma, Q.W., Xu, P., 2020. Analysis of shadowing effect of propeller-ice milling conditions with peridynamics. Ocean engineering 195, 106591.
[26] Murat, J., Lainey, L., 1982. Some experimental observations on the Poisson's ratio of sea-ice. Cold Regions Science and Technology 6 (2), 105-113.
[27] Chunyu GUO, Kang HAN, Chao W, et al. Numerical modelling of the dynamic ice-milling process and structural response of a propeller blade profile with state-based peridynamics [J]. Ocean Engineering, 2022, 264: 112457.
[28] Wang, J., 2007. Prediction of propeller performance on a model podded propulsor in ice (propeller-ice interaction). Memorial University of Newfoundland.

On suitable CFD setups for cavitation dynamics

Youjiang Wang[1,2]

[1]Key Laboratory of Marine Intelligent Equipment and System, Ministry of Education, P.R. China

[2]Shanghai Jiao Tong University, Dongchuan Road 800, 200240 Shanghai, China

E-mail: `youjiang.wang@sjtu.edu.cn`

Abstract. In the current work, cavitating simulations were carried out with different setups including different mesh resolutions, time steps, RANS and IDDES turbulence models to investigate their influence on cavitation dynamics. The simulations were conducted for the Delft Twist hydrofoil, for which comprehensive experimental data is available. The open-source CFD solver OpenFoam is adopted because further model improvement is straightforward. The hydrodynamics, cavity volume, and cavitation shedding frequency are compared between different simulations. It is found that a good mesh for hydrodynamics is by far not enough to obtain appropriate cavitation dynamics. The turbulence model plays a prominent role on cavitation dynamics. The current work provides a valuable reference for choosing suitable setups to simulate cavitation dynamics.

1. Introduction

To minimize shipping's impact on the environment, the underwater radiated noise of merchant ships got increasing attention in recent years. Duarte et al identified 538 peer-reviewed papers that considered how changes in anthrophony, geophony, and biophony affect marine fauna[1]. A summary of these investigations shows clearly that noise compromises hearing ability and induces physiological and behavioural changes in marine animals. The shipping noise is emitted between 10 Hz and 20 kHz and is audible to a wide range of taxa.

The issue of underwater noise and its impact on marine mammals was first raised at IMO in 2004. Following instruction from the MEPC in 2021, the IMO Sub-Committee on Ship Design and Construction reviewed the 2014 Guidelines and would submit new recommendations for the reduction of underwater noise to MEPC 80 in 2023 [2]. The EU has defined the Marine Strategy Framework Directive 2008/56/EC to achieve good environmental status, which includes reduced underwater noise[3]. In 2022, the first ever EU-wide mandatory limits for underwater noise were passed [4]. The new limits mean that no more than 20% of a given marine area can be exposed to continuous underwater noise over a year. Similarly, no more than 20% of a marine habitat can be exposed to impulsive noise over a given day, and no more than 10% over a year.

Much, if not most, of the underwater noise is caused by propeller cavitation [5]. Cavitation happens on marine propellers when they cause low pressure in water. To maintain the propulsion efficiency of propellers in ships' wakefields, cavitation is an inevitable phenomenon. Thus, it is highly demanded to develop marine propellers that have both low cavitating noise and high propulsion efficiency. The accurate numerical method to predict the cavitating noise of marine propellers will in no doubt promote this development.

A prerequisite to predicting cavitating noise satisfactorily is the appropriate simulation of the cavitation dynamics. Both potential and viscous methods could be used to model cavitation dynamics. However, it was found that the potential method can only give satisfactory results for low tonal frequencies. The viscous CFD technique must be used for broadband and high-frequency dynamics. While with modern CFD solvers, setting the prism layers correctly and choosing a second order of discretization scheme can easily lead to hydrodynamics prediction within an engineering acceptable error (5%), it is still a difficult task without general rules to obtain the appropriate transient cavitation dynamics.

Currently, one consensus is that the RANS model fails to capture the transient cavitation dynamics, and corrections or RANS-LES hybrid models must be applied to reproduce the physical dynamics [6]. However, the guidance to choose the suitable mesh resolution, time step and turbulent models is not mature. Lots of researchers use hydrodynamic forces as the criteria to verify the simulation setups, and then use them to predict cavitation dynamics and hydrodynamic noise. Such a way lacks strictness and could very probably result in large errors or even wrong conclusions.

Thus, new knowledge about the correct setup of cavitation dynamics simulation is required. The influence of insufficient mesh resolution and time step should be clarified, as the best choice cannot always be guaranteed. For this reason, cavitating simulations were carried out with different setups including different mesh resolutions, time steps, RANS and IDDES turbulence models to investigate their influence on cavitation dynamics. The hydrodynamics, cavity volume, and cavitation shedding frequency are compared between different simulations. Such work provides a valuable reference for choosing suitable setups to simulate cavitation dynamics.

2. Theory

The open-source CFD solver OpenFOAM is used for the simulations in the current work. The underlaid theories are described briefly in this section.

2.1. RANS, LES, and IDDES

The governing equations of the fluid flow are the NS (Navier-Stokes) equations. To avoid solving the NS equations directly, which requires resolving eddies down to the Kolmogorov scale, the RANS, LES, or their combination called DES (including DDES and IDDES) methods are normally used for engineering simulations.

The RANS (Reynolds Averaged Navier-Stokes) method is based on ensemble averaging. Each fluid flow variable is separated into its averaged quantity and fluctuation. By averaging the NS equation, one can obtain the RANS equations, which govern the variation of the averaged quantities and have also some terms related to the fluctuated quantities. The latter is called the Reynolds stresses. Turbulence models are used to relate the Reynolds stresses to the averaged quantities so that the equations are closed. Turbulence models are mostly based on the Boussinesq hypothesis, which assumes that the Reynolds stresses have a similar diffusion effect as the molecular viscosity. Turbulence models give then an estimation of the local eddy viscosity. Popular turbulence models include the SA (Spalart-Allmaras) model[7], $k-\omega$ SST model[8], realizable and RNG $k-\epsilon$ models[9, 10].

The LES (Large Eddy Simulation) method is based on low-pass filtering. A spatial-temporal low-pass filtering is applied to the fluid flow variables and the NS equations. Only low frequencies (large scales) are resolved and solved. The effect of small scales being filtered out is modelled through subgrid-scale models. For an LES simulation, a much more refined mesh is required than for the RANS simulation. The filter size, normally the mesh size, should lie in the inertial subrange, where the turbulence energy spectra exhibit a constant -5/3 decay with the wave number. For practical engineering problems, fulfilling this requirement in the boundary layer is still not feasible.

As a fact, the equations for RANS and LES look identical. They differ from each other in the physical meaning of the variables. In RANS, they are averaged quantities and fluctuations, while in LES, they are filtered quantities and subgrid-scale quantities. The effects of turbulence models in RANS and subgrid-scale models in LES are also similar. With this observation, the hybrid method has been established, where RANS and LES methods are used for different flow regions based on the relation between the local turbulent length scale and mesh size[11]. The expected effect is that the RANS method is used in the boundary layer, while LES is used in the other region. Such a technique is called Detached Eddy Simulation (DES), and later improvements include Delayed DES (DDES) and improved DDES (IDDES)[12]. All of them are realized by the smooth interconversion between turbulence models and subgrid-scale models. The turbulence models having been adjusted to have such functionality includes the SA model and $k-\omega$ SST model.

The formulas for RANS, LES, and turbulence models are available in various literature and websites and are not given here for their lengthiness. The reader can refer to [13, 14] for more details. In this work, the RANS and IDDES simulations will be carried out, using both the SA and $k-\omega$ SST turbulence models.

2.2. The cavitation model
To simulate the cavitating flows, the volume of fluid approach (VOF) and mass transfer model is used. With the VOF approach, the mixture of water and vapour is modelled as a single-phase flow. A new variable, that is the volume fraction of liquid, is used to describe the spatial distribution of liquid and vapour. The volume fraction of liquid is also used to determine the density and viscosity of the mixture fluid. The mass transfer model is used to model the conversion between water and vapour and serves as a source term for the volume fraction of liquid. In the current work, the Schnerr-Sauer model [15] is adopted for mass transfer. The default values in OpenFOAM are used for setups, that is 1.6×10^{13} for the nuclei density, 2×10^{-6}m for the diameter of initial nuclei, 1 for the vaporisation and condensation coefficients.

3. Numerical setups
3.1. Simulation case
The well-documented cavitating flow on the Delft Twist hydrofoil is chosen as the simulation case in the present work. This is a benchmark case being made open-access on the SMP11 conference[6]. The foil has a chord length of 150mm and is twisted so that it has a larger angle of attack in the middle than in the tip. Such a geometry concentrates the cavitating phenomenon near the middle of the foil. The readers can refer to [6] and [16] for more details of the Delft Twist foil. For the simulated case, the foil has an angle of attack of -2 degrees and the inflow velocity is 6.97m/s. The pressure on the outlet is 29kPa and the saturated vapour pressure is 2.97kPa. Thus, the cavitation number is 1.07. Both the fully wet and cavitating situations were simulated by various institutes during the workshop of SMP11. The results observed in the experiment and obtained in the previous simulations are listed in table 1.

3.2. Meshes
We have adopted the computational domain recommended in [6] to facilitate comparisons with others. The domain extends 2c ahead of the leading edge and 4c behind the trailing edge, where c is the chord length of the hydrofoil. The cross-section has a size of 0.15m x 0.3m. The symmetrical condition is used in the y direction to reduce the domain size and mesh numbers. The computational domain is depicted in figure 1.

Four sets of meshes were generated systematically. As listed in Table 2, all meshes have the same topology and are refined from M0 to M3 gradually with the factor $\sqrt{2}$. The boundary layer mesh is kept the same, where the first cell height is 5e-3mm, and the expansion rate is

IOP Conf. Series: Materials Science and Engineering 1288 (2023) 012053 doi:10.1088/1757-899X/1288/1/012053

Table 1. The hydrodynamics and cavitating characteristics observed in the experiment and obtained in simulations presented on SMP11[6, 17, 18].

	Cl(wet)	Cd(wet)	Cl mean (cavitating)	cavity volume range	cavity shedding frequency
LR	0.4268	0.0151	0.4275	4.5 - 11×10^{-6}	35±2.5 Hz
MARIN	0.4279	0.01447	-	-	38Hz
Charlmers	-	-	0.42 - 0.45	4.0 - 10×10^{-6}	30 - 34Hz
Experiment	0.461	-	0.51	-	32.2Hz

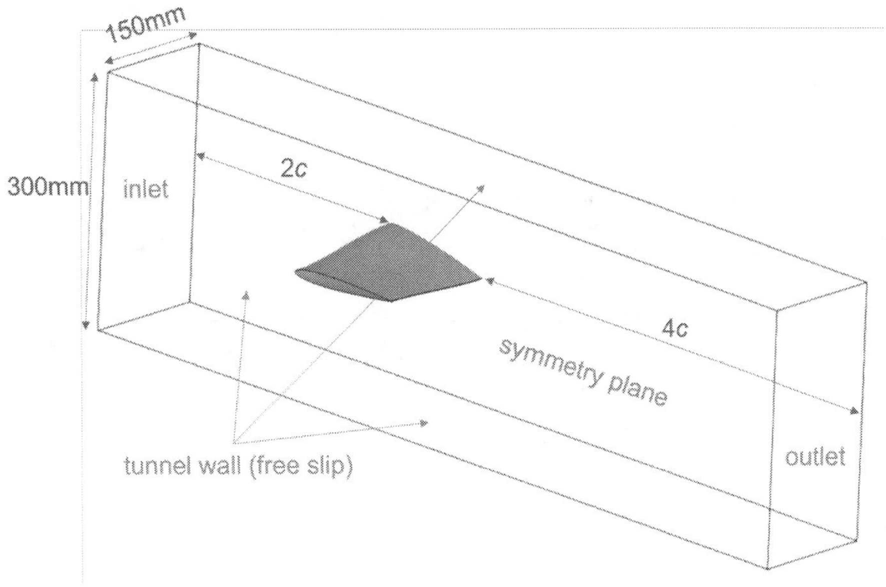

Figure 1. The computation domain and boundary conditions.

1.2. This results in a y+ value of around 1. The mesh M3 is shown in Figure 2, in the form of surface mesh on the symmetry plane and the foil surface.

Table 2. Details of the meshes used in the current work.

Mesh ID	cell size on leading edge	cell size on tunnel walls	number of nodes for the chord	total number of cells
M0	2.8mm	28mm	59	772K
M1	2.0mm	20mm	73	1.25M
M2	1.4mm	14mm	100	1.95M
M3	1.0mm	10mm	135	3.40M

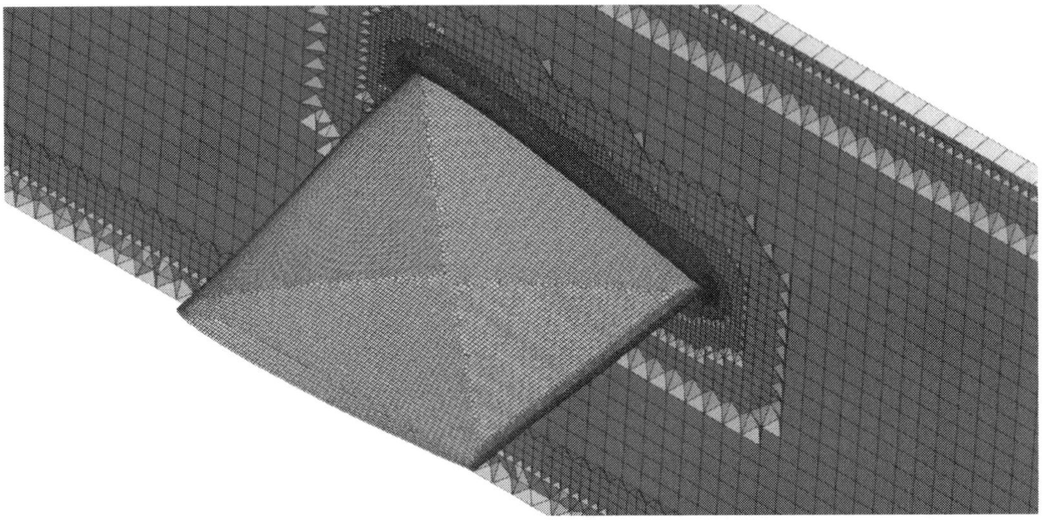

(a) mesh on the foil surface

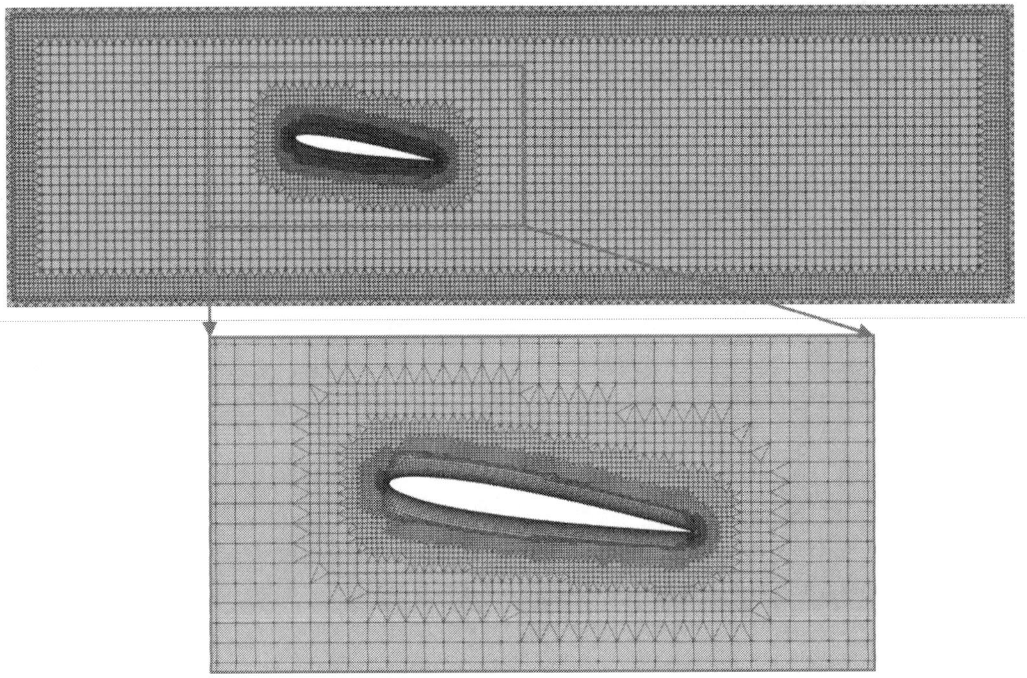

(b) mesh on the symmetrical plane

Figure 2. The mesh M3. The meshes M0, M1, M2 have the same topology, but different refinement levels.

3.3. Boundary conditions and numerical schemes

For the boundary conditions, the recommendation in [6] is adopted. On the tunnel walls, the free slip boundary condition is applied. On the hydrofoil surface, the no-slip boundary condition is applied. As for numerical schemes, the second-order upwind scheme is used for the convection of velocity, turbulence kinematic energy k and the variable $\tilde{\nu}$ in the SA turbulence model. First order upwind scheme is used for the convection of the variable ω in the k- ω type turbulence models to keep the simulation stable. The second-order implicit scheme called backward is used for the temporal advancing and the central difference scheme is used for the diffusion term. In the PIMPLE scheme, 5 outer correction iterations and 3 inner correction iterations are used. The dynamically adjustable time step with the max Courant number being 10 is applied if not otherwise stated.

4. Results

Steady and transient RANS fully wet simulations, transient RANS and IDDES cavitating simulations are carried out for different meshes. The openFOAM solvers simpleFoam, pimpleFoam and interPhaseChangeFoam were used for these simulations. The transient RANS simulation is also named uRANS in the following text.

4.1. RANS simulation for fully wet case

RANS simulations with the k-ω SST turbulence model are carried out for all meshes. For the mesh M2, another simulation with the SA model is carried out. The lift coefficient C_l and drag coefficient C_d are listed in table 3. The variation of hydrodynamic forces with the mesh refinement is given in figure 3. Both C_l and C_d are in the convergence range, that is they approach a value with decreasing error monotonously as the mesh is refined. Besides, the difference between the hydrodynamics obtained with RANS and uRANS is negligible (¡0.1%). The results obtained in the current work correlate well with the previous numerical simulations (see table 1), while all simulations exhibit an underestimation when compared to the measured value.

Table 3. The lift and drag coefficient obtained in different RANS simulations. M0 M3 means different meshes. SST and SA means k-ω SST and SA turbulence model, respectively.

case	C_l	C_d	C_l(uRANS)
M0_RANS_SST	0.4265	0.01617	0.4268
M1_RANS_SST	0.4286	0.01480	0.4286
M2_RANS_SST	0.4296	0.01460	0.4297
M3_RANS_SST	0.4298	0.01449	0.4297
M2_RANS_SA	0.4288	0.01639	-

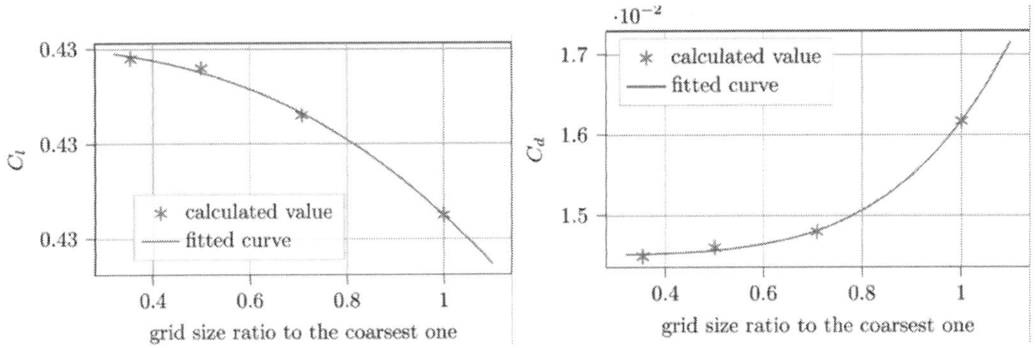

Figure 3. Variation of the lift coefficient and drag coefficient with the grid size.

With the Richardson extrapolation theory [19], the error and mesh size have the following relationship.

$$e_h = y_h - y_{acc} = Ch^p = (Ch_0^p)(h/h_0)^p \tag{1}$$

where y_{acc} is the accurate value, y_h is the value obtained with grid size being h, p is the convergence order, h_0 is an arbitrary basic grid size and C is a constant coefficient. Applying the least square fitting for the results of RANS simulations, the convergence properties are obtained and listed in table 4. The fitting curves are shown in figure 3. The lift coefficient has generally a smaller uncertainty than the drag coefficient. Even with the coarsest mesh M0, the uncertainty of the lift coefficient is as low as around 1%. With mesh finer than M2, the uncertainty of lift and drag are both smaller than 1%.

Table 4. Convergence properties for Cl and Cd wrt. mesh for RANS simulations.

	C_l	C_d
accurate value y_{acc}	0.4301	0.01451
oder of convergence p	2.65	4.90
uncertainty (M0)*	1.05%	12.85%
uncertainty (M1)*	0.431%	2.47%
uncertainty (M2)*	0.139%	0.795%
uncertainty (M3)*	0.081%	0.148%

*For the uncertainty, the safety factor of 1.25 is used.

With the SA model (case M2_RANS_SA), the accuracy of the hydrodynamics is a little lower than that with the k-ω SST model. The C_l of case M2_RANS_SA is quite similar to that of case M1_RANS_SST and the C_d is even worse than that of M0_RANS_SST.

4.2. uRANS simulation for cavitating case

The unsteady RANS simulations for the cavitating case are carried out. The cases and the averaged hydrodynamics and cavity volume are listed in table 5. The cavity volume is calculated as the volume-weighted integral of the vapour volume fraction. When compared with the fully wet case, the lift decreases, and the drag increases. Besides, the hydrodynamic forces show more sensitivity to the mesh. Furthermore, the results do not show a clear convergence trend. As shown in figure 4, the lift and drag do not approach a specific value gradually as the mesh is refined. The convergence order p obtained with the Richardson extrapolation theory is less than 1, which is abnormal. This may suggest that the cavitation model consisting of the VOF and mass transfer methods is mesh dependent.

The variation of the lift coefficient and cavity volume are shown in figure 5. As expected, the fluctuation of the cavity volume and the lift force are both small. For the cavity volume, the fluctuation amplitude is less than 5e-8 m^3, which is far less than the mean cavity volume (¡1%). The fluctuation amplitude of the lift coefficient is also within 1% of the mean value. The inability of RANS to model the cavitation dynamics is now a trivial observation and is verified again here, with either the SA or the k-ω SST turbulence model.

4.3. IDDES simulation for cavitating case

The k-ω SST IDDES and the SA IDDES models are utilized for the cavitating simulations. The quantitative results including the average lift coefficient ($\bar{C}_l$), cavity volume ($\bar{V}_C$), the main fluctuation frequency $f_{V_C,1}$ and the corresponding amplitude of the cavity volume $A_{V_C,1}$ are

Table 5. The average hydrodynamics and cavity volume for cavitating uRANS simulations. M0 - M3 means different mesh. SST and SA means k-ω SST and SA turbulence model, respectively.

case	C_l	C_d	cavity volume (m^3)
M0_uRANS_SST	0.4172	0.02055	7.680e-6
M1_uRANS_SST	0.4100	0.01963	6.890e-6
M2_uRANS_SST	0.4005	0.01904	6.893e-6
M3_uRANS_SST	0.3892	0.01841	5.414e-6
M2_uRANS_SA	0.4108	0.01929	8.470e-6

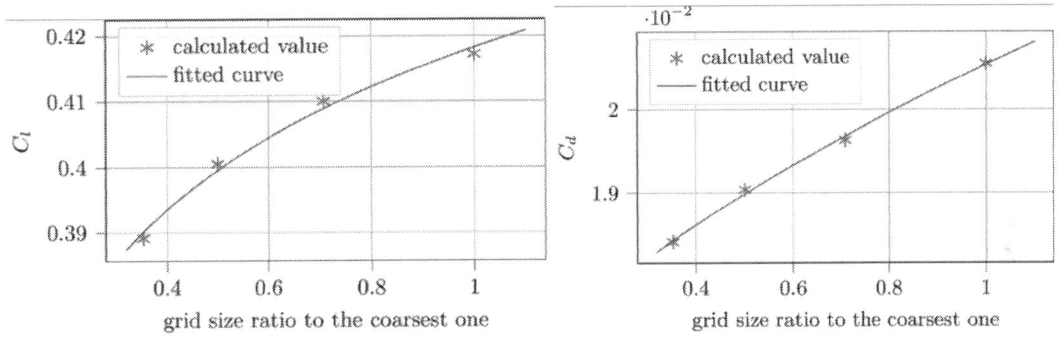

Figure 4. Variation of the lift and drag coefficients with the grid size for cavitating uRANS simulations.

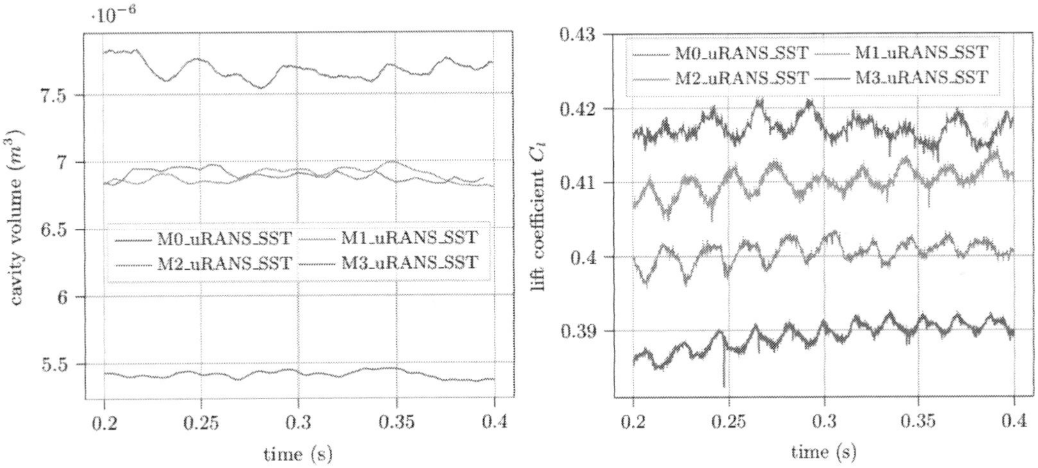

Figure 5. The temporal variation of cavity volume and lift coefficient in uRANS cavitating simulations.

listed in table 6. The temporal variations of the lift coefficient and cavity volume are shown in figure 6. Significant cavity volume fluctuation and cavity shedding phenomenon are observed in the simulations. The cavity volume fluctuation amplitude varies in each period, i.e. the shedding is not very regular. The value of $A_{V_C,1}$ given in Table 5 is obtained using FFT.

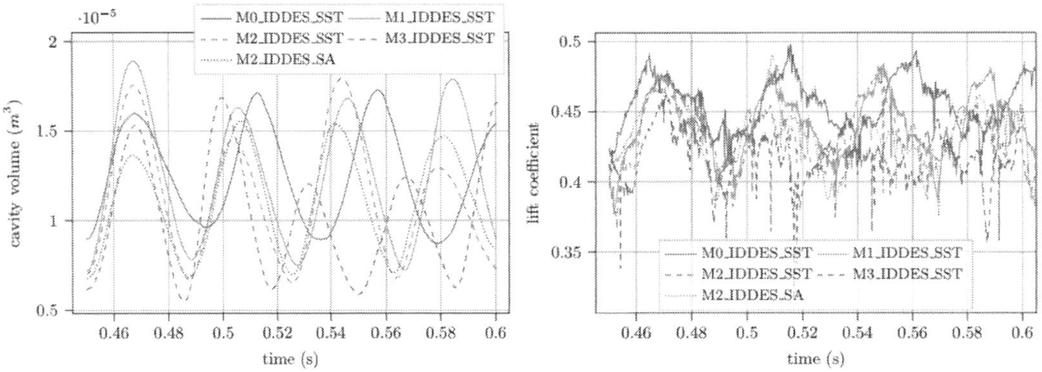

Figure 6. Temporal variation of cavity volume and lift coefficient in IDDES simulations.

The averaged lift coefficient and cavity volume decrease with the mesh refinement except for abnormal behaviour of the mesh M2. The cavity shedding frequency being represented by $f_{V_C,1}$ increases with the mesh refinement. Only to the mesh M3, the shedding frequency begin to correlate well with previous simulations and the experimental measurements (30-35Hz). The cavity volume variation range is roughly between 6 18$10^{(}-6)m^3$, which is larger than the previous simulations presented on SMP11 (see table 1). Compared to the k-ω SST IDDES model, the SA IDDES model leads to smaller values of lift coefficient, cavity volume, and cavity volume fluctuation amplitude. However, the differences are insignificant, and the shedding frequency is at the same level.

Table 6. Quantitative results of IDDES cavitating simulations. M0 M3 means different mesh. SST and SA means k-ω SST IDDES and SA IDDES models, respectively.

case	$\bar{C_l}$	$\bar{V_C}$ (m^3)	$f_{V_C,1}$	$A_{V_C,1}$ (m^3)
M0_IDDES_SST	0.4513	1.128e-5	22.5Hz	3.73e-6
M1_IDDES_SST	0.4384	1.119e-5	25.0Hz	4.01e-6
M2_IDDES_SST	0.4328	1.213e-5	27.5Hz	4.27e-6
M3_IDDES_SST	0.3153	0.946e-5	30.9Hz	4.10e-6
M2_IDDES_SA	0.4254	0.979e-5	26.7Hz	3.56e-6

The method to calculate the cavity volume may have an influence on the results. Except for the volume-weighted integral of vapour volume fraction, the summation of cell volumes where the vapour volume fraction α_v is larger than a threshold value is also utilized to check the influence. Different threshold values including 0.3, 0.5, and 0.7 are tested. The results for a specific time are given in table 7. The result of the volume-weighted integral method is similar to that of the threshold method with the threshold value being 0.5. With the threshold method, the cavity volume decreases with the threshold value. However, the difference is not sufficient to explain the deviation of the current cavity volume range to those presented on SMP11. Considering its implication in hydrodynamic noise, the volume-weighted integration method is preferred in the current work.

To check the influence of timestep, simulations with different max Courant numbers are conducted with mesh M1 and the k-ω SST IDDES model. The results are given in table 8. As

IOP Conf. Series: Materials Science and Engineering 1288 (2023) 012053 doi:10.1088/1757-899X/1288/1/012053

Table 7. The cavity volume calculated with different methods.

method	cavity volume (m^3)
volume weighted integral of vapour volume fraction	9.54e-6
sum of cell volumes with $\alpha_v > 0.3$	1.21e-5
sum of cell volumes with $\alpha_v > 0.5$	1.01e-5
sum of cell volumes with $\alpha_v > 0.7$	7.72e-6

the max Courant number decrease from 10 to 2, the average lift coefficient decreases by 0.5%, the average cavity volume increases by 4%. The shedding frequency increases a little. Generally, a smaller timestep leads to better results, but the influence is insignificant considering that five times of computational resources are consumed.

Table 8. Results of simulations with different max Courant numbers. C10, C5, and C2 means that the max Courant number is 10, 5, and 2, respectively. O10 means that the number of pimple outer corrector iterations is 10.

case	$\bar{C}_l$	$\bar{V}_C$ (m^3)	$f_{V_C,1}$	$A_{V_C,1}$ (m^3)
M1_IDDES_SST_C10	0.4384	1.119e-5	25.0Hz	4.01e-6
M1_IDDES_SST_C5	0.4366	1.158e-5	25.0Hz	3.95e-6
M2_IDDES_SST_C2	0.4366	1.165e-5	26.0Hz	3.22e-6
M3_IDDES_SST_O10	0.4395	1.139e-5	25.0Hz	3.77e-6

The PIMPLE iteration steps affect the final residual in each timestep and thus influence the simulation accuracy. Thus, the influence of the number of outer corrector iterations is also investigated. Another simulation using the M1 mesh with 10 outer corrector iterations is carried out. The results are given in table 8. The temporal variations of cavity volume are shown in figure 7. It should be pointed out that the computational resources consumed by M1_IDDES_SST_C5 and M1_IDDES_SST_O10 are roughly the same. While decreasing timestep and increasing the number of pimple outer corrector iterations both increase the shedding frequency, the latter has a greater effect. Besides, increasing the number of pimple outer corrector iterations also makes the fluctuation more regular. However, the difference is too small to explain the deviation of current simulations from previous ones.

5. Conclusion

In the current work, cavitating simulations were carried out with different setups including different mesh resolutions, time steps, RANS and IDDES turbulence models, for the delft twist 11 hydrofoil. The hydrodynamics, cavity volume, and cavitation shedding frequency are compared between different simulations. By comparing the results, the following points are found:

(i) With the RANS technique the cavitation dynamics cannot be captured, while with the IDDES technique the cavitation shedding is successfully simulated.

(ii) The Spalart-Allmaras and k-ω SST IDDES models lead to similar results, although the Sparlart-Allmaras model shows a slightly lower accuracy.

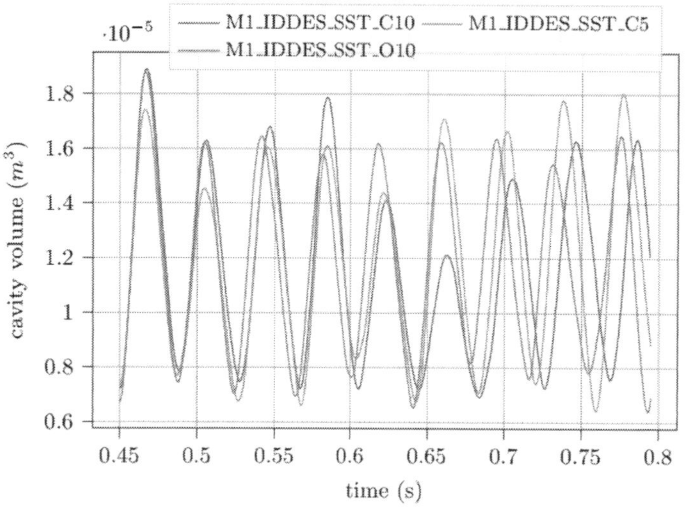

Figure 7. Temporal variation of cavity volume for different timesteps and pimple outer corrector iterations.

(iii) From mesh M0 to M3 (the mesh refinement ratio is $(\sqrt{2})^3$), while the lift force changes less than 1% in fully wet case, it changes around 10% in the cavitating case. The shedding frequency also changes from 22.5Hz to 30Hz. Thus, a verified computational mesh and setup in the fully wet case is not always suitable for cavitating simulations.

(iv) In the current work, the cavitation shedding frequency increases almost linearly with mesh refinement. No clear convergence has been observed. This may due to the mesh-dependent nature of IDDES models.

(v) The timestep and number of pimple outer iterations influence the cavitation shedding frequency to some extent, but not significantly.

Acknowledgments

The work is supported by the Fundamental Research Funds for the Central Universities.

References

[1] Duarte C M, Chapuis L, Collin S P, Costa D P, Devassy R P, Eguiluz V M, Erbe C, Gordon T A, Halpern B S, Harding H R, Havlik M N, Meekan M, Merchant N D, Miksis-Olds J L, Parsons M, Predragovic M, Radford A N, Radford C A, Simpson S D, Slabbekoorn H, Staaterman E, Van Opzeeland I C, Winderen J, Zhang X and Juanes F 2021 *Science* **371** ISSN 10959203

[2] IMO 2022 Noise 2022 accessed on: 2023-01-31 URL `https://www.imo.org/en/MediaCentre/HotTopics/Pages/Noise.aspx`

[3] European-Commission 2019 Our oceans, seas and coasts accessed on: 2020-07-31 URL `https://ec.europa.eu/environment/marine/good-environmental-status/descriptor-11/index_en.htm`

[4] EU 2022 Zero pollution and biodiversity: First ever eu-wide limits for underwater noise 2022 accessed on: 2023-01-31 URL `https://environment.ec.europa.eu/news/zero-pollution-and-biodiversity-first-ever-eu-wide-limits-underwater-noise-2022-11-29_en`

[5] Cruz E, Lloyd T, Lafeber F, Bosschers J, Vaz G and Djavidnia S 2022 *22nd International Symposium on Nonlinear Acoustics* **48** 070021

[6] Hoekstra M, Van Terwisga T and Foeth E 2011 *Proceedings of the 2nd International Symposium on Marine Propulsors, Hamburg, Germany* (Citeseer)

[7] Spalart P and Allmaras S 1992 *30th aerospace sciences meeting and exhibit* p 439

[8] Menter F 1993 *23rd fluid dynamics, plasmadynamics, and lasers conference* p 2906

[9] Yakhot V, Orszag S, Thangam S, Gatski T and Speziale C 1992 *Physics of Fluids A: Fluid Dynamics* **4** 1510–1520

[10] Shih T H, Liou W W, Shabbir A, Yang Z and Zhu J 1994 A new k-epsilon eddy viscosity model for high reynolds number turbulent flows: Model development and validation Tech. Rep. CMOTT-94-6 NACA

IOP Conf. Series: Materials Science and Engineering 1288 (2023) 012053 doi:10.1088/1757-899X/1288/1/012053

[11] Spalart P R 1997 *Proceedings of the First AFOSR International Conference on DNS/LES, 1997* pp 137–147
[12] Gritskevich M S, Garbaruk A V, Schütze J, Menter F R *et al.* 2012 *Flow Turbulence and Combustion* **88** 431
[13] Ferziger J H, Street R L and Peric M 2019 *Computational methods for fluid dynamics* (Springer International Publishing AG)
[14] CFD Online Turbulence modeling accessed on: 2023-04-07 URL `https://www.cfd-online.com/Wiki/Turbulence_modeling`
[15] Schnerr G H and Sauer J 2001 *Fourth international conference on multiphase flow* vol 1 (ICMF New Orleans)
[16] van Terwisga T Explanation with the experimental data for the delft twist 11 foil test case URL `http://maritimetechnology.tudelft.nl/SHS/`
[17] Bensow R E 2011 *SMP'11*
[18] Whitworth S 2011 *SMP'11* pp 15–17
[19] Roache P J 1997 *Annual review of fluid Mechanics* **29** 123–160

Multi-scale characteristics investigation of sheet/cloud cavitation

B C Tian[1], L M Li[2], Q Wu[1], B Huang[1], and G Y Wang[1]
[1]Beijing Institute of Technology, School of Mechanical Engineering, Beijing, China
[2] Zhejiang Sci-Tech University, Zhejiang Hangzhou, China
Email: wuqin919@bit.edu.cn; huangbiao@bit.edu.cn

Abstract: Sheet and cloud cavitation involve the formation of large-scale cavities and microbubbles across a wide range of length scales. In this study, a two-way coupling Eulerian-Lagrangian algorithm was utilized to simulate the cavitating flow around a NACA66 hydrofoil at an incidence of 8 degrees and a cavitation number of 1.2. The large eddy simulation (LES) and volume of fraction (VOF) methods are used to capture the large scale vapor structures in Eulerian frame. Meanwhile, the discrete bubble model (DBM) and simplified Rayleigh-Plesset equation are implied to solve the dynamic of Lagrangian microbubbles smaller than local grid scale. The results indicate that large-scale cavities are periodically shed downstream, influenced alternatively by the re-entrant jet and shock wave mechanisms, resulting in periodic variations of both the number and Sauter mean diameter of microscale bubbles. During the quasi-periodic evolution process of sheet/cloud cavitation, the microbubbles scatter around the large scale cavities where have strong turbulence intensity and high vorticity, and the probability density function of discrete bubble diameter similarly conforms to Gamma distribution with the dominant bubble size between 30 and 70μm. During the growth of an attached cavity, the number density of bubbles follows a power-law scaling with a value of -5/3 on radius. For the other stages, the bubble size spectrum of smaller microbubbles exhibits a -2/3 power-law scaling when the diameter of the microbubble is less than 200 μm. For larger bubbles with a diameter greater than 300 μm, the bubble density is proportional to the bubble radius to the power of -6.

1. Introduction

Cavitation is an unsteady, compressible, complex two-phase flow phenomenon that often occurs when the pressure in a fluid is reduced to a certain extent. It can gradually develop into a multiphase, multiscale state that includes small-scale cavity growth and collapse behavior, as well as large-scale cavitation evolution behavior. Meanwhile, cavitation typically results in vibration[1], erosion[2], and noise[3], which can significantly impede the advancement of technology and equipment in high-speed hydrodynamic fields such as underwater weapons, underwater launch, aerospace, fluid machinery, and national defense. Therefore, accurately capturing the bubble structure at various scales and analyzing the transformation mechanism between bubbles at different scales are critical for understanding and controlling cavitation behavior.

Based on high-speed water tunnels and high-speed camera technology, numerous experimental studies have shown that cavitation patterns gradually develop from incipient to sheet, cloud, and super-cavitation as the Reynolds number increases or the cavitation number decreases. Moreover, sheet/cloud cavitating flow exhibits typical periodic evolution characteristics, including vapor cavity fluctuation, fracture, and shedding downstream, caused by the re-entrant jet[4], Kelvin-Helmholtz waves[5], and condensation shockwaves[6]. Arabnejad et al.[7] investigated the shedding mechanism of cloud cavitation around NACA0009. The results showed that there are two mechanisms for the periodic shedding of cloud cavitating flow: the re-entrant jet mechanism, which results in large-scale vapor cloud shedding, and the shockwave mechanism, which generates small-scale horseshoe vortices. Kubota et al.[8] investigated the unsteady flow field structure of cloud cavitation. Their findings indicated the presence of two scales of velocity disturbances in the flow field: large-scale disturbances formed by the evolution of bubble clouds, and small-scale disturbances generated by bubble collapse. Additionally, the detached cloud vapor clusters contain numerous microscale bubbles. Russell et al.[9] used diffused laser shadowgraphy technology to measure the size and concentration of microbubble populations in the size range of 5 to 300μm in the far wake of turbulent cavitating flow. They discovered that the dominant bubble size was between 10 and 15μm. Additionally, they observed that the microbubble concentration increased with increasing Reynolds number, and the microbubble size increased with decreasing microbubble number. However, due to the refraction of the vapor-liquid interface, the overlap of vapor structures, and resolution limitations, cavitation experiments based on high-speed camera technology cannot provide all the information on bubbles at various scales, which restricts our comprehension of the multi-scale characteristics of cavitating flow.

The advancement of computational fluid dynamics (CFD) technology has made it a trustworthy and effective way to comprehend the physical mechanisms of turbulent cavitating flows. Nearly all numerical methods for modeling cavitation in the Eulerian frame treat water and vapor as a single homogeneous mixture and use different cavitation models to simulate the phase change process. To analyze cavity structures, the volume of fluid (VOF) method[10] and level-set (LS) model[11] are widely used to capture the two-phase interface by calculating the Navier-Stokes equations. However, the homogeneous mixture model and interface capture method of the two-phase interface can only simulate large-scale vapor structures, and subgrid-scale discrete bubbles cannot be captured. In recent years, the Euler-Lagrange method has become an effective way to simulate multi-scale cavitating flows. This method comprehensively uses the Eulerian mixture model to model large cavity structures and the Lagrangian approach to trace sub-grid scale structures and discrete bubble clusters. Chahine et al.[12] developed the discrete bubble model (DBM) to investigate the movement of bubbles in two-phase flows. Hsiao et al.[13] proposed an E-L method by coupling VOF and Discrete Singularity Model (DSM), and applied the model to inception cavitation, growth of nuclei in vortex flow and sheet/cloud cavitation around a hydrofoil. Ghahramani et al.[14-16] applied the coupled method of Eulerian model and Lagrangian model to overcome the limitation of the transport equation model on tracking the microbubble clusters. This hybrid solver was used to analyze the multiscale cavitating flow around various objects, such as hydrofoils, semi-circular cylinders, and blunt bodies. Li et al. [17-18]developed a hybrid and multiscale cavitation model by coupling DBM with VOF, and applied it to study the formation of discrete bubbles and the evolution of large-scale cavity structures in a convergent-divergent channel. This hybrid method has also been successfully used to simulate inception cavitation, cloud cavitation, tip vortex cavitation, and the multiscale evolution of vortex shedding and discrete microbubble clusters. However, while the research on multiscale cavitating flow mainly focuses on observing and studying the transformation process of vapor with different scales, there is less attention given to the influence of the continuous phase flow field on discrete bubble and the quantitative analysis of micro-scale bubbles.

In present work, the previously proposed Eulerian-Lagrangian method was used to simulate the cloud cavitation around NACA66 hydrofoil.[30] Meanwhile, the evolution process of multi-scale cavitating flow and the influence of different shedding mechanisms on the spatiotemporal distribution characteristics of microscale bubble was discussed.

2. Numerical Methodology

2.1. Governing equations and Large Eddy Simulation

The homogeneous equilibrium flow model (HEM) and Large Eddy Simulation (LES) method were applied for the simulation of cavitating flows in Euler frame. The mass and momentum conservation equations of Euler flow as follows:

$$\frac{\partial(\rho_v\alpha_v)}{\partial t}+\frac{\partial\left(\alpha_v\rho_v\bar{u}_j\right)}{\partial x_j}=\dot{m}^{+}+\dot{m}^{-}+S_b \tag{1}$$

$$\frac{\partial(\rho_m\bar{u}_i)}{\partial t}+\frac{\partial(\rho_m\bar{u}_i\bar{u}_j)}{\partial x_j}=-\frac{\partial\bar{P}}{\partial x_i}+\frac{\partial}{\partial x_j}\left[\mu_m\left(\frac{\partial\bar{u}_i}{\partial x_j}+\frac{\partial\bar{u}_j}{\partial x_i}\right)\right]+\rho_m g_i-\frac{\partial\tau_{ij}}{\partial x_j}+f+F_s \tag{2}$$

where u is the velocity, p is the pressure, ρ is the density, and α denotes the volume fraction. The subscripts (i, j, k) represent the directions of the Cartesian coordinates. And the subscripts (l, v) denote the materials water and vapor, respectively. Besides, the VOF technique was used to capture the interface of water and vapor.

The mixture density ρ_m and dynamic viscosity μ_m are defined as

$$\rho_m=\rho_l\alpha_l+\rho_v\alpha_v \tag{3}$$

$$\mu_m=\mu_l\alpha_l+\mu_v\alpha_v \tag{4}$$

$$\alpha_l+\alpha_v=1 \tag{5}$$

The f is the momentum source term and the S_b is the mass source term, which describe the effect of the Lagrangian bubbles on the continuous fluid were defined in Sec.2.4. Where F_s represents the surface tension[19]. The source term m^+, and the sink term m^-, in equation (1) represent the condensation and evaporation rates, respectively.

According to the Boussinesq hypothesis, the SGS stress τ_{ij} in equation (2) can be calculated as

$$\tau_{ij}=\rho_m\left(\overline{u_iu_j}-\bar{u}_i\bar{u}_j\right) \tag{6}$$

$$\tau_{ij}-\frac{1}{3}\tau_{kk}\delta_{ij}=-2\mu_t\bar{S}_{ij} \tag{7}$$

The wall-adapted local eddy-viscosity (WALE) model is formulated locally to solve the eddy viscosity

$$\mu_t=\rho L_S^2\frac{\left(S_{ij}^dS_{ij}^d\right)^{3/2}}{\left(\bar{S}_{ij}\bar{S}_{ij}\right)^{5/2}+\left(S_{ij}^dS_{ij}^d\right)^{5/4}} \tag{8}$$

$$S_{ij}^d=\frac{1}{2}\left(\bar{g}_{ij}^2+\bar{g}_{ji}^2\right)-\frac{1}{3}\delta_{ij}\bar{g}_{kk}^2 \tag{9}$$

$$\bar{g}_{ij}=\frac{\partial\bar{u}_i}{\partial x_j} \tag{10}$$

$$L_s=\min\left(\kappa d,C_SV^{1/3}\right) \tag{11}$$

where $\bar{S}_{ij}$ is the strain rate tensor for the resolved scale, $S_{ij}^dS_{ij}^d$ is a tensor invariant, L_s is the mixing length for sub-grid scale, k is von Karman constant, d is the distance to the closest wall, with the WALE constant $Cs = 0.5$. $V^{1/3}=\left(\Delta x\cdot\Delta y\cdot\Delta z\right)^{1/3}$ represent the local grid size, where Δx, Δy and Δz are the mesh sizes in each direction.

2.2. Cavitation model

In Schnerr–Sauer cavitation model,[20] the growth and collapse of the bubble clusters are assumed to be governed by the simplified Rayleigh–Plesset equation,[21]neglecting the second-order term and the surface tension force. The source and sink terms in equation (3) are used to predict the mass transfer rate in the turbulent cavitating flow and are governed by

$$\begin{cases}\dot{m}^+=C_V\dfrac{3\alpha_v(1-\alpha_v)}{R_B}\dfrac{\rho_v\rho_l}{\rho}\sqrt{\dfrac{2}{3}\dfrac{P_v-P}{\rho_l}},P<P_v\\[2ex]\dot{m}^-=C_c\dfrac{3\alpha_v(1-\alpha_v)}{R_B}\dfrac{\rho_v\rho_l}{\rho}\sqrt{\dfrac{2}{3}\dfrac{P-P_v}{\rho_l}},P\geq P_v\end{cases} \tag{12}$$

According to Schnerr-Sauer cavitation model, the connection between the vapor volume fraction and nucleation expressed by

$$\alpha_v=\frac{n\frac{4}{3}\pi R_b^3}{1+n\frac{4}{3}\pi R_b^3} \tag{13}$$

where R_B is the bubble radius, p_v is the saturated liquid vapor pressure, and p is the local fluid pressure. In present work, the assumed model constants are C_v=1, C_c=1 and the bubble number density $n=10^{13}$. In addition, the bubble radius can be calculated as

$$R_B=\left(\frac{\alpha_v}{1-\alpha_v}\frac{3}{4\pi n}\right)^{1/3} \tag{14}$$

2.3. Discrete bubble model

The discrete bubble model (DBM) for microscale bubble is applied to achieve a multiscale simulation for cavitating flow. Each bubble is tracked individually using Newton's second law. The kinematics equation and dynamics equation of discrete bubble can be written as

$$\frac{dx_b}{dt}=u_b \tag{15}$$

$$m_b\frac{du_b}{dt}=F_{buo}+F_D+F_{VM}+F_{PG} \tag{16}$$

where x_b is bubble displacement, m_b is the bubble mass and u_b is the bubble velocity. The terms on the right-hand of equation(16) represent the forces acting on microscale bubble, presenting the buoyancy force, the drag force,[22] the virtual mass force,[23] and the pressure gradient force,[24] respectively. Which are defined as

$$F_{buo}=V_b g\left(\rho_l-\rho_b\right) \tag{17}$$

$$F_D=V_b\frac{18\mu}{d_b^2}\frac{C_D\,\mathrm{Re}}{24}\left(u_l-u_b\right) \tag{18}$$

$$F_{VM}=C_{VM}\rho_l V_b\left(u_b\nabla u_l-\frac{du_b}{dt}\right) \tag{19}$$

$$F_{PG}=V_b\rho_l u_l\nabla u_l \tag{20}$$

where the subscripts l and b represent the Euler fluid and the Lagrangian bubble in those equations, d_b is the bubble diameter and the relative Reynolds number Re defined as

$$\mathrm{Re}=\frac{\rho_l d_b\left|u_b-u_l\right|}{\mu_l} \tag{21}$$

The drag coefficient C_D in equation (18) is determined by the spherical drag law, and the virtual-mass force coefficient C_{VM} in equation (19) equal to 0.5.

In addition, bubbles would grow and collapse rapidly depending on the local pressure after nucleation in cavitating flow. In present work, the cavitation nuclear is injected randomly in the global computational region and the bubble radius is calculated as

$$\frac{dR_b}{dt}=\sqrt{\frac{2}{3}\left(\frac{\left|1.25P_v-P\right|}{\rho_l}\right)} \tag{22}$$

Referring to the previous research, the initial cavitation nuclear density is assumed to 2×10^8, [25-26] the initial bubble radius is set as 1×10^{-6} m.[27] On account of the contact time and the Weber number of microscale bubble extremely smaller than the relevant film drainage time and the critical Weber number in turbulent cavitating flow, the aggregation and breakage of discrete bubble can be neglected in this work.[28-29]

2.4. Transition between Eulerian and Lagrangian frames

The pressure, velocity, and vapor volume fraction (α_v) are obtained by solving the continuity equation and momentum equations of the continuous flow field in the Eulerian frame. Once the volume fraction exceeds 1, the cavitation vapor is treated as a large-scale vapor structure and presented through the VOF method. Otherwise, the small-scale cavity will be transformed into the Lagrangian frame as a discrete bubble. The cavitation nuclei will be inserted randomly in the fluid domain as soon as the Euler field iterative calculation is completed, and the bubbles will be tracked in the Lagrangian frame. In this algorithm, discrete bubbles will collapse and be removed immediately in the region where the local pressure is higher than the saturated vapor pressure. In the lower pressure area, the bubble sizes and locations will be updated by solving the motion and dynamic equations in the Lagrangian frame. If the discrete bubble is large enough to be resolved by enough computational cells, such that the relative volume of bubbles exceeds the local discretized domain grid, it is calculated through the VOF method. Otherwise, it is tracked as microscale dispersed bubbles in the Lagrangian frame using the DBM method. The continuous field and discrete phase are then solved alternately until the resolutions meet the convergence requirements before advancing to the next time step.

It should be noted that the governing equation needs to be modified by updating the source terms in equation (1) and equation (2) before each continuous field iterative in order to ensure the mass and momentum conservation in cavitating flow simulation

The mass source S_b in equation (1) presents the effect of removed bubble on Euler continuous field, which can be derived as

$$S_b = \frac{V_B}{V_{cell}\Delta t} \cdot \rho_v \tag{23}$$

Additionally, the cavitation vapor is categorized as Eulerian structures and Lagrangian bubbles to be tracked in the corresponding framework. Therefore, the vapor volume fraction for the next continuous field iterative solution should be modified as

$$\alpha_v' = \alpha_v + \frac{\sum V_{b,j}}{V_{cell,i}} \tag{24}$$

Where the α_v' denotes the vapor volume value for the next iterative in Euler frame.

The interaction force f in equation (2) is the resultant of buoyancy force, drag force, virtual-mass force and pressure gradient force. Which is integrated and averaged in each control volume for momentum exchange to achieve the two-way Euler-Lagrangian coupling simulation, as described

$$f_i = -\frac{1}{V_{cell,i}} \sum \left(F_{D,j} + F_{buo,j} + F_{VM,j} + F_{PG,j} \right) \tag{25}$$

Where the subscript i denotes the serial number of control volume and the subscript j denotes the serial number of discrete bubbles in volume i.

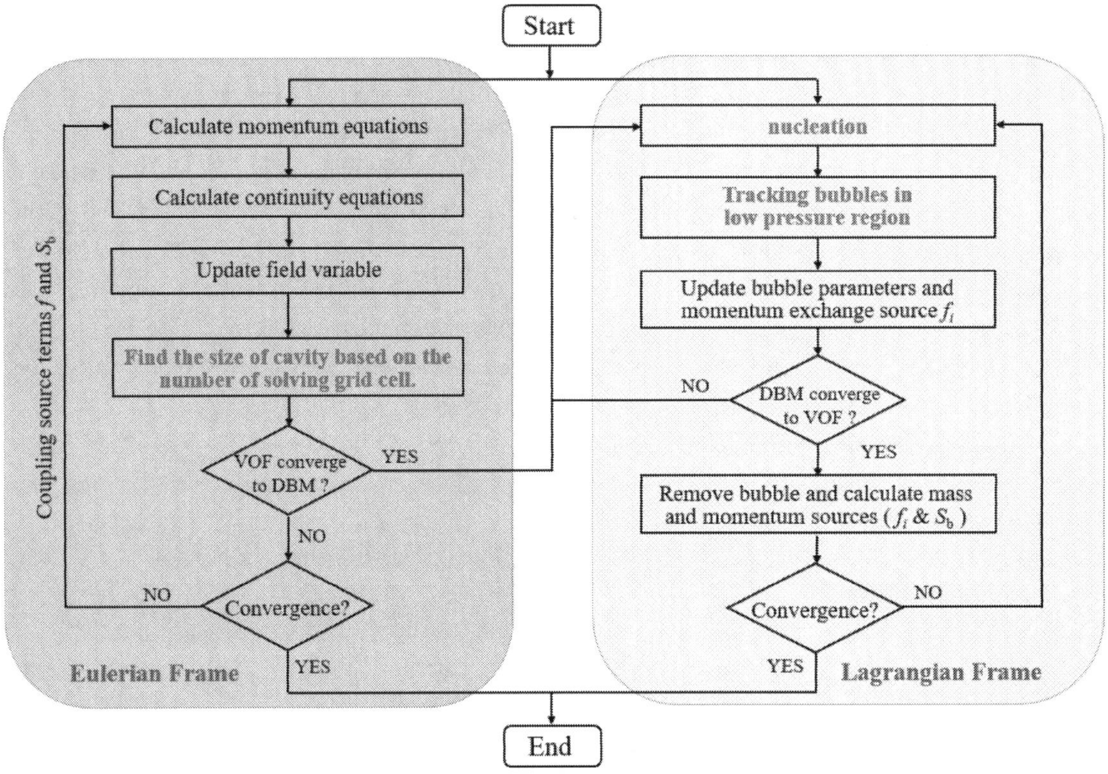

Figure 1. Flow diagram of the coupled Euler-Lagrangian simulation.

2.5. computational domain and mesh generation

As shown in figure 2, the length and height of the computational domain are 9c and 2.5c, respectively. The width of the computational domain is set to 0.3c, which is equal to the hydrofoil span width and more than double the thickness of the hydrofoil, to ensure both space economy and computational efficiency in accurately resolving the stream-wise vortex. The hydrofoil is located at the center of the computational domain with a fixed angle of attack (AOA) of 8°. The leading edge of the hydrofoil is 4c downstream from the inlet of the domain and 1.25c below the top surface of the domain. The top and bottom boundaries, as well as the surfaces of the hydrofoil, are set as no-slip boundary conditions. The front and back boundaries of the tunnel are symmetric. The left boundary is set as velocity inlet with a fixed velocity $U_\infty = 10m/s$, which is corresponding to the Reynolds number $\mathrm{Re} = 10^6$. In addition, the reference pressure of outlet p_∞ varies with the cavitation number, which is defined as $\sigma = (p_\infty - p_v)/(0.5\rho_l U_\infty^2)$. Where the saturated vapor pressure $p_v = 3169pa$. The structured grid is given in figure 3 for computational domain meshing with the encryption around the leading edge, trailing edge, and wake zone of the hydrofoil for ensuring that the grid resolution satisfy the $\Delta y{+}{=}u_\tau\Delta y/v{\sim}1$ condition along the surface of the hydrofoil, where u_τ is the friction velocity on the wall, and v is the kinematic viscosity of the liquid. Finally, the fine 3D fluid mesh which is composed of more than 4.5 million is conducted to capture the detail structure of cavitating flow.

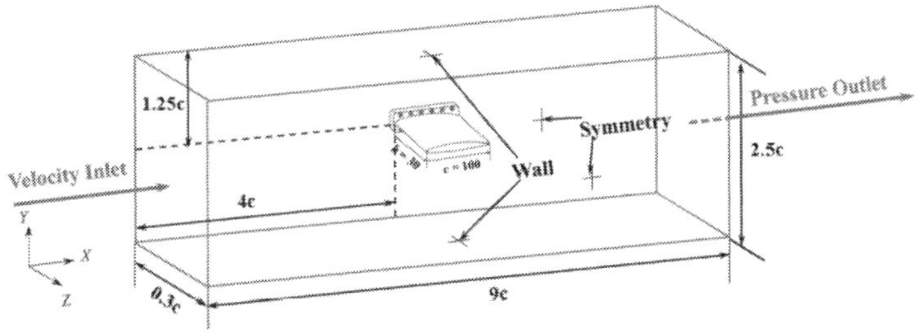

Figure 2. Computational domain and boundary condition

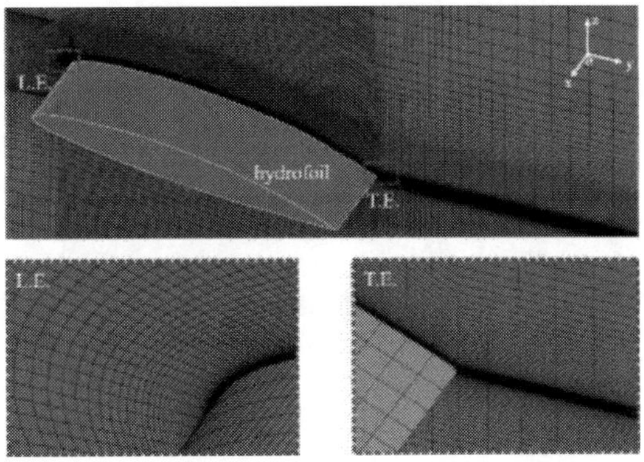

Figure 3. Structured meshes around the hydrofoil.

3. Results and Discussion

3.1 Evolution of multiscale cavity topologies

The Eulerian-Lagrangian cavitation model is applied to simulate cloud cavitating flow. Figure 4 shows the variation of the normalized length of the large-scale cavity over time in one cycle. Under the condition of U=10 m/s and σ=1.2, the period of cloud cavitation is about T_{ref}=56ms. The cavity length shows two different variation features in one cycle. From t=t0 to $t=t_0+0.39T_{ref}$, the large-scale cavity gradually grows to 0.85c and instantly detaches from the suction surface under the effect of the re-entrant jet. Then, from $t=t_0+0.39T_{ref}$ to $t=t_0+0.69T_{ref}$, the large-scale cavity grows to c at a faster rate while gradually shortening until it falls off completely under the influence of the shock wave.

As shown in figure 5 and figure 6, several numerical snapshots of vapor cavity structures at different shedding mechanism in a typical cycle of cloud cavitation. From $t=t_0$ to $t=t_0+0.21T_{ref}$, the sheet cavity initially occurs at the hydrofoil leading edge and gradually grows to the middle of hydrofoil suction surface with a few sub-grid scale bubbles distributing on the surface and the rear region of sheet cavity. As depicted at $t=t_0+0.21T_{ref}$, the re-entrant jet gradually generates from the attached cavity trailing edge and moves upstream along the hydrofoil suction surface due to the reverse pressure gradient between the vapor field and water field. Under the influence of re-entrant jet, the surface of sheet cavity fluctuates more violently and the discrete bubble number increases obviously in the disturbed region of attached cavity. Then at $t=t_0+0.39T_{ref}$ the re-entrant jet reaches the hydrofoil leading edge causing the attached cavity detach from the suction surface of hydrofoil. At this instant, the microscale bubbles scatter in the whole detached sheet cavity. From $t=t_0+0.39T_{ref}$ to $t=t_0+0.69T_{ref}$, as shown in figure 6, under the combined effect of mainstream and the re-entrant flow, the detached cavity shrinks, converges, and then sheds downstream. During this process, the shedding cavity quickly changes from a large-scale cavity into numerous microscale bubbles which finally collapse in the high-pressure region at the trailing edge of hydrofoil. Meanwhile, the largescale cavity regenerates from the leading edge of hydrofoil again and reaches the trailing edge at $t=t_0+0.69T_{ref}$. From $t=t_0+0.69T_{ref}$ to $t=t_0+T_{ref}$, microscale bubble cluster collapses rapidly at the tail edge of the hydrofoil resulting shock wave propagates upstream along hydrofoil causing the largescale break into numerous bubbles. When the shock wave propagates to the leading edge of the hydrofoil $t=t_0+T_{ref}$, the largescale attached cavity disappears and cavitation enters the next period.

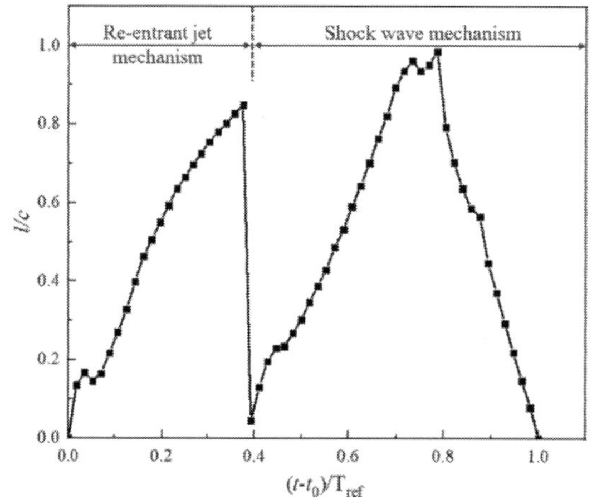

Figure 4. Normalized lengths (l/c) of largescale cavity for one cycle.

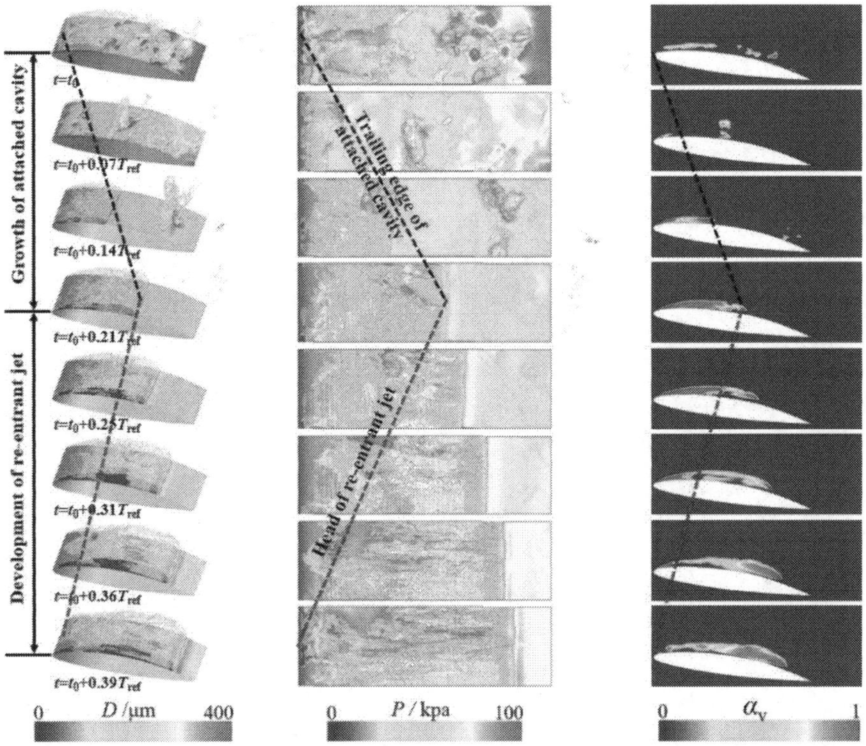

Figure 5. Evolution process of vapor structure under re-entrant jet mechanism

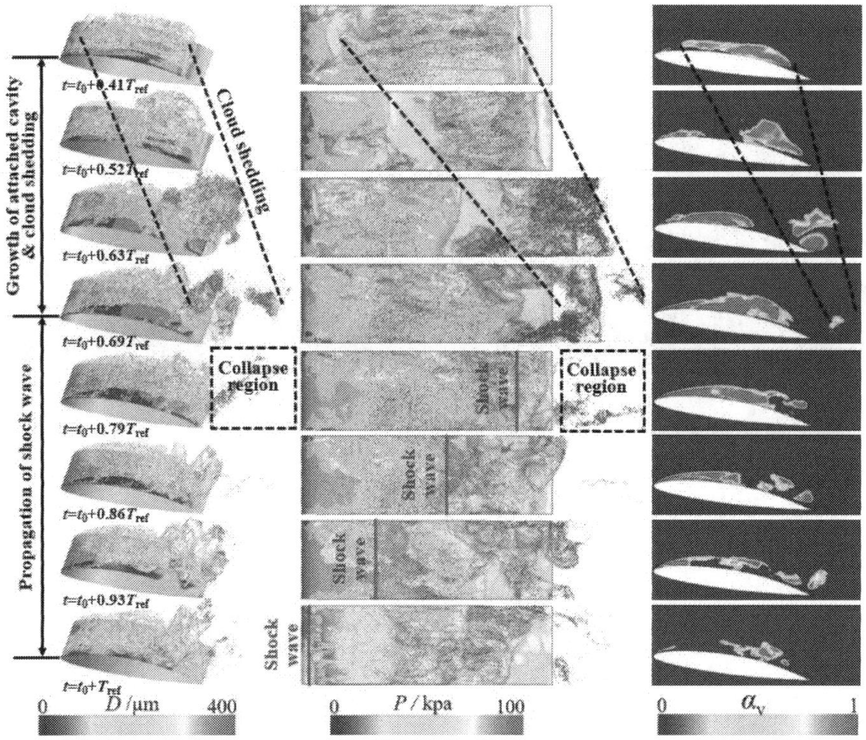

Figure 6. Evolution process of vapor structure under shock wave mechanism

Figure 7 presents power spectrum density of velocity at x=0.7c on suction surface of the mid-section of hydrofoil. Cavitating flow exhibits two typical frequency characteristics. For the low-frequency range below 100Hz, the energy spectrum has high amplitude and low-frequency fluctuation characteristics. The dominant frequency is 17.8Hz, which corresponds to the cavitation evolution cycle. Additionally, the second-order main frequency of 37.6Hz and third-order main frequency of 69.32Hz correspond to the largescale shedding caused by shock wave and re-entrant jet, respectively. In contrast, for frequencies higher than 100Hz, the amplitude of the power spectral density is relatively small and follows a power-law distribution with a slope of -5/3. This suggests that the cavities gradually break into smaller ones due to the vortex cascade.

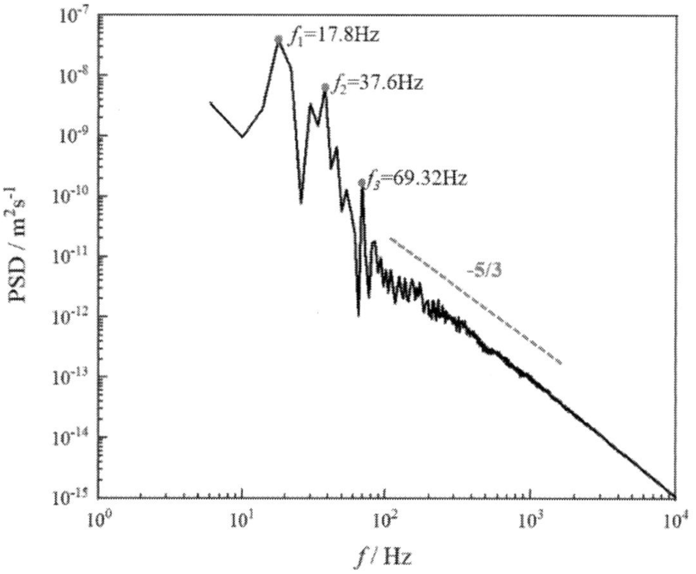

Figure 7. Power spectrum density of velocity at x=0.7c of suction surface on the mid-section of hydrofoil

3.2 Transient temporal-spatial distribution of microbubbles

According to abovementioned discussion, cloud cavitating flow around NACA66 hydrofoil with 8°attack angle is characterized by re-entrant jet and shock wave alternatively under the condition of σ=1.2. The transient evolution process of large-scale cavity contains four stages: (a) the growth of attached cavity, (b) development of re-entrant jet, (c) cloud shedding and (d) propagation of shock wave. In addition to, the periodic variation of largescale cavity is accompanied by the complex dynamic evolution of a microscale bubbles.

Figure 8 display the variation of the number and Sauter mean diameter of microbubbles over time in one cycle of cloud cavitating flow. The Sauter mean diameter (SMD) can quantitatively characterize the mean size feature of microscale bubble groups,[31] which defined as

$$d_{32} = \sum_{i=1}^{n} d_i^3 / \sum_{i=1}^{n} d_i^2 \tag{26}$$

From $t=t_0$ to $t=t_0+0.21T_{\rm ref}$, the attached cavity on the leading edge gradually grows toward the trailing edge. The number and size of microbubbles in the flow field remain relatively stable. From $t=t_0+0.21T_{\rm ref}$

to $t=t_0+0.39T_{ref}$, the adverse pressure gradient promotes a re-entrant jet formed at the trailing edge of the sheet cavity. Then the re-entrant flow moves upstream and enhances the attached cavity fluctuations, causing the rupture of the attached cavity and the mixture of water and vapor. At $t=t_0+0.39T_{ref}$, the re-entrant jet reaches the hydrofoil leading edge and the attached cavity breaks off from the suction surface instantly. During the development of re-entrant jet, both the number and SMD of bubbles increase gradually. From $t=t_0+0.39T_{ref}$ to $t=t_0+0.69T_{ref}$, due to the impact of the mainstream, the detached cavity first converges into a large-scale cavity and sheds downstream along with the hydrofoil. The cavity then breaks into numerous discrete bubbles, leading to a rapid increase in the number and Sauter mean diameter (SMD) of microbubbles in the flow field. The maximum number of discrete bubbles is around 70,000, and the maximum SMD is around 350 μm. Finally, from $t=t_0+0.69T_{ref}$ to $t=t_0+T_{ref}$, the bubbles quickly collapse and result in a shock wave, accompanied by a sharp reduction in the number and SMD of bubbles. Additionally, the number and size remain relatively stable during the propagation of the shock wave along the suction surface of the hydrofoil.

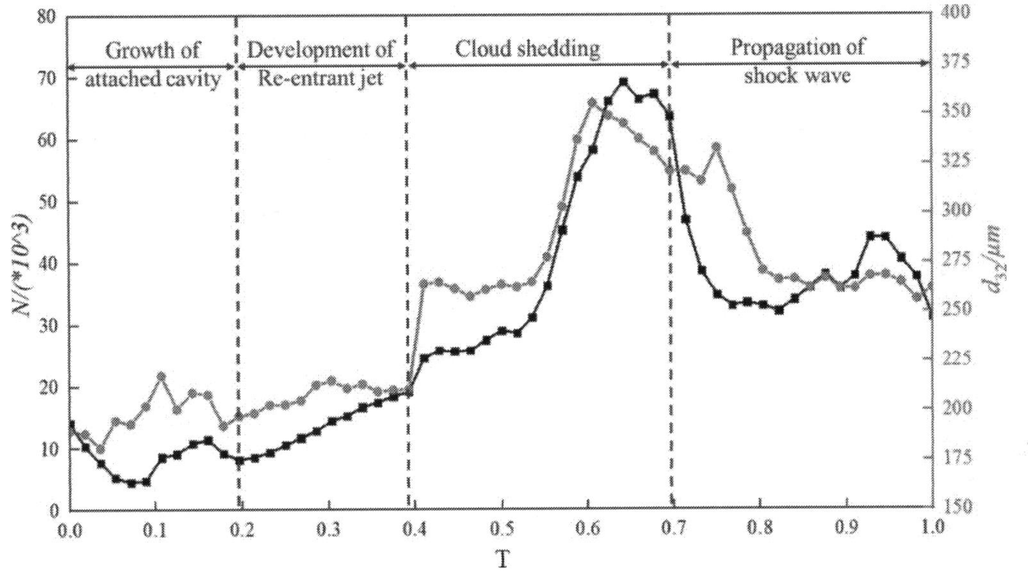

Figure 8. Variation of the number and Sauter mean diameter of discrete bubble in one cycle

To comprehensively investigate the distribution characteristics of discrete bubbles along the hydrofoil, figure 9 displays the time-evolution number density of microbubbles for four stages of cloud cavitation. The color scale denotes the log-scale of the number of microbubbles, N, with a diameter interval of $\Delta d=10\mu m$ and a displacement interval of $\Delta x=0.05c$. The contour consists of solid lines in black, blue, brown, and grey, which denote isolines of number density with values of 300, 500, 700, and 900, respectively. The results indicate that during the growth of the attached cavity, microbubbles mainly distribute in the middle and rear part of the hydrofoil, as well as the wake region, where the large-scale cavity of the previous cycle breaks into microbubbles with sizes concentrated in the range of 0-120μm. During the development of the re-entrant jet, microbubbles mainly distribute in the disturbance region of the re-entrant jet, with a diameter of microbubble less than 200μm. In the stage of cloud shedding, microbubbles mainly distribute in the middle and rear of the hydrofoil, as well as the wake area of the flow around the hydrofoil, with microbubbles concentrated in the range of 0-300μm. In the process of shock wave propagation, the attached cavity is disturbed by the shock wave, causing rapid collapse, and the number density of microbubbles at various scales is relatively large, with concentration in the range of 0-300μm

In order to furtherly illustrate the characteristics of microbubble clusters, figure 10 shows the averaged statistical character of microscale bubble number density. And the probability density functions (PDF) of bubble diameter conform to Gamma distributions.

$$f\left(d_b^*,\alpha,\beta\right)=\frac{\alpha^\beta}{\Gamma(\beta)}d_b^{*\beta-1}e^{-\alpha d_b^*} \tag{27}$$

where α is inverse scale parameter, β is shape parameter. The non-dimensional bubble diameter is defined as

$$d_b^*=d_b/\overline{d_{32}} \tag{28}$$

As shown in figure 10, the number density of bubbles increases rapidly with increasing bubble diameter and then gradually decreases. The dominant sizes of microbubbles are consistently in the 30-70μm range across all stages of cloud cavitation.

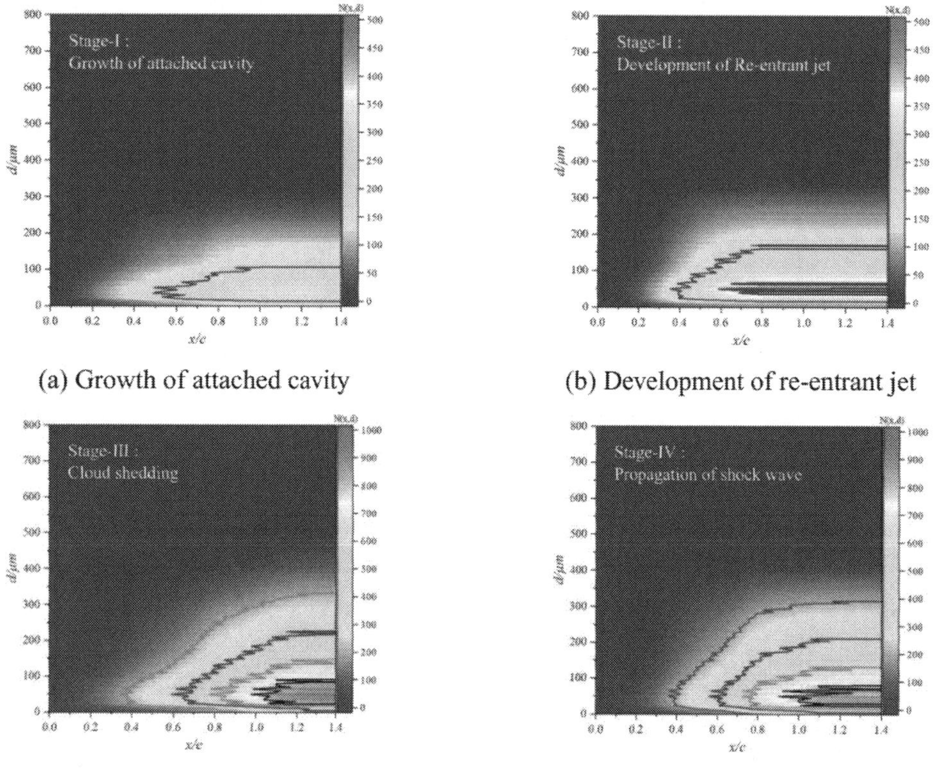

(a) Growth of attached cavity (b) Development of re-entrant jet

(c) Cloud shedding (d) Propagation process of shock wave

Figure 9. Time-averaged distribution along chord wise of various scale bubble for different stage of cloud cavitation

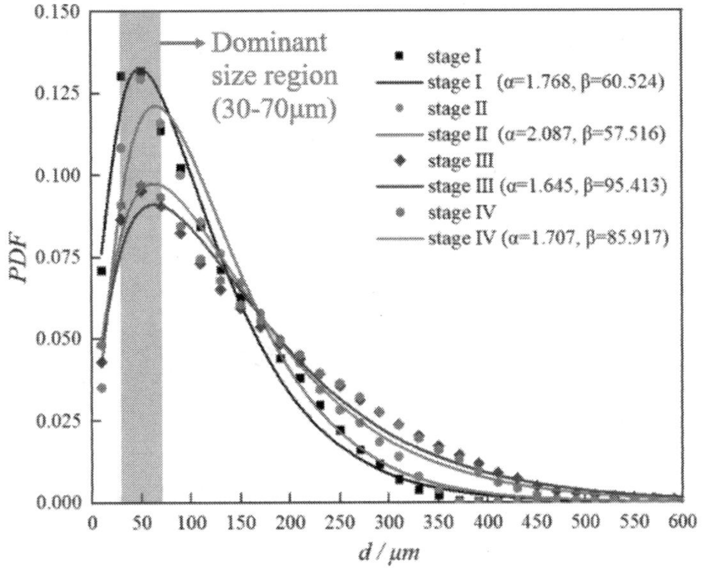

Figure 10. probability density of various scale bubble for different stage of cloud cavitation.

3.3 Effect of turbulence on dynamic characteristics of microbubble

Figure 11 illustrates the spatial characteristics of microbubbles at typical instants during different stages of cloud cavitation. The white solid line represents the iso-line of the vapor interface with a value of α=0.1, while the contour indicates the number density of microbubbles of all sizes at different locations along the hydrofoil. The results indicate that discrete bubbles are primarily scattered in the rear part region of the attached cavity interface and the rupture region of the largescale cavity. This is due to the complex vortex structures that exist in the tail edge of the largescale cavity and the cloud cavity area, which causes strong turbulent pulsation. The energy dispersion effect of the turbulent flow is significant, promoting the rupture of the cavity interface and the generation of discrete bubbles,[32] as illustrated in figure 12.

To investigate the underlying mechanism behind the model for the bubble size spectrum in cloud cavitation, figure 13 illustrates the logarithmic distribution of microbubbles with varying diameters. The bubble size distribution was estimated by calculating the average number of microbubbles across various scales for the four different stages of cloud cavitation. During the growth of the attached cavity, the turbulence of the cavitating flow is relatively weak, and the number density of bubbles follows a power-law scaling on the radius, with a value of -5/3. For the stages of high-intensity turbulence, such as the development of the re-entrant jet, cloud shedding, and propagation of shock waves, the bubble spectrum reveals two distinct scaling laws. The bubble size spectrum of smaller microbubbles exhibits a -2/3 power-law scaling when the diameter of the microbubble is less than 200μm. For larger bubbles with a diameter larger than 300μm, the bubble density is proportional to the bubble radius to the power of -6.

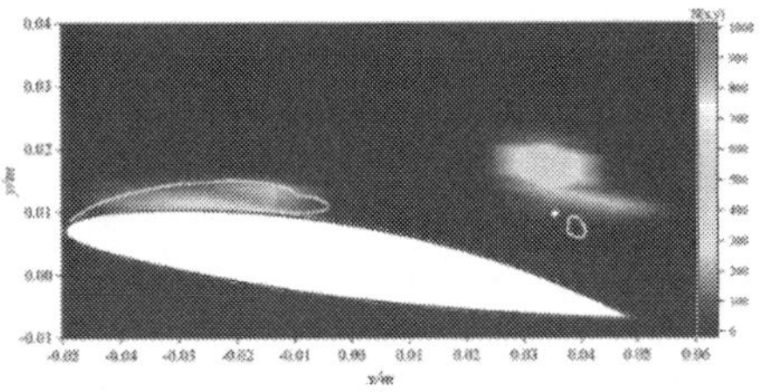

(a) typical instant in the growth of attached cavity t=t_0+0.16T_{ref}

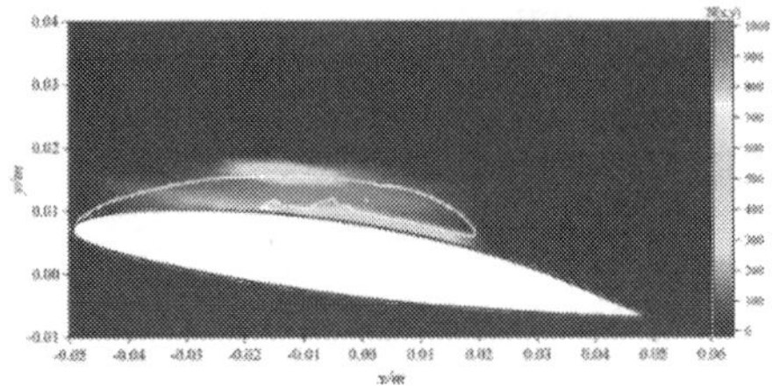

(b) typical instant in the development of re-entrant jet t=t_0+0.27T_{ref}

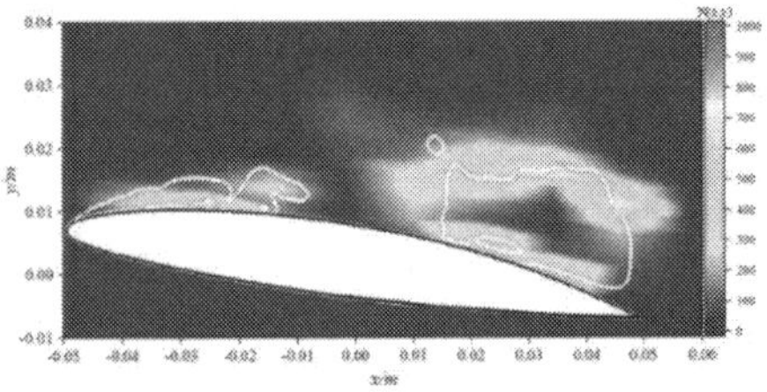

(c) typical instant in the cloud shedding t=t_0+0.55T_{ref}

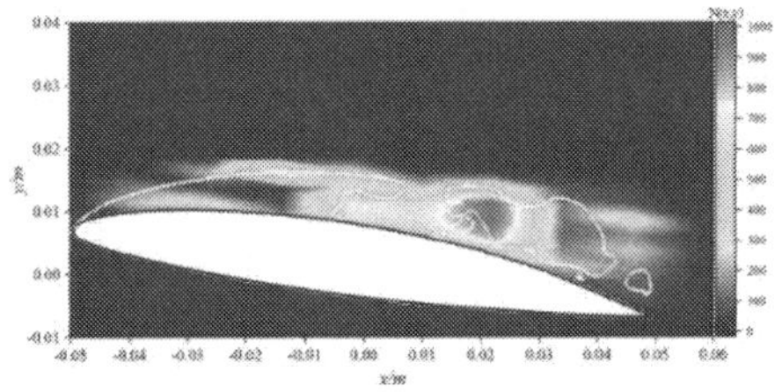

(d) typical instant in the propagation of shock wave t=t_0+0.88T_{ref}

Figure 11. Spatial characteristics of microbubble for typical instants of different stage of cloud cavitating flow.

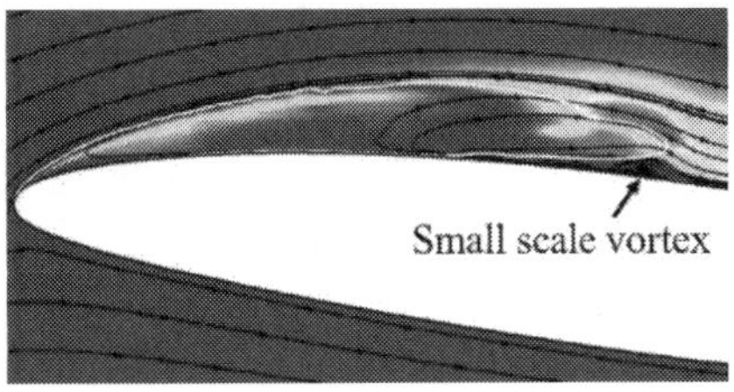

(a) typical instant in the growth of attached cavity t=t_0+0.16T_{ref}

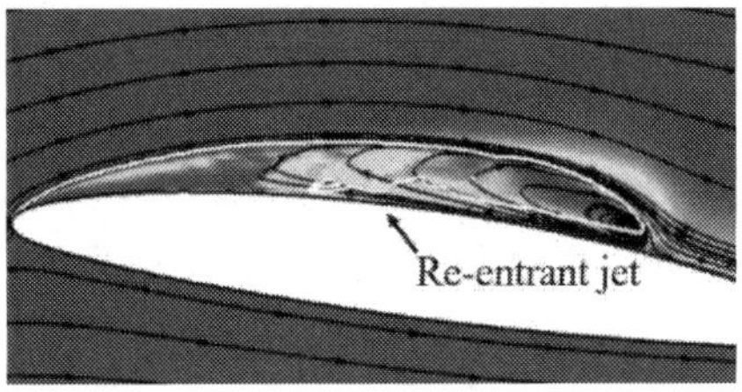

(b) typical instant in the development of re-entrant jet t=t_0+0.27T_{ref}

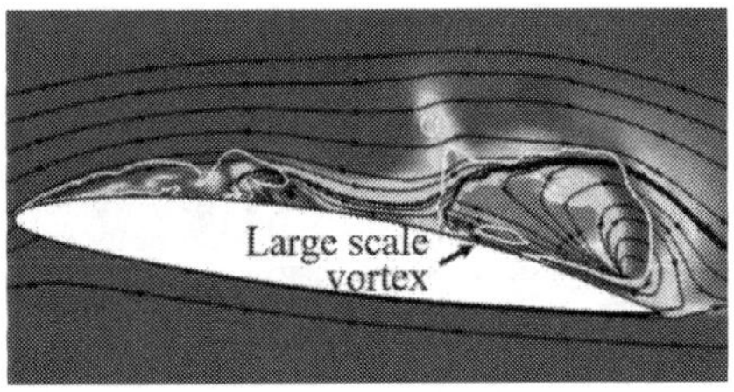

(c) typical instant in the cloud shedding t=t_0+0.55T_{ref}

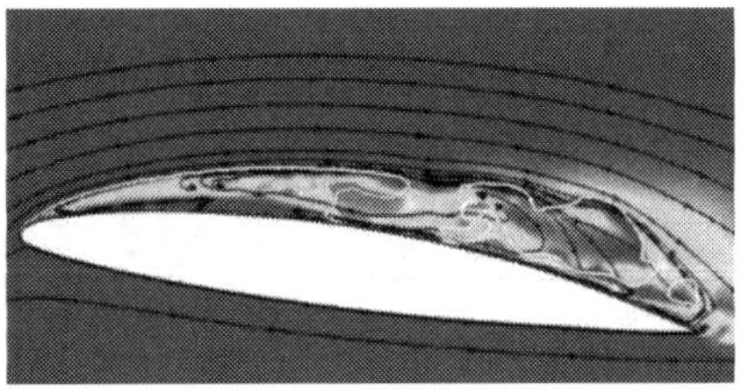

(d) typical instant in the propagation of shock wave t=t_0+0.88T_{ref}

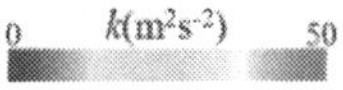

Figure 12. Spatial characteristics of turbulent kinetic energy for typical instants of different stage of cloud cavitating flow.

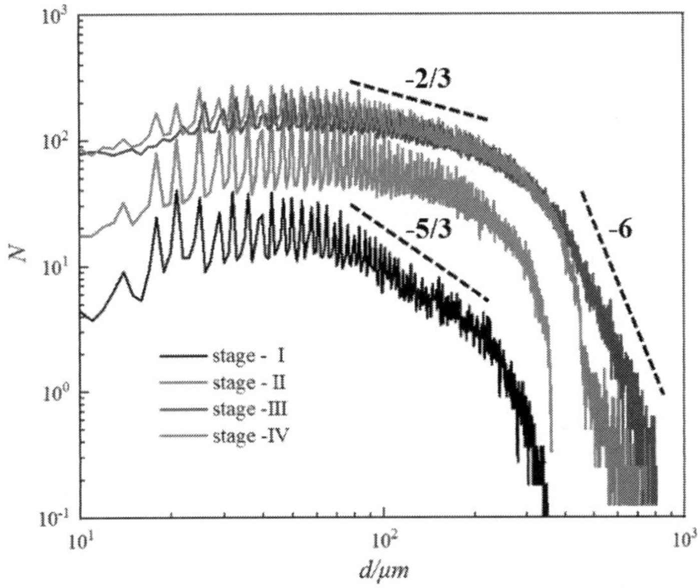

Figure 13. Bubble-size spectrum of different stage of cloud cavitating flow.

4. Conclusion

In the present study, a two-way coupled Eulerian-Lagrangian method was utilized to investigate the multiscale characteristics of cavitating flow around NACA66 under the conditions of U=10m/s and σ=1.2. The large-scale cavity was captured using the VOF method, and the discrete bubbles were tracked using the DBM model. Detailed investigations were conducted on the complex evolution of cavity topologies, statistical characteristics, and the effect of turbulent flow on the dynamics of microbubbles. The primary findings are:

(1) The previously proposed Eulerian-Lagrangian method can effectively simulate the turbulent multiscale cavitating flow characteristics under the conditions of U=10m/s and σ=1.2. One period of multiscale cavitating flow lasts approximately 56ms and includes four stages: growth of the attached cavity, development of the re-entrant jet, shedding of the cloud, and propagation of the shock wave. The microbubbles are mainly distributed at the interface and the breaking region of the large-scale cavity, where high-intensity turbulence occurs.

(2) As cloud cavitation develops, the number and Sauter mean diameter of microscale discrete bubbles in the cavitation field gradually increase, eventually reaching their peak value at the end of the cloud shedding stage when the large-scale cavity breaks into microbubbles. Subsequently, the microbubbles rapidly collapse in the trailing edge region of the hydrofoil, leading to shock waves that propagate upstream along the suction surface. This process results in a reduction in the length of the attached cavity and the transformation of the large-scale cavity into microbubbles.

(3) The probability density of the bubble radii follows similar Gamma distributions at all stages, with the dominant size of microbubbles ranging between 30-70μm. During the relatively stable growth stage of the attached cavity, the number density of bubbles follows a power-law scaling on radius, with a value of -5/3. For the other stages, including the development of the re-entrant jet, cloud shedding, and propagation of the shock wave, the bubble spectrum reveals two distinct scaling laws. The size spectrum of smaller microbubbles exhibits a -2/3 power-law scaling when the diameter is less than 200μm. For

larger bubbles with diameters greater than 300μm, the bubble density is proportional to the bubble radius to the power of -6.

References

[1] Feng J, Liu B, Luo B, Zhu G, Li K, and Wu G, Experimental investigation on characteristics of cavitation-induced vibration on the runner of a bulb turbine, 2022 *Mech. Syst. Signal Process.* **189** 110097

[2] Wang C, Wang G, Zhang M, and Huang B, Experimental investigation of wall-pressure fluctuations in compressible turbulent cavitating flows with emphasis on non-Gaussian features 2022 *Exp. Therm. Fluid Sci.* **139** 110726

[3] Kuroyama T, Ogasawara H, and Mori K, Characterization as measurement sound source of acoustic cavitation noise from bubble clusters under ultrasonic horn 2023 pp. 0–6, 2023.

[4] Brunhart M, Investigation of cavitation and vapor shedding mechanisms in a Venturi nozzle, 2020 *Phys. Fluids*, **32** 8

[5] K. Yoon, S. Qin, S. Shao, and J. Hong, "Internal flows of ventilated partial cavitation," *Exp. Fluids*, vol. 61, no. 4, 2020, doi: 10.1007/s00348-020-2931-y.

[6] Wang C, Liu Y, Chen J, Zhang Y, Huang B, and Wang G, Cavitation vortex dynamics of unsteady sheet/cloud cavitating flows with shock wave using different vortex identification methods, 2019 *J. Hydrodyn.* **31** 3

[7] Arabnejad M, Amini A, Farhat M, and Bensow R, Numerical and experimental investigation of shedding mechanisms from leading-edge cavitation, 2019 *Int. J. Multiph. Flow*, **119**

[8] Kubota A, Kato H, Yamaguchi H, and Maeda M, Unsteady structure measurement of cloud cavitation on a foil section using conditional sampling technique, 1989 *J. Fluids Eng. Trans. ASME*, **111**, 2

[9] Russell P, Giosio D, Venning J, Pearce B, and Brandner P, Microbubble Generation from Condensation and Turbulent Breakup of Sheet Cavitation, 2016 *31st Symp. Nav. Hydrodyn.* **9** 11–16

[10] Hirt C and Nichols B, Volume of fluid (VOF) method for the dynamics of free boundaries, 1981 *J. Comput. Phys.* **39** 1

[11] Osher S and Sethian J, Fronts propagating with curvature-dependent speed: Algorithms based on Hamilton-Jacobi formulations, 1988 *J. Comput. Phys.* **79** 1

[12] CHAHINE G, Numerical simulation of bubble flow interactions, 2009 *J. Hydrodyn.*, **21** 3

[13] Hsiao C, Ma J, and Chahine G, Multiscale tow-phase flow modeling of sheet and cloud cavitation, 2017 *Int. J. Multiph. Flow* **90** 102–117

[14] Ghahramani E, Arabnejad M, and Bensow R, Realizability improvements to a hybrid mixture-bubble model for simulation of cavitating flows 2018 *Comput. Fluids* **174** 135–143

[15] Ghahramani E, Arabnejad M, and Bensow R, A comparative study between numerical methods in simulation of cavitating bubbles 2019 *Int. J. Multiph. Flow* **111** 339–359

[16] Ghahramani E, Ström H, and Bensow R, Numerical simulation and analysis of multi-scale cavitating flows, 2021 *J. Fluid Mech.* **922** 1–54

[17] Li L, Wang Z, Li X, Wang Y, and Zhu Z, Very Large Eddy Simulation of Cavitation from Inception to Sheet/Cloud Regimes by A Multiscale Model, 2021 *China Ocean Eng.* **35** 3

[18] Li L, Jiang B, Wei G, Li X, and Zhu Z, Multiscale multiphase flow simulations using interface capturing and Lagrangian particle tracking, 2022 *Phys. Fluids*, **34** 12

[19] Brackbill J, Kothe D, and Zemach C, A continuum method for modeling surface tension, 1992 *J. Comput. Phys.* **100** 2

[20] Schnerr G and Sauer J, Physical and Numerical Modeling of Unsteady Cavitation Dynamics, 2001 *4th Int. Conf. Multiph. Flow*, **6**

[21] Athavale M, Li H, Jiang Y, and Singhal A, Application of the full cavitation model to pumps and inducers,2002 *Int. J. Rotating Mach.* **8** 1

[22] Morsi S and Alexander A, An investigation of particle trajectories in two-phase flow systems, 1972 *J. Fluid Mech.* **55** 2

[23] Stokes G, On the effect of the Internal friction of fluids on the motion of pendulums - Section III,1850 *Trans. Cambridge Philos. Soc.*, **IX**

[24] Duan H, Ren Y, and Zhang L, Effects of Interphase Forces on Fluid Flow in Gas-Stirred Steel Ladles Using the Eulerian–Lagrangian Multiphase Approach, 2018 *JOM*, **70** 10

[25] Hsiao C, Ma J, and Chahine G, Multiscale tow-phase flow modeling of sheet and cloud cavitation, 2017 *Int. J. Multiph. Flow*, **90**

[26] Li L, Huo Y, Wang Z, Li X, and Zhu Z, Large eddy simulation of tip-leakage cavitating flow using a multiscale cavitation model and investigation on model parameters, 2021 *Phys. Fluids*, **33** 9

[27] Zwart P, Gerber A, and Belamri T, A two-phase flow model for predicting cavitation dynamics,2004 *5th Int. Conf. Multiph. Flow*

[28] Chesters A, Modelling of coalescence processes in fluid-liquid dispersions. A review of current understanding, 1991 *Chem. Eng. Res. Des.* **69** 4

[29] Lau Y, Bai W, Deen N, and Kuipers J, Numerical study of bubble break-up in bubbly flows using a deterministic Euler-Lagrange framework, 2014 *Chem. Eng. Sci.*, **108**

[30] Tian B, Li L, Chen J, Huang B, and Cao J, Numerical Study of Multiscale Cavitating Flow Around a Hydrofoil, 2022 *Lixue Xuebao/Chinese J. Theor. Appl. Mech.* **54** 1557–1571

[31] Kowalczuk P and Drzymala J, Physical meaning of the Sauter mean diameter of spherical particulate matter, 2016 *Part. Sci. Technol.* **34** 6

[32] Yu X, Hendrickson K, and Yue D, Scale separation and dependence of entrainment bubble-size distribution in free-surface turbulence, 2019 *J. Fluid Mech.* **885** 1–12

Numerical simulations of submarine self-propulsion flows near the free surface

Enkai Guo[1], Liushuai Cao[1*], Zhiben Shen[2], Yun Wang[2]

[1] Computational Marine Hydrodynamics Lab (CMHL), School of Naval Architecture, Ocean and Civil Engineering, Shanghai Jiao Tong University, Shanghai, China.
[2] Wuhan Second Ship Design and Research Institute, Wuhan, China.
[*] Corresponding author: liushuaicao@sjtu.edu.cn

Abstract. Submarine needs to complete more tasks near the free-surface area in modern war. When sailing near the free surface, the self-propelled characteristics of the submarine and its nearby flow field will be affected by the free surface. This paper presents the research on the self-propulsion submarine near the free surface with computational fluid dynamics (CFD) method. We take the generic Joubert BB2 submarine and Marine 7371R propeller models as the research objects. The RANS method and SST k-ω turbulence model are selected to solve the turbulent flows. The volume of fluid (VOF) method is used to capture the free surface. Grid convergence study is performed based on three sets of grids with different resolution. Numerical simulations are performed under the self-propulsion cases with different submergence depths and velocities. The proportional integral (PI) controller method is proposed to obtain the self-propulsion point. The results are more accurate compared with the traditional one. It is found that with the decrease of the submergence depths, the rotating speed of the self-propelled point and the corresponding thrust increase. The study also explores the free surface effect on flow field characteristics, wave-making characteristics and vortex structures. The scientific findings provide some useful results for the self-propulsion submarines near the free surface and support the further research in the future.

1. Introduction

Modern warfare requires submarines to perform a variety of tasks in near-free-surface conditions, such as navigating in near-free-surface formations, coordinating with air forces, launching missiles, and entering or leaving ports for maintenance and resupply. Additionally, submarine listening, surveillance, and special operations capabilities in shallow coastal waters are becoming increasingly important. Modern operations require that submarines not only stick to deep-water operations, but also have near-surface combat capabilities.

Research has shown that the hydrodynamic performance and flow field characteristics of submarines change significantly when sailing near the free surface. Different research methods have been used to investigate this phenomenon, including theoretical analysis, model experiments, and computational fluid dynamics (CFD).

Theoretical analysis method is an important method. Some early scholars used slice theory and potential flow method to solve the hydrodynamic performance of the near-surface submarines [1]. Gourlay and Dawson [2] adopted the Havelock-source panel method to describe the potential flow around submarines near the free surface. It is found that the method can accurately calculate the wave resistance, pressure drag, vertical force, and trim moment of DARPA SUBOFF and Joubert submarines compared with the experimental data. The method also shows advantages in computation

IOP Conf. Series: Materials Science and Engineering 1288 (2023) 012055 doi:10.1088/1757-899X/1288/1/012055

speed, simplicity and robustness. However, the theoretical method is often limited by conditions and experience, such as the selection of empirical formula, which is insufficient to deal with complex flow field problems. The results of theoretical analysis often need to be verified by experimental data.

The advantage of the model experiment method is that it can create a real physical environment, therefore the results are highly reliable. Wang et al. [3] determined the hydrodynamic performance and wake characteristics of the submarine propeller when it is self-propelled near the free surface by using stereoscopic particle image velocimetry (SPIV) method. However, the model experiment method also has some shortcomings, such as the limitation of test conditions and the difficulty of flow field measurement. So, there are few documents and articles on the study self-propulsion of submarine near the free surface by using the model experiment method, which lacks sufficient data support.

With the rapid development of CFD technology, simulation software based on CFD can quickly analyze and simulate the flow characteristics of submarines. Many researchers have investigated the relative problems with different CFD methods. Lungu [4] used the detached eddy simulation (DES) method to study the flow around the self-propelled SUBOFF submarine near the free surface and mainly discussed the wave variation under different conditions. Li et al. [5] coupled the Reynolds-Averaged Navier-Stokes (RANS) equation with k-ω turbulence model to conduct the self-propulsion simulation and analyze the influence of the free surface on the performance of propeller. Carrica [6] used a general purpose CFD code REX to conduct near-surface self-propulsion simulations and study the hydrodynamics performance of submarine and propeller both in calm water and waves. Dong [7] used RANS solver to simulate the impact of long-crested waves on the self-propulsion submarines. Huang et al. [8] also used RANS method to analyze the wave-making characteristics of the near-surface submarines.

This paper presents the research on the self-propulsion submarines near the free surface. The study in this paper takes the standard Joubert BB2 submarine and Marine 7371R propeller as the research object. The RANS equation and SST k-ω turbulence model is selected. The VOF method is used to establish the free surface. Numerical simulations are conducted under the cases with different submergence depths and velocities. The linear interpolation method and the PI controller method are applied to obtain the self-propulsion points. There is no significant difference between their results. The results are compared to explore the influence of and submergence depth on submarines self-propulsion performance, flow field characteristics, wave-making characteristics and vortex structures. All of the simulations are conducted based on the CFD software STAR-CCM+.

2. Methodology

2.1. Submarine and propeller geometries

The Joubert BB2 submarine model and six-bladed propeller 7371R model were used for the self-propulsion study. They were both designed by Maritime Research Institute Netherlands (MARIN). Figure 1 shows the entire self-propulsion model and the appendages on it. The BB2 model consists of main hull, sail, sailplanes and X rudders. Table 1 shows the main parameters of the submarine and propeller models at model scale (λ=18.348) and full scale.

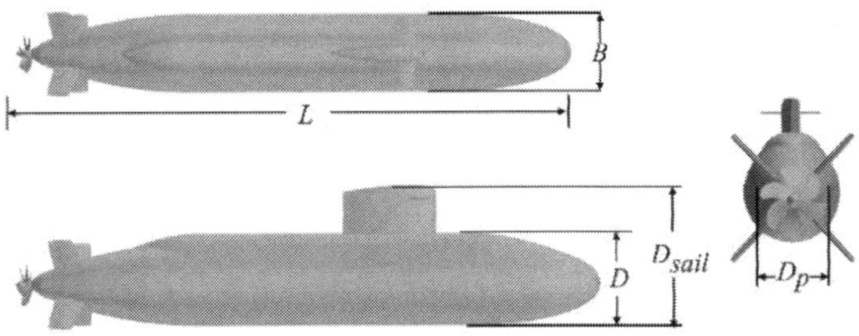

Figure 1. The submarine and propeller model.

Table 1. The main parameters of the submarine and propeller models.

Description	Symbol	Full scale value	Model scale value	Unit
Length	L	70.20	3.826	m
Beam	B	9.60	0.5232	m
Drought to deck	D	10.60	0.5777	m
Drought to sail	D_{sail}	16.20	0.8829	m
Diameter of propeller	D_p	5	0.273	m
Pitch ratio	P/D_p	0.966	0.966	-

2.2. Governing equations and turbulence model

Simulations in this study were conducted by solving Reynolds-Averaged Navier-Stokes (RANS) equations in STAR-CCM+ software. The flow is incompressible and turbulent. The governing equations of RANS method can be expressed as follows:

$$\frac{\partial \bar{u}_i}{\partial x_i} = 0 \tag{1}$$

$$\frac{\partial \bar{u}_i}{\partial t} + \bar{u}_j \frac{\partial \bar{u}_i}{\partial x_j} = \bar{f}_i - \frac{1}{\rho}\frac{\partial \bar{p}}{\partial x_i} + \nu \frac{\partial^2 \bar{u}_i}{\partial x_j \partial x_j} - \frac{\partial \overline{u_i' u_j'}}{\partial x_j} \tag{2}$$

where $\bar{u}_i$, $\bar{f}_i$ and $\bar{p}$ are the time averaged values of velocity component, force and pressure.

In this study, the SST k-ω turbulence model is selected to close the solving equations, which is proposed by Menter [9]. It uses k-ω model to deal with the flow in the boundary layer and k-ε to deal with the flow of free fluid. Its transport equations are given as:

$$\frac{\partial}{\partial t}(\rho k) + \frac{\partial}{\partial x_i}(\rho k u_i) = \frac{\partial}{\partial x_j}\left(\Gamma_k \frac{\partial k}{\partial x_j}\right) + G_k - Y_k + S_k \tag{3}$$

$$\frac{\partial}{\partial t}(\rho \omega) + \frac{\partial}{\partial x_i}(\rho \omega u_i) = \frac{\partial}{\partial x_j}\left(\Gamma_\omega \frac{\partial \omega}{\partial x_j}\right) + G_\omega - Y_\omega + S_\omega \tag{4}$$

where Γ_k, Γ_ω represent the effective diffusivity; G_k, G_ω represent the turbulent kinetic energy; Y_k, Y_ω represent the energy dissipation; S_k, S_ω represent the source item.

2.3. Free surface model

Volume of fluid (VOF) method is a moving interface tracking method based on Euler grid, which is proposed by Hirt and Nichols [10]. Its basic principle is to simulate the moving interface by defining a VOF function of the ratio between the volume of the target fluid and the grid, and calculating the function value on each cell element. VOF method is widely used in the simulation of multiphase transition. The advantage of VOF method is that it has good mass conservation characteristics and leads to high accuracy.

The free surface separates the air domain and the water domain. The volume fraction α is introduced to represent the ratio of a kind of fluid in the grid. The α=0 represents the air and α=1 represents the water. The equation of calculating the volume fraction α can be expressed as:

$$\frac{\partial \alpha}{\partial t} + \vec{u} \cdot \nabla \alpha = 0 \tag{5}$$

2.4. Computational domain and boundary conditions

The computational domain consists of two regions. The outer region is the static region and the inner region is the rotating region. The rotating region is a cylinder with a diameter of $1.32D_p$ and a height of $0.26D_p$. It completely encloses the blades and hub of the propeller. The boundary of the rotating region is defined as interface. The boundary of the static region is a cuboid. The inlet boundary is $3.18L$ away from the front nose tip of submarine which is defined as velocity inlet. The outlet boundary is 4.18L away from the tail tip of submarine which is defined as pressure outlet. The top boundary is 2.09L away from the submarine's surface. The bottom boundary is 4.18L away from the submarine's surface. The bilateral boundaries are 4.18L away from the port or starboard. They are all defined as symmetry plane. The surfaces of body (including hull, appendages, propeller blades and shaft) are defined as no-slip wall.

As shown in Figure 2, the free surface is established parallel to the submarine's base plane (z=0 plane). The distance between the base plane and the free surface is defined as the submergence depth H. The static region is divided into air domain and water domain by the free surface.

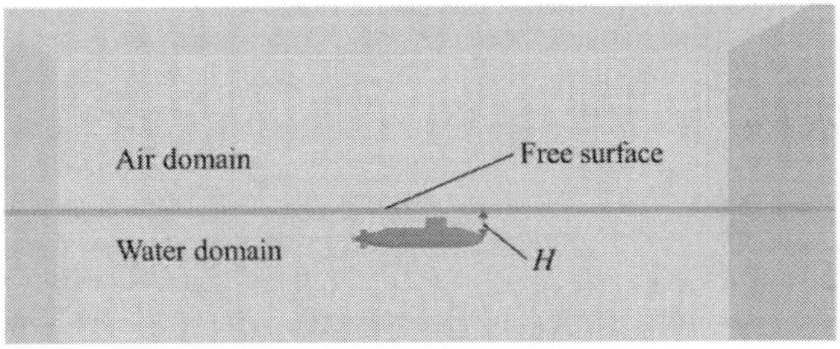

Figure 2. The computational domain.

2.5. *Self-propulsion simulation method*

In this study, two different methods were used to obtain self-propulsion point in order to verify the accuracy of simulation.

The first one is a traditional method which was proposed by Yang et al. [11]. The key to this method is to adjust rotating speed in order to balance the thrust of propeller and total resistance of submarine under fixed velocity. The process is as follows.

(1) There is a uniform flow with fixed speed in the computational domain. We estimate an initial rotating speed through propeller's open water performance.

(2) The numerical simulation is conducted at fixed speed and rotating speed. The thrust (T) value and resistance (D) value are monitored and noted when they are convergent.

(3) If $T>D$, reduce the rotating speed; If $T<D$, increase the rotating speed. Then we conduct the simulation again and repeat the steps above until the thrust value is equal to the resistance value.

(4) When $T=D$, the corresponding rotating speed represents the self-propulsion point at that speed. The wake fraction, thrust deduction coefficient and other self-propulsion factors can be calculated according to the resistance curve of the hull and the open water performance curve of the propeller.

However, in the realistic operating process, it is often difficult to directly obtain the self-propulsion point where the thrust value is exactly equal to the resistance value. Therefore, simulations are conducted at several rotating speeds near the self-propulsion point and the balancing point of thrust and resistance can be acquired through linear interpolation.

The other one is using a proportional-integral (PI) controller to find the self-propulsion point. A PI controller is designed to drive the propeller's rotating speed based on the error between the submarine's current speed and target speed. During the simulation, the rotating speed is not fixed but changing dynamically. The process of every step is given in Figure 3.

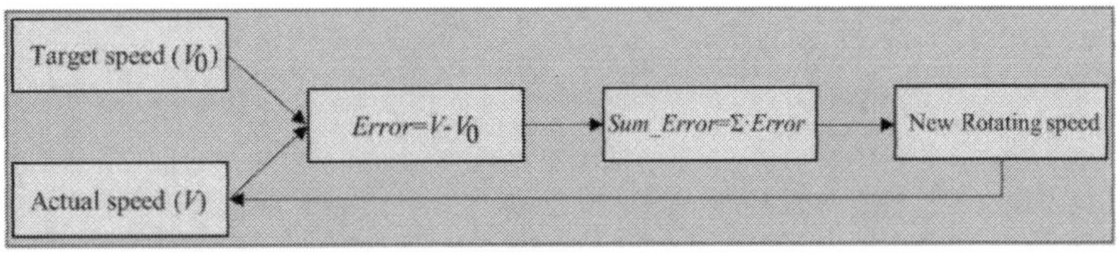

Figure 3. The process of PI controller simulation.

The general formula for the PI controller can be expressed as follows:

$$n(t)=n_0\left(1+K_p\cdot error+K_i\cdot\sum error\right) \tag{6}$$

where $n(t)$ is the rotating speed at different times; n_0 is the initial value of rotating speed; error is the error at certain time which means submarine's target speed minus current speed here. K_p is the gain for the proportional component of the controller; K_i is the gain for the integral component of the controller.

3. Mesh generation and validation

In this study, mesh generation is mainly based on the Trimmed Mesher in both static region and rotating region. The Prism Layer Mesher is used in the boundary layer around the submarine and the propeller to simulate the fluid flow in the boundary layer more accurately. The mesh is refined in the turbulence wake region of rudders and sail. Meanwhile the mesh in far field is relatively coarser.

To verifying the mesh dependence and the accuracy of simulation, we totally established three different grids and classify them as coarse grid, medium grid and fine grid based on their basic cell sizes. Then all of the grids are used to simulate at deep depth (submergence depth H is big enough) in a fixed speed of 1.19m/s. The results of rotating speed and thrust at self-propulsion point are compared with the results from Kim et al. [12] as shown in Table 2. It can be noticed that all the values of other cases are well below 4% which is acceptable. The results of the case with the finest grid also don't show much superiority compared with others. So there is no need to further increase the grid size. The medium grid is finally chosen to be used for the following simulation which has an approximate 5.8 million cells. The details of the mesh in the midship section are shown in Figure 4.

Table 2. The results of mesh dependence validation.

Grid	cells (million)	points (million)	rotating speed (rpm)	δ_1 (%)	thrust (N)	δ_2 (%)
coarse	3.8	4.2	276.6	2.07	26.9	3.07
medium	5.8	6.3	275.2	1.55	26.6	1.92
fine	9.7	10.3	271.9	0.33	25.9	-0.77
Kim (2017)	-	-	271	-	26.1	-

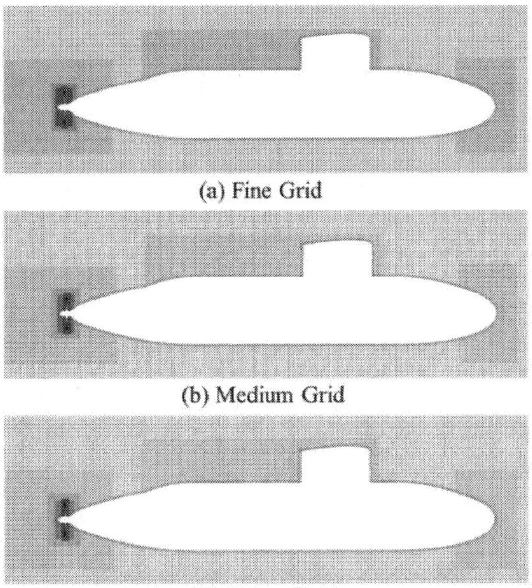

(a) Fine Grid

(b) Medium Grid

(c) Coarse Grid

Figure 4. The details of the mesh in the midship section.

4. Results and discussions

4.1. Hydrodynamic forces

Simulations were conducted in totally 8 different cases including 2 submarine velocities (1.201m/s and 1.6309m/s) and 4 submergence depth (0.5m, 0.845m, 1.0m and infinite) as is given in Table 3. In every case, both of the linear interpolation method and the PI controller method were used to obtain the self-propulsion point.

Table 3 shows the results of rotating speed (n) and thrust (T) at the self-propulsion point in corresponding cases. Obviously at the same submergence depth, higher submarine velocity corresponds to higher rotating speed and thrust. At the same velocity, deeper submergence depth corresponds to lower rotating speed. The results of the linear interpolation method and the PI controller method have the same trend. There is no significant difference between them, which proves that the simulations' accuracy.

Table 3. The results of mesh dependence validation.

case	Velocity (m/s)	Depth (m)	Traditional method		PI-controller method	
			n (r/min)	T (N)	n (r/min)	T (N)
1	1.201	0.5	310.3	37.9	310.8	38.8
2	1.201	0.845	280.7	28	282.9	29.1
3	1.201	1	278.9	27.6	280	28.1
4	1.201	infinity	277.7	27.1	277.6	27
5	1.6309	0.5	443.3	85.4	443.2	85.2
6	1.6309	0.845	386.4	54.7	386.4	54.6
7	1.6309	1	376.7	50.2	377.9	50.7
8	1.6309	infinity	372.2	47.7	372.5	47.8

4.2. Wave-making characteristics

Figure 5 shows the height of the free surface relative to the initial position under different depth conditions, that is, the wave height. The same color bar scale is used at the same velocity, which reflects the wave making characteristics at the free surface. The wave on the free surface resembles the Kelvin wave pattern well, which consists of transverse wave system perpendicular to the direction of motion and divergent wave system diagonal to the direction of motion.

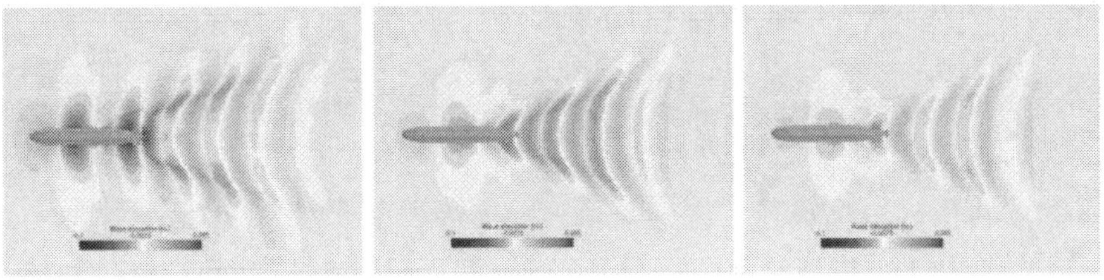

(a) Case 1 (b) Case 2 (c) Case 3

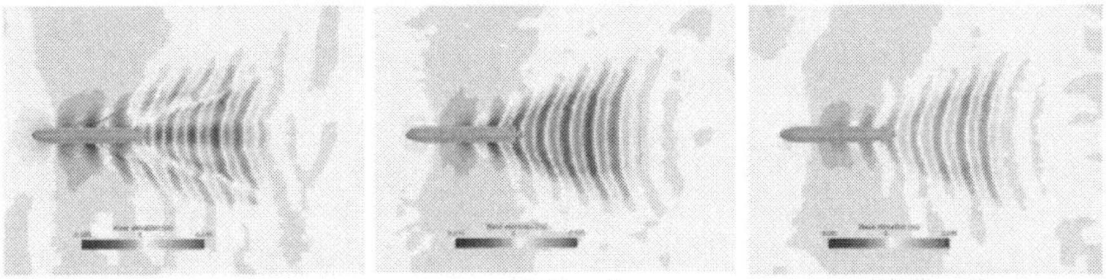

(d) Case 5 (e) Case 6 (f) Case 7

Figure 5. The waves on the free surface in different cases.

In case 1 and case 5, where the submergence depth is 0.5m, clear Kelvin wave system characteristics can be observed, especially steep divergent waves. As part of the sail and sailplanes protruding from the water, the free surface interacts with the hull to form a wave with large amplitude. In case 2 and 6, where the submergence depth increases to 0.845m. Kelvin wave system features become weak and divergent waves cannot be clearly observed. In case 3 and 7, where the submergence depth increases to 1.0m. Divergent wave system features become further weaker and the Kelvin angle decreases. In addition, due to the lower amplitude, there are some clutter wave interferences.

4.3. Flow field characteristics

Figure 6 shows the velocity nephograms at the midship section of the submarine at the velocity of 1.6309 m/s at various depths. Color bar scale is represented by dimensionless velocity. When the free surface exists, the bow flow field gradually diffuses from the front stagnation point of the submarine to the free surface in a circle, forming a wave crest; The high-velocity flow formed above the sail and sailplanes diffuses upward and backward to the free surface, forming a wave trough with large amplitude; The flow behind the sail diffuses to the rear of the submarine and merges with the flow behind the stern rudders and the flow formed by the propeller rotation, which diffuse upward to the free surface, forming a wave crest with large amplitude. They correspond to the wave-making characteristics in Section 4.2. Compared the cases (b), (c), (d) with the case (a), it is found that the flow field below the submarine is similar, while the flow field above it changes greatly. Due to Bernoulli's principle, the flow velocity increases in the area below the wave trough, forming a high-speed area. The flow velocity decreases in the area below the wave crest respectively. In addition, the existence of free surface aggravates the disturbance of the flow field near the propeller, increasing the wake velocity of the propeller and making the two wake lines generated by the propeller shift downward.

When the submergence depth is shallow, the flow near the sail interacts with the transverse waves on the free surface, generating divergent waves and other wave systems. When the submergence depth is deep, while the distance from the free surface is far, the hull can hardly affect the waveform. Therefore, the transverse waves are the main waveform.

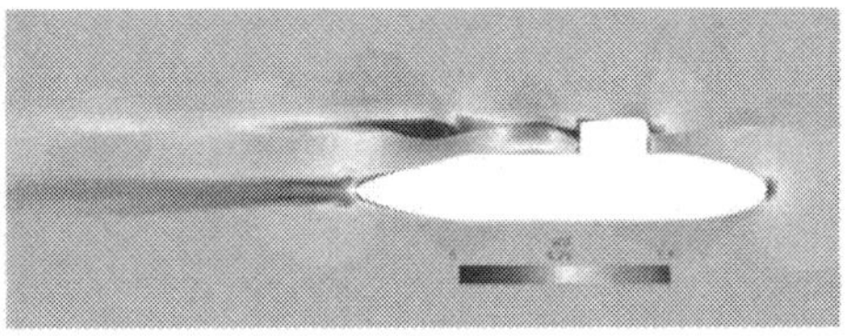

(a) Case 5

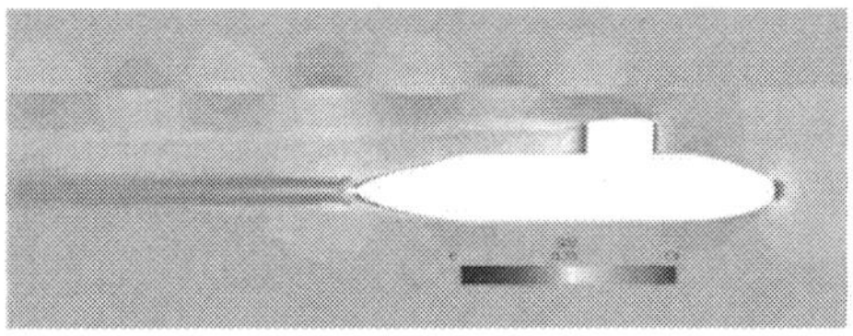

(b) Case 6

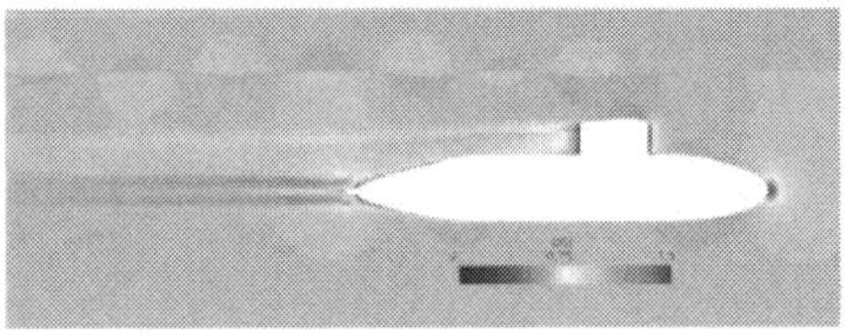

(c) Case 7

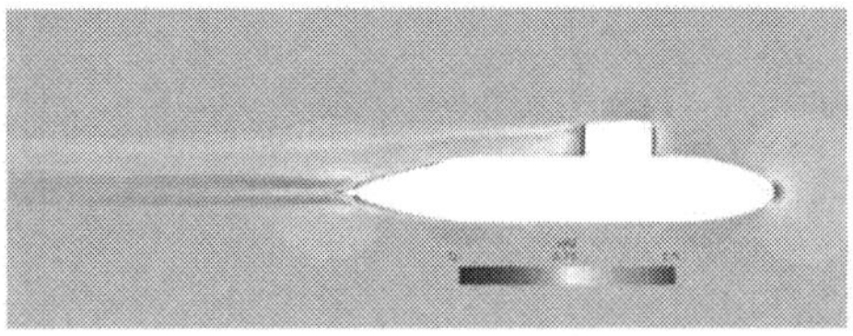

(d) Case 8

Figure 6. The velocity nephograms at the midship section in different cases.

4.4. Vortex structures

After the analysis of flow field, Q-criterion is used to analyze the vortex structure around the submarine and propeller. Q-criterion is widely used for discriminating and visualizing vortex tubes. The basic expression of Q is as follows:

$$Q = \frac{1}{2}\left(\|\Omega\|^2 - \|S\|^2\right) \tag{7}$$

where $\Omega = 0.5\left[\nabla v - (\nabla v)^T\right]$ is the antisymmetric vorticity tensor, $S = 0.5\left[\nabla v + (\nabla v)^T\right]$ is the symmetric strain tensor.

Figure 7 shows the vortex structures visualized by Q-iso-surface under three different submergence conditions. The Q values of all the conditions are the same. The pressure field function is used for coloring. In Figure 7 (a) where the sail and sailplanes protruding from the free surface, it can be seen that the vortex structure formed behind the sail merges with the free surface, and then tends to approach the upper surface of the hull. The tip vortex behind the rudders is affected by the free surface, and the vortex tube becomes longer and thicker. The vorticity on the free surface is high. In Figure 7 (b) the tip vortex tube behind the sail tends to approach the free surface. In Figure 7 (c) the vortex structure is influenced little by the free surface.

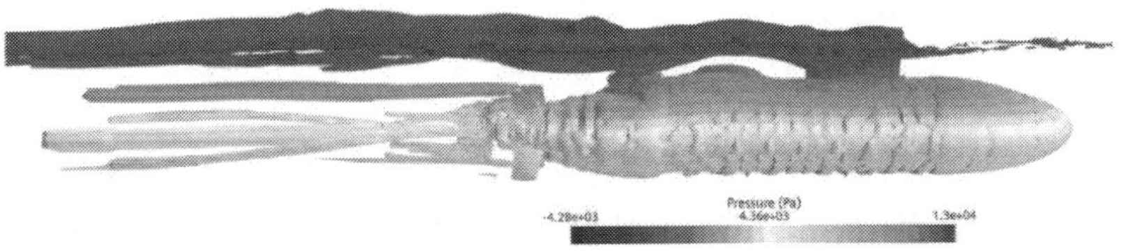

(a) Case 5

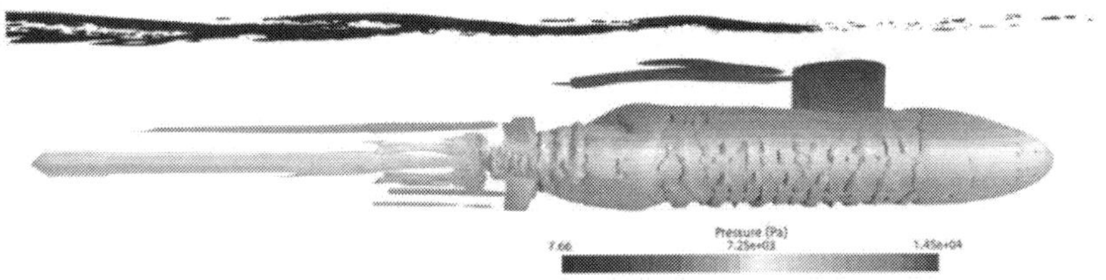

(b) Case 6

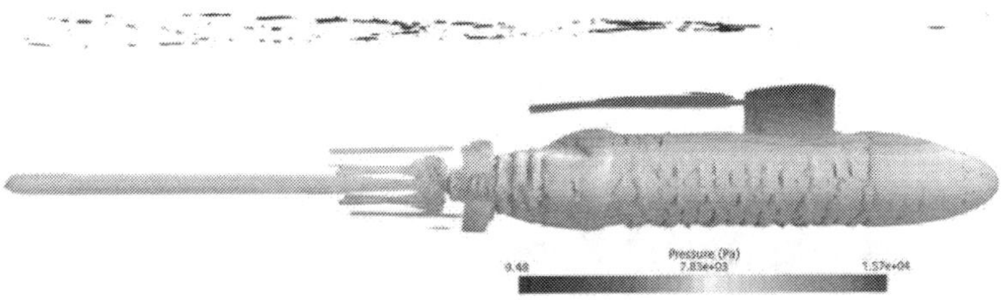

(c) Case 7

Figure 7. The vortex structures in different cases.

5. Conclusions

This study conducted the self-propulsion numerical simulations of Joubert BB2 submarine and Marine 7371R propeller through both linear interpolation method and PI controller method based on a commercial software STAR-CCM+. The main findings are described below:

(1) The self-propulsion results calculated by the PI-controller method is similar to the results calculated by the traditional method. With the decrease of the submergence depth, the rotating speed and corresponding thrust at the self-propulsion points will increase.

(2) The wave-making characteristics on the free surface are similar to Kevin waves. With the increase of submergence depth, divergent wave features become weaker and only transverse waves can be observed. Under the same velocity, the case with deeper submergence depth has the smaller amplitude, while at the same submergence depth, the case with higher velocity has greater amplitude and longer wavelength. A wave crest appears at the bow, a wave trough with large amplitude appears at the middle of the hull and a wave crest with large amplitude appears behind the propeller.

(3) The existence of the free surface changes the flow field between the free surface and the upper surface of the hull. The flow near the sail spread upward to the free surface. The interaction between the flow field and the free surface aggravates the inhomogeneity of the wake velocity behind the rudders and the propeller.

(4) When the sail protrudes from the free surface, a high vorticity field will be formed on the free surface around it. The vortex structure formed behind the sail merges with the free surface, and then tends to approach the upper surface of the hull. When the sail is below the free surface, the tip vortex behind the sail extends longer and tends to diffuse to the free surface.

This paper presents some basic research results of submarine self-propulsion in straight near the free surface. In the future, complex maneuverability tests such as steady turning tests can be investigated as further research.

Acknowledgements

This work was supported by the National Natural Science Foundation of China (52001210, 52131102), and the National Key Research and Development Program of China (2019YFB1704200), to which the authors are most grateful. The authors specially thank to Prof. Decheng Wan for his kind suggestions and supervision.

References

[1] Feng X, Jiang Q, Miao Q and Kuang X 2002 *Journal of Ship Mechanics* **02** 1-14
[2] Gourlay T and Dawson E 2015 *J. Mar. Sci. Appl.* **14** 215-224
[3] Wang L, Martin J E, Felli M and Carrica P M 2020 *Ocean. Eng.* **206** 107304
[4] Lungu A 2022 *Ocean. Eng.* **244** 110358
[5] Li P, Wang C, Han Y, Kuai Y and Wang S 2021 *Chinese Journal of Theoretical and Applied Mechanics* **53(9)** 2501-2514
[6] Carrica P M, Kim Y and Martin J E 2019 *Ocean. Eng.* **183** 87-105
[7] Dong K, Wang X, Zhang D, Liu L and Feng D 2022 *J. Mar. Sci. Eng.* **10(1)** 90
[8] Huang F, Meng Q, Cao L and Wan D 2022 *Ocean. Eng.* **250** 111062
[9] Menter F R 1994 *AIAA. J.* **32(8)** 1598-1605
[10] Hirt C W and Nichols B D 1981 *J. Comput. Phys.* **39(1)** 201-225
[11] Yang Q, Wang G, Zhang Z, Feng D and Wang X 2013 *Chinese Journal of Ship Research* **8(2)** 22-27
[12] Kim H, Ranmuthugala D, Leong Z Q and Chin C 2018 *Ocean. Eng.* **150** 102-112

Cavitation of some copper alloys for naval propellers: electrolyte effect

M V Biezma Moraleda[1], L Merino Galván[1], P Linhardt[2]

[1]Dpto. de Ciencia e Ingeniería del Terreno y de los Materiales, Universidad de Cantabria, Dique de Gamazo 1, 39004 Santander, Spain
[2]Institute for Chemical Technologies and Analytics, Technische Universität Wien, Getreidemarkt 9, 1060 Vienna, Austria

biezmav@unican.es

Abstract. Efficient shipping is becoming more challenging, not only because of the increasing demand for this means of transport, the performance requirements and the optimisation of resources, but also because of the environmental constraints. Cavitation is a harmful phenomenon that can affect different systems of a ship such as pumps, valves, impellers, pipes, etc., but is particularly known for its effect on propellers, associated with loss of power and even impairment of structural integrity. The mechanical effect of cavitation on metallic materials is in general superimposed by corrosion and may thus lead to a synergistic degradation phenomenon. In this work, the cavitation behaviour and the potential synergistic effect with corrosion was evaluated for three widely used copper-based alloys, NAB (Nickel Aluminium Bronze), MAB (Manganese Aluminium Bronze), and a brass (low lead content) using three types of water: natural seawater, synthetic seawater and synthetic brackish water. Experiments were carried out in an ultrasonic bath, followed by structural surface characterisation. The main result is that MAB is most susceptible to both, general surface damage and deep localized attack. This is attributed to its poor ability in regenerating the protective layer, the area ratio of their phases and the high hardness of the κ-phase, in combination with high mechanical stresses during impacts affecting the grain boundaries. Moreover, natural seawater was found to increase cavitation, attributable to the dispersion of micro-particles, while brackish water with its content of sulfidic species was found to promote the strongest synergy between corrosion and cavitation.

1. Introduction

The phenomenon of cavitation, i. e. the growth and collapse of vapor bubbles in a liquid due to local pressure fluctuations, has been and still is extensively studied within the field of marine engineering. One of its undesired consequences are the mechanical impacts on solid surfaces which arise from the collapsing bubbles, possibly leading to damages at components. The attempt to achieve greater efficiency of systems and navigation, resulting in the design of larger propellers operating at higher revolutions, increases the likelihood of cavitation and in combination with the corrosive environment the potential for damages due to this phenomenon has increased [1]. In particular, considering the need to extend the service life of propellers, to save material and manufacturing costs, to extend the operating regime, etc., an exhaustive study of cavitation is required for design optimization [2],[3],[4]. Moreover, environmental conditions must be taken into account, as they can cause corrosion problems and aggravate the phenomenon of cavitation, while, on the other hand, cavitation may also accelerate the corrosion processes. This appears critical in marine conditions due to the hostile environment of rather complex chemical composition, possibly aggravated by the presence of contaminants [5, 6].

In this study, the cavitation behaviour of three copper alloys (nickel-aluminium bronze, manganese-aluminium bronze, and a duplex brass) as are used for naval propellers, were investigated with respect to the interaction with different electrolytes, i.e. different water compositions: artificial seawater, artificial brackish water, and natural seawater.

2. Experimental

2.1. Materials

Alloys most commonly used for propellers, impellers and valves due to their good mechanical properties and corrosion resistance were studied in their as-cast state: manganese aluminium bronze (MAB), nickel aluminium bronze (NAB), and a brass (CB773S). The brass may be considered an *ecobrass*, having a low Pb content and appearing attractive in view of future environmental restrictions. The chemical composition of the materials was provided by the suppliers and is summarized in Table 1. Their hardness was assed previously [7] by 205 HB (NAB), 207 HB (MAB), and 158 HB for the brass.

Table 1. Chemical composition of the alloys (%wt).

Alloy	Cu	Al	Ni	Fe	Mn	Zn	Pb	Sn	Si
NAB	79.80	9.31	4.57	4.92	1.23	0.0492	0.0056	0.0058	0.0393
MAB	70.9	7.12	2.06	4.59	12.63	2.35	0.0126	0.0059	0.0345
CB773S	57.97	0.17	0.01	0.15	-	≈ 40	0.09	0.03	-

The specimens were machined to 20 x 15 mm with one corner rounded by a radius of ca. 13 mm. The thickness of these coupons was 3 mm. In view of a desired surface roughness for propellers of 0.6 μm, the specimens received a surface finish by grinding with SiC paper up to # 1200 and final polishing with diamond paste from 6 down to 0.25 μm.

The microstructures of the materials as observed after polishing are presented in figure 1. NAB consists of α phase, β phase and four rather small κ-phases (formed by Fe3Al and NiAl, ca. 5 μm). MAB is a bronze that consists of α-phase (rich in copper), β-phase and relatively large κ-phases (mainly rich in Fe, Mn, ca. 40 μm wide) [8]. The brass exhibits a duplex structure consisting of α-(rich in copper) and β-phase (rich in zinc).

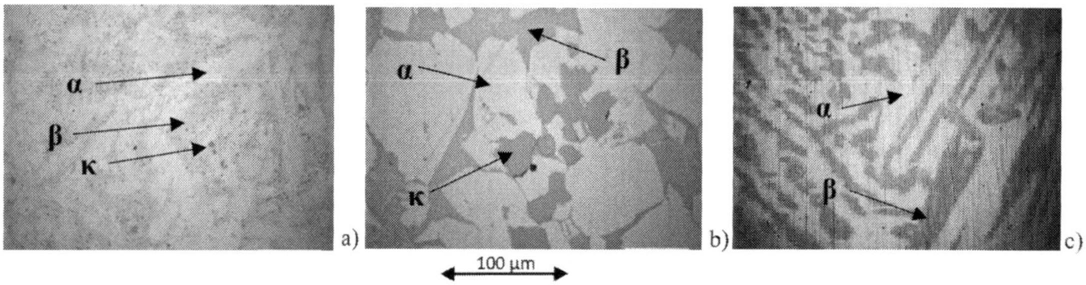

Figure 1. Microstructures of tested materials. a) MAB, b) NAB, c) brass.

2.2. Test media

Three test media were selected to reflect different waters that propellers may exposed to during service conditions. Consequently, they also represent the electrolytes, which are relevant for the electrochemical corrosion processes, which act on the metals in general, and which get modified by cavitation.

IOP Conf. Series: Materials Science and Engineering 1288 (2023) 012056 doi:10.1088/1757-899X/1288/1/012056

Artificial Seawater (ASW) and Artificial Brackish Water (ABW) were prepared by dissolving the relevant salts in deionized water, based on a standard [9]. Table 2 presents their composition. Santander Bay Water (SBW) was collected in the Bay of Santander, Spain, and its typical physicochemical characteristics, as provided by the Instituto Oceanográfico de Santander, are listed in Table 3.

Table 2. Salts used for preparing 1 L of ASW and ABW.

Chemical compounds	ASW	ABW
sodium chloride, 28.0 g $NaCl$	X	X
magnesium chloride, 5.0 g $MgCl_2 \cdot 6\ H_2O$	X	X
calcium chloride, 2.4 g $CaCl_2 \cdot 6\ H_2O$	X	X
magnesium sulphate, 7.0 g $MgSO_4 \cdot 7\ H_2O$	X	X
sodium bicarbonate, 0.20 g $NaHCO_3$	X	-
tri-sodiumcitrate, 1.0 g $C_6H_5Na_3O_7 \cdot 2\ H_2O$	-	X
thioacetamide, 0.50 g CH_3CSNH_2.	-	X

Table 3. Physicochemical properties of SBW.

Parameter	Quantity
Density	1.025 g/mL
pH	8.2
Acidity	0.06-0.18 mEq/L
Alkalinity	2.36-2.72 mEq/L
Hardness	6.4-6.8 g $CaCO_3$/L
Ammonium	0-50 µg N-NH_4^+/L
Calcium	320-680 mg Ca/L
Nitrite	0-25 µg N-NO_2^-/L
Nitrate	100-125 mg N-NO_3^-/L
Sulphate	720-900 mg S- SO_4^{2-}/L
Phosphate	0-23 g P-PO_4^{3-}/L
Chloride	17.6-19.2 g Cl^-/L

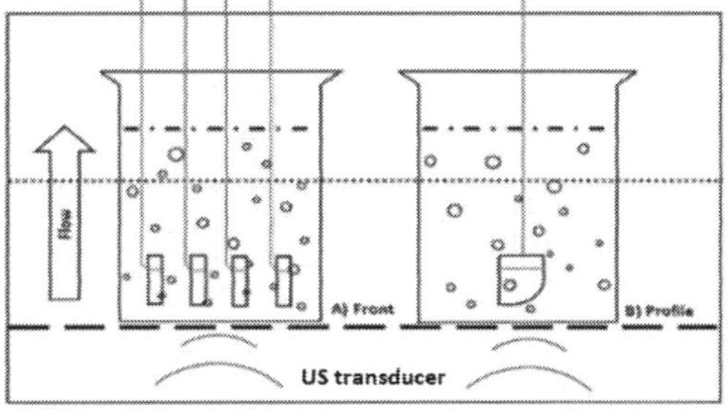

Figure 2. Position of specimens vertically inside beakers in the ultrasonic bath.

IOP Conf. Series: Materials Science and Engineering 1288 (2023) 012056 doi:10.1088/1757-899X/1288/1/012056

2.3. Cavitation testing

An Emerson Branson 2800 ultrasonic bath, operating at 40 kHz at a power of 100 W and filled with tap water was used to perform the cavitation tests. The test media were filled in beaker glasses placed in the bath and the specimens were introduced vertically by hanging them in, ca. 0.5 mm above the bottom. figure 2 provides a schematic view of the test setup.

The test time was 45 min. or 90 min. (in two intervals of 45 min. each), allowing to observe the surface changes in the early stage, and at some more severe state of destruction. The temperature in the bath was kept below 30 °C during testing. All tests were done in triplicate at minimum.

2.4. Evaluation of tested specimens

To observe type and level of cavitation damage and for finding a correlation with the microstructural features, the tested specimens were investigated using a Leica EZ4 HD stereo microscope and a Leica DM 4000 - U optical microscope.

For quantification of the damage, each side of the specimen was subdivided by a mesh (40x40) and each surface element was evaluated by 0 (no damage) and 1 (damage found), yielding a value for the Affected Zone as the percentage of damaged surface elements (% AZ). Moreover, the mean pit diameter (Φm, μm) was calculated by averaging the diameter of the pits in different areas. It should be noted that weight loss has not been considered for evaluation since the amount of material loss during the tests was found insufficient for this method.

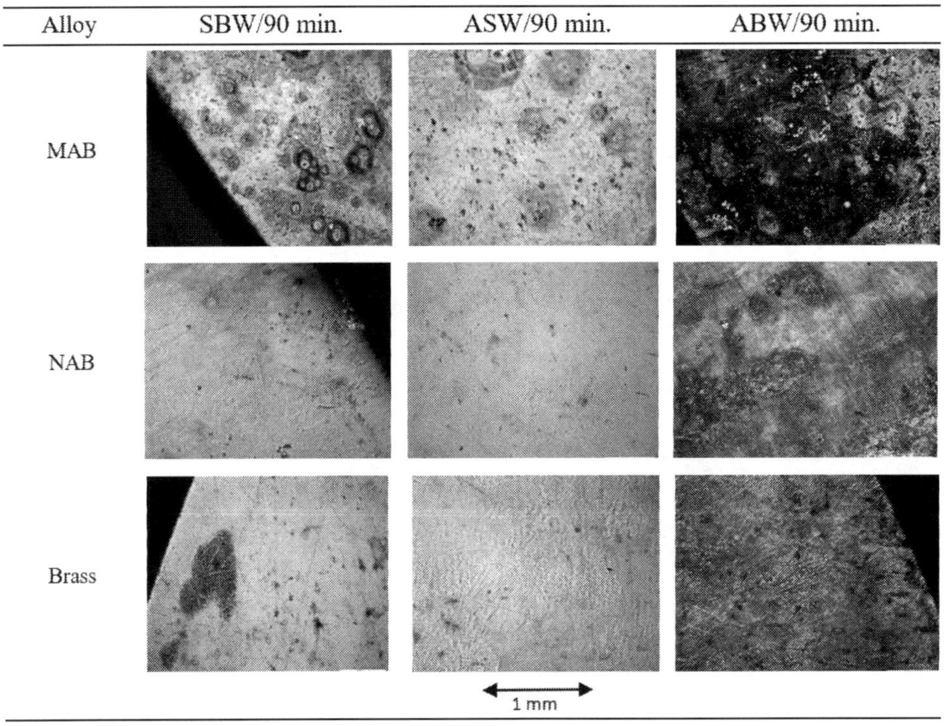

Figure 3. Surface views of specimens after 90 min. testing.

3. Results

Figure 3 presents views on the surface of tested specimens after 90 min. of testing and Table 4 summarizes the results of cavitation damage evaluation after 45 min and 90 min test time in SBW and ASW. For ABW, this method of quantification was not successful since a well adhering crust of corrosion products prevented from assessing the data accurately This may be anticipated by figure 3 and particularly for MAB in figure 4 at larger scale, comparing the effect of the three test media.

Table 4. Affected zone (AZ) and mean pitting diameter (Φ_m) for MAB, NAB and brass in SBW and ASW.

	AZ (%)						Φ_m (μm)					
	45 min			90 min			45 min			90 min		
	MAB	NAB	Brass	MAB	NAB	Brass	MAB	NAB	Brass	MAB	NAB	Brass
SBW	46	24	28	82	39	47	45	10	15	90	24	40
ASW	35	15	15	57	24	26	25	3	7	48	12	16

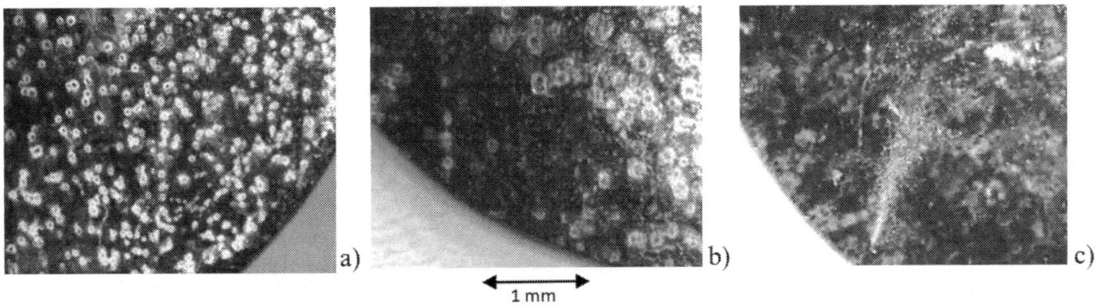

Figure 4. MAB tested for 90 min. in a) SBW, b) ASW, c) ABW.

The data in Table 4 indicate that the damage increases with time for both parameters, density of damage (AZ) and size of damaged spots (Φm). There appears a systematic ranking for the susceptibility of the materials to cavitation damage in this test: MAB>brass>NAB, which is more pronounced for Φm than for AZ. E.g. we find for ASW after 90 minutes for Φm 48:12 μm (MAB:NAB) and for AZ 57:24 % (MAB:NAB), and similar ratios were gained with SBW. However, damage parameter values for SBW appear significantly higher (about double) than those for ASW.

For the microscopic inspection of the surfaces, the specimens tested for 45 min. were found more suitable since the degree of damage was sufficiently low to reveal the consequences of impacts on the different materials. Figure 5 presents typical observations made with ASW tested specimens. For brass and NAB, mainly effects of plastic deformation from the cavitation impacts were observed, with more diffuse appearance (brass) up to distinct impact craters (NAB). For MAB, however, damages appeared preferentially as deep pits related to locations where the kappa phase is present, which was removed in part (attack at the boundary to surrounding alpha), and sometimes it erodes completely, leaving some kind of brownish halo, presumably a thin layer of corrosion products, on the surrounding surface.

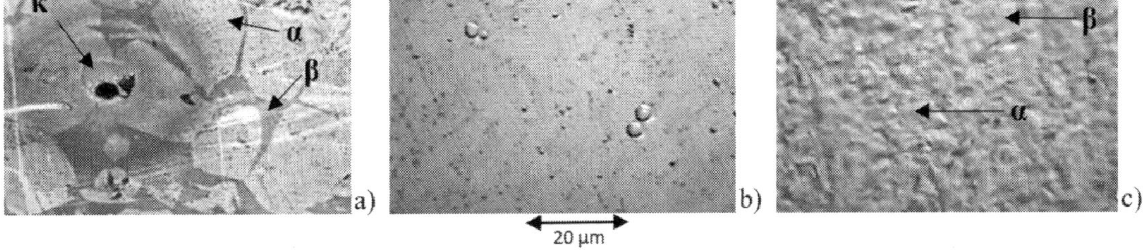

Figure 5. Typical surface damage phenomena at a) MAB, b) NAB, and c) brass, tested in ASW.

With SBW, the phenomena were similar to those observed with ASW but there were indications for the impacts having been more severe: figure 6a presents an example from brass, where some material appears eroded in an irregular manner, affecting mainly the alpha phase. figure 6b provides an example from NAB, where rather deep pits have appeared, accompanied by massive brown halos. Their relation to a certain microstructural feature is not traceable any more, but the color of the halo indicates the presence of iron and thus indicates the dissolution of some kappa phase, mainly those with higher content of this chemical element, denominated κ_I and κ_{II}.

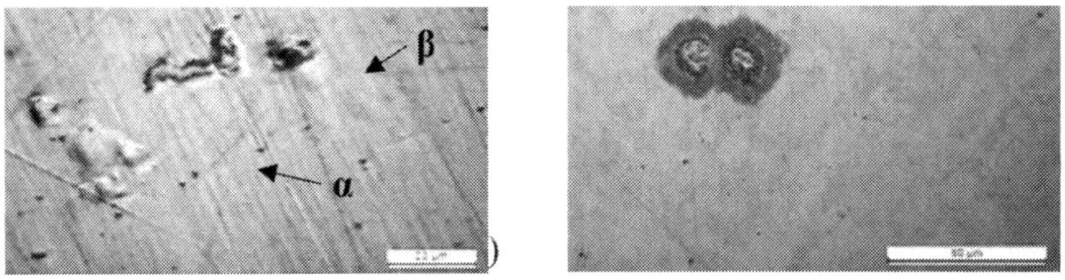

Figure 6. Typical surface damage phenomena at a) brass and b) NAB, tested in SBW.

With ABW it was generally hard to identify distinct locations of damage due to the accumulated corrosion products (figure 3) being darker for MAB in relation to NAB and brass. After their careful removal, the surface features summarized in figure 7 could be observed. The attack at MAB (figure 7a) occurred preferentially at the interface between κ- and α-phase, or by complete removal of κ, accompanied by intense halos of corrosion products, indicating a significant contribution of corrosion to the removal of material. This phenomenon is similar that observed with ASW. With NAB (figure 7b), preferential localized removal of β-phase was observed, leading to a dendritic spreading of the material loss by corrosion, emerging from impact sites. A similar pattern was found for brass (figure 7c), where also the β-phase gets preferentially removed in a very localized manner.

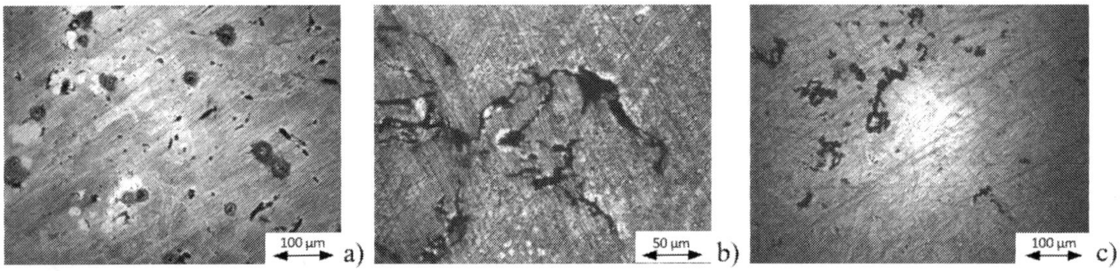

Figure 7. Typical surface damage phenomena at a) MAB, b) NAB, and c) brass, tested in ABW.

4. Discussion

The data in Table 4 and the impressions from microscopic inspection (figures 3, 5, and 6) indicate natural seawater, i.e. SBW, acting more aggressive on the tested materials under cavitation conditions compared to the synthetic medium (ASW). However, this increased aggressiveness appears not related to corrosiveness (the appearance of corrosion products is rather limited in both media), but rather to more frequent and more intense impacts from cavitation occurring in SBW. This may be explained by the suspended solids which are present in natural waters and which modify substantially the cavitation behavior of the water. As reported previously [10], these microscopic particles may serve as nucleation sites for the cavitation-caused vapor bubbles, increasing the probability of cavitation impacts.

Under cavitation conditions, surface inhomogeneity may induce the formation of pits, as was observed on MAB and NAB. A particular role in this context play the κ-phases, i.e. intermetallic compounds rich in Fe, Al and Ni, and this observation is in agreement with results from previous studies [11, 12]. κ-phases of MAB and NAB are different in chemical composition, size, shape and distribution and may induce the nucleation of bubble implosion, driving their erosion due to large dislocations between grain boundaries [13], in particular at the α/κ-interface. Such a surface inhomogeneity induces a continued erosion process since it acts as preferential site for bubble implosions causing an accumulative effect of damage. Consequently, the material is unable to recover its protective layer, enhancing the synergistic interaction between cavitation and corrosion [14]. Besides, the turbulent flow conditions under cavitation conditions enhance the mass transport of oxidant and aggressive species to the metal surface by eliminating diffusion limitation, thus accelerating the corrosion processes [11].

By contrast, brass suffers mainly plastic deformation affecting preferentially the α-phase, which contains less Zn than the β-phase and therefore is less hard [15].

ABW turned out to create a strong additional corrosive effect to the mechanical cavitation damaging. This may be related to its content of sulfidic species, which is well known to act detrimental on copper and its alloys by retarding the formation of the regular, copper oxide-based protective layer [2, 16]. Moreover, the corrosion products on the surface may have influenced the further nucleation of impacts, resulting in a significant change in the damaging pattern, particularly for NAB and brass (Figure 7b, c).

To summarize, a significant effect of the composition of the test medium on cavitation damage was observed. Natural seawater and synthetic seawater, although of very similar chemical composition, were found to induce different degrees of cavitation damage, attributed to some dispersed solid phase in the natural water. On the other hand, artificial brackish water was found to add a substantial contribution of corrosion to the cavitation process. Its contribution is related to the presence of dissolved sulfidic species, which negatively affects the ability of the material to develop a new protective layer after an impact event.

5. Conclusions

Cavitation corrosion is a rather complex phenomenon: cavitation, as a mechanical phenomenon, erodes the most brittle and hard phases of the material, which are frequently the intermetallic compounds. Corrosion, as an electrochemical phenomenon, attacks the most reactive phases and controls the formation of a more or less protective layer, depending on the chemical composition of the electrolyte. A continuous cavitation process removes corrosion products and impacts until the brittle phases are eliminated.

There appears a systematic ranking for the susceptibility of the investigated materials to cavitation damage in this test, independent of the test medium: MAB>brass>NAB. This behaviour is attributable to the chemical composition, distribution and particular large size of the rather hard and more brittle κ-phase of MAB that acts as nucleation site for impact from bubble collapse.

Artificial brackish water, ABW, is the most aggressive test medium, causing a strong corrosion contribution to the cavitation effect. This is attributed to the presence of sulfidic species which reacts with copper to non-protective corrosion products.

Natural sea water, SBW, was found to act more aggressive under cavitation conditions compared to the synthetic medium, ASW. This may be explained by the suspended solids present in the natural water as these microscopic particles may serve as nucleation sites during the cavitation process.

Acknowledgement: to JC Navalips, Maliaño, Spain, and Latones del Carrión, Córdoba, Spain, for providing bronzes and brass, respectively, and to Gobierno de Cantabria, for financial support of project SUBVTC-2021-0024.

References

[1] Ivan Richardson 2016 Guide to Nickel Aluminium Bronze for Engineers (ed. Carol Powell), *Copper Development Association Publication No 222*, Copper Development Association
[2] Zhao X, Qi. Y, Wang J, Peng TX, Zhang Z, Li K 2020 *Metals* **10** 1227
[3] Doijode P S, Hickel S, van Terwisga T, Visser K 2022 *Applied Ocean Research* **124** 103174

[4] Peters A, Lantermann U, Moctar O 2018 *Wear* **408–409** 1–12
[5] De Sanchez S R, Schiffrin D . 1982 *Corrosion Science* **22** 585-607
[6] Song Q N, Xu N, Bao Y F, Jiang Y F, Gu W, Zheng Y, Qiao Y X 2017 *Acta Metall. Sin.* (Engl. Lett.) **30** 712–720
[7] Cobo I, Biezma M V, Linhardt P 2022 *Mater. Corr.* **73** 1788-1799
[8] Mota N M, Tavares S S M, Nascimento A, Zeeman G, Biezma-Moraleda M V 2021 *Eng. Failure Analysis* **129** 105732
[9] DIN 50905-4 2018 Deutsches Institut für Normung e. V. (Beuth Verlag GmbH, Berlin, Germany)
[10] Kunpeng S, Jianhua W, Dingkang X 2020 *Ultrason Sonochem* **68** 105214
[11] Biezma M V, Linhardt P 2022 *11th Iberian Conference on Tribology 6-7 October 2022 – Setúbal, Portugal, Book of Abstracts* 96-98
[12] Linhardt P, Kührer S, Ball G, Biezma M V 2018 *Mater. Corros.* **69** 358-364
[13] Karimi A, Martin J L 1986 *International Metals Reviews* **31** 1–26
[14] Linhardt P, Biezma, M V, Strobl S, Haubner R 2023 *Solid State Phenomena* **341** 25-30
[15] Moriarty M, Wu Y, Murray T, Hutchinson C 2021 *Corros. Sci.* **184** 109366
[16] Schüssler A, Exner H E 1993 *Corros. Sci.* 34 1793-1802

Rudder hydrodynamics behind a propeller rotating ahead and astern

Suli Lu[1,*], **Djahida Boucetta**[1], **Wim Van Hoydonck**[2], **Evert Lataire**[1], **Guillaume Delefortrie**[1]

[1]Maritime Technology Division, Ghent University, Technologiepark 60, 9052 Ghent, Belgium
[2]Flanders Hydraulics, Berchemlei 115, 2140 Antwerp, Belgium

E-mail: suli.lu@ugent.be

Abstract. Ships sailing in harbour environments will experience four-quadrant manoeuvres, based on the direction of ship velocity and propeller rate. A better understanding of ship hydrodynamics in such manoeuvres will contribute to the navigation safety. Aimed at ships equipped with a conventional single propeller and single rudder, the hydrodynamic performance of a rudder with the engine ahead and astern is studied. The KP505 propeller and NACA 0018 semi-balanced rudder (from the KCS benchmark ship) are selected for numerical studies for extended experimental data published. After validating numerical methods with open-water test data for the single propeller and the single rudder, CFD simulations based on RANS methods are conducted for different advance ratios ahead and astern, with rudder deflections ranging from 0 to 15 degrees. To understand propeller impact on lift and drag forces of the rudder in different working conditions, inflow to the rudder in propeller slipstreams are analysed by extracting the flow field data in different profiles along the rudder span. Streamlines around the rudder and pressure distributions on the rudder surfaces, along with the turbulent kinematic energy distributions and the vortex structures visualised by Q-criterion, are compared to reveal propeller-rudder interaction mechanisms in ahead and astern conditions. The study shows that the propeller rotation mode has a significant impact on rudder inflow patterns, and induced rudder forces change in different trends with propeller loading variations.

1. Introduction

As the main steering device in ship manoeuvres, rudders play a significant role in the course-changing of ships. Hydrodynamic forces acting on the rudder, therefore, are key factors to be determined in ship manoeuvring modelling. For conventional vessels, rudders operate in the turbulence flow generated by the hull wake and the propeller slipstream. The propeller generally rotates ahead during sailing, while in exceptional cases such as harbour manoeuvres, the propeller can rotate astern. The rudder behind an astern-rotating propeller can exhibit different hydrodynamic performance compared with widely studied ahead ones, leading to different trends in motion parameters' variations.

Denoted by the signs of the hull longitudinal velocity and the propeller rate, the four-quadrant motions of the ship can be defined. In this article, the first and the second quadrant of the single-propeller-single-rudder (SPSR) system are studied. Many researchers have paid attention to flow mechanisms of SPSR systems in the first quadrant. Posa [1] investigated the load distributions along the rudder spanwise directions in the propeller slipstream, and it is found that the core wake shed by the propeller blades has a much larger influence on the rudder surface pressure fields than the hub vortex and the tip vortices. Hu [2] performed CFD simulations on the rudder surface pressure fluctuations within the propeller wake, and numerical results reveal that the propeller tip vortex can enhance the fluctuation periodicity while

IOP Conf. Series: Materials Science and Engineering 1288 (2023) 012057 doi:10.1088/1757-899X/1288/1/012057

the hub vortex enhances the fluctuation randomness. Wang [3] conducted Detached Eddy Simulation (DES) on wake dynamics of the propeller before a rudder, showing that heavy loading conditions and finite rudder angles lead to more complex vortex topological structures.

For the propeller-rudder system working in the second quadrant, limited studies are conducted experimentally and numerically compared with the first quadrant. Extensive wind tunnel tests by Molland [4] are executed for four quadrant conditions, and rudder forces data was provided under different propeller loadings and rudder angles. As for studies based on Computational Fluid Dynamics (CFD) methods, some simulations are performed for the propeller in the reversely-rotating mode. Kroll [5] studied a ducted propeller in crashback with LES methods, and the obtained unsteady force coefficients show that the tip-leakage flow leads to high rotor blade tip loading, while it is mainly from the vortex ring in open water conditions. Jang [6] performed Large Eddy Simulation (LES) on a reverse rotating propeller and presented the unsteady flow physics by analysing the unsteady vortex ring separations and propeller forces variations. Verma [7] compared the hydrodynamic performance of a reverse rotating propeller with and without a submarine hull in two advance coefficients, and the case with a higher negative advance ratio shows larger differences in the recirculation zone area and vortex ring location due to hull impacts.

Though there are many publications on ships' steering and propulsion computations, the modelling of the rudder behaviour behind the propeller in four-quadrant wake fields is rarely done. The overall objectives of this paper are to enhance the prediction of the flow field details and analyse the rudder hydrodynamics behind the propeller rotating ahead and astern. Following the introduction, Section 2 demonstrates the geometrical information and numerical settings involved in CFD simulations. Section 3 shows the validation procedures for the SPSR system. In Section 4, flow field details are analysed to show rudder hydrodynamics behind the propeller rotating ahead and astern, and corresponding conclusions are summarized in Section 5.

2. Computational details

The five-blade and right-handed KP505 propeller, which are equipped on the KCS benchmark ship, are modelled. The rudder configuration consists of two components: a fixed horn part and a movable rudder blade part which rotates with respect to the rudder stock, relatively positioned in accordance with the KCS stern configuration. The geometries of the propeller and the rudder are shown in Figure 1, and the main particulars are presented in Table 1.

Table 1. Main particulars of KP505 propeller and NACA 0018 semi-balanced rudder.

Propeller			Rudder		
Particular	Symbol	Value	Particular	Symbol	Value
Diameter (m)	D	0.250	Mean chord length (m)	c_m	0.174
Area ratio ($-$)	A_E	0.448	Span (m)	B	0.313
Hub ratio ($-$)	r_h	0.150	Aspect ratio ($-$)	Λ_G	1.799
Pitch ratio ($-$)	$(P/D)_{0.7R}$	1.000	Area of the movable part (m^2)	S	0.054
Scale ratio ($-$)	λ	1/31.6	Scale ratio ($-$)	λ	1/31.6

Vyšohlíd [8] stated that RANS is unable to adequately predict propeller crashback because of the pervasive large-scale unsteadiness. However, considering that RANS methods can still yield reasonable results for the propeller rotating astern compared with more accurate numerical strategies such as DES (10% $\sim$ 20% deviations according to Guo [9]) with relatively lower computational cost, the RANS

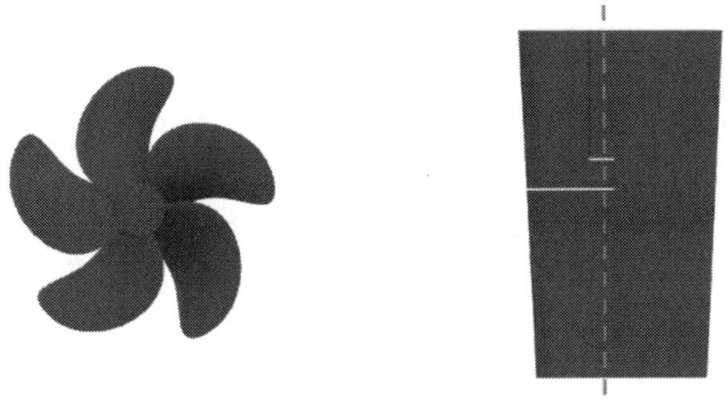

(a) KP505 propeller (b) NACA 0018 semi-balanced rudder

Figure 1. KP505 propeller and NACA 0018 semi-balanced rudder geometry (the red dashed line denotes the rudder stock).

method is applied for CFD simulations in this paper, and the governing equations for incompressible viscous flow are as follows:

$$\frac{\partial \bar{u}_i}{\partial x_i} = 0\,,$$

$$\rho\frac{\partial \bar{u}_i}{\partial t} + \rho\bar{u}_j\frac{\partial \bar{u}_i}{\partial x_j} = f_i - \frac{\partial \bar{p}}{\partial x_i} + \frac{\partial}{\partial x_j}\left(\mu\frac{\partial \bar{u}_i}{\partial x_j} - \rho\overline{u_i'u_j'}\right), \quad (i = 1,2,3)\,,$$

where $\bar{u}_i$ is the mean velocity component in the x direction, ρ and μ are the density and viscosity coefficient of the fluid, $\bar{P}$ is the mean pressure, f_i is the body force, while $\rho\overline{u_i'u_j'}$ is the Reynolds stress tensor. $k-\omega$ SST turbulence model is used to close the RANS equation in this paper for its advantages in capturing flow separations, and the overPimpleDyMFoam solver is selected for following simulations with the rotating propeller.

Open water simulations and the propeller-rudder interaction simulations are conducted in a cubic calculation domain, which extends 6 D upstream and 8 D downstream, with the lateral distance of 3.5 D, as shown in Figure 2. The overset meshing strategy is utilised to simulate the rotating propeller, with the propeller wrapped in a cylinder rotation domain and then merged with the static background mesh. The rudder is positioned in the static region with fixed angles of attack for propeller-rudder interaction simulations. All the meshes are generated in OpenFOAM v2106 with blockMesh and snappyHexMesh.

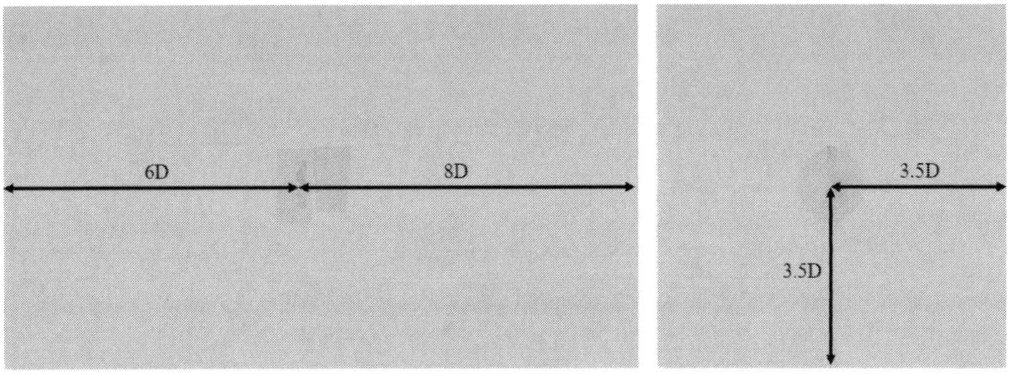

Figure 2. Calculation domain for the open water and propeller-rudder interactions simulations.

3. Grid convergence study

Grid convergence studies are performed for the open water propeller simulations firstly with the time step $\Delta t = 2.315 \times 10^{-4}s$, corresponding to the time the propeller rotates for 1 degree. Three sets of meshes with the coarse, medium and fine level are generated by a refinement ratio $\sqrt{2}$ with the number of cells ranging from $7.08 \times 10^5 \sim 4.18 \times 10^6$, as shown in Figure 3, where the red parts denote for the overset regions. For the case with the refinement level r_i, the first layer cell height Δs_i and the boundary layer growth ratio r_{bl_i} is adjusted according to Crepier [10]:

$$\Delta s_i = \Delta s_0 \frac{1-r_{bl_0}^{\frac{1}{r_i}}}{1-r_{bl_0}}$$

$$r_{bl_i} = r_{bl_0}^{\frac{1}{r_i}}, \quad (i=0,1,2)$$

where $i = 0$ denotes the coarsest mesh. Wall functions are applied to model the boundary layer flow, and y^+ is set as around 30. Obtained thrust force T and torque Q are expressed in dimensionless forms by:

$$K_T = \frac{T}{\rho n^2 D^4}$$

$$K_Q = \frac{Q}{\rho n^2 D^5}$$

$$\eta = \frac{K_T}{K_Q} \cdot \frac{J}{2\pi}$$

where K_T is the thrust coefficient, K_Q is the torque coefficient, η is the propeller efficiency, n is the propeller rotation rate ,and D is the propeller diamter. $J = 0.6$ and $n = 12\ rps$ are selected as the working conditions in this case, and the calculation lasts for $1s$ in physical time. Obtained K_T and K_Q are processed by a low-pass filter, and the average values are formulated by averaging filtered data in the last two revolutions. Grid Convergence Index (GCI) method from [11] is adopted for the convergency evaluation, and estimated uncertainty results are shown in Table 2. K_T shows the monotonic convergence while K_Q is oscillatory, probably due to fluctuations of viscous components in calculations. Considering that the magnitudes of the grid uncertainty values are small, the medium mesh settings are used for following studies.

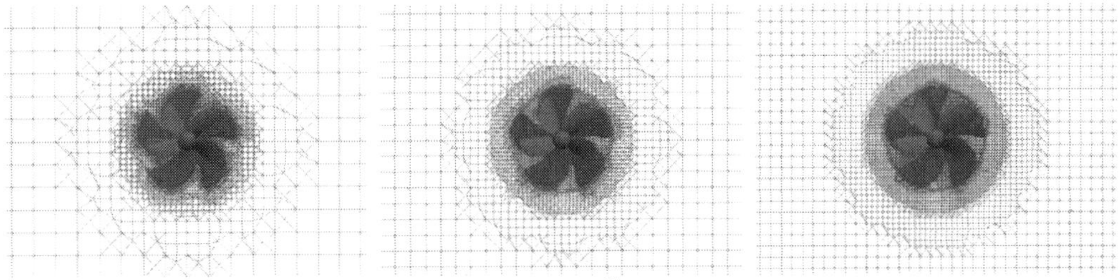

(a) $r_0 = 1$ (b) $r_1 = \sqrt{2}$ (c) $r_2 = 2$

Figure 3. Meshes with different refinement ratios around the propeller.

Besides $\Delta t = 2.315 \times 10^{-4}s$, another two times steps $\Delta t = 1.157 \times 10^{-4}s$ and $\Delta t = 4.630 \times 10^{-4}s$ are selected for time step independence studies, and simulation results are shown in Table 3. Thrust and torque coefficients are the same with the simulation interval $\Delta\theta$ changing from 0.5° to 2°, partially because of the adjustTimeStep scheme activated in OpenFOAM. Therefore, the medium $\Delta t = 2.315 \times 10^{-4}s$ is adopted from the perspective of balancing the accuracy and the efficiency of numerical simulations.

Table 2. Thrust and torque coefficients obtained from meshes with different refienment ratios.

	K_T	$10K_Q$
ε_{21}	-3.295×10^{-3}	6.632×10^{-3}
ε_{32}	-1.363×10^{-2}	-2.005×10^{-2}
$\varepsilon_{21}/\varepsilon_{32}$	0.242	-0.331
Convergence	Monotonic	Oscillatory
p	4.096	3.192
$U_{G_{21}}$	1.135×10^{-3}	2.167×10^{-3}
$U_{G_{32}}$	4.764×10^{-3}	6.437×10^{-3}

By changing the inflow velocity while keeping $n = 12\ rps$ fixed, exactly as experimental settings in Fujisawa [12], RANS simulations are conducted for open water characteristics of KP505 propeller. The comparison with the experimental data in Figure 4 shows that deviations between CFD and EFD results for K_T and K_Q are within 5% for $J = 0.5 \sim 0.7$, while η show larger deviations for higher J. Simulation results for lighter loading conditions are not as accurate as heavier ones, and one possible reason is that the propeller is assumed to experience fully developed turbulence flows, so transitions from laminar to turbulent states existing in model tests are not captured accurately. Following propeller-rudder interaction simulations for ahead conditions in Section 4 will be conducted for $J = 0.5 \sim 0.7$, and current numerical methods are validated for such applications.

Table 3. Thrust and torque coefficients obtained from the medium mesh with different time steps.

Δt (s)	$\Delta\theta$ ($^\circ$)	K_T (-)	$10K_Q$ (-)
1.157×10^{-4}	0.5	0.2269	0.3845
2.315×10^{-4}	1.0	0.2269	0.3845
4.630×10^{-4}	2.0	0.2269	0.3845

Similar strategies are applied for the rudder blade simulation in open water conditions, with meshes of different refinement ratios presented in Figure 5. The y^+ value is around 30 with wall functions utilized, and the numbers of cells range from $4.14 \times 10^5 \sim 3.25 \times 10^6$. The fixed rudder horn part is excluded in the convergence study since the experimental data to be compared with in Van Hoydonck [13] only takes the movable part into account. Lift and drag coefficients C_L and C_D of the rudder can be expressed as:

$$C_L = \frac{F_L}{0.5\rho U^2 S}$$
$$C_D = \frac{F_D}{0.5\rho U^2 S}$$

where F_L is the lift force, F_D is the drag force, U is the inflow velocity, and S is the rudder surface area. Grid convergence results for rudder simulations with $\Delta t = 2.315 \times 10^{-4} s$ in Table 4 show that C_L and C_D both achieve monotonic convergence, and the medium level mesh is adopted. C_L and C_D for $\alpha = 0^\circ \sim 25^\circ$ are compared with experimental results in Figure 6. Lift coefficients agree with experimental data better for $\alpha > 15^\circ$ than smaller angles. Drag coefficients also coincide better under higher rudder deflections, but show larger relative errors than lift coefficients. From the perspective of forces components, drag forces are dominated by the frictional shear stress, the viscous part, while lift forces are mainly contributed by inviscid pressure differences on rudder surfaces [14]. Therefore,

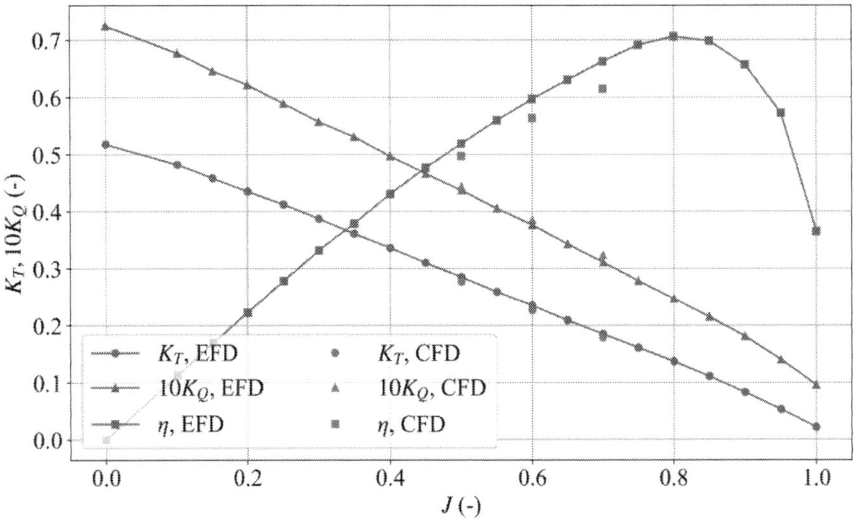

Figure 4. Comparison of open water characteristics of KP505 propeller with experimental data.

larger deviations in drag coefficients are partially related to some inherent defects of RANS methods in predicting viscous effects and the wall function model applied, and the existing accuracy is presently acceptable considering the smaller contributions of the drag force to the forces relevant for manoeuvring (sway force and yaw moment).

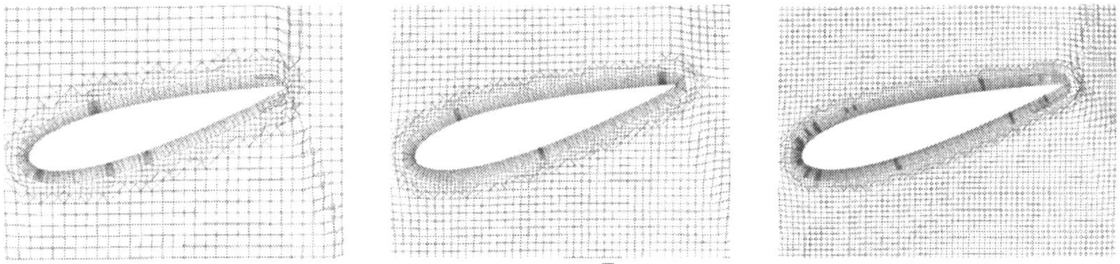

(a) $r_0 = 1$ (b) $r_1 = \sqrt{2}$ (c) $r_2 = 2$

Figure 5. Meshes with different refinement ratios around the rudder.

4. Results and discussions

With the numerical methods validated with experimental data for the propeller and the rudder in open water conditions in Section 3, RANS simulations for propeller-rudder interactions in the first and the second quadrant are performed. Section 4.1 presents rudder hydrodynamic results behind a propeller rotating ahead, and flow details are compared under different advance ratios and angles of attack. The rudder in the wake of a reverse rotating propeller is studied in Section 4.2, while only two working conditions ($J = -0.5/\alpha = 15°$ and $J = -0.7/\alpha = 15°$) are simulated considering significantly longer physical time needed for the propeller crashback simulations, and such results are analysed and compared with the first quadrant results.

Table 4. lift and drag coefficients obtained from meshes with different refienment ratios.

	C_L	C_D
ε_{21}	-5.817×10^{-3}	2.068×10^{-3}
ε_{32}	-2.194×10^{-2}	7.671×10^{-3}
$\varepsilon_{21}/\varepsilon_{32}$	0.265	0.270
Convergence	Monotonic	Monotonic
p	3.830	3.782
$U_{G_{21}}$	1.743×10^{-3}	1.316×10^{-3}
$U_{G_{32}}$	6.703×10^{-3}	4.811×10^{-3}

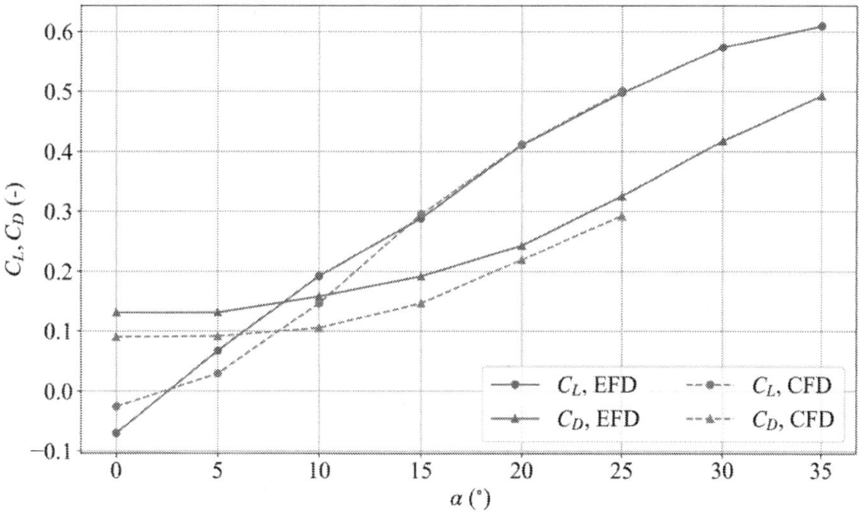

Figure 6. Comparison of open water characteristics of KCS rudder with experimental data.

4.1. Rudder behind a propeller rotating ahead

Vortex structures around the propeller and the rudder are visualised by the isosurface of $Q = 300s^{-2}$ in Figure 7. Vortices shed by the propeller blades and the hub, along with those by rudder ends, constitute the wake of the propeller-rudder system. Interacting with the downstream rudder, the propeller jet flow is blocked and the vortices break down, demonstrating different continuity characteristics from the free evolution in single-propeller simulations. In the far field, vortices are highly skewed and dissipated, while such a phenomenon is partially due to deficiencies in RANS methods and insufficient mesh resolutions. When $\alpha = 0°$, vortices near the rudder are helix-shaped, but the evolution pattern is disrupted in larger angles of attack. For the case with $\alpha = 15°$, the flow separation leads to more complicated turbulence. In lighter loading conditions, the vortex merging tends to delay, and more stable vortex evolutions can be captured.

Axial velocity contours in *xoz* plane are shown in Figure 8. Regions influenced by propeller tip vortices are in high velocities while velocities of flows between the propeller hub and the rudder leading edge are low because of the blockage effect and shedded hub vortices. Periodic high-velocity regions in the propeller blade wake as a result of the projection of 3D vortices on 2D profiles can be observed, and the development of such regions is unstable in the far field. For $\alpha = 0°$ cases, velocities show symmetries with respect to the central line of the rudder profile. Larger angles, however, deflected

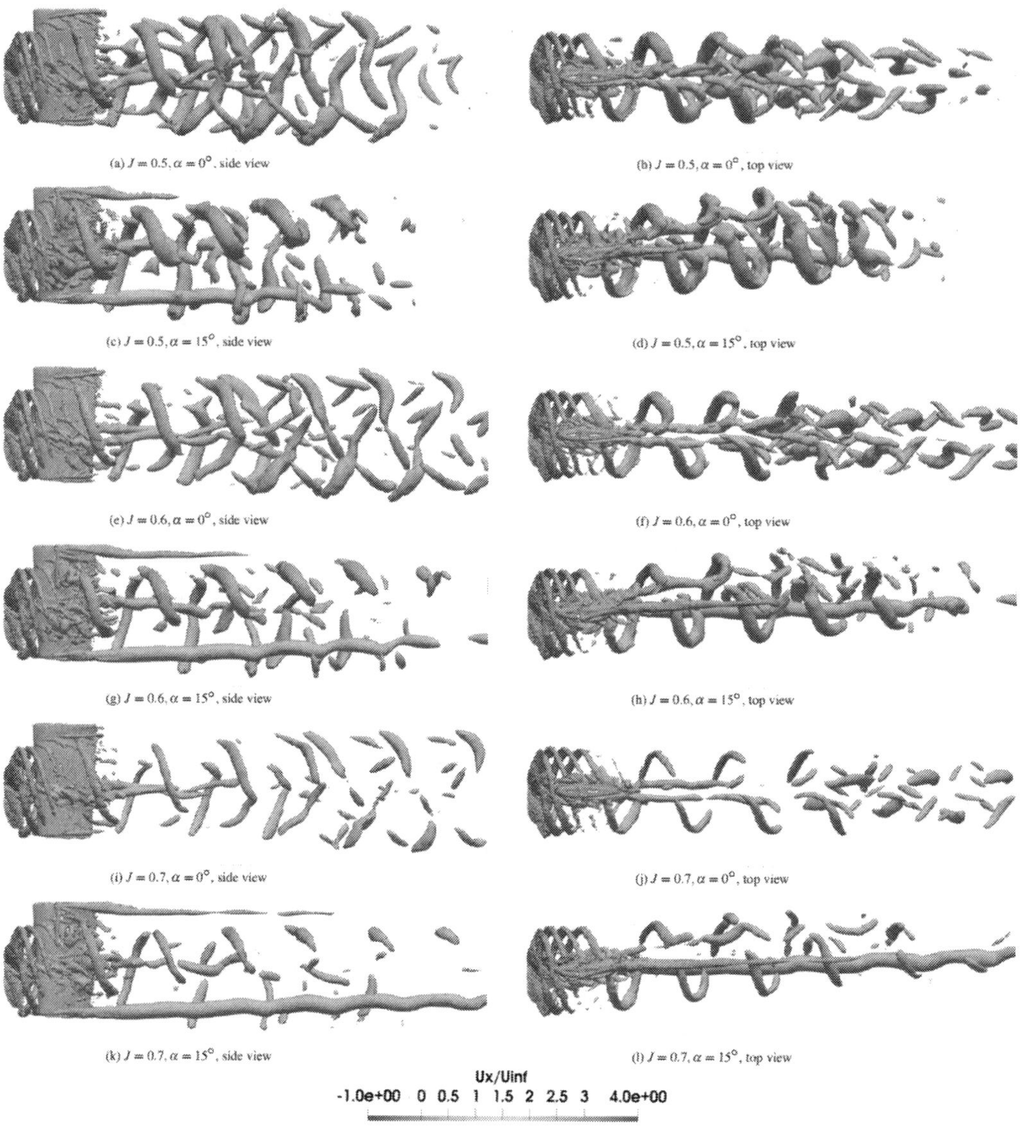

Figure 7. Isosurface of $Q = 300s^{-2}$ around the propeller-rudder system in ahead conditions, $t = 1$ s.

the development of the high-velocity regions and contributed to the expansion of the low-velocity area near the leeward surface for $\alpha = 15^\circ$ due to backflows. Average velocity magnitudes are higher for larger advance coefficients for larger inflow velocities, and the high-velocity area patterns are similar for $J = 0.5 \sim 0.7$.

Figure 9 shows pressure distributions on rudder surfaces from $\alpha = 0^\circ \sim 15^\circ$ in ahead conditions. The approximately mirror-symmetrical locations of the high-pressure regions on both sides of the rudder when $\alpha = 0^\circ$ show the impact of the evolving tip vortex. The high-pressure areas on the starboard surface, together with low-pressure areas on the port side counterpart, increase with larger angles of attack. Heavy loadings produce more remarkable pressure differences on rudder surfaces, and increased lift coefficients can be obtained.

Lift and drag coefficients of the rudder behind the propeller under different advance coefficients are

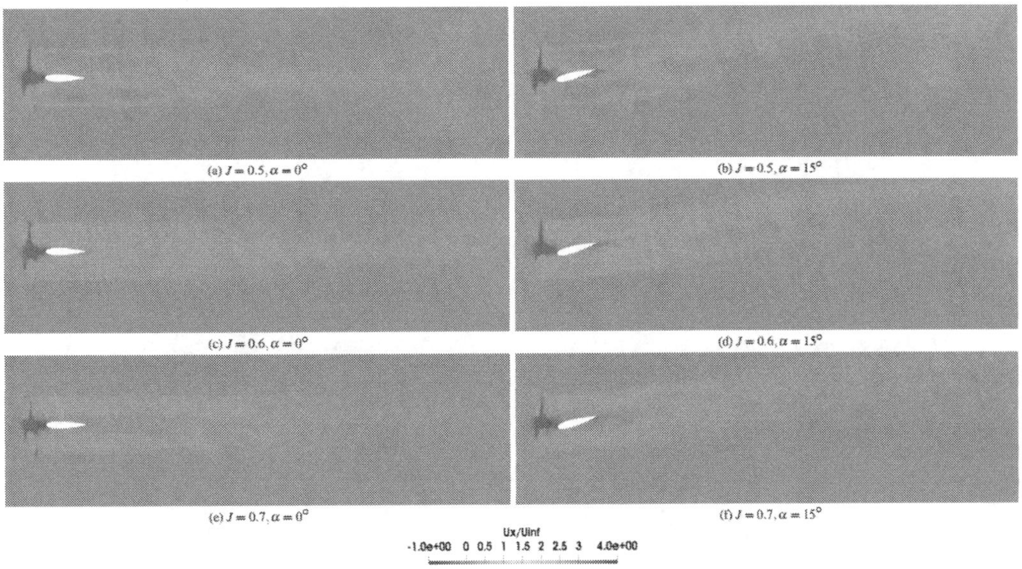

Figure 8. Axial velocity contours around the propeller-rudder system in ahead conditions, $t = 1$ s.

shown in Figure 10. Although the fixed rudder horn is also included in propeller-rudder interactions simulations, C_L and C_D in Figure 10 refer to forces acting on the rudder blade and are made non-dimensional by the area of the movable part. The acceleration effects of the propeller in the first quadrant enhance the inflow velocities to rudder surfaces, and resulted C_L are larger compared with open water conditions. Magnitudes of C_L and its increment ratios versus rudder angles both decrease with advance coefficients, demonstrating more significant propeller impact on rudder lift forces under heavy loadings. For zero rudder angles, incremental positive magnitudes of C_L for three loading conditions are derived from spatially asymmetrical slipstreams near the port side and starboard. Smaller C_D values for $\alpha = 0° \sim 10°$ than open water rudders, apart from the effect of the horn, are related to strong swirl components of propeller jets, which shift local rudder deflections and may result in negative drag forces [15].

4.2. *Rudder behind a propeller rotating astern*

Two working conditions of the propeller rotating astern, $J = -0.5$ and $J = -0.7$, are simulated with $\alpha = 15°$ for the physical time of 8 s, corresponding to 96 propeller revolutions. Time histories of K_T, K_Q, C_L, and C_D in second quadrant conditions are shown in Figure 11. All the forces show large fluctuations even after 60 propeller revolutions, which is attributed to the unsteady nature of the propeller jet in counterflows. The oscillation frequencies of K_T and K_Q are very similar to each other, and the similarity is in accordance with LES results on a crashback propeller by Jang [6]. The rudder experienced the reversed flow which is in opposite to inflow directions, so the starboard rudder angle provides the negative lift forces.

For the propeller rotating astern, the isosurfaces for $Q = 300s^{-2}$ around the propeller-rudder system for $J = -0.5$ and $J = -0.7$ in $t = 8s$ are shown in Figure 12. A completely different vortex evolution compared with the first quadrant is formulated due to the opposite directions of the uniform inflow and the propeller jet, and the highly unsteady vortex ring wrapping the propeller-rudder system is captured. The reverse vortex induced by blades forms in the near front of the propeller, and the trends of moving forward are blocked by incoming flows, leaving the location of the vortex ring core relatively fixed. However, the vortex shedding from the top and the bottom of the ring indicates that the ring patterns are

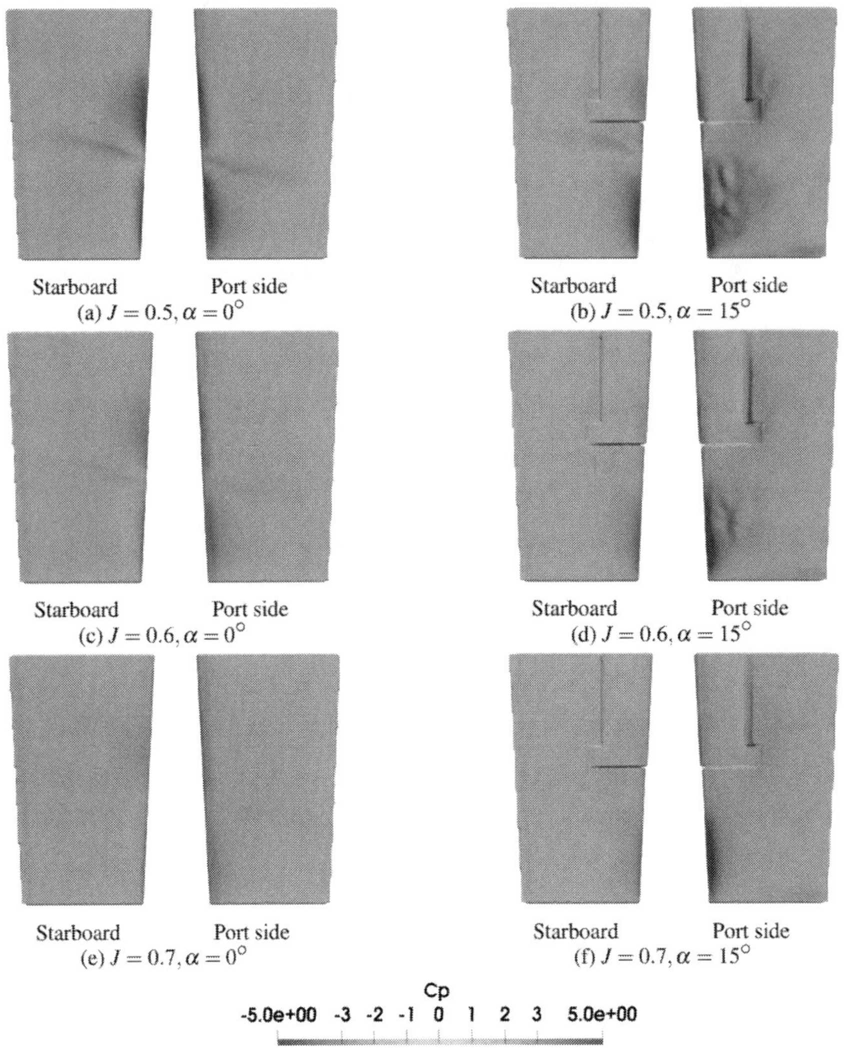

Figure 9. Pressure distributions on rudder surfaces behind a propeller rotating ahead, $t = 1$ s.

still evolving.

Figure 13 shows axial velocity distributions in astern conditions. The rudder is located in negative axial velocities due to the pumping of the propeller, and the zero U_x regions are formed in front of the propeller hub. Near the outer edge of the negative U_x zone, positive U_x with higher magnitudes than inflow velocities emerge, which explains the conservation of mass flow rates across the section.

Pressure distributions on rudder surfaces for astern conditions are shown in Figure 14. No alternating regions with the high and low pressure are observed, and port side and starboard surfaces near the leading edge are both in low pressure as the result of propeller slipstream interactions with the inflow. Rudder surfaces in the vicinity of the trailing edge show slightly larger pressure owing to being away from the reverse flows, and such pressure distributions cause negative C_L in Figure 11.

3D Streamlines around the propeller-rudder system are presented in Figure 15(a)~(b), and are compared with the first quadrant conditions in Figure 15(c)~(d). The uniform inflow passing through the propeller disk, instead of being accelerated and flowing straightforward to the rudder, enter a

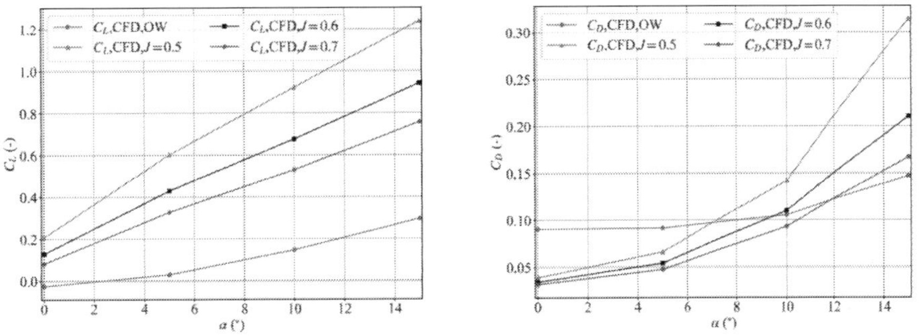

(a) C_L (b) C_D

Figure 10. Lift and drag coefficients of the rudder behind a propeller rotating ahead.

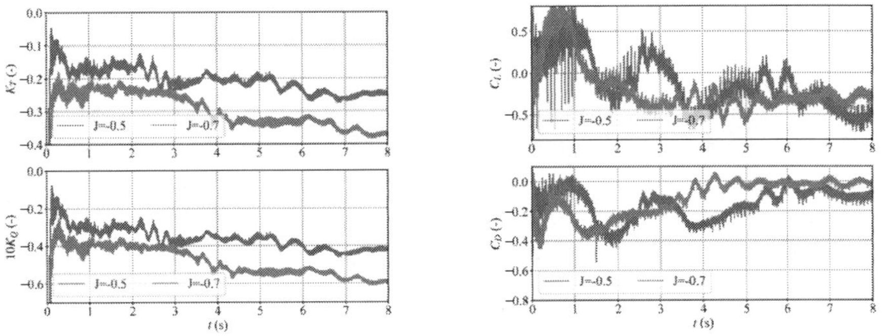

(a) Propeller forces (b) Rudder blade forces

Figure 11. Time histories of the propeller and the rudder blade forces in second quadrant conditions.

recirculation zone stated by Jang [6]. The strong reverse flows because of the pumping effect of the propeller significantly decrease inflow velocities to the rudder, corresponding to large low-velocity areas in Figure 14. For the $J = -0.5$ case, stronger pumping effects lead to larger scale motions of reverse flows compared to $J = -0.7$, so the vortex ring evolves with a larger radius.

Axial velocity distributions of rudder inflows in ahead and astern conditions are shown in Figure 16 (a)~(d). A section between the propeller hub and the rudder leading edge is selected to demonstrate inflow velocities in the near front of the rudder, and dimensionless axial velocities along the span direction are extracted in three profiles, i.e. $z = -0.08\ m$, $z = 0\ m$, and $z = 0.08m$. The dashed pink lines in the coordinate system denote the position of the rudder projection area. In ahead conditions, far ahead inflows are accelerated within an approximately circular region of the propeller diameter, except for some flows shedded from the propeller hub. Standing in sharp contrast, the rudder in the second quadrant condition is in reverse U_x for the whole rudder span length, and heavier loadings make the reverse flow more violent inside a larger recirculation zone.

Turbulent kinematic energy (TKE) distributions around the propeller-rudder system are shown in Figure 17 (a)~(d). As the measurement of 3D unsteadiness and turbulence in the flow [7], TKE k can be expressed as:

$$k = \frac{1}{2}\left(\overline{(u')^2} + \overline{(v')^2} + \overline{(w')^2}\right),$$

IOP Conf. Series: Materials Science and Engineering 1288 (2023) 012057 doi:10.1088/1757-899X/1288/1/012057

(a) $J = -0.5, \alpha = 15^{\circ}$, front view

(b) $J = -0.5, \alpha = 15^{\circ}$, side view

(c) $J = -0.7, \alpha = 15^{\circ}$, front view

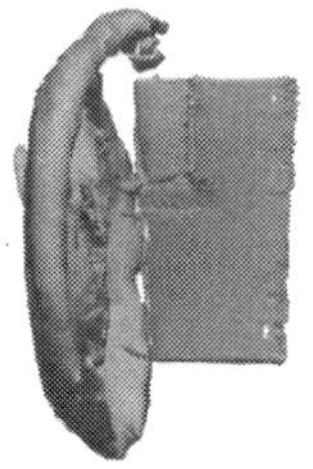

(d) $J = -0.7, \alpha = 15^{\circ}$, side view

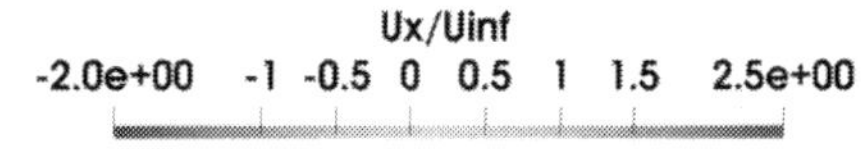

Figure 12. Isosurface of $Q = 300s^{-2}$ around the propeller-rudder system in astern conditions, $t = 8$ s.

(a) $J = -0.5, \alpha = 15^{\circ}$ (b) $J = -0.7, \alpha = 15^{\circ}$

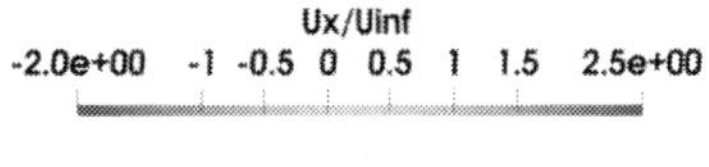

Figure 13. Axis velocity contours around the propeller-rudder system in astern conditions, $t = 8$ s.

where u', v', and w' are the differences between the instantaneous and the average velocity components. k is nondimensionalized to k_0 with the inflow velocity in this paper:

$$k_0 = \frac{k}{\frac{1}{2}U_{inf}^2},$$

TKE in ahead conditions show small magnitudes near the propeller for stable flow regimes, and flow separations close to the rudder port side slightly increase the value. Large TKE inside the vortex ring region for astern cases indicate that the rudder experiences highly unsteady flow conditions behind the reverse rotating propeller, which significantly extend to both lateral sides of the propeller-rudder system.

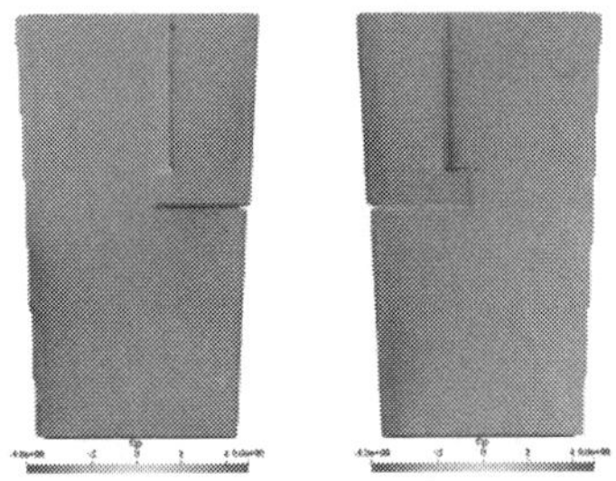

Starboard Port side
(a) $J = -0.5, \alpha = 15^\circ$

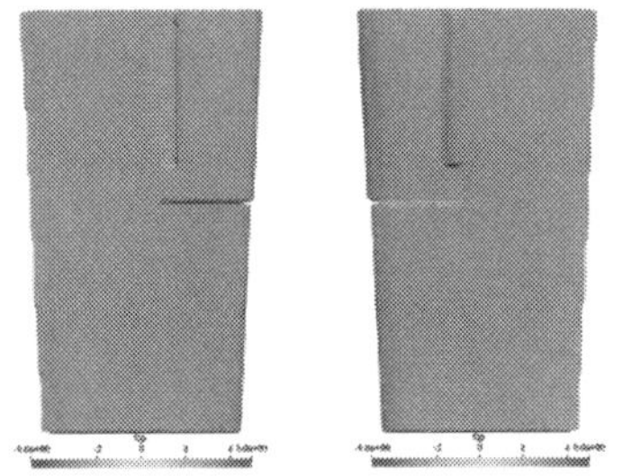

Starboard Port side
(b) $J = -0.7, \alpha = 15^\circ$

Figure 14. Pressure distributions on rudder surfaces behind a propeller rotating astern, $t = 8$ s.

(a) $J = -0.5, \alpha = 15^\circ, t = 8$ s

(b) $J = -0.7, \alpha = 15^\circ, t = 8$ s

(c) $J = 0.5, \alpha = 15^\circ, t = 1$ s

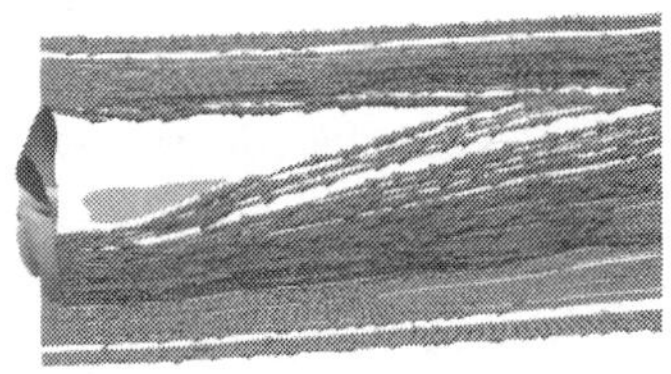

(d) $J = 0.7, \alpha = 15^\circ, t = 1$ s

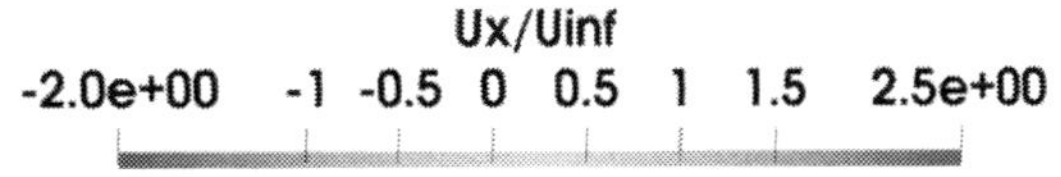

Figure 15. 3D Streamlines around the propeller-rudder system in ahead and astern conditions.

5. Conclusions

RANS simulations of propeller-rudder interactions in the first and the second quadrant are conducted in this paper, and rudder hydrodynamics behind the propeller rotating ahead and astern are analysed based on numerical results. In the first quadrant, the propeller accelerates and distorts the flow field

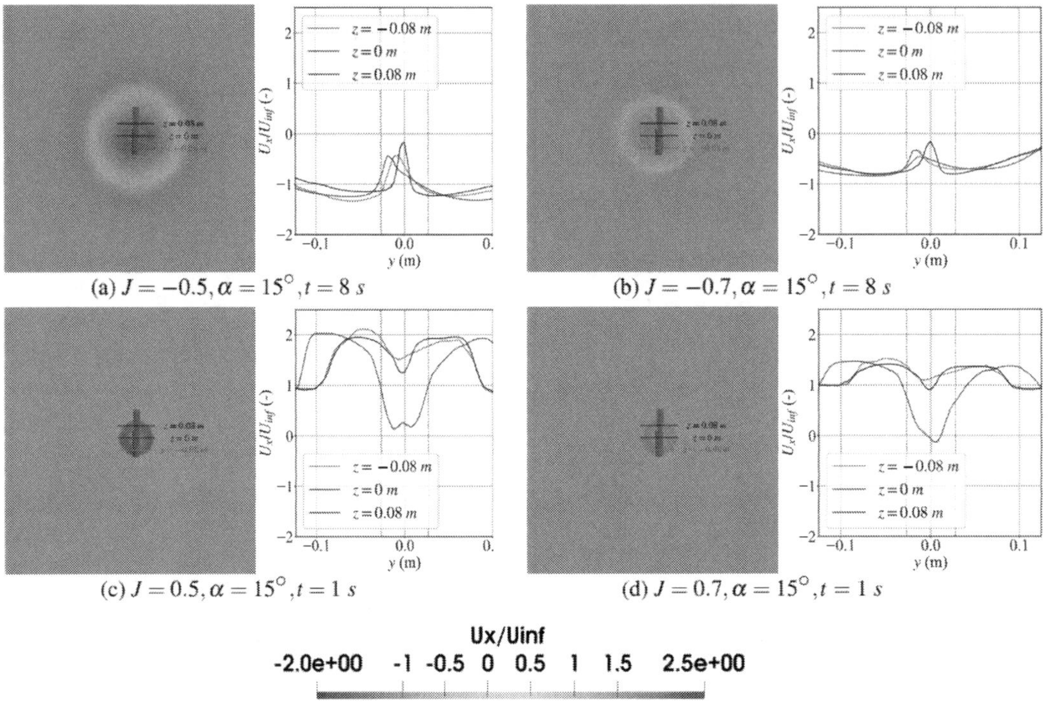

Figure 16. Axial velocity distributions of rudder inflows in ahead and astern conditions.

to the rudder, and such flow straightening effects lead to enlarged rudder lift forces and alternating high-pressure regions on rudder surfaces, which are more significant under heavier loading conditions and larger rudder deflections. Compared with the propeller rotating ahead around design conditions, the propeller working in the second quadrant conditions brings highly unsteady characteristics for the propeller-rudder system. The interaction between the uniformly incoming flow and the propeller jet in opposite directions produces the time-dependent vortex ring, so forces acting on the propeller and the rudder all undergo low-frequency fluctuations. Surrounded by low-velocity recirculation regions, rudders in such flow conditions can produce negative lift forces due to reverse inflows, while the impact of advance ratios and rudder deflections needs to be concluded with more working conditions in future research.

Apart from performing more simulations, methodological investigations on numerical studies of four-quadrant ship motions are also important. As a time-averaged numerical strategy, RANS may not be the optimal method for resolving the propeller slipstream in the crashback mode, and other advanced numerical strategies such as LES and DDES should be evaluated. The final aim of the research is to investigate the interaction between the rudder and propeller behind the hull in four quadrants.

Acknowledgement
The first author Suli Lu is financed by the China Scholarship Council under Grant 202206950018. The computational resources (Stevin Supercomputer Infrastructure) and services used in this work were provided by the VSC (Flemish Supercomputer Center), funded by Ghent University, FWO and the Flemish Government - department EWI.

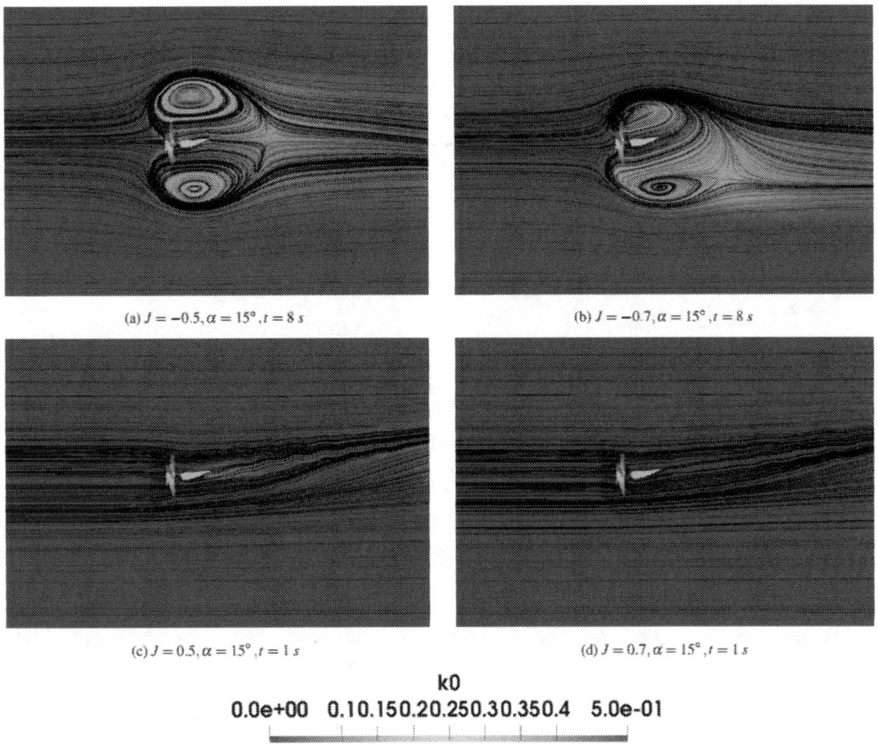

Figure 17. Turbulent kinematic energy distributions around the propeller-rudder system in ahead and astern conditions.

References

[1] Posa A and Broglia R 2022 *Ocean Eng.* **264** 112542
[2] Hu J, Zhang W, Sun S and Guo C 2019 *Ocean Eng.* **190** 106446
[3] Wang L, Guo C, Xu P and Su Y 2019 *Ocean Eng.* **188** 106250
[4] Molland A F and Turnock S R 1993 *Wind tunnel tests on the influence of propeller loading on ship rudder performance: four quadrant operation, low and zero speed operation*
[5] Kroll T B and Mahesh K 2022 *Flow.*
[6] Jang H and Mahesh K 2013 *J. Fluid Mech.* **729** 151–179
[7] Verma A, Jang H and Mahesh K 2012 *J. Fluid Mech.* **704** 61–88
[8] Vyšohlíd M and Mahesh K 2005 *Proceedings of Flow Induced Unsteady Loads and the Impact on Military Applications*
[9] Guo H, Li G and Du L 2022 *7th International Symposium of Marine Propulsors*
[10] Crepier P 2017 *Proceedings of the VII International Conference on Computational Methods in Marine Engineering* pp 365–376
[11] Celik I B, Ghia U, Roache P J and Freitas C J 2008 *J. Fluids Eng.-Trans. ASME* **130**
[12] Fujisawa J, Ukon Y, Kume K and Takeshi H 2000 *Local velocity field measurements around the KCS model in the SRI 400m towing tank*
[13] Van Hoydonck W, Delefortrie G, De Maerschalck B and Vantorre M 2018 *32nd Symposium on Naval Hydrodynamics*
[14] Roberts T W 1990 *Prospects for Drag Prediction Using Computational Fluid Dynamics*
[15] Molland A F and Turnock S R 2011 *Marine rudders and control surfaces: principles, data, design and applications* (Elsevier)

Numerical Investigation of the Propulsive Efficiency of a Containership under Load Variation

T H Yeh[1], S H Chen[1], T Y Lin[1] and S W Chau[1]

[1]Department of Engineering Science and Ocean Engineering, National Taiwan University, No. 1, Sec. 4, Roosevelt Rd., Taipei 106216, Taiwan (R.O.C.)

E-mail: tylin110@ntu.edu.tw

Abstract. The International Towing Tank Conference (ITTC) suggests that determining the variations of propulsion efficiency can be achieved by model tests that may be costly and inefficient. This is the reason why this paper investigates the variation of propulsion factors under various ship loading conditions by computational fluid dynamics (CFD). This study proposed a non-iterative method for load variation tests that reduced the computational time. The CFD software Star-CCM+ was used to solve the transient free-surface Reynolds-averaged Navier–Stokes equations. To further reduce the computational time, the propeller was modeled using the actuator disk method. The resistance augmentation, wake fraction, and propeller efficiency of a container ship model KCS with propeller KP505 were studied. These values were used to calculate the load variation coefficient of the shaft revolution speed ζ_N and load variation coefficient of the delivered power ζ_P at full scale. The results showed that as the resistance increased, the rotation rate correspondingly increased to maintain the service speed. As the ship resistance increased, the hull and open water efficiencies decreased, and when the actuator disk model assumed a relative rotative efficiency of 1.0, the quasi-propulsion efficiency also decreased.

1. Introduction

1.1. Ship propulsion test

Ship propulsion tests estimate the horsepower and propulsion efficiency of full-scale ships during navigation. A propulsion test includes driving a ship model with its propulsion system, generally a propeller, using a towing carriage and servomotor. The towing force, propeller force, and torque are measured to predict the overall performance of the ship. Using the bare hull resistance and open water performance, the hull–propeller interaction is characterized to establish the total propulsion efficiency. To analyze the complex hydrodynamic interactions between the ship hull and the propeller, the ITTC recommends dividing the total propulsion efficiency into several components, such as the thrust deduction factor, effective wake fraction, relative rotative efficiency, and open water efficiency. Through model ship correlations, namely, scale effect corrections, we can estimate the propeller rotation rate, thrust, and power at full scale.

Self-propulsion tests are typically conducted for prediction under calm water conditions; this resolve only the ideal performance in still water. Real ships operate in unstable sea conditions, where sea states, operating conditions, and hull surface roughness dramatically change the ship resistance. To maintain the same ship speed, the propeller rotation rate must be adjusted to account for the changes in resistance.

In addition to the self-propulsion test, load variation tests are used to understand the effects of changes in ship resistance on performance metrics such as efficiency, rotation rate, and propulsion power to quantify the correlations between these parameters under different load conditions.

The load variation test procedures are detailed in the ITTC 2017 guidelines [1]. The implementation of load variation tests requires additional propulsion tests rather than self-propulsion; hence, several times works by either model tests or computational simulations are required. That is, self-propulsion is a special case of load variation. The major challenge in load variation tests is determining the specified towing force, which is the difference between the ship resistance and the propeller thrust, by tuning the propeller rate. In the British propulsion method, that is, constant speed, self-propulsion is interpolated using at least three propeller rates to determine the target towing force. In this regard, a few examples of load variation tests have been reported by naval academies.

This study proposes a better approach to address the significant effort required in the traditional load variation procedure. The main idea is to drop the target towing force and modify the loading factor as a dependent variable on the propeller rate. This method avoids the iteration of the force balance and thus reduces the simulation time.

1.2. CFD for ship propulsion

Computational fluid dynamics (CFD) is a mature tool for analyzing ship hydrodynamics to determine a ship's resistance and propulsion performance [2]. CFD software solves fluid dynamics problems based on the given conditions and obtains objective quantities such as ship resistance, propeller performance, seakeeping, and maneuverability. A typical CFD procedure includes computer-aided design modelling, governing equations, flow field discretization, boundary conditions, fluid properties, turbulence models, convergence criteria, and numerous solver parameters.

Simulating the operation of a propeller in a fluid using CFD requires fine spatial discretization and extremely small time intervals, resulting in considerable computational effort [3]. The actuator disk model simplifies the rotating propeller as a rotor disk, on which the force and torque are determined by an external solver or simply by open water performance data. This model, also called the propeller body force method, enables simulation of the complex hydrodynamic phenomena of a propeller and its interactions with fluids and other objects [4]. Because the propeller geometry is not resolved, the number of grids can be reduced; this decreases computational complexity while ensuring more stable results.

This study employed the propeller body force method for the ship propulsion simulations. The selected KCS hull and KP505 propeller were widely studied both experimentally and numerically. [5] First, the ship resistance and open-water performance were simulated at the model scale. Subsequently, self-propulsion and four load variations were performed using this new approach. The propulsive factors were predicted at full scale using the ITTC-recommended procedure. Finally, the load variation coefficient of the shaft revolution speed and delivered power were calculated.

2. Method

2.1. Governing equations

The commercial software Star-CCM+ is used to solve the unsteady Reynolds-averaged Navier–Stokes equations (RANS) for the turbulent flow problems. RANS is a time-averaging simulation method that decomposes the pressure and fluid velocity in the Navier–Stokes equation into time-averaged and fluctuating quantities and then averages the equation over time to obtain the mean flow field [6]. Reynolds stress tensor is modelled using the $k - \varepsilon$ turbulence model. The free surface is modelled using the volume of fluid (VOF) method [7]. The density of fresh water is set 997.3 kg/m^3, and dynamic viscosity is set 8.9×10^{-4} Pa-s. The density of air is set 1.18 kg/m^3, and dynamic viscosity of air is set 1.86×10^{-5} Pa-s.

IOP Conf. Series: Materials Science and Engineering 1288 (2023) 012058 doi:10.1088/1757-899X/1288/1/012058

2.2. Propeller body force

The present study uses open water propeller characteristics as the input to resolve the body force on the actuator disk without actual propeller geometry. The position and orientation of the disk are defined in the propeller plane. The rotation rate is specified. By providing propeller performance curves, the mean flow on the disk is retrieved from the RANS domain and processed as the inflow for the propeller. The flow field around the propeller and its force can be iteratively solved using the inlet flow velocity and body forces acting on the actuator disk [8,9].

To calculate the body forces, non-dimensional propeller thrust coefficient K_T, torque coefficient K_Q, and open water efficiency η_O in terms of non-dimensional advance ratio J are required. Advance ratio J is defined as follows:

$$J = \frac{V_a}{n \times D} \tag{1}$$

where V_a denotes the inflow velocity of the propeller, n denotes the rotation rate, and D denotes the diameter of the propeller. After specifying n, the body force distribution on the actuator disk can be obtained through the following steps. The inner iteration is shown in figure 1 [10].

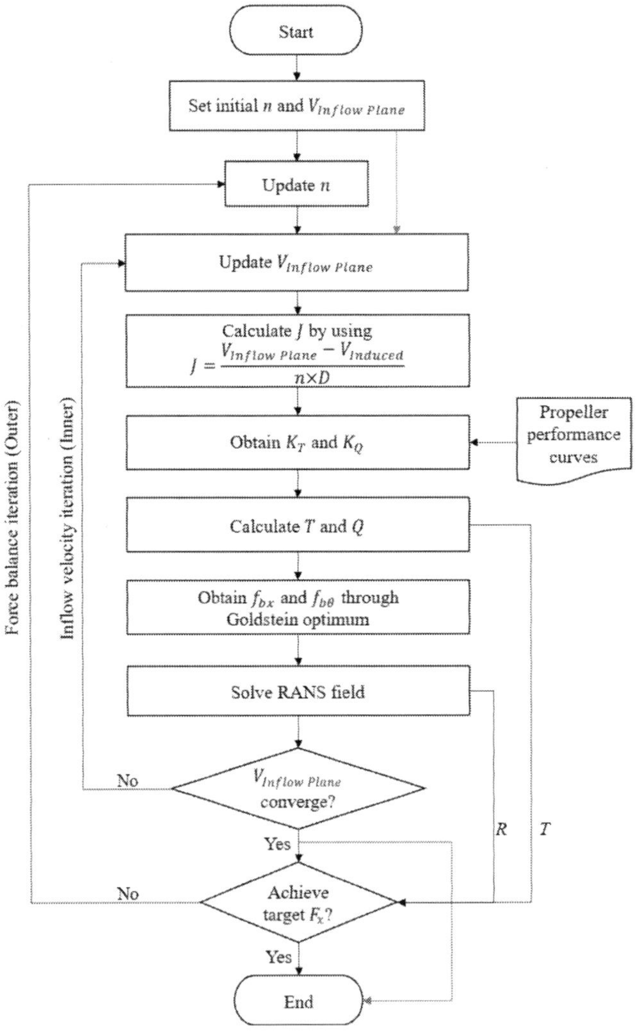

Figure 1. Flow chart of propulsion simulation.

1. V_a is obtained by subtracting induced velocity $V_{Induced}$ from mean axial velocity $V_{Inflow\ Plane}$ of the flow on the propeller plane, and J can be calculated using the following equation:

$$J = \frac{V_{Inflow\ Plane} - V_{Induced}}{n \times D} \tag{2}$$

2. The propeller thrust and torque coefficients are obtained by interpolating the propeller performance curves.
3. Propeller thrust T and torque Q are calculated as follows:

$$T = K_T \times \rho \times n^2 \times D^4 \tag{3}$$

$$Q = K_Q \times \rho \times n^2 \times D^5 \tag{4}$$

4. The axial and tangential components of the volume force, f_{bx} and $f_{b\theta}$, can be obtained through Goldstein optimum in the radial direction [10,11,12].
5. The RANS field is solved, and $V_{Inflow\ Plane}$ is updated. Return to the first step and continue iterating until $V_{Inflow\ Plane}$ and ship resistance R converge.

2.3. Propulsion factors
For a given rotation rate, the propulsive factors are defined as follows. The wake fraction (w) can be calculated from the inflow velocity (V_a) and model ship speed (V_m) as follows:

$$w = 1 - \frac{V_a}{V_m} \tag{5}$$

The thrust deduction factor (t) can be calculated using R, T, and the bare hull resistance (R_T).

$$t = \frac{R - R_T}{T} \tag{6}$$

Towing force F_X can be obtained through the following equation:

$$F_X = R - T \tag{7}$$

Hull efficiency (η_H) can be obtained as follows:

$$\eta_H = \frac{1 - t}{1 - w} \tag{8}$$

Because T and Q of the actuator disk are directly obtained from the propeller performance curves, the relative rotative efficiency is 1.0. Therefore, the quasi-propulsive efficiency (η_D) and delivered horsepower (P_D) can be calculated using the following equations:

$$\eta_D = \eta_H \times \eta_O \tag{9}$$

$$P_D = 2\pi \times Q \times n \tag{10}$$

2.4. Load variation factor
A load variation test is conducted at a selected draft and at least one ship speed; this was the design speed in this study. The load variation test must include at least four propulsive tests, each conducted at a different towing force while maintaining the same ship speed. In other words, resistance variation is implemented by varying the towing force. Equation (11) defines load factor A as the ratio of the resistance variation to the bare hull resistance. Self-propulsion is achieved when A is zero. The ITTC recommends A between -10% and +20%, specifying target F_X, for a load variation setup [2]. The ITTC procedure is depicted as an outer iteration in figure 1. To predict the full-scale values, skin friction F_D is corrected according to the ITTC formula.

$$A = \frac{\Delta R}{R_{TS}} \tag{11}$$

$$\Delta R = (F_D - F_X)\lambda^3 \frac{\rho_s}{\rho_M} \tag{12}$$

Referring to the resistance test and to satisfy equations (11,12), towing force F_X can be expressed as follows:

$$F_X = F_D - A \times (R_{TM} - F_D) \tag{13}$$

Unlike the ITTC procedure, our method specifies n, and F_X is computed using equation (7). As indicated by the red arrows in figure 1, no outer iteration is required. Following equations (13,11), A and ΔR are obtained, respectively. After conducting load variation tests for a range of A, R_T, t, and w are corrected to full scale to calculate full-scale resistance R_{TS}, rotation rate n, delivered power P_D, and quasi-propulsive efficiency η_D. By setting the self-propulsion point as the reference, the deviations in the above variables are obtained. Finally, under the linear assumption of deviations, equations (14,15) compute the load variation coefficient of the delivered power ζ_P and load variation coefficient of the shaft revolution speed ζ_N, respectively.

$$\frac{\eta_{D\_LV}}{\eta_{D\_SP}} = \zeta_P \frac{\Delta R}{R_{TS}} + 1 \tag{14}$$

$$\frac{\Delta n}{n} = \zeta_N \frac{\Delta P_D}{P_D} \tag{15}$$

3. Computational setup

3.1. KCS containership and propeller

The target ship was KCS (KRISO Container Ship) with KP505 propeller, established by the Korea Research Institute for Ship and Ocean Engineering (KRISO). The MOERI model [13,14] was selected as the computational model, and its basic dimensions are listed in table 1. The propeller performance curves for the KP505 are shown in figure 2.

3.2. Computational domain and boundary conditions

The Cartesian coordinate system was used in the computational domain. The origin was located at the intersection of the after perpendicular (AP) and design draft. The domain box was a hexahedron, with the boundaries of X extending 3.2*L* in the positive direction and 2.24*L* in the negative direction, boundaries of Y extending 2.24*L* in the positive and negative directions, and boundaries of Z extending 0.96*L* in the positive direction and 1.6*L* in the negative direction. The boundary conditions were set as follows. The front, top, bottom, left, and right sides of the hexahedron were specified as velocity inlets whereas the back side as the pressure outlet. The hull surfaces were no-slip walls, as shown in figure 3. Dynamic sinkage and trim were not considered in this study.

Table 1. Main particulars of the KCS model and KP505.

Length (*L*)	Beam	Depth	Draft	Displacement	Wet Surface	Speed
7.28 m	1.02 m	0.60 m	0.342 m	1.649 m^3	9.5441 m^2	2.196 m/s

Hub ratio	Diameter (m)	P/D	Z	Expanded area ratio	Rotation direction
0.18	0.25	0.9967	5	0.8	Right-handed

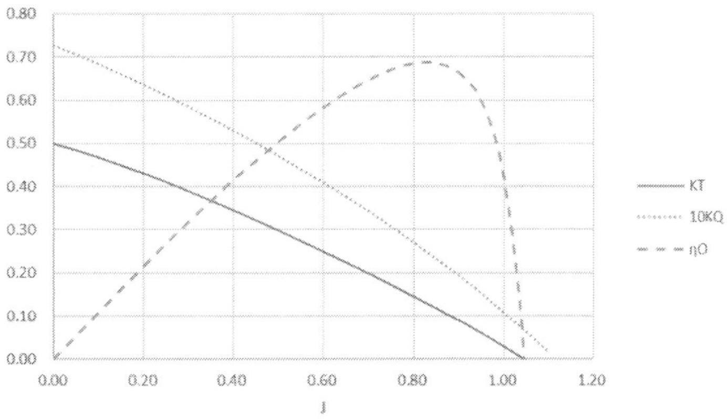

Figure 2. Propeller performance curves.

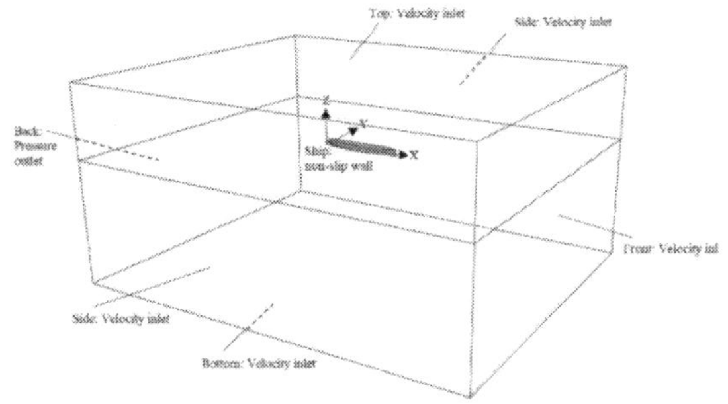

Figure 3. Flow domain and boundaries.

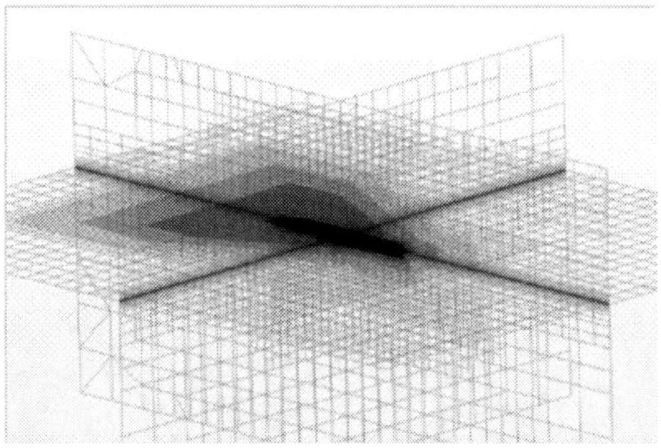

Figure 4. Mesh settings.

3.3. Mesh settings

Spatial discretization of the flow field was performed using the trimmer mesh and prismatic layer algorithms. Mesh refinement was performed on surfaces that changed significantly in the bow and stern, free surface, and wake field to analyze the flow field changes and force distribution around the propeller. Anisotropic mesh refinement was also applied in the region of divergence and transverse waves around the free surface. By adjusting the total thickness, growth rate, and number of prism layers in the boundary layer mesh, the averaged dimensionless wall distance (y+) was controlled to approximately 90, and the total number of mesh was 3.18 million. The mesh configuration is shown in figure 4.

3.4. Computational validation

The resistance coefficient of the bare hull was obtained through resistance tests (static water condition) at the same speed and draft; this was 3.55×10^{-3} [15,16]. The calculated result was 3.52×10^{-3}, with a difference of -0.85%. The calculated resistance was smaller because the dynamic sinkage and trim were not considered.

For self-propulsion (SP), the CFD propulsion factors were compared with the experimental results, and the errors were calculated, as listed in table 2. The maximum error was +10.53% for w that led to a slight overestimation of η_H because its denominator was 1-w. In terms of the propeller inflow velocity, 1-w, the error remained at only 2% and marginally impacted subsequent calculations. The thrust deduction factor and quasi-propulsion efficiency were considerably accurate. The coefficient with the largest difference was the wake fraction that was overestimated because of the overestimation of the inflow velocity, resulting in an underestimated rotation rate and overestimated thrust coefficient. Overall, the errors in the CFD were acceptable.

3.5. Open water calibration

In the propeller body force method, the diameter and position of the inflow plane should be provided for computing $V_{Inflow\ Plane}$. The inflow plane is ideally identical with the propeller. Because $V_{Inflow\ Plane}$ in RANS field contains the induced velocity, which is analytically calculated from the momentum theory, the diameter of the inflow plane is suggested to be slightly larger than the propeller diameter, and its axial offset should be at a distance in front of the rotor disk [10]. Without tuning these two parameters, open water performance cannot be recovered.

In this study, an open water test was simulated by fixing the inflow velocity and varying n. To reduce the number of variables, the radius of the inflow plane was set to be the same as that of the actuator disk, and induced velocity correction was enabled. By changing the offset of the inflow plane, the error between simulated J and set J was calculated using equation (2). The calibration of the offset minimized the error, and hence recovered the original open water performance. The results are summarized in table 3. Notably, when the inflow plane offset was set to 0.02 m, that was 8% of the propeller diameter, the error was within 3%, and the error around self-propulsion point speed (J = 0.728) was less than 1%. Therefore, an offset of 0.02 m was applied to the inflow plane.

Table 2. Self-propulsion validation.

	CFD	EXP [17,18]	Error (%)
n (rps)	9.288	9.5	-2.23
K_T	0.184	0.172	+6.74
w	0.230	0.208	+10.53
t	0.146	0.147	-0.41
η_H	1.109	1.077	+2.92
η_O	0.660	0.662	-0.30
η_D	0.732	0.713	+2.66

Table 3. Inflow plane offset calibration.

J (set)	J (CFD)	Error (%)
1.098	1.0997	+0.15
0.976	0.9744	-0.16
0.878	0.8760	-0.27
0.798	0.7972	-0.17
0.732	0.7329	+0.12
0.676	0.6797	+0.59
0.627	0.6350	+1.20
0.586	0.5970	+1.94
0.549	0.5643	+2.79

4. Results and discussion

4.1. Dependencies of propulsive factors

In this study, we did not use a -10% to +20% load factor for the load variation tests, as recommended by the ITTC for two reasons. First, designating the value of variable load required iterative calculation of the rotation rate until the condition was met, requiring more time. Second, larger load range better demonstrated nonlinear ζ_P and ζ_N. Therefore, the present CFD simulation was directly performed with four propeller rotation rates: 8, 9, 10, and 11 rps. The results are listed in table 4. Figures 5–8 show the results plotted against the propeller rotation rate. The vertical green line in each figure represents the self-propulsion condition (SP).

As the rotation rate increased, the resistance of the hull and propeller thrust increased. Because the increased speed of the thrust was faster than that of the resistance, the towed resistance exhibited a decreasing trend. The figure shows that when the resistance of the hull increased, the propeller speed correspondingly increased to maintain the same velocity of the ship.

For the propeller, an increase in the rotation rate increased the load and thrust coefficients and decreased the advance ratio. A decrease in the advance ratio decreased the open water efficiency, whereas an increase in the thrust coefficient increased the thrust to counteract resistance. Regarding the propulsive coefficient, as the rotation rate increased, the wake fraction coefficient decreased, that is, the inflow velocity of the propeller increased. The thrust deduction factor increased slightly but was almost unaffected by an increase in the rotation rate. The combined effects of w and t led to a decrease in the hull efficiency. With a decrease in the hull and open water efficiencies, the quasi-propulsion efficiency also decreased.

Overall, when the hull resistance increased, the propeller had to increase its rotation rate to generate more thrust, consuming more transmitted horsepower and sacrificing greater propulsive efficiency.

4.2. Flow field analysis
The variation in the quasi-propulsion efficiency originates from the hull and open water efficiencies. The hull efficiency can be calculated using the thrust deduction factor and wake fraction. Therefore, we can probe the physical mechanisms of these two coefficients through a flow field analysis.

The axial velocity distribution at the plane section of the propeller was calculated and contour-plotted, and the radial and tangential velocities were represented as vector fields. Figure 9 shows the wake fields at various rotation rates: bare hull (a), 8 (b), 9 (c), 10 (d), and 11 rps (e). The colors indicate the normalized axial velocities. The velocity obtained from the flow field was the sum of the effective wake and propeller-induced velocities. As the rotation rate increased, the effective wake velocity increased, that is, w decreased; simultaneously, the propeller load increased, increasing the induced velocity. The simultaneous velocity change in these two parameters increased the total wake velocity. The flow velocity was lower in the 12 o'clock direction because it was close to the stern plate. In addition, larger velocity vectors were observed in the radial and tangential directions, indicating a greater propeller efficiency loss.

Table 4. Propulsive factors of load variation at model scale.

n (rps)	8	9	10	11
R (N)	85.389	88.774	92.473	96.3595
T (N)	33.047	54.516	79.245	107.257
Q (N · m)	1.575	2.426	3.392	4.479
F_X (N)	52.342	34.258	13.230	-10.898
K_T	0.133	0.173	0.203	0.228
J	0.823	0.749	0.691	0.644
w	0.251	0.233	0.214	0.194
t	0.138	0.146	0.147	0.145
η_H	1.150	1.114	1.085	1.061
η_O	0.685	0.670	0.641	0.613
η_D	0.788	0.746	0.696	0.650
P_D (W)	79.163	137.189	213.119	309.563

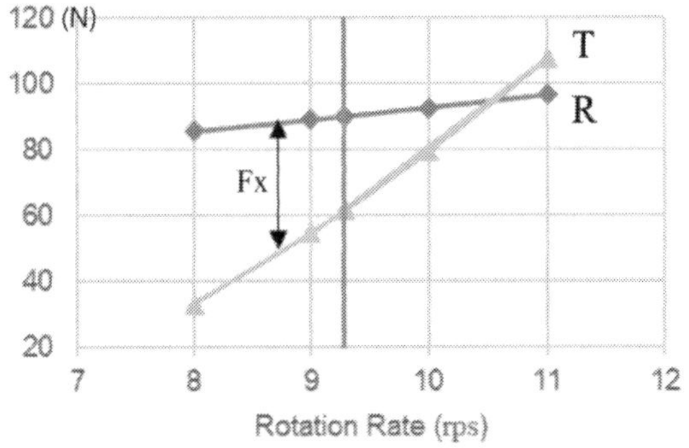

Figure 5. R and T at different rotation rates.

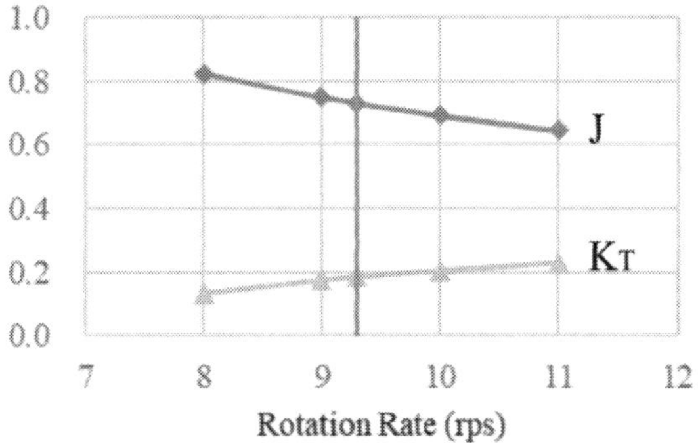

Figure 6. J and K_T at different rotation rates.

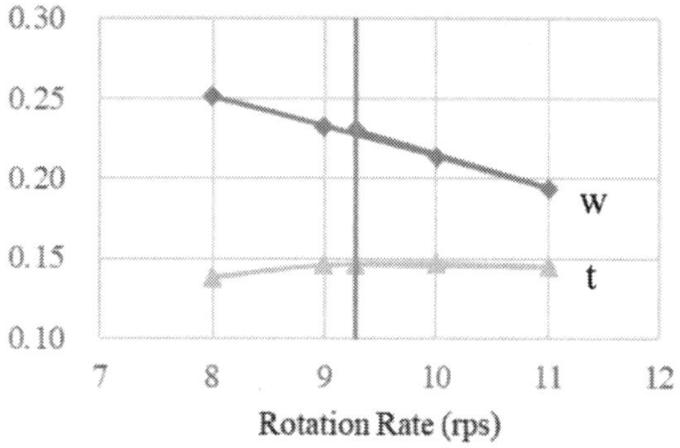

Figure 7. w and t at different rotation rates.

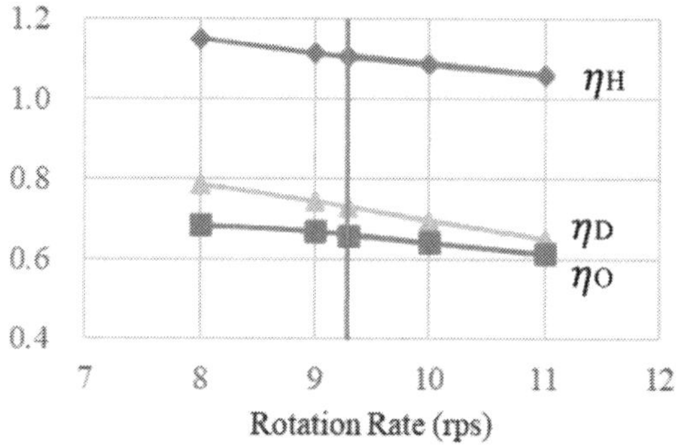

Figure 8. η_H, η_O, and η_D at different rotation rates.

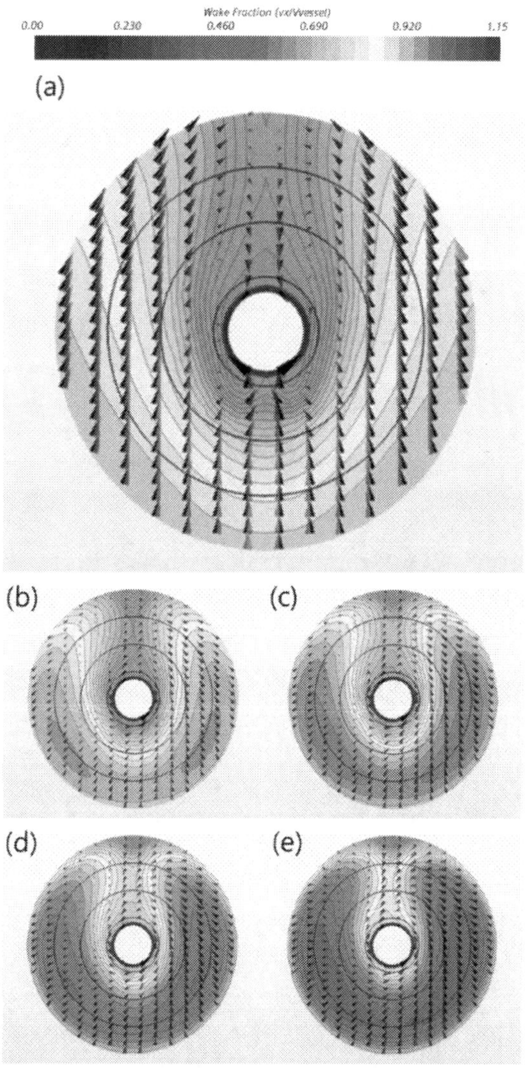

Figure 9. Wake field on the propeller plane. (a) nominal (b) 8 rps (c) 9 rps (d) 10 rps (e) 11 rps.

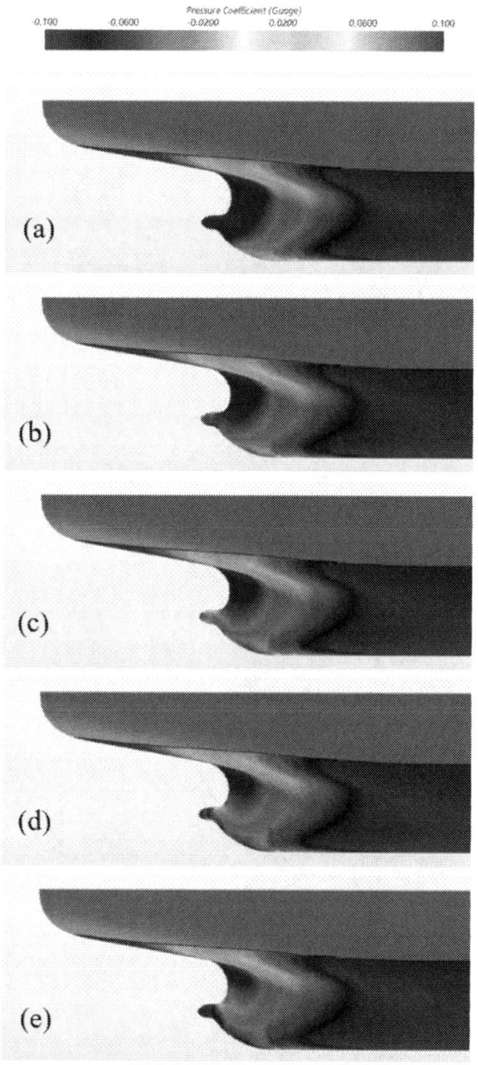

Figure 10. Pressure coefficient at the stern. (a) bare hull (b) 8 rps (c) 9 rps (d) 10 rps (e) 11 rps.

IOP Conf. Series: Materials Science and Engineering 1288 (2023) 012058 doi:10.1088/1757-899X/1288/1/012058

The thrust deduction factor was derived from the increase in resistance caused by the propeller-induced flow around the hull. Figure 10 shows the contours of the dimensionless stern dynamic pressure, with the color indicating the magnitude of the hull pressure coefficient. From top to bottom, the conditions were the bare hull, and with the propeller at 8, 9, 10, and 11 rps. The ideal flow predicts the stern to be a high-pressure stagnation point, with the pressure decreasing toward the bow. Under the influence of the propeller, the low pressure on the suction side in front of the actuator disk decreased the stern pressure, as shown in figure 10, as the white region became larger. As the rotation rate increased, the high-pressure region was neutralized by the low pressure on the suction side, resulting in the lower stern pressure and higher pressure resistance, that is, the higher resistance. Notably, the thrust deduction factor was defined as the ratio of the towing force (increment of the hull resistance) to the propeller thrust. As shown in figure 5, the increases in thrust and resistance were proportional and did not affect the hull efficiency.

4.3. Correlations of load variation

The load variation coefficient of the delivered power, denoted as ζ_P, was calculated using equation (14). Resistance, efficiency, delivered power, and rotation rate were predicted at full scale according to the ITTC guidelines, as shown in table 5, given the density of sea water as 1023.6 kg/m^3 and dynamic viscosity as 9.65×10^{-4} Pa-s. Figure 11 plots η_{D_LV}/η_{D_SP} against $\Delta R/R_T$. As the load variation test was referenced to the SP, the line passed through (0,1). By fitting a quadratic trendline, the slope of the tangent line at the origin was determined to be $\zeta_P = -0.071$, implying that when resistance doubled, the propulsive efficiency decreased by 7%.

Similarly, load variation coefficient of the shaft revolution speed ζ_N was calculated using equation (15). Figure 12 plots $\Delta n/n$ against $\Delta P_D/P_D$. By fitting a quadratic trendline, ζ_N was determined by the slope of the tangent at the origin, whose value $\zeta_N = 0.223$. This implied that when the delivered horsepower doubled, the rotation rate increased by 22%.

In addition, figure 13 plots $\Delta n/n$ against $\Delta R/R_T$, showing that as the resistance increased, the rotation rate also correspondingly increased to maintain the speed. The coefficient of the resistance–RPM correlation, that is $\zeta_N/(1+\zeta_P)$, is 0.239.

Table 5. Load variation of propulsion values.

n (rpm)	85.94	95.94	106.17	116.52
$\Delta R/R_{TS}$	-0.515	-0.130	0.318	0.831
η_{D_LV}/η_{D_SP}	1.020	1.012	0.962	0.908
$\Delta P_D/P_D$	-0.524	-0.140	0.369	1.017
$\Delta n/n$	-0.131	-0.030	0.073	0.178

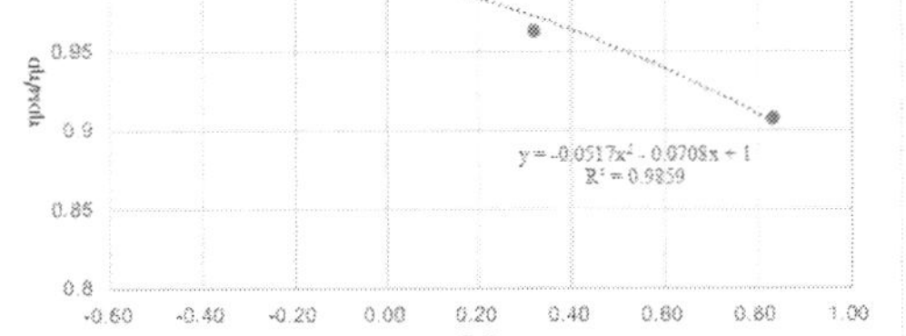

Figure 11. Correlation between quasi-propulsion efficiency and resistance.

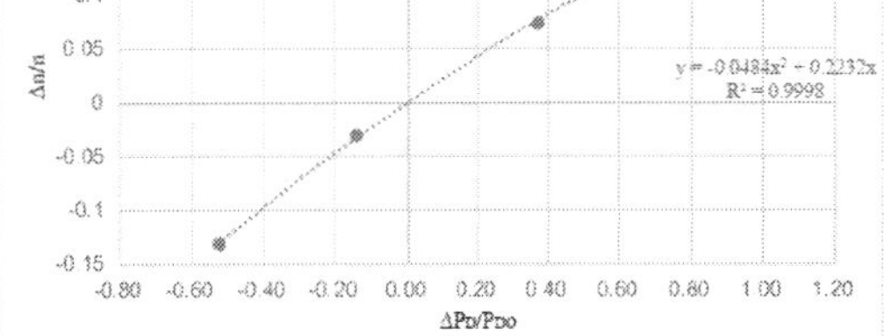

Figure 12. Correlation between propeller rate and power.

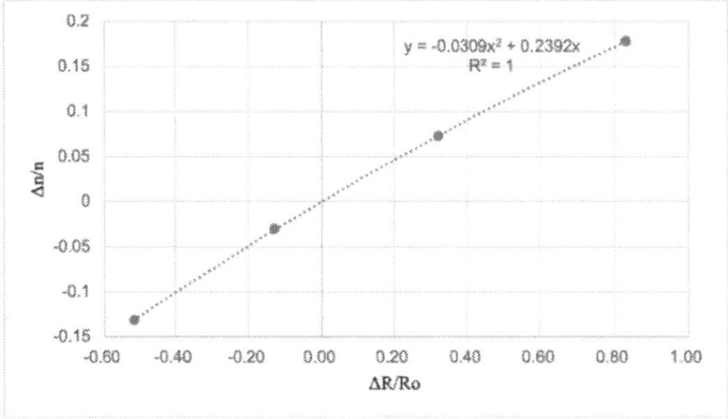

Figure 13. Correlation between propeller rate and resistance.

5. Conclusions

This study proposed a new approach for load variation tests for ship propulsion. A ship model KCS containership was used, and the propeller was modelled using an actuator disk through CFD analysis. Two sets of loads that were higher and lower were calculated with reference to self-propulsion conditions. In the new approach, propeller rotation rates were directly set to avoid the iterative process and demonstrate ζ_P and ζ_N better than that using the load factor recommended by ITTC for load variation tests.

Notably, as the hull resistance increased, the rotation rate increased to maintain a constant ship velocity, thus increasing the propeller thrust and thrust coefficient and decreasing the advance ratio and wake fraction coefficient, whereas the thrust deduction factor remained unaffected. With an increase in hull resistance, the hull and open water efficiencies decreased, reducing the quasi-propulsive efficiency. The load variation coefficient of the delivered power was -7%, and the load variation coefficient of the shaft revolution speed was 22%.

In future studies, multiple ship speeds can be analyzed under variable loads to obtain the load variation coefficient of the ship speed that can be further applied to EEDI speed calculations.

6. References

[1] International Towing Tank Conference 2017 ITTC-Recommended Procedures and Guidelines 1978 ITTC Performance Prediction Method 7.5-02-03-01.4
[2] Yasser M. Ahmed, O. B. Yaakob, M. F. A. Rashid and A. H. Elbatran 2015 Determining ship resistance using computational fluid dynamics (CFD) *J. Transport Syst. Eng.* **2** 20–25
[3] Dadhich V and Kumar R 2019 Numerical analysis of wing-tip mounted propeller interaction using actuator disk model *21st Annual CFD Symp. (Bangalore)*
[4] Conway JT 1998 Exact actuator disk solutions for non-uniform heavy loading and slipstream contraction *J. Fluid Mech.* **365** 235–267
[5] Alireza Nadery, Hassan Ghassemi, and Leszek Chybowski 2021 The effect of the PSS configuration on the hydrodynamic performance of the KP505 propeller behind the KCS *Ocean Eng.* **234** 109310
[6] Giancarlo Alfonsi July 2009 Reynolds-averaged navier-stokes equations for turbulence modeling *Appl. Mech. Rev. DOI: 10.1115/1.3124648*
[7] Hirt CW and Nichols BD 1981 Volume of fluid (VOF) method for the dynamics of free boundaries *J. Comput. Phys.* **39** 201–225
[8] Ping-Chen Wu 2022 CFD body force propeller model with blade rotational effect *Appl. Sci.* **12** 11273 *https://doi.org/10.3390/app122111273*
[9] Hsin CY, Lin CT, Lee CP, Lin CC and Chen JH 2016 Study of the wave effects on propeller performance by computations *J. Taiwan Soc. Nav. Archit. and Mar. Eng.* **35** 185–197

[10] Siemens Star-CCM+ user manual, ver. 2020.1
[11] Sydney Goldstein 1929 On the vortex theory of screw propellers *Proc. R. Soc. Lond.* A **123** 440–465 *http://doi.org/10.1098/rspa.1929.0078*
[12] Hrvoje Jasak, Vuko Vukčević, Inno Gatin and Igor Lalovič 2019 CFD validation and grid sensitivity studies of full scale ship self propulsion *Int. J. Nav. Architect. Ocean Eng.* **11** 33–43
[13] MOERI Container Ship (KCS) http://www.simman2008.dk/KCS/kcs_geometry.htm
[14] Stefango Gaggero, Diego Villa and Michele Viviani 2015 The KRISO container ship (KCS) test case: An open source overview *6th Int. Conf. on Computational Methods in Marine Engineering MARINE 2015*
[15] Dakui Feng, Bin Ye, Zhiguo Zhang and Xianzhou Wang 2020 Numerical simulation of the ship resistance of KCS in different water depths for model-scale and full-scale *J. Mar. Sci. Eng.* **8** 745 *https://doi.org/10.3390/jmse8100745*
[16] Ludwig Kerner 2014 KCS resistance calculation *Benchmark Report 2014-09-19 FM Resistance-KCS, https://www.numeca.com/docs/finemarine kcs g2010 2-1-2-3 18092014.pdf*
[17] Kume K, Ukon Y, Fujisawa J, Hori T, Tsukada T and Haruya T 2000 Uncertainty analysis for the KCS model tests in the SRI 400 m towing tank *Ship Performance Division Report No. 00-008-1*
[18] Fujisawa J, Ukon Y, Kume K and Haruya T 2000 Local velocity field measurements around the KCS model in the SRI 400 m towing tank *Ship Performance Division Report No. 00-003-2*

Acknowledgments
This study was supported by the National Science and Technology Council, Project NSTC-111-2222-E-002-008-, Taiwan.

IOP Conf. Series: Materials Science and Engineering 1288 (2023) 012059 doi:10.1088/1757-899X/1288/1/012059

Analysis of the Hydrodynamic Performance of a Gate Rudder System

Serkan Turkmen[1], Li Wang[2], Stefanos Raftopoulos[1], Chang Li[1] and Rose Norman[1]

[1]School of Engineering, Marine, Offshore and Subsea Technology Group, Newcastle University, NE1 7RU, Newcastle upon Tyne, United Kingdom
[2]School of Naval Architecture Ocean and Energy Power Engineering, Wuhan University of Technology. Wuhan, Hubei, P.R.China.

serkan.turkmen@ncl.ac.uk

Abstract: An alternative twin rudder system known as the Gate Rudder© has been applied to newbuild vessels and an existing ship with the aim to improve their hydrodynamic performance and reduce greenhouse gas emissions. The system consists of two asymmetric cross-sections and L-shaped blades located either side of a propeller. The blades generate thrust (duct effect) from the propeller and ship-included flow. Initially, forces and moments of a single blade were measured using a 6-axis strain gauge sensor in the towing tank without a propeller. An alternative measuring device, consisting of a Piezoelectric Macro-Fiber Composite Actuator, was then attached to the blade to investigate flow-induced forces. Finally, Computational Fluid Dynamic (CFD) simulations were compared with the experimental results to identify the source of the oscillating forces. The results show that flow-induced vibrations may occur on the gate rudder blades at both shedding frequencies and the structure's natural frequency.

1.Introduction

The International Maritime Organization (IMO) strategy on reducing ship induced greenhouse gases (GHGs) by at least 50% by 2050 has been the priority for shipping, ports, engineering, and leisure marine industries [1]. To achieve this target, technology is a key factor influencing nations' and organisations' prosperity and competitiveness. Emission reduction solutions can be categorised based on propulsion systems, alternative fuels, renewables and operation [2]. Ducted propulsion systems, or propellers, are hydrodynamically based energy-saving devices (ESD) that can improve total thrust [3]. There are several ducted systems which have a different effect on the Quasi-Propulsive Coefficient (QPC). The most commonly utilised systems are a Wake Equalizing Duct (WED), Super Stream Duct (SSD), Sumitomo Integrated Lammeren Duct (SLID), Mewis Duct or Weather Adopted Duct (WAD) [4,5]. A new rudder system called "Gate Rudder©" acts like a flow accelerating type of ducted propeller. A Gate Rudder is a twin rudder system, consisting of two asymmetric rudders located either side of the propeller [6,7]. A proportion of the total thrust is taken by the blades. Apart from the increased QPC, the steerable rudders can provide manoeuvrability, and seakeeping ability [8]. It has been applied to both new buildings and an existing vessel as shown in Figure 1.

Figure 1. The first Gate Rudder application on a new Building 400TEU Container Vessel (left) and the first retrofitted system on 4500DWT Dry Cargo Vessel (right)

The Gate Rudder System was developed using the "Fine powering performance prediction technology" in the Emerson Cavitation Tunnel (ECT) [6]. This method enhances understanding of the interaction between the hull, propeller, and appendages. For instance, to investigate forces generated by the Gate Rudder and their impact on the thrust deduction, the segmented stern of a dummy hull and rudder blades are connected to various load cells. The results obtained from load cell measurements can be compared with the torque and thrust measurements to determine propulsive performance.

Both model-scale and full-scale rudder tests show that this unconventional rudder improves propulsive performance and reduces GHG emissions [9]. Investigators examined the effects of force generated by the Gate Rudder System on maneuvering and found advanced maneuverability for different operation modes since each of the twin rudder blades can be steered independently [8]. Extensive research was carried out on underwater radiated noise performance, with investigations conducted in the ECT [10], however, no previous study exists which has presented dynamic structural performance of the rudder blades. A cavitation tunnel test showed that it is possible that the rudder blades could not withstand loadings under certain test conditions. Figure 2 shows a flaw on a Gate Rudder blade. This result is likely to be related to the accelerated flow seen by the propeller. Pressure fluctuations on the surface also occur as vortices are shed from the body. Vortex shedding can excite the structure to vibrate and generate acoustic sound [11]. Hence, an alternative explanation of damage occurrence is that the blade and its shaft were exposed to forces originating from both flow and propeller pressure pulses. This phenomenon is commonly called Fluid-Structure Interaction (FSI).

Figure 2. A Damaged blade due to the accelerated flow

This research seeks to address the structural dynamic performance of the Gate Rudder by conducting a simplified test campaign for a single blade without a propeller using a towing tank and computational methods. The results support improvements to the rudder design and can be utilised to avoid critical operation conditions to reduce the risk of failure and improve the fatigue life cycle.

2. Towing Tank Experiment Setup and Conditions

A Gate Rudder model with a geometrical scale of 1/14 was tested in the Newcastle University towing tank. The structure is composed of three parts: the rudder blades, rudder shaft and upper installation. The materials used for each are PET-CF15, steel and aluminum alloy. The parameters of the materials are shown in Table 1.

The dimensions of the water tank are 37x3,7x1.25m and the carriage can operate at a maximum speed of 3m/s. A 6-axis strain gauge sensor was installed on the carriage to measure the forces (see Figure 3). The numbers indicated in Figure 3 correspond to 1) Water tank, 2) Carriage, 3) 6-axis strain gauge sensors, 4) PC processor, 5) GoPro camera, 6) Connection shaft, and 7) Model Gate Rudder.

The 6-axis strain gauge sensor was calibrated between -40N and +40N with accuracy across the full-scale equal to 0.05N, which is ±0.125% of the value of maximum force. Then, appropriate software was used to collect and analyse the voltage signal produces by the sensor. A shaft was used to connect the underwater model rudder to the strain gauge sensor placed out of the water. This shaft was able to rotate allowing multiple angles to be set in the experimental tests. The tests were performed for the same conditions (water temperature, humidity, level of water) at speeds 0.5, 0.75, 1 and 1.5m/s and angle of attack (AoA) 0, ±15, ±30 degrees.

Table 1. Gate Rudder structure Material parameters.

Material name	Density (kg/m^3)	Young's modulus (Pa)	Passion ratio
PET_CF15	1160	1.85e9	0.04
Steel	7850	2e11	0.3
Aluminium alloy	2770	7.1e10	0.33

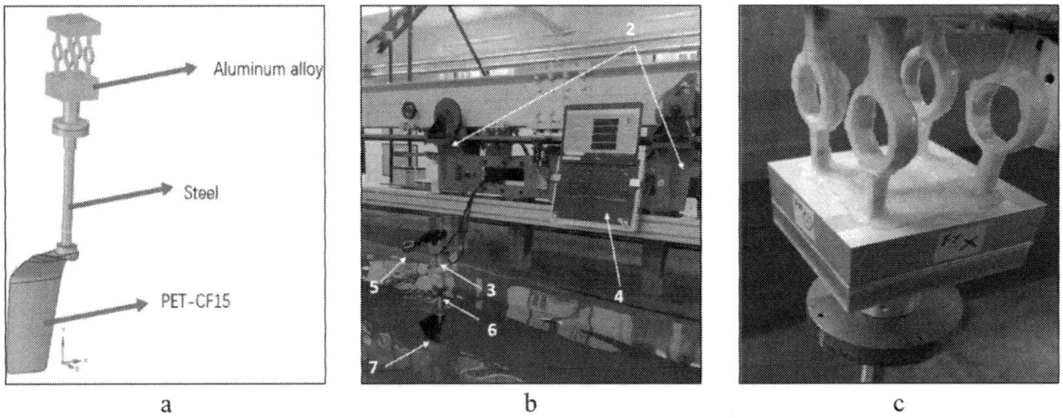

Figure 3. (a) Port side blade of a Gate Rudder System, (b) Towing tank test equipment set-up, (c) the 6-axis strain gauge sensor apparatus connected to the rudder shaft

During the experimental tests, a part of the shaft was immersed in the water, as shown in Figure 4, left hand side four pictures. This created additional resistance due to the shaft. Thus, by running the experiment with the same conditions with the shaft standalone, as presented in Figure 4, right hand side four pictures, the drag of the rudder shaft could be measured and subtracted from the total results of drag. It can be also seen from Figure 4, a strong flow disturbance occurs at 1.5 m/s due the shaft of the rudder. The flow separation is less severe at a speed of 1.0 m/s or below.

The results of the strain gauge sensors for drag and lift were not static but fluctuating. This was a result of all the phenomena that take place in the towing tank and affect the measurements of the rudder, such as vortices, waves and wake. A Fast Fourier Transform (FFT) was used for the separation of each vibration force from the desired drag and lift force, which are static forces acting on the rudder for steady conditions.

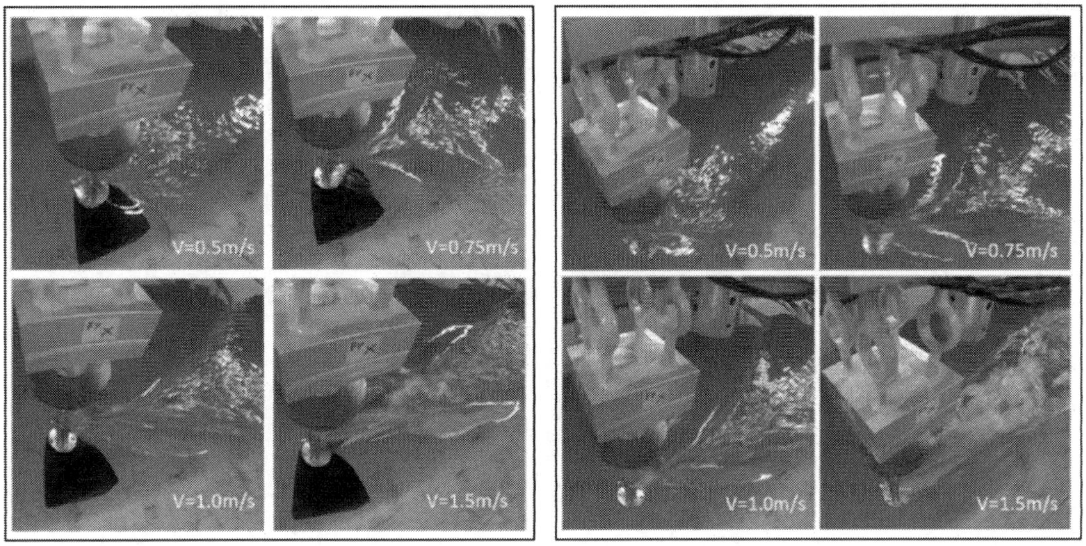

Figure 4. (a) the model rudder at Zero AoA , (b) the shaft standalone

3.CFD Simulation Setup

CFD simulations were performed for the same conditions of speed and angle for validation and flow field investigation purposes. The CFD model work was established on NUMECA. Because the flow is incompressible and turbulent, the governing equations are the continuity equation and the momentum equations. The developed turbulent flow is quantitatively described using RANS equations and the time-averaged continuity equation. The turbulent model that was implemented in the present CFD study was k-ω Shear Stress Transport (SST). The advantage of the k-ω SST model is that it provides a better prediction of flow separation than most RANS models. Furthermore, it performs better for boundary layers under adverse pressure gradients.

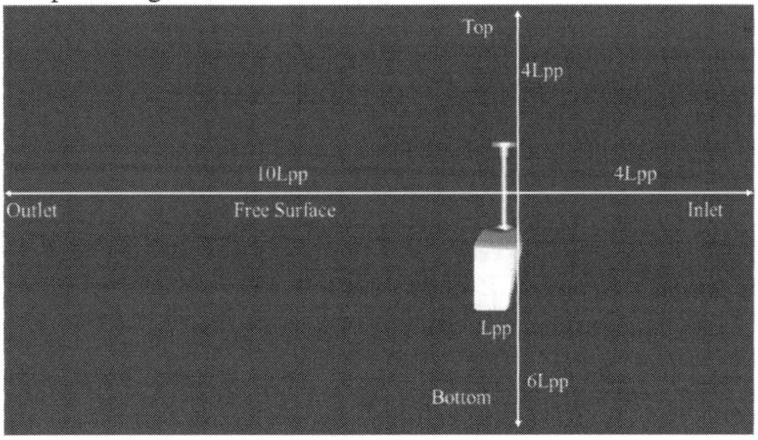

Figure 5. CFD domain of the GRS rudder

The simulations of resistance tests were also conducted at the same conditions as the experiments. Figure 5 indicates the dimensions and the boundary conditions of the computational domains with the free surface for the resistance simulations. The dimensions of the domain through CFD were decided according to the ITTC recommended procedure [12] for solving resistance analysis. The surfaces of the model were defined as no-slip wall type boundary conditions. The top and bottom were defined as zero-gradient conditions. The inlet, outlet and port-starboard sides were defined as velocity far-field boundaries. The mesh generation was done by using a full hexahedral mesh. The grid near the free

IOP Conf. Series: Materials Science and Engineering 1288 (2023) 012059 doi:10.1088/1757-899X/1288/1/012059

surface was refined to capture the wave elevation precisely. There are three layers of boundary-layer mesh near the model surfaces and the average wall function y+ value was kept at 50.

4.Experimental vs CFD Results

Figure 6 shows flow field around a single blade of the GRS rudder at different AoA. When the GRS rudder angle of attack is increasing (Figure 6 b and d), the amount of flow separation increases which results in eddy formation becoming noticeable. As the angle keeps increasing (Figure 6 c and e), the formation of eddies increases rapidly. At a rudder angle of ±30°, strong flow separation occurred.

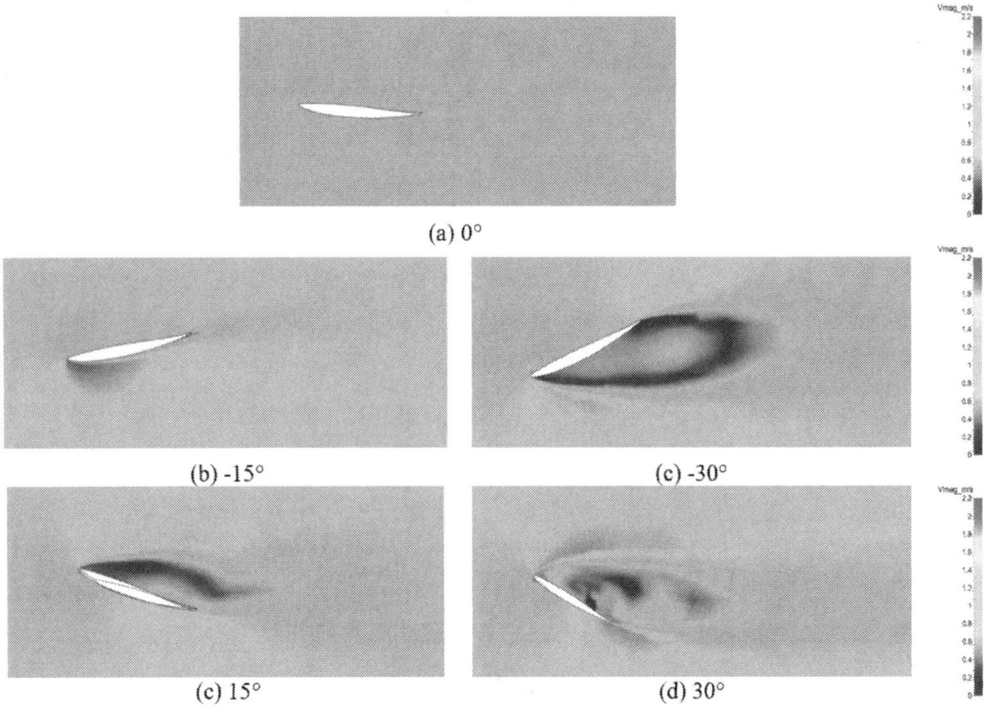

Figure 6. Fluid separation from the Gate Rudder under towing tests

The graphs in Figure 7 show the experimental measurements of lift and drag forces and their comparison with the CFD calculations for different speed and AoA. Both methos indicate that drag forces increased for a greater angle of rudder and for higher speeds. There are slight differences between positive and negative angles, with the former showing higher force values, especially at high speeds.

The Lift force diagram follows the theoretical trend of rudders. The rudder generates zero lift force at 0° angle. Specifically, for angles starting from 0° to around 20° the Lift force increases. After 20° the lift force remains constant or starts to decrease. The lift force is negative for negative angles and positive for positive angles. This is an expected result since the Gate Rudder's blade acts as a conventional rudder without a propeller. the Gate Rudder blade will be able to produce higher lift force to generate thrust with a further contribution comes from the accelerated the flow by the propeller [10].

IOP Conf. Series: Materials Science and Engineering 1288 (2023) 012059 doi:10.1088/1757-899X/1288/1/012059

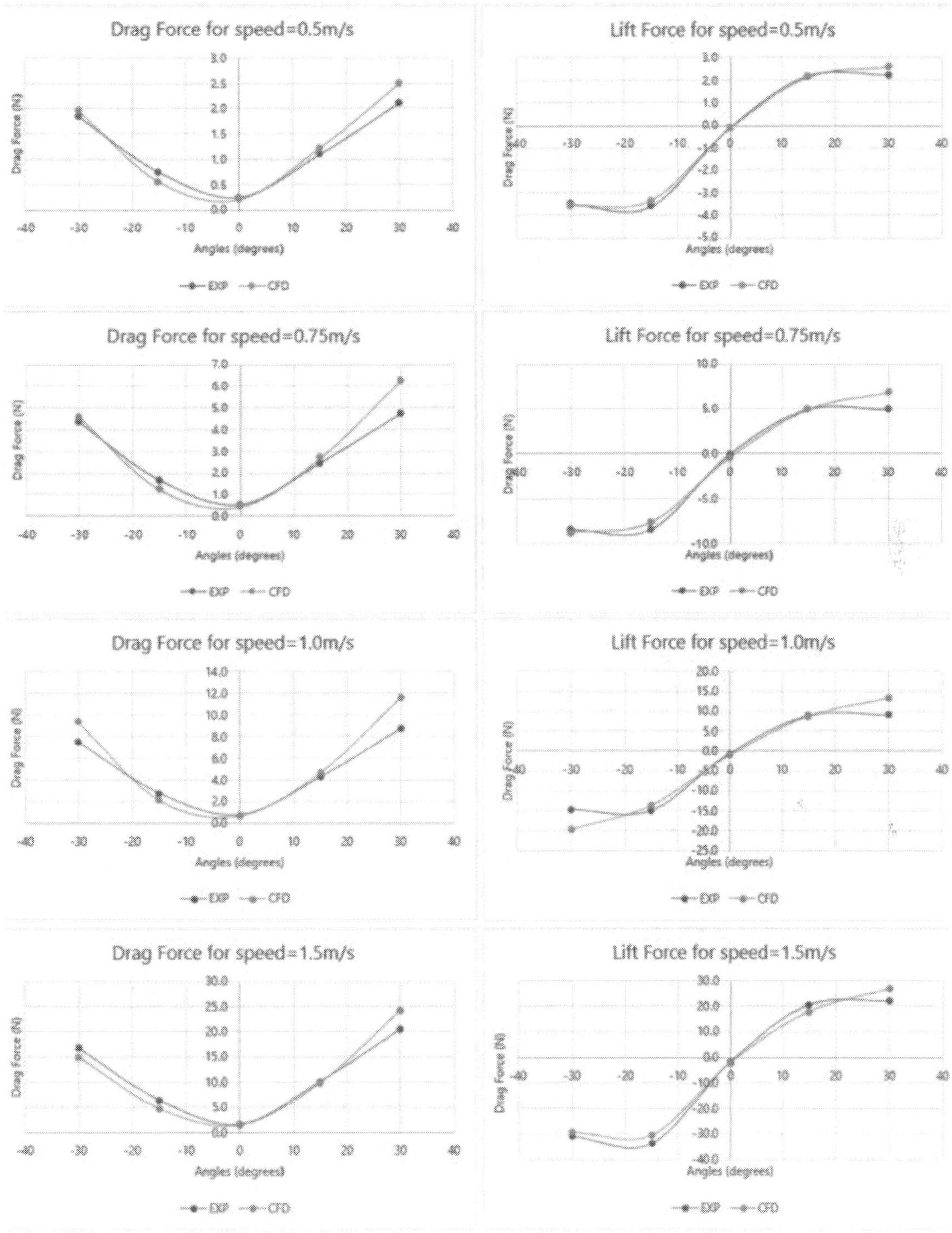

Figure 7. Drag and lift force measurements and simulations

IOP Conf. Series: Materials Science and Engineering 1288 (2023) 012059 doi:10.1088/1757-899X/1288/1/012059

5.Fluid Structure Interaction

Before carrying out the numerical simulation to investigate modal analysis of the fluid-solid coupling of a Gate Rudder, a modal experiment was conducted in the air and water to verify the modal analysis results. A series of vibration experiments was carried out in dry and wet conditions. The method was that a Piezoelectric Macro-Fiber Composite (PMFC) patch was installed on the back of the Gate Rudder Blade, as shown in Figure 8. The PMFC gives an output signal as a voltage which is measured using a digital oscilloscope. To receive a signal, the rudder blade is hit using a hammer and the oscilloscope displays the vibration signal of the rudder blade. Following this, FFT is performed on the signal to obtain the frequency spectrum. The peak value of the frequency spectrum can be considered as a certain modal frequency of the Gate Rudder.

5.1. Gate Rudder free modal experiment

The single rudder blade, shown in Figure 3 (left), is suspended from the shaft by an elastic rope to investigate the natural frequencies of the blade. A hammer is used to hit to the centre of the rudder blade. The time signal and frequency spectrum signal obtained from PMFC are shown in Figure 8. Two distinct responses appeared at 138Hz and 355 Hz.

The finite element free mode analysis was carried out on a single rudder blade, and the corresponding mode shape and frequency compared with the experimental results. Numerical simulation results, given in Table 2, are consistent with the experimental values.

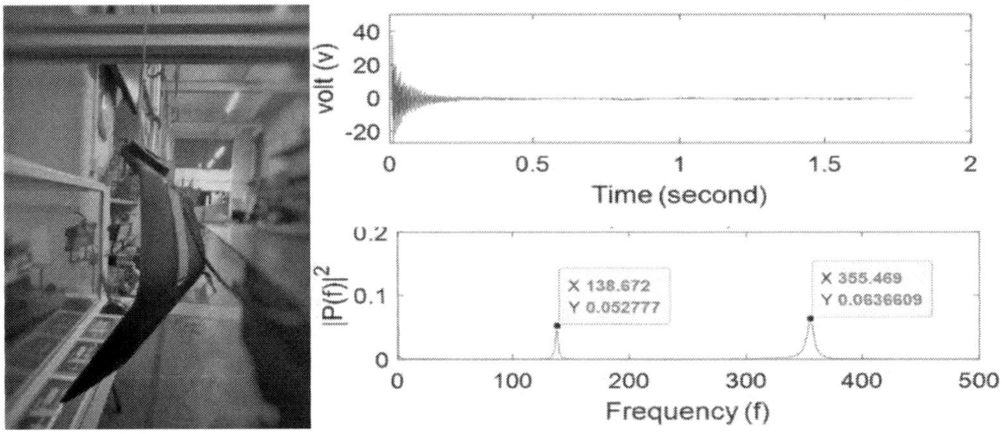

Figure 8. Rudder blade free modal experiment installation and its vibration signal

Table 2. A rudder blade free modal analysis.

	First (Hz)	Third (Hz)
	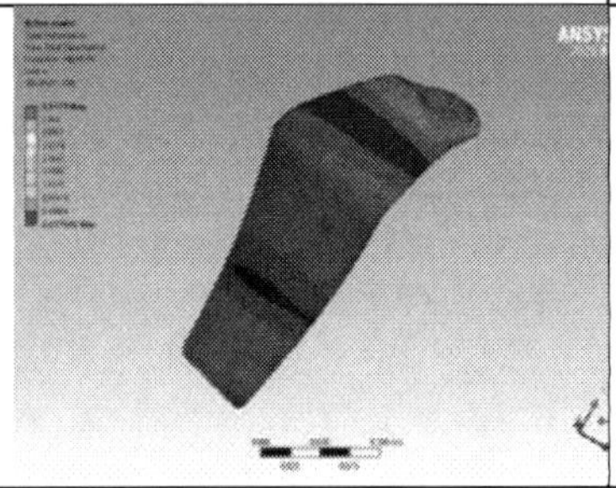	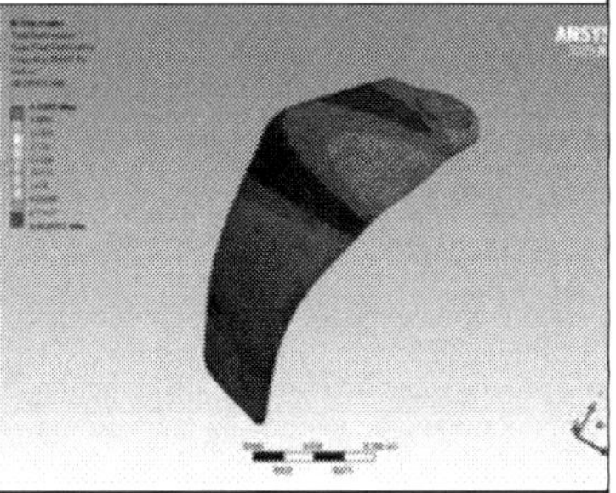
FEA	140.7	354.9
EXPERIMENT	138.7	355.5

IOP Conf. Series: Materials Science and Engineering 1288 (2023) 012059 doi:10.1088/1757-899X/1288/1/012059

5.2 The blade constrained modal experiment stationary condition in air

The rudder blade was constrained by a fixed degree of freedom at the position where it is connected to the rudder shaft, and the center of the rudder blade was hit with the hammer. The time signal and frequency spectrum from the PMFC are shown in Figure 9.

The finite element analysis of a single rudder blade under constrained boundary conditions is compared with the experimental results. The first-order modal frequency is highly consistent with the peak spectrum of the strain gauge as shown in Table 3.

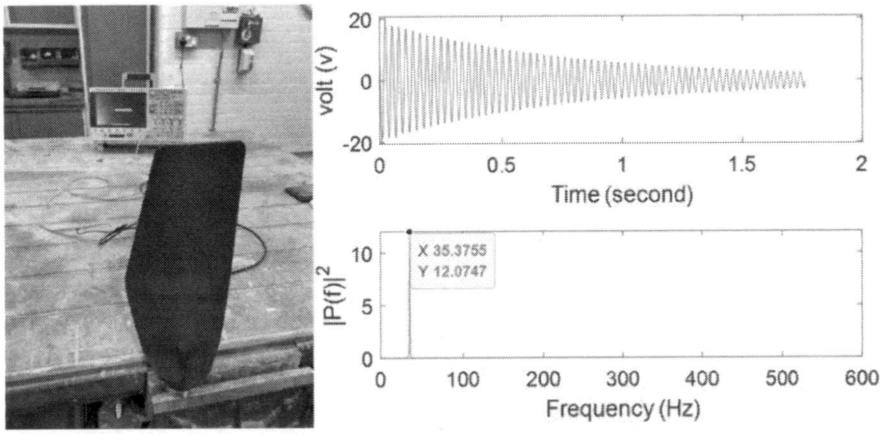

Figure 9. Rudder blade constrained modal experiment installation and its vibration signal

Table 3. The rudder blade constrained modal analysis.

	First (Hz)	Third (Hz)
	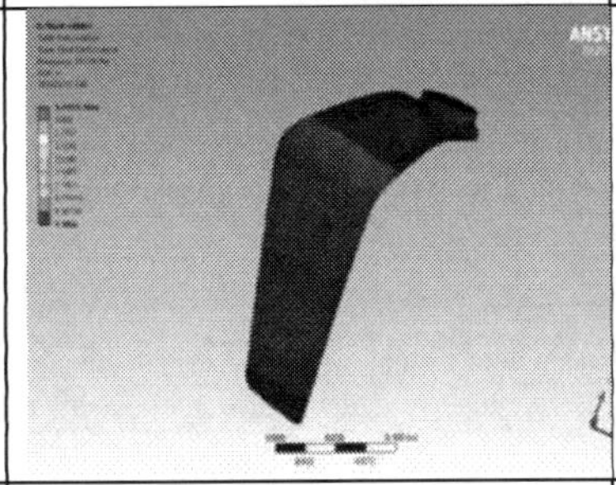	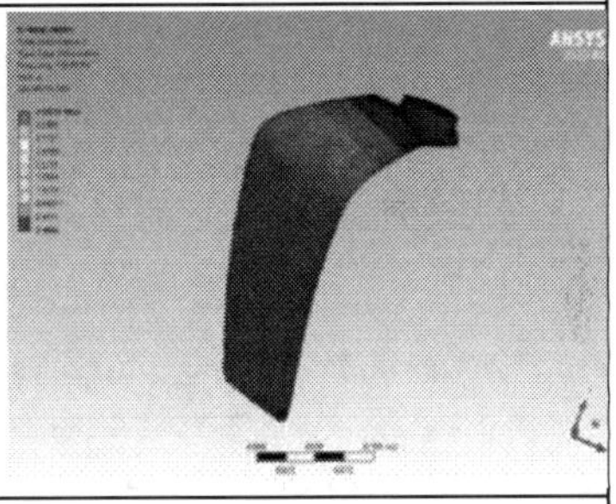
FEA	35.7	139.7
EXPERIMENT	35.4	None

5.3 The rudder and shaft constrained modal experiment

The rudder blade is connected to the rudder shaft, and a fixed displacement boundary condition is applied to the end of the rudder shaft. Again, the center of the rudder blade is hit by the hammer. The time signal and frequency spectrum gauge are shown in Figure 10.

The finite element modal analysis of the rudder blade and shaft under constrained displacement boundary conditions were again compared with the experimental results. The first-order modal frequency is consistent with the peak spectrum of the strain gauge. Table 4. shows that the FEA produced a higher modal frequency than the experiment. A possible explanation for this might be that the numerical model does not include the bolts and interface between the rudder and structure.

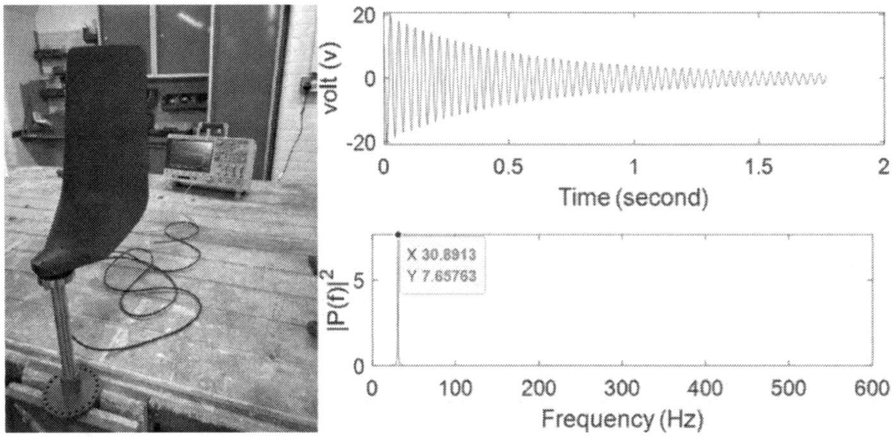

Figure 10. The rudder shaft constrained modal experiment

Table 4. The rudder and shaft constrained modal analysis in the air.

	First (Hz)	Third (Hz)
	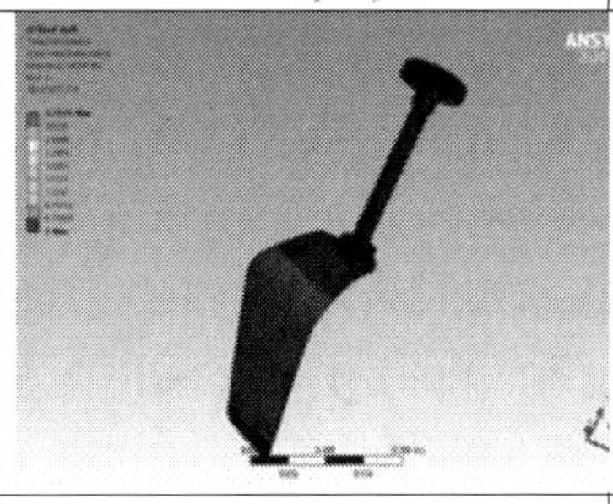	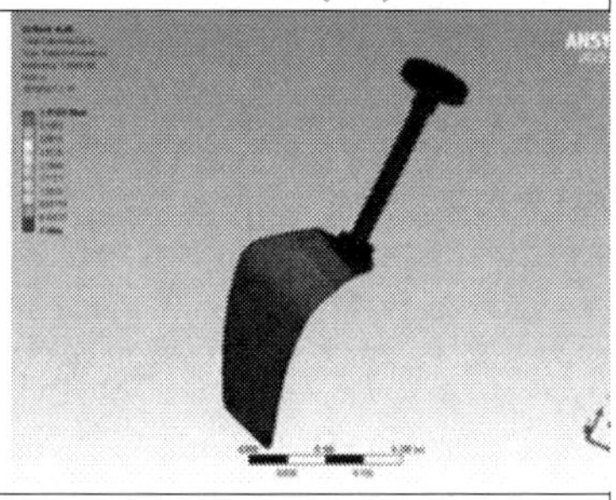
FEA	34.29	129.83
EXPERIMENT	30.89	None

5.4 Modal experiment of the rudder in the water

Pressure pulses generated by the propeller have an impact on the rudder blade during operation. Since the rudder itself moves in the water, it may also produce forces which is a phenomenon similar to the Karman vortex street. As a result, the rudder deforms and vibrates when the frequency of the external excitation force is close to or consistent with the natural frequency of the Gate Rudder structure. If the rudder is operated at this critical frequency the structure will resonate and be damaged. Therefore, the Gate Rudder structure should not be regarded as a rigid body for structural analysis, instead it should be regarded as a flexible body. Analysing the inherent characteristics of the Gate Rudder structure is very important for the design and safe use. Figure 11 shows the structural form of the Gate Rudder and its installation in water.

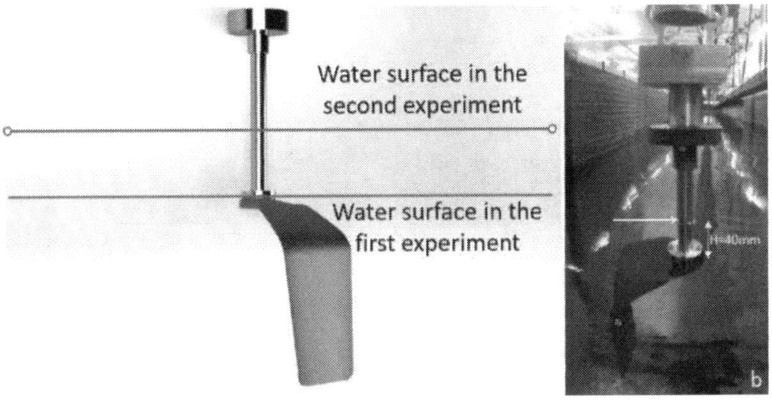

Figure 11. Gate Rudder structure and its installation in water

The assembly of the Gate Rudder structure and the water body is shown in Figure 12, where the water body is represented by a semicircular sphere. The modal analysis of the rudder blade immersed in water was done in the modal acoustic module of ANSYS. In this modal acoustic module, the Gate Rudder is defined as the structural domain, The topmost surface of the Gate Rudder is the DOF constrained surface. The water body is defined as the acoustic domain, the truncated surface of the water body is a semicircular spherical surface which is set as a non-reflecting radiation boundary, and the surface in contact with air is defined as the acoustic impedance boundary. Acoustic impedance is defined as:

$$Z = \rho C \tag{1}$$

where, Z is the acoustic impedance, ρ is the acoustic material density, C is the acoustic material acoustic speed. For water, the density is 1000 [kg/m^3], C is 1480 m/s, then the water acoustic impedance is 1.48e6 [kg/m^2s].

The mesh model of the Gate Rudder is shown in Figure 12. The mesh size of the Gate Rudder is 0.005 m, and the mesh size of the water body is 0.01 m. For the frequency range of 0-200Hz, in the acoustic analysis, there must be at least 6 elements in one wavelength. At a frequency of 200Hz, the wavelength of sound waves in water is 7.4 meters, and the maximum element length is: 7.4/6=1.23 meters, which is far greater than 0.01 meters. The mesh size therefore meets the requirements of acoustic analysis. Overall, there are 12978 nodes and 45449 elements, all of which are linear tetrahedral elements.

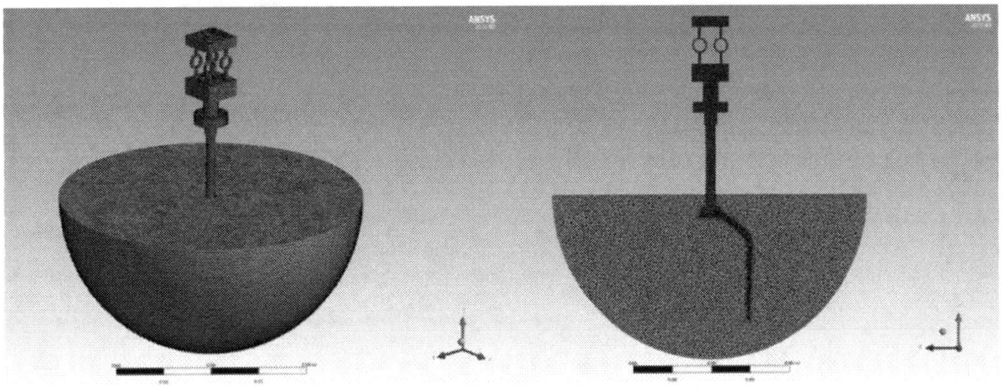

Figure 12. The Gate Rudder and water body mesh

The modal analysis results of the Gate Rudder both in air and in water are shown in Table 5. For the Gate Rudder studied in this paper, its mode shapes in air and water are basically the same. However, the corresponding modal frequencies are different. The modal frequencies in water are significantly lower than those in air, which shows that water has a damping effect on the vibration of the Gate Rudder.

Table 5. The Gate Rudder modal analysis comparison in the water and in the air.

	In the water		In the air	
	Frequency (Hz)	Modal shape	Frequency (Hz)	Modal shape
1	33.788	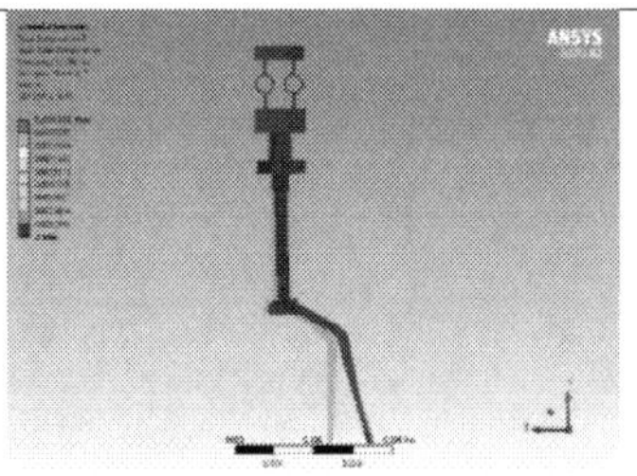	35.601	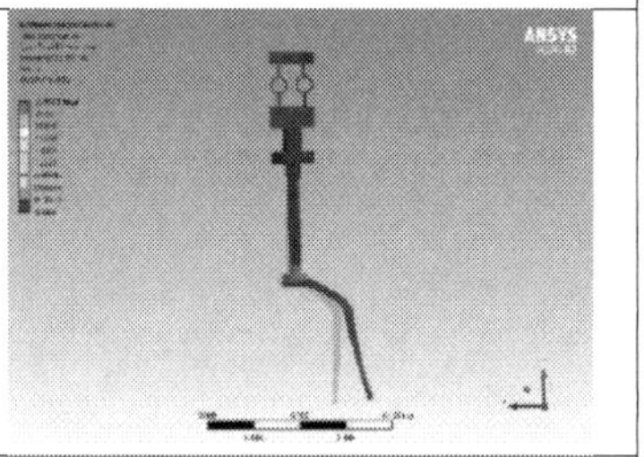
2	44.901	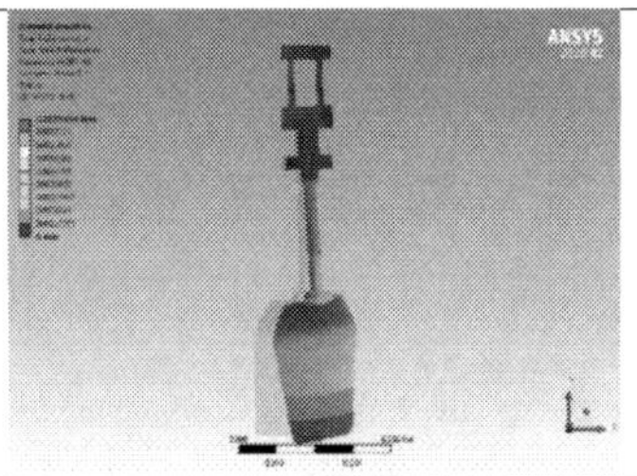	46.07	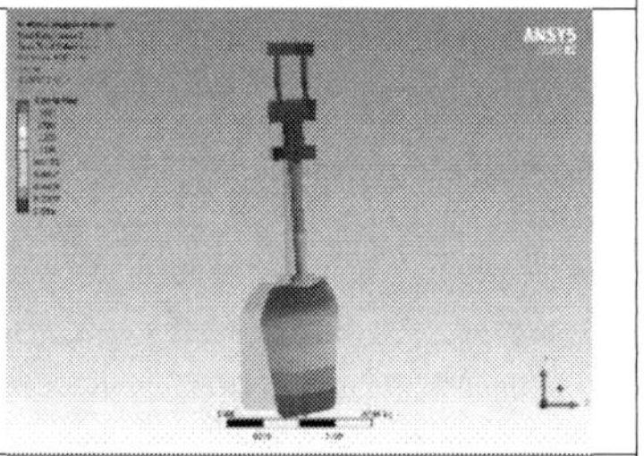
3	65.852	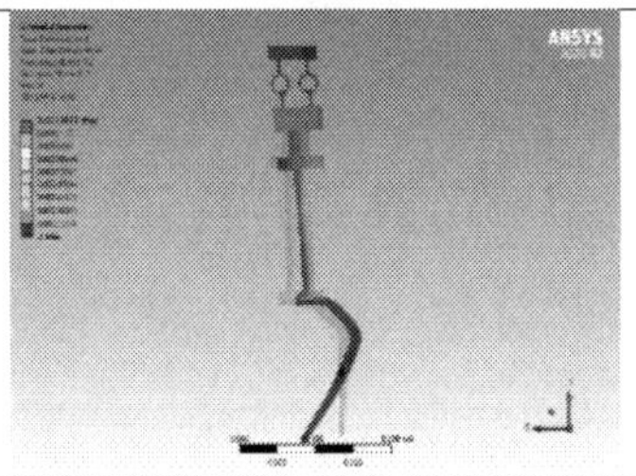	68.79	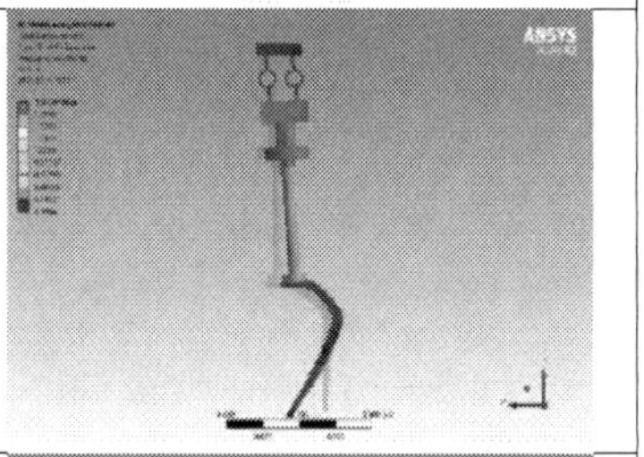
4	96.763	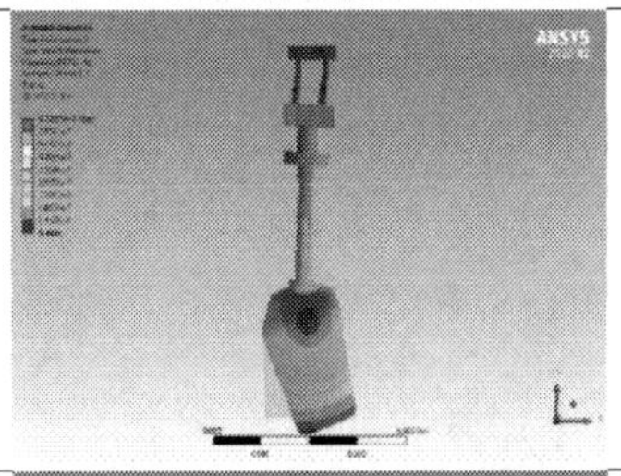	101.6	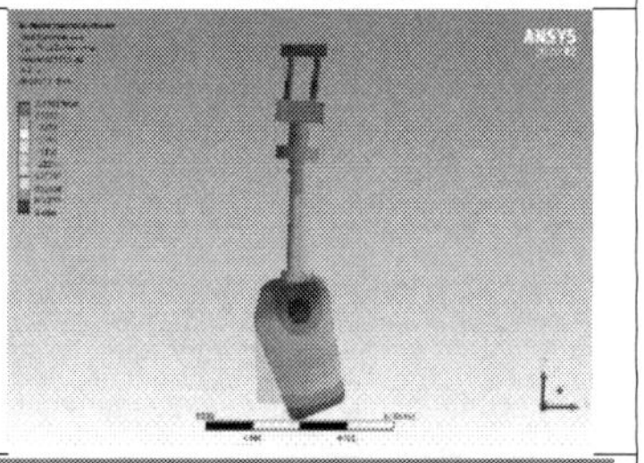
5	145.02	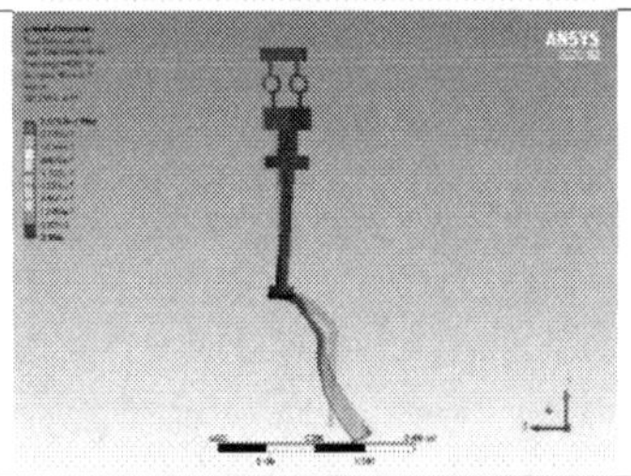	154.71	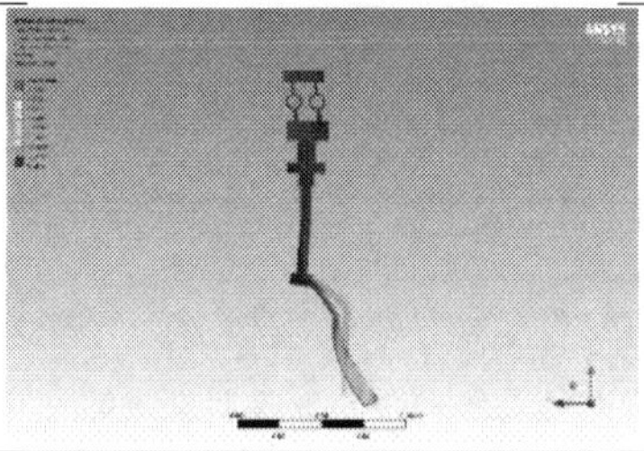
6	157.27	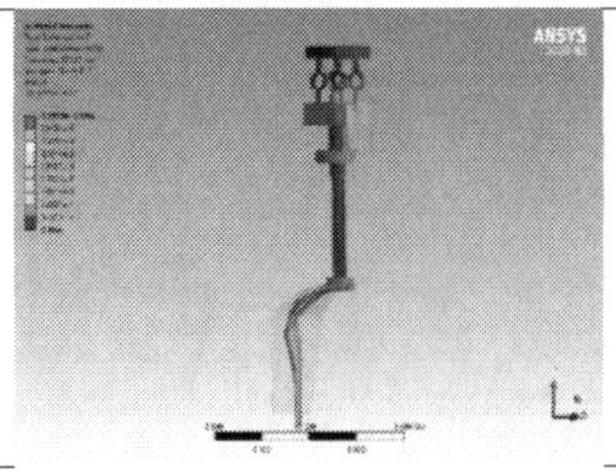	166.2	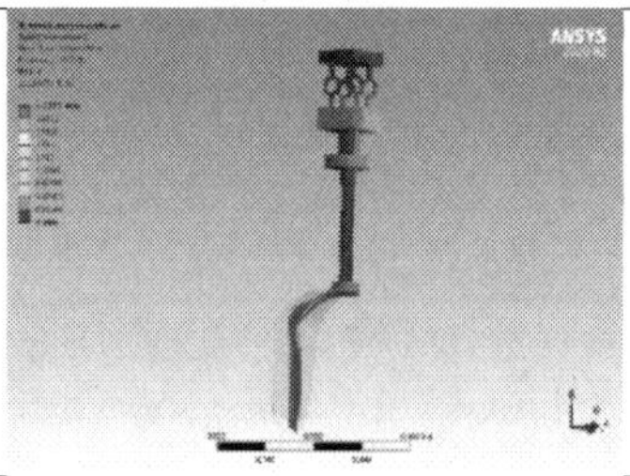

IOP Conf. Series: Materials Science and Engineering 1288 (2023) 012059 doi:10.1088/1757-899X/1288/1/012059

5.5 Vibration analysis of the towing tank test

It is likely that vortex shedding occurs due to the cylindrical shape of the rudder and shaft. The main dimensions are given in Figure 13. The vortex shedding frequencies (*f*) can be calculated by the Strouhal number (*St*) given as

$$St = \frac{fD}{v} \tag{2}$$

where f the frequency (Hz), D the diameter of the submerged body (m), and V the velocity of the water flow (m/s). The Strouhal number is equal to 0.2 according to Blevins [11] for this range of Reynolds number (Re ≈ 1.5x105).

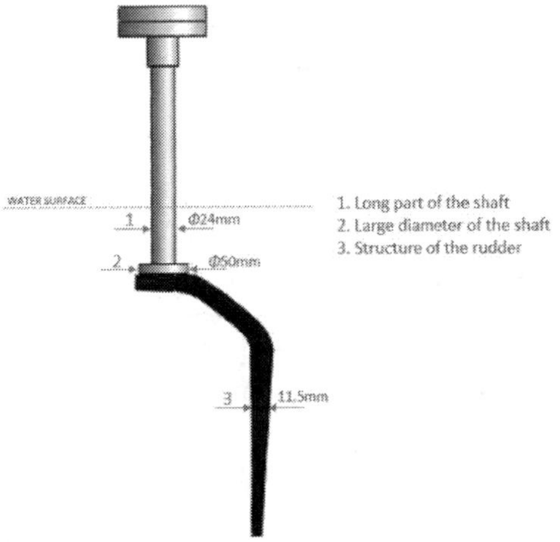

Figure 13. Thickness of each part of shaft-rudder configurations, (1) long part of the shaft, (2) large diameter of the shaft, (3) structure of the rudder

Both experimental and numerical analysis confirm the vibration sources summarized in Table 6. The oscillation due to vortex shedding at a frequency of 5.9 Hz, 12.8 Hz, and 25.5 Hz are originated from the large diameter of the shaft (2), the long part of the shaft (1), and the rudder (3) as Figure 13 depicts. The structure's natural frequencies are also measured at 29.8 and 43.4 Hz. These frequencies correspond to the first and second modes of the complete structure.

Table 6. Measured frequencies using Piezoelectric Macro-Fibber Composite at 1.5m/s and 30° AoA

Frequency (Hz)	Source	Origin
5.9	Vortex	2-large shaft diameter
12.8	Vortex	1- long shaft
25.5	Vortex	3- blade of the rudder
29.8	Natural Frequency	Blade's 1st Mode
43.4	Natural Frequency	Blade's 2nd Mode

6. Conclusions

The main goal of the current study was to identify the oscillating forces acting on the rudder using experimental and numerical methods. Firstly, the experimental results show that this Gate Rudder creates lift and drag forces similar to a conventional rudder. The results demonstrate a clear and consistent association between experimental and numerical simulations. Thus, CFD set-up and simulations are correct and can be used for different angles/speed conditions or for a full-scale rudder, giving a complete picture of its performance. The modal comparison between the blade in the air and immersed in the water shows that the corresponding mode shape does not change. However, its modal frequency is significantly reduced in water. This result indicates that water has a suppressive effect on the vibration which is more obvious at higher frequencies than at low frequencies. It was also found that vortex shedding originates from the rudder shaft and further contributes the oscillation. Thus, defining shaft geometry should be part of the design process.

Although the study has successfully demonstrated the flow and rudder interaction using both experimental fluid dynamics and computational fluid dynamics, it has certain limitations in terms of representing the entire Gate Rudder System. A further study using the Piezoelectric Macro-Fiber Composite Actuator on the rudder blade in a cavitation tunnel and on a full-scale vessel could assess the long-term effects of propeller induced vibration on rudder structure.

Acknowledgments
The work presented in this paper is part of the activities in the GATERS (Gate Rudder System as a Retrofit for the Next Generation Propulsion and Steering of Ships) project which is sponsored by the EC H2020 Programme initiative under the grant agreement no: 860337. The Authors kindly acknowledge support received from colleagues of the Newcastle University Marine Hydrodynamic Laboratory and the Emerson Cavitation Tunnel.

References

[1] MEPC72 I M O 2018 Resolution MEPC. 304 (72) *Initial IMO Strateg. Reduct. GHG Emiss. from Ships*

[2] Joung T-H, Kang S-G, Lee J-K and Ahn J 2020 The IMO initial strategy for reducing Greenhouse Gas (GHG) emissions, and its follow-up actions towards 2050 *J. Int. Marit. Safety, Environ. Aff. Shipp.* **4** 1–7

[3] Narita H, Yagi H, Johnson H D and Breves L R 1981 Development and full-scale experiences of a novel integrated duct propeller *Trans. SNAME* **89** 319–46

[4] Lewis E V 1988 Principles of naval architecture second revision *Jersey: Sname* **2** 152–7

[5] Carlton J 2018 *Marine propellers and propulsion* (Butterworth-Heinemann)

[6] Turkmen S, Carchen A and Atlar M 2015 A new energy saving twin rudder system-Gate Rudder *SCC 2015 International Conference on Shipping in Changing Climates*

[7] Sasaki N, Atlar M and Kuribayashi S 2016 Advantages of twin rudder system with asymmetric wing section aside a propeller: The new hull form with twin rudders utilizing duct effects *J. Mar. Sci. Technol.* **21** 297–308

[8] Carchen A, Turkmen S, Piaggio B, Shi W, Sasaki N and Atlar M 2021 Investigation of the manoeuvrability characteristics of a Gate Rudder system using numerical, experimental, and full-scale techniques *Appl. Ocean Res.* **106** 102419

[9] Fukazawa M, Turkmen S, Marino A and Sasaki N 2018 Full-scale Gate Rudder performance obtained from voyage data *Proceedings of the A. Yücel Odabası Colloquium Series: 3rd International Meeting-Progress in Propeller Cavitation and Its Consequences: Experimental and Computational Methods for Predictions, Istanbul, Turkey* pp 15–6

[10] Turkmen S, Fukazawa M, Sasaki N and Atlar M 2018 Cavitation Tunnel tests and full-scale review of the first Gate Rudder system installed on the 400TEU container ship *A. Yücel Odabaşı Colloquium Series 3rd International Meeting on Progress in Propeller Cavitation and its Consequences: Experimental and Computational Methods for Predictions* (Istanbul Turkey) pp 29–39

[11] Blevins R D 1977 Flow-induced vibration *New York*

[12] ITTC 2014 27 th ITTC Full Conference *27th ITTC Conf.* **51** 23–30

Estimation of the entrained water effect of tandem propellers based on numerical method

Jianzheng Wang*[1,2,3], Ruiping Zhou[1], Yukuo Guo[1], Xiaoyuan Luo[2,3]

[1] School of Naval Architecture, Ocean and Energy Power Engineering, Wuhan University of Technology, Wuhan, 430063, China
[2] National Engineering Laboratory for Marine and Ocean Engineering Power System, Shanghai, 201108, China
[3] Shanghai Marine Diesel Engine Research Institute, Shanghai, 201108, China

E-mail: hnwjz060@126.com

Abstract. The objective of the paper is to propose the empirical formulas of the entrained water effect based on numerical method to estimate the virtual mass and the virtual moment of inertia for tandem propellers. The fluid-solid interaction of combined propellers in the flow field has been taken into consideration in the calculation. The axial added mass and the added moment of inertia of the entrained water are calculated for a single propeller using the finite element method and the existing empirical formulas to verify the feasibility and accuracy of numerical method. Consequently, the numerical calculation of tandem propellers is conducted to obtain the novel empirical formulas of entrained water properties, which are available to evaluate the influence of the blade diameter, the expanded area ratio and the pitch/diameter ratio. The corresponding correction coefficients in the empirical formulas considering the entrained water effect of tandem propellers are given. The empirical formulas are beneficial to relevant practical engineering calculation of entrained water properties for combined propellers.

1. Introduction

The entrained water effect caused by propellers in the flow field is a very important parameter in the shaft alignment and vibration calculations. The fluid-solid interaction of the propeller rotating in the flow field causes the propeller to generate axial acceleration and circumferential angular acceleration, namely the axial entrained water mass and the circumferential added entrained water moment of inertia. In the engineering calculation, the entrained water effect of a single propeller is usually obtained by multiplying the mass and moment of inertia of the propeller by empirical coefficients. To date, the combined propeller which exhibits higher propulsion efficiency and better energy saving effect are favored by ship workers[1]. However, few investigations have been carried to calculate the entrained water effect of combined propellers. It would be therefore worthwhile exploring the relevance empirical formulas of added water mass and moment of inertia for the tandem propeller.

The water effect of a single propeller is widely studied and a variety of empirical formulas are proposed to estimate the virtual mass and the virtual moment of inertia. Burrill[2] conducted various propeller model tests and obtained the longitudinal added mass formula and the corresponding coefficients. Schwanecke[3] derived the formula of the added moment of inertia of entrained water for propellers of merchant ships. Brooks[4] studied the longitudinal entrained water mass by theoretical methods and experiments. Parsons[5]made a statistical analysis of the numerical calculation of B series propeller with the lifting line and lifting surface theories, and obtained the formulas and corresponding coefficients of the axial added mass and the added moment of inertia of the entrained water for a single propeller. Harrington[6] developed a formula for calculating the entrained water effect based on the blade thickness and width of propellers. Donald[7]presented an overview analysis of the previous estimation methods of the entrained water effect of propellers, and improved the Burrill's estimation formula. The accuracy is verified in comparison with the Burrill's experimental data. Korotkin[8] investigated the entrained water effect of propellers for ship structures theoretically. Hutchison[9] calculated the added mass of the entrained water for ducted propellers. None of the aforementioned studies has yet been exploited to deal with the entrained water effect of combined propellers.

Motivated by the current limitation in the literature, the novel empirical formulas of the entrained water effect based on numerical method will be proposed in this work to estimate the virtual mass and the virtual moment of inertia for tandem propellers. Firstly, the solid model and FEM model of a single propeller is established by commercial software NX and ANSYS, respectively. The axial added mass and the circumferential added moment of inertia of the entrained water of the single propeller are calculated to verify the correctness of the numerical method, compared to the results of the existing empirical formula. On this basis, numerical calculation of the entrained water effect of a variety of tandem propellers with different parameters is conducted and it is available to evaluate the influence of the blade diameter, the expanded area ratio (EAR), and the pitch/diameter ratio (P/D). Finally, the novel empirical formulas for estimating the entrained water effect of tandem propellers are derived by using multiple regression model.

2. Numerical method

To consider the fluid-solid interaction of the combined propeller in the flow field, both the propeller and the fluid domain are modeled by finite element method. The vibration of the fluid and the propeller can be analyzed by harmonic analysis. Here, the whole system concluding the propeller and the fluid domain is simplified as a single-degree-of-freedom forced vibration system, which will generate acceleration response under the external excitation force. The vibration differential equation of the whole system is

$$m\ddot{x} + c\dot{x} + kx = F\sin\omega t \tag{1}$$

Ignoring the effects of damping term and stiffness term (relatively small compared with the inertia term), the vibration equation can be simplified as

$$m\ddot{x} = F\sin\omega t \tag{2}$$

The mass of the single-degree-of-freedom system is

$$m = \frac{F}{\ddot{x}} \tag{3}$$

where $m = m_p + m_f$, m_p is the propeller mass, and m_f is the mass of the entrained fluid.

As soon as the mass of the whole system is calculated by equation (3), the mass of the entrained water can be obtained by subtracting the mass of the propeller. It is noted that the frequency responses of the acceleration in vacuum and in water are calculated simultaneously to obtain the mass of the entrained water. By replacing the mass with the moment of inertia and the external force with the torque, the moment of inertia of the entrained water can also be obtained.

3. Validation of numerical method

In this section, the mass and moment of inertia of the entrained water of a single propeller are calculated by the finite element method and the existing empirical formula to verify the feasibility and accuracy of the numerical method.

IOP Conf. Series: Materials Science and Engineering 1288 (2023) 012060 doi:10.1088/1757-899X/1288/1/012060

The geometry and material properties of a single propeller model for a multi-purpose ship are given in Table 1.

Table 1. Geometry and material properties of a single propeller.

Geometric parameters	Number of blades	Blade diameter (m)	Pitch/diameter ratio	Expanded area ratio	
	4	5.15	0.604	0.55	
Material parameters	Mass (kg)	Moment of inertia (kg/m^2)	Density(kg/m^3)	Elastic modulus (MPa)	Poisson's ratio
	11560	14110	7500	117700	0.34

The three-dimensional solid model of the single propeller is established using NX software. After importing it into ANSYS software, the three-dimensional model of the water domain is established. Then the single propeller-water coupling finite element model is obtained after meshing. The characteristics of this model are as follows: (1) The radius of the water domain model is 6 times of the radius of the propeller and a non-reflection boundary condition is set for the outermost surface of the water domain; (2) The entrained water effect is ignored between the hub and the water, and the cylindrical surface of the hub is set as a constrained boundary condition; (3) The inner surface of the hub is coupled with a reference point which is used to apply the external force and torque; (4) The elements connected to the blade surface and the water are bonded to simulate the contact between them.

The SOLID45 tetrahedral element type and the FLUID30 acoustic tetrahedral element type are used for the propeller and the water, respectively. The FLUID130 acoustic surface element type is covered on the external surface of the water. The meshing of the contact region between the blade and the water is consistent. The meshing is dense inside the water and sparse outside. It should be noted that the maximum mesh size is 0.2m for the propeller and covers the frequency up to 350Hz.

The single propeller-water coupling finite element model is shown in Figure 1. The propeller and the water domain model, the constrained boundary condition of the propeller hub, the internal coupling reference point, the water-propeller contact, and the external non-reflection boundary condition are presented, respectively.

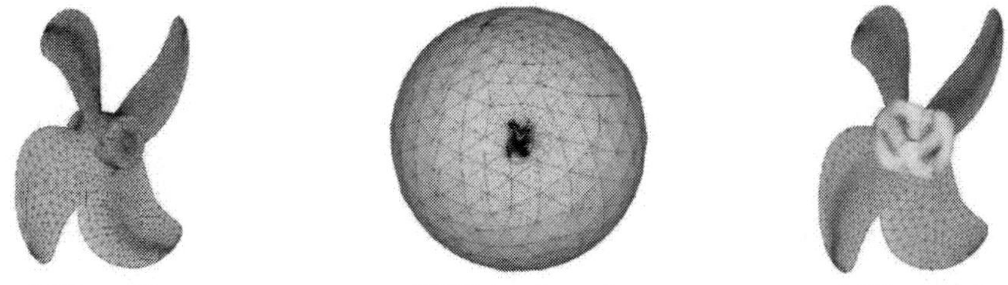

(a)Propeller (b)Water domain (c)Restraint

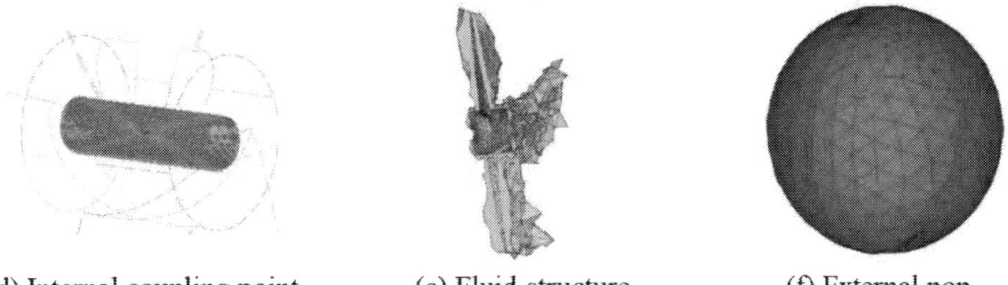

(d) Internal coupling point of the propeller hub (e) Fluid-structure contact (f) External non-reflection boundary

Figure 1. The single propeller-water coupling finite element model.

IOP Conf. Series: Materials Science and Engineering 1288 (2023) 012060 doi:10.1088/1757-899X/1288/1/012060

A 10 N exciting force in the axial direction is applied to the coupling point of the propeller and the harmonic response analysis is carried out in vacuum and water, respectively. The acceleration frequency responses at the excitation reference point are obtained, as shown in Figure 2. It can be seen form Figure 2 that the entrained water effect results in the reduction of the propeller's resonance frequency, namely the natural frequency, which is in accordance with the actual vibration condition.

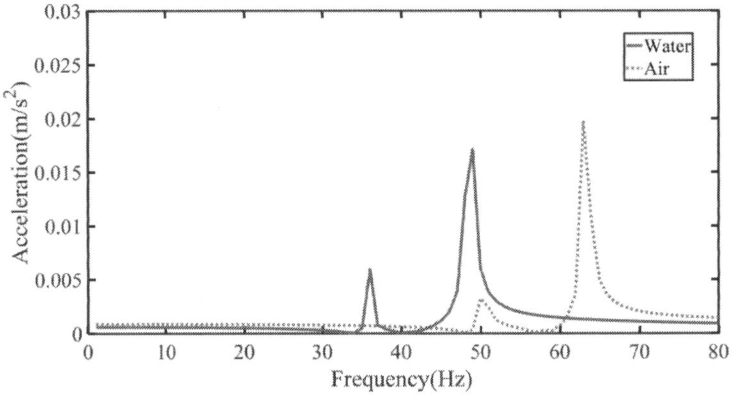

Figure 2. Accelerations at the excitation reference point.

The angular accelerations are calculated by the same procedure while the propeller torque at the excitation point is 10 N m, as shown in Figure 3.

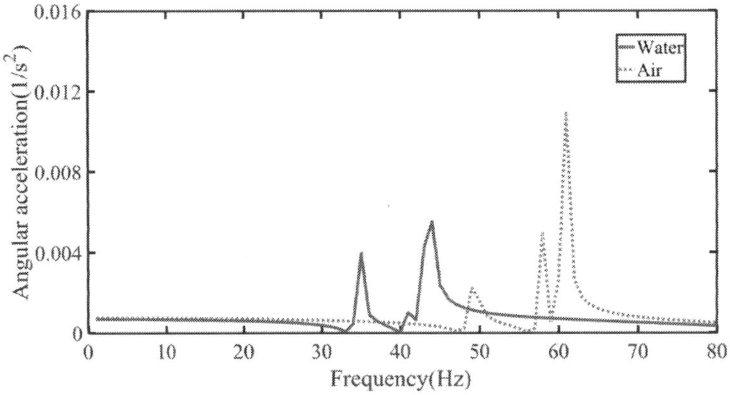

Figure 3. Angular accelerations at the excitation reference point.

The acceleration in the frequency range far from the resonant frequency can be used to calculate the entrained water mass. The reason is that the results near the resonance frequency is very large, resulting in the failure of the method. It is recommended to choose the frequency points in the low frequency region which is also far away from the resonance region.

Table 2. Mass and moment of inertia of the entrained water for the single propeller.

Frequency (Hz)	Exciting force (N)	Acceleration in vacuum (m/s^2)	Acceleration in water (m/s^2)	Added mass (kg)	Donald/Burrill formula	Relative deviation
10	10	0.000870188	0.000487258	9031.25	9360.84	3.52%
Frequency (Hz)	Excitation torque (Nm)	Angular acceleration in vacuum (1/s^2)	Angular Acceleration in water (1/s^2)	Added moment of inertia(kg/m^2)	Donald/Burrill formula	Relative deviation
10	10	0.000743953	0.000649312	1959.21	2011.87	2.62%

Here, the frequency response at 10 Hz is selected for comparative analysis and calculation, and the mass and moment of inertia of the entrained water for the single propeller are finally calculated, as shown in Table 2. For the purpose of comparison, the added mass and moment of inertia of the single propeller estimated by Donald/Burrill empirical formula are also listed.

As can be seen from Table 2, the maximum relative deviation between the numerical calculation results in this work and those calculated by Donald/Burrill empirical formula is 3.52%, both of which are within 5%. It indicates that the numerical calculation method has very good accuracy and high precision property, which can be used to conduct the following analysis.

4. Empirical formulas of the entrained water effect for tandem propellers

In this section, the proposed numerical method is employed to analyze the response of tandem propellers having various forms of geometrical parameters. Table 3 shows the mass and moment of inertia of the entrained water for tandem propellers using ANSYS with different geometric parameters. In Table 3 the design variables of empirical formulas of the entrained water effect are selected as follows: the blade diameter, the expanded area ratio and the pitch/diameter ratio are independent variables, and the axial added mass and the added moment of inertia of the entrained water are dependent variable.

Table 3. Mass and moment of inertia of the entrained water for tandem propellers with different geometric parameters.

No.	Front Propeller			Rear Propeller			Added mass (kg)	Added moment of inertia (kg/m^2)
	EAR	P/D	Diameter (m)	EAR	P/D	Diameter (m)		
1	0.4	0.4	0.25	0.4	0.5	0.25	0.698871713	0.000241817
2	0.4	0.5	0.25	0.4	0.6	0.25	0.659472773	0.000321032
3	0.4	0.6	0.25	0.4	0.7	0.25	0.606512106	0.000411519
4	0.4	0.7	0.25	0.4	0.8	0.25	0.570597737	0.00050301
5	0.4	0.8	0.25	0.4	0.9	0.25	0.573467103	0.000637306
6	0.4	0.9	0.25	0.4	1	0.25	0.55037697	0.000786075
7	0.4	0.6	0.25	0.3	0.6	0.25	0.476048115	0.000274836
8	0.5	0.6	0.25	0.4	0.6	0.25	0.791480859	0.000450101
9	0.6	0.6	0.25	0.5	0.6	0.25	1.173832155	0.000659229
10	0.7	0.6	0.25	0.6	0.6	0.25	1.532889211	0.000850416
11	0.5	0.4	0.35	0.5	0.4	0.25	1.968682595	0.000847396
12	0.5	0.4	0.25	0.5	0.4	0.35	1.901528736	0.000839746
13	0.5	0.4	0.35	0.5	0.4	0.35	2.761666046	0.001405341
14	0.5	0.4	0.25	0.5	0.4	0.25	1.12784578	0.000292486

Some FEM models of tandem propellers are shown in Figure 4-Figure 6. Here, at least 6 or so elements per wavelength is satisfied for the numerical calculation as a rule-of-thumb of conventional FEA.

(a) Front view (b) Left view (c) Isometric view

Figure 4. FEM model of No.1 tandem propeller.

(a) Front view (b) Left view (c) Isometric view

Figure 5. FEM model of No.11 tandem propeller.

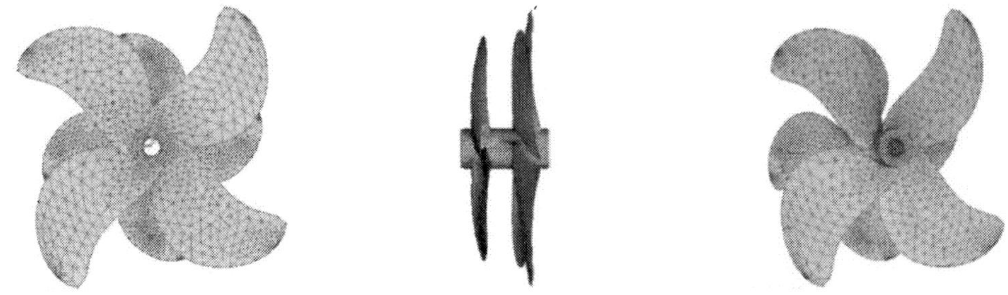

(a) Front view (b) Left view (c) Isometric view

Figure 6. FEM model of No.12 tandem propeller.

4.1. Multiple linear regression model

Considering the existing empirical formula of the entrained water effect, the following multivariate linear regression model is established

$$M_w = \rho_w D_f^{\,3}\left(\alpha_{0f} + \alpha_{1f}\frac{A_f}{A_{0f}} + \alpha_{2f}\frac{P_f}{D_f}\right) + \rho_w D_b^{\,3}\left(\alpha_{0b} + \alpha_{1b}\frac{A_b}{A_{0b}} + \alpha_{2b}\frac{P_b}{D_b}\right) \tag{4}$$

where M_w is the entrained water mass, ρ_w is the water density, $\frac{A_f}{A_{0f}}$ and $\frac{A_b}{A_{0b}}$ are respectively the EARs of front and rear blades, $\frac{P_f}{D_f}$ and $\frac{P_b}{D_b}$ are respectively the pitch/diameter ratios of front and rear blades, D_f and D_b are the blade diameters of front and rear blades, and $\alpha_{0f}, \alpha_{1f}, \alpha_{2f}, \alpha_{0b}, \alpha_{1b}, \alpha_{2b}$ are the undetermined coefficients.

The multivariate linear regression model for the entrained water moment of inertia is

$$J_w = \rho_w D_f^{\,5}\left(\beta_{0f}+\beta_{1f}\frac{A_f}{A_{0f}}+\beta_{2f}\frac{P_f}{D_f}\right)+\rho_w D_b^{\,5}\left(\beta_{0b}+\beta_{1b}\frac{A_b}{A_{0b}}+\beta_{2b}\frac{P_b}{D_b}\right) \quad (5)$$

where J_w is the entrained water moment of inertia and $\beta_{0f}, \beta_{1f}, \beta_{2f}, \beta_{0b}, \beta_{1b}, \beta_{2b}$ are the undetermined coefficients.

Analyzing the data in Table 3 by the least square method in Matlab software, the coefficients of the entrained water effect of tandem propellers are obtained, as shown in Table 4.

By substituting the coefficients of Table 4 into equation (4) and equation (5), the entrained water mass and moment of inertia under different geometric parameters in Table 3 can be recalculated as shown in Figure 7 and Figure 8.

Table 4. Coefficients of the entrained water effect for tandem propellers in the multiple linear regression model.

M_w	α_{0f}	α_{1f}	α_{2f}	α_{0b}	α_{1b}	α_{2b}
	-0.016926551	0.114155101	-0.010719309	-0.021852898	0.113228875	-0.008232098
J_w	β_{0f}	β_{1f}	β_{2f}	β_{0b}	β_{1b}	β_{2b}
	-0.000599241	0.001726523	-0.000296979	-0.000553728	0.000255805	0.001397515

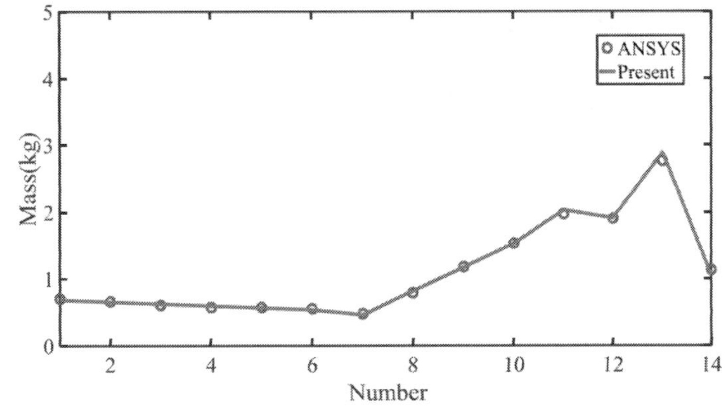

Figure 7. Comparison of the entrained water mass calculated by FEM and empirical formula for the multivariate linear regression model.

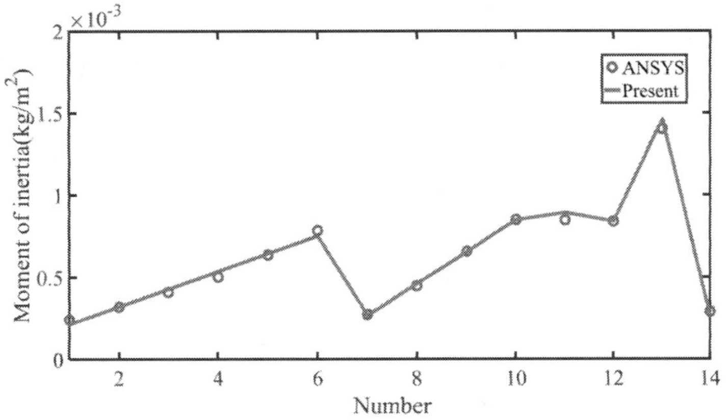

Figure 8. Comparison of the entrained water moment of inertia calculated by FEM and empirical formula for the multivariate linear regression model.

As shown in Figure 7 and Figure 8, the results of the entrained water mass and moment of inertia obtained by empirical formulas for the multivariate linear regression model are basically consistent with FEM results in Table 3. The relative error of inertia of No.6, No.11 and No.13 is slightly high.

4.2. Multiple quadratic regression model

The following is the form of the multiple quadratic regression model for empirical formula of the entrained water effect

$$M_w = \rho_w D_f^3\left(\alpha_{0f}+\alpha_{1f}\frac{A_f}{A_{0f}}+\alpha_{2f}\frac{P_f}{D_f}+\alpha_{3f}\left(\frac{A_f}{A_{0f}}\right)^2+\alpha_{4f}\left(\frac{P_f}{D_f}\right)^2+\alpha_{5f}\frac{A_f}{A_{0f}}\frac{P_f}{D_f}\right)+ \rho_w D_b^3\left(\alpha_{0b}+\alpha_{1b}\frac{A_b}{A_{0b}}+\alpha_{2b}\frac{P_b}{D_b}+\alpha_{3b}\left(\frac{A_b}{A_{0b}}\right)^2+\alpha_{4b}\left(\frac{P_b}{D_b}\right)^2+\alpha_{5b}\frac{A_b}{A_{0b}}\frac{P_b}{D_b}\right) \tag{6}$$

where $\alpha_{0f}, \alpha_{1f}, \alpha_{2f}, \alpha_{3f}, \alpha_{4f}, \alpha_{5f}, \alpha_{0b}, \alpha_{1b}, \alpha_{2b}, \alpha_{3b}, \alpha_{4b}, \alpha_{5b}$ are the undetermined coefficients.

The multivariate quadratic regression model for the entrained water moment of inertia is

$$J_w = \rho_w D_f^5\left(\beta_{0f}+\beta_{1f}\frac{A_f}{A_{0f}}+\beta_{2f}\frac{P_f}{D_f}+\beta_{3f}\left(\frac{A_f}{A_{0f}}\right)^2+\beta_{4f}\left(\frac{P_f}{D_f}\right)^2+\beta_{5f}\frac{A_f}{A_{0f}}\frac{P_f}{D_f}\right)+ \rho_w D_b^5\left(\beta_{0b}+\beta_{1b}\frac{A_b}{A_{0b}}+\beta_{2b}\frac{P_b}{D_b}+\beta_{3b}\left(\frac{A_b}{A_{0b}}\right)^2+\beta_{4b}\left(\frac{P_b}{D_b}\right)^2+\beta_{5b}\frac{A_b}{A_{0b}}\frac{P_b}{D_b}\right) \tag{7}$$

where $\beta_{0f}, \beta_{1f}, \beta_{2f}, \beta_{3f}, \beta_{4f}, \beta_{5f}, \beta_{0b}, \beta_{1b}, \beta_{2b}, \beta_{3b}, \beta_{4b}, \beta_{5b}$ are the undetermined coefficients.

Analyzing the data in Table 3 by the least square method in Matlab software, the coefficients of the entrained water effect for tandem propellers are obtained, as shown in Table 5.

Table 5. Coefficients of the entrained water effect for tandem propellers in the multiple quadratic regression model.

M_w	α_{0f}	α_{1f}	α_{2f}	α_{3f}	α_{4f}	α_{5f}
	1114978.459	-4553537.872	170225.6563	3830078.681	851128.6376	0
	α_{0b}	α_{1b}	α_{2b}	α_{3b}	α_{4b}	α_{5b}
	-800060.8902	3787522.3	0	-3830078.611	-851128.6028	0
J_w	β_{0f}	β_{1f}	β_{2f}	β_{3f}	β_{4f}	β_{5f}
	5415.811122	0	-12923.12959	-17854.32073	26356.38154	0
	β_{0b}	β_{1b}	β_{2b}	β_{3b}	β_{4b}	β_{5b}
	2516.609249	-23720.74258	0	17854.32114	-26356.38059	45486.0139

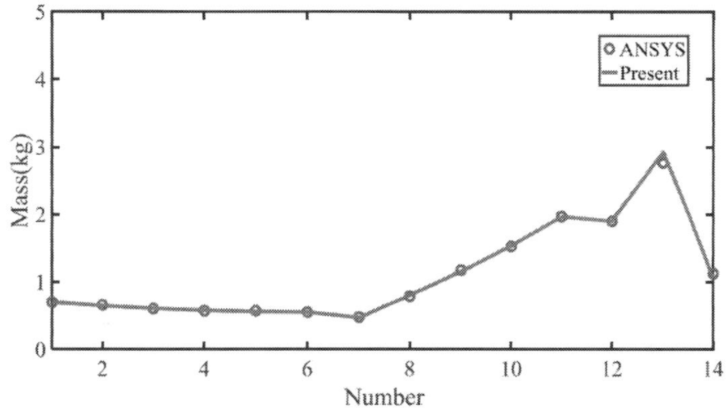

Figure 9. Comparison of the entrained water mass calculated by FEM and empirical formula for the multivariate quadratic regression model.

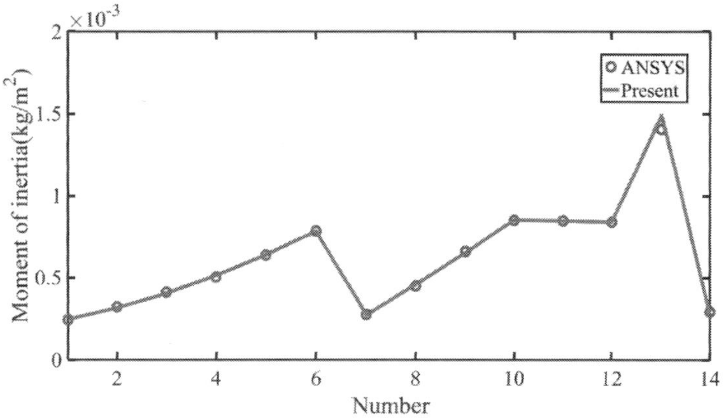

Figure 10. Comparison of the entrained water moment of inertia calculated by FEM and empirical formula for the multivariate quadratic regression model.

By substituting the coefficients of Table 5 into equation (6) and equation (7), the entrained water mass and moment of inertia under different geometric parameters in Table 3 can be recalculated as shown in Figure 9 and Figure 10.

As shown in Figure 9 and Figure 10, the results of the entrained water mass and moment of inertia obtained by empirical formulas for the multivariate quadratic regression model are consistent with FEM results in Table 3. Compared with Figure 7 and Figure 8, the results of the entrained water mass and moment of inertia calculated by empirical formulas of the multivariate quadratic regression model improve greatly.

5. Conclusion

The novel empirical formulas of the entrained water effect for tandem propellers are presented in this work. The validation of numerical method is verified by the single propeller. Based on numerical method, the empirical formulas of entrained water properties for multivariate linear/quadratic regression models of tandem propellers are proposed. It is concluded that:

(1) Excellent accuracy property can be observed for the numerical method used to calculate the entrained water mass and moment of inertia for tandem propellers.

(2) The empirical formulas of the entrained water effect for tandem propellers are given, considering the influence of the blade diameter, the expanded area ratio and the pitch/diameter ratio.

(3) Compared with the multivariate linear regression model, the accuracy of the entrained water mass and moment of inertia calculated by the multivariate quadratic regression model is higher.

The empirical formulas of tandem propellers obtained by numerical method to calculate the entrained water effect in this work effectively solve the lack of relevant practical engineering calculation.

Acknowledgments
This study is supported by the Natural Science Foundation of China (No. 51839005) and the Development of High-efficient Tandem Propeller (No. Z2021SFC-11/DZ).

References
[1] Nikoozadeh A, Behzadmehr A and Payan S 2019 *Journal of Thermal Analysis and Calorimetry* **140** pp 1029–44
[2] Burrill L C and Robson W 1962 *Trans. North East Coast Institution of Engineers and Shipbuilders* **78** pp 325–360
[3] Schwanecke H 1963 *Sonderdruck aus Jahrbuch der Schiffbautechnischen Gesellschaft* B **57** 252
[4] Brooks J E 1976 *David W Taylor Naval Ship Research and Development Center* (Maryland: DTNSRDC) pp 01–11
[5] Parsons M G and Vorus W S 1981 *Propellers Symposium* (Virginia: Trans. SNAME) p 016
[6] Harrington R L 1992 *Marine engineering* (Virginia: Trans. SNAME)
[7] MacPherson D M, Puleo V R and Packard M B 2007 *SNAME New England Section* (Virginia: Trans. SNAME) pp 01–11
[8] Korotkin A I 2009 *Added Masses of Ship Structures* (Berlin: Springer) pp 333–353
[9] Hutchison S, Steen S and Sanghani A 2013 *Proc. of the 3rd International Symposium on Marine Propulsors* (Launceston: SMP) pp 69–77

AUTHOR INDEX